PRINCIPLES OF MATERIALS CHARACTERIZATION AND METROLOGY

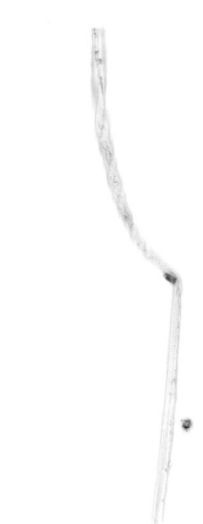

Principles of Materials Characterization and Metrology

Kannan M. Krishnan

University of Washington, Seattle

OXFORD

UNIVERSITY PRESS

OXFORD

UNIVERSITY PRESS

Great Clarendon Street, Oxford, OX2 6DP,
United Kingdom

Oxford University Press is a department of the University of Oxford.
It furthers the University's objective of excellence in research, scholarship,
and education by publishing worldwide. Oxford is a registered trade mark of
Oxford University Press in the UK and in certain other countries

First Edition published in 2021

Impression: 1

Published in the United States of America by Oxford University Press
198 Madison Avenue, New York, NY 10016, United States of America

British Library Cataloguing in Publication Data
Data available
Library of Congress Control Number: 2020952840

ISBN 978–0–19–883025–2 (hbk.)
ISBN 978–0–19–883026–9 (pbk.)

DOI: 10.1093/oso/ 9780198830252.001.0001

Printed and bound by
CPI Group (UK) Ltd, Croydon, CR0 4YY

To

Amma,

Appa,

MN,

and my students—past, present, and future

Preface

Materials science and engineering (MSE) is a multidisciplinary field, impacting every aspect of our technological society today. At the heart of MSE is understanding the relationship between structure and properties of materials. In fact, it is now well established that, by optimizing composition and structure ranging from the macroscopic to atomic dimensions, the properties of materials can be not only well controlled but also tailored for any specific application. In this endeavor, *materials characterization* and *analysis* involving a range of *diffraction, imaging,* and *spectroscopy* methods, at relevant length scales, has enabled the structure–property–processing–performance tetrahedron that epitomizes the field.

Traditionally, an *undergraduate* curriculum in MSE emphasizes the practical application of optical microscopy and spectroscopy, imparts a working knowledge of X-ray diffraction and, where resources are available, scanning and transmission electron microscopy, and atomic force microscopy. However, recent advances in developing materials for a wide range of applications, emphasizing atomic-scale tailoring of microstructure and exploiting size-dependent properties, require an interdisciplinary approach to materials development where a judicious use of available characterization methods becomes important. This requires a coherent discussion of the underlying *physical principles* of materials characterization and metrology using the wide range of electrons, photons, ions, neutrons, and scanning probes.

Following a broad introduction (§1), this book lays the foundations of characterization, analysis, and metrology and builds on concepts that should be familiar to an upper-division student in any branch of science or engineering. Starting with atomic structure, we develop spectroscopy methods based on intra-atomic electronic transitions (§2), followed by bonding and the electronic structure of molecules and solids motivating a number of spectroscopy methods (§3). We then discuss the periodic arrangement of atoms and develop principles of crystallography (§4), which leads to an introduction to diffraction in both real and reciprocal space. Next, we address different probes and present relevant details of the generation and use of photons, electrons, ions, neutrons, and scanning probes (§5), followed by a presentation of ion-based scattering methods (§5). A concise introduction to optics, optical microscopy, polarization of light, and ellipsometry follows (§6). The second part of the book includes a comprehensive discussion of diffraction and imaging methods that emphasize techniques widely used in the characterization and analysis of materials. This includes X-ray (§7), electron (§8), and neutron (§8) diffraction, as well as transmission and analytical electron (§9), scanning electron (§10), and scanning probe (§11) microscopies. Throughout the text, the characterization techniques are also used to introduce and illustrate fundamental properties and materials science concepts encountered in a wide range of materials. The book is generously illustrated throughout with figures, data tables, comparison between related methods, worked examples, and concludes (§12) with three unique and comprehensive tables summarizing the salient points of all the spectroscopy, diffraction, and imaging methods presented.

To keep the overall extent of the book to a manageable length, I have mainly emphasized *probe-based techniques*. Other methods, such as thermal characterization and property measurements, including mechanical testing, are deliberately not included. However, I do include numerous applications of materials and structures in fields ranging from science, technology, art, and biology. Each chapter also includes a number of worked examples to help tie together the concepts introduced therein, an extensive set of test-your-knowledge questions to help readers consolidate understanding of the subject matter based on the text, problem sets to further deepen learning, a summary highlighting the key concepts and ideas presented in each chapter, and an extensive bibliography for further reading.

While specialized books do exist for most of the techniques discussed here, including encyclopedias of materials characterization, a coherent textbook on materials characterization and metrology at the undergraduate or early graduate level, emphasizing fundamental physical principles, such as this one, is both lacking and highly desirable. Combining discussions of the underlying principles with practical examples, and detailed sets of exercises, this book will ideally serve as a text for a *year-long course* at the undergraduate and/or early graduate level. Completion of such a course should give students an entry into the interdisciplinary field of Materials Science and Engineering, and a solid foundation in characterization and analysis methods, with the ability to select and apply the appropriate technique for any characterization problem at hand. Alternatively, the book can be adapted to a *semester-length course*, taking a more traditional approach by devoting about five weeks each to spectroscopy (Chapters 2, 3, 9, and 10), diffraction (Chapters 4, 7, and 8), and imaging (Chapters 6, 9–11). If one is further constrained in time to a *10-week quarter or term*, as in many US universities, and which is the case for the course I have been teaching (syllabus available on request) for many years at UW—where UG students take this course after prior exposure to X-ray diffraction and the electronic structure of solids—this book may be adopted by teaching, selectively, the essential concepts of Chapters 1–6 at the approximate pace of one chapter per week, and in the last four weeks, covering scanning electron (Chapter 10) and scanning probe (Chapter 11) microscopies. Assigning term paper topics for self-study, involving more specialized techniques and their applications, including transmission electron microscopy (Chapter 9), would further strengthen student learning.

For those students from different disciplines other than MSE, this book provides what they will essentially need to know in materials characterization, including additional background, at an early stage of their study. Overall, this book is expected to potentially have a wide readership and academic relevance for teaching a course on characterization, analysis, and metrology, across multiple disciplines of engineering, physics, chemistry, geology, biology, art conservation, etc. The examples in the book are selected to reinforce this breadth of disciplines. Finally, even though this textbook is tailored for the teaching of upper-division undergraduate or early-stage graduate students, it is also written for self-study by experienced researchers, including those in industry, who realize that, to deliver a program/product satisfactorily, they need to know more about the microstructure of their materials than they currently do!

In writing this book, I benefitted from discussions with numerous colleagues and teachers who generously shared their knowledge in multiple disciplines with me over the last four decades. Some of them also reviewed sections of this manuscript at various stages of development. Alphabetically they include: S. D. Bader, P. Blomqvist, S. Brück, J. N. Chapman, D. E. Cox, U. Dahmen, C. J. Echer, R. Egerton, M. Farle, P. J. Fischer, E.E. Fullerton, C. Hetherington, F. Hofer, W. Grogger, R. Gronsky, R. Kilaas, C.A. Lucas, R. K. Mishra, Y. Murakami, C. Nelson, S. Paciornik, S. J. Pennycook, L. Rabenberg, P. Rez, D. Shindo, I. K. Schuller, S.G.E. te Velthuis, N. Thangaraj, S. Thevuthasan, G. Thomas, M. Varela, D. O. Welch, T. Wen, and T. Young. I also offer a special note of thanks to my former students, Eric Teeman and Ryan Hufshmid, who have independently created, for the exercises in the book, a solution manual (available from OUP to those who adopt this book for teaching a course), and the anonymous reviewer who provided a chapter-by-chapter review of the entire book. Also, I benefitted immensely from interactions with many generations of graduate students and post-doctoral fellows at both UCB and UW who, driven by their own curiosity and interests, provided me the motivation to learn and apply a wide range of characterization methods in our research. The list is too long to acknowledge them individually here, but many of their contributions are reflected in this book. Finally, over the past many years, students of my course on Principles of Materials Characterization (MSE333) at UW have used draft chapters of this book as it has evolved over time with subsequent revisions. Their constructive feedback and relentless criticisms have significantly improved the book, making it more accessible and tailored to student teaching and learning. I am deeply indebted to all of the people mentioned here; however, I am entirely responsible for any remaining omissions, errors, or mistakes, and if they are brought to my attention, I will be more than happy to address them in subsequent revisions. This book has been many years in the making and parts of it were written during multiple residencies in a number of places. I am particularly beholden to the Whitely Center, an idyllic writing retreat at Friday Harbor, the Brahm Prakash Visiting Professorship at the Indian Institute of Science, Bangalore, the JSPS Senior Fellowship at the University of Tohoku, and the Humboldt Career Research Award at the University of Duisburg-Essen, all of which provided the right atmosphere to make substantial progress in writing this book.

Kannan M. Krishnan
Seattle, August 2020

Contents

Introduction to Materials Characterization, Analysis, and Metrology

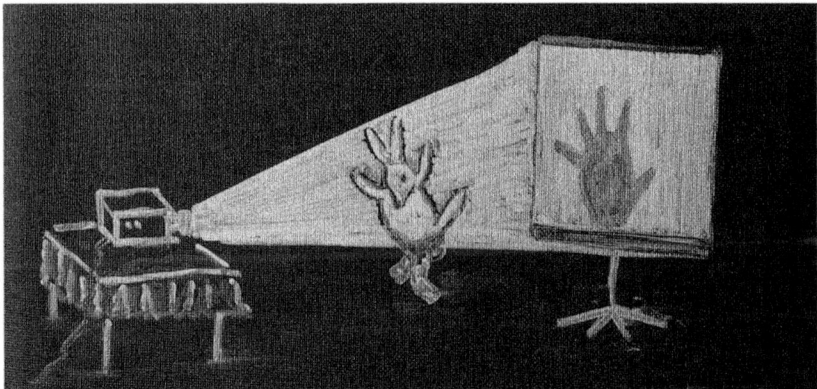

This illustration, by the author, based on a cartoon by John O'Brien (*The New Yorker*, February 25, 1991), succinctly describes the challenges in materials characterization. We are often called upon to describe the material microstructure (rabbit) based on the measured signals (hand) in diffraction, spectroscopy, or imaging methods. Needless to say, a poor understanding of the fundamental principles underlying the characterization methods generally lead to bad experimental design, hasty interpretations, and/or erroneous conclusions.

Principles of Materials Characterization and Metrology. Kannan M. Krishnan, Oxford University Press (2021).
© Kannan M. Krishnan. DOI: 10.1093/oso/9780198830252.003.0001

1.1 Microstructure, Characterization, and the Materials Engineering Tetrahedron

Materials science and engineering (MSE) is an enabling and multidisciplinary field, impacting nearly every aspect of society today. The reach of MSE is enormous—advanced semiconductors have stretched the limit of high-performance computers; optical fibers have dramatically increased the bandwidth and speed of intercontinental data transmission; magnetic materials in data storage have revolutionized information access, including the proliferation of the Internet; light-weight metals, polymers, and composites have transformed aircraft design and fuel efficiency; novel batteries and fuel-cell materials power things from cell phones to public buses; and increasingly, innovation in materials is at the heart of biomedicine. As the late William Baker, past president of Bell Laboratories, put it so elegantly: "everything is made of something and always will be!" In other words, MSE exemplifies the use-inspired fundamental studies of "Pasteur's Quadrant" (Stokes, 1997). The dramatic societal impact of the work of materials scientists and engineers can be illustrated by numerous examples; a few of them, included in Figure 1.1.1, are adapted from a special report by a distinguished panel of members of the U.S. National Academy of Engineering [1].

When we engineer any material, we tailor its properties for a specific application. This requires that it perform in a predictable and reliable manner when it is fabricated in the desirable shape or *form*. The latter may be a bulk material, a composite, a coating, a thin film or heterostructure, a wire or rod, a nanoparticle or its dispersion in a matrix, a surface, a nanoscale structure, a lithographically patterned element or array, etc. In other words, we make materials with sizes ranging from the atomic to the macroscopic, and dimensionality ranging from zero to three. Sometimes, the critical feature of interest in the material may be deep inside; an example is the buried interface in many modern semiconductor, magnetic, or photonic devices that are designed and fabricated in the form of thin film heterostructures.

Characterization and analysis of materials is central to the practice of materials sciences and engineering. The properties of all materials are determined by their *structure*, by which we broadly mean the composition, electronic structure, thermodynamic state/phase, and the arrangements of their internal components. The structure of materials can be described at various *length scales* or levels of detail. At the *atomic* level, it describes the bonding and organization of atoms or molecules relative to one another. At the *mesoscopic* level, it refers to an intermediate-length scale, between the atomic and microscopic, where material properties are different from the bulk, often determined by quantum mechanics, and dominated by surface effects. This is also the length scale of particular interest in nanoscience and nanotechnology (Owens and Poole, 2008). At

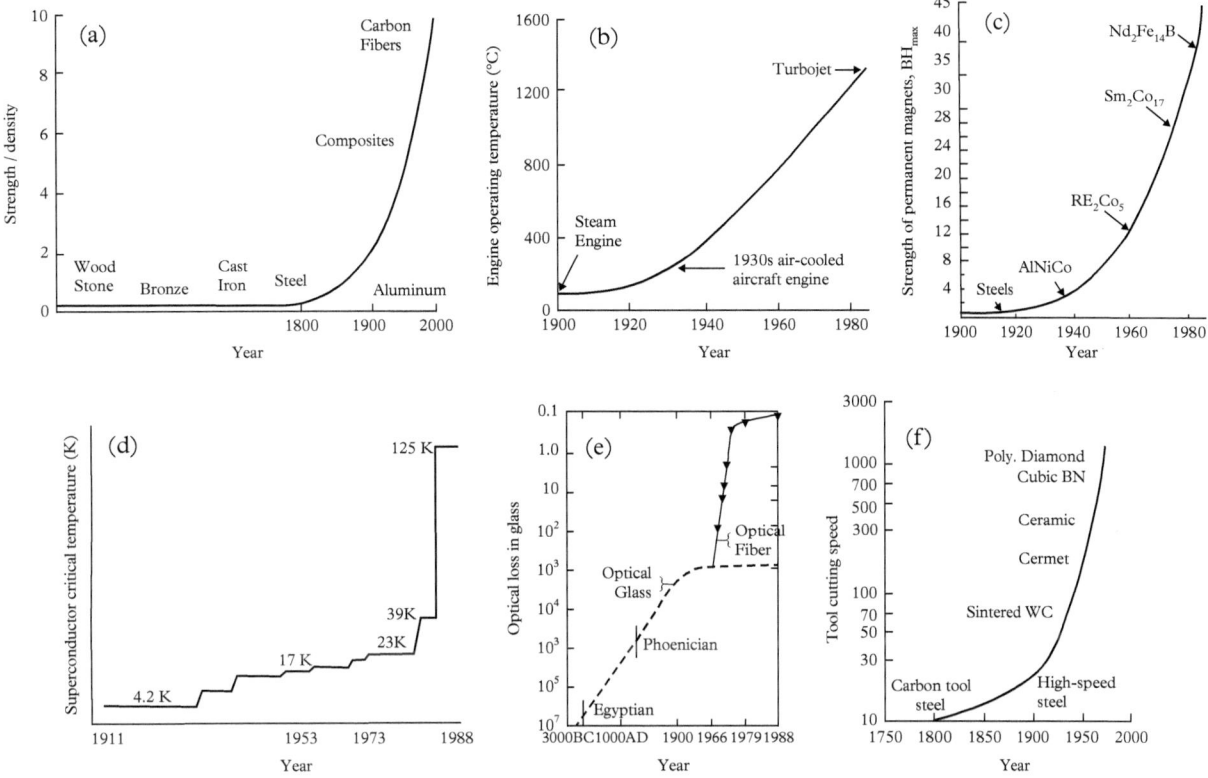

Figure 1.1.1 The societal impact of materials science and engineering is illustrated by a few representative examples. (a) Strength-to-density ratio of structural materials has increased fiftyfold, compared to cast iron used two centuries ago, with application ranging from light-weight eye glasses to composite airplanes. (b) The design of high-efficiency engines with reduced environmental impact requires materials that are strong at high temperature—superalloys and specialty ceramics can operate at temperature as high as 1,100–1,400°C, with theoretical efficiencies of ~80%. (c) The strength of a permanent magnet, given by its energy product, determines the design of smaller and more powerful motors—a 100-fold increase from the 1930s is evident. (d) Progress in the critical temperature of superconducting materials. (e) Optical fibers are now 100 times more transparent than they were in the 1960s. (f) New hard abrasive materials have increased cutting tool speeds by a factor of 100 from the early twentieth century, making manufacturing processes cheaper and more efficient.

Adapted from [1].

the *microscopic*—that which can be observed by some type of microscope—level, it refers to the arrangements of larger groups of atoms, such as grains and thermodynamic phases, including their morphology, chemistry, and crystallographic relationships. As materials scientists/engineers, when we use

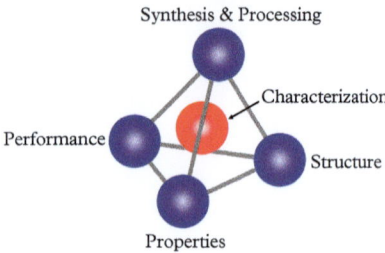

Synthesis & Processing

Characterization

Performance

Structure

Properties

Figure 1.1.2 The materials engineering tetrahedron, where characterization plays a central role.

the term *microstructure*, we generally mean <u>*all*</u> relevant structural details described here, from the atomic up to the microscopic-length scale. Lastly, at the *macroscopic* level, we refer to structure that can be viewed with the naked eye.

The core activities of a materials scientist/engineer can be represented by a tetrahedron (Fig. 1.1.2). Naturally, we begin with the *synthesis* of any material and then *process* it to achieve the desired *structure*, which in turn, determines its *properties* and its required *performance* in an economical and socially acceptable manner. Note that *characterization*, or *evaluating the microstructure at the appropriate length scale*, plays a critical role in tailoring the synthesis and processing of materials to achieve the desired properties and performance of any engineering component. So, we conveniently place characterization in the center of this tetrahedron. Note that characterization, as discussed in this book, does not include measurements of properties—mechanical, magnetic, electrical, thermal, etc.—that can be found in other specialized textbooks.

A range of characterization methods is required to elucidate the *processing–structure–property–performance* tetrahedron exhibited by the wide variety of materials that are tailor-made for specific functionality today. The list is long and includes metals/alloys, ceramics, polymers, amorphous materials and glasses that do not express any long-range crystallographic order, semiconductors, biological or biomimetic materials, composites, and natural materials like wood and paper. Alternatively, we can also classify materials in terms of their application, i.e. where structural, mechanical, functional (electrical, magnetic, optical), nuclear, biocompatibility properties, either individually or in some combination, become important. It may well be a soft material that is susceptible to damage when probed. Irrespective of the class of material that may be of interest, understanding the role of microstructure and tailoring it to optimize its properties is central to MSE. The microstructure of interest may be a combination of chemical, electronic, structural, crystallographic, or magnetic (*domains*) features, and has to be elucidated at the appropriate length scale that describes the behavior of the material. The characterization methods are many and the one to be applied for a specific problem has to be chosen judiciously among those readily available. Table 1.1.1 provides a list of *probe-based* characterization methods discussed in this book, especially those commonly identified by acronyms.

Broadly, the methods of characterization using different *probes* are classified as *diffraction*, *spectroscopy*, and *imaging*. The physical principles underlying these methods are the foundations of this book. Various techniques follow from these principles and are presented, with varying detail, as appropriate for this comprehensive presentation; naturally, detailed discussions of individual techniques abound in more advanced texts, including encyclopedias of materials characterization (see Further Reading). For working engineers, the American Society of Testing and Materials, now known as ASTM International, provides detailed standards[1] for a variety of materials characterization

[1] https://www.astm.org/studentmember/metallurgybycommittee.html

Table 1.1.1 Probe-based characterization methods, including acronyms where appropriate, discussed in this book, with sections indicated. Many of these techniques are mentioned in this chapter.

Technique	Acronym/Abbreviation	Section
Analytical electron microscopy	AEM	§9.4
Auger electron spectroscopy	AES	§2.6.2
Atomic force microscopy	AFM	§11.8
Atom probe field ion microscopy	AP-FIM	§1.4.3
Back-scattered electron imaging		§10.3.3
Biological force spectroscopy	BFS	§11.8.4.5
Cathodoluminescence	CL	§10.6.2
Confocal scanning optical microscopy	CSOM	§6.8.4
Convergent beam electron diffraction	CBED	§8.6.3
Dip-pen nanolithography	DPN	§11.8.5
Energy dispersive X-ray spectrometry	EDXS	§2.5.1.1
Electron back-scattered diffraction	EBSD	§10.4
Electron energy-loss spectroscopy	EELS	§9.4.2
Electron holography		§9.3.7
Electron probe microanalysis	EPMA	§2.5.2.2
Electron tomography		§9.5.1
Ellipsometry		§6.9
Energy filtered imaging in a TEM	EFTEM	§9.4.2.6
Environmental scanning electron microscopy	ESEM	§10.7.1
Extended X-ray absorption fine structure	EXAFS	§3.9.2
Field ion microscopy	FIM	§1.4.3
Focused ion beam milling	FIB	§9.6.5
Fourier transform infrared spectroscopy	FTIR	§3.5.2
High-angle annular dark field imaging	HAADF	§9.3.4

continued

Table 1.1.1 Continued

Technique	Acronym/Abbreviation	Section
High-resolution electron microscopy	HREM	§9.3.5
Inductively coupled plasma mass spectrometry	ICP-MS	§5.4.4
Inductively coupled plasma optical emission spectrometry	ICP-OES	§5.4.4
Interference contrast microscopy		§6.8.3.3
Inverse photoemission spectroscopy	IPES	§3.8
Lateral force microscopy	LFM	§11.5.2
Local electrode atom probe	LEAP	§1.4.3
Lorentz microscopy		§9.3.6
Low-energy electron diffraction	LEED	§8.4.1
Low-energy ion scattering spectroscopy	LEISS	§5.4.2
Magnetic force microscopy	MFM	§11.8.2
Metallography		§6.8.6
Neutron scattering		§8.8
Optical microscopy		§6.8
Particle induced X-ray emission	PIXE	§5.4.5
Photoemission spectroscopy	PES	§3.8
Raman spectroscopy		§3.6
Rayleigh scattering		§3.4.2
Reflection high-energy electron diffraction	RHEED	§8.4.3
Rutherford back-scattering spectroscopy	RBS	§5.4.1
Scanning electron microscopy	SEM	§10
Scanning electron microscopy with polarization analysis	SEMPA	§10.5.2
Scanning force microscopy	SFM	§11.4
Scanning probe microscopy	SPM	§11
Scanning thermal microscopy	SThM	§11.8.3
Scanning transmission electron microscopy	STEM	§9.2.8
Scanning tunneling microscopy	STM	§11.2
Secondary ion mass spectroscopy	SIMS	§5.4.3
Selected area diffraction	SAD	§8.6.1
Transmission electron microscopy	TEM	§9
Ultraviolet and visible spectroscopy	UV-Vis	§3.4.2
Ultraviolet photoelectron spectroscopy	UPS	§3.8
Wavelength dispersive (X-ray) spectroscopy	WDS	§2.5.1.2
X-ray absorption near edge structure	XANES	§3.9.2
X-ray absorption spectroscopy	XAS	§3.9.1
X-ray diffraction	XRD	§7
X-ray fluorescence spectroscopy	XRF	§2.5.2.1
X-ray magnetic circular dichroism	XMCD	
X-ray photoelectron spectroscopy	XPS	§2.6.5
X-ray transmission microscopy	XTM	§6.6.7
Z-contrast imaging		§9.3.4

methods encountered in their professional practice; familiarity with them and adhering to these standard practices will help advance the career of those in industry.

First and foremost, this book is not a catalog of characterization techniques, but, as already mentioned, it emphasizes the physical principles underlying each of the measurement techniques. The study of measurements, broadly known as *metrology*, is also of much industrial relevance with many companies having separate metrology departments to monitor the production and characterization of materials, devices, and components. The contents of this book are also relevant to *failure analysis*, an interdisciplinary engineering subject that is discussed at length in other specialized textbooks (Brooks and Choudhury, 2001). Finally, any reader who is convinced of the importance and breadth of materials characterization may skip this introductory chapter, but mastery of the basic concepts introduced here are critical for making progress through the rest of the book.

1.2 Examples of Characterization and Analysis

Materials scientists and engineers are called upon to work with and characterize a wide range of materials (Callister and Rethwisch, 2009). These include metals/alloys, ceramics, polymers, semiconductors, composites, amorphous or glassy materials, wood, paper, etc. An alternative way to classify materials of interest is in terms of their *application or function*; for example, these include structural or mechanical properties, functional (electronic, optical, magnetic) behavior, biomaterials optimized for specific *in vivo* applications that raise additional issues of biocompatibility and toxicity, "smart" materials that are able to sense changes in their environment and respond to them in a predetermined manner, nanomaterials, etc.

In general, the microstructural questions to be resolved are particular to the morphology of the material. For a bulk material, we may be interested in identifying its crystal structure (*point or space group*, Bravais lattice, unit cell dimensions, distribution of atoms in the unit cell; all described in §4), composition, chemical homogeneity, grain size and distribution, and if multi-phase, the orientation relationship between the matrix phases and the distribution of secondary phases, if any, especially at grain boundaries. We may also be interested in identifying the nature, extent, and distribution of defects (§4:1.9), be they planar, line, or point, and their effect on the properties of interest. In addition, in thin films, multilayers, and superlattices, we may wish to investigate the crystallographic relationships between the layers, i.e. texture and *epitaxy*.[2] Questions about the nature of the

[2] The nature or artificial growth of crystalline materials on single crystal substrates determining their orientation.

interfaces, especially the buried ones, such as roughness, compositional mixing, and in the case of magnetic materials, the spin lattice at the interface, may also be of interest. Further, elucidation of their growth mode—e.g. columnar, layer-by-layer—could also be relevant (§8.4.4). For surfaces, additional questions of the surface electronic structure and changes in crystallography due to surface reconstruction also arise (§8.3.1). For zero-dimensional objects like *nanoparticles*, resolving their crystallography and defects, compositional homogeneity, phase purity, size, size-distribution, and shape, especially when they are on the nanoscale, is an ongoing challenge (Fig. 9.3.14). Last, but not least, for glassy or amorphous materials without long-range order, identifying a good descriptor or a method to quantify such structures is an enduring question.

It is really impossible to find a *single* technique or an *Eierlegende Wollmilchsau*,[3] to address these wide range of microstructural questions. To be effective, we have to identify the best technique, or set of techniques, to resolve the microstructural questions at hand. Sometimes, the optimal technique may not be readily available and one may have to make do with a less-effective alternative.

Here, we begin with some typical examples of characterization and analysis of materials used in a variety of fields including engineering, biology, art, and geology. Each of these examples requires a judicious selection of characterization methods. Then, in subsequent chapters, we elucidate the fundamental principles behind these methods of *spectroscopy*, *diffraction*, and *imaging*, building on elementary concepts that should be familiar to the reader, i.e. atomic structure (§2), then moving to molecules (§3) and their vibration modes, and finally to crystalline solids (§4).

1.2.1 Ni-Based Superalloys: Ultrahigh Temperature Materials for Jet Engines

The efficiency and performance of jet engines are strongly dependent on the highest temperature attainable in their high-pressure turbine section (Fig. 1.2.1). Higher thrust requires higher operating temperatures, and for higher efficiency the engine must be made lighter without loss of thrust. Even though Ni-based superalloys in single crystal form have the required properties—high melting point ($\sim$1,650°C), good thermal conductivity, low density, and intrinsic corrosion resistance—their *microstructure* and the ensuing thermo-mechanical properties depend on the alloying elements, their concentrations, and their processing conditions.

Ni-based superalloys usually have a dual-phase microstructure (Fig. 1.2.2c) consisting of a $L1_2$ ordered γ' phase, existing in cuboidal shapes with {100} faces, separated by narrow channels of the FCC γ phase in between, and creating a coherent $\gamma|\gamma'$ interface (Fig. 1.2.3c). The chemically ordered $L1_2$ structure

[3] This is an old term in German for an imaginary animal, which provides everything for every purpose. A literal translation is oviparous-wooly-milking-sow. Here it means an animal that can lay eggs, give milk, and provides wool and meat. Needless to say, such an animal does not exist.

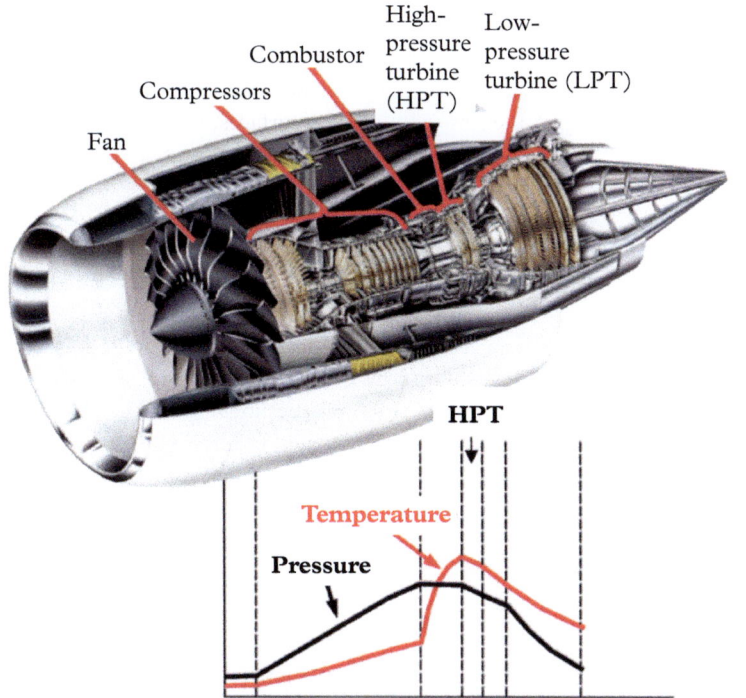

Figure 1.2.1 Illustration of the various components of a GE 90-115B jet engine. The variation of temperature and pressure from the back to the front of the engine is shown below.

Adapted from [34].

(Fig. 1.2.3j) of the γ' phase renders it highly rigid with low dislocation tolerance; hence the dislocations are confined to the γ channels, providing the required strength and high temperature creep properties. However, when subject to thermo-mechanical loads, the microstructure evolves—forming dislocation networks, coarsening the γ' cuboids (Fig. 1.2.2d–f), and precipitating topologically close-packed phases—and deteriorates the creep resistance of the alloy. Various alloying elements and heat treatments are used to control the microstructure and its evolution upon thermo-mechanical loading: precipitation elements (Al, Ta, Ti, Re) stabilize the γ' phase, solid-solution elements (Cr, Co, Mo) strengthen the γ phase by increasing the solidus temperature and resistance to dislocation movement, grain boundary elements (C, B) form carbides and borides along the grain boundaries to prevent casting pores and strengthen low-angle grain boundaries, and oxidation resistance elements (Al) form a protective Al_2O_3 surface layer.

Figure 1.2.3 shows a typical microstructure analysis carried out on a single crystal superalloy after heat treatment; in practice, this is correlated with mechanical behavior like creep as described in [3]. However, in the context of characterization it is important to point out that the analysis outlined here

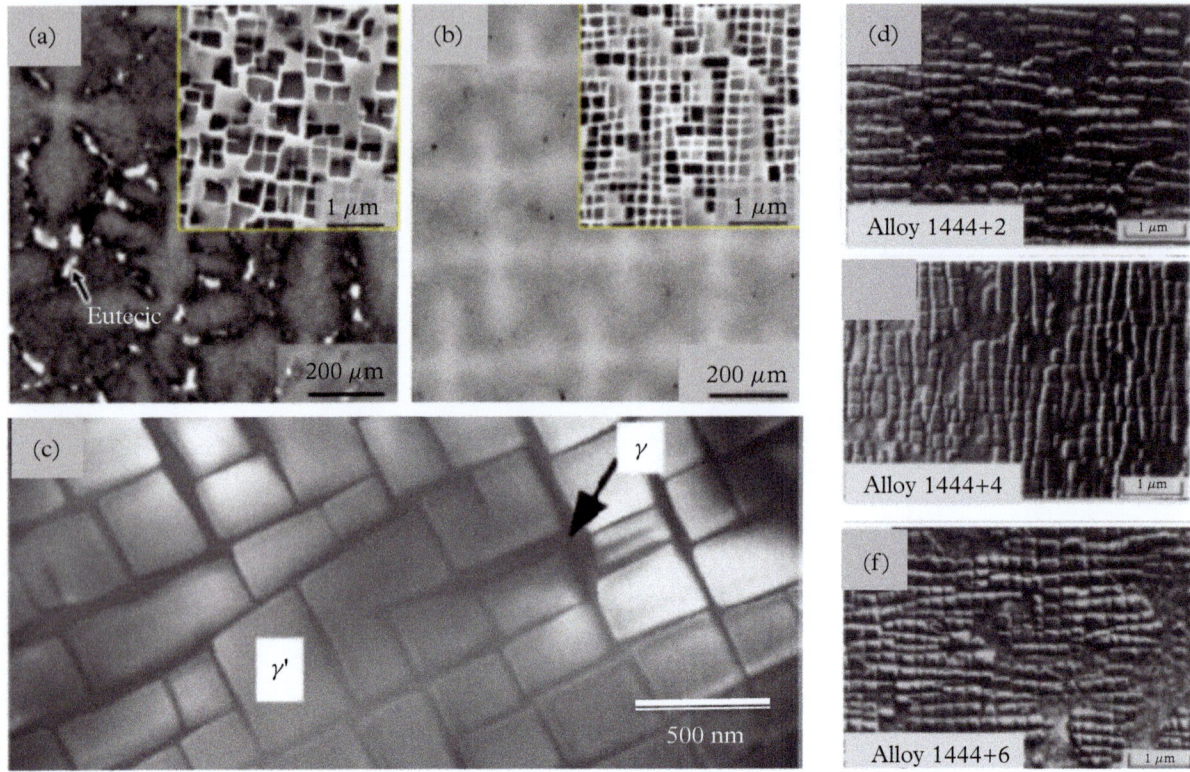

Figure 1.2.2 Microstructure of Ni-based single crystal superalloys (a) in the as-cast state, and (b) after the formation of the γ' phase following heat treatment, with the morphology of the two phases shown in greater detail (c). The rate of coarsening of the γ' phase, when the samples are aged for 2 hours at 1,065°C, is slowed down by the addition of Re, as shown in (d) 2% Re, (e) 4% Re, and (f) 6% Re. These images were obtained using transmission electron microscopy (TEM), a technique discussed further in §9.

Adapted from [2].

emphasizes the use of electron-based imaging, diffraction (§8), and spectroscopy methods in both transmission (§9) and scanning (§10) geometries.

1.2.2 Unraveling the Structure of Deoxyribonucleic Acid (DNA)

"It has not escaped our notice that the specific pairing we have postulated immediately suggests a possible copying material for this genetic material". With this characteristic understatement, the first paper [6] describing the structure of the genetic building blocks of life was published. The model of DNA proposed by

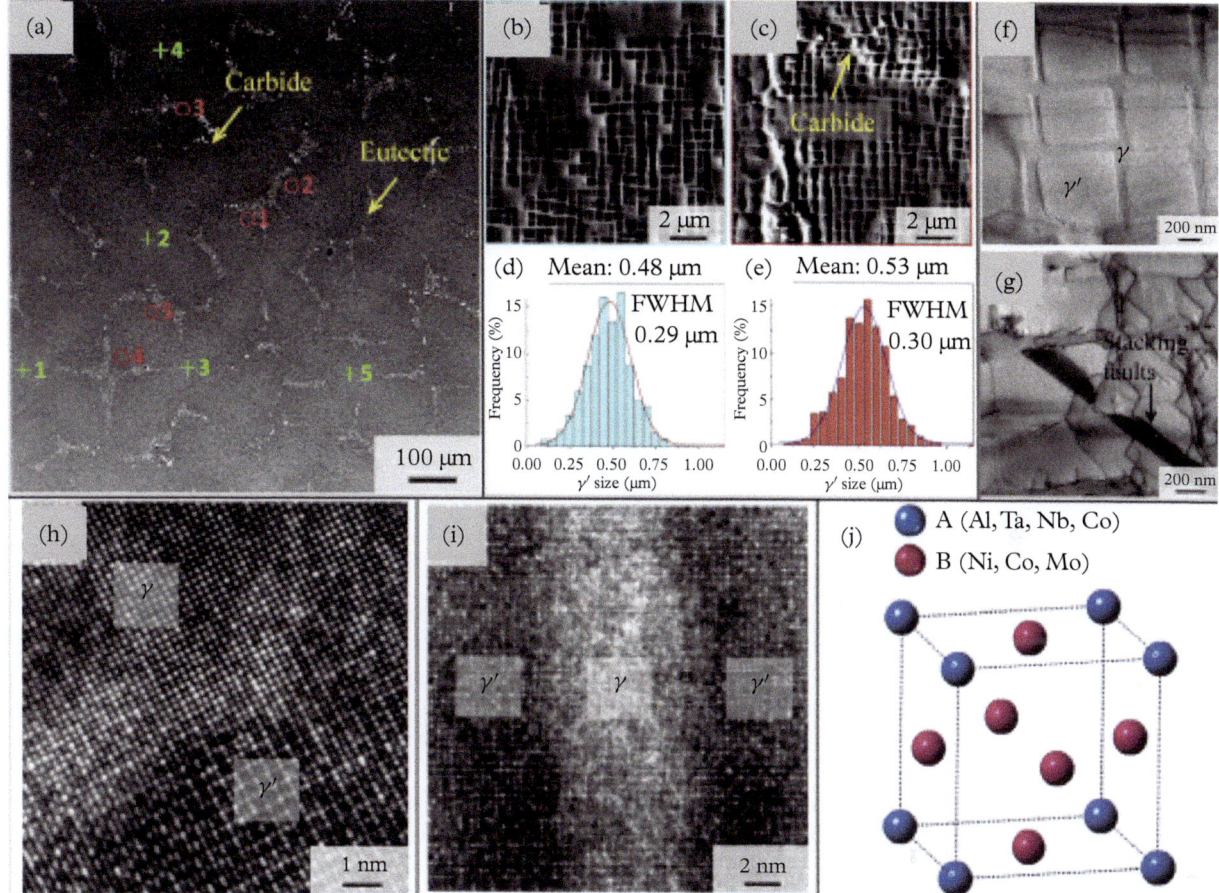

Figure 1.2.3 Typical microstructure analysis of a superalloy after heat treatment showing (a) secondary electron image of the microstructure seen from the transverse directions, identifying dendritic (green, b) and interdendritic (red, c) regions, where the γ' precipitate size distribution was determined, as shown in (d) and (e), respectively. The bright field TEM image of the alloy in different regions is shown in (f) and (g). (h) A high-resolution TEM, and (i) a high-angle annular dark field scanning TEM, HAADF-STEM image of the $\gamma \mid \gamma'$ interface illustrating the high level of coherency. (j) Schematic of the $L1_2$ ordered unit cell of the γ' phase. These methods are discussed in later chapters of electron diffraction (§8), TEM (§9), and SEM (§10).

Adapted from [3].

Watson and Crick (Fig. 1.2.4a) shows two phosphate-sugar, right-handed, helical chains, each coiled around the same axis, with the horizontal rods indicating the pairs of bases that holds the chains together. In particular, the bases are on the inside of the helix and the phosphates on the outside.

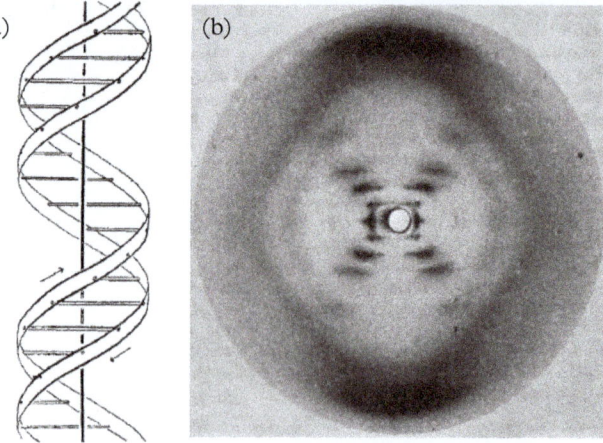

Figure 1.2.4 (a) The Watson–Crick model of DNA proposed in 1953. (b) The crucial X-ray diffraction (XRD, §7) photograph that was key to identifying the helical structure of DNA.

Adapted from [7].

The crucial experiment that provided key evidence for the correctness of the Watson–Crick[4] model of DNA was the X-ray diffraction (XRD) photograph (Fig. 1.2.4b) published in the same issue of the journal [7]. The specimen is a fiber with high water content and containing many millions of DNA strands aligned along the fiber axis; supposedly, this is the form of DNA in living cells. The X-ray beam is incident normal to the fiber and the X-ray photograph shows, in a striking manner, the characteristic features of helical structures. The key features of the diffraction pattern are four diamond-shaped outlines of fuzzy diffraction halos and separated by two arms of spots radiating from the center. The two arms are characteristic of helical structures, and the angle between the arms is proportional to the ratio of the width of the molecule (20 Å) to the repeat period (34 Å) of the helix. Further, careful study of the sequence of spots along the arms indicates an absence of the fourth spot, which confirms that there are only two intertwined helices involved in the structure.

1.2.3 Characterizing a Picasso Painting Reveals Hidden Secrets

Characterization and analysis play a significant role in art conservation, and are often used for authentication and to rule out forgery of paintings and sculptures when their provenance is questionable. Sometimes they reveal hidden secrets. Figure 1.2.5a shows Picasso's *La Miséreuse accroupie,* painted early in his Blue Period era of work. Using X-ray fluorescence (§2.5.2.1) measurements in a specialized set up (Fig. 1.2.5b) to scan such large paintings, it was discovered that the current Picasso painting was painted over a landscape painting (Fig. 1.2.8c) by another artist.

---

[4] F. H. C. Crick, J. D. Watson and M. H. F. Wilkins shared the 1962 Nobel Prize in Physiology in Medicine, and were cited "for their discoveries concerning the molecular structure of nucleic acids and its significance for information transfer in living material."

 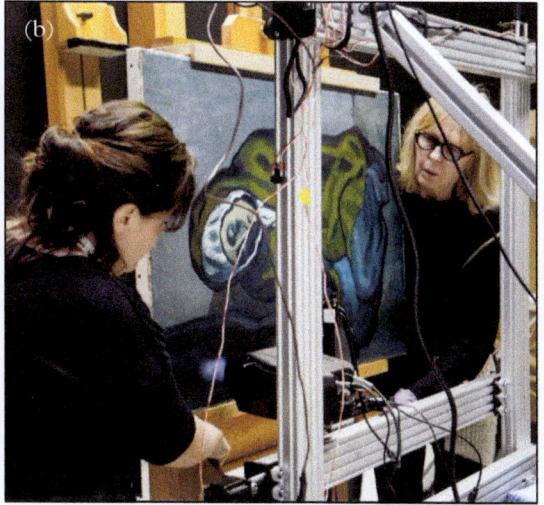 

Figure 1.2.5 (a) The current Picasso *La Miséreuse accroupie*. (b) Mounting of the painting for element-specific, spatially-resolved X-ray fluorescence measurements. (c) The hidden landscape painting buried underneath the current painting detected by element-specific X-ray fluorescence (XRF, §2) images.

Adapted from the *New York Times*, February 21, 2018.

In X-ray fluorescence, the incident probe of X-rays is absorbed by the various elements in the pigments, which then re-emit their characteristic X-rays (or fluoresce) at specific wavelengths (§2.5.2). This element-specific X-ray fluorescence can be locally excited and mapped spatially to obtain their two-dimensional distribution in the painting. Maps of iron (Fig. 2.Frontispiece.d) representing the use of Prussian blue, which is an iron-based pigment, and chromium (Fig. 2.Frontispiece.e), which is used in yellow pigments, matches the structure of the painting as seen today. However, the distributions of cadmium (Fig. 2.Frontispiece.f) used in multiple colored pigments, including red, yellow, and orange, and lead (Fig. 2.Frontispiece.g), used as a white pigment, clearly shows a different painting underneath.

1.2.4 Failure Analysis: Metallurgy of the *RMS Titanic*

In the early part of the twentieth century, passengers and mail between Europe and North America crossed the Atlantic Ocean by passenger steamship. One of the most luxurious steamships to be built for this purpose was the *RMS Titanic* (Fig. 1.2.6a), which, on its maiden voyage in 1912, struck an iceberg that damaged its hull and broke the ship in two. Within three hours, it sank, and more than 1,500

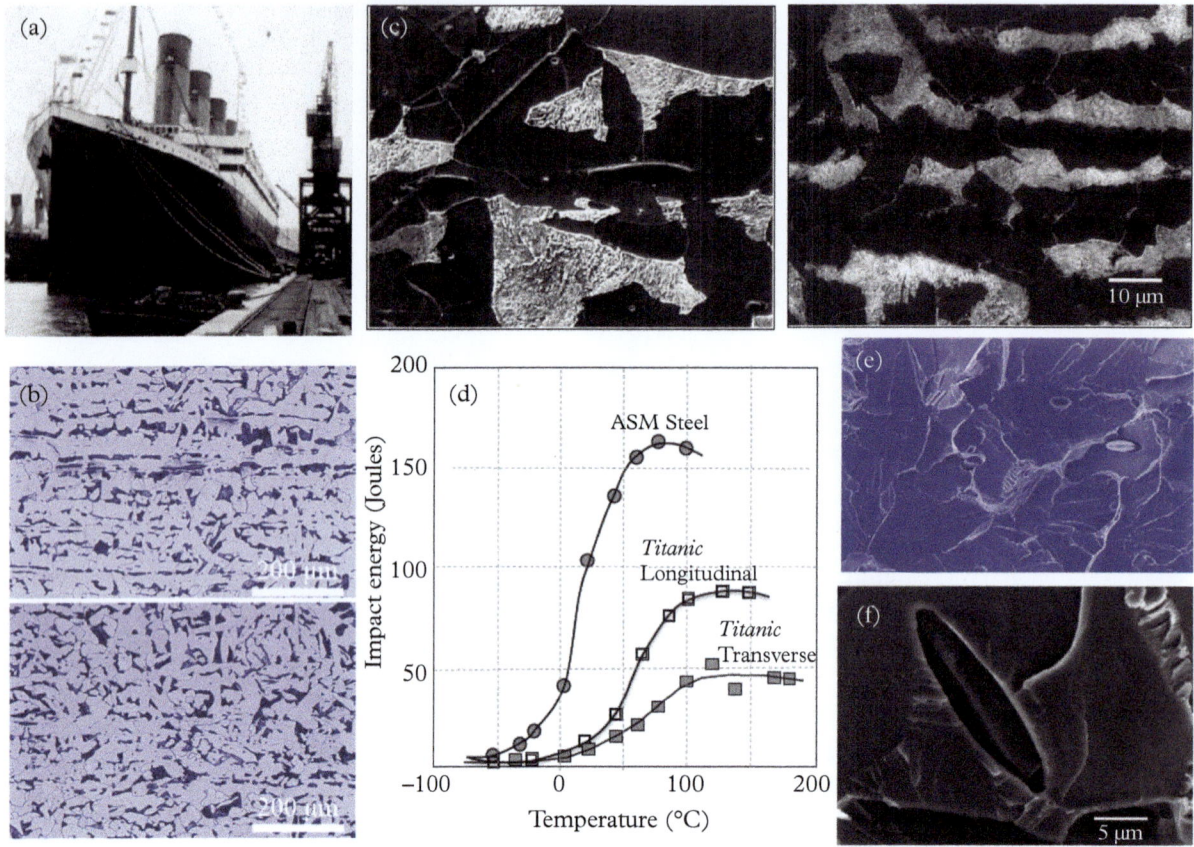

Figure 1.2.6 (a) The *HMS Titanic* before its maiden voyage. (b) Optical micrographs (§6) of the steel from the *Titanic* hull in the longitudinal (top) and transverse (bottom), showing the banding with elongated pearlite colonies and MnS precipitates. (c) An SEM micrograph of the *Titanic* hull plate (longitudinal section) and the ASTM standard. (d) Charpy impact energy as a function of temperature for longitudinal and transverse specimens of the *Titanic* hull and ASTM A36 standard. (e) An SEM micrograph of a Charpy impact fracture surface newly created at 0°C shows cleavage planes with ledges and MnS particles; the latter is shown magnified in (f).

Adapted from [8].

of its passengers died. An oft-cited culprit for this disaster was the quality of the steel used in its construction. A metallurgical analysis of the hull steel recovered from the wreckage provides interesting insight into its failure.

The first analysis of the steel looking at its overall composition Table 1.2.1 revealed a low nitrogen content, indicating that the steel was not made by the Bessemer process, which was known to render the steel against being brittle, especially at freezing temperatures. Instead, it was made by the then-alternative

Table 1.2.1 The composition (at %) of steels from the *Titanic*, a lock gate of the same era, and ASTM A36 steel

	C	Mn	P	S	Si	Cu	O	N	Mn:S ratio
Titanic Hull Plate	0.21	0.47	0.045	0.069	0.017	0.024	0.013	0.0035	6.8:1
Lock gate	0.25	0.52	0.01	0.03	0.02	-	0.018	0.0035	17.3:1
ASTM A36	0.20	0.55	0.012	0.037	0.007	0.01	0.079	0.0032	14.9:1

open-hearth process, as suggested by the relatively high oxygen and low silicon content. In addition, it contained a higher than normal phosphorus content, a very high sulfur content, and a low manganese content; the ratio of Mn:S was 6.8:1, a very low ratio by modern standards. The overall composition suggests that this steel was prone to embrittlement, or loss of ductility, especially at the freezing conditions encountered by the ship on that fateful night.

Metallographic preparation for optical microscopy, consisting of grinding, polishing, and etching with 2% Nital (§6.8.5), revealed the microstructure of the steel by optical microscopy (Fig. 1.2.6b). The steel is clearly banded in the longitudinal direction (top) with an average grain size of ∼60 μm, and the pearlite phase cannot be resolved. A scanning electron micrograph (§10) reveals the pearlite colonies, ferrite grains, small non-metallic inclusions, and MnS particles (Fig. 1.2.6c) identified by energy-dispersive X-ray spectrometry (EDXS, §2.5.1.2), elongated in the direction of banding. The Charpy impact test, performed from −55°C to 179°C (Fig. 1.2.6d) shows that the ductile–brittle transition temperature is 32°C for the hull steel, and −27°C, for the comparable ASTM A36 steel. The sea water temperature at the time of the collision was −2°C! Note that the ASTM A36 standard has a higher Mn:S ratio, and a substantially lower phosphorous content, both of which lead to reduced ductile-brittle transitions. A scanning electron microscope (SEM) image (Fig. 1.2.6e) shows a fractured lenticular MnS particle that protrudes edge-on from the fractured surface; further, slip lines radiating away from the MnS particle can be seen. Based on such analysis [8] it can be concluded that, even though the hull steel used was the best available in 1909–1911—when the *Titanic* was built—it would not meet the standards for plate steel used in ship construction today.

1.2.5 Beneath Our Feet: Microstructure of Rocks and Minerals

Common materials used in engineering are sourced from common minerals, which in turn are the major constituents of common rocks. Earth is a series of "shells"; it has a liquid core composed mainly of iron (and some nickel), an intermediate mantle (solid rock rich in oxygen, silicon, iron, and magnesium, in the form of silicates), and an outer crust (which averages around 30 km in thickness) that is made mostly of aluminosilicates, alkali elements, and calcium. The most abundant chemical element in the Earth's crust is oxygen (47 wt%, 94 at%), followed by silicon (28 wt%, 1% at%), and aluminum (8 wt%, 0.5 at%). Needless to say, most metals and ceramics are extracted from this outer crust.

The most common minerals are chemical compounds of silicon, aluminum, and oxygen, with small amounts of other elements distributed in them. Silicates dominate the minerals in the crust, the most abundant (58%) being feldspars (Orthoclase—$KAlSi_3O_8$, Albite—$NaAlSi_3O_8$, Anorthite—$CaAl_2Si_2O_8$), followed (13%) by pyroxenes and amphiboles (Diopside—$CaMgSi_2O_6$, Enstatite—MgSiO3, Tremolite), and to a lesser extent, $\sim$10–11% each, by quartz (SiO2) and mica (Muscovite—$KAl_2(AlSi_3O_{10})(OH)_2$, Kaolinite—$Al_2Si_2O_3(OH)_4$), respectively.

Petrologists study the mineralogical and chemical details of rocks, and structural geologists study the structural aspects of minerals and rocks, especially from the viewpoint of deformation processes. Detailed studies of rocks at the microscopic level provide a link between these two areas of study (Vernon, 2004). Microstructures of standard thin or polished sections of rocks are routinely observed in optical microscopes (§6.8.3) using polarized light (Fig. 1.2.7a). Cathodoluminescence (CL, §10.6.2) is used (Fig. 1.2.7b) to reveal the internal microstructure of grains, especially from those minerals (quartz, feldspar, and calcite) that appear colorless in light microscopes. Moreover, CL arises from imperfections in the crystal lattice (§4.1.9), e.g. impurity atoms, vacancies, and dislocations that are produced during the formation and growth of the minerals. Higher resolution images can be obtained in an SEM (§10), using secondary or back-scattered electrons (Fig. 1.2.7c) the latter with element sensitivity, and are used to reveal detailed microstructures of small grains, fine-grain aggregates, and intergrowths. Transmission electron microscopes (§9) can provide further details of finer features, e.g. exsolution lamellae. Computed tomography (§9.5.1) maps the variation of X-ray attenuation within a solid, along multiple directions [10]. The attenuation varies with the amount of each mineral present, and a series of cross-section or 2D images, produced along different directions, are computed to provide a 3D representation of the grains and aggregates in the rock (Fig. 1.2.7d,e). Finally, electron probe microanalysis (EPMA, §2.5.2.2) and mapping produces maps of compositional distribution, particularly in fine-grained aggregates of minerals (Fig. 1.2.7c). In general, the microstructure of rocks is a product of a complicated sequence of geological events and processes. As such, its microstructural analysis, including its chemical information, applying many of the techniques discussed in this book, can provide insight into the rock's formation, its geological history, and mineral value.

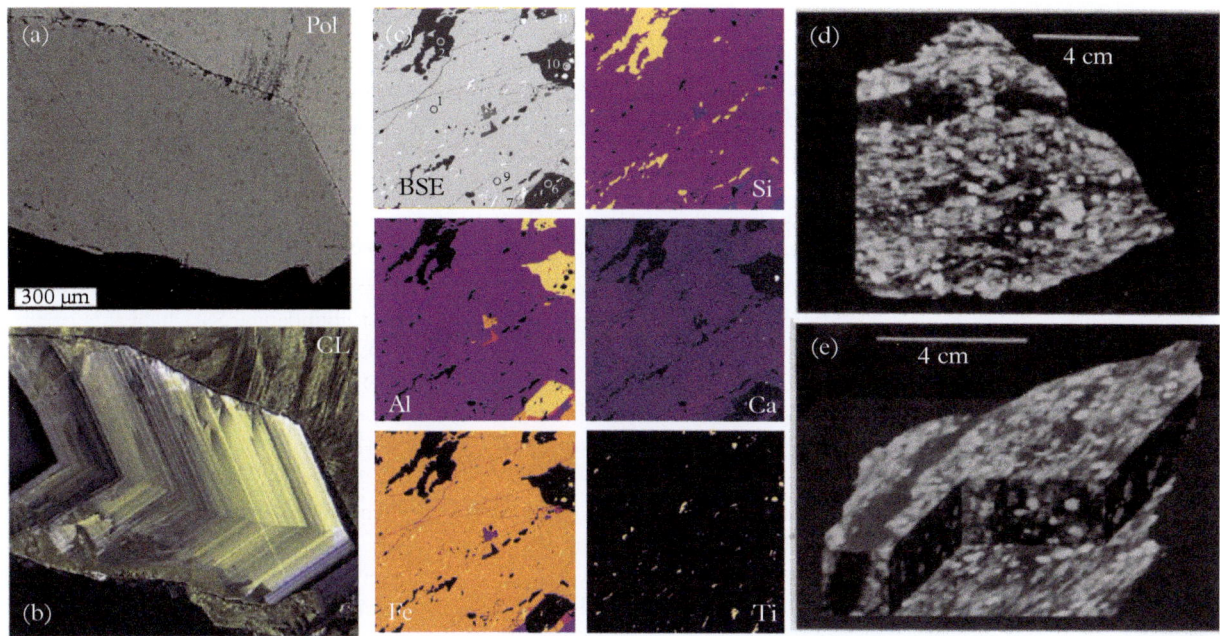

Figure 1.2.7 A specimen of hydrothermal quartz imaged by (a) polarizing and (b) cathodoluminescence (CL, §10) microscopy. Note that CL reveals internal structures and growth zoning that is not visible in the former. Adapted from [9]. (c) Back-scattered electron image (§10) of a specimen of peletic schist from Antarctica, along with composition maps for the principal elements. X-ray tomographic images of garnetiferous metamorphic rocks. (d) A single slice of a peletic schist—garnets are light gray to white ovals, kainite appears as medium gray laths, and dark gray to black regions are rich in quartz, feldspar, and muscovite. (e) A perspective view of a 3D density map. The single slice at the bottom of the stack locates the cutaway block in the interior of the specimen.

Adapted from [10].

1.2.6   Ceramic Materials: Sintering and Grain Boundary Phases

Ceramic materials are typically heterogeneous, multiphase materials, often containing crystalline and glassy (non-crystalline) phases with unique properties that make them suitable for high-temperature structural applications. As mentioned in §1.1, microstructure that may be too small to be seen with the naked eye, plays an important factor in the final property of a material. For ceramics, the microstructure is made up of small crystals called grains (Fig. 4.1.1), and in general, the smaller the grain size, the stronger and denser is the ceramic material.

Typically, ceramic materials are prepared by sintering, often in the presence of additives to introduce a liquid phase during sintering to overcome the poor solid-state diffusion, and achieve high densities. For example, in the case of silicon nitride, the liquid phase is introduced by oxide additives, which form a low-temperature eutectic[5] liquid with the oxidized surface layers of the silicon nitride powder precursors. However, the glassy intergranular layer is often retained after sintering, causing a deterioration in the mechanical properties. An understanding of the intergranular layer and its structure is required to improve the performance of ceramic materials. The characterization of the microstructure, including the grain boundary phases, is carried out using multiple techniques, such as X-ray microanalysis (§2.5.2), scanning and transmission electron microscopy (§9 and §10), and X-ray (§7) and electron (§8) diffraction, which is then correlated with rigorous mechanical testing.

Now, we briefly introduce the microstructural evaluation of the sintering of silicon nitride with small additions of La_2O_3, Y_2O_3, and SrO [4]. A polished specimen is first prepared and observed in an SEM (Fig. 1.2.8a) and its composition analyzed by EDXS (Fig. 1.2.8b). The polished surface reveals Si_3N_4 grains (90%) with a distribution of a boundary phase at grain boundaries and multi-grain junction regions (pockets); the latter can be estimated to be $\sim$10% of the volume.

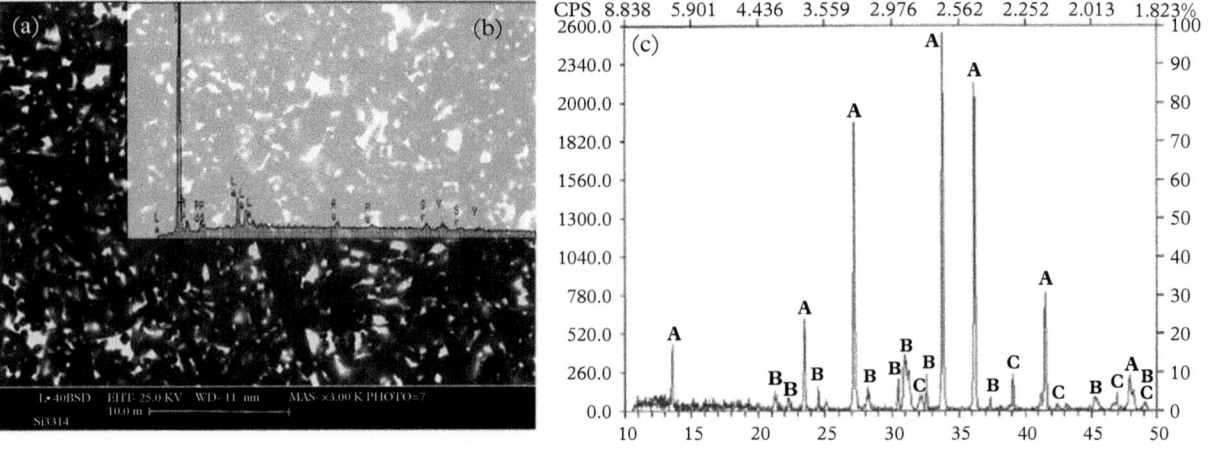

Figure 1.2.8 (a) SEM micrograph of a polished surface of a silicon nitride specimen showing the presence of the boundary phase (white), especially in the multiple grain junctions. (b) X-ray microanalysis (§2.5.2) of the polished surface shows peaks of Si and La, with very small intensities for Y and Sr. (c) X-ray diffraction (§7) patterns showing peaks that can be indexed for β-Si_3N_4 (A), $La_5Si_3O_{12}N$ (B), and $Y_5Si_3O_{12}N$ (C).

Adapted from [4].

The EDXS spectrum reveals the presence of the major sintering aid (La_2O_3) with La peak intensities substantially more intense than that for Y and Sr. Further, routine XRD θ-2θ scans (§7.9.2, Fig. 1.2.8c) reveal the presence of a majority β-Si_3N_4 and minority/boundary $La_5Si_3O_{12}N$ and $Y_5Si_3O_{12}N$ crystalline phases; however, it is not easy to say if the crystalline boundary phases were formed during sintering or during subsequent heat treatments.

Figure 1.2.9a is a lower magnification, bright-field transmission electron microscope (TEM) image of a two-grain region that shows a near-uniform dark contrast at the grain boundary. High-resolution electron microscopy (HREM) using phase contrast (§9.2.7) reveals further details of the microstructure at the atomic level. The HREM image from the two-grain region (Fig. 1.2.9b) shows that the β-Si_3N_4 grains are separated by a uniform amorphous phase $\sim$1.8 nm in thickness. However, a HREM image (Fig. 1.2.9c) of a three-grain junction shows evidence of a crystalline phase, with two sets of orthogonal lattice fringes (0.32 nm and 0.33 nm) present in the pocket. The fringe spacing(s) is in agreement with the interatomic spacing for the (210) and $(1\bar{1}2)$ lattice planes of silicon

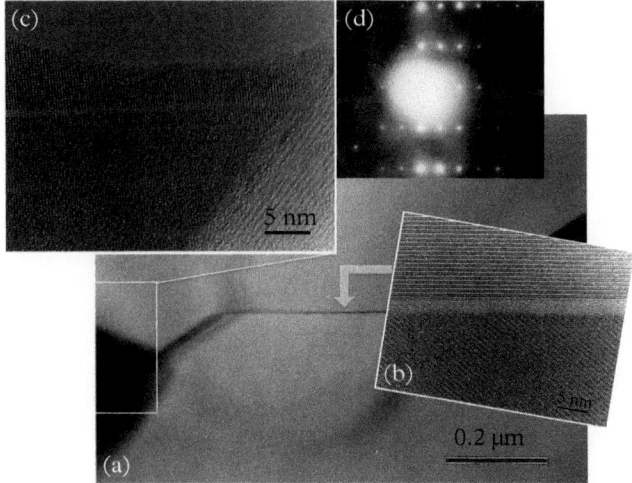

Figure 1.2.9 (a) A lower magnification bright-field TEM image showing a straight two-grain boundary. (b) An HREM micrograph of a two-grain boundary region. Using the lattice plane spacing of β-Si_3N_4 as an internal calibration, the amorphous phase thickness can be estimated to be $\sim$1.8 nm. (c) The multigrain junction (pocket) shows two-dimensional lattice fringes, with orthogonal spacing of 0.32 nm and 0.33 nm, respectively, corresponding to $La_5Si_3O_{12}N$. A residual glassy phase between $La_5Si_3O_{12}N$ and the β-Si_3N_4 grain is also found. (d) The crystallized material at the grain pocket can be indexed as the <012> diffraction pattern of $La_5Si_3O_{12}N$.

Adapted from [4].

lanthanum oxynitride ($La_5Si_3O_{12}N$), which is further confirmed by the selected area electron diffraction (Fig. 1.2.9d). Detailed analysis of this microstructure [4] shows that it is consistent with models [5] of sintering that predict an equilibrium thickness ($\sim$1 nm) at two-grain boundaries. Further, the grain pockets still show the presence of an amorphous phase, suggesting that it is hard to achieve a complete recrystallization of the grain boundary phase in silicon nitride. However, the partial recrystallization of the grain boundary phase can be correlated with the superior high-temperature strength of these ceramic materials.

1.2.7   Microstructure and the Properties of Materials: An Engineering Example

We now present an engineering example that illustrates a number of microstructural features; these are of particular relevance to materials characterization. Consider a dual-phase steel alloy of iron and chromium, doped with other elements, and bulk composition $Fe_{64}Cr_{25.3}Ni_{4.7}Mn_{1.34}Mo_{1.48}C_{0.04}$, typically used in automobile body panels, wheels, and bumpers. A number of *specimens*[6] of this alloy, all of the same bulk composition, were prepared and first annealed for one hour at 1,300°C. Then, each of them was subsequently annealed again at a different temperature, T_a, over the range, 400°C < T_a < 1200°C. Their mechanical properties were measured and the ultimate tensile strength (UTS) of the alloys varied with T_a (Fig. 1.2.10a) with a minimum at T_a = 900°C. Now, since all the alloys have the same chemical composition, the variation of mechanical properties can only be explained in terms of the *microstructure*. Figure 1.2.10b shows a typical microstructure observed in a transmission electron microscopy (§9) bright field image common to all these alloys. Clearly it has two *phases* with distinctly different features, i.e. dark particles distributed in a light matrix.

Careful *crystallographic structure* analysis by transmission electron diffraction (§8) shows that the *structural arrangement* of the *unit cells* in the two phases is different—the dark particles have a face-centered cubic (FCC) structure (Fig. 4.1.12), which is referred to as the γ-austenite phase, and they are distributed in a body-centered cubic (BCC), α-ferrite matrix phase. The volume fraction, f_γ, of the γ-austenite phase, is obtained by analyzing these images and it varies from 5% to 40% as a function of T_a, (Fig. 1.2.10c) with a maximum for T_a = 900°C. It is clear that the ultimate tensile strength depends *inversely* on the content, f_γ, of the γ- phase. Now, it is easy to understand the mechanical behavior. Compared to the α-ferrite matrix (BCC) the γ-austenite (FCC) particles have a lower resistance to plastic deformation, a larger ductility, and lower values of the UTS. This is not the end of the story. If we carefully plot the value of UTS as a function of f_γ, we see not one but two distinct curves (Fig. 1.2.10d). It turns out that these two curves correspond to different *compositions* of the particle and matrix phases, as confirmed by appropriate *chemical analysis*, using EDXS (§2.5) in a TEM

[6] In materials characterization, the terms *specimen* and *sample* are used interchangeably. However, in this book we will endeavor to distinguish the two. The sample is generally the overall subject of the study, and the specimen is specifically what is used or prepared to carry out a measurement or examination. For example, a chemically *inhomogeneous* sample may have a certain bulk/average composition; however, different specimens prepared from different parts of the sample may have different compositions because of the inhomogeneity.

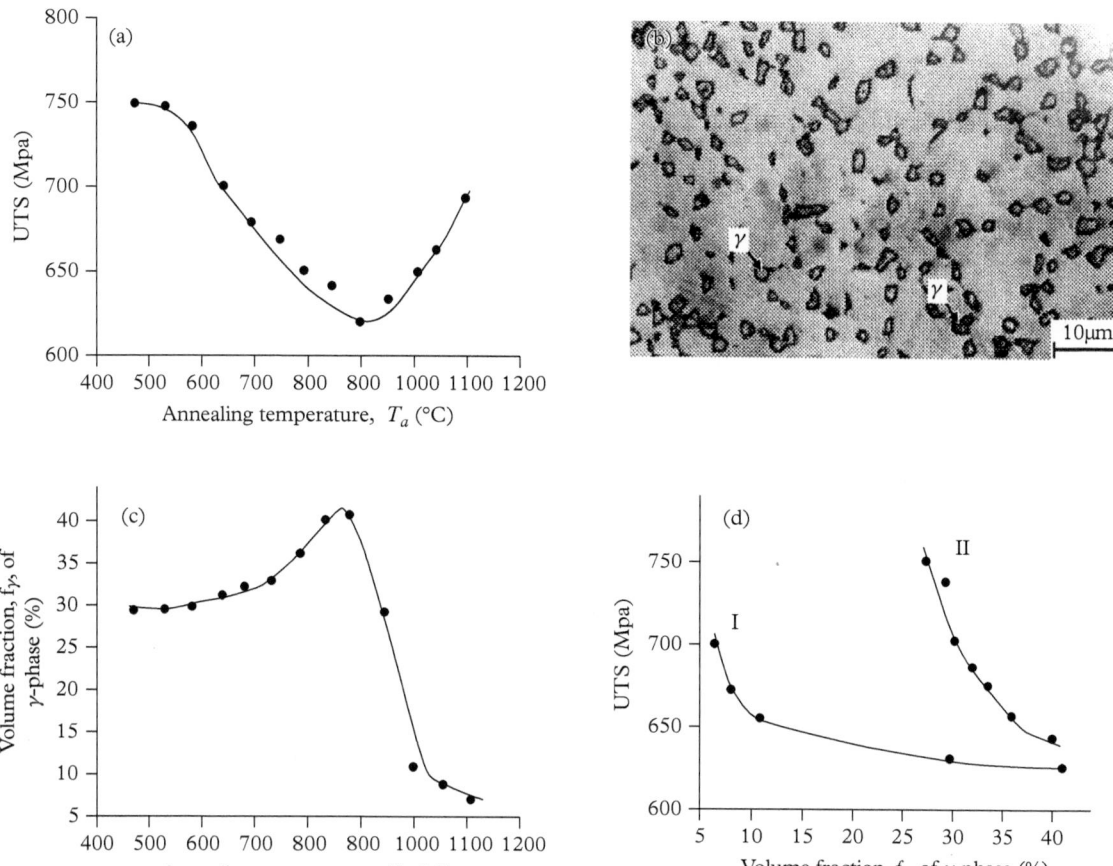

Figure 1.2.10 An example of the relationship between microstructure and properties in dual-phase steel. (a) Variation of the ultimate tensile strength (UTS) with annealing temperature. (b) The two-phase microstructure of γ-austenite (FCC) precipitates in an α-ferrite (BCC) matrix. (c) The γ-phase volume fraction, f_γ, as a function of annealing temperature. Note the inverse correlation with UTS. (d) The variation of UTS with f_γ further depends on the composition.

Adapted from Kurzydlowski and Ralph (1995).

(§9.4.3). Thus, we can see that these two phases not only differ in their mechanical properties but also their alloying content. From the perspective of this book, we can see that the microstructure involves *structural* (crystallography), *chemical* (composition), and *morphological* (volume fraction) aspects, among others, that need to be understood to optimize the mechanical properties of these dual-phase steel alloys.

1.3 Probes for Characterization and Analysis: An Overview

1.3.1 Probes and Signals

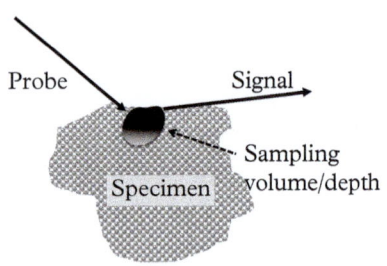

Our general approach to materials characterization is illustrated in Figure 1.3.1. We typically use an incident radiation or *probe* to interrogate the material and then detect a *signal* that can be used to infer or interpret the microstructure of interest. The interaction of the probe with the specimen can be *elastic* or *inelastic*. By elastic, we mean that there is no difference in the energy of the probe and signal, and applies predominantly to scattering or diffraction methods. By inelastic interactions, we mean a change in energy and such dispersions of signal intensity, as a function of energy, form the basis of spectroscopy techniques. However, signals from both elastic and inelastic interactions can be spatially resolved or mapped, and such spatial intensity distributions form the basis of imaging methods.

Figure 1.3.1 A general representation of materials characterization using probe and signal radiations to infer details of the microstructure of materials. In general, the probe and signal may be the same type of radiation, or they may be different.

We emphasize that the probe and signal may or may not be the same; in other words, they can be different. For example, in X-ray photoelectron spectroscopy (XPS, §2.6.3), we use X-rays or photons as the incident probe, and detect the photoelectrons as the signal. In this case, it is also important to recognize that, even though the *penetration depth* of the incident photons or X-rays may vary, the signal will be restricted to the *escape depth* of the photoelectrons, which is largely limited to the surface. A counter-example would be EDXS, where the incident probe is a high-energy or "fast" electron beam in an SEM (§10) or TEM (§9), and the detected signals are the characteristic X-rays emitted from elements constituting the specimen. Alternatively, electron energy-loss spectroscopy (EELS) in a TEM (§9.4.2), which is the electron analog of X-ray absorption spectroscopy (XAS, §3.9.1), uses fast electrons as the incident probe but then detects the energy dispersion of the inelastically scattered electrons, typically in the forward direction, after passing through the specimen.

1.3.2 Probes Based on the Electromagnetic Spectrum and their Attributes

The electromagnetic spectrum, extending from radio waves to γ-rays (Fig. 1.3.2) and including the relationship between energy (E), wavelength (λ), and frequency (f), is a good starting point to look at various probes used in materials characterization and analysis. The visible spectrum ranging from $\lambda \sim 380 - 700$ nm, with red ($\lambda = 650$ nm), green ($\lambda = 530$ nm), and blue ($\lambda = 470$ nm) indicated prominently, is used in optical imaging/microscopy and Raman spectroscopy. At larger wavelengths, we encounter infrared (IR) radiation used in probing vibration modes in molecules, followed by microwaves and radio waves; the latter, includes nuclear magnetic resonance (NMR) of wide use in chemistry and medical imaging. At wavelengths shorter than the visible, we find ultraviolet (UV)

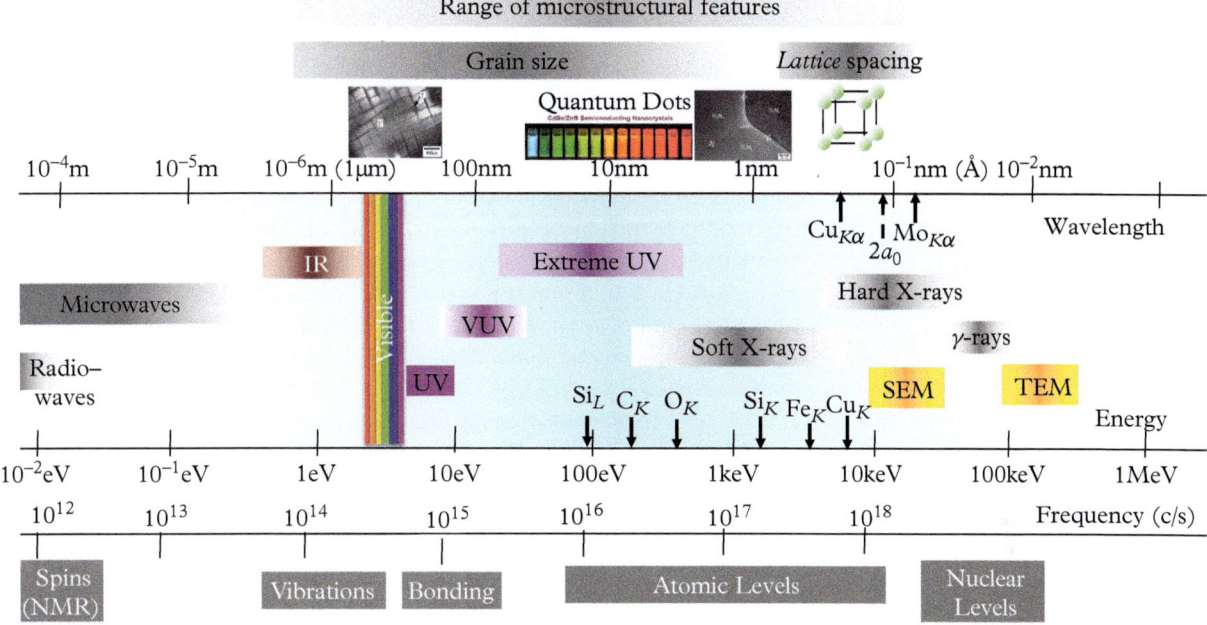

Figure 1.3.2 The electromagnetic spectrum illustrating the range of probes that can be used in materials characterization. In addition, we must also include scanning probes as well as ions and neutrons.

radiation, useful in probing bonding states of molecules, and further down we encounter extreme UV (EUV) radiations, followed by soft X-rays,[7] hard X-rays, and γ–rays, in that order. Soft X-rays overlap in energy with the atomic levels of core electrons; for reference, the Si_L (99.2 eV, $\lambda = 12.5$ nm), C_K (284 eV, $\lambda = 4.4$ nm), O_K (532 eV, $\lambda = 2.3$ nm), Si_K (1840 eV, $\lambda = 0.674$ nm), Fe_K (7112 eV, $\lambda = 0.174$ nm), and Cu_K (8980 eV, $\lambda = 0.138$ nm) absorption edges are shown. The $Cu_{K\alpha}$ (8.05 keV, $\lambda = 0.154$ nm) and $Mo_{K\alpha}$ (17.48 keV, $\lambda = 0.071$ nm) radiations, often used as laboratory sources for X-ray diffraction, are also indicated. The radius of the $n = 1$ orbit in the Bohr model (§2.2) of the hydrogen atom is $a_0 = 0.53$ Å. Twice this Bohr radius, or the diameter, $2a_0$, is a critical dimension that represents the spatial extent within which most of the charge for all atoms is contained; this value $2a_0 = 1.06$ Å, is also indicated. However, in atoms with multiple (Z) electrons, the inner shells have radii of the order of a_0/Z because they are not shielded by the outer electrons and experience the full Coulombic interactions of the nuclear charge ($+Ze$).

Most commercial SEMs (§10.2) and TEMs (§9.2) operate in the range of 1–30 kV and 100–300 kV, respectively. The energy ranges of these "fast" electrons straddle those of γ-rays, and are also included in Figure 1.3.2. However, we

[7] Wilhelm C. Röntgen (1845–1923) was the German physicist who discovered X-rays, also known as Röntgen rays, in 1895. He received the *first* Nobel prize in Physics in 1901 "in recognition of the extraordinary services he has rendered by the discovery of the remarkable rays subsequently named after him."

should not forget that electrons are charged particles and that Coulomb forces are very strong. Microstructural features of interest in materials characterization range from the interatomic lattice spacing ($\sim$1 Å), to intergranular phases ($\sim$1 nm), to grains and precipitates (10 nm $\sim$ 1 μm). Representative microstructural examples (§1.2) of γ and γ' phases in bulk Ni-based super alloys, semiconductor quantum dots, and intergranular amorphous phase in a sintered silicon nitride ceramic, illustrate the spatial range of materials encountered in characterization and are included for comparison in Figure 1.3.2.

Note that in Figure 1.3.2, energy is not expressed in the standard SI or MKS unit of joule (J), but in unit of electron volts (eV), which is the kinetic energy gained by an electron accelerated from rest through a potential difference of 1 V. It is generally recognized that the unit of joule is so large that it is inconvenient for use in routine materials characterization. Using the electron charge ($e^- = -1.602 \times 10^{-19}$ Coulomb), and the fact that a joule is equal to a Coulomb-Volt, we can readily show that 1 eV = 1.602×10^{-19} J. Commonly, multiples of the eV, i.e. keV (10^3 eV) and MeV (10^6 eV), are also used. Further, based on the dispersion relation in vacuum, $c = f\lambda$, where $c = 299{,}792{,}458$ m/s, is the velocity of light, f is the frequency, and λ is the wavelength, one can write the energy, E, of the photon[8] as

$$E = hf = \frac{hc}{\lambda} = \frac{12.4 \text{ (keV)}}{\lambda(\text{Å})} \tag{1.3.1}$$

where Planck's[9] constant, $h = 4.136 \times 10^{-15}$ eV s = 6.626×10^{-34} Js, and λ is in units of Å($= 10^{-10}$ m).

1.3.3 Wave–Particle Duality

We have described probes within the electromagnetic spectrum in terms of their unique attributes of energy, wavelength, and frequency. In materials characterization we also tend to view the probe and signal radiations as discrete particles, e.g. electrons, photons, ions, and neutrons. The wave–particle duality—a concept central to modern physics—is the key to understanding this dichotomy. Historically, the behavior of electrons and photons provided key insight into the wave and particle nature of matter. Two important classical experiments are discussed here.

The photoelectric effect: When light is incident on a clean metal surface it ejects electrons (known as photoelectrons). The intensity of the light determines the number of photoelectrons emitted but the energy of the photoelectrons depends only on the frequency, f, of the light. This is impossible to reconcile with a wave description of light because it requires that the photoelectrons be emitted

[8] The word *photon* was introduced by G. N. Lewis in *Nature*, December 18, 1926.

[9] Max Planck (1858–1947) was a German physicist who was awarded the Nobel prize in Physics in 1918 "in recognition of the services he rendered to the advancement of Physics by his discovery of energy quanta."

with a velocity proportional to the intensity of light. Einstein[10] explained the photoelectric effect by considering the incident light as a beam of particles or photons, each with a *quantum* of energy, hf, such that a single photon would eject an electron from the metal surface with velocity, v, given by

$$\frac{1}{2}m_e v^2 = hf - \Phi \tag{1.3.2}$$

where, m_e is the mass of the electron, h is the Planck constant, and Φ is the surface work function (§3.7) required to remove a photoelectron from the solid. In this then-radical theory, increasing the intensity of light increases the number of incident photons, and leads to an increase in the number of ejected electrons without changing their velocities. In contrast, diffraction of light and X-rays, the latter from the planes of atoms in crystals, arises from interference (§6.6) and confirms their wave-like nature.

Example 1.3.1: A beam of photons illuminates a metallic surface (work function = 3.45 eV) and ejects electrons with a velocity of 765 km/s. What is the wavelength of the incident photon? What part of the electromagnetic spectrum does this photon correspond to?

Solution: First we convert the work function into SI units:

$$\Phi = 3.45 \text{ eV} = 5.527 \times 10^{-19} \text{ J}$$

Then, applying (1.3.2) and using the values of the fundamental constants, we get the frequency of the incident photon:

$$f = \frac{\frac{1}{2}m_e v^2 + \Phi}{h} = 1.236 \times 10^{15} \text{s}^{-1}$$

From (1.3.1), we get

$$\lambda = c/f = 2.43 \times 10^{-7} \text{m} = 243 \text{ nm},$$

where

$$c = 3 \times 10^8 \text{m/s}$$

is the velocity of light.

Thus, the wavelength of the incident photon is 243 nm, and falls in the UV region of the electromagnetic spectrum (Fig. 1.3.2).

---

[10] Albert Einstein (1879–1955) received the Nobel prize in Physics in 1921 for "his services to theoretical physics and especially for his discovery of the law of the photoelectric effect."

Electron diffraction: Electrons are deflected in electric and magnetic fields consistent with a classical particle-like behavior. In fact, such deflections are used in the design of electron spectrometers (§2.6.1.2) and systems for the magnetic imaging of domains (§9.3.6). However, one can associate a wavelength, λ, and a momentum, $\mathbf{p}$, where $|\mathbf{p}| = p$, with the motion of the electrons as postulated by de Broglie[11] (1924), in his principle of *wave-particle duality:*

$$\lambda = h/p \tag{1.3.3}$$

where h is the Planck constant. In fact, such wave-like behavior of electrons was verified almost immediately by Davisson[12] and Gerner (1925), who demonstrated the diffraction of electrons (§8) from the surface of nickel single crystals.

Thus, diffraction studies of surfaces require electrons with energies of the order of 100 eV, and such surface techniques are termed low energy electron diffraction (LEED, §8.4.1) or microscopy (LEEM).[13] See Example 1.3.2. However, to probe the internal crystal structure of materials, electrons with substantially higher energy (100–200 keV), e.g. in TEMs, with wavelengths in the range (0.037–0.0251 Å) are used (§9). Such high-energy electrons can also probe the structure of surfaces in reflection mode (RHEED, §8.4.3). In addition, scattering of protons (H^+) and He^+ ions of energy $\sim 1.0 - 2.0$ MeV, are also used in materials characterization (§5.4); typically, a 2.0 MeV He^+ ion has a wavelength of 10^{-5} nm.

[11] L. de Broglie (1892–1987) was a French physicist who was awarded the Nobel prize in Physics in 1929 and was cited for his "his discovery of the wave nature of electrons."

[12] C. J. Davisson (1881–1958) was an American physicist who won the Nobel prize in Physics in 1937 and was cited for "the experimental discovery of the diffraction of electrons by crystals."

[13] If, in addition, spin polarized (SP) electrons are used for imaging, the resulting SPLEEM microscope can be used to image surface magnetic domain structures. See Krishnan (2016), Ch. 8.

Example 1.3.2: What is the kinetic energy of an electron suitable for electron diffraction of crystalline materials?

Solution: The distances between lattice planes in a crystal are of the order of 0.1 nm (1 Å) and for diffraction of electrons their wavelengths should be of comparable magnitude. Thus, for a wavelength of 0.1 nm, the velocity of the electron is, from (1.3.3), $\lambda = \frac{h}{p} = \frac{h}{m_e v}$.

Thus,

$$v = \frac{h}{m_e \lambda} = \frac{6.626 \times 10^{-34}}{(9.109 \times 10^{-31})(0.1 \times 10^{-9})} = 7.27 \times 10^6 \text{ m/s} \tag{1.3.4}$$

and its kinetic energy is

$$E = \frac{1}{2} m_e v^2 = 2.41 \times 10^{-17} \text{ J} \sim 150.4 \text{ eV} \tag{1.3.5}$$

where 1 eV $= 1.602 \times 10^{-19}$ J.

1.3.4   Nature and Propagation of Electromagnetic Waves

We briefly review the laws governing the propagation of electromagnetic waves (see §6 and/or Hecht (2002) for more details). We state here, without a detailed discussion, that Maxwell[14] equations, which apply to all electromagnetic phenomena, require that the electric, **E**, and magnetic, **H**, fields that form the components of any electromagnetic radiation propagate as waves in free space. Both **E** and **H** vary along the direction of propagation, with the variation being harmonic in time, and the related displacements being always along the propagation direction. Further, the waves have only transverse and no longitudinal components. Figure 1.3.3 shows such an electromagnetic wave, propagating in the *z*-direction, at a particular instant of time, with **E** confined to the *xz*-plane, and **H** confined to the *yz*-plane. For this wave, the plane of vibration of the electric field, **E**, is the *xz*-plane and its direction of vibration is the *x*-direction. Such plane waves are said to be *linearly* or *plane polarized*.

For a plane-polarized radiation, with the coordinates chosen such that **E** is parallel to the *x*-axis, assuming that the variation of **E** with both position, *z*, and time, *t*, are sinusoidal, we can express the electric field as

$$E_x = E_{x0} \sin 2\pi \left(\frac{z}{\lambda} - ft + \theta_x \right) \qquad (1.3.6)$$

where, E_{x0} is the amplitude of the wave, λ is its wavelength, f is the frequency, and θ_x is the phase angle. Naturally, the frequency and wavelength are related, i.e. $c = f\lambda$. Note that any electromagnetic radiation, such as an X-ray beam, carries energy, and the rate of flow of this energy per unit area perpendicular to the

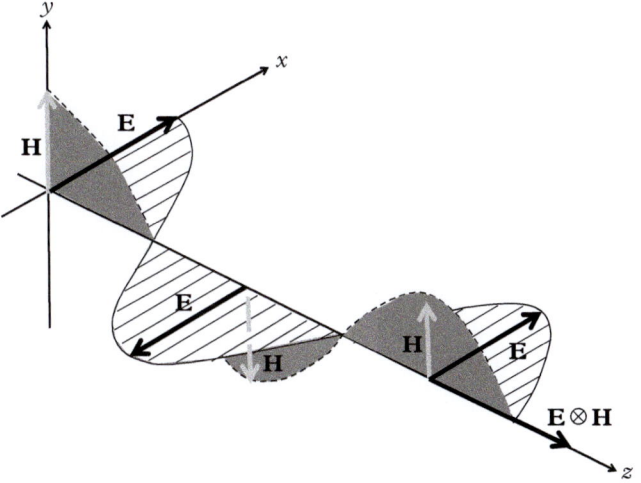

Figure 1.3.3 The variation of **E** and **H**, for a plane polarized electromagnetic wave propagating along the positive *z*-direction.

[14] James Clerk Maxwell (1831–1879) was a Scottish physicist best known for his formulation of electromagnetic theory.

direction of propagation is given by its intensity that is proportional to the square of its amplitude, E_{x0}^2. The unit of intensity is $J/m^2/s$.

The associated magnetic field, $\mathbf{H}$, is always perpendicular to $\mathbf{E}$, and in this case varies along the y-axis. Thus

$$H_y = H_{y0} \sin 2\pi \left(\frac{z}{\lambda} - ft + \theta_y \right) \tag{1.3.7}$$

where all terms have been defined earlier. If we are only concerned with the variations in time, at a specific location, z, of $\mathbf{E}$ (or, for that matter, $\mathbf{H}$) we can simplify (1.3.6) for the plane wave as

$$E_x = E_{x0} \sin 2\pi \left(ft + \theta_x \right) \tag{1.3.8}$$

Further, if the phase angle, θ_x, is not of interest it can be set to zero.

Waves with other states of polarization do exist. A simple way to visualize them is to consider the combination of two waves of the same frequency, propagating in the same direction, (say, z-axis), but with the direction of vibration of the electric vector in the x-direction for one wave and the y-direction for the other. Then the resultant polarization will depend on the combination of their amplitudes, E_{x0} and E_{y0}, and their phases, θ_x and θ_y, which is given by

$$E_x = E_{x0} \sin 2\pi \left(\frac{z}{\lambda} - ft + \theta_x \right) \tag{1.3.9a}$$

and

$$E_y = E_{y0} \sin 2\pi \left(\frac{z}{\lambda} - ft + \theta_y \right) \tag{1.3.9b}$$

Following some mathematical manipulations, at any given position along the z-axis, we can show that the locus of positions with coordinates of E_x and E_y are, in general, an ellipse (§6.2.6). The characteristics of this ellipse depend on the amplitudes, E_{x0} and E_{y0}, and the phase difference, $\Delta\theta = \theta_x - \theta_y$, of the two waves. The minor, y', and major, x', axes of this ellipse do not lie along the original x- and y-direction, but make an angle, ψ, with the x-direction such that

$$E_{x'} = E_x \cos \psi + E_y \sin \psi$$
$$E_{y'} = -E_x \sin \psi + E_y \cos \psi \tag{1.3.10}$$

where E_x and E_y can be substituted from (1.3.6). This *elliptically polarized* wave is shown in Figure 1.3.4. Further, when $E_{x0} = E_{y0}$, the ellipse simplifies to a circle and the wave is called *circularly polarized*. We discuss the application of polarized light in ellipsometry in §6.9.

A word on the nature of photons is now in order. Its nature can be illustrated with reference to Figure 1.3.5, which shows its electric field component, E, propagating in the z-direction. What we see is a pulse, or wave packet, moving with

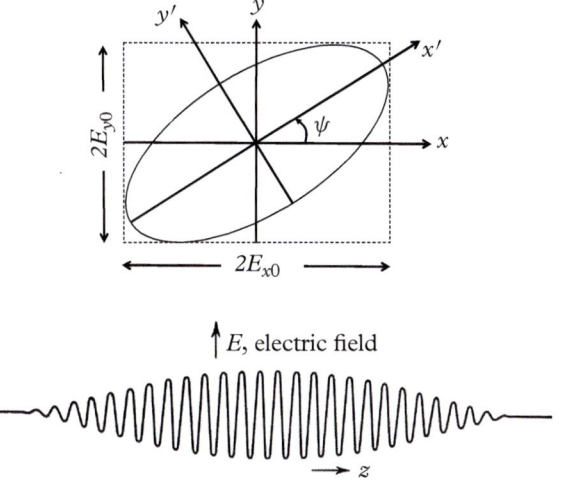

Figure 1.3.4 The vibrational ellipse of the electric vector, **E,** of an elliptically polarized wave propagating in the z-direction (out of the plane of the paper. Note that when $E_{x0} = E_{y0}$, the ellipse becomes a circle and the radiation is called circularly polarized. Also see Figure 6.2.7.

Figure 1.3.5 A photon can be considered as a wave packet, with its electric field component, $E(z)$, shown at any instant of time, t. Note that the entire wave packet moves with the velocity, c, of light.

Adapted from Sproul (1963).

the velocity, c, of light. Even though only a small number of oscillations are shown for clarity, in reality a typical photon emitted by an atom following a transition (Fig. 2.3.1) would have millions of oscillations. The motion of this photon or *wave packet*, would also follow all the properties of waves (wavelength, diffraction, polarization, etc.). Both, when light is emitted or absorbed, photons are created and absorbed as single indivisible units. Its energy is hf, where $f = c/\lambda$, and h is the Planck constant.

Example 1.3.3: How many photons of wavelength, λ(nm), are required to deliver 1 J of energy?

Solution: The energy of each photon is $E = hf = hc/\lambda$. Thus, the number (#) of photons required to deliver 1 J of energy is

$$\# = 1/E = \lambda/hc = \lambda \times 10^{-9}/(6.63 \times 10^{-34} \times 2.9979 \times 10^{8}) = 5.03 \times 10^{15}\lambda.$$

1.3.5 Interactions of Probes with Matter and Criteria for Technique Selection

Any characterization technique involving a probe will interact and invariably perturb the material (§5.3). Such perturbation has the potential to cause damage. Hence, the probe should be selected carefully to give maximum information with minimum damage. Of all the available probes, electromagnetic radiation in the visible spectrum (light, in common parlance) causes the least damage, but at the same time the information it provides is restricted to the surface, and even that, only with modest resolution. Thus, initial observations of materials should be made with optical methods (§6) using either/both elastic and inelastic scattering strategies; this includes both imaging and spectroscopic methods. Subsequently, higher-energy X-rays can be used to probe the material at greater depths to provide crystallographic (§7) and chemical information (§2.5). If imaging at higher resolution is desired, it can be accomplished by either using an SEM with higher energy (5–30 keV) electrons (§10), or using a TEM (§9) with electrons of energy > 100 keV to obtain information at the highest spatial resolution, provided electron transparent thin foil specimens representative of the behavior of interest can be made (§8 and §9). Before resorting to electron-based probes, e.g. SEM and TEM, when available, scanning probe methods (§11) may also provide useful information from the *specimen surface*, albeit at small areal (typically, 10 μm × 10 μm) coverage. Finally, the materials can be investigated with ions (§5.4), or neutrons (§8.8), to provide a variety of complementary information. Chapter 5 discusses in detail in the various probes available (photons, electrons, ions, neutrons), the nature of their interactions—both elastic and inelastic—with the specimen, their generation, and factors to consider in their selection and use; where relevant the discussion continues in subsequent chapters dealing with specific techniques. Here, in brief, is a preview of some important points to be considered.

1.3.5.1 *Penetration Depth and Mean Free Path Length*

These define different distances travelled in the material being characterized by the probe radiation. *Penetration depth* is a measure of how deep the electromagnetic radiation can penetrate into a material; often, it is defined as the depth at which the intensity falls to $1/e$ (37%) of its original value. *Mean free path length* is the average distance traveled by a moving particle in a material between successive impacts/collisions that modify its direction or energy. For any technique, if the probe and signal radiations are not the same and have different mean free path lengths in the material, the volume analyzed (or, the *sampling depth*) will be determined by the radiation—either probe or signal—with the smaller mean free path length. In general, it is impossible to provide a single, comprehensive description of the penetration depth of the radiation across the entire electromagnetic spectrum. Instead, it is only possible to discuss it in terms of some specific wavelengths of importance in materials characterization. For example, in the vicinity of the visible spectrum, IR radiation is used to probe materials based on how they are absorbed (Fourier transform IR (FTIR) spectroscopy, see §3.5.2), visible light is used to examine the specimen surface (optical microscopy, §6.8), and UV radiation is used to resolve the surface electronic structure (UV photoelectron spectroscopy, UPS, §3.8).

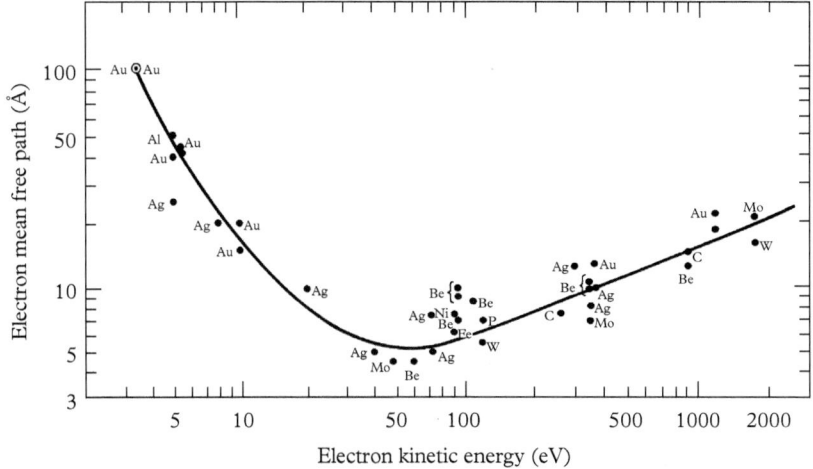

Figure 1.3.6 The mean free path of electrons as a function of energy for various metals. A universal curve, with a minimum of 4.0 Å for energies in the range ~50–100 eV, of relevance for Auger Electron Spectroscopy (AES, §2.6.2) and X-ray photoelectron spectroscopy (XPS, §3.8), is observed.

Adapted from [35].

Higher-energy probes, particularly X-ray radiation, have a uniform and predictable behavior in all materials. The absorption of X-rays, is defined by the attenuation coefficient, μ, (see Table 2.4.1) which increases with the average atomic number of the material, and determines the depth of penetration as

$$I = I_0 \exp\left(-\mu t\right) \tag{1.3.11}$$

where, I_0 is the intensity of the incident X-ray probe, and I is its intensity after being transmitted through the material of thickness, t (§2.4.2). The intensities of γ-rays show the same exponential dependence on thickness as X-rays, but with their much higher energies (~50 keV – 50 MeV) they penetrate much larger distances.

For electron probes, the mean free path length varies dramatically with their energy and the (average) atomic number of the material. For low energy (~0–2000 eV) electrons, the mean free path length is of the order of a few Å, and curiously, for all materials, satisfy a universal curve (Fig. 1.3.6) as a function of energy. This behavior, in simple terms, can be explained as follows.

For the range of energies of interest, the electrons in the solid can be approximated as a free electron gas. Then, the plasma frequency, which is a function of the mean electron-electron distance, r_s, determines the loss function. The inverse of the mean free path length, λ^{-1}, for electron propagation, is determined by r_s, which, to first order, is the same for all materials. For high-energy ($E \geq 5$ keV) electrons, the penetration depth, d_P, behaves as $d_P \propto E^{1.7}/Z$, where Z is the average atomic number of the material.

Neutrons have ~1,000 times the mass of electrons but do not have an electric charge. As a result, neutrons penetrate much greater distances than electrons and X-ray photons. Precise details of the penetration of neutrons in materials depend on the specific atomic species; in fact, the scattering lengths of neutrons show an erratic variation in magnitude and sign with atomic number (§8.8).

The interaction of high-energy ions with materials is also complex. At low (∼eV) energy they are reflected back from the surface, following simple rules of conservation of energy and momentum; at higher energies, they interact with the material, causing atomic displacements, formation of clusters and sputtering, or the removal of atoms, ions, and electrons from the specimen. In this context, it is customary to define a stopping distance, rather than a mean-free path, in the material for the ion. Further details are in §5.4.

1.3.5.2   *Resolution*

In the direction of the incident beam, i.e. *depth resolution*, the resolution is equivalent to the penetration depth already described in the previous section. However, the resolution in a direction normal to the direction of incidence, also known as the *spatial resolution*, depends on the diameter of the incident beam, its wavelength, and its mean free path length in the material. It also depends on the mode of imaging or signal collection. If the image is formed using an imaging system with "lenses", utilizing either the transmitted or reflected radiation, then the spatial resolution will depend on the wavelength of the radiation, the quality of the imaging system, including the "aberrations" inherent in the lenses, and the coherence of the source. Microscopes—conventional optical, electron in diffraction or phase contrast modes, certain X-ray, and some ion—operate in this manner. To first order, the Rayleigh criterion (Fig. 1.4.10) provides a good working rule to determine the spatial resolution of these imaging modes (also see §6.8.1).

The alternative method of imaging is to focus a narrow beam of radiation onto the specimen surface and again to detect the transmitted or reflected radiation. The image is formed by rastering the incident probe on the specimen surface and recording any changes in the radiation either due to elastic or inelastic interactions. Now, the spatial resolution is determined by the diameter and wavelength of the incident radiation, and the nature of the interaction, including the degree of localization of the interaction event that constitutes the signal of interest. Confocal light microscopy using lasers (§6.8.5), SEMs (§10), and ion-based optical imaging systems are examples of this mode of imaging. It is important to recognize that the spatial resolution of a technique using such focused probes of high-intensity radiation can also be limited by the potential damage the beam can cause to the material being analyzed; this is discussed briefly in the next section and in detail in §5.3.

Complementing this, *temporal resolution*, defined as the precision of a measurement with respect to time, is a very important criterion for designing *in situ* and dynamic experiments for the study of growth, morphological evolution, and response of materials to various applied stimulus (§1.4.5). Further details of the achievable *depth*, *spatial*, and *temporal* resolution, and the factors that influence them, are discussed for specific techniques in future chapters.

1.3.5.3 *Damage*

Damage to the specimen can be caused by the transfer of energy and momentum from the probe (see §5.3 for a detailed discussion). For photons, the transfer of energy in the form of heat largely causes the damage. The degree and spatial extent of the damage will be determined by the penetration of the radiation in the material, and the energy and flux of the incident photon. Specifically, for photons in the IR, visible, and UV portions of the electromagnetic spectrum, their momentum is quite small. However, higher-energy photons in the X-ray range, particularly when focused, can cause significant damage if the flux density is high, such as that obtained in synchrotron sources (§2.4.3) using zone-plate "lenses" (§6.6.7).

Electrons can behave both as particles and waves (§1.3.3), and when accelerated through several hundred keV in a TEM, they can cause significant damage by breaking interatomic bonds, particularly in polymeric materials (note: bond-breaking is not so common in inorganic or metallic materials). However, if the electrons are accelerated through higher voltages (~1 MeV), such as in high-voltage electron microscopes (HVEMs), the momentum transferred to the atomic nuclei by elastic large-angle scattering is sufficient to cause significant atomic displacements even in inorganic materials and alloys (Fig. 1.3.7). Similar atomic displacements are also caused by ions, and the extent of the damage is determined by the incident ion flux. For low flux, areas of displacement damage are isolated from each other, but at high flux the displacement damage is uniformly distributed in the specimen, resulting in the formation of an amorphous region even in crystalline materials. Note that an HVEM can be a powerful tool for *in situ* studies of radiation processes and the kinetics of defect formation; in fact, if it can be combined with an ion beam source such as in a tandem facility,[15] where if necessary, conditions similar to that experienced in a high-flux nuclear reactor can be approximated and the ensuing damage in the material can be studied at high resolution [12].

1.3.5.4 *Specimen Preparation or Requirements*

These are also to be considered in probe and technique selection. While these requirements are technique specific and are described in appropriate sections throughout the book, some typical considerations are outlined here. Optical metallography (§6.8.5) often requires the polishing of surfaces, with specialized etchants (Table 6.8.1) to provide adequate contrast to resolve the features of interest. FTIR and Raman spectroscopy are quite flexible, require no vacuum, and can investigate liquids, gases, and solids. SEM requires minimal preparation, as long as the specimen is vacuum compatible, but if it is an insulator, it will require a thin (~10 nm) coating of a conducting layer of carbon, gold, or other metal to prevent image degradation due to charging effects (§10.8). The most important requirement for carrying out TEM experiments is the ability to prepare high-quality, electron transparent thin foil specimens, which are representative of the

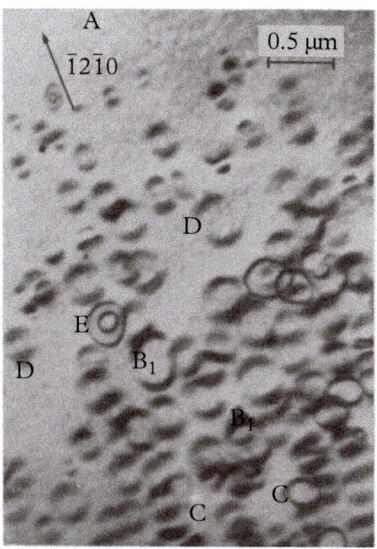

1.3.7 Complex damage, showing a variety of dislocation loops, produced by 400 keV electrons in zinc irradiated to 10^{-3} displacements per atom (d.p.a) at room temperature. The electron beam direction was [0001].

Adapted from [11].

[15] Intermediate Voltage Electron Microscopy (IVEM)—Tandem Facility, Argonne National Laboratory, USA.

sample microstructure. The same holds true for X-ray transmission microscopy (XTM). For atomic resolution scanning tunneling microscopy (STM, §11.3), an atomically flat specimen that is also conducting to establish a tunneling current, is required; however, scanning force microscopy (SFM), commonly known as atomic force microscopy (AFM) does not require a conducting surface, but both STM and AFM require solid surfaces that are somewhat rigid to avoid deformation during scanning. Ion scattering techniques (§5.4) require a solid specimen that is vacuum compatible.

1.4 Methods of Characterization: Spectroscopy, Diffraction, and Imaging

Three basic processes underpin the foundations of most characterization methods using probes and signals. The first, *spectroscopy*, is generally described by considering the probe radiation as a particle and involves absorption and emission processes. The second, *scattering* and/or *diffraction*, is described by considering the radiation as a wave, and monitoring its intensity in different directions. *Imaging*, a third way to characterize materials, largely follows from the first two by recording the signal in a spatially resolved fashion. Images can be formed by various contrast mechanisms arising from variations in mass, chemical composition, diffraction, and phase. In scanning probe techniques (§11), contrast is obtained by mapping various forces or registering a tunneling current between the specimen and a tip as it is moved across the surface.

1.4.1 Spectroscopy: Absorption, Emission, and Transition Processes

We start with a simple Bohr model of the atom (§2.2.1), represented by electrons orbiting around a nucleus of charge $+Ze$. In standard X-ray notation, the electron orbits are labeled as $K\,(n = 1)\,, L\,(n = 2)\,, M\,(n = 3)\ldots$, where n is the principal quantum number of the electron. Further details of the electronic structure of atoms (§2) and molecules and solids (§3), as they pertain to various materials characterization methods, are discussed later.

Now, consider a multi-electron atom (Fig. 1.4.1). When a primary electron (probe) with sufficient energy, E_P, to overcome the binding energy, E_B, of the inner-shell electron, is incident on the atom, it removes or ejects the *core electron*. In this process, the primary electron is scattered in some new direction with reduced energy, $E_{P'}$. Note that not all primary electrons have a close encounter with one of the core electrons to result in such an ejection. The probability of such a close encounter is given by the *cross-section* or the probability of the occurrence of such a collision and ejection. Alternatively, the average distance that the primary electron

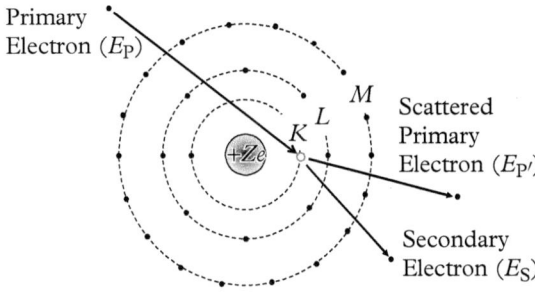

Figure 1.4.1 Interaction of a primary electron with the atom resulting in a core hole, a scattered primary electron, and a secondary electron.

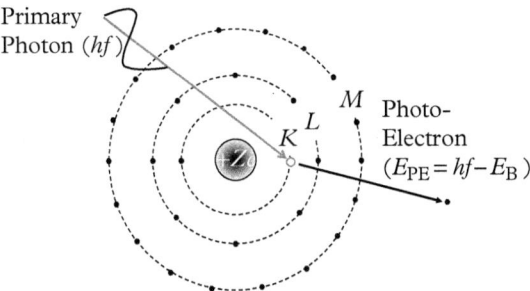

Figure 1.4.2 An incident photon causing an inner-shell ionization and the ejection of a photoelectron.

travels between such collisions in the material is called the *mean free path length* (see §5.3.5.1 for related definitions). The core electron that is removed is now free and is referred to as the *secondary electron*. Further, the primary electron may impart some kinetic energy, E_S, to the secondary electron. By conservation of energy we can easily see that the energy lost by the primary electron, $E_P - E_{P'} = E_B + E_S$, is sensitive to the binding energy of the core electron, and by measuring this loss of energy accurately, the electronic structure of the atom/material can be probed. This technique, EELS, can be implemented either to probe a surface or the bulk— the latter using a TEM (§9.4). It is also possible to map the distribution of such secondary electrons as the primary electron beam is rastered along the specimen surface, and this technique forms the basis of SEM (§10).

Alternatively, instead of the primary electron if a *photon* of sufficient energy, $E_P = hf$, is incident, the related process of photoionization can be realized (Fig. 1.4.2). The photon can be absorbed by the atom and a *photoelectron* with kinetic energy, $E_{PE} = hf - E_B$, can be emitted. As the K-shell electrons are bound with more energy than the L-shell electrons, for the same incident photon, they will emerge with lower kinetic energy. Atomic electron binding energies are tabulated in Table 2.2.2. However, even though, to first order, binding energies of core electrons are impervious to the nature of atomic bonding, they can be perturbed by the electronic states of the outer electrons; these include electronic bonding in molecules and the valence/conduction bands in solids (see §2.6.3, §3.7). Thus, studies of the energy distribution of photoelectrons for photon incidence, i.e. XPS, is a powerful technique to study chemical states in materials. Naturally, as the K-shell electrons are more tightly bound and shielded, compared to the L-shell electrons, they are less commonly probed and L-shell photoelectrons are preferred in photoemission studies (§2.6).

In both ionization processes, following the ejection of the core electron, the atom is left in an excited state with an inner-shell vacancy and can rearrange its electronic configuration to minimize its energy. This is accomplished by the transition of an outer-shell electron of higher energy, aided by the strong nuclear Coulombic attractive potential, to the vacancy created by the core hole. This can be accomplished by two competing processes.

1.4.1.1 Characteristic X-Ray Emission

In this process of atomic rearrangement (Fig. 1.4.3), the transition of the electron is accompanied by the fluorescent emission of an X-ray photon with characteristic energy equal to the difference in energy between the initial and final atomic levels. As expected, these characteristic X-ray emission energies (Table 2.3.1), are element specific and their detection by energy or wavelength dispersion form the basis of elemental analysis by energy (EDXS) or wave-length dispersive X-ray spectrometry (WDS) methods. The semiconductor detectors for energy dispersion are rather compact (§2.5.1.2) and are widely implemented in the narrow confines of both SEMs and TEMs. The alternative, wavelength dispersion analysis (§2.5.1.1) requires a rather bulky crystal detector, and is implemented in dedicated instruments called electron microprobe analyzers (§2.5.2.2). Commonly referred to as microprobes, they are popularly used by geologists to analyze mineral samples (Fig. 1.2.8b).

1.4.1.2 Non-Radiative Auger Emission

In this competing process (Fig. 1.4.4), the atomic rearrangement is accompanied by a non-radiative emission of a second Auger electron, again with energies characteristic of the atom (Table 2.3.2). Since three atomic levels are involved, the emitted Auger electron is labeled with three capital letters—the first represents the shell where the original vacancy is created, the second represents the shell from which the vacancy is filled, and the third represents the shell from which the Auger electron is ejected. Figure 1.4.4 shows two representative Auger electrons, *KLL* and *KLM*.

Which of these two competing processes is favored? Figure 1.4.5 shows X-ray fluorescence and Auger electron yields, as a function of atomic number. It can be clearly seen that, between fluorescence and non-radiative Auger emission, the former (X-rays) is preferred for high atomic number (Z) elements, but the latter (Auger electrons) is favored by low atomic number elements.

In molecules, additional transition between rotational, vibrational, and electronic levels can also be probed (Fig. 3.4.2). If a monochromatic radiation of frequency, f_I (usually from a laser, but in the first experiments a high-intensity mercury arc lamp was used), irradiates a specimen (Fig. 1.4.6) it can be both

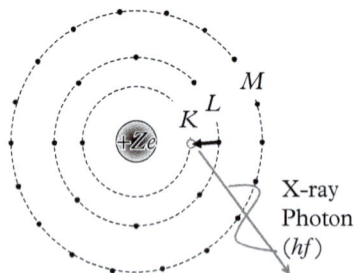

Figure 1.4.3 Characteristic X-ray fluorescence following the creation of a core hole. Note that the emitted photon would have the characteristics of a wave packet illustrated in Figure 1.3.5.

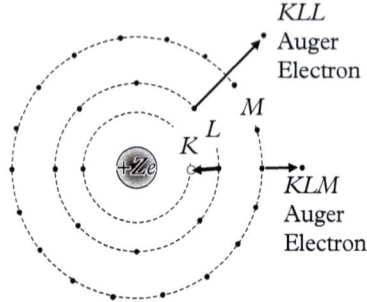

Figure 1.4.4 Atomic relaxation process by non-radiative Auger electron emission. Two alternative scenarios, to illustrate that the Auger electron need not be emitted from the same shell, are shown.

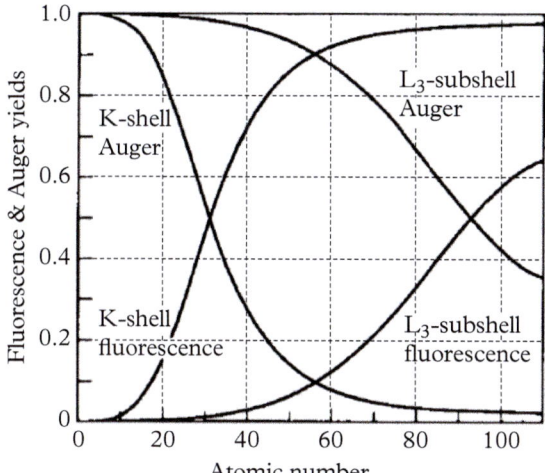

Figure 1.4.5 Auger electron and X-ray fluorescence yields, for K- and L-shell ionization, as a function of atomic number. It is clear that lower atomic number elements favor Auger emissions. See also Example 2.3.2.

Adapted from [13].

elastically (no loss of energy) or inelastically (change in energy) scattered. The scattered radiation has both intensity and polarization (§1.3.4) characteristics different from the incident radiation, and both of these attributes will also depend on the direction of observation.

If the frequency content of the scattered radiation is analyzed, it will show not only the original frequency, f_I (referred to as Rayleigh scattering, §3.4.2), but also pairs of additional frequencies, $f_S = f_I \pm f_M$ (referred to as Raman[16] scattering), where f_M is an internal frequency that depends on molecular, rotational, and electronic transitions (Fig. 3.4.2). In molecular crystals, the Raman spectrum is further dependent on the crystal symmetry, or point group (§4.3), and the intermolecular interactions, which can result in further splitting of the Raman band. Finally, understanding Raman scattering from crystalline solids requires detailed application of group theory and lattice phonon dynamics; however, in practice, a key challenge is to avoid damage (§5.7) of the specimen under irradiation by the intense laser beam. Such details are beyond the scope of this book (see Sherwood, 1972 for details); it suffices to say that for specific materials, such as diamond, the Raman modes are very good "finger prints" for the confirmation of the structure. Basic principles of Rayleigh scattering and Raman spectroscopy are introduced in §3.6.

1.4.2 Scattering and Diffraction

An electromagnetic radiation can be considered as a continuous wave or a discrete quantum of energy, called photons. Even the quanta are discrete bundles of waves

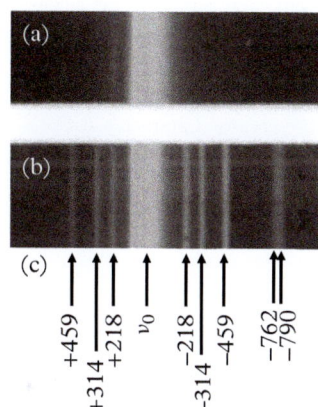

Figure 1.4.6 One of the first published Raman spectra of carbon tetrachloride liquid. (a) Spectrum of the mercury arc lamp in the region of ($\lambda = 435.9$ nm, or wavenumber $\upsilon_0 = 22938$ cm^{-1})[17] used as the incident probe. (b) Rayleigh and Raman spectrum from liquid CCl_4. (c) The principal lines from the spectrum of CCl_4 are indexed in wave numbers (cm^{-1}) with respect to $\upsilon_0 = 22938$ cm^{-1}

Adapted from [14].

---

[16] Sir C. V. Raman received the Nobel prize in Physics in 1930 and was cited for "his work on the scattering of light and for the discovery of the effect named after him."

[17] It is common in optical spectroscopy to specify the radiation in inverse wavelength or wavenumber, $\upsilon = 1/\lambda$

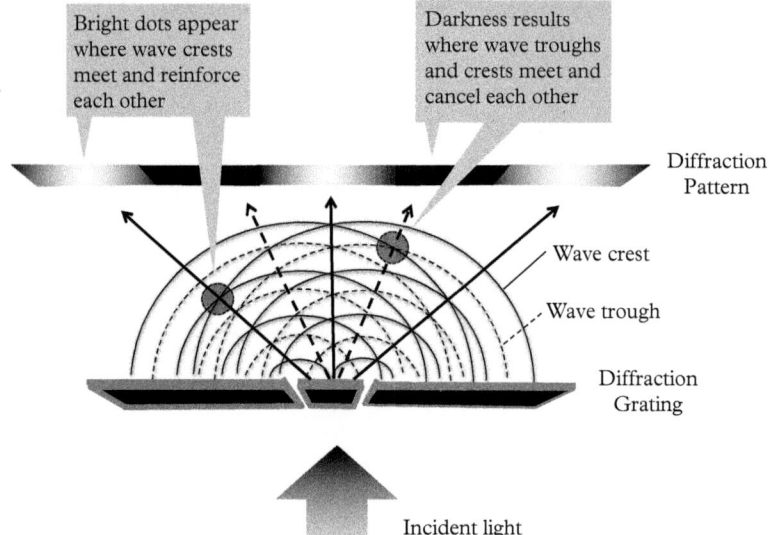

Figure 1.4.7 Interference from a diffraction grating consisting of two slits. Adapted from *https://www.nobelprize.org/prizes/chemistry/2011/press-release/*. For further details of diffraction gratings see §6.6. In the case of crystalline materials, the slits are replaced by a periodic arrangement of atoms.

(Fig. 1.3.5) and their wavelength determines their energy, and in the case of the visible spectrum, their color. When multiple quanta reach the same point in space, they either interfere constructively (become brighter) or destructively (become darker). Simply put, the results of the interference between two waves depend on their phase difference. In the extreme case, if the two waves of the same wavelength are completely in phase (phase difference $= m\pi$, where m is 0 or an even integer) they interfere constructively (addition of amplitudes). Alternatively, if the two waves are out of phase (phase difference $= m\pi$, where m is an odd integer), they interfere destructively (subtraction of amplitudes).

Such an interference effect is illustrated in the high-school physics experiment involving the passage of monochromatic light through two slits: a diffraction grating (Fig. 1.4.7). As the light waves travel through the grating, the waves from the two slits interfere and an alternating dark and light pattern on a screen positioned behind the grating is formed. This diffraction pattern arises, in the general sense, whenever a wave motion (incident radiation) encounters an ordered array of scatterers (such as slits in the grating), which then redirects the incident radiation into relatively well-defined directions. The only requirement for such diffraction to occur is that the wavelength of the incident radiation should be of the same magnitude as the distance between the scattering centers.

In crystalline materials, instead of slits, the gratings are made up of a periodic arrangement of atoms in three dimensions (see discussion at the end of §6.6.2). The repeat distance, or the period, d, in crystalline materials is of the order of 1 Å (0.1 nm), and so X-rays (Fig. 1.3.2) with similar wavelengths are ideally suited for such diffraction experiments (see Examples 1.4.1 and 1.4.2). Alternatively, electrons subject to an acceleration potential (high energy, $\sim$100 keV, for bulk studies in a transmission electron microscope, or low energy, $\sim$100 eV, for surface studies), or neutrons can also be used. The periodic array of atoms in the crystal form well-defined planes that coherently scatter the incident radiation along specific directions, given by Bragg's[18] law

$$\lambda = 2\,d\,\sin\,\theta \qquad (1.4.1)$$

where λ is the wavelength of the incident radiation, d is the interplanar spacing, and θ is measured from the reflecting planes (Fig. 1.4.8). The periodic arrangement of atoms in a 3D crystal describe many such planes with different interplanar spacings, and hence, diffraction peaks or positive interference occurs along several directions (Fig. 1.4.9a). In contrast, amorphous materials show a broad peak (Fig. 1.4.9b) and quasicrystal (§4.4) show unusual fivefold symmetry (Fig. 1.4.9c), a rotational symmetry inconsistent with lattice translations (see §4.1). Diffraction, particularly in reciprocal space is introduced in §4, and discussed in detail, starting in §7 for X-rays and continuing in §8 for electrons and neutrons.

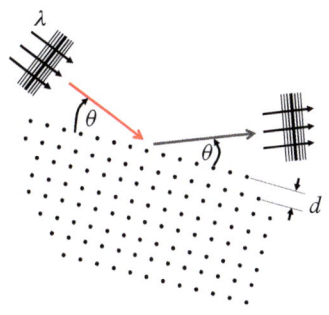

Figure 1.4.8 Diffraction from a periodic arrangement of scattering sites, such as atoms in crystalline materials.

Example 1.4.1: A XRD measurement for NaCl (Fig. 7.9.5) shows the most intense Bragg peak for $d_{hkl} = 2.82$ Å at $2\theta = 31.69°$. What is the wavelength of the X-ray radiation?

Solution: Applying (1.4.1), we get $\lambda = 2*2.82$ Sin (15.89) = 1.54 Å, which is the Cu Kα radiation, often used for XRD in the laboratory.

(a)

(c)

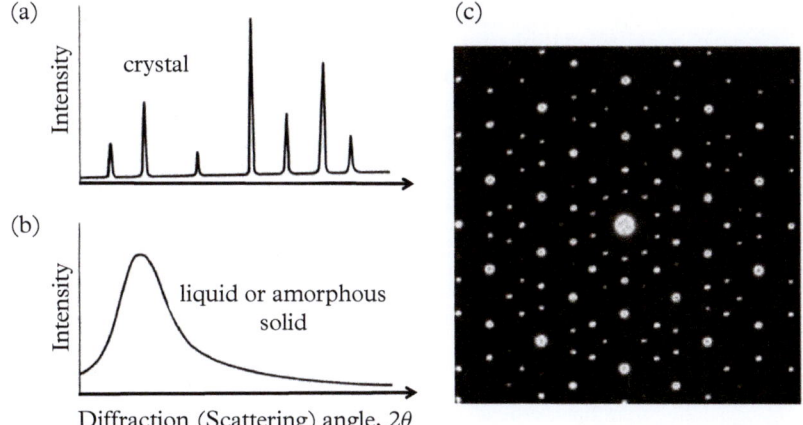

(b)

Diffraction (Scattering) angle, 2θ

Figure 1.4.9 X-ray diffraction pattern ($\theta - 2\theta$ scan) as a function of angle, 2θ, in (a) crystalline, and (b) amorphous materials. (c) An electron diffraction pattern from a quasicrystalline material, originally identified by Professor Schechtman.[19] Notice the unusual tenfold rotation symmetry—a rotational axis that is inconsistent with lattice translations. Such crystals are mathematically regular but do not repeat themselves periodically.

[18] Sir Lawrence Bragg, at age 25, was the youngest Nobel prize (1915) winner in Physics, and was cited for "services in the analysis of crystal structure by means of X-rays."
[19] Dan Shechtman won the Nobel prize in Chemistry in 2011, and was cited for "the discovery of quasicrystals."

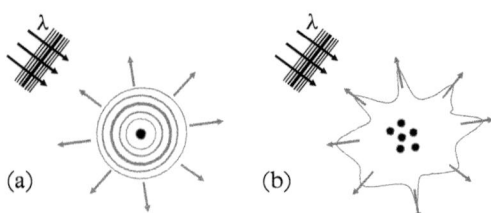

Figure 1.4.10 Scattering of incident radiation by (a) point scatterer isotropically in all directions and (b) partially ordered material scattering non-isotropically.

(a)   (b)

In addition to diffraction, incident radiation can also be redirected over a wide angular range by a point scatterer, such as an individual atom (Fig. 1.4.10a), and by a partially ordered material or a rough surface (Fig. 1.4.10b). The angular distribution of such scattering process is related to the spatial periodicities of the scattering object—obtained from the Fourier transform of its charge density distribution function. Thus, a point object scatters radiation evenly in all directions. In contrast to crystalline materials, in amorphous solids or liquids the atoms are tightly packed together with a preference, statistically speaking, for a specific interatomic spacing but with a general lack of periodicity. Figure 1.4.9 shows representative X-ray scattering from periodic crystals and amorphous solids.

It is worth mentioning here that if a collimated beam of monoenergetic incident particles (typically ^{4}He ions with MeV energy) are used to probe a solid, the ions are kinematically scattered (§5.3.5.3) following simple rules of conservation of energy and momentum both parallel and perpendicular to the direction of incidence. It can be easily shown that the energy of the ions after scattering is determined by the masses of the particle (^{4}He ions) and the target atoms in the material, as well as the scattering angle. For direct back-scattering, the energy ratio of the scattered to the incident ion has the lowest value and this geometry is often used in a technique called Rutherford back-scattering spectrometry (RBS, §5.4.1) to determine composition profiles, with depth, of a wide variety of materials. RBS is particularly popular in the analysis of semiconductor heterostructures and further details are discussed in §5.4.

Example 1.4.2: In a low-energy electron diffraction experiment, a peak is observed at $2\theta = 24.42°$ for a specimen with an interplanar spacing of 2.9 Å. What is the voltage used to accelerate the electron in this experiment?

Solution: Applying Bragg's law, (1.4.1), we solve for the wavelength of the incident electrons:

$$\lambda = 2\ d\ sin\theta = 2{*}2.9{*} \sin\left(12.21°\right) = 1.227 \text{ Å}$$

From the de Broglie relation, (1.3.3), we have

$p = m_e v = h/\lambda$. Thus, the velocity of the of the incident electron is

$$v = \frac{h}{\lambda m_e} = 5.93 \times 10^6 \text{m/s}$$

Its kinetic energy is

$$E = \frac{1}{2}m_e v^2 = 1.60 \times 10^{-17}\,\text{J}$$

Assuming that all this kinetic energy is due to the acceleration potential, V, we get

$V = E/e = 100$ V, where $e = 1.6 \times 10^{-19}$C is the charge of the electron.

1.4.3 Imaging and Microscopy

Complementing spectroscopy and diffraction/scattering, imaging is the third principal component of materials characterization. Optical microscopy/metallography (§6.8.5) was the first technique developed by Sorby[20] to reveal the microstructure of metallic surfaces. From these early observations of steels by optical methods [15], it became apparent that materials not only had structure, but more importantly, that the defects in steels could be related to their properties. This gave rise to the technique of metallography, including related surface preparation (Table 6.8.1) and laid the foundations of structure–property correlations in metallurgy and materials sciences.

To coordinate with the excellent characteristics of data collection and image formation of the human eye (§6.7), optical imaging is optimized for its sensitivity range ($\lambda = 380 - 700$ nm), peaking in the green ($\lambda = 560$ nm) portion of the visible spectrum. Hence, green filters are commonly used to focus optical microscopes and the viewing screens of TEMs are coated with a green phosphor. Now, the resolution, δ, of a microscope is given by the Rayleigh criterion (§6.8.1):

$$\delta = \frac{0.6\lambda}{n \sin \alpha} \tag{1.4.2}$$

where, λ is the wavelength of the radiation, n is the refractive index of the material, and 2α is the angle that a point source subtends at the lens (Fig. 1.4.11). For a given geometry, fixing 2α, the resolution can be improved by going to smaller wavelengths, λ. Figure 1.4.12 shows a typical example that illustrates the effect of wavelength on resolution. The optical micrograph (a) at 500×, of the Toluca iron meteorite is compared with an image (b) of the same meteorite surface taken with a SEM at 2000×, with the latter showing substantially greater details. Nevertheless, optical microscopy (§6.8) is a rapid and efficient technique that is the mainstay of materials characterization; further details, including the use of crossed polarizers for non-cubic materials, are discussed in §6.8.

Based on the Rayleigh criterion of resolution, (1.4.2), it is indeed tempting to consider using X-rays of shorter wavelengths than the visible spectrum to improve

[20] Henry Clifton Sorby, 1826–1908.

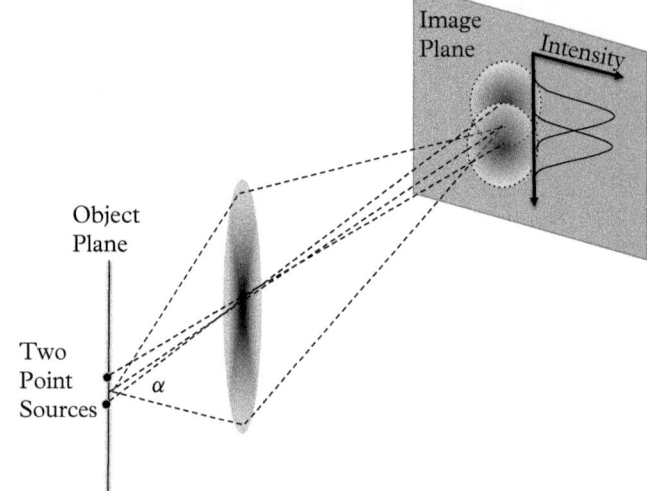

Figure 1.4.11 The Rayleigh criterion, defined by the separation in the image of two point-sources, such that the maxima of one coincides with the first minima of the other.

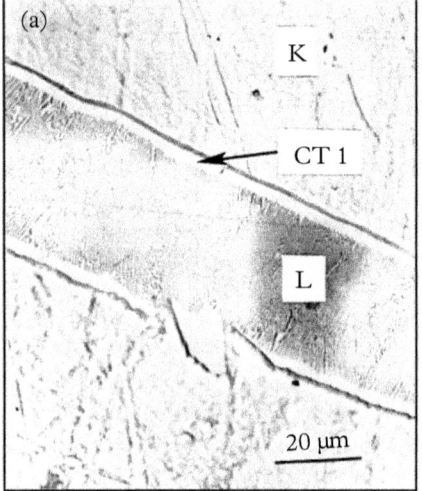 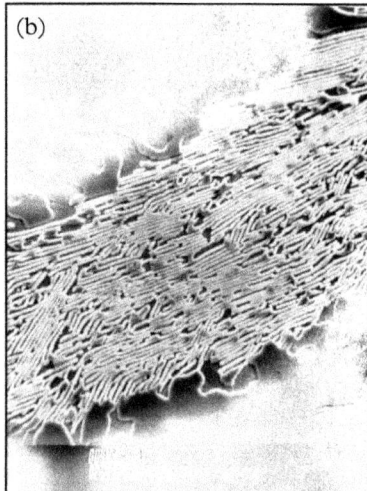

Figure 1.4.12 (a) The lamellar (L) structure in a high-Ni Taenite (T) phase of the Toluca iron meteorite observed using an optical microscope at 500x magnification. The grey region (CT1) surrounding the lamellar structure is a high-Ni ordered FeNi phase. (b) The same lamellar structure observed in an SEM at 2000×; now the details of the lamellar structure, composed of aligned K and T phases, are clearly resolved.

Adapted from Williams, Pelton, and Gronsky (1991).

resolution. However, optical lenses are based on the simple idea of refraction, i.e. the bending of radiation at the interface between materials of different refractive indices (§6.5). Unfortunately, for EUV and soft X-rays wavelengths, significant refraction cannot be accomplished within a single absorption length. As a result, real images using refraction (lenses) at X-ray wavelengths are not practical. Instead, diffraction techniques using Fresnel zone plate lenses (§6.6.7) are employed for high-resolution soft X-ray ($\lambda = 0.4-4.4$ nm) microscopy. There are two versions of such microscopes: (a) the *full-field* soft X-ray microscope (Fig. 6.6.12), which uses the zone plate lens after the specimen to form a complete image, point by point, much like a common optical microscope, and (b) the *scanning* X-ray microscope, which uses the zone plate to focus a spot of X-ray radiation on the specimen, which is then scanned and the radiation transmitted through the specimen is used to construct its image, pixel by pixel. In addition to recording the absorption as a function of position, it can also record element specific signals, such as emission or fluorescence, giving chemical information as a function of position. Alternatively, using circularly polarized light (Fig. 1.3.4) and X-ray magnetic circular dichroism (XMCD) for magnetic contrast, the magnetic microstructure, or the domain structure, can also be imaged (Fig. 1.4.13). Further discussion of various magnetic imaging methods can be found in Krishnan (2016). XTM, using synchrotron radiation, is introduced in §6.6.7.

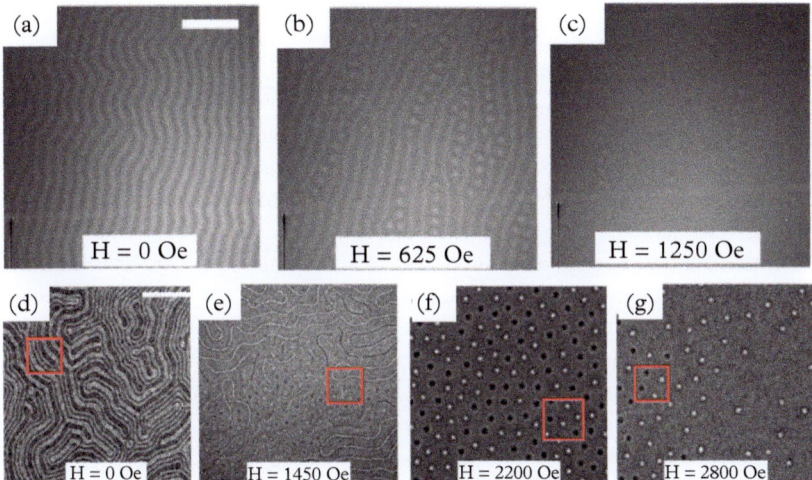

Figure 1.4.13 Magnetic images of $\{Fe_{0.34nm}Gd_{0.4nm}]_{80}$ multilayers under magnetic fields applied perpendicular to the film as indicated. (*a-c*) Images taken using a full-field soft X-ray transmission microscope (XTM), at the Fe_{L3} edge, with contrast sensitive to the out of plane magnetization. Starting with a stripe domain (a), the domains are pinched to form cylinders in the same space (b), followed by final dissipation (c) of the cylindrical domains. (d-f) Lorentz microscopy images (§9.3.6), recorded at room temperatures, in a FEI Titan microscope equipped with an aberration corrector. Transitions from stripe domains (d), to a magnetic skyrmion[21] lattice (f), and subsequently to disordered skyrmions (g) are observed. Lorentz microscopy is sensitive to the in-plane magnetic induction only.

Adapted from [16].

[21] A magnetic skyrmion is a quasiparticle that defines the smallest perturbation to a uniform magnetic field and is visualized as a point-like region of reversed magnetization surrounded by a whirling twist of spins.

Attempts to improve upon the resolution of optical microscopy by using shorter wavelength radiation, such as soft X-rays, have been successful but not easily accessible as they require synchrotron radiation sources (§2.4.3) to generate the X-ray radiation. For routine laboratory use, a more successful optical approach is to reduce the size and intensity of the light source to a sub-micron scale, i.e. using a laser, to generate image signals from individual microscopic spots on the specimen surface. Further, apertures are used to eliminate all light in the image from any plane in the specimen outside the plane of focus. In other words, this method improves resolution and contrast by employing spatial filtering methods to eliminate scattered or reflected light from planes that are out of focus. Two-dimensional images are formed by rastering the spot on the specimen surface; 3D images can be constructed by changing the image plane in the vertical direction. Such confocal scanning optical microscopes (CSOM), described in §6.8.5, have found much use in imaging biological materials (Sheppard and Shotton, 1997).

Following (1.4.2), electrons accelerated at higher voltages (∼100 kV), are a good alternative to achieving much smaller wavelengths, and thus superior resolutions. However, the electromagnetic lenses required for electron microscopy suffer from various lens aberrations (§9.2.2) that limit the collection angle, α, of electrons to 0.05–0.5° (10^{-2}–10^{-3} rad). In vacuum, the refractive index, $\mu = 1$, and the Rayleigh criterion for small angles α, gives a resolution $\delta = 1.2\lambda/\mu \sin\alpha = 1.2\lambda/\alpha$. Interatomic distances in solids are of the order of 0.1 nm (see §5.3.5.1 for a discussion of sizes and dimensions); thus, atomic resolution should be attainable for electrons microscopes at ∼100 kV. Further, taking into consideration the aberrations of electromagnetic lenses, and being mindful of the need to penetrate electron transparent thin foils representative of the bulk material being investigated, atomic resolution was typically achieved in microscopes operating only at higher voltages. However, recent developments in the design and availability of aberration corrected microscopes have made sub-Å resolution in transmission electron microscopy routinely achievable. Figure 1.4.14 provides a very convincing case of the advancement made in image resolution, with successive generations of high-resolution TEMs, using an example of the structural determination of β-SiAlON.

From the first demonstration of a transmission electron microscope in 1933 by Ernst Ruska,[22] electron microscopy and diffraction have undergone spectacular development over the years. Further details of this versatile technique, where a single instrument incorporating multiple detectors can provide comprehensive information on the physical, chemical, and magnetic microstructure (Fig. 1.4.13) of the specimen, all at unmatched spatial resolution (Fig. 9.1.2) through diffraction, imaging, and spectroscopy, are discussed in §8 and §9.

Complementing TEM, the SEM (§10), in which the resolution depends on the size of a finely focused probe incident on the specimen, is a versatile instrument

[22] Ernst Ruska (1906–1988) was a German physicist who received a much-belated Nobel prize in Physics in 1982; he was cited for "his fundamental work in electron optics, and for the design of the first electron microscope."

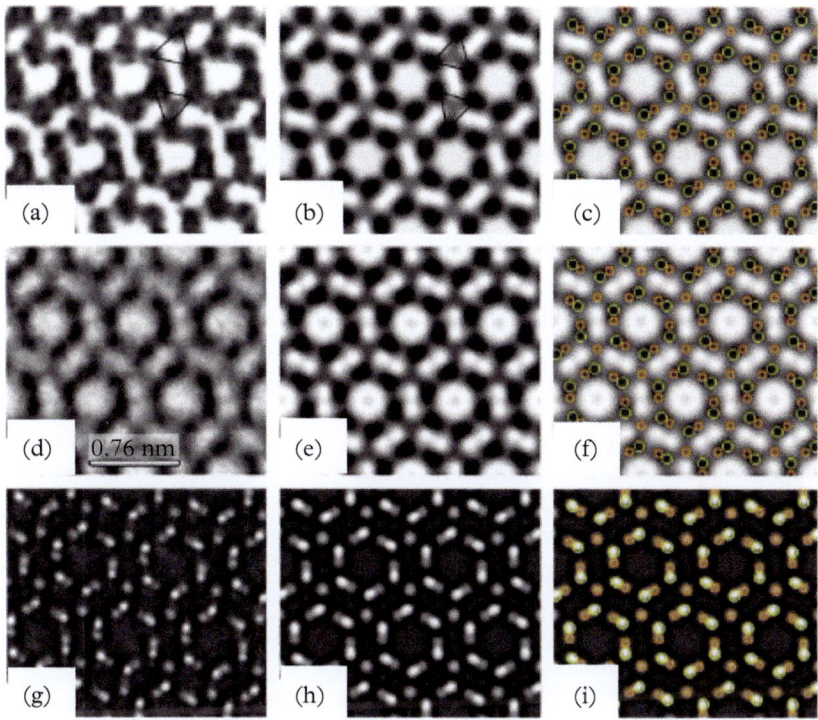

Figure 1.4.14 High resolution atomic structure images of a ceramic material, β′-SiAlON, taken with three generations of transmission electron microscopes at the National Center for Electron Microscopy, Berkeley. Top: The Atomic Resolution Microscope (ARM) designed to achieve atomic resolution using high voltage (1 MeV, λ = 0.00087 nm) and a spherical aberration coefficient, C_s = 2 mm. Middle: The next generation One Ångstrom Microscope (OAM), uses a 300 kV, Schottky emission gun source and a much smaller C_s = 0.6 mm. Bottom: The most recent Transmission Electron Aberration-corrected Microscope (TEAM) uses hardware corrections of the aberrations allowing C_s to be tuned at will, and designed for a resolution limit of 0.05 nm. For each microscope, the left column shows the image recorded at Scherzer defocus (§9.2.7.4), with the specimen viewed along the (0001) zone axis of this hexagonal material. The middle column shows a simulation of the structure and the right column shows an overlay of the simulated image on the atomic structure. While all three images (a, d, g) show sixfold symmetry, only the TEAM image shows a direct relationship to the atomic structure. TEM is described in detail in §9.

Adapted from [17].

providing various modes for imaging using secondary or back-scattered electrons. It provides topographic (Fig. 1.4.12b), compositional (Fig. 1.4.15), voltage, and magnetic contrast as well as crystallographic information through channeling patterns (§10.4).

Another elegant way to obtain microstructural information at atomic resolution (Fig. 1.4.16) is to apply a large electric field to a specimen of the material, shaped in the form of a needle, such that the potential barrier, V, for the electrons to leave the specimen can be locally overcome at the tip. Such field emission (§5.2.2) was used to image local variations in the work function. Alternatively, by admitting a small quantity of gas into the chamber and reversing the voltage on the needle (in the first experiments, the material happened to be tungsten) the gas close to the needle surface can be ionized. As the electric field is increased, these ionized gas atoms are repelled from the tip surface towards a fluorescent screen, producing a magnified image of the needle specimen (Fig. 1.4.16) that can clearly delineate the individual atoms. This technique, called field ion microscopy (FIM) [18, 19], has made significant contributions to the study of the structure of interfaces and grain boundaries on the atomic scale [20], but will not be discussed further in this book. A serious limitation of the FIM has been its inability to identify the chemical

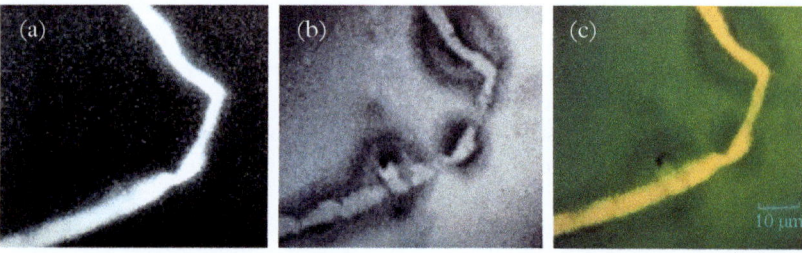

Figure 1.4.15 SEM compositional maps, using characteristic X-rays, of a Cu–Zn specimen showing diffusion-induced grain boundary migration. (a) Zn-map, with Zn in white, (b) Cu-map, with 90% Cu being black, and (c) a pseudo-color superposition image with Cu in green and Zn in yellow.

Adapted from Gronsky, Pelton, and Williams (1991).

Figure 1.4.16 (a) Schematic of an atom probe field ion microscope (AP-FIM) and (b) FIM image of a Tungsten tip. Note that individual atoms are resolved. FIM was invented by E. R. Müller.

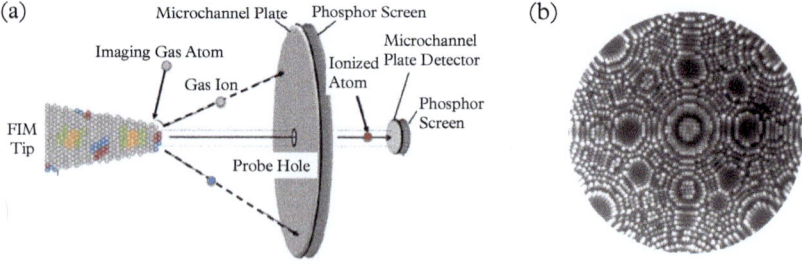

nature of the individually imaged atoms. To overcome this limitation, Müller[23] conceived and built the atom-probe FIM (Fig. 1.4.16a), which is a combination of a probe-hole FIM with a mass spectrometer (§5.4.3) having single particle sensitivity [21]. During observation, the observer selects an atomic site of interest by placing it over a probe hole in the image screen. Pulsed field evaporation sends the chosen particle through the hole and into the spectrometer section. The AP-FIM (see Miller et al., 1996 for a detailed discussion) has undergone numerous major improvements, including the imaging atom probe [22] and the position-sensitive atom probe [23]. The most common geometry currently used is a local electrode atom probe (LEAP) [24], which first appeared commercially in 2002. The current status of atom probe tomography is discussed in [19].

If such a "needle" reverses its role, i.e. it is used as a probe, rather than as the specimen, and is mounted on a flexible cantilever, which is then brought very close—within a few atomic distances—to the specimen surface, a tunneling current can be established between the specimen and the tip. If the tip is now scanned, the specimen surface in vacuum can be imaged at atomic resolution, either by monitoring the tunneling current at constant height, or by monitoring the height at constant tunneling current (§11.2). Such an instrument is called an STM and was invented by Binnig and Rohrer.[24] Moreover, the sharp tip can serve as a fine probe to measure a variety of physical properties on the local scale by using a variety of spectroscopic methods. Note that the STM resolves individual atoms on conducting surfaces (Fig. 1.4.17). For resolving individual atoms on insulating surfaces the alternative atomic force microscope (AFM) was introduced [26]. Here, instead of the tunneling current, atomic forces between the needle and the specimen surface, as the tip is scanned, can also be measured

Figure 1.4.17 Large-scale, atomic resolution, topographical STM image of the Si(100)2 × 1 surface showing two kinds of steps and various point defects. This is technologically most important as integrated circuits are made on this silicon surface and understanding the local structure on the atomic scale is important for processing and manufacturing of semiconductor microprocessors.

Adapted from [25].

---

[23] E. R. Müller (1911–1977), German–American physicist.
[24] G. Binning and H. Rohrer shared the 1986 Nobel prize in Physics with E. Ruska, and they were cited for "their design of the scanning tunneling microscope."

(§11.4). The forces between the tip and surface can have various origins, such as electrostatic, van der Waals, magnetic, etc., giving rise to various variants of these AFMs. These scanning probe instruments and their applications in materials analysis (Weisendanger, 1994; Meyer, Hug, and Bennewitz, 2003) are discussed in §11.

1.4.4 Digital Imaging

As electronic acquisition and analysis of images is now routinely used in materials characterization, it is important to have at least a basic knowledge of digital image acquisition, storage, processing, and analysis methods. While there are excellent textbooks (Gonzales and Woods, 2018; Russ, 1992) and review articles [27–29] that an interested reader can consult for a detailed treatment of the subject, a very brief introduction is included here for completeness.

The sequence of procedures followed in digital imaging *is acquisition* (Fig. 1.4.18a), followed by *processing, analysis,* and *output* (Fig. 1.4.18b). Typically, an image acquired from a materials characterization source, such as an optical, electron, or scanning probe microscope, is digitized and then stored. By digitization we mean the generation of a digital image, $a[m,n]$, in a 2D *discrete* space that is derived from an analog image, $a(x,y)$, in a 2D *continuous* space through a *sampling* process. The analog image is divided into N rows and M columns and the intersection of a specific row and column, with integer coordinates $[n,m]$, is called a picture element or *pixel*. The value, $a[m,n]$, assigned to each pixel in the digital image is the average brightness in the pixel, rounded to the nearest integer value, with L different gray levels in a process called amplitude

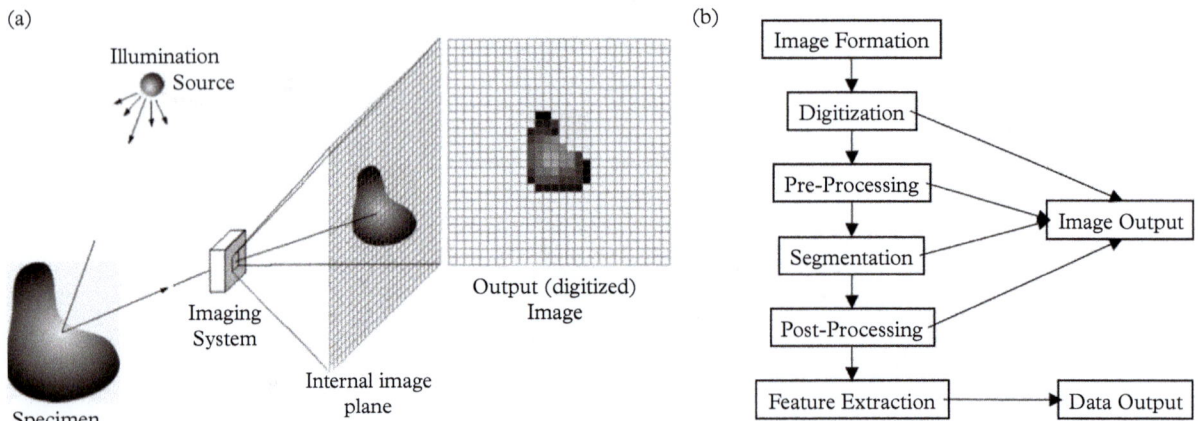

Figure 1.4.18 (a) Schematic illustration of the digital image acquisition process, which include illumination, the specimen, the imaging system and the digitized image. (b) Steps in digital imaging, shown as a flowchart.

(a) Adapted from Gonzalez and Woods (2002). (b) Adapted from [27].

quantization. Usually, $L = 2^B$, where B is the number of bits used to represent the brightness levels in the image; if $B = 1$, we get a binary (black and white) image, and if $B > 1$, we get a gray-scale image.

While the acquired image can itself be the output, it often suffers from defects such as the inclusion of electronic noise, uneven illumination over the field of view, and/or specimen drift. These defects are corrected in the *pre-processing* step. Further analysis of the image for quantitative interpretation requires a sophisticated and complex *segmentation* step, where features of interest in the image are identified and discriminated from the background. Often, the segmentation result requires some *post-processing* correction and then the image, containing the distribution of the desired object(s), is analyzed in the *feature extraction* step to obtain a number of quantitative parameters, including size, position, texture, and (when possible and necessary) appropriate statistical analysis. Here, we introduce only the first three important steps of general interest in digital imaging; the final two steps of post-processing and feature extraction are specific to the particular imaging methods and the characterization problem at hand, and as such are best left for more detailed discussions, readily available in specific textbooks cited earlier.

1.4.4.1 *Image Acquisition, Digitization, and Storage*

A typical SEM image of a cleaved silicon surface (Fig. 1.4.19) can be used to illustrate the effect of sampling in creating the digital image. The rate at which the signal intensity changes in space is called the spatial frequency; typical features in the image such as edges and small particles represent high spatial frequencies and are best sampled at shorter intervals to improve the quality of the digital image. The sampling frequency in the spatial axis is also known as the *resolution*; in the intensity axis it is known as *quantization*. Figure 1.4.20 shows the effects of resolution and quantization on the quality of the digitized image. While it is obvious that the best-quality image is obtained for largest values of resolution and quantization, this choice often leads to very large data files, which reduces its attractiveness.

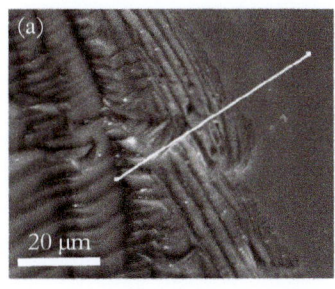
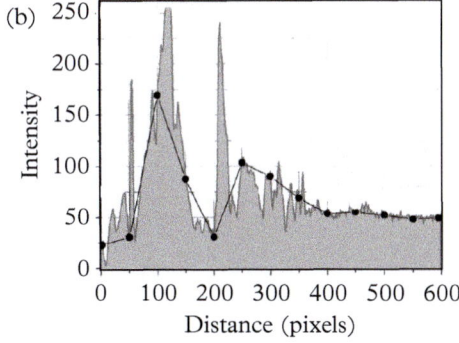

Figure 1.4.19 (a) A scanning electron micrograph of a cleaved silicon wafer, and (b) the intensity trace along the white line shown in the image (a). The image is typically sampled at a specific number of points (black dots) and the digital approximation is shown by the black line.

Adapted from [27].

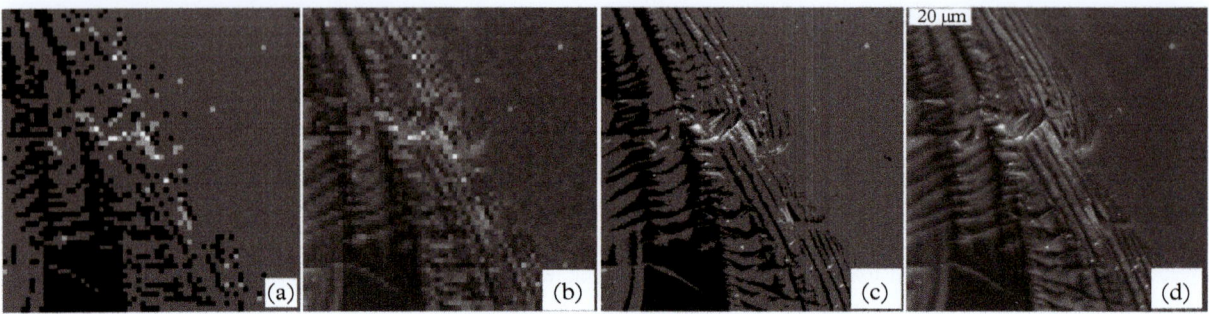

Figure 1.4.20 The effect of resolution and quantization on the digital image. The image from Figure 1.4.19, now shown with (a) 64 × 64 pixels and four gray levels, (b) 64 × 64 pixels and 256 gray levels, (c) 512 × 512 pixels and four gray levels, and (d) 512 × 512 pixels and 256 gray levels.

Adapted from [27].

The image file size, *IFS*, is given by

$$IFS = N_x N_y B_{pp} \qquad (1.4.3)$$

where N_x and N_y are the number of pixels in the two orthogonal axes, and B_{pp}, which depends on the quantization, is the number of bytes occupied by each pixel. By definition, a byte is composed of 8 binary digits (bits) and a full byte represents 256 values. For examples, a typical gray-scale image with 256 levels corresponds to a digital value ranging from 0 (black) to 255 (white), will require 1 byte/pixel. Such digital images or arrays of numbers can be stored in different file formats, with different compression protocols developed to reduce the file size. TIFF (Tagged Image File Format) is the most flexible format that can be easily exchanged between different computer platforms, and which accepts different kinds of lossless compression methods. It is also the format of choice for many characterization applications. Alternatively, JPEG (Joint Photograph Expert Group) is another standard and commonly used file format that presents images of good visual quality, but often with lossy compression (Fig. 1.4.21). It allows for several levels of data compression, and the higher the level, the smaller the file (more efficient storage) and the greater the loss of information. For scientific and technical purposes, it is important to retain the precision and details of the acquired image and therefore it is best to avoid the JPEG format since the compressions compromise the data.

1.4.4.2   *Pre-Processing: Look-Up Tables, Histogram Equalization, Point, and Kernel Operations*

An image displayed on a computer monitor need not be a direct mapping of the original image. Generally, a look-up table (LUT) is used to map the image

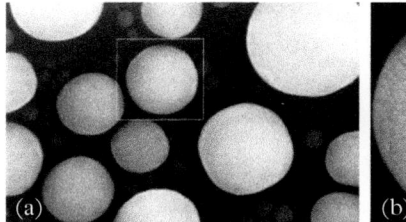 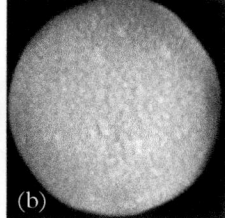 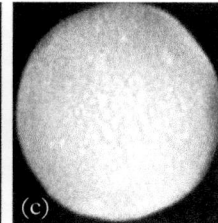

Figure 1.4.21 Storing SEM images of micrometer-size tin spheres. (a) A .TIF file requiring 306 kB of storage. The same file enlarged by a factor of four (4×) using the original image stored as (b) .TIF, and (c) .JPEG requiring only 42 kB of storage. Note that the .JPEG image is degraded and pixelated compared to the .TIF image.

Adapted from Goldstein et al. (2003).

intensity values to the brightness values in the display (Fig. 1.4.22a). If the LUT is linear with unit slope and zero intercept the image is directly mapped to the display. The most common example of a LUT operation is the one used to control the brightness (Fig. 1.4.22b) and contrast (Fig. 1.4.22c) in the displayed image. The functional form of the linear LUT is $g(u) = u + b$ with image values $u \in [0, 1]$. When b > 0 (< 0), we get a brightening (darkening) of the image. In addition, the contrast can also be manipulated using a LUT of the form $g(u) = mu + b$, where $b = (1 - m)/2$. Then the contrast is enhanced with a large slope, $m (\to \infty)$, and negative intercept, $b (\to -\infty)$; alternatively, the contrast is reduced with a small slope, $m (\to 0)$ and positive intercept, $b \to 1/2$. Note that the image is inverted when $m = -1$, $b = 1$. Finally, a nonlinear LUT of the form $g(u) = u^{\gamma}$, also known as the gamma correction, is often used (Fig. 1.4.22d). Now, when $\gamma < 1$, the function approximates a logarithmic LUT with contrast expansion in the dark region, and when $\gamma > 1$, it approximates an exponential LUT with contrast reduction in the dark region.

In practice, the intensity values of a typical image, plotted as a histogram, are often clustered around the mid-gray value and fall off on either side (Fig. 1.4.21e, top). Such a clustered histogram indicates that the contrast of the image is not maximized. This image is normally transformed so that the distribution of intensity values is such that all intensity values are equally represented in the image. This is known as *histogram equalization* (Fig. 1.4.21e, bottom), and helps optimize image contrast. Modern image acquisition software includes a live window that includes the image and its histogram, and it can be optimally adjusted by changing the illumination and/or exposure time. Needless to say, it can also be adjusted off-line, post image acquisition.

In general, the LUT operations discussed so far are classified as *point operations*, where the intensity, $I_O[m,n]$, of the output image at any pixel with coordinates

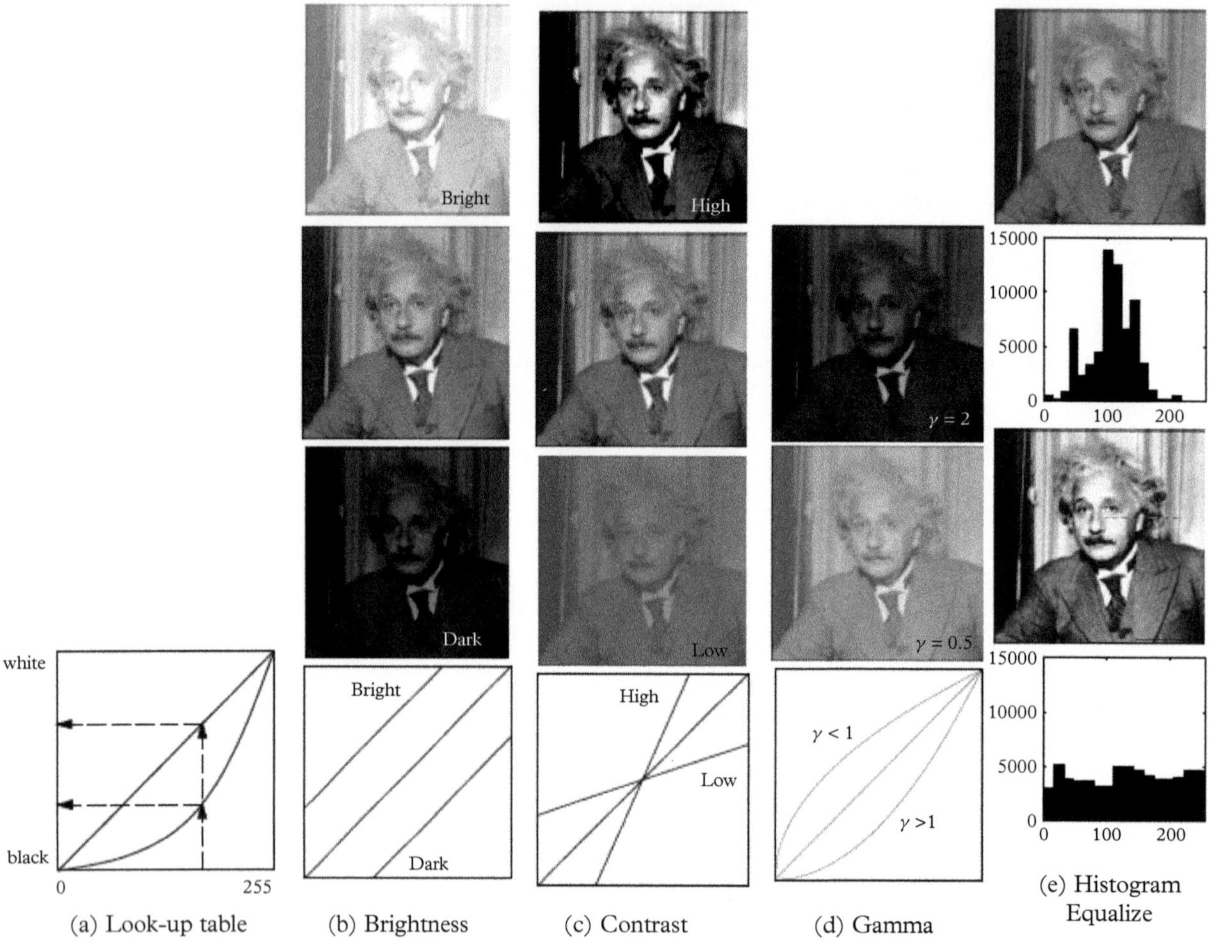

Figure 1.4.22 (a) A look-up table maps the image intensities to the display intensities. A linear LUT can be used to manipulate (b) brightness and (c) contrast in the image. (d) A nonlinear LUT, such as the gamma correction can be used to increase ($\gamma > 1$) or decrease ($\gamma < 1$) contrast. (e) A histogram of the number of pixels in the image at every pixel intensity (0 to 255) is typically clustered (top) around mid-value. By equalizing or stretching the histogram over the entire intensity range, typically improves the contrast in the image.

Adapted from Farid (2010).

$[m,n]$ is dependent only on the input image, $I_I[m,n]$, at the same coordinates. i.e. the two are related by the function, F, such that

$$I_O[m, n] = F(I_I[m, n]) \qquad (1.4.4)$$

Such point operations readily lend themselves to algebraic and logic operations on two or more images, in such a way that

$$I_O[m, n] = F(I_1[m, n], I_2[m, n]) \qquad (1.4.5)$$

where the function, F, relates the intensities at the same pixel position $[m,n]$ in the two input images. The simplest algebraic functions are addition, subtraction, multiplication, and division.

Addition is often used to improve the signal-to-noise ratio (SNR): several fields of the same image are acquired and added together to give an improvement of SNR by the square root of the number of fields. Alternatively, subtraction is used to eliminate contributions from noise or uneven background as may be expected for poor or uneven illumination in an optical microscope (Fig. 1.4.23). Multiplication is used to change pixel intensities using another image as a mask, and division is sometimes used to normalize the response of a CCD camera.

Neighborhood operations differ from point operations, in that their output intensity at any pixel position depends not only on the input intensity at the same pixel position but also on the intensities of its neighbors. Also known as *kernel operations*, they are critically important for applications, which include noise filtering, background subtraction, and edge detection. The example, Figure 1.4.24(i), uses a neighborhood—which is always odd-sided and whose weights are represented as a matrix of 3×3 pixels. The intensity of each pixel is multiplied by a certain weight (kernel), the results are summed together and divided by the total weight for all pixels, and the result is then plotted as the output for the neighborhood center.

Kernels are of two types: (a) those with only positive values, called *low-pass filters*, because they reduce the high-frequency component of the images, and result in reducing *noise* and *blur* in the image, and (b) those with mixed positive and negative values, called *high-pass filters*, because they increase high spatial frequencies in the image, and result in sharpening the image but at the same time increasing the noise. These two filters are also illustrated in Figure 1.4.24(ii), and their effects on an image, i.e. the blurring effect of a low-pass filter, and

Figure 1.4.23 Background subtraction. (a) An optical micrograph of hypereutectic cast iron at 200× BF, 1300 ×1030 pixels that is unevenly illuminated. (b) The estimated background image intensity. (c) The background subtracted image. Adapted from [27].

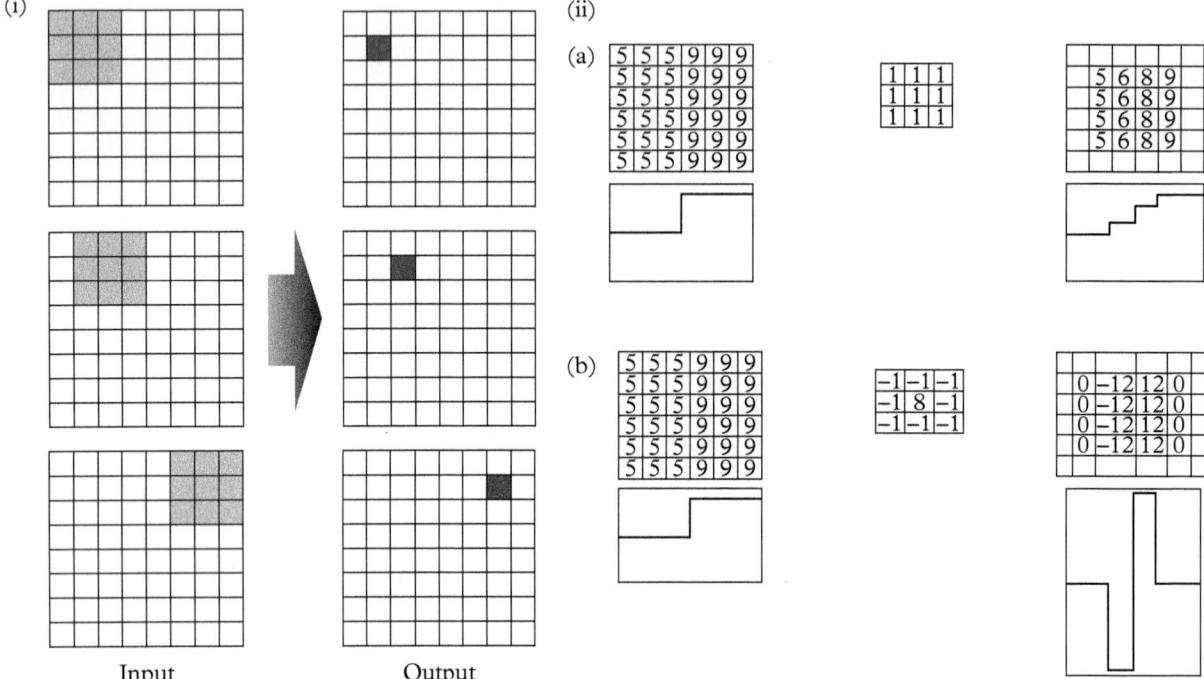

Figure 1.4.24 (i) Sequence used in a neighborhood operation, with the neighborhood analyzed on the left and the calculated output pixel on the right. The process starts at the top and is carried out, column by column, across a line and then line by line, until the entire image is processed. (ii) Application of low-pass and high-pass filter to a simple image. (a) Original image and line profile (left), the low-pass kernel (center), and the output image and line profile (right). (b) Original image and line profile (left), the high-pass kernel (center) and the output image and line profile (right).

Adapted from [27].

the sharpening (edge enhancement) of the high-pass filter, are illustrated in Figure 1.4.25.

Now, consider the filters (kernels) shown in Figure 1.4.26a, which are directional filters that have different effects in different image directions. As it turns out, these kernels are low-pass filters in one direction, and high-pass filters in the orthogonal direction. Hence, they enhance edge directionality, and approximate the derivative of the image in x- and y-directions. In fact, they are the basis of the *Sobel edge detector*, which is often used in segmentation, and corresponds to the intensity gradient in the image. It is given by

$$\text{Sobel}\left[I\left(x,y\right)\right] = \sqrt{\left(\frac{\partial I\left(x,y\right)}{\partial x}\right)^2 + \left(\frac{\partial I\left(x,y\right)}{\partial y}\right)^2} \tag{1.4.6}$$

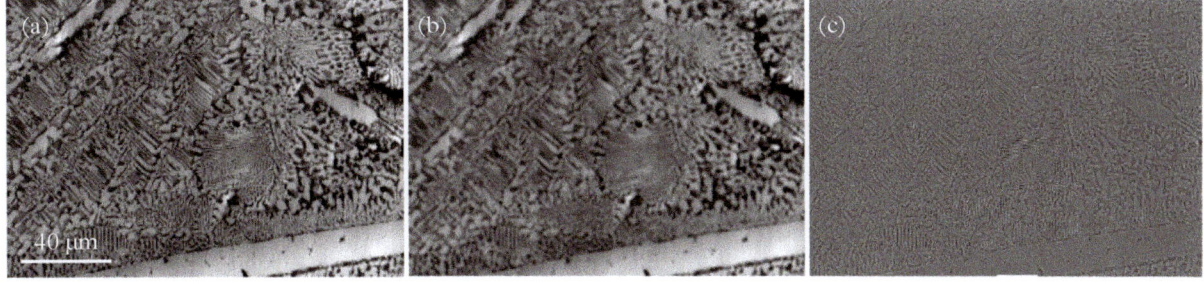

Figure 1.4.25 The effect of low-pass and high-pass filters. (a) The original image. The effect of (b) a 9 × 9 kernel low-pass filter, and (c) a 3 × 3 kernel high-pass filter.

Adapted from [27].

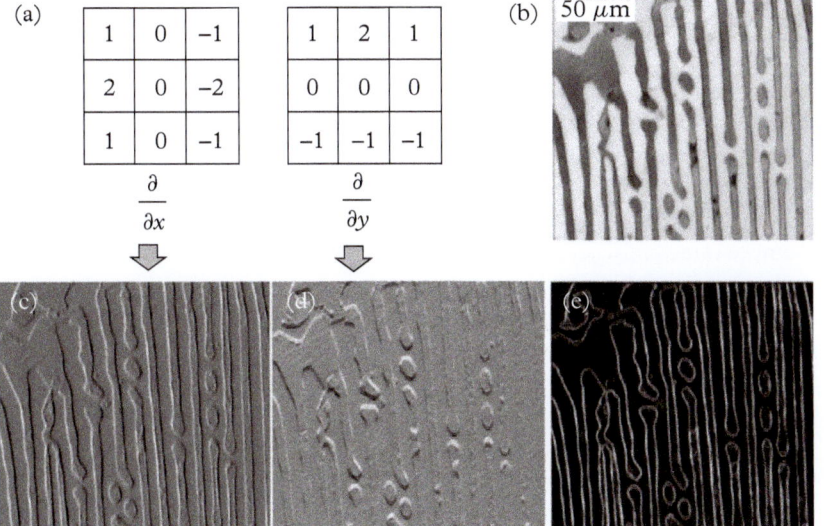

Figure 1.4.26 (a) The kernels for partial *x*- and *y*-derivatives. (b) The original image. (c) *x*-partial derivative image. (d) *y*-partial derivative image, and (e) Magnitude of the image after the Sobel operation.

Adapted from [27].

where the partial derivatives correspond to the two kernels in Figure 1.4.26a. The effect of these operations on an image is shown in Figure 1.4.26b–e; in particular, the Sobel operator enhances the edges while the uniform regions become black, and can be used in the segmentation step to discriminate objects in an image.

1.4.4.3   Image Segmentation

This is the most complex step in the digital imaging flow-chart illustrated in Figure 1.4.18b, because it requires the computer to perform cognitive functions similar to the human brain. The latter performs a rapid processing of numerous inputs—specific shape, texture, brightness, boundaries—in association with previous experience, but unfortunately, computers do not have such associative

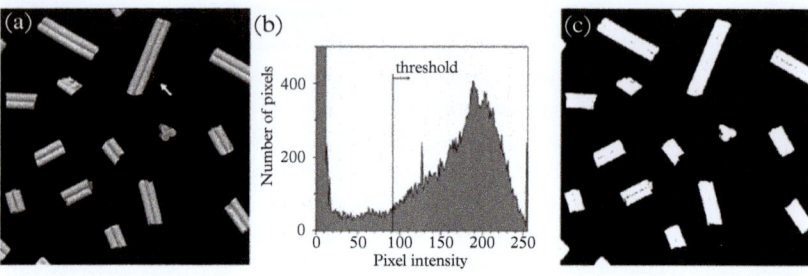

Figure 1.4.27 (a) An image of catalyst particles. (b) The histogram of intensities and a highlighted threshold level. (c) The segmentation obtained from the thresholding operation.

Adapted from [27].

power, and the classification of an object is done by classifying each pixel of the image as belonging to the object, or not.

The simplest way to segment is to work with the intensities of the pixels; it is considered to be part of the object if its brightness is above a certain level. In the simplest process of intensity thresholding, the choice of the threshold, T, is based on the image histogram, i.e. if $I(x,y) > T$, then the pixel at (x,y) belongs to the object class (Fig. 1.4.27). Alternatively, segmentation can also be based on the contours, and the simplest method is to apply the Sobel edge detector. If such edges form closed boundaries, they can be clearly identified with the object, but if the boundaries or contours are incomplete, it limits the ability to detect the objects of interest. Several other contour-based methods, such as Marr-Hildreth [30] and Canny [31] methods, have been developed to overcome these limitations; however, further details are beyond the introductory nature of this discussion.

The practice of image processing cannot be defined in general terms, as so much of it relates to the actual tasks in mind. For instance, one can include in image processing the task of modeling the background and then subtracting it, which is routinely done in EELS spectroscopy (§9.4.2.3). Examples include the coloring of an image using a well-defined scheme (Fig. 6.8.12), or according to the atomic species in quantitative chemical mapping using EDXS (Fig. 1.4.15). On the other hand, techniques such as particle measurements, closing, erosion, area, moment, counting, and so on are incredibly important at lower resolution, but rarely (if ever) used in high-resolution TEM (§9.3.5). Then, there are the techniques of filters and convolutions, both in real and reciprocal space, which have been introduced, but again, in practice, must be tailored to the actual "question" in mind. Finally, image processing has been covered in all its detail in specialized textbooks (Gonzales and Woods, 2002), but here our interest is limited to providing basic information of general use in characterization.

1.4.5 *In Situ* Methods Across Spatial and Temporal Scales

The emphasis of this book is to describe principles and applications of characterization methods that are broadly applied to studying static and motionless states of materials. These studies include structural and analytical methods that have provided incredible details of the structure, form, and function of matter, nature, and life. However, matter reacts, undergoes transformations, and routinely exhibits changes in its thermodynamic state. These changes in equilibrium are often unstable and transient in nature. As a result, there is substantial interest in "seeing" how the building blocks of nature, if possible at the atomic length scale, react to external forces, e.g. changes in the thermodynamic state of the material. In other words, a fourth dimension, *time*, now becomes important in characterizing the dynamic behavior of materials. Even though time-resolved studies are not new, and date back to the study of galloping horse in the nineteenth century [36], recent developments permit *in situ* studies at the atomic-length scale and even at femtosecond time scales. In practice, there are many hurdles to carry out these dynamic studies even at slower time scales, much of which revolves around the nature of the specimen, their preparation, and the appropriate specimen environment. We provide very brief introductions to these dynamic methods of materials characterization in this book for X-ray (§7.11.5), neutron (§8.8.3), and electron (§9.5.5) probes, but for those interested in further details, the monograph edited by Ziegler et al. (2014) gives a good overview of time-resolved and *in situ* methods, providing a snapshot of these rapidly developing techniques with particular emphasis on parameters that limit their application.

1.5 Features of Materials Used for Characterization

In general, two principal features of materials, i.e. their electronic structure, including their atomic mass, and crystallography, are used for characterization and are highlighted throughout this text.

For spectroscopy, we probe different aspects of the electronic structure of the material; this can be done starting with the energy levels of its constituent atoms (§2). The core levels, or the inner-shell electron energy levels, are barely perturbed by atomic bonding, and continue to provide valuable chemical information, even in bulk form, for most materials. Techniques that are based on probing such energy levels and related transitions, including the design of simple detectors for measuring X-rays and electrons, are described first (§2). The electronic structure of materials, especially the outer levels, change when the atoms bind into molecules and solids. In addition, molecules exhibit vibrational and rotational modes, whereas crystalline solids develop a bulk band structure, conveniently described by an itinerant or delocalized electron model (Sutton, 1993). The solid also develops acoustic or phonon modes of excitation. Following a conceptual introduction to the relevant electronic structure of molecules and solids, various techniques that are based on probing their rich electronic structures are introduced in §3.

Whenever electromagnetic radiation interacts with matter, some form of scattering always takes place. In particular, the type of scattering that we consider can be simply thought of as arising from the absorption of the incident radiation followed by its re-emission. This model lends itself to both elastic (Rayleigh) and inelastic (Stokes and anti-Stokes) scattering for molecules and atomic arrangements in solids (§3).

An electromagnetic wave, such as an X-ray, can be described by a propagating electric field, which varies sinusoidally with time (Fig. 1.3.4). At any given location, when such an oscillating electric field encounters a charged particle, e.g. an electron in an atom, it will set it also in oscillatory motion about its mean position. An oscillating or accelerating electric charge will emit an electromagnetic radiation. Thus, the atomic electron will not only absorb the incoming radiation but will also re-emit it with the same frequency (f) and wavelength (λ). This process is called scattering, and in X-ray diffraction it is characterized by the re-emitted radiation being coherent with the incident wave. By coherent, we mean a constant phase difference, ϕ, between the incident and scattered radiation, such that in the case of X-rays, $\phi = \lambda/2$. Even if the incident radiation is a parallel beam, the scattered radiation is re-emitted in all directions, with intensities dependent on the angle of scattering (we discuss such Thomson scattering in §7.2.1). Note that in this simple model, the contributions from the positively charged nucleus can be ignored because its mass is much larger ($\sim 1{,}000$ times) than that of the electron, and as such, it cannot be oscillated to any significant degree by the incident wave. The collection of all electrons in an atom will thus scatter the incident wave in all directions (Fig. 1.4.10). However, for a crystalline material, where atoms are arranged in a well-defined periodic array, their collective scattered radiations interfere constructively only along certain well-defined direction determined by Bragg law, (1.4.1), but destructively interfere for all other directions (§7.4). Further, if the crystal is not perfect, i.e. it includes defects, then the requirement for complete destructive interference is not met, and intensities can be found at angles other than the Bragg angle.

From this brief discussion, it is important to recognize that before we present the physics of diffraction, we need to develop a technical vocabulary to describe the symmetry and periodic arrangement of atoms in crystals as well as their defects. Thus, in §4, we begin with an introduction to crystallography and the general principles of diffraction. We then discuss details of the different kinds of probes and scattering of ions and ion-based characterization methods (§5), and present an introduction to optics, microscopy, and ellipsometry (§6). In the second half of the book, we discuss diffraction in detail for X-rays (§7), and electrons and neutrons (§8). Finally, we discuss imaging with electrons in transmission (§9) and scanning (§10) modes, as well as scanning probe methods (§11), and conclude with a comprehensive set of tables (§12) comparing the different spectroscopy, imaging, and diffraction methods discussed throughout the book.

Summary

Optimizing materials microstructures is central to materials development, enabling new technologies, and in a broad sense, impacting every aspect of our life today. Central to this exercise is materials characterization, which helps relate synthesis and processing of materials to their structure, properties, and performance. The reach of materials characterization is indeed broad and includes, for example, optimization of materials microstructures for advanced technologies, unraveling the structure of biological molecules and complexes, failure analysis, understanding the morphology of rocks and geological processes, and even in the analysis, conservation, and authentication of sculptures or paintings.

Materials characterization typically uses probe and signal radiations to interrogate a specimen. The probe and signal radiations may or may not be the same, and the interactions between the probe and the specimen may be elastic or inelastic, coherent or incoherent. The characterization techniques can be broadly classified as spectroscopy (involving absorption, emission, and/or transitions processes), diffraction and scattering, and imaging and microscopy. The probes are broadly based on the electromagnetic spectrum, and their characteristics (energy, wavelength, momentum, polarization) define their interaction with matter, and determine the nature, scope, and details of the possible methods of characterization. In addition, the probes or signals can also be ions or neutrons. Principal features of the materials, i.e. details of their electronic structure, including atomic mass, composition, and crystallography, contribute to the observable signals and define possible characterization methods that are explored systematically in later chapters.

This introductory chapter provides some basic background, including definitions, on the underlying physics of spectroscopy, diffraction and imaging, interactions of radiations with matter, and a rudimentary introduction to digital imaging. Further, it provides numerous motivational examples of characterization in a variety of different contexts.

. . . .

FURTHER READING

Brandon, D., and W. D. Kaplan. *Microstructural Characterization of Materials.* Chichester: Wiley, 2008.

Brooks, C., and A. Choudhury. *Failure Analysis of Engineering Materials.* New York: McGraw-Hill, 2001.

Brundle, C. R., C. A. Evans and S. Wilson (eds.). *Encyclopedia of Materials Characterization.* Stoneham: Butterworth-Heinemann, 1992.

Callister, W. D., Jr., and D. G. Rethwisch. *Materials Science and Engineering: An Introduction.* New York: Wiley, 2009.

Farid, H. *Fundamentals of Image Processing*. Hanover: Dartmouth College Press, 2010.

Feldman, L. C., and J. W. Mayer. *Fundamentals of Surface and Thin Film Analysis*. New York: North-Holland, 1986.

Flewitt, P. E. J., and R. K. Wild. *Physical Methods of Materials Characterization*. Bristol: IoP Press, 2003.

Goldstein, J., D. Newbury, D. Joy, C. Lyman, P. Echlin, E. Lifshin, L. Sawyer, and J. Michael, *Scanning Electron Microscopy and X-Ray Microanalysis*. New York: Springer, 2003.

Gonzalez, R. C., and R. E. Woods. *Digital Image Processing*. Upper Saddle River: Prentice-Hall, 2002.

Hecht, E. *Optics*. Reading: Addison Wesley, 2002.

Hüfner, S. *Photoelectron Spectroscopy*. Berlin: Springer-Verlag, 2003.

Krishnan, K. M. *Fundamentals and Applications of Magnetic Materials*. Oxford: Oxford University Press, 2016, Chapter 8.

Kurzydlowski, K. J., and B. Ralph. *The Quantitative Description of the Microstructure of Materials*. Boca Raton: CRC Press, 1995.

Meyer, E., J. J. Hug, and R. Bennewitz. *Scanning Probe Microscopy—The Lab on a Tip*. Berlin: Springer-Verlag, 2003.

Miller, M. K., A. Cerezo, M. G. Hetherington, and G. D. W. Smith. *Atom Probe Field Ion Microscopy*. Oxford: Oxford University Press, 1996.

Owens, F. J., and C. P. Poole Jr., *The Physics and Chemistry of Nanosolids*. Wiley, 2008.

Russ, J. C. *The Image Processing Handbook*. Boca Raton: CRC Press, 1992.

Sheppard, C. J. R., C. Sheppard, and D. Shotton. *Confocal Laser Scanning Microscopy*. Abigndon: BIOS Scientific Publishers, 1997.

Sherwood, P. M. A. *Vibrational Spectroscopy of Solids*. Cambridge: Cambridge University Press, 1972.

Somorjai, G. A. *Chemistry in Two Dimensions: Surfaces*. New York: Cornell University Press, 1981.

Stokes, D. E. *Pasteur's Quadrant: Basic Science and Technological Innovation*. Washington, DC: Brookings Institution Press, 1997.

Sutton, A. P. *Electronic Structure of Materials*. Oxford: Oxford University Press, 1993.

Vernon, R. H. *A Practical Guide to Rock Microstructure*. Cambridge: Cambridge University Press, 2004.

Vickerman, J. C., ed. *Surface Analysis—The Principal Techniques*. Chichester: Wiley, 1997.

Wiesendanger, R. *Scanning Probe Microscopy and Spectroscopy*. Cambridge: Cambridge University Press, 1994.

Williams, D., A. R. Pelton, and R. Gronsky, eds. *Images of Materials*. Oxford: Oxford University Press, 1991.

Woodruff, D. P, and T. A. Delchar. *Modern Techniques of Surface Analysis*. Cambridge: Cambridge University Press, 1986.

Ziegler, A., H. Graafsma, X. F. Zhang, and J. W. M. Frenken, eds. *In-Situ Materials Characterization Across Spatial and Temporal Scales*. New York: Springer, 2014.

....

REFERENCES

[1] *Materials Science and Engineering for the 1990s: Maintaining Competitiveness in the Age of Materials*, Washington, DC: The National Academy Press, 1989. A pdf copy of the report is available from http://nap.edu/758

[2] Long, H., S. Mao, Y. Liu, Z. Zhang, and X. Han. "Microstructural and compositional design of Ni-based single crystalline superalloys — A review." *Journal of Alloys and Compounds* 743 (2018): 203–20.

[3] Xiang, S., S. Mao, H. Wei, Y. Liu, J. Zhang, Z. Shen, H. Long, H. Zhang, X. Wang, Z. Zhang, and X. Han "Selective Evolution of Secondary $\gamma\prime$ Precipitation in a Ni-Based Single Crystal Superalloy Both in the γ Matrix and at the Dislocation Nodes." *Acta Materialia* 116 (2016): 343–53.

[4] Liu, M., and S. Nemat-Nasser. "The Microstructure and Boundary Phases of *In-Situ* Reinforced Silicon Nitride." *Materials Science and Engineering A* 254, (1998): 242–52.

[5] Clarke, D. R. "Grain Boundaries in Polycrystalline Ceramics." *Annual Review of Materials Science* 17 (1987): 57.

[6] Watson, J. D., and F. H. C. Crick. "Molecular Structure of Nucleic Acids: A Structure for Deoxyribose Nucleic Acid." *Nature* 171 (1953): 737–8.

[7] Franklin, R. E., and R. G. Gosling. "Evidence for 2-Chain Helix in Crystalline Structure of Sodium Deoxyribonucleate." *Nature* 172 (1953): 156–7.

[8] Felkins, K., H. P. Leighly, and A. Jankovic. "The Royal Mail Ship Titanic: Did a Metallurgical Failure Cause a Night to Remember?" *Journal of Metals* 1998: 12–17.

[9] Götze, J. "Potential of Cathodoluminescence (CL) Microscopy and Spectroscopy for the Analysis of Minerals and Materials." *Analytical and Bioanalytical Chemistry* 374 (2002): 703.

[10] Carson, W. D., and C. Denison. "Mechanisms of Porphyroblast Crystallization: Results from High-Resolution Computed X-ray Tomography." *Science* 257 (1992): 1236.

[11] Whitehead, M. E., A. S. A. Karim, M. H. Loretto and R. E. Smallman. "Electron Radiation Damage in H.C.P. Metals—II. The Nature of the Defect Clusters in Zn and Cd Formed by Irradiation in the HVEM." *Acta Metallurgica* 26, no. 6 (1977): 983–93.

[12] Li, M., M. A. Kirk, P. M. Baldo, D. Xu and B. D. Wirth. "Study of Defect Evolution by TEM with *In Situ* Ion Irradiation and Coordinated Modelling." *Philosophical Magazine* 92, no. 16 (2012): 2048–78.

[13] Krause, M. O. "Atomic Radiative and Radiationless Yields for K and L Shells." *Journal of Physical and Chemical Reference Data* 8 (1979): 307.

[14] Raman, C.V., and K. S. Krishnan. "The Production of New Radiations by Light Scattering. Part I." *Proc. Roy. Soc.* 122, no. 789 (1929): 23–35.

[15] Nutall, R. H. "The First Microscopes of Henry Clifton Sorby." *Technology and Society* 22 (1981): 275–80.

[16] Montoya, S.A., S. Couture, J. J. Chess, J. C. T. Lee, N. Kent, D. Henze, S. K. Sinha, M.-Y. Im, S. D. Kevan, P. Fischer, B. J. McMorran, V. Lomakin,

S. Roy, and E. E. Fullerton. "Tailoring Magnetic Energies to Form Dipole Skyrmions and Skyrmion Lattices." *Physical Review B* 95 (2017): 024415.

[17] Thorel, A., J Ciston, T. P. Bardel, C. -Y. Song, and U. Dahman. "Observation of the Atomic Structure of ß'-SiAlON Using Three Generations of High-Resolution Electron Microscopes." *Philosophical Magazine A* 93(2013): 1172–81.

[18] Müller, E. R., and K. Bahadur. "Field Ionization of Gases at a Metal Surface and the Resolution of the Field Ion Microscope." *Physical Review* 102 (1956): 624.

[19] Miller, M. K., T. F. Kelly, K. Rajan, and S. P. Ringer. "The Future of Atom Probe Tomography." *Materials Today* 15, no. 4 (2012): 158–65.

[20] Amouyal, Y., and G. Schmitz. "Atom Probe Tomography—A Cornerstone in Materials Characterization". *MRS Bulletin* 41 (2016): 13–18.

[21] Muller, E. W., J. A. Panitz, and S. B. McLane. "The Atom-Probe Field Ion Microscope." *Review of Scientific Instruments* 39 (1968): 83.

[22] Panitz, J.A. "The 10 cm Atom Probe." *Review of Scientific Instruments* 44 (1973): 1034.

[23] Cerezo, A., T. J. Godfrey and G. D. W. Smith. "Application of a Position-Sensitive Detector to Atom Probe Microanalysis." *Review of Scientific Instruments* 59 (1988): 862.

[24] Kelly, T. F., T. T. Gribb, J. D. Olson, R. L. Martens, J. D. Shepard, S. A. Wiener, T. C. Kunicki, R. M. Ulfig, D. R. Lenz, E. M. Strennen, E. Oltman, J. H. Bunton, and D. R. Strait. "First Data from a Commercial Local Electrode Atom Probe (LEAP)." *Microscopy and Microanalysis* 13 (2004): 373–83.

[25] Hamers, R. J., R. M. Tromp, and J. E. Demuth. "Electronic and Geometric Structure of Si(111)-(7 × 7) and Si(001) Surfaces." *Surface Science* 181 (1987): 346–55.

[26] Binnig, G., C. F. Quate and Ch. Gerber. "Atomic Force Microscope." *Physical Review Letters* 56 (1986): 930.

[27] Paciornik, S., and M. H. P. Mauricio. "Digital Imaging." In: *ASM Handbook, Metallography and Microstructures*, edited by G. F. Vander Voort, Vol. 9, 368–402. Materials Park: ASM International, 2014.

[28] Russ, J. C. "Fundamentals of Image Processing and Measurement." In *Encyclopedia of Computer Science and Technology*, 2nd ed., edited by P. A. Laplate. Boca Raton: Taylor & Francis Group, 2016.

[29] Russ, J. C. "Image Analysis of Foods." *Journal of Food Science* 80, no. 9 (2015): 1974–87.

[30] Marr, D., and E. Hildreth. "Theory of Edge Detection." *Proceedings of the Royal Society of London. Series B, Biological Sciences* 207, no. 1167 (1980): 187–217.

[31] Canny, J. "A Computational Approach to Edge Detection." *IEEE Transactions on Pattern Analysis and Machine Intelligence* 8 (1986): 679–698.

[32] Feynman, R. P. "There's Plenty of Room at the Bottom (Data Storage)." *Journal of Microelectromechanical Systems* 1, no. 1 (1992): 60–66. doi:10.1109/84.128057. This is a transcript of the talk by Richard Feynman.

[33] Bednorz, J. G., and K. A. Müller. "Possible High T_c Superconductivity in the Ba−La−Cu−O System." *Zeitschrift für Physik B Condensed Matter* 64 (1986): 189–93.

[34] Zhao, J. C., and J. H. Westbrook. "Ultrahigh-Temperature Materials for Jet Engines." *MRS Bulletin* 28, no. 9 (2003): 622–30.

[35] Seah, M. P., and W. A. Dench. "Quantitative Electron Spectroscopy of Surfaces: A Standard Data Base for Electron Inelastic Mean Free Paths in Solid." *Surface and Interface Analysis* 1, no. 1 (1979): 2–11.

[36] Eadweard Muybridge collections, Stanford University Libraries. Available at: https://library.stanford.edu/collections/eadweard-muybridge-photographs.

....

EXERCISES

A. Test Your Knowledge

There may be more than one, or no, correct answer.

1. Metrology is
   (a) the study of measurements.
   (b) the study of weather patterns.
   (c) not relevant for good materials characterization.
2. Microstructure
   (a) includes atomic, mesoscopic, and microscopic length scales.
   (b) is related to materials properties and function.
   (c) can be observed by the naked eye.
3. Materials characterization involves
   (a) probes and signals.
   (b) probes that may damage some materials.
   (c) probes and signals that are always different radiations.
4. Resolution in the context of materials characterization is
   (a) a measure of how determined you are to get the job done.
   (b) only given by the penetration depth of the probe.
   (c) divided into spatial, depth, and temporal categories.
5. The interaction of a probe with the specimen is
   (a) elastic if it does not lose any energy.
   (b) coherent if probe and signal have the same phase.
   (c) coherent if all the signal is uniformly of the same phase.
6. The electromagnetic spectrum
   (a) provides a rich source of probes for materials analysis.
   (b) includes scanning probes.
   (c) can be displayed in terms of energy, frequency, or wavelength.
   (d) matches with critical length scales that describe the microstructure of materials.

7. Light
   (a) is <u>always</u> unpolarized.
   (b) is a transverse wave with **E** and **H** propagating in orthogonal planes.
   (c) can be linearly, elliptically or circularly polarized.

8. A photon is a quantum of light that
   (a) behaves as a particle.
   (b) behaves as a wave packet.
   (c) can be divided into smaller units of energy.

9. Interaction of a probe with matter
   (a) will generally not perturb the material.
   (b) can cause damage.
   (c) is characterized by a penetration depth and a mean free path length.

10. The Planck constant
    (a) is used to strengthen your abs.
    (b) is a very large number.
    (c) relates the energy of a photon to its frequency.
    (d) can be used to calculate the wavelength of an object if its velocity is known.

11. Spectroscopy can involve
    (a) absorption, emission, and transition processes.
    (b) non-radiative processes.
    (c) energy levels associated with molecular vibrations and rotations.

12. X-ray diffraction
    (a) of crystals involves Bragg law.
    (b) is dominated by the interaction only with atomic electrons.
    (d) includes contributions from the electrons, protons and neutrons of the atoms.

13. The resolution of a microscope
    (a) is given by the Rayleigh criterion.
    (b) can be improved by reducing the wavelength of the probe.
    (c) can be improved by increasing the refractive index of the medium.

14. Images with *atomic resolution* can be obtained by _____ microscopy.
    (a) optical
    (b) scanning <u>probe</u>
    (c) scanning tunneling
    (d) high resolution transmission electron
    (e) field ion
    (f) scanning electron

15. A *digitized* image
    (a) generates an image in *discrete* space.
    (b) is often derived from an image in continuous space.
    (c) typically involves a *sampling* process.

16. A *pixel*
    (a) stands for a picture element.
    (b) has an *averagebrightness* value.
    (c) has binary values (1 or 0) for a grey scale image.
17. In a digital image, the *sampling frequency*
    (a) in the spatial axis gives the resolution.
    (b) in the intensity axis is known as quantization.
    (c) depends on the number of specimens imaged.
18. *File formats* in digital imaging include
    (a) TIFF, which is flexible and involves loss-free compression.
    (b) JPEG, which is common but has lossy compression.
    (c) no other formats because they are not discussed here.
19. *A look-up table*
    (a) maps the image intensities to the display intensities.
    (b) if linear, can routinely manipulate contrast and brightness.
    (c) can be non-linear.
    (d) operation, in general, is NOT a *point operation*.
20. *Kernel* operations in digital imaging
    (a) are neighborhood operations.
    (b) only allow low-pass filters.
    (c) can produce derivatives of images.

B. Problems

1. What do you understand by the term "microstructure of materials?" Write a brief (50–100 word) definition of "microstructure."
2. What is meant by "materials characterization?" In your own words, briefly (50–100 word) define "materials characterization."
3. What is meant by "sensitivity" or "detection limit" of a characterization technique?
4. Is it possible to damage a material in the process of characterizing it? Explain, in your own words, using appropriate examples.
5. What is the difference between *spatial* and *temporal* resolution?
6. A dust particle of mass 1 µg travels at a velocity of 25,000 kph. Calculate its wavelength.
7. What is the kinetic energy in Joules of an electron accelerated from rest through a potential of 5 kV?
8. What is the de Broglie wavelength of
    (i) hydrogen atoms ($m = 1.67 \times 10^{-27}$ kg) moving with velocity = 10^3 m/s?
    (ii) electrons accelerated by 5,000 V in a SEM?
    (iii) tennis balls (100 g) traveling with a velocity of 20 m/s?
    (iv) $^4\text{He}^+$ ions accelerated through 1 MV for RBS experiments?

9. If the wavelength of an electron is infinite, the electron must be stationary. True or false? Justify your answer.

10. Please do some general reading and complete as much of the following table as possible.

Characterization Technique	Probe in	Probe out	Resolution, sensitivity	Depth probed	Highlights or comments
Light Microscopy					
Scanning electron microscopy (SEM)					
Transmission electron microscopy (TEM)					
Energy dispersive X-ray spectrometry (EDS)					
Auger electron spectrometry (AES)					
X-ray photoelectron spectroscopy (XPS)					
Fourier transform infrared spectroscopy (FTIR)					
Raman spectroscopy					
Ultraviolet and visible spectrometry (UV-Vis)					
Rutherford back-scattering spectroscopy (RBS)					
X-ray diffraction (XRD)					
X-ray fluorescence (XRF)					
Scanning tunneling microscopy (SPM)					
Atomic force microscopy (AFM)					
Nuclear magnetic resonance (NMR)					
Secondary ion mass spectrometry (SIMS)					
Inductively coupled plasma mass spectrometry (ICP-MS)					
Field ion microscopy					

11. For $Cu_{K\alpha}$ radiation, how many photons are required to deliver 1 Joule of energy? What is the flux (photons/s) required to deliver a power of 1 watt?

12. X-rays typically used in diffraction have a wavelength of 1.5 Å. What is the *energy* and *momentum* of this X-ray photon? What would be the velocity of a corresponding (i) electron, and (ii) proton, that has the same momentum?

13. Read Richard Feynman's famous article "*There is Plenty of Room at the Bottom*" [32], considered by many as the first lecture on the promise of nanoscience and nanotechnology. In this talk, he first outlines how all the world's data can be stored on a speck of dust. He then makes the case for developing the finest electron microscopes to "see" atoms. He then, with characteristic certainty, says: "It would be very easy to make an analysis of any complicated chemical substance; all one would have to do would be to look at it and see where the atoms are." Now, examine the case of high-T_c oxide superconductors [33], for which its discoverers, J. G. Bednorz and K. A. Müller, received the Nobel Prize in 1987, and address the following questions:

    (i) Is the structure, including the position of atoms in the perovskite unit cell, for these high-T_c $YBa_2Cu_3O_7$ oxides well known?

    (ii) Has this led to an understanding of why these materials are superconducting?

    (iii) Does this prove or refute the arguments of Professor Feynman? Please explain.

2

Atomic Structure and Spectra

(d) Iron   (e) Chromium   (f) Cadmium   (g) Lead   (h) Barium   (i) Zinc

Characterization and analysis also play a significant role in art conservation, and are often used for authentication, or help rule out forgery when the provenance of the work is questionable. Sometimes such analysis reveals hidden secrets, as illustrated here in Picasso's Blue Period painting *La Miséreuse accroupie*. (a) The painting as it is seen today. (b) Mounting the painting for element-specific, spatially-resolved X-ray fluorescence measurements described in §2.5.2.1. (c) Using such core-level spectroscopy revealed, underneath the Picasso painting, a hidden landscape painting by another artist. Element-specific X-ray fluorescence images of the pigments (d) Iron (blue), (e) chromium (yellow), (f) cadmium (red, orange), (g) lead (white), (h) barium, and (i) zinc.

Adapted from the *New York Times*, February 21, 2018.

Principles of Materials Characterization and Metrology. Kannan M. Krishnan, Oxford University Press (2021).
© Kannan M. Krishnan. DOI: 10.1093/oso/9780198830252.003.0002

2.1   Introduction

Section 1.4.1 presented a brief discussion of absorption, emission, and radiation processes, based on a simple model of the atom to motivate various spectroscopy methods used in materials characterization. This chapter develops the Bohr[1] model further and then presents, in a concise form, a quantum- or wave-mechanical model of atomic structure. This model overcomes the major flaw of the Bohr model, i.e. electrons in perpetual acceleration emitting no radiation, a behavior contradicting classical radiation physics.

Then, using the latter as a working model of the atomic structure, we revisit in further detail the set of radiative and non-radiative processes introduced in §1 for both electron and X-ray incidence. We discuss the emission, transmission, and detection of characteristic X-rays, and then develop a formulism for their use in quantitative microanalysis. We look at two non-radiative processes: Auger electron spectroscopy (AES) with electron incidence, and X-ray photoelectron spectroscopy (XPS) with X-ray incidence, developing a conceptual understanding of these methods. We also discuss the instrumentation required for the detection of X-rays and electrons.

Section §2.2 deals with elementary ideas of atomic structure. Those familiar with these concepts, including *spin–orbit coupling*, may skip it and move directly to §2.3.

2.2   Atomic Structure

2.2.1   **Bohr–Rutherford–Sommerfeld Model**

Our early understanding[2] of the atomic structure, especially the nuclei, come from scattering experiments; the classical experiment carried out by Rutherford,[3] using incident α particles, showed the existence of a positively-charged nucleus, surrounded by electrons in circular orbits.[4] At that time, it was also known, from extensive spectroscopic measurements, that atoms emit characteristic but narrow lines of radiation with well-defined wavelength (or frequency), and most importantly, in prescribed numerical sequences. Further, the principle of wave-particle duality (§1.3.3), where the electron wavelength, λ, and momentum, p, are related through the de Broglie[5] relation, $\lambda = h/p$, was well accepted.

Incorporating these conceptual ideas, Bohr proposed the early quantum mechanical model of the atom by first equating the Coulombic attractive force, $\frac{1}{4\pi\varepsilon_0}\frac{Ze^2}{r^2}$, experienced by the electron (charge, $e = -1.602 \times 10^{-19}$ C; rest mass, $m_e = 9.11 \times 10^{-31}$ kg) in the atom due to the nuclear charge, $+Ze$, with the centripetal force, $m_e v^2/r$. He also introduced the concept of stationary orbits ($2\pi r = n\lambda$) and included the principle of wave-particle duality, to show that the circular electron orbits are not continuous but quantized in angular momentum, $m_e v r = nh/2\pi$, ($n = 1, 2, 3, \ldots.$), total energy

[1] Niels H. D. Bohr (1885–1962) was a Danish physicist who received the 1922 Nobel prize in Physics for "his services in the investigation of the structure of atoms and of the radiation emanating from them."

[2] Arnold J. W. Sommerfeld (1868–1951) was a German physicist.

[3] Ernest Rutherford received the 1908 Nobel prize in Chemistry for "his investigations into the disintegration of the elements, and the chemistry of radioactive substances."

[4] It is now well established that the scattering of high-energy α particles from different nuclei is distinctly different, and the measurement of their intensity and energy distribution, after being scattered by a material, also provides a direct method to determine the elemental composition of the specimen (see §5.4 for a discussion of the technique of Rutherford backscattering spectroscopy, or RBS).

[5] Louis de Broglie (1892–1987) was a French physicist who received the 1929 Nobel prize in Physics for "his discovery of the wave nature of electrons."

$$E_n = -\frac{m_e Z^2 e^4}{8\varepsilon_0^2 h^2}\frac{1}{n^2} = -13.606\frac{Z^2}{n^2}\ (\text{eV}) \qquad (2.2.1)$$

and radius

$$r_n = \frac{\varepsilon_0 h^2}{\pi m_e Z e^2} n^2 = \frac{a_0 n^2}{Z}(\text{Å}) \qquad (2.2.2)$$

where, $m_e e^4/8\varepsilon_0^2 h^2 = 13.606$ eV, is known as the Rydberg constant, $\varepsilon_0 = 8.85 \times 10^{-12}$ F m^{-1} is the permittivity of free space, and $h = 4.135 \times 10^{-15}$ eV s, is Planck's constant. Note that the radius is commonly expressed in terms of the first Bohr radius, $r_1 = a_0 = 0.529$ Å, of the hydrogen atom ($Z = 1$).

The significant problem with the Bohr model is that it contradicts classical radiation physics as the electron in orbit, in spite of its continuous acceleration, is assumed to not emit any radiation. However, the Bohr model does allow for the emission of radiation when the electron transitions from one energy level (n_i) to another (n_f), derived from (2.2.1) as:

$$hf = E_i - E_f = \frac{m_e Z^2 e^4}{8\varepsilon_0^2 h^2}\left(\frac{1}{n_f^2} - \frac{1}{n_i^2}\right) = -13.6 Z^2 \left(\frac{1}{n_f^2} - \frac{1}{n_i^2}\right)(\text{eV}) \qquad (2.2.3)$$

Now, as both n_i and n_f are integers, numerical sequences are possible for different combinations of n_i and n_f , and various observed series, such as Balmer ($n_i = 3, 4, 5, \ldots$, and $n_f = 2$), Lyman ($n_i = 2, 3, 4, \ldots$, and $n_f = 1$), Paschen ($n_i = 4, 5, 6 \ldots$, and $n_f = 3$) etc., can be explained by (2.2.3). Sommerfeld improved upon the work of Bohr by introducing non-circular elliptical orbits with the use of a second azimuthal quantum number to define the degree of ellipticity, and explained additional fine structure observed in the spectra. However, not all the predicted emission lines were observed, suggesting that of all the possible quantum states, only some were preselected and determined the allowed transitions.

The currently accepted wave-mechanical version of the quantum theory, developed in a decade of unusual creativity with contributions from many scientists, including Heisenberg,[6] Schrodinger,[7] and Dirac,[7] provided a way to go beyond the limitations of the Bohr model. It accurately predicts and matches observed atomic behavior, including observed transitions and emissions, of particular interest to our discussion. Details are found in any text on quantum mechanics (e.g. Rae, 1992; Sutton, 1993), but here we provide a brief summary highlighting the essential physics involved in electron transitions in atoms and related emissions.

2.2.2   Quantum Mechanical Model

In the quantum- or wave-mechanical model, the electrons are described in terms of a probabilistic wave-function, $\Psi(\mathbf{r}, t)$, such that $|\Psi(\mathbf{r}, t)|^2 d\mathbf{r}$ gives the probability

[6] Werner Heisenberg received the 1932 Nobel prize in Physics for "the creation of quantum mechanics, the application of which has, inter alia, led to the discovery of the allotropic forms of hydrogen."

[7] Erwin Schrödinger and P. A. M. Dirac shared the 1933 Nobel prize in Physics for "the discovery of new productive forms of atomic theory."

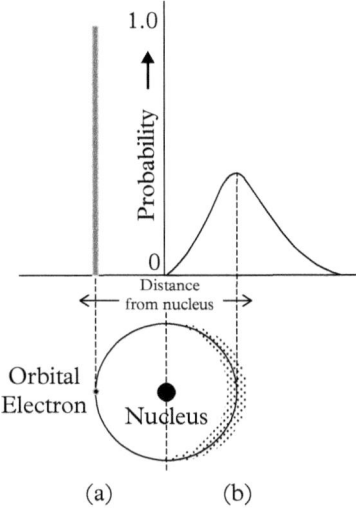

(a) (b)

Figure 2.2.1 A pictorial representation of the (a) Bohr and (b)
wave-mechanical model of the spatial
distribution of electrons in the atom.

Adapted from Callister and Rethwisch
(2010).

of finding the electron, at any time, t, in a volume $d\mathbf{r} = dx.dy.dz$, located at the position $\mathbf{r} = (x, y, z)$. In this model, any electron in the atom no longer occupies a discrete orbital, as in the Bohr model, but rather its position is described by the probability function (Fig. 2.2.1). In classical mechanics, the path or motion of a particle of mass, m_e, is prescribed by the Newton law of motion; in quantum mechanics the corresponding equation of motion of the electron is the Schrodinger equation, which describes the time evolution of the wave-function, $\Psi(\mathbf{r}, t)$, under a potential, $V(\mathbf{r}, t)$, as

$$-\frac{\hbar^2}{2m_e}\nabla^2\Psi(\mathbf{r}, t) + V(\mathbf{r}, t)\,\Psi(\mathbf{r}, t) = i\hbar\frac{\partial\Psi(\mathbf{r}, t)}{\partial t} \qquad (2.2.4)$$

where, to avoid writing 2π often, $\hbar = \frac{h}{2\pi}$, where $\hbar$ is the reduced Planck constant, m_e is the mass of the electron, and ∇ is the vector gradient. This time-dependent Schrodinger equation, (2.2.4), can be solved by the method of separation of variables, $\Psi(\mathbf{r}, t) = \psi_E(\mathbf{r})\mathrm{T}(t)$, where $\mathrm{T}(t)$ can be shown to be $\mathrm{T}(t) = e^{\left(\frac{-iEt}{\hbar}\right)}$. Thus

$$\Psi(\mathbf{r}, t) = \psi_E(\mathbf{r})\exp(-iEt/\hbar) \qquad (2.2.5)$$

and the spatially-dependent term, $\psi_E(\mathbf{r})$, is an *eigenfunction* of the time-independent Schrodinger equation

$$-\frac{\hbar^2}{2m_e}\nabla^2\psi_E(\mathbf{r}) + V(\mathbf{r})\psi_E(\mathbf{r}) = E\psi_E(\mathbf{r}) \qquad (2.2.6)$$

For electrons in a spherically symmetric Coulombic potential, $V(\mathbf{r}) = -\frac{Ze^2}{4\pi\varepsilon_0 r}$, experienced due to nuclear charge, $+Ze$, in the atom, the solution to the Schrodinger equation, (2.2.6), also reflect the spherical symmetry, and takes the form

$$\psi_E(\mathbf{r}) = \psi(r, \theta, \phi) = \mathrm{R}_{nl}(r)\mathrm{Y}_l^{m_l}(\theta, \phi) \tag{2.2.7}$$

where r, θ, ϕ are spherical coordinates, $\mathrm{Y}_l^{m_l}(\theta, \phi)$ are spherical harmonics, and θ is measured from the z-axis. The first term, $\mathrm{R}_{nl}(r)$, depends only on the distance of the electron from the nucleus, and the second term, $\mathrm{Y}_l^{m_l}(\theta, \phi)$, takes care of the spherical symmetry. Further, the requirement that this solution be finite, continuous, single valued, and normalized, introduces three quantum numbers, n, l, *and* m_l, i.e. one for each coordinate axis, which characterize the *eigenfunctions*[8] of the electron.

Here, n is the *principal quantum number*, associated with the radial coordinate, $\mathrm{R}_{nl}(r)$, and has allowed integer values, $n = 1, 2, 3 \ldots$. Note that this quantum number, and only this one, is associated with the earlier Bohr model.

l is the *orbital quantum number* that determines the magnitude of the electron angular momentum, $|L| = \sqrt{l(l+1)}\hbar$, with integer values, $l = 0, 1, 2, 3, \ldots$, $n-1$, denoting the sub-shell. The letters $s(l = 0)$, $p(l = 1)$, $d(l = 2)$, $f(l = 3)$, ..., are also commonly used.

m_l is the *magnetic quantum number*, with $m_l = 0, \pm 1, \pm 2, \ldots, \pm l$, that determines the z-component of the angular momentum $L_z = m_l\hbar$.

Finally, a relativistic treatment completes the quantum mechanical description of the electrons, by introducing a fourth quantum number, $s = \frac{1}{2}$, such that the magnitude of the spin angular momentum is $|S| = \sqrt{s(s+1)}\hbar = \frac{\sqrt{3}}{2}\hbar$ and its components along the z-axis are given by $m_s\hbar$, where $m_s = \pm 1/2$. This quantum number, m_s, describes the component of the spin angular momentum, which in simple terms, must be oriented with z-component either up (1/2) or down (–1/2). The occupation of electrons in different orbitals defined by these set of four quantum numbers, n, l, m_l, and m_s, must also satisfy the Pauli[9] exclusion principle, which states that no two electrons can have the same set of four *identical* quantum numbers.

The constraints on the four quantum numbers enumerated above, combined with the exclusion principle, imposes a limit on the number of electrons in each orbital. Thus, the first shell $(n = 1)$ can hold two electrons, the second shell $(n = 2)$ can hold two electrons in the $l = 0$ (s) sub-shell and six electrons in the $l = 1$ (p) subshell, for a total of eight electrons. The third shell can hold 18 electrons with two in the $3s$ ($n = 3, l = 0, m_l = 0, m_s = \pm 1/2$), six in the $3p$ ($n = 3, l = 1, m_l = \pm 1, 0, m_s = \pm 1/2$), and 10 in the $3d$ ($n = 3, l = 2, m_l = \pm 2$, $\pm 1, 0, m_s = \pm 1/2$) sub-shells. For the atom with $Z = 36$ (Kr), where all the shells up to the third are completely filled, its electronic structure can be written as $1s^2, 2s^2, 2p^6, 3s^2, 3p^6, 3d^{10}, 4s^2, 4p^6$. Note that the available electrons fill states

[8] A parameter-dependent equation with non-vanishing solutions only for particular values (*eigenvalues*) of the parameter is an eigenvalue equation, and the associated solutions are the *eigenfunctions*.

[9] Wolfgang Pauli received the 1945 Nobel prize in Physics for "the discovery of the Exclusion Principle, also called the Pauli Principle."

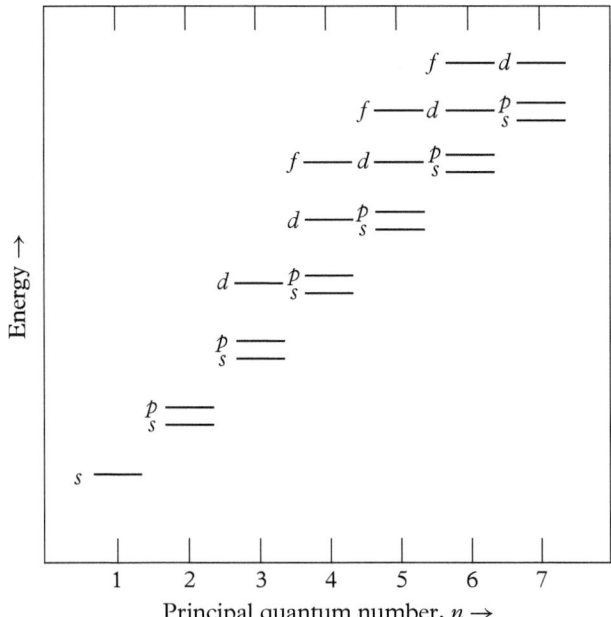

Figure 2.2.2 The relative energies of the electrons in the atom occupying various shells and sub-shells. Further, note that for $l > 1$, the spin–orbit coupling leads to further splitting of the sub-shells as discussed in the text, and in Figure 2.2.6.

Adapted from Callister and Rethwisch (2010).

of the lowest energy; such a configuration is also called the *ground* or unexcited state of the atom. The energies of the electrons are determined by the principal quantum number, n, and to a lesser extent by the orbital quantum number l (Fig. 2.2.2). The periodic table of the elements is built by assigning electrons, as described above, and Table 2.2.1 summarizes the ground state electronic configuration of the first 36 common elements.

The orbitals that we have discussed thus far also have an angular character, which is central to directional bonding in molecules and covalent crystals (to be discussed further in §3). This angular character is determined by the angular dependence of the wave functions or the appropriate spherical harmonic in (2.2.7), with its shape determined by the probability distribution, $\left|Y_l^{m_l}(\theta, \phi)\right|^2$. In other words, the angular dependence of the probability density, $\left|\psi_E(\mathbf{r})\right|^2$, with shapes corresponding to the s, p, and d orbitals are shown in Figure 2.2.3. For those interested, further details can be found in Pettifor (1995) or Borg and Dines (1992).

Figure 2.2.4 shows the energies of the valence electrons, defined as those that occupy the outermost shell, and principally involved in any bonding for all s and p levels. Unlike the hydrogen atom, where the energy levels (2.2.1) are determined by $E_n \propto n^{-2}$, for atoms with multiple electrons, states with the same principal quantum number, n, but different orbital quantum number, l, have non-degenerate, or different, energy levels. This is because the presence of additional

Table 2.2.1 Electronic structure of the atom, including the first ionization energy, for the first 36 elements.

Principal Quantum Number, n			1	2	2	3	3	3	4	4
Orbital Quantum Number, l			0	0	1	0	1	2	0	1
Letter Designation of State			$1s$	$2s$	$2p$	$3s$	$3p$	$3d$	$4s$	$4p$
Z	Symbol	Element	Ionization Energy (eV)							
1	H	Hydrogen	13.60	1						
2	He	Helium	24.58	2						
3	Li	Lithium	5.39		1					
4	Be	Beryllium	9.32		2					
5	B	Boron	8.30		2	1				
6	C	Carbon	11.26	He	2	2				
7	N	Nitrogen	14.54	Core	2	3				
8	O	Oxygen	13.61		2	4				
9	F	Fluorine	17.42		2	5				
10	Ne	Neon	21.56		2	6				
11	Na	Sodium	5.14				1			
12	Mg	Magnesium	7.64				2			
13	Al	Aluminum	5.98		Ne		2	1		
14	Si	Silicon	8.15		Core		2	2		
15	P	Phosphorus	10.55				2	3		
16	S	Sulfur	10.36				2	4		
17	Cl	Chlorine	13.01				2	5		
18	Ar	Argon	15.76				2	6		
19	K	Potassium	4.34							1
20	Ca	Calcium	6.11							2
21	Sc	Scandium	6.56						1	2
22	Ti	Titanium	6.83						2	2
23	V	Vanadium	6.74						3	2
24	Cr	Chromium	6.76				Ar		4	2
25	Mn	Manganese	7.43				Core		5	2
26	Fe	Iron	7.90						6	2
27	Co	Cobalt	7.86						7	2
28	Ni	Nickel	7.63						8	2
29	Cu	Copper	7.72						10	1
30	Zn	Zinc	9.39						10	2
31	Ga	Gallium	6.00						10	2	1
32	Ge	Germanium	7.88						10	2	2
33	As	Arsenic	9.81						10	2	3
34	Se	Selenium	9.75						10	2	4
35	Br	Bromine	11.84						10	2	5
36	Kr	Krypton	14.00						10	2	6

Adapted from [1].

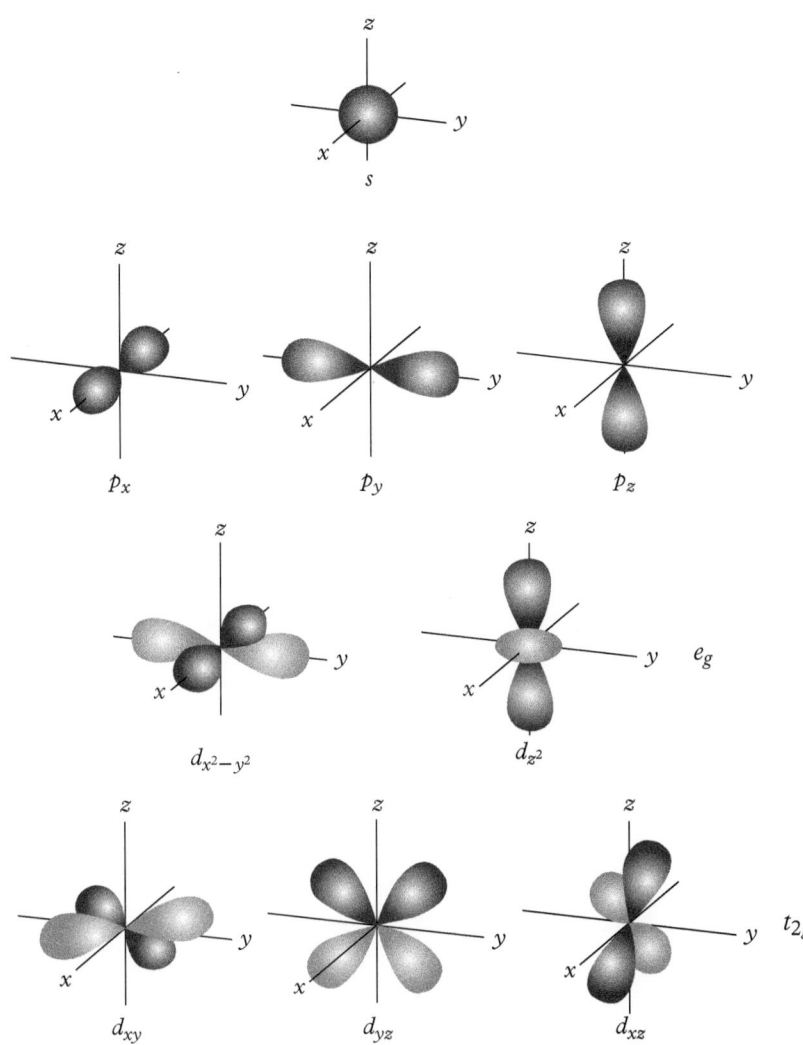

Figure 2.2.3 The angular component of the probability distribution corresponding to the *s*, *p*, and *d* orbitals.

Adapted from Krishnan (2016).

(more than one) electrons outside the nucleus, perturbs the potential, $V(\mathbf{r})$, and it no longer shows a simple r^{-2} dependence.

Three important observations can be made from Figure 2.2.4. First the $2s$ levels are all well below the $2p$ levels for the first-row elements, B through Ne. Second, both E_p and E_s depend linearly on the atomic number, Z, and not as Z^2 as expected from (2.2.1)—but why? Both of these observations can be explained in terms of the change in the potential due to the presence of other electrons. Finally, the energy difference, $E_p - E_s$, decreases as one goes from the right to the left of the periodic table. This is significant because as the difference becomes small, it will

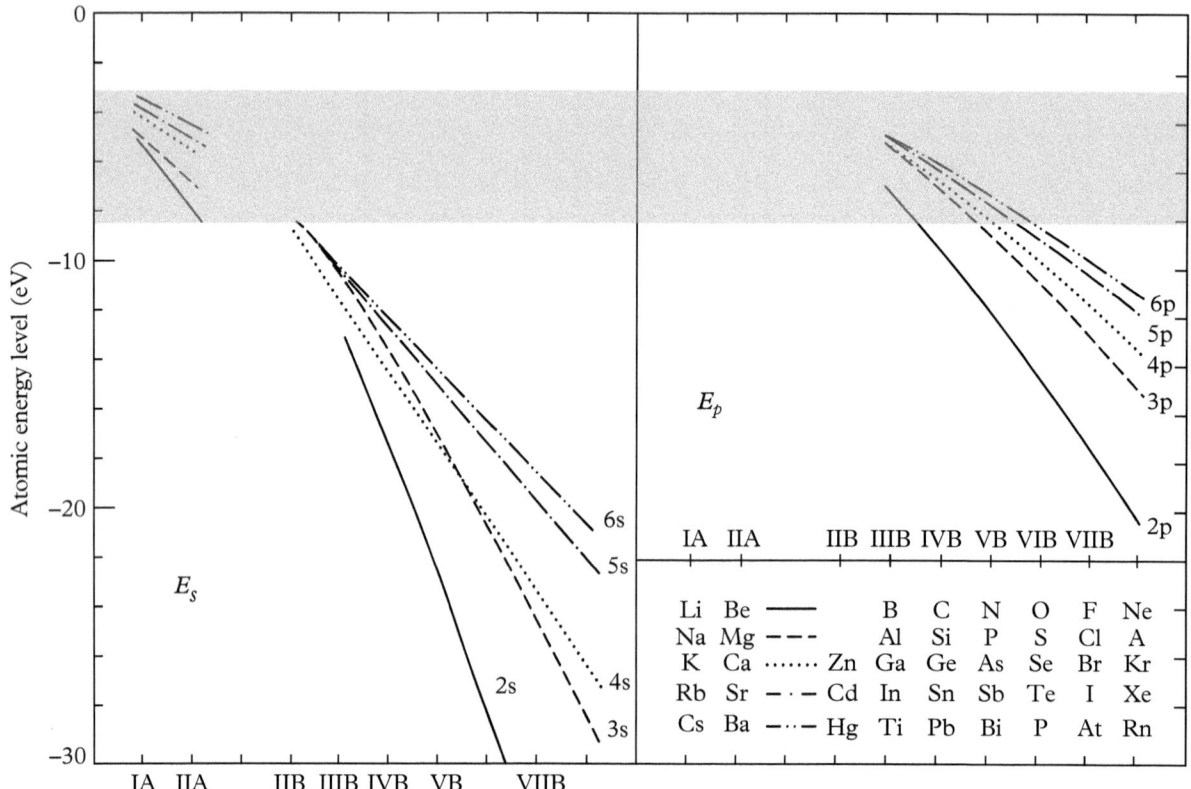

Figure 2.2.4 The valence *s* and *p* energy levels, E_s and E_p, respectively.

Adapted from Pettifor (1995).

enable the valence *s* and *p* electrons to "hybridize" and form common *sp* orbitals (Fig. 2.2.5). As Chapter 3 shows, these hybridized orbitals and their bonding configurations provide signature differences in their unoccupied levels that can be probed with absorption spectra using either electron or X-ray incidence (see Problem 3.6).

Lastly, it is helpful to briefly describe the concept of spin–orbit interaction or coupling. By changing the coordinate frame in an atom (Fig. 2.2.6), rather than considering the electron in orbit around the nucleus, we can keep the electron stationary and consider the nucleus in orbit around it with a radius equivalent to the Bohr radius. The orbiting nuclear charge, $+Ze$, produces a well-defined field/induction (think of a solenoid) at the position of the stationary electron. The *intrinsic spin magnetic moment* of the electron interacts with this field with its energy lower (higher) when they are both aligned parallel (antiparallel). Thus, we can have an *effective coupling* between the spin and the orbit of the electron.

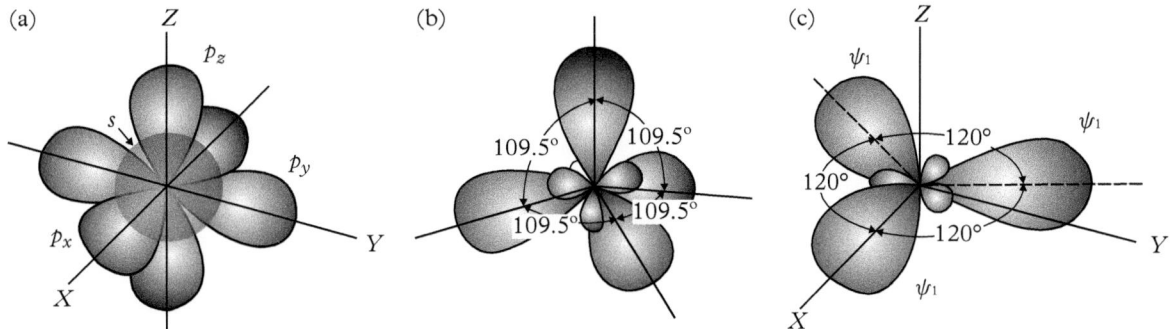

Figure 2.2.5 One *s* and three *p* orbitals shown in (a) can hybridize and form four *sp*³ orbitals (b) along the directions of a tetrahedron. These *sp*³ orbitals are key to the bonding in diamond. (c) If one *s* and two *p* orbitals hybridize, they form three *sp*² orbitals, all lying in a single plane. These *sp*² orbitals are key to the bonding in graphite.

Adapted from Borg and Dienes (1992).

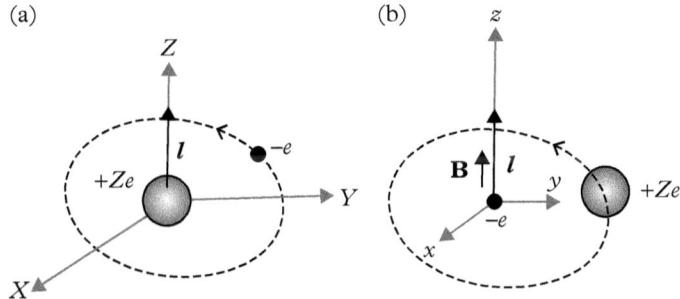

Figure 2.2.6 The spin–orbit interaction can be explained by changing the frame of reference from the nucleus (a), to the orbiting electron (b). In (b) the orbiting nuclear charge, $+Ze$, generates an effective magnetic field/induction at the electron that can couple with its intrinsic spin with different energies depending on their relative orientation.

Adapted from Krishnan (2016).

It is common practice to include this effective interaction between the spin and the magnetic field generated by the electron orbit, or *spin–orbit coupling*, by combining the spin and orbital contribution to the angular momentum for each electron (see, for example, Krishnan (2016), Sect. 2.8, for a more detailed discussion). Then, the appropriate set of quantum numbers are $n = 1, 2, 3 \ldots$, $l = 0, 1, 2, \ldots n - 1$, and two new quantum numbers, the total angular momentum quantum number $j = l \pm s$, and $m_j = \pm 1/2, \pm 3/2, \cdots \pm j$ that gives its z-component, where both j and m_j are, by definition, half integers. In this scenario, the shells are further split in energy to give a configuration such as $1s, 2s, 2p_{1/2}, 2p_{3/2}, 3s, 3p_{1/2}, 3p_{3/2}, 4s, 3d_{3/2}, 3d_{5/2}, \ldots$, where the additional subscript denotes the total angular momentum quantum number, $j = l \pm s$, which accounts for the spin–orbit splitting. The difference in energies between the $2p_{1/2}$ and $2p_{3/2}$ levels, or the $2p_{1/2} - 2p_{3/2}$ splitting as it is called, is small for small Z (~ 1 eV for Al), but increases as Z increases (~ 23 eV for Zn, and ~ 75 eV for Y). As we show in the next sections, this splitting is important and is included in the nomenclature of emission spectra.

Table 2.2.2 summarizes the electron binding energies for all elements in their natural form. Note that the energies are given in eV with respect to the *vacuum level* for $H_2, N_2, O_2, F_2,$ and Cl_2, and the rare gases, but relative to the *Fermi level* (Fig. 3.2.3) for metals, and relative to the top of the valence band for semiconductors (Fig. 3.7.1).

2.3   Atomic Spectra: Transitions, Emissions, and Secondary Processes

Three important characterization or analysis techniques discussed in this chapter are based on atomic electron transitions. These include two-transition processes such as the emission of characteristic X-rays, which involves the creation of a *core hole* (first transition) and the decay of an outer electron (second transition) along with the emission of an X-ray photon. This process is the basis of the techniques of X-ray fluorescence (XRF; X-rays in, X-rays out; §2.5.2.1) and electron probe microanalysis (EPMA; electrons in, X-rays out; §2.5.2.2), or microanalysis in a TEM/SEM (energy dispersive X-ray spectrometry, EDXS; electrons in, X-rays out; §2.5.1.1, §9.4.3, and §10.6.1). It is also the basis of particle-induced X-ray emission (PIXE, high-energy particle in, X-rays out), discussed later in §5.4.5. Similarly, AES (electrons in, electrons out; §2.6.2), involves the creation of a core hole (first transition), followed by the Auger decay (second transition). Note that the emission of characteristic X-rays and Auger electrons are competing processes and the former (latter) is favored by high-Z (low-Z) elements (Fig. 1.4.5). An example of an analysis method consisting of a single transition is XPS (X-rays in and electrons out), which involves the simultaneous formation of a core hole and the emission of an energetic photoelectron. Of the many possibilities, only some transitions of electrons in atoms are observed; the rules governing these allowed transitions, as you may have surmised, depend on quantum mechanics, and they are discussed in the next section.

2.3.1   Dipole Selection Rules and Allowed Transitions of Electrons in Atoms

We consider a radiative decay process consisting of a transition between two stationary states, *i.e.* the initial, higher energy state, ψ_i, and the final, lower energy

Table 2.2.2 Electron binding energies (eV) of the elements in their natural form.

Element	K 1s	L₁2s	L₂2p₁/₂	L₃2p₃/₂	M₁3s	M₂3p₁/₂	M₃3p₃/₂	M₄3d₃/₂	M₅3d₅/₂	N₁4s	N₂4p₁/₂	N₃4p₃/₂
1 H	13.6											
2 He	24.6*											
3 Li	54.7*											
4 Be	111.5*											
5 B	188*											
6 C	284.2*											
7 N	409.9*	37.3*										
8 O	543.1*	41.6*										
9 F	696.7*											
10 Ne	870.2*	48.5*	21.7*	21.6*								
11 Na	1070.8†	63.5†	30.65	30.81								
12 Mg	1303.0†	88.7	49.78	49.50								
13 Al	1559.6	117.8	72.95	72.55								
14 Si	1839	149.7*b	99.82	99.42								
15 P	2145.5	189*	136*	135*								
16 S	2472	230.9	163.6*	162.5*								
17 Cl	2822.4	270*	202*	200*								
18 Ar	3205.9*	326.3*	250.6†	248.4*	29.3*	15.9*	15.7*					
19 K	3608.4*	378.6*	297.3*	294.6*	34.8*	18.3*	18.3*					
20 Ca	4038.5*	438.4†	349.7†	346.2†	44.3†	25.4†	25.4†					
21 Sc	4492	498.0*	403.6*	398.7*	51.1*	28.3*	28.3*					
22 Ti	4966	560.9†	460.2†	453.8†	58.7†	32.6†	32.6†					
23 V	5465	626.7†	519.8†	512.1†	66.3†	37.2†	37.2†					
24 Cr	5989	696.0†	583.8†	574.1†	74.1†	42.2†	42.2†					
25 Mn	6539	769.1†	649.9†	638.7†	82.3†	47.2†	47.2†					
26 Fe	7112	844.6†	719.9†	706.8†	91.3†	52.7†	52.7†					
27 Co	7709	925.1†	793.2†	778.1†	101.0†	58.9†	59.9†					
28 Ni	8333	1008.6†	870.0†	852.7†	110.8†	68.0†	66.2†					
29 Cu	8979	1096.7†	952.3†	932.7	122.5†	77.3†	75.1†					
30 Zn	9659	1196.2*	1044.9*	1021.8*	139.8*	91.4*	83.6*	10.2*	10.1*			
31 Ga	10367	1299.0*b	1143.2†	1116.4†	159.5†	103.5†	100.0†	18.7†	18.7†			
32 Ge	11103	1414.6*b	1248.1*b	1217.0*b	180.1*	124.9*	120.8*	29.8	29.2			
33 As	11867	1527.0*b	1359.1*b	1323.6*b	204.7*	146.2*	141.2*	41.7*	41.7*			
34 Se	12658	1652.0*b	1474.3*b	1433.9*b	229.6*	166.5*	160.7*	55.5*	54.6*			

continued

Table 2.2.2 Continued

Element	K $1s$	L_1 $2s$	L_2 $2p_{1/2}$	L_3 $2p_{3/2}$	M_1 $3s$	M_2 $3p_{1/2}$	M_3 $3p_{3/2}$	M_4 $3d_{3/2}$	M_5 $3d_{5/2}$	N_1 $4s$	N_2 $4p_{1/2}$	N_3 $4p_{3/2}$
35 Br	13474	1782*	1596*	1550*	257*	189*	182*	70*	69*			
36 Kr	14326	1921	1730.9*	1678.4*	292.8*	222.2*	214.4	95.0*	93.8*	27.5*	14.1*	14.1*
37 Rb	15200	2065	1864	1804	326.7*	248.7*	239.1*	113.0*	112*	30.5*	16.3*	15.3*
38 Sr	16105	2216	2007	1940	358.7†	280.3†	270.0†	136.0†	134.2†	38.9†	21.3	20.1†
39 Y	17038	2373	2156	2080	392.0*b	310.6*	298.8*	157.7†	155.8†	43.8*	24.4*	23.1*
40 Zr	17993	2532	2307	2223	430.3†	343.5†	329.8†	181.1†	178.8†	50.6†	28.5†	27.1†
41 Nb	18986	2698	2465	2371	466.6†	376.1†	360.6†	205.0†	202.3†	56.4†	32.6†	30.8†
42 Mo	20000	2866	2625	2520	506.3†	411.6†	394.0†	231.1†	227.9†	63.2†	37.6†	35.5†
43 Tc	21044	3043	2793	2677	544*	447.6	417.7	257.6	253.9*	69.5*	42.3*	39.9*
44 Ru	22117	3224	2967	2838	586.1*	483.5†	461.4†	284.2†	280.0†	75.0†	46.3†	43.2†
45 Rh	23220	3412	3146	3004	628.1†	521.3†	496.5†	311.9†	307.2†	81.4*b	50.5†	47.3†
46 Pd	24350	3604	3330	3173	671.6†	559.9†	532.3†	340.5†	335.2†	87.1*b	55.7†a	50.9†
47 Ag	25514	3806	3524	3351	719.0†	603.8†	573.0†	374.0†	368.3	97.0†	63.7†	53.3†
48 Cd	26711	4018	3727	3538	772.0†	652.6†	618.4†	411.9†	405.2†	109.8†	63.9†a	63.9†a
49 In	27940	4238	3938	3730	827.2†	703.2†	665.3†	451.4†	443.9†	122.9†	73.5†a	73.5†a
50 Sn	29200	4465	4156	3929	884.7†	756.5†	714.6†	493.2†	484.9†	137.1†	83.6†a	83.6†a
51 Sb	30491	4698	4380	4132	946†	812.7†	766.4†	537.5†	528.2†	153.2†	95.6†a	95.6†a
52 Te	31814	4939	4612	4341	1006†	870.8†	820.0†	583.4†	573.0†	169.4†	103.3†a	103.3†a
53 I	33169	5188	4852	4557	1072*	931*	875*	630.8	619.3	186*	123*	123*
54 Xe	34561	5453	5107	4786	1148.7*	1002.1*	940.6*	689.0*	676.4*	213.2†	146.7	145.5*
55 Cs	35985	5714	5359	5012	1211*b	1071*	1003*	740.5*	726.6*	232.3*	172.4*	161.3*
56 Ba	37441	5989	5624	5247	1293*b	1137*b	1063*b	795.7†	780.5*	253.5†	192	178.6†
57 La	38925	6266	5891	5483	1362*b	1209*b	1128*b	853*	836*	274.7*	205.8	196.0*
58 Ce	40443	6549	6164	5723	1436*b	1274*b	1187*b	902.4*	883.8*	291.0*	223.2	206.5*
59 Pr	41991	6835	6440	5964	1511	1337	1242	948.3*	928.8*	304.5	236.3	217.6
60 Nd	43569	7126	6722	6208	1575	1403	1297	1003.3*	980.4*	319.2*	243.3	224.6
61 Pm	45184	7428	7013	6459	–	1471	1357	1052	1027	–	242	242
62 Sm	46834	7737	7312	6716	1723	1541	1420	1110.9*	1083.4*	347.2*	265.6	247.4
63 Eu	48519	8052	7617	6977	1800	1614	1481	1158.6*	1127.5*	360	284	257
64 Gd	50239	8376	7930	7243	1881	1688	1544	1221.9*	1189.6*	378.6*	286	271

65 Tb	51996	8708	8252	7514	1968	1768	1611	1276.9*	1241.1*	396.0*	322.4*	284.1*
66 Dy	53789	9046	8581	7790	2047	1842	1676	1333	1292.6*	414.2*	333.5*	293.2*
67 Ho	55618	9394	8918	8071	2128	1923	1741	1392	1351	432.4*	343.5	308.2*
68 Er	57486	9751	9264	8358	2207	2006	1812	1453	1409	449.8*	366.2	320.2*
69 Tm	59390	10116	9617	8648	2307	2090	1885	1515	1468	470.9*	385.9*	332.6*
70 Yb	61332	10486	9978	8944	2398	2173	1950	1576	1528	480.5*	388.7*	339.7*

Element	$N_4\,4d_{3/2}$	$N_5\,4d_{5/2}$	$N_6\,4f_{5/2}$	$N_7\,4f_{7/2}$	$O_1\,5s$	$O_2\,5p_{1/2}$	$O_3\,5p_{3/2}$	$O_4\,5d_{3/2}$	$O_5\,5d_{5/2}$	$P_1\,6s$	$P_2\,6p_{1/2}$	$P_3\,6p_{3/2}$
48 Cd	11.7†	10.7†										
49 In	17.7†	16.9†										
50 Sn	24.9†	23.9†										
51 Sb	33.3†	32.1†										
52 Te	41.9†	40.4†										
53 I	50.6	48.9										
54 Xe	69.5*	67.5*	–	–	23.3*	13.4*	12.1*					
55 Cs	79.8*	77.5*	–	–	22.7	14.2*	12.1*					
56 Ba	92.6†	89.9†	–	–	30.3†	17.0†	14.8†					
57 La	105.3*	102.5*	–	–	34.3*	19.3*	16.8*					
58 Ce	109*	–	0.1	0.1	37.8	19.8*	17.0*					
59 Pr	115.1*	115.1*	2.0	2.0	37.4	22.3	22.3					
60 Nd	120.5*	120.5*	1.5	1.5	37.5	21.1	21.1					
61 Pm	120	120	–	–	–	–	–					
62 Sm	129	129	5.2	5.2	37.4	21.3	21.3					
63 Eu	133	127.7*	0	0	32	22	22					
64 Gd	–	142.6*	8.6*	8.6*	36	28	21					
65 Tb	150.5*	150.5*	7.7*	2.4*	45.6*	28.7*	22.6*					
66 Dy	153.6*	153.6*	8.0*	4.3*	49.9*	26.3	26.3					
67 Ho	160*	160*	8.6*	5.2*	49.3*	30.8*	24.1*					
68 Er	167.6*	167.6*	–	4.7*	50.6*	31.4*	24.7*					
69 Tm	175.5*	175.5*	–	4.6	54.7*	31.8*	25.0*					
70 Yb	191.2*	182.4*	2.5*	1.3*	52.0*	30.3*	24.1*					

continued

Table 2.2.2 Continued

Element	K 1s	L$_1$2s	L$_2$2p$_{1/2}$	L$_3$2p$_{3/2}$	M$_1$3s	M$_2$3p$_{1/2}$	M$_3$3p$_{3/2}$	M$_4$3d$_{3/2}$	M$_5$3d$_{5/2}$	N$_1$4s	N$_2$4p$_{1/2}$	N$_3$4p$_{3/2}$
71 Lu	63314	10870	10349	9244	2491	2264	2024	1639	1589	506.8*	412.4*	359.2*
72 Hf	65351	11271	10739	9561	2601	2365	2108	1716	1662	538*	438.2†	380.7†
73 Ta	67416	11682	11136	9881	2708	2469	2194	1793	1735	563.4†	463.4†	400.9†
74 W	69525	12100	11544	10207	2820	2575	2281	1872	1809	594.1†	490.4†	423.6†
75 Re	71676	12527	11959	10535	2932	2682	2367	1949	1883	625.4†	518.7†	446.8†
76 Os	73871	12963	12385	10871	3049	2792	2457	2031	1960	658.2†	549.1†	470.7†
77 Ir	76111	13419	12824	11215	3174	2909	2551	2116	2040	691.1†	577.8†	495.8†
78 Pt	78395	13880	13273	11564	3296	3027	2645	2202	2122	725.4†	609.1†	519.4†
79 Au	80725	14353	13734	11919	3425	3148	2743	2291	2206	762.1†	642.7†	546.3†
80 Hg	83102	14839	14209	12284	3562	3279	2847	2385	2295	802.2†	680.2†	576.6†
81 Tl	85530	15347	14698	12658	3704	3416	2957	2485	2389	846.2†	720.5†	609.5†
82 Pb	88005	15861	15200	13035	3851	3554	3066	2586	2484	891.8†	761.9†	643.5†
83 Bi	90524	16388	15711	13419	3999	3696	3177	2688	2580	939†	805.2†	673.8†
84 Po	93105	16939	16244	13814	4149	3854	3302	2798	2683	995*	851*	705*
85 At	95730	17493	16785	14214	4317	4008	3426	2909	2787	1042*	886*	740*
86 Rn	98404	18049	17337	14619	4482	4159	3538	3022	2892	1097*	929*	768*
87 Fr	101137	18639	17907	15031	4652	4327	3663	3136	3000	1153*	980*	810*
88 Ra	103922	19237	18484	15444	4822	4490	3792	3248	3105	1208*	1058	879*
89 Ac	106755	19840	19083	15871	5002	4656	3909	3370	3219	1269*	1080*	890*
90 Th	109651	20472	19693	16300	5182	4830	4046	3491	3332	1330*	1168*	966.4†
91 Pa	112601	21105	20314	16733	5367	5001	4174	3611	3442	1387*	1224*	1007*
92 U	115606	21757	20948	17166	5548	5182	4303	3728	3552	1439*b	1271*b	1043†

Element	$N_4 4d_{3/2}$	$N_5 4d_{5/2}$	$N_6 4f_{5/2}$	$N_7 4f_{7/2}$	$O_1 5s$	$O_2 5p_{1/2}$	$O_3 5d_{3/2}$	$O_4 5d_{3/2}$	$O_5 5d_{5/2}$	$P_1 6s$	$P_2 6p_{1/2}$	$P_3 6p_{3/2}$
71 Lu	206.1*	196.3*	8.9*	7.5*	57.3*	33.6*	26.7*					
72 Hf	220.0†	211.5†	15.9†	14.2†	64.2†	38*	29.9†					
73 Ta	237.9†	226.4†	23.5†	21.6†	69.7†	42.2*	32.7†					
74 W	255.9†	243.5†	33.6*	31.4†	75.6†	45.3*b	36.8†					
75 Re	273.9†	260.5†	42.9*	40.5*	83†	45.6*	34.6*b					
76 Os	293.1†	278.5†	53.4†	50.7†	84*	58*	44.5†					
77 Ir	311.9†	296.3†	63.8†	60.8†	95.2*b	63.0*b	48.0†					
78 Pt	331.6†	314.6†	74.5†	71.2†	101.7*b	65.3*b	51.7†					
79 Au	353.2†	335.1†	87.6†	84.0	107.2*b	74.2†	57.2†					
80 Hg	378.2†	358.8†	104.0†	99.9†	127†	83.1†	64.5†|	9.6†	7.8†			
81 Tl	405.7†	385.0†	122.2†	117.8†	136.0*b	94.6†	73.5†	14.7†	12.5†			
82 Pb	434.3†	412.2†	141.7†	136.9†	147*b	106.4†	83.3†	20.7†	18.1†			
83 Bi	464.0†	440.1†	162.3†	157.0†	159.3*b	119.0†	92.6†	26.9†	23.8†			
84 Po	500*	473*	184*	184*	177*	132*	104*	31*	31*			
85 At	533*	507	210*	210*	195*	148*	115*	40*	40*			
86 Rn	567*	541*	238*	238*	214*	164*	127*	48*	48*	26		
87 Fr	603*	577*	268*	268*	234*	182*	140*	58*	58*	34	15	15
88 Ra	636*	603*	299*	299*	254*	200*	153*	68*	68*	44	19	19
89 Ac	675*	639*	319*	319*	272*	215*	167*	80*	80*	–	–	–
90 Th	712.1†	675.2†	342.4†	333.1†	290*a	229*a	182*a	92.5†	85.4†	41.4†	24.5†	16.6†
91 Pa	743*	708*	371*	360*	310*	232*	232*	94*	94*	–	–	–
92 U	778.3†	736.2†	388.2*	377.4†	321*ab	257*ab	192*ab	102.8†	94.2†	43.9†	26.8†	16.8†

Adapted from *X-ray Data Booklet*, Published by the Center for X-ray optics and Advanced Light
Source: LBNL/PUB-490 Rev-2 (2001).

state, ψ_f, of the atom. During a transition from ψ_i to ψ_f, the average position of the electron oscillates between these two states, with a frequency $f_{if} = (E_i - E_f)/h$, given by the difference in their energies (Fig. 2.3.1a). This oscillation between these two states, referred to in quantum mechanics as a *mixed state*, goes on for a finite period of time, known as the *lifetime* of the transition. During this lifetime, the probability of finding the electron in the higher energy state, ψ_i, decreases gradually from unity to zero, while the probability of finding the electron in the lower energy state, ψ_f, increases gradually from zero to unity (Fig. 2.3.1b). The lifetime is sufficiently long that during the transition period the electron executes many millions of oscillations, back and forth, between ψ_i and ψ_f. Moreover, in quantum mechanical terms, the larger the number of oscillations in the period defined by the lifetime of the transition, the better defined is the wavelength of the emitted photon or wave packet (Fig. 1.3.5); this also translates into a narrow line width in the observed spectrum.

In quantum mechanics the transition probability between states, ψ_i and ψ_f, is proportional to the square of the dipole matrix element

$$-e\int \psi_i \mathbf{r} \psi_f d\mathbf{r} = -e\mathbf{r}_{if} \tag{2.3.1}$$

A detailed discussion of (2.3.1), which provides a method to compute the atomic transition probabilities between the two states, is beyond the scope of this book, but can be found in Quantum Mechanics textbooks, e.g. Rae (1992). However, suffice it to say that the transition from ψ_i to ψ_f occurs only when the matrix element gives a finite oscillation amplitude, $\mathbf{r}_{if}(t)$. If the wave functions, ψ_i and ψ_f, are such that the integral, (2.3.1), is zero, the transition is not allowed as the oscillations are non-existent. By simple examination of the integral, one can see that $\mathbf{r}$ is an odd function of the coordinates (replacing r by $-r$ changes its sign)

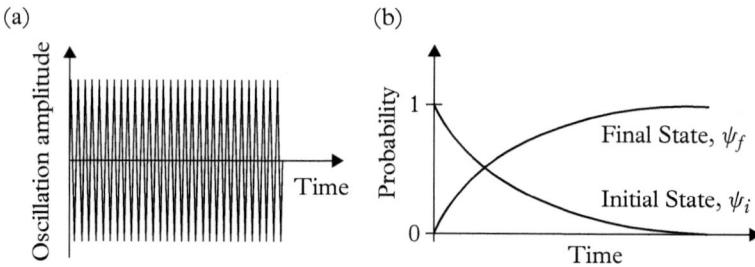

Figure 2.3.1 A radiative decay process consists of (a) the mixed atomic state oscillating between the initial, ψ_i, and final, ψ_f, states, with the probability (b) of the initial state decaying to zero during the *lifetime*—involving millions of such oscillations—of the transition.

Adapted from Atwood (2000).

and hence ψ_i and ψ_f should be of opposite parity (thus one of them should be odd, and the other even, in terms of the coordinates) to give a nonzero value of the integral. The parity of the wave functions alternates with increasing orbital quantum number, l, giving the first *dipole selection rule* for an observable or allowed transition in a single electron atom as

$$\Delta l = \pm 1 \tag{2.3.2a}$$

In addition, the total angular momentum quantum number, $j = l \pm s$, must also satisfy

$$\Delta j = 0, \pm 1 \tag{2.3.2b}$$

A physical way to understand the selection rules is to consider the emitted photon as a particle with an angular momentum of unity. Thus, to conserve angular momentum and energy, transitions where angular momentum changes by one unit, *i.e.* $\Delta l = \pm 1$ and $\Delta j = 0, \pm 1$, are only allowed (Fig. 2.3.4).

2.3.2 Characteristic X-Ray Emissions and their Nomenclature

Figure 2.3.2 shows the nomenclature used to describe X-rays generated from specific atomic transitions. To understand this nomenclature, consider the electron energy-level diagrams for two representative elements, sodium and molybdenum (Fig. 2.3.3). Here the energy of the entire atom is plotted, with zero indicating the neutral atom in its ground state.

Any excitation, say the removal of a particular core electron, leaving behind a hole or a vacant quantum state, leads to an increase in energy of the atom (excited state). For example, the energy of 1070.8 eV associated with $1s$ level of Na (Fig. 2.3.3) indicates the energy of the excited state of the Na atom when

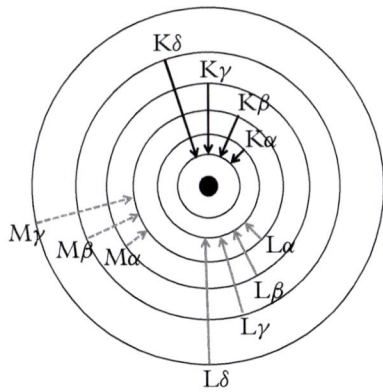

Figure 2.3.2 The nomenclature used to describe different X-ray lines arising from specific atomic transitions.

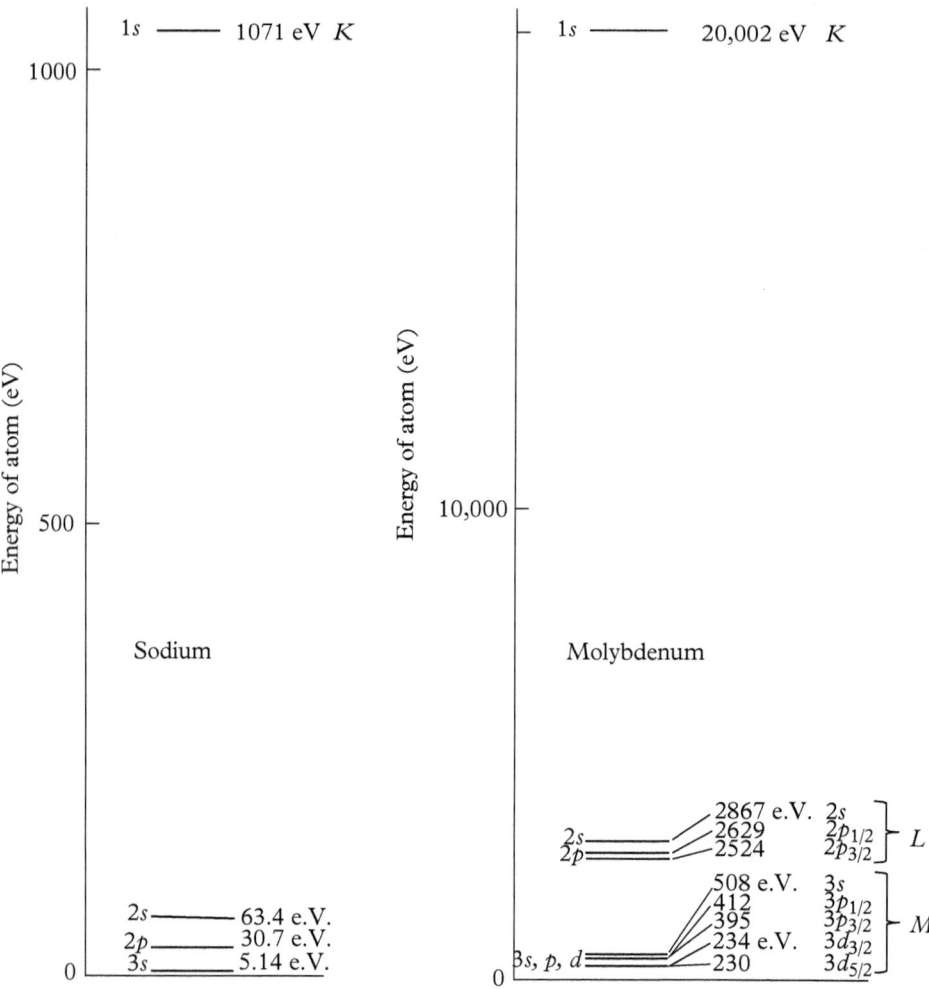

Figure 2.3.3 Electron energy level diagram for Na and Mo. Excited states of the atom, upon the removal of specific core electrons, as indicated, are shown.

Adapted from Sproul (1980).

the 1s electron from the K-shell ($n = 1$) has been removed. In this scenario, a 2p electron, following the dipole selection rule, (2.3.2a), can make a transition to 1s level, filling the K-shell, and then emitting an X-ray photon. This X-ray photon, called the $K\alpha$ spectral line (or K-line), would have a characteristic energy of 1040.1 eV (= 1070.8 − 30.7). The selection rule would also allow a $3p \rightarrow 1s$ transition ($K\beta$ line), but this is not observed for Na as the $3p$ state is unoccupied. We shall illustrate these concepts further with the example of Mo.

Example 2.3.1: Calculate the energies of the characteristic X-ray emissions for Mo.

Solution: The binding energies for Mo (Fig. 2.3.3) are $E_{1s} = 20,002$ eV, and since the $2p$ levels are split due to spin–orbit coupling we have $E2p_{3/2} = 2,524$ eV and $E2p_{1/2} = 2,629$ eV.

The resulting $K\alpha$ X-ray line is also split into two:

$$K\alpha_1 = 20,002 - 2524 = 17,478 \text{ eV}$$

and

$$K\alpha_2 = 20,002 - 2629 = 17,373 \text{ eV}.$$

In addition, the $K\beta$ lines are also split into two: $K\beta_1 = 1s - 3p_{3/2} = 20002 - 395 = 19607$ eV and $K\beta_3 = 1s - 3p_{1/2} = 20002 - 412 = 19590$ eV Alternatively, if the initial core hole were to be in the $2p$ (L –shell) level, we observe three X-ray lines

$$L\alpha_1 = 2p_{3/2} - 3d_{5/2} = 2524 - 230 = 2294 \text{ eV}$$

$$L\alpha_2 = 2p_{3/2} - 3d_{3/2} = 2524 - 234 = 2290 \text{ eV}$$

and

$$L\beta_1 = 2p_{1/2} - 3d_{3/2} = 2629 - 234 = 2395 \text{ eV}.$$

An alternative convention, which we favor in this book, is to plot the electron binding energies (negative) in all the shells in the multi-electron atom (Fig. 2.3.4).

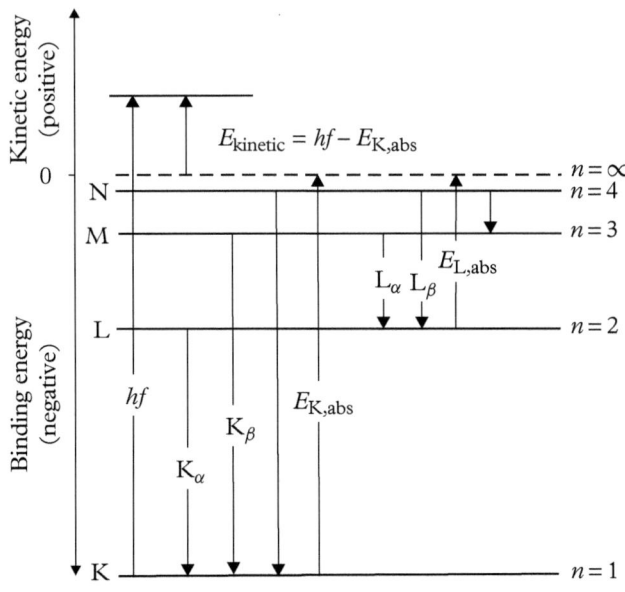

Figure 2.3.4 Energy levels for a multi-electron atom, showing the allowed transitions and the K-shell absorption edge.

Adapted from Atwood (2000).

The transitions, following the allowed dipole selection rules, that show narrow emission lines, *i.e.* $K\alpha$ ($n = 2 \rightarrow 1$), $K\beta$ ($n = 3 \rightarrow 1$), $L\alpha$ ($n = 3 \rightarrow 2$), etc., are also indicated. Further, the energy, $E_{K,abs}$, required to take an electron from the K-shell to the continuum ($n = \infty$) is called the *absorption edge*. If a photon (or electron) of energy greater than the absorption edge, $hf > E_{K,abs}$, is incident, it can remove the K-shell electron beyond the continuum limit, to a state with positive kinetic energy. As an L-shell electron has a lower binding energy compared to the K-shell electron, if it is also lifted to the continuum by a similar photon, it would have a larger kinetic energy. These transitions are also shown in the same figure.

Figure 2.3.5 shows a detailed energy level diagram for copper, for all the allowed transitions satisfying the selection rules, $\Delta l = \pm 1$ and $\Delta j = 0, \pm 1$, and including the spin–orbit coupling. The energy levels are not to scale. The reader is encouraged to study this figure carefully and understand each of the allowed

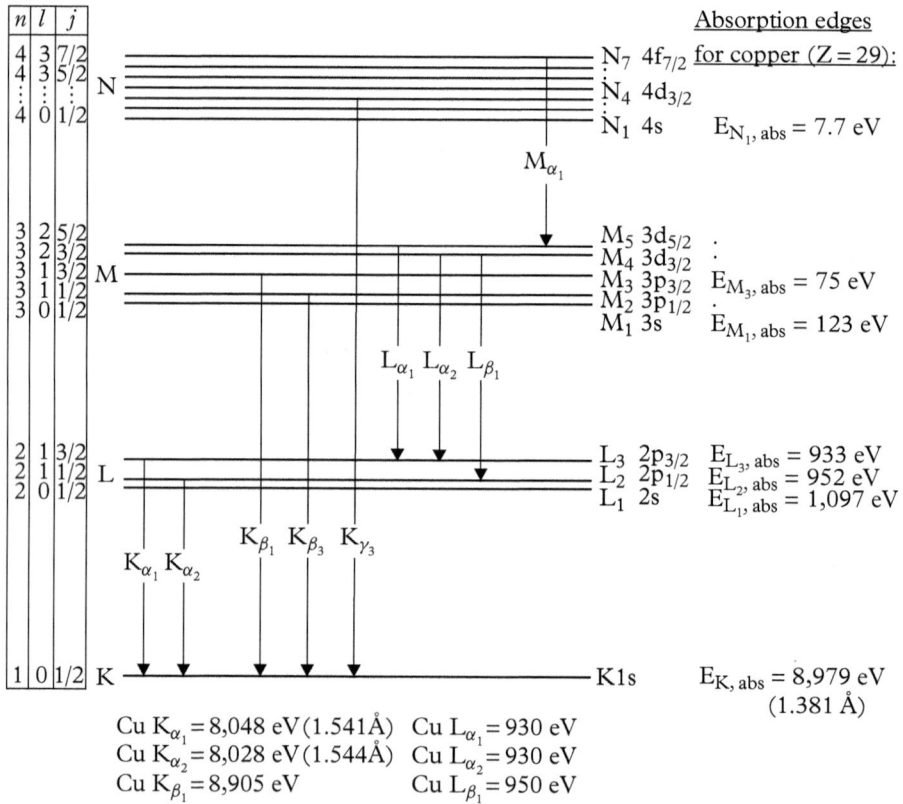

Cu K_{α_1} = 8,048 eV (1.541Å)   Cu L_{α_1} = 930 eV
Cu K_{α_2} = 8,028 eV (1.544Å)   Cu L_{α_2} = 930 eV
Cu K_{β_1} = 8,905 eV   Cu L_{β_1} = 950 eV

Figure 2.3.5 Energy level diagram for copper showing all the transitions allowed by the dipole selection rules and including the spin–orbit coupling.

Adapted from Atwood (2000).

transitions and the nomenclature used to describe it, before attempting Problem 2.1. The characteristic X-ray photons generally have long mean free path lengths in solids (§2.4.3). If used as an exciting probe, they are also deeply penetrating (see §5.3). As a result, techniques that use X-rays as a probe (XRF) or signal (EPMA), give rise to *bulk analytical tools*. In contrast, the non-radiative Auger electron emission and photoelectron emission are both *surface sensitive* signals (Fig. 1.3.5) and are discussed in the next section.

Table 2.3.1, adapted from [9], lists the energies of the principal X-ray emission lines for all elements.

Table 2.3.1 Energies (eV) of X-ray emission lines [9]. The corresponding wavelength in Å is given by $\lambda = 12398/E$, where E is in eV.

Element	$K\alpha_1$	$K\alpha_2$	$K\beta_1$	$L\alpha_1$	$L\alpha_2$	$L\beta_1$	$L\beta_2$	$L\gamma_1$	$M\alpha_1$
3 Li	54.3								
4 Be	108.5								
5 B	183.3								
6 C	277								
7 N	392.4								
8 O	524.9								
9 F	676.8								
10 Ne	848.6	848.6							
11 Na	1,040.98	1,040.98	1,071.1						
12 Mg	1,253.60	1,253.60	1,302.2						
13 Al	1,486.70	1,486.27	1,557.45						
14 Si	1,739.98	1,739.38	1,835.94						
15 P	2,013.7	2,012.7	2,139.1						
16 S	2,307.84	2,306.64	2,464.04						
17 Cl	2,622.39	2,620.78	2,815.6						
18 Ar	2,957.70	2,955.63	3,190.5						
19 K	3,313.8	3,311.1	3,589.6						
20 Ca	3,691.68	3,688.09	4,012.7	341.3	341.3	344.9			
21 Se	4,090.6	4,086.1	4,460.5	395.4	395.4	399.6			
22 Ti	4,510.84	4,504.86	4,931.81	452.2	452.2	458.4			
23 V	4,952.20	4,944.64	5,427.29	511.3	511.3	519.2			
24 Cr	5,414.72	5,405.509	5,946.71	572.8	572.8	582.8			
25 Mn	5,898.75	5,887.65	6,490.45	637.4	637.4	648.8			
26 Fe	6,403.84	6,390.84	7,057.98	705.0	705.0	718.5			
27 Co	6,930.32	6,915.30	7,649.43	776.2	776.2	791.4			
28 Ni	7,478.15	7,460.89	8,264.66	851.5	851.5	868.8			
29 Cu	8,047.78	8,027.83	8,905.29	929.7	929.7	949.8			
30 Zn	8,638.86	8,615.78	9,572.0	1,011.7	1,011.7	1,034.7			

Table 2.3.1 Continued

Element	$K\alpha_1$	$K\alpha_2$	$K\beta_1$	$L\alpha_1$	$L\alpha_2$	$L\beta_1$	$L\beta_2$	$L\gamma_1$	$M\alpha_1$
31 Ga	9,251.74	9,224.82	10,264.2	1,097.92	1,097.92	1,124.8			
32 Ge	9,886.42	9,855.32	10,982.1	1,188.00	1,188.00	1,218.5			
33 As	10,543.72	10,507.99	11,726.2	1,282.0	1,282.0	1,317.0			
34 Se	11,222.4	11,181.4	12,495.9	1,379.10	1,379.10	1,419.23			
35 Br	11,924.2	11,877.6	13,291.4	1,480.43	1,480.43	1,525.90			
36 Kr	12,649	12,598	14,112	1,586.0	1,586.0	1,636.6			
37 Rb	13,395.3	13,335.8	14,961.3	1,694.13	1,692.56	1,752.17			
38 Sr	14,165	14,097.9	15,835.7	1,806.56	1,804.74	1,871.72			
39 Y	14,958.4	14,882.9	16,737.8	1,922.56	1,920.47	1,995.84			
40 Zr	15,775.1	15,690.9	17,667.8	2,042.36	2,039.9	2,124.4	2,219.4	2,302.7	
41 Nb	16,615.1	16,521.0	18,622.5	2,165.89	2,163.0	2,257.4	2,367.0	2,461.8	
42 Mo	17,479.34	17,374.3	19,608.3	2,293.16	2,289.85	2,394.81	2,518.3	2,623.5	
43 Te	18,367.1	18,250.8	20,619	2,424	2,420	2,538	2,674	2,792	
44 Ru	19,279.2	19,150.4	21,656.8	2,558.55	2,554.31	2,683.23	2,836.0	2,964.5	
45 Rh	20,216.1	20,073.7	22,723.6	2,696.74	2,692.05	2,834.41	3,001.3	3,143.8	
46 Pd	21,177.1	21,020.1	23,818.7	2,838.61	2,833.29	2,990.22	3,171.79	3,328.7	
47 Ag	22,162.92	21,990.3	24,942.4	2,984.31	2,978.21	3,150.94	3,347.81	3,519.59	
48 Cd	23,173.6	22,984.1	26,095.5	3,133.73	3,126.91	3,316.57	3,528.12	3,716.86	
49 In	24,209.7	24,002.0	27,275.9	3,286.94	3,279.29	3,487.21	3,713.81	3,920.81	
50 Sn	25,271.3	25,044.0	28,4860	3,443.98	3,435.42	3,662.80	3,904.86	4,131.12	
51 Sb	26,359.1	26,110.8	29,725.6	3,604.72	3,595.32	3,843.57	4,100.78	4,347.79	
52 Te	27,472.3	27,201.7	30,995.7	3,769.33	3,758.8	4,029.58	4,301.7	4,570.9	
53 I	28,612.0	28,317.2	32,294.7	3,937.65	3,926.04	4,220.72	4,507.5	4,800.9	
54 Xe	29,779	29,458	33,624	4,109.9	—	—	—	—	
55 Cs	30,972.8	30,625.1	34,986.9	4,286.5	4,272.2	4,619.8	4,935.9	5,280.4	
56 Ba	32,193.6	31,817.1	36,378.2	4,466.26	4,450.90	4,827.53	5,156.5	5,531.1	
57 La	33,441.8	33,034.1	37,801.0	4,650.97	4,634.23	5,042.1	5,383.5	5,788.5	833
58 Ce	34,719.7	34,278.9	39,257.3	4,840.2	4,823.0	5,262.2	5,613.4	6,052	883
59 Pr	36,026.3	35,550.2	40,748.2	5,033.7	5,013.5	5,488.9	5,850	6,322.1	929
60 Nd	37,361.0	36,847.4	42,271.3	5,230.4	5,207.7	5,721.6	6,089.4	6,602.1	978
61 Pm	38,724.7	38,171.2	43,826	5,432.5	5,407.8	5,961	6,339	6,892	—
62 Sm	40,118.1	39,522.4	45,413	5,636.1	5,609.0	6,205.1	6,586	7,178	1,081
63 Eu	41,542.2	40,901.9	47,037.9	5,845.7	5,816.6	6,456.4	6,843.2	7,480.3	1,131
64 Gd	42,996.2	42,308.9	48,697	6,057.2	6,025.0	6,713.2	7,102.8	7,785.8	1,185
65 Tb	44,481.6	43,744.1	50,382	6,272.8	6,238.0	6,978	7,366.7	8,102	1,240
66 Dy	45,998.4	45,207.8	52,119	6,495.2	6,457.7	7,247.7	7,635.7	8,418.8	1,293
67 Ho	47,546.7	46,699.7	53,877	6,719.8	6,679.5	7,525.3	7,911	8,747	1,348
68 Er	49,127.7	48,221.1	55,681	6,948.7	6,905.0	7,810.9	8,189.0	9,089	1,406
69 Tm	50,741.6	49,772.6	57,517	7,179.9	7,133.1	8,101	8,468	9,426	1,462
70 Yb	52,388.9	51,354.0	59,370	7,415.6	7,367.3	8,401.8	8,758.8	9,780.1	1,521.4

Table 2.3.1 Continued

Element	Kα_1	Kα_2	Kβ_1	Lα_1	Lα_2	Lβ_1	Lβ_2	Lγ_1	Mα_1
71 Lu	54,069.8	52,965.0	61,283	7,655.5	7,604.9	8,709.0	9,048.9	10,143.4	1,581.3
72 Hf	55,790.2	54,611.4	63,234	7,899.0	7,844.6	9,022.7	9,347.3	10,515.8	1,644.6
73 Ta	57,532	56,277	65,223	8,146.1	8,087.9	9,343.1	9,651.8	10,895.2	1,710
74 W	59,318.24	57,981.7	67,244.3	8,397.6	8,335.2	9,672.35	9,961.5	11,285.9	1,775.4
75 Re	61,140.3	59,717.9	69,310	8,652.5	8,586.2	10,010.0	10,275.2	11,685.4	1,842.5
76 Os	63,000.5	61,486.7	71,413	8,911.7	8,841.0	10,355.3	10,598.5	12,095.3	1,910.2
77 Ir	64,895.6	63,286.7	73,560.8	9,175.1	9,099.5	10,708.3	10,920.3	12,512.6	1,979.9
78 Pt	66,832	65,112	75,748	9,442.3	9,361.8	11,070.7	11,250.5	12,942.0	2,050.5
79 Au	68,803.7	66,989.5	77,984	9,713.3	9,628.0	11,442.3	11,584.7	13,381.7	2,122.9
80 Hg	70,819	68,895	80,253	9,988.8	9,897.6	11,822.6	11,924.1	13,830.1	2,195.3
81 Tl	72,871.5	70,831.9	82,576	10,268.5	10,172.8	12,213.3	12,271.5	14,291.5	2,270.6
82 Pb	74,969.4	72,804.2	84,936	10,551.5	10,449.5	12,613.7	12,622.6	14,764.4	2,345.5
83 Bi	77,107.9	74,814.8	87,343	10,838.8	10,730.91	13,023.5	12,979.9	15,247.7	2,422.6
84 Po	79,290	76,862	89,800	11,130.8	11,015.8	13,447	13,340.4	15,744	—
85 At	81,520	78,950	92,300	11,426.8	11,304.8	13,876	—	16,251	—
86 Rn	83,780	81,070	94,870	11,727.0	11,597.9	14,316	—	16,770	—
87 Fr	86,100	83,230	97,470	12,031.3	11,895.0	14,770	14,450	17,303	—
88 Ra	88,470	85,430	100,130	12,339.7	12,196.2	15,235.8	14,841.4	17,849	—
89 Ac	90,884	87,670	102,850	12,652.0	12,500.8	15,713	—	18,408	—
90 Th	93,350	89,953	105,609	12,968.7	12,809.6	16,202.2	15,623.7	18,982.5	2,996.1
91 Pa	95,868	92,287	108,427	13,290.7	13,122.2	16,702	16,024	19,568	3,082.3
92 U	98,439	94,665	111,300	13,614.7	13,438.8	17,220.0	16,428.3	20,167.1	3,170.8
93 Np	—	—	—	13,944.1	13,759.7	17,750.2	16,840.0	20,784.8	—
94 Pu	—	—	—	14,278.6	14,084.2	18,293.7	17,255.3	21,417.3	—
95 Am	—	—	—	14,617.2	14,411.9	18,852.0	17,676.5	22,065.2	—

Adapted from *X-ray Data Booklet*, Published by the Center for X-ray optics and Advanced Light Source: LBNL/PUB-490 Rev-2 (2001).

2.3.3 Non-Radiative Auger Electron Emission

Auger electron emission (§1.4.1) was discovered independently by Lise Meitner (in 1922) and Pierre Auger (in 1923), but named only after the latter! When an atom is ionized with the formation of a core hole by either an incident photon or electron of sufficient energy, the excited atom can lower its energy by filling the core hole with an electron from an outer level, along with the emission of energy. The emitted energy can be in the form of characteristic X-ray photons, following dipole selection rules (§2.3.2). Alternatively, the de-excitation process can be accompanied by a nonradiative emission of electrons (Fig. 2.3.6), known as Auger emissions. Note that in the process of de-excitation and Auger electron emission, the atom is left in its final state with two vacancies, or holes. Further, in the case of

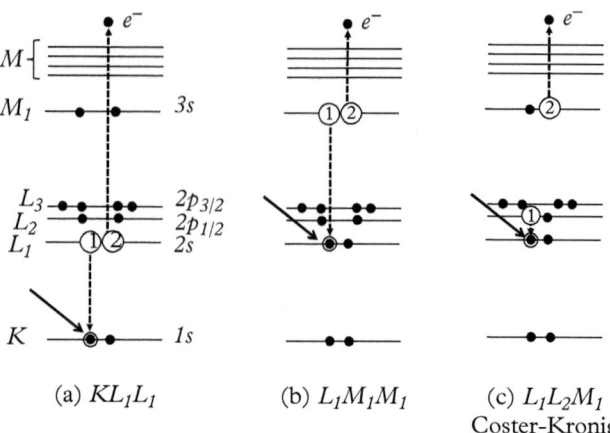

(a) KL_1L_1 (b) $L_1M_1M_1$ (c) $L_1L_2M_1$ Coster–Kronig

Figure 2.3.6 Schematic representation of three different two-electron (ABC) de-excitation processes. (a) The KL_1L_1 transition, starting with a core hole in the K-shell (step A), which is filled with an L_1 electron (step B); the other L_1 electron is ejected to the continuum (step C). Table 2.3.2 gives the energies of all KLL lines. (b) A similar L_1M_1M Auger process but with L-shell core hole. (c) The $L_1L_2M_1$ line, known as a Coster–Kronig process. Note that in each case, eventually two holes, as indicated by numbers 1 and 2, are formed.

Auger electron emission, the dipole selection rules are not followed. Figure 2.3.6 shows three cases, i.e. KL_1L_1, $L_1M_1M_1$, and $L_1L_2M_1$, of Auger emission. The $L_1L_2M_1$ line, specifically known as a Coster–Kronig process, where the initial and final vacancies are in the same shell (but not in the same sub-shell), tends to have a higher probability of occurring than the other Auger lines.

The kinetic energy of a specific Auger electron, in the general case involving three levels, A, B, and C, is given by

$$E_{ABC} = E_A - E_B - E_C \qquad (2.3.3)$$

where, E_A, E_B, and E_C are one-electron binding energies, and E_{ABC} is the kinetic energy of the emitted Auger electron. In the case of the KL_1L_1 line shown in Figure 2.3.6, (2.3.3) can be specifically written as

$$E_{KL_1L_1} = E_{B,K} - E_{B,L_1} - E_{B,L_1} \qquad (2.3.3a)$$

Note that, typically, one core level (E_A) that is characteristic of the atomic species and remains unchanged in any bonding, is always involved in the Auger emission. Hence this is a form of element-specific, core-level spectroscopy. This relationship, (2.3.3), though reasonable, is not a very accurate energy description of the Auger electron, for it does not take into account that the true energy is the difference between the binding energies of a one-hole and two-hole state in the atom. Thus, for a proper description of the kinetic energy of the Auger electron, the common practice is to rewrite (2.3.3) as

$$E_{ABC} = E_A - E_B - E_C - U \tag{2.3.4}$$

where, $U = H - P$, with H being the hole-hole interaction energy and P takes care of the extra atomic screening or relaxation effects in the presence of the holes. Typically, U, is a parameter that is experimentally determined. However, the kinetic energy of the Auger electron for an element, Z, can also be empirically written as

$$E_{ABC}^Z = E_A^Z - E_B^Z - E_C^Z - \frac{1}{2}\left(E_C^{Z+1} - E_C^Z + E_B^{Z+1} - E_B^Z\right) \tag{2.3.4a}$$

Recall that the emission of characteristic X-rays and Auger electrons are complementary processes (Fig. 1.4.5). The yields of X-rays (W_X) and Auger electrons (W_A) are also given semi-empirically as a function of atomic number, Z:

$$\frac{W_X}{W_A} = \left(-a + bZ - cZ^3\right)^4 \tag{2.3.5}$$

where $a = 6.4 \times 10^{-2}, b = 3.4 \times 10^{-2}$, and $c = 1.03 \times 10^{-6}$. A typical Auger electron spectrum (Fig. 2.6.6) is a plot of the intensity versus kinetic energy and further details of this spectroscopy technique for surface chemical analysis are discussed in §2.6.2.

Example 2.3.2:

   (a) Estimate the energy of the KL_1L_2 Auger electrons for Ni.

   (b) Also calculate the fraction of Auger electron yield for Ni.

Solution:

   (a) We shall apply (2.3.4a) and determine this empirically. The electron binding energies (keV) can be obtained from Table 2.2.2. They are $E_K^{Ni} = 8.333, E_{L_1}^{Ni} = 1.008, E_{L_2}^{Ni} = 0.872, E_{L_2}^{Cu} = 0.951, E_{L_1}^{Cu} = 1.096$ Then, from (2.3.4a), we get

$$\begin{aligned} E_{KL_1L_2}^{Ni} &= E_K^{Ni} - E_{L_1}^{Ni} - E_{L_2}^{Ni} - \tfrac{1}{2}\left(E_{L_2}^{Cu} - E_{L_2}^{Ni} + E_{L_1}^{Cu} - E_{L_1}^{Ni}\right) \\ &= 8.333 - 1.008 - 0.872 - \tfrac{1}{2}(0.951 - 0.872 + 1.096 - 1.008) \\ &= 6.369 \text{ keV} \end{aligned}$$

   The observed value (Table 2.3.2) 6.384 keV, is in reasonably good agreement.

Table 2.3.2 KLL Auger lines (eV).

		$2s^0 2p^6$	$2s^1 2p^5$				$2s^2 2p^4$			
		$^1S_0$ KL_1L_1	$^1P_1$ KL_1L_2	$^3P_0$ KL_1L_2	$^3P_1$ KL_1L_3	$^3P_2$ KL_1L_3	$^1S_0$ KL_2L_2	$^1D_2$ KL_2L_3	$^3P_0$ KL_3L_3	$^3P_2$ KL_3L_3
C	6	243	252	258	258	258	265	266	267	267
N	7	356	362	369	369	369	373	375	377	377
O	8	474	486	495	495	495	504	507	509	509
F	9	610	627	638	638	638	650	654	657	657
Ne	10	761	781	794	794	794	808	813	816	816
Na	11	928	952	967	967	967	984	989	993	993
Mg	12	1 105	1 135	1 151	1 151	1 151	1 172	1 179	1 183	1 183
Al	13	1 301	1 336	1 354	1 354	1 354	1 379	1 387	1 392	1 392
Si	14	1 516	1 554	1 574	1 574	1 575	1 602	1 611	1 616	1 617
P	15	1 742	1 784	1 805	1 806	1 806	1 835	1 845	1 851	1 852
S	16	1 982	2 034	2 057	2 058	2 059	2 096	2 107	2 114	2 115
Cl	17	2 249	2 305	2 329	2 330	2 331	2 370	2 382	2 389	2 391
A	18	2 527	2 586	2 612	2 613	2 614	2 656	2 669	2 677	2 679
K	19	2 815	2 881	2 909	2 910	2 912	2 959	2 973	2 981	2 984
Ca	20	3 122	3 195	3 224	3 225	3 227	3 279	3 294	3 303	3 306
Sc	21	3 456	3 533	3 563	3 504	3 567	3 622	3 638	3 647	3 651
Ti	22	3 799	3 886	3916	3 919	3 922	3 985	4 002	4011	4 016
V	23	4 168	4 259	4 290	4 293	4 298	4 362	4 381	4 391	4 397
Cr	24	4 557	4 651	4 683	4 687	4 692	4 757	4 778	4 788	4 795
Mn	25	4 956	5 056	5 089	5 094	5 100	5 169	5191	5 202	5 211
Fe	26	5 374	5 480	5 514	5 519	5 527	5 598	5 622	5 634	5 644
Co	27	5 808	5 923	5 957	5 964	5 972	6 049	6 075	6 088	6 099
Ni	28	6 264	6 384	6 419	6 426	6 436	6 514	6 542	6 556	6 568
Cu	29	6 732	6 861	6 896	6 905	6 916	7 000	7 030	7 045	7 059
Zn	30	7 214	7 348	7 384	7 394	7 407	7 493	7 526	7 543	7 558
Ga	31	7 712	7 852	7 888	7 900	7 915	8 000	8 037	8 057	8 073
Ge	32	8 216	8 365	8 401	8 416	8 433	8 523	8 563	8 586	8 603
As	33	8 749	8 903	8 939	8 957	8 975	9 063	9 107	9 133	9 152
Se	34	9 283	9 447	9 483	9 504	9 524	9 616	9 665	9 695	9 715
Br	35	9 840	10 014	10 049	10 074	10 096	10 189	10 244	10 279	10 300
Kr	36	10 412	10 594	10 630	10 658	10 682	10 777	10 837	10 877	10 899
Rb	37	10 995	11 186	11 221	11 255	11 280	11 376	11 442	11 487	11 511
Sr	38	11 595	11 795	11 830	11 870	11 897	11 992	12 066	12 118	12 143

	Z									
Y	39	12 213	12 422	12 457	12 503	12 532	12 626	12 708	12 767	12 793
Zr	40	12 851	13 069	13 104	13 157	13 188	13 279	13 370	13 437	13 464
Nb	41	13 505	13 731	13 766	13 827	13 860	13 948	14 049	14 125	14 153
Mo	42	14 179	14 414	14 449	14 519	14 554	14 639	14 750	14 836	14 865
Te	43	14 867	15 111	15 146	15 226	15 263	15 343	15 466	15 563	15 593
Ru	44	15 574	15 827	15 862	15 952	15 991	16 066	16 202	16 310	16 341
Rh	45	16 298	16 560	16 595	16 697	16 738	16 806	16 956	17 077	17 109
Pd	46	17 040	17 312	17 347	17 462	17 504	17 505	17 729	17 864	17 897
Ag	47	17 797	18 078	18 113	18 242	18 286	18 339	18 519	18 668	18 702
Cd	48	18 568	18 857	18 892	19 037	19 082	19 125	19 322	19 488	19 523
In	49	19 354	19 653	19 688	19 849	19 896	19 930	20 144	20 327	20 364
Sn	50	20 157	20 465	20 501	20 680	20 728	20 750	20 984	21 185	21 223
Sb	51	20 977	21 295	21 331	21 529	21 579	21 588	21 844	22 066	22 104
Te	52	21 814	22 142	22 179	22 398	22 449	22 444	22 722	22 965	23 005
I	53	22 668	23 006	23 043	23 284	23 338	23 316	23 618	23 884	23 925
Xe	54	23 527	23 879	23 916	24 182	24 237	24 201	24 530	24 822	24 863
Cs	55	24 426	24 783	24 820	25 111	25 167	25 109	25 463	25 781	25 823
Ba	56	25 330	25 697	25 735	26 053	26 111	26 033	26 416	26 762	26 805
La	57	26 251	26 631	26 669	27 018	27 077	26 978	27 393	27 769	27 813
Ce	58	27 201	27 590	27 628	28 009	28 069	27 945	28 393	28 802	28 847
Pr	59	28 171	28 572	28 610	29 024	29 086	28 936	29 420	29 863	29 909
Nd	60	29 163	29 574	29 612	30 063	30 126	29 947	30 468	30 948	30 905
Pm	61	30 170	30 592	30 631	31 120	31 184	30 976	31 537	32 056	32 104
Sm	62	31 199	31 631	31 671	32 200	32 266	32 024	32 627	33 186	33 235
Eu	63	32 247	32 690	32 730	33 303	33 370	33 092	33 740	34 345	34 395
Gd	64	33 315	33 769	33 809	34 429	34 497	34 182	34 877	35 528	35 579
Tb	65	34 402	34 868	34 909	35 576	35 646	35 291	36 036	36 736	36 788
Dy	66	35 512	35 988	36 029	36 749	36 820	36 421	37 220	37 972	38 025
Ho	67	36 640	37 127	37 169	37 944	38 016	37 570	38 425	39 234	39 287
Er	68	37 788	38 287	38 329	39 162	39 236	38 740	39 655	40 522	40 576
Tm	69	38 958	39 469	39 512	40 406	40 481	39 934	40 911	41 840	41 895
Yb	70	40 151	40 674	40 716	41 675	41 752	41 149	42 192	43 186	43 242
Lu	71	41 361	41 897	41 940	42 967	43 045	42 383	43 496	44 559	44 617
Hf	72	42 589	43 137	43 181	44 280	44 359	43 635	44 821	45 967	46 015

Table 2.3.2 Continued

Element	Z	$2s^0 2p^6$	$2s^1 2p^5$				$2s^2 2p^4$			
		$^1S_0$ KL_1L_1	$^1P_1$ KL_1L_2	$^3P_0$ KL_1L_2	$^3P_1$ KL_1L_3	$^3P_2$ KL_1L_3	$^1S_0$ KL_2L_2	$^1D_2$ KL_2L_3	$^3P_0$ KL_3L_3	$^3P_2$ KL_3L_3
Ta	73	43 831	44 391	44 436	45 611	45 691	44 900	46 164	47 377	47 436
W	74	45 097	45 671	46 715	46 971	47 053	46 193	47 538	48 831	48 891
Re	75	46 385	46 972	47 018	48 357	48 440	47 507	48 938	50 315	50 376
Os	76	47 690	48 291	48 337	49 767	49 851	48 839	50 361	51 830	51 892
Ir	77	49 022	49 636	49 682	51 205	51 291	50 195	51 812	53 375	53 437
Pt	78	50 375	51 003	51 050	52 672	52 759	51 575	53 292	54 954	55 017
Au	79	51 752	52 393	52 440	54 167	54 255	52 978	54 801	56 568	56 633
Hg	80	53 149	53 802	53 849	55 685	55 774	54 397	56 330	58 206	58 272
Tl	81	54 554	55 227	55 275	57 225	57 316	55 840	57 890	59 882	59 948
Pb	82	55 992	56 677	56 726	58 799	58 891	57 302	59 476	61 591	61 658
Bi	83	57 451	58 155	58 205	60 402	60 495	58 799	61 098	63 338	63 406
Po	84	58 918	59 640	59 690	62 026	62 120	60 299	62 739	65 118	65 187
At	85	60 427	61 163	61 213	63 689	63 784	61 836	64 416	66 935	67 005
Rn	86	61 980	62 720	62 771	65 392	65 489	63 397	66 124	68 789	68 860
Fr	87	63 523	64 286	64 337	67 114	67 212	64 983	67 868	70 690	70 762
Ra	88	65 103	66 887	65 939	68 879	68 978	66 604	69 654	72 640	72 712
Ac	89	66 720	67 509	67 562	70 673	70 774	68 232	71 453	74 611	74 684
Th	90	68 341	69 153	69 207	72 498	72 600	69 898	73 302	76 040	76 714
Pa	91	70 016	70 842	70 896	74 373	74 476	71 599	75 190	78 714	78 789
U	92	71 704	72 550	72 604	76 280	76 384	73 327	77 116	80 839	80 916
Np	93	73 437	74 297	74 351	78 236	78 342	75 085	79 086	83 019	83 096
Pu	94	75 204	76 080	76 135	80 237	80 344	76 884	81 103	85 254	86 332
Am	95	77 060	77 930	77 985	82 317	82 425	78 727	83 177	87 558	87 837
Cm	96	78 867	79 590	79 646	84 386	84 495	80 240	85 099	89 888	89 968
Bk	97	80 594	81 528	81 585	86 408	86 518	82 388	87 331	92 204	92 284
Cf	98	83 286	84 187	34 245	89 453	89 565	85 017	90 348	95 607	95 688
Es	99	85 219	86 146	86 204	91 701	91 814	86 997	92 617	98 165	98 248
Fm	100	87 205	88 144	88 203	93 998	94 113	89 006	94 926	100 774	100 857
Md	101	89 221	90 192	90 251	96 356	96 471	91 085	97 315	103 472	103 556
No	102	91 267	92 260	92 320	98 763	98 880	93 173	99 744	106 240	106 325
Lr	103	93 373	94 388	94 448	101 250	101 368	95 322	102 252	109 108	109 194
Ku	104	95 518	96 555	96 615	103 796	103 915	97 510	104 820	112 055	112 142

Adapted from Siegbahn et al. (1967).

(b) To calculate the fractional Auger electron yield for Ni ($Z = 28$) we apply (2.3.5) and get

$$\frac{W_X}{W_A} = \left(-6.4 \times 10^{-2} + 3.4 \times 10^{-2} \times 28 - 1.03 \times 10^{-6} \times 28^3\right)^4$$
$$= 0.8654$$

But $W_X + W_A = 1$. Hence $W_A = 0.5361$ or 53.6%.

2.3.4 Electron Photoemission

For all practical purposes, the emission of electrons from atoms upon photon incidence is a very simple process involving a single transition (Fig. 1.4.2). An electron with a specific binding energy, E_B, absorbs a photon of energy, hf, and then the same electron emerges from the atom with a kinetic energy, $E_K = hf - E_B$. Thus, the observed energy distribution of all photo-emitted electrons is simply related to the binding energies of all electrons in the atom by a simple shift of the energy scale by hf. In a solid, this simple picture is valid only if the probability of the photon being absorbed by all the electronic states in the solid is the same; unfortunately, this assumption is seldom true. Further in a solid, the choice of incident photon is simple; as long as hf also exceeds the *work function*, Φ, or the energy required for the electron to escape from the surface of the material, it can be used. In practice, soft X-ray Al Kα (1486.7 eV) and Mg Kα (1253.6 eV) photons, readily available using laboratory sources, are employed.

Strictly speaking, the work function of the specimen and spectrometer should be accounted for in the energy balance. If the specimen and spectrometer are in electrical contact and thus in thermodynamic equilibrium, their Fermi levels are equal and then the measured kinetic energy (Fig. 2.3.7) $E_{K,meas} = hf - E_B - \Phi_{spect}$. XPS is introduced further in §2.6.3 and discussed in detail in §3.8.

2.4 X-Rays as Probes: Generation and Transmission of X-Rays

In principle, there are two ways of generating X-rays using high-energy electrons to bombard a metal surface such as in an X-ray tube (Fig. 2.4.2). The electrons ionize the atoms on the surface of the target metal, and in the subsequent relaxation process, following dipole selection rules, produce characteristic X-ray photons (§2.3.1). In addition, all charged particles emit electromagnetic radiation when they are accelerated or decelerated. Thus, the high-energy electrons, on deceleration in the target metal, generate a continuous spectrum of radiation known as breaking radiation, or *bremsstrahlung* in German. A typical X-ray

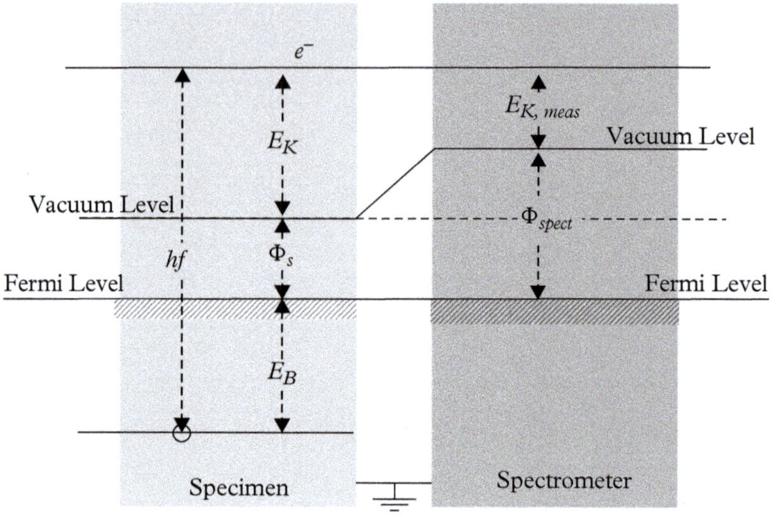

Figure 2.3.7 Schematic of the energy levels of the specimen and spectrometer in thermodynamic equilibrium. The measured kinetic energy of the photoemitted electron is now $E_{K,meas} = hf - E_B - \Phi_{spect}$, where all energy terms are marked in the figure.

Adapted from Feldman and Mayer (1986).

spectrum Fig. 2.4.1 produced by the bombardment of a metal (Mo) target is a combination of these two processes. A continuous band of bremsstrahlung X-rays, starting with a ridge corresponding to the energy of the incident electrons, λ_{SWL}, and continuing through much larger wavelengths (lower energies) is observed as the background. Superimposed on this background are the characteristic X-ray peaks.

Accelerating charged particles (electrons or positrons), such as those that are made to travel in curved trajectories using the Lorentz force (5.2.10) generated by a magnetic field, also emit X-rays of much higher intensities. In addition, if the charged particles were to move at relativistic speeds, as in a synchrotron, the emitted radiation is directed tangentially outward with respect to the particle motion in the shape of a narrow cone.

Details of X-ray generation in laboratory-scale X-ray tubes and synchrotron sources, with access available to the latter as national user facilities in many countries, are discussed later.

2.4.1 Laboratory Sources and Methods of X-Ray Generation

Figure 2.4.2 shows a schematic drawing of an X-ray tube. It contains a filament heated to high temperatures as the source of electrons, a set of two electrodes—one of which is the target (anode)—across which a high voltage, V, is applied to accelerate the electrons rapidly towards the water-cooled target (anode). All these components are sealed in a vacuum tube. X-rays produced at the point of impact are guided out of the tube through windows. Since only 1% of the electron

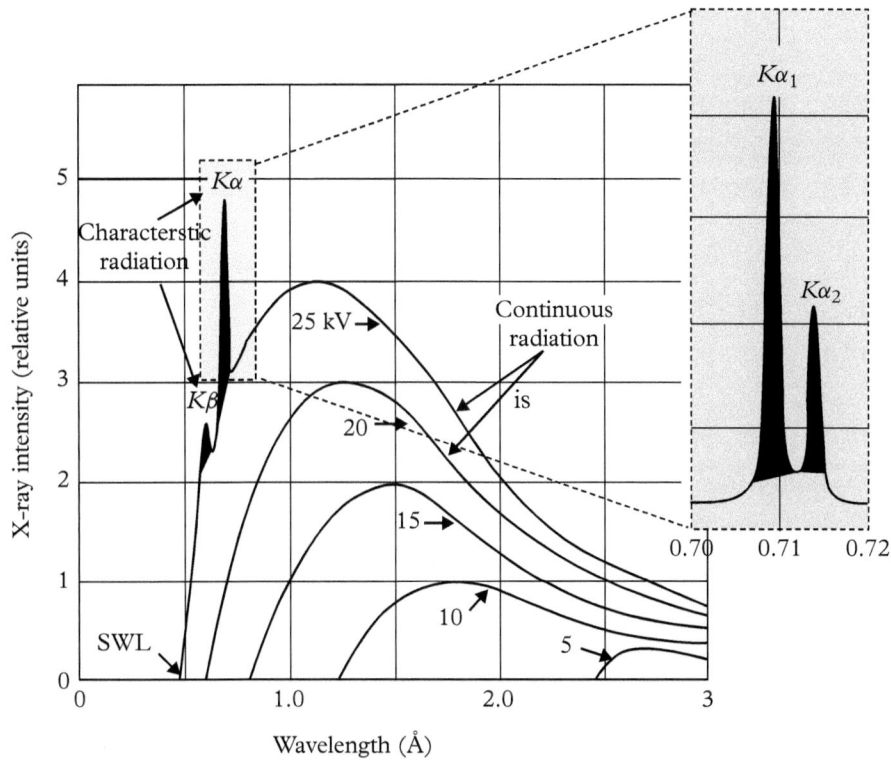

Figure 2.4.1 X-ray intensity as a function of wavelength for a Mo X-ray tube, as a function of applied voltage (kV). The natural line widths ($K\alpha_1 \sim 2$ eV, $\lambda_{\mathrm{FWHM}} \sim 0.001$Å) are not to scale. Inset shows details of the $K\alpha$ emission split into $K\alpha_1$ and $K\alpha_2$ emissions due to spin–orbit coupling.

Adapted from Cullity (1977).

energy is converted to X-rays—the remaining 99% is converted to heat—extensive cooling of the target is required.

Figure 2.4.1 shows how the intensity of the X-rays varies with wavelength. It is zero up to a certain wavelength, λ_{SWL}, called the short wavelength limit (SWL), and corresponds to the energy of the incident electrons, and then increases rapidly to a maximum before decreasing gradually as the wavelength increases. Electrons in the tube that are stopped in one impact by the target, generate photons with the maximum energy (or, minimum wavelength, λ) by transferring all their energy (eV) to the photon $\left(hf_{\mathrm{max}}\right)$. Thus

$$eV = hf_{\mathrm{max}} = hc/\lambda_{\mathrm{min}}, \text{ or } \lambda_{\mathrm{min}} = \lambda_{\mathrm{SWL}} = \frac{hc}{eV} = \frac{12.39 \times 10^3}{V}(\text{Å}) \qquad (2.4.1)$$

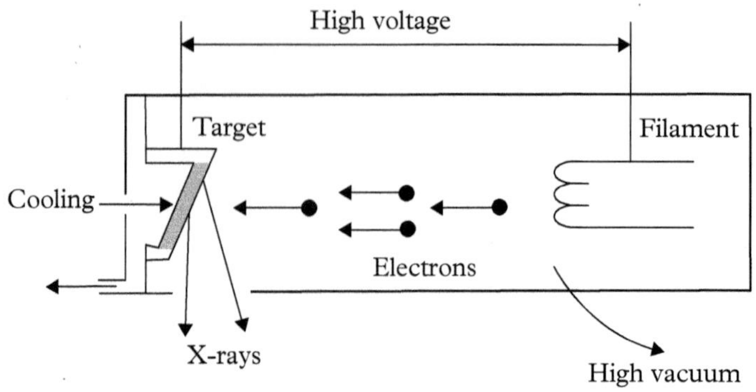

Figure 2.4.2 A schematic drawing of an X-ray tube showing the principal components.

Note that λ_{SWL} decreases as the high voltage in the tube, V, increases. If the electron does not lose all its energy in a single impact, but continues to move in the metal target with reduced velocity after the impact, then the emitted photon has energy less than hf_{max}, and its wavelength is $\lambda > \lambda_{SWL}$. The bremsstrahlung reflects the total contribution of all such electrons that undergo partial transfer of energy to the photons. As the applied voltage in the tube is increased, both the average energy of the photons and the total number of photons increase, explaining the observed variation with voltage in Figure 2.4.1. The total intensity, I_B, of the X-ray background, given by the area under any one curve, is a function of the atomic number, Z, of the target, the voltage, V, of the tube, and the current, i. It can be written as $I_B \propto iZV^m$, where the exponent, $m \sim 2$.

Superimposed on the background are the characteristic X-ray lines. For, Mo, with a binding energy, $E_B \sim 20.002$ keV, for the $1s$ electron, a K-line only appears if the excitation voltage, V, of the electron is larger than this value; thus, these characteristic emission lines do not appear in the lower energy curves in Figure 2.4.1. The intensity of the K-line, I_{K-line}, is given by $I_{K-line} \propto i(V - E_B)^n$, where $1 < n < 2$. The narrow width of the K-line, defined by its full width at half maximum, $\lambda_{FWHM} < 0.001$ Å, makes high resolution XPS and diffraction from crystals possible.

From the dipole selection rules discussed in §2.3.1, a K-shell ($1s$) vacancy can produce $K\alpha_1$, $K\alpha_2$, or $K\beta$ radiation. In principle, one ionized atom may produce a $K\alpha_1$ photon, whilst another ion may produce a $K\beta$ photon. In practice, it is more probable that a K-shell ($1s$) vacancy is filled by a L-shell ($2p$) electron ($K\alpha$ radiation) than by a M-shell ($3p$) electron ($K\beta$ radiation); thus, the former shows a much higher intensity. Finally, the $1s$ electrons (K-shell) in the atom have a larger binding energy compared to the $2s$ (L-shell) or $3s$ (M-shell) electrons. Thus, if the high voltage electron generated in the X-ray tube has sufficient energy to ionize a K-shell, it certainly will have more than enough energy to ionize the outer L- and M-shell. Hence, the characteristic K-radiation is always accompanied by the lower energy (higher wavelength) L- and M-radiation.

Most spectroscopy and diffraction experiments require *monochromatic* X-ray photons to enhance energy resolution and avoid interference effects. Absorption methods are used to filter out all the emissions other than the one that is of interest; this is discussed in the next section.

From a practical point of view, the X-ray tube is less than 1% efficient in producing X-rays, as most of the energy of the high-voltage electrons, impinging on the target, is converted to heat. One solution to increase the fraction of X-ray production is to have a rotating anode. By constantly rotating the anode, it always presents a fresh surface of the target metal, thereby increasing the amount of power that can be applied without generating too much heat. The rotating geometry provides additional challenges in delivering cooling water without breaking vacuum, and such machines with 5–10 times more power compared to X-ray tubes, are commercially available.

Example 2.4.1: Design an X-ray tube using Ag as the target material.

(a) Assuming that the Ag $K\alpha$ line has 2/3 contribution from the $K\alpha_1$ line and 1/3 contribution from the $K\alpha_2$ line, what is its wavelength?

(b) What is the difference in energies between the $K\alpha$ and $K\beta$ photons of Ag?

(c) What is the excitation voltage required for Ag?

(d) What would be the short wavelength limit for such a tube if the applied voltage is 10 kV greater than the excitation voltage?

Solution:

(a) From Table 2.3.1 the energies are $E(K\alpha_1) = 22,163$ eV, and $E(K\alpha_2) = 21,990$ eV.
The corresponding wavelengths ($\lambda \sim 12,400/E$) are $\lambda(K\alpha_1) = 0.5595$ Å, and $E(K\alpha_2) = 0.5662$ Å. Hence, the Ag $K\alpha$ would be observe at a wavelength of $\lambda(K\alpha) = 0.667 * 0.5595 + 0.333 * 0.5662 = 0.5619$Å. This corresponds to an energy $E(K\alpha) = \sim 12,400/\lambda = 22,068$ eV

(b) From Table 2.3.1, $E(K\beta) = 24,942$ eV.
Thus, the difference in energy is $24,942 - 22,068 = 2874$ eV.

(c) From Table 2.2.2, the binding energy for Ag, K_{1s} is 25,515 eV, which is also the excitation voltage required for this Ag tube.

(d) Therefore, the applied tube voltage is $25,515 + 10,000 = 35,515$ V. Hence the short wavelength limit for this Ag tube would be $\lambda_{SWL} = 12,400/35,515 = 0.3491$ Å.

2.4.2   X-Ray Absorption and Filtering

X-ray photons are both transmitted and absorbed by a material depending on its thickness. It has been shown experimentally that the fractional decrease, $-dI/I$, of X-ray intensity is proportional to the distance, dx, travelled, i.e. $-dI/I = \mu\, dx$, where μ is the linear absorption coefficient of the material. By integrating this simple relationship, we get

$$I_x = I_0 \exp\left(-\mu x\right) \tag{1.3.11}$$

where, I_0, is the initial intensity, and I_x is the intensity of the X-ray beam after traveling a distance, x, in the material. It is common practice to define a *mass absorption coefficient*, μ/ρ, where ρ is the density, such that it is independent of the physical state (liquid, solid, gas) of the material, and rewrite (1.3.11) as

$$I_x = I_0 \exp -\left(\frac{\mu}{\rho}\right)\rho x \tag{2.4.2}$$

In multi-element materials, with weight fractions, w_i, for each one, the effective mass absorption coefficient is $\frac{\mu}{\rho} = \sum_i w_i\left(\frac{\mu}{\rho}\right)_i$.

The mass absorption coefficients, μ/ρ, are tabulated in the literature [10], and adapted for many common elements and X-ray wavelengths in Table 2.4.1. In general, μ/ρ increases with increasing atomic number, Z, but decreases with increasing energy (decreasing wavelength) of the photon (Fig. 2.4.3). However, superimposed on the monotonically decreasing behavior of μ/ρ with wavelength, sharp jumps at well-specified wavelengths are seen, and marked in the figure as L_I, L_{II}, L_{III}, and K. These are wavelengths (energies) at which the incident X-ray photon ionizes the material and generates a photoelectron (§2.3.4). Thus, each of these steps (or jumps), or absorption edges, in the mass absorption coefficient marks the wavelength (or energy) when the X-ray photon has the right energy to eject a specific L, M, or N photoelectron from the atom. Between these edges, the observed monotonic behavior can be expressed mathematically as

$$\frac{\mu}{\rho} = k\lambda^3 Z^3 \tag{2.4.3}$$

where k is a different constant for different branches of the curve between the edges.

In practice, the absorption features can be used to filter and narrow the energy width of the X-ray radiation (Fig. 2.4.4). We can choose a material with an absorption edge wavelength (energy) located slightly lower (higher) than that of the $K\alpha$ radiation of interest. Then the $K\beta$ radiation and all radiations including the background up to the absorption edge of the filter material can be removed (Fig. 2.4.4). Note that the thicker the filter–typically, only a thin foil is used–the greater the ratio of $K\alpha$ to $K\beta$ in the transmitted beam. Table 2.4.2 shows specific

Table 2.4.1 Mass absorption coefficients (cm^2/g) and densities ($g\ cm^{-3}$) for select elements and X-ray radiations.

Absorbing Element	Density ($g\ cm^{-3}$)	Mo Kα 0.711 Å	Mo Kβ 0.632 Å	Cu Kα 1.542 Å	Cu Kβ 1.392 Å	Co Kα 1.790 Å	Co Kβ 1.621 Å	Cr Kα 2.291 Å	Cr Kβ 2.085 Å
Be	1.85	0.2451	0.2216	1.007	0.7742	1.522	1.152	3.183	2.388
C$_{graphite}$	2.27	0.5348	0.4285	4.219	3.093	6.683	4.916	14.46	10.76
O	1.33×10^{-3}	1.147	0.8545	11.03	8.062	17.44	12.85	37.19	27.88
Na	0.966	2.939	2.098	30.30	22.23	47.34	35.18	98.48	74.66
Mg	1.74	3.979	2.825	40.88	30.08	63.54	47.38	130.8	99.62
Al	2.7	5.043	3.585	50.23	37.14	77.54	58.08	158.0	120.7
Si	2.33	6.533	4.624	65.32	48.37	100.4	75.44	202.7	155.6
Ca	1.53	19.00	13.56	171.4	129.0	257.4	196.4	499.6	389.3
Ti	4.51	23.25	16.65	202.4	153.2	300.5	231.0	571.4	449.0
V	6.09	25.24	18.07	222.6	168.0	332.7	254.7	75.1	501.0
Cr	7.19	29.25	20.99	252.3	191.1	375.0	288.1	85.71	65.79
Mn	7.47	31.86	22.89	272.5	206.7	405.1	311.2	96.08	73.75
Fe	7.87	37.74	27.21	304.4	233.6	56.25	345.5	113.1	86.77
Co	8.8	41.02	29.51	338.6	258.7	62.86	47.7	124.6	96.1
Ni	8.91	47.24	34.18	48.83	282.8	73.75	56.05	145.7	112.5
Cu	8.93	49.34	35.77	51.54	38.74	78.11	59.22	155.2	119.5
Zn	7.13	55.46	40.26	59.51	45.3	88.71	68.00	171.7	133.5
Ga	5.91	56.9	41.69	62.13	46.65	94.15	71.39	186.9	144.0
Ge	5.32	60.47	44.26	67.92	51.44	102.0	77.79	199.9	154.5
As	5.78	65.97	48.57	75.65	57.01	114.0	86.76	224.0	173.3
Y	4.48	97.56	72.57	127.1	96.19	190.2	145.4	368.9	286.9
Zr	6.51	16.1	75.2	136.8	103.3	204.9	156.6	398.6	309.7
Nb	8.56	16.96	81.22	148.8	112.3	222.9	170.4	431.9	336.4
Mo	10.22	18.44	13.29	158.3	119.7	236.6	181.0	457.4	356.5
Pd	12.0	24.42	17.63	205.0	155.6	304.3	234.0	580.9	455.1
Ag	10.5	26.38	19.1	218.1	165.8	323.5	248.9	617.4	483.5
Nd	7.0	53.28	38.88	417.9	319.8	531.7	475.9	271.3	213.4
Sm	7.54	57.96	42.4	453.5	346.6	411.8	446.3	295.0	231.5
Ta	16.67	89.51	66.07	161.5	123.9	238.3	183.9	454.7	355.0
W	19.25	95.75	70.57	170.5	131.5	249.7	193.7	470.4	369.1
Pt	21.44	108.6	80.23	198.2	151.2	295.2	226.4	571.6	443.9
Au	19.28	111.3	82.33	207.8	160.6	303.3	235.7	568.0	446.7
Pb	11.34	122.8	90.55	232.1	178.6	340.8	263.8	644.5	504.9

Adapted from Cullity (1977).

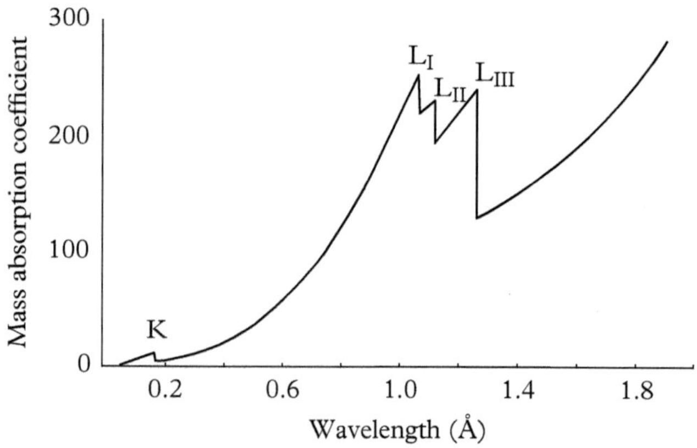

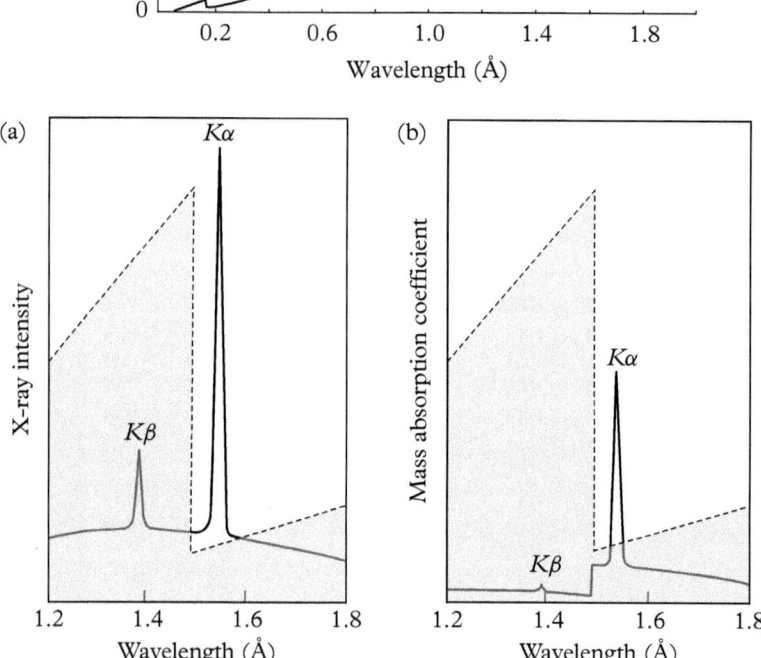

Figure 2.4.3 Mass absorption coefficients, μ, for Ba showing the K- and L-absorption edges superimposed on the monotonically decreasing background.

Adapted from Leng (2013).

Figure 2.4.4 Schematic representation of X-ray radiation from a Cu tube (a) before and (b) after filtering through a thin foil of Ni. The shaded (and dashed) curve is the mass absorption coefficient of Ni.

Adapted from Cullity (1977).

Table 2.4.2 Filters for suppression of $K\beta$ radiation.

Target	$K\alpha$ (eV)	$K\beta$ (eV)	Filter	Absorption edge (eV)	Incident beam $I(K\alpha)/I(K\beta)$	Filter thickness for $I(K\alpha)/I(K\beta) = 500$
Mo	17,479	19,608	Zr	17,998	5.4	0.117 mm
Cu	8,048	8,905	Ni	8,333	7.5	0.020 mm
Co	6,930	7,649	Fe	7,112	9.4	0.018 mm
Fe	6,404	7,058	Mn	6,539	9.0	0.018 mm
Cr	5,415	5,947	V	5,465	8.5	0.015 mm

filter materials that are used to suppress $K\beta$ radiation for a variety of common X-ray sources.

> **Example 2.4.2:** For a Cu X-ray tube, show that the Ni filter thickness required for obtaining
> $I(K\alpha)/I(K\beta) = 500$ is 0.020 mm. Use data from Table 2.4.1 and Table 2.4.2.
>
> **Solution:** From Table 2.4.2, for the incident beam $I_0(K\alpha)/I_0(K\beta) = 7.5$
> Assume that the Ni filter is x cm thick. Let $I_0(K\alpha) = 1$. Then $I_0(K\beta) = 1/7.5 = 0.1333$
> Applying (2.4.2), and using the data in Table 2.4.1, we get
>
> $$I_x(K\alpha) = I_0(K\alpha) \exp - (48.83 * 8.91) x$$
>
> $$I_x(K\beta) = I_0(K\beta) \exp - (282.8 * 8.91) x$$
>
> Now, $I_x(K\alpha)/I_x(K\beta) = 500 = [I_0(K\alpha)/I_0(K\beta)]\exp[(282.8 - 48.8) * 8.91x]$
> $= 7.5 \exp(234 * 8.91x)$
> Thus $x = 0.002$ cm or 0.020 mm (as shown in Table 2.4.2).

2.4.3 Synchrotron Sources of X-Ray Radiation

Accelerating or decelerating charged particles emit radiation. For charged particles traveling in a curved path at relativistic velocities, v, defined by the Lorentz factor $\gamma = 1/\sqrt{(1 - v^2)/c^2}$, where c is the velocity of light, the emitted radiation is directed tangential to the path with an emission angle of $1/\gamma$, in radians (Fig. 2.4.5a). Typical values of γ, for two synchrotron radiation sources, are $\gamma = 3,720$ (Advanced Light Source (ALS), Berkeley, USA, optimized for soft X-rays) and $\gamma = 13,700$ (Advanced Photon Source (APS), Argonne, USA, optimized for hard X-rays). Magnets are used in synchrotron rings to control the trajectory of the charged particles and tailor their emissions. They come in three categories: (a) bending magnets, (b) undulators, and (c) wigglers, with each one producing a radiation beam with different spatial and energy characteristics (Fig. 2.4.5). The bending magnets cause a uniform curved trajectory of the charged particle, with centripetal acceleration, and as they go around the bend, they produce a fan of radiation. Undulators are insertion devices consisting of periodic magnetic structures of alternating polarity, with weak magnetic fields, placed in sections where the charged particles move in straight lines. The resultant angular excursions of the charged particle are smaller than $1/\gamma$ and the undulator produces a bright, partially coherent X-ray beam with a very narrow frequency spread. A wiggler is also an insertion device, where at least in one plane the

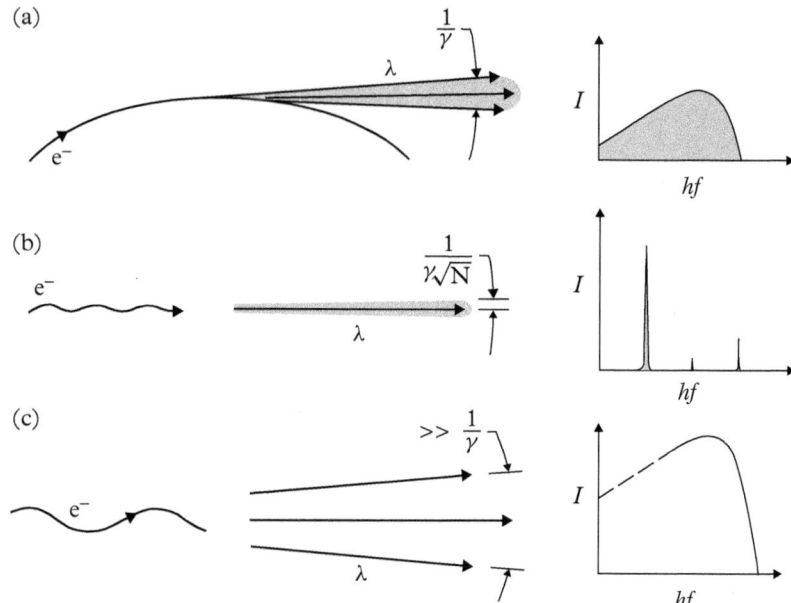

Figure 2.4.5 Magnets used for synchrotron radiation and the X-ray intensities they produce as a function of energy: (a) bend magnets, (b) undulators with small angular excursions, and (c) wigglers with angular excursions larger than $1/\gamma$.

Adapted from Atwood (1999).

magnetic field excursions are larger than $1/\gamma$; now, the accelerations are stronger and the wiggler produces a higher photon flux and energy peak, with a broad radiation spectrum, similar to a bend magnet. Last but not least, bending magnet radiation is linearly polarized in the horizontal plane of acceleration. Above and below this plane, the radiation is elliptically polarized—a feature that has also been used to probe magnetic properties and image magnetic domains. See Figure 1.3.4 or Figure 6.9.4 for a description of elliptical polarization.

Historically, high-energy physicists considered synchrotron radiation as a nuisance for it was a source of energy loss in electron storage rings. At first, parasitic ports were built in such storage rings for scientific experiments using X-ray photons. Subsequent second-generation synchrotron sources were dedicated facilities for X-ray photons built with bend magnets. Third-generation synchrotrons, of which the ALS[10] (Fig. 2.4.6) and APS[11] are good examples, have many straight sections optimized for the insertion of undulators and wigglers to produce soft and hard X-rays, respectively. The excellent text by Atwood (1999) has further details of synchrotron radiation and its wide range of applications.

[10] https://als.lbl.gov/
[11] https://www.aps.anl.gov/

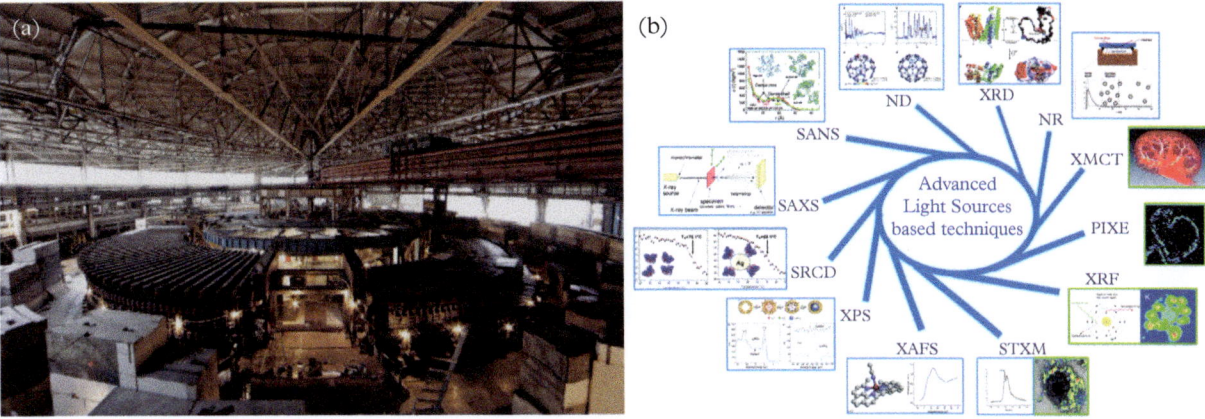

Figure 2.4.6 The Advanced Light Source (ALS) at the Lawrence Berkeley National Laboratory in Berkeley, CA. (a) The interior of the synchrotron, showing the beam lines in the periphery. (b) Schematic illustration of the different beam lines available and the wide range of X-ray based characterization methods implemented at the ALS. See https://als.lbl.gov/ for specific details of the ring and different end-stations.

2.5 X-Rays as Signals: Core-Level Spectroscopy with X-Rays

An inner-shell vacancy in the atom can be created with X-ray photons, high-energy electrons, or accelerated particles. The core-shell vacancy can, in turn, be occupied by an outer shell electron and accompanied by the spontaneous emission of an element-specific X-ray photon characteristic of the atom. The core-levels are unperturbed by bonding or the immediate environment of the atom (§3.3), and hence detection of these characteristic X-ray lines helps unequivocally identify the constituent atoms in the material. Moreover, their intensities can be interpreted in terms of the composition of the specimen. Such microanalysis using characteristic X-ray emission can be carried out with X-ray, electron, or particle incidence (probes). Before we discuss these methods—XRF, EPMA, and PIXE—we must first understand how we detect X-ray photons and measure their intensities.

2.5.1 Instrumentation for Detecting X-Rays

X-ray photons or wave packets are characterized by both their energy and wavelength. For example, Cu $K\alpha_1$ radiation corresponds to a wavelength of 1.541 Å or an energy of 8.048 keV. Thus, we can detect them and measure their intensities using either wavelength or energy dispersion methods.

2.5.1.1 *Wave Length Dispersion Spectrometer (WDS)*

Figure 2.5.1a shows a WDS, which consists of an analyzing/diffracting crystal of finite size, typically 25x10 mm, that diffracts X-rays of selected wavelength by

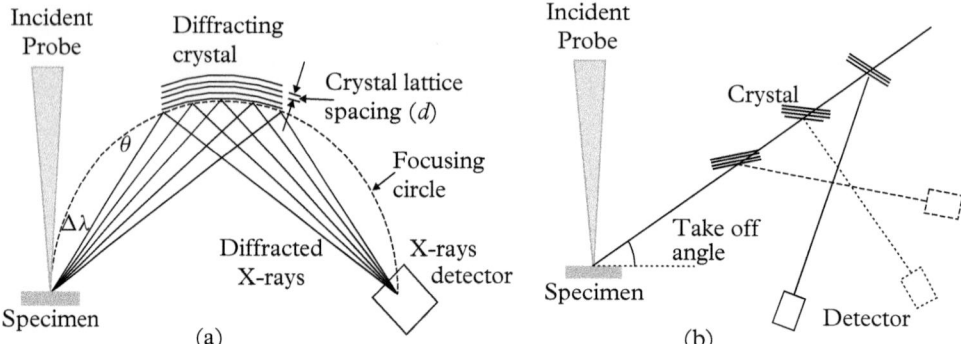

Figure 2.5.1 A wavelength dispersive spectrometer. (a) Schematic of the apparatus illustrating the diffracting crystal and detector geometry to select a narrow wavelength range that is diffracted. (b) As the take-off angle of the X-ray photons is fixed, the detector and crystal geometry have to be varied in concert to retain a focused condition at the detector for a range of X-ray wavelengths.

Bragg law, $n\lambda = 2d\sin\theta$, (1.4.1), and a *photo-detector* for measuring their intensity. Thus, for a given diffracting crystal with well-defined inter-planar spacing, d, and a given angle of incidence, θ, the wavelength of the unknown characteristic X-ray line can be determined. In addition, because of its finite size, the detecting single crystal is curved (ground or bent) to a given radius, focusing the diverging X-ray beam from the specimen to a fixed spot on the detector. Further, since the take-off angle of X-rays from the specimen surface is fixed, the crystal and detector are moved together to maintain focus, as the angle of the detector is changed to cover a range of wavelengths (Fig. 2.5.1b).

Normally, the maximum achievable θ in WDS is about 73°. From Bragg's law, the maximum wavelength of X-rays that can be diffracted/analyzed by the crystal is $1.91d$. To detect light elements with long wavelengths (low energy), crystals with large inter-planar spacing are required; in fact, synthetic multilayered materials, with artificial spacing, d, such that $25 < d < 60$ Å, are prepared specially for detecting light elements ($4 < Z < 9$). Further, the resolution of WDS can be obtained by differentiating Bragg's law, i.e.

$$\frac{d\theta}{d\lambda} = \frac{n}{2d\cos\theta} \tag{2.5.1}$$

Thus, using crystals with smaller d-spacing gives better resolution. On the other hand, using a small d-spacing, will restrict the range of wavelengths that can be analyzed. Hence, in practice, a crystal with a larger d-spacing is used to scan the specimen to identify characteristic X-ray peaks of interest; then, a crystal with smaller d-spacing is used to obtain a high-resolution spectrum over a narrower range of wavelengths. Note that the geometry of WDS restricts a specific crystal with a well-defined set of lattice planes, or d-spacing, to detect only a small range

Table 2.5.1 Diffracting crystals used in wavelength dispersive spectrometers and their range of applicability.

Crystal	Plane	Spacing, d (Å)	Atomic range (Z) K-lines	Atomic range L-lines
LiF	(220)	1.424	> Ti ($Z=22$)	> La ($Z = 57$)
LiF	(200)	2.014	K(19)–Br(35)	> Cd ($Z = 48$)
NaCl	(200)	2.82	S(16)–Rb(37)	> Mo ($Z = 42$)
Pentaerythritol	(002)	4.371	Al(13)–K(19)	
W-Si Multilayer (LSM)		variable	Be(4)–F(9)	Be(4)–N(7)

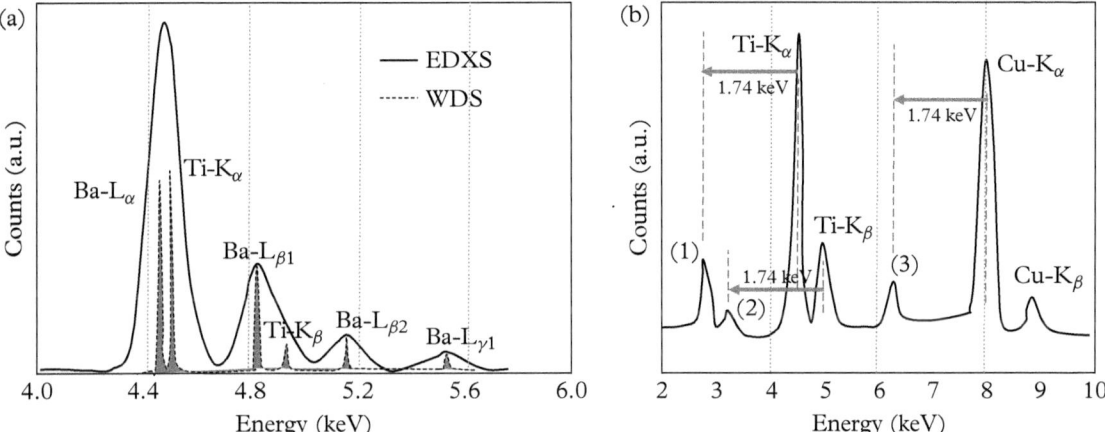

Figure 2.5.2 (a) Comparison of wave length and energy dispersive X-ray spectra of $BaTiO_3$ over identical energy ranges. Notice the much better energy resolution in WDS compared to EDXS. The natural line width of an X-ray peak is ~2 eV. Measured values are 3.0 eV (WDS) and 150 eV (EDXS) for Ti-K_α. (b) Energy dispersive X-ray spectrum of a Cu–Ti alloy, showing the escape peaks from (1) Ti-K_α, (2) Ti-K_β, and (3) Cu-K_α X-ray emissions.

of wavelengths, typically of the order of a few Angstroms, covering only a small range of elements. Therefore, spectrometers are built to accommodate multiple diffracting crystals, which can be easily interchanged to cover the broadest range of wavelengths and elements. Table 2.5.1 is a partial list of crystals used in WDS and the range of elements whose characteristic X-rays that each one can detect.

WDS is characterized by high resolution and high signal to background (Fig. 2.5.2). Its principal limitations are its bulkiness, as multiple analyzing crystals need to be available in the same spectrometer to cover the energy range required for a complete analysis. Further, it involves moving parts that are susceptible to breakdown; moreover, the data acquisition is sequential and very slow. It also suffers from the possibility of higher order Bragg peaks overlapping with other characteristic X-ray lines, complicating the interpretation. An oft-cited example is S Kα ($n = 1$) at $\lambda = 5.372$ Å, overlapping with Co Kα ($n = 3$) at $\lambda = 1.789$ Å.

2.5.1.2   *Energy Dispersive X-Ray Spectrometer (EDXS)*

EDXS uses a solid-state semiconductor crystal to simultaneously detect all the X-rays emitted from a specimen. In this way, they are able to overcome a significant limitation of WDS, which involves sequential detection as a function of wavelength, and often requires multiple analyzer crystals to cover the broad range necessary for chemical analysis. Energy dispersive detectors have the advantage of being compact, which allows them to be installed even in the small confines available between electromagnetic lenses in a TEM/SEM column (Fig. 2.5.3a).

Figure 2.5.3b shows a schematic arrangement for EDXS. It includes a semi-conductor detector, processing electronics, and a multichannel analyzer (MCA) for the display. The semiconductor detector can be either a Li-drifted silicon (Si–Li) or a germanium (Ge) crystal. In the case of an Si–Li detector, the Li is ion-implanted to uniformly compensate any intrinsic defects in silicon. To optimize the performance of the semiconductor detector (Fig. 2.5.4), it is often cooled to liquid nitrogen (L-N_2) temperatures. Since the specimen to be probed by electron

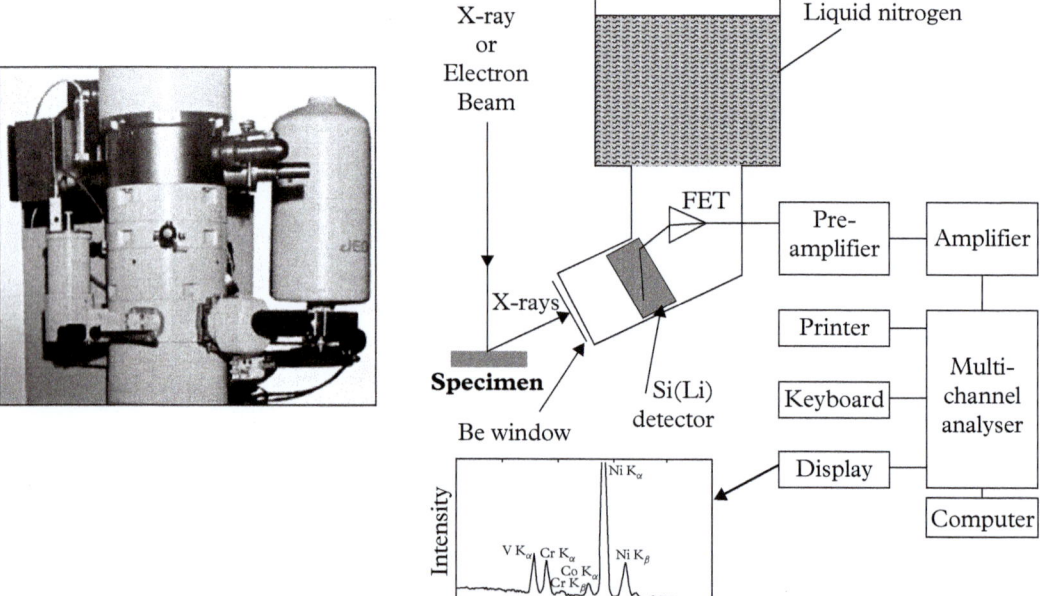

Figure 2.5.3 (a) An EDXS detector installed in a TEM column; only the liquid-N_2 Dewar (seen) is outside the column. (b) A schematic representation of an EDXS detection system, including signal processing and display.

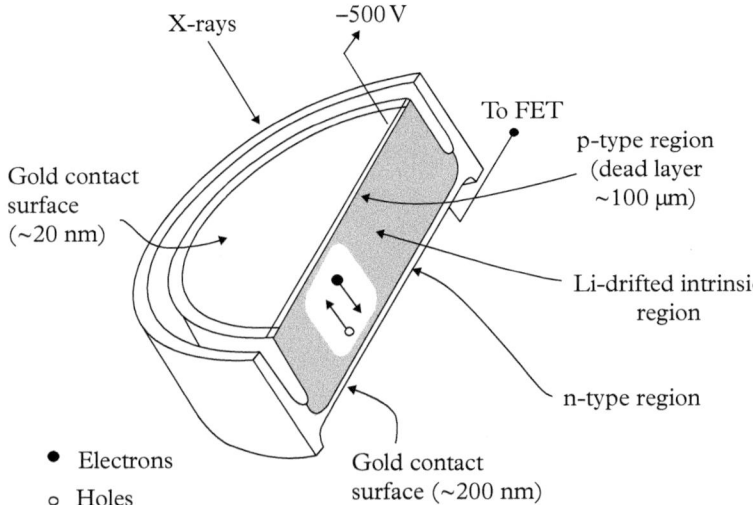

X-rays

−500 V

To FET

p-type region
(dead layer
~100 μm)

Gold contact
surface
(~20 nm)

Li-drifted intrinsic
region

n-type region

• Electrons

∘ Holes

Gold contact
surface (~200 nm)

Figure 2.5.4 An EDXS detector showing all the principal components.

Adapted from Williams and Carter (1996).

or X-ray incidence is kept in vacuum, the cooled detector often ends up having water vapor or hydrocarbon contamination on its surface. Thus, to protect the detector, it is isolated from its environment by sealing it with a window that allows substantial X-ray intensity to be transmitted through. The standard window is made of a thin foil (~7 μm) of beryllium (Be), but even this thin Be window filters X-rays of energy < 1 keV ($Z\sim11$, Na Kα). Thus, to detect low atomic number elements, such as B, C, N, and O, of importance in polymer and ceramic materials, an ultrathin window (~100 nm of polymer +Al, or diamond-like carbon film) that can transmit X-rays down to ~150 eV (i.e. detect B Kα = 192 eV) or even a windowless detector, provided the "column" has a good enough differential pumping system to prevent contamination, is employed.

When X-ray photons interact and transfer energy to a semiconductor crystal, they produce electron–hole (e–h) pairs by transferring electrons from the valence to the conduction bands. For Si–Li crystals at L-N_2 temperatures, the energy required to create a *single e–h* pair is $E_{e-h} \approx 3.8$ eV, which is a statistical average and much larger than the energy of the band gap. Therefore, a single characteristic X-ray photon of energy, E_K, produces $N_{e-h} = E_K/E_{e-h}$ electron-hole pairs; in other words, the number, N_{e-h}, of e–h pairs is proportional to the characteristic energy of the X-ray photon. To detect these e–h pairs, the Si–Li crystal is sandwiched between two ohmic contacts (Au coatings), producing a p-type layer on one surface and an n-type layer on the other, with an intrinsic region of the Si–Li crystal in between. A reverse bias applied to this p-i-n junction type crystal detector, separates the e–h pairs and a charge pulse, with magnitude proportional to the number of e–h pairs generated by the specific X-ray photon, is registered. The current signal is amplified, converted to a voltage, and then sent to the MCA to display the counts as a function of energy.

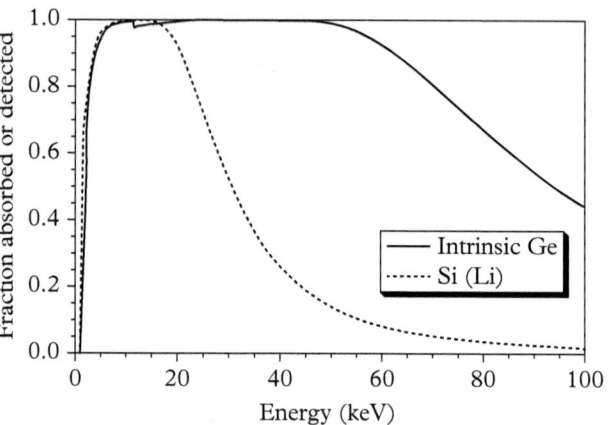

Figure 2.5.5 Efficiency of detection as a function of energy for Si–Li and intrinsic Ge detectors.

Adapted from Williams and Carter (1996).

Si–Li detectors drop off in efficiency at 20 keV, as higher energy X-ray photons are simply transmitted undetected through the thin detector without creating e–h pairs. However, if an intrinsic Ge crystal of much higher atomic number replaces the Si–Li crystal, it can detect X-rays up to ~80 keV (Pb Kα) with good efficiency (Fig. 2.5.5). In addition, $E_{e-h} = 2.9$ eV, for intrinsic Ge and so it produces more e–h pairs for the same X-ray photon. The EDXS signal is also dependent on two parameters of the signal processing electronics: time constant and dead time. The time constant, τ, is the time window (10–50 μs) the processor allots to evaluate the magnitude (N_{e-h}) of the signal pulse. If τ is large, the system can better assign an energy to the pulse, improving resolution, but at the expense of processing fewer pulses in a given time. However, if τ is small, it will help with the statistics (more total counts per second), but at the expense of resolution. To discriminate between pulses, the electronics switches off the detection for a short period of time, called the dead time, immediately after a pulse is detected. If the number of photons arriving at the detector is large, then the dead time would be unacceptably high; in that case, it is best to change the incident probe (X-ray or electron) intensity to reduce the total characteristic X-ray emission.

Figure 2.5.2a compares WDS and EDXS spectra from the same specimen of BaTiO₃. The theoretical resolution of a Si–Li detector is ~110 eV for a Mn Kα (5.9 keV) radiation. In practice, the best EDXS detectors have achieved ~130–140 eV (Si–Li) and 115 eV (*i*Ge) for the same Mn Kα radiation; this is very poor compared to WDS (~3 eV). Two important artifacts observed in EDXS are a sum peak (two X-ray photons arriving simultaneously, but detected as one with twice their energy) and in a Si–Li detector, an escape peak at $E_k - 1.74$ keV, which is caused by the incoming X-ray photon first fluorescing a Si Kα X-ray photon (Kα~1.74 keV) in the detector before being detected with a correspondingly reduced energy (Fig. 2.5.2b).

2.5.2 Chemical X-Ray Microanalysis

Characteristic X-rays can be produced either by photon or electron incidence on the specimen. However, the underlying approach for quantitative chemical micro-analysis of materials, by measuring the emitted characteristic X-ray intensities, is common to both methods.

2.5.2.1 *Photon Incidence: X-Ray Fluorescence Spectroscopy (XRF)*

XRF is a non-destructive method for chemical analysis of materials. It irradiates a specimen with a high-energy primary X-ray beam that, in turn, fluoresces characteristic X-ray radiation representative of the chemical composition of the specimen. The characteristic X-ray emissions can be analyzed by either EDXS (Fig. 2.5.3) or WDS (Fig. 2.5.1) detection (§2.5.1). A laboratory XRF system uses an X-ray tube of high power (0.5–3.0 kW) and voltage (30–50 kV) as the source to ensure that the primary X-ray beam (the probe) has sufficient energy to excite the characteristic X-ray photons of interest (the signal) from the specimen. Typically, the target metals in the sealed tube are W, Cu, Mo, Cr, Rh, and Ag, and the tube voltage is adjusted accordingly.

XRF does not require a vacuum, although it is preferred for best analysis, and can also function in atmospheres of air and helium. In particular, the ability to detect low-Z elements is affected by the atmosphere, e.g. using EDXS detectors, XRF can detect $Z > 6$ (C) in vacuum, $Z > 11$ (Na) in helium, and $Z > 13$ (Al) in air. Specimens for XRF can be liquids, powders, or bulk; the main requirement to obtain an analysis that is truly representative of its composition is a physically flat surface, and good chemical homogeneity, particularly near the surface, as the primary X-ray beam illuminates a large surface area ($\sim$5 cm^2) and penetrates $\sim$50 μm into the specimen.

Strictly speaking, the penetration depth of the primary X-ray beam can be controlled by the angle of incidence. At total internal reflection (shallow angles of incidence) it is a few nm, but increases to several μm at large angles of incidence. XRF analysis can be quantitative (§2.5.2.3) and at larger angles of incidence can detect concentrations $> 0.1\%$. For trace element analysis, a grazing incidence angle (few tenths of a degree) with detection limit down to 10^{13} atoms/m^2, called total reflection XRF (TRXF) [2] has been implemented (Fig. 2.5.6a).

TRXF is routinely used as a monitoring tool in semiconductor fabrication to satisfy the stringent demands of surface purity (metallic contamination) on semiconductor (typically, silicon and to a lesser extent, gallium arsenide) wafer surface. The grazing angle of incidence is kept below the critical angle, ϕ_c, given by

$$\phi_c = 3.72 \times 10^{-11} \sqrt{n_e}/E \tag{2.5.2}$$

for total internal reflection, where n_e is the electron density (cm^{-3}) in the specimen, and E is the energy (keV) of the incident X-ray photon. For Mo Kα,

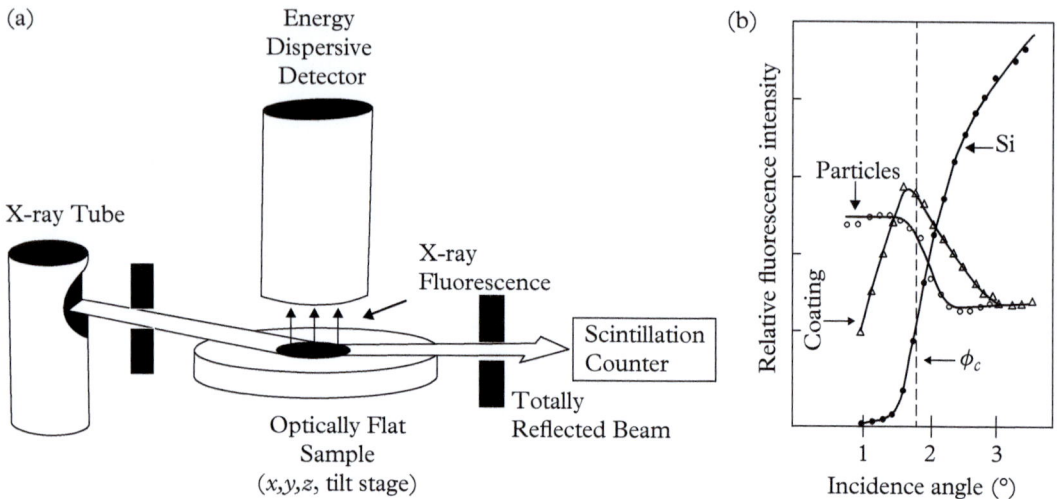

Figure 2.5.6 (a) A schematic arrangement for total reflection X-ray fluorescence. (b) Analysis of a silicon surface with a thin layer of nickel, both as a uniform coating and dispersed as fine particles, as a function of angle around the critical angle for total internal reflection.

Adapted from Leng (2013).

$E = 17.48$ keV, and a silicon surface, $\phi_c \sim 1.8$ mrad. The angular variation of fluorescence around the critical angle will also depend on the chemical nature of the impurity (Fig. 2.5.6b) and hence a precise control of the angle of incidence is important for good TRXF analysis. Further details of TRXF can be found in appropriate ASTM standards [3].

2.5.2.2 *Electron Incidence: Electron Probe Microanalysis (EPMA)*

An EPMA, originally developed by Castaing [4], is similar to XRF, but instead of X-rays it uses high-energy electrons, with focusing electron-optics, as the probe to generate and measure characteristic X-ray photons with good spatial resolution ($\sim$100 nm–1 µm). In principle, EMPA instruments are similar to standard scanning electron microscopes (SEMs) equipped with EDXS detectors, but they typically generally have one or more WDS detectors as well. Therefore, the underlying physical principles and quantification methods presented here also applies to microanalysis with SEMs (see §10.6.1).

While the interaction of focused electron probes with solid materials is discussed in detail later (§5.3), here it is important to point out that the volume (radius and depth; in the first approximation we will assume they are the same) of the specimen producing characteristic X-ray photons is larger than the diameter, d_p, of the incident electron probe. The X-rays are typically generated from a pear-shaped volume under the beam (Fig. 2.5.7). The depth, R_x (μm), of characteristic X-ray generation is given by

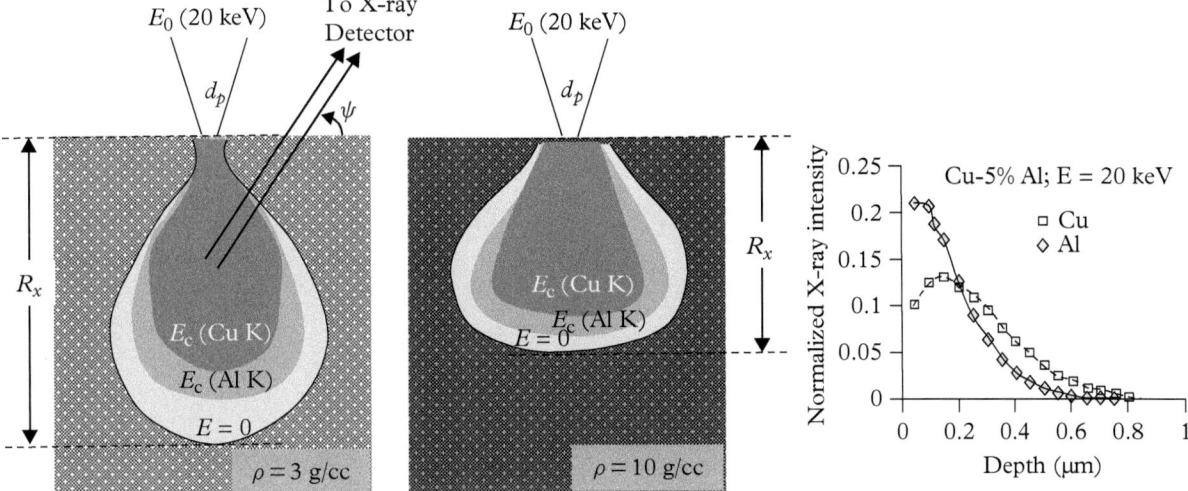

Figure 2.5.7 Schematic comparison (left) of X-ray production with depth in two specimens of different densities. The critical volumes for Cu K and Al K emissions are also shown, which gives an idea of the depth/spatial resolution. The take-off angle, ψ, determines the path length of the characteristic X-rays in the specimen. The plot (right) shows X-ray intensities with depth, but now including absorption. Clearly Al K, even though it is generated deeper in the specimen, is absorbed more and is attenuated earlier.

Adapted from Leng (2013).

$$R_x = \frac{0.0276A}{Z^{0.89}\rho} \left(E_0^{1.67} - E_x^{1.67}\right) \qquad (2.5.3)$$

where A is the atomic weight (g/mole), Z is the atomic number, and ρ is the density (g/cc) of the matrix, E_0 is the incident beam energy in keV, and E_x is the energy in keV required to excite the characteristic X-ray line of interest.

Another important factor that determines the accuracy of microanalysis, particularly for low-Z elements, is the take-off angle, ψ, determined by the position of the detector with respect to the specimen. The absorption path length in the specimen for any X-ray photon generated at a depth, z, is $z\,\csc\psi$, and such absorption can be minimized by making ψ as large as possible, within the constraints of the instrument. Note that, in an EPMA or SEM, the ability to focus and scan the primary beam over the surface, makes it also possible to produce composition maps of the specimen (Fig. 1.4.14).

In summary, EPMA is a quantitative method for elemental analysis of materials in a nondestructive manner. The analysis can be typically carried out for micron-sized volumes at the ppm level. The more sophisticated EPMA systems (Fig. 2.5.8) are capable of simultaneous X-ray (WDS and/or EDXS detectors) analysis, SEM (§10), and back-scattered electron (§10.3) imaging.

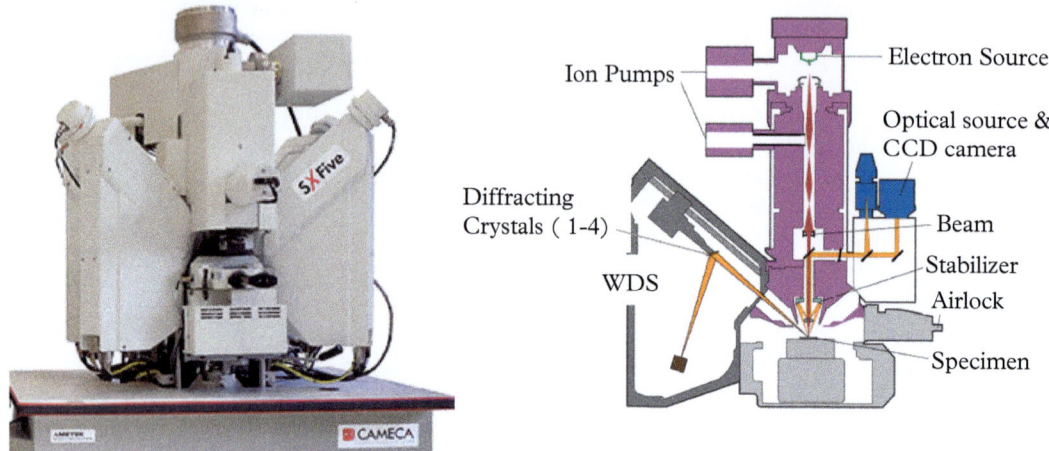

Figure 2.5.8 A commercial EPMA machine (CAMECA, SXFive) and a schematic cutout illustrating its principal components.

2.5.2.3 *Quantitative X-Ray Microanalysis*

X-ray spectra can provide qualitative analysis of the elements present in the specimen with relative ease. Characteristic X-ray emission lines (K, L, M, . . .) for all elements are tabulated (Table 2.3.1), and using the software for analysis provided by the manufacturer is a trivial exercise to match and label the observed peaks. Even if the peaks overlap, the relative positions and intensities of the α and β lines can help to identify the elements present. However, we are often interested in the composition or the concentration of the different elements present in the specimen and hence methods for quantitative analysis of the X-ray spectrum are required.

In general, the concentration, C_A, of an element A of interest is related to its observed characteristic X-ray peak intensity, I_A, corrected for instrument, X, and specimen or matrix, M, factors, i.e. $C_A \propto I_A X\, M$. The instrument factors include details of the source (intensity or current), the geometrical arrangement of the detector (takeoff angle, collection angle) and detector characteristics (efficiency of detection, detector dead time).

The matrix factors are threefold:

1. *Atomic number effect* (Z): This effect relates to how the beam changes with depth as it propagates in the specimen. In XRF, the main concern is the primary absorption of the incident X-ray beam before it reaches the atom to be excited. For electrons there is a similar effect that involves the stopping power of the electron in the specimen; in addition, electrons are back-scattered, and those that are do not engage in characteristic X-ray production. For example, a significant fraction (~15% in Al, ~30% in Cu, and ~50% in Au) of the incident beam is backscattered without

generating X-rays. This loss in X-ray generation increases with increasing atomic number of the specimen.

2. *Absorption effect* (A): As described, X-rays are generated over a range of depths and as they propagate towards the detector, they are absorbed or scattered in the specimen itself. The intensity of the characteristic X-ray, produced at depth, z, leaving the specimen follows (2.4.2), and varies with the distance given by $z \csc \psi$, i.e.

$$\frac{I}{I_0} = \exp\left(-\frac{\mu}{\rho}\right) \rho z \csc \psi \tag{2.5.4}$$

Figure 2.5.7 also shows a plot of the sampling depth of Cu $K\alpha$ and Al $K\alpha$ radiation in a $Cu_{95}Al_5$ alloy. Even though the Al $K\alpha$ is produced deeper in the specimen, the self-absorption of the X-rays by the specimen, makes the sampling depth of Cu $K\alpha$ to be larger than that of Al $K\alpha$, because of the difference in their mass absorption coefficients. In short, this effect is also dependent on the specimen composition.

3. *Fluorescence effect* (F): This arises from the secondary fluorescence of the characteristic X-rays of interest by the characteristic X-rays of other higher atomic number elements, if present in the specimen. Further, higher-energy bremsstrahlung radiation can also contribute to this secondary fluorescence and the overall effect is dependent on the specimen composition.

Quantitatively, X-ray intensities, I, can be related to the concentration, C, by comparing a standard, S, of known composition, with the unknown specimen, U, using the relationship

$$\frac{C_U}{C_S} = \frac{I_U^G}{I_S^G} \tag{2.5.5}$$

where G refers to the generated X-ray intensity, corrected for only the instrument factors but not the matrix effects. However, what is measured in an experiment are the actual intensities, I_U^E and I_S^E, and the ratio $k_{US}^E = I_U^E/I_S^E$, also known as the k-factor, which would be different from the generated intensities in (2.5.5), due to matrix effects. Interactions of high-energy X-rays or electrons with solid materials is well understood, and analytical expressions for the Z, A, and F correction factors exist. Therefore, a quantitative ZAF correction is incorporated in all X-ray microanalysis software; it is important to re-emphasize that all three corrections depend on the specimen composition and can be calculated if that is known. Including these corrections, and modifying (2.5.5), we can write the composition of the unknown as

$$C_U = Z\,A\,F\left(\frac{I_U^E}{I_S^E}\right)C_S = Z\,A\,F\,k_{US}^E C_S \tag{2.5.6}$$

Now, to perform a quantitative analysis, an approximate composition of the unknown specimen is assumed. Based on the assumed composition, the ZAF correction factors are calculated, and a first value, k_{US}^1, for the k-factor is predicted. The difference $\Delta k_{US}^1 = k_{US}^1 - k_{US}^E$ is computed and in this manner the composition is iterated until the difference, Δk_{US}^n, is below an acceptable value to converge on the quantitative analysis. Since the k-value in (2.5.6) is a ratio of intensities for the same X-ray energy, in the unknown and standard, it is not instrument-dependent, nor does it require detailed knowledge of ionization cross-sections or fluorescence yields.

Note that microanalysis with X-rays becomes much simpler when we can ignore the ZAF corrections, such as in a TEM where we use very thin electron transparent specimens (§9.4.3). In that case, k-factors for the elements are measured or calculated with respect to Si, i.e. k_{ASi}. For example, in the case of a binary alloy, A_xB_y, we know that $C_A + C_B = x + y = 1$. Further, it is easy to relate the measured characteristic X-ray intensities (I_A, I_B) to the specimen concentrations (C_A, C_B), simply as follows:

$$\frac{I_A}{I_B} = k_{AB}\frac{C_A}{C_B} = \frac{k_{ASi}}{k_{BSi}}\frac{C_A}{C_B} \tag{2.5.7}$$

Thus, if the k-factor, k_{AB}, is known or can be experimentally determined, by measuring the characteristic X-ray intensities (I_A, I_B), the composition, (C_A, C_B), can be determined (all the other factors cancel out). This approach is discussed in Example 2.5.1, and further in §9.4.3.

Example 2.5.1: Microanalysis using k-factors. You are trying to determine the composition of an unknown *binary alloy* specimen, A_xB_y, in *thin film form*. You measure the following X-ray characteristic peak intensities: $I_A = 10,000$ and $I_B = 12,000$ counts. You also measure the characteristic X-ray intensities of two available standards, $A_{50}Si_{50}$, and get $I_A = 10,000$ and $I_{Si} = 1,000$ counts, and for the other, $B_{0.67}Si_{0.33}$, you measure $I_B = 60,000$ and $I_{Si} = 10,000$ counts.

(a) Determine the k-factors, k_{ASi}, k_{BSi}, and k_{AB}.

(b) What is the composition of the alloy specimen, A_xB_y?

Solution: We neglect the ZAF corrections as the alloy is in thin film form. Then we apply (2.5.7) to the two standards.

For $A_{50}Si_{50}$, $I_A/I_{Si} = k_{ASi}C_A/C_{Si} = 10000/1000 = k_{ASi}(50/50)$. Thus, $k_{ASi} = 10$.

Similarly, for $B_{0.67}Si_{0.33}$, $\frac{I_B}{I_{Si}} = k_{BSi}\frac{C_B}{C_{Si}} = \frac{60000}{10000} = k_{BSi}\frac{67}{33}$. Thus, $k_{BSi} = 3$.

Further, $k_{AB} = k_{ASi}/k_{BSi} = 10/3 = 3.33$

Now $I_A/I_B = k_{AB}C_A/C_B = 10000/12000 = 3.33(C_A/C_B)$, which gives $C_A/C_B = 1/4$.

But $C_A + C_B = 1$. Hence, $C_A = 0.2 = x$ and $C_B = 0.8 = y$, and the compound is $A_{0.2}B_{0.8}$

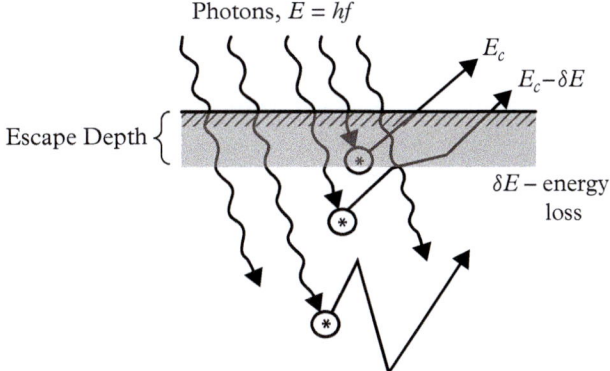

Figure 2.6.1 Energetic photons incident on the surface of a specimen ionize atoms and create characteristic photoelectrons deep inside. Only those photoelectrons within the escape depth of the surface, emerge with their energy, E_c, intact. Those from further down suffer an energy loss, δE, and emerge with energy $E_c - \delta E$.

2.6 Surface Analysis: Spectroscopy with Electrons

We now introduce two of the principal surface analysis techniques, AES and XPS, which involve the detection of electrons with energy in the range of 5–2000 eV. A more extensive discussion of these two methods can be found in Watts and Wolstenholme (2003) and Briggs and Seah (1990). In this energy range the electrons have a high probability of inelastic scattering, with an associate loss of energy, δE, as their mean free path lengths, in most solids, are less than 100 Å (Fig. 1.3.5). Thus, if we are able to detect electrons of energy, E_c, which is known to be unchanged as it emerges from the specimen, we can safely assume that it originates within a shallow layer (called the *escape depth*) of the specimen surface. In other words, these techniques are surface sensitive or even *surface specific* (Fig. 2.6.1). Second, as both the incident electrons and X-ray photons used as the probe in AES and XPS, respectively, have reasonably high energies ($\sim$1.5 keV), they can penetrate and interact with the inner shell electrons of the atoms and, as such, are useful for elemental identification, albeit only at the surface of the specimen. Note that if lower-energy electrons or photons are used as the probe they interact with the outermost, less-tightly bound electrons that are associated with the chemical bonding and are not necessarily associated with a specific atom. Third, since these techniques analyze the surface, any surface contamination, including gas molecules, would adversely affect the signal intensities. Hence, these measurements require clean surfaces, and are carried out in a chamber held at ultra-high vacuum, 10^{-6}–10^{-8} Pa (10^{-8}–10^{-10} mbar). Finally, all such techniques, including XPS and AES, require some form of electron energy analyzer, and an understanding of the basic principles involved in the construction and use of such analyzers (§2.6.1.2) would help significantly in experiment design and improve metrology.

2.6.1 Instrumentation for Surface Analysis with Electron Spectroscopy

2.6.1.1 Vacuum Chamber and Other Components

Figure 2.6.2 shows a schematic of a modern instrument combining both AES and XPS in a single vacuum chamber for comprehensive surface analysis. Figure 2.6.3 shows a commercial XPS system with its principal components. In principle, an SEM column (§10.2) could also be added to image the area of analysis. Moreover, the electron gun used for exciting Auger electrons can also be focused and used to scan the specimen surface. In this manner, maps of the surface Auger electron distribution in two dimensions can be made (Fig. 2.6.9); the technique is then referred to as scanning Auger microscopy. Finally, a sputter ion gun using Argon ions, can be employed either to clean the specimen surface of any contamination and/or to remove very thin layers of the material between the generation of Auger maps. These maps can also be digitally combined to create a three-dimensional Auger electron map of the specimen.

Ultrahigh vacuum (UHV, 10^{-6}–10^{-8} Pa or 10^{-8}–10^{-10} mbar) is required for surface electron spectroscopy to keep the surfaces free of any contamination. It also prevents the low-energy photoelectrons and Auger electrons from being scattered by residual gas molecules in the chamber, thus enhancing their probability of reaching the analyzer. Adsorption of gas molecules on the surface of the specimen is very rapid (a monolayer accumulates in less than 1 second, compared to typical AES/XPS data acquisition time of several minutes) and a major issue of concern for any surface science experiment. To avoid gas contamination the chamber is baked at 250-300°C; this helps dislodge most gas molecules attached to the chamber walls to be removed by the pumping system.

The X-ray gun, typically for XPS, is similar to an X-ray tube (Fig. 2.4.2) and common sources are Al Kα (1.4866 keV) or Mg Kα (1.2536 keV), with narrow line widths of the order of ~1eV. The sputter ion gun, with energies in the range 0.5–5.0 keV, in addition to cleaning the surface, can also be used for secondary ion mass spectrometry (see §5.5 for further details on SIMS). The electron gun used for AES is similar to those used in electron microscopy; both lanthanum hexaboride, LaB_6, filaments and field emission sources are now common, and are discussed later (§5.2.2 and §9.2.1).

2.6.1.2 Electrons as Signals: Electron Energy Spectrometers and Analyzers

A common and desirable principle for measuring the number of electrons in a narrow energy window is to design a spectrometer as a band-pass analyzer. This is achieved by creating a dispersing field, which can either be electrostatic or magnetic. When the electrons pass through this field they are deflected, and the degree of deflection would depend on their velocity or kinetic energy. Electrostatic deflection analyzers are more common for surface analysis as they are compact, UHV compatible, do not generate any significant fields outside the analyzer, and

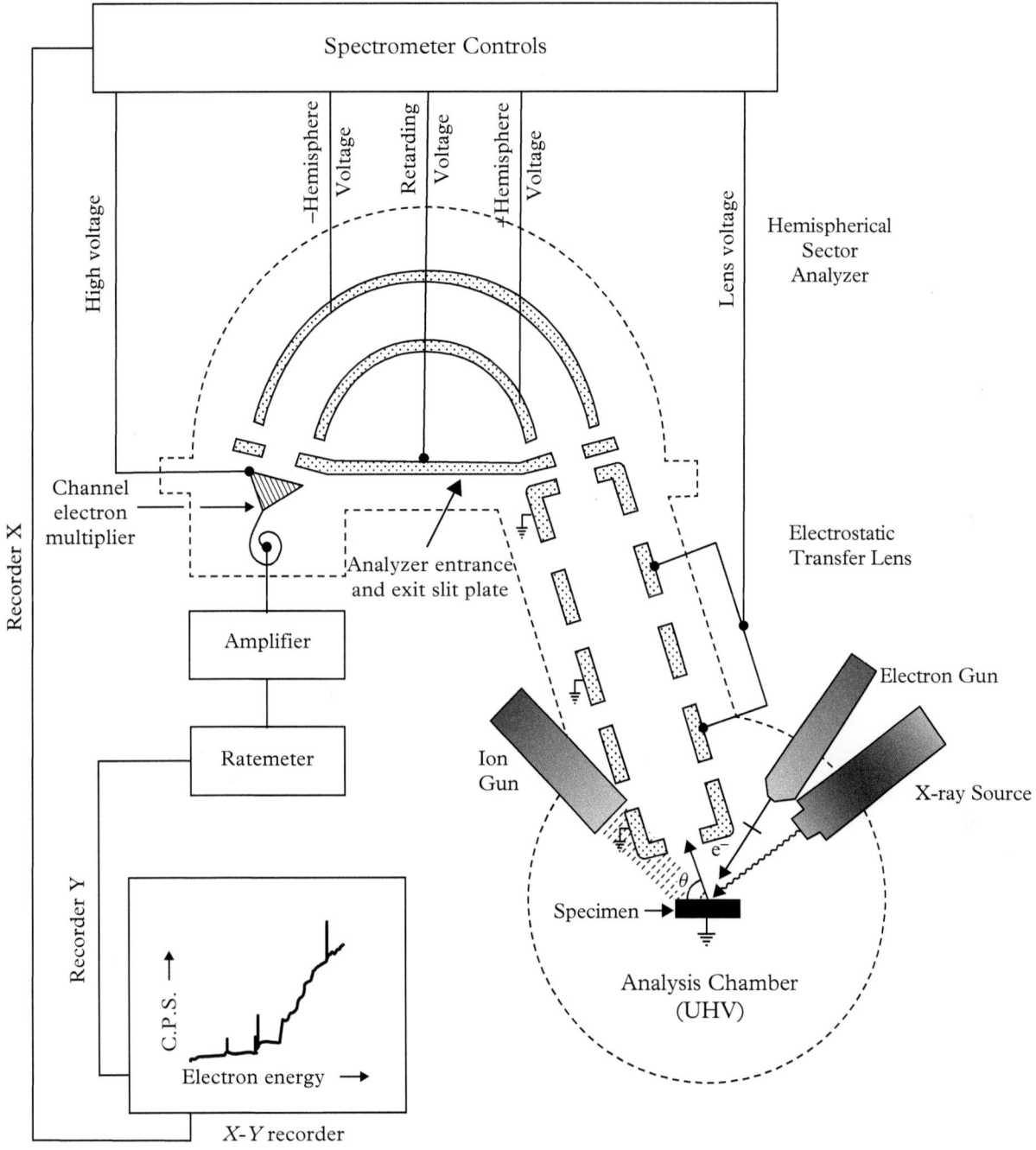

Figure 2.6.2 A schematic drawing of a compound AES and XPS system.

Adapted from Watts (1990).

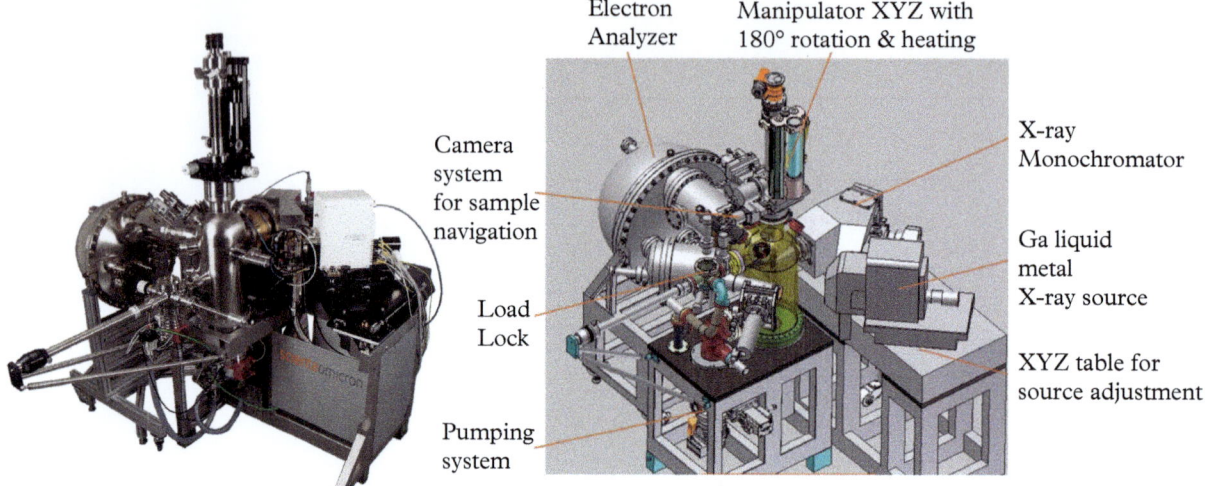

Figure 2.6.3 A commercial Omicron-HAXPES system. A schematic of this hard X-ray photoelectron system, with its principal components is shown on the right.

are optimal for the low energy (<1400 eV) electrons encountered in AES and XPS. Magnetic sector analyzers (Fig. 5.2.6) generate high fields and are used for dispersing high-energy electrons (>100 keV), such as in an analytical TEM, and are discussed later (§9.4.2.4) in the context of electron energy-loss spectroscopy.

The principles governing the design of an electrostatic analyzer can be understood by considering the simple case of two parallel plates set at different potentials (Fig. 2.6.4a), producing an electric field, E_x. If electrons of different energy are directed between these plates along the z-direction, they will be deflected by the field with the magnitude of deflection being inversely proportional to their energies. If an aperture or opening is inserted into one of the plates, by adjusting the field as well as the aperture size, electrons of a specific energy can be made to emerge though the aperture and be counted. However, Figure 2.6.4a shows how the degree of deflection of the electrons would also be affected by their angle of entry into the analyzer. A good spectrometer should be capable of focusing electrons of the same energy, but entering the analyzer over a finite range of incident angles, at the same exit aperture. Figure 2.6.4b shows how, in the parallel plate analyzer, this can be achieved by injecting the electrons at a fixed angle.

Based on these principles, two designs of analyzers are used in surface science: the concentric hemisphere analyzer (CHA), with a mean deflection of 180° (Fig. 2.6.4c), and a cylindrical mirror analyzer (CMA), consisting of two concentric cylinders (Fig. 2.6.4d). Generally, the aperture for the CHA would be such that it accepts a total solid angle of 10^{-2} steradians; in contrast, the CMA would accept ~1 steradian (~100× larger than CHA) giving higher collection

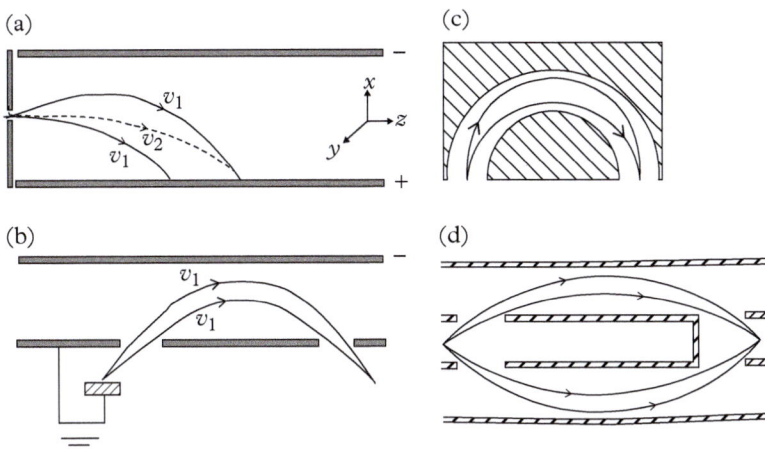

Figure 2.6.4 Principle of electrostatic electron analyzers based on trajectories of electrons in a parallel plate capacitor. (*a*) Electrons introduced parallel to the plates with velocities, $v_2 > v_1$, strike the plate at different positions. However, another electron with velocity, v_1, may arrive at the same point as v_2, if it is injected axially. (b) The same parallel plates, in a plane mirror focusing geometry. Electrons of the same velocity/energy, injected at different angles are brought to the same focal point. (c) A concentric hemispheric analyzer (CHA) with a 180° spherical sector, and (d) a concentric mirror analyzer (CMA) with cylindrical symmetry about the axis.

Adapted from Woodruff and Delchar (1986).

efficiencies and signal to noise characteristics. Further details of the design of these analyzers are beyond the scope of this book and can be found elsewhere (Woodruff and Delcher, 1986), but here we can summarize their main attributes. A typical CMA with an outer cylinder diameter of 10–15 cm and no retardation, can operate at a resolving power ($E_0/\Delta E$) of ~200 and a working distance of ~5 mm; in other words, a CMA can be a high collection efficiency, low-resolution analyzer. A CHA of comparable size can have a resolving power of 1,000–2,000 and a working distance 25–50 mm; thus, it is a low collection efficiency, high-resolution analyzer. Figure 2.6.2 shows a CHA as part of the surface analysis system.

2.6.2 Auger Electron Spectroscopy

AES is an electrons-in (probe), electrons-out (signal) technique conducted in vacuum. The Auger process (§2.3.3) involves the ionization of core levels by the relatively high-energy (>1.5 keV) incident electrons and the detection of Auger electrons of discrete energy, following the non-radiative decay (Fig. 2.3.6) of the core hole. Auger electrons typically have energies <1.5 keV (Fig. 2.6.5) for all

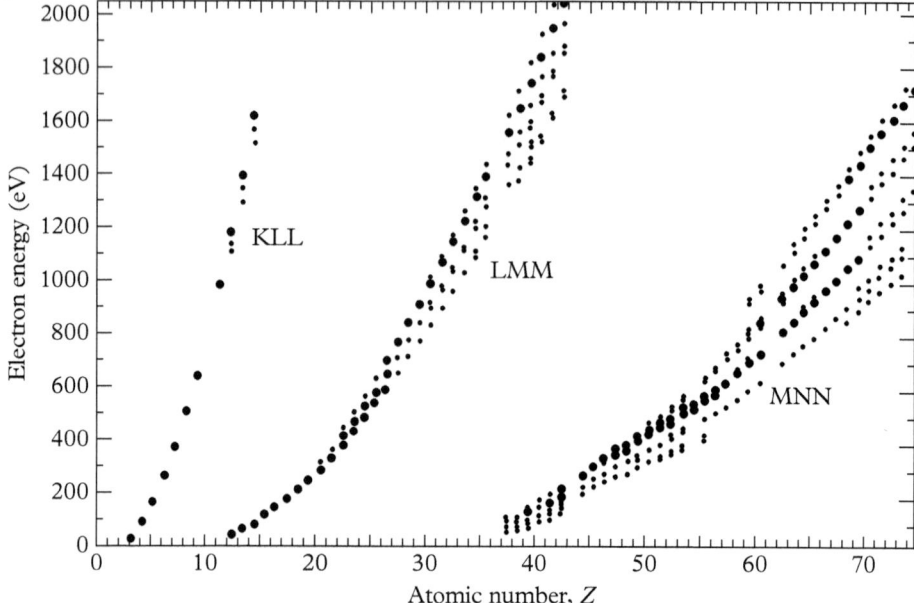

Figure 2.6.5 Energies of principal Auger lines, KLL, LMM, and MNN for the elements indicated by dots. Large dots indicate the principal peaks in the Auger spectrum. See also Table 2.3.2.

Adapted from Watts (1990).

the elements. Note that the mean free path length for core ionization is rather long (∼100 nm). Thus, the surface sensitivity of AES is determined not by the incident electrons, but rather by the inelastic scattering of the emerging Auger electrons.

Typically, when a surface is irradiated with high-energy electrons (Fig. 2.6.6), with primary energy, $E_p = 2$ keV, the spectrum of electrons detected consists of both elastically (no energy loss) and inelastically scattered electrons. The spectrum can be broadly classified into three regions: (I) elastically scattered, (II) inelastically scattered, and (III) secondaries. In addition to the elastic or zero-loss peak, the right side (I) of the figure includes electrons that have lost energy in a single interaction due to some discrete quantum excitation. These include inter-band transitions (∼few eV loss), plasmon (both bulk and surface) excitations (10–30 eV loss) and any inner shell ionization (element specific, and ranging from a few 10s to 1,000 eV loss) (see also Fig. 9.4.1 and related discussion).

True secondaries are produced by a cascade process involving the energy-loss of the high-energy primaries, and lie at the extreme left end (region III) in the range of 0–50 eV. In between is the inelastically scattered region (II)—note that it is difficult to truly distinguish them from the secondaries—where we also find electrons arising from other decay processes. The most prominent decay process

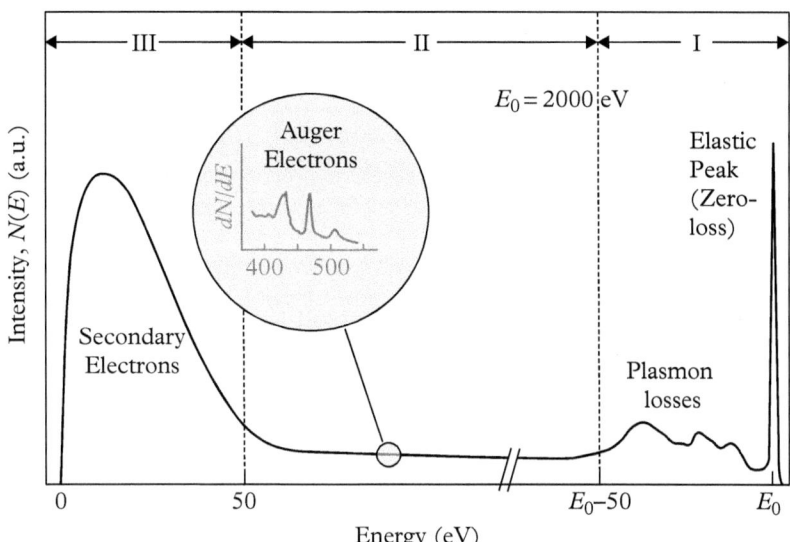

Figure 2.6.6 Spectrum of 2 keV electrons (probe) scattered from a surface showing three distinct regions—(I) elastically scattered, (II) inelastically scattered, and (III) secondaries. The energy scale is nonlinear. The inset shows a portion of the inelastically scattered region (II) where the Auger spectrum in differential form is magnified.

Adapted from Feldman and Mayer (1986).

here is the non-radiative Auger electron emission, which appears as weak features superimposed over the inelastic background. Thus, rather than plot the direct spectrum, $N(E)$, to reveal these weak features the derivative spectrum, $dN(E)/dE$ is displayed. Figure 2.6.7 shows a typical AES spectrum, including both $N(E)$ and $dN(E)/dE$, to demonstrate the use of the derivative technique for a cobalt specimen.

As discussed earlier (2.3.4), the binding energy of electrons in atoms in a charged state can be different from the theoretical values of neutral atoms. Similarly, when two elements combine to form a compound, small changes in the binding energy can occur. Lastly, in the case of hybridization (Fig. 2.2.5), new orbital configurations and energies may contribute to the bonding and also to the related Auger process. Since three energy levels are involved in Auger emission, a chemical shift may occur for all three of the levels; thus, for a KLM Auger process, the observed chemical shift is $\Delta E_{shift} = \Delta E_K - \Delta E_L - \Delta E_M$. For example, the chemical shifts in the $C(1s)$ binding energy between carbides and CF_4 can extend over 12 eV. Moreover, the Auger peak shape can vary with its chemical state (Fig. 2.6.8a). The spectra show variations in line shapes due to differences in the chemical environments of the carbon atoms in the four samples. Although the data clearly show that bonding information can be obtained from Auger spectroscopy, interpreting chemical state and chemical environment information from Auger spectra is challenging because the Auger process involves three energy levels. More detailed electronic structure information (e.g. hybridization, electron delocalization, screening effects) may be obtained from quantitative analysis of Auger spectral line shapes; in practice, this is a daunting task that is best left to the more advanced practitioners.

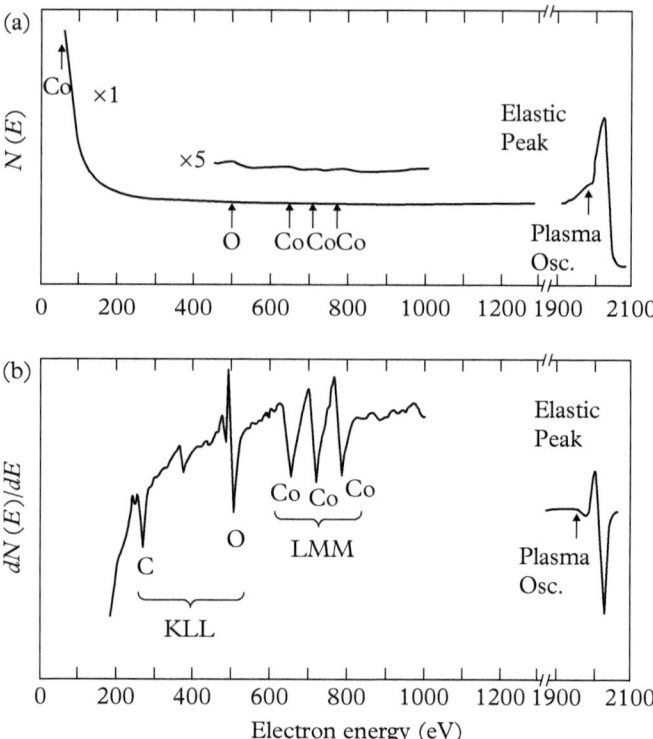

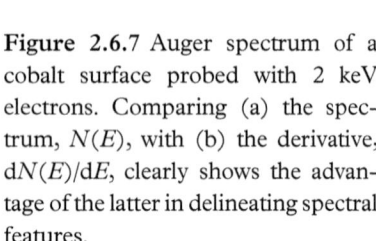

Figure 2.6.7 Auger spectrum of a cobalt surface probed with 2 keV electrons. Comparing (a) the spectrum, $N(E)$, with (b) the derivative, $dN(E)/dE$, clearly shows the advantage of the latter in delineating spectral features.

Adapted from Feldman and Mayer (1986).

It is possible to obtain spatially resolved chemical state information using a scanning AES system (Fig. 2.6.9). In this example of a Ti–Al superalloy specimen, the spatial information in the Auger maps, combined with the chemical shift, have been used to distinguish between microstructural features where nitrogen is bound to aluminum or to titanium.

Furthermore, using a sputter gun, the specimen can be slowly etched away, or sectioned, and AES spectra can be obtained as a function of *depth*. Figure 2.6.10 shows such data from the interface region of a Ta–Si film grown on polycrystalline silicon (p-Si). Clearly, the presence of a native oxide layer on the surface of p-Si can be resolved.

> **Example 2.6.1:** Al Kα X-rays ($\lambda = 8.3$ Å) are used to probe a specimen containing calcium.
>
> (a) Which innermost shell of the element calcium can this photon eject?
>
> (b) What Auger electron peaks are potentially visible? At what energies?
>
> (c) If you are analyzing a compound A_xCa_y, where A is an unknown element, which technique, AES or XRF, would be appropriate to identify the element, A, most easily? Why?

Solution:

(a) The energy of the X-ray photon is $E(\text{Al K}\alpha) = 12,400/\lambda = 1494$ eV.

From Table 2.2.2 of binding energies, these photons can only ionize the $2s$ electrons of Ca.

(b) Two potential Auger electrons that may be visible are the $L_1 M_1 M_1$, and the $L_1 L_2 M_1$ lines. The binding energies for Ca are $E(2s) = 438.4$ eV, $E(2p_{1/2}) = 349.7$ eV, and $E(3s) = 44.3$ eV. Applying (2.3.3), we get

$E(L_1 M_1 M_1) = 438.4 - 44.3 - 44.3 \text{ eV} = 349.8 \text{ eV}$

$E(L_1 L_2 M_1) = 438.4 - 349.7 - 44.3 \text{ eV} = 44.7 \text{ eV}$

(c) Since Auger electron emission involve three electronic levels, they are more accurate in identifying elements, compared to characteristic X-rays, which can sometimes be interpreted as multiple elements.

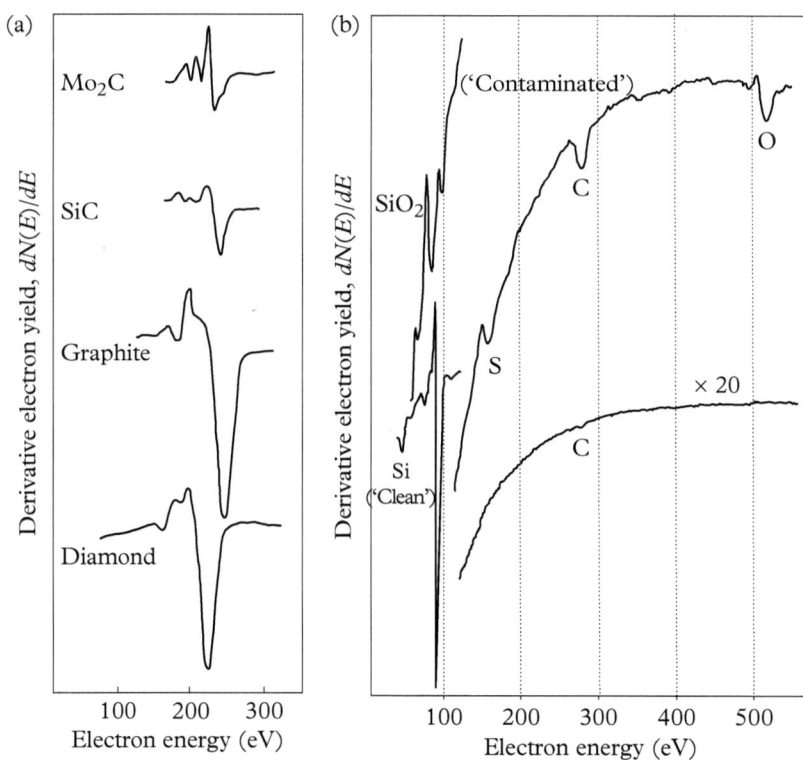

Figure 2.6.8 (a) The derivative of the electron-yield in C KVV Auger electron spectra for two carbides, graphite, and diamond. (b) AES from a clean and contaminated silicon surface—both $N(E)$ (right) and dN(E)/dE (left) are shown. There is a significant difference in the Si $L_{2,3}$ LVV Auger transitions. The contaminated surface shows a line shape characteristic of SiO_2. See Figure 2.6.14 of XPS data from a similar surface.

Adapted from Woodruff and Delchar (1986).

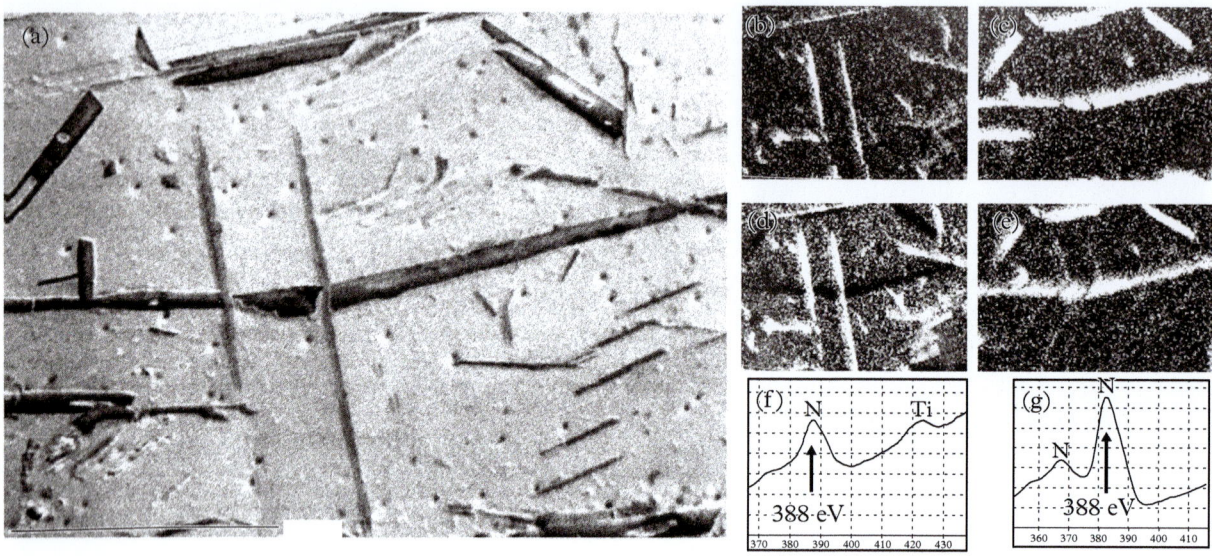

Figure 2.6.9 Secondary electron image (a), along with Auger maps (b–e), of a coated Ti–Al superalloy. The Ti map (b) is closely associated with the N map (d) at 388 eV. Similarly, the Al map (c) is associated with the N map (e) at 383 eV. The N peaks in the Auger spectrum from the Ti (f) and Al (g) rich regions of the superalloy are also shown.

Adapted from [5].

Figure 2.6.10 Sputter depth profiling with AES of a Ta–Si film deposited on a poly-Si substrate. The focus is on the buried interface region where the native oxide layer can be clearly resolved from the oxygen signal.

Adapted from [6].

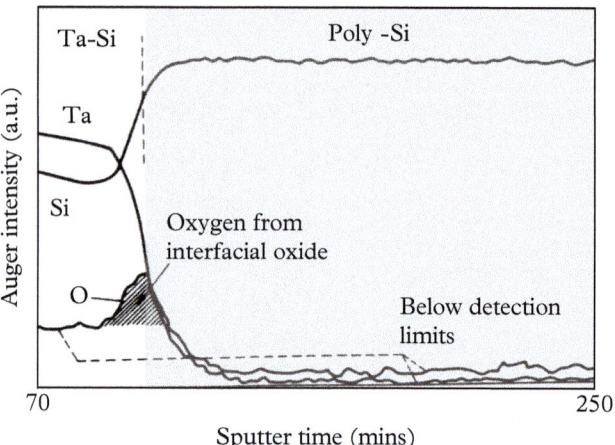

2.6.3   X-Ray Photoelectron Spectroscopy

Photoelectron spectroscopy, as the name implies, is a photon-in (probe), electron-out (signal) technique and is, in principle, a simple process. For atoms and molecules, a photon (hf) penetrates it and ejects an electron of binding energy, E_b, which then emerges as a photoelectron with energy, $E_{pe} = hf - |E_b|$. In the

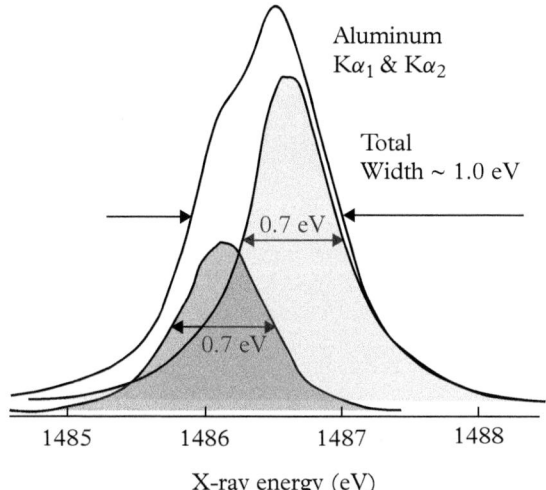

Aluminum
Kα₁ & Kα₂

Total
Width ~ 1.0 eV

0.7 eV

0.7 eV

1485   1486   1487   1488

X-ray energy (eV)

Figure 2.6.11 The components of the Kα spectrum of Al. Each of the sub-peaks, Kα1 and Kα2, have a FWHM ~0.7 eV, but they overlap, giving a FWHM ~1.0 eV for the total Kα peak.

Adapted from [7].

case of a solid (§3.8), we set the energy scale such that $E_b = 0$, at $E = E_F$, and then we have to include the work function, Φ, i.e. the photoelectron will now have an energy, $E_{pe} = hf - |E_b| - \Phi$. Thus, for a solid, any photon with energy greater than the work function, i.e. $hf > \Phi$, can be used as the probe, which eliminates near-ultraviolet, visible, and higher wavelengths.

In general, photoelectron spectroscopy is implemented using two types of photon sources available in a laboratory: either a soft X-ray source, $1200 < hf < 1400$ eV, used specifically for XPS, or a gas discharge lamp (§5.2.1), $10 < hf < 40$ eV, typically used for ultraviolet photoelectron spectroscopy (UPS). The escape energies (E_{pe}) of photoelectrons are 500–1,400 eV in XPS, and ~17 eV for UPS, if an He I resonance line ($hf = 21.2$ eV) is used. From Figure 1.3.5, we can infer that for the case of UPS, the energies correspond to the deep drop in the mean free path with energy for inelastic scattering, and hence, the surface sensitivity would also be significantly dependent on the material. The large gap in energy between these two radiations can be filled, if necessary, by synchrotron radiation (§2.4.3), which provides a large band of tunable radiation from the soft ultraviolet to hard X-rays ($hf > 10$ keV); using a suitable monochromator, photoelectron spectroscopy can be performed at any energy within this range.

To gain useful information about the binding energies of electrons in the specimen, the photons from the X-ray source should be as monochromatic as possible. Further, a highly conducting metal source, to minimize cooling in the UHV chamber, is preferred for the target. These choices lead to the two materials already mentioned, Al (Kα = 1.4866 keV) or Mg (Kα = 1.2536 keV). Strictly speaking, the Kα line is composed of the Kα₁ ($2p_{3/2} \rightarrow 1s$) and Kα₂ ($2p_{1/2} \rightarrow 1s$) lines (Fig. 2.6.11), but even when combined, for Al they are concentrated to give a FWHM ~1.0 eV. The corresponding FWHM, taking such spin–orbit coupling into consideration for the Kα lines, for other common metal sources are

Figure 2.6.12 (a) An XPS spectrum of a partially oxidized surface of Al with some contamination. Notice that the Delpha photoelectron energy (E_{PE}) measured and the binding energy (E_b) are simply related as $E_{PE} = hf - E_b$, neglecting the work function, which is indicated in the plot separately for Al $K\alpha$ and Mg $K\alpha$ in the abscissa. (b) The lower binding energy range (0–200 eV) is enlarged. Compare with the binding energies for these elements in Table 2.2.2.

Adapted from Woodruff and Delchar (1986).

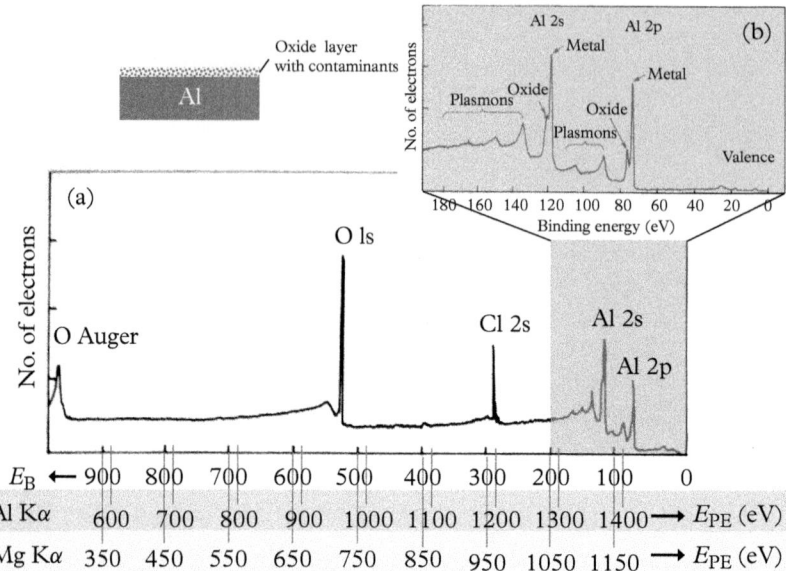

Mg ~ 0.8 eV, Cr (~ 2 eV), Cu (~ 2 eV), and Mo (~ 6 eV). Thus, Al $K\alpha$ or Mg $K\alpha$ emission lines are narrow enough for most analysis and are commonly used. If higher energy sources are used, or a narrower energy resolution is required, a crystal diffraction monochromator (§7.9.5) can be used, but this will generally result in a loss of intensity of the incident probe radiation.

Figure 2.6.12a shows a typical photoelectron spectrum for a partially oxidized and partially contaminated Al film, using monochromatic Al $K\alpha$ irradiation. The spectrum is dominated by a number of sharp peaks associated with the core states of the atoms, from which they emerge without undergoing any further energy loss. Each of the sharp peaks is followed by higher-energy "tails," following the original photo-excitation of the core holes. Whilst we have argued that the inelastic mean free path of the outgoing photoelectrons defines the surface sensitivity of XPS, the depth of photoionization is dependent on the propagation and absorption of the incident photons. Thus, any resultant loss of energy of the photon, prior to the photoionization event, is reflected in the inelastic tails observed. Figure 2.6.13 shows the variation in binding energy, E_b, of the electrons in the atom as a function of atomic number, Z. As expected (2.2.1), the energies increase as Z^2. The accessible energy levels for Al $K\alpha$ and Mg $K\alpha$ are indicated and for $Z > 30$ only the outer M and N shells can be ionized. On the higher energy side, additional energy loss peaks in Figure 2.6.12b, corresponding to plasmon resonances—collective excitations of the valence band—by the outgoing photoelectron are observed. For further details on plasmons (§9.4.2.2) and inter-band transitions, see Rather (2013).

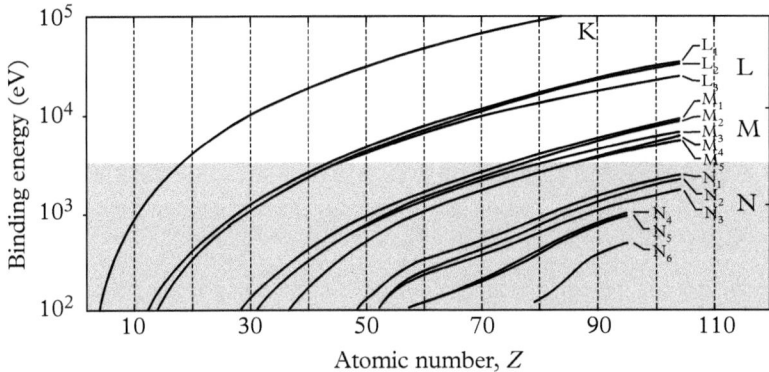

Figure 2.6.13 Binding energy (log scale) as a function of atomic number for all the elements. The region normally relevant for XPS, using laboratory sources, is shaded.

Adapted from Feldman and Mayer (1986).

If we look at the photoelectron spectrum on a finer scale (Fig. 2.6.12b) and focus on the Al peaks, in addition to the presence of multiple plasmon losses, we notice a chemical shift in the peaks from the metal to the oxide. When bonding takes place between elements to form a solid (§3.3), electrons are transferred (ionic) or shared (covalent) between the atoms. This changes the chemical environment of a specific element and a spatial redistribution of the valence electron charges around a specific atom. In other words, the potential around a specific atom is slightly modified and their binding energy changes. These shifts in binding energy of core electrons for the silicon $2p$ line, upon oxidation, and carbon $1s$ lines in various molecules are shown in Figure 2.6.14a and b, respectively.

Example 2.6.2: The C_{1s} binding energies measured in XPS for organic samples shows the following values, depending on the ligand: C–H (285.0 eV), C–N (286.0 eV), C–O (286.5 eV), C–F (287.8 eV), F–C–F (292.0 eV). Can you explain these trends (hint: consider electronegativity values from the periodic table)?

Solution: It is reasonable to assume, as a first approximation, that the initial state effects are only responsible for the observed chemical shifts. Thus, as shown in Figure 2.6.14, as the oxidation state of the element increases the binding energy of the photoelectron increases. Similarly, in these compounds, the chemical shift will depend on the electronegativity of the element that binds to carbon. Pauling electronegativity values are H = 2.1, C = 2.5 eV, N = 3.0, O = 3.5, and F = 4.0. In other words, there will be more electron transfer from C, equivalent to increased oxidation, as the electronegativity of the other element increases, which explains the observed trend.

There are other fine structures in an XPS spectrum for which a more involved explanation of the photoelectron emission process is in order. Strictly speaking,

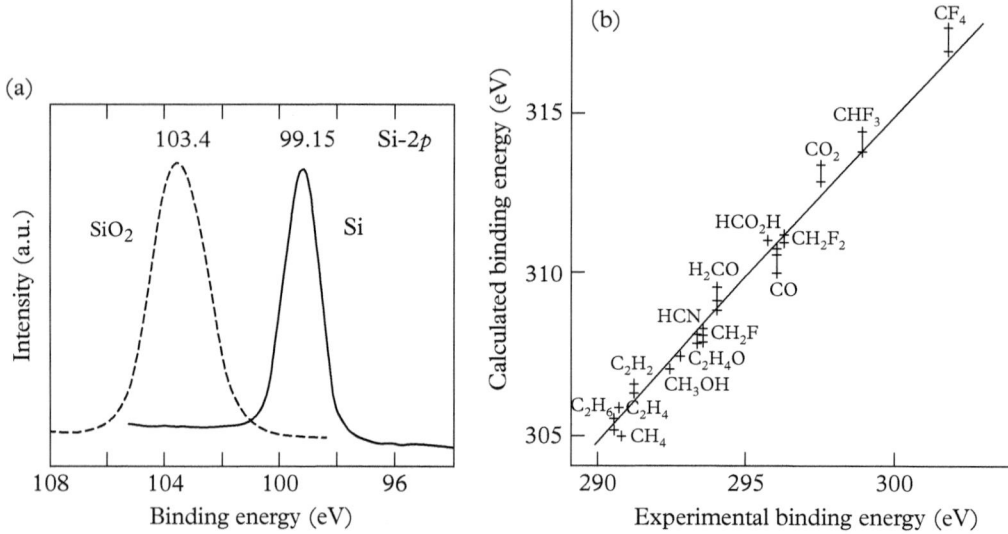

Figure 2.6.14 Examples of chemical shifts observed in XPS. (a) The shift in Si-$2p$ between elemental Si and SiO_2 depends on the oxidation state, and is ~4.25 eV. (b) Experiment and theoretical values of chemical shifts for C-$1s$ in various compounds.

the ideal energy, $E_{\rm PE} = hf - |E_b|$, of the photoelectron is not what is observed. In an atom, when the core hole is created, the other electrons readjust their energy to lower energy states to screen the core hole, and this additional energy, E_a, may be transferred to the photoelectron. As a result of this intra-atomic relaxation, the energy of the photoelectron is modified as $E_{\rm PE} = hf - |E_b| + E_a$. This is what one would expect if the photoionization and photoemission are stable equilibrium processes. In reality, the process is much too fast and the atomic rearrangement is switched on rather suddenly. Then the energy associated with this relaxation may excite a valence electron to an empty higher level—a process known as "shake-up." The energy for this shake-up may come from the emerging photoelectron, which would then appear as a peak with lower energy. The peaks are also known as "shake-off" if the valence electron is excited to the continuum. Shake-up peaks are particularly strong for some transition metals, rare earths, and aromatic organic systems. In addition to the change in peak position, in elements with valence or core shells containing unpaired electrons, the spin–orbit coupling can produce states with different energies, resulting in additional fine structure, called *multiplet splitting*, in the photoelectron spectra. Such multiplet splitting is small and would require the use of monochromators to be resolved. Figure 2.6.15 shows examples of shake-up peaks and multiplet splitting, and further details can be found in more advanced texts such as Briggs and Seah (1990) or Watts and Wolstenholme (2003).

Example 2.6.3: For the two morphologies of specimens, (a) a surface layer of atom B over atom A, and (b) a homogeneous ordered alloy of atoms A and B, as shown in the following image, what would the intensity ratio I_A/I_B in XPS/ESCA look like as a function of the X-ray incident angle, θ?

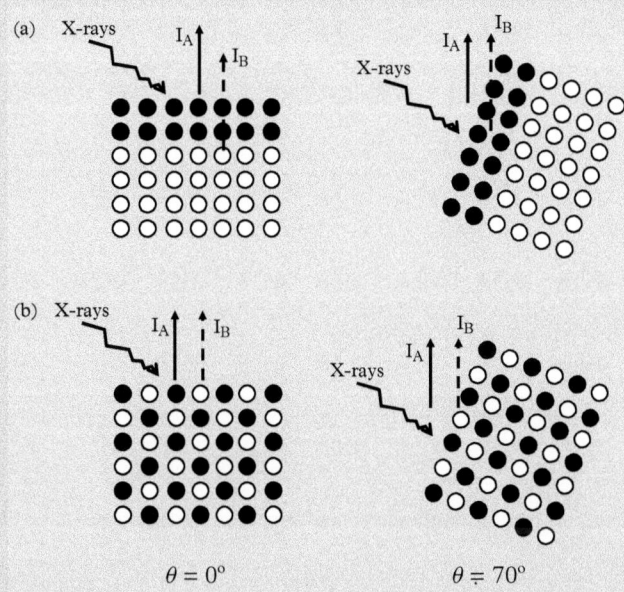

$\theta = 0°$ $\theta \doteq 70°$

Solution: The XPS or ESCA signal is mainly detected from the surface of the specimen. Thus for (a) we expect I_B to drop off and I_A/I_B to increase rapidly with the angle, θ. However, for (b), a homogeneous specimen, there will be no change in I_A/I_B with angle, θ.

Representative plots illustrating this variation are shown in the following figure.

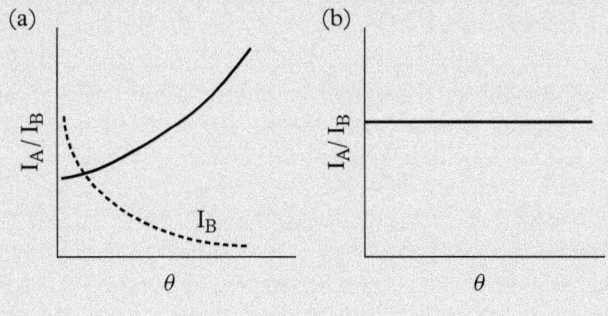

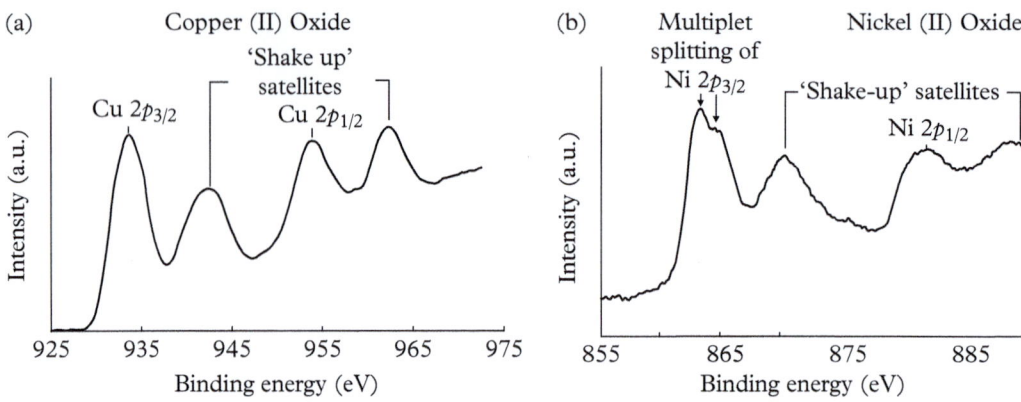

Figure 2.6.15 (a) Shake-up peaks observed in copper (II) oxide and (b) multiplet splitting of Ni $2p_{1/2}$ observed in nickel (II) oxide.

Adapted from Leng (2013).

2.6.4 Surface Compositional Analysis with AES and XPS

It is possible to perform quantitative chemical analysis, similar to X-rays (§2.5.2.3) with both AES and XPS.

In the case of AES, the yield, dN_x, of Auger electrons for the three-step ABC transition (Fig. 2.3.6), at any thickness, dt, and depth, t, for primary electrons with incident energy, E_p, can be written as

$$dN_x = I_P(t)n_x(t)\sigma_A(t)R_B\left(E_p, t\right)\omega_x p_{no-loss}(t)d\Omega\eta \qquad (2.6.1)$$

where $I_P(t)$ is the incident electron flux, n_x is the number of atoms of the element, x, of interest per unit volume, $\sigma_A(t)$ is the ionization cross-section of energy level A, $R_B(E_p, t)$ is the back-scattering factor for electrons of energy, E_p, ω_x is the fluorescence yield or probability of the decay of A to give the specific ABC transition, $p_{no-loss}(t)$ is the probability of no-loss escape from depth, t, $d\Omega$ is the acceptance solid angle of the analyzer, and η is the instrument detection efficiency. Thus, the total electron flux at depth, t, is $I(t) = I_P(t)\left[1 + R_B(t)\right]$. The first two terms in (2.6.1) are the incident flux and the atomic concentration to be determined. The last two terms are dependent on the specific instrumentation. The third term is the ionization cross-section, which can be either theoretically calculated or experimentally measured [7]. Typically, the general trend in electron ionization cross-section as a function of incident electron energy, E_p, is for it to rise rapidly above the threshold (given by the core binding energy, E_b) to a peak value at $E_p \sim 3$–4 E_b, and then falling off slowly at higher energies. Here, the fluorescence yield is the probability that, following an inner shell ionization, an Auger electron

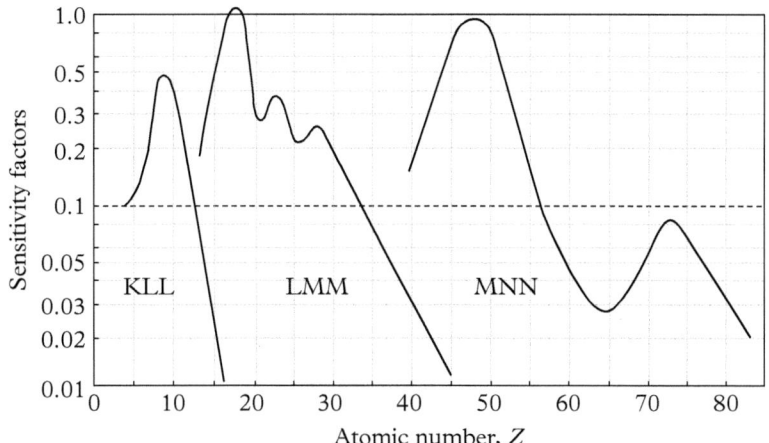

Figure 2.6.16 AES sensitivity factors for quantitative analysis for 5 keV electron probes.

Adapted from Flewitt and Wild (2003).

(instead of the production of an X-ray photon) is observed. It is given by

$$\omega = 1 - \frac{1}{1 + aZ^{-4}} \tag{2.6.2}$$

where Z is the atomic number, and a is a constant with large values of $a = 1.12 \times 10^6$ for K-shells and $a = 6.4 \times 10^7$ for L-shells. Thus, for low atomic number elements, Auger emission is favored, but as Z becomes large (>35), characteristic X-ray production starts to dominate (Fig. 1.4.5). The back-scatter factor, R_B, depends on both electron energy, E_p, and the atomic number, Z. In practice, because of this very complicated dependence of Auger yields on a rich variety of factors, quantitative analysis of Auger spectra is carried out using known standards to obtain so-called sensitivity factors (S_i), for any element, i. For a well-defined beam current, the Auger signal from an element is measured and normalized with respect to the signal from a standard, the ratio being the sensitivity factor, S_i. For an unknown specimen, an Auger spectrum is measured and after the peaks are identified, appropriate sensitivity factors are applied to give the concentration C_i as

$$C_i = \frac{I_i/S_i}{\sum I_i/S_i} \tag{2.6.3}$$

The sensitivity factors are published in the literature (McGuire, 1979), and obtained either from instrument manufacturers (Fig. 2.6.16) or measured for the specific instrument. The last method is rather tedious, but it is the most accurate. In any case, compositional analysis using this method assumes that the specimen to be analyzed is spatially homogeneous, which is seldom the case in practice.

Quantitative analysis of XPS data, a technique also referred to as electron spectroscopy for chemical analysis (ESCA), follows along the same lines. Similar

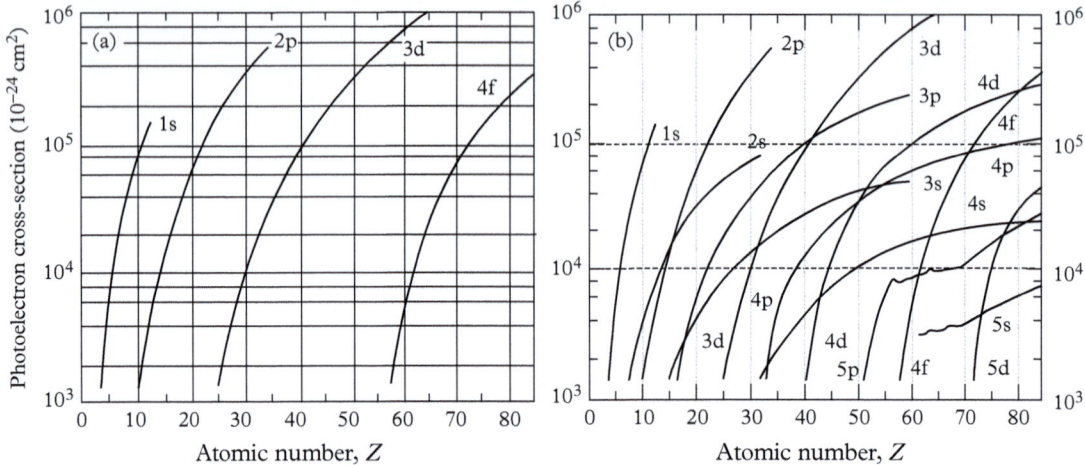

Figure 2.6.17 Photoelectron cross-section calculated for different sub-shells for all the elements: (a) the shells most used in XPS, and (b) the complete set of subshells.

Adapted from [8].

to AES, the yield, dN_i, for a photoelectron peak, at any thickness, dt, and depth, t, for primary X-rays with incident energy, E_p, can be written as

$$dN_i = I_x(t)n_i(t)\sigma_i(t)p_{no-loss}(t)d\Omega\eta \qquad (2.6.4)$$

where $I_x(t)$ is the X-ray flux, $n_i(t)$ is the number of atoms of element, i, per unit volume, $\sigma_i(t)$ is the photoelectric cross-section of element, i, $p_{no-loss}(t)$ is the probability of no-loss escape of photoelectrons from depth, t, which depends on escape depth, λ, and the analyzer angle, $d\Omega$ is the acceptance solid angle of the analyzer, and η is the instrument detection efficiency.

Note that, for all practical purposes, the X-ray flux is not attenuated over the escape depth of the photoelectrons. Even though the photoelectric cross-sections have been calculated for different subshells (Fig. 2.6.17), the absolute quantification of XPS data is complicated. Here again, a simpler approach is to use pure standards and determine the sensitivity factor, and the concentrations, C_i, as discussed earlier for AES.

2.6.5   Comparison of AES and XPS

AES and XPS are two of the most widely used surface analysis techniques, as they both have relatively uniform sensitivity to all elements other than H and He. It is also relatively much easier to produce a high-intensity electron beam as opposed to intense X-ray sources in the laboratory (synchrotron sources are exceptions) and so AES is faster in data acquisition and more sensitive than XPS (the appropriate cross-sections have similar values). On the other hand, high electron beam intensities may cause more surface damage in AES; in contrast, XPS is relatively benign. However, as electrons can be easily focused using electro-optical lenses, AES (unlike XPS) can provide high spatial resolution, including 2D maps of surface composition. The quantification process to obtain quantitative surface microanalysis in both cases is complex, but relative concentrations can be obtained with both methods with reasonable degree of accuracy (∼5%) using standards and sensitivity factors. Lastly, the analysis of chemical effects has been highly developed for XPS, and this technique is used widely to probe chemical states of surfaces.

2.7 Select Applications

2.7.1 XRF Analysis of Dental and Medical Specimens

It is well known that human tissues contain many kinds of minerals and trace elements that act as catalysts or structural components [11]. Quantitative analysis, including determining the spatial distribution and chemical state of these trace elements, is important for the analysis of metabolic processes. Sometimes these foreign elements lead to lesions; therefore, their analysis can impact diagnosis. For these materials, XRF is particularly suited for elemental analysis and imaging their spatial distribution for the following reasons: X-rays are easily transmitted in air, they seldom damage specimens, including biological tissue, and no pre-treatment, such as fixing, dehydrating, or coating with an electrical conductor, is required. Moreover, XRF can analyze wet and thermally low-resistive materials. Finally, microanalysis using XRF often requires a capillary focusing optics (Fig. 2.7.1a), instead of optical lenses, since the refractive index of glass for X-rays is almost equal to one.

The XRF spectrum of a paraffin-embedded lung biopsy specimen confirms the presence of tungsten in a patient suspected of having tungsten carbide pneumoconiosis (Fig. 2.7.1b). Synchrotron radiation sources have intensities greater than laboratory sources (X-ray tubes) by several orders of magnitude. Moreover, laboratory sources have high background, whereas synchrotrons produce highly monochromatized X-rays with low background. As a result, XRF spectra using synchrotron radiation shows negligible background, with the possibility of detecting trace (∼ppm) quantities (Fig. 2.7.1c,d).

In summary, teeth, soft tissues, and pathology specimens are regularly examined by XRF. For example, such studies have shown correlation between Ca content and tooth demineralization, as well as Zn, S, and Pb content in teeth with the levels of environmental pollution. Further details can be found in this excellent review [11] and the references therein.

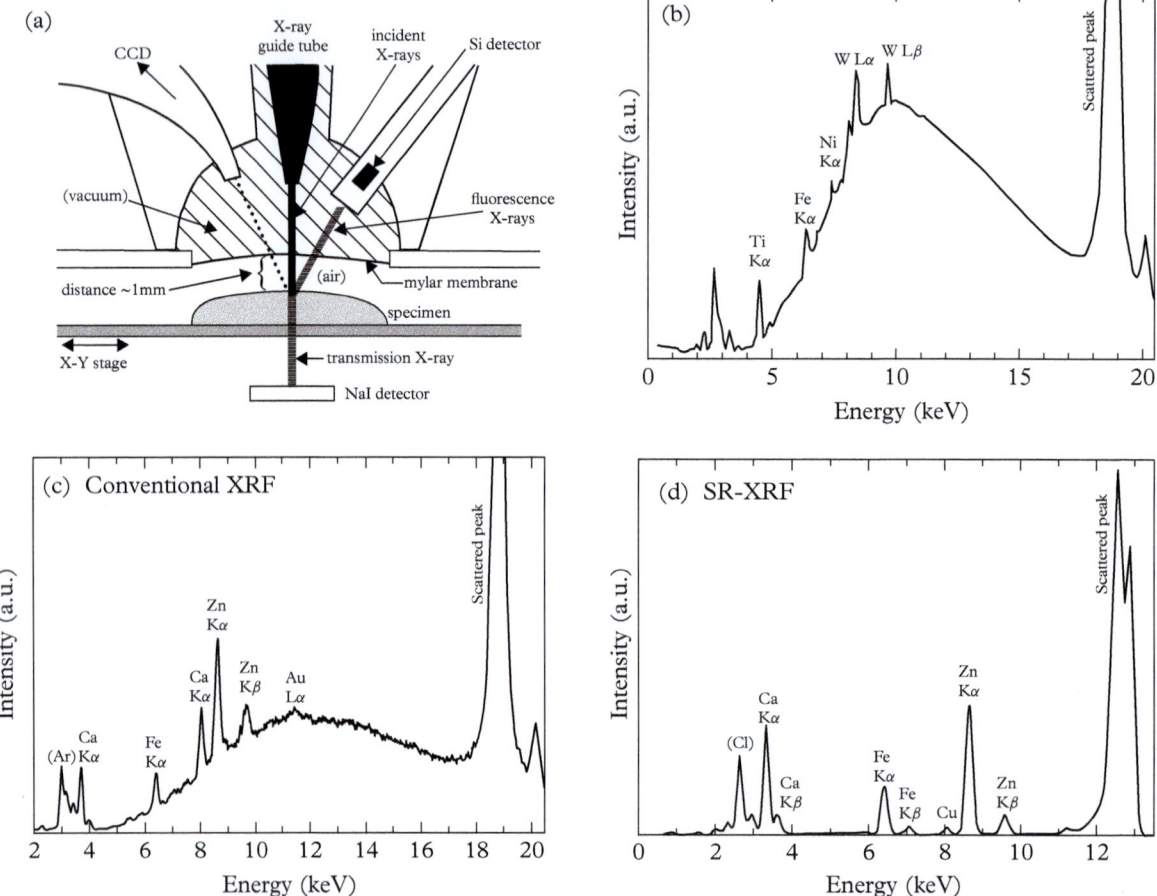

Figure 2.7.1 (a) Schematic diagram of a micro-focused X-ray fluorescence system using a capillary tube to focus X-rays. The inner surface of the capillary tube is designed to be a paraboloid of revolution and the total internal reflection from the inner surface guides the X-rays to focus (b) X-ray spectrum of a paraffin-embedded lung tissue specimen (experimental conditions: 50 kV, 1 mA, Rh target, and 600s/point). Comparison of X-ray fluorescence spectra from identical specimens using (c) a conventional laboratory X-ray tube, and (d) monochromatized synchrotron radiation.

Adapted from [11].

2.7.2   Environmental Science: Contamination in Ground Water Colloids

There is now substantial agreement that mobile colloids in ground water are important for transporting contaminants in subsurface environments [12]. Combining synchrotron radiation and energy dispersive X-ray spectrometry, elemental

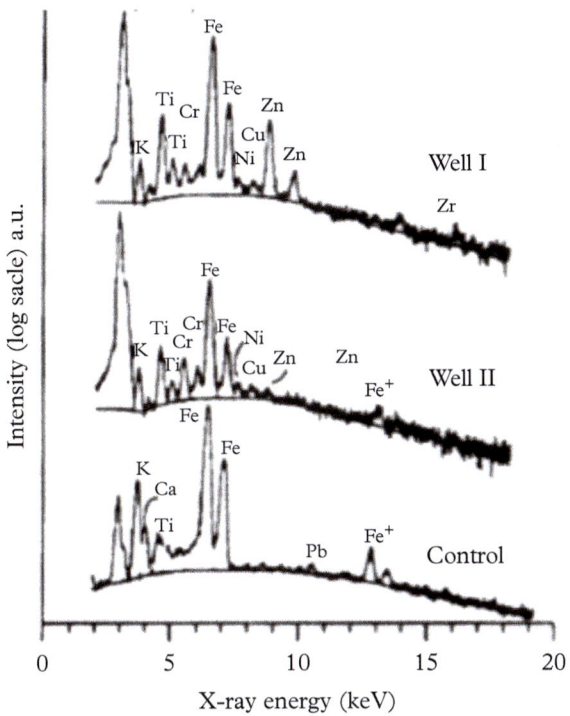

Figure 2.7.2 Synchrotron radiation X-ray fluorescence spectra of suspended colloids collected from two contaminated wells and a control. In all cases, the background intensity is consistently ~100 counts.

Adapted from [13].

composition of individual colloids, and the identity of the contaminants associated with these colloids have been determined (Fig. 2.7.2). Such analysis of colloids, including controls, showed that Al, Si, K, Ti, and Fe elements were primarily of natural origin, but Ca, Cr, Ni, Cu, Zn, Pb, and Zr, and especially Zn, with large peaks were observed in contaminated wells. Further details of these analysis can be found in [13].

Summary

We began with a brief description of the electronic structure of the atom, including quantum numbers that describe the state of the electron, hybridization, and spin–orbit coupling. We also introduced the dipole selection rules that govern single electron transitions, and put all this together to explain the emission of characteristic X-rays, following inner shell ionization, and the nomenclature used to describe them. Alternatively, de-excitation processes can also be non-radiative, with the emission of Auger electrons. Auger processes involve three atomic energy levels (as reflected in their nomenclature), leave the atom in its final state with two vacancies or holes, and need not follow dipole selection rules. A third consequence of photon incidence is the emission of characteristic photoelectrons, and their energy distribution can be simply related to the binding energies of the electrons in the solid.

X-rays are generated in the laboratory by the bombardment of a metal target by high-energy electrons, and typically show characteristic peaks superimposed on a continuous background. High-intensity X-rays are also generated in synchrotrons. X-rays are detected by wavelength dispersion using single crystals of well-defined lattice spacing and applying Bragg's law of diffraction. To cover the broad range of wavelengths required, multiple crystals are employed. Alternatively, they can also be detected by energy dispersion using semiconductor detectors, albeit with poorer energy resolution. However, unlike the bulky single crystal WDS detectors, the semiconductor EDXS detectors are compact, and are often included even in the narrow confines of electron optical columns, such as in TEMs and SEMs.

Characteristic X-rays are used for X-ray microanalysis, and are generated in instruments either with photon (XRF spectrometers) or electron (EPMA) incidence. The intensities of characteristic X-ray peaks can be related to the specimen composition, provided appropriate correction for instrument (details of the source, geometrical arrangement of the detector, efficiency of detection, etc.) and matrix/specimen (atomic number—Z, absorption—A, and fluorescence—F) factors are made. Alternatively, using standards with well-known composition and a method of ratios, or k-factors, allows for routine chemical analysis.

The complementary AES (typical energies <1.5 keV) and XPS (energies in the range 5–2,000 eV) methods are mainly used for surface analysis. They require clean surfaces and are often carried out in UHV (10^{-6}–10^{-8} Pa) environments. AES is an electrons-in (probe), electrons-out (signal) technique and such non-radiative de-excitation process is favored over characteristic X-ray emission for lower atomic number elements. The Auger electron spectrum appears as a weak feature superimposed over the inelastic background and, as such, is best resolved in the derivative spectrum. Further, Auger spectra show variations in line shapes due to differences in the chemical state and environment, but such spectral features are hard to interpret. Scanning AES instruments provide spatially resolved information, and when combined with a sputter gun, can provide three-dimensional compositional information.

Finally, we briefly introduced XPS, a photons-in (probe), electrons-out (signal) method; this technique is discussed in greater detail in §3, starting with §3.7, after we introduce bonding in solids (§3.2, §3.3), and then discuss the vibrational spectra of molecules (§3.4), including infrared (§3.5) and Raman (§3.6) spectroscopy.

. . . .

FURTHER READING

Attwood, D. *Soft X-Rays and Extreme Ultraviolet Radiation.* Cambridge: Cambridge University Press, 2000.

Borg, R. J., and G. J. Dines. *The Physical Chemistry of Solids*. Boston: Academic Press, 1992.

Briggs, D. A., and M. P. Seah. *Practical Surface Analysis with Auger and X-Ray Photoelectron spectroscopy*. New York: Wiley, 1990.

Callister, W. D., and D. G. Retwisch. *Materials Science and Engineering: An Introduction*, 8th ed. New York: Wiley, 2010.

Cullity, B. D. *Elements of X-Ray Diffraction*, 2nd ed. Reading: Addison-Wesley, 1977.

Feldman, L. C., and J. W. Mayer. *Fundamentals of Surface and Thin Film Analysis*. Amsterdam: North-Holland, 1986.

Flewitt, P. E. J., and R. K. Wild. *Physical Methods of Materials Characterization*. Boca Raton: IoP Publishing, 2003.

Krishnan, K. M. *Fundamentals and Applications of Magnetic Materials*. Oxford: Oxford University Press, 2016.

Leng, Y. *Materials Characterization: Introduction to Microscopic and Spectroscopic Methods*. Weinheim: Wiley-VCH, 2013.

McGuire, G. E. *Auger Electron Spectroscopy Reference Manual*. New York: Plenum Press, 1979.

Pettifor, D. *Bonding and Structure of Molecules and Solids*. Oxford: Oxford University Press, 1995.

Rae, A. I. M.*Quantum Mechanics*. Boca Raton: IOP Publishing, 1992.

Raethr, H. *Excitations of Plasmons and Interband Transitions by Electrons*. New York: Springer, 2013.

Rohrer, G. *Structure and Bonding in Crystalline Materials*. Cambridge: Cambridge University Press, 2001.

Siegbahn, K. *ESCA, Atomic, Molecular and Solid-State Structure Studied by Means of Electron Spectroscopy*. Uppsala: Almqvist & Wiksells, 1967.

Somorjai, G. A. *Chemistry in Two Dimensions: Surfaces*. New York: Cornell University Press, 1981.

Sproull, R. L., and W. A. Phillips, *Modern Physics*, 3rd ed. New York: Wiley, 1980.

Sutton, A. P. *Electronic Structure of Materials*. Oxford: Oxford University Press, 1993.

Watts, J. F. *An Introduction to Surface Analysis by Electron Spectroscopy*. Oxford: Oxford University Press, 1990.

Watts, J. P., and J. Wolstenholme. *An Introduction to Surface Analysis by XPS and AES*. New York: Wiley, 2003.

Williams, B. D., and C. B. Carter. *Transmission Electron Microscopy*, Vol IV. New York: Plenum Press, 1996.

Woodruff. D. P., and T. A. Delcher. *Modern Techniques of Surface Science*. Cambridge: Cambridge University Press, 1986.

....

REFERENCES

[1] Moore, C. E. *Atomic Energy Levels*, vol II. Washington, DC: US Government Printing Office, 1952.

[2] Boston, M. A., R. Klockenkaemper, J. Knoth, A. Prange, and H. Schwenke. "Total-Reflection X-Ray Fluorescence Spectroscopy." *Analytical Chemistry* 64, no. 23 (1992): 1115A–23A.

[3] Eichinger, P., H. J. Rath and H. Schwater. In *Semiconductor Fabrication: Technology and Metrology*, ASTM STP 990, edited by D.C. Gupta, 305. West Conshohocken, PA: ASTM International, 1989.

[4] Castaing, R. "Electron Probe Microanalysis." *Advances in Electronics and Electron Physics* 13 (1960): 317.

[5] Heard, P. J., J.C.C. Day and R. K. Wild. *Microscopy and Analysis* May (2000): 9.

[6] Pawlik, D., H. Oppozler and T. Hillner. "Characterization of Thermal Oxides Grown on TaSi2/Polysilicon Films." *Journal of Vacuum Science and Technology B* 3 (1985): 492.

[7] Bishop, H. E., and J.C. Riviere. "Estimates of the Efficiencies of Production and Detection of Electron-Excited Auger Emission." *Journal of Applied Physics* 40 (1969): 1740.

[8] Scofield, J. H. "Hartree–Slater Subshell Photoionization Cross-Sections at 1254 and 1487 eV." *Journal of Electron Spectroscopy and Related Phenomena* 8, no. 2 (1976): 129–37.

[9] Bearden, J. A. "X-Ray Wavelengths." *Reviews of Modern Physics* 39 (1967): 78.

[10] Creagh, D. C., and J. H. Hubbell. "X-Ray Absorption (or Attenuation) Coefficients" (Sec. 4.2.4). In *International Tables for Crystallography*, Vol. C, edited by A. J. C. Wilson, 189–206. Dordrecht: Kluwer Academic Publishers, 1992.

[11] Uo, M., T. Wada and T. Sugiyama. "Applications of X-Ray Fluorescence Analysis (XRF) to Dental and Medical Specimens." *Japanese Dental Science Review* 51, no. 1 (2015): 2–9.

[12] McCarthy, J. F., and C. Degueldre. "Sampling and Characterization of Colloids and Particles in Groundwater for Studying Their Role in Contaminant Transport." *Environmental Particles. Part II. Sampling and Characterization of Particles of Aquatic Systems*, edited by J. Buffle and H.P. van Leeuwen, 247–315. Boca Raton: Lewis Publishers, 1993.

[13] Kaplan, D. I., D. B. Hunter, P. M. Bertsch, S. Bajt, and D. C. Adriano. "Application of Synchrotron X-Ray Fluorescence Spectroscopy and Energy Dispersive X-Ray Analysis to Identify Contaminant Metals on Groundwater Colloids." *Environmental Science & Technology* 28 (1994): 1186–9.

....

EXERCISES

A. Test Your Knowledge

Which of the following (may be more than one, or none) is correct?

1. How many quantum numbers (EXCLUDING SPIN) are needed to describe an electron moving in TWO dimensions?
   (i) One
   (ii) Two
   (iii) Three
   (iv) Four

2. When an electron "jumps" from a higher energy level to a lower one, the energy difference is usually
   (i) absorbed by the nucleus.
   (ii) emitted as heat.
   (iii) emitted as light.
   (iv) emitted as a continuous electromagnetic wave.
   (v) emitted as a photon.

3. If the atomic number of an element is Z and its atomic weight is A, the number of protons in its nucleus is
   (i) Z
   (ii) $A - Z$
   (iii) A
   (iv) $Z - A$

4. The maximum number of electrons in the L-shell ($n = 2$) is
   (i) 4
   (ii) 6
   (iii) 20
   (iv) 8
   (v) 14

5. The characteristic feature of a transition elements is
   (i) a partly filled valence subshell.
   (ii) an empty inner subshell.
   (iii) an unfilled outer subshell.
   (iv) a partly filled inner subshell.

6. The element with the configuration $1s^2, 2s^2, 2p^6, 3s^2, 3p^6, 3d^8, 4s^2$ is a transition metal.
   (i) True
   (ii) False

7. The electrostatic nature of ionic bonding makes it
    (i) non-directional.
    (ii) weak.
    (iii) applicable only to Groups 1 and 2.
8. The electronegativity of an element is a measure of
    (i) the excess of protons over electrons.
    (ii) the number of electrons in the valence shell.
    (iii) the strength with which electrons are attracted to the atom.
9. (a) The principal quantum number, n, may have values
        (i) $0,1,2\ldots$.
        (ii) $1,2,3,\ldots$.
        (iii) $\pm1,\pm2,\pm3,\ldots$.
    (b) The angular momentum quantum number, l, may have values
        (i) $0,1,2,3,\ldots(n-1)$.
        (ii) $0,1,2,3,\ldots n$.
        (iii) $1,2,3,\ldots n$.
        (iv) $1,2,3,\ldots(n-1)$.
    (c) The magnetic quantum number, m_l, may have values
        (i) $0,\pm1,\pm2,\pm3,\ldots\pm l$.
        (ii) $0,\pm1,\pm2,\pm3,\ldots\pm n$.
        (iii) $0,\pm1,\pm2,\pm3,\ldots+(l\text{-}1)$.
        (iv) $0,\pm1,\pm2,\pm3,\ldots+(n\text{-}1)$.
10. H and He cannot be detected by AES because
    (i) No one has tried to do so.
    (ii) It is too expensive to make suitable specimens.
    (iii) Their electronic structure does not allow such non-radiative decay.

B. Problems

1. **Characteristic X-rays**: Consider the element, Mo:
    (a) What is its electronic distribution ($1s^2,\ldots$)
    (b) What are the binding energies (in eV) of the different atomic levels (look this up)?
    (c) What are energies of the allowed transitions following the ionization of inner shell electrons?
    (d) Draw the energy levels and label these transitions.
    (e) If each of these electronic transitions lead to the emission of characteristic *X-rays*, what would be their wavelengths?
    (f) The FWHM of a Mo K_β line is 0.0008 Å. What is its energy width?
2. For a Mo X-ray tube, you wish to obtain I($K\alpha$)/I($K\beta$) =500, using a **filter**.
    (a) What element will you use as the filter?
    (b) What is the filter thickness required?

(c) What would be the transmission factor for Mo K_α and Mo K_β radiations when this filter is used?

3. **Mosley Diagram**: Consider five elements, C, Na, Mg, Cu, and Mo.
   (a) What is the energy, E_K, of each of their K_α photons?
   (b) Plot $(E_K)^{1/2}$ as a function of (Z) where Z is their atomic number.
   (c) Draw the best-fit curve to this data. Is it a straight line? Explain.

4. **XPS**: You are using Mo K_α X-rays to probe a specimen of steel (which is predominantly iron) by XPS or ESCA. Assume that the binding energies are unchanged from the atom in the steel.
   (a) At what values of energies will you see peaks in the XPS/ESCA spectrum.
   (b) Instead of Mo K_α you use CuK_α radiation. Now, where will the peaks show up?

5. **AES**: In Problem 4, what Auger peaks will you observe? Indicate the transitions (KLL, etc.) and the energies involved. Use the table of binding energies.

6. **AES**: What are the energies for the KL_1L_2 and $L_1M_1M_1$ Auger electrons of Fe? Use (2.3.4a).

7. **Electron Spectrometer**: You are designing a simple magnetic spectrometer to detect electrons 20–50 eV in energy. You want to build a spectrometer that is not larger than 10 cm in radius. What value of the uniform magnetic induction, **B**, do you need to generate to have a functioning spectrometer that fits this dimension?

8. **X-ray Shielding**: You are working for a dental instrument manufacturer. They want you to build a simple lead shield to put on patients to protect them from *X-ray* exposure. Assume they use CuK_α radiation for imaging teeth. How thick should the lead shielding be to ensure that the patients are protected in case of accidental exposure? State your assumptions. Note $I = I_0 e^{-At}$, where I_0 is the incident intensity, A is the absorption coefficient, and t is the thickness.

9. **X-ray analysis**: The chemical composition of a mixture of barium-oxide and chromium-carbide in equal proportions is measured using energy dispersive X-ray spectrometry/microanalysis over the energy range 0–40 keV.
   (a) If you are using a **windowless** Si(Li) detector, plot the spectra that you will see, including the two principal artefacts: a Si escape peak and a sum peak. Identify and label all the peaks carefully.
   (b) What will this spectrum look like if you are using a **thin Be window** to protect your semiconductor detector?
   (c) What will this spectrum look like if you are using **a Ge detector** instead of a Si(Li) detector?
   (d) What will this spectrum look like if you are using a **wavelength dispersive** spectrometer?

10. **WDS analysis**: You are asked to analyze a specimen containing tungsten and carbon using wavelength dispersive X-ray analysis. You decide to analyze the Kα radiation for both elements.

   (a) What *lattice spacing* for the crystal analyzers are required to carry out the analysis for C Kα and W Kα_1 lines?

   (b) From Table 2.5.1, which crystal(s) will you use for the analysis?

   (c) If you decide to analyze W Lα_1 instead of W Kα_1, what crystal spacing is required?

   (d) Which would you prefer to analyze W Kα_1 or W Lα_1? Why?

11. In the **EDXS microanalysis** of an unknown binary alloy, $A_x B_y$, the following intensities (counts) are measured $I_A = 10,000$ & $I_B = 2,500$. For two standards $A_{50}Si_{50}$ and B_2Si_1, the intensities measured are $I_A = 20,000$ & $I_{Si} = 10,000$, and are $I_B = 16,000$ & $I_{Si} = 2,000$, respectively. What is the composition of the binary alloy? State your assumption(s), if any.

12. **Auger Electron Spectroscopy** analysis of the alloy, Tungsten Nitride (WN_2).

   (a) For which of the two elements, W and N, do you expect to see Auger peaks?

   (b) For that element, estimate the KL_1L_2 Auger electron energy. Hint: Use (2.3.4a) and Table 2.2.2.

   (c) For that element, what is the fraction of Auger electron yield? Hint: Use (2.3.5).

Bonding and Spectra of Molecules and Solids

3

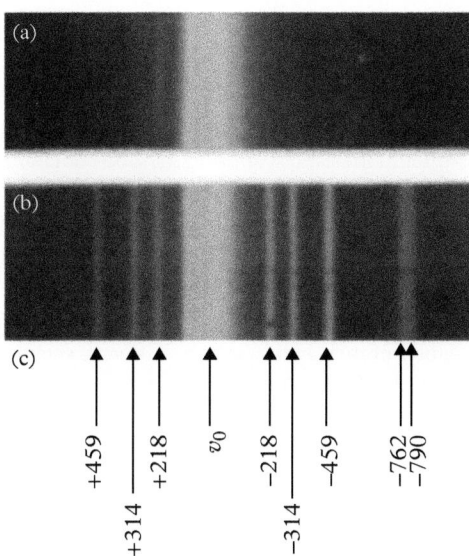

One of the first published Raman[1] spectra from carbon tetrachloride liquid. (a) Spectrum of the mercury arc lamp in the region of (λ = 435.9 nm, or wavenumber υ_0 = 22,938 cm^{-1}) used as the incident probe. (b) Rayleigh and Raman spectrum from liquid CCl_4. (c) The principal lines from the spectrum of CCl_4 are indexed in wave numbers (cm^{-1}) with respect to υ_0 = 22,938 cm^{-1}.

Adapted from Raman, C. V., and K. S. Krishnan. "The Production of New Radiations by Light-Scattering. Part I." *Proceedings of the Royal Society of London A* **122**, (1929): 23–35.

[1] Sir C. V. Raman (1888–1970) was an Indian physicist who was awarded the Nobel prize in Physics in 1930, and was cited for "his work on the scattering of light and for the discovery of the effect named after him."

3.1 Introduction

We discussed atomic structures and the periodic table of the *elements*, based on a systematic filling of atomic energy levels by electrons, following the Pauli exclusion principle and Hund rules (§2.2). We then discussed atomic transitions, emissions, secondary processes (§2.3), and related spectroscopy methods, all involving inner-shell electrons in one form or the other. In addition to these core-level electrons, which occupy filled shells with energies barely perturbed by bonding, there are also the valence electrons that occupy the outer energy levels, participate in bonding and the formation of molecules and solids, and determine many of the physical, chemical, and structural properties of materials.

This chapter begins with a brief overview of the electronic structures of molecules, involving the bonding of a small number of atoms. These molecular energy levels are also modulated by their vibration and rotation modes; we provide a simple model to describe these modulations in order that we can discuss absorption and two-photon inelastic scattering methods of characterization. The chapter then covers bulk solids, described by the bonding of a very large number, N_A, of atoms, where N_A is the Avogadro number. The four principal types of interatomic bonds (ionic, covalent, metallic, and secondary) describe, either singly or in combination, the wide range of engineering materials we encounter today (Fig. 3.1.1). In general, the most direct way to probe the energy levels of molecules and solids, and the nature of their bonding, is by different spectroscopy methods; this is the main subject of this chapter. In the next one (§4), we will move away from spectroscopy methods, and introduce the spatial arrangement of atoms in solids (symmetry and crystallography) that allows us to probe their structure by

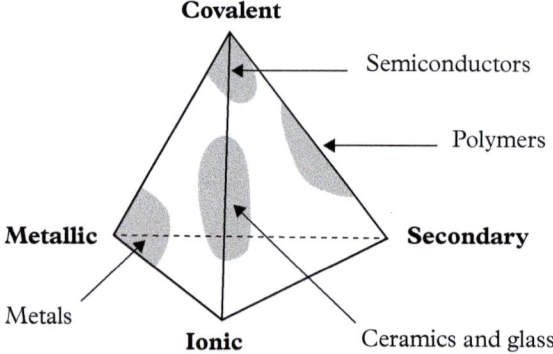

Figure 3.1.1 A tetrahedron with the principal bonding types forming its apexes, and illustrating their relative contribution to the important classes of engineering materials that are encountered in materials characterization.

Adapted from Shackelford (1996).

diffraction and scattering. Scattering of ions (§5), and diffraction of X-rays (§7) and electrons and neutrons (§8), are discussed in detail in future chapters.

In §3.2 and §3.3 we introduce the basics of bonding in molecules, the formation of bands, and the types of bonding in solids. The first is important in understanding various molecular spectroscopic methods, and the latter two are required to appreciate methods that probe the electronic structure of solids. Needless to say, those familiar with these elementary concepts can move directly to §3.4.

3.2 Bonds and Bands

As a first approximation, the molecular orbitals can be considered as a *linear combination of atomic orbitals* (LCAO). The simplest case to consider is a diatomic molecule, such as H_2 (Fig. 3.2.1), where the $1s$ orbitals with wavefunctions, ψ_1 and ψ_2, are combined linearly to form molecular orbitals. Two such molecular orbitals, bonding, $\psi_b = \psi_1 + \psi_2$ and antibonding, $\psi_{ab} = \psi_1 - \psi_2$, that satisfy the symmetry requirement that the environment around each hydrogen atom in the molecule be the same are possible. The bonding molecular orbital, ψ_b, with increased electron density between the two hydrogen nuclei, strongly stabilizes the molecular bond by lowering its overall energy. In the antibonding molecular orbital, ψ_{ab}, the electrons avoid the inter-nuclear region and raise their overall energy. If heavier elements are involved, molecular orbitals are constructed in the same way with atomic orbitals, provided only the valence electrons of similar energy and the correct relative symmetry of the orbitals are included.

For polyatomic molecules, the linear combination of atomic orbitals (Fig. 3.2.2 a,b) can also generate molecular orbitals by extending over all the constituent atoms. However, the *total* number of molecular orbitals, including the bonding (sometimes referred to as the highest occupied molecular orbitals,

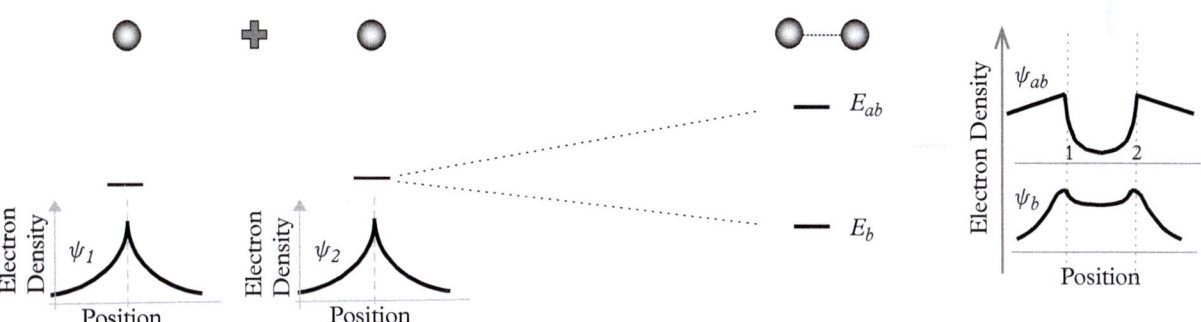

Figure 3.2.1 Bonding in a simple divalent molecule, such as H_2, involving two atomic wavefunction of the valence electrons. The two atomic levels (*s*-orbital wavefunctions shown in the left-bottom) combine to form bonding, E_b, and antibonding, E_{ab}, levels, with two different energies. Note that when two atomic orbitals combine, they conserve the number of orbitals and give rise to two possible molecular orbitals.

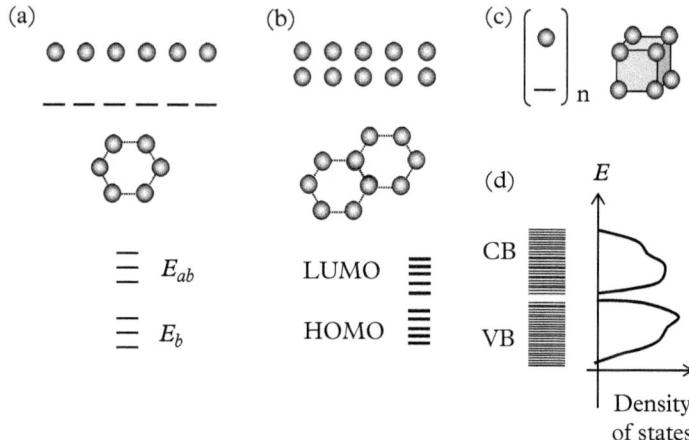

Figure 3.2.2 Orbital energies for (a) a small molecule of six atoms, giving rise to three bonding and three antibonding levels; (b) a larger molecule with ten atoms—note that the number of energy levels is always conserved in the bonding; and (c) a solid involving a very large number ($\sim N_A$) of atomic orbitals. (d) The density of states (DOS), $N(E)$, for the valence (VB) and conduction (CB) bands shown in (c). In each case, the top row indicates the atomic levels, and the bottom row the molecular or solid electronic levels. Note that the energy gap decreases with the size of the molecule or solid.

Adapted from Krishnan (2016).

HOMO) and antibonding (referred to as the lowest unoccupied molecular orbital, LUMO) levels, is exactly the same as the total number of atomic orbitals involved in the formation of the molecule. For larger molecules, the number of molecular orbitals increases and their spacing in energy becomes closer (Fig. 3.2.2b). When we reach the size of a solid—a very, very large molecule, indeed—the orbitals extend throughout the crystal, and these crystal energy levels are now so closely spaced that they form a band of energy (Fig. 3.2.2d). Note that in a finite solid, the number of crystal orbitals is equal to the total number of atomic orbitals involved and can be quite large ($\sim N_A$, the Avogadro number, 6.02×10^{23} mol^{-1}). Hence, the energy levels in a band are so closely spaced that they are considered to be essentially continuous in energy.

In the solid, the crystal energy levels may *not* be distributed *uniformly* over all energies; in fact, their distribution vary as a function of energy and are concentrated at some energies. In addition, there are also regions in energy called *band gaps*, where there are no levels present. This distribution of energy levels is described by the *density of states* (DOS) function, $N(E)$, which is defined as the number of energy levels, per unit volume, present in the energy range, $E - (E + \Delta E)$; note that $N(E) = 0$, in the forbidden band gap. Even though $N(E)$ may have some complicated form (Fig. 3.2.2d), they are often represented by simple blocks of band energies. The width of the bands depends on the strength of the interactions between the neighboring atoms in the solid. For valence s and p electrons, as in the elements on the left-hand side of the Periodic Table, the orbital overlaps are strong and give bands that are very broad in energy (many eVs wide). On the other hand, as emphasized earlier, the core orbitals contract and interact weakly to give very narrow bands (width ~0.1 eV). The latter contribute very little to the bonding in solids, but retain their atomic identity, unaffected by the bonding—a fact exploited by the various core-level spectroscopies discussed in §2, for unequivocal chemical identification and microanalysis.

The electronic properties of materials depend on the energies of the bands, their distribution and widths, as well as the gaps between them. In the case of metals, the density of states, $N(E)$, within the conduction band can be calculated to first order, at 0 K, by the free electron model (Kittel, 1986):

$$N(E) = 4\pi \left(\frac{2m_e}{h^2}\right)^{3/2} E^{1/2} \tag{3.2.1}$$

where E is the energy, and all other physical constant have been defined earlier. Figure 3.2.3(b) shows this density of states function, $N(E)$, with $E^{1/2}$ dependence

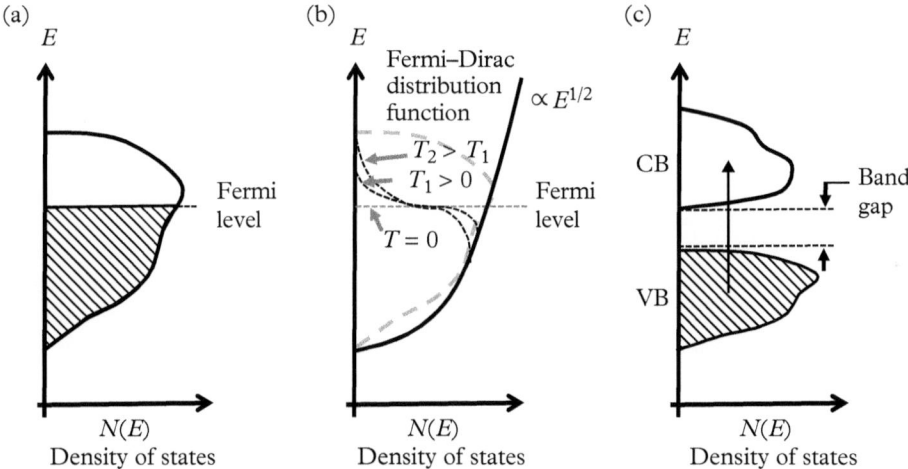

Figure 3.2.3 Density of states (DOS) in (a) metal, showing the highest occupied state called the Fermi level, E_F. (b) DOS in the free electron model for a metal, including the Fermi–Dirac distribution showing the occupancies of the levels at 0 K and at higher temperatures, $T_1 > 0$ and $T_2 > T_1$. (c) DOS in a non-metal, showing the filled valence band and the empty conduction band separated by a band gap. In (a) and (c), the shading shows the occupied levels. The filled levels in materials are typically probed in X-ray photoemission spectroscopy (XPS) and the unfilled levels in X-ray absorption spectroscopy (XAS) and inverse photoemission spectroscopy (IPES).

(3.2.1). The photoelectron spectra (§3.8) of many simple metals follow this free-electron relationship; the calculated maximum filled state or Fermi level, E_F, for example, Al $\sim$11.7 eV, Mg $\sim$7.1 eV, and Na $\sim$3.2 eV, matches the photoemission electron spectroscopy (PES) measurements (Al $\sim$10.7 eV [1], Mg $\sim$6.1 eV [2], Na $\sim$2.5 eV) reasonably well. At higher temperatures, the thermal energy excites electrons to higher energy levels beyond E_F, and the Fermi–Dirac distribution gives the fractional occupation of levels as:

$$f(E) = \frac{1}{1 + e^{\left(\frac{E - E_F}{k_B T}\right)}} \tag{3.2.2}$$

Figure 3.2.3(b) also shows this function. In fact, the electron distribution around the Fermi level, as a function of temperature, measured by ultraviolet photoemission spectroscopy (UPS) is in very good agreement with the Fermi–Dirac function (Fig. 3.8.8).

In its ground state, a solid insulator or semiconductor will display a filled band known as the valence band, with a band gap, E_g, separating it from the empty conduction band. Simple optical absorption measurements (Fig. 3.2.4) can be used to determine the band gap, which vary from 0.1 eV in semiconductors to $\sim$12 eV in some ionic insulators (Fig. 3.3.2). A metal, on the other hand, shows a partially filled conduction band and since there is no gap, their optical properties are quite different; in fact, because of the strong interaction of optical radiation with the free electrons in the solid, metals show high reflectivity. Moreover, metallic materials conduct down to low temperatures, but the conductivity decreases with increasing temperature; in contrast, non-metallic materials are non-conducting at low temperatures with some small increase in conductivity as the temperature is increased. Generally, the electrical conductivity is understood in terms of the filling of energy levels in the bands (Fig. 3.2.3). This rudimentary summary of the electronic structure of materials is sufficient to understand related methods of characterization but, for those interested, further details can be found with a physics flavor in Kittel (1986) and Pettifor (1995), or with a chemistry flavor in Cox (1987) and Borg and Dines (1992).

This chapter presents throughout a variety of spectroscopy methods (Fig. 3.7.1 offers a summary) used to measure the electronic structure of solids and surfaces, including the occupied and unoccupied levels. However, before we discuss these methods, we need to briefly review the principal types of bonding and how they affect the electronic structure of materials.

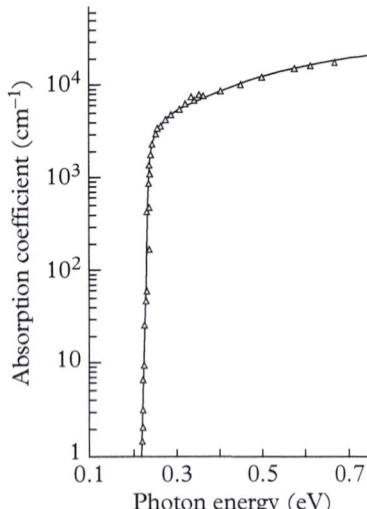

Figure 3.2.4 The band gaps in semiconductors are typically measured by optical absorption spectroscopy. Such transitions from the valence band to unoccupied states in the conduction band, measured in Indium antimonide, InSb, a semiconductor with a band gap of $\sim$0.23 eV at 0 K is shown. See also Figure 3.7.1.

Adapted from Kittel (1986).

3.3 Interatomic Bonding in Solids

The primary bonding in solids (Fig. 3.3.1) is classified into three categories: *ionic*, *covalent*, and *metallic*. For each type, the bonding necessarily involves the sharing

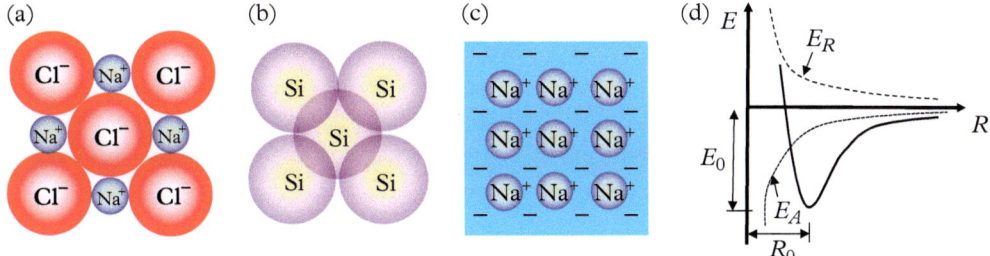

Figure 3.3.1 Schematic representation of (a) an ionic bond, such as in NaCl, with local charge transfer from the cation to the anion; (b) a covalent bond, such as in an elemental semiconductor, Si, with charge sharing to stabilize the bond; and (c) a metallic bond, such as in Na, with a "sea" of delocalized electrons holding the positively charged ion cores together in the solid. (d) The general shape of the bond energy curve as a function of interatomic separation, including the attractive, E_A, and repulsive, E_R, energy components that stabilize the bond. This curve, generally applicable to all bond types, shows the equilibrium bond energy, E_0, and bond length, R_0. See also Figure 3.4.1.

or transfer of the outermost electrons, and specifically, the bonding determines the final electronic structure of the solid. In addition, weaker secondary bonds resulting from dipolar interactions, without any change in the electronic structure of the atoms or molecules that form the solid, are also observed. In practice, such secondary bonds contribute to the stability of polymeric solids (Fig. 3.1.1); however, here we introduce only the three primary bonds, and further details on bonding can be found in Pettifor (1995).

3.3.1 Ionic Bonding

The ionic bond (Fig. 3.3.1a) is readily found in compounds made of metallic and nonmetallic elements, where valence electrons are readily transferred from the metal to the nonmetal. In this manner, all the atoms acquire a stable inert gas configuration, forming either positively (cations) or negatively (anions) charged ions. The energy of interaction between the ions is due to isotropic Coulombic forces, with an attractive component, $E_A \propto -A/R$, and a repulsive component, $E_R \propto B/R^n$ (Fig. 3.3.1d). Here, A is a constant that depends on the charge of the cation and anion, B is an empirical constant, R is the inter-ionic distance, and the exponent, n, is a constant in the range of 6–9 that reflects the short range of the repulsive interaction. The $1/R$ dependence in the attractive potential is long range and hence all interactions in the solid, in addition to the nearest neighbor one, must be considered. Thus, depending on the specific crystallographic arrangements in the solid, the summation of all Coulombic interactions, both attractive (typically next-neighbor interactions of opposite charges) and repulsive (next next-neighbor interactions of like charges), results in the constant A being

Figure 3.3.2 The valence and conduction band energies of NaCl can be derived by starting with (a) free ions, then including (b) the Madelung potential, followed by correcting for (c) the electrostatic polarization arising from adding/removing an electron, and finally (d) the inclusion of the bandwidth from orbital overlap. The band gap is ~7 eV, the width of the valence band (Cl $3p$) is ~2 eV, and the conduction band (Na 3s) is ~6 eV. Different spectroscopy measurements discussed in this chapter allow for a detailed evaluation of both occupied and unoccupied states arising in materials as a result of their bonding.

Adapted from Cox (1986).

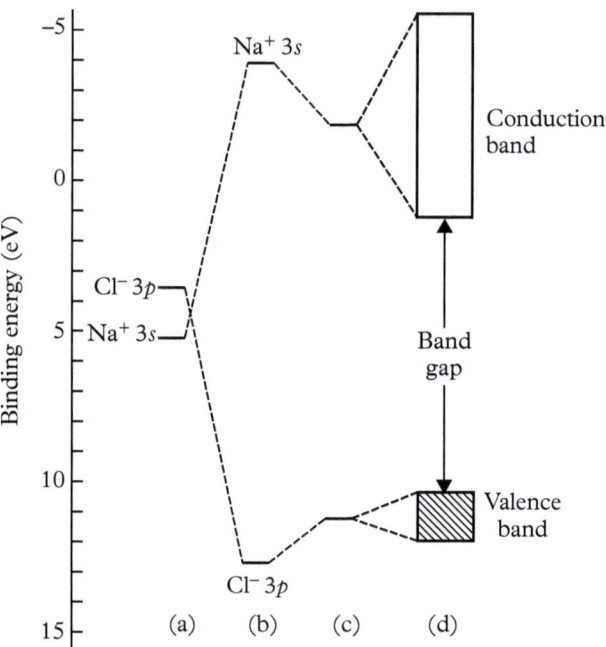

replaced by the *Madelung[2] constant,* A_M, for the solid. Typically, the lowest energy structures have the largest value of A_M, thus favoring arrangements with the largest coordination (consistent with the notion of a non-directional, isotropic ionic bond). Alternatively, from space filling considerations, the coordination number is determined by the radius ratios of the two ions.

The model compound, NaCl, illustrates the electronic structure, i.e. the filled and empty levels, of ionic compounds. The fact that each ion has a closed–shell configuration ensures that the associated bands are either empty or completely filled. The filled band (VB) is thus expected to be the Cl $3p$ orbital with an additional electron, whereas the empty band (CB) is the Na $3s$ orbital from which the electron has been removed. The final band structure (Fig. 3.3.2d), then involves adjustments for the Madelung potential of the lattice, and the energy arising from the orbital overlaps.

3.3.2   Covalent Bonding

Covalent bonding is central to geology (silicates), biology (DNA, proteins, etc.), and the technologically important family of inorganic semiconductors that includes the elements Si and Ge, as well as the compounds GaAs, InSb, and SiC. In solids with covalent bonding (Fig. 3.3.1b), stable electronic configurations are

[2] Erwin Madelung (1881–1972) was a German theoretical physicist.

achieved by the sharing of electrons. The bonding in the solid is not different from that in small molecules; for example, the bond energy and bond length of C–C bond in diamond is similar to that found in many alkanes. The number of covalent bonds that is possible for a particular atom is given by the 8–N rule, where N is the number of valence electrons. The covalent bond is directional, i.e. it is formed between specific atomic orbitals placed along well-defined directions.

Covalent bonding in solids, including the additional constraints imposed by the crystallographic symmetry (§4.1.7), results in the "band" structure. Simply put, the *band structure* is the distribution of the energy levels in the covalent crystal along different directions of the *reciprocal lattice* (which is simply related to the real lattice; see §4.2 for details). Measurements of the band structure of semiconductors, which typically show a filled valence band separated from an empty conduction band by the band gap, are a topic of both academic and technological interest. Figure 3.3.3 shows a detailed measurement of the band structure of GaAs, involving the techniques of *photoemission and inverse photoemission spectroscopy*, as well as *angle-resolved* PES measurements. Note the direct band gap of ~1.4eV in GaAs is observed in the measurements. Photoemission spectroscopy to probe the occupied levels was introduced in §2.6 and further details of this technique, and its complement, or inverse photoemission spectroscopy, to probe unoccupied energy levels, will be discussed in §3.8.

In practice, many materials have mixed bonding character. The concept of *electronegativity*, defined as a measure of the ability of the atom to attract electrons, helps quantifying the fraction of ionic character in an A-B bond as

$$\% \text{ of ionic character} = 100 \left[1 - e^{\frac{-1}{4}(x_A - x_B)^2} \right] \qquad (3.3.1)$$

where x_A and x_B are the Pauling[3] electronegativities (see Periodic Table of the elements, page 848 for values) of elements A and B, respectively.

Example 3.3.1: Calculate the percent ionic character in (a) Na–Cl, (b) Al–N, (c) Mg–O, (d) Si–Ge, and (e) Cu–Zn. Use Pauling electronegativity values.

Solution: We look up a table of Pauling electronegativities for the elements and applying (3.3.1) we solve in the form of a table.

	Na–Cl	Al–N	Mg–O	Si–Ge	Cu–Zn
x_A	0.9	1.5	1.2	1.8	1.9
x_B	3.0	3.0	3.5	1.8	1.6
% Ionic character	66.8%	43.02%	73.35%	0%	2.2%

[3] Linus Pauling (1901–1994) was an American chemist who received the Nobel prize in Chemistry in 1954 for "his research into the nature of the chemical bond and its application to the elucidation of the structure of complex substances." He was also awarded the Nobel Peace prize in 1962!

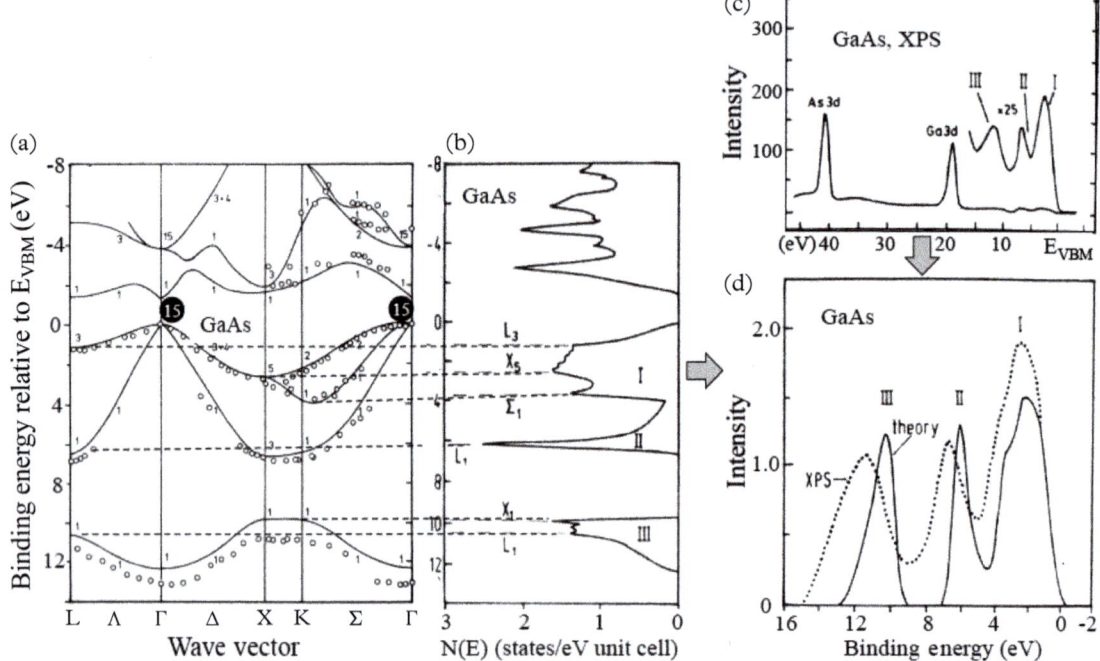

Figure 3.3.3 (a) The band structure of GaAs derived from both photoemission spectroscopy (PES) and inverse photoemission spectroscopy (IPES), above the Fermi level (labeled as 15), obtained by angle-resolved measurements (not discussed in this book). The band structure is simply the distribution of the energy levels in the crystal along different directions (indicated by the symbols L, Λ, Γ, Δ, X, K, and Σ) of the reciprocal lattice. The solid lines are the result of calculations. (b) Calculated density of states. (c) Measured XPS data, and (d) comparison of the valence region, between experimental data, after background subtraction, and calculated density of states.

Adapted from Hüfner (2003).

3.3.3   Metallic Bonding

A relatively simple model of metals involves delocalized valence electrons that are shared by many atoms. In general, metallic solids have a relatively larger number of orbitals than the available number of electrons, making the sharing of the latter energetically favorable. Thus, in metals, a "sea of electrons" holds together the positively charged ion cores distributed periodically throughout the solid (Fig. 3.3.1c). Furthermore, the electrons shield the positively charged ion cores from mutual electrostatic repulsion, further lowering the energy. A detailed description of metals requires an understanding of band theory, and can be found in standard condensed matter texts such as Kittel (1986) and Sutton (1993).

Figure 3.3.4a shows a calculated density of states for the three elemental ferromagnetic metals, Fe, Co, and Ni. Notice that the density of states is spin-split, i.e. it is different for spin-up and spin-down electrons, because the *exchange interaction*—fundamental to the origin of ferromagnetism—-causes a difference in the population of spin-up and spin-down bands for these three elemental ferromagnets. For both Co and Ni, the spin-up band is completely full but it is not so for iron. Hence, even though iron has a larger magnetic moment, it is considered a weak ferromagnet. The degree of *exchange splitting*, δE_x, is proportional (~ 1 eV/μ_B) to the magnetic moment [3], and can be measured by spin-polarized photoemission (Fig. 3.3.4b), which gives a value $\delta E_x \sim 2.2$ eV for Fe. We mention this here only for completeness—further details of such measurements can be found in some excellent reviews [4,5] and in Krishnan (2016).

We revisit photoemission and related spectroscopic methods in §3.8 and §3.9, but first, we discuss the vibration modes of molecules, and the related absorption and inelastic scattering methods used for their analysis.

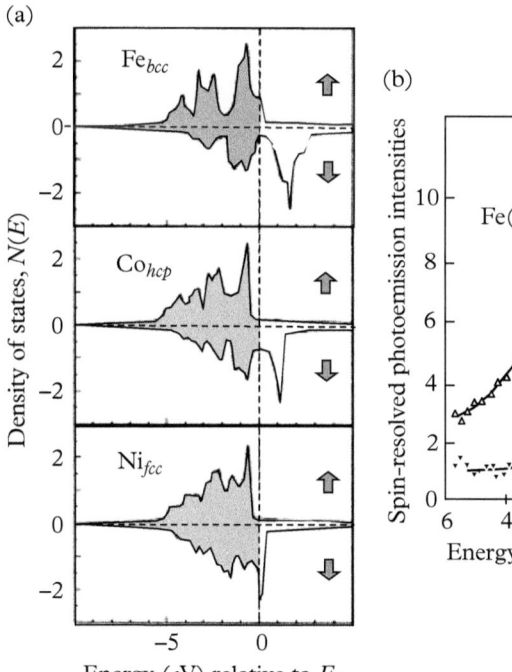

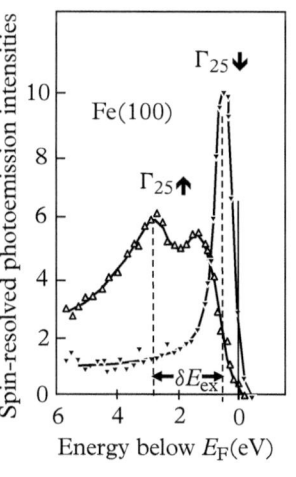

Figure 3.3.4 (a) Calculated spin-split density of states for metallic Fe, Co, and Ni. (b) Ferromagnetic exchange splitting, between majority and minority bands of iron, measured by *spin-polarized* photoemission (not discussed here). Further details of such measurements can be found in Krishnan (2016).

(a) Adapted from Papaconstantopoulos (1980). (b) Adapted from Krishnan (2016).

3.4 Molecular Spectra

The next three sections introduce different optical spectroscopic methods to identify different molecular species by probing their electronic, vibrational, and rotational characteristics. This is followed by methods to probe the occupied and unoccupied states of solids (§3.7–§3.9). Much of the latter falls broadly in the realm of surface analysis (Watts, 1990) and such measurements also require ultra-high vacuum.

3.4.1   Vibrational and Rotational Modes

Figure 3.3.1d shows how materials are composed of atoms or molecules bonded together with an equilibrium bond energy, E_0, and bond length, R_0, arising from a balance between attractive and repulsive forces. However, molecules vibrate about their equilibrium position, R_0, and such vibrations of a diatomic molecule can be understood by considering its *energy-displacement curve*. If the two atoms are separated by a distance, $\pm \Delta R$, slightly larger or smaller than R_0, the energy is increased to $E + \Delta E$, and this, in effect, acts as a restoring force. We can fit a parabola to the actual $E(R)$ curve (Fig. 3.4.1a), which serves as a good approximation for small vibrations, $\pm \Delta R$, about R_0.

The restoring force, F_R, of this *harmonic oscillator*, in the classical sense, is then

$$F_R = -\left(\frac{d^2E}{dR^2}\right)(R - R_0) = -A(R - R_0) \tag{3.4.1}$$

where, the *force constant*, A, is a measure of the *bond strength*. The energy, E_{vib}, of this harmonic oscillator, classically, is given by

$$E_{vib} = \frac{1}{2}A(R - R_0)^2 \tag{3.4.2}$$

Figure 3.4.1 (a) Vibrational energy levels of a *diatomic molecule*. The depth of the well is $E_0 = |E_d| + hf_{vib}/2$. The number of vibrational levels increases for heavier molecules. (b) Rotational energy levels for a diatomic molecule with moment of inertia, B.

Adapted from Sproul and Philips (1980).

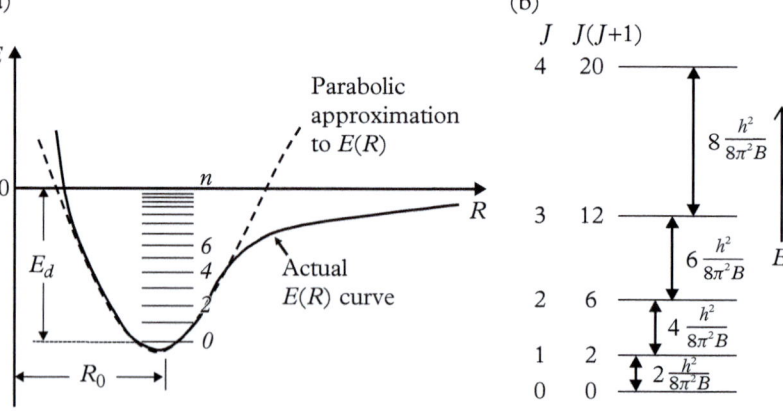

and its classical vibrational frequency, f_{vib}, is

$$f_{vib} = \frac{1}{2\pi\mu^{1/2}} \left[\left(\frac{d^2E}{dR^2} \right)_{R=R_0} \right]^{1/2} \tag{3.4.3}$$

where $\mu = \frac{M_1 M_2}{M_1 + M_2}$, the *reduced mass* of the two-particle vibrating system, simplifies the problem of two particles (mass, M_1 and M_2) vibrating about a common center of mass, to a single particle of mass, μ, vibrating about a fixed point. In quantum mechanics, this vibration energy of a diatomic molecule is quantized, such that

$$E_{vib,n} = (n + 1/2) hf_{vib} \tag{3.4.4}$$

where $n = 0, 1, 2, 3, \ldots$.

All $E_{vib,n}$ are measured in relation to $E = 0$, when the system is at rest at $R = R_0$, and f_{vib} is the classical oscillation frequency given by (3.4.3). Thus, even at 0 K, when $n = 0$, the atoms are vibrating with energy, $E_{vib} = 1/2\, hf_{vib}$, the so called *zero-point energy*.

Example 3.4.1: Consider the linear molecule H_2, with $f_{vib} = 1.28 \times 10^{14}$ Hz.

(a) Calculate its zero-point energy.

(b) Compare *qualitatively* the vibration energy levels of H_2, HD, and D_2, where D is deuterium, the heavy isotope $_1H^2$ of hydrogen.

(c) What is the difference in the dissociation energy of H_2 and D_2?

Solution:

(a) The zero-point energy of the H_2 molecule is

$$E_{vib} = 1/2\, hf_{vib} = 4.24 \times 10^{-20} \text{ J} = 0.265 \text{ eV}.$$

(b) The reduced mass for the molecules is $\mu(H_2) = \frac{1}{2}$, $\mu(HD) = 2/3$, and $\mu(D_2) = 1$ From (3.4.3), $f_{vib} \propto 1/(\mu)^{1/2}$. Thus, their vibrational energies can be qualitatively related as $E_{vib}(H_2) = (2/3^{1/2}) E_{vib}(HD) = 2^{1/2} E_{vib}(D_2)$.

(c) The difference in dissociation energy is assumed to be the difference in the zero-point energy of the two molecules. Thus, it is given by

$$\left(1 - 1/2^{1/2} \right) E_{vib}(H_2) = 0.293 \times 0.265 \text{ eV} = 0.078 \text{ eV}.$$

In fact, structural details of proteins are determined by carefully substituting deuterium for hydrogen in specific locations of the molecule and monitoring the change in vibrational frequency of that specific bond (see §3.10.1).

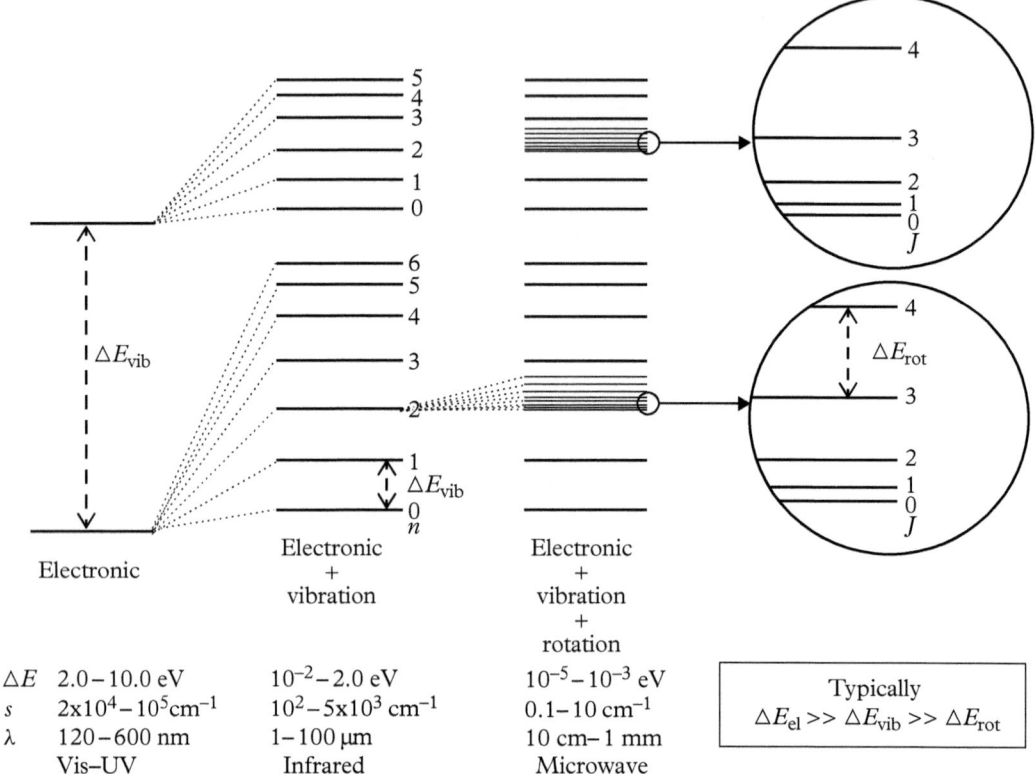

$$
\begin{array}{llll}
\Delta E & 2.0-10.0 \text{ eV} & 10^{-2}-2.0 \text{ eV} & 10^{-5}-10^{-3} \text{ eV} \\
s & 2 \times 10^4 - 10^5 \text{cm}^{-1} & 10^2 - 5 \times 10^3 \text{ cm}^{-1} & 0.1-10 \text{ cm}^{-1} \\
\lambda & 120-600 \text{ nm} & 1-100 \ \mu\text{m} & 10 \text{ cm} - 1 \text{ mm} \\
& \text{Vis-UV} & \text{Infrared} & \text{Microwave}
\end{array}
$$

Typically
$\Delta E_{el} \gg \Delta E_{vib} \gg \Delta E_{rot}$

Figure 3.4.2 Molecular energy levels comprised of electronic, vibrational, and rotational contributions. The rotation levels are only shown for vibrational $n = 2$ states. The transitions between electronic, vibrational, and rotational levels are observed in Vis-UV, infrared, and microwave regions, respectively, of the electromagnetic spectrum.

Typically, for various vibrating molecules, we observe that $f_{vib} = 6 \times 10^{12} - 1.2 \times 10^{14}$ Hz, a frequency in the *mid-infrared*. In solids, one also encounters *lattice vibrations* or *phonons*—a synchronized movement of all atoms in a crystal, typically at lower frequencies ($f_{vib} = 6 \times 10^{11} - 9 \times 10^{12}$ Hz) compared to molecular vibrations. However, lattice vibrations when compared to molecular vibrations, are more sensitive to changes in temperature. In addition to vibrations, the kinetic energy of molecules is also affected by rotations about their common *center of mass*. In quantum mechanics, by solving the Schrodinger equation for such rotations of diatomic molecules, we get quantized energy levels, in terms of J, a rotational quantum number:

$$E_{rot, J} = \frac{h^2}{8\pi^2 B} J (J + 1) \tag{3.4.5}$$

where the moment of inertia, B, is a constant given by $B = \frac{M_1 M_2}{M_1 + M_2} R_0^2 = \mu R_0^2$. The first few rotational energy levels are plotted in Figure 3.4.1b. The total energy of a molecule is a combination of the electronic, vibrational and rotational contributions. To give a sense of scale, energy levels are separated by 2–10 eV for electronic energies, 10^{-2}–2.0 eV for vibrational energies, and 10^{-5}–10^{-3} eV for rotational energies. Figure 3.4.2 schematically illustrates these energies for a diatomic molecule, with the rotational contribution exaggerated for clarity.

The number of vibration modes in a molecule is related to its number of *degrees of freedom*. Thus, a nonlinear molecule with N atoms, has $3N$ degrees of freedom and $3N-6$ vibration modes (a linear molecule would have one less, i.e. $3N-5$ vibrations modes). In complicated molecules, with atoms bonded to more than one atom, and including nonlinear chains of atoms, the vibration modes can be more complex (Fig. 3.4.3). Vibrations along the bond direction are called stretching modes, while those vibrations perpendicular to the bond direction are known as bending modes. These vibration and related modes can be probed by the absorption of infrared (IR) radiation or by inelastic scattering of light, known as Raman scattering, which we discuss in the next section. Note that IR and Raman spectroscopy are normally discussed in terms of wavenumbers ($s = 1/\lambda$) in units of cm^{-1}. Figure 3.4.4 shows two examples of vibrations modes in H_2O and CO_2.

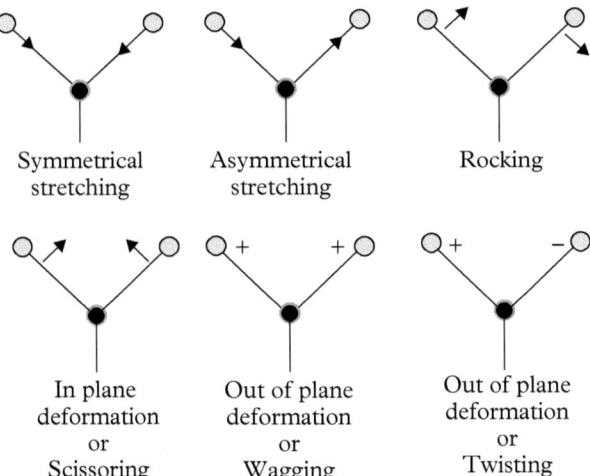

Symmetrical stretching

Asymmetrical stretching

Rocking

In plane deformation or Scissoring

Out of plane deformation or Wagging

Out of plane deformation or Twisting

Figure 3.4.3 Possible molecular vibration modes for two identical atoms bonded to a different, third atom.

Adapted from Flewitt and Wild (2003).

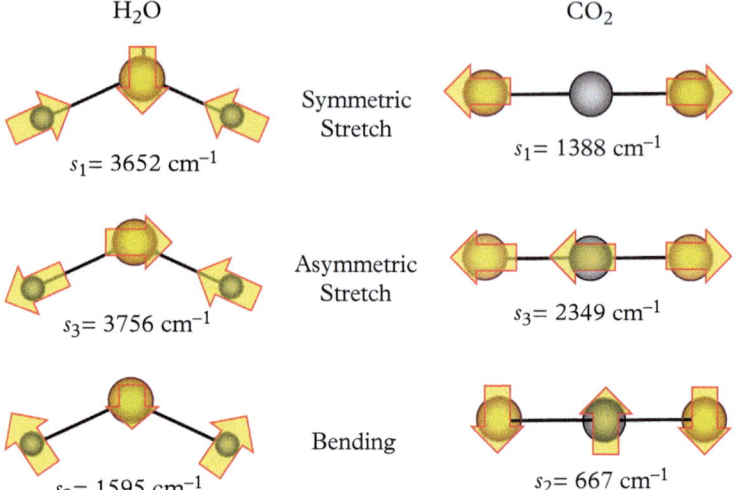

H_2O

CO_2

Symmetric
Stretch

$s_1 = 3652$ cm^{-1}

$s_1 = 1388$ cm^{-1}

Asymmetric
Stretch

$s_3 = 3756$ cm^{-1}

$s_3 = 2349$ cm^{-1}

Bending

$s_2 = 1595$ cm^{-1}

$s_2 = 667$ cm^{-1}

Figure 3.4.4 Two examples of different vibrations modes, observed in H_2O and CO_2. The associated wavenumber for each mode is also indicated. See also Figure 3.4.12.

Example 3.4.2: The molecule NaH undergoes a rotational transition from $J = 0$ to $J = 1$, when it absorbs a photon of 1.2×10^{-3} eV energy. What is the equilibrium separation between Na and H in the molecule?

Solution: From (3.4.5) we can see that for the $J = 0$ to $J = 1$ transition the energy $\Delta E = \frac{h^2}{8\pi^2 B}(2 - 0) = \frac{h^2}{4\pi^2 B} = \frac{h^2}{4\pi^2 \mu R_0^2} = 1.2 \times 10^{-3}$eV $= 1.2 \times 10^{-3} \times 1.6021 \times 10^{-19} = 1.9480 \times 10^{-22}$J. The reduced mass of the molecule is $\mu = \frac{m_{Na} m_H}{m_{Na} + m_H} = \frac{(22.989)(1.0078)}{22.989 + 1.078} = 0.9655$ *a.m.u* $= 1.55 \times 10^{-27}$ *kg*. Thus, the equilibrium separation,

$$R_0 = \left(\frac{h^2}{4\pi^2 \mu \Delta E}\right)^{\frac{1}{2}} = \left(\frac{\left(6.6257 \times 10^{-34}\right)^2}{4\pi^2 \left(1.55 \times 10^{-27}\right)\left(1.948 \times 10^{-22}\right)}\right)^{\frac{1}{2}}$$

$$= 1.9 \times 10^{-10} \ m = 1.9 \ \text{Å}.$$

3.4.2　Ultraviolet and Visible Spectroscopy (UV-Vis)

The absorption of light in the visible ($\lambda \sim 380 - 700$ nm) and ultraviolet (UV, $\lambda \sim 180 - 380$ nm) regions of the electromagnetic spectrum (Fig. 1.3.2) results from excitations of certain electronic states (Fig. 3.4.2) and forms the basis of UV-Vis spectroscopy, which measures how strongly a specimen absorbs different wavelengths of light. In practice, a UV-Vis spectrophotometer sends monochromatic photons of UV or visible light through a specimen, measures the light transmitted (not absorbed) through it, compares it to a reference cuvette, and plots the absorbance as a function of wavelength. Such absorbance may be measured in some specimens of solid phase, but the majority of UV-Vis tests are run in solution. Hence, a typical specimen is dissolved in a solvent and contained in a cuvette made of quartz, glass, or plastic. It is important that the selected solvent and cuvette material be also transparent to the wavelength being studied. Moreover, the concentration of the solution (specimen) is limited by the Beer–Lambert law, which relates the absorbance, A, of the molecule to its concentration, C, and its molar extinction coefficient, ε, as

$$A = \log\left(\frac{I_0}{I}\right) = \varepsilon C l \tag{3.4.6}$$

where I_0 is the intensity measured after the light passes through a reference cuvette, I is the intensity of the light transmitted through the specimen, and l is the path length (standardized to 1 cm for most cuvettes) of the light traveling through the specimen. Here, it is assumed that the solution is sufficiently dilute to avoid any absorbing molecule being in the shadow of another one, such that all molecules present in the cuvette contribute to the absorption signal. Needless to say, the Beer–Lambert law breaks down at higher concentrations. From (3.4.6), for any molecule in solution, we can determine either its concentration (if the molecular extinction coefficient is known) or molecular extinction coefficient (if its concentration is known), all other parameters being measured.

UV and visible photons have sufficient energy to excite certain electrons and provide details of the electronic structure of molecules (Fig. 3.4.2). Typically, UV-Vis spectra display one main peak (λ_{max}) characteristic of the molecular specimen, with the highest absorbance and may also include some smaller peaks. As discussed in Chapter 2, probing transitions between atomic levels require energies in the X-ray region. However, when atoms form molecules, the orbitals hybridize, and create σ and π bonding orbitals, σ^* and π^* anti-bonding orbitals, n nonbonding orbitals with lone pairs of electrons. Moreover, in molecules the separation between the HOMO and the LUMO is significantly smaller than in the case of isolated atomic transitions, making it possible for it to absorb UV-Vis light. In addition, σ-bonded electrons are tightly bound and require higher energies to excite them, compared to the loosely bound π and n electrons. Thus, only lower energy transitions involving π, σ^*, π^*, and n molecular orbitals, i.e. $\pi \rightarrow \pi^*$, $n \rightarrow \pi^*$, and $n \rightarrow \sigma^*$, contribute to the UV-Vis spectrum (Fig. 3.4.5). In addition, as more atoms are added to the molecule, the separation between the HOMO and LUMO levels decreases (Fig. 3.2.2); thus, more *conjugated molecules* absorb at lower energies or longer wavelengths. In addition, molecules with heteroatoms, such as oxygen, nitrogen, and sulfur, contribute to molecular resonances, increase the effective conjugation of the molecule, and absorb light of lower energies.[4] In practice, simple molecules absorb at $\lambda \sim 200$ nm, and larger dye molecules absorb in the visible region.

[4] The parts of the molecule that absorb visible light, and are responsible for its color, are called chromophores.

(a) (b)

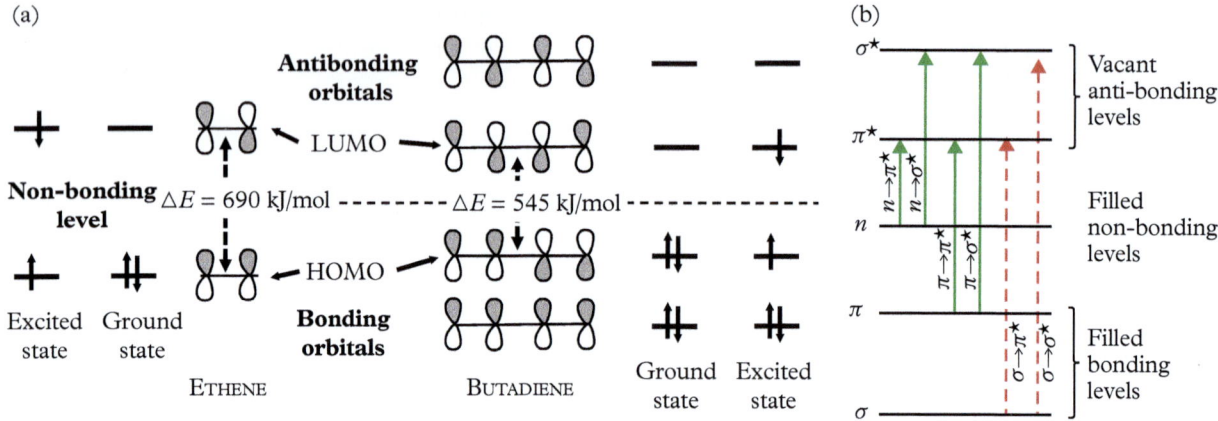

Figure 3.4.5 (a) Ground and excited states for isolated (left) and bonded (right) orbitals, and the magnitude of the HOMO–LUMO gap for two different molecules. (b) Relative energies of the bonding, non-bonding, and antibonding levels in molecules, with transitions included in (blue solid, observed) and outside (red hatched, not observed) the range of UV-Vis spectroscopy.

Adapted from Harwood and Claridge (2003), and Duckett and Gilbert (2000), respectively.

Example 3.4.3: Show that the HOMO–LUMO transition in butadiene can be probed by UV-Vis spectroscopy.

Solution: From Figure 3.4.5, the HOMO–LUMO energy gap, $\Delta E = 545$ kJ/mol for butadiene. The energy gap per molecule is $E = \Delta E / N_0 = 545{,}000/6.022 \times 10^{23}$ J. $= 9.05 \times 10^{-19}$ J

The wavelength, λ, of the EM radiation required to probe this transition is

$$\lambda = h\,c/E = 6.626 \times 10^{-34} \text{ J.s.} \times 299{,}792{,}458 \text{ m s}^{-1}/9.05 \times 10^{-19} \text{ J}$$
$$= 2.19 \times 10^{-7} \text{ m} = 219 \text{ nm.}$$

This is clearly in the range of UV-Vis spectroscopy (Fig. 3.4.2).

In addition to concentrations obtained by applying Beer–Lambert law, the location (λ) of the absorption peak in a UV-Vis spectrum gives general information about the structure of a molecule (Fig. 3.4.6). In general, the advantage of UV-Vis is the speed with which it can analyze a specimen and newer models, which forego cuvettes and use fiber-optic probe systems inserted into the specimen solution, are incredibly fast with capabilities of scanning through the entire spectral range in less than three seconds. However, the resolution of the UV-Vis spectrum is too poor to resolve the structure of the molecule or the energies of specific bonds. As

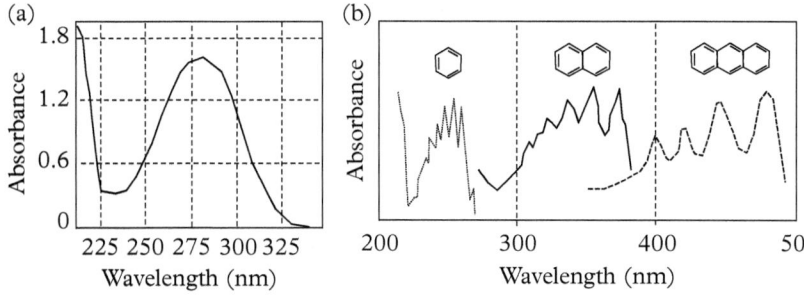

Figure 3.4.6 Typical UV-Vis absorption spectra of (a) propane in hexane solvent, and (b) benzene, naphthalene, and anthracene, superimposed to illustrate the effect of increasing conjugation.

Adapted from Duckett and Gilbert (2000), and Harwood and Claridge (2003), respectively.

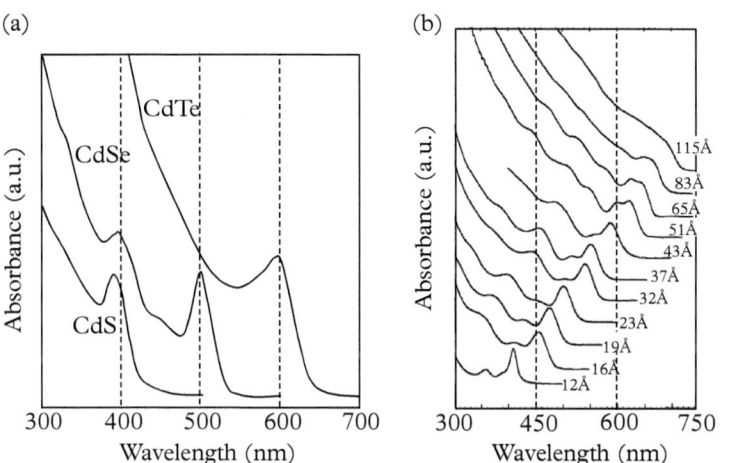

Figure 3.4.7 UV-Vis spectra of quantum dots shown for (a) three different semiconductors of 20-30 Å dia. with dramatic shifts in the absorption peaks, and (b) for CdSe as a function of diameter.

Adapted from [17].

such, it is not a particularly useful spectroscopic tool for stand-alone diagnostics of molecular composition and structure, but it can be used to characterize optical properties of materials such as bandgaps of semiconductor nanocrystals as a function of their size (Fig. 3.4.7).

Example 3.4.4: In a UV-Vis experiment with a path length of 1 cm, we observe concentration and absorbance values as shown in the following table:

Absorbance	0.14	0.24	0.48	0.70	1.04	1.4	1.65
Concentration (μM)	24	41.2	82.3	120	178	257	321

(a) What is the physical significance of the molar extinction coefficient?
(b) From the above data, determine the molar extinction coefficient? What are its units? (c) At what concentration does the Beer–Lambert law fail?

Solution:

(a) The Beer–Lambert law (3.4.6) can be rearranged as $\varepsilon = A/Cl$. Thus, the molar extinction coefficient, ε, physically represents the amount of light absorbed per unit concentration.

(b) We plot the absorbance data as a function of concentration as shown in the image

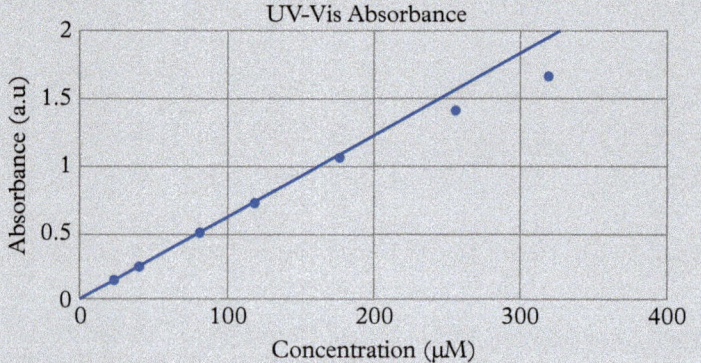

From the slope of the linear portion of the data, we determine the molar extinction coefficient to be $\varepsilon = 5.83 \times 10^{-3}\mathrm{M^{-1}cm^{-1}}$.

(c) The Beer–Lambert law fails when the concentration approaches ~ 250 μM.

Readers interested in learning more about UV-Vis spectroscopy are referred to Harwood and Claridge (2003), and Duckett and Gilbert (2000).

3.4.3 Classical Model of Rayleigh and Raman Scattering

When light of frequency, f_I, usually from a laser (§5.2.1.2), is incident (probe) on a specimen, it can be scattered (signal). The scattering can be *elastic*, with the frequency of the scattered light remaining unchanged (referred to as Rayleigh scattering), or *inelastic*, resulting in some shifted frequency, $f_S = f_I + f_{\mathrm{int}}$ (referred to as Raman scattering). The frequency, f_{int}, can correspond to some internal vibrational, rotational, or electronic transition, as discussed in the previous section. However, the most important one is the vibrational Raman effect, which we discuss here. Typically, in Raman scattering (Fig. 3.4.8), if the scattered frequency is lower (red shifted), it is referred to as Stokes scattering, and if the frequency is

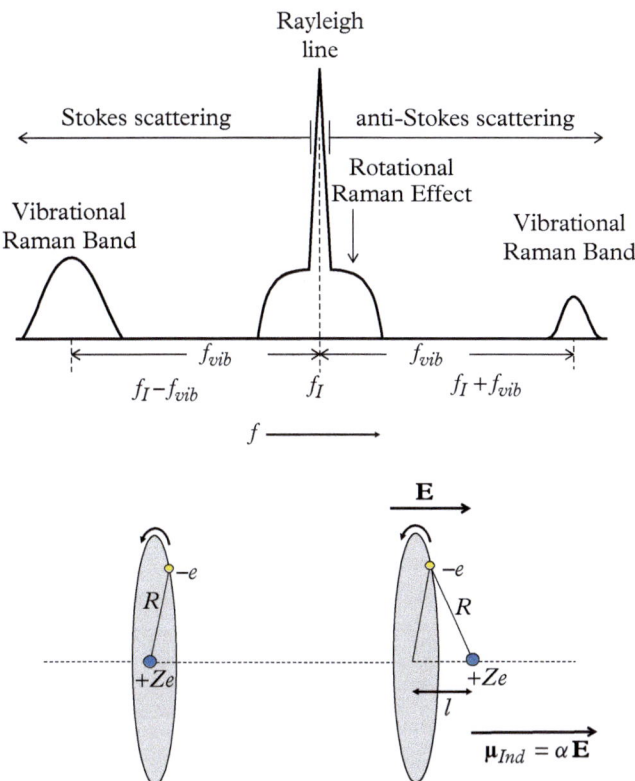

Figure 3.4.8 Schematic illustration of the Raman effect, including vibrational and rotational contributions.

Figure 3.4.9 Schematic of the induced dipole, μ_{Ind}, in an atom/molecule on the application of an electric field, **E**. Here R is the radius of the electron orbit, and l is the displacement of the centroid of the orbiting electron from the nuclear charge, $+Ze$.

higher (blue shifted) it is known as anti-Stokes scattering. Raman scattering can be understood physically using a classical model.

When a molecule is subject to an electric field, **E**, its electrons and nuclei respond by moving in opposite directions (Fig. 3.4.9); thus, the applied field induces a dipole moment, μ_{Ind}. In the molecule, and for small fields, the relationship between the induced dipole moment and the applied field is linear, i.e.

$$\mu_{Ind} = \alpha \mathbf{E} \qquad (3.4.7)$$

Here, the proportionality constant, α, called the *polarizability*, is characteristic of the molecule. Note that this assumes that the molecule is highly symmetric and the induced dipole is along the same directions as the applied field. In the general case, less-symmetric molecules, **E** and μ_{Ind}, are vector quantities and can be along different directions. In that case, the polarizability would become a second-rank tensor, α_{ij}, relating the two vectors **E** and μ_{Ind}. To keep matters simple, we assume a linear relationship for the rest of this discussion.

If the electrical field oscillates with time, such as in an electromagnetic radiation (Fig. 1.3.3), then the induced dipole moment will also oscillate with time. Now, the *intensity* of the scattered light is proportional to the square of the amplitude of the oscillating dipole. Furthermore, any internal motion of the molecule, such as vibration and rotation, can modulate this induced dipole moment, resulting in the appearance of additional frequencies. In a classical description, the polarizability now has a static term, α_0, and a sinusoidal oscillating term with amplitude, α_1, where

$$\alpha = \alpha_0 + \alpha_1 \cos(2\pi f_{\text{int}} t) \tag{3.4.8}$$

Further, for any specific vibrating mode, V_i, we can assume

$$\alpha_1 = \left. \frac{\partial \alpha}{\partial V_i} \right|_{V=0} V_i \tag{3.4.9}$$

such that when the polarizability does not change with the vibration, i.e. if $\partial \alpha / \partial V_i |_{V_i=0} = 0$, the Raman scattering effect is not observed. Basically, this means that the induced dipole moment oscillates at a frequency other than that of the incident wave, f_I, because of the oscillating polarizability. Now, for an electromagnetic wave incidence, we can represent $\mathbf{E} = E_0 \cos(2\pi f_I t)$, and rewrite (3.4.7) as

$$\begin{aligned} \mu_{Ind} = \alpha E &= \alpha E_0 \cos(2\pi f_I t) = [\alpha_0 + \alpha_1 \cos(2\pi f_{\text{int}} t)] \, E_0 \cos(2\pi f_I t) \\ &= \alpha_0 E_0 \cos(2\pi f_I t) + \alpha_1 E_0 \cos(2\pi f_{\text{int}} t) \cos(2\pi f_I t) \\ &= \alpha_0 E_0 \cos(2\pi f_I t) + \tfrac{\alpha_1 E_0}{2} \left[\cos 2\pi \, (f_I - f_{\text{int}}) \, t + \cos 2\pi \, (f_I + f_{\text{int}}) \, t \right] \end{aligned}$$
$$\tag{3.4.10}$$

The first term is not shifted in frequency and corresponds to elastic Rayleigh scattering (Fig. 3.4.8) of the wave at the same frequency, f_I. Now, specifically for vibrational Raman effect, we can set $f_{vib} = f_{\text{int}}$, and then the lower-frequency term, $(f_I - f_{vib})$, is the Stokes scattering and the higher-frequency term, $(f_I + f_{vib})$, is the anti-Stokes scattering. Note that in this classical formulism, the intensities of Stokes and anti-Stokes scattering is predicted to be the same, but this is not usually observed and Stokes scattering has a higher intensity (Fig. 3.4.8). Details on why this is the case can be found in more specialized texts, e.g. Cothup, Daly, and Wiberley (1990).

Figure 3.4.10 shows the energy level diagram corresponding to this *two-photon process*; both elastic and inelastic scattering can be understood in terms of the energy transfer from the incident photons to the molecules. When light is incident, certain resonance frequencies are absorbed, raising the molecule to some *virtual excited state*. If the molecule decays to the original level, a photon at the same frequency is emitted; this is the elastic Rayleigh scattering. The molecule may contain energy levels, corresponding to the vibration modes, at energies higher and lower than the energy level to which it was initially excited. Some of these adjacent energy levels may be unfilled because they may have also been

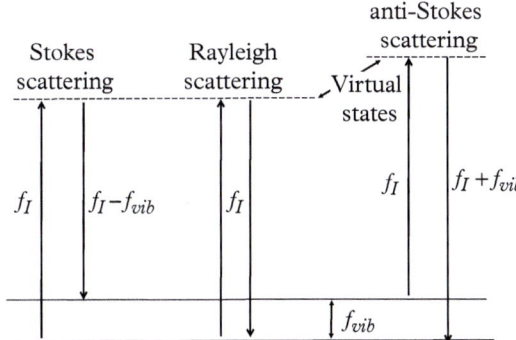

Figure 3.4.10 Energy level diagram of a molecule showing Rayleigh, Stokes, and anti-Stokes scattering.

excited by the incident photons. The excited atom may decay into one of these adjacent levels, with frequencies larger (anti-Stokes) or smaller (Stokes) than the Rayleigh line.

Before we present practical details of IR absorption spectroscopy and Raman scattering, we briefly discuss the complementarity of these two methods. For readers interested in the physics of Raman scattering, further details can be found in Bernath (1995) and Long (1977).

Example 3.4.5: A Raman spectroscopy experiment is conducted with incident light of wavelength, $\lambda = 532$ nm, and Stokes scattering corresponding to a wavenumber $s = 806$ cm^{-1}, is observed. At what wavelength will this signal photon be observed?

Solution: The frequency of the incident light, $f_I = c/\lambda = 3 \times 10^8 / 532 \times 10^{-9} = 5.6 \times 10^{14}$ Hz. The frequency of the internal vibration is

$$f_{\text{vib}} = c/\lambda_{\text{vib}} = c\,s = 3 \times 10^{10} \times 806 = 2.42 \times 10^{13} \text{ Hz}.$$

Thus, the Stokes scattering will be observed at a frequency

$$f_{\text{Stokes}} = f_I - f_{\text{vib}} = 5.6 \times 10^{14} - 2.42 \times 10^{13} = 5.36 \times 10^{14} \text{ Hz}.$$

This corresponds to a wavelength $\lambda_{\text{Stokes}} = c/f_{\text{Stokes}} = 3 \times 10^8 / 5.36 \times 10^{14}$ m $= 560$ nm, for the observed photon.

3.4.4 Selection Criteria for Infrared and Raman Activity

IR spectroscopy is based on the *absorption* or *transmittance* of electromagnetic radiation in the IR region by molecular vibration. If a particular frequency, f_{IR},

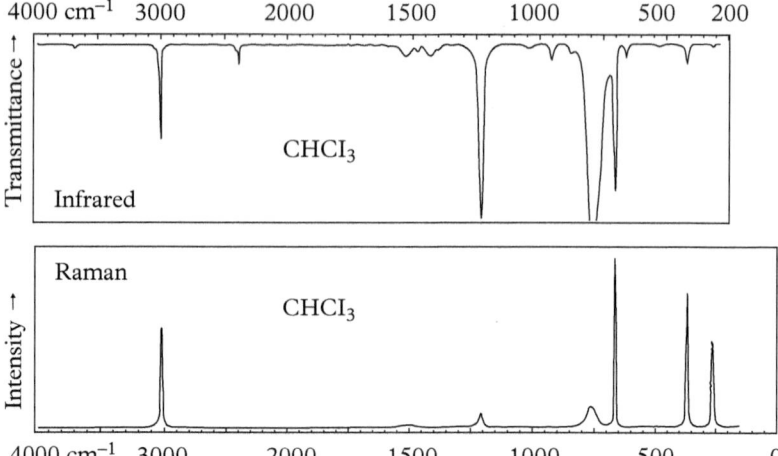

Figure 3.4.11 Complementary infrared transmittance and Raman spectral intensity of CHCl₃ as a function of wavenumber ($s = 1/\lambda$). Compare the two spectra at 1,210, 750, and 370 cm⁻¹.

Adapted from Leng (2013).

matches the vibrational frequency, f_{vib}, of a molecule, then the *photon* can be absorbed and raise the molecule to an excited state; typically, the absorption would be dominated by transition from the $n = 0$ to $n = 1$ state in Figure 3.4.1a. Raman spectroscopy, on the other hand, is a *two-photon event*, involving the absorption of a photon, accompanied by the excitation of the molecule to a virtual state, followed by the decay and emission of a "scattered" photon, modulated by the vibration modes of the molecule (Fig. 3.4.10).

In practice, IR absorption and Raman activity are complementary, as shown for CHCl₃ molecules in Figure 3.4.11. It turns out that the *selection rules* for Raman scattering are different from those for IR (and microwave) absorption spectroscopy; some transitions can be observed only through Raman scattering, some only through IR spectroscopy, some through both, and some through neither.

There are some broad guidelines to predict if we can expect IR spectroscopy or Raman scattering for a given molecule. For a molecule to be IR active, the specific vibration mode must cause a change in the dipole moment, $\mu = ql$, of the molecule; here, q is the electric charge, positive and negative, at the charge centers of the molecule, and l is their separation. The dipole moment can be intrinsic or induced. Mathematically, IR activity requires that the derivative of the magnitude of the dipole moment, with respect to the relevant vibration parameter, ϑ, proportional to $(R-R_0)$ in (3.4.1), at equilibrium ($\vartheta = 0$), is nonzero, i.e.

$$\left(\frac{\partial \mu}{\partial \vartheta}\right)_{\vartheta=0} \neq 0 \qquad (3.4.11)$$

In contrast, for a molecule to be Raman active, the derivative of the polarizability, α, defined in (3.4.7), with respect to the vibration parameter, ϑ, should be nonzero, i.e.

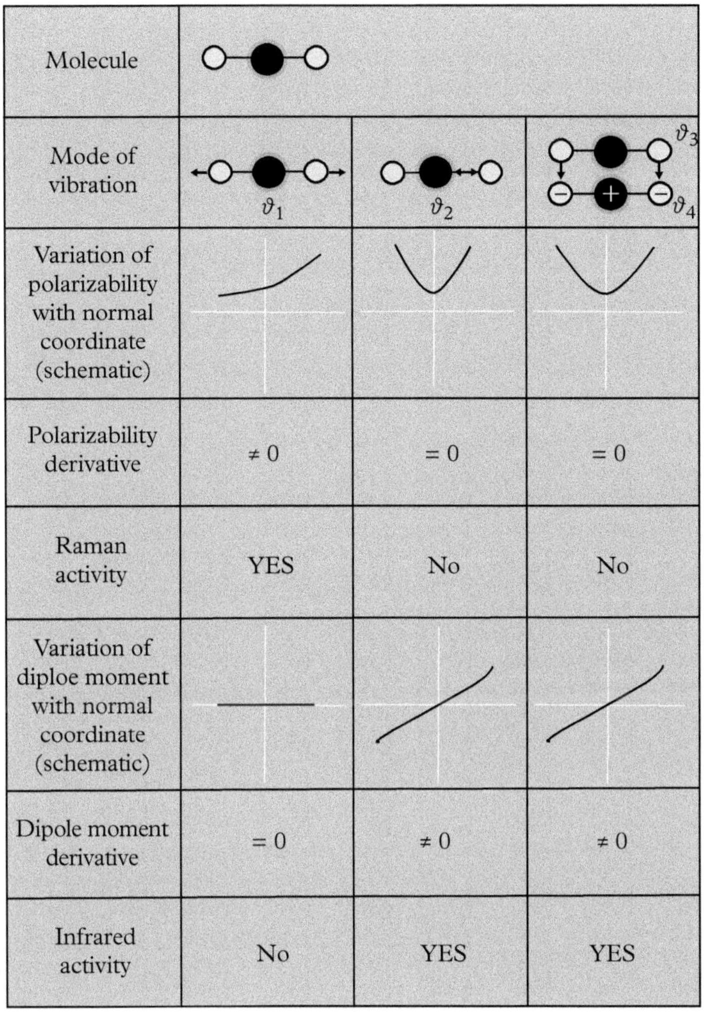

Molecule			
Mode of vibration	ϑ_1	ϑ_2	ϑ_3 ϑ_4
Variation of polarizability with normal coordinate (schematic)			
Polarizability derivative	$\neq 0$	$= 0$	$= 0$
Raman activity	YES	No	No
Variation of diploe moment with normal coordinate (schematic)			
Dipole moment derivative	$= 0$	$\neq 0$	$\neq 0$
Infrared activity	No	YES	YES

Figure 3.4.12 Polarizability and the variation of the dipole moment for CO_2, a linear molecule, in the neighborhood of the equilibrium position for Raman and IR activities. For each of the four possible vibration modes (row 2), a schematic variation of the polarizability (row 3), and dipole moment (row 6), as a function of the spatial coordinate is shown. If the derivative of the polarizability is nonzero (row 4), such as for mode ϑ_1, Raman activity is expected. On the other hand, if the derivative of the dipole moment is nonzero (row 7), such as for modes ϑ_2, ϑ_3, and ϑ_4, IR activity results.

Adapted from Long (1977).

$$\left(\frac{\partial \alpha}{\partial \vartheta}\right)_{\vartheta=0} \neq 0 \qquad (3.4.12)$$

Figure 3.4.12 illustrates these criteria with the vibration modes of CO_2, which has two stretching modes, ϑ_1 (symmetric) and ϑ_2 (antisymmetric), as well as two bending modes, ϑ_3 and ϑ_4. Figure 3.4.12 shows the derivative of the polarizability and dipole moment, as a function of the vibration parameter, for each mode. Clearly, only the *symmetric stretching* mode, ϑ_1, where each bond is polar but the vector sum is always zero, is Raman active (the other, ϑ_2, *asymmetric stretch*, where one bond is compressed and the other one stretched, creates different bond polarities and results in a dipole, is not), and only mode ϑ_1 is not IR active (the other three are). In practice, broadly speaking, we can expect in molecules with relatively low symmetry all vibration modes to be both IR and Raman active; however, for high-symmetry molecules, especially with a center of symmetry, the IR and Raman activity are mutually exclusive. Further, vibrations of molecules with ionic bonds, such as OH^-, show strong IR activity, but those with covalent bonds, such as C=C, show strong Raman activity. In the next section(s), we discuss practical details of IR and Raman spectroscopy.

3.5   Infrared Spectroscopy

All IR spectroscopy measurements use a radiation source with a broad range of frequencies, and after interaction with the molecule, plot some form of intensity as a function of wave number. The IR portion of the electromagnetic spectrum (Fig. 1.3.2) stretches from the long wavelength end of the visible spectrum ($\lambda = 0.7\,\mu m$; wavenumber, $s = 1/\lambda = 14000$ cm^{-1}) to the short wavelength limit of the microwave spectrum ($\lambda = 1000\mu m, s = 10$ cm^{-1}). The near IR ($\lambda = 0.70 - 2.5\mu m, s = 14285 - 4000$ cm^{-1}) region is not of much interest for vibrational spectroscopy. The mid-IR region ($\lambda = 2.5 - 50\mu m$, $s = 4000-200$ cm^{-1}) is the region where many useful absorptions occur and contains two important regions: (a) in the "group frequency" region, $\lambda = 2.5-8.0\mu m$, $s = 4000 - 1250$ cm^{-1}, many of the strongest absorption bands observed can be associated with vibrations of a portion, or functional group, of the molecule, with minimal contributions from the rest of the molecule; (b) in the fingerprint region, $\lambda = 8.0 - 15.4\mu m$, $s = 1250 - 650$ cm^{-1}, the common features are associated with single-bond stretching and bending vibrations similar to that shown for CO_2 in Figure 3.4.4. Finally, absorption in the far IR region, $\lambda = 50 - 1000\mu m$, $s = 200 - 10$ cm^{-1}, is associated largely with other bending vibrations. In practice, the IR absorption features are measured and then the problem is to determine what molecular species has contributed to the observed spectra. Generally, the practice is to compare the observed spectra to a catalog of recorded spectra for various modes of known molecules; such reference spectra can be found in Cothup et al. (1990), and Table 3.5.1 gives typical values for some common molecules or bonds.

Table 3.5.1 Typical values of stretching and bending vibrations in wave numbers (cm^{-1})

Molecule or bond	Stretching mode	Bending mode
C–H	2,800–3,000	
N–N	3,300–3,500	
H_2O	3,500–3,800	1,600
C=O	1,700	
C=C	1,600	

3.5.1 Instrumentation for Raman and IR Spectroscopy

There are two kinds of instrumentation—*dispersive* and *interferometry*—for IR and Raman spectroscopy measurements. In dispersive systems, the radiation from the IR source is dispersed as a function of wavelength using prisms and gratings. A diffraction grating (§6.6.3), as shown in Figure 3.5.1, is used to disperse light according to its wave number. The incident beam (for the sake of illustration, only two rays, A and A′, are shown) is dispersed by the grating in a discrete direction (rays B and B′), based on Bragg's law, the wavelength of the radiation, and the period of the grating. Typically, the grating is rotated to change the angle of incidence, and in this way, one grating is used to cover a range of approximately 1,000 wave numbers. From the figure it is easy to see that when the path difference between rays B and B′ satisfies the condition

$$d(\sin \theta_i - \sin \theta_r) = m\lambda \qquad (3.5.1)$$

where m is an integer, and d is the period of the grating, constructive interference is observed and a particular wavelength is detected (optical gratings are discussed further in §6.6). To define a spectral element, slits are used to select a window of energy by wavelength. The spectrum is acquired sequentially with a measurement time interval to record each spectral element as defined by the dispersive component and slit, to finally give the intensity as a function of wavelength. Table 3.5.2 summarizes the components used in IR spectroscopy, for different ranges of wavelength.

In addition to requiring a long time to acquire a complete spectrum, dispersive instrumentation have many moving parts that need synchronized movement with good accuracy (e.g. slit widths must be closely matched and varied with wavelength). To overcome these limitations, a different approach is used (Fourier transform IR (FTIR) spectroscopy), where all spectral information is contained in an *interferogram* produced by scanning a Michelson interferometer. FTIR has

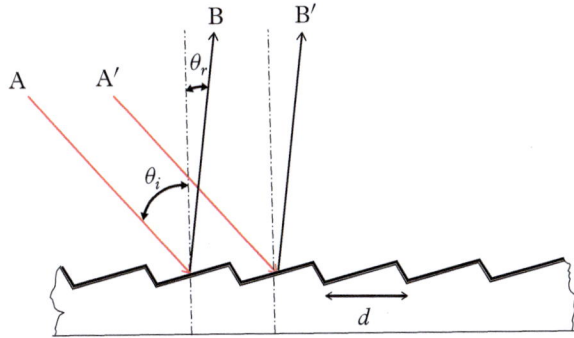

Figure 3.5.1 Schematic representation of a diffraction grating with periodic grooves separated by a distance, *d*. See §6.6 for a further discussion on gratings.

Table 3.5.2 Components used in IR spectroscopy.

	Near IR	Mid IR	Far IR
Wavenumber $(1/\lambda)$, cm^{-1}	12,500–4,000	4,000–200	200–10
Wavelength range, μm	0.8–2.5	2.5–50	50–1,000
IR source (see §5.2.1 for details)	Tungsten filament lamp	Nernst glower, globar, or nichrome coil	High-pressure mercury arc lamp
Optical setup	1–2 quartz prisms or prism grating double monochromator	2–4 diffraction gratings with either a foreprism monochromator or IR filters	Double-prism grating up to 700 μm; interferometric spectrometers up to 1000 μm
Detector	Lead sulfide photoconductor	Thermopile, transistor, or pyroelectric	Golay or pyroelectric

Adapted from Willard et al. 1988.

three main advantages—it records all wavelengths at the same time, has much higher signal collection speeds with possibility of repeat scans, and high signal-to-noise ratios—over dispersive methods.

3.5.2 Michelson Interferometer and the Fourier Transform Infrared (FTIR) Method

In practice, to work in the mid-IR (finger printing region) of wavelength, $\lambda = 10$ μm, which corresponds to a frequency, $f_{IR} = c/\lambda = 3 \times 10^8 / 10 \times 10^{-6} = 3 \times 10^{13}$ Hz, requires specialized detectors and electronic circuits that are fast enough to rapidly sense at the frequencies involved. However, as we later demonstrate, when the signal is modulated using a Michelson interferometer the frequency at which it is to be measured is practically in a more reasonable range (400 Hz).

Figure 3.5.2 shows the optical layout of an FTIR setup, with the principal Michelson interferometer component. The incident electromagnetic radiation (note, only the electric component, E, that is of interest, is considered) is given by

$$E(s) = A(s) \sin 2\pi \, (f_{IR}t - sx) \tag{3.5.2}$$

where, $s = 1/\lambda = f_{IR}/c$, and $A(s)$ is the amplitude for wavenumber, s. The *intensity* of the beam is

$$I(s) = E^2(s) = A^2(s)\sin^2 2\pi \, (f_{IR}t - sx) \tag{3.5.3}$$

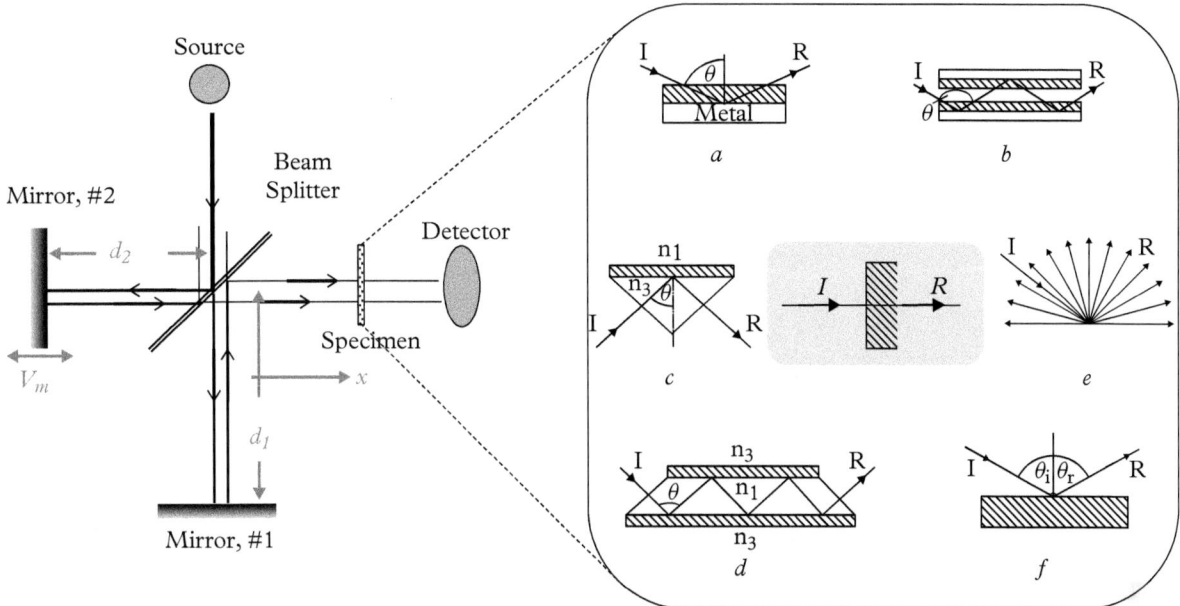

Figure 3.5.2 The optical layout for a FTIR set up using a Michelson interferometer. The incident beam is split into two by the beam splitter (note, intensity is indicated by the widths of the beam) and the mirrors, with one of them moving at a fixed speed, control the path length of the individual beams impinging on the specimen. The interferogram is then interpreted in terms of its Fourier component. The inset shows various specimen geometries, (a–f), other than transmission: (a) reflection on a metal surface, (b) multiple reflection, taking advantage of (d) total internal reflection are some of the ways to enhance the FTIR signal.

and its *average value*, over many cycles, is

$$I_{avg}(s) = A^2(s)/2 \qquad (3.5.4)$$

where the average, $< \sin^2\theta > = 1/2$.

At the beam splitter, half of the intensity goes into each of the two beams traveling towards the two mirrors. After reflection by the mirror, the two beams encounter the splitter again, and now half of their intensities (i.e. ¼ of the original) is transmitted to the specimen and detector. In other words, the amplitude of each of the two beams going towards the detector is half the original amplitude, i.e., ½ $A(s)$. Let d_1 (fixed) and d_2 (variable) be the distances of the two mirrors from the beam splitter; in addition, mirror no. 1 is fixed but mirror no. 2 is moved with a velocity, V_m, as shown. At the point, x, where the specimen is placed (as shown) in front of the detector, the electromagnetic radiation for the two beams are

$$E_1(s) = \frac{1}{2}A(s) \sin 2\pi \left[f_{IR}t - s(x + 2d_1) \right] \qquad (3.5.5)$$

and

$$E_2(s) = \frac{1}{2}A(s) \sin 2\pi [f_{IR}t - s(x + 2d_2)] \tag{3.5.6}$$

The total radiation, $E(s) = E_1(s) + E_2(s)$, using the relationship, $\sin a + \sin b = 2 \sin \left(\frac{a+b}{2}\right) \cos \left(\frac{a-b}{2}\right)$, is

$$E(s) = A(s) \sin 2\pi [f_{IR}t - sx + s(d_1 + d_2)] \cos 2\pi s(d_1 - d_2) \tag{3.5.7}$$

Setting the path difference, δ, between the two beams to be $\delta = 2(d_1 - d_2)$, the intensity of the combined beam is obtained by averaging the $\sin^2\theta$ term at a fixed point, x, over a long time as

$$I(\delta) = \frac{1}{2}A^2(s)\cos^2(\pi s\delta) \tag{3.5.8}$$

Using $\cos^2\theta = (1 + \cos 2\theta)/2$, we can rewrite, $I(\delta)$ as

$$I(\delta) = \frac{1}{4}A^2(s) [1 + \cos(2\pi s\delta)] \tag{3.5.9}$$

Now taking the average intensity, as δ is varied by moving the mirror no. 2, gives

$$I_{avg}(\delta) = A^2(s)/4 \tag{3.5.10}$$

which is consistent with the initial beam being split into two. If the intensity, $I(\delta)$, is recorded as a function of time, as δ is varied the total signal, over all wavelengths, is given by the Fourier cosine transform of the intensity:

$$I_{avg,Total}(\delta) = \int I_{avg}(s) \cos(2\pi f_{IR}s) \, ds \tag{3.5.10}$$

In other words, by measuring the $I_{avg,Total}(\delta)$, and taking its inverse Fourier transform, we can obtain the intensity, $I_{avg}(s)$, transmitted at each wavenumber, s.

Example 3.5.1: In a typical FTIR instrument the mirror moves with a speed of 0.2 cm/s. Assuming that the wavelength of IR radiation involved is 10 μm, will the frequency to be measured be within practical detector capabilities?

Solution: Let us assume that the mirror no. 2 moves with a speed, $V_m = 0.2$ cm/s. Then for a given time interval, Δt, the change in path difference is $\Delta\delta = 2V_m\Delta t$. If the change $\Delta\delta = \lambda_{IR}$, we can expect constructive interference. Thus, the frequency, f_M, to be measured is $f_M = 1/\Delta t = 2V_m/\lambda_{IR} = 2V_m f_{IR}/c = 2V_M s$. Thus, $f_M = \frac{2V_m}{\lambda_{IR}} = \frac{2\times0.2\times10^{-2}}{10\times10^{-6}} = 400$ Hz, which is well within modern detector capabilities.

3.5.3 Practice and Application of FTIR

Figure 3.5.3 shows the implementation of FTIR spectroscopy for a specimen of polystyrene [6] using the Michelson interferometer. First, the interferogram, $I_{avg,Total}$ (δ), with no specimen (a), and the polystyrene specimen (c), is recorded. Fourier transform (3.5.11) of these interferograms produces the IR spectra for the measurement without (b) and with (d) the specimen. Already, the IR characteristics of polystyrene can be seen in (d). By dividing the spectrum I of polystyrene (d) by the spectrum I_0 of (b), we obtain the transmittance, T, defined as $T = I/I_0$, as shown in (e). Note that the IR spectrum can also be plotted as the absorbance (A), which is defined as

$$A = -\log T. \tag{3.5.11}$$

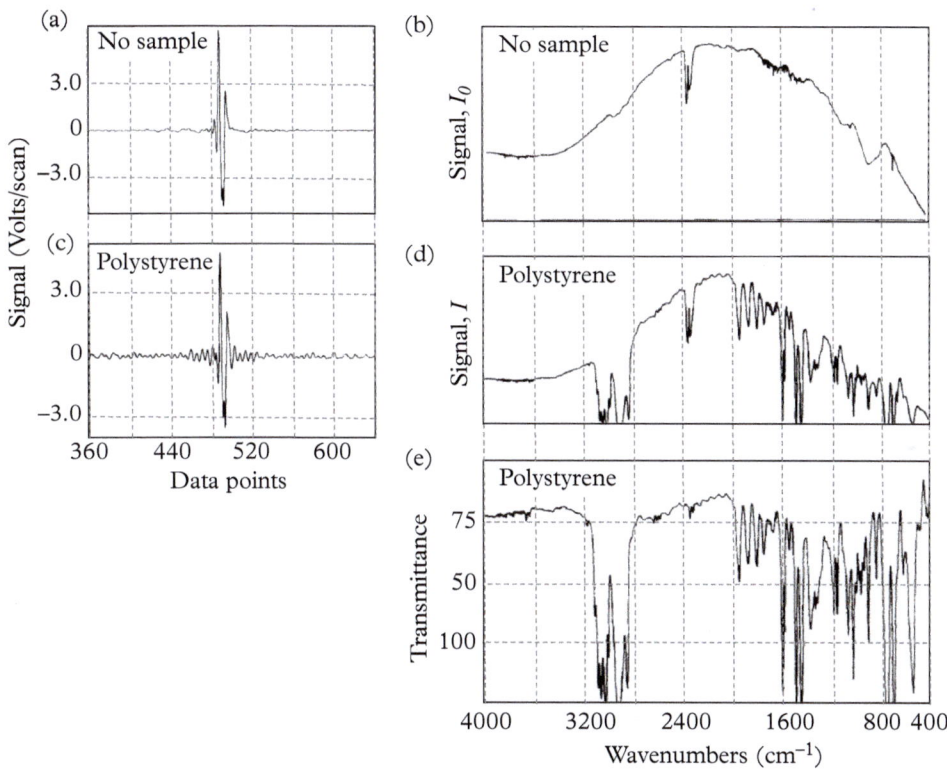

Figure 3.5.3 Interferograms obtained without (a) and with a polystyrene (c) specimen, with the corresponding FTIR spectra, as a function of wave number, s, determined after the inverse Fourier transform, are shown in (b) and (d), respectively. The transmittance spectrum of polystyrene is obtained by dividing spectrum (d) by spectrum (b).

Adapted from [6].

The range of frequencies, over which the interferometer can be used is determined by the spacing between data points in (a) and (c). In practice, for a helium–neon laser, this usually works out to 400–4,000 cm^{-1}, as shown in (b), (d), and (e). Further, the wavenumber resolution, Δs_{res}, of the IR spectrum, depends inversely on the maximum path length difference, δ_{max}, i.e. $\Delta s_{res} \sim 1/\delta_{max}$. Thus, a typical FTIR set up, with $\delta_{max} = 10$ cm, gives $\Delta s_{res} \sim 0.1$ cm^{-1}, which is more than adequate for most polymeric materials.

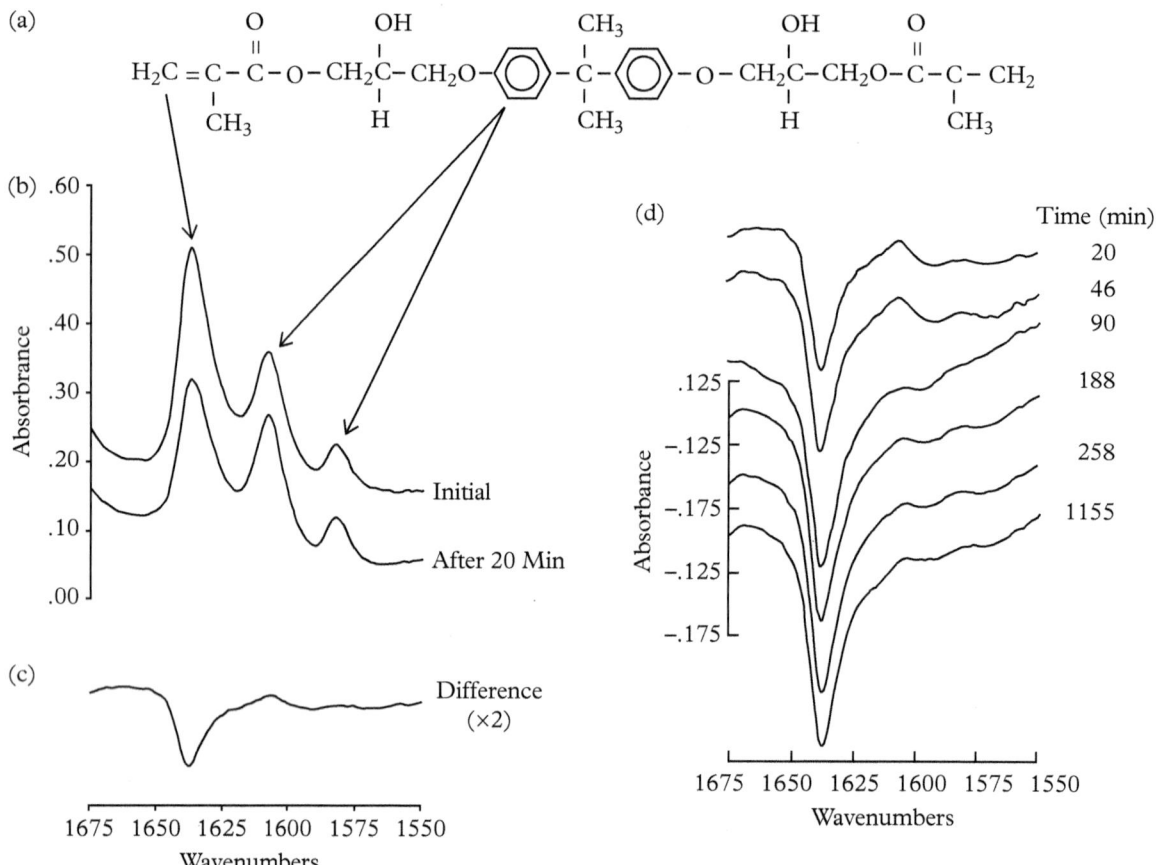

Figure 3.5.4 FTIR can monitor the curing of polymers used in fabricating polymer-matrix composites, such as bisphenol (a). The FTIR spectrum (b) shows a clear stretching mode of the double-bond at 1640 cm^{-1} and two modes of ring vibrations, at 1,608 cm^{-1} and 1,582 cm^{-1}, to serve as internal standards. After processing, the polymer is loaded and any continued cross-linking of the polymer can be monitored (c) by the reduction in the double-bond stretching mode intensity. (d) This feature, monitored over time, shows that curing continues even for many days after the processing.

Adapted from [6].

Recent developments in fabricating aircraft structures using polymer-matrix composite materials, has made FTIR spectroscopy into an essential industrial tool. It is used to evaluate the post-cure evolution, or the cross-linking of pre-polymers used in such fabrication, as a function of the type and amount of initiator used and the molecular flexibility of the pre-polymers. An example of such a polymer is bisphenol (Fig. 3.5.4a), which has a double-bond stretching mode at 1,640 cm^{-1}, and which will decrease in intensity as cross-linking takes place. In addition, two neighboring bands, at 1,608 cm^{-1} and 1,582 cm^{-1}, assigned to benzene ring vibrations, serve as internal standards. The polymers are cured at elevated temperatures, and their initial spectra measured immediately (b). The polymers are then placed in a simulated environment similar to their final use, removed after specific periods of time, and their FTIR spectra measured (b). Additional cross-linking, taking place in the simulated environment even after the processing step, is confirmed by the difference between the two spectra (c). Further, time-dependent evaluation (e) of the polymer cross-linking, shows that the 1,640 cm^{-1} band continues to increase, suggesting that the curing continues for several days after the initial process.

3.6 Raman Spectroscopy

3.6.1 Raman, Resonant Raman, and Fluorescence

Normally, in Raman spectroscopy (§3.4; Fig. 3.4.4; Fig. 3.4.6), the incident photon is of a wavelength not specifically related to any electronic excitation line (Fig. 3.6.1a) of interest in the material. However, if the incident laser light is tunable, its wavelength or energy can be made to match that of a specific electronic transition (Fig. 3.6.1b) and then the resulting Raman spectrum is further resonance-enhanced with orders of magnitude increase in signal. In addition to signal enhancement, the resonance signal may even be different, with the possibility of the appearance of new bands. An obvious, but important, application of resonance Raman spectroscopy is the ability to work with lower concentrations of the material. Figure 3.6.2 is a good illustration of how the resonance Raman spectrum differs from the normal Raman spectrum for K_2CrO_4. The normal Raman spectrum shows only the four bands characteristic of the CrO_4^{2-} ion; in contrast, the resonance Raman spectrum shows a progression of ten harmonics of the symmetric stretching mode (853 cm^{-1}), i.e. increasing bandwidth but with a reduction in intensity.

Many colored specimens, with impurities, can absorb the incident laser light and re-emit it as a fluorescence (Fig. 3.6.1c). Unfortunately, the intensity of the fluorescence can be higher by as much as a factor of 10^4 than the weaker Raman signal, and can completely mask the signal of interest. In fact, fluorescence is considered a major drawback in the application of resonance Raman spectroscopy. Various approaches to overcome this limitation include (a) prolonged irradiation

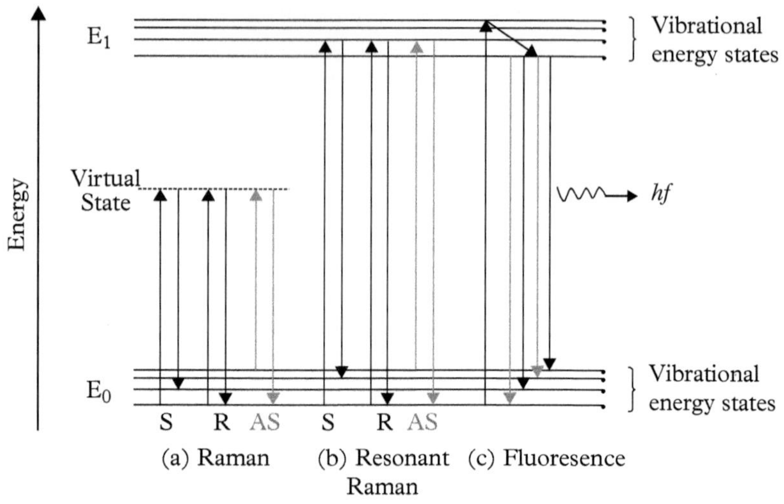

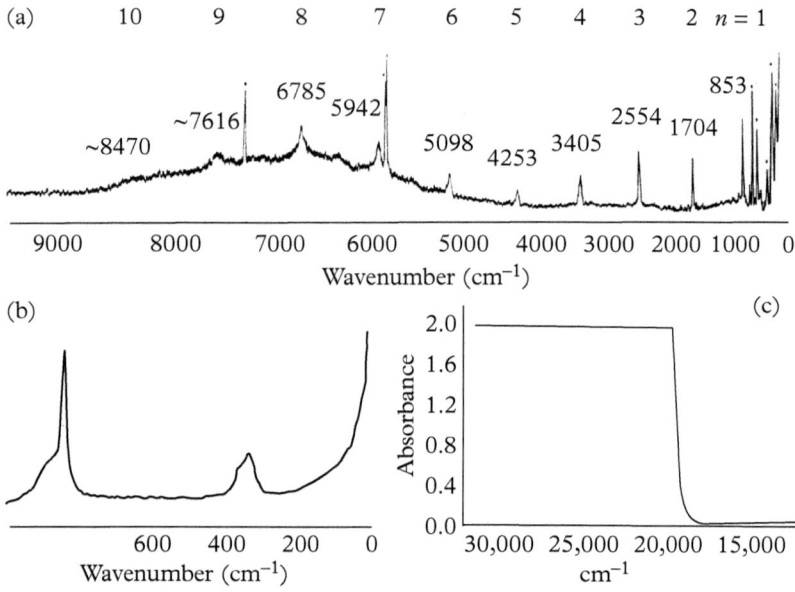

Figure 3.6.1 Schematic illustration of the Stokes, Rayleigh, and anti-Stokes transitions for (a) Raman and (b) resonant Raman scattering. Often, (c) fluorescence from the specimen dominates and swamps the Raman spectrum.

Figure 3.6.2 (a) Resonance Raman spectrum of K_2CrO_4 at 363.8 nm excitation, compared with (b) a normal Raman spectrum at 632.8 nm excitation. (c) Absorption spectrum of aqueous K_2CrO_4 solution.

Adapted from Long (1978).

of the specimen, prior to the Raman experiment, to bleach out the fluorescence signal; (b) using a larger wavelength laser with lower excitation energy, and (c) using a pulsed laser to take advantage of the shorter lifetime (1–100 ps) of Raman scattering, compared to that of fluorescence (1–100 ns), combined with electronic time-gating to discriminate between the two contributions.

3.6.2   Instrumentation for Raman Spectroscopy and Imaging

A simple laser is used as a source of monochromatic light and a dispersion grating (§6.6.2) arrangement for Raman spectroscopy (Fig. 3.6.3a). In principle, there are two ways to increase the spectral resolution, i.e. the ability to resolve spectral features of such a system. One is by increasing the focal length (Fig. 3.6.3);

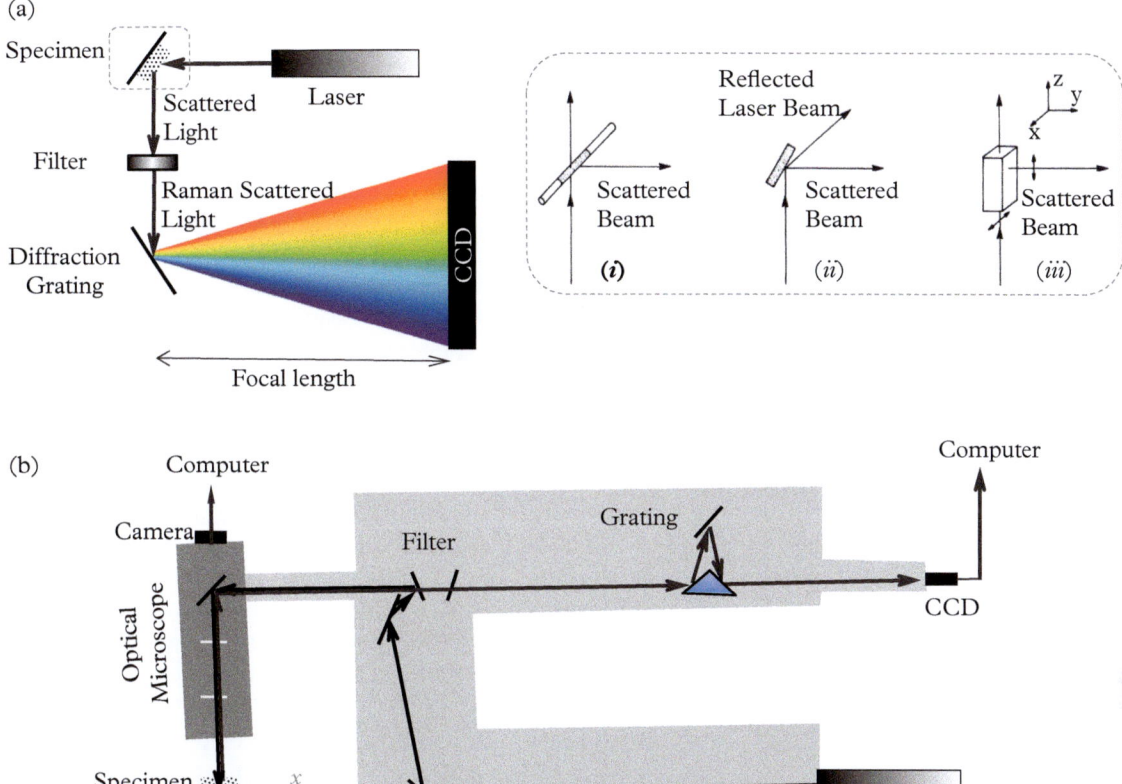

Figure 3.6.3 (a) Raman scattering using a filter and diffraction grating, for different specimen arrangements, shown in inset include (i) liquid capillary tube, (ii) powder pellets, and (iii) single crystals. The spectral resolution depends on the focal length and the density of gratings. (b) An imaging, micro-Raman arrangement combined with a standard optical microscope. The laser beam can be focused on a specific feature of a specimen, and the Raman backscattered light is filtered to eliminate the Rayleigh line and dispersed with a grating. The *x-y* specimen stage can be moved to obtain data at points across the specimen. The arrows indicate the beam path from the source (laser) to the detector (CCD) after interaction with the specimen.

approximately, doubling the focal length also doubles the spectral resolution. Alternatively, the grating can be changed, and similarly, doubling the density of gratings gives twice the spectral resolution, but higher density gratings have restricted working range; for example, IR radiation is incompatible with gratings of line density larger than ~2,000 lines/mm. Moreover, as discussed, to minimize fluorescence from swamping out the Raman signal, a longer wavelength laser may be necessary. On the other hand, higher-energy UV lasers might be required to ensure good specimen penetration, especially if fluorescence is not a problem. From a practical point of view, it may be preferable to work with lasers in the visible spectrum. As a result of all these competing factors, a good system will either have a *tunable laser*, or a number of lasers, that are switchable, along with different filters to eliminate the Rayleigh line, and matched gratings for dispersion and detection. In addition, Figure 3.6.3a shows scattering geometries appropriate for (i) liquids in capillary tubes or glass fibers, (ii) powder pellets or films on substrates, and (iii) oriented single crystals, which gives the maximum information.

Direct imaging involves examining the whole specimen for characteristic shifts, e.g. of a single compound. This generates an image showing the chemical distribution of that compound. In contrast, using a focused laser probe, Raman spectra can be taken at points across the specimen, helping identify multiple compounds and their spatial distribution. Generally, such a micro Raman imaging system (Fig. 3.6.3b) is combined with an optical microscope to observe the specimen and identify microstructural features of interest, with connections for the incoming laser light and the outgoing scattered light. Further, a filter is installed in the return optical path after Raman scattering, to remove stray light, specifically the Rayleigh line, and a grating is used to disperse the Raman scattered light into a spectrum. Incidentally, such data acquisition for Raman imaging requires lot of time, computer power, and data storage. Last but not least, Fourier transform Raman spectroscopy, similar to what we have discussed for FTIR, has been developed; details can be found elsewhere (Hendra, Jones, and Warnes, 1991).

3.6.3   Application of Raman Spectroscopy in Chemical and Materials Analysis

Table 3.6.1 summarizes the application of Raman spectroscopy to identify common chemical functional groups.

Figure 3.6.4a shows a typical application of Raman spectroscopy in materials research. $Si_{1-x}Ge_x$ is a semiconductor synthesized over a wide range of compositions, and is used for heterojunction bipolar transistors or as a strain-inducing layer for CMOS transistors. The Raman spectrum from this compound semiconductor shows four characteristic vibrational bands related to Si–Si, Ge–Ge, Si–Ge, and the Si–Si$_{substrate}$ modes. Moreover, the position of the Si–Si vibration band, depends linearly on the Ge concentration, x, as shown in

Table 3.6.1 Raman frequencies of common functional groups.

Functional Group	Position (cm^{-1})	Remarks
>S–S<	500–550	
C–C	~1,060 and 1,127	Polyethylene
C–C	~77–1260	Highly mixed in complex molecule
Aromatic ring	~1,000	Monosubstituted; 1,3 disubstituted; 1,3,5 trisubstituted
Aromatic ring	~860	1,4 disubstituted
CH_3 umbrella mode	~1,375	
CH_3 and CH_2 deformations	1,410–1,460	
>C=C<	~1,650	
>C=C<	~1,623	Ethylene
>C=O mixed with NH deformation	1,620–1,690	Amide
>C=O	1710–1745	Changes for aldehyde, ketone, and ester
C=C	2,100–2,300	
SH	2,540–2,600	
$>CH_2$	2,896 and 2,954	Ethane
$>CH_2$	2,845 and 2,880	Polyethylene
CH_3	2,870 and 2,905	Polypropylene
CH	2,900	Cellulose
CH	3,015	Olefinic CH
CH	3,065	Aromatic CH
CH	3,280–3,340	Acetylenic CH
NH	3,150–3,340	Broadened and shifted by H bonding
OH	3,000–3,600	Broadened and shifted by H bonding

the *calibration plot* (Fig. 3.6.4b). Thus, by measuring the line position (510 cm^{-1}) of the Si–Si Raman peak in the unknown alloy, ~120 nm thick, and using the calibration plot, we can determine its composition to be $Si_{0.7}Ge_{0.3}$.

3.6.4 Surface-Enhanced Raman Spectroscopy (SERS)

It is well-known that the Raman signal for many molecules, using laser excitation in the visible or near-IR (NIR), is orders of magnitude larger when they are adsorbed on metal surfaces compared to isolated molecules or in solution (see Ferraro, Nakamura, and Brown, 2003). This *surface-enhanced* Raman spectroscopy (SERS) is a promising development as an *in situ* vibrational probe for studying liquid–solid, gas–solid, and solid–solid interfaces. Most common substrates used

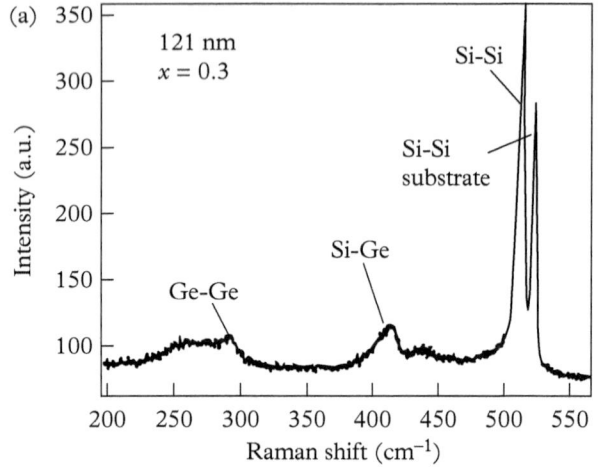
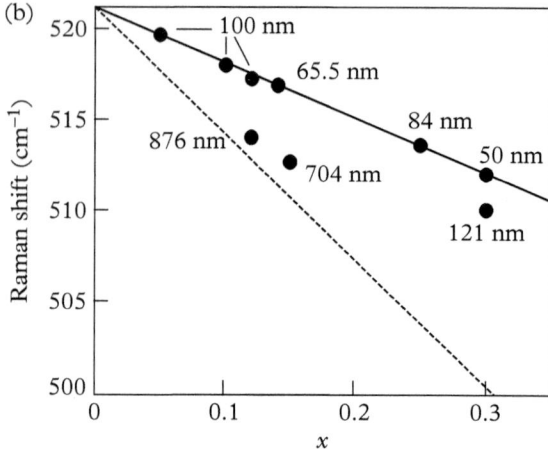

Figure 3.6.4 Application of Raman spectroscopy in materials research. (a) The Raman spectrum from a 121 nm thick specimen of $Si_{1-x}Ge_x$. (b) A calibration plot showing that the position of the Si–Si vibration line varies linearly with the concentration of Ge in the alloy. From the calibration plot, and the observed position of the Si–Si line ($\sim$510 cm^{-1}), the composition of the unknown alloy is determined to be $x = 0.3$.

Adapted from Leng (2013).

for SERS studies are colloidal nanoparticles of gold and silver, as well as electro-chemically roughened silver electrodes.

The fundamental mechanism underlying the SERS effect is still poorly under-stood. The Raman signal is proportional to the square of the induced dipole moment, $\mu_{Ind} = \alpha \mathbf{E}$. Thus, the enhancement can occur either through changes in the molecular polarizability, α, called the chemical effect, or through the electric field, $\mathbf{E}$, called the electromagnetic effect. The electric field, $\mathbf{E}$, may be significantly changed in the vicinity of a metal surface, particularly when fine particles with significant curvature or rough surfaces are involved. The laser light also collectively excites the valence band—known as plasmon resonance—of the metal, causing the roughness feature of the metal to be polarized and enhancing the local electromagnetic field from that of the applied radiation, leading to Raman signal enhancement. The chemical effect is attributed to a charge transfer or bond formation between the metal and the absorbing molecule, thereby increasing its polarizability.

In summary, Figure 3.6.5 compares the relative strengths of Raman scattering, FTIR, and NIR absorption, that can help plan and identify the application of these complementary techniques.

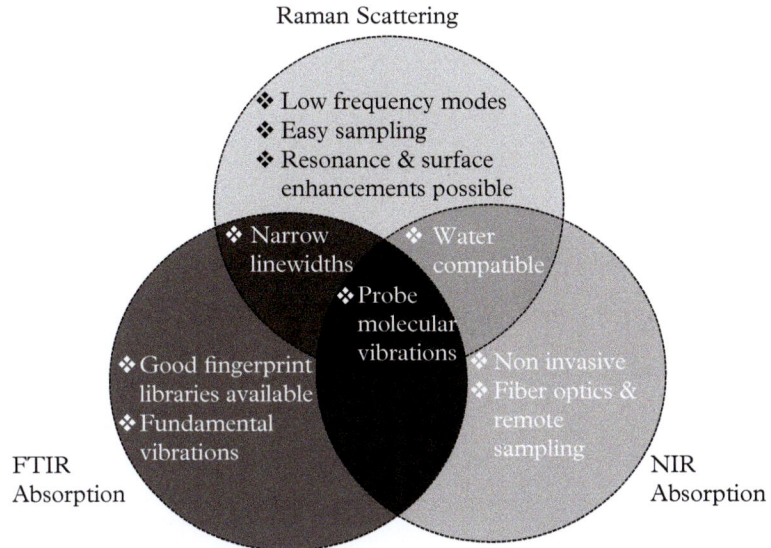

Raman Scattering

❖ Low frequency modes
❖ Easy sampling
❖ Resonance & surface
  enhancements possible

❖ Narrow   ❖ Water
  linewidths   compatible
      ❖ Probe
        molecular
        vibrations
❖ Good fingerprint   ❖ Non invasive
  libraries available ❖ Fiber optics &
❖ Fundamental   remote
  vibrations   sampling

FTIR   NIR
Absorption   Absorption

Figure 3.6.5 Comparison of Raman, FTIR, and NIR methods in materials characterization.

Adapted from McCreery (2000).

3.7   Probing the Electronic Structure of Solids

Spectroscopy methods routinely provide information about the energy levels in a solid (Fig. 3.7.1). In particular, details of the structure of the valence band and the unoccupied states in the conduction band (in addition to the core levels already discussed; see §2) are of interest. The simplest and commonly used method is that of optical absorption in the UV and visible range, referred to as UV-Vis spectroscopy. Such UV-Vis spectra (Fig. 3.7.1a) of solids shows absorption peaks corresponding to transitions of electrons from filled valence bands to empty states in the conduction band. Specifically, in nonmetallic materials (e.g. semiconductors), this allows for the determination of the band gap (Fig. 3.2.4). However, detailed interpretation of such optical transitions depends on the complex details of both the *filled* and *empty* bands, and it is often difficult to resolve them separately. On the other hand, there are other spectroscopic techniques, using X-ray photons or high-energy electrons, that offer opportunities for a straightforward interpretation of the spectrum in terms of the energy levels and their dispersion, i.e. band structure. We explore these techniques in the remaining sections of this chapter.

Figure 3.7.1 summarizes the principal spectroscopic methods with photons and electrons, schematically illustrating the techniques and the energy levels involved. Note that we define the *vacuum level*, E_V, as the energy at which an electron is able to escape completely from the electric potential of the solid, and define our energy scale such that $E_V = 0$. However, inside the solid, the normal reference point for energy of the electrons is the Fermi level, E_F, already defined in

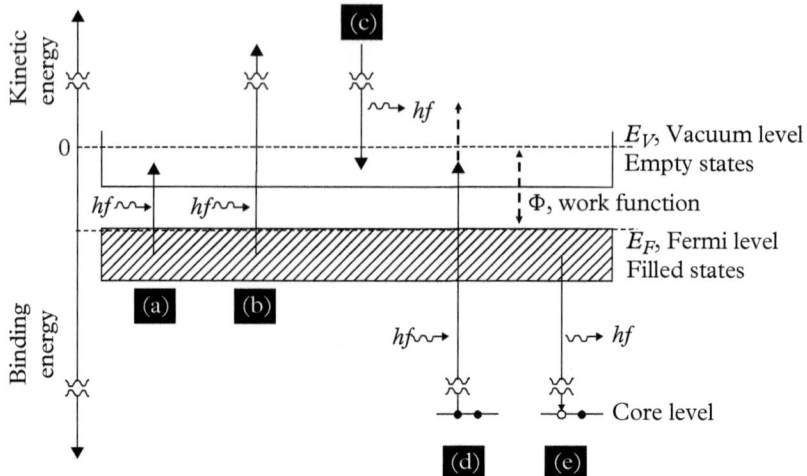

Figure 3.7.1 This is a key figure that schematically represents the important spectroscopic techniques used to probe the electronic structure of a solid, including (a) optical spectroscopy, see Figure 3.2.4; (b) photoemission, (c) inverse photoemission, see §3.8; (d) X-ray absorption, §3.9, where the core electron can be excited either to the continuum or some empty state, and (e) X-ray emission (discussed in §2). Energy levels of interest, and the work function, are also indicated. Electrons with energies greater than the vacuum level can either leave (b) or enter (c) the solid. Note that in this plot the energy scale is set such that $E = 0$ at $E = E_V$.

Adapted from Cox (1987).

Figure 3.2.2. The difference between these two energy levels is the *work function*, Φ, which is simply given by the difference as

$$\Phi = E_V - E_F \tag{3.7.1}$$

Thus, the work function is the energy required to remove an electron from the solid into the vacuum (Fig. 5.2.4), which in a metal is typically ~2–5 eV. In practice, the work function is important for many surface analysis methods, and depends on various surface effects, such as adsorption of different molecules, and surface crystallography (which is introduced in §4.1.7 and discussed further in §8.3). From a practical point of view, work function is also important for thermionic emissions from solids (§5.2.2). Such thermionic emitters release electrons upon heating and are used as electron sources or guns, e.g. in electron microscopes (see §9.2.1 and §10.2.1). The best thermionic emitting materials have low work functions, and the metallic compound lanthanum hexaboride (LaB_6) is widely used.

The energy scale is defined such that the electrons bound in the solid have negative values and those that are free, having left the solid with sufficient energy to overcome the work function, are assigned positive values (Fig. 3.7.1). The best way to probe the energy levels and the dispersion of electrons in filled levels of solids, especially in the valence band, is by (b) photoelectron spectroscopy. The complementary, inverse photoelectron spectroscopy, as shown in (c), probes the unoccupied, or empty, states. Information about empty states is also obtained from X-ray absorption spectroscopy (XAS) (d). Here, the *fine structure* seen at the onset of characteristic absorption edges, known as X-ray absorption near edge structure (XANES), and long-range oscillations, known as extended X-ray absorption fine structure (EXAFS), are commonly investigated (see §3.9.2).

X-ray photoelectron spectroscopy (XPS) and XAS are complementary methods (Fig. 3.7.2). In XPS, a bound electron is excited to a free state in vacuum with a well-defined kinetic energy; such transitions, shown for inner-shell excitation

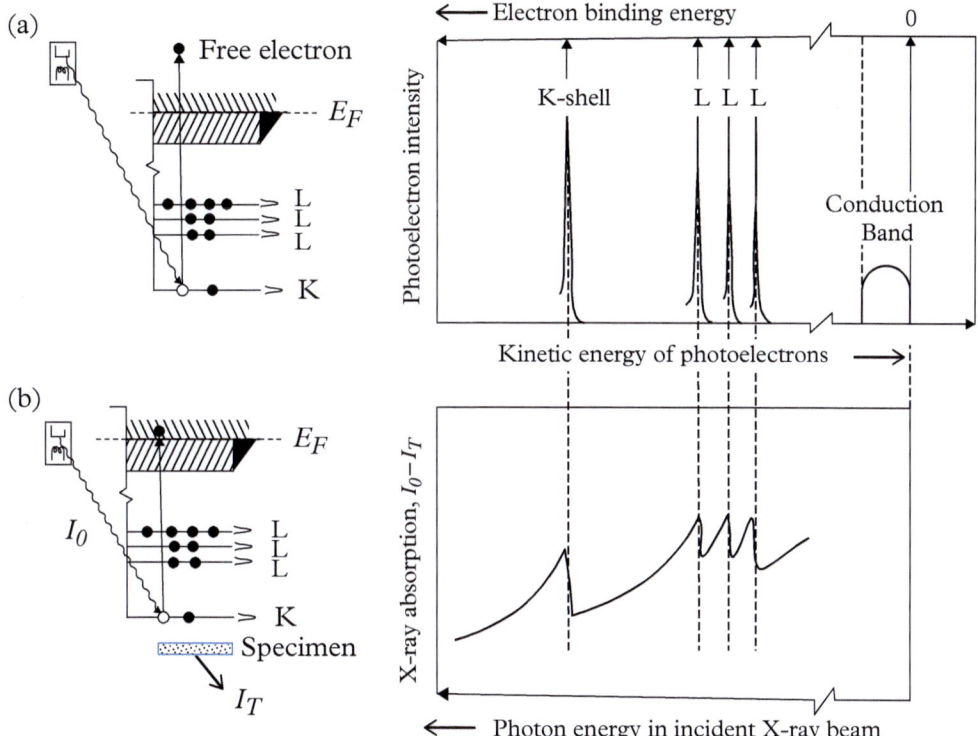

Figure 3.7.2 Complementary nature of (a) X-ray photoelectron (XPS) and (b) X-ray absorption spectroscopies (XAS).

Adapted from Siegbahn et al. (1967).

(K- and L-shells) as sharp lines, can also probe the valence or conduction (shown here for a metal) bands. Note that a constant energy photon probe is used and the kinetic energy of the photoelectron signal is measured. In XAS (Fig. 3.7.2b), an edge occurs when an X-ray photon is absorbed and a bound electron is excited, following dipole selection rules, to the first available unoccupied state above the Fermi level, displaying some fine structure based on the unoccupied density of states; some electrons are excited beyond the vacuum level, with probability decreasing with increasing energy transfer, and appearing as a gradually decreasing tail in the absorption spectrum. In XAS, the absorption is measured as a function of photon energy.

We first revisit photoelectron spectroscopy as a probe of the occupied electronic levels in a solid, followed by an introductory description of absorption methods.

3.8   Photoemission and Inverse Photoemission from Solids

Section 2.6 introduced the physical principles of photoemission and some details, including instrumentation, of the related technique of photoelectron spectroscopy using incident X-rays (XPS) or ultra-violet radiation (UPS). In principle, all photoemission experiments are performed as described in Figure 2.6.2. In addition to an X-ray tube, and a UV gas discharge lamp (§5.2.1.1), synchrotron radiation sources (§2.4.3) with higher intensities and tunability of the radiation are also used. The light impinges on a specimen and ejects a photoelectron, which is analyzed both in terms of its kinetic energy, E_{kin}, and its momentum, $\mathbf{p}$. Thus, in experiments, the kinetic energy of the emitted photoelectrons and angle-resolved measurements that specify the collection geometry (Fig. 3.8.1a), and thereby the direction of their momentum, are measured. In a solid specimen, the kinetic energy and momentum of the photoelectron are simply related to the binding energy, E_B, (with $E_B = 0$ at E_F) and the work function, Φ,

$$E_{kin} = hf - \Phi - | E_B | \tag{3.8.1}$$

and

$$p = \sqrt{2m_e E_{kin}} \tag{3.8.2}$$

with the direction of $\mathbf{p}$ being determined by the polar, θ, and azimuth, φ, angles (Fig. 3.8.1a) along which the electrons are collected by the spectrometer after they leave the specimen. In this specialized case, where the photoelectrons are collected as a function of direction, the technique is known as angle-resolved photoemission spectroscopy (ARPES).[5]

[5] Pioneered by Kai Siegbahn (1918–2007) and colleagues in Sweden. Professor Siegbahn shared the 1981 Nobel prize in Physics and was cited for "his contribution to the development of high-resolution electron spectroscopy."

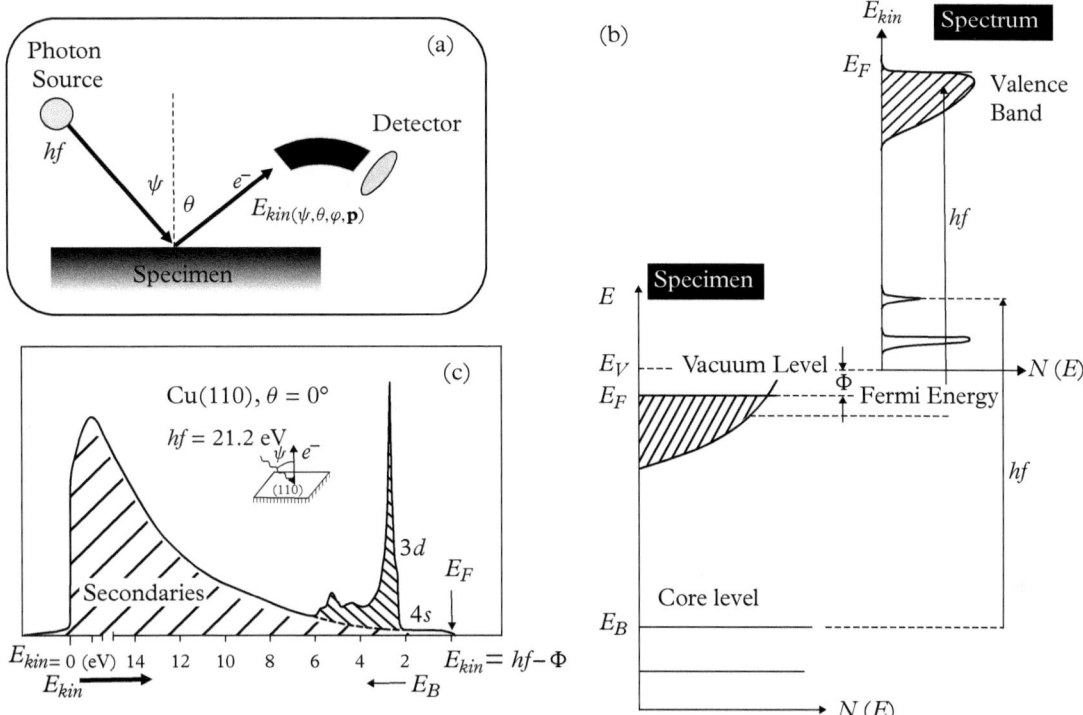

Figure 3.8.1 (a) Experimental arrangement for photoelectron spectroscopy (see also Figure 2.6.2). The photon source can either be a UV discharge lamp, an X-ray tube, or a synchrotron storage ring. The photoelectrons are detected by using an electrostatic analyzer described in §2.6.1.2. (b) Relationship between the energy levels in the solid and the distribution of energies observed in photoemission when probed with photons of energy, hf. The kinetic energy of the photoelectrons is measured with respect to the vacuum level. Experimental measurements are often displayed in terms of the binding energy of the electrons in the solid with the Fermi level set to zero. (c) UPS spectrum ($hf = 21.2$ eV) from the surface of Cu(110), with normal emission. The work function ($\Phi = 4.5$ eV) can also be obtained from such measurements.

Adapted from [18].

Figure 3.8.1b shows the relationship for a metal between the kinetic energy distribution of the photoelectrons and their binding energy in the solid, when initially occupying the core levels or the valence band. The Fermi level, E_F, is at the top of the valence band, and is separated from the vacuum level, E_V, by the work function, Φ, as shown. The binding energy, E_B, is defined as $E_B = 0$, at $E = E_F$ in solids, but in isolated free atoms or molecules, $E_B = 0$, at $E = E_V$; in both cases $E_B < 0$ (Fig. 3.7.1). However, in the photoemission literature, it is common to assign positive values for E_B (which creates some difficulties for inverse photoemission, but we discuss this later), and we follow this convention for this section. Further, we can see the simple relationship between the energy levels in the solid and the kinetic energy of the photoelectrons emitted using an incident photon of energy, hf (Fig. 3.8.1b). Moreover, in practice, the variation of intensity is plotted as a function of E_B (Fig. 3.8.1c). Here, to investigate the valence band of a Cu(110) surface, lower-energy ($hf = 21.2$ eV) UV photons are used and UPS data is acquired normal ($\theta = 0°$) to the specimen. The onset of the photoemission signal, with $E_{kin} = hf - \Phi$, is at E_F, followed in terms of increasing binding energy, E_B, by the shallow $4s$ valence band (~ 2 eV wide), and then between 2–6 eV, the structured $3d$ band. Another important feature is the large background of secondaries, arising from those electrons that have lost some energy due to inelastic scattering processes.

It is important to re-emphasize that the features seen in a photoemission spectrum are truly interpretable in terms of the electronic structure of the solid only if they emerge from the shallow escape depth (Fig. 1.3.6), of the order of ~ 10 nm of the specimen where the photoemission originates without loss of energy. In practice, the surface sensitivity of XPS can be further enhanced by changing the angle of incidence of the photons (Fig. 3.8.2), which clearly shows enhanced sensitivity of the Al $2s$ signal to surface oxidation at shallow angles of detection ($\theta = 82.5°$ from the surface normal).

Photoelectron spectroscopy experiments can be conveniently interpreted using a three-step model (Fig. 3.8.3). Even though photoemission takes place as a

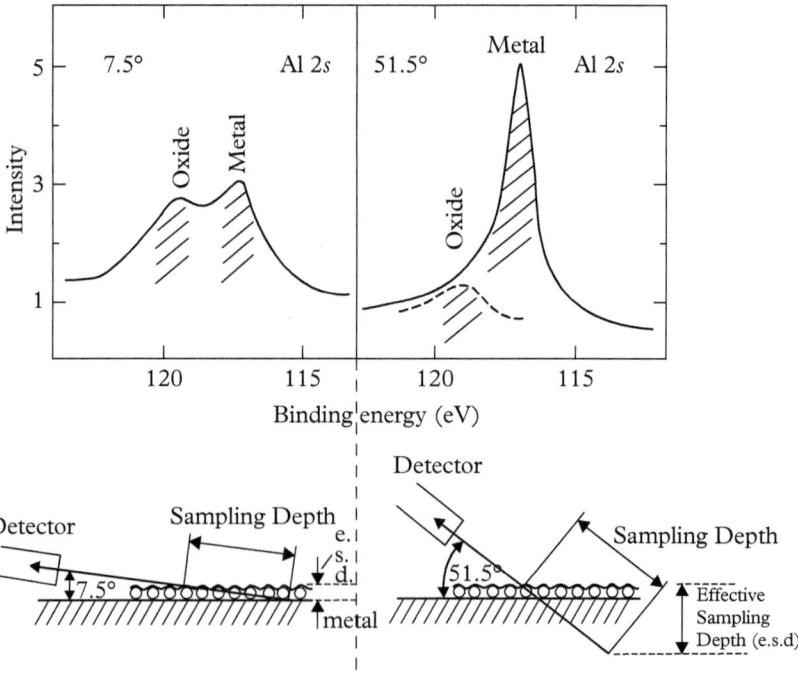

Figure 3.8.2 Experiment demonstrating the surface sensitivity of XPS by changing the angle of detection relative to the surface of Al that is slightly oxidized. At 7.5° ($\theta = 82.5°$), the Al $2s$ signal is of the same magnitude for both metal and oxide, but at 51.5° ($\theta = 38.5°$) only the metal signal is visible.

Adapted from Hüfner (2003).

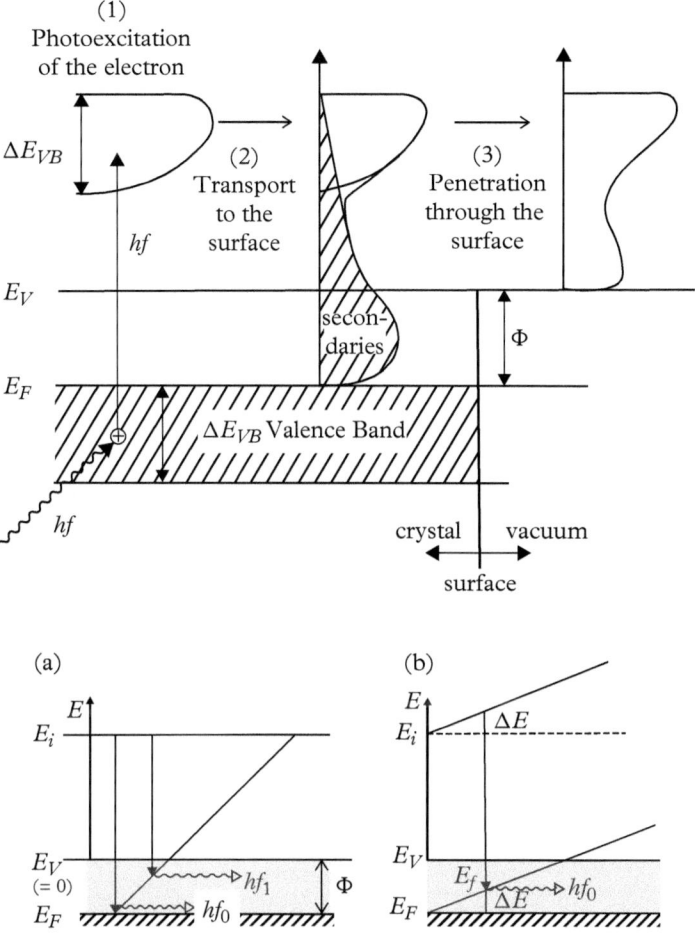

Figure 3.8.3 Photoemission as a three-step process: (1) photoexcitation, (2) transport to the surface of the solid, and (3) penetration and escape to the vacuum.

Adapted from Hüfner (2003).

Figure 3.8.4 Inverse photoemission (IPES) experiments can be performed in two different ways; either (a) with fixed electron energy, E_i, and tunable photon detector, or (b) with a photon detector of fixed energy, hf_0, and variable incident electron energy, E_i. The latter (b) is also known as bremsstrahlung isochrome spectroscopy.

Adapted from Hüfner (2003).

single process, for ease of understanding it can be divided into three steps: (1) photoexcitation of electrons, (2) transport of the electrons to the surface with simultaneous production of secondaries, and (3) penetration through the surface layer and escape into the vacuum, where it is collected by the spectrometer.

Inverse photoemission spectroscopy (IPES) measurements (Fig. 3.7.1c) can be carried out in two different ways (Fig. 3.8.4a,b). In the first method (a), the specimen is probed with electrons of a fixed energy, E_i, and the range of emitted photons, produced by the bremsstrahlung process, as the electron is rapidly brought to a stop inside the material, is measured using a *tunable* X-ray detector. For example, photons measured with energy, hf_1, as shown, correspond to an unoccupied level, $E_1^u = \Phi - (hf_1 - E_i)$, above E_F. The alternative method of bremsstrahlung isochrome spectroscopy (BIS) (Fig. 3.8.4b) probes the surface with electrons of variable energy, $E_i + \Delta E_i$, and the photons of a specific energy, hf_0, are detected. The corresponding energy of the unoccupied level would be $E_1^u = \Delta E_i$ (also, above E_F). Note that in PES, the energy range between E_F and E_V is not accessible, but in IPES/BIS the inaccessible range is for energies less than E_F, and it is the unoccupied states, $E_F < E < E_V$, that are truly resolved. In other words, the two methods are complementary (Fig. 3.8.5).

We can compare the photoemission spectra, using UV light, of the same hydrocarbon material in the gas and solid phases (Fig. 3.8.6) for benzene (C_6H_6). In the gas phase (a), the photoemission describes the occupied molecular orbitals (MO), starting with the top of the filled π MO as a band around $E_B \sim 9$ eV. The higher energy bands reflect further changes in molecular geometries as the electrons are removed from different MOs. Overall, the energy levels in the molecular solid, (b), closely reflect the energy levels of the MOs in the individual molecules of C_6H_6. However, the bands appear broader in the solid compared to the gas; this is attributed to the crystal vibration modes that are likely excited in the photoionization process. Moreover, the energy bands are shifted to lower energies in the solid by ~ 1.15 eV. This shift, as discussed earlier (§2.6.3), is the

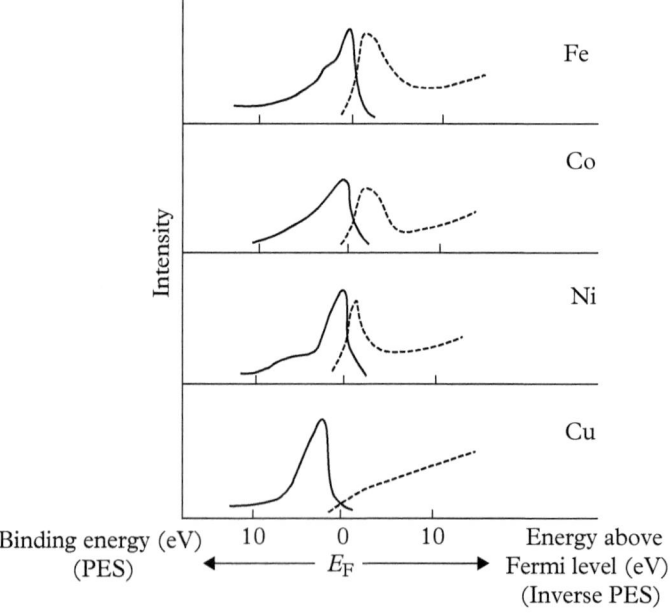

Figure 3.8.5 X-ray photoelectron spectra [8], continuous line, and inverse photoemission [9], dashed line, obtained at a photon energy of 530 eV, for the $3d$ band in the transition elements, Fe, Co, Ni, and Cu. Notice the complementarity of the two techniques.

Fe

Co

Ni

Cu

Intensity

Binding energy (eV)   10   0   10   Energy above
(PES) $\longleftarrow$ E_F $\longrightarrow$ Fermi level (eV)
(Inverse PES)

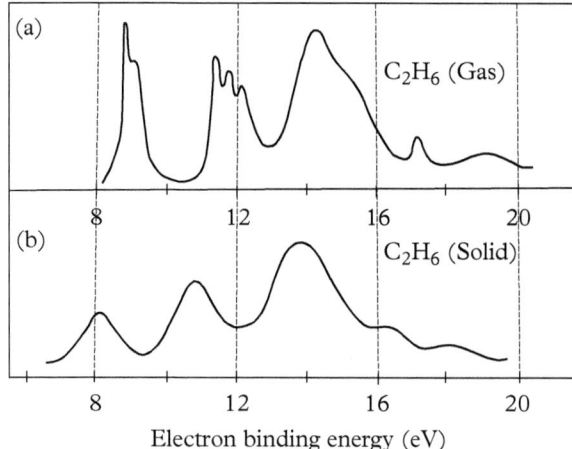

Figure 3.8.6 UPS spectra of C_6H_6 (benzene) in (a) the gas phase, and (b) as a molecular solid. Other than the broadening and a relaxation shift of ~ 1.15 eV, in the solid the two spectra are not very different.

Adapted from [10].

result of the response of the other electrons in the solid to the ionization process. The positive hole left behind at the site of ionization will instantaneously polarize the electrons in the surrounding molecules, which will affect the energy required to produce the hole! Such polarization energy has been calculated (see Cox, 1987) to be of the order of 1 eV, in agreement with the experiment.

The effect of ligands in photoemission studies can be understood further by considering the XPS spectra in three different copper dihalides (Fig. 3.8.7). In the ground state, these three dihalides have Cu in the $3d^9$ configuration and a filled ligand shell, L, which is $2p$ for F$^-$, $3p$ for Cl$^-$, and $4p$ for Br$^-$. Thus, the electronic configuration for dihalide can be written as $3d^9L$. The photoexcitation of the Cu $2p$ levels can lead to two possible final states: either the $3d^9L$ configuration remains unchanged, or after the core hole is created, one electron is transferred from the ligand shell to the Cu d-shell, resulting in a final $3d^{10}L^{-1}$ configuration. If we now recall the dipole selection rules (§2.3.1) and the spin-orbit coupling (§2.2.2), we can assign the main peaks observed in the XPS spectrum (Fig. 3.8.7) as $2p_{3/2}^{-1}3d^{10}L^{-1}$ and $2p_{1/2}^{-1}3d^{10}L^{-1}$. The two satellites are $2p_{3/2}^{-1}3d^9L$ and $2p_{1/2}^{-1}3d^9L$, respectively. In this first approximation, the main line energy shows a strong correlation with the ligand as the photoionization energy includes the energy required to produce a hole in the ligand valence orbital, which in turn depends on the nature of the ligand, and explains the observed shift in the position of the main lines.

Figure 3.8.8a shows the UPS measurement for polycrystalline Ag at room temperature, using $hf = 21.2$ eV incidence, for the energy region around E_F (compare with Figure 3.2.3b). The instrument used had an energy resolution of 25 meV and the fit of the data to the Fermi–Dirac distribution (3.2.2) is in good agreement. The width of the Fermi–Dirac distribution is ~ 100 meV or $4k_BT$, which is larger than the resolution, and determines the shape of the curve. Measurements around the Fermi "edge" at lower temperatures are a good way to determine the energy

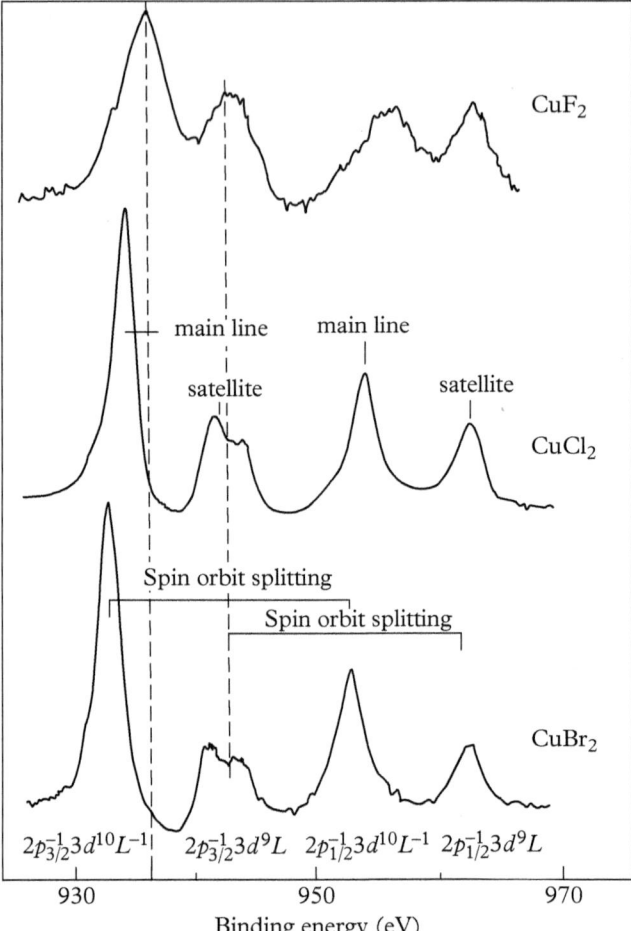

Figure 3.8.7 XPS spectra of copper dihalides, CuF_2, $CuCl_2$, and $CuBr_2$. The main lines are assigned as $2p_{3/2}^{-1}3d^{10}L^{-1}$ and $2p_{1/2}^{-1}3d^{10}L^{-1}$, respectively, and their energies vary because of the difference in the binding energies of the ligands.

Adapted from [11].

resolution of an UPS instrument. Figure 3.8.8b illustrates a similar measurement using a very high-resolution spectrometer for polycrystalline Ag at $T = 8$ K. As we can see, using a 3 meV Gaussian resolution function to convolute the Fermi–Dirac distribution, gives a very good fit to the experimental data.

Figure 3.8.9 compares XPS and UPS measurements of a complex oxide, $Na_{0.7}WO_3$. In XPS, the X-ray photons used have sufficient energy to ionize the core levels, including the $Na1s$, $O1s$, $W4d$, and $W4f$ levels. Additional Auger transitions can also be observed. The small shift in binding energies, from that of the pure elements, is consistent with the oxidation states. The valence energy levels are better probed by UPS, because the inherently smaller energy line widths, lead to better energy resolution. The valence spectra show two bands, assigned to the $W5d$ and $O2p$ levels, and the relative intensities of the two bands is consistent

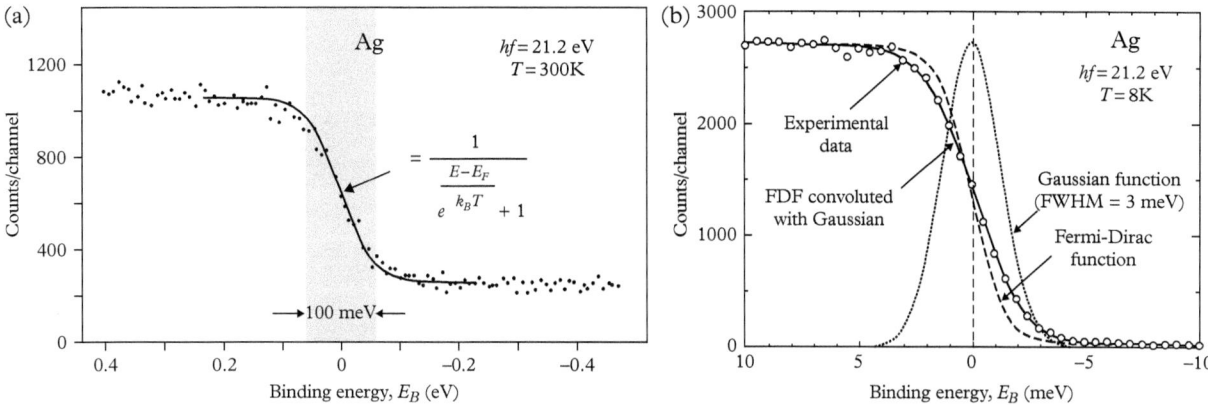

Figure 3.8.8 UPS spectrum ($hf = 21.2$ eV) of polycrystalline Ag at (a) $T = 300$ K and FWHM 25 meV, and (b) $T = 8$ K and FWHM $\sim$3 meV, showing the energy distribution around the Fermi level. As expected, the data fits very well to the Fermi–Dirac distribution.

Adapted from Hüfner (2003).

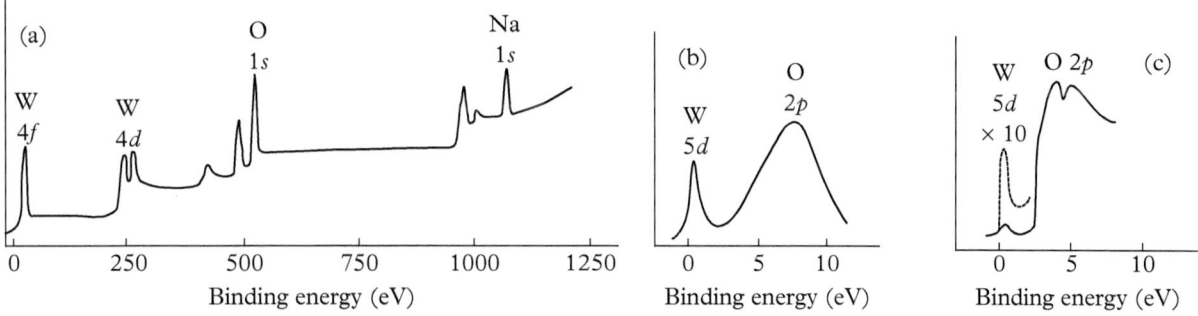

Figure 3.8.9 PES spectra of $Na_{0.7}WO_3$. (a) A wide scan showing the core levels. (b) XPS of the valence band and (c) a higher resolution UPS of the valence band.

Adapted from Cox (1987).

with the difference in their ionization cross-sections, as a function of the incident photon energy.

This section ends with a discussion of a practical application of photoemission spectroscopy in materials research [12]. CaF_2 has a crystal lattice structure that matches very well to Si(111), and together they are a model system to study insulator–semiconductor epitaxial[6] behavior by photoemission spectroscopy. Moreover, the PES can be acquired in situ, if the spectrometer is combined with a UHV growth chamber, to monitor the evolution of the electronic structure as a function of epitaxial growth. Figure 3.8.10b shows three such PES spectra corresponding to different stages of the growth of CaF_2; starting with a

[6] Artificial growth of crystals on a crystalline substrate with a well-defined crystallographic orientation relationship between them.

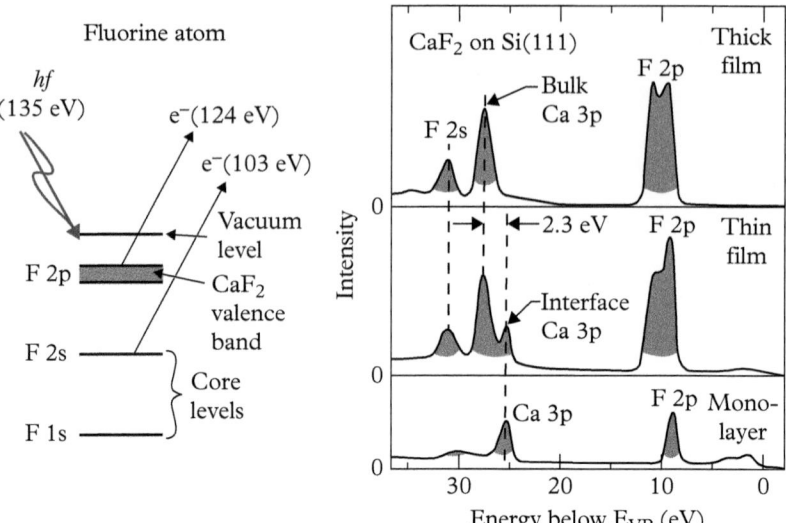

Figure 3.8.10 Practical application of PES in materials research. Here it is used to study the evolution of the bonding of CaF_2 (insulator) to a Si(111) surface. The energy levels (left) and three PES spectra for a monolayer (bottom), a thin 1.1 nm layer (middle), and a thick 5 nm film (top), all of CaF_2, are shown. Further details are discussed in the text.

Adapted from [12].

single monolayer or a molecule thick film of CaF_2 (bottom), to a relatively thick (∼5 nm) film (top), and a thin (∼1.1 nm) film in between. Incident photon energy was 135 eV, and the experiments were performed using a synchrotron source. The plots show all energies in eV below the top of the valence band.

In the monolayer spectrum, one observes narrow atom-like sharp features at 8.3 eV (F $2p$) and 25.6 eV (Ca $3p$). In the case of Ca, this feature is associated with bonding to the surface. The role of fluorine is minimal, as the Ca–Si bonding is preferred, and Ca retains a larger fraction of the shared electron, as reflected in the lower binding energy. In the thin film spectrum, the F $2p$ peak broadens and has characteristics of the bulk valence band; in addition, the Ca $3p$ now develops a split peak, at 25.6 eV and 27.9 eV, with features of the bulk and surface valence bands. One can assign the 2.3 eV difference to the binding energy of Ca in the bulk and surface environments. In addition, a core F $2s$ feature is also seen. Finally, in thick films, both F $2p$ and Ca $3p$, show fully developed bulk features; now, the Ca outer $4s^2$ electrons are strongly attracted by the surrounding eight highly electronegative fluorine atoms, and thus increasing the binding energy of the remaining $3p$ electrons.

It is clear from this discussion that PES has great utility in chemical identification and analysis, as discussed in §2, and also in elucidating bonding in different chemical environments. The only requirement is a specimen with a clean surface that is representative of the bulk structure. The examples of PES in this section are all *angle-integrated* spectra; by this we mean that electrons emitted over a wide solid angle are collected and displayed. In practice, the energy distribution of electrons photoemitted from a solid is dependent on the direction of both the incoming photon and the outgoing photoelectron. If data were to be acquired in an angle-resolved manner, detailed information on the 3D band structure (Fig. 3.3.3) of the solid or the 2D band structure of a surface adsorbate layer, including the orientation of such adsorbed molecules, can be obtained. In practice, to get a full appreciation of the full range of capabilities of this technique, it requires variation of the photon energy (using synchrotron radiation) and the polarization of the incident radiation. Further details of PES can be found in many texts and monographs, including Hüfner (2003), emphasizing fundamentals, and Watts (1990) for a good introduction from a materials applications point of view.

Example 3.8.1: For Al Kα radiation (1,486.7 eV) incident on a Cu film, determine the ratio of 2s to 3s photoelectron yields based only on cross-sections (Fig. 2.6.17) and escape depths (Fig. 1.3.6).

Solution: The probability, P_{ie}, per incident photon to create a photoelectron, simplified (2.6.4), is given by $P_{ie} = \sigma^k Nt$, where Nt is the number of atoms/cm^2 in a layer of thickness t, and σ^k is the cross-section (Fig. 2.6.17) for ejecting a photoelectron from the k-orbital. However, the number of electrons that can escape from the solid and be detected decreases with depth as $\exp(-x/\lambda)$, where λ is the mean free path length (Fig. 1.3.6). Thus, the probability, P_d, of an incident photon producing a detectable photoelectron is then $P_d = \sigma^k N\lambda$. We ignore instrument efficiencies of detection.

From Table 2.2.2, the binding energies for Cu, $E_b^{2s} = 1096.7\ eV$, $E_b^{3s} = 122.5\ eV$

From Figure 2.6.17, the corresponding cross-sections for Al Kα radiation are

$$\sigma^{2s} = 8 \times 10^4, \sigma^{3s} = 1.5 \times 10^4.$$

For incident Al Kα radiation, the photoelectrons have energies $E_{PE}^{2s} = 1486.7 - 1096.7 = 390\ eV$ and $E_{PE}^{3s} = 1486.7 - 122.5 = 1364.2\ eV$. The corresponding mean free path lengths (Fig. 1.3.6) are $\lambda^{2s} = 10$ and $\lambda^{3s} = 20$ Å.

Thus $\dfrac{P_d^{2s}}{P_d^{3s}} = \dfrac{\sigma^{2s}N\lambda^{2s}}{\sigma^{3s}N\lambda^{3s}} = \dfrac{8\times10^4 N10}{1.5\times10^4 N20} \approx 2.67$

3.9 Absorption Spectroscopies–Probing Unoccupied States

3.9.1 X-Ray Absorption Spectroscopy (XAS)

Figure 3.7.1 summarizes electronic transitions that form the basis of experimental studies of the electronic structure of both occupied and unoccupied states in materials. Section 3.8 described photoemission spectroscopy (PES) techniques and discussed how XAS complements PES methods (Fig. 3.7.2). In XAS, the lowest

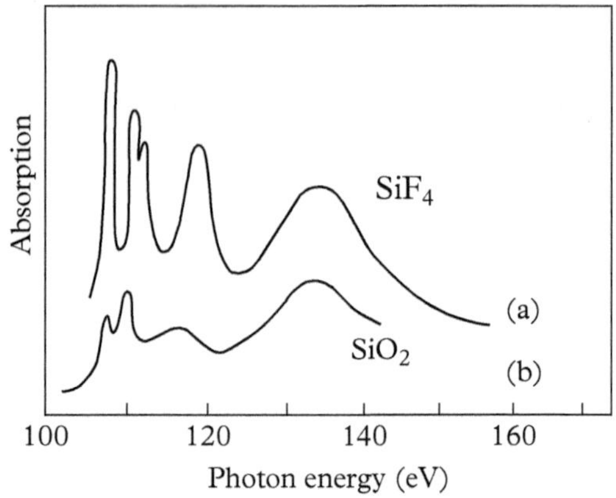

Figure 3.9.1 X-ray absorption L-edge spectra of tetrahedrally coordinated silicon in (a) gas phase SiF_4, and (b) solid SiO_2.

Adapted from [13].

energy excitations are from occupied to unoccupied states in the valence band; here the shape of the spectrum is a convolution of the single particle densities of states of the initial and final levels, and since its interpretation is complicated, it is not popularly used. However, higher-energy excitations correspond to local transitions from a core level to an unoccupied state in the valence band, and subject to dipole selection rules, (2.3.2), as discussed in §2.3.1. The interpretation of the spectra, in most cases, is simple as it relates directly to the final *unoccupied* density of states.

Such a single-electron transfer picture of XAS gives data that is related to the specific site and symmetry-selected density of states. To illustrate this point, consider the Si-*L* edge absorption spectra (Fig. 3.9.1) for a gas (SiF_4) and a solid (SiO_2, quartz); in each case, four highly electronegative atoms tetrahedrally coordinate the silicon. The two spectra are very similar as they reflect transitions of electrons from the Si $2p$ levels to unoccupied states, confirming the importance of local symmetry in the XAS spectra of both gas and solid phases. Note that, at this introductory level, we ignore the interaction between the excited electron and the core hole, the possible formation of *excitons*,[7] and the broadening of the XAS spectra because of the short lifetime (Fig. 2.3.1) of the core hole due to their decay either by fluorescence or Auger electron emission. Furthermore, XAS is similar to inverse photoemission, introduced in the last section, but with one difference. Unlike XAS, inverse photoemission samples all the unoccupied states and is not limited by the dipole selection rules.

In practice, XAS studies are best carried out using synchrotron radiation, where monochromatic X-rays suitable for XAS, with energies >800 eV, are generated by Bragg diffraction using a double-crystal *monochromator* (§7.9.5.2). However, for lower-energy (50–800 eV) X-rays, diffraction gratings (Fig. 3.5.1) are preferred. The absorption of monochromatic X-rays can be measured in two different ways (Fig. 3.9.2). In the transmission mode (a), the spectrum is the ratio of the beam intensity before and after it passes through the specimen. In general,

[7] An *exciton* is an electrically neutral quasiparticle arising from the bound state of an electron and a hole, attracted to each other by the electrostatic Coulomb force. It typically exists in insulators, semiconductors, or in some liquids.

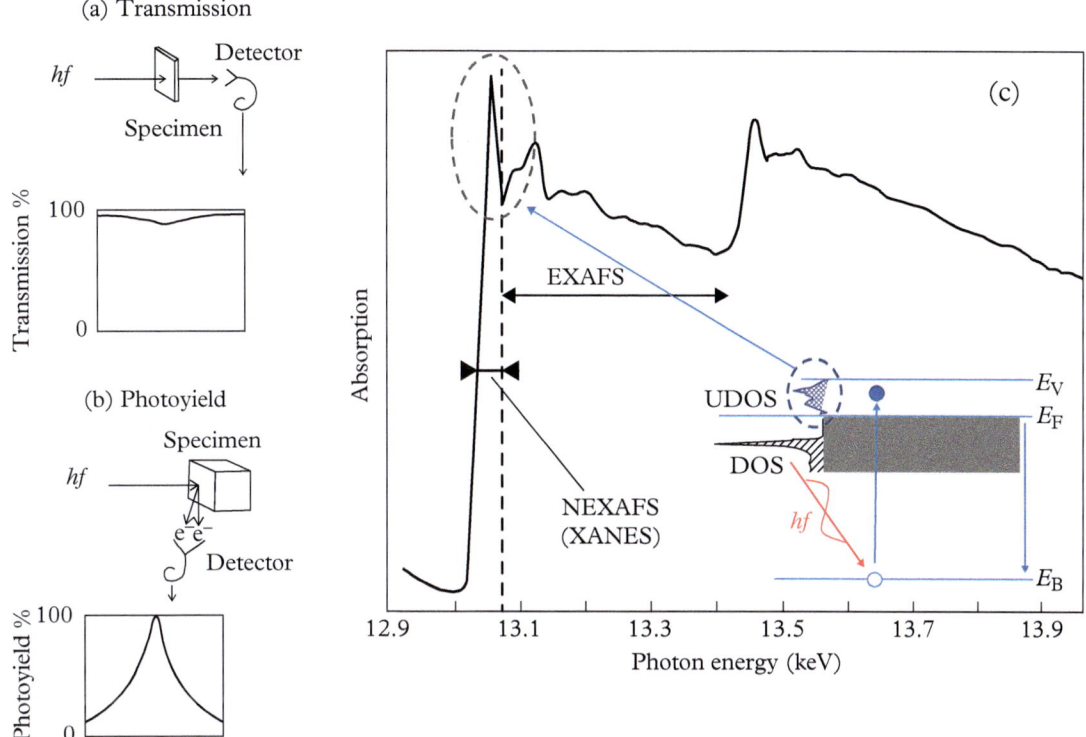

Figure 3.9.2 Schematic diagram of XAS measurements in (a) transmission mode through a thin film specimen, and in (b) the photo-yield mode. (c) A typical absorption L_3 edge of Pb in an oxide specimen, showing energy ranges of interest for NEXAFS and EXAFS. The inset shows an idealized energy-level diagram illustrating how the transitions to levels in the *unoccupied* density of states, subject to dipole selection rules, contribute to the near-edge structure.

this method is used to study deeply bound core levels, but for energies <800 eV, the absorption *cross-sections* (§5.3.5.1) increase rapidly with decreasing energy, and very thin specimens (~2–10 nm thick) without any pin holes are required. The alternative XAS measurements (b) is to use the photo-yield method, which measures the current of electrons and fluorescent radiation from the specimen. Here, it is generally assumed—and experiments bear this out—that the mean absorption depth of the photon is much larger than the mean escape depth of the photoelectrons (Fig. 1.3.6), and thus the photo-yield signal is proportional to the fraction of incident radiation absorbed within the surface. Moreover, since there is no energy selection, such yield measurements of XAS are more sensitive than PES. Furthermore, XAS has greater depth (>10 nm) sensitivity than PES (0.5–2 nm), because all electrons, including those that have suffered inelastic scattering events before escaping the specimen are collected and included in the

XAS signal. Recent developments in transmission electron microscopy (TEM) hardware, including field emission sources and monochromators, make it possible to obtain measurements equivalent to XAS with comparable energy resolution using electron energy-loss spectroscopy (EELS), but with much better spatial resolution (~1 nm). The underlying physical principles of XAS and transmission EELS in the forward scattering direction are the same, and we discuss this further in §9.4.

3.9.2   Near-Edge and Extended X-Ray Absorption Fine Structure (NEXAFS and EXAFS)

Figure 3.9.2c shows a typical X-ray absorption spectrum of the L_3 edge of Pb from a solid specimen of $BaPb_{1-x}Bi_xO_3$, illustrating two types of fine structure superimposed on the generic X-ray absorption edge profile (§2.4.2). The first one, typically extending ~50 eV above the edge, referred to either as near edge X-ray absorption fine structure (NEXAFS), or as XANES, is determined by the final density of states, the transition probability, and many body effects. The analysis of NEXAFS is quite complex, often requiring substantial theoretical effort (see Fuggle and Inglesfield, 1992). Therefore, rather than discuss this complicated topic, we restrict our discussion to the example of the "white lines" observed in $3d$ transition metals (Fig. 3.9.3) where the interpretation of the fine structure can be based on a simple one-electron picture. In addition to the near-edge structure, a second fine structure extending from ~50 eV above the absorption edge for several hundreds of eV, due to the interference effects of the wave function of the excited electrons propagating in the solid, is also observed. This is called the extended X-ray absorption fine structure (EXAFS). After a brief introduction to the white line transitions observed in NEXAFS, we conclude this chapter with a discussion of EXAFS.

Figure 3.9.3 X-ray absorption spectra, recorded by total electron yield detection, of the L_3 and L_2 edges of transition metals (Fe, Co, Ni, and Cu) with the gradual filling of their d-orbitals. Note that the white lines are absent in Cu because of its filled d shell.

Adapted from Schlachter and Wuilleumier (1994).

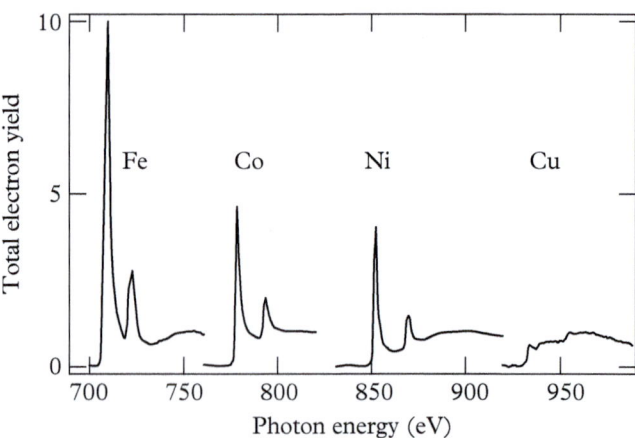

Figure 3.9.3 shows L-edge ($p \rightarrow d$) XAS spectra, satisfying dipole selection rules and obtained by total electron yield, for the ferromagnetic $3d$ transition metals (Fe, Co, Ni, and Cu). We observe separate L_3 ($2p_{3/2} \rightarrow 3d_{unoccupied}$) and L_2 ($2p_{1/2} \rightarrow 3d_{unoccupied}$) transition features, because of the strong spin-orbit splitting of the $2p$ levels (§2.2) in the edges for all elements except Cu (absent, because of the $3d^{10}$ configuration).

The $p \rightarrow d$ transition, which is divided into three different cases depending on how the d valence states are treated, can be understood in the one-electron picture by referring to Figure 3.9.4. In the first case, all 10 d states are assumed to be degenerate, with one of them being empty, and an electron is excited from either the $2p_{3/2}$ or $2p_{1/2}$ spin-orbit split core states into the d-hole, giving rise to the sharp L_3 and L_2 features in the XAS spectra. In the second case, we have assumed that the d states are also split into spin-up and spin-down states as a result of exchange interactions in the ferromagnetic elements. If an external magnetic field, H, is applied such that the spin-up states are lower in energy, but with a hole in the spin-down state that can accommodate a photoelectron, then the transition is allowed, as shown in (b). The third case (c) corresponds to the spin-orbit splitting of both the final $3d$ and initial $2p$ states. Now, the selection rule that applies is $\Delta j = 0, \pm 1$, and the $1/2 \rightarrow 5/2$ transition is not allowed. As a result of these various factors, in general, the ratio of the L_3/L_2 white line intensity ratio can change dramatically along the $3d$ transition metal series (Fig. 3.9.5). Further, the

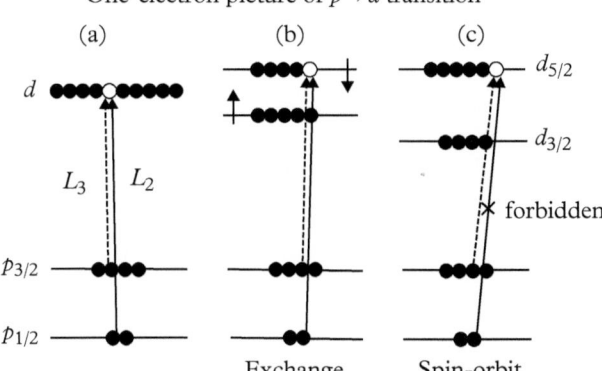

Figure 3.9.4 One-electron picture of white line transitions, $2p \rightarrow 3d$, in transition metals for three different consideration. (a) Spin-orbit interaction only in the $2p$ core level. (b) In addition to (a), the exchange interactions in the d shell, causing a splitting of the spin-up and spin-down states for the ferromagnetic elements is included. (c) The exchange interaction in the d shell is replaced by the spin-orbit interaction with splitting into $d_{5/2}$ and $d_{3/2}$ levels.

Adapted from Schlachter and Wuilleumier (1994).

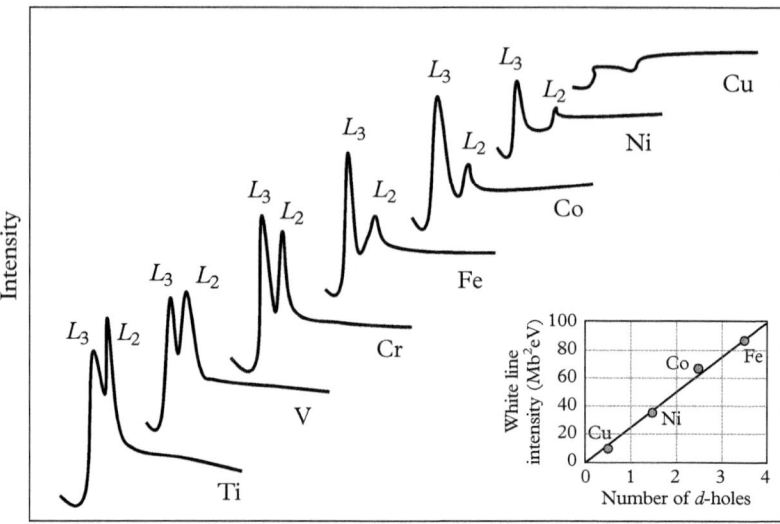

Figure 3.9.5 The ratio of the white line intensity, L_3/L_2, varies linearly with atomic number along the $3d$ transition metal series. Adapted from electron energy-loss measurements [14]; note that the photon energy is not to scale. Inset shows that the total white line intensity is linear with the number of d-holes.

Adapted from [15].

sum of the intensities of L_3 and L_2 white lines is proportional to the number of d-holes (Fig. 3.9.5, inset) and the white line ratio is also sensitive to the oxidation state of the transition metal (see §9.4).

The second, long-range oscillations (EXAFS) (Fig. 3.9.2c) have been developed into a useful, non-destructive, *element-specific* technique to directly probe the atomic environment of a particular X-ray absorber atom. It provides chemical bonding information (nearest neighbor distances and coordination number) for both crystalline and amorphous materials, and also for nonsolids such as molecules, surface adsorption species, etc. In principle, it complements diffraction (to be introduced in Chapter 4), as a tool for structural analysis.

The basic principles of EXAFS, using either inner (K, L) or outer (M, N) shell ionization, are quite easy to understand, even though its practical implementation requires sophisticated data analysis. As the X-ray photon energy approaches the binding energy characteristic of a core level of an element, absorption occurs and a sharp onset of the absorption edge is observed. As the photon energy is increased above the absorption edge, the core electron has sufficient energy to be excited to the continuum. The ionized photoelectron propagates as a wave (recall the wave-particle duality discussed in §1.3.3) and is scattered by the neighboring atoms (Fig. 3.9.6a). The outgoing wave from the absorber atom and the back-scattered wave from the neighboring atoms interfere, both constructively and destructively, to give the modulated EXAFS spectrum.

The EXAFS spectrum depends on the local atomic structure distribution in the immediate environment of the absorber atom (Fig. 3.9.6b); in particular, the frequency of EXAFS oscillations depends on the *distances* between the absorber and backscattering atoms, and its amplitude depends on the *number, type, and*

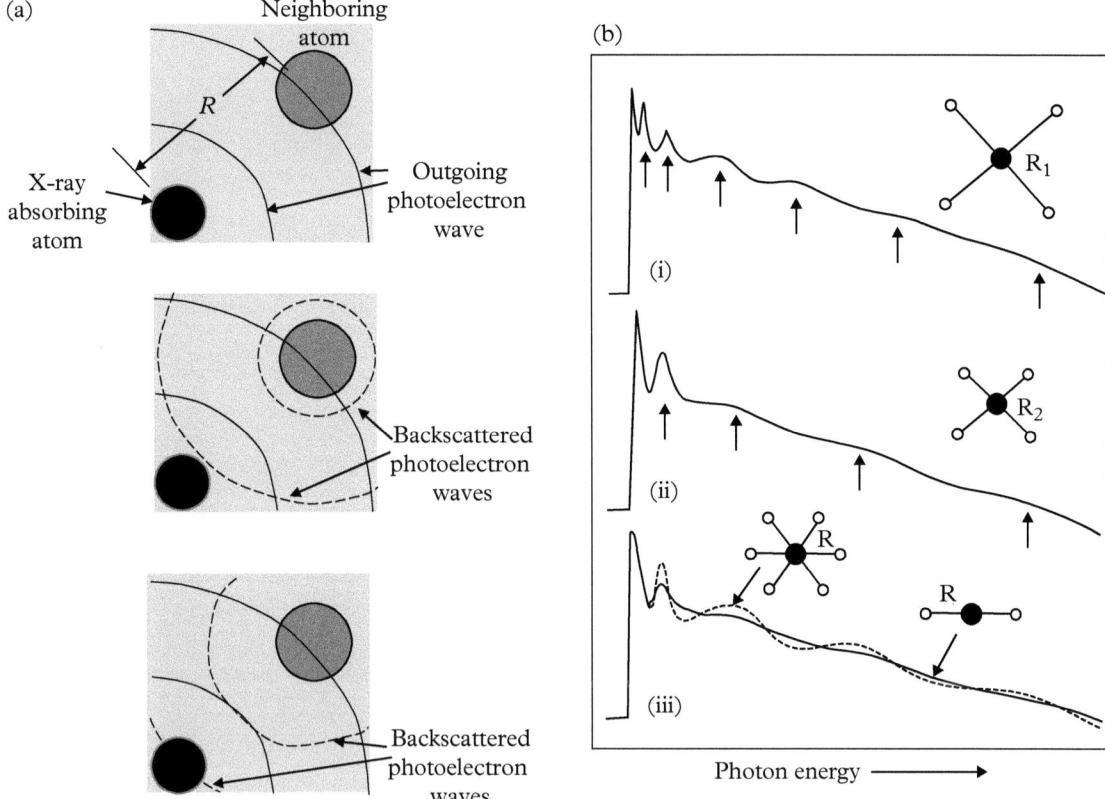

Figure 3.9.6 (a) A schematic illustration of the phenomenon of EXAFS. Top: The outgoing photoelectron wave from the absorber atom encounters a nearest neighbor atom, at a distance, R, in the crystal. Middle: Destructive interference of the backscattered and outgoing photoelectron waves. Bottom: Constructive interference at the site of the absorber atom. (b) The effect of nearest neighbor distance, (i) and (ii), and coordination number, (iii), on EXAFS. The frequency of the oscillations is related to the distance, R; when $R_2 < R_1$, we observe a longer EXAFS period. The amplitude of EXAFS is related to the coordination number; when the number of backscattering atoms increases, the larger is the EXAFS amplitude (here, 6 and 2 nearest neighbors are compared).

Adapted from Stohr (1981).

arrangement of back-scattering atoms. It is also possible to identify the back-scattering atoms chemically, as they have a unique way of altering the phase of the back-scattered wave.

The analysis of EXAFS data can be understood following Figure 3.9.7, which shows the absorption spectrum for the Mo-K edge. First the shape of the edge is fitted to a polynomial function ($\ln I_0/I_t$ versus E, in Figure 3.9.7a) to obtain the modulatory function, $\mu(E)$. The smooth absorption background function,

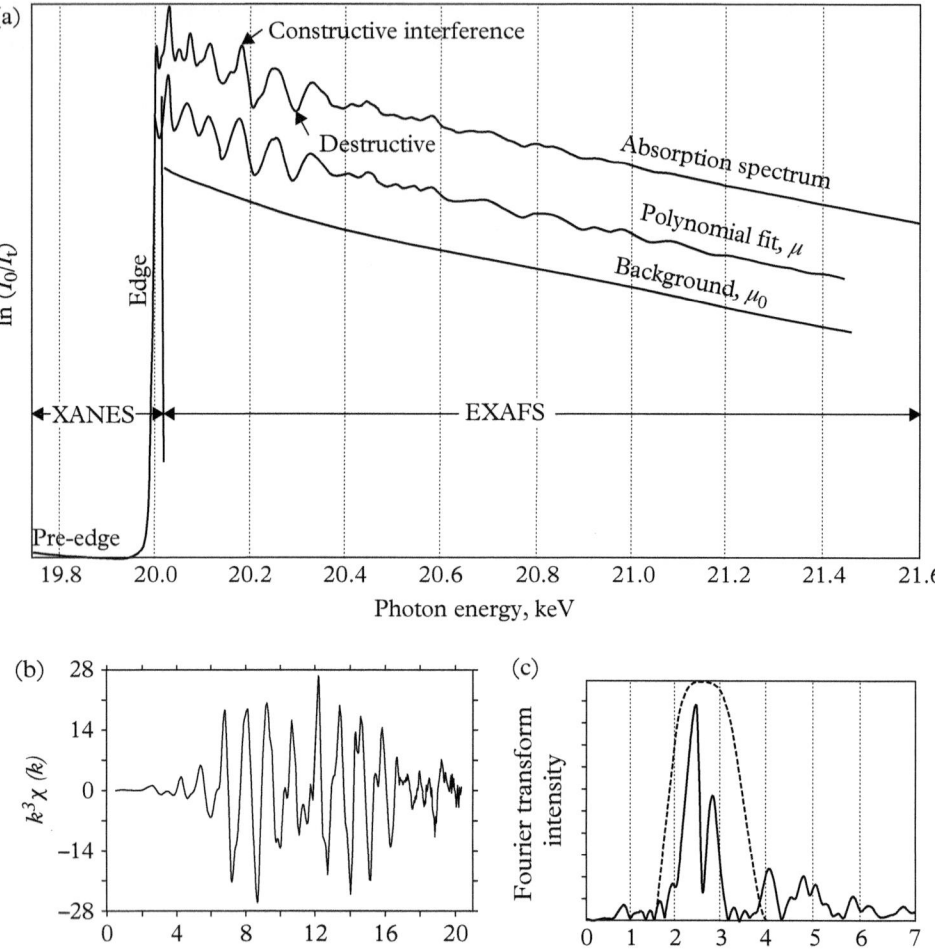

Figure 3.9.7 (a) Mo K-edge absorption spectrum obtained by transmission through a thin metal foil at 77 K with synchrotron radiation. The fitted polynomial modulatory function, $\mu(E)$, and the smooth background function, $\mu_0(E)$, both shifted vertically for clarity are also shown. (b) Background subtracted and k^3 weighted EXAFS spectrum, $\chi(k)k^3$ versus k, based on the Mo K-edge data in (a). (c) Fourier transform of (b), as a function of the effective interatomic separation, R', observed; this does not include the phase shifts. Two large peaks, observed at 2.38 Å and 2.78 Å, and two small peaks observed at 4.04 Å and 4.77 Å, are consistent with interatomic spacing in Mo, provided a phase shift of 0.37 ± 0.2 Å is introduced for all backscattering events.

$\mu_0(E)$, is then subtracted from $\mu(E)$ to give the EXAFS oscillations as a function of energy

$$\chi(E) = (\mu(E) - \mu_0(E)) / \mu_0(E) \tag{3.9.1}$$

The energy (eV) is now converted to a wave-vector (k) scale, where

$$k = 2\pi \frac{\sqrt{2m(E - E_0)}}{h} = 0.263(E - E_0)^{1/2} \tag{3.9.2}$$

to give $\chi(k)$, the oscillating portion of the absorption coefficient as a function of the wave-vector. Note that E_0 is the energy threshold at which the onset of the K-edge is observed (for Mo-K it is 20,002 eV), chosen to define the origin of the EXAFS spectrum. In other words, $k = 0$, when the incident X-ray photons have energy $E = E_0$, and the photoelectrons have no kinetic energy. Figure 3.9.7b shows EXAFS data, $\chi(k)$, which are then normalized by multiplying by k^n to compensate for amplitude attenuation as function of k (the value of $n = 1, 2$, or 3, and is determined empirically, based on best fit to the data), and $\chi(k)\,k^3$, for the Mo-K edge. Finally, the data is Fourier transformed to provide a pseudo-radial distribution function, R', of the atoms around the absorber atom (Fig. 3.9.7c). In general, when compared to the true distances in the solid, the peaks observed in the Fourier transform are shifted by a small distance. This is because of the element specific phase shift introduced in the backscattering process. For the case of Mo, the observed distances, R', given by the peaks in Figure 3.9.7c, are at 2.38 Å, 2.78 Å, 4.04 Å, and 4.71 Å. The actual Mo–Mo distances, R, in the metal are 2.73 Å, 3.15 Å, 4.45 Å, and 5.22 Å. Applying a uniformly small phase shift of 0.37 ± 0.2 Å gives a good fit between experiment and all true nearest neighbor distances.

An important example of the application of XAS is in understanding the photosynthesis reaction [16]. It helps establish the relationship between structure and function of metalloproteins, such as Mn, Ca, and possibly Cl, which are the metal sites in the oxygen-evolving complex (OEC) involved in photosynthesis. In particular, the role of the Mn_4Ca cluster in photosynthesis (PS II) is of special interest. Figure 3.9.8 shows the Mn K-edge from spinach PS II, highlighting the EXAFS region of the spectrum. Following the analysis described in Figure 3.9.7, the first observed oscillatory EXAFS contribution, $k^3\chi(k)$, is shown in Figure 3.9.8a (inset), which is then Fourier transformed to obtain the apparent distances (Fig. 3.9.8b) between the atoms of interest. The EXAFS is interpreted in real space as shells at 1.8 Å (Peak I) attributed to O or N atoms, and a second shell at 2.7 Å (Peak II) attributed to Mn–Mn interactions. Further, an additional shell for Mn and Ca is also observed at ~3.35 Å (Peak III). By tracking the EXAFS changes [16] as a function of the photosynthesis reaction, such measurements showed that the Mn_4Ca complex (OEC) undergoes structural changes that are triggered by the changes in oxidation state and protonation or deprotonation events.

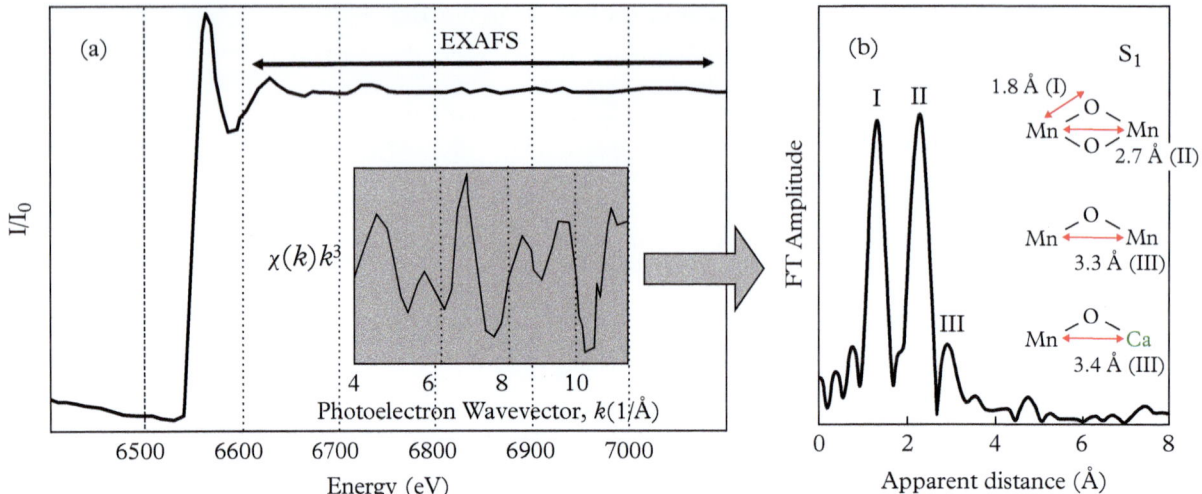

Figure 3.9.8 (a) Mn K-edge EXAFS spectrum from a spinach specimen used for studying the role of metal clusters in photosynthesis. The inset shows the spectrum in *k* space. (b) The Fourier transform of the *k* space EXAFS spectrum, gives the inter atomic distance of the other elements (Ca and O) with respect to Mn.

Adapted from [16].

Further details of the use of XAS in understanding photosynthesis can be found in [16].

It is clear from this example that EXAFS also provides useful information to biochemists and structural biologists. In particular, it provides *element-specific* structural information (distance between absorber and back-scatter atoms up to a distance of ~5 Å, with a precision of 0.02 Å) without interference from a periodic arrangement of atoms, that is independent of the state of the sample (the sample can be a powder, a solution, or specifically a frozen solution for biological materials), and at intermediate stages of a chemical process. In materials science, EXAFS is a versatile method that complements X-ray diffraction with unique advantages in studying *in situ* processes on materials surfaces, such as catalysis and electrochemical reactions. Further details can be found in specialized texts and monographs; for example, see Teo and Joy (1981).

3.10 Select Applications

3.10.1 Structure of Proteins Resolved by FTIR

We can generalize the result of Example 3.4.1 and state that in a protein if any specific atom is replaced by its isotope, the vibrational frequency of that specific bond will change proportional to $1/(\mu)^{1/2}$, where μ is the reduced mass of the two atoms involved in the bond. Thus, by substituting ^{1}H in proteins with deuterium, the structure of proteins can be resolved [19]. Further, in these experiments [19] the specific carbonyl group involved was substituted with ^{13}C and ^{18}O, such that the specific ^{13}C=^{18}O bond will give a specific signature (wavenumber) different from other C=O bonds.

The specific question addressed was the hydrogen bond between different amide groups in the protein (Fig. 3.10.1a–d). On deuterating the protein, the IR peak shifted to lower wavenumbers as the effective mass increased. By comparing with density functional theory calculations, it was found that the shift in the IR absorption peak to lower wavenumbers (Fig. 3.10.1e) for the ^{13}C=^{18}O bond was caused by the deuteration of positions (B) shown in red, but not in positions (A) shown in blue. In other words, noncovalent bonds, e.g. the protons in positions A, do not affect the C=O bonding, but the covalent bond (position B) influences the vibration frequency. Furthermore, they also analyzed the in-plane bending of the N–H bond at position B, under the influence of an external electric field; the interested reader is referred to [19] and the related supplementary information.

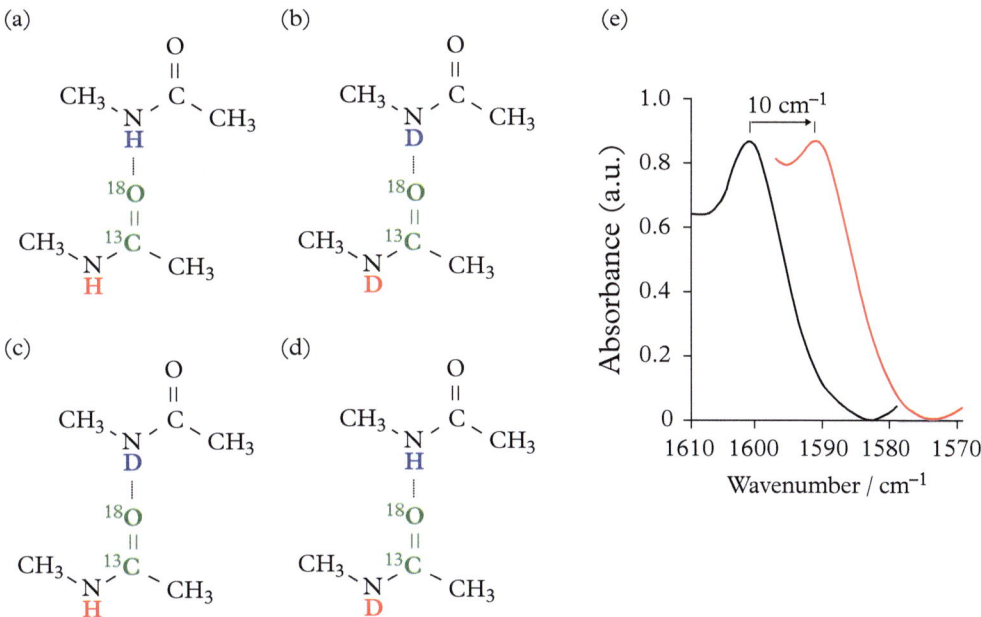

Figure 3.10.1 (a–d) Possible arrangement for the deuteration of the protein. Each system has two amide carbonyl groups, with the isotopically labeled one, participating in the hydrogen bond, shown in green. Further the hydrogen-bonding hydrogen (Position B) is shown in blue, and the amide hydrogen (Position A) is shown in red. (e) shows the change in IR absorption between (a) and (b).

Adapted from [19].

3.10.2 Analysis of Catalytic Particles by XAS and XPS

The structure, chemistry, and oxidation states of catalytic particles are key to their functionality. To purify automobile exhaust, Pt-based three-way catalyst (TWC) nanoparticles are used. However, above 800°C, these Pt nanoparticles sinter, decreasing their surface area available for reactions in oxidative environments. XAS was used to study mechanisms to inhibit such sintering [20]. Pt/cereia-based catalysts (referred to as CZY) did not sinter at 800°C, but Pt/Al$_2$O$_3$ did. The Pt L$_3$ edge was studied, and both near-edge structure (XANES) and extended fine structure were analyzed. The white lines observed at the onset of the Pt L$_3$ edge (Fig. 3.10.2a) reflects the vacancy of the $5d$ orbitals of Pt, and its intensity can be correlated with the Pt oxidation state. Thus, we can conclude that Pt/Al$_2$O$_3$, with white line features similar to Pt foil, is in the Pt^0 (metallic) state after aging. On the other hand, Pt/CZY, shows white line features similar to PtO$_2$, thus indicating that Pt^{2+} and Pt^{4+} species were present in these aged samples. If the white line intensity is linear with oxidation state, the aged Pt/CZY catalysts can be expected to have an average oxidation state of 3.53. Figure 3.10.2b shows EXAFS data, after Fourier transformation, for the same set of samples. The peak at 2.76 Å for the Pt foil is assigned to the Pt–Pt bond, and the peaks at 2.04 Å and 3.10 Å for the PtO$_2$ powder are assigned to the Pt–O and Pt–O–Pt bonds, respectively. Again, the Pt/Al$_2$O$_3$ data (FT spectrum) agrees with that of the Pt foil. On the other hand, the aged Pt/CZY specimen shows data different from both Pt foil and PtO$_2$ powder. The peak at 2.02 Å is close to the Pt–O bond length in PtO$_2$ powder, but the peak at 3.01 could only be fit to Pt–O–Ce bond after considerable simulation [20], giving an indication of how/why sintering is inhibited in Pt/CZY catalysts.

In other metal catalysts it is important to maintain the metal oxidation state, even upon exposure to air, as the catalyst facilitates redox reactions. XPS was used to analyze the stability of Co(Ni)MoS catalysts before and after exposure to air for three minutes (Fig. 3.10.2c,d). It is clear from the figure(s) that the relative amount of CoMoS decreased, while the oxide phases, Co$_9$S$_8$, and Co(II)-oxide increased. Further details can be found in the original manuscript [21].

Summary

When atoms form molecules and solids, their electronic structure are significantly altered, particularly the energy levels of those outer electrons involved in the bonding. In addition, the vibrational and rotational degrees of freedom in molecules modulate their electronic structure. As a result, molecules and polymers exhibit a combination of electronic (2–10 eV, UV-Vis), vibrational (10^{-2}–2 eV, IR), and rotational (10^{-5}–10^{-3} eV, microwave) energy levels that can be probed by appropriate spectroscopy methods.

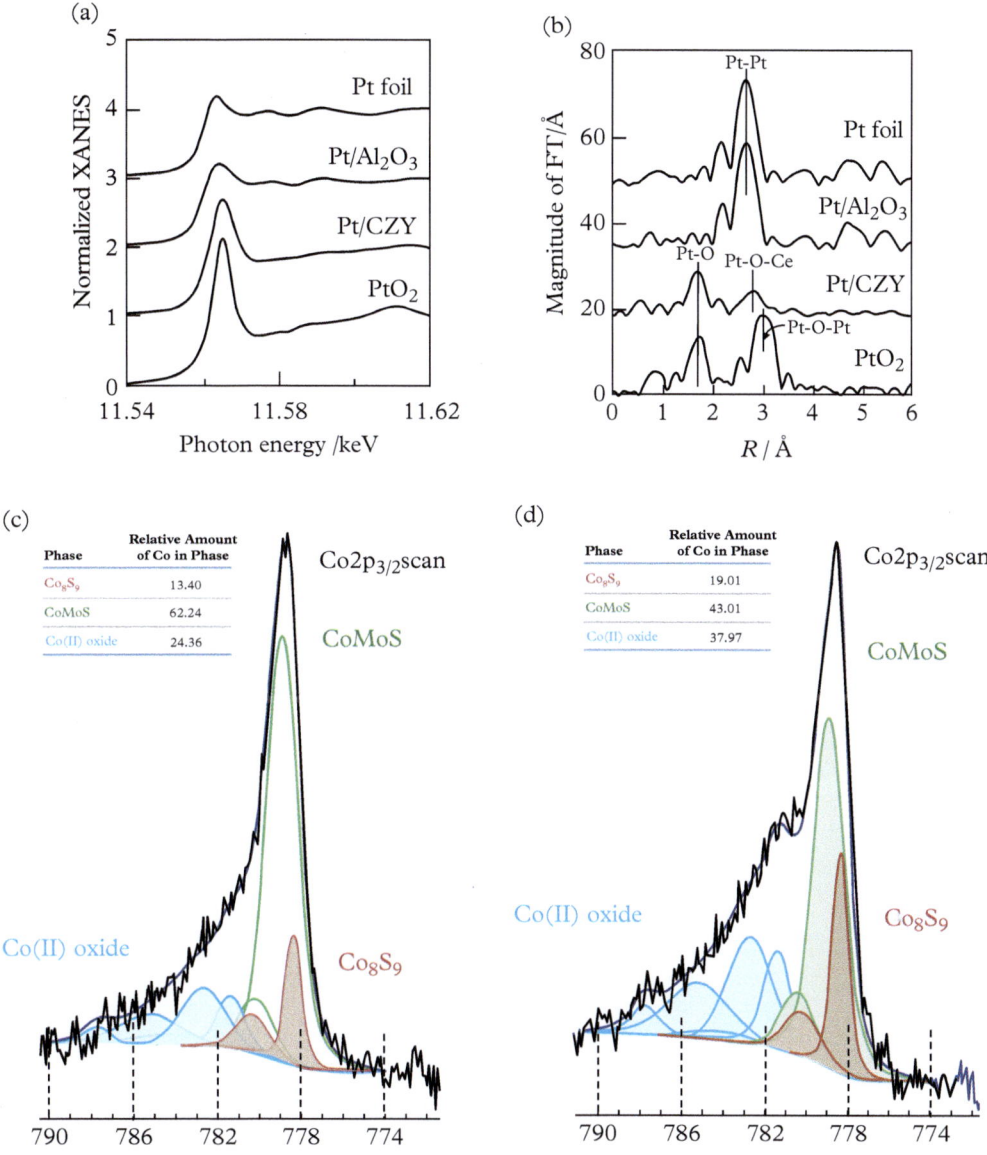

Figure 3.10.2 (a) Pt L_3 XANES and (b) Fourier transformed k^3x data for Pt L_3 EXAFS, for Pt supported catalysts aged at 800°C in air, together with standards for Pt metal and PtO_2. Co $2p_{3/2}$ XPS spectrum of (c) fresh and (d) exposed to air, catalyst of CoMoS particles. Note that the Co $2p_{3/2}$ peak was deconvoluted into component peaks, which requires some prior knowledge of the composition of the specimen.

(a, b) Adapted from [20]. (c, d) Adapted from [21].

Light, especially in the IR region, incident on a molecule or molecular solid can be either absorbed or scattered. The scattering can either be elastic (Rayleigh scattering) or inelastic (Raman scattering), the latter resulting in a signal that exhibits either a smaller (Stokes scattering) or larger (anti-Stokes scattering) frequency. Unlike the single photon involved in an IR absorption measurement, Raman scattering is a two-photon process. First, the probing photon is absorbed raising the molecule to a virtual excited state; this is followed by a decay and emission of a second photon of either the same (Rayleigh) or altered (Raman) frequency. The change in frequency arises primarily from the internal modes of quantized vibration and secondarily from the rotation of the molecule. Further, IR absorption and Raman scattering are complementary in nature. For a given molecule, if the derivative of its dipole moment with respect to the vibration mode is nonzero, it can show IR activity; alternatively, if the derivative of its polarizability with respect to the vibration mode is nonzero, it can show Raman activity. Signals in the IR region of the electromagnetic spectrum can be detected by dispersion using gratings or interferometry methods. The latter detects all wavelengths at the same time (parallel detection) with much higher signal collection speeds. These two spectroscopy methods find wide use in a range of technological applications involving polymers, and include the ability to monitor the curing of fiber-reinforced polymer composites increasingly used in the manufacture of fuselages of commercial airplanes.

A different class of spectroscopy methods provides details of the electronic structure of a crystalline solid using either X-rays or electrons as probes. In particular, they include X-ray or UV photoemission (PES, probing occupied levels) and inverse photoemission (probing unoccupied levels) spectroscopy. The energies of the probes or signals involved make these methods particularly suited for surface analysis. In addition, if the photoelectron intensity in PES is also measured in an angle-resolved manner (ARPES), both its energy and momentum can be determined simultaneously, allowing the complete determination of the band structure of the solid.

Alternatively, the unoccupied electronic states of the material are probed by XAS that follows dipole selection rules. The fine structure at the onset of the characteristic X-ray edge XANES provides information on the final density of unoccupied states, the transition probabilities, and many body effects. Further, the long-range oscillations extending from $\sim$50 eV for hundreds of eV (extending X-ray absorption fine structure, EXAFS) from the onset of the characteristic absorption edge, arise from interference effects of the wave scattered by the absorber atom with those back-scattered from neighboring ones. Thus, analyzing EXAFS data provides important *element-specific* structural information, e.g. nearest neighbor distances and their coordination number. A significant advantage of EXAFS over diffraction methods (§4, §7, and §8) is that it does not require a periodic crystal, and as such finds particular use in dynamic measurements of reactions and processes in structural biology, biochemistry, catalysis, and electrochemistry.

....

FURTHER READING

Bernath, P. F. *Spectra of Atoms and Molecules*. Oxford: Oxford University Press, 1995.

Borg, R. J., and G. J. Dines. *The Physical Chemistry of Solids*. New York: Academic Press, 1992.

Callister, W. D. Jr., and D. G. Rethwisch. *Materials Science and Engineering*. Hoboken: Wiley, 2010.

Cox, P. A. *The Electronic Structure and Chemistry of Solids*. Oxford: Oxford University Press, 1987.

Cothup, N. B., L. H. Daly, and S. E. Wiberley. *Introduction to Infrared and Raman Spectroscopy*. New York: Academic Press, 1990.

Duckett, S., and B. C. Gilbert. *Foundations of Spectroscopy*. Oxford: Oxford University Press, 2000.

Egerton, R. F. *Electron Energy-Loss Spectroscopy in the Electron Microscope*. Boston: Plenum Press, 1996.

Ferraro, J. R., K. Nakamura, and C. W. Brown. *Introductory Raman Spectroscopy*. New York: Academic Press, 2003.

Flewitt, P. E. J., and R. K. Wild. *Physical Methods for Materials Characterization*. Boca Raton: IOP, 2003.

Fuggle, J. C., and J. E. Inglesfield, eds. *Unoccupied Electronic States: Fundamentals for XANES, EELS, IPS and BIS*. Berlin: Springer-Verlag, 1992.

Harwood, L. M., and T. D. W. Claridge. *Introduction to Organic Spectroscopy*. Oxford: Oxford University Press, 2003.

Hendra, P., C. Jones, and G. Warnes, *Fourier Transform Raman Spectroscopy*, New York: Ellis Horwood, 1991.

Hüfner, S. *Photoelectron Spectroscopy: Principles and Applications*. Springer, 2003.

Ibach, H., and H. Lüth. *Solid-State Physics*. Berlin: Springer, 1991.

Kittel, C. *Introduction to Solid-State Physics*. Chichester: Wiley, 1986.

Krishnan, K. M. *Fundamentals and Applications of Magnetic Materials*. Oxford: Oxford University Press, 2016.

Leng, Y. *Materials Characterization*. New York: Wiley-VCH, 2013.

Long, D. A. *Raman Spectroscopy*. New York: McGraw–Hill, 1977.

McCreery, R. L. *Raman Spectroscopy in Chemical Analysis*. New York: Wiley, 2000.

Papaconstantopoulos, D. A. *Handbook of Band Structures of Elemental Solids: From $Z = 1$ To $Z = 112$*, 2nd ed. New York: Springer, 1986.

Pettifor, D. *Bonding and Structure of Molecules and Solids*. Oxford: Oxford University Press, 1995.

Schlachter, A. S., and F. J. Wuilleumier, eds. *New Directions with Third-Generation Soft X-Ray Synchrotron Radiation Sources*. Dordrecht: Kluwer Academic Publishers, 1994.

Shackelford, J. F. *Introduction to Materials Science for Engineers*. New York: Prentice Hall, 1996.

Siegbahn, K. *ESCA: Atomic, Molecular, and Solid-State Structure Studied by Means of Electron Spectroscopy*. Uppsala: Almqvist & Wiksells, 1967.

Stohr, J. "EXAFS and Surface EXAFS: Principles, Analysis, and Applications." In *Emission and Scattering Techniques*, edited by P. Day. Dordrecht: D. Reidel Publishing, 1981.

Sproul, R. L., and W. A. Philips. *Modern Physics*. Hoboken: Wiley, 1980.

Sutton, A. P. *Electronic Structure of Materials*. Oxford: Oxford University Press, 1993.

Teo, B. K., and D. C. Joy. *EXAFS Spectroscopy*. New York: Springer, 1981.

Willard, H. H., L. L. Merritt Jr., J. A. Dean, and F. A. Settle Jr. *Instrumental Methods of Analysis*. Belmont: Wadsworth Publishing, 1988.

Watts, J. F. *An Introduction to Surface Analysis by Electron Spectroscopy*. Royal Society Microscopy Handbook 32. Oxford: Oxford University Press, 1990.

....

REFERENCES

[1] Jensen, E., and E. W. Plummer. "Experimental Band Structure of Na." *Physical Review Letters* 55 (1985): 1912.

[2] Plummer, E. W. "Deficiencies in the Single-Particle Picture of Valence-Band Photoemission." *Surface Science* 152/153 (1985): 162.

[3] Himpsel, F. J. "Exchange Splitting of Epitaxial fcc Fe/Cu(100) versus bcc Fe/Ag(100)." *Physical Review Letters* 67 (1991): 2363.

[4] Himpsel, F. J., E. Ortega, G. J. Mankey, and R. F. Willis. "Magnetic Nanostructures." *Advances In Physics* 47 (1998): 511–97.

[5] Schneider, C. M., and J. Kirschner. "Spin- and Angle-Resolved Photoelectron Spectroscopy from Solid Surfaces with Circularly Polarized Light." *Critical Reviews in Solid State and Materials Sciences* 20, (1995): 179–283.

[6] Fanconi, B. M. "Fourier Transform Infrared Spectroscopy of Polymers—Theory and Application." *Journal of Testing and Evaluation* 12, no. 1 (1984): 33–9.

[7] Brosseau, C. L., F. Casadio, and R. P. van Duyne. "Revealing the Invisible: Using Surface-Enhanced Raman Spectroscopy to Identify Minute Remnants of Color in Winslow Homer's Colorless Skies." *Journal of Raman Spectroscopy* 42, (2011): 1305–10.

[8] Hüfner, S., and G. K. Wertheim. "X-Ray Photoemission Studies of the 3d Metals from Mn to Cu." *Physics Letters A* 47, no. 5 (1974): 349–50.

[9] Turtle, R. R., and R. J. Liefeld. "Densities of Unfilled One-Electron Levels in the Elements Vanadium and Iron through Zinc by Means of X-Ray Continuum Isochromats." *Physical Review B* 7, (1973): 3411.

[10] Yu, K. J., J. C. McMenahim, and W. E. Spicer. "UPS Measurements of Molecular Energy Level of Condensed Gases." *Surface Science* 50 (1975): 149–56.

[11] van der Laan, G., C. Westra, C. Haas, and G. A. Sawatzky. "Satellite Structure in Photoelectron and Auger Spectra of Copper Dihalides." *Physical Review B* 23 (1981): 4369.

[12] Olmstead, M. A., R. I. G. Uhrberg, R. D. Bringans, and R. Z. Bachrach. "Photoemission Study of Bonding at the CaF_2-on-Si(111) Interface." *Physical Review B* 35 (1987): 7526.

[13] Bianconi, A. "Core Excitons and Inner Well Resonances in Surface Soft X-Ray Absorption (SSXA) Spectra." *Surface Science* 89 (1979): 41–50.

[14] Pearson, D. H., C. C. Ahn, and B. Fultz. "White Lines and d-Electron Occupancies for the $3d$ and $4d$ Transition Metals." *Physical Review B* 47, (1993): 8471.

[15] Stohr, J. "Exploring the Microscopic Origin of Magnetic Anisotropies with X-Ray Magnetic Circular Dichroism (XMCD) Spectroscopy." *Journal of Magnetism and Magnetic Materials* 200, (1999): 470–97.

[16] Yano, J., and V. Yachandra. "X-Ray Absorption Spectroscopy." *Photosynthesis Research* 102, no. 2–3 (2009): 241–54.

[17] Murray, C. B., D. J. Norris, and M. G. Bawendi. "Synthesis and Characterization of Nearly Monodisperse CdE (E = S, Se, Te) Semiconductor Nanocrystallites." *Journal of the American Chemical Society* 115 (1993): 8706–15.

[18] Hüfner, S., S. Schmidt, and F. Reinert. "Photoelectron Spectroscopy—An Overview." *Nuclear Instruments and Methods in Physics Research A* 547, no 1 (2005): 8–23.

[19] Brille, E., and I. Arkin. "Site-Specific Hydrogen Exchange in a Membrane Environment Analyzed by Infrared Spectroscopy." *The Journal of Physical Chemistry Letters* 9 (2018): 4059–65.

[20] Nagai, Y., T. Hirabayashi, K. Dohmae, N. Takagi, T. Minami, H. Shinjoh, and S. Matsumoto. "Sintering Inhibition Mechanism of Platinum Supported on Ceria-Based Oxide and Pt-Oxide–Support Interaction." *Journal of Catalysis* 242, no. 1 (2006): 103–9.

[21] Gandubert, A. D., C. Legens, D. Guillaume, S. Rebours, and E. Payen. "X-ray Photoelectron Spectroscopy Surface Quantification of Sulfided CoMoP Catalysts – Relation Between Activity and Promoted Sites – Part I: Influence of the Co/Mo Ratio." *Oil & Gas Science and Technology* 62, no. 1 (2007): 79–89.

. . . .

EXERCISES

A.   Test Your Knowledge

For each statement, identify ALL the correct possibilities.

1. Core and valence electrons
    (i) are the same; they are indistinguishable.
    (ii) are not affected by bonding.
    (iii) behave differently on bonding. Core electron energies are barely perturbed while valence electrons participate actively in the bonding.
2. For polyatomic molecules, the total number of molecular orbitals
    (i) depends on the type of bonding.
    (ii) depends on the temperature.
    (iii) is the same as the total number of valence orbitals of all the atoms involved in the bonding.
3. Energy bands in solids
    (i) are groups of electrons that make quantum music.
    (ii) have widths that depend on the strength of interactions between neighboring atoms.
    (iii) determine electronic properties of materials.
4. The density of states of a solid
    (i) determines whether it will float or sink in water.
    (ii) describes the distribution of electronic energy levels in the material.
    (iii) in the free electron model shows a $E^{1/2}$ dependence.
5. The three primary types of bonds
    (i) are ionic, covalent, and van der Waals.
    (ii) necessarily involve transfer or sharing of electrons.
    (iii) largely influence materials properties.
6. All bonding types
    (i) involve attractive and repulsive forces.
    (ii) have an attractive potential proportional to $1/R$.
    (iii) consider only short-range interatomic interactions.
7. Covalent and ionic bonds
    (i) are both directional bonds.
    (ii) are determined by optimal space-filling considerations.
    (iii) differ in both directionality of the bonds and the importance of space-filling considerations.
8. In molecules and solids
    (i) electronic, vibrational, and rotational energy levels are separated by the same magnitude.

(ii) the vibrational modes can be approximated by a classical harmonic oscillator model.

(iii) even at $T = 0$ K, the molecules vibrate with the zero-point energy.

9. Vibration modes in a molecule

(i) are related to the number of degrees of freedom.

(ii) involve stretching, bending, rocking, and scissoring.

(iii) are sensitive to temperature.

10. Molecular vibrations

(i) lead to Raman and infrared activity in ALL molecules.

(ii) with a derivative of the dipole moment lead to Rayleigh scattering.

(iii) with a derivative of the dipole moment lead to IR activity.

(iv) with a derivative of the polarization lead to Stokes scattering.

(v) with a derivative of the polarization lead to Raman scattering.

(vi) give the same intensity for both Stokes and anti-Stokes scattering.

11. Using a Michelson interferometer in FTIR allows

(i) signal modulation and a measurable frequency consistent with detector response.

(ii) higher signal collection speeds with improved signal-to-noise ratio.

(iii) allows for setting up the experiment inexpensively.

12. Resonance enhancement in Raman scattering

(i) increases the signal by orders of magnitude.

(ii) often produces additional bands in the signal.

(iii) in practice is affected by competing fluorescence effects.

13. In practical implementation of Raman spectroscopy

(i) doubling the focal length of the grating doubles the spectral resolution

(ii) lasers in the visible spectrum are avoided for safety reasons.

(iii) Rayleigh line is NOT a problem, and can be ignored.

(iv) the surface on which the molecules are adsorbed does not influence the signal.

14. The band gap of semiconductors can be determined by

(i) optical spectroscopy.

(ii) PES.

(iii) XAS.

15. Important energy levels in a solid for spectroscopy are

(i) vacuum and Fermi levels INSIDE the solid.

(ii) vacuum and Fermi levels OUTSIDE the solid.

(iii) the Fermi level inside and the vacuum level outside the solid.

16. The work-function

(i) defines the work involved in doing any spectroscopy measurement.

(ii) is the difference in energy between the vacuum and Fermi levels of a solid.

(iii) is of the order of 2–5 eV in most metals.

(iv) is of no practical importance, especially in electron guns.

17. It is best to use
    (i) PES to probe filled levels in the valance bands.
    (ii) inverse PES to probe unfilled levels in the valence band.
    (iii) XAS to probe unoccupied levels.
18. The convention for binding energy, E_B, is
    (i) arbitrary.
    (ii) such that $E_B = 0$, at $E = E_F$ in solids.
    (iii) such that $E_B = 0$, at $E = E_V$ in molecules.
    (iv) such that $E_B > 0$, for PES.
19. In PES, it is best to
    (i) use UV light to investigate core levels.
    (ii) use UV light to investigate the valence band.
    (iii) use UV light only if an appropriate X-ray source is not available.
    (iv) set the onset of the signal to be at $E = E_F$.
    (v) assume that there is no background from secondaries.
    (vi) use a shallow angle of incidence to enhance surface sensitivity.
    (vii) ignore the effects of ligands in interpreting the spectrum.
    (viii) interpret the data as sensitive only to the surface.
20. Inverse photoemission is
    (i) implemented in only one well-defined way.
    (ii) best used to study occupied states with $E < E_F$.
    (iii) best used to study unoccupied states with $E_F < E < E_V$.
    (iv) is complementary to PES.
    (v) probes only those levels allowed by dipole selection rules.
21. The Fermi–Dirac distribution
    (i) can be used to measure the resolution of a PES spectrometer.
    (ii) is only of academic interest to physicists.
    (iii) gives the probability of the occupancy of electronic levels as a function of T.
22. In XAS
    (i) dipole selections determine allowed transitions.
    (ii) data interpretation requires knowledge of BOTH initial and final states.
    (iii) data interpretation can be based on final unoccupied DOS.
23. XAS
    (i) is exactly the same as inverse PES.
    (ii) is similar to electron energy-loss spectroscopy.
    (iii) often requires a synchrotron radiation source.
    (iv) can only be measured in transmission through a very thin specimen.
    (v) has no depth sensitivity and is a surface analysis technique.
24. The fine structures in XAS
    (i) are of two kinds: near-edge and extended.
    (ii) cannot give any information on the bonding.

(iii) can give information on the nearest neighbor distances and coordination numbers.

B. Problems

1. **Classic harmonic oscillator**: for the harmonic oscillator describing the vibrations of two atoms, what will happen to the vibration frequency
   (a) if one of the atoms is replaced with one of heavier mass?
   (b) if the bonding becomes weaker, but keeping the masses of the two atoms the same?

   In addition, what is the reduced mass for
   (c) a homonuclear diatomic molecule, with atomic mass, m?
   (d) the molecule HX, where $m_X \gg 1$?

2. The **potassium hydride** molecule, K–H, has an equilibrium bond length of 2.32 Å. It undergoes a rotational transition from $J = 1$ to $J = 2$. What is the frequency of the photon that it needs to absorb to make this transition possible?

3. **X-ray sources**: The $2p$ peak in XPS for a given element using Al K_α (1487 eV) is at a kinetic energy of 700 eV. Instead, if Mg K_α (1,254 eV) were to be used, where will the same peak show up in the spectrum?

4. The **photoelectron energy spectrum** of an unknown specimen, using Mg $K\alpha$ irradiation at 1.250 keV is shown in Figure 3.Pr.4. The energy axis is $E_B = hf - E_{kin}$, and the vertical axis is the observed intensity in arbitrary units. Two regions of the spectrum have been enhanced for clarity.

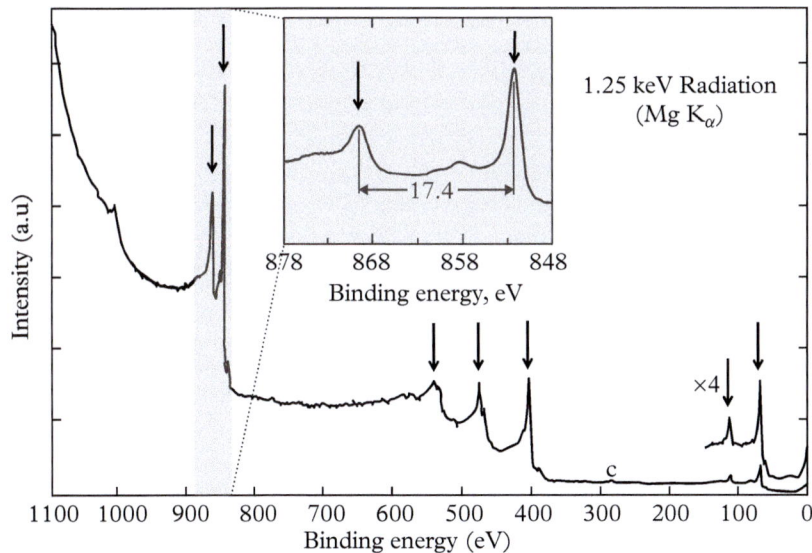

(a) Index all the major features indicated by arrows.

(b) Now, identify the material.

5. **FTIR with a Michelson interferometer**: your FTIR detector electronics can respond up to a maximum frequency of 1 kHz. The mirror in the interferometer can be moved within a velocity range of 0.1–0.75 cm/s.

(a) What is the range of incident photon wavelengths that can be used in this set-up?

(b) Will this instrument be capable of covering the mid-IR from the group-frequency to thefinger-printing regime?

6. The **X-ray absorption spectrum** (actually, this is EELS data, but for your purposes they are the same) of three forms of carbon (amorphous, graphite, and diamond) are shown in Figure 3.Pr.6.

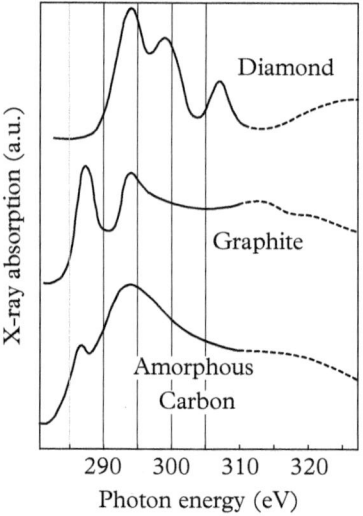

(a) Identify the edge (K, L, etc.).

(b) Draw a simple molecular orbital (MO) energy diagram of sp^2 and sp^3 hybridized carbon in graphite and diamond, respectively.

(c) See if you can associate the main features in the absorption spectrum with the levels in your MO picture.

7. Using the plot (Fig. 3.Pr.7a) that gives various possible characteristic bands in vibrational spectra for diatomic stretching vibrations, see if you can identify any of the features in the **IR** and **Raman spectra** of

(a) $CHCl_3$ in Figure 3.4.11

(b) styrene/butadiene rubber shown in Figure 3.Pr.7b.

If not, look up any reference "finger print" data, e.g. Colthup et al. (1990), and see if you can do better.

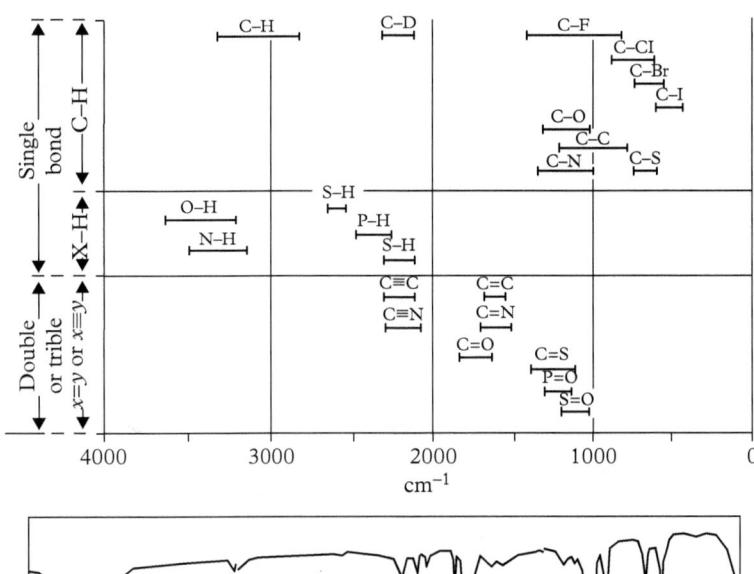

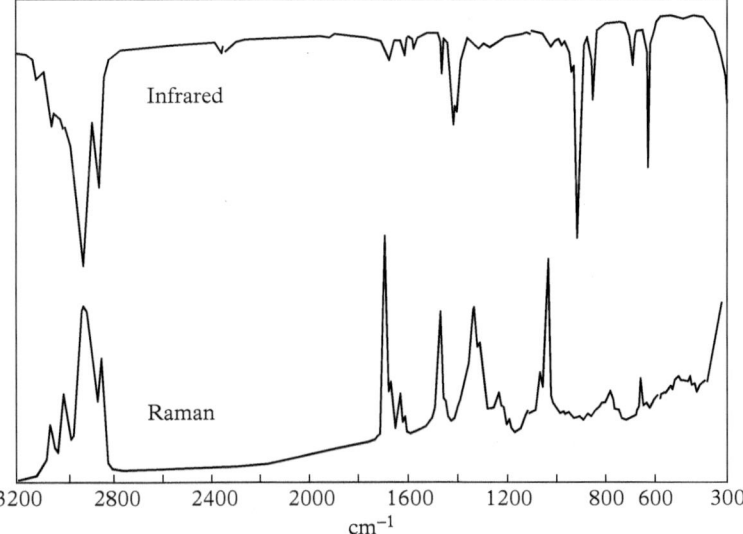

8. **X-ray photoemission spectroscopy** using Al Kα radiation. In your spectrum you see peaks at 942 eV, 778 eV, 765 eV, and 643 eV. What are the binding energies of these electrons? Identify the two elements present in your specimen. Hint: Use Table 2.2.2.

9. **Michelson interferometer**: An He–Ne laser based FTIR detector has a mirror that can be moved with a velocity in the range of 0.2–0.8 cm/s. What maximum response frequency is required for this device if it has to detect wavenumbers in the range 400–4,000 cm^{-1}? Will this device cover the mid-IR region, from the group frequency to the finger-printing regime?

10. Can **IR spectroscopy** detect a transition with an energy of 325 kJ/mol?

4

Crystallography and Diffraction

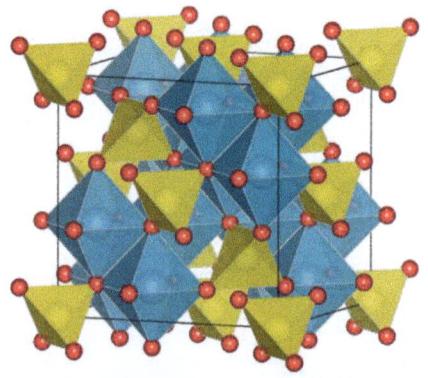

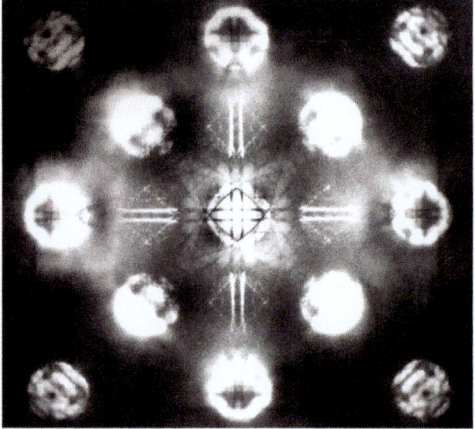

The structure and diffraction pattern of a spinel ($MgAl_2O_4$) crystal.

Top: Drawing of the cubic unit cell, in *real space*, where red spheres represent the oxygen atoms, Mg atoms sit within the yellow tetrahedral sites, and Al atoms sit within the blue octahedral sites. A unit cell contains eight formula units. The space group of this crystal is $Fd3m$.

Bottom: A transmission electron diffraction pattern, in *reciprocal space*, obtained with a convergent beam (CBED) incident along the [001] direction of the spinel crystal. Such CBED patterns are discussed further in §8.6.3.

Principles of Materials Characterization and Metrology. Kannan M. Krishnan, Oxford University Press (2021).
© Kannan M. Krishnan. DOI: 10.1093/oso/9780198830252.003.0004

Chapter 3 introduced different types of bonding in materials and developed spectroscopic methods to analyze them. We saw that the type of bonding determines the local arrangements of atoms including their coordination number and interatomic distances. The bonding also determines the magnitude of a given property of the material, and the local symmetry or point group. The latter, in turn, defines the components of the tensor that relates the response of the materials to an applied stimulus (see Nye, 2000). An example is the second-rank, polarizability tensor, α_{ij}, relating the applied field, $\mathbf{E}$, and the dipole moment, $\boldsymbol{\mu}$, introduced in §3.4.2.

In some materials, the arrangement of the atoms determined by the bonding is limited to their immediate environment; materials with such short-range order are commonly referred to as *amorphous*, or formless, materials. In many other materials, the local atomic arrangement is spatially repetitive in a periodic way, and extends over the macroscopic extent of the crystal. Materials with such long-range order are traditionally called crystalline materials, but recent discovery of quasicrystalline materials (see §4.4) has led to a redefinition of what makes a material crystalline. Nevertheless, it is important to be able to rigorously describe the entire class of periodic crystalline materials, i.e. be familiar with the field of classical crystallography, as it is the foundation of structural characterization of materials by diffraction. Note that, even though it may initially appear that a very large number of periodic crystal structures are to be found in nature or synthesized as engineering materials, they can be classified in three dimensions into a small number of groups based on local, or point group symmetry (32 in number), and the patterns, or space groups (230 in total), they generate by periodic repetition.

Crystallography is a highly developed subject that is discussed in detail in a number of excellent books; see, among others, Allen and Thomas (1998), Borchard-Ott (1995), Buerger (1970), Hammond (2006), and Sands (1975). In the context of materials characterization and analysis, our interest and scope here is modest. We only wish to use crystallography as a language to describe crystalline materials in the context of developing diffraction/scattering methods to elucidate their structure and symmetry in appropriate detail.

In this chapter, after an overview of periodic crystal structures, we discuss the important concept of a *reciprocal* lattice. Then, we introduce the basic principles of diffraction in both real and reciprocal space and discuss it qualitatively for both X-ray and electron incidence. We conclude this chapter with a very brief introduction to an interesting family of materials, called quasicrystals, discovered by Professor Dan Schechtman.[1] Detailed quantitative treatment of diffraction, and different scattering geometries follows for X-rays in §7, and electrons and neutrons in §8.

[1] Dan Shechtman (1941–) won the Nobel prize in Chemistry in 2011, and is cited for "the discovery of quasicrystals."

4.1   The Crystalline State

We begin with some simple definitions.

A *crystal* is a homogeneous body composed of a *periodic arrangement* of atoms, ions, or molecules, in three dimensions with anisotropic properties. By anisotropic properties we mean that the properties of a crystal are different when measured along different directions. However, this is by no means universal, as there also exist isotropic crystals, where the physical property is the same when measured in any direction.

If the periodic arrangement in a solid is perfect, and extends uniformly throughout the entire sample in all directions, we refer to it as a *single crystal*. Such single crystals are common in nature, and high quality, defect-free ones are also synthesized artificially in the laboratory. In materials research we also encounter them as thin films grown *epitaxially* on other single crystal substrates. However, most engineered crystalline solids used in practice are made up of small units of single crystals, or *grains*, on the micrometer length scale. These materials are termed *polycrystalline*, and they are characterized by random crystallographic orientations; the regions where different grains meet in a solid are called *grain boundaries* (Fig. 4.1.1; §4.1.9.3). In some cases, the grains in a polycrystalline material exhibit a preference for a particular crystallographic orientation (these

Figure 4.1.1 Schematic representation of the stages in the solidification of a polycrystalline material; each square grid represents a unit cell: (a) Small nuclei. (b) Early stages of growth, with obstruction of some grains. (c) After solidification with the formation of irregular-shaped grains; the size of the grains determines the extent of *long-range order*, and also affects the full-width at half maximum (FWHM) of diffraction peak intensities. (d) Grain boundaries of (c) visible in an optical microscope. Further, the composition and structure of individual grains and grain boundaries can be characterized, at atomic resolution, by various transmission electrons microscopy (§9) methods.

Adapted from Callister and Rethwisch (2009).

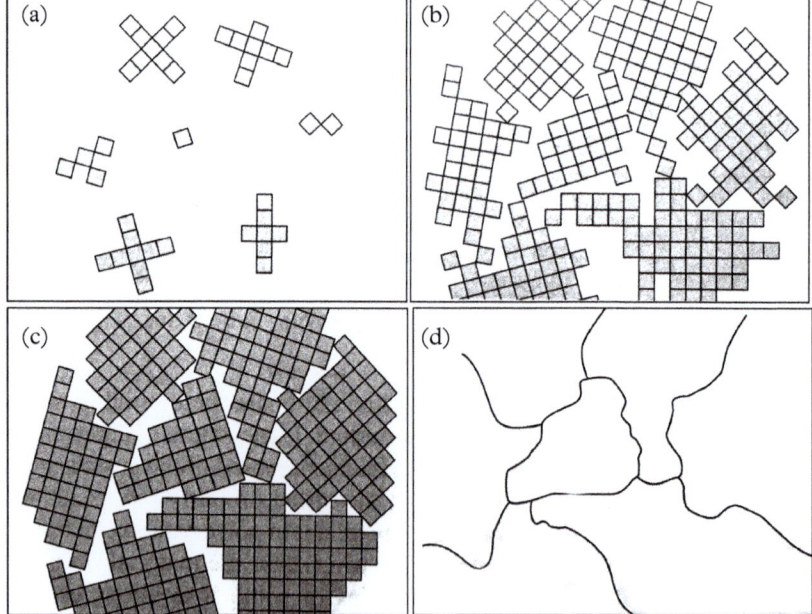

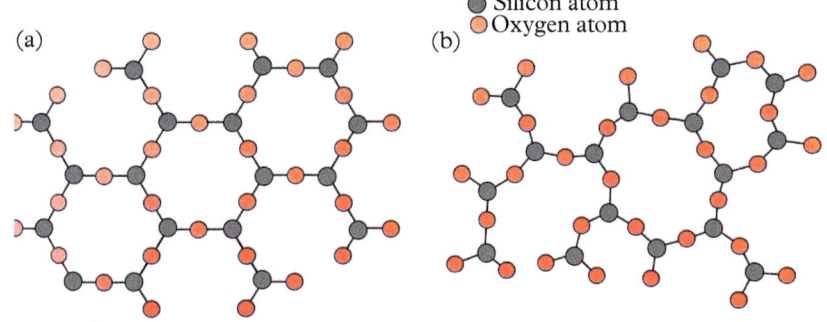

Figure 4.1.2 A schematic representation of SiO_2 in (a) crystalline form, and (b) non-crystalline form. Note that the nearest neighbor coordination in both cases are the same.

Adapted from Callister and Rethwisch (2009).

terms will be defined in §4.2), and such materials are said to have a *texture*. As we will see, diffraction methods probe the long-range order (LRO) in the solid, and the grain size, which defines the spatial extent of the long-range order, will determine the broadening (FWHM) of the diffraction peaks (§7.8).

If the temperature is low enough, and given sufficient time, all materials will eventually crystallize, since the ordered crystalline state is the one with the lowest energy. However, some solids, such as *glass*, never reach the lowest energy state; thus, they have higher energy than the equivalent crystalline state and are referred to as *noncrystalline* or *amorphous* materials. In such amorphous materials, the building blocks defined by the bonding are retained (Fig. 4.1.2) and the short-range order (SRO) is preserved, but the crystallinity or long-range order is not developed. Such SRO (§7.11.5) has its own signature in the form of diffuse intensities in diffraction patterns (Warren, 1990).

4.1.1 Lattices

Regularity and repetition are the central features of a periodic crystal structure. To appreciate these factors, the atoms in the crystal can be replaced by points, such that each point in space has a fixed relationship to the atoms in the crystal. In this manner, we can generate a *space* lattice, defined as an *array of points arranged in space with identical surroundings*; Figure 4.1.3a shows it in its most general form. This lattice can also be visualized as a repeating set of *unit cells*, in the form of parallelepipeds (Fig. 4.1.3b), which are identical in size, shape, and orientation with respect to their neighbors. Now, there are two ways to describe the size and shape of the unit cell. One way is to describe it using a set of three, non-coplanar vectors (**a**, **b**, and **c**) drawn from one corner of the unit cell and defined as the crystallographic axes of the cell. Further application of the lattice translation to these vectors, i.e. $P\mathbf{a} + Q\mathbf{b} + R\mathbf{c}$, where P, Q, and R, are integers, will generate all the points of the space lattice. Alternatively, the repeating unit cell geometry is completely defined in terms of six parameters, i.e. the three edge lengths (a, b, and c), and the three inter-axial angles (α, β, and γ) between them (Fig. 4.1.3b).

Figure 4.1.3 (a) A space lattice with one unit-cell highlighted (b) The highlighted unit cell with all the six parameters indicated.

Note: a two-dimensional or surface lattice can be defined by only three variables, a, b, and γ. See Figure 4.1.15 and §8.3.1.

Adapted from Cullity (1978).

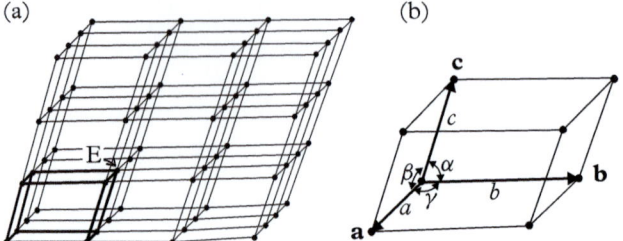

Note that every unit cell has eight identical vertices and six faces. However, a lattice point at any one of the vertices (say E, in Fig. 4.1.3a), is shared by eight unit-cells that meet at that point. Thus, only one eighth of the lattice point at any of the eight vertices of the parallelepiped may contribute to a specific unit cell. Since, there are eight vertices or corners, N_C, each contributing one eighth, in effect the unit cell contains a single lattice point and is defined as a single or *primitive unit cell*, with the crystallographic symbol, P.

Now, we generalize this concept to classify all crystals in three dimensions into fourteen Bravais lattices and seven crystal systems.

4.1.2   Generalized Crystal Systems and Bravais Lattices

The unit cell geometry, completely defined by the six parameters (a, b, and c, and α, β, and γ) in Figure 4.1.3, gives rise to seven possible combinations, each of which represents a distinctly different crystal system. These crystal systems, in decreasing order of symmetry, where cubic[2] ($a = b = c$ and $\alpha = \beta = \gamma = 90°$) is the highest, and triclinic ($a \neq b \neq c, \alpha \neq \beta \neq \gamma$) is the lowest, are listed in Table 4.1.1. Note that the rhombohedral (R) system, can be considered as a subset of the hexagonal system; in this case, the number of crystal systems reduces from seven to six.

By periodically repeating the unit cells of the seven crystal systems, we can generate seven different primitive lattices (if rhombohedral (R), is excluded, then we have six primitive lattices). However, as defined, the primary requirement of a space lattice is that each point has identical surroundings. Now, additional variations (nonprimitive) of these seven lattices can be generated, but it turns out that there are only seven more nonprimitive possibilities. Thus, in total we have fourteen Bravais[3] lattices (Fig. 4.1.4). In addition to the simple or primitive (P) lattice, already discussed in the last section, additional face centered (F), body centered (I), and base centered (A, B, or C) lattices are introduced. The F-lattice includes additional lattice points, N_F, occupying face centers and shared by two unit-cells, and the I-lattice includes an additional lattice point, N_I, at the exact center of the unit cell that is not shared with any other. In general, the total number of lattice point, N, per unit cell is given by

[2]   In any system, if $a = b = c$, then they are all given the same symbol, a, to indicate the equality.

[3]   A. Bravais (1811–1863) was a French mathematician and crystallographer.

Table 4.1.1 Crystal systems (7) and Bravais lattices (14), their characteristic symmetries, and point groups (32), in three-dimensions.

System	Bravais lattice	Axial lengths and angles	Characteristic (minimum) symmetry	Noncentrosymmetric point groups	Centrosymmetric point groups
Cubic	Simple, P Body centered, I Face centered, F	$a = b = c$ $\alpha = \beta = \gamma = 90°$	**4** *threefold* rotation (triad) axes at 109.47°	23 432 $\overline{4}3m$	$m\overline{3}$ $m\overline{3}m$
Tetragonal	Simple, P Body centered, I	$a = b \neq c$ $\alpha = \beta = \gamma = 90°$	**1** *fourfold* rotation★ (tetrad) axes	4 422 $\overline{4}$ $4mm, \overline{4}2m$	$4/m$ $4/mmm$
Orthorhombic	Simple, P Body centered, I Face centered, F Base centered, C	$a \neq b \neq c$ $\alpha = \beta = \gamma = 90°$	**3** *twofold* rotation★ (diad) axes at 90° (set of **3** orthogonal mirrors)	222 mm2	*mmm*
Rhombohedral (Trigonal)	Simple	$a = b = c$ $\alpha = \beta = \gamma \neq 90°$	**1** *threefold* rotation★ (triad) axis	3 32 3m	$\overline{3}$ $\overline{3}m$
Hexagonal	Simple	$a = b \neq c$ $\alpha = \beta = 90°,$ $\gamma = 120°$	**1** *sixfold* rotation★ (hexad) axis	6 622 $\overline{6}$ 6mm $\overline{6}m2$	$6/m$ $6/mmm$
Monoclinic	Simple, P Base centered, C	$a \neq b \neq c$ $\alpha = \gamma = 90°$ $\neq \beta \geq 90°$	**1** *twofold* rotation★ (diad) axes	2 m	$2/m$
Triclinic	Simple	$a \neq b \neq c$ $\alpha \neq \beta \neq \gamma \neq 90°$	None	1	1

★ with or without inversion

Adapted from Hammond (2006).

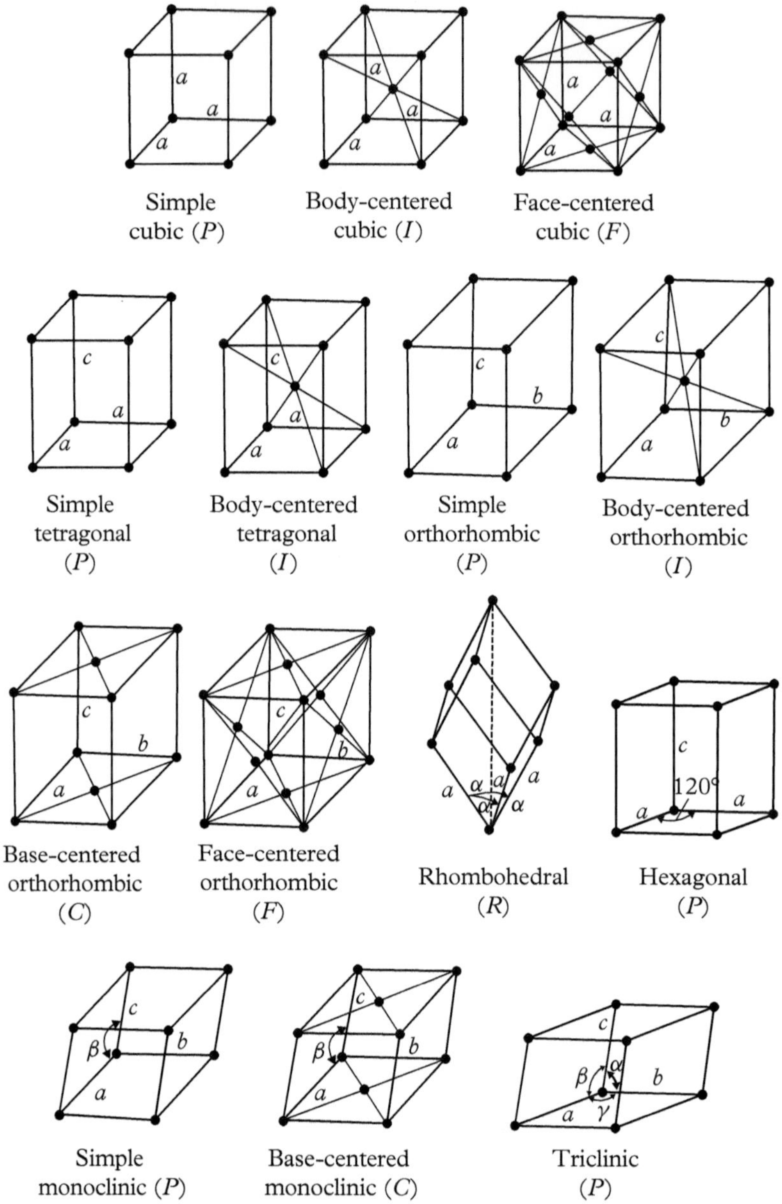

Simple cubic (P)

Body-centered cubic (I)

Face-centered cubic (F)

Simple tetragonal (P)

Body-centered tetragonal (I)

Simple orthorhombic (P)

Body-centered orthorhombic (I)

Base-centered orthorhombic (C)

Face-centered orthorhombic (F)

Rhombohedral (R)

Hexagonal (P)

Figure 4.1.4 The seven crystal structures that give rise to the fourteen Bravais lattices as presented in Table 4.1.1.

Simple monoclinic (P)

Base-centered monoclinic (C)

Triclinic (P)

$$N = N_C/8 + N_F/2 + N_I \qquad (4.1.1)$$

where N_C, as mentioned, is the total number of points on the corners of the unit cell.

In addition to I and F, possible nonprimitive arrangements (Fig. 4.1.4; Table 4.1.1) are the A, B, and C type, base centered lattices, where additional lattice points are centered on any one pair of faces (note, the A face is defined by the *b* and *c* axis, and the others are defined cyclically). Needless to say, the periodic application of the unit-cell vectors (**a**, **b**, and **c**) to all points, wherever they are located in the cell, can also extend to all the nonprimitive unit cells through space.

Real crystal structures are generated from the Bravais lattices by having atoms, ions, or molecules, occupy all the points of the space lattice. The arrangement of atoms associated with every lattice point is called a *basis*. Then, by applying the lattice translations that replicate the atoms throughout the lattice, a crystal structure is generated. In other words, we have

$$Lattice + basis = crystal \tag{4.1.2}$$

We should also point out that any of the non-primitive Bravais lattices can be referred to an equivalent primitive unit cell; Figure 4.1.5 shows an example for the face centered cubic (FCC) structure. This will be important later in §7 and §8 as we "index" diffraction patterns of materials.

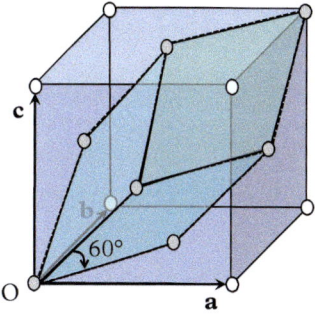

Figure 4.1.5 The face centered cubic (FCC) lattice referred to cubic (multiple) and rhombohedral (unit) cells.

4.1.3 Lattice Points, Lines, Directions, and Planes

Every lattice point in the crystal can be uniquely defined, with respect to the origin of the space lattice, by the vector, $\boldsymbol{\tau} = u\mathbf{a} + v\mathbf{b} + w\mathbf{c}$, where **a**, **b**, and **c**, are unit cell vectors with lengths, *a*, *b*, and *c* (Fig. 4.1.6). Thus, only the coordinates, u v w, are required to specify a point; if u, v, and w, are integers, they represent points on the primitive or P-lattice. Alternatively, u,v,w can also have fractional values of ½, ¼, 1/3, or 2/3; in such cases, they represent points within the unit cell.

Mathematically, in any coordinate system, two points can specify a line. In crystallography, the direction of any line in a lattice is described by drawing a line parallel to it but passing through the origin, and then using the coordinates of any lattice point on the line. Thus, if the line passes through 000 and uvw, where the latter need not be integers, then [uvw] are indices of the direction of the line

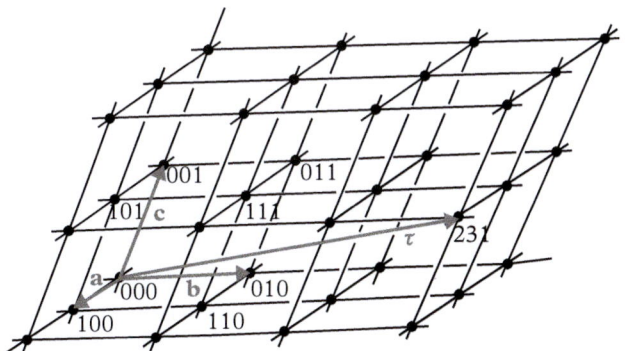

Figure 4.1.6 Lattice points are defined by the vector from the origin to the point, uvw. Here, τ defines the lattice point 231.

Adapted from Borchardt–Ott (1995).

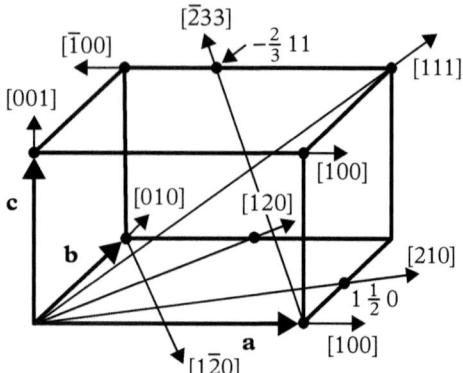

Figure 4.1.7 Indices, [uvw], for different directions.

Adapted from Cullity (1978).

(Fig. 4.1.7). In practice, the values of [uvw] are converted to a set of smallest integers; hence, [1/2 ¼ 1], [214], and [1½2], all refer to the same line, but [214], with all integers, is preferred. Negative indices are indicated with a line or bar on top, i.e. [ūvw]. Many lines are generated by the symmetry of the crystal, and those ones that are related by symmetry are represented by angular brackets; hence, in a cubic system, <110> represents the family of six face diagonals—[110], [101], [011], [$\bar{1}$10], [10$\bar{1}$], [0$\bar{1}$1].

The orientation of a plane in a lattice can be represented symbolically as (*hkl*), based on its intercepts with the three principal axes. Since a plane parallel to a specific crystallographic axis will only "intercept" it at infinity, the convention is to use the reciprocal of its intercept. Thus, we define the Miller[4] indices, (*hkl*), for a lattice plane as the reciprocal of its intercept with the three crystallographic axes. Again, in practice, the convention is to choose (*hkl*) such that they are the smallest integral multiples of the reciprocals of the plane intercepts (Fig. 4.1.8).

> **Example 4.1.1:** Calculate the Miller indices of the planes shown in Figure 4.1.8a.
>
> **Solution:** The intercepts, PQR, of the planes are 111 (A), and 2 2.5 3.5 (B)
>
> The equation of the plane is given by $\frac{x}{P} + \frac{y}{Q} + \frac{z}{R} = 1$.
> Multiplying by PQR, we get $xQR + yPR + zPQ = PQR = xh + yk + zl$.
> Substituting the values of P, Q, and R, we get
>
> for Plane A: $x + y + z = 1$, which gives its Miller index as $h = k = l = 1$, or (111)
> for Plane B: $\frac{x}{2} + \frac{y}{2.5} + \frac{z}{3.5} = 1$, or $8.75x + 7y + 5z = 17.5$, and multiplying by 4 to set all coefficients as the smallest integers, we get $35x + 28y + 20z = 70$, giving the Miller index of (35 28 20).

[4] W. H. Miller (1801–1880), who first used this notation, was an English mineralogist and crystallographer.

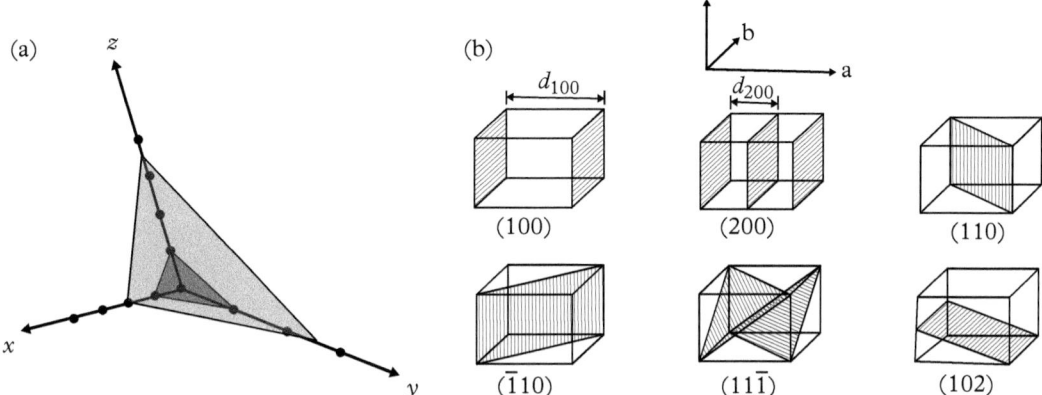

Figure 4.1.8 Miller indices of planes. (a) Two planes with intercepts of 111 and 2 2.5 3.5 are shown. Their indices are (111) and (35 28 20). (b) Miller indices of some lattice planes and the inter-planar spacing, d_{hkl}. See Table 4.1.2.

In reality, for any plane in any lattice, there must exist a whole set of parallel planes that are equidistant, one of which passes through the origin. Also, planes with Miller indices ($nh\ nk\ nl$) are parallel to (hkl), but with $1/n$th the inter-planar spacing. The inter-planar spacing, d_{hkl}, in crystals is important in diffraction, as we will see in §4.2, §7, and §8. They are tabulated, in terms of the unit cell parameters, for all seven crystal systems in Table 4.1.2. Finally, it must be mentioned that only in cubic systems, the direction [hkl] is always normal to the plane (hkl).

For the hexagonal system, consistent with its symmetry, planes can also be indexed by a different system, the Miller–Bravais indices, ($hkil$), using the inverse intercepts on all the four crystallographic axes, $\mathbf{a}_1$, $\mathbf{a}_2$, $\mathbf{a}_3$, and $\mathbf{c}$ (Fig. 4.1.9). However, since $\mathbf{a}_1$, $\mathbf{a}_2$, and $\mathbf{a}_3$, are coplanar and separated by 120°, it is easy to see that $\mathbf{a}_1 + \mathbf{a}_2 = -\mathbf{a}_3$, and $h + k = -i$. Hence, the index, i, is redundant and is replaced by a dot, and the index of the plane is written as ($hk.l$); often, for further simplicity, the dot is not even used!

As for the directions, [UVW], in the hexagonal system, they are expressed in terms of the three axes, $\mathbf{a}_1$, $\mathbf{a}_2$, and $\mathbf{c}$. They can be inter-related to the indices [uvtw], in the four axes, $\mathbf{a}_1$, $\mathbf{a}_2$, $\mathbf{a}_3$, and $\mathbf{c}$ system, by the simple relationships

$$
\begin{aligned}
U &= u - t & u &= (2U - V)/3 \\
V &= v - t & v &= (2V - U)/3 \\
& & t &= -(u + v) = -(U + V)/3 \\
W &= w & w &= W
\end{aligned} \tag{4.1.3}
$$

Table 4.1.2 Interplanar spacing, d_{hkl}, in terms of unit cell parameters for all crystals systems.

Crystal system and their unit cell parameters		Interplanar spacing, d_{hkl}
Cubic	$a = b = c$ $\alpha = \beta = \gamma = 90°$	$\frac{1}{d^2} = \frac{h^2+k^2+l^2}{a^2}$
Tetragonal	$a = b \neq c$ $\alpha = \beta = \gamma = 90°$	$\frac{1}{d^2} = \frac{h^2+k^2}{a^2} + \frac{l^2}{c^2}$
Orthorhombic	$a \neq b \neq c$ $\alpha = \beta = \gamma = 90°$	$\frac{1}{d^2} = \frac{h^2}{a^2} + \frac{k^2}{b^2} + \frac{l^2}{c^2}$
Hexagonal	$a = b \neq c$ $\alpha = \beta = 90°; \gamma = 120°$	$\frac{1}{d^2} = \frac{4}{3a^2}\left(h^2 + hk + k^2\right) + \frac{l^2}{c^2}$
Rhombohedral	$a = b = c$ $\alpha = \beta = \gamma < 120° \neq 90°$	$\frac{1}{d^2} = \frac{1}{a^2}\frac{(1+\cos\alpha)\left\{\left(h^2+k^2+l^2\right)-\left(1-\tan^2\frac{\alpha}{2}\right)(hk+kl+lh)\right\}}{1+\cos\alpha-2\cos^2\alpha}$
Monoclinic	$a \neq b \neq c$ $\alpha = \gamma = 90° \neq \beta$	$\frac{1}{d^2} = \frac{h^2}{a^2\sin^2\beta} + \frac{k^2}{b^2} + \frac{l^2}{c^2\sin^2\beta} - \frac{2hl\cos\beta}{ac\sin^2\beta}$
Triclinic	$a \neq b \neq c$ $\alpha \neq \beta \neq \gamma$	$\frac{1}{d^2} = \frac{1}{V^2}\left(s_{11}h^2 + s_{22}k^2 + s_{33}l^2 + 2s_{12}hk + 2s_{23}kl + 2s_{31}lh\right)$ where $V^2 = a^2b^2c^2\left(1 - \cos^2\alpha - \cos^2\beta - \cos^2\gamma + 2\cos\alpha\cos\beta\cos\gamma\right)$ and $s_{11} = b^2c^2\sin^2\alpha$, $s_{22} = c^2a^2\sin^2\beta$, $s_{33} = a^2b^2\sin^2\gamma$, $s_{12} = abc^2\left(\cos\alpha\cos\beta - \cos\gamma\right)$, $s_{23} = a^2bc\left(\cos\beta\cos\gamma - \cos\alpha\right)$ $s_{31} = ab^2c\left(\cos\gamma\cos\alpha - \cos\beta\right)$

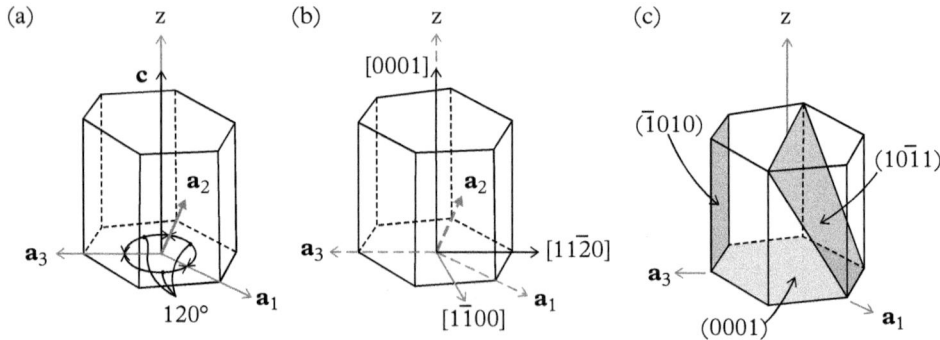

Figure 4.1.9 Hexagonal crystals system. (a) Unit cell with the principal axes. Miller–Bravais indices for select (a) directions and (b) planes.

Adapted from Callister and Rethwisch (2009).

Example 4.1.2: Convert the following directions in a hexagonal crystal from the three-index to a four-index notation, applying (4.1.3): [123], [1$\bar{2}$2], and [310].

Solution: (a) [123] = [UVW], thus u = 0, v = 1, t = −1, w = 3; thus [uvtw] = $\left[01\bar{1}3\right]$

(b) $\left[1\bar{2}2\right]$ = [UVW], thus u = 4/3, v = −5/3, t = 1/3, w = 2; thus [uvtw] = $\left[4\bar{5}16\right]$

(c) [310] = [UVW], thus u = 5/3, v = −1/3, t = −4/3, w = 0; thus [uvtw] = $\left[5\bar{1}40\right]$

4.1.4 Zonal Equations

In general, the equation of any plane may be written as

$$\frac{x}{m} + \frac{y}{n} + \frac{z}{p} = 1 \tag{4.1.4}$$

where, xyz are the coordinates of any point lying on the plane, and m, n, and p, are the intercepts of the plane with the principal crystallographic axes. We have already defined (hkl) as the reciprocal of the intercepts, and so we can rewrite (4.1.4), in its general form as,

$$hx + ky + lz = C \tag{4.1.5}$$

where C is an integer. Now, (4.1.5) represents a family of parallel planes, and for all $h, k, l > 0$, $C = 1$ defines the plane nearest the origin in the positive direction of a, b, and c. Similarly, $C = -1$, defines the plane nearest the origin, but in the negative a, b, and c directions. When $C = 0$, it defines the plane in the same family that passes through the origin and has the equation

$$hx + ky + lz = 0 \tag{4.1.6}$$

If any specific point, xyz, on the plane is defined by the lattice line, [uvw], we rewrite (4.1.6) as

$$h\mathrm{u} + k\mathrm{v} + l\mathrm{w} = 0 \tag{4.1.7}$$

and refer it to as the *zonal equation*. Any two non-parallel planes, $(h_1k_1l_1)$ and $(h_2k_2l_2)$, will have a common intersection along a lattice line [uvw], given by the solution of $h_1\mathrm{u} + k_1\mathrm{v} + l_1\mathrm{w} = 0$, and $h_2\mathrm{u} + k_2\mathrm{v} + l_2\mathrm{w} = 0$, which is the cross product

$$\begin{aligned} \mathrm{u} &= k_1 l_2 - k_2 l_1 \\ \mathrm{v} &= l_1 h_2 - l_2 h_1 \\ \mathrm{w} &= h_1 k_2 - h_2 k_1 \end{aligned} \tag{4.1.8}$$

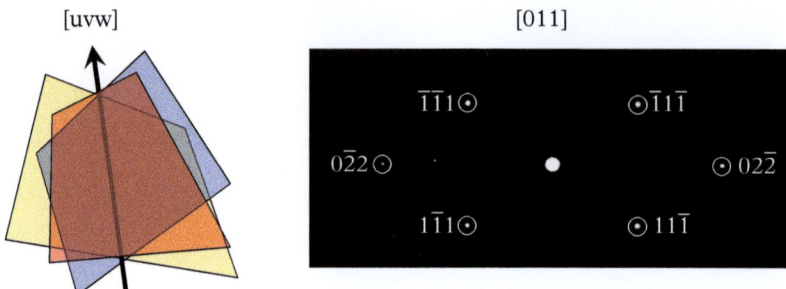

Figure 4.1.10 A common line, or zone axis, [uvw], at the intersection of different planes (only three planes are shown). On the right is an actual electron diffraction pattern from a single crystal of silicon. In this case, each diffraction spot represents a family of planes, (*hkl*), as indexed. All of them share a common line, [uvw] = [011], referred to as the zone axis that, here, is also the incident electron beam direction. Note that each of the planes satisfies the condition, $uh + vk + wl = 0$. See §8.6 for further details on transmission electron diffraction.

The line [uvw] is referred to as the *zone axis* of $(h_1 k_1 l_1)$ and $(h_2 k_2 l_2)$. It is easy to see that a set of planes, consistent with the Bravais lattice, or the symmetry of the crystal, can be part of the same zone axis, provided the planes satisfy the condition (4.1.6). The concept of zone axis is of practical use in *electron diffraction* (Fig. 4.1.10; Fig. 8.6.3).

Example 4.1.3: The following figure is a projection of a lattice along the *c*-axis, onto the *a-b* plane. The red lines labeled (A) and (B) are the traces of the planes parallel to the *c*-axis.

(a) Index planes (A) and (B).

(b) Calculate [uvw], the line common to the planes (A) and (B).

(c) Draw the trace of the planes (320) and $(1\overline{2}0)$ on the same projection.

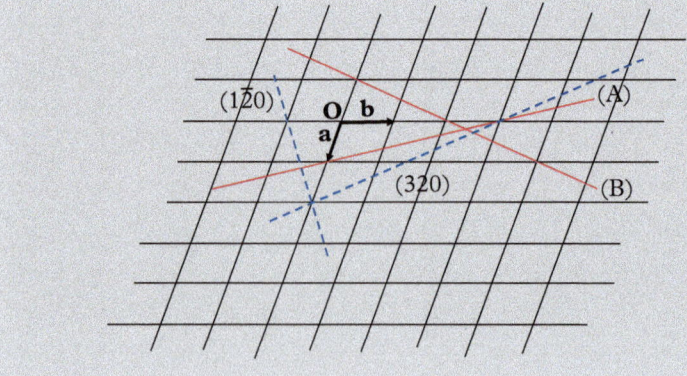

Solution: The intercepts of the two planes are (A) 1 3 ∞, and (B) –1 2 ∞. Their indices, from (4.1.4) are (A) (310), and (B)($\bar{2}$10). The common line is along the z-axis (normal to the plane shown). For (320) the intercepts are 2 3 ∞, and for (1$\bar{2}$0), the intercepts are 2 –1 ∞. These are also plotted in the figure.

4.1.5 Atomic Size, Coordination, and Close Packing

As shown previously, the structure of actual physical crystals can be related to the space lattice by placing basis atoms either at, or in some specific relation to, the points of the Bravais lattice. The process of combining a lattice and basis (4.1.2) generates a periodic arrangement of atoms in three dimensions to form a crystal. Representative examples of such structures are presented in §4.1.6.

Here, we briefly introduce the additional concept of close packing in crystal structures, where the atoms/ions are considered as hard spheres that pack together to satisfy three important criteria: (i) fill space most efficiently, (ii) achieve an environment of the highest symmetry, and (iii) achieve maximum interaction with the largest coordination number of nearest neighbors. Note that the representation of atoms as hard spheres is not intended to show anything specific about their physical or chemical nature. Their diameters merely represent how closely they approach their neighbors, and depend on how they are packed together in a solid. Thus, the hard sphere model is not meant to be a representation of the structure of atoms or ions, but instead, a way to interpret their structure, mainly size, when they are packed together in a solid.

In two dimensions, atoms can be arranged in a close-packed hexagonal or honeycomb pattern (Fig. 4.1.11a), which is the most compact way possible. We refer to this as the A layer. Note that there are two possible interstices or troughs, marked as B and C, where a second layer can be stacked. Without loss of generality, the second layer is stacked such that atoms are placed over the interstice B, and each atom touches three atoms in the layer below (Fig. 4.1.11b). For the third layer, there are two choices. They can sit in the interstices directly above the atoms in the first layer (Fig. 4.1.11c) giving the sequence ABABAB . . . , and form a hexagonal close packed (HCP) structure, with the unit cell shown in Figure 4.1.11e. Common metals with the HCP structure include Be, Mg, α–Ti, Co, and Zn. Alternatively, the third layer may not sit above the first layer, but instead sits over the interstices marked as C (Fig. 4.1.11d) giving a repeat sequence ABCABCABC . . . , and form a face centered cubic (FCC) structure, (also, known as cubic close-packed) with the unit cell shown in Figure 4.1.11f. FCC metals include Al and its alloys, γ–Fe, Ni, Cu, Pb, Ag, Au, Pt, etc. Note that both HCP and FCC structures are close-packed and equally dense.

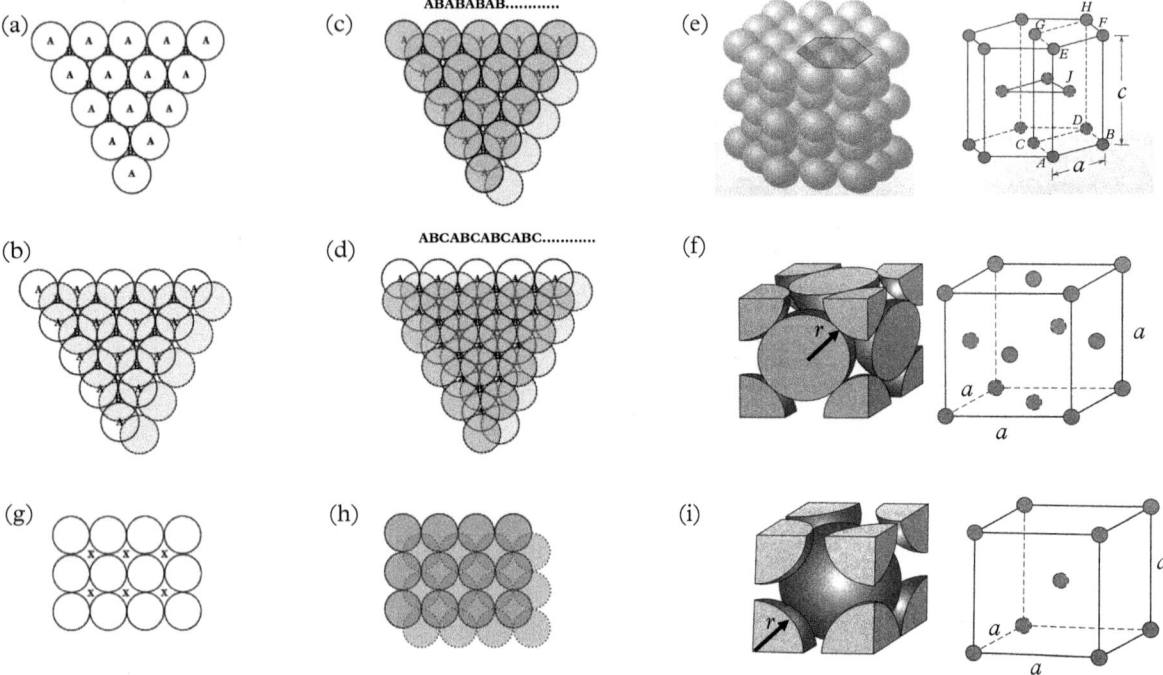

Figure 4.1.11 (a) A single layer of a hexagonal close-packed plane of atoms, A; on the surface of this plane there are two sets, B and C, of equivalent troughs of low energy where the next layer of atoms can be placed. (b) The next layer of atoms placed in position, B. (c) The third layer can be placed on top of the first layer, A, and this sequence can be repeated to generate an ABAB stacking and a hexagonal close packed, HCP, structure. (d) Alternatively, the third layer can be placed at position, C, and the sequence repeated, to give an ABCABC . . . stacking and a face centered cubic, FCC, close packed structure. (e) The unit cell of the HCP structure. (f) The unit cell of the FCC structure. Notice that the face diagonal, $4r = \sqrt{2}a$, where r is the radius of the atom. (g) Instead of the close-packed hexagonal array we can start with a less efficient, in terms of space-filling, square array of atoms. A second array can be placed exactly on top of this layer to form a simple cubic crystal (not shown). (h) A second plane can be placed in the positions of the troughs shown in (g), and the third layer can be placed back in position A. This gives the body centered cubic, BCC, structure. (i) The unit cell of the BCC structure. Note that now the body diagonal is $4r = \sqrt{3}a$.

Alternatively, in two dimensions, the first layer can form a square lattice arrangement (Fig. 4.1.11g) where there are a large number of interstices (marked as X). Now, the second layer can be stacked such that atoms are on top of each other and form a simple cubic structure (not shown in the figure). The other possibility is for the second layer to sit on the array of interstices (Fig. 4.1.11h) to form a body centered cubic (BCC) structure, with unit cell shown in Figure 4.1.11i. Note that neither the simple cubic, nor the body centered cubic, structure are close packed. However, a number of metals, notably Cr, Mo, α–Fe, and W, are found in the BCC structure.

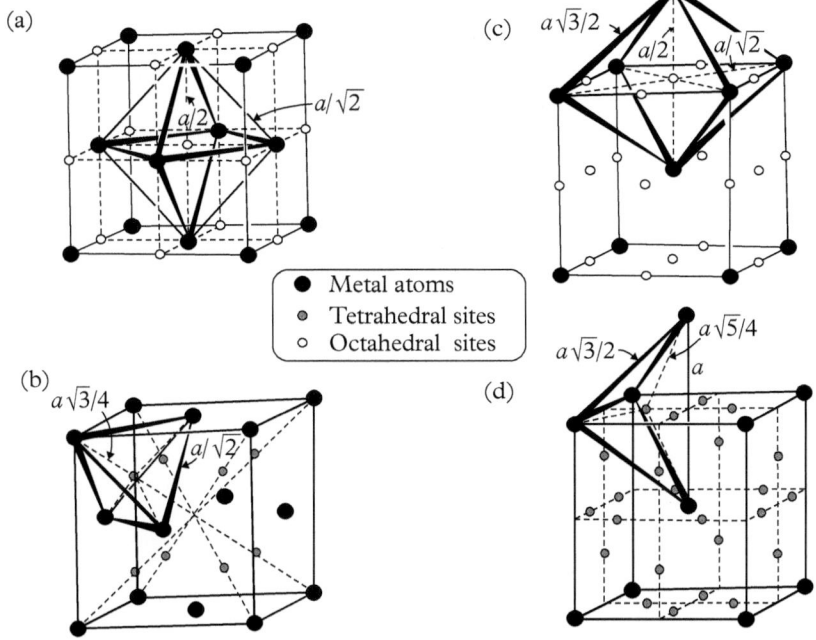

(a)

(b) $a\sqrt{3}/4$

(c) $a\sqrt{3}/2$ $a/2$ $a/\sqrt{2}$

(d) $a\sqrt{3}/2$ $a\sqrt{5}/4$ a

- ● Metal atoms
- ◉ Tetrahedral sites
- ○ Octahedral sites

Figure 4.1.12 Unit cell of the FCC structure showing (a) octahedral sites with $r_X/r_A = 0.414$, and (b) tetrahedral sites with $r_X/r_A = 0.225$. Similarly, for the BCC structure, (c) octahedral sites with $r_X/r_A = 0.154$, and (d) tetrahedral sites with $r_X/r_A = 0.292$, are shown. In general, in ionic compounds, we expect threefold, trigonal planar coordination for $r_X/r_A \geq 0.15$, tetrahedral coordination for $r_X/r_A \geq 0.22$, sixfold octahedral coordination for $r_X/r_A \geq 0.41$, and eightfold coordination for $r_X/r_A \geq 0.73$.

Adapted from Hammond (2006).

These stacking sequences not only describe the crystal structures of different elements, but they also describe the structure of a wide range of compounds with two or more elements. Specifically, they describe compounds where smaller atoms (cations) fit into the interstitial spaces between larger atoms (anions). The available interstitial sites, with fourfold coordination, called *tetrahedral* sites, and sixfold coordination, called *octahedral* sites, in FCC and BCC crystals are shown in Figure 4.1.12a–d. From simple space filling considerations, assuming that the atoms are hard spheres, the radius ratios, r_X/r_A, determine the size, X, of the second atom/ion that can be accommodated in these interstitial sites. In the FCC structure, $r_X/r_A = 0.225$ and $r_X/r_A = 0.414$, for the tetrahedral and octahedral sites, respectively.

It is easy to calculate the radius ratios for these sites based on the hard sphere model. Consider the tetrahedral site in the FCC structure shown in Figure 4.1.12b. The face diagonal of the cube along which the atoms are in contact gives $4r_A = \sqrt{2}a$; the interstitial atom is located at $1/4^{\text{th}}$ the body diagonal of the cube ($\sqrt{3}a$) and this distance is the sum of the two radii, i.e. $r_X + r_A = \sqrt{3}a/4$. Thus $r_X/r_A = 0.225$. In the BCC structure, the octahedral sites are located both at the center of the faces and at midpoints between the edges (Fig. 4.1.12c), whereas the tetrahedral sites (Fig. 4.1.12d) are located halfway in between on a line joining the centers of the faces and the midpoints of the edges. Moreover, the octahedron is somewhat squashed in one direction as shown. Thus, in the

BCC structure, $r_X/r_A = 0.291$ and $r_X/r_A = 0.154$, for the tetrahedral and octahedral sites, respectively. The sizes of the interstices are practically important; for example, in carbon steels, carbon occupies the tetrahedral site in BCC iron, but since it is much larger ($r_C/r_{Fe} > 0.29$), it distorts the structure and strengthens the steel.

4.1.6   Describing Crystal Structures—Some Examples

In general, we are interested in describing the structures of materials or compounds of different elements; in particular, we wish to specify the coordinates of the basis atoms in the Bravais lattice as a way to calculate the intensities of diffraction patterns (discussed further in §7.4). In order to do this correctly, the arrangements of atoms in the unit cell must follow two simple rules:

1) First, if the crystal structure has a body, face, or base centering translation, it must begin and end on the same atomic species (see later discussion of CsCl). For example, such translations are (000, ½ ½ ½), referred to as *body centering*, in BCC crystals, and (000, ½ ½ 0, ½ 0 ½, 0 ½ ½), referred to as *face centering* in FCC crystals. In simple terms, the Bravais lattice must be populated in such a way that these centering translations move one atom, A, to another atom, A, of the same kind in the unit cell.

2) Second, the Bravais lattices and the actual crystals generated from them, possess various kinds of symmetry. By symmetry, we mean that there exist certain operations, such as reflection, rotation, inversion, translation, etc., called *symmetry operators*, that when applied to the crystal renders its structure unchanged. Further, each crystal system possesses a characteristic symmetry element (Table 4.1.1, column 4); we discuss symmetry in more detail in the next section (§4.1.7). However, it is important that each atom type in the crystal must independently satisfy the symmetry requirements of the entire crystal (see later discussion of NaCl). Simply put, any symmetry operation particular to the crystal system must move an atom, A, into coincidence with another atom, A. In other words, the symmetry operation cannot move an atom, A, to a different atom, B.

We apply these two rules by first considering the crystal structures of Cu (FCC) and V (BCC). These two structures are cubic, with atomic positions in the unit cell, given by 4 Cu at 000, ½ ½ 0, ½ 0 ½, and 0 ½ ½, or, 4 Cu at 000 + face centering translations and 2 V at 000 and ½ ½ ½, or 2 V at 000 + body centering translations.

Next, we consider the structures of the two alkyl halides, CsCl and NaCl. The unit cell of CsCl is shown in Figure 4.1.13a, with Cl⁻ ions at 000, and Cs⁺ ions at ½ ½ ½. The Bravais lattice is simple cubic, and not BCC, because in the latter,

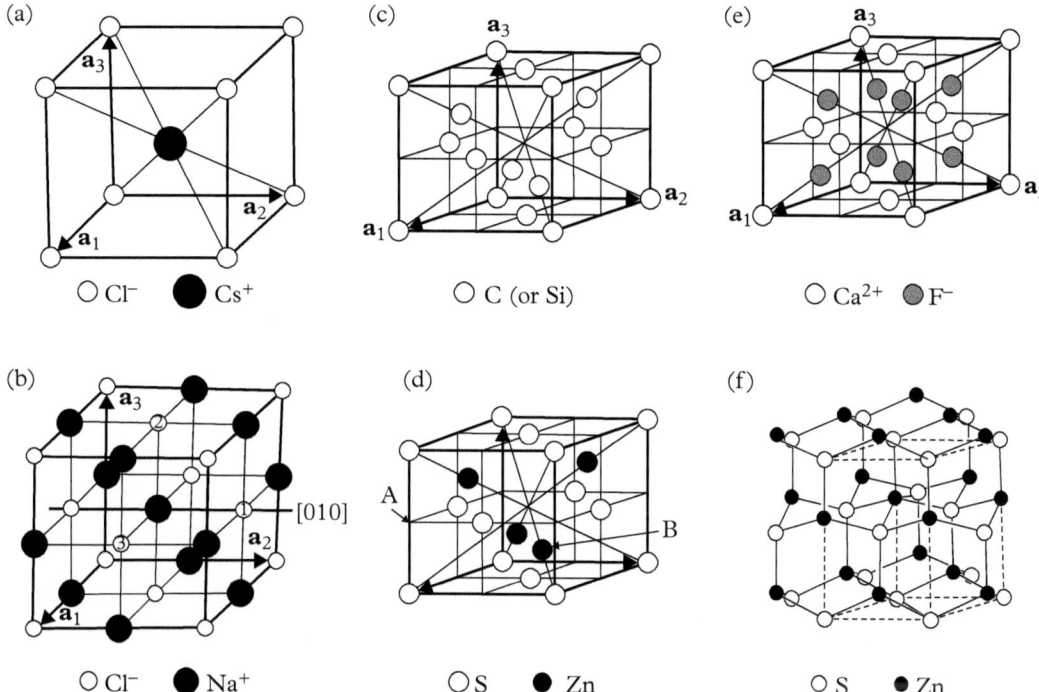

(a) ○ Cl⁻   ● Cs⁺

(b) ○ Cl⁻   ● Na⁺

(c) ○ C (or Si)

(d) ○ S   ● Zn

(e) ○ Ca²⁺   ⦿ F⁻

(f) ○ S   • Zn

Figure 4.1.13 Unit cells of typical crystal structures. (a) CsCl with two interpenetrating simple cubic unit cells (b) NaCl with two interpenetrating FCC unit cells. (c) The diamond cubic structure of carbon and common to elemental semiconductors (Si, Ge,). (d) The zinc-blende structure common to many compound semiconductors, including GaAs, ZnS, etc. (e) The CaF_2 structure, which is a variant of the diamond cubic structure. (f) The hexagonal Wurtzite structure.

the body centering translation would violate our first rule, and take a Cl^- ion to a Cs^+ ion. In the language of crystallography, we have a simple cubic lattice with a basis of two ions, Cl^- ion at 000 and a Cs^+ ion at ½ ½ ½. Similarly, one can interpret the NaCl, or the rock salt structure, shown in Figure 4.1.13b. Here it is easy, following (4.1.1), to see that the unit cell contains 4 Cl^- ions at 000, ½ ½ 0, ½ 0 ½, and 0 ½ ½, and 4 Na^+ ions at ½ ½ ½, 00 ½, 0 ½ 0, and ½ 00.

The Cl^- ions are face centered cubic, as are the Na^+ ions, but with interpenetrating lattices displaced by ½ ½ ½. Thus, the overall Bravais lattice of NaCl is FCC, and the positions of the ions in the unit cell can also be written as 4 Cl^- ions at 000 + face centering translations, and 4 Na^+ ions at ½ ½ ½ + face centering translations.

Further, in agreements with the second rule, the application of the characteristic symmetry operation for the cubic system, i.e. the threefold rotation axes, or 120° rotation, along <111>, as shown, leaves the unit cell unchanged (for example, look at Cl^- ions, marked as 1, 2, 3, in the figure). Note that many monoxides, MO, where M^{2+}, a metal ion, replaces the Na^+, and O^{2-} replaces the Cl^- in the rock salt structure, such as MgO, FeO, NiO, etc., are well known.

Semiconductor materials come in crystal structures that are variants of the FCC structure; in fact, the structures of all the well-known covalently bonded semiconductors can be derived from a parent FCC lattice with the cations filling the available tetrahedral (and octahedral) sites [1] (see also Exercises: Problem 13). Two structures commonly observed in semiconductors—diamond cubic (e.g. Si, Ge) (Fig. 4.1.13c) and zinc-blende (e.g. GaAs, SiC, ZnS, HgI) (Fig. 4.1.13d)—are both derived from the FCC structure.

The structure of diamond, with eight atoms per unit cell (Fig. 4.1.13c) is similar to Cu, discussed earlier, but with two atoms forming the basis, with coordinates (for carbon) of 4 C at 000 + face centering translations, and 4 C at ¼ ¼ ¼ + face centering translations.

Notice that ¼ ¼ ¼ is a tetrahedral site in the FCC unit cell described in Figure 4.1.12b.

The zinc-blende (ZnS) structure, also known as *sphalerite*, is similar to diamond cubic, but with a different atom on the two sites: 4 S at 000 + face centering translations and 4 Zn at ¼ ¼ ¼ + face centering translations.

ZnS is also found in the *wurtzite* structure (Fig. 4.1.13f), with a hexagonal primitive lattice and atom positions: 2 Zn at 00 ½ + z, and 2/3 1/3 z, and 2 S at 000, and 2/3 1/3 1/2.

Here, z, is a variable parameter with a value of z ~1/8.

Another variant of the diamond cubic structure is the CaF_2 structure (Fig. 4.1.13e). Here the Ca^{2+} ions form an FCC unit cell, where the F^- ions occupy ALL the eight available tetrahedral sites, with the positions 4 Ca^{2+} at 000 + face centering translations; 4 F^- at ¼ ¼ ¼ + face centering translations, and 4 F^- at ¼ ¼ ¾ + face centering translations.

One complex-oxide structure observed in technologically important materials, built on the cubic lattice, is *perovskite* (ABO_3), where A and B are cations occupying the body center ($N_I = 1$) and corners ($N_C = 1$), respectively, and the oxygen occupies the face centers ($N_F = 3$), consistent with its stoichiometry. Other complex-oxide unit cells of interest are *spinel* structure compounds, (MO. Fe_2O_3), introduced at the beginning of this chapter (frontispiece), where M is a divalent metal, with eight formula units or 56 atoms per unit cell, and *garnets* $\left(M_3^{3+}Fe_2^{3+}Fe_3^{3+}O_{12}\right)$, with 160 atoms per unit cell! The spinel structure is also based on an FCC Bravais lattice, and hence the number of atoms per unit cell for each element, as expected, is a multiple of four. Moreover, in spinels, two formula units of 14 atoms are associated with each point of the FCC Bravais lattice.

Example 4.1.4: An oxide of Cu has the structure, $a_0 = b_0 = c_0 = 4.27$ Å, $\alpha = \beta = \gamma = 90°$, with basis atoms Cu: ¼ ¼ ¼, ¾ ¾ ¼, ¾ ¼ ¾, ¼ ¾ ¾ and O: 000 and ½ ½ ½.

(a) Draw a projection of the unit cell on the *a,b* plane, as well as the unit cell.

(b) What is the stoichiometry (chemical formula) of this compound? How many formula units are in a unit cell?

(c) What is the shortest Cu–O distance?

(d) What is the density of this compound?

Solution: (a)

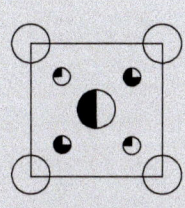
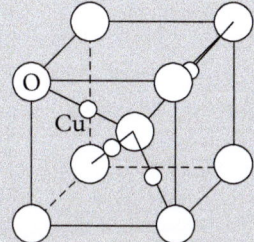

(b) Cu_2O; there are two formula units in the unit cell.

(c) From the figure, the shortest distance is $\frac{\sqrt{3}a_0}{4} = 1.85$ Å.

(d) The density of the compound is $\rho = \frac{2(2 \times 63.54 + 16)1.674 \times 10^{-27}}{(4.27 \times 10^{-10})^3} =$ 6,153 kg/m³.

4.1.7   Symmetry and the International Tables for Crystallography

So far, we have described crystal structures in terms of a lattice and basis of atoms, within the constraints of simple rules of space filling. There is an alternative, more rigorous, and systematic way to describe how repetitive patterns can fill space. By assigning atoms, or molecules, or groups of atoms and molecules to these patterns, subject to symmetry constraints, we can arrive at a general description of crystal structures. We present a brief description of this approach, beginning with the essential ideas in two dimensions; the extension to three dimensions will follow logically, but only the basic principles are introduced in this chapter. Our goal here is twofold; to introduce the characteristic (minimum) symmetries of the seven different crystal systems (Table 4.1.1, column 4) and the *International Tables for Crystallography*,[5] of use in describing all the periodic crystal structures and interpreting diffraction patterns (§7 and §8).

We define an *asymmetric object* as one with no remaining symmetry, such as the shape of a hand, or the letter, R. By subjecting the asymmetric object to two symmetry elements or operations, i.e. a mirror, *m*, and different rotation axes, we can evaluate the patterns that are generated. In conventional crystallography, a fivefold rotation axis is not encountered, as it is not consistent with lattice translations (see Fig. 4.4.1). Neither, do we encounter rotation axes greater than sixfold. Thus, the rotation axes of interest are onefold or rotation by 360° (represented by the symbol, 1), twofold (diad) or rotation by 180° (represented as 2), threefold (triad) or rotation by 120° (3), fourfold (tetrad) or rotation by 90° (4), and sixfold (hexad) or rotation by 60° (6). Each one is also represented by the number, *n*, where 360°/*n* represents the rotation angle. The combination of a mirror with each of these rotation axes gives rise to 10 crystallographic point groups in two dimensions (also known as plane point groups) (Fig. 4.1.14). They are called point groups because their rotation axis, perpendicular to the page, and the mirror line(s) pass through a single point. Furthermore, the point groups correspond to the two-dimensional motifs, or projections of molecules in two dimensions (Fig. 4.1.14).

What kinds of lattices are consistent with these point groups, or how can we extend the symmetry elements of the motifs in a two-dimensional pattern? Rotation axes with *n* = 1, 2, can be repeated in a pattern consisting of a general parallelogram (*p*1 and *p*2 in Fig. 4.1.15). Since the mirror symmetry must extend through the whole pattern, when a mirror is introduced, the parallelogram is reduced either to a simple (*pm*) or centered (*cm*) rectangular lattice. To be consistent with a fourfold rotation axis (*n* = 4), the lattice has to be a square (*p*4), and *n* = 3, 6 generate an equilateral triangular (*p*3 and *p*6) lattice. In all, for two-dimensional crystals, we have the possibility of five lattices (Fig. 4.1.15). In addition, the periodic patterns, combining the allowed rotation axes (1, 2, 3, 4, and 6) and the translations, must be generated from only these five possibilities. It turns out that there are only 10 periodic patterns in total. Now, the addition of point groups containing mirrors with the lattice translations gives rise to a new symmetry element—the glide mirror, which is combination of a mirror and a translation. When the glide is combined with the five lattices, it generates seven more periodic patterns, giving a total of seventeen two-dimensional patterns, or plane groups (also known as two-dimensional space groups; Fig. 4.1.15). To summarize then, in two dimensions, we have 10 crystallographic point groups or motifs (Fig. 4.1.14), five possible lattices, and 17 different plane groups (two-dimensional space groups) or patterns.

To extend this analysis to three dimensions, we have to introduce three additional symmetry elements: inversion (which transforms all points *xyz* to $\overline{xyz}$), roto-inversion (combination of rotation and inversion), and screw (combination

[5] A revised version of the original publication, *The International Tables of X-ray Crystallography*, by N. F. M. Henry and K. Londsdale (1952).

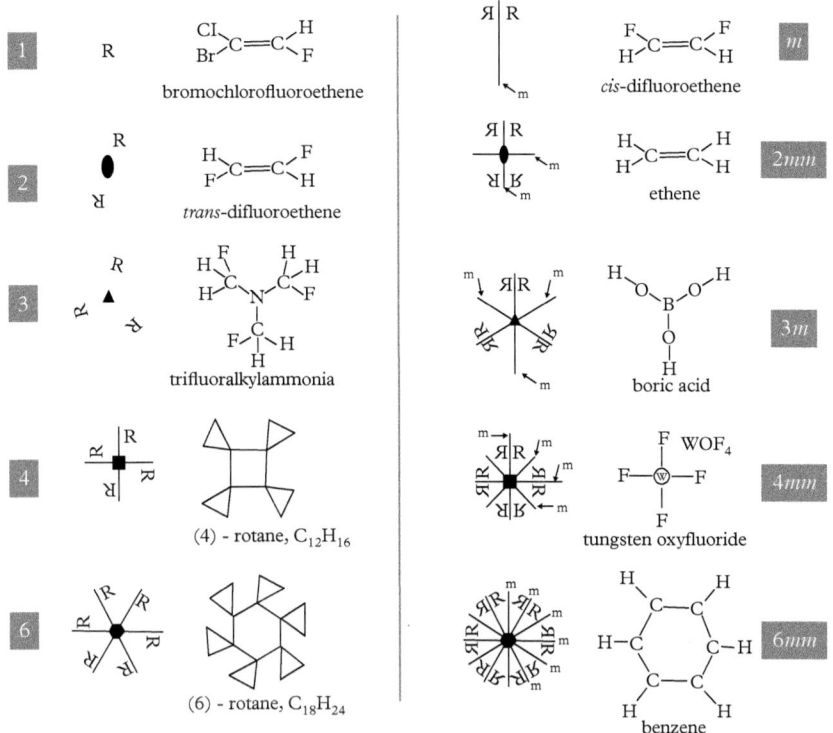

Figure 4.1.14 The 10 crystallographic point groups in two dimensions or plane point groups, including examples of molecules with the same symmetry. On the left are the pure rotations and on the right are possible combinations of rotations and mirrors. Note that a mirror (*m*), combined with a rotation axis generates an *additional* mirror for the twofold (*2mm*), fourfold (*4mm*), and sixfold (*6mm*), but does not for threefold (*3m*) rotations.

Adapted from Hammond (2006).

of rotation and parallel translation). In addition, combinations of two rotation axes generate a third rotation axis passing through the same point. However, to be compatible with lattice translation, the generated axis can only be one of the five possible rotation axes (1, 2, 3, 4, and 6). The latter restricts the rotation axes combinations to the monoaxial group (simple rotations of 1, 2, 3, 4, and 6), dihedral group or sets of two twofold rotation axes combined with a third (222, 223, 224, 226), the isometric groups, 233 (tetrahedral) or 234 (octahedral), and the special case, $\bar{4}$ (an inversion tetrad axis). Furthermore, in three dimensions, combinations of mirrors and translations give rise to a number of possible glide planes, and combinations of rotations and translations give screw axes, not encountered in two dimensions.

Including these symmetry elements in three dimensions, we find a total of 32 crystallographic point groups (Table 4.1.1), some without (column 5) and some with (column 6) a center of symmetry. This is an important distinction as a diffraction pattern is always *centro-symmetric*, even if the crystal does not possess a center of symmetry; this phenomenon is known as the *Friedel law* (to be discussed further in §7.6.2). Furthermore, the five possible plane lattices in two dimensions expand to the 14 Bravais lattices presented in Figure 4.1.4. Moreover, the 32 point-groups when mapped into the seven crystal systems,

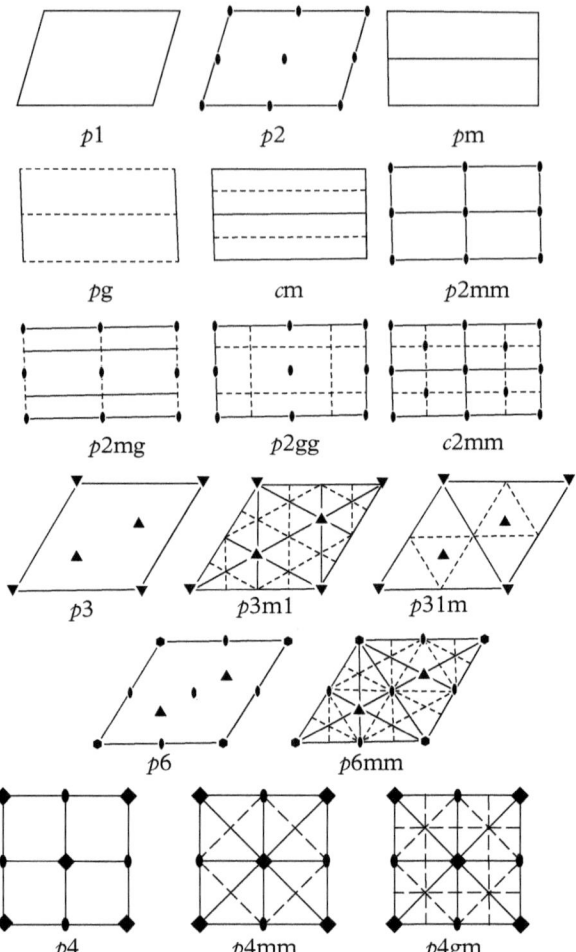

Figure 4.1.15 The 17 two-dimensional space groups (also known as the plane groups) of importance in *surface electron diffraction* (see §8.3). Note that these are based on the five possible plane lattices: the oblique *p*-lattice (*p*1, *p*2) with $a \neq b, \gamma \neq 90°$; the rectangular *p*-lattice (*pm*, *pg*, *p2mm*, *p2mg*, *p2gg*) with $a \neq b, \gamma = 90°$; the rectangular *c*-lattice, allowing for an additional centering translation and creating a multiple cell (*cm*, *c2mm*); the square *p*-lattice (*p*4, *p4mm*, *p4gm*) with $a = b, \gamma = 90°$; and the hexagonal *p*-lattice (*p*3, *p*6, *p3m1*, *p31m*, *p6mm*) with $a = b, \gamma = 120°$.

Adapted from Hammond (2006).

reveal important characteristic symmetries associated with each crystal system (Table 4.1.1, column 4). For example, the characteristic symmetry for the cubic system is a set of four triad ($n = 3$) axes, equally inclined at 109.47° along the body diagonal. Similarly, the orthorhombic system, is characterized by three diads ($n = 2$), equally inclined at 90° or equivalently, a set of three orthogonal mirrors.

Finally, the 17 plane groups or patterns in two dimensions, increase to 230 space groups in three dimensions, arising from the combination of the fourteen Bravais lattices with the 32 point-groups. These 230 space groups are numbered, beginning with triclinic (lowest symmetry) and ending with cubic (highest symmetry), drawn systematically, and described in the *International Tables for Crystallography* (ITC). Note that to be proficient in materials characterization, it only requires an ability to read and interpret these tables.

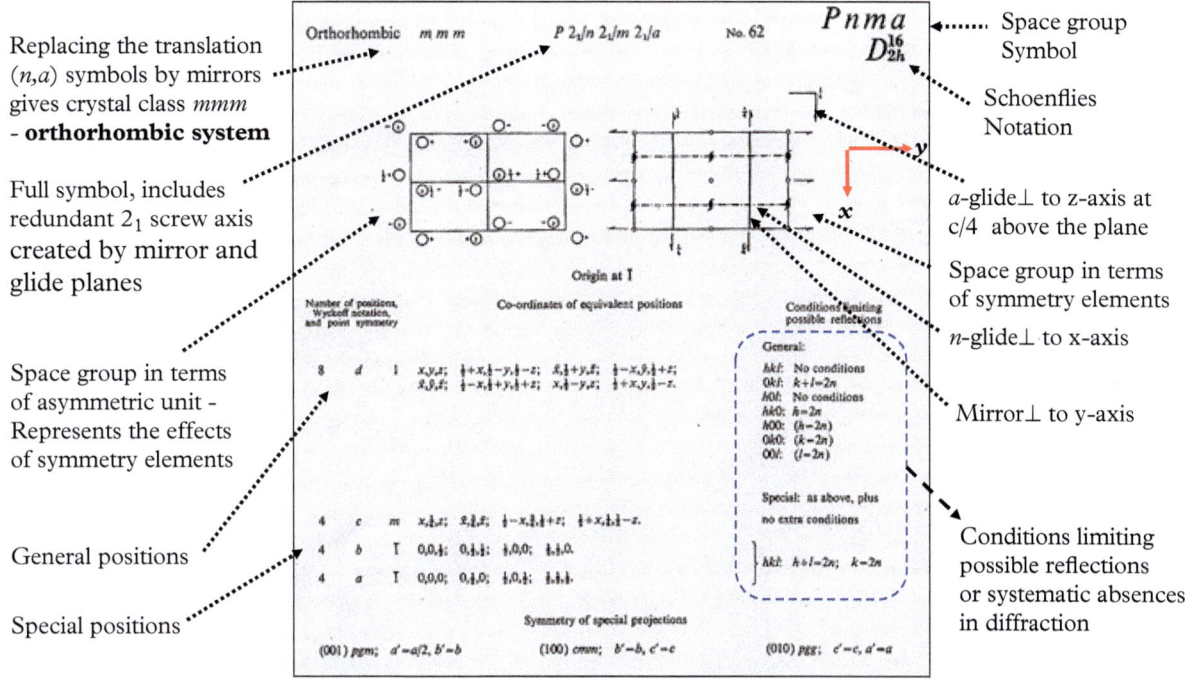

Replacing the translation (*n,a*) symbols by mirrors gives crystal class *mmm* - **orthorhombic system**

Full symbol, includes redundant 2_1 screw axis created by mirror and glide planes

Space group in terms of asymmetric unit - Represents the effects of symmetry elements

General positions

Special positions

Space group Symbol

Schoenflies Notation

a-glide⊥ to z-axis at *c*/4 above the plane

Space group in terms of symmetry elements

n-glide⊥ to x-axis

Mirror⊥ to y-axis

Conditions limiting possible reflections or systematic absences in diffraction

Figure 4.1.16 An annotated page from the *International Tables for Crystallography*, Space group #62, *Pnma*. Adapted from Allen and Thomas (1999).

Figure 4.1.16 shows an annotated example of a page from the ITC, for space group *Pnma* (#62). Each space group includes two drawings of the unit cell, typically projected in the *x-y* plane, one representing all the symmetry elements, and the other, the equipoints in the cell generated by applying the symmetry elements (shown in the second drawing) to an asymmetric object (○)—defined as any object with no remaining symmetry—placed in the asymmetric unit. Note, that a mirror image is indicated by ⊙. Further, this is an orthorhombic crystal system as it involves the characteristic symmetry of three orthogonal mirrors, *mmm*.

The coordinates of the general positions, as well as their number (8) per unit cell, the "Wycoff letter" identifying the position (*d*), and the symmetry (1) of the position are also shown under the drawings. If the asymmetric object were to be placed in a special position, such as the mirror, a simpler pattern would result. The eight asymmetric units would merge into four, and occupy the special equivalent position, and described as 4*c*. Similarly, the special positions 4*a* and 4*b*, both with multiplicity of four in the unit cell, represent sites with inversion symmetry ($\bar{1}$). Note that any atom or molecule occupying a special position must at least possess the minimum symmetry of the site. Last, but most importantly, from the point of diffraction, the table on the bottom right lists the Laue conditions, or when one is expected to observe systematic absences of certain reflections in X-ray diffraction intensities. We discuss this more in §7, where we cover X-ray diffraction intensities in a quantitative manner.

We now revisit some of the structures discussed in the last two sections and assign their space groups and numbers from the ITC. Simple cubic metals are $Pm\bar{3}m$ (#221), FCC metals are $Fm\bar{3}m$ (#225), BCC metals are $Im\bar{3}m$ (#229), and HCP crystals are $P6_3/mmc$ (#194). CsCl is also $Pm\bar{3}m$, but with both Cs and Cl occupying special positions. The structure has a center of symmetry, as indicated by $\bar{3}$, the triad roto-inversion axis. The zinc blende structure of the semiconductors ZnS and GaAs is the space group, $F\bar{4}3m$ (#216), with atoms in special positions; note the absence of the center of symmetry or no inversion. The diamond cubic structure is $Fd\bar{3}m$ (#227).

Only a very small number of space groups are commonly observed, and many of the 230 possible space groups do not correspond to any actual crystal structures. Most of the elements (70%) belong to five space groups ($Fm\bar{3}m$, $Im\bar{3}m$, $Fd\bar{3}m$, $F\bar{4}3m$, and $P6_3/mmc$), and 60% of the organic/inorganic compounds belong to six space groups ($P2_1/c, C2/c, P2_1, P\bar{1}, Pbca, P2_12_12_1$)! Not surprisingly, these space groups are the ones that lead to close packed crystals.

4.1.8   The Stereographic Projection

Crystals are three-dimensional objects and in any written text, such as this one, we need to represent them on a two-dimensional page. While this can be done in combinations of plan and elevation drawings, the angular relationships between lattice planes and directions are better represented in the stereographic projection. Figure 4.1.17 shows the principles of this projection. The crystal of interest is placed at the center of the sphere, normals to all planes of interest are drawn, and their intersections with the surface of the sphere, called *poles*, are recorded (Fig. 4.1.17a). Note that the angle between the poles, ϕ, is different from the dihedral angle, Φ, between the faces (Fig. 4.1.17c); however, $\phi + \Phi = 180°$, is always true. Moreover, for any two poles on the surface of the sphere, there always exists a great circle that passes through both of them. To complete the stereographic projection, lines are drawn from the poles to the south pole (Fig. 4.1.17b), and the intersection of these lines with the equatorial plane constitutes the stereographic projection (Fig. 4.1.17b).

The stereographic projection (Fig. 4.1.18a) has some specific properties:

(a) Any circle on the sphere also appears as a circle on the plane of projection. However, the centers of the two circles, generally, do not coincide.

(b) A great circle on the sphere appears as a circular arc on the projection. Further, it cuts the basic circle of the projection at two, diametrically opposite, endpoints.

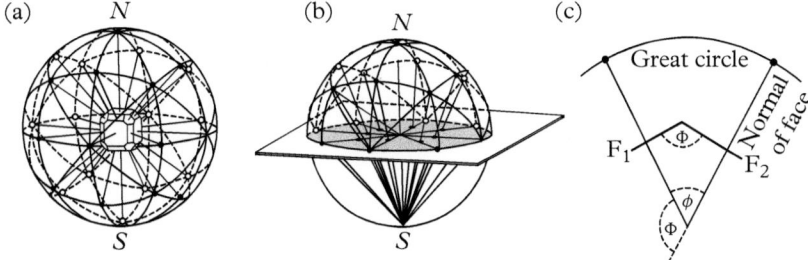

Figure 4.1.17 Basics of the stereographic projection. (a) The crystal (in this case, Galena, the cubic form of PbS) is placed at the center of a unit sphere. The normal to the crystallographic planes are extended and their points of intersection, called poles of the planes, are recorded. (b) Lines are drawn from each pole to the south pole, S, of the sphere as shown. The intersection of this second set of lines with the equatorial plane (grey) is the stereographic projection. A detailed example, for a complete cubic crystal, is shown in Figure 4.1.18. (c) Schematic illustrating the angle, ϕ, between the normals, as well as the dihedral angle, Φ, of two planes, F_1 and F_2.

    (c) Angles are preserved in this projection. Thus, it is possible, to measure the angle of intersection between great circles on the projection. This is done with a Wulff net (see Fig. 4.1.18b and the discussion that follows).

    (d) Unfortunately, areas are not preserved in the projection.

    In practice, the stereographic projection is used in conjunction with a Wulff net (Fig. 4.1.18b). The planes and directions are first plotted in a stereographic projection and then this is overlaid on a Wulff net, such that the center of the stereographic projection and the Wulff net are superimposed and they can rotate relative to one another. Now, with this arrangement it is possible to do the following:

    (a) Measure the angle between two crystallographic planes or directions. The Wullf net is rotated such that the stereographic projection of the two direction of interest lie on the same great circle. Then the angles are simply read off the Wulff net.

    (b) To rotate the projection about an axis, align the axis of rotation with the N–S axis of the Wulff net. Then each point of the projection is moved by the required angle of rotation along the small circle of the Wulff net on which it lies.

    (c) To index unknown crystallographic directions, compare the angle between them and known crystallographic directions, with the angle between indexed directions in a standard projection.

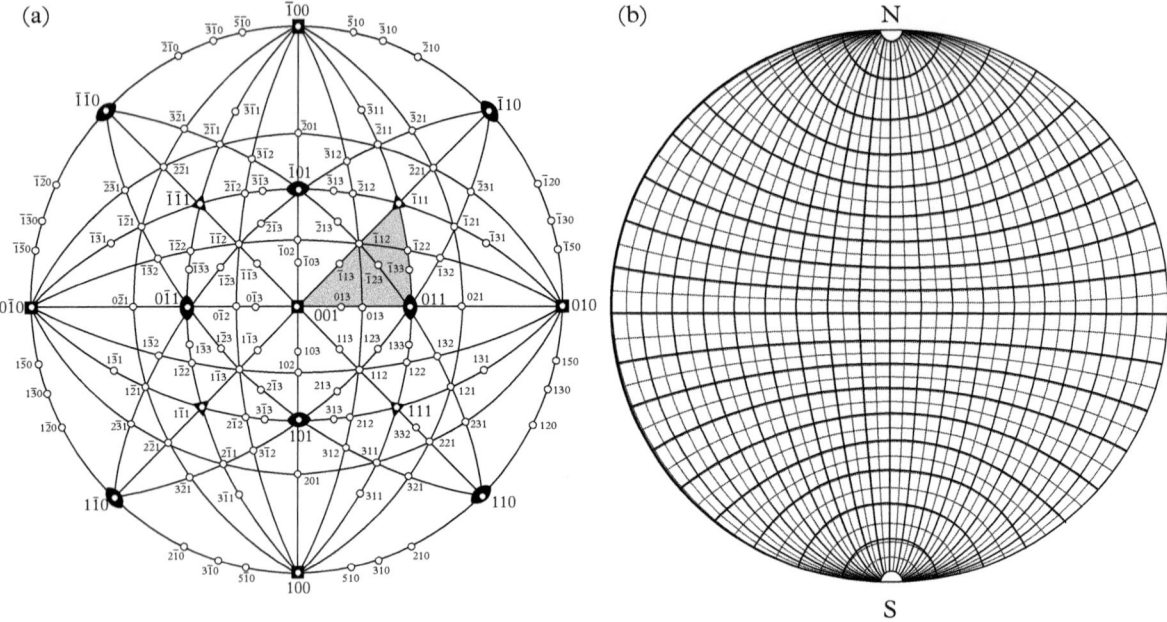

Figure 4.1.18 (a) A complete stereographic projection of a cubic crystal. (b) A Wulff net used for measuring angles (see text for details).

Finally, for those interested in reviewing the use of stereographic projections and Wulff nets in diffraction, it is worth their while looking up the detailed treatment of the subject in Cullity (1978).

Example 4.1.5: Construct the stereographic projection for the point group 4*mm* shown in Figure 4.1.14.

Solution: We construct it in parts as follows:

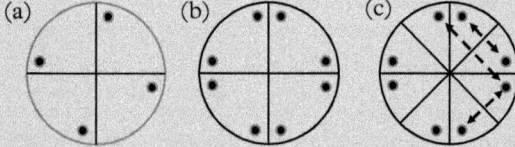

(a) Fourfold rotation axes showing the equipoints.

(b) Mirror planes, shown as bold lines, perpendicular to [100]. The four-fold rotation axes also act on the mirrors, such that two additional mirrors perpendicular to [110] are also generated.

(c) However, the new mirrors perpendicular to [110] do not generate any additional equipoints in the stereographic projection.

4.1.9 Imperfections in Crystals

So far, we have discussed ideal crystalline materials; in practice, ideal crystals are seldom found, and real crystals show distinct features, called *defects*, that deviate from the ideal. Such defects determine materials behavior as many of their properties are profoundly affected by deviations from the ideal crystalline structure. Hence, evaluating such imperfections is central to materials characterization and analysis. Imperfections are classified based on lattice irregularities as point, line, and planar defects. There is a vast literature on the subject, including some classic textbooks (Nabarro, 1967; Weertman and Weertman, 1992), but we only present an introduction that is relevant for a description of their analysis in later chapters (§9.5.2).

4.1.9.1 Point Defects

The simplest or *point defect* is a vacancy in a lattice site originally occupied by an atom. Vacancies are found in all crystals (Fig. 4.1.19). Alternatively, a self-interstitial is an atom from the regular crystal that is found elsewhere at an interstitial site; the latter, typically of a smaller volume than a regular site. Often, in metals, such interstitials lead to large distortions and strain in the surrounding lattice. If a second element, either an impurity or a deliberate addition, as in alloying, is included, it can either be a substitutional atom, forming a solid solution, or occupy an interstitial position. The incorporation of an additional element as a solid solution will be determined by (a) atomic size ratios to minimize distortion of the lattice, (b) crystal structure—-preferably the same for the element and alloy, (c) electronegativity difference—to be as small as possible, and (d) their valence. Accurate characterization of point defects is difficult and challenging, but most success has been obtained by *positron annihilation spectroscopy* (PAS), a technique that is beyond the scope of this book. However, for those interested, an excellent introduction to the subject of PAS, emphasizing its applications to metals and alloys, can be found in this accessible review [2].

4.1.9.2 Line Defects

This type of linear defect, called a dislocation, is often associated with local atoms that are misaligned in some way. An *edge dislocation* (Fig. 4.1.20a) involves an extra half-plane of atoms, referred to as the dislocation line, which terminates within the crystal. The extra half-plane of atoms leads to a displacement of a part of the crystal with respect to the rest. For an edge dislocation, this displacement vector,

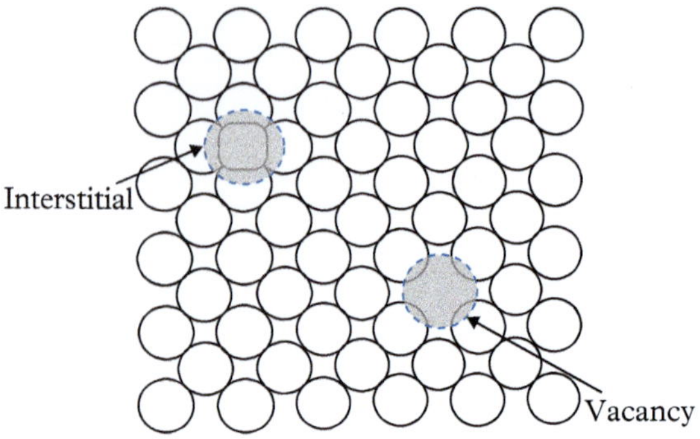

Figure 4.1.19 Point defects are of two kinds, interstitial or vacancy, as shown.

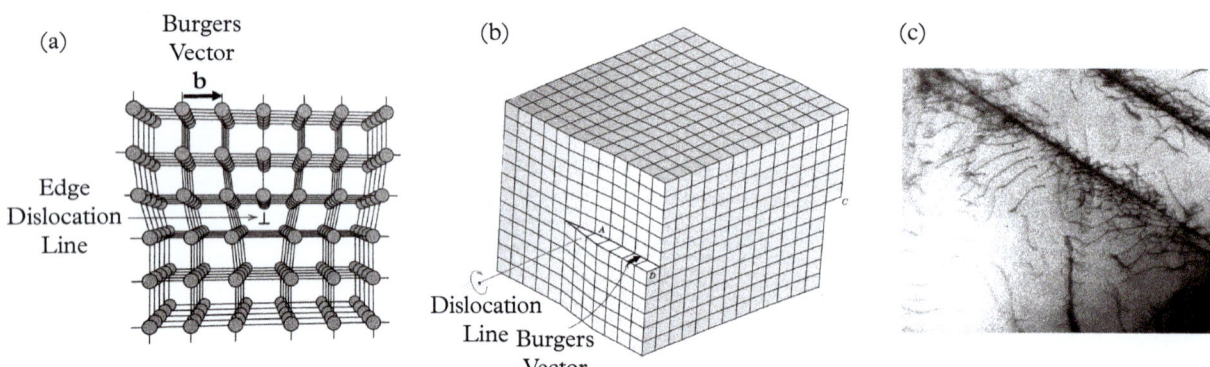

Figure 4.1.20 Schematic illustration of (a) edge and (b) screw dislocations; note that, in the latter, all the atoms are displaced parallel to the Burgers vector, **b**. See §8.5.4 and §9.5.2. In reality, dislocations tend to have partial edge and partial screw character. (c) One of the earliest TEM images of dislocations in cold-rolled stainless steel.

Adapted from [5].

known as the Burgers vector, **b**, is normal to the dislocation line. A second type of dislocation, called the *screw dislocation*, arises from shear stress to produce a linear displacement of atoms (Fig. 4.1.20b). Screw dislocations are characterized by the Burgers vector, **b**, being parallel to the dislocation line, and play an important role in crystal growth.

Edge and screw dislocations are end-members of a continuum, as most dislocations found in crystalline materials are mixed, having both edge and screw components. From a practical point of view, deformation of crystalline materials involves the motion or pinning of dislocations, and as such, they play a very important role in plastic deformation in materials. Dislocations are observed and their Burgers vector determined by electron microscopy techniques (Fig. 4.1.20c) and further details are in §9.5.2.

4.1.9.3 Planar Defects

Figure 4.1.1 introduced the concept of grain boundaries. Figure 4.1.21a schematically shows the atomic mismatch between two crystalline grains, and Figure 4.1.21b shows an atomic resolution image of the same. A small-angle grain boundary involves a small mismatch and can be described by an array of dislocations. If the grain boundary is formed by a series of edge dislocation (Fig. 4.1.21c), it is known as a *tilt boundary*. If the grain boundary is formed by a series of screw dislocations, it is known as a *twist boundary*. The interatomic distances at a grain boundary are typically larger than that of an ideal crystal of the same material; as such, they are higher-energy arrangements of the atoms and the magnitude of their *interfacial energy* is proportional to the degree of misalignment.

A *stacking fault* is typically found in an FCC crystal, where the ABCABC . . . stacking of the close-packed planes (§4.1.5) is interrupted and locally altered to an ABC<u>ABAB</u>ABC . . . stacking. Similar changes can also be expected to occur in HCP crystals, i.e. local changes from ABABAB . . . to AB<u>ABC</u>ABA . . . stacking.

A *twin boundary* is a special kind of grain boundary arising from the growing together of two crystals with well-defined orientation relationships. A common twin element is a mirror (Fig. 4.1.21d), such that atoms on one side of the boundary are located in positions that are mirror images of the atoms on the other side. Twins are formed during growth or through mechanical deformation, and occur on well-defined crystallographic planes; the characteristic twin planes are {111} and {112} for FCC and BCC, respectively (see §8.7.6 for further discussion of twinning).

An *antiphase boundary* (APB), is the interface between two domains in an ordered alloy. Such interfaces are typically coherent and across the boundary the lattice planes are continuous but the site occupancy is different. A good example is the APB in B2 ordered alloys with a BCC unit cell, where on one side atoms

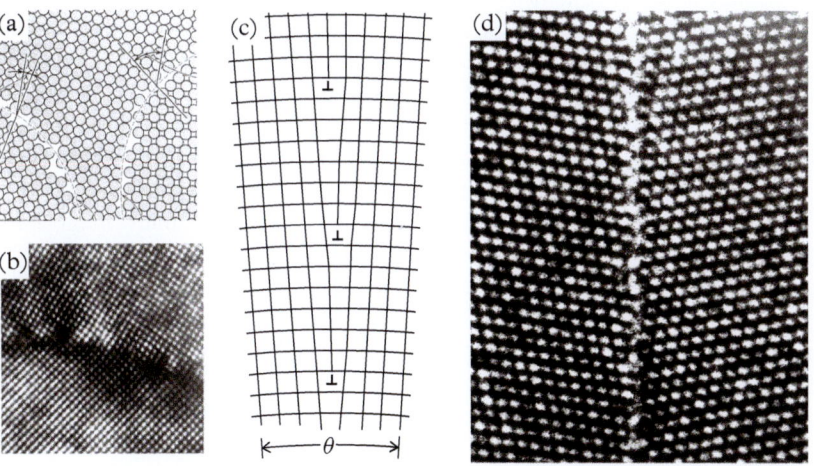

Figure 4.1.21 (a) Schematic representation of a grain boundary (b) An atomic resolution microscopy image of a GB in Mo. The white dots are separated by 0.23 nm. (c) Schematic of a small angle tilt boundary consisting of a series of edge dislocations. (d) A grain boundary in Al. The white dots are separated by 0.2 nm.

Adapted from Williams, Pelton and Gronsky (1991).

of type A occupy the cube corners and atoms of type B occupy the body centers, and on the other side of the APB the positions of atoms A and B are reversed.

Defects are analyzed by various imaging techniques that include transmission electron diffraction (§8.6) and microscopy (§9.5.2). They also do alter positions of atoms in the crystal, and this in turn, affects the intensities and positions of observed reflections in diffraction. We discuss this further in §8.

The problem of crystal structure determination by diffraction can be largely divided into two parts. The first part is to obtain the size and shape of the unit cell, including the lattice parameters, from the geometry of the diffraction pattern. The second, much more challenging, part is to obtain the lattice type, and the positions of the atoms in the unit cell.

Section 4.3 introduces the concept of diffraction as a way to provide an overview of the first part, and leave the details, as well as the second part, to later chapters, i.e. §7 and §8. But first, to understand diffraction, we now describe the important concept of the *reciprocal lattice*.

4.2   The Reciprocal Lattice

X-ray diffraction was originally interpreted as an interference effect of X-rays reflected by a parallel set of lattice planes that depends on both their *orientation* and their *interplanar spacing*[2]. From this perspective, the *reciprocal lattice* construction is an excellent aid, as each of its lattice points represents an entire *family* of lattice planes, (*hkl*), at a given orientation and a fixed interplanar spacing, d_{hkl}. Needless to say, the reciprocal lattice has a fixed relationship to the real lattice, from which it is derived. As such, the reciprocal lattice can be constructed using either a physical or a mathematical (vector) approach, and we now begin with the former.

Consider the three families of planes, with inter-planar spacing d_1, d_2, and d_3, perpendicular to the plane of the paper (Fig. 4.2.1a). We draw normals to the planes (Fig. 4.2.1b) and set the vectors (Fig. 4.2.1c) proportional to the inverse of their inter-planar spacing, i.e. $|\mathbf{d}_i^*| = K/d_i$, where the proportionality constant, $K = 2\pi$, is assumed.[6] The vectors, $\mathbf{d}_i^*, i = 1, 2, 3$, are called the reciprocal lattice vectors, as their dimensions are 1/length (in general, Å^{-1}). Notice that $\mathbf{d}_3^* = \mathbf{d}_1^* + \mathbf{d}_2^*$, and these three vectors, can indeed generate a space lattice.

A similar exercise can be done with the *x-z* projection of a primitive monoclinic unit cell, with the **b** axis normal to the page (Fig. 4.2.2a). Figure 4.2.2b shows the corresponding reciprocal lattice vectors for many of the planes, and Figure 4.2.2c shows the projected unit cell of the reciprocal lattice. It is clear that the magnitudes of the reciprocal lattice vectors are $|\mathbf{a}^*| = d_{100}^* = 2\pi/d_{100}$ and $|\mathbf{b}^*| = d_{010}^* = 2\pi/d_{010}$. We can extend this approach to the third dimension and generate the complete unit cell of the monoclinic P crystal in reciprocal space (Fig. 4.2.5a).

---

[6]   The factor 2π is used for the proportionality constant, K, to be consistent with our definition of the magnitude of the wave vector, $|\mathbf{k}| = k = 2\pi/\lambda$ (see §6 on optics and optical methods, and specifically, §6.2 where k is defined). Alternatively, we can set $K = 1$, but then we have to redefine the wave vector $|\mathbf{k}| = k = 1/\lambda$, which is the practice in some textbooks. Note that $K = 1$ is commonly used in the interpretation of high-resolution electron microscopy. See §9.2.7.3 and specifically (9.2.17).

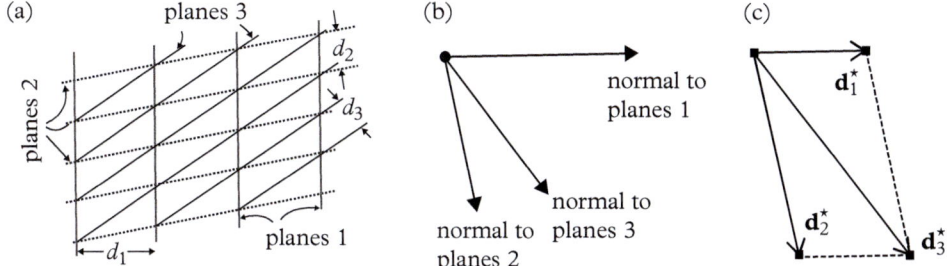

Figure 4.2.1 The reciprocal lattice construction. (a) A schematic of three sets of lattice planes in real space for a monoclinic crystal. (b) Directions of the normal to each of the sets of planes. (c) The reciprocal lattice vectors, $\mathbf{d}_i$, with magnitudes inversely related to their inter-planar spacing. Note that $\mathbf{d}_3^* = \mathbf{d}_1^* + \mathbf{d}_2^*$.

Adapted from Hammond (2006).

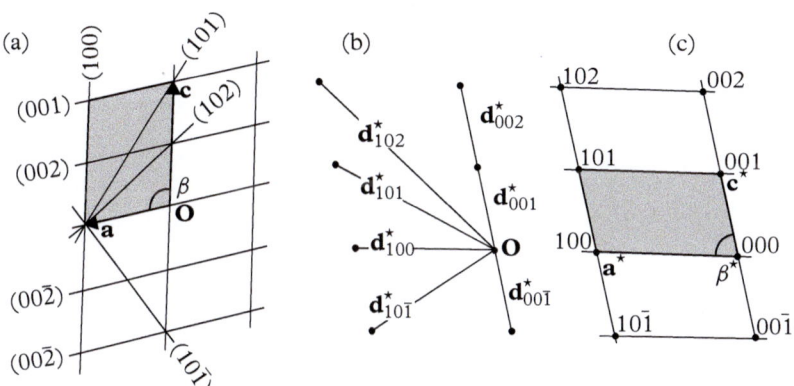

Figure 4.2.2 (a) A monoclinic unit cell, with the **b** axis normal to the page (*x-z* projection), and some planes are indexed as shown. (b) The reciprocal lattice vector for some of the planes shown in (a). Note that the reciprocal lattice vectors are oriented normal to the planes and their lengths are related inversely to their inter-planar spacing. (c) The unit cell of the reciprocal lattice.

Adapted from Hammond (2006).

Now that we have a physical idea of the reciprocal lattice and how it is generated, we can take a look at the mathematical approach that is easier to generalize.

We start with the most general case, or the lowest symmetry triclinic unit cell defined by the vectors, **a**, **b**, and **c** (Fig. 4.2.3). Then the volume, V, of the unit cell is the area of the parallelogram, OACB, multiplied by the height of the cell, OP. In vector notation, we can write this as $V = \mathbf{a} \cdot (\mathbf{b} \otimes \mathbf{c}) = \mathbf{b} \cdot (\mathbf{c} \otimes \mathbf{a}) = \mathbf{c} \cdot (\mathbf{a} \otimes \mathbf{b})$. Then the reciprocal lattice vectors, **a★**, **b★**, and **c★** (Fig. 4.2.3) are defined as

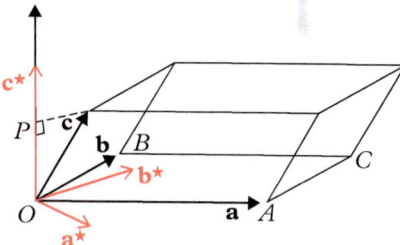

Figure 4.2.3 The unit cell of a primitive triclinic unit cell showing the lattice vectors in real space as well as the reciprocal lattice vectors.

$$\mathbf{a}^* = 2\pi\,(\mathbf{b} \otimes \mathbf{c})/V$$
$$\mathbf{b}^* = 2\pi\,(\mathbf{c} \otimes \mathbf{a})/V \qquad (4.2.1)$$
$$\mathbf{c}^* = 2\pi\,(\mathbf{a} \otimes \mathbf{b})/V$$

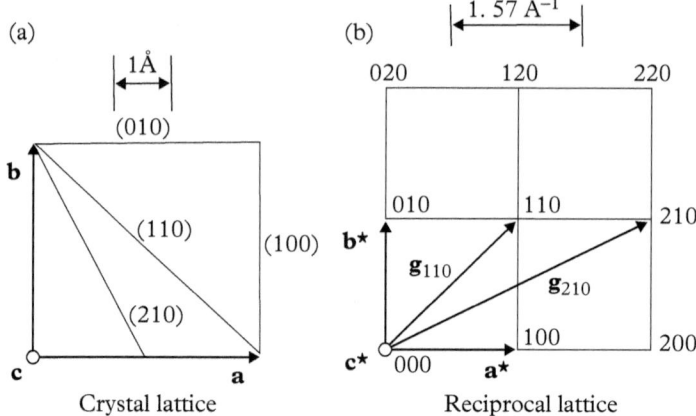

Figure 4.2.4 A simple cubic crystals showing (a) the real lattice with *c*-axis normal to the page, and (b) its reciprocal lattice. The unit cell parameter for the real crystal, $a = 4$ Å.

Adapted from Cullity (1978).

It is easy to see in the figure that $\mathbf{c}^\star$ is normal to the plane of $\mathbf{a}$ and $\mathbf{b}$, and its length is

$$|c^*| = \frac{2\pi \,|\mathbf{a} \otimes \mathbf{b}|}{V} = \frac{2\pi \cdot \text{area of parallelogram, OACB}}{(\text{area of parallelogram, OACB}) \times (\text{height of cell, OP})}$$

$$= \frac{2\pi}{\text{OP}} = \frac{2\pi}{d_{001}}, \tag{4.2.2}$$

which is simply proportional (by a factor of 2π) to the inverse of the inter-planar spacing of the (001) planes of the real crystal. Similarly, we can show that $\mathbf{a}^\star$ is normal to the (100) planes, with magnitude, $|\mathbf{a}^*| = 2\pi/d_{100}$, and $\mathbf{b}^\star$ is normal to the (010) planes, with magnitude, $|\mathbf{b}^*| = 2\pi/d_{010}$. We can extend this approach to all the planes of the crystal lattice and build the reciprocal lattice by repeated translations of the reciprocal lattice vectors, $\mathbf{a}^\star, \mathbf{b}^\star$, and $\mathbf{c}^\star$. We then label the array of points generated by this lattice in terms of the basis vectors, i.e. the end of $\mathbf{a}^\star$ is labeled as 100, the end of $\mathbf{b}^\star$ is labeled as 010, and the end of $\mathbf{c}^\star$ is labeled as 001. Hence, any vector, $\mathbf{g}_{hkl}$, in the reciprocal lattice, is *normal* to the set of planes with the Miller indices (*hkl*). Thus, $\mathbf{g}_{hkl}$ can be written in terms of the unit cell vectors of the reciprocal lattice as

$$\mathbf{g}_{hkl} = h\mathbf{a}^* + k\mathbf{b}^* + l\mathbf{c}^* \tag{4.2.3a}$$

In addition, its length[7] is

$$|\mathbf{g}_{hkl}| = 2\pi/d_{hkl} \tag{4.2.3b}$$

As a first exercise, Figure 4.2.4 illustrates the reciprocal lattice generated by the application of (4.2.1) to the unit cell of a simple cubic crystal. Note that $\mathbf{g}_{200}$ is, as expected, parallel *to* $\mathbf{g}_{100}$, but is twice as long. Also, as mentioned earlier, planes

[7] If we set the proportionality constant, $K = 1$, we have $|\mathbf{g}_{hkl}| = g_{hkl} = 1/d_{hkl}$.

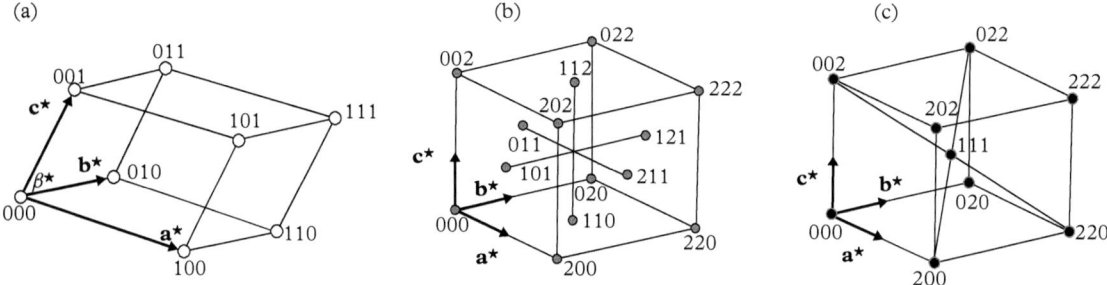

Figure 4.2.5 Reciprocal lattice unit cells for the real (a) Monoclinic (b) BCC, and (c) FCC unit cells. Note that the reciprocal unit cell for the real BCC is FCC, and for the real space FCC the reciprocal unit cell is BCC. Further for the real BCC, the reciprocal unit cell satisfies the condition $h + k + l =$ even, and for the real FCC structure, hkl must be either all even or all odd. The latter has implications for the intensities, including absences of specific reflections, observed in diffraction patterns (see §7.5).

$(nh\ nk\ nl)$, where n is an integer, are parallel to (hkl), but with $1/n^{\text{th}}$ the inter-planar spacing. Thus $|\mathbf{g}_{nh\ nk\ nl}| = n|\mathbf{g}_{hkl}|$. The reciprocal unit cells of the cubic (I) unit cell (BCC), and the cubic (F) unit cell (FCC), are shown in Figure 4.2.5b and Figure 4.2.5c, respectively.

The cubic unit cell in real space is based on mutually perpendicular basis vectors, **a**, **b**, and **c**. In such cases, which also includes the orthorhombic and tetragonal crystals, the reciprocal lattice vectors, **a**⋆, **b**⋆, and **c**⋆, are also mutually perpendicular. For crystals, such as monoclinic and triclinic, discussed earlier, where the unit cell vectors in real space are not mutually perpendicular, the construction of the reciprocal lattice involves complementary angles. As an example, Figure 4.2.6 shows the real and reciprocal unit cells of the hexagonal lattice.

Note that in general, since **c**⋆ is normal to both **a** and **b**, its scalar product with either one should be zero, i.e.

$$\mathbf{c}^* \cdot \mathbf{a} = \mathbf{c}^* \cdot \mathbf{b} = 0$$
$$\mathbf{c}^* \cdot \mathbf{c} = 2\pi \tag{4.2.4}$$

It is now easy to prove the validity of the zonal equation, (4.1.7), we discussed in §4.1.4. By definition, the planes of a zone are all parallel to a common line, [uvw], called the zone axis, and the normal, $\mathbf{g}_{hkl}$, to the plane must be coplanar and normal to [uvw]. Thus, the two vectors, the zone axis, $\mathbf{Z} = u\mathbf{a} + v\mathbf{b} + w\mathbf{c}$, and the reciprocal lattice vectors, $\mathbf{g}_{hkl} = h\mathbf{a}^* + k\mathbf{b}^* + l\mathbf{c}^*$, must be perpendicular with a scalar product of zero. Thus,

$$\mathbf{Z} \cdot \mathbf{g}_{hkl} = (u\mathbf{a} + v\mathbf{b} + w\mathbf{c}) \cdot (h\mathbf{a}^* + k\mathbf{b}^* + l\mathbf{c}^*) = 0$$
$$uh + kv + lw = 0 \tag{4.2.5}$$

as illustrated in Figure 4.1.10.

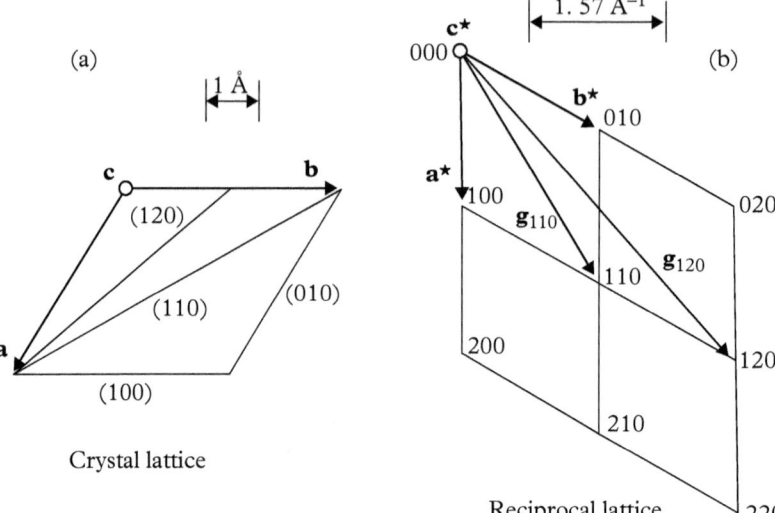

Figure 4.2.6 The hexagonal crystal showing (a) the real lattice and the unit cell with *c*-axis normal to the page. The unit cell parameter for the real crystal, $a = 4$ Å.(b) Its reciprocal lattice with some reciprocal lattice vectors indicated.

Adapted from Cullity (1978).

Example 4.2.1: Consider an **orthorhombic** crystal with lattice parameters

$$a = 2\,\text{Å}, \quad b = 3\,\text{Å} \quad \text{and} \quad c = 4\,\text{Å}\,(\alpha = \beta = \gamma = 90°).$$

(a) Draw the unit cell.

(b) Calculate the inter-planar spacing, d_{hkl}, using equation in Table 4.1.2, for the (100), (002), (030), (011), (111), (310), (311) and (312) planes.

(c) What is the common direction for the (011) and (111) planes? Would either (310) or (311) planes, or both, share this common direction (i.e. form part of the same zone)?

(d) Calculate the unit cell dimensions of the reciprocal lattice? Draw this in the same figure as (a) but with different scale (note: units are different).

Solution: (a)　The unit cell is shown in the following figure.

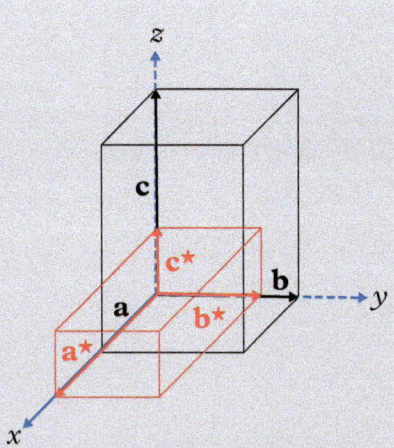

(b) The interplanar spacing can be calculated (Table 4.1.2), using $\frac{1}{d_{hkl}^2} = \frac{h^2}{a^2} + \frac{k^2}{b^2} + \frac{l^2}{c^2}$, as shown in the following table

Plane	(100)	(002)	(030)	(011)	(111)	(310)	(311)	(312)
d_{hkl} (Å)	2.0	2.0	1.0	2.4	1.54	0.65	0.64	0.62

(c) The common direction is given by the cross product, (4.1.8). Thus, for (011) and (111), we have $[uvw] = [01\bar{1}]$. (311) would share the same common direction (dot product = 0), but (310) would not (dot product $\neq$ 0).

(d) The reciprocal lattice vectors are calculated using (4.2.1). Thus

$$\mathbf{a}^* = 2\pi \, (\mathbf{b} \otimes \mathbf{c})/V = \pi \boldsymbol{i}$$

$$\mathbf{b}^* = 2\pi \, (\mathbf{c} \otimes \mathbf{a})/V = 2\pi/3 \, \boldsymbol{j}$$

$$\mathbf{c}^* = 2\pi \, (\mathbf{a} \otimes \mathbf{b})/V = \pi/2 \, \boldsymbol{k}$$

The reciprocal unit cell is also shown in the preceding figure (See Example 4.3.2).

We are now ready to introduce diffraction, both in real space and in reciprocal space formulations.

4.3 Diffraction

In 1912, von Laue[8] made the earliest suggestion that X-rays might be "diffracted" by crystals. At that time, it was commonly understood that the geometry of macroscopic crystals could be explained by a three-dimensional arrangement of atoms with interatomic distances of the order of 1–2 Å. By comparing the relationship between visible light and the repeat distances of the rulings in gratings (Fig. 3.5.1) used in optical diffraction (gratings are discussed further in §6.6), with the wavelengths of X-rays (Fig. 1.3.2) and the periodicities in crystals, von Laue suggested that crystals could serve as three dimensional gratings for X-ray diffraction. Indeed, this prediction was immediately confirmed by the work of Friedrich and Knipping [3]. However, the simple and elegant picture of W. H. Bragg and W. L. Bragg quickly overshadowed this approach of considering the crystal as a three-dimensional grating. In the Bragg picture [4], the crystal is considered to be made up of planes of atoms that specularly[9] reflect the X-rays. When the path difference between reflections from successive planes of the crystal of the "reflected" beam at a specific X-ray incidence angle (Fig. 4.3.1), is some integral number of the X-ray wavelength, they suggested that an enhancement in intensity, or "strong" beams are produced. Even though this picture of

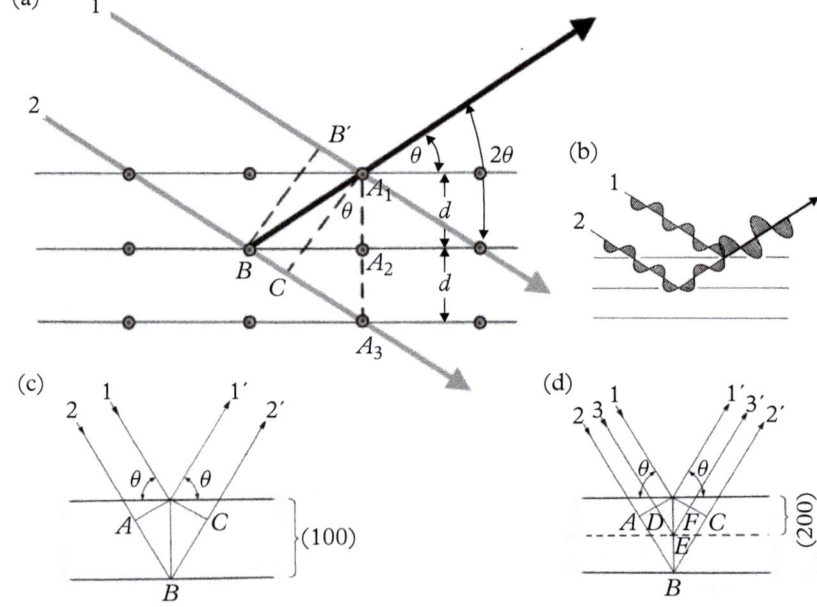

Figure 4.3.1 (a) Schematic illustration of Bragg's law of diffraction showing a simple scheme to calculate the path difference. (b) When the path difference for waves "reflected" from successive planes is an integral multiple of the wavelength, constructive interference, as shown, is expected. (c) A second-order diffraction from the (100) planes is equivalent to a first order diffraction (d) from the set of (200) planes.

Adapted from Cullity (1978).

[8] Max von Laue (1879–1960) was a German physicist who was awarded the 1914 Nobel prize in Physics for "his discovery of the diffraction of X-rays by crystals."
[9] Like a mirror.

mirror-like reflection is, strictly speaking, physically incorrect, the geometry of scattering is correctly described by the simple Bragg's law of diffraction.[10]

4.3.1 Bragg's Law: Interpreting Diffraction in Real Space

Consider a parallel, monochromatic beam of X-rays, incident at an angle, θ, on the surface of a crystal, with lattice planes separated by a distance, d (Fig. 4.3.1a). We assume that all three—the incident beam, the normal to the surface, and the diffracted beam, are always *coplanar*. Further, the diffraction angle, 2θ, between the incident and diffracted beams, typically measured in experiments, is always twice the incidence angle, θ. For diffraction or positive interference (Fig. 4.3.1b) the path difference between the two beams, 1 and 2, should be some integral number of wavelengths, $n\lambda$. The path difference, PD, between beams 1 and 2, in Figure 4.3.1a, is given by $PD = BA_1 - B'A_1 = BA_3 - BC = CA_3 = 2d \sin \theta$. Thus

$$n\lambda = 2d \sin \theta \tag{4.3.1}$$

where the integer, n, determines the order of the diffraction. We can rewrite (4.3.1) as

$$\lambda = 2 \left(\frac{d_{hkl}}{n} \right) \sin \theta \tag{4.3.2}$$

However, as we have seen before, $d_{hkl} = nd_{nh\ nk\ nl}$. Hence, n is not written explicitly, for we can consider a diffraction of nth order from any set of planes, as a first order diffraction from a parallel set of planes with $1/n^{\text{th}}$ the spacing (Fig. 4.3.1c,d). Thus, Bragg's law can be written in general terms as

$$\lambda = 2d \sin \theta \tag{4.3.3}$$

and we will use (4.3.3) throughout the book. However, note that $\sin \theta < 1$, and hence, $\lambda < 2d$, or $0 < 1/d < 2/\lambda$. Since, in most crystals, $d < 3$ Å, the useful range of X-ray wavelengths for diffraction is $\lambda < 6$ Å. Further, if the incident beam is along the x-axis, and the diffracting crystal is placed at the origin, O (Fig. 4.3.2), then all the allowed values of $g_{hkl} (= 2\pi/d_{hkl})$ in reciprocal space, which have the potential to diffract X-rays with wavelength, λ, would be enclosed in a limiting sphere of radius, $4\pi/\lambda$.

In general, the interatomic spacing, d_{hkl}, of the planes to be used in Bragg's law, is a function of their Miller indices and the parameters $(a, b, c, \alpha, \beta, \gamma)$ of the crystal lattice, as given in Table 4.1.2. As an exercise, we can start with the general equations for the triclinic crystal given in Table 4.1.2. Thus, for a tetragonal crystal, with axes $a = b$, and c, and $\alpha = \beta = \gamma = 90°$, we get, $V = abc = a^2c$,

[10] For those interested, *Fifty Years of X-Ray Diffraction* by Ewald et al. (1962) is an invaluable sourcebook covering the fascinating early beginnings of X-ray diffraction.

$s_{11} = b^2c^2 = a^2c^2 = s_{22}, s_{33} = a^2b^2 = a^4$ and $s_{12} = s_{23} = s_{31} = 0$. Substituting these values, we derive the expression in Table 4.1.2, for a tetragonal unit cell, and combine it with (4.3.3) to get

$$\frac{1}{d^2} = \frac{h^2 + k^2}{a^2} + \frac{l^2}{c^2} = \frac{4\sin^2\theta}{\lambda^2} \tag{4.3.4}$$

or

$$\sin^2\theta = \frac{\lambda^2}{4}\left(\frac{h^2 + k^2}{a^2} + \frac{l^2}{c^2}\right) \tag{4.3.5}$$

Similarly, we can show for an orthorhombic crystal (left as an exercise for the reader) that

$$\sin^2\theta = \frac{\lambda^2}{4}\left(\frac{h^2}{a^2} + \frac{k^2}{b^2} + \frac{l^2}{c^2}\right) \tag{4.3.6}$$

To summarize, the diffraction directions, 2θ, can be obtained purely from the *size and shape* of the unit cell.

Example 4.3.1: Consider a powder diffraction experiment conducted on a polycrystalline specimen of cubic KI (density 3130 kg/m^3) using Cu Kα ($\lambda = 1.54$ Å) radiation. The first eight lines were measured at the following values of 2θ:

21.75, 25.19, 36.05, 42.45, 44.5, 51.7, 56.75, and 58.5.

(a) Index these diffraction lines and calculate the d-values.
(b) What is the lattice parameter, a_0?
(c) How many formula units, Z, are in the unit cell?
(d) What structure do you expect for KI, given $r_{K+} = 1.33$ Å, $r_{I-} = 2.20$ Å?

Solution:

(a) Applying Bragg's law, (4.3.3), we get 4.08 Å (111), 3.53 Å (200), 2.5 Å (220), 2.13 Å (311), 2.04 Å (222), 1.76 Å (400), 1.621 Å (311), and 1.58 Å (420).
(b) From $d_{400} = 1.76$ Å, we get the lattice parameter $a_0 = 4\,d_{400} = 7.04$ Å.
(c) Comparing the experimental and theoretical density based on the measured lattice parameter, we get $Z = 4$.
(d) For $Z = 4$, there are two possible structures, sphalerite or rock salt. But the radius ratio (~0.61) suggests the NaCl (rock salt) structure (see Fig. 4.1.12).

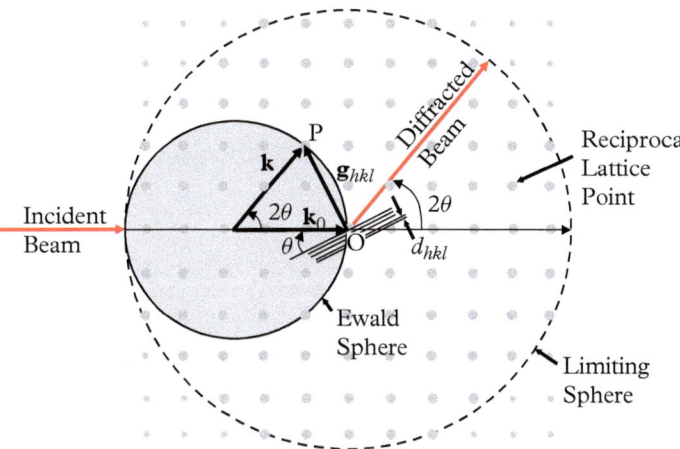

Figure 4.3.2 Interpreting diffraction in reciprocal space. The Ewald (reflecting) sphere construction for a set of planes, d_{hkl}, at the exact Bragg angle; the vector, g_{hkl}, satisfies the diffraction condition, $\mathbf{k} = \mathbf{k}_0 + \mathbf{g}_{hkl}$. The limiting sphere of radius, $4\pi/\lambda$, that includes all the possible reciprocal lattice points that can be diffracted is also shown. Note that the reciprocal lattice is fixed at the center, O.

4.3.2 The Ewald Construction: Interpreting Diffraction in Reciprocal Space

We can also represent Bragg's law in reciprocal space through the Ewald[11] sphere construction. The incident, and diffracted beams are defined by their wave vectors, $\mathbf{k}_0$ and $\mathbf{k}$, respectively, such that $| \mathbf{k}_0 |=| \mathbf{k} |= 2\pi/\lambda$ (Fig. 4.3.2). The origin of the reciprocal lattice is fixed at O, on the surface of the actual crystal. The Ewald, or reflecting, sphere, with radius, $2\pi/\lambda$, lies within the limiting sphere of radius, $4\pi/\lambda$, centered at O, and is positioned such that it just touches the limiting sphere at one end, and the origin, O, is at the other end, as shown. Note that $\mathbf{k}$ is parallel to the diffracted beam, making an angle, 2θ, with $\mathbf{k}_0$, and intersects the Ewald sphere at a point, P. Bragg's law is satisfied if, and only if OP is a reciprocal lattice vector, $\mathbf{g}_{hkl}$, of the crystal. Thus

$$\mathbf{k} = \mathbf{k}_0 + \mathbf{g}_{hkl} \qquad (4.3.7)$$

where, $\mathbf{g}_{hkl}$, is perpendicular to the diffracting planes (*hkl*) with interplanar spacing, d_{hkl}. This construction, attributed to Ewald, is the vector form of Bragg's law (also known as the Laue criterion for diffraction) and can indeed be very useful.

We can make three important conclusions of practical consequence from the Ewald construction:

1. The radius ($4\pi/\lambda$) of the limiting sphere is inversely related to the wavelength of the X-rays. If we reduce the wavelength, the radius of the limiting sphere increases, and larger reciprocal lattice vectors, or smaller inter-planar distances, are potentially included.

---

[11] P. P. Ewald (1890–1985) was an English physicist.

2. For the same incident beam direction, if the crystal is rotated about the origin, the reciprocal lattice of the crystal sweeps through the Ewald sphere. When a reciprocal lattice point intersects the Ewald sphere, Bragg's law is satisfied, and diffraction with enhanced intensity is observed. We discuss the magnitudes of the diffracted intensity in §7.4. Further, even larger portions of the three-dimensional reciprocal lattice can be sampled by rotating the crystal about two orthogonal axes. We discuss experimental methods of X-ray diffraction in §7.9.

3. The reciprocal lattice points have a finite size determined by the physical size and perfection of the real crystals (§7.8; Fig. 8.5.4). Thus, diffraction can be observed even if the Ewald sphere does not pass exactly through a reciprocal lattice point, but in close proximity to it. This is discussed further in the chapter on electron diffraction (§8.5.3).

These three rules assume purely coherent scattering without any loss in energy of the X-rays. Further, it is assumed that the incident beam is monochromatic and perfectly parallel. In practice, two principal deviations from these two conditions can be observed. Figure 4.3.3 shows the changes, using the Ewald sphere construction for (a) two different wavelengths of X-ray incidence, and (b) for poor collimation leading to beam divergence.

In practice, Bragg's law of diffraction is applied in two different ways. By using a crystal of known spacing, d_{hkl}, and measuring the angle, θ, we can determine the wavelength, λ, of the X-ray radiation. This is the principle of

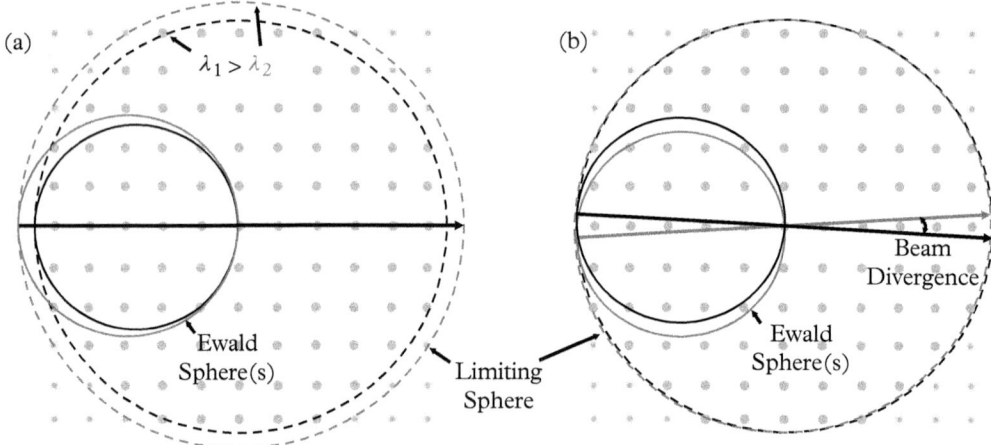

Figure 4.3.3 The Ewald and limiting sphere construction illustrating (a) two different X-ray wavelengths, and (b) poor collimation and related beam divergence.

wave-length dispersive spectroscopy as discussed in §2.5. Alternatively, it is used for crystal structure analysis where λ is known, θ is measured, and d_{hkl} of various planes (hkl) in the crystal are to be determined. This is done by a number of diffractions methods using powder samples (Debye–Scherrer and diffractometer methods), single crystals (Laue and rotating-crystal methods), or thin films (X-ray reflectivity). We discuss these methods and other experimental details of X-ray diffraction in further detail in §7.9. However, in all cases, atomic positions within unit cells can only be established by measuring the diffracted intensities. In §7, we also develop a formulism to determine the diffraction intensities by first considering how X-rays are scattered by single electrons, then by an atom, and finally by the unit cell of the periodic crystal. Beyond §7 and §8, interested readers should consult Cullity (1978) and Hammond (2006) for further details of diffraction; for a more advanced perspective, see Woolfson (1997), Warren (1990), and Cowley (1975).

Example 4.3.2: For the orthorhombic crystal in Example 4.2.1:

(a) Apply Bragg's law and calculate the angle, θ, of diffraction for all the planes using (i) Cu Kα, and (ii) Mo Kα, radiations. If you are interested in the (310) reflection, which radiation is better for this measurement?

(b) Plot the two-dimensional real and reciprocal lattices (only **a***-**b*** plane) for this crystal.

(c) Draw the incident and diffracted vectors, and the Laue condition for diffraction of the (310) planes. Draw all angles accurately.

Solution:

(a) The wavelengths for the two radiations are $\lambda_{Cu,\,K\alpha} = 1.54$ Å, and $\lambda_{Mo,\,K\alpha} = 0.709$ Å.

We apply Bragg's law, (4.3.3), and solve it as a table:

Plane	(100)	(002)	(030)	(011)	(111)	(310)	(311)	(312)
d_{hkl} (Å)	2.0	2.0	1.0	2.4	1.54	0.65	0.64	0.62
θ (Cu K$_\alpha$)	22.64°	22.64°	50.35°	18.71°	30.0°	–	–	–
θ (Mo K$_\alpha$)	10.2°	10.2°	20.76°	8.5°	13.31°	33.05°	33.63°	34.78°

Based on the preceding table, it is not possible to use Cu K$_\alpha$ radiation to measure the (310); reflection and Mo K$_\alpha$ are preferred.

(c) For $\theta = 33.05°$

(b)

4.3.3   Comparison of X-Ray and Electron Diffraction

The wave particle duality, corrected for relativistic effects, gives the wavelength of electrons accelerated through a voltage, V, in volts, as

$$\lambda = \frac{h}{p} = \frac{h}{\sqrt{2m_e e V \left(1 + \frac{eV}{m_e c^2}\right)}} \tag{4.3.8}$$

where m_e is the rest mass of the electron, e is its charge, and c is the speed of light. Typically, in a transmission electron microscope (§9), electrons are accelerated through a voltage in the range 100 kV $< V <$ 400 kV, with corresponding wavelengths, λ, including relativistic corrections in the range 0.0037 nm $> \lambda >$ 0.00164 nm. Note that, when relativistic effects are ignored, (4.3.8) can be simplified as

$$\lambda = \frac{h}{p} = h(2m_e e V)^{-1/2} \sim \left(\frac{150}{V}\right)^{1/2} \tag{4.3.8a}$$

where V is in volts and λ is in Å.

Assuming 100 keV electron incidence, for diffraction from crystals with typical inter-planar spacing, $d_{hkl} \sim 0.2$ nm, the Bragg scattering angle, θ, is expected to be $< 1°$. At such small angles, $\sin\theta = \theta$, and Bragg's law of diffraction simplifies to $\lambda = 2d\theta$. We can use this simplified form in the Ewald sphere construction for electron diffraction obtained in transmission electron microscopes (TEMs) using thin electron transparent specimens. The very short wavelength of electrons, compared to the inter-planar spacing, d_{hkl}, in crystals, can modify the Laue equation, (4.3.7), i.e. $|\mathbf{k}_0| = |\mathbf{k}| = 2\pi/\lambda \gg \mathbf{g}_{hkl}$, and affect the Ewald sphere construction (Fig. 4.3.4). The electron diffraction pattern, obtained in transmission, contains all the reciprocal lattice points intersected by the Ewald sphere. Because of the extremely shallow curvature, or very large radius of the Ewald sphere when compared to the reciprocal lattice, multiple regions

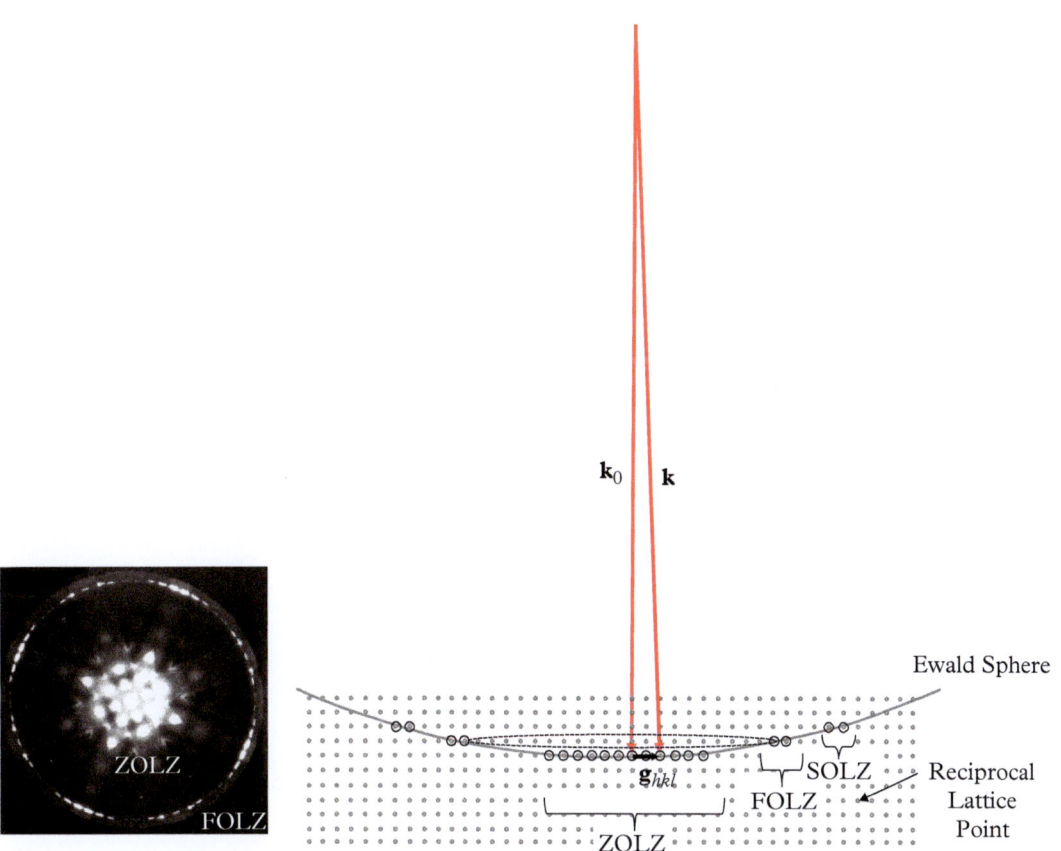

Figure 4.3.4 The Ewald sphere construction for electron diffraction, typically encountered in a transmission electron microscope. Because of the very short wavelength of the electrons the Ewald sphere radius is very large with a very flat surface, which intersects the reciprocal lattice in different zones and satisfying different zonal conditions. For a beam along [UVW], the lower image shows the zero order Laue zone, ZOLZ, with $hU + kV + lZ = 0$, and the ring of the first order Laue zone, FOLZ, with $hU + kV + lZ = 1$. The specimen is a single crystal of cubic spinel viewed along [001].

of diffraction spots, called Laue zones, can be simultaneously excited and seen in electron diffraction patterns (also see §8.6.3). First, a set of diffraction spots around the transmitted beam, 000, incident along a direction [UVW], called the zero order Laue zone (ZOLZ) is observed. The ZOLZ is so called because it also includes the origin, 000, in reciprocal space. Recall the zonal equation, (4.2.5), illustrated in Figure 4.1.10, and the related discussions. Thus, all the diffracted beams in the ZOLZ satisfy the zonal equation $hU + kV + lZ = 0$. In addition, because of its shallow curvature, the Ewald sphere can intersect reciprocal lattice points from planes with normal that are *not parallel* to the electron beam. These are called higher order Laue zones (HOLZ) and they are given specific names, i.e. first-order Laue zone (FOLZ) if they satisfy $hU+kV+lZ = 1$, second-order Laue zone (SOLZ), if they satisfy $hU+kV+lZ = 2$, and so on. Figure 4.3.4 includes an example of a diffraction pattern from a single crystal spinel ($MgAl_2O_4$) specimen, illustrating the ZOLZ and FOLZ.

It is instructive to compare typical scattering geometries for X-ray and transmission electron diffraction. We use an oriented single crystal thin film with an orthorhombic unit cell as an example (Fig. 4.3.5). For the θ-2θ X-ray scattering geometry shown (a), the measured lattice parameters would correspond to the family of d_{0k0} planes. On the other hand, if the same film were to be studied with a TEM, with plan view specimens, as shown in (b), the measured lattice parameters would be different and correspond to the d_{h00} family of planes. In other words,

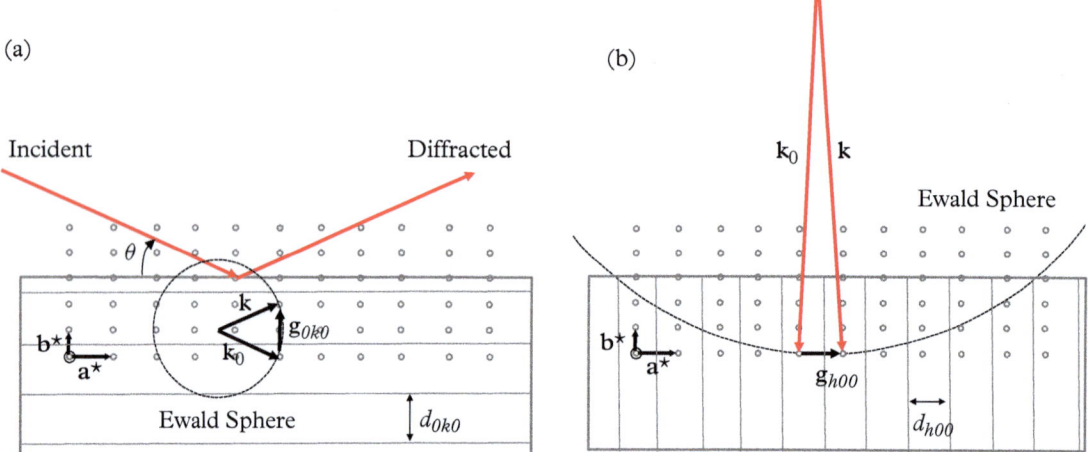

Figure 4.3.5 (a) X-ray diffraction, including the Ewald sphere construction, for a typical θ-2θ scan of a single crystal thin film. Note that in this scattering geometry the inter-planar spacing of $(0k0)$-type planes are measured. (b) The same film measured by transmission electron diffraction. The scattering geometry, including the Ewald sphere, represents the diffraction condition shown and measures the $(h00)$-type planes.

the two techniques measure inter-planar spacing of completely different families of planes, with significant consequence in the analysis of noncubic crystals.

Electron diffraction is discussed further in §8 and §9, as well as in pedagogically excellent books by Williams and Carter (1996), Cowley (1984), and Spence and Zuo (1992). Here, from this brief introduction, the important point to take away is that in typical transmission electron diffraction the wavelength of the electrons is much smaller than the inter-planar spacing in crystals, i.e. $\lambda_e << d_{hkl}$, and this flattens the Ewald sphere and affects the Laue interpretation of diffraction.

Example 4.3.3: X-ray diffraction, using **Cu Kα** ($\lambda = 0.154$ nm) radiation, gives a peak for the (111) planes of polycrystalline Aluminum at $2\theta = 38.42°$.

(a) What is the interplanar spacing of the (111) planes?

(b) What is the lattice parameter for cubic aluminum?

(c) When 100 kV electrons are used, instead of X-rays, the peak for the same (111) planes is now observed at $2\theta = 0.9°$. What is the wavelength of the high-energy electrons?

(d) Calculate the de Broglie wavelength of 100 kV electrons? Compare with the result of (c) and comment on the difference.

Solution: (a) The interplanar spacing can be calculated from Bragg's law, (4.3.3), as

$$d_{111} = \lambda / (2 \; \mathrm{Sin}\theta) = 0.154 / (2 \; \sin \; 19.21) = 0.234 \; \mathrm{nm}.$$

(b) For a cubic crystal, from Table 4.1.2, $d_{111} = a_0/(3)^{1/2}$. Thus, the lattice parameter,

$$a_0 = (3)^{1/2} d_{111} = 0.405 \; \mathrm{nm}.$$

(c) Again, applying (4.3.3), $\lambda_{\mathrm{electron}} = 2 \; d_{111} \; \mathrm{Sin}\theta = 2 \times 0.234 \times$ Sin $0.45 = 0.0037$ nm.

(d) The energy of electrons accelerated through 100 kV is $E = 100,000 \times 1.6 \times 10^{-19}$ J.

If we assume all this energy is kinetic, then $E = p^2/2m_e$, where p is the momentum and m_e is the rest mass of the electron.
   The de Broglie wavelength

$$\lambda = h/p = h/(2m_e E)^{1/2} = 6.626 \times 10^{-34}/(2 \times 9.1 \times 10^{-31} \times 1.6 \times 10^{-14})^{1/2}$$

$$= 0.00388 \; \mathrm{nm}$$

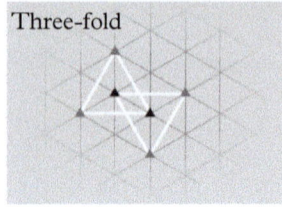 Three-fold

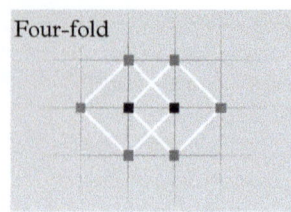 Four-fold

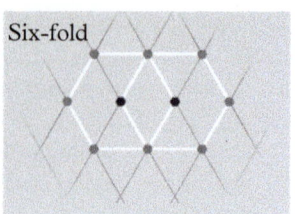 Six-fold

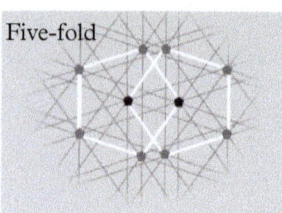 Five-fold

Figure 4.4.1 Lattice translations and rotational axes. Note that a fivefold rotation axes is not consistent with lattice translations.

The results of (c) and (d) are close, and the difference can be attributed to the need to use relativistic corrections, (8.2.4), for the mass of the electron when it is accelerated though 100 kV potential.

4.4 Quasicrystals and the Definition of a Crystalline Material

Section 4.1.7 showed that the 230 space groups cataloged in the ITC represent all the possible combinations of symmetry elements, and thus, all possible patterns in three dimensions that can be generated by periodic repetition. One of the cardinal rules of classical crystallography is that a fivefold rotational symmetry is inconsistent with lattice translations (Fig. 4.4.1).

However, Shechtman discovered crystals with diffraction patterns that could be rotated by 36° (tenfold rotation) without any change in the pattern (see Fig. 1.4.8c and Fig. 4.4.2a–c). The discrete diffraction pattern suggests that the atoms in the materials are packed into an ordered crystal. Such materials with fivefold symmetry [6], subsequently named as quasicrystals, are characterized by interatomic distances that are regular, in that they correlate with the Fibonacci sequence.[12] A fascinating aspect of quasicrystals is the appearance of the golden ratio[13], Φ (144/89 =1.61803 . . .), in the ratios of various inter-planar distances in the material. However, the regularity is not the same as when the crystal is periodic. In actuality, perfect quasicrystals appear in the form of an icosahedrons, a solid with close-packing on each of its twenty triangular faces [7] and a point group symmetry of 235.

Since its discovery in 1982, many quasicrystals have been synthesized in the laboratory, and in 2009, a naturally occurring mineral, *icosahedrite*, with characteristic symmetries of the icosahedron (Fig. 4.4.2a–c), was also found [8]. Figure 4.4.2d shows an example of a stable bulk quasicrystal, $Al_{65}Cu_{20}Fe_{15}$ alloy, showing pentagonal dodecahedral faces [9].

[12] A sequence of numbers, where each one is the sum of the preceding two, i.e. 1, 2, 3, 5, 8, 13, 21, etc.

[13] The golden ratio is an irrational number that appears in mathematics, art, and nature. It can be seen in the spirals of mollusk shells, pineapple segments, fir cones, and the arrangement of seeds in a sunflower!

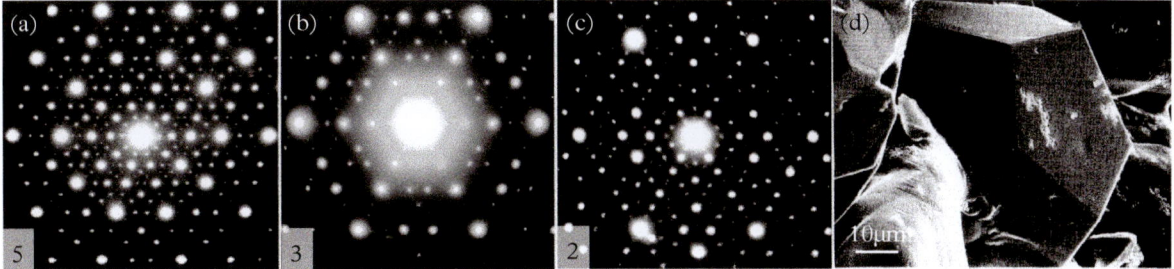

Figure 4.4.2 Electron diffraction patterns from a naturally occurring mineral icosahedrite with characteristic point group symmetry, 235, of an icosahedron. (a) Fivefold, (b) threefold, and (c) twofold. (d) Scanning electron micrograph of a $Al_{65}Cu_{20}Fe_{15}$ quasicrystal prepared by conventional rapid solidification and annealing.

(a–c) Adapted from [6]. (d) Adapted from [8].

Before Schechtman's discovery of *quasicrystalline* materials, a crystalline material was defined as "a substance in which the constituent atoms, molecules, or ions are packed in a regular, ordered, repeating, three-dimensional pattern". As a result of his discovery, the International Union of Crystallography changed the definition of a crystalline material to "any solid having an essentially discrete diffraction pattern".

So far, we have motivated our development of spectroscopy and diffraction by emphasizing their importance in the context of bonding and structure of materials. In Chapter 5 we make a detour to systematically look at specific characteristics of different probes, including electrons, photons, neutrons, and ions, some of which have already been introduced. In the process, we look particularly at ion-based characterization methods (§5.4) in greater detail and discuss related techniques. After that, in §6 we visit the important topic of optics and optical diffraction, which serves as a good entry point to more sophisticated methods of diffraction (§7, §8) and microscopy (§9, §10) discussed in subsequent chapters.

Summary

Most of the materials—-be they metals, alloys, ceramics, semiconductors, and even some polymers and biomaterials—-encountered in characterization have a periodic arrangement of atoms and exhibit long-range order. The description of such a crystalline state starts with the definition of a space lattice, which is then used to develop useful concepts of Bravais lattices, crystal systems, the nomenclature to prescribe its various features (lattice points, lines, directions, and planes), and the practically useful concepts of interplanar spacing and zonal equations for interpreting diffraction patterns (see §7 and §8 for more).

Crystals can be described simply in terms of space filling and close packing. However, the most comprehensive and systematic way to describe periodic crystals is to understand how repetitive patterns can fill spaces. Thus, by assigning atoms/molecules to these patterns, subject to various symmetry operations, we arrive at a general description of crystal structures. This requires an understanding of symmetry, symmetry elements, and point and space groups. These concepts were developed in two dimensions, then extended to three dimensions, and followed with an introduction to the ITC, which provides a comprehensive description of all the 230 possible space groups. An annotated page was used to describe a typical entry for the space group *Pnma* (#62), and the reader is advised to review this carefully. Finally, practical materials are not ideal crystals as they harbor a wide range of imperfections. These are classified as point, line, and planar defects, and their identification and enumeration are an important part of characterization, for defects significantly affect a wide range of materials properties.

The concept of reciprocal lattice is the key to our understanding of diffraction. It has a fixed and well-defined relationship to the real lattice from which it is derived using either a physical or a mathematical approach. Both approaches were introduced, along with a number of examples.

Crystallography and diffraction are closely related. In fact, diffraction provides the methodology—some may even argue that this may be the only method—to reveal the structure of crystals. The inter-relationship between crystallography and diffraction was introduced in this chapter and will be developed further in dedicated chapters using X-rays (§7) and electrons (§8). Interpretation of diffraction patterns is done either in real space (Bragg's law), or in reciprocal space (Laue criterion and Ewald sphere construction); however, one must hasten to add that the two are equivalent, but the latter is most general and finds particular use in electron diffraction and microscopy.

Finally, recent discovery of quasicrystal materials exhibiting fivefold rotational symmetry, which is inconsistent with lattice translations, has resulted in redefining a crystalline material as "any solid having an essentially discrete diffraction pattern."

. . . .

FURTHER READING

Allen, S., and E. L. Thomas. *The Structure of Materials*. Hoboken: Wiley, 1999.

Borchardt-Ott, W. *Crystallography*. Berlin: Springer, 1995.

Buerger, M. *Elementary Crystallography*. Cambridge: MIT Press, 1970.

Callister, W. D. Jr., and D. G. Rethwisch. *Materials Science and Engineering—An Introduction*. Hoboken: Wiley, 2009.

Cowley, J. M. *Diffraction Physics*. Amsterdam: North-Holland, 1984.

Cullity, B. D. *Elements of X-Ray Diffraction*, 2nd ed. Reading: Addison-Wessley, 1978.

Ewald, P. P. (ed). *Fifty years of X-Ray Diffraction*. Utrecht: International Union of Crystallography, 1962.

Hahn, T. (ed). *International Tables for Crystallography, Vol A: Space-Group Symmetry*, 3rd ed. Dordrecht: D. Reidel Publishing Company, 1993.

Hammond, C. *The Basics of Crystallography and Diffraction*. Oxford: Oxford University Press, 2006.

Nabarro, F. R. N. *Theory of Crystal Dislocations*. Oxford: Oxford University Press, 1967.

Nye, J. F. *Physical Properties of Crystals*. Oxford: Oxford University Press, 2000.

Sands, D. E. *Introduction to Crystallography*. Wilmington: Dover Publications, 1975.

Spence, J. C. H., and J. M. Zuo. *Electron Microdiffraction*. New York: Plenum Press, 1992.

Warren, B. E. *X-Ray Diffraction*. Dover, 1990.

Weertman, J., and J. R. Weertman. *Elementary Dislocation Theory*. Oxford: Oxford University Press, 1992.

Williams, D. B., and C. B. Carter. *Transmission Electron Microscopy, Vol II. Diffraction*. New York: Plenum Press, 1996.

Williams, D.B., A.R. Pelton, and R. Gronsky (eds). *Images of Materials*. Oxford: Oxford University Press, 1991.

Woolfson, M.F. *An Introduction to X-Ray Crystallography*. Cambridge: Cambridge University Press, 1997.

....

REFERENCES

[1] Mooser, E. "Semiconducting Compounds." *Science*, 132, no. 3436 (1960): 1285–91.

[2] Siegel, R.W. "Positron Annihilation Spectroscopy." *Annual Review of Materials Science* 10 (1980): 393–425.

[3] von Laue, M. "Kritische Bemerkungen zu den Deutungen der Photogramme von Friedrich und Knipping." *Physikalische Zeitschrift* 14, no. 10 (1913): 421–3.

[4] Bragg, W.L. "The Diffraction of Short Electromagnetic Waves by a Crystal." *Proceedings of the Cambridge Philosophical Society* 17 (1913): 43–57.

[5] Whelan, M.J., P.B. Hirsch, R.W. Horne, and W. Bollmann. "Dislocations and Stacking Faults in Stainless Steel." *Proceedings of the Royal Society A: Mathematical, Physical a nd Engineering Sciences* 240, no. 1223 (1957): 524–38.

[6] Shechtman, D., I. Blech, D. Gratias, and J.W. Cahn. "Metallic Phase with Long-Range Orientational Order and No Translational Symmetry." *Physical Review Letters* 53, (1984): 1951.

[7] Mackay, A.L. "A Dense Non-Crystallographic Packing of Equal Spheres." *Acta Crystallographica* 15, no. 9 (1962): 916–18.

[8] Bindi, L., P.J. Steinhardt, N. Yao, and P.J. Lu. "Icosahedrite, $Al_{63}Cu_{24}Fe_{13}$, The First Natural Quasicrystal." *American Mineralogist* 96 (2011): 928–31.

[9] Tsai, A., A. Inoue, and T. Matsumoto. "A Stable Quasicrystal in Al-Cu-Fe System." *Japanese Journal of Applied Physics* 26, no. 9A (1987): L1505.

[10] Bancel, P.A., P.A. Heiney, P.W. Stephens, A.I. Goldman, and P.M. Horn. "Structure of Rapidly Quenched Al-Mn." *Physical Review Letters* 54 (1985): 2422.

. . . .

EXERCISES

A. Test Your Knowledge

Complete each statement with all the possible (maybe none, or more than one) correct answers.

1. A polycrystalline material **always** contains
   a) crystals of different chemical composition.
   b) crystallites of the same composition but different structures.
   c) crystallites with different orientations.

2. A material containing grain boundaries is
   a) grainy.
   b) metallic.
   c) a polymer.
   d) polycrystalline.

3. A crystal is a homogeneous body with
   a) isotropic properties.
   b) cubic symmetry.
   c) a discrete diffraction pattern.

4. A lattice is
   a) an array of points in space with identical surroundings.
   b) a repeating set of unit cells.
   c) defined by six parameters.

5. Cubic symmetry is characterized by
   a) one tetrad axis.
   b) three orthogonal tetrad axes.
   c) four triad axes along the body diagonals.

6. The area of the (101) plane in a *bcc* structure is
   a) $4\sqrt{5R^2}$
   b) $(16/3)\,R^2$
   c) $\left(16\sqrt{2}/3\right)R^2$

7. [1/3 ½ 1] and [2 3 6] refer to
    a) two *different* planes.
    b) the *same* line.
    c) a *family* of parallel lines passing through the origin.
8. [*hkl*] is normal to (*hkl*)
    a) in *all* crystal systems.
    b) in *cubic* crystals.
    c) for *select* planes in the orthorhombic and tetragonal crystals.
9. Diffraction from crystals can be described by
    a) Bragg's law.
    b) the Ewald construction applying the Laue condition.
    c) both (a) and (b).
10. When we rotate a crystal for a diffraction experiment
    a) the reciprocal lattice remains fixed/unchanged in the laboratory frame of reference.
    b) the reciprocal lattice attached to the crystal rotates in the laboratory frame of reference.
    c) the angle between the incident beam and any reciprocal lattice vector does not vary.
11. The lattice spacing of any *general* crystal structure depends on
    a) only the indices, (*hkl*), identifying the plane of interest.
    b) only the unit cell parameters, a, b, c.
    c) only the unit cell angles, α, β, γ.
    d) all of the above considered together.
12. The zonal equation identifies
    a) the zone where crystals are normally oriented.
    b) a zone containing any set of planes.
    c) a direction that is contained and common to a specific set of lattice planes.
13. In the Bragg equation, $\lambda = 2d \sin\theta$, the angle, θ, is between the
    a) incident and diffracted X-ray beams.
    b) incident beam and the *normal* to the diffracting planes.
    c) incident beam and the diffracting planes.
14. An FCC crystal has a
    a) ABCABC . . . stacking sequence.
    b) BCC unit cell in the reciprocal lattice.
    c) lattice parameter, $a = 2\sqrt{2}R$, where R is the radius of the atom.
15. CsCl has a
    a) BCC Bravais lattice.
    b) structure that is made of two interpenetrating simple cubic unit cells.
    c) body centering translation.

16. A fivefold rotational symmetry axes
    a) can be found in molecules.
    b) cannot be found in crystals with translational symmetry.
    c) has been found in actual crystals.
17. The International Tables of Crystallography
    a) had its name changed from *International Tables of X-Ray Crystallography* because it was found to be not relevant for X-ray diffraction.
    b) is a good source of useful information on diffraction conditions.
    c) has entries for 230 space groups, but this is subject to change.
18. A Burgers vector
    a) is very tasty and filling.
    b) defines the displacement associated with a dislocation.
    c) is always along or normal to a dislocation line.
19. Bragg's law of diffraction
    a) applies *only* to single crystals.
    b) depends on the parameters of the unit cell.
    c) depends *only* on the Miller indices of the diffracting planes.
    d) is different from the Ewald sphere construction.
    e) is equivalent to the Laue criterion for diffraction in reciprocal space.

B. Problems

1. Consider the **tetragonal** crystal with lattice parameters $a = 3$ Å, $b = 3$ Å and $c = 4$ Å ($\alpha = \beta = \gamma = 90°$).
    a) Draw the unit cell.
    b) Calculate the inter-planar spacing, for the first ten planes.
    c) What is the common direction for the (011) and (111) planes (Hint: Use cross-product)? Would either (310) or (311) planes, or both, share this common direction (i.e. form part of the same zone)?
    d) Calculate the unit cell dimensions of the reciprocal lattice? Draw this in the same figure as (a) but with different scale (note: units are different).
2. Apply **Bragg's law** and calculate the angle, θ, of diffraction for the first 10 planes of the same tetragonal crystal in Problem 1, using **Cu** K_α and **Mo** K_α radiations. Which radiation would be preferred if you are interested in the (310) reflection?
3. Using the results of Problems 1 and 2, plot
    a) the two-dimensional reciprocal lattice (only the $\mathbf{a^\star}$-$\mathbf{b^\star}$ plane) for this crystal.
    b) then draw the incident and diffracted wave vectors, as described in the text, and illustrate the Laue condition for diffraction of the (310) planes. Draw all angles accurately.
4. The **angle**, ρ, between two vectors, $\mathbf{a}$ and $\mathbf{b}$, is given by $\rho = \cos^{-1}\left(\frac{\mathbf{a}\cdot\mathbf{b}}{|\mathbf{a}||\mathbf{b}|}\right)$

a) Show that the angle, ρ, between normals of planes $(h_1 k_1 l_1)$ and $(h_2 k_2 l_2)$ in an orthorhombic crystal is given by

$$\cos \rho = \frac{\left(\frac{h_1 h_2}{a^2}\right) + \left(\frac{k_1 k_2}{b^2}\right) + \left(\frac{l_1 l_2}{c^2}\right)}{\sqrt{\left[\frac{h_1^2}{a^2} + \frac{k_1^2}{b^2} + \frac{l_1^2}{c^2}\right]\left[\frac{h_2^2}{a^2} + \frac{k_2^2}{b^2} + \frac{l_2^2}{c^2}\right]}}$$

b) What will be the expression for the angle, ρ, for cubic crystals?
c) Using the expression derived in (b), for a cubic crystal, determine the angle between (111) and (110) plane normals?
d) Using the stereographic projection and Wulf net in Figure 4.1.18, see if you can determine the same angle.

5. Draw the unit cell of the **diamond structure.**
   Determine the Miller indices of a plane containing three atoms that are *nearest neighbors* in the diamond lattice.
6. Derive the [111] **reciprocal lattice section** for the cubic-F and cubic-I lattices.
7. Complete the following table:

Element	Latticeparameter	Structure	Distance between centers of neighboring atoms	Atomic radius
Na	0.424 nm	bcc		
K	0.462 nm	bcc		
Cu	0.361 nm	fcc		
Ag	0.408 nm	fcc		

8. For the **NaCl structure**, what is the direction common to the planes $(\bar{1}11)$ and (002)?
9. You have a single crystal of NaCl ($a_0 = 0.564$ nm) cut (atomically clean) with a (001) normal. If you grow a film of Au ($a_0 = 0.4065$ nm) on top, how will the film grow epitaxially (if you can tolerate a strain of 3%). Note, best **epitaxy** is achieved if the lattice planes of the substrate and film, have a well-defined orientation relationship, and overlap with minimum strain. Hint: calculate inter-planar distances for the principal planes in the film and substrate, and compare.
10. An electron diffraction **zone axis pattern** along the [111] direction for a BCC crystal (similar to Figure 4.3.4) shows a FOLZ ring. Is this possible?
11. Following Example 4.1.5, draw the stereographic projection for the point group $\frac{4}{m} mm$.
12. An X-ray diffraction pattern for a **quasicrystalline** Al–Mn alloy, obtained in a synchrotron, is in Figure 4.Pr.12. Note that the intensity

is plotted as a function of Q ($= 2\pi/d$), and the intensities of the four strongest peaks have been truncated. Adapted from [8].

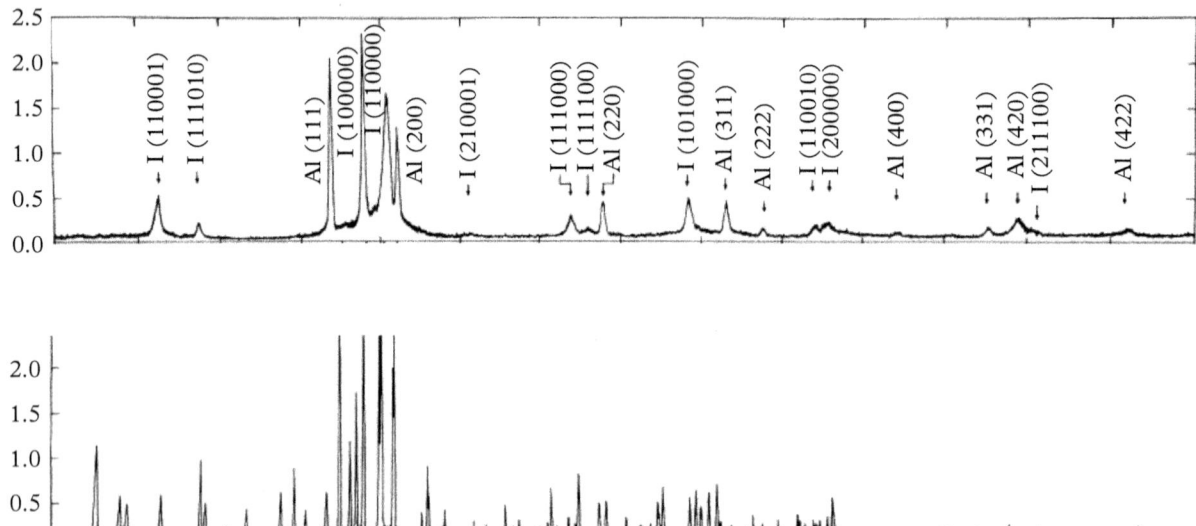

a) Make a table of all the observed interplanar spacing from the X-ray scan (Fig. 4.Pr.12).
b) In this experiment, the quasicrystalline phase is mixed with pure Al.
c) Calculate or look up the *dhkl* spacing of aluminum and identify the peaks that correspond only to the quasicrystal phase.
d) Show, by taking ratios, that the inter-planar spacing observed for the quasicrystalline phase satisfy the golden rule.

13. Crystal structure of **semiconducting compounds**: see Figure 4.Pr.13, adapted from [1].

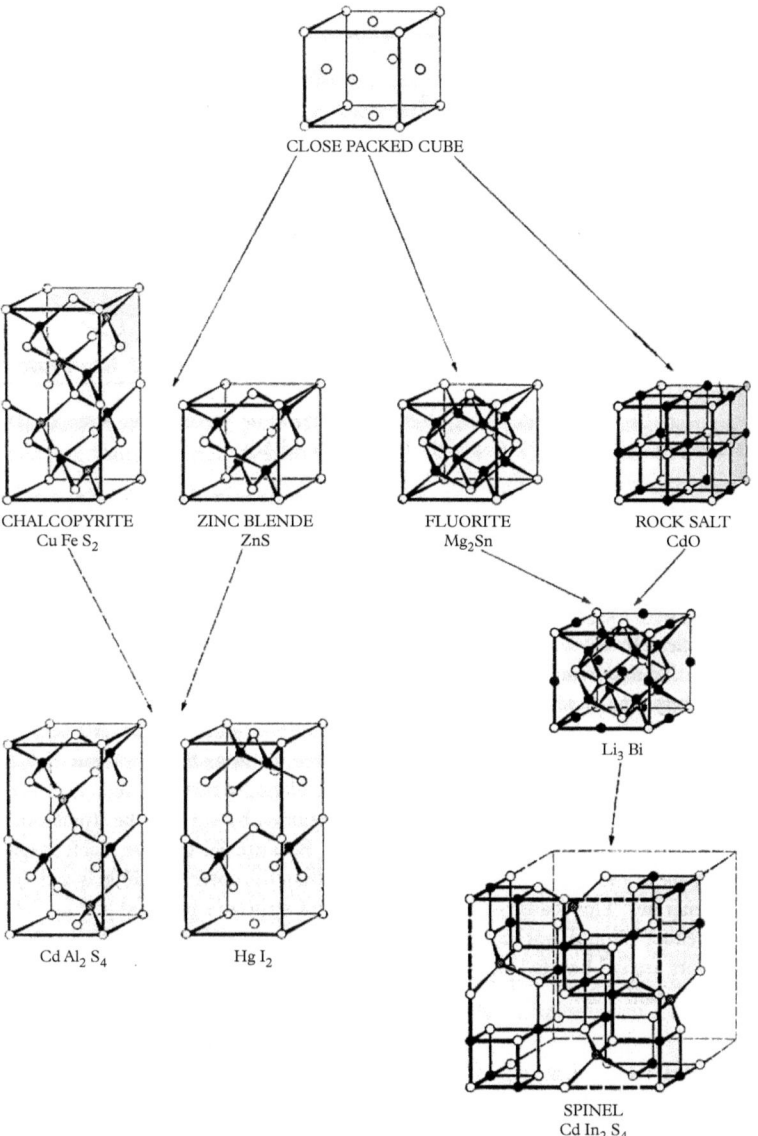

a) Why do we expect tetrahedral and octahedral coordination in typical semiconductor compounds?

b) Draw a unit cell of the FCC structure and identify ALL the tetrahedral and octahedral sites.

c) The figure (adapted from [1]) shows some of the semiconductor structures derived from the close-packed cubic structure.

Based on an inspection of these structures, give a short explanation of how the unit cells of each of these semiconductor compounds (chalcopyrite, zinc blende, fluorite, rock salt, and spinel) can be derived from the parent FCC structure?

14. In a θ-2θ **X-ray scan** for a *cubic* material with lattice parameter, $a_0 = 4.8$ Å, using Cu Kα radiation, three peaks are observed at $2\theta = 26°, 32°$, and $38°$.

   a) Calculate the interplanar spacing, d_{hkl}, for each peak.
   b) Index or identify the crystal plane (hkl) for each case.
   c) Is this a FCC, BCC, or simple-cubic crystal?

15. Draw the standard (010) **stereographic projection** for a cubic crystal, with the (100) pole pointing south, and label the Miller indices for all major planes.

16. Gold has an **FCC structure** as shown in Figure 4.1.11f. (a) How many atoms of Au are there per unit cell? What is the relationship between the radius, R, of a gold atom and the lattice parameter, a, of the unit cell? What is the atomic packing factor (APF) defined as the ratio of the volume of the atoms in the unit cell to the unit cell volume? Given the atomic mass of gold is 196.97 g/mol and the lattice parameter $a = 0.407$ nm, what is the density of a gold crystal? What is the atomic concentration or the number of gold atoms per unit volume?

Probes: Sources and Their Interactions with Matter

5

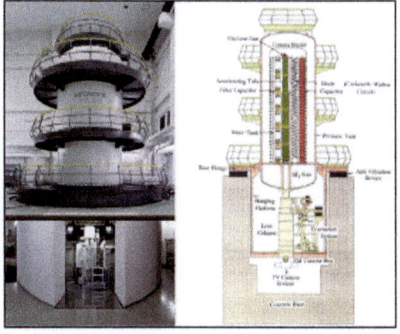

Probes can be generated in the laboratory or in large-scale user facilities. The latter provide probes (photons, ions, electrons, and neutrons) with unique capabilities for materials characterization and analysis. *Top*: The Advanced Photon Source at Argonne National Laboratory (Illinois, USA) is a 7-GeV synchrotron that provides ultra-bright, high-energy, storage ring-generated X-ray beams for research in almost all scientific disciplines. The diameter of the circular experimental hall, where numerous photon beam lines for specific experiments are located, is 373.4 m. *Middle*: An ion beam facility at Lawrence Berkeley National Laboratory (California, USA) producing He^{2+} ions, with variable energies (0.6–5.0 MeV) and used for materials analysis. The principal ion-beam based characterization methods (RBS, SIMS, PIXE) are discussed later in this chapter (§5.4). *Bottom*: An ultra-high-voltage (3 MV) transmission electron microscope at Osaka University (Japan). Such machines are being replaced by aberration-corrected electron microscopes (§9.2.9) with superior resolution achieved at much lower voltages. See also Figure 5.2.6, which shows a spallation neutron source.

Principles of Materials Characterization and Metrology. Kannan M. Krishnan, Oxford University Press (2021).
© Kannan M. Krishnan. DOI: 10.1093/oso/9780198830252.003.0005

5.1 Introduction

In any probe-based materials characterization or analysis technique, the main goal is to maximize information while minimizing any damage to the specimen.

To this end, we first consider commonly available probes (photons, electrons, neutrons, and ions), describe ways to generate them in a controlled manner, emphasize different types of practical sources, and discuss their principal characteristics. Next we discuss the interactions of these probes with matter, including the types of damage we can expect in the course of materials characterization. The emphasis in this chapter is on electrons and ions; however, optics and optical methods are discussed in §6, photons and X-rays in §7, neutrons in §8, electrons in §8–10, and scanning probes in §11.

The second half of this chapter is devoted to ion-based methods. We begin with a discussion of the kinematics of ion–atom collisions and use the physical principles of such scattering to discuss ion-based characterization methods (§5.4). In particular, principles and applications of Rutherford back-scattering spectroscopy (RBS), low-energy ion-scattering spectroscopy (LEISS), and secondary ion mass spectroscopy (SIMS), followed by induction-coupled plasma mass spectrometry (ICP-MS) and particle-induced X-ray emission spectrometry (PIXE) are presented.

5.2 Probes and Their Generation

5.2.1 Photons: Lamps and Lasers

Photon sources for materials characterization and analysis across the electromagnetic spectrum (Fig. 1.3.2) include X-ray tubes (Fig. 2.4.2), synchrotron radiation (§2.4.3), incandescence (heated filaments) and plasma discharge (arc) lamps, and simulated emission of a gas or solid (lasers). The first two were discussed in §2.4, and this section discusses the last three, of particular importance in molecular (§3.4), infrared (IR; §3.5) and Raman (§3.6) spectroscopy, as well as optical microscopy and metallography (§6.8).

5.2.1.1 *Light Sources for Optical Microscopy*

Incandescent light sources originally included tungsten and carbon filaments. Now, they are largely dominated by tungsten-halogen lamps and are used reliably in optical microscopy (§6.8). When heated by an electric current, they generate a continuous spectrum of light, ranging from the central ultraviolet (UV) ($\lambda = 200 - 380$ nm), through the visible ($\lambda = 380 - 700$ nm), and into the IR region ($\lambda = 700 - 1000$ nm; Fig. 5.2.1). Typically, in tungsten-halogen lamps, the filaments are heated to $\sim$2,000–2,500 K, and hence, the longer wavelengths predominate when compared to the emission spectrum of sunlight.[1] The emission profile of tungsten-halogen lamps can be made to shift to shorter wavelengths

---

[1] The sun being an ideal black body radiator at $\sim$5,500°C.

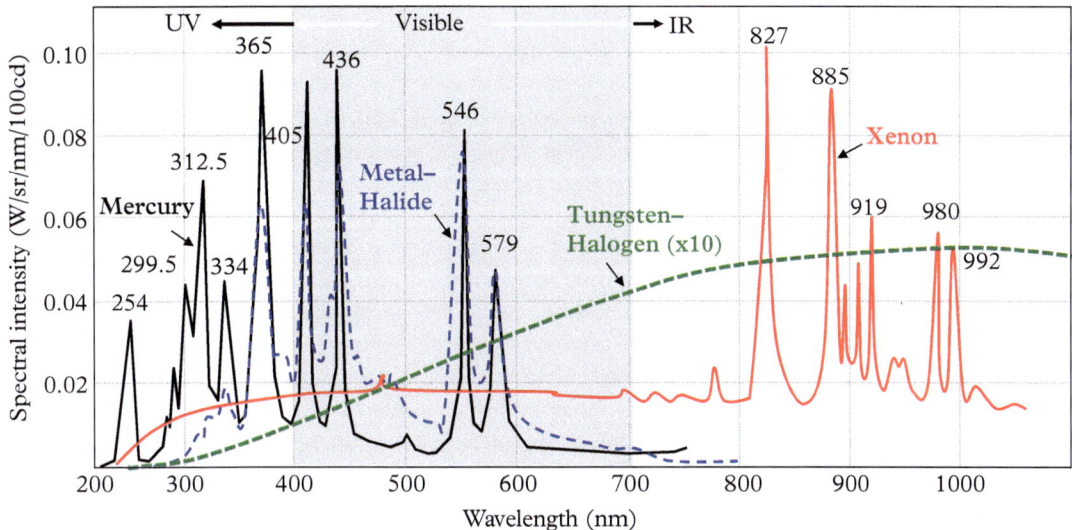

Figure 5.2.1 Emission characteristics of various optical sources and lamps. For each one, principal emission lines and their relative intensities are shown. The emission profile of a mercury arc lamp shows distinct peaks at 365 nm (near UV with 10.7% of the total intensity), 405 nm (violet, 4%), 436 nm (dark blue, 12.6%), 546 nm (green-yellow, 7%), and 579 nm (yellow, double band, 8%). The last two are at the center (550 nm) of the highlighted region of the visible spectrum.

by increasing the temperature. At temperatures closer to the melting point of tungsten ($T_m = 3,965$K), the fraction of the visible spectrum in the radiation increases substantially. Needless to say, such high operating temperatures are not a very practical solution for extended time of use. However, one advantage of the tungsten-halogen lamp is that once it reaches its operating temperature its emission is very stable.

Alternatively, if two metal rods are connected to a source of direct current and then they are slowly drawn apart, a bright arc of light is generated between them. In early experiments, with Fe, Cu, or Al, used as the metal rods, it was seen that the radiation came mainly from a gas of metal vapor that traversed the arc. This effect is particularly enhanced for low melting point metals, such as sodium or mercury. In early designs, the sodium metal was enclosed in a glass tubing or envelope and filled with argon or neon at low pressure. In this case, the discharge is initiated in the rare gas, and as the temperature in the tube rises, it eventually melts the sodium and vaporizes it into the arc. In short order, the radiation from the ionized Na atoms replaces the rare-gas spectrum with emission in the yellow region of the visible spectrum ($\lambda = 589.2$ nm).

Modern versions of these discharge lamps come in three flavors—mercury, metal-halide, and xenon arc lamps—each with its distinct spectral characteristics (Fig. 5.2.1). The green-yellow line ($\lambda = 546$ nm) of the mercury lamp is very sharp and is used as a universal reference for calibrating wavelengths in various optical experiments, particularly in biological experiments. This arc lamp has about four times the relative power output of the incandescent tungsten-halogen lamps. Note that the mercury arc lamp delivers a third of its output in the visible and half of its output in the UV, the remaining being lost as heat. However, mercury arc lamps produce significant fluctuations in intensity primarily because the gas plasma is unstable as both stray magnetic fields and the erosion of the electrode over time affect the performance of the lamp.

The metal-halide lamp has a spectrum identical to that of the mercury lamp. In addition, the metal-halide arc lamps produce higher output levels between the peaks, i.e. in the continuous regions, making them more useful for exciting fluorophores[2] that may lack an absorption band exactly at the characteristic spectral lines. Unlike the mercury lamp, the xenon lamp has a low but uniform and continuous output in the entire visible range; however, most of its energy is concentrated in the IR, with wavelengths larger than 800 nm.

5.2.1.2 *Simulated Emissions and Lasers*

Consider a gas of free atoms enclosed in a container with a number of possible energy levels, at least one of which is metastable. Shining white light on this gas causes the atoms to be pumped from the ground state to higher energy states by the absorption of radiation. When the atoms relax back to the ground state, we expect some of them to be trapped in the metastable state mentioned. If the intensity of the white light is sufficiently large, we can envision a scenario where the number of atoms in the metastable state is much larger than the number in the ground state. This scenario is called a *population inversion*.

It is logical to expect that any atom that is part of this population inversion will eventually return from the metastable to the ground state by the emission of a photon of specific energy, *hf*. This phenomenon is called a *phosphorescent radiation*. This photon, as it passes by a nearby atom that is also in the same metastable state can, by resonance, stimulate the radiation of a photon with exactly the same frequency, and return the second atom also to its ground state (Fig. 5.2.2a). Curiously, under certain conditions the second photon resulting from the stimulated emission can have exactly the same attributes as the first; in other words, the two photons are spatially (same frequency, direction, and polarization) and temporally (same phase and velocity) coherent! Subsequently, both photons can be considered as primary waves and the process of stimulation continues on and on, ad infinitum, as long as the primary white light maintains the population inversion. To create a laser beam, the stimulated radiation must be collimated, which is done by building a cavity with reflecting mirrors, and where the waves are used again and again for stimulated radiation. If the cavity is properly designed, the photons will move back and forth in the cavity in a self-sustaining fashion and the system will oscillate spontaneously. This is called *lasing*. Further, if there are openings or windows in the cavity, the escaping photons can be used as the source of visible radiation (Fig. 5.2.2b). Lasing can also be achieved in a

[2] A fluorescent chemical compound that can re-emit light upon excitation by light.

(a)

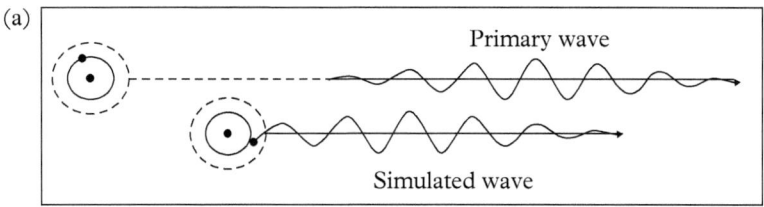

(b)

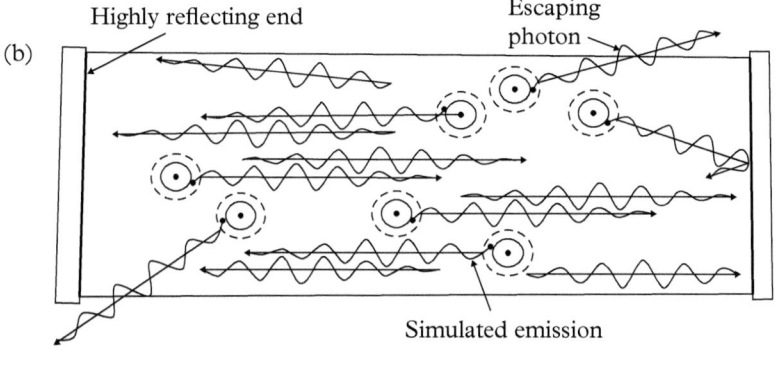

Figure 5.2.2 (a) The principle of simulated emission from an atom where the primary and simulated waves are spatially and temporally coherent. (b) A "cavity" with reflecting coatings at either ends showing multiple stimulated emissions and the "escape" of photons to form a laser.

Adapted from Jenkins and White (1976).

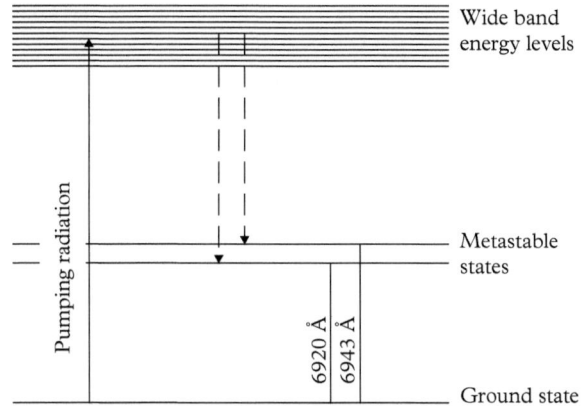

Figure 5.2.3 Energy levels of ruby that forms a solid-state laser.

Adapted from Jenkins and White (1976).

solid material as long as it has the possibility of a population inversion. Such a crystal is ruby (Al_2O_3), whose energy levels are shown in Figure 5.2.3, with high probability for a population inversion and a subsequent phosphorescent radiation.

Example 5.2.1: A typical HeNe laser has an output of 5 mW, and operates using a dc voltage of 2.5 kV and a current of 8 mA. What is the efficiency of this laser?

Solution: Efficiency, η, is defined as output power/input power. Thus

$$\eta = (5 \times 10^{-3}\ W)/(2.5 \times 10^3\ V \times 8 \times 10^{-3}\ A) = 2.5 \times 10^{-4} = 0.025\%$$

Table 5.2.1 Some common gas and solid-state lasers with their emission wavelengths in nanometers.

Laser	Medium	UV	Blue	Green	Red	IR
He–Cd	Gas	325	441.6	537.8	636	–
Ar	Gas	351.1, 356.4	457.9	514.5	–	–
Kr	Gas	350.7, 356.4	461.9	–	676.4	753, 799
Ar–Kr	Gas	–	467.5	–	676.4	–
He–Ne	Gas	–	–	543	632.8	1150, 3390
Ruby	Solid	–	–	–	694.3	–
GaAs	Solid–diode	–	–	–	–	904
Nd	Solid	–	–	–	–	1060
Nd–YAG	Solid	266	–	–	–	–

Lasers were invented by Townes[3] and first used to determine atomic energy levels with great precision by Schawlow.[4] Lasers find a variety of uses in materials science and engineering, including thin-film deposition (laser ablation), imaging (confocal microscopy, §6.8.5, scanning probe microscopy §11), chemical analysis (laser ionization mass spectrometry), and magnetic characterization (magneto optic Kerr effect). Some common types of lasers encountered in practice are summarized in Table 5.2.1. Note that lasers, typically, have low power efficiency (Example 5.2.1).

5.2.2   Electrons: Thermionic and Field-Emission Sources

Electron guns, found in transmission (§9) and scanning (§10) electron microscopes, are generated using two kinds of sources (Fig. 5.2.4): thermionic, which produces electrons when heated, and field emission, which generates electrons by applying an intense electric field. In the former, the kinetic energy of the electrons in the solid increases to the point that they can overcome the energy barrier, or the work function, Φ, that keeps the electrons within the solid (see §3.7). In the latter, the energy barrier, Φ, is modified such that electrons can tunnel (§11.2) through and escape the solid that forms the cathode (Fig. 5.2.4c). In both cases, the emission current is established by keeping the cathode at a negative potential, such that the potential difference between the cathode and anode draws the emitted electrons away from the solid (Fig. 5.2.5).

In thermionic emission, the current density, j_c, defined as the current emitted per unit area at the cathode is given by Richardson's[5] law, and is exponentially related to the work function, Φ, and the cathode operating temperature, T_c (K), as

$$j_c = AT_c^2 \exp\left(\frac{-\Phi}{k_B T_c}\right) \tag{5.2.1}$$

[3] C. H. Townes (1915–2015) was a US physicist who received the 1964 Nobel prize in Physics for "fundamental work in the field of quantum electronics, which has led to the construction of oscillators and amplifiers based on the maser-laser principle."

[4] A. L. Schawlow (1921–1999) was a US physicist who received the 1981 Nobel prize in Physics for "contribution to the development of laser spectroscopy."

[5] O. W. Richardson (1879–1959) was a UK physicist who received the 1928 Nobel prize in Physics and was cited for "work on the thermionic phenomenon and especially for the discovery of the law named after him."

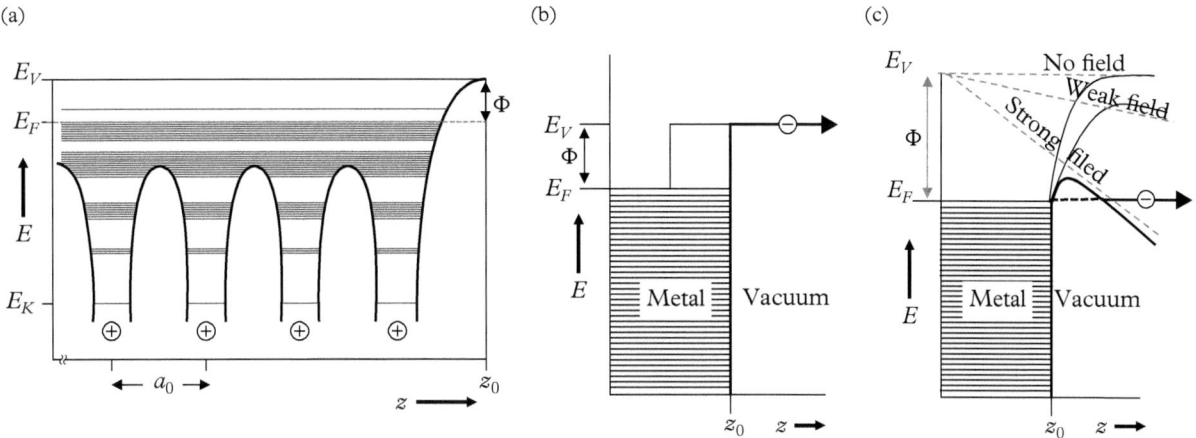

Figure 5.2.4 (a) Energy bands in a periodic potential of a crystal. At the crystal surface, z_0, the symmetry is broken, the potential reaches its vacuum value, E_V, and exceeds the Fermi level, E_F, by the work function, Φ. (b) Sufficient thermal energy, as in thermionic emission shown here, can allow the electrons to overcome the work function, Φ, and be emitted into the vacuum. (c) Alternatively, a strong electric field can reduce the effective work function and allow electrons to be emitted either by weak thermal excitation or quantum mechanical tunneling.

Adapted from Fuchs, Oppolzer, and Rehme (1990).

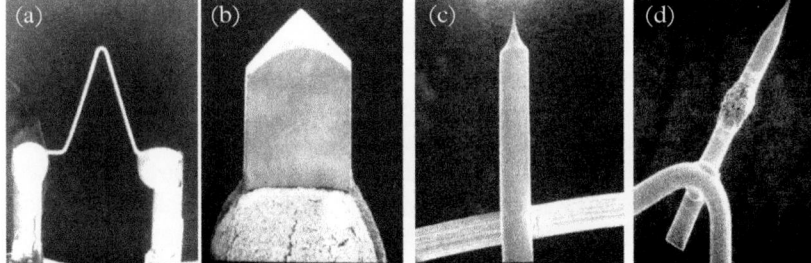

Figure 5.2.5 Electron sources. (a) Conventional thermionic filament—a tungsten hairpin wire spot-welded to support posts. (b) Lanthanum hexaboride single crystal block mounted on support. (c) Cold field emission tip—a tungsten <310> single crystal tapered to a sharp point (radius <100 nm) spot-welded to a tungsten wire, and (d) A Schottky emission source.

Adapted from Goldstein et al. (2003).

where, k_B, is the Boltzmann constant, and A is a material-dependent constant (see Table 5.2.2). As Φ appears in the exponent, it has a particularly strong influence on the current density. The alkali earth materials, such as cesium (Cs), have a low work function, but they are unstable in air even at room temperature. The only materials that can survive the operating temperatures required for

Table 5.2.2 Parameters of commonly used thermionic and field-emission sources/cathodes. See also §9.2.1.

	Thermionic emission	Thermionic emission	Field emission (cold)	Schottky emission
Cathode material	W	LaB_6	W	W/ZrO_2
Work function, Φ (eV)	4.5	2.7	4.5	1.7
Richardson constant, A $(A/cm^2/K^2)$	75–120	30	–	–
Working temperature, T_c (K)	2,800	1,400–2,000	1,000	1,350–1,450
Vacuum required (Pa)	10^{-2}–10^{-3}	10^{-3}–10^{-5}	10^{-8}	10^{-7}
Lifetime (hours)	25–50	150–200	1,000	5,000
Total current, I (µA)	5–100		1–10×10^{-3}	100–200×10^{-3}
Current density, j_c (A/m^2)	1–3×10^4	~ 2.5–10×10^5	10^8–10^{10}	10^7–10^8
Angular current distribution $dI/d\Omega$ (A/sr)	$\sim 10^{-2}$		$\sim 10^{-4}$	$\sim 10^{-4}$
Energy width (eV)	1–3	1–2	0.1–0.5	0.3–0.9
Source brightness, β $(A/m^2/sr)$	5×10^8 $(E = 10 \text{ keV})$ 1–5×10^9 $(E = 100 \text{ keV})$	2×10^9 $(E = 10 \text{ keV})$	5×10^{11}–5×10^{12} $(E = 20 \text{ keV})$ 2×10^{12}–2×10^{13} $(E = 100 \text{ keV})$	10^{13} $(E = 3 \text{ keV})$
Cross-over diameter, d_c (µm)	20–40	10–20	5–10×10^{-3}	10^{-2}

Adapted from Reimer (1993).

the electrons to overcome the work function, without melting and vaporizing, are the high melting point refractory metals, such as tungsten ($T_m = 3965$ K, $\Phi = 4.5$ eV). Alternatively, lanthanum hexaboride, LaB_6, has an intrinsically lower work function ($\Phi \sim 2.5$ eV, for <100> single crystal); thus, W and LaB_6, are the two principal materials used for thermionic emission. However, the higher operating temperatures ($\sim 2,800$ K) required for W, lead to a higher vaporizing rate, and restrict the lifetime of the filament to 25–50 hours with current densities of $\sim 10^4$ A/m^2. On the other hand, LaB_6 ($T_m = 2,210°C$) sources operate at much lower temperatures (1,400–2,000 K), with extended lifetimes (150–200 hours) and modest increase in current density ($\sim 10^5$ A/m^2). In the latter case, the preferred material is a single crystal rod with conical points to increase current density, and exposed low-index crystallographic planes to enhance the emission (Figure 5.2.5c). Important characteristics of thermionic and field emission sources when used as electron "guns" are summarized in Table 5.2.2, and discussed further in chapters on transmission (§9) and scanning (§10) electron microscopy.

The alternative, field emission source, is based on the quantum mechanical tunneling effect (see §11.2). When an electric field of the order of 10^6 V/m is applied, it reduces the potential barrier at the metal surface allowing the electrons to escape by tunneling (Fig. 5.2.4c). To achieve the high fields required, fine tungsten wires with the favorable <310> orientation and a sharp tip with radius of curvature, $r \sim 0.1$ μm are used. Tungsten is chosen as the cathode material for its ability to also withstand the large mechanical stresses placed on the tip in very high electrical fields. The current density, j_{FE}, of field emission can be estimated from the Fowler–Nordheim formula [1], as

$$j_{FE} = \frac{k_1 |\mathbf{E}|^2}{\Phi} \exp\left(-\frac{k_2 \Phi^{\frac{3}{2}}}{|\mathbf{E}|}\right) \tag{5.2.2}$$

where $|\mathbf{E}| = U/r$, U is the potential applied between the tip and the first anode, r is the radius of the tip, and the constants, k_1 and k_2, depend very weakly on $|\mathbf{E}|$. In addition, to ensure a uniform tunneling of electrons, the tip surface must be pristine with no contamination. At room temperature this can be achieved at ultra-high vacuum ($<10^{-8}$ Pa) conditions, and is called cold field emission. Alternatively, the W tip can be heated to operate under relatively poorer vacuum conditions ($<10^{-7}$ Pa); this is the thermal field emission process, which can be further improved by treating the W surface with ZrO_2 (Schottky emitters). In reality, the field strength in this type of ZrO/W source is not sufficient to emit electrons purely by the quantum mechanical tunneling effect. Instead, the lowering of the potential barrier, Φ, from 4.5 eV to 1.7 eV by the applied field, now allows the electrons to overcome the barrier by their thermal energy. Hence, strictly speaking, Schottky emitters are still thermionic sources albeit with a lower energy barrier and emitting electrons with a narrower energy spread.

Note that all electrons from these types of sources/cathodes do not emerge with the same energy. Typically, the energy distribution of electrons has a full width at half maximum (FWHM) of $\sim$1–3 eV for thermionic and 0.1–0.5 eV for field emissions. Further, the two sources have important differences in how the current is generated. The current density, $j_c = 1 - 3 \times 10^4$ Am^{-2} for thermionic emission, but orders of magnitude larger values, $j_{FE} = 10^8$–10^{10} Am^{-2}, can be obtained for field emission. On the other hand, the total current, I, for thermionic emission (5–100 μA) is much larger than that generated by field emission (0.1–1 μA). This is because the electrons in field emission sources are emitted from an extremely small virtual area of the order of 10 nm in diameter (d_{co}) into a very large aperture, typically over an angle of 0.1 rad ($\sim$11.5°). As a result, the current per unit solid angle, $dI/d\Omega$, in field emission is of the order of 10^{-4} A/sr, and about 100 times smaller than that for thermionic sources.

To use the electrons emitted from the cathode and form a useful probe, the electrons must be accelerated and focused. This requires an electron-optical system, also called an electron "gun" (Fig. 5.2.6), which include the cathode (source), a Wehnalt electrode to form a small disc-like cross-over image (a virtual source) in front of the cathode, with a diameter, d_{co}, and an anode to propel the emitted electrons in the forward direction. The performance of this electron gun is characterized by its *brightness*, β_{co}, defined as current density per unit solid angle, $\Delta\Omega = \pi\alpha_{co}^2$, where α_{co} is defined by the aperture half-angle (Fig. 5.2.6). Thus

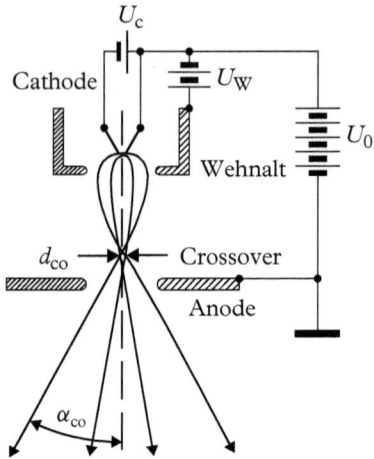

Figure 5.2.6 An electron "gun", comprised of the cathode, the Wehnalt electrode and the anode. The emitted electrons are focused to a diameter, d_{co}, (virtual source) with a subsequent divergence semi-angle of α_{co}. Both d_{co}, and α_{co} can be controlled by changing the voltages, particularly U_W, of the Wehnalt electrode.

$$\beta_{co} = \frac{j_{co}}{\Omega_{co}} = \frac{j_{co}}{\pi \alpha_{co}^2} \tag{5.2.3}$$

The current density at the cross-over, j_{co}, depends on the current density at the cathode, j_c. Further, assuming that the energy and momenta of the emitted electrons satisfy the Maxwell–Boltzmann distribution, it can be shown for thermionic emission, with a cathode operating temperature, T_c, that the maximum gun brightness, $\beta_{\max}$, is given by

$$\beta_{\max} = \frac{j_c e U_0}{\pi k_B T_c} \tag{5.2.4}$$

where U_0 is the anode voltage, referred to in electron microscopy as the acceleration voltage (see §9 and §10, for further discussions), and j_c is given by (5.2.1). Note that, the brightness increases linearly with U_0.

Example 5.2.2: A thermionic source uses a LaB$_6$ filament, operating at 2,000 K, and an acceleration voltage of 10 kV.

(a) What is the current density and the maximum brightness of the source?

(b) If we intend to use a W source instead, what must be its operating temperature to achieve the same current density? Is this feasible?

Solution: (a) The work function for a LaB$_6$ source (Table 5.2.2) is

$$\Phi = 2.7 \text{ eV} = 2.7 \times 1.6 \times 10^{-19} \text{ J}.$$

Using the data in Table 5.2.2, the current density (5.2.1) is

$$j_c = AT_c^2 \exp\left(\frac{-\Phi}{k_B T_c}\right) = 30(2000)^2 \exp\left(\frac{-2.7 \times 1.6 \times 10^{-19}}{1.38 \times 10^{-23} \times 2000}\right)$$

$$= 19.12 \text{A cm}^{-2} = 1.912 \times 10^5 \text{A m}^{-2}.$$

The brightness of the source (5.2.4) is

$$\beta_{max} = \frac{j_c e U_0}{\pi k_B T_c} = \frac{1.912 \times 10^5 \times 1.6 \times 10^{-19} \times 10000}{3.14 \times 1.38 \times 10^{-23} \times 2000}$$

$$= 3.5 \times 10^9 \text{ A/m}^2/\text{sr}.$$

(b) Using the data in Table 5.2.2, for a W source to generate the same current density the operating temperature (T) must satisfy (5.2.1), i.e. $100 \, (T)^2 \exp\left(\frac{-4.5 \times 1.6 \times 10^{-19}}{1.38 \times 10^{-23} \times T}\right) = 19.12$ A cm^{-2}.

We solve this numerically to give $T \sim 2{,}960$ K. This is feasible and the operating temperature is not too high, compared to its melting point ($T_m = 3{,}965$ K), to achieve a meaningful operating lifetime for the source.

5.2.3 Neutrons

Fission reactors and spallation sources are used to produce neutrons. In the case of fission reactors, slow neutrons are absorbed by metastable uranium (^{235}U). When this excited state decays, it produces a cascade of fission products and, on average, produces ~2.5 neutrons for the fission of each ^{235}U atom. The decay is moderated in a reactor with neutrons selectively allowed to escape, and subsequently they are separated according to their wavelength using large single crystal monochromators. The production of neutrons is continuous and steady state, i.e. it does not vary with time.

Recently developed spallation sources generate pulses of neutrons using high-energy protons, generated by an accelerator, to bombard metals such as uranium, tungsten, lead, or mercury. The neutrons produced in this manner have high energies and they are then slowed down by placing hydrogenous moderators, such as water, around the source to have wavelengths suitable for diffraction and scattering from materials. Typically, pulses of neutrons of a few microsecond duration and frequency of ~50 Hz, depending on the proton source beam and the moderator, are produced.

As of this book being written, the Spallation Neutron Source (SNS) at Oak Ridge National Laboratory (Fig. 5.2.7) produces the most intense neutron beam to date. Here, negatively charged hydrogen (H$^-$) ions, consisting of a proton orbited by two electrons, produced by an ion source are injected into a linear

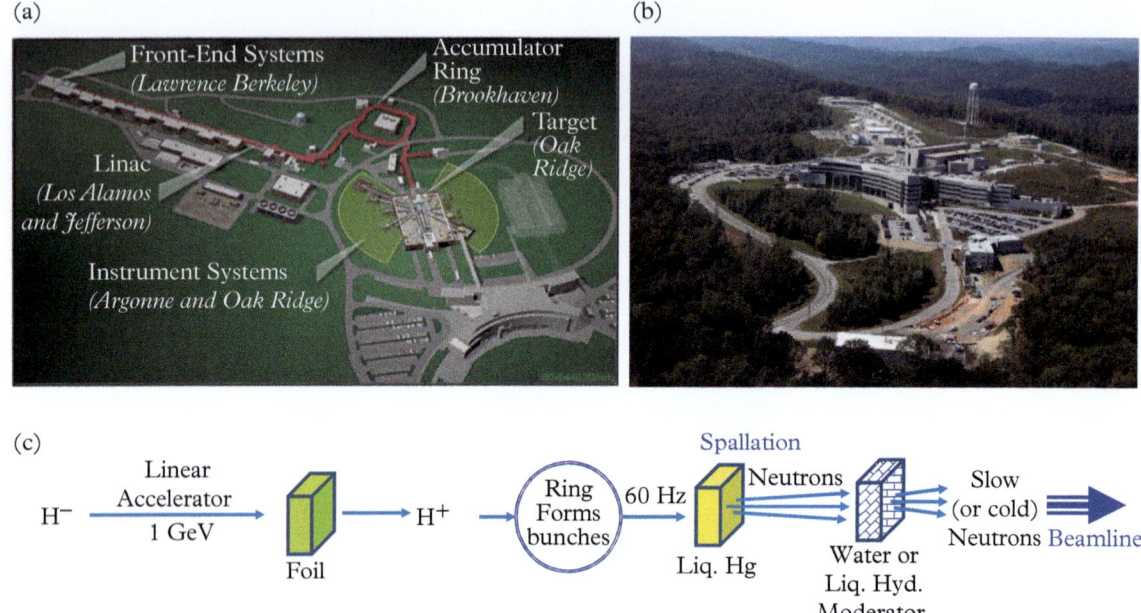

Figure 5.2.7 (a) The layout and (b) photograph of the be controlled by changing the voltages, particularly spallation neutron source (SNS) at Oak Ridge National Laboratory (USA) that presently (2019) produces the most intense neutron beams. Different US national laboratories are responsible for different parts of the facility. (c) A schematic illustration of the different components of the SNS at ORNL. For additional details, see text as well as https://neutrons.ornl.gov/content/how-sns-works, from where this photograph is used with permission.

particle accelerator (Linac), which accelerates them to high energies ($\sim$1 GeV). Then, they are passed through a foil, which strips the two electrons from H^- and converts it to a proton (H^+). The protons then pass into a ring-shaped structure (accumulator ring) where they first form bunches. Then, one at a time, these bunches are released from the ring at a rate of 60 Hz to strike a heavy metal (liquid mercury). This generates pulses of neutrons (spallation process), which are slowed down in moderators (using water to achieve room temperature, or liquid hydrogen at 20 K, to produce cold neutrons) and guided through beam lines to specialized instruments for experiments. (See §8.8.3 for *in situ* kinetic studies of a ubiquitous material, cement, using neutrons).

5.2.4 Ions

5.2.4.1 Guiding and Focusing

For neutral atoms, the number of negatively charged electrons exactly compensates the positive charge of the nucleus; thus, by adding or removing an electron a negatively or positively charged ion can be produced. Typically, the charge of an ion is equivalent to either an elementary electron charge (e^-) or its integral multiples. However, the mass of an ion is essentially the nuclear mass and for typical atoms, $m_{ion} \sim 10^5 m_e$, i.e. the mass of the ion is much larger than the mass of an electron, and this difference leads to considerably different physical effects.

We can use electric or magnetic fields to guide and focus any charged particle. In general, the force, **F**, and kinetic energy, E_{KE}, of a particle of charge, q, subject to an electric field, **E**, or an applied potential, U_0, is

$$\mathbf{F} = q\,\mathbf{E} \tag{5.2.5}$$

$$E_{KE} = qU_0 = mv_0^2/2 \tag{5.2.6}$$

From the preceeding, it follows that its velocity, v_0, is given by

$$v_0 = \sqrt{\frac{2qU_0}{m}} \tag{5.2.7}$$

Note that for electrons accelerated through typical potentials in electron microscopes relativistic corrections, as in (4.3.8), must be applied. However, the key point to remember is that, even though ions have similar charges as electrons, because of their substantially larger masses, their velocities ($\propto 1/\sqrt{m}$), for the same potential, U_0, are significantly smaller.

Charged particles, including ions, can be deflected by both electric and magnetic fields. Consider a particle charge, q, entering a capacitor of length, l, with an electric field, $E_c\,(= U_c/d)$, after being accelerated by a potential, U_0 (Fig. 5.2.8a). The velocity, v, of the particle is given by (5.2.7), and its perpendicular component, $v_\perp$, after traveling a distance, l, is

$$v_\perp = a_\perp t = a_\perp l/v = q\frac{E_c}{m}\frac{l}{v} = \frac{q}{m}\frac{U_c}{d}\frac{l}{v} \tag{5.2.8}$$

(a) (b) (c)

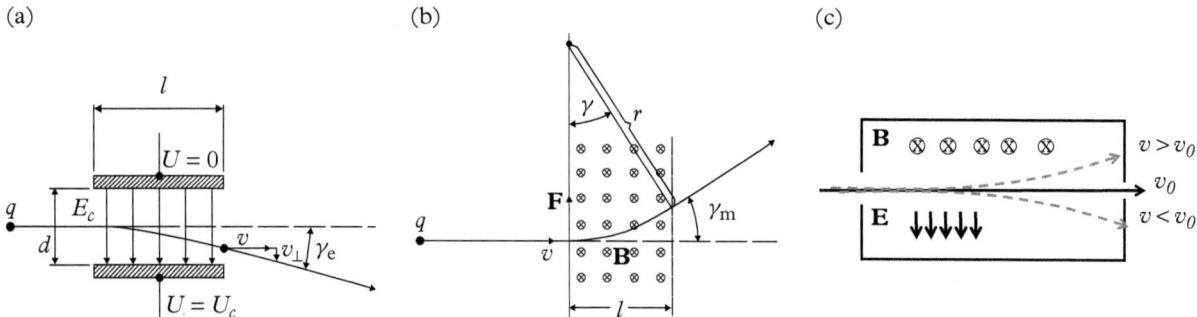

Figure 5.2.8 Deflection of a charged particle in (a) electric field, E_c, as in a capacitor, and (b) a uniform magnetic field, with flux density, **B**. (c) A Wien filter combines variable magnetic and electric fields, applied in an *orthogonal* fashion, to select a charged particle of a well-defined velocity, v_0, to emerge from the filter.

and the deflection angle, γ_e, assumed to be small, is given by

$$\tan \gamma_e = \frac{v_\perp}{v} = \frac{q}{m} \frac{U_c}{d} \frac{l}{v^2} = \frac{1}{2} \frac{U_c}{U_0} \frac{l}{d} \simeq \gamma_e \qquad (5.2.9)$$

where (5.2.7) has been substituted for v. Since there is no charge or mass dependence in the deflection angle, γ_e, both electrons and ions require the same field, E_c, for a given deflection.

On the other hand, consider a charged particle entering a region of uniform magnetic field with flux density, $\mathbf{B}$ (Fig. 5.2.8b). It will now experience a Lorentz force

$$\mathbf{F}_L = q\mathbf{v} \otimes \mathbf{B} \qquad (5.2.10)$$

such that it is balanced by the centripetal force, i.e. $F_L = qvB = mv^2/r$, leading to a circular trajectory of radius, r. The deflection, γ_m ($<< 1$), for traversing a field of length, l, is given by

$$\gamma_m \approx \frac{l}{r} = \frac{lqB}{mv} = \sqrt{\frac{q}{m}} \frac{lB}{\sqrt{2U_0}} \qquad (5.2.11)$$

which, unlike the case of an applied electric field, depends on the ratio of the charge to mass of the particle. Since the mass of an ion is much larger than that of an electron, by about five orders of magnitude, to observe the same deflection an ion will require much higher values of the magnetic flux density.

These arguments, based on deflection, also apply to focusing of electron beams (§9.2.2). Thus, it is easy to see that *electrostatic systems* are favored for guiding and focusing of ions; however, electrostatic lenses suffer from larger aberrations and nowadays are not used in electron microscopes.

Example 5.2.3:

(a) Show that a Wien filter (Fig. 5.2.8c) selects particles based only on their velocity. How is this done in practice?

(b) Such a filter is used to detect electrons incident at 1keV that have lost 284 eV of energy. If your filters applies an electric field of 1 kV/m, what will be the magnetic flux density (B) required to detect those specific energy-lost electrons?

Solution:

(a) A Wien filter applies electrical and magnetic fields in orthogonal directions, with anti-parallel forces applied on the moving particle of charge, q, and velocity, v.

Then, it satisfies $F_E = qE = F_B = q\,\mathbf{v} \otimes \mathbf{B} = qvB$. Thus, the velocity selected is $v = E/B$, or by tuning the ratio of the electric to magnetic field.

(b) The energy of the electrons to be detected, $E_e = 1000 - 284 = 716$ eV. Assuming that all its energy is kinetic, $E_e = mv^2/2$, where m is the mass of the electron.

Thus, $v = (2E_e/m)^{1/2} = E/B$. Hence $B = E/(2E_e/m)^{1/2} = 1000/(2 \times 716 \times 1.6 \times 10^{-19}/9.1 \times 10^{-31})^{1/2} = 6.3 \times 10^{-5}$ Tesla.

5.2.4.2 Ion Sources

There are a number of ion sources producing different beam currents and different energy distributions of ions. Our main interest is to discuss the generation of protons (H^+) and alpha ($^4He^+$) particles, used in a number of ion-based techniques for materials characterization (§5.4).

The simplest RF plasma ion source (Fig. 5.2.9a) contains a cylindrical discharge space where the electrons emitted from a filament, F, at the cathode, C_1, are guided by the magnetic field generated by the coils, M. The electrons undergo a helical path towards the second repeller electrode, C_2, before finally ending up at the anode, A. In the meantime, the electrons collide with the gas entering the chamber at G, ionizing them to generate positive ions. The ions are extracted from the discharge cylinder, as a beam, through the narrow opening of the electrode, E.

A further development of the RF plasma source is the duoplasmatron (Fig. 5.2.9b). Now an intermediate electrode, I, is placed between the cathode, C, and anode, A. Both I and A act as pole-pieces for the magnet, M, generating an inhomogeneous magnetic field before the emission aperture. The gas enters at G, is ionized by the cathode, then the plasma is concentrated in front of the emission aperture of A, and finally extracted by the electrode, to form the ion beam, B.

Ion beams extracted from such sources are typically of low energy (20 keV). To produce a beam of sufficient energy (MeV) to be useful in materials analysis, the ions are injected into a megavolt electrostatic van de Graaf accelerator or tandem accelerator. At the positive terminal of a van de Graaf accelerator, a positive ion beam of charge, $+q$, and voltage, $+V$, is created and accelerated

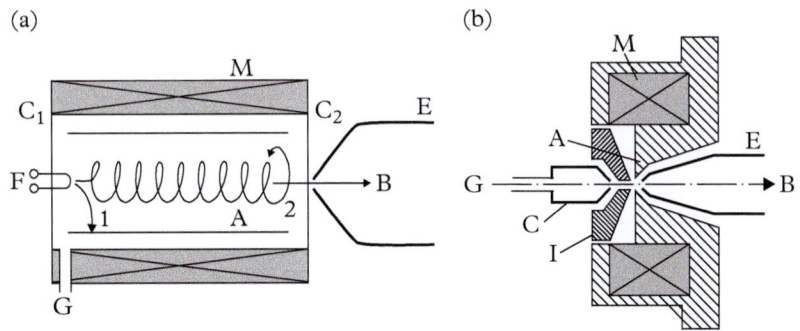

Figure 5.2.9 Schematic diagram of (a) plasma ion source, and (b) duoplasmatron ion source. See text for details.

Adapted from Fuchs, Oppolzer, and Rehme (1990).

towards a ground potential, where it emerges with energy, qV. On the other hand, in a tandem accelerator, a positive ion beam from the source is first changed into a negative beam for injection into the accelerator. It is then accelerated towards a positive terminal of voltage, $+V$; then, the charge is stripped from the negative ion and turned into a positive ion beam of charge, $+q$, which then accelerates away from the positive terminal with an energy of $(1+q)V$. Thus, the terminal voltage required in a Tandem accelerator is less than that required in a van de Graaf generator. In addition, the ion source being outside (inside) in a tandem (van de Graaf) accelerator, it is easier (harder) to maintain.

Ion beams are normally focused with electrostatic lenses to perform experiments. Further details on ion sources can be found in [2], and in the excellent handbook edited by Tesmer and Nastasi (1995).

5.3 Interactions of Probes with Matter, Including Damage

To characterize a material and discern its microstructure, it is often necessary to interrogate it with some form of probe radiation (Fig. 1.3.1). The process of interrogation may itself alter the specimen, and our goal in selecting a probe is to minimize any such damage, while maximizing the information that can be obtained. To this end, where possible, the sample is first probed with light, although in some situations it may not be as benign[6] as it appears; related details of optics, optical diffraction, and microscopy are discussed in §6. If higher resolution is desired, a scanning electron microscope (SEM; §10) with electrons of higher energy (5–30 keV) is used. These electrons have sufficient energy to penetrate deeper into the material and may also cause minor materials damage. Alternatively, instead of visible light, one may start with X-rays to probe the crystal structure (§7), followed by electrons, and then finally move to ions. Which set of techniques to choose, and the sequence in which to apply them, depend on the available resources, the microstructural questions of interest, the interactions of a specific probe with the material, including the penetration depth and volume analyzed, and any damage that may ensue.

Some of the basic ideas—penetration depth, mean free path length, resolution, damage, and specimen preparation requirements—were introduced in §1.3.5. Recall the difference between the penetration depth of a probe and the sampling or information depth, which depends on the depth of origin of the detected signal (may be the same or may be different from the probe) associated with a specific technique. Here, for the four different probes (photons, electrons, neutrons, and ions) introduced earlier (§5.2), we specifically discuss the nature of their interactions with matter, emphasizing the penetration depth, as well as the damage that they may cause to the specimen in the course of analysis.

[6] For example, a photographic plate or film comes to mind, but this technology is rapidly giving way to digital recording and processing of information (see §1.4.4).

5.3.1 Photons

The electromagnetic spectrum (Fig. 1.3.2) of interest in materials characterization ranges from microwaves ($\lambda \sim 10^{-4}$ m, $E \sim 10^{-2}$ eV) to γ-rays ($\lambda \sim 10^{-10}$ m, $E \sim 10^2$ keV), both in terms of their wavelength and energy. Also, recall the concept of wave-particle duality (§1.3.3), which was used to describe light as discrete quanta of electromagnetic radiation (Fig. 1.3.5) satisfying the relationship (1.3.1) between their wavelength, λ, frequency, f, and energy, E.

Given the wide range of energies and wavelengths in the electromagnetic spectrum, it is not surprising that the penetration of photons in different materials shows a large range or variation. It is not necessary to catalog the penetration depth of all the components of the electromagnetic spectrum, so we restrict our discussion to those wavelengths of importance in materials characterization and analysis. As already seen, IR radiation is used in absorption spectroscopy (§3.5) to characterize technologically important materials (Fig. 3.5.3), visible light is used in microscopy (§6.8) and metallography (§6.8.6) to obtain a visual image of the surface of a material (Fig. 1.4.4), higher-energy or shorter-wavelength UV light is used to obtain electronic structure information of the surface atoms (§3.8), and even higher-energy X-ray photons are used to obtain details of the crystallographic structure by diffraction (§7). It is also well known that some materials are transparent to visible light, some reflect, and some are opaque. However, even the most opaque or reflective materials will allow photons to penetrate a distance of the order of a wavelength beneath the surface. Typically, in the middle of the visible spectrum ($\lambda \sim 500$ nm), the penetration depth is on average between 50–300 nm, and for most materials, with inter-planar spacing of ~ 0.2 nm (see §4.3.3), this corresponds to 250–1,500 atomic layers below the surface.

Table 5.3.1 shows the penetration depth of soft X-ray photons, electrons, and ions of 1,000 eV energy. The X-ray photons penetrate considerably deeper into the specimen, whereas electrons and ions of similar energy penetrate far less and stay close to the surface. In contrast, since electrons and ions deposit all their energy near the surface, they create more damage, including sputtering (§5.4) of the surface atoms, when compared to photons.

We have discussed the absorption of X-rays in materials (2.4.2). The mass absorption coefficient, μ/ρ, for X-ray photons, which increases with the average

Table 5.3.1 Penetration depth of electrons, photons and ions of the same energy.

Probe	Energy	Penetration Depth
X-ray photons	1,000 eV	1,000 nm
Electrons	1,000 eV	20 nm
Ions	1,000 eV	10 nm

atomic number of the material, and decreases with increasing energy of the photon, determines the depth of penetration. Similarly, γ-rays also decrease in intensity with distance traveled in a material, but because of their high energy (50 keV–50 MeV), they are not practically used in materials characterization and analysis as they will go through most bulk materials of interest.

It is not possible to probe a material using any radiation without perturbing it in some way, but in the case of photons this damage can be minimal. Photons in the visible, UV, and IR range have very small momenta and as a result they do not displace atoms in a solid. Generally, the damage caused by photons is due to local heating and the degree to which this occurs depends on their depth of penetration, the energy of the radiation, and the photon flux. These factors can become particularly acute when lasers are used, for they can cause holes or ablate materials by heating to instantaneous melting temperatures. However, the ablated species can also be analyzed by appropriate mass spectrometers (see §5.4.3) to obtain spatially resolved chemical analysis by scanning the laser over the specimen surface. This technique is known as laser-induced mass analysis (LIMA).

In summary, visible light can cause the least damage and, where possible, should be used first for materials characterization and analysis.

5.3.2 Electrons

The interaction of electrons with materials is more complicated than photons. In addition to the broadening of the electron beam in the specimen, depending on their energy they also cause displacements of the atoms in the material, including their sputtering, as well as heating, electrostatic charging, and hydrocarbon contamination of the specimen. We discuss these in further detail.

5.3.2.1 *Beam Broadening*

When electrons are used as probes, after hitting the solid surface, they penetrate into the material and interact with the constituent atoms. The primary interactions, governed by *electrostatic* Coulomb forces, are twofold: collisions of the probe electrons with the positively charged and much heavier nuclear core, or with the negatively charged electrons in the outer atomic shells. The collision of the probe electrons with the nucleus is mainly elastic because of the significantly larger mass of the latter, and therefore, it only changes their direction with negligible change in their energy.

As a first approximation, we neglect the effects of the atomic shells and treat the nucleus as a point positive charge, $q = +Ze$, where Z is the atomic number. The resulting small angle deflection, θ (Fig. 5.3.1a) is given by

$$\theta = \frac{q}{dU_0} = \frac{Ze}{dU_0} \tag{5.3.1}$$

where the electron accelerated through a potential, U_0, before the collision, passes at a distance, d, from the nucleus. It is also possible for the electrons to collide

(a)   (b)

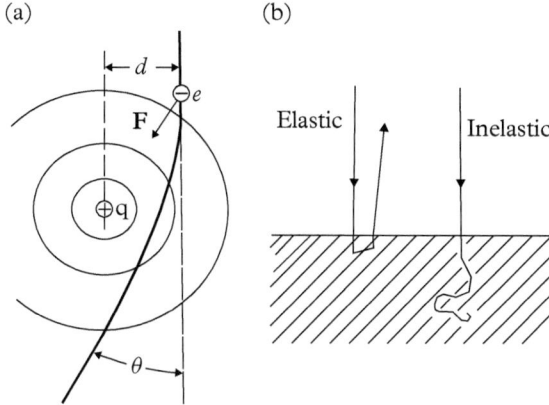

Figure 5.3.1 (a) Collision of an electron with the nucleus leading to elastic
scattering and angular deflection, θ,
in the forward direction. (b) Head-on
collision with the nucleus (left, elastic)
may lead to back scattering and even
ejection from the specimen. Alternatively, collisions with the outer-shell
electron lead to loss of energy (right,
inelastic) and a change in the direction
of propagation. See also Figure 9.3.1.

head-on with the nucleus; in this case, the electron is scattered backward and may
even leave the solid (Fig. 5.3.1b).

When the electrons of the probe and those occupying the atomic shells collide,
since they have identical mass, energy is transferred from the incident electrons
to those in the atomic shells. If the energy transferred is sufficiently high, an
inner-shell electron of the atom can be ejected and it can subsequently relax
by the emission of characteristic X-rays and/or Auger electrons (§2.3). In such
inelastic collisions, the electrons in the probe lose energy, change direction through
small scattering angles (Fig. 5.3.1b), and are slowed down, ultimately coming to
rest in the bulk solid. Generally, both processes (elastic scattering by the nuclei
and inelastic scattering by the electronic shell) occur simultaneously in time and
spatially side by side (for different incident electrons) in the specimen. As a result,
a sharply focused incident electron probe broadens into a much larger volume in
the material. The shape and size of this volume (§10.2.7), which depends on the
material (Z), the energy (U_0e), and angle of incidence of the probe (Fig. 5.3.2),
determines the penetration depth, d_p, of the electron and the broadening of the
beam in the specimen. It also determines the spatial resolution of any technique
that uses a focused electron probe for generating signals of interest.

The mean free path of low-energy electrons in the range of 0–2 keV
(Fig. 1.3.6) follows a universal curve, and varies over a rather small range
(0.4–300 nm) in depth. However, the penetration depth of high-energy electrons,
such as in SEMs (§10) and transmission electron microscopes (TEMs, §9), varies
strongly with the energy of the electrons as well as the average atomic number of
the specimen. Figure 5.3.3 shows the penetration depth, d_p, of electrons, based
on both theoretical simulations and experimental measurements, plotted in units
of ρd_p, where ρ is the density of the material, such that it is independent of its
state (amorphous or crystalline) for a large range of materials. It is clear that
most materials exhibit a similar dependence (shown to be $\sim E^{1.7}$) on the electron
energy.

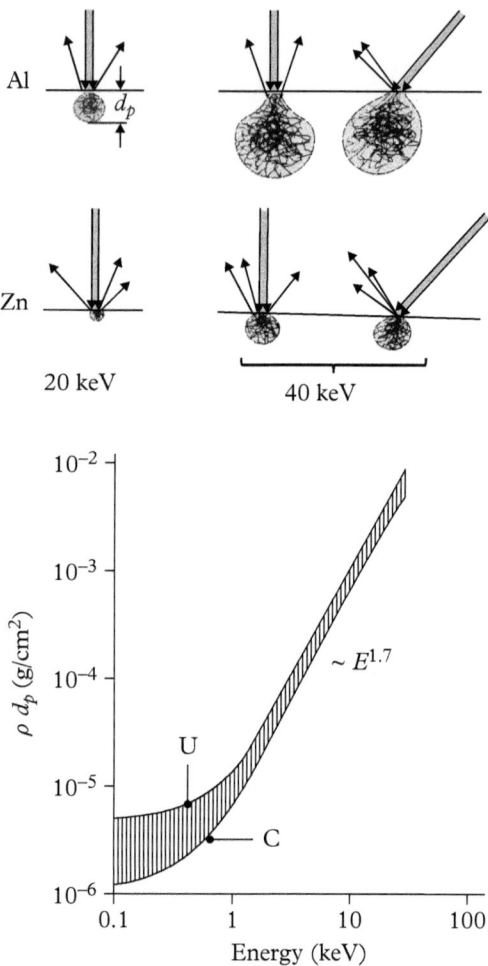

Figure 5.3.2 Broadening of an electron probe due to a series of successive inelastic scattering events depends on both the average atomic number, Z, of the specimen, its energy and the angle of incidence. Both penetration depth and resolution are affected by the broadening. See also Figures 10.2.13–15.

Adapted from Fuchs, Oppolzer, and Rehme (1990).

Figure 5.3.3 The product, ρd_p, of the penetration depth, d_p, and the density, ρ, as a function of incident electron energy (E) shows an $E^{1.7}$ dependence. The behavior in most materials falls within the shaded area bounded by carbon (C) and uranium (U).

The elastic interaction of electrons with atoms leads to back-scattering, which is used to measure the average atomic number of the specimen in an SEM (§10.3.3), and also to electron diffraction (§4.3.3) in periodic crystals. The inelastic scattering of the probe electrons also leads to the emission of secondary electrons from the surface layers, which is used to generate SEM images (§10.3.1). In this case, in addition to depth, the spatial resolution in imaging the secondary electrons is also affected by the spatial distribution of the incident electrons, which depends on the average atomic number of the specimen. Figure 5.3.4 shows the scattered secondary electron distribution for two different probes diameters incident on two materials with very different atomic numbers.

In addition to secondary electrons, the inelastic scattering leads to emission of characteristic X-rays and Auger electrons, generation of electron–hole pairs,

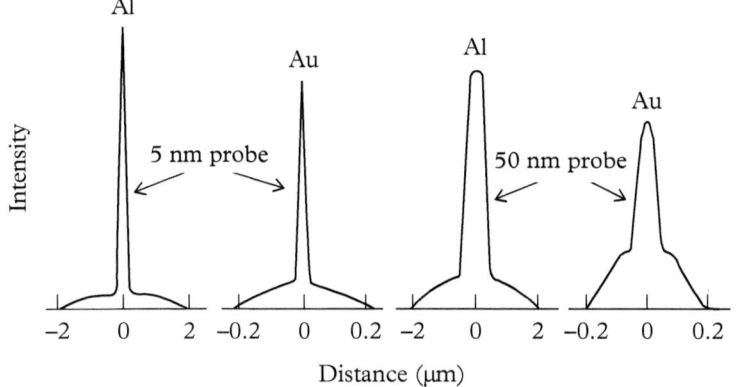

Figure 5.3.4 Spatial distribution of *secondary electrons* from two different materials (Al, Ag) and two different incident probe sizes (5 nm and 50 nm).

Adapted from Flewitt and Wild (2003).

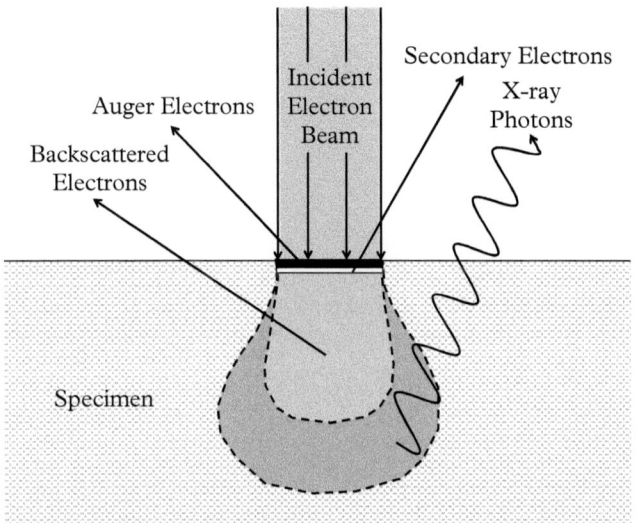

Figure 5.3.5 Multiple signals and their depth dependence or excitation volume, obtained from a specimen upon electron incidence. See also Figure 10.2.6.

and collective excitations of electrons in the valence band (plasmon resonances); these topics are discussed further in §9 (TEM/AEM) and §10 (SEM). Moreover, the volume in the material from which each of these signals emerge are different (Fig. 5.3.5). Since the back-scattered electrons can have higher energy than both Auger and secondary electrons, they can originate much deeper in the specimen and still be detected. In addition, they are scattered with a momentum component normal to the incident beam and hence have a larger excitation volume in the specimen. The characteristic X-ray photons can be generated from wherever the scattered electrons reach, and because of their sufficiently large energy, they can generally make it out of the specimen. This leads to the largest excitation volume for X-ray emission as shown in the same figure.

We now discuss damage, permanent or temporary, caused by electrons to the surface or bulk structure of a specimen in a TEM or SEM (Fig. 5.3.6).

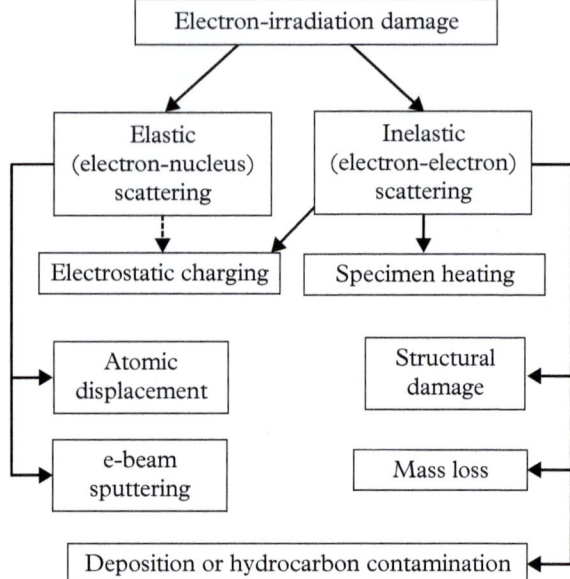

Figure   5.3.6 Electron   irradiation damage due to elastic and inelastic scattering processes in a TEM and SEM, classified according to the effects produced in the specimen.

Adapted from [3].

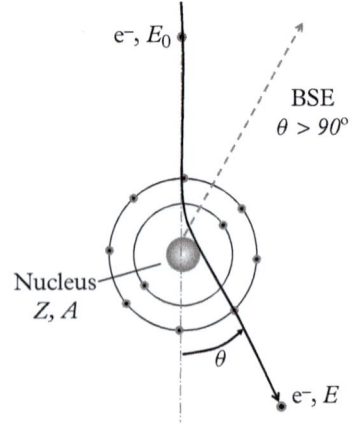

Figure 5.3.7 Scattering of electrons by the nucleus of atomic number, Z, and atomic weight, A. Incident electrons of energy, E_0, are scattered with energy, E, as shown. For elastic scattering, $E \sim E_0$.

---

[7] Defined as the molecular decomposition of a substance by ionizing radiation.

The damage can be broadly divided into two categories: (a) *elastic* interactions of the electron probe with the Coulomb field of the atoms that can, under favorable circumstances, lead to *atomic displacements* and *electron beam heating*, and (b) *inelastic* interactions of the electron probe with the electronic shell causing radiolysis[7] effects that can produce mass loss or, if ambient hydrocarbons are present, produce local contamination leading to image deterioration.

5.3.2.2   *Atomic Displacements*

We consider the deflection, θ, of the transmitted electron beam as a result of scattering by the nucleus of an atom of atomic weight, A (Fig. 5.3.7). For an incident electron of energy, E_0 (eV), by applying the conservation of energy and momentum, the energy transfer, ΔE, can be shown to be

$$\Delta E = E_0 - E = E_{max}\sin^2\left(\theta/2\right) \tag{5.3.2}$$

where

$$E_{max} = E_0 \frac{1.02 + E_0/10^6}{465.7A} \tag{5.3.3}$$

If we make a first order approximation of elastic scattering, i.e. that there is no change in the energy of the transmitted electron beam, the scattering angle, θ, is very small, and the energy transfer, ΔE, is negligible ($\Delta E << 1$ eV). However, if the electron is back-scattered ($\theta > 90°$), and E_0 is sufficiently high, ΔE can be a few eV. Further, if the collision is head on ($\theta = 180°$), the energy transfer, $\Delta E = E_{max}$, can be several eV for high E_0 and low atomic weight (A) of the target atoms. Now, if $\Delta E = E_d$, where the displacement energy, E_d, is a function of the atomic weight, bond strength, and the crystal structure of the material, then a sufficiently large energy transfer can displace the atoms in the material. Such displacement damage can result in squeezing the atoms into interstitial positions (Fig. 4.1.19), and under favorable conditions can be observed as interstitial loops in TEM micrographs (Fig. 5.3.8).

Displacement energies for various materials have been determined experimentally and the corresponding threshold energies, $E_0 = E_{Td}$, for which atomic displacements are observed, can be calculated (Table 5.3.2). Typical SEMs operate at a much lower voltage (5–30 kV) to observe any displacement damage, but in higher-voltage TEMs (>300 kV), it can be significant.

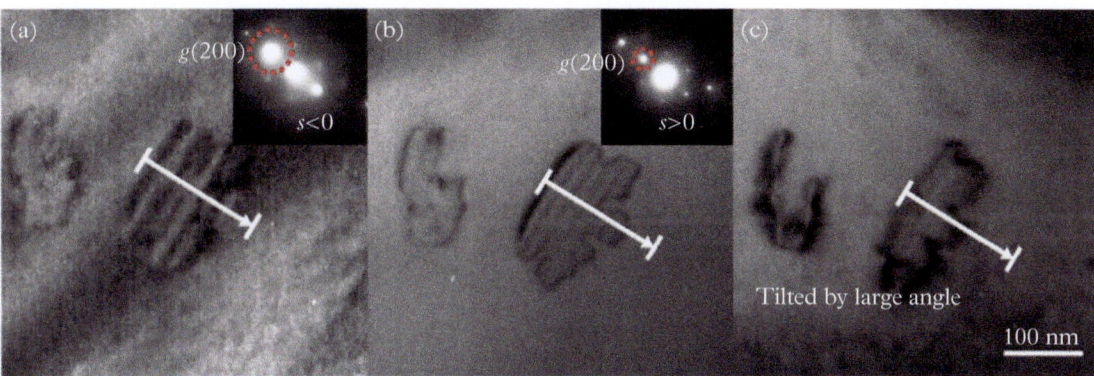

Figure 5.3.8 Dislocation loops in an austenitic stainless steel. (a–c) Interstitial loops formed under electron beam irradiation at 1000 keV in a TEM, observed with different angles of specimen tilt.

Adapted from [4].

Table 5.3.2 Displacement energy, E_d, and the corresponding threshold energy, $E_0 = E_{Td}$, for common materials.

Material	E_d (eV)	E_{Td} (keV)
Graphite	30	140
Diamond	80	330
Aluminum	17	180
Copper	20	420
Gold	34	1320

Adapted from [5].

Example 5.2.4: A 200 keV incident electron beam is back-scattered by an angle $\theta = 170°$. Will it be able to displace atoms in a specimen of (a) Al ($A = 26.98$), (2) Cu ($A = 63.55$), and (3) diamond ($A = 12.01$)?

Solution: Here, $E_0 = 200\text{keV} = 200 \times 10^3 \text{eV}$, and $\sin^2(\theta/2) = 0.9924$

For each element, from (5.3.3), $E_{max} = 523.9/A$ and from (5.3.2), $\Delta E = 519.96/A$.

We calculate ΔE for each element, and compare with their displacement energy, E_d (Table 5.3.3). Thus:

$\Delta E\,(\text{Al}) = 519.96/26.98 = 19.27$ eV (> 17 eV)—Yes, displacement is possible
$\Delta E\,(\text{Cu}) = 519.96/63.55 = 8.18$ eV (< 20 eV)—No displacement
$\Delta E\,(\text{Diamond}) = 519.96/12.01 = 43.29$ eV (< 80 eV)—No displacement.

5.3.2.3　*Sputtering*

Atoms on the surface of a specimen, subject to high-angle elastic scattering, satisfy the conditions (5.3.2) and (5.3.3). However, the energy, E_s, required for the displacement of surface atoms is much lower than that of the bulk, i.e. $E_s \ll E_d$. Rather than being squeezed as an interstitial atom, surface atoms are free to leave the solid, and such ejection of surface atoms is referred to as sputtering. In the transmission (TEM) geometry, the momentum transfer from the electron probe to the atoms is largely in the forward direction and hence it is expected that the sputtering would take place, with the formation of a "crater", largely in the *exit surface* of a thin electron-transparent specimen [6]. As a first approximation, the sputtering energy, E_s, is assumed to be the sublimation energy, E_{sub}, of the material, and applying (5.3.3), the expected threshold values for sputtering,

Figure 5.3.9 Threshold energy for sputtering, $E_0 = E_{Ts}$, as a function of atomic number, Z, using the sublimation energy as the displacement energy for a surface atom.

Adapted from [3].

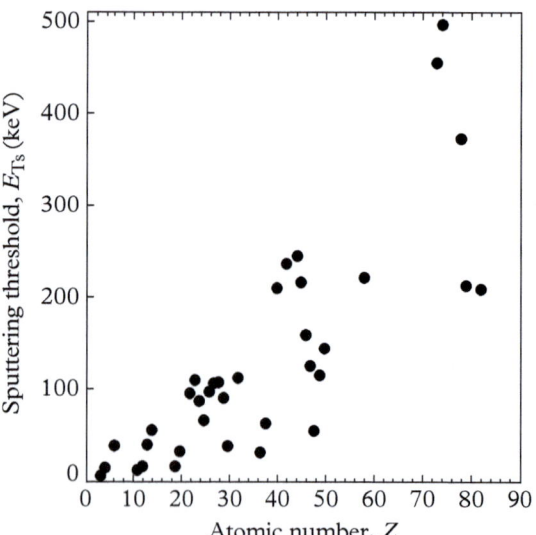

$E_0 = E_{Ts}$, are plotted in Figure 5.3.9; clearly, electron beam sputtering is mainly a concern for low Z elements, even if they are present within a high Z matrix material.

The sputtering rate, S, in units of monolayers per second is given by

$$S = \left(\frac{J}{e}\right)\left(\frac{Z^2}{AE_0}\right)\left(\frac{1}{E_s} - \frac{1}{E_{max}}\right) 3.54 \times 10^{-17} \ \text{cm}^2 \qquad (5.3.4)$$

where J/e is the incident current density in units of electrons/cm^2/s, and the rest of the expression is a sputtering cross-section, with E_s in eV, and E_{max} is a function of E_0 as described by (5.3.3).

To avoid sputtering, it is best to restrict the radiation dose (product of current density and time of exposure). Alternatively, for TEM, coating the *exit surface* of the specimens with a high atomic number (Z) element, such as Au, could avoid mass loss by sputtering.

5.3.2.4   *Beam Heating*

Appreciable energy can be transferred between the high-energy electron probe and atomic electrons as a result of inelastic collisions between these particles of equal mass. Much of the energy transferred ends up as heat within the specimen resulting in a local temperature rise, $\Delta T = T - T_0$.

Assume a mean free path length, λ_{ie}, for inelastic scattering of electrons in a specimen. Then the number, n_{ie}, of inelastic collisions in a specimen of thickness, t, is $n_{ie} = t/\lambda_{ie}$. For a beam of current, I, diameter, d, and average energy loss, $\langle E \rangle$, in Joules, for inelastic collisions per incident electron, the heat deposited, q, per second in the specimen is given by

$$q = \frac{I}{e}\langle E \rangle n_{ie} = I\langle E(eV)\rangle\frac{t}{\lambda_{ie}} \qquad (5.3.5)$$

Under steady-state conditions, the heat generated by the beam is balanced by the conductive heat loss over a radius, R_0, in the material of thermal conductivity, κ, such that [3]

$$I\langle E(eV)\rangle\frac{t}{\lambda_{ie}} = \frac{4\pi\kappa t (T - T_0)}{0.58 + 2\ln\left(\frac{2R_0}{d}\right)} + \frac{\pi d^2}{2}\varepsilon\sigma\left(T^4 - T_0^4\right) \qquad (5.3.6)$$

In practice, the second radiative term on the right of (5.3.6) can be neglected, and the temperature rise, $\Delta T (= T - T_0)$, is then independent of the specimen thickness.

In a TEM, the specimen (§9.6) is rather thin and the heat conduction is essentially in two dimensions, with beam heating a problem when high incident currents are used. However, for organic materials with poor thermal conductivity, κ (typically $\kappa = 0.2 - 2$ W/m/K compared to metals with $\kappa > 100$ W/m/K), beam heating can be a problem as they are susceptible to melting, even for low currents.

In the case of SEM (§10), the heat conduction is radial in three dimensions, and results in a much smaller temperature rise. When the beam diameter, d, is smaller than the electron range, R, (Fig. 5.3.14) the temperature rise in a stationary probe is

$$\Delta T = \frac{1.5\pi \, (I + E_0)}{\kappa R} \tag{5.3.7}$$

where I is the probe current and E_0 is the acceleration voltage. For typical values, $I = 1$ nA, $E_0 = 20$ keV, we get $\Delta T < 0.1$ K for metals, but ΔT is larger (a few degrees) for polymers.

5.3.2.5 *Electrostatic Charging*

Typically, charging is given by the difference between the current entering the specimen through the incident beam and the current leaving in the form of back-scattered or secondary electrons. We consider the case of incident electrons with back-scattering coefficient, η, and a secondary electron yield, $\delta\,(\varepsilon)$, where ε is the kinetic energy of the secondary electrons (see also Fig. 10.3.4 and related discussion). Under steady-state conditions, a current balance between the current entering and leaving the specimen is observed

$$I + \frac{V_s}{R_s} = I\eta + I\delta\,(V_s) = I\eta + I \int_{eV_s}^{E_0 - eV_s} \frac{d\delta}{d\varepsilon} d\varepsilon \tag{5.3.8}$$

where V_s is the surface potential developed in the beam and R_s is an effective resistance between the irradiated and surrounding areas in the specimen. The terms on the left represent the current entering the specimen and those on the right are the losses of electrons due to back-scattering and secondary emissions.

At high incident energy, E_0, most secondary electrons are generated deep in the specimen and do not escape into the vacuum. Thus $\delta\,(V_s)$ is very small, and a negative V_s is required on the left to achieve a current balance (Fig. 5.3.10). For highly insulating specimens, R_s is large and $|V_s|$ may be several kV, causing the incident beam to be repelled and giving rise to unstable or distorted SEM images (Reimer, 1998).

At low E_0, the beam penetrates a shallow distance (a few nm or less) and most of the secondary electrons are able to escape the specimen. Thus, $\delta\,(V_s)$ is very large, and a positive V_s is required to achieve a current balance. When the beam hits the specimen, V_s increases and $\delta\,(V_s)$ decreases until the current balance, (5.4.7), is satisfied. For highly insulating materials, with large R_s, the V_s/R_s term can be insignificant and the balance can be achieved at a potential, V_s, which prevents the low energy secondaries from leaving. In practice, the operating voltage, $E_0 = E_2$, is chosen such that $V_s = 0$, to prevent the loss of secondary electron signal. Typical values for E_2 (keV) are: low-density resist ~ 0.55, PMMA

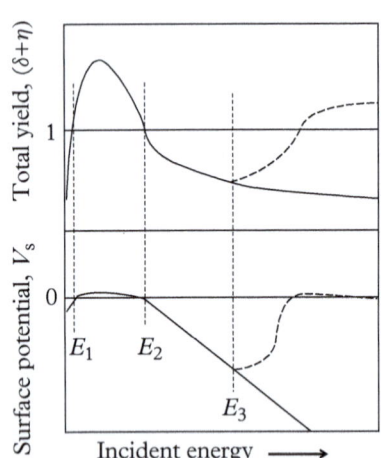

Figure 5.3.10 Total electron yield and surface potential for a poorly conducting specimen in bulk (continuous) and thin film (dashed) form, as a function of incident beam energy, E_0.

Adapted from [3].

resist $= 1.6$, NaCl $= 2.0$, SiO$_2$ (quartz) $= 3.0$, and GaAs $= 2.6$. Further details for different materials and related information can be found in [7].

5.3.2.6   *Hydrocarbon Contamination*

In a TEM or SEM, any hydrocarbon molecules present on the surface of the specimen can be polymerized upon beam irradiation. These polymers have a low vapor pressure and poor surface mobility and over time of continued irradiation they increase in thickness on the specimen. It is known that surface diffusion of hydrocarbons can also be a source of such contamination and various strategies to combat this problem, primarily by keeping the specimen surface and the chamber vacuum sufficiently free of hydrocarbons have been implemented [3].

5.3.3   Neutrons

There are two significant differences between neutrons and electrons. First, neutrons are much heavier than electrons, i.e. $m_n = 1.675 \times 10^{-27}$ kg $= 1,839\, m_e$. Moreover, since the de Broglie wavelength, (1.3.3), of the particle is proportional to $m_n^{-1/2}$, neutrons show wavelengths of the order of inter-planar spacing in crystals suitable for diffraction (§8.8). Second, neutrons have no charge and, as such, their interaction with atoms is only through the nucleus and is not affected by the electrons in the orbitals. Thus, neutrons are able to penetrate much more deeply in materials compared to electrons or even X-ray photons; in most materials the penetration depth of neutrons is of the order of several millimeters. As a result, a substantial amount of material is required for neutron scattering experiments. In some cases, such as in ultra-thin films, difficulties in preparing sufficient quantities of materials may be a limitation to neutron-based scattering analysis.

5.3.4   Protons

Protons are similar to electrons, but they have a positive charge and their mass, m_p, is substantially larger, i.e. $m_p \sim 1,836\, m_e$. In characterization and analysis, protons with MeV energy are used, and they have substantially larger momentum than electrons that are typically used with 10–200 keV energy. Protons impinging on a solid lose a small fraction of their momentum with each atomic collision and go a long way without deviating from their path in a specimen. Figure 5.3.11a is a schematic comparison of the scattering of a proton and electron beam. The proton beam undergoes only a slight broadening and penetrates several tens of micrometers into the specimen. The electron beam is typically stopped after 5–10 μm of penetration and its profile is strongly broadened (in the shape of a tear drop) due to inelastic scattering by the atomic electrons (Fig. 5.3.2).

The range of travel and the depth of penetration of protons (or any other ion) in a material is correlated with its stopping power,[8] S, which typically decreases with increasing average atomic number of the specimen and increasing proton energy

[8] The stopping power defined as the rate of loss of energy with thickness, $(-dE/dt)$, depends both on the material and the energy of the ion. To avoid problems with the state of matter, the energy loss is divided either by the density and S defined in units of [keV mg^{-1} cm^2], or by the atom density (§5.4.1), to give S in units of [eV/(atoms/cm^2)].

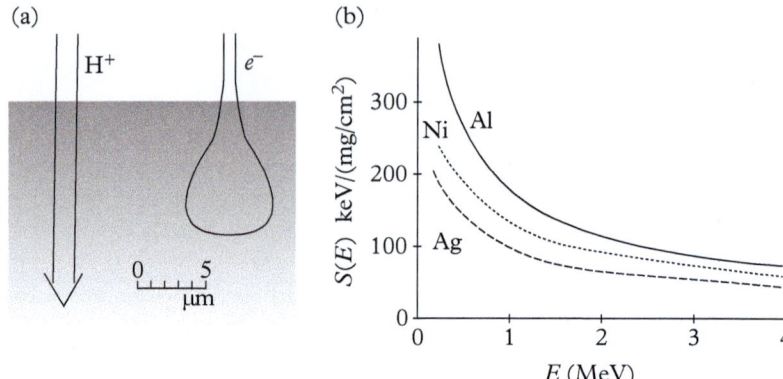

Figure 5.3.11 (a) Comparison of beam broadening of MeV protons and keV electrons in a specimen. See also Figure 5.3.2. (b) The stopping power, S, of MeV protons in different elements.

Adapted from Johansson, Campbell, and Malmqvist (1995).

(Fig. 5.3.11b and Table 5.4.1). For a 2.5 MeV proton, penetration depths are 55 μm ($S = 123 \, \text{keV} \, \text{mg}^{-1} \, \text{cm}^2$) in carbon, and 28 μm ($S = 56 \, \text{keV} \, \text{mg}^{-1} \, \text{cm}^2$) in silver. High-energy protons ($\sim$MeV) are also used to excite characteristic X-rays, which are then analyzed by energy dispersive detectors (§2.5.1.2) to obtain sample chemistry; this technique (§5.4.5) is known as particle induced X-ray emission (PIXE).

5.3.5 Ions

5.3.5.1 Dimensions, Sizes, and Cross-Sections

The interactions of ions with atoms in a specimen involve sizes and distances over many orders of magnitude. The radius of single atoms, (2.2.2), is of the order of 10^{-10} m, with a nucleus of the order of 10^{-14} m, and a number of electrons occupying various electron shells (§2.2.2). The innermost, K- and L-shells, have a typical mean radius of the order of 10^{-12} m. Isolated atoms are electrostatically neutral and can approach each other until the Coulomb interaction affects their outermost electrons. In a crystalline solid, as described in §4.1, we find a periodic arrangement of atoms with average separation of 10^{-9}–10^{-10} m, sharing outer electrons among which the atomic nuclei are distributed; the latter only occupy a small volume ($\sim$10%) of the solid. In addition, atoms are not stationary but vibrate about their equilibrium position with a mean amplitude, at room temperature, of about one-tenth of the average spacing (10^{-10}–10^{-11} m). We are interested in understanding how a projectile ion, $\sim$$10^{-12}$ m in radius, interacts with this complex arrangement in a solid.

As a starting point, the wave-particle duality (§1.3.3) can be applied to the projectile ion. For example, a proton with energy, $E_0 > 10$ eV, has a wavelength much less than the inter-atomic spacing, and can easily penetrate a solid. Further, the de Broglie wavelength, (1.3.3), is inversely proportional to the square root of the atomic mass; thus, heavier atoms and the nuclei of such ions can more easily penetrate a solid. However, in ion–solid interaction, the important parameter is not

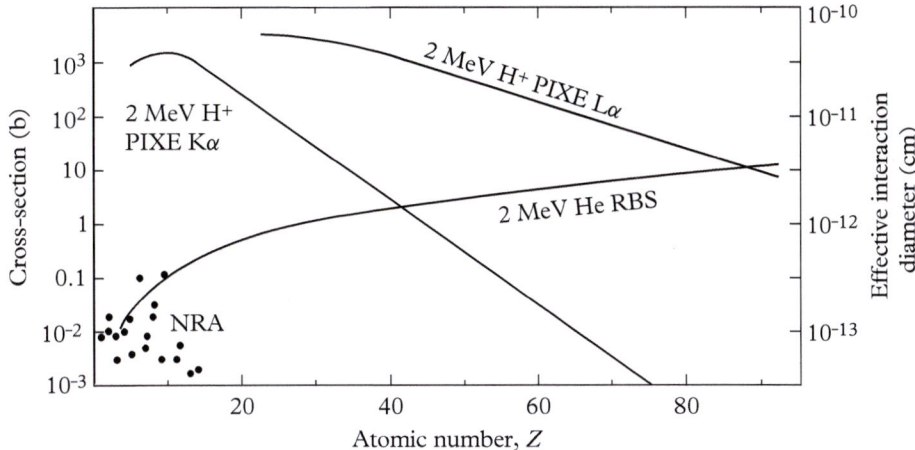

Figure 5.3.12 Cross-sections in units of barns (1b $= 10^{-24}$ cm^2/atom) for the interactions of protons and Helium ions as a function of atomic number. Scattered data points for nuclear reaction analysis (NRA) are also shown.

Adapted from Bird and Williams (1989).

the distance the ion travels in a solid, but the number of atoms the ion encounters, as defined by the atomic areal density (number of atoms/cm^2), ρ_A. In addition, the probability of a particular interaction between the ion and the atom is given by the *cross-section*, σ, defined as the effective area of the atom or nucleus seen by the ion. Thus, the probability, P, that a projectile ion will have a specific interaction is given by the product, i.e. $P = \rho_A \sigma$. Figure 5.3.12 shows cross-sections for the interactions of high-energy ions and atoms, as a function of atomic number, relevant for materials analysis.

The average distance between collisions is defined as the *mean free path length* (§1.3.5.1), and is inversely proportional to the cross section. If the cross-section is high, the mean free path length is low, and the ion beam is significantly attenuated as it propagates in the solid. The attenuated beam flux is given by the same equation that we used to describe X-ray absorption (2.4.2), but with the parameter, μ, representing the attenuation coefficient for the ion beam. Note that the same equation, (2.4.2), applies to all cases of photons, ions, and neutrons, that are attenuated in a solid without undergoing any energy loss.

5.3.5.2   *Ion–Solid Interactions*

Figure 5.3.13 schematically illustrates the interaction of ions with a specimen. Broadly speaking, it is possible to divide the ion–specimen interaction into two categories. *Kinematic collisions* between the primary ion (mass, M_1, and initial energy, E_0) and the atoms (mass, M_2, and initial energy, $E = 0$, i.e. at rest) of the specimen, leading to their scattering by angles, θ_1 and θ_2, with energies, E_1 and E_2,

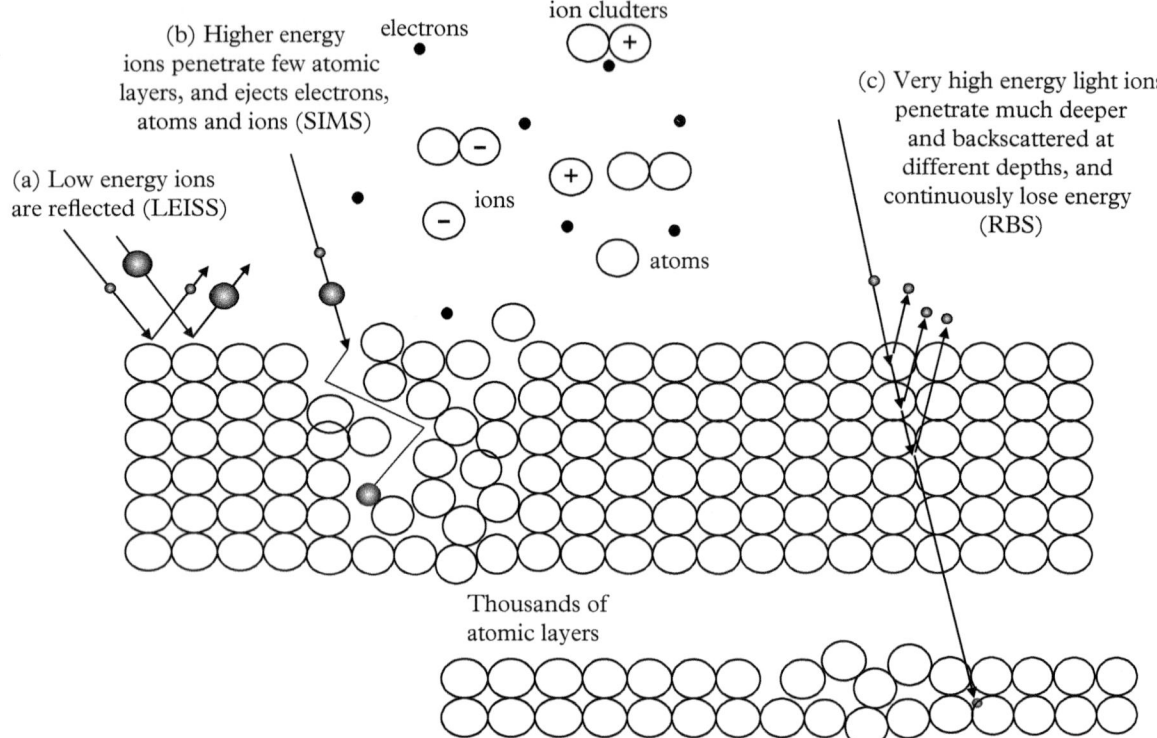

Figure 5.3.13 Ion–solid interactions: (a) Low energy primary ions are reflected, and form the basis of low-energy ion-scattering spectroscopy (b) Higher energy ions penetrate, at most, a few atomic layers, undergoing many major interactions and causing collision cascades. In the process, electrons, atoms and ions ejected from the solid, can be analyzed in a technique known as secondary ion mass spectrometry (c) Very high-energy light ions penetrate many thousands of atomic layers, occasionally undergoing major interactions and being back-scattered; however, they also suffer inelastic scattering and continuously lose energy as they propagate in the solid. Analysis of the energy distribution of the back-scattered ions forms the basis of Rutherford back-scattering spectroscopy (RBS).

respectively (Fig. 5.3.15). In this process, very high-energy (MeV) primary ions are back-scattered, even from substantial depths in the solid. In addition, both incoming and scattered ions continuously lose energy by inelastic scattering processes, which has to be accounted for in the scattering kinematics (Fig. 5.3.17; Fig. 5.4.2).

Alternatively, when higher-energy ions strike a specimen, they penetrate it, and following a number of collisions with atoms are slowed down and brought to rest. Now, the kinetic energy of the primary ion is distributed over several atoms, which in turn, strike other atoms, causing a collision cascade. As a result, the primary ion travels in an irregular path before coming to rest in the specimen (Fig. 5.3.13b).

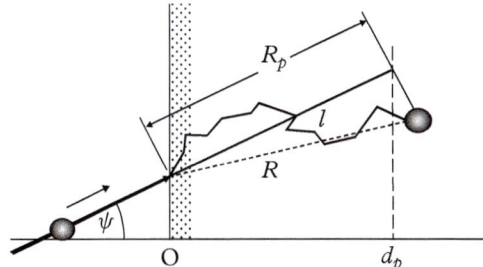

Figure 5.3.14 Schematic drawing defining the projected range, R_p, depth of penetration, d_p, and the path, l, of an ion in a solid, and defining its range, R, which is a straight line from the point of entry to its final resting point.

This process is also known as *ion implantation*. The distance from the point of entry traveled by the ion in the solid before it comes to rest is called its *range*, R, and its projection along the direction of incidence is called the *projected range*, R_p (Fig. 5.3.14). Note that the actual path, l, travelled may be larger than R. For normal incidence, $\psi = 0°$, the projected range is the same as the *penetration depth*, i.e. $d_p = R_p$.

The atoms in the specimen struck by the collision cascade are also displaced, more or less in an isotropic angular distribution from the original sites. Atoms on the surface of the specimen may acquire sufficient energy in the collision cascade to overcome the surface binding forces and be ejected from the specimen. This process is called *sputtering*, and it is sometimes used as a way to remove surface layers, atomic layer by atomic layer. Further, after a layer of the specimen is removed by sputtering, the specimen can be chemically analyzed using a surface analytical technique, such as Auger spectroscopy, to give chemical information as a function of depth (Fig. 2.6.9). Last, but not least, secondary ions, and clusters of ions and atoms are also ejected from the specimen surface (Fig. 5.3.13). These secondary ions can be detected and their mass analyzed by a technique called secondary ion mass spectrometry or SIMS (§5.4.3), a destructive method that is nevertheless used routinely in surface analysis.

5.3.5.3 Physics of Kinematic Collisions

The scattering between an energetic primary ion and the atoms in a specimen can be described as a classical two-body elastic collision, interacting through a centrosymmetric potential. Figure 5.3.15 shows the trajectories of the two masses, M_1 and M_2, before and after collision, in the laboratory frame. M_1 is the projectile/primary ion with energy E_0, and the target atom, M_2, is initially at rest, i.e. $E = 0$. After collisions, their energies are E_1 and E_2, and the scattering angles are θ_1 and θ_2, respectively.

The distance of closest approach, D_c, for a head-on collision, D_0, defined for a zero-impact parameter (Fig. 5.3.15) is important to determine the *nature* of the ion-atom collision. It is given by

$$D_0 = 1.44 \times 10^{-9} \, (Z_1 Z_2) \, (M_1 + M_2) / (M_2 E_0) \tag{5.3.9}$$

Figure 5.3.15 Kinematic scattering geometry between a primary/projectile ion and a stationary atom in the specimen. The impact parameter, D_1, which is the separation between the ion and atom (center to center) perpendicular to the initial trajectory of the ion, and the distance, D_c, of closest approach that determines the nature of the ion-atom interaction is shown. Typical value for keV ions (LEISS, §5.4.2) with $\lambda \sim 10^{-3}$ nm is $D_c \sim 0.05$ nm, and for MeV ions (RBS, §5.4.1) with $\lambda \sim 10^{-5}$ nm, is $D_c \sim 0.001$ nm.

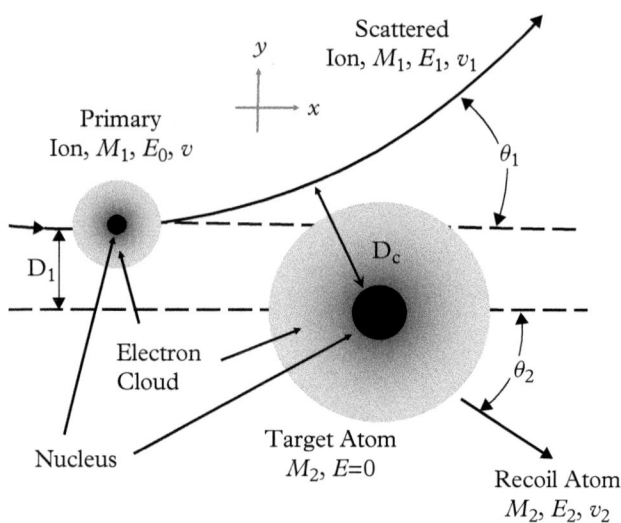

where E_0 is in eV, M is the atomic mass in amu, Z is the atomic number, and D_0 is in meters (Fig. 5.3.15). If D_0 is less than the sum of the atomic or ionic radii, *excitations* can take place, and the classical kinematic collision described here must be replaced by a quantum mechanical description.

Example 5.3.1: Are the collisions with a K^+ ion (radius ~ 0.138 nm) for a He^+ ion (atomic radius ~ 0.031 nm) accelerated at (1) 2 MeV (RBS) and (2) 2 keV (LEISS), kinematic?

Solution: We apply (5.3.9) with $Z_1 = 2$, $M_1 = 4$, $Z_2 = 19$, $M_2 = 39.1$ and determine D_0.

(1) For 2 MeV He^+ ion, $D_0 = 3.0 \times 10^{-14}$ m $= 3 \times 10^{-5}$ nm, which is smaller than the sum of the two diameters.

(2) For 2 keV He^+ ions, $D_0 = 3.0 \times 10^{-11}$ m $= 3 \times 10^{-2}$ nm, which is also smaller than the sum of the two diameters.

In both cases, it will cause excitation (or inelastic scattering, where the incident ion loses energy) and a simple kinematical collision model cannot be applied. Nevertheless, we shall do so, but assume that the inelastic scattering gives rise to a continuous back-scattered energy spectrum up to a maximum energy dictated by the kinematic collision for the heaviest atomic mass in the specimen (see RBS spectrum, Fig. 5.3.17).

We define the dimensionless, kinematic factor, $K_1 = E_1/E_0 = v_1^2/v^2$, for the projectile/primary ion. From a conservation of energy, we get

$$\frac{1}{2}M_1 v^2 = \frac{1}{2}M_1 v_1^2 + \frac{1}{2}M_2 v_2^2 \tag{5.3.10a}$$

Now, conserving the momentum along the direction of motion (x) of the primary ion we get

$$M_1 v = M_1 v_1 \cos\theta_1 + M_2 v_2 \cos\theta_2 \tag{5.3.10b}$$

and similarly, for conservation of momentum perpendicular to the direction of motion (y) gives

$$0 = M_1 v_1 \sin\theta_1 - M_2 v_2 \sin\theta_2 \tag{5.3.10c}$$

Eliminating θ_2 first, and then v_2, from (5.3.10 a–c) we get

$$\frac{v_1}{v} = \frac{\pm\left(M_2^2 - M_1^2\sin^2\theta_1\right)^{1/2} + M_1\cos\theta_1}{M_1 + M_2} \tag{5.3.10d}$$

from which we can determine the dependence of K_1 as a function of the scattering angle, θ_1, as

$$K_1(\theta_1) = \frac{E_1}{E_0} = \left(\frac{\cos\theta_1 \pm \left(A^2 - \sin^2\theta_1\right)^{1/2}}{A + 1}\right)^2 \tag{5.3.11}$$

Note that the kinematic factor, K_1, only depends on the mass ratio, $A = M_2/M_1$, and the scattering angle, θ_1. *The positive sign holds for $A > 1$,* and both signs for $A < 1$. The corresponding equation for the recoiling target atom is

$$K_2(\theta_2) = \frac{E_2}{E_0} = \frac{4A}{(1 + A)^2}\cos^2\theta_2 \tag{5.3.12}$$

with a maximum value when $\theta_2 = 0°$.

Figure 5.3.16 plots the function, $K_1(\theta_1)$, based on (5.3.11), for different values of A. Two specific cases are of interest in materials analysis. For $\theta_1 = 90°$, $K_1(\theta_1)$ is particularly simple, i.e.

$$K_1(90°) = \frac{E_1}{E_0} = \frac{A - 1}{A + 1} = \frac{M_2 - M_1}{M_2 + M_1} \tag{5.3.13}$$

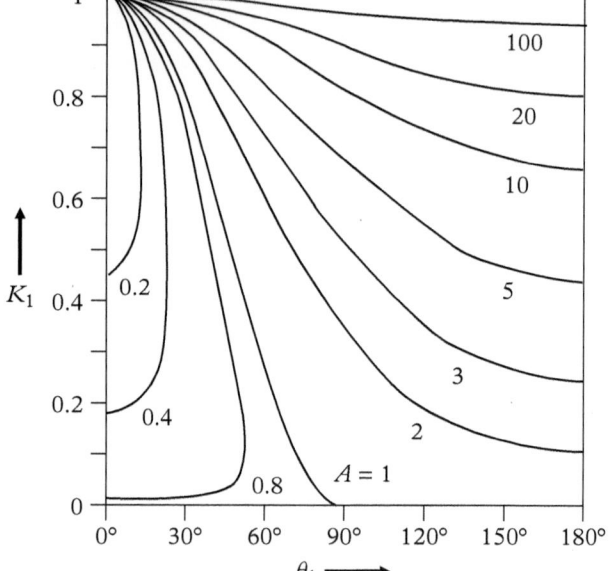

Figure 5.3.16 The kinematic factor, K_1, as a function of the scattering angle, θ_1, in the laboratory frame of reference. The parameter, $A = M_2/M_1$, is the ratio of the scattering atomic mass (M_2) to the primary/projectile ion mass (M_1).

Adapted from Bird and Williams (1989).

which can be rearranged as

$$M_2 = -M_1 \left(\frac{E_1 + E_0}{E_1 - E_0} \right) \tag{5.3.14}$$

Thus, if M_2, or the atomic mass of the element(s) in the specimen, is unknown, it can be determined by measuring the energy of the projectile after collision. Similarly, for $\theta_1 = 180°$

$$K_1 (180°) = \left(\frac{A-1}{A+1} \right)^2 = \left(\frac{M_2 - M_1}{M_2 + M_1} \right)^2 = \frac{E_1}{E_0} \tag{5.3.15}$$

The relationships (5.3.11) and (5.3.12) can be used to identify the scattering atoms from ion energy spectra. In fact, they are sufficiently sensitive to be able to resolve individual isotopes of light elements with MeV ^{4}He$^+$ ions. Figure 5.3.17 shows typical energy spectra for scattering of ^{4}He$^+$ ions at $\theta_1 = 140°$, from a specimen containing ^{108}Ag, ^{28}Si, and ^{16}O, for two different energies, $E_0 = 1$ keV (LEISS, §5.4.2) and $E_0 = 1$ MeV (RBS, §5.4.1).

Example 5.3.2: For the two cases discussed in Example 5.3.1, calculate the kinematic factor assuming that the scattering angle is 170°.

Solution: We use (5.3.11), with $A = M_2/M_1 = 39.1/4 = 9.785$, Cos $\theta_1 = -0.985$, Sin $\theta_1 = 0.174$. Then

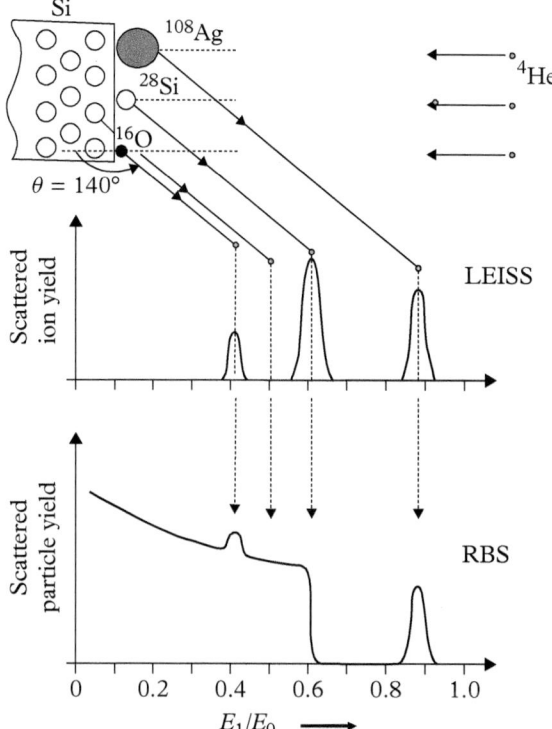

Figure 5.3.17 Schematic representation of the scattering of $^4\mathrm{He}^+$ ions, at an angle of 140°, from bulk silicon with atoms of Ag, Si and O, on its surface. Two representative spectra for low-energy, $E_0 = 1$ keV (LEISS, upper), and high-energy, $E_0 = 1$ MeV (RBS, lower) ions are shown.

Adapted from Vickerman and Gilmore (2009).

(1) $E_0 = 2$ MeV, and $K_1 = [\{-0.985 + (9.785^2 - 0.03)^{1/2}\}/(9.785 + 1)]^2 = 0.6655$

Note that this agrees with Figure 5.3.16.

(2) For $E_0 = 2$ keV, $K_1 = 0.6655$. Clearly, the kinematic factor does not depend on E_0.

For light ions, such as $^4\mathrm{He}^+$, penetrating a solid, both the incident and scattered ions lose energy primarily through excitations and ionizations of atomic electrons. These inelastic collisions are microscopically discrete, but macroscopically, we assume that the moving ions lose energy continuously. Thus, any inelastic scattering that occurs at some depth below the surface, gives rise to a continuous backscattered energy spectrum up to a maximum energy dictated by the kinematic collision for the heaviest atomic mass in the specimen. This can be seen clearly for silicon in the RBS spectrum (Fig. 5.3.17). Further details of LEISS and RBS, and the interpretation of the respective spectra, are discussed in §5.4.

5.3.5.4   *Specimen Damage*

If primary ions with sufficient kinetic energy strike a specimen, apart from the small fraction (~1%) that is reflected from the surface, they generally penetrate it and are slowed down by elastic (nucleus) and inelastic (electrons) collisions. The former results in a collision cascade with a number of possible consequences:

1. The primary ion will traverse a path, l, and range, R (Fig. 5.3.14) and will eventually come to a stop inside the specimen. This is referred to as *ion implantation* and can lead to a local change in composition at some depth in the interior of the specimen. In the semiconductor industry such ion implantation is used to dope wafers at specific depth locations.

2. The collision cascade displaces the atoms in the solid from their original crystallographic sites. The resultant change in atomic configuration can lead to a complete loss of long-range order (amorphization) and/or *atomic mixing*, including the implanted primary ions, in the specimen.

3. A very small number of atoms in the specimen may experience direct impact from the primary ions. Such atoms experience an energy transfer larger than in the cascade and are pushed in the forward direction. This leads to so-called *recoil mixing*, and results in local changes, most significantly, in buried interfaces. Similarly, surface contaminants and adsorbates, if present, can also experience a direct impact and be pushed forward into the interior of the specimen; this process is called *recoil implantation*.

4. If the specimen is highly insulating, it can become charged upon primary ion irradiation. Then, an electric field extending far into the interior of the specimen can be formed, which can promote the motion of charged ions. This process, known as *electromigration*, has been observed for alkali ions in SiO_2.

5. The atoms on the surface (uppermost monolayers) of the specimen, stuck by the collision cascade may possess sufficient kinetic energy to escape the solid by overcoming the surface binding forces. This process is known as ion-beam *sputtering* (see §5.3.5.5).

These processes, are covered in detail in more specialized books on ion-materials interactions (see Bird and Williams, 1989). From a materials analysis point of view, ion-beam sputtering (5), is of particular interest as it forms the basis of depth profiling (systematic removal of surface layers, one monolayer or so at a time, followed by chemical analysis, such as Auger electron spectroscopy). Alternatively, the sputtered ions are directly mass-analyzed, as in SIMS (§5.4.3), to provide chemical information; in this case, the yield of secondary ions is affected by (1–4), which are collectively known as *matrix effects*.

5.3.5.5 Ion-Beam Sputtering

The rate of sputtering, dN_i/dt, for atoms of element, i, in the collision cascade model (Fig. 5.3.13) defined as the number of sputtered atoms per unit time, is proportional to the primary ion flow rate, $dN_0/dt = I_0/q$ and given by

$$dN_i/dt = Y_i \, dN_0/dt \qquad (5.3.16)$$

where I_0 is the current of ions with charge, q, and the proportionality factor, Y_i, is called the *sputtering yield*. In the case of a single element specimen, the yield, $Y_i^e(E_0)$, will depend on how much energy the primary ion will deposit close to the surface, $(dE/dz)_n|_{z=0}$, and the binding strength of the surface atoms. Thus, for a specific ion–specimen combination, we have

$$Y_i^e(E_0) \sim (dE/dz)_n|_{z=0}/E_{sub} \qquad (5.3.17)$$

where the numerator is the specific loss, due to collisions with the nuclei on the surface, and the binding energy of the surface atoms is approximated by the sublimation energy, E_{sub}, of the material. For example, Figure 5.3.18 shows the values for the sputtering yield of Al, as a function of the primary ion energy, E_0. While the sputtering yield will depend on the crystallinity and topography of the specimen, it can be seen that the yield (Fig. 5.3.18) is a maximum for $E_0 \sim 10$ keV; a typical value often used in practice for sputtering. Moreover, for fixed E_0, the sputtering yield will vary by a factor of 3–5 across the periodic table.

For a multi-element specimen, (5.3.17) is modified to

$$Y_i^s = X_i^s Y_i^e \qquad (5.3.18)$$

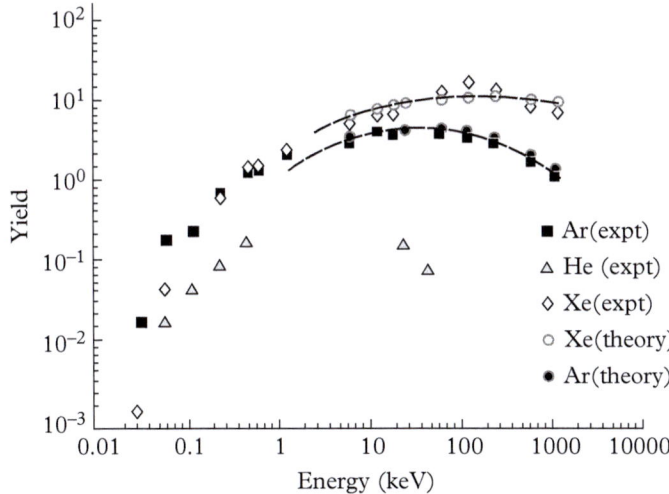

Figure 5.3.18 Sputter yield data for aluminum using different primary ions as a function of ion energy. Notice the peak at ~ 10 keV.

Adapted from Vickerman and Gilmore (2009).

where, X_i^s is the fractional surface concentration of element, i, in the specimen. In practice, due to the matrix effects, (1–4), discussed earlier in §5.3.5.4, the specimen composition is affected by the ion–specimen interactions, and the sputtering yield is altered from Y_i^e to Y_i^s. Thus, for a multi-element specimen, the total sputtering yield, Y, is

$$Y = X_i^s Y_i^s = \frac{dN/dt}{dN_0/dt} \tag{5.3.19}$$

Further, the number of ions produced by sputtering is given by the *secondary ion yield*

$$Y_i^\pm = P_i^\pm Y_i^s \tag{5.3.20}$$

where, $P_i^\pm$ is the ionization probability. In practice, the instrument detection parameters, β_i, are included and the secondary ion intensity is written as

$$I_i^\pm = \beta_i P_i^\pm Y_i^s (dN_0/dt) \tag{5.3.21}$$

Since the ionization process is affected by the electronic state of the surface, the secondary ion yields vary by several orders of magnitude across the periodic table (Fig. 5.3.19). They are also dependent on the chemical state of the surface; thus, we can expect the ion yields to be different for a metal and its oxide.

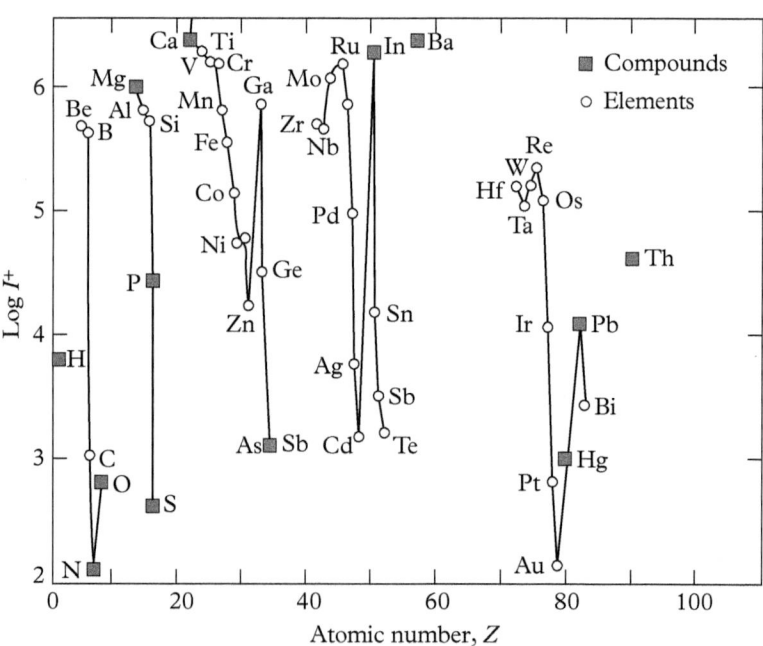

Figure 5.3.19 Positive ion yield as a function of atomic number for elements (open circles) and compounds (filled squares). A primary current of 1 nA, 13.5 keV, O-ions were used.

Adapted from [8].

5.4   Ion-Based Characterization Methods

A high-energy (MeV) primary ion penetrating a solid will undergo energy transfer by a variety of collision processes (Fig. 5.3.13). The most common process is inelastic scattering with the valence electrons, and involves energy losses of the order of 10s of eV per collision, but with insignificant deviation in the trajectory (i.e. no deflection) of the primary ion beam. The *cross-section* for such processes is high and of the order of 10^{-16} cm^2, with correspondingly large impact parameters of ~1 Å, i.e. of the order of typical lattice parameters in materials. At smaller impact parameters, collisions leading to inner-shell ionization and subsequent de-excitation by the emission of characteristic X-rays are observed. Such chemical analysis of materials by PIXE is discussed in §5.4.4. If such a collision leads to back-scattering ($\theta_1 \sim 140°–180°$; Fig. 5.3.15), then the final energy of the scattered ion will depend on the elastic nuclear collision at a given depth combined with a number of inelastic collisions, each with a small energy-loss, as the primary ion travels into and the back-scattered ion propagates out of the specimen. Thus, measuring the energy of the scattered ion not only gives the mass of the scattering atom in the specimen but also its depth location within the specimen. This is the basis of RBS, discussed in §5.4.1. It is worth emphasizing here that the kinematic collision and the associated scattering cross-section are independent of the chemical bond, and hence the back-scattered spectrum is independent of the nature of the bonding or the electronic structure of the specimen. Compared to RBS, which yields an edge feature (Fig. 5.4.3) in the spectrum, LEISS, as in Figure 5.3.17, is characterized by sharp peaks for each atomic species found on the specimen surface. In LEISS, only ions back-scattered from the surface that retain their original charge, i.e. not neutralized, are detected with appropriate electrostatic analyzers. LEISS is discussed in §5.4.2. Last, but not least, the secondary ions produced upon primary ion bombardment can be detected and their mass analyzed; this technique of SIMS is discussed in §5.4.3.

5.4.1   Rutherford Back-Scattering Spectroscopy (RBS)

One parameter of importance for RBS is the energy loss of the primary ions per unit length—dE/dz—due to the inelastic scattering in the specimen. Commonly it is expressed in terms of the stopping power, S (eV/(atoms/cm^2)) where

$$S = -\frac{dE}{dz}\frac{1}{N} \tag{5.4.1}$$

and N is the atom density (number of atoms/cm^3). The stopping power data for all elements has been tabulated (Zeigler, 1977), and (5.4.1) is used extensively in RBS analysis to calculate the energy loss per unit length. Figure 5.4.1 shows a typical plot of the stopping power for ^{4}He$^+$ in Ni. The broad peak at ~1 MeV is normally used as the operating energy for RBS.

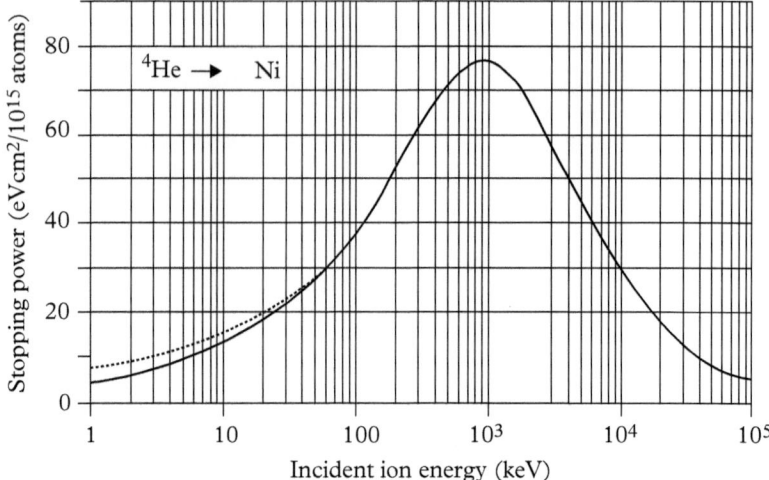

Figure 5.4.1 Stopping power of $^4\text{He}^+$ ions in Ni as a function of incident ion energy. Notice the broad peak at $\sim$1 MeV.

Adapted from Vickerman and Gilmore (2009).

For compounds and mixtures, a simple additive rule of energy loss may be used. Thus, the stopping power for a compound, $A_m B_n$, is given by

$$S_{A_m B_n} = m S_A + n S_B \tag{5.4.1a}$$

where $m + n$ is normalized to unity. The stopping power for ^4He and ^1H are tabulated for select elements and energies in Table 5.4.1.

> **Example 5.4.1:** Calculate the stopping power for Si_3N_4 for 3 MeV ^4He ions.
>
> **Solution:** Table 5.4.1 gives S^{Si} (3 MeV) = 39.17 eV cm^2/10^{15} atoms, and S^{N} (3 MeV) = 25.01 eV cm^2/10^{15} atoms.
>   From (5.4.1a), we get $S_{A_m B_n}$ = (3 × 39.17 + 4 × 25.01) /7 = 31.08 eV cm^2/10^{15} atoms.

Table 5.4.1 Proton (^{1}H) and Helium (^{4}He) stopping power (eVcm^2/10^{15} atoms) for select elements and energies.

		^{1}H					^{4}He				
Z	Element	100 keV	1 MeV	2 MeV	3 MeV	4 MeV	100 keV	1 MeV	2 MeV	3 MeV	4 MeV
1	H	4.690	1.125	0.656	0.474	0.374	7.66	11.87	7.68	5.62	4.5
2	He	6.123	1.906	1.116	0.808	0.638	9.14	17.42	12.48	9.43	7.64
3	Li	8.120	2.662	1.638	1.208	0.965	13.51	22.02	16.29	12.82	10.67
4	Be	10.044	3.296	2.064	1.530	1.226	22.92	24.24	19.15	15.63	13.22
5	B	13.653	4.045	2.488	1.834	1.465	23.14	33.75	24.58	19.44	16.22

Table 5.4.1 Continued

Z	Element	¹H 100 keV	1 MeV	2 MeV	3 MeV	4 MeV	⁴He 100 keV	1 MeV	2 MeV	3 MeV	4 MeV
6	C	14.428	4.602	2.866	2.131	1.714	23.62	37.78	27.83	22.05	18.46
7	N	16.222	5.154	3.209	2.404	1.943	26.91	45.21	32.54	25.01	20.67
8	O	16.145	5.737	3.602	2.700	2.183	26.70	46.17	34.92	27.54	23.01
9	F	13.434	6.170	4.740	3.343	2.934	22.80	43.28	36.34	29.41	24.75
11	Na	20.975	7.680	4.826	3.600	2.904	40.46	54.99	44.49	36.41	30.81
12	Mg	20.700	7.605	4.854	3.646	2.949	44.33	55.61	44.11	35.95	30.52
13	Al	20.460	7.786	4.994	3.760	3.049	40.64	53.18	43.99	36.56	31.25
14	Si	26.428	8.271	5.288	4.002	3.264	41.84	63.71	48.33	39.17	33.20
15	P	27.108	8.773	5.634	4.270	3.482	45.21	70.21	51.56	41.53	35.21
16	S	27.706	8.182	5.313	4.081	3.355	45.49	73.02	50.69	39.20	32.84
17	Cl	30.424	9.596	6.219	4.773	3.922	53.23	85.27	59.38	45.98	38.52
19	K	32.604	11.067	6.879	5.159	4.193	47.14	85.90	66.59	53.15	44.41
20	Ca	27.434	11.140	7.197	5.520	4.537	50.77	82.55	66.4	53.13	44.72
22	Ti	30.923	11.771	7.528	5.687	4.627	49.40	86.81	68.49	55.71	47.27
23	V	32.362	12.043	7.691	5.865	4.812	45.77	96.76	73.21	57.72	48.35
24	Cr	29.221	12.065	7.767	5.911	4.845	43.02	83.70	69.18	56.93	48.46
25	Mn	27.737	12.153	7.864	6.000	4.926	40.06	82.76	69.20	57.20	48.82
26	Fe	29.619	12.358	8.039	6.156	5.066	43.68	85.12	70.29	58.09	49.65
27	Co	27.314	12.358	8.105	6.231	5.141	41.33	82.50	69.48	57.85	49.65
28	Ni	25.010	12.557	8.366	6.456	5.335	39.97	76.11	68.03	58.08	50.47
29	Cu	23.737	12.616	8.524	6.593	5.445	34.95	75.15	67.14	57.89	50.72
30	Zn	24.168	12.869	8.657	6.729	5.589	33.71	77.22	69.50	59.41	51.73
31	Ga	25.262	12.848	8.798	6.886	5.728	44.73	81.60	70.18	59.22	51.66
32	Ge	27.976	13.602	9.413	7.377	6.137	51.38	83.05	72.06	62.06	54.70
33	As	32.061	13.326	9.145	7.172	5.977	62.62	88.62	72.76	61.35	53.59
34	Se	31.135	13.657	9.420	7.387	6.157	49.33	87.64	73.06	62.48	54.93
39	Y	41.148	15.751	10.694	8.386	6.992	73.67	116.25	90.23	73.56	63.34
41	Nb	42.281	17.371	11.361	8.741	7.223	76.22	119.39	98.62	81.60	69.85
42	Mo	28.632	16.170	10.673	8.256	6.847	55.02	109.86	91.09	75.70	65.03
46	Pd	36.105	17.511	11.714	9.096	7.558	68.64	109.38	95.79	81.22	70.46
47	Ag	34.727	16.982	11.471	8.998	7.524	56.37	111.17	94.98	79.14	68.33
48	Cd	39.929	17.530	11.754	9.153	7.623	57.80	114.85	96.66	81.40	70.54
49	In	38.637	17.837	11.972	9.323	7.762	56.85	116.53	98.45	82.82	71.78
50	Sn	40.480	17.627	11.817	9.212	7.676	64.87	119.46	98.61	82.14	70.94
56	Ba	56.156	20.675	13.817	10.740	8.926	87.15	147.60	116.41	98.48	83.23
57	La	54.859	20.686	13.512	10.446	8.680	79.39	148.21	119.16	97.64	83.25
78	Pt	36.452	20.522	14.915	15.040	10.234	49.79	117.60	104.40	92.07	82.81
79	Au	35.401	20.968	15.094	12.217	10.414	53.96	122.53	110.71	95.39	84.59
82	Pb	44.030	21.859	15.664	12.667	10.784	72.95	139.14	118.03	99.98	88.19

Note: values for intermediate energies can be linearly interpolated.

Adapted from Tesmer and Nastasi (1995).

Example 5.4.2: A binary compound, Na_xCl_{1-x}, with a density of 8×10^{22} atoms/cm^3, is analyzed by RBS, using 2 MeV ^{4}He ions. The ions lose energy at the rate of 4.16×10^{39} eV/cm.

(a) Calculate the stopping power for this compound.

(b) What is the composition of the binary alloy?

Solution:

(a) The stopping power, (5.4.1), for the compound can be calculated as
$S_{BC} = \left(4.16 \times 10^{39}\right) / \left(8 \times 10^{22}\right) = 5.19 \times 10^{16}$ eVcm^2/atom $= 51.9$ eV cm^2/10^{15} atoms.

(b) From Table 5.4.1, the stopping power for the two elements are
$S_{Na} = 44.49$ eV cm^2/10^{15} atoms, and $S_{Cl} = 59.38$ eV cm^2/10^{15} atoms

From (5.4.1a), we have $S_{BC} = x\,S_{Na} + (1-x)\,S_{Cl}$, which gives $51.9 = x\,44.49 + (1-x)\ 59.38$.
Thus, $x = 0.5$, and the compound is $Na_{0.5}Cl_{0.5}$ or NaCl.

5.4.1.1   *Energy Width in Back-Scattering Spectroscopy*

We now determine the energy width in the RBS spectrum corresponding to the finite layer thickness of a specimen and illustrate it with a simple example (Fig. 5.4.2). The total energy loss of the primary ion with energy, E_0, as it penetrates a thin film of thickness, Δt, is

$$\Delta E_{in} = \int_0^{\Delta t} \frac{dE}{dz} dz = \left.\frac{dE}{dz}\right|_{in} \Delta t \tag{5.4.2}$$

where, $\left.\frac{dE}{dz}\right|_{in}$ is an average value between E_0 and $E_0 - \Delta t \left.\frac{dE}{dz}\right|_{in}$. Then, at any depth, t, the energy of the ion in the specimen is

$$E(t) = E_0 - t\left.\frac{dE}{dz}\right|_{in} \tag{5.4.3}$$

After large angle (θ_1) back-scattering, the energy of the ion just after back-scattering is $K_1 E(t)$, where K_1 is the kinematic factor, (5.3.11). In the outward

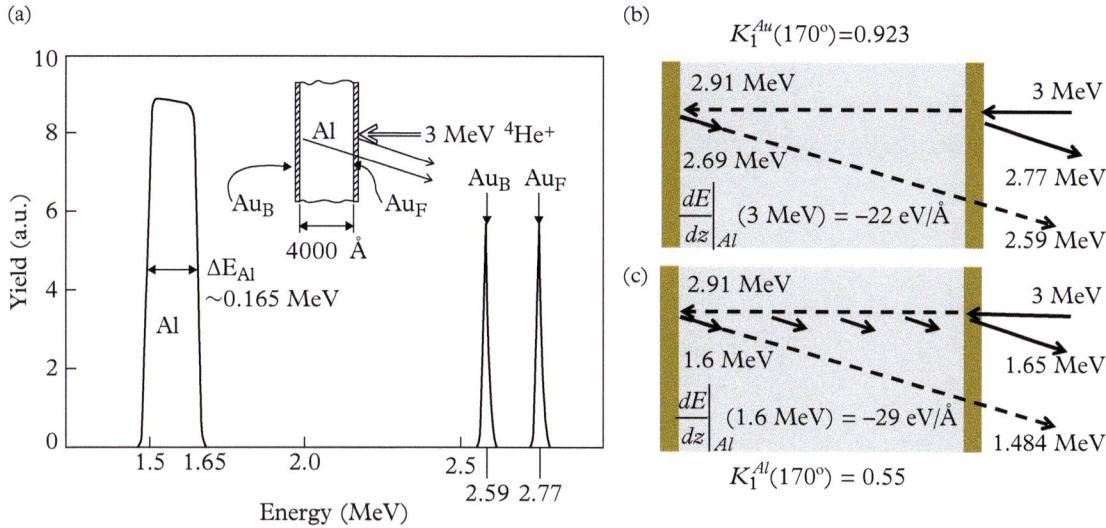

Figure 5.4.2 (a) Back-scattering spectra of 3 MeV ^{4}He$^+$ ions of a 400 nm thick Al film with two Au surface markers. (b) For the Au$_F$ layer one sees a sharp peak at 2.77 MeV; the incident ion losses energy (0.088 MeV) traversing the Al film and arrives at the Au$_B$ layer with an energy of 2.91 MeV. Upon back-scattering it has an energy of $\sim$ 2.69 MeV, loses energy ($\sim$ 0.1 MeV) in the Al film, and emerges with an energy of 2.59 MeV, appearing as a sharp peak. (c) On encountering the front surface of the Al layer, the He ion is back-scattered at an energy of 1.65 MeV. It loses energy in the Al layer and arrives at the back surface with an energy $\sim$ 2.91 MeV. Upon back-scattering at the very back layer it has an energy of $\sim$ 1.6 MeV, loses energy (0.116 MeV) in the Al film, and emerges with an energy of 1.484 MeV. Note that as the ion traverses into the film it is back-scattered at various depths and as a result, the back-scattered yield for Al has a broad shaped peak with a width of 0.165 MeV, as shown.

Adapted from Feldman and Mayer (1986).

(back-scattered) path, the ion continues to lose energy and its energy, $E_{out}(t)$ at any depth, t, is

$$E_{out}(t) = K_1 E(t) - \frac{t}{|\cos\theta_1|} \left.\frac{dE}{dz}\right|_{out} = -t\left[K_1 \left.\frac{dE}{dz}\right|_{in} + \frac{1}{|\cos\theta_1|}\left.\frac{dE}{dz}\right|_{out}\right] + K_1 E_0$$

$$(5.4.4)$$

Thus, for a film of thickness, Δt, the energy width, ΔE, in the RBS signal is

$$\Delta E = \Delta t\left[K_1 \left.\frac{dE}{dz}\right|_{in} + \frac{1}{|\cos\theta_1|}\left.\frac{dE}{dz}\right|_{out}\right] = \Delta t\,[S] \qquad (5.4.5)$$

where, [S] is the back-scattering energy-loss factor. In the surface energy approximation, for typical films with $\Delta t \leq 100$ *nm*, assuming that the change in energy along the path of the ion is negligible, (5.4.5) simplifies to

$$\Delta E_0 = \Delta t \left[K_1 \frac{dE}{dz}\bigg|_{E_0} + \frac{1}{|\cos\theta_1|} \frac{dE}{dz}\bigg|_{K_1 E_0} \right] = \Delta t \, [S_0] \qquad (5.4.6)$$

The implications of (5.4.6) are illustrated in Figure 5.4.2a, which shows an RBS spectrum, using 3 MeV $^4\mathrm{He}^+$ ions, for a standard test film of Al, 400 nm thick, with Au markers on the top and bottom surface. Features of the RBS spectrum, for the Au markers (Fig. 5.4.2b) and the Al layer (Fig. 5.4.2c) are also explained. For Al, $\frac{dE}{dz}\big|_{in} \sim 22$ eV/Å for 3 MeV ions (stopping power, $S = 36.56$ eV/$(10^{15}$ atoms/cm^2)) and $N_{Al} = 6 \times 10^{22}$ atoms/cm^3) back-scattered at 170°, and $\frac{dE}{dz}\big|_{out} \sim 29$ eV for ~ 1.5 MeV ($S = 48.34$ eV/$(10^{15}$ atoms/cm^2)), with K_1(Al) ~ 0.55; note that the energy loss rate is determined from known values of stopping power, S, as in (5.4.1) and known values of the atom density, N. Substituting these values in (5.4.5), we get ΔE (Al) ~ 165 keV for the film. This is very close to the observed separation (~ 175 keV) of Au peaks, but the difference is understandable as K_1(Au) is used in (5.4.5).

5.4.1.2   *Shape of the Back-Scattering Spectrum*

Is there a well-defined shape to an ideal spectrum in RBS data?

To answer this question, we consider a back-scattering experiment from an infinitely thick Au target, a detector with a solid angle, Ω, and N_0 incident $^4\mathrm{He}^+$ particles of 1.4 MeV energy (Fig. 5.4.3). Then the yield, $Y(t)$, from a thin layer of thickness, Δt, with $N\Delta t$ target atoms/cm^2 in the layer is

$$Y = \sigma\,(\theta)\ \Omega\,N_0\,N\Delta t \qquad (5.4.7)$$

where $\sigma(\theta)$ is the scattering cross-section for elements of atomic numbers, Z_1 (primary particle) and Z_2 (matrix atom), given by

$$\sigma\,(\theta) \simeq \left(\frac{Z_1 Z_2 e^2}{4E_1(t)} \right)^2 \qquad (5.4.8)$$

and $E_1(t)$ is the energy of the particle at depth, t.

The ratio, A, of the energy lost in the outward path, $\Delta E_{out} = K_1 E(t) - E_1$, and the inward path, $\Delta E_{in} = E_0 - E(t)$, where $E_1(t)$ is the measured energy, is given by

$$A = \frac{\Delta E_{out}}{\Delta E_{in}} = \frac{K_1 E(t) - E_1}{E_0 - E(t)} = \frac{(dE/dz)_{out}}{(dE/dz)_{in}} \qquad (5.4.9)$$

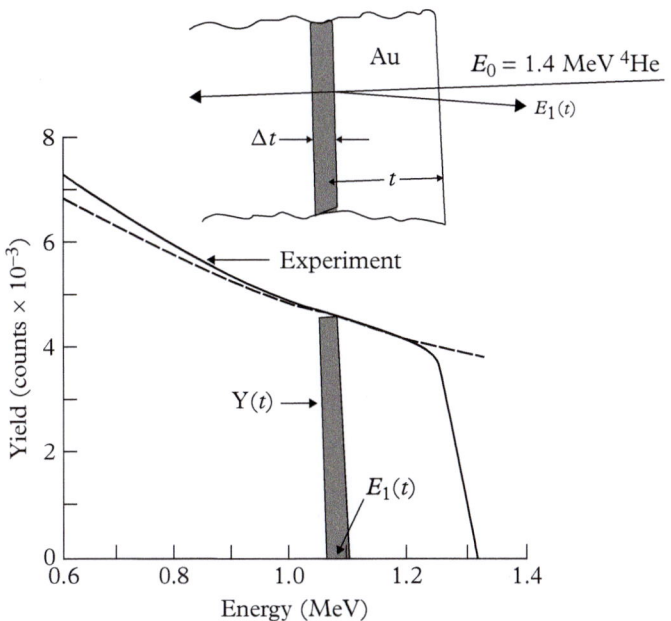

Figure 5.4.3 The back-scattering yield for 1.4 MeV ^{4}He$^+$ ions incident on a thick gold specimen. The dashed line is calculated using (5.4.11) and normalized to the experiment.

Adapted from Feldman and Mayer (1986).

For MeV electrons, A is a constant, and then by rearranging terms we get

$$E(t) = \frac{E_1 + AE_0}{K_1 + A} \tag{5.4.10}$$

For medium to heavy mass target atoms, $K_1 = 1$ and $A = 1$, and thus

$$Y(E_1) \propto \left(\frac{1}{E_0 + E_1}\right)^2 \tag{5.4.11}$$

which accounts for the typical spectral shape in RBS spectra (Fig. 5.4.3).

5.4.1.3 Ni Thin Films Grown on Silicon: A Technological Example

A routine problem in silicon semiconductor films is the formation of metal silicides at the interface of a metal contact layer by the reaction of metal films when grown on silicon. We now illustrate how this thin film materials microstructure is analyzed by RBS.

Figure 5.4.4a shows the 170° back-scattered spectrum of a 1000 Å thick film of Ni, deposited on a silicon substrate, probed with ^{4}He$^+$ primary particles of energy, $E_0 = 2$ MeV. The penetration depth of the 2 MeV ^{4}He$^+$ beam in Ni is many microns. The particles back-scattered, with energy E_1, from the front surface, satisfy the kinematic equation, (5.3.11), which gives $E_1 = K_1 E_0 = 1520$ keV, with $K_1 = 0.76$ for Ni, and $K_1 = 0.57$ for Si. The 2 MeV particles traveling through

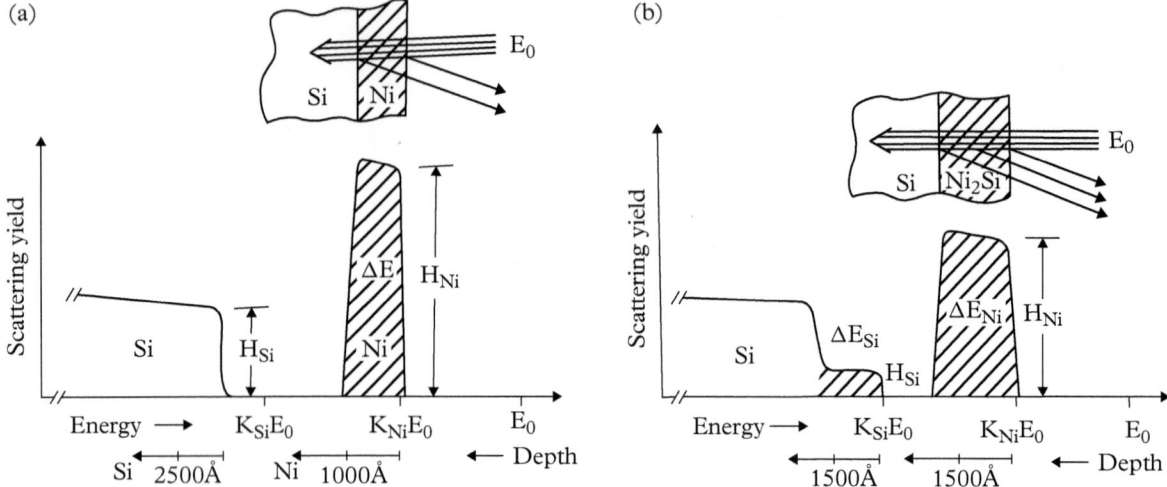

Figure 5.4.4 Back-scattering yield for MeV ^{4}He$^+$ ions from (a) 100 nm thick Ni film grown on a silicon substrate, and (b) a reacted Ni$_2$Si (silicide) layer. The energy scale is converted to depth as shown.

Adapted from Feldman and Mayer (1986).

the nickel layer lose energy continuously along the path at the rate of 64 eV/Å (stopping power, $S = 70.04$ eV/(10^{15} atoms/cm^2), and ideal atom density ($N_{Ni} = 9.17 \times 10^{22}$ atoms/cm^3). Assuming the energy loss is linear with the thickness, the 2 MeV particles will lose 64 keV in energy while penetrating through the nickel layer. Just at the interface, the particles back-scattered by Ni atoms will have an energy $K_{Ni} (E_0 - 64)$ keV, and as they propagate outward they will continue to lose energy at the rate of 69 eV/Å (the difference from 64 eV/Å is due to the small energy dependence of the stopping power, S), and emerge with an energy of 1402 keV. Thus, the energy difference, ΔE_{Ni}, of the particles emerging after back-scattering from the Ni surface and the Ni/Si interface is, $\Delta E_{Ni} \sim 1520 - 1402 = 118$ keV.

On the other hand, if the Ni were to react with silicon and form a Ni$_2$Si compound (Fig. 5.4.4b), we can see that the energy width, ΔE_{Ni}, is broader due to the presence of Si in the reacted layer. In addition, the back-scattering spectrum for Si shows a step corresponding to its presence in the Ni$_2$Si layer. Further, the product of the respective step heights, H_{Ni} and H_{Si}, and energy widths, ΔE_{Ni} and ΔE_{Si}, are proportional to the composition of the silicide layer, i.e.

$$\frac{N_{Ni}}{N_{Si}} = \frac{H_{Ni}}{H_{Si}} \frac{\Delta E_{Ni}}{\Delta E_{Si}} \frac{\sigma_{Ni}}{\sigma_{Si}} = \frac{H_{Ni}}{H_{Si}} \frac{\Delta E_{Ni}}{\Delta E_{Si}} \left(\frac{Z_{Si}}{Z_{Ni}}\right)^2 \tag{5.4.12}$$

Example 5.4.3: A thin film, 100 nm thick, of $YBa_2Cu_3O_7$ is grown on a single crystal $SrTiO_3$ substrate (adapted from Vickerman, 2000).

(a) What would the RBS spectrum look like if 2.5 MeV $^4He^{2+}$ primary ions are used?

(b) What would the individual components of the $YBa_2Cu_3O_7$ spectrum look like after background subtraction?

(c) Derive a simple expression for the compositional ratio of two elements, assuming that their energy widths are the same.

Solution: (a) The expected RBS spectrum is plotted in the following figure.

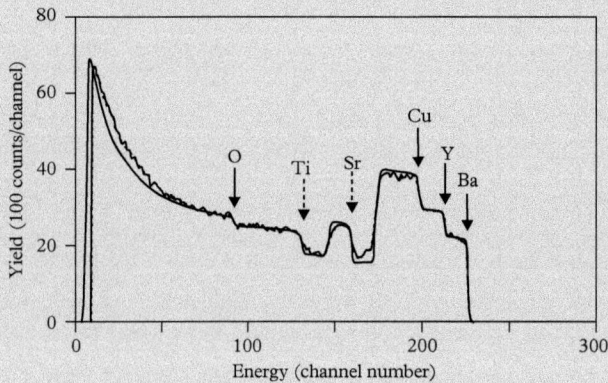

(b) The following figure shows what, after background subtraction, the individual component of the $YBa_2Cu_3O_7$ RBS spectrum would look like.

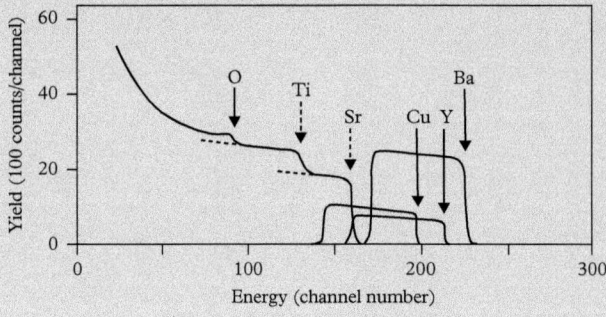

(c) The stoichiometry of the film can be calculated by taking ratios of any two species, i and k, as in (5.4.12) to give

$$\frac{N_i}{N_k} = \frac{H_i}{H_k} \frac{\Delta E_i}{\Delta E_k} \frac{\sigma_i}{\sigma_k} = \frac{H_i}{H_k} \frac{\Delta E_i}{\Delta E_k} \left(\frac{Z_k}{Z_i}\right)^2 \tag{5.4.13}$$

As a first approximation, we can assume that $\Delta E_i = \Delta E_k$, and obtain

$$\frac{N_i}{N_k} = \frac{H_i}{H_k} \left(\frac{Z_k}{Z_i}\right)^2 \tag{5.4.14}$$

as an estimate of the stoichiometry.

5.4.1.4   Ion Channeling

In single crystal materials, a high-energy ion incident along certain crystallographic orientation encounters far fewer atoms, or scattering centers, when compared to other incident directions. This geometric effect is termed channeling and is exhibited by all incident probes—photons, electrons, neutrons, and ions.

If the film is highly monocrystalline, and the collimated primary ion beam is incident along a low-index crystallographic direction, then it is steered in between the atomic columns (Fig. 5.4.5). This is called channeling, and in this case, the back-scattering yield is substantially reduced when compared to the case of

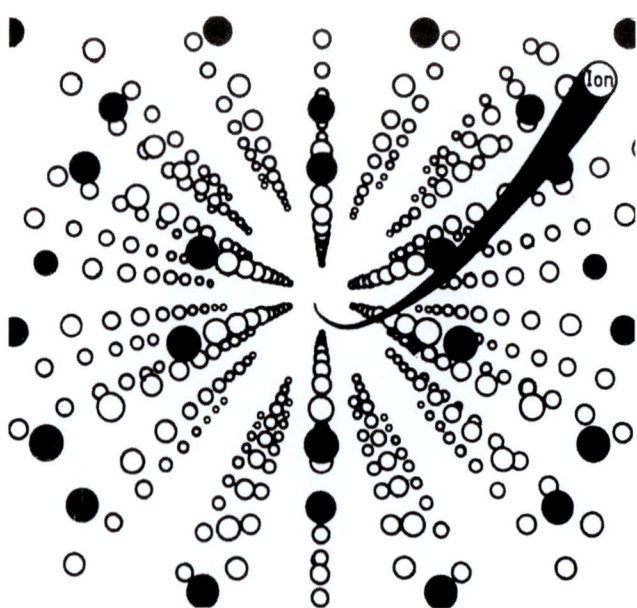

Figure 5.4.5 A schematic representation of the path of an ion when channeled in a diamond cubic lattice.

Adapted from Bird and Williams (1989).

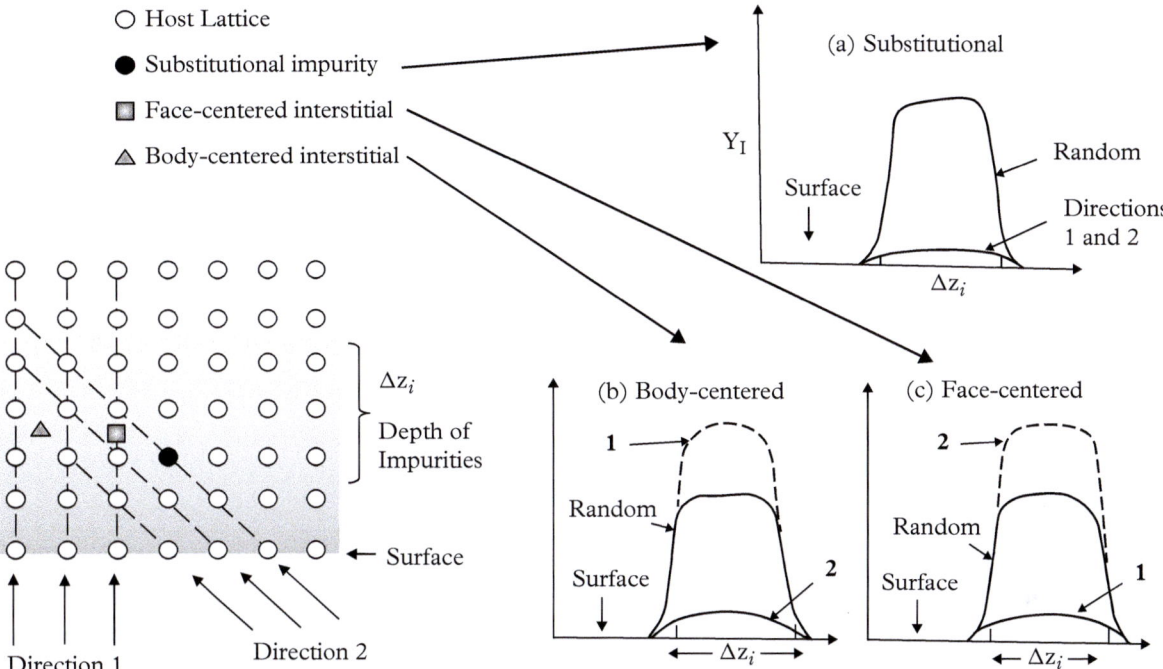

Figure 5.4.6 Determination of the position of foreign atoms in a lattice by the channeling and blocking of the primary high-energy ion beam incident along different directions. For the structure shown on the left and ions incident along the two directions, 1 and 2, the depth profile is shown for impurities on (a) substitutional sites, (b) BCC interstitial sites, and (c) FCC interstitial sites.

Adapted from Bird and Williams (1989).

"random" incidence. However, if there are any deviations from this ideal crystal lattice, such as the presence of interstitial atoms, the back-scattered flux intensity and its angular distribution can vary. By appropriate triangulation, the position of foreign atoms in the lattice can be determined (Fig. 5.4.6).

5.4.2 Low-Energy Ion Scattering Spectroscopy (LEISS)

A low-energy ion scattering spectroscopy experiment is similar to RBS but is carried out at low energies ($\sim$keV). It also satisfies the kinematic relationship between energy and mass of the ions as described in (5.3.11) and (5.3.12). However, the scattered ions are detected using an *electrostatic analyzer and therefore only positively charged ions are detected*. For low-energy ions, only those particles back-scattered from the top surface monolayer have a high probability of surviving the scattering without their charge being neutralized. Thus, since the electrostatic analyzer requires a positive charge, LEISS is a *highly surface-sensitive technique*.

Further, LEISS exhibits a sharp peak for each element present on the surface (Fig. 5.3.17); in comparison, a sharp edge in the spectrum for each element is observed in RBS.

The sensitivity to neutralization is most pronounced for noble gas (He^+, Ne^+, Ar^+) ion scattering, and typically, the probability, P, for surviving as an ion after scattering is ~5% for the first atomic monolayer, and at least one order of magnitude lower (~0.5%) for deeper levels; hence, the surface sensitivity. However, the ion survival probability, P, is not well-known; typical values are ~10% for 1 keV $^4He^+$, ~5% for 1 keV $^{20}Ne^+$. Thus, quantitative analysis of LEISS data is not straightforward largely because of poor knowledge of the probability, P. Additionally, there is also substantial uncertainty in the absolute scattering cross-sections for low-energy ions.

Figure 5.4.7a shows a typical LEISS spectrum for an Al_2O_3 surface covered with a monolayer of Rh. It is compared with an RBS spectrum (Fig. 5.4.7b) of the same film using 1 MeV $^4He^+$ ions. Notice the sensitivity of LEISS to the surface monolayer of Rh.

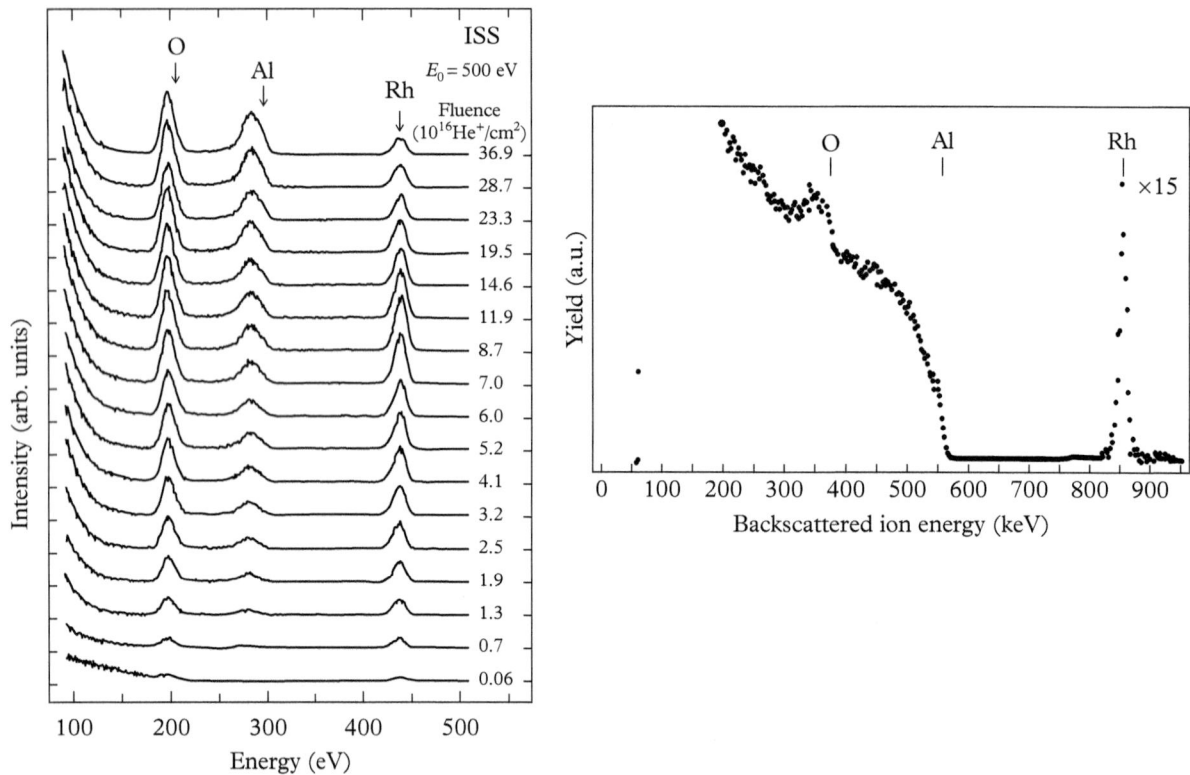

Figure 5.4.7 (a) LEISS and (b) RBS spectra from the same specimen of Al_2O_3 with a surface monolayer of Rh.

Adapted from Vickerman and Gilmore (2009).

In summary, LEISS can be used to determine (1) the chemical species on the surface, (2) the relative concentration of different elements on a surface, and (3) where the surface atom (impurity, adsorbate) is located relative to the host lattice (how?).

Example 5.4.4:   Consider the **ion scattering experiment** shown in the following figure.

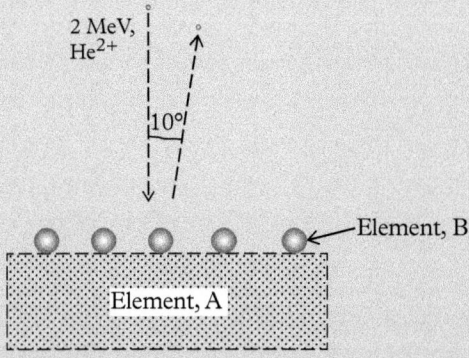

(a) Calculate the kinematical scattering factors for $_{79}Au^{197}$, $_{14}Si^{28}$.

(b) If necessary, make a very simple assumption that $\left.\frac{dE}{dz}\right|_{Si} = 20eV/\overset{\circ}{A}$, and $\left.\frac{dE}{dz}\right|_{Au} = 40eV/\overset{\circ}{A}$ for both the incoming and outgoing ions. Now:

  (i) If element A is gold (Au) and the atoms of element B, dispersed on its surface are silicon (Si), plot its RBS spectrum.

  (ii) If the situation is reversed, i.e. element A is silicon (Si) and the atoms of element B, dispersed on its surface are gold (Au), plot the new RBS spectrum.

 (iii) Which of the two, (i) or (ii), would be better to resolve in an RBS experiment? Why?

 (iv) For (i) and (ii), if we perform a LEISS measurement with 2 keV incident ions, what would the corresponding spectra look like?

  (v) When would you prefer to use LEISS over RBS?

Solution:

(a) Kinematic factors, (5.3.11), for $\theta = 170°$, with $A_{Au} = 197/4 = 49.25$ and $A_{Si} = 28/4 = 7$, are $K_{Au} = 0.923$ and $K_{Si} = 0.565$.

(b) Thus, for $E_0 = 2$ MeV, $E_1(Au) = 1.85$ MeV and $E_1(Si) = 1.13$ MeV.

(i) Here Au is a thick film and its RBS spectrum would look like Figure 5.4.3, with an onset at 1.85 MeV. Si will have a sharp peak at 1.13 MeV, similar to Au in Figure 5.4.2. The overall RBS spectrum will be a sum of the two, with a small Si peak superimposed on the Au, as shown in the following figure.

(ii) Si is the thick film and Au is on the surface. The onsets will be the same, but now Au and Si features are separately resolved.

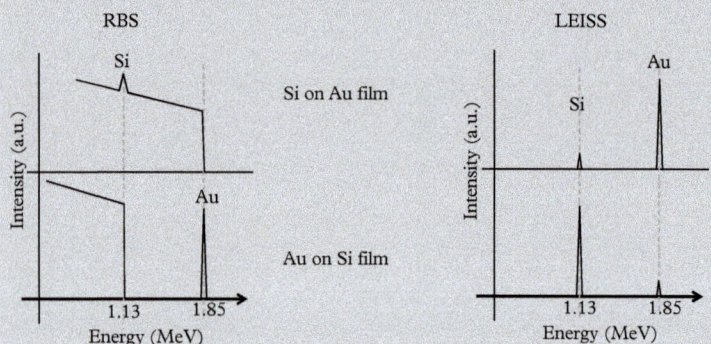

(iii) Clearly, RBS will resolve Au atoms on Si film better than Si atoms on Au.

(iv) LEISS mainly probes the surface atoms with intensity proportional to the concentration. The corresponding spectra for the two cases are shown in the figure on the right.

(v) For Si on Au film, we prefer to use LEISS over RBS.

5.4.3 Secondary Ion Mass Spectrometry (SIMS)

When a solid is bombarded by ions, a number of interactive effects take place (Fig. 5.3.13), including changes in local chemistry (implantation) and structure (atomic displacements and sputtering). If the primary ions are incident normal to the surface of a specimen, the structural and chemical alteration of the solid extends to a depth equivalent to the projected range, R_p, of the ion (Fig. 5.3.14). After the passage of a certain period of time, corresponding to the sputtering of twice the projected range (depth $= 2R_p$) of the specimen, an equilibrium of sorts between implantation and sputtering is established. From this point on, the composition of the materials removed by sputtering, as secondary ions, resembles closely the composition of the specimen, and can be analyzed to give useful information. However, the fraction of secondary ions in the sputtered material is quite small ($<5\%$), and in practice, also depends on matrix effects (see §5.3.5.4).

In equilibrium, the number, $I_i^\pm$, of positive or negative secondary ions of an element, i, recorded per unit time is given by modifying (5.3.19) as

$$I_i^\pm = \beta_i P_i^\pm Y X_i \frac{dN_0}{dt} \tag{5.4.15}$$

where X_i is the mole fraction of element, i, in the specimen, Y is the total sputter yield, and all other terms have been defined earlier. The SIMS analysis, using (5.4.15), is quite straightforward and can resolve all elements as well as their isotopes. Further, SIMS is surface sensitive to ∼1 nm in depth, since only the atoms on the surface layers are ejected from the specimen. As discussed, successive sputtering and mass analysis can provide depth-profiling information. Finally, since ions can be focused using electrostatic lenses, such information can be obtained with spatial resolution ranging from 50 nm to 2 μm.

Figure 5.4.8 shows the basic SIMS set up. The primary ion is generated by a source (IS) and is focused using two or more lenses (BL and FL) to have independent control over both the probe diameter and probe current. The velocity of the primary ion is further selected using a Wien filter (WF; Fig. 5.2.6c); such filtering controls not only the energy of the primary ion but also its mass, thus avoiding any contaminant ions. The secondary ions generated from the specimen are detected using highly sensitive mass spectrometers, of which there exist two principal designs.

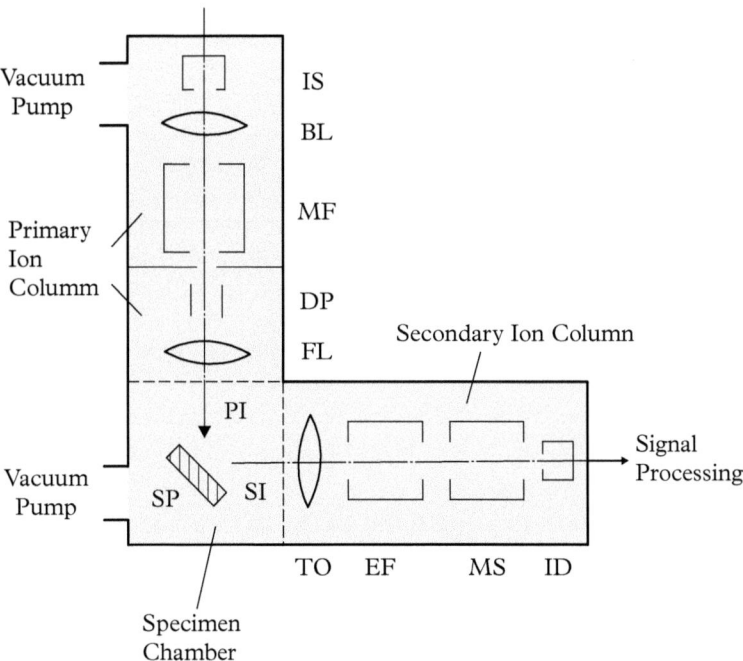

Figure 5.4.8 The basic secondary ion mass spectrometry (SIMS) set-up showing both primary ion (PI) and secondary ion (SI) columns. IS is the ion source, BL and FL are lenses used in focusing the ions, SP is the specimen, MF is the mass filter, EF is an energy filter, typically of the Wien type, ID is the ion detector, and MS is the mass spectrometer.

Adapted from Fuchs, Oppolzer, and Rehme (1990).

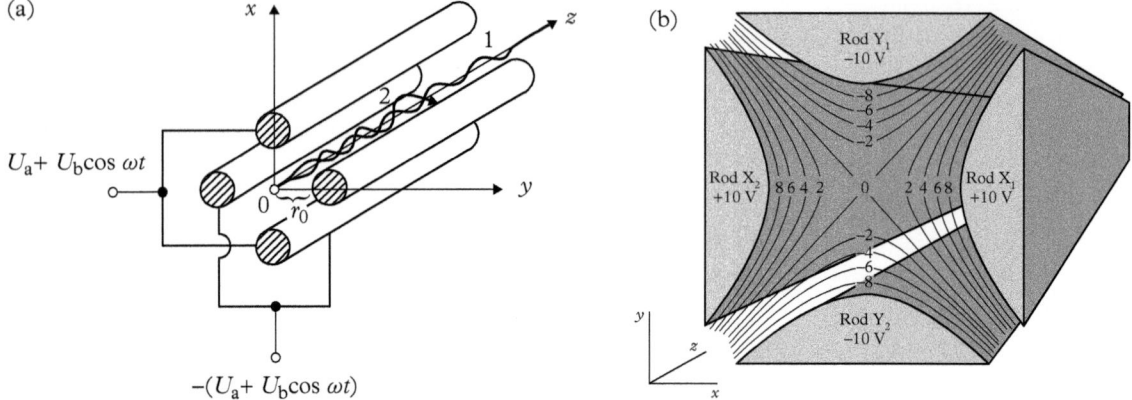

Figure 5.4.9 (a) A schematic drawing of a quadrupole mass spectrometer and (b) cross-section showing the equipotentials when a voltage of $+10$ V is applied to X_1 and X_2, and -10 V is applied to Y_1 and Y_2.

A *quadrupole mass spectrometer* (Fig. 5.4.9) uses a combination of DC (U_a) and RF ($U_b \cos \omega t$) electric fields applied to the four rods to separate the ions according to their mass-to-charge (m/q) ratio. A combined DC and RF potential, $U_a + U_b \cos \omega t$, is applied to one pair of rods and an equal but opposite voltage, $-(U_a + U_b \cos \omega t)$, is applied to the other pair. The rapid switching of the potential sends most of the ions into an unstable oscillation, striking the rods and preventing their transmission. However, ions of *specific* mass-to-charge ratio follow a stable trajectory and are transmitted through to the detector. By changing the field conditions (U_a, U_b, ω), ions of different m/q ratio can be selected and transmitted through for detection. Note that a quadrupole mass spectrometer transmits only a small fraction ($<1\%$) of ions; moreover, it detects the ions sequentially (all others being discarded) and is highly inefficient.

The alternative *time of flight* (TOF) *mass spectrometer* is relatively simple in concept (Fig. 5.4.10). Pulses of secondary ions are accelerated to a given potential (3–8 kV), such that all the ions have the same kinetic energy; after that, they enter a field-free space of a well-defined length, called a drift tube, and are then detected at the other end. Even though they all enter the drift tube with the same kinetic energy, heavier ions travel more slowly and take more time, t, to travel the length, L, of the drift tube before being detected. Thus, t can be used to determine the m/q ratio of the secondary ion, for a given voltage, V, applied before entering the drift tube, as

$$t = L \left(\frac{m}{2qV} \right)^{1/2} \tag{5.4.16}$$

The TOF-SIMS technique requires a highly accurate clock to measure the time of flight, t, and considerable computing power for data acquisition and

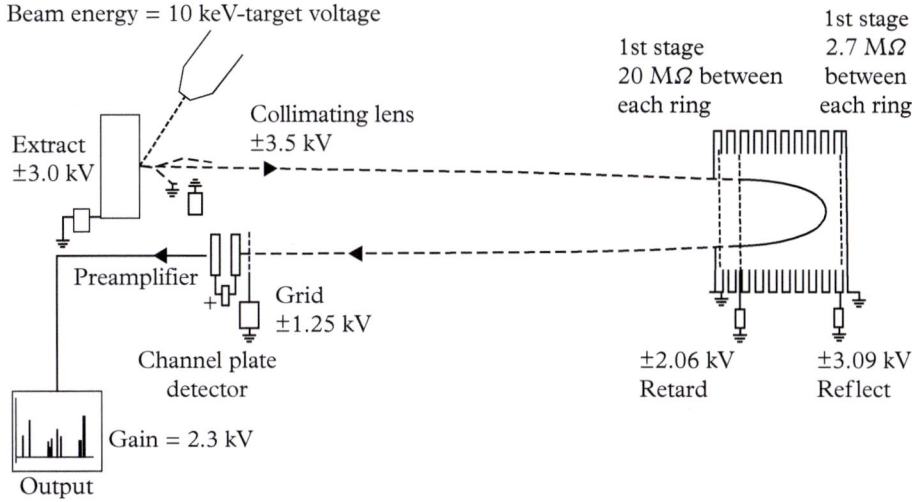

Beam energy = 10 keV-target voltage

1st stage 20 MΩ between each ring

1st stage 2.7 MΩ between each ring

Extract ±3.0 kV

Collimating lens ±3.5 kV

Preamplifier

Grid ±1.25 kV

Channel plate detector

±2.06 kV Retard

±3.09 kV Reflect

Gain = 2.3 kV

Output

Figure 5.4.10 A time of flight (TOF) mass spectrometer.

Adapted from Vickerman and Gilmore (2009).

analysis. In practice, the flight times for many ions are measured electronically, and using (5.4.16) related to their mass. TOF-SIMS detects 10–50% of the secondary ions in parallel, and as such is quite efficient. Figure 5.4.11 shows a typical TOF-SIMS spectrum.

In practice, there are two main modes of SIMS analysis. Static SIMS uses a *low-dose* primary ion beam to generate atomic and molecular ions from the top surface (MLs) of a specimen, with focus on molecular characterization of surfaces. In contrast, dynamic SIMS is a destructive technique that uses *high-dose* primary ion beams, which sequentially erode the surface of a specimen to provide elemental compositional analysis with ppb sensitivity for many elements as a function of depth ranging from nanometers to tens of micrometers.

5.4.4 Induction-Coupled Plasma Mass Spectrometry (ICP-MS)

This method is used to obtain the composition of a sample with ppb sensitivity, and finds wide application in the analysis of natural materials, including rocks, minerals, soils, sediments, water, air, and plant/animal tissues. It is a destructive method as the specimen is ionized at high temperatures in an inductively coupled plasma, and subsequently transported to a quadrupole mass spectrometer, with a sensitivity of ~1 amu, to be analyzed. All the elements in the periodic table can be detected/analyzed, provided they do not interfere with components of the plasma (for example, when Argon plasma is used, ArC^+ at 52 amu, $Ar^{13}C^+$ at 53

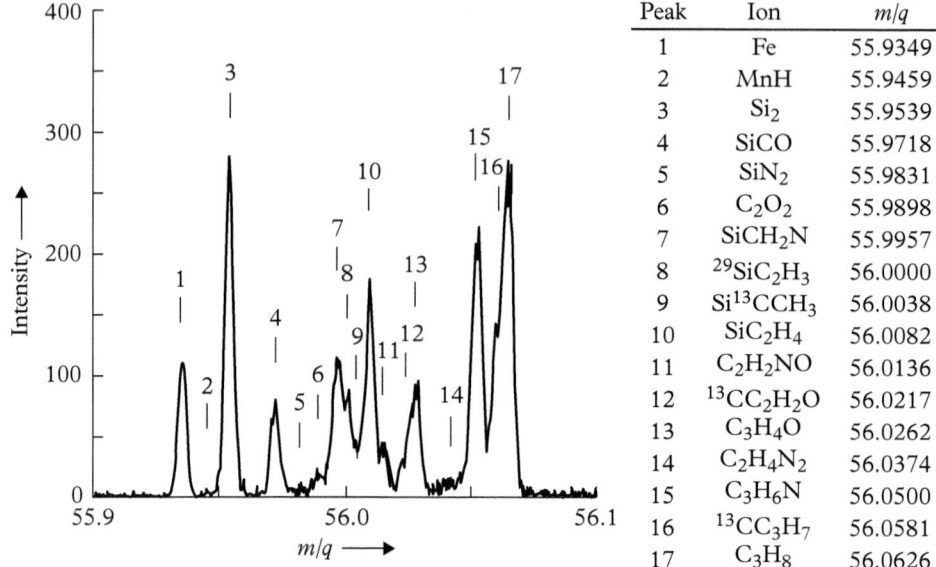

Peak	Ion	m/q
1	Fe	55.9349
2	MnH	55.9459
3	Si_2	55.9539
4	SiCO	55.9718
5	SiN_2	55.9831
6	C_2O_2	55.9898
7	$SiCH_2N$	55.9957
8	$^{29}SiC_2H_3$	56.0000
9	$Si^{13}CCH_3$	56.0038
10	SiC_2H_4	56.0082
11	C_2H_2NO	56.0136
12	$^{13}CC_2H_2O$	56.0217
13	C_3H_4O	56.0262
14	$C_2H_4N_2$	56.0374
15	C_3H_6N	56.0500
16	$^{13}CC_3H_7$	56.0581
17	C_3H_8	56.0626

Figure 5.4.11 A TOF-SIMS spectrum of a contaminated silicon wafer showing the possibility of resolving different charge to mass ratios characteristic of the contaminant molecules listed in the table.

Adapted from Vickermann (2000).

amu, ArN^+ at 54 amu, ArO^+ at 56 amu, and $ArOH^+$ at 57 amu can be sources of interference). Typically, specimens are introduced into the plasma as aqueous solutions using a nebulizer. The role of the nebulizer is to transform the liquid sample into an aerosol to be carried into the plasma by the flowing argon.

The preparation of aqueous solutions of the materials to be analyzed is a major challenge in ICP-MS analysis. Generally, the sample is dissolved in an acid; the simplest one to work with is nitric acid, as it has minimal spectral interference. Organic polymers may be analyzed by ashing—oxidation of the carbon-containing matrix, leaving only an inorganic residue—and then dissolving them in the acid. Direct sampling of solid samples may be carried out using laser ablation, where a high-powered laser is used to vaporize the sample, which is then carried into the plasma for ionization. In addition to not requiring any dissolution of the sample, laser ablation ICP-MS has a spatial resolution of $\sim$25–50 μm, and a depth resolution of $\sim$1–10 μm per laser pulse.

Figure 5.4.12 shows a schematic of the ICP-MS. The plasma torch is aimed at the mass spectrometer, and the ions are transported physically into the spectrometer through an interfacial region. The 6,000°C plasma is coupled to the spectrometer, and the ions are transported from atmospheric pressure of the plasma to the 10^{-1} Pa base pressure in the spectrometer, through an expansion

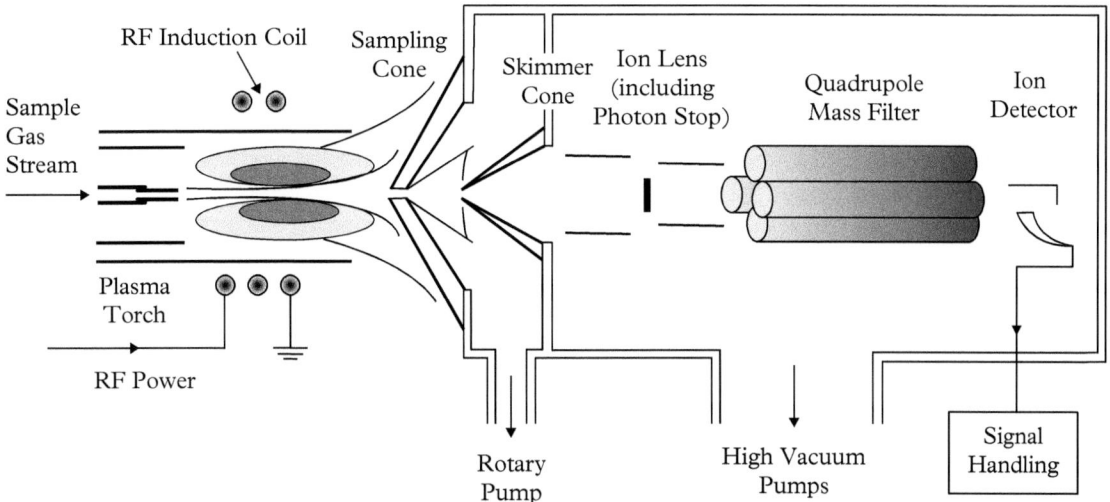

Figure 5.4.12 Schematic of an ICP-MS system, indicating the various regions of interest.
Adapted from Brundle, Evans, and Wilson (1992).

chamber at intermediate pressure consisting of two cones. One is the sampling cone where the plasma flame impinges and the other is a skimmer cone, with the intermediate region being continuously pumped. The skimmer cone has a smaller aperture, which creates a pressure of ~1 Pa in the intermediate region. A set of ion lenses focuses the sample ions on to the quadrupole spectrometer, and the overall sensitivity of the instrument is dependent on the quality of the design of these ion optics. In high-end instruments the quadrupole is replaced by a magnetic-sector mass spectrometer.

Figure 5.4.13 shows a typical ICP-MS spectrum. In practice, composition analysis of a specimen is dependent on the matrix and instrument conditions. However, the response of the mass spectrometer for the small mass range of the elements required for elemental analysis can be easily determined using internal or external standards. Typical accuracies are in the range of 5–10%. For complete quantitation, the preparation of a standard solution is paramount.

A variant of this analytical method is the ICP-OES, which combines an optical detection system, instead of a mass spectrometer, to the plasma. The excited atoms and ions in the plasma emit light at characteristic wavelength in the UV or visible part of the spectrum (Table 5.4.2), where the intensity of each line is proportional to the concentration of the element in the specimen. In general, ICP-OES is less sensitive than ICP-MS because the latter has two orders of magnitude better/lower detection limits, but the former is more common because of its quantitative accuracy (using standards), accessibility, rapid analysis, and substantially lower (50%) cost.

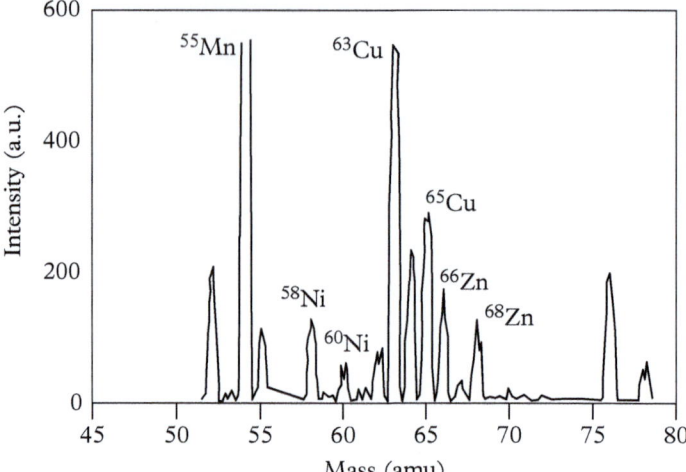

Figure 5.4.13 A low-resolution ICP-MS spectrum from a multi-element alloy.

Table 5.4.2 Recommended spectral lines (nm) for ICP-OES analysis of select elements arranged in alphabetical order.

Element	λ (nm)	Element	λ (nm)	Element	λ (nm)	Element	λ (nm)
Ag	328.07	Cr	267.72	Mo	313.26	Sn	189.98
Al	308.22	Cu	324.75	Na	588.99	Sr	407.77
As	193.7	Fe	259.94	Nb	316.34	Ta	226.23
Au	242.8	Ga	287.42	Ni	231.6	Ti	337.28
B	249.67	Ge	265.12	P	213.62	V	290.88
Ba	455.4	Hf	277.34	Pb	220.35	W	224.88
Be	313.04	Hg	194.23	Rb	780.02	Y	371.03
Bi	306.77	K	766.49	Sb	206.83	Zn	213.86
Ca	317.93	Li	670.78	Sc	361.38	Zr	339.20
Cd	226.5	Mg	383.83	Se	196.03		
Co	228.6	Mn	257.61	Si	288.16		

Adapted from Thompson and Walsh (1988).

5.4.4.1   *Application of ICP-MS in the Pharmaceutical Industry*

ICP-MS is widely used in the pharmaceutical industry to detect inorganic impurities, particularly heavy metals (Cd, Cu, Hg, and Cr) that can pose high risk to patients at even small concentrations, and satisfy the stringent impurity level requirements in drugs and medications. Traditionally, precipitation methods were used for such chemical analysis, but often they produced inconclusive results, which are now resolved by ICP-MS analysis. Figure 5.4.14 highlights those elements that can be detected by ICP-MS, including their detection limits, which

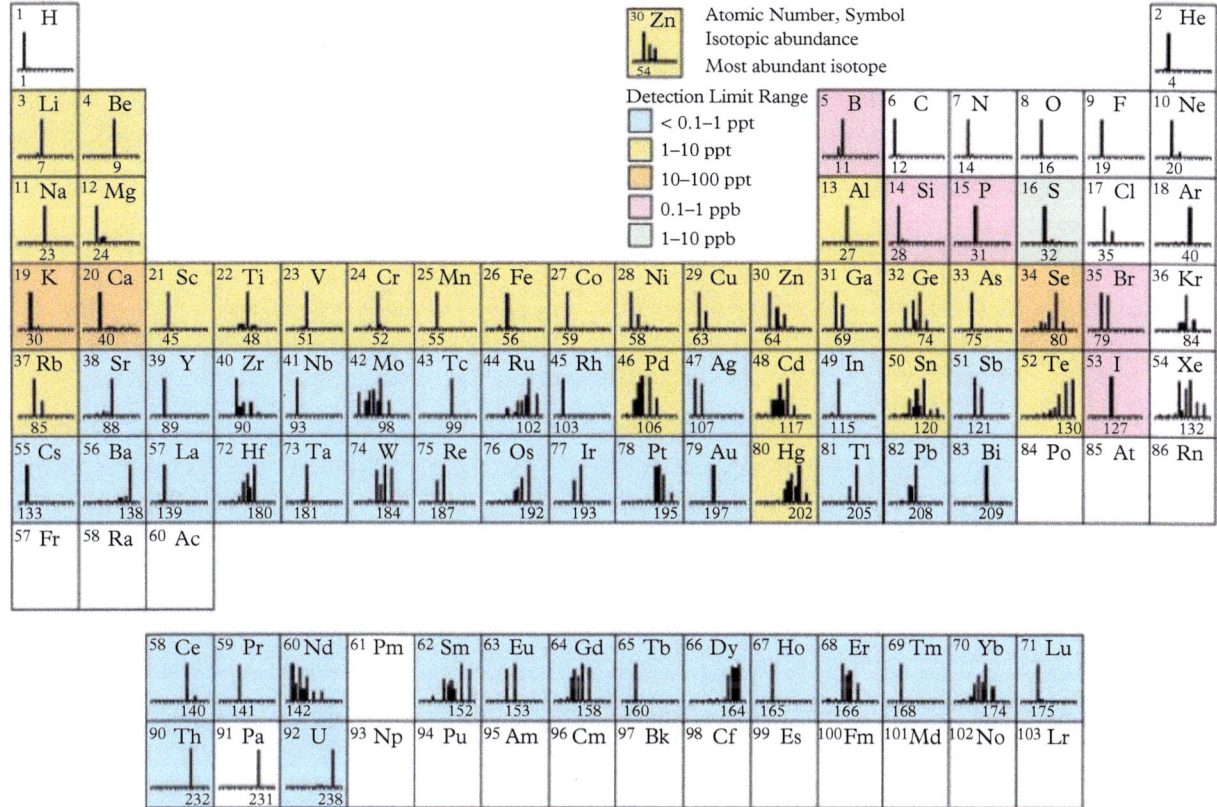

Figure 5.4.14 A modified periodic table of the elements showing their detection limits in ICP-MS spectroscopy. Notice the significant decrease in elemental detection limits with increasing atomic number.

Adapted from Agilent.

decrease substantially with increasing atomic numbers, indicating that this method is particularly suited for detecting heavy metals.

The first step in such analysis is to prepare a suitable specimen of the drug or medication. If the sample is a liquid, then the elements must be stabilized in ionic form, often using nitric or hydrochloric acid. In addition, the carbon matrix is oxidized (ashed) to leave behind an inorganic residue. Solid samples are dissolved either in acids or in hydrogen peroxide. Alternatively, laser/spark ablation or electrochemical vaporization is used, and the vaporized sample is transported to the plasma using an inert gas such as argon. Typically, the elemental composition is then compared to the expected composition, and the purity of the drug is determined (Fig. 5.4.15) for the screening of the isotopes of Cd in an antiviral drug.

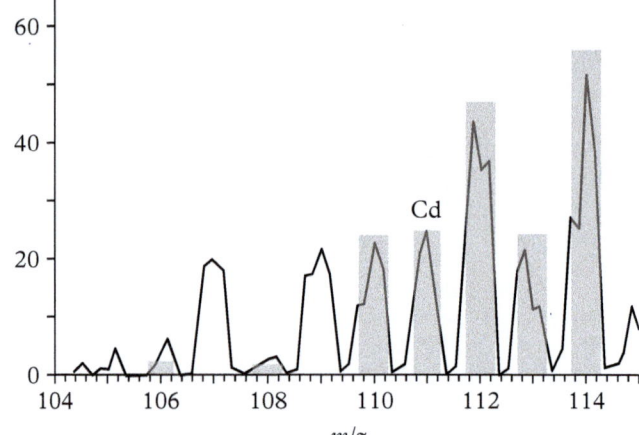

Figure 5.4.15 The ICP-MS spectrum of an antiviral drug to screen for Cd, measured at 5 ppt. The overlay shows expected isotope ratios normalized to *m/z*.

Adapted from [9].

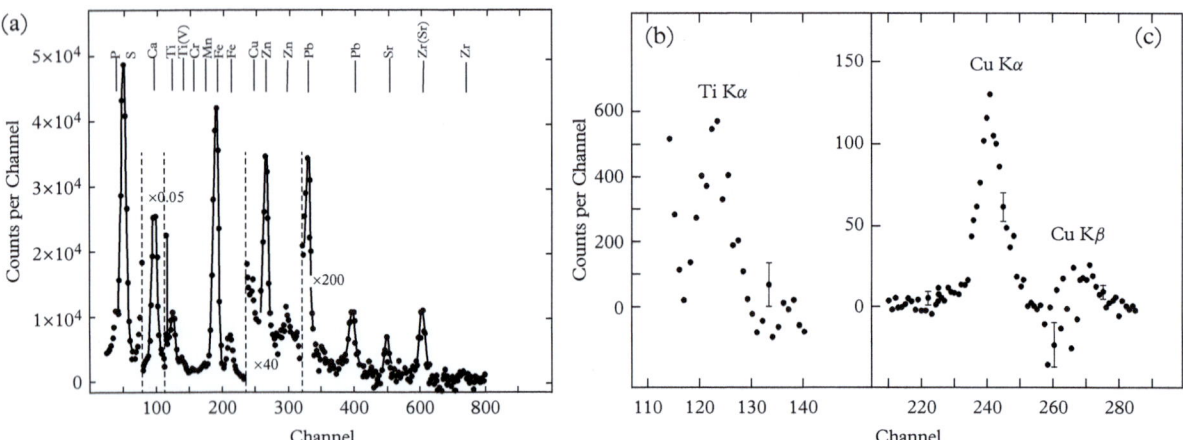

Figure 5.4.16 PIXE spectrum from (a) air pollution collected on a carbon grid using 2 MeV protons and a Si(Li) detector. (b) The Kα spectrum from 4×10^{-11} g of Ti, and (c) 8×10^{-10} g of Cu.

Adapted from Johansson, Campbell, and Malmqvist (1995).

5.4.5   Particle-Induced X-Ray Emission (PIXE)

In this method, the specimen is irradiated with a high-energy (MeV) ion and following inner-shell ionization, the characteristic X-rays emitted from the specimen are analyzed using solid-state energy-dispersive semiconductor detectors (§2.5.1.1). The technique has high sensitivity (Fig. 5.4.16a) and a detection limit of 8×10^{-10} g (Cu in Fig. 5.4.16c). There is also very little beam broadening in the specimen for high-energy particles (Fig. 5.3.11a). As a result, for thin specimens (1–10 μm) a focused primary ion beam, or micro-PIXE, gives two orders of magnitude lower detection limit compared to an electron microprobe (§2.5.2.2), while operating under similar spatial resolution conditions.

Summary

Materials characterization requires probes and techniques that maximize information while minimizing damage to the specimen. All probes—electrons, photons, ions, and neutrons—are judiciously applied to satisfy this maxim. The first three are generated using either laboratory sources or in large user facilities; the latter, particularly synchrotrons, include unique characteristics and capabilities. However, neutron beams are only produced in fission reactors or in spallation sources, both requiring specialized facilities.

Sources of light for microscopy and spectroscopy include incandescence (heated filaments) and plasma discharge (arc) lamps with spectral features ranging from the UV to the IR. On the other hand, electron beams are generated by thermionic (heating) or field-emission sources, with a wide range of characteristics (Table 5.2.2) of which total current, current density, and energy spread are most important for characterization. X-rays (§2.4) are generated using tubes in the laboratory or by synchrotron radiation. Lastly, RF plasma sources generate ions, principally He^+ and H^+, at low voltages that are subsequently accelerated to high energies (MeV) using van de Graff generators or tandem accelerators. Being charged particles, ions can be guided or focused by electric or magnetic fields, or a combination of the two (Wien filter); however, electrostatic systems are preferred because of the low (compared to electrons) charge-to-mass ratio of the ions.

For proper use of any probe in characterization, its interaction with matter, including damage processes, must be well understood. From this point of view, where possible, specimens should be probed first with light, as it generally does not cause any damage. Electrons interact with matter in a more complicated manner, causing beam broadening that depends on the average atomic number of the specimen, and the energy and angle of incidence of the electron beam. Further, elastic interaction of electrons can lead to atomic displacements, sputtering, and local heating, whereas their inelastic interactions cause radiolysis leading to mass loss and local contamination. Neutrons, being substantially heavier than electrons and with neutral charge, penetrate much more deeply in materials and require substantially more sample material for any scattering experiment. Protons have a positive charge but are heavier than electrons (by a factor of 1836) and high-energy protons go a long way in the specimen without significant deviation. Typical penetration depths for 2.5 MeV protons range from 55 μm in carbon to 28 μm in silver.

The interactions of ions with solids can be broadly classified into two categories. Kinematic collisions of high-energy primary ions, arising from a conservation of energy and momentum, lead to their back-scattering even from substantial depths in the solid. Typically, this is accompanied by inelastic scattering processes, causing the primary ion to continuously lose energy both in its inward and outward paths in the specimen. When this is properly accounted for, as in RBS, it is possible to infer not only the mass of the scattering atom, but also its depth location and distribution in the specimen. Alternatively, the primary ion can penetrate and create a collision cascade in the specimen. This can lead to ion implantation in the specimen. Further, the collision cascade displaces the atoms in the specimen, and if the displaced atoms have sufficient energy to overcome the surface-binding forces, they may be ejected from the specimen. Such species sputtered from the surface can be detected and their mass analyzed by SIMS. If specimen destruction is not an issue, an aqueous dispersion of the specimen can be ionized in an ICP and analyzed either in an MS or in an OES to obtain its elemental composition with ppb sensitivity. Finally, irradiation of a specimen by high-energy ions can also lead to inner-shell ionization and characteristic X-ray radiation, which can be analyzed (§2.5) with high sensitivity (PIXE) to provide compositional information of the specimen.

....

FURTHER READING

Bird, J. R., and J. S. Williams. *Ion Beams for Materials Analysis*. New York: Academic Press, 1989.

Brundle, C. R., C. A. Evans, and S. Wilson. *Encyclopedia of Materials Characterization*. Oxford: Butterworth-Heinemann, 1992.

Feldman, L.C., and J. W. Mayer. *Fundamentals of Surface and Thin Film Analysis*. Amsterdam: North-Holland, 1986.

Flewitt, P. E. J., and R. K. Wild. *Physical Methods of Materials Characterization*. Bristol: IoP Press, 2003.

Fuchs, E., H. Oppolzer, and H. Rehme. Particle Beam Microanalysis. *New York: VCH*, 1990.

Goldstein, J., D. Newbury, D. Joy, C. Lyman, P. Echlin, E. Lifshin, L. Sawyer, and J. Michael. *Scanning Electron Microscopy and X-Ray Microanalysis*. New York: Springer, 2003.

Jenkins, F. A., and H. E. White. *Fundamentals of Optics*. Aukland: McGraw Hill, 1976.

Johansson, S. A., J. L. Campbell, and L. G. Malmqvist. *Particle-Induced X-Ray Emission Spectrometry (PIXE)*. New York: Wiley-Interscience, 1995.

Reimer, L. *Transmission Electron Microscopy*. Berlin: Springer Verlag, 1993.

Reimer, L. *Scanning Electron Microcopy*. Berlin: Springer Verlag, 1998.

Tesmer, J. R., and M. Nastasi, eds. *Handbook of Modern Ion Beam Materials Analysis*. Pittsburgh: MRS Press, 1995.

Thompson, M., and J. N. Walsh. *Handbook of Inductively Coupled Plasma Spectrometry*. New York: Chapman and Hall, 1988.

Vickerman, J. C., ed. *Surface Analysis: The Principle Techniques*. Chichester: Wiley, 2000.

Vickerman, J. C., and I. S. Gilmore. *Surface Analysis: The Principle Techniques*, 2nd ed. Chichester: Wiley, 2009.

Zeigler, J. F. *Helium Stopping Power and Ranges in All Elemental Matter*. New York: Pergamon Press, 1977.

. . . .

REFERENCES

[1] Swanson, L. W., and L. C. Crouser. "Total-Energy Distribution of Field-Emitted Electrons and Single-Plane Work Functions for Tungsten." *Physical Review* 163 (1967): 622.

[2] Breese, M. B. H., G. W. Grime, and F. Watt. "The Nuclear Microprobe." *Annual Review of Nuclear and Particle Science* 42 (1992): 1–38.

[3] Egerton, R. F., P. Li, and M. Malac. "Radiation Damage in the TEM and SEM." *Micron* 35, (2004): 399–409.

[4] Yang, Z., N. Sakaguchi, S. Watanabe, and M. Kawai. "Dislocation Loop Formation and Growth under *In Situ* Laser and/or Electron Irradiation." *Scientific Reports* 1 (2011): 190.

[5] Hobbs, L. W. "Murphy's Law and the Uncertainty of Electron Probes." *Scanning Microscopy Supplement* 4 (1990): 171–83.

[6] Crozier, P. A., M. R. McCartney and D. J. Smith. "Observation of Exit Surface Sputtering in TiO2 Using Biased Secondary Electron Imaging." *Ultramicroscopy* 237 (1990): 232–40.

[7] Joy, D. C., and C. S. Joy. "Low Voltage Scanning Electron Microscopy." *Micron* 27 (1996): 247–63.

[8] Storms, H. A., K. F. Brown, and J. D. Stein. *Anal. Chem.* 49 (1977): 2023.

[9] Santamaria-Fernandez, R., S. Merson, and R. Hearn. "Semiquantitative Screening of Pharmaceutical Antiviral Drugs using the Agilent 7500ce ICP-MS in Helium Collision Mode." Agilent Technologies Application Notes 5989-9443EN2. Santa Clara: Agilent Technologies, Ltd., 2017.

. . . .

EXERCISES

A. **Test Your Knowledge**

Which of the following is correct?

1. Tungsten-halogen lamps
   (i)   are ideal black body radiators.
   (ii) typically operate at filament temperatures of ~2,000–2,500 K.

(iii) have an emission profile that can shift to shorter wavelengths at lower temperatures.

(iv) operated close to the melting point of tungsten have a higher fraction of the visible spectrum.

2. A population inversion is when we have
   (i) more refugees than natives.
   (ii) more electrons occupying metastable states than their ground state.
   (iii) electrons occupying only metastable states.

3. A Schottky emitter is
   (i) just a thermionic emitter.
   (ii) is the emitter portion of a transistor.
   (iii) is a thermally enhanced field emission source.

4. A thermal emitter produces
   (i) a higher total current than a field emitter.
   (ii) a higher current density than a field emitter.
   (iii) heat inside a microscope for *in situ* annealing experiments.

5. To focus ions, it is
   (i) better to use electrostatic rather than electromagnetic lenses.
   (ii) better to use electromagnetic rather than electrostatic lenses.
   (iii) not important what kind of lenses, electromagnetic or electrostatic, are used.

6. To decide on the most optimal probe to solve a given characterization problem we must consider
   (i) its penetration depth.
   (ii) the damage it will cause to the material.
   (iii) both (i) and (ii).

7. When we determine the penetration depth for a given technique, we consider the mean free path length for
   (i) the probe radiation.
   (ii) the signal radiation.
   (iii) the shorter of the two mean free path lengths.

8. The ZAF correction
   (i) is unrelated to materials characterization.
   (ii) describes only the absorption of X-rays.
   (iii) is used to correct the analysis of materials using X-rays.

9. The range of wavelength of photons used for characterization
   (i) should match the microstructural features of interest.
   (ii) varies from $\sim 10^{-10}$ cm to 10^{-2} cm.
   (iii) can be anything.

10. In an SEM, a focused beam broadens in a way that
    (i) is independent of the material.
    (ii) depends on the average atomic number of the material.
    (iii) is over a larger area but with reduced intensity in Al compared with Au.

11. Displacement damage for electron incidence is significant
    (i) in SEMs under normal operation.

(ii) in TEMs operating with $E_0 > 300$ keV.

(iii) only for light elements.

12. The penetration depth of neutrons is
  (i) smaller than electrons because it has neutral charge.
  (ii) larger than protons because of its neutral charge.
  (iii) same as protons because their masses are similar compared to electrons.

13. Photons
  (i) generally, cause the least damage in materials.
  (ii) in the form of lasers, can cause instantaneous heating to melting temperature.
  (iii) do not penetrate any materials and are reflected.

14. High-energy electrons incident on a material can cause
  (i) sputtering only on the exit surface of a thin foil.
  (ii) atomic displacements due to elastic interactions.
  (iii) break bonds by inelastic scattering mechanisms.

15. The temperature rise (e^- beam heating) in a TEM is
  (i) always independent of thickness.
  (ii) varies linearly with thickness.
  (iii) independent of thickness if radiation losses are neglected.

16. A quadrupole mass spectrometer detects ions
  (i) based on their charge.
  (ii) based on their charge-to-mass ratio.
  (iii) in a very efficient manner.

17. The mean free path length is
  (i) defined as the average distance between collisions.
  (ii) unrelated to scattering cross-section.
  (iii) inversely proportional to the scattering cross-section.

18. LEISS is a highly surface-sensitive technique because it
  (i) uses low-energy ions.
  (ii) uses electrostatic analyzers.
  (iii) detects only neutral atoms.

19. High-energy electrons incident on a material can cause charging
  (i) in metals.
  (ii) in oxides.
  (iii) in nonconducting polymers.

B. Problems

1. The **displacement energies** for Al and Cu in an Al-Cu alloy are 18 eV and 20 eV, respectively. Calculate the threshold value of the incident energy of electrons in a high-voltage TEM that can cause displacement damage?

2. Derive equation (5.3.11) based on **kinematical scattering** conditions for two ions as described in the text.

3. Based on actual data in Example 5.4.3, and applying (5.4.12) calculate the **composition** of the superconducting film? Is it different from the expected stoichiometry?

4. Calculate the **stopping power** for Al_2O_3, for (a) ^4He of 4 MeV and (b) ^1H of 100 keV incidence.

5. Applying the **kinematic scattering** theory (5.3.11) to Figure 5.3.17, calculate the positions of the peaks and edges in the LEISS and RBS spectra, respectively.

6. **SIMS and Auger** can probe a specimen as a function of depth by sputtering and plotting the signal for different elements as a function of sputtering time. Compare the strengths and weaknesses of these two techniques.

7. A **mass spectrometer** can use either a time of flight or quadrupole analyzer. Which one would you prefer, and under what circumstances?

8. A tri-layer thin film specimen, shown in Figure 5.Pr.8, is subject to an **RBS experiment** with the detector at 140°.

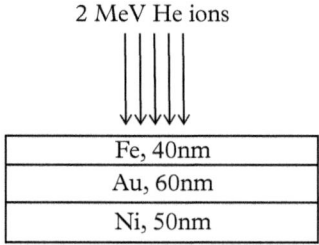

Calculate the expected RBS spectrum and plot it similar to Figure 5.4.2. Use Table 5.4.1 for stopping power data. Make any reasonable assumption and state it clearly.

9. Now repeat the same calculation of the **RBS spectrum** for the thin film structure shown in Figure 5.Pr.9, for 4 MeV He ions with the detector at 170°. Use Table 5.4.1 for stopping power data.

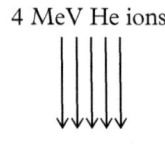

10. In an **RBS experiment**, 2 MeV He ions are used to probe a silver target.

   (a) What is the distance of closest approach, D_0, in this back-scattering experiment?

   (b) Compare this with the Bohr radius, and the radius of the Ag K-shell. Is the kinematic scattering approximation valid?

   (c) The scattering cross-section is related to D_0 and the scattering angle, θ_1, as

$$\sigma\,(\theta) = \frac{(D_0/4)^2}{sin^4\,(\theta_1/2)}$$

   Calculate the cross-section in Barns for a 180° back-scattering experiment.

   (d) Now, repeat (a)–(c), for a forward scattering PIXE experiment using 2 MeV protons and a silicon specimen.

11. Consider the following **RBS experiment** (Fig. 5.Pr.11).

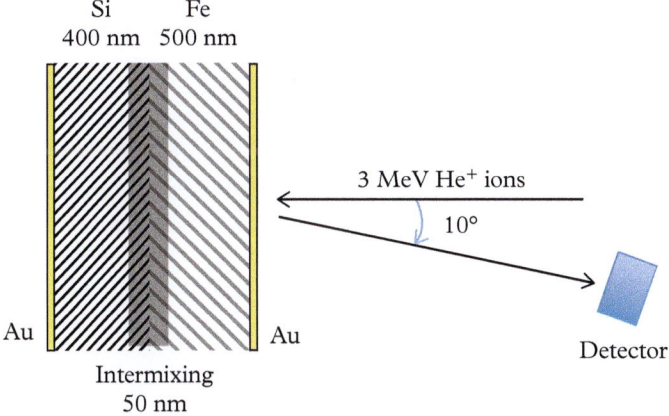

a) Use Figure 5.3.16 or calculate the kinematic scattering factors

b) Make a very simple assumption that $\frac{dE}{dz}\big|_{Si} = 20\text{eV/Å}$ for BOTH the incoming and outgoing ions

c) Similarly, make a very simple assumption that $\frac{dE}{dz}\big|_{Fe} = 30\text{eV/Å}$ for BOTH the incoming and outgoing ions

Now draw the expected RBS spectra for the structure assuming that the Si/Fe interface is

(i) ATOMICALLY sharp

(ii) INTERMIXED forming a 50-nm thick iron–silicide layer.

12. A thin film (100 nm) of Ni is grown on InP (Fig. 5.Pr.12a). It is the annealed at 250°C for 30 min, which forms an alloy of InNiP, 50 nm thick, at the interface (Fig. 5.Pr.12b).

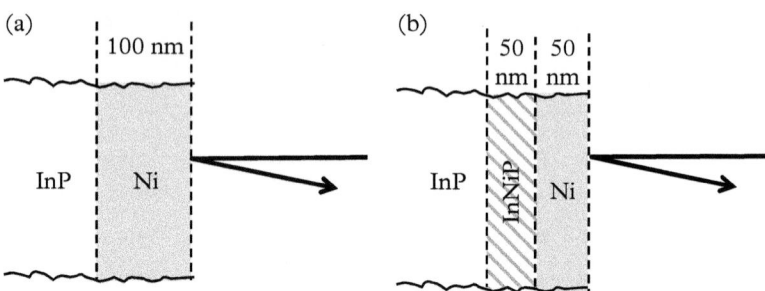

a) Calculate and plot the **RBS spectrum** for these two cases, using 2 MeV He⁺ ions at 170°.

b) Instead of RBS, you decide to probe the specimen as a function of depth by **sputter Auger electron spectroscopy**. Now, plot the AES signal for each of the three elements as a function of sputtering time for these two specimens.

13. In a **LEISS experiment**, 0.5 keV He ions are used to probe a specimen of Au. Determine the distance of closest approach for a head-on collision. Will this collision be kinematic? For this interaction, what is the cross-section in barns, for a scattering angle of 140°? Where will the peaks appear for a specimen containing both Ag and Ni?

Optics, Optical Methods, and Microscopy

6

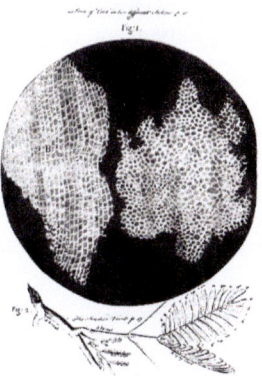

Robert Hooke (1635–1703) published the first book, *Micrographia* (1665), on optical imaging based on his observations using a microscope with a variety of lenses and illumination conditions. Hooke is believed to have used this microscope (right, top) manufactured by Christopher White of London. Numerous meticulous drawings, including the one on the right (bottom) of the structure of a plant (cork), and for which he coined the term "cell", now central to biology, formed the basis of his book.

Principles of Materials Characterization and Metrology. Kannan M. Krishnan, Oxford University Press (2021).
© Kannan M. Krishnan. DOI: 10.1093/oso/9780198830252.003.0006

6.1 Introduction

This chapter discusses fundamental principles of optics, Fraunhofer and Fresnel diffraction, optical microscopy, and ellipsometry. It describes the propagation of waves in a medium, and then presents details of the related phenomena of interference, scattering, polarization, reflection, refraction, and diffraction, which form the basis of many characterization techniques. Specifically, Fraunhofer diffraction from a one-dimensional grating with multiple slits provides a good foundation for understanding the diffraction of X-rays and electrons by the periodic arrangement, in three dimensions, of atoms in crystalline materials. The chapter also addresses details of optical gratings used in various spectroscopies (§3), zone plates used in X-ray microscopy, optical microscopy, metallography, and the measurement of the optical constants of materials using ellipsometry. Note that an introduction to the propagation and polarization of electromagnetic waves (§1.3.4), and scattering and diffraction (§1.4.2 and §4), were presented earlier; the reader should review those sections as appropriate. Further, those familiar with elementary concepts of optics and waves can skip the first five or six introductory sections of this chapter.

6.2 Wave Equation for Simple Harmonic Motion

Light waves are transverse waves (Fig. 1.3.3), where all points on the wave vibrate in the same plane (Fig. 6.2.1), and have displacements, y (or x), normal to the direction, z, of propagation. Mathematically, they are described as

$$y = a\sin\left(\frac{2\pi z}{\lambda}\right) = a\sin kz \tag{6.2.1}$$

where λ is the spatial period or wavelength, a is the amplitude or maximum disturbance, and $k = 2\pi/\lambda$, is the magnitude of its wavevector. For the wave to propagate *without changing its shape* in the positive z-direction with velocity, v, we introduce the time, t, explicitly as

$$y = a\sin\frac{2\pi}{\lambda}(z - vt) \tag{6.2.2}$$

The period, T, for one complete vibration, i.e. starting and returning to the same specific point, and its reciprocal, the frequency, f, are given by

$$T = \lambda/v = 1/f \tag{6.2.3}$$

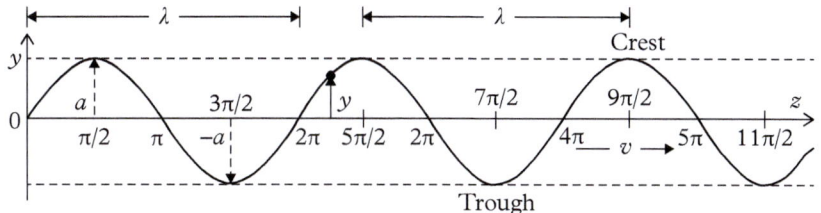

Figure 6.2.1 A transverse wave, vibrating in the plane of the page, showing the displacement, y, wavelength, λ, amplitude, a, and velocity, v. Note that one wavelength, λ, corresponds to a phase change of 2π radians.

Substituting (6.2.3) in (6.2.2), we get

$$y = a \sin 2\pi \left(\frac{z}{\lambda} - ft\right) = a \sin (kz - \omega t) \qquad (6.2.4)$$

where $\omega = 2\pi f$ is the angular frequency.

Example 6.2.1: The peak sensitivity of the human eye corresponds to green light ($\lambda = 550$ nm). How many wavelengths of this light will fit into a thin film (thickness, $t = 110$ μm)? If we use radio waves ($f = 1$ MHz, $v = 3 \times 10^8$ m/s) instead, how far will the same number of waves extend?

Solution: Number of waves $= t/\lambda = 110 \times 10^{-6}/\left(550 \times 10^{-9}\right) = 200$
For radio waves, $\lambda = v/f = 3 \times 10^8/10^6 = 300$ m, which will extend to 60 km!

6.2.1 The Phase Angle

We can also display the instantaneous displacement and direction of propagation of the wave, (6.2.4), on a circle (Fig. 6.2.2) of radius, a, at any time, t, where the angle, $\theta = \omega t$, measured counter-clockwise from the z-axis is defined as the phase angle. Thus, $y = a \sin \theta = a \sin \omega t$. If we now explicitly include the initial phase ($\theta = \alpha$, at $t = 0$), we can write the displacement as $y = a \sin (\omega t + \alpha)$. To include propagation, we include the wave vector, $k = 2\pi/\lambda$ and write

$$y = a \sin (kz - \omega t + \alpha) \qquad (6.2.5)$$

The constant, α, can be eliminated by rotating the coordinate frame by the same angle to give

$$y = a \sin (kz - \omega t) \qquad (6.2.6)$$

In the general case, (6.2.5), the phase of the wave, $\varphi (z, t)$, is given by

$$\varphi (z, t) = kz - \omega t + \alpha \qquad (6.2.7)$$

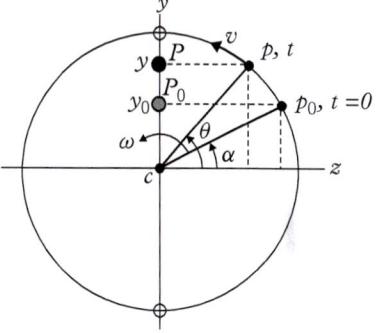

Figure 6.2.2 A simple harmonic motion along the y-axis, shown for the circle of reference with radius equal to the amplitude, a, the initial phase angle, α, the angular frequency, ω, the initial position, p_0, at $t = 0$, and position, p, at any time, t.

The partial derivative of $\varphi(z, t)$, with respect to time, t, holding z constant is the angular frequency, or the *rate of change of phase with time*, i.e.

$$\left|\left(\frac{\partial\varphi}{\partial t}\right)_z\right| = \omega \tag{6.2.8a}$$

and the *rate of change of phase with distance*, holding t constant, is the wave vector, k, i.e.

$$\left|\left(\frac{\partial\varphi}{\partial z}\right)_t\right| = k \tag{6.2.8b}$$

Then, the velocity, v, of a point on the wave of constant phase, such as the crest with the maximum displacement, called the *phase velocity*, is given by

$$\left(\frac{\partial z}{\partial t}\right)_\varphi = \frac{-(\partial\varphi/\partial t)_z}{(\partial\varphi/\partial z)_t} = \pm\frac{\omega}{k} = \pm v \tag{6.2.9}$$

and carries a positive (negative) sign for propagation along the direction of increasing (decreasing) z.

Example 6.2.2: Consider the following waves:

(a) $y = 5\sin 2\pi(0.4z - 2t)$
(b) $y = \frac{\sin(12z + 4t)}{3}$
   where z is in meters and t is in seconds.

For each case determine its (i) frequency, (ii) wavelength, (iii) period, (iv) amplitude, (v) phase velocity, and (vi) direction of motion.

Solution: Comparing with (6.2.4), i.e. $y = A\sin(kz - \omega t)$, we get for (a):
(i) $\omega = 4\pi = 2\pi f$, thus $f = 2$ Hz. (ii) $k = 2\pi/\lambda = 0.8\pi$, $\lambda = 2.5$ m, (iii) $T = 1/f = 0.5$s, (iv) $A = 5$ m, (v) $v = \omega/k = 2/0.4 = 5$ m/s. (vi) positive z direction.
Similarly, we get for (b):
(i) $\omega = 4 = 2\pi f$, thus $f = 2/\pi$ Hz. (ii) $k = 2\pi/\lambda = 12$, $\lambda = \pi/6$ m, (iii) $T = 1/f = \pi/2$ s, (iv) $A = 1/3$ m, (v) $v = \omega/k = 4/12 = 0.33$ m/s. (vi) negative z direction.

6.2.2   The Superposition Principle

Later in this chapter we discuss the phenomena of polarization, interference, and diffraction. Conceptually, they are related to the same question—what happens when two or more waves overlap in some region of space? The answer to this question will, of course, depend on the precise circumstances of the overlap or superposition. In general, we are interested in how the specific properties (amplitude, frequency, and phase) of the constituent waves affect the resultant wave.

We begin by considering the addition of two simple harmonic waves (SHW)[1] of the same angular frequency, ω, but with different amplitudes and initial phase angles, traveling along the same line, and given by

$$y_1 = a_1 \sin (\omega t - \alpha_1) \qquad (6.2.10a)$$

$$y_2 = a_2 \sin (\omega t - \alpha_2) \qquad (6.2.10b)$$

By the principle of superposition, the resultant displacement, y, of the two waves combined is given by their algebraic sum:

$$
\begin{aligned}
y = y_1 + y_2 &= a_1 \sin (\omega t - \alpha_1) + a_2 \sin (\omega t - \alpha_2) \\
&= (a_1 \cos \alpha_1 + a_2 \cos \alpha_2) \sin (\omega t) - (a_1 \sin \alpha_1 + a_2 \sin \alpha_2) \cos (\omega t)
\end{aligned}
\qquad (6.2.11)
$$

Since a_1, a_2, α_1, and α_2 are constants, we can rewrite (6.2.11) as

$$y = A \cos \theta \sin (\omega t) - A \sin \theta \cos (\omega t) = A \sin (\omega t - \theta) \qquad (6.2.12)$$

where the constant, A, is given by

$$A^2 = a_1^2 + a_2^2 + 2a_1 a_2 \cos (\alpha_1 - \alpha_2) \qquad (6.2.13)$$

In other words, the resultant intensity is not the sum of the component intensities, but has an additional interference term, $2a_1 a_2 \cos (\alpha_1 - \alpha_2)$. Further, the angle, θ, satisfies

$$\tan \theta = \frac{a_1 \sin \alpha_1 + a_2 \sin \alpha_2}{a_1 \cos \alpha_1 + a_2 \cos \alpha_2} \qquad (6.2.14)$$

Hence, the *linear* superposition, or the algebraic sum of two SHWs, produces a new SHW with a different amplitude, A, and a new phase angle, θ. Thus, when we bring two beams of light together, as is done in a Michelson interferometer (Fig. 3.5.2), the intensity, I, of the light at any point is given by the square of the resultant amplitude, i.e.

$$I \approx A^2 = 2a^2 (1 + \cos \delta) = 4a^2 \cos^2 (\delta/2) \qquad (6.2.15)$$

where, it is assumed that $a_1 = a_2 = a$, and the phase difference, $\delta = \alpha_1 - \alpha_2$. If $\delta = n\pi$, where n is an even integer, then $I = 4a^2$ or four times the intensity of either beam; alternatively, if n is an odd integer, $I = 0$. For any intermediate value of the phase difference, δ, the intensity, I, varies between these two limits as the square of the cosine of the phase difference. Such modification of the intensities

[1] Simple harmonic motion, such as in light waves, is defined as periodic motion in which the restoring force is proportional to the displacement.

of the waves, when combined, is known as *interference effects* and is often observed in experiments (see §6.4 for further discussion of interference).

6.2.3 Phasor Representation and the Addition of Waves

The resultant amplitude (6.2.13) and phase (6.2.14) of the superposition of two waves of the same angular frequency, ω $(= 2\pi f)$, but different amplitudes and phases can be derived from a simple geometrical construction (Fig. 6.2.3). Here, at any given time, t, each wave is represented by an arrow of length corresponding to its amplitude, making an angle, ωt, with respect to the reference axis. This representation combining the amplitude and the phase angle is referred to as a *phasor* and tells us everything that is necessary about the harmonic wave.

For the triangle formed by a_1, a_2, and A, it is easy to show, from the law of cosines, that

$$A^2 = a_1^2 + a_2^2 - 2a_1a_2 \cos{[\pi - (\alpha_1 - \alpha_2)]} \tag{6.2.16}$$

which is the same as (6.2.13). This graphical construction can be readily generalized for the superposition of many waves. An example for four waves with the same amplitude, a, and phase difference, δ, is shown in Figure 6.2.4a; a more general case is shown in Figure 6.2.4b.

6.2.4 Complex Representation of a Simple Harmonic Wave

To avoid geometrical constructions (Fig. 6.2.4), especially when many waves are superposed, we use an analytical treatment involving complex numbers, which is a representation that is mathematically simple to process. The wave amplitude and

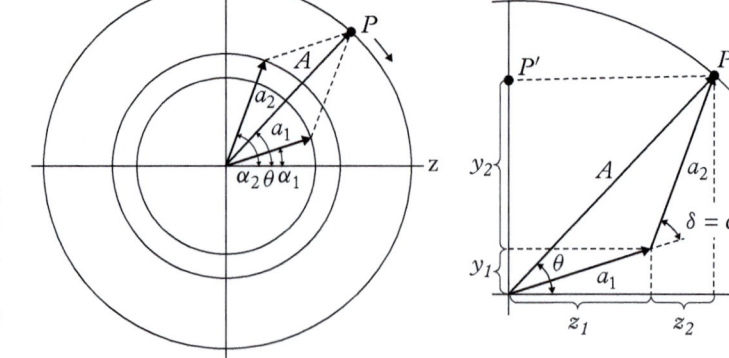

Figure 6.2.3 (a) Graphical representation for the addition of two waves of the same angular frequency, ω, but different amplitudes a_1 and a_2, and initial phase angles, α_1, and α_2, respectively. (b) A magnified view of (a) indicating the specific parameters of interest.

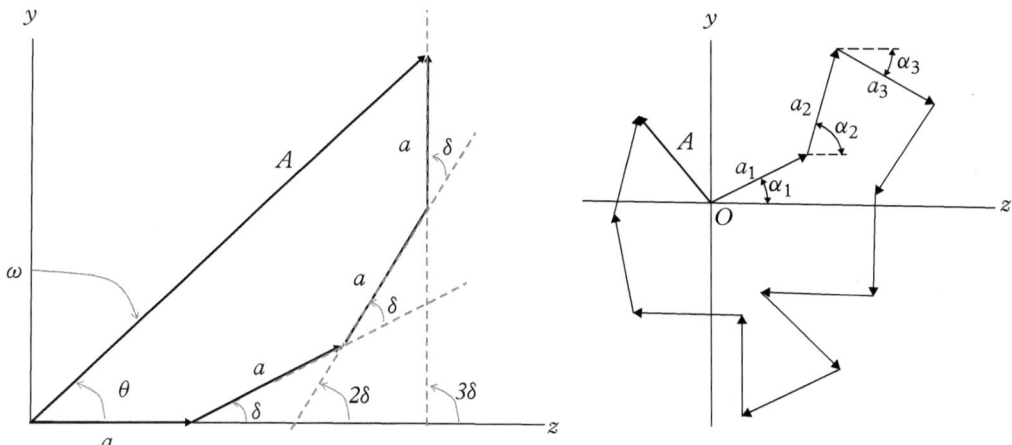

Figure 6.2.4 Phasor additions of (a) four waves with the same amplitude, a, and phase difference, δ, and (b) twelve waves of different amplitudes and phases.

phase are represented in complex space (Fig. 6.2.5) with the analytical expression for the wave given as a complex number, known as the Euler formula[2], as

$$a \exp(i\phi) = a \cos \phi + i a \sin \phi \tag{6.2.17}$$

The intensity of the wave is proportional to the square of its amplitude. Thus

$$|a \exp(i\phi)|^2 = a \exp(i\phi)\, a \exp(-i\phi) = a^2 \tag{6.2.18}$$

and many waves, in the complex notation, can be simply added as

$$A = \sum_n a_n \exp(i\phi_n) \tag{6.2.19}$$

a method that is used, e.g. (7.4.7), to calculate structure factors for scattering of X-rays by all atoms in a unit cell.

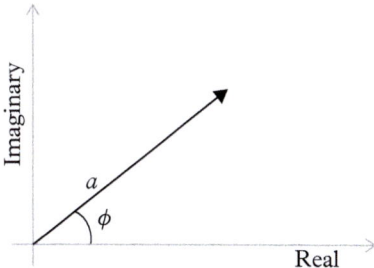

Figure 6.2.5 Representing a wave as a complex number in terms of its real and imaginary components. This is also known as an Argand diagram.

6.2.5 Superposition of Two Waves of the Same Frequency

We now consider the combination of two waves of the same frequency and amplitude, traveling in the same direction, $+z$, but with one wave propagating slightly ahead of the other by a small distance, d. In this case, the two waves are written mathematically as

$$y_1 = a \sin(\omega t - kz) \tag{6.2.20a}$$

[2] $e^{i\theta} = \cos\theta + i\sin\theta$. Also $\cos\theta = \dfrac{e^{i\theta}+e^{-i\theta}}{2}$, and $\sin\theta = \dfrac{e^{i\theta}-e^{-i\theta}}{2i}$.

$$y_2 = a \sin [\omega t - k (z + d)] \tag{6.2.20b}$$

Using the principle of superposition, the resultant wave, y, is given by

$$\begin{aligned} y = y_1 + y_2 &= a \sin (\omega t - kz) + a \sin [\omega t - k (z + d)] \\ &= 2a \cos (kd/2) \sin [\omega t - k (z + d/2)] \end{aligned} \tag{6.2.21}$$

We now have a new wave with the same frequency, but with a different amplitude. When $d \ll \lambda$, the new amplitude, A, is $\sim 2a$, but if $d = \lambda/2, A = 0$. Note that if the two waves propagate in opposite directions,

$$y_1 = a \sin (\omega t - kz) \tag{6.2.22a}$$

$$y_2 = a \sin (\omega t + kz) \tag{6.2.22b}$$

the resultant wave is

$$y = y_1 + y_2 = 2a \cos (-kz) \sin \omega t \tag{6.2.23}$$

This is commonly referred to as a *standing wave*; thus, at any position, z, we observe a simple harmonic motion, but with amplitude, A, dependent on z. When $kz = \pi/2, 3\pi/2, 5\pi/2, ..., A = 0$, and the disturbance will be zero at all times; these points are known as *nodes*. However, when $kz = 0, \pi, 2\pi, ..., A = 2a$, has its maximum value, and these points are known as *antinodes*.

Example 6.2.3: Two sources at positions, z_1 and z_2, *incoherently* emit light of the same frequency. What would be the amplitude and intensity of the resulting interference pattern?

Solution: The two waves can be described by (6.2.5) as

$$y_1 = A_1 \sin (kz_1 - \omega t - \alpha)$$

$$y_2 = A_2 \sin (kz_2 - \omega t)$$

where α is an *additional phase difference that varies randomly with time*.
   Then the phase difference is $\delta = 2\pi (z_1 - z_2) /\lambda + \alpha$ with the resulting amplitude at the point of interference given by

$$A^2 = A_1^2 + A_2^2 + 2A_1 A_2 \cos [2\pi (z_1 - z_2) /\lambda + \alpha]$$

where A is not a constant because α varies randomly with time. So, we find the average value, $\langle A^2 \rangle$. However, because of random fluctuations of α, we can safely assume that $\cos [2\pi (z_1 - z_2) /\lambda + \alpha] = 0$
and hence

$$\langle A^2 \rangle = A_1^2 + A_2^2$$

Or, since the intensity is proportional to the square of the amplitude, the average of the intensity $\langle I \rangle = I_1 + I_2$, is the sum of the individual intensities, with no fluctuation observed as a function of time, and is the same at all points!

We can generalize this and say that the intensity of the superposition of n identical waves, with random phases, is n times the intensity of a single wave.

In other words, the resultant amplitude, A, for a large number of waves added in random in Figure 6.2.4b, will not average to zero, but increases in length as n increases!

6.2.6   Addition of Waves on Orthogonal Planes and Polarization

So far, we have established that light is a transverse electromagnetic wave. We have also assumed that the light is *linearly* or *plane* polarized, by which we mean that the orientation of the electric field vector, **E**, always remains in the same plane, although its magnitude and sign varies with time. In the last two sections we have considered the superposition of waves whose electric field vectors are collinear. Now, we consider the alternative case of the combination of two light waves with electric field directions that are mutually perpendicular. As we will see, this results in light waves that may or may not be linearly polarized (Fig. 6.2.6). Note that measuring the changes in polarization of light forms the basis of the technique of ellipsometry (§6.9) used to determine the optical constants of materials and thicknesses of thin films.

Consider two simple harmonic (sine) waves of the same frequency, but with displacements in orthogonal directions, x and y, and described by the equations

$$y = a_1 \sin(\omega t - \alpha_1) \tag{6.2.24a}$$

$$x = a_2 \sin(\omega t - \alpha_2) \tag{6.2.24b}$$

To add these two waves, according to the principle of superposition, we eliminate the variable, t, from the two equations

$$y/a_1 = \sin \omega t \cos \alpha_1 - \cos \omega t \sin \alpha_1 \tag{6.2.25a}$$

$$x/a_2 = \sin \omega t \cos \alpha_2 - \cos \omega t \sin \alpha_2 \tag{6.2.25b}$$

by multiplying (6.2.25a) by $\sin \alpha_2$ and (6.2.25b) by $\sin \alpha_1$, and subtracting the first from the second we get

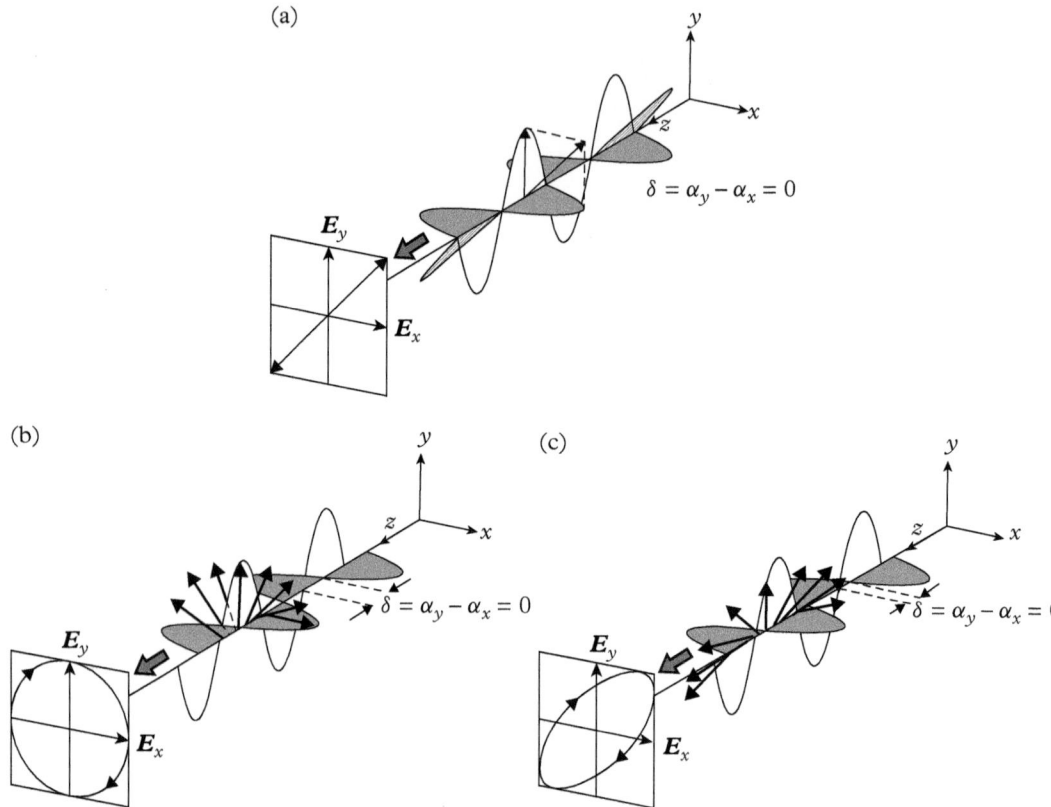

Figure 6.2.6 Schematic illustration of the addition of two simple harmonic waves of the *same amplitude* propagating in the same direction but vibrating on orthogonal planes. Three cases are shown: (a) *linearly* polarized with phase difference, $\delta = \alpha_y - \alpha_x = 0$, (b) *circularly* polarized, $\delta = \alpha_y - \alpha_x = \pi/2$, with constant scalar amplitude of the electric field vector, and (c) *elliptically* polarized light, $\delta = \alpha_y - \alpha_x = \pi/4$, with time varying amplitude of **E**.

Adapted from Fujiwara (2007).

$$-\frac{y}{a_1} \sin \alpha_2 + \frac{x}{a_2} \sin \alpha_1 = \sin \omega t \, (\cos \alpha_2 \sin \alpha_1 - \cos \alpha_1 \sin \alpha_2) \qquad (6.2.26)$$

Similarly, by multiplying (6.2.25a) by $\cos \alpha_2$ and (6.2.25b) by $\cos \alpha_1$, and subtracting the first from the second gives

$$\frac{y}{a_1} \cos \alpha_2 - \frac{x}{a_2} \cos \alpha_1 = \cos \omega t \, (\cos \alpha_2 \sin \alpha_1 - \cos \alpha_1 \sin \alpha_2) \qquad (6.2.27)$$

Eliminating, t, from (6.2.26) and (6.2.27) we get

$$\sin^2 (\alpha_1 - \alpha_2) = \left(\frac{y}{a_1}\right)^2 + \left(\frac{x}{a_2}\right)^2 - \frac{2yx}{a_1 a_2} \cos (\alpha_1 - \alpha_2) \qquad (6.2.28)$$

If we set the phase difference, $\delta = \alpha_1 - \alpha_2$, the equation for the resultant wave is

$$\left(\frac{y}{a_1}\right)^2 + \left(\frac{x}{a_2}\right)^2 = 1 \tag{6.2.29}$$

when $\delta = \pi/2, 3\pi/2, ...$ This equation represents an ellipse with the principal axes aligned along the x and y directions. Such waves, in general, are called *elliptically polarized*. Physically, this means that the electric field vector, **E**, will change its magnitude and direction (rotate) over time as well.

Now, if $a_1 = a_2 = a$, and $\delta = \pi/2, 3\pi/2, ...$, the ellipse becomes a circle and the wave is referred to as *circularly* polarized; furthermore, the scalar amplitude, E, of the electric field vector, **E**, remains constant over time. If the electric field is rotating clockwise (counter-clockwise) when the light is moving towards the observer, the convention is to call it right (left) circularly polarized. Note that by combining two oppositely polarized circular waves of equal amplitude a linearly polarized light can be generated. We leave this as an exercise to the reader (Problem 12). Finally, if $\delta = 0, \pi, 2\pi, ...$, then

$$\left(\frac{y}{a_1}\right)^2 + \left(\frac{x}{a_2}\right)^2 - \frac{2yx}{a_1 a_2} = 0 \tag{6.2.30}$$

or

$$\left(\frac{y}{a_1} - \frac{x}{a_2}\right)^2 = 0 \quad \text{or} \quad y = \frac{a_1}{a_2} x \tag{6.2.31}$$

which represents a straight line. Now, the resultant wave is considered to be *plane polarized*. Figure 6.2.7 shows plots of the resultant electric field vectors for various values of the phase difference, δ.

Example 6.2.4: The addition of two waves on orthogonal planes (Fig. 6.2.6; Fig. 6.2.7b) with different amplitudes and a phase difference, can be written vectorially as

$$\mathbf{E}(z,t) = \mathbf{E}_x(z,t) + \mathbf{E}_y(z,t) = \hat{\mathbf{i}}\, a_x \sin(kz - \omega t) + \hat{\mathbf{j}}\, a_y \sin(kz - \omega t + \alpha).$$

Then, for each of the following waves, describe the complete state of polarization:

(a) $\mathbf{E}(z,t) = \hat{\mathbf{i}}\, a_0 \sin(kz - \omega t) + \hat{\mathbf{j}}\, a_0 \sin(kz - \omega t)$.

(b) $\mathbf{E}(z,t) = \hat{\mathbf{i}}\, a_x \sin(kz - \omega t) + \hat{\mathbf{j}}\, a_y \sin(kz - \omega t + \pi/4)$.

(c) $\mathbf{E}(z,t) = \hat{\mathbf{i}}\, a_0 \cos(\omega t - kz) + \hat{\mathbf{j}}\, a_0 \cos(\omega t - kz - \pi/2)$.

Solution:

(a) The amplitudes of the two initial waves are equal.
Thus, $\mathbf{E}(z, t) = \hat{\mathbf{i}}\, a_0 \sin(kz - \omega t) + \hat{\mathbf{j}}\, a_0 \sin(kz - \omega t + \pi)$.
The phase difference is π, and so the resultant wave is linearly polarized at an angle of 135° or 45°, because $a_x = a_y$.

(b) The amplitudes of the two initial waves are different. E_x leads E_y by $\pi/4$. This is an elliptically polarized wave. The ellipse is tilted by an angle $\theta = Tan^{-1}(a_y/a_x)$.

(c) The amplitudes of the two initial waves are equal. We can rewrite the resultant wave as

$$\mathbf{E}(z, t) = \hat{\mathbf{i}}\, a_0 \sin(kz - \omega t + \pi/2) + \hat{\mathbf{j}}\, a_0 \sin(kz - \omega t).$$

This is a circularly polarized wave.

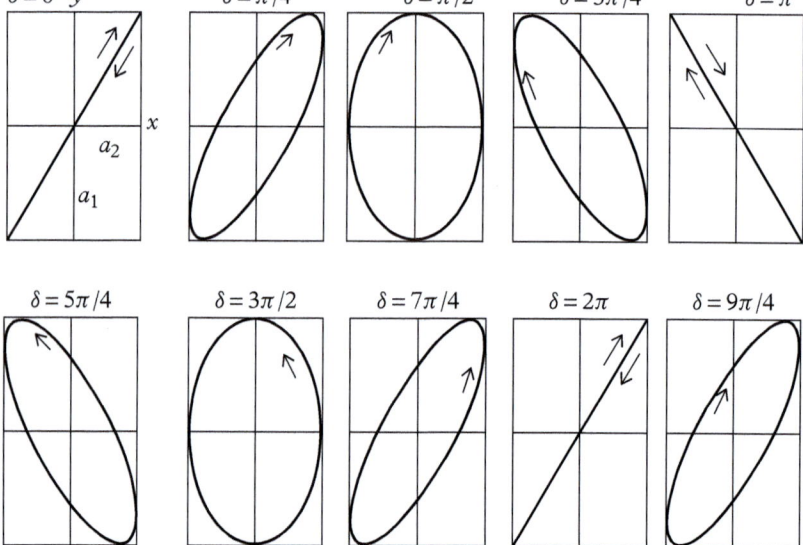

Figure 6.2.7 Superposition of two simple harmonic waves of the same frequency but different amplitudes and phases, as indicated, and propagating in orthogonal planes. For $\Delta = 0, \pi, 2\pi, \ldots$ the resultant wave is *plane* polarized; all others are *elliptically* polarized. Note that if $a_1 = a_2$, for $\Delta = \pi/2, 3\pi/2$, the light wave is *circularly* polarized.

6.3 Huygens' Principle

If an *opaque object* is placed between a source of light and a screen, the shadow that it casts is found to have smeared edges, and unlike what would be expected from a straightforward application of geometrical optics (§6.8.3). This observation is related to the phenomenon of *diffraction*, which is best illustrated by the spreading

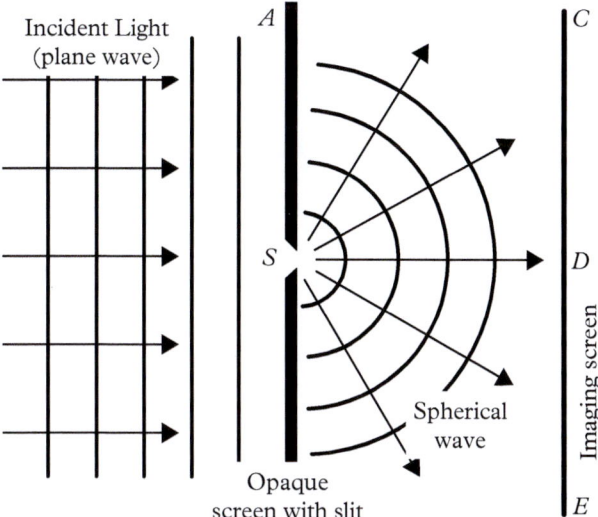

Incident Light
(plane wave)

A

C

S

D

Imaging screen

Spherical
wave

Opaque
screen with slit

E

Figure 6.3.1 The propagation of a plane wave through a pin hole illustrating the Huygens' principle.

of a wave after passing through a small aperture, such as a pin hole[3] or narrow slit (Fig. 6.3.1).

Typically performed with light as shown and assuming that light travels in straight lines, in the experiment one would expect to see a narrow patch of light at position D on the screen. However, in practice, as the slit is made narrower, the patch of light at D becomes broader, and if this process is continued to the point where the slit is smaller than the wavelength of the light used, the patch broadens sufficiently to produce a nonzero intensity even at an angle approaching 90° from the direction of propagation. Huygens[4] explained this "bending" of light more than three centuries ago by proposing that "every point on an advancing wave front is a new source of spherical wavelets, and the resultant waveform at some other point is the envelope of these wavelets propagating with the same speed and frequency as the primary wave." Thus, the pin hole becomes the source of a secondary spherical wavelet, and explains the detected intensity at large angles from the direction of propagation.

6.4 Young Double-Slit Experiment

We now discuss the phenomenon of *optical interference*. In simple terms, optical interference is the interaction of two or more light waves yielding a resultant wave that deviates from a simple sum of the components. The superposition of waves was discussed in §6.2, but light is a vector phenomenon, involving the propagation of electric and magnetic fields. Appreciating this difference is important for an intuitive understanding of optics and diffraction.

[3] Defined as a wave whose surfaces of constant phase are infinite parallel planes normal to the direction of propagation.

[4] C. Huygens (1629–95) was a Dutch scientist who developed the wave theory of light, as described in his *Traite de Lumiere* (1690).

We begin this discussion with the classic Young[5] double-slit experiment (Fig. 6.4.1). A plane wave (sunlight) is allowed to pass through a pin hole, S, and then after traveling a considerable distance, again pass through two pin holes, S_1 and S_2, separated by a distance, d, to interfere with each other. The resultant symmetric pattern is detected on the screen, positioned at a distance, D, where $D >> d$. We shall derive a simple expression for the intensity at any point, P, on the screen, at a distance, z, as shown. The two spherical harmonic waves, emanating from S_1 and S_2, travel with a path difference, l_p $(= S_2P—S_1P)$ and are then superimposed with a phase difference, Δ, given by

$$\Delta = l_p \frac{2\pi}{\lambda} \tag{6.4.1}$$

The intensity, (6.2.15), at P is then

$$I_P = 4a^2\cos^2(\Delta/2) \tag{6.4.2}$$

where a is the amplitude of the two spherical waves. Note that $D >> d$ and $D >> z$. Under these conditions, the path difference

$$l_p \sim d\sin\theta = d\frac{z}{D} \tag{6.4.3}$$

Thus, the intensity, (6.4.2), has a maximum value of $4a^2$ whenever Δ is an integer multiple of 2π. From (6.4.1), these maxima will occur when $l_p \approx m\lambda$, where m is an integer, and bright fringes are observed when

$$z = m\lambda\frac{D}{d} \tag{6.4.4}$$

or

$$m\lambda = \frac{dz}{D} = d\sin\theta \tag{6.4.5}$$

Similarly, we can show that the intensity is a minimum with a value of zero, whenever the phase difference $\Delta = \pi, 3\pi, 5\pi \ldots$, and for these values corresponding to

$$z = \left(m + \frac{1}{2}\right)\lambda\frac{D}{d} \tag{6.4.6}$$

we observe dark fringes. Figure 6.4.1c,d shows such an *interference pattern* for a double-slit experiment. One of the principal requirements for the interference of two beams to produce a stable pattern is that they must have (very nearly) the same frequency. If the frequency difference is large, it would result in a time varying phase difference, which would then cause the resultant intensity to average to zero.

[5] Thomas Young (1773–1829) revived the Huygens wave theory. Also known for his work on elastic behavior of solids (Young modulus).

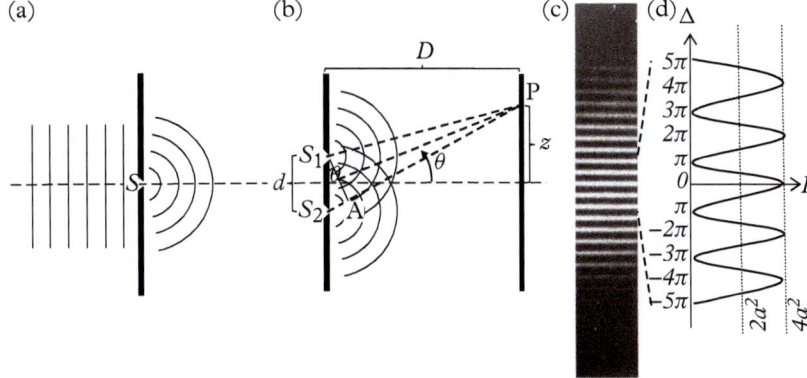

(a)   (b)   (c)   (d)

Figure 6.4.1 Young double-slit experiment: (a) A plane wave passes through a pin hole, S, and then after a certain distance passes though (b) two pin holes and the resultant projected on a screen. The path difference, $d \sin \theta$, between the waves from S_1 and S_2, leads to constructive and destructive interference as a function of position, P, on the screen. (c) The observed *fringe* pattern and (d) the intensity, I, variation as a function of the phase difference, Δ, for the first few fringes are also shown.

Example 6.4.1: In a Young double-slit experiment, the two slits separated by 1.0 mm are illuminated with green light ($\lambda = 550$ nm), and the interference pattern is observed on a screen placed at a distance of 1.0 m from the slits. What is the separation between two successive bright or dark fringes?

Solution: The separation between the bright fringes (dark fringes will be the same) are given by (6.4.5). Here, $d = 1.0$ mm, $D = 1.0$ m, $\lambda = 550$ nm, and fringes will be observed for

$$z = m\lambda \, D/d = m \, 550 \times 10^{-9}/10^{-3} = m \, 5.5 \times 10^{-4} \text{ m} = 0.55m \text{ mm}.$$

Thus, for $m = 1$, the separation between successive fringes will be 0.55 mm, and may require magnifying glasses to be observed by the eye.

6.5   Reflection and Refraction

Huygens' principle can be readily applied to well-known laws of reflection and refraction of light (Fig. 6.5.1) by simply considering point sources on the interface of mediums 1 and 2, and then deriving the resultant wave in the new medium (refraction) or direction (reflection). We first consider reflection (Fig. 6.5.1).

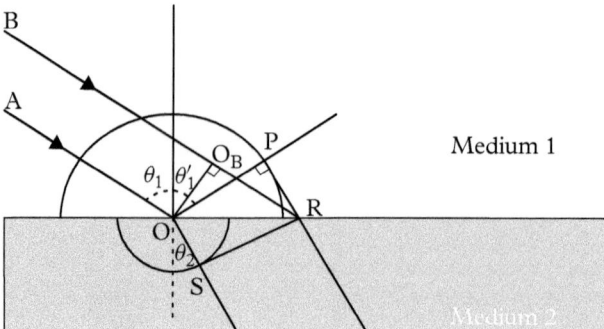

Figure 6.5.1 Reflection of light from a surface separating Medium 1 and Medium 2.

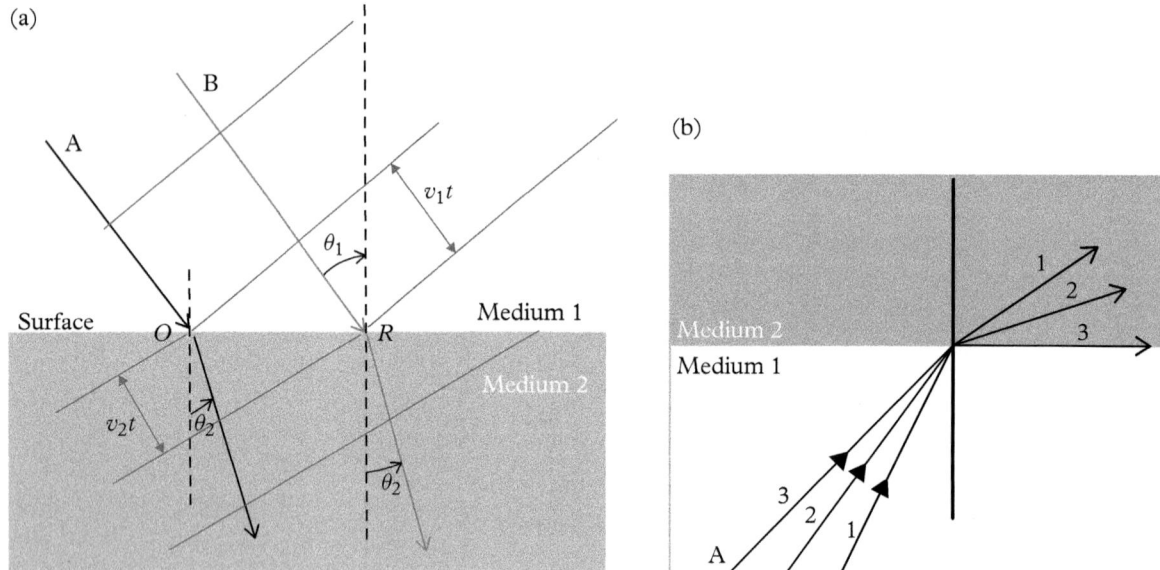

Figure 6.5.2 (a) Refraction of light at the interface between two media, and (b) total internal reflection (of ray 3).

At time, $t = 0$, the ray AO reaches the surface of medium 2, but the wave at B will reach R later at a time $t_B = \frac{O_B R}{v_1} = \frac{OR \sin \theta_1}{v_1}$. The spherical wave leaving O, will reach P, at a time $t_P = \frac{OP}{v_1} = \frac{OR \sin \theta_1'}{v_1}$. If the two times are equal, i.e. $t_B = t_P$, for the two waves to be in phase, then

$$\sin \theta_1 = \sin \theta_1' \tag{6.5.1}$$

or the angle of incidence is equal to the angle of reflection, i.e. $\theta_1 = \theta_1'$.

Generally, a fraction of the light will penetrate medium 2, from medium 1, and be refracted (Fig. 6.5.2a); the rest (not shown) is reflected at the interface. Here, the velocity of light in the two media are, v_1 and v_2, respectively. In medium 1, ray

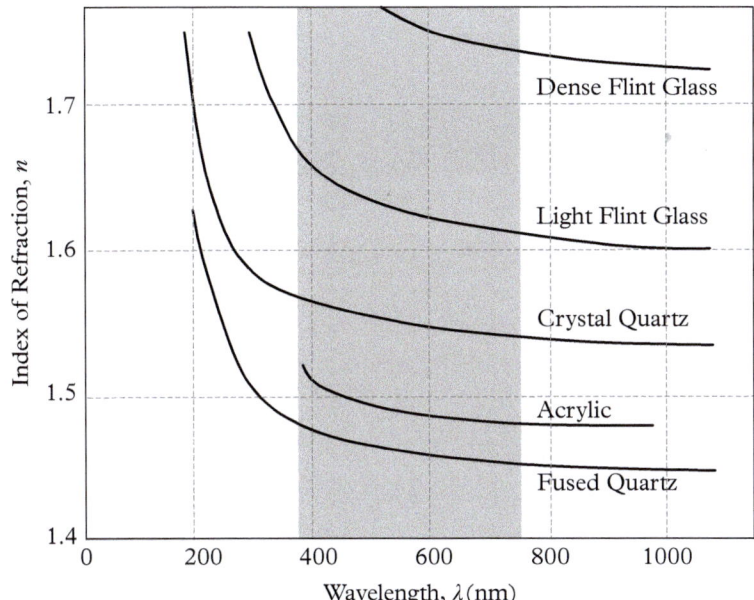

Figure 6.5.3 The change in index of refraction, n, as a function of the wavelength of light for various materials. The visible region is highlighted.

Adapted from Hecht (2001).

A will reach point O first, and then after a time, t, ray B will reach R. In this time, ray B travels a distance $v_1 t = OR \sin \theta_1$ in medium 1, and ray A travels a distance $v_2 t = OR \sin \theta_2$ in medium 2. Thus, we get the Snell[6] law of refraction:

$$\frac{\sin \theta_1}{\sin \theta_2} = \frac{v_1}{v_2} = \frac{n_2}{n_1} \qquad (6.5.2)$$

where we define the refractive index, n, of the medium as the ratio between the speed of light in vacuum to the speed of light in the medium. Note that the velocity of light in a specific medium also depends on its wavelength (Fig. 6.5.3) for various optical materials, including the commonly used fused quartz.

Further, if light is traveling from a medium of higher refractive index to a medium of lower refractive index, i.e. $n_1 > n_2$, it is bent away from the surface normal towards the plane of the surface. Thus, at some incident angle, $\theta_1 = \theta_c$, we reach a point where $n_1 \sin \theta_c = n_2 \sin (\pi/2) = n_2$, or $\theta_c = \sin^{-1} (n_2/n_1)$ and all light is reflected (no refraction) for $\theta_1 > \theta_c$ (Fig. 6.5.2b). This is known as *total internal reflection* and is used in many applications. In particular, it is used in fiber optic cables to transmit light signals without any loss in signal intensity.

6.6 Diffraction

When light passes through a narrow slit or pin hole it is diffracted (Fig. 6.3.1). In fact, this is a general characteristic of wave phenomena and occurs whenever a portion of a wave front is obstructed in some way. Figure 6.6.1 shows a common example of this phenomenon that deviates significantly from geometric optics.

[6] Attributed to the Dutch mathematician, W. Snellius (1580–1626), but historical research shows that this law already was formulated by the Persian mathematician and optician Ibn Salah in 984 CE!

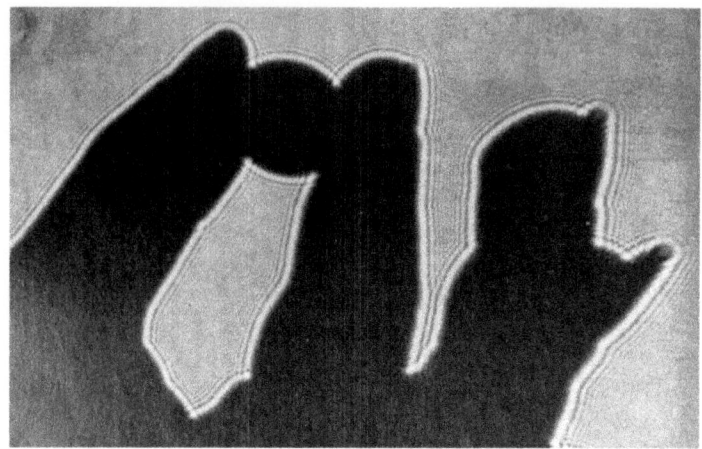

Figure 6.6.1 The shadow of a hand holding a coin illustrating the effect of diffraction.

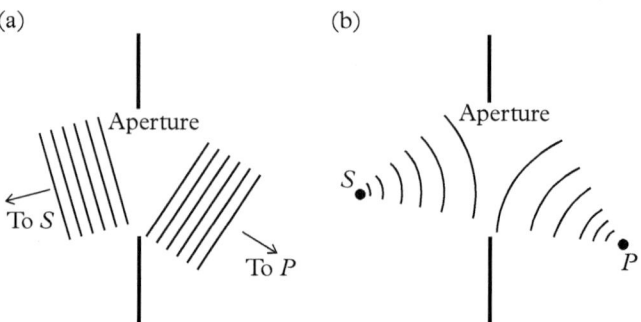

Figure 6.6.2 Diffraction by an aperture or slit, illustrating the conditions for (a) Fraunhofer and (b) Fresnel diffraction. Here, S is the source and P is the point of detection.

Adapted from Fowles (1989).

In the detailed treatment of such diffraction, it is common practice to distinguish between two general cases known as Fraunhofer[7] and Fresnel[8] diffraction (Figure 6.6.2). Fraunhofer diffraction is important to understand the behavior of gratings, introduced in Figure 3.5.1 in the context of infrared spectroscopy, and is discussed further in §6.6.3. It is expected when the incident and diffracted light are essentially plane waves; in practice, this is the case when the distances from the source to the diffraction aperture and from the aperture to the screen are sufficiently far apart to neglect the curvatures of the incident and diffracted waves (Fig. 6.6.2a). On the other hand, if either the source or the detector is close enough to the diffraction aperture, such that the curvature of the wave front is significant, we have Fresnel diffraction (Fig. 6.6.2b). A practical application of Fresnel diffraction is the design of zone plates used routinely to focus high energy X-rays (see §6.6.7). We shall now discuss specific examples of Fraunhofer and Fresnel diffraction by various types of apertures.

[7] Joseph Fraunhofer (1787–1826).

[8] Augustin J. Fresnel (1788–1827).

6.6.1 Fraunhofer Diffraction from a Single Slit

Figure 6.6.3a shows the experimental arrangement for Fraunhofer diffraction from a single slit of width, w. Figure 6.6.3b shows the geometrical construction to obtain the intensity distribution on a screen at a distance, D (where $D >> w$), illuminated by a plane wave. For analysis, the slit is divided into infinitesimally small elements of width, ds, at a distance, s, from the origin, O, *each of which is a source of secondary wavelets.* For the wavelet emitted by the element, ds, at the origin, its amplitude at point. P, on the screen will be directly proportional to the element width, ds, and inversely proportional to the distance, x. Thus, the displacement, dz_0, it produces at the point, P is

$$dz_0 = \frac{a\,ds}{x} \sin(\omega t - kx) \tag{6.6.1}$$

For the element, ds, at the position $+s$, the difference in path-length, l_p, is given by $l_p = s \sin \theta$. The corresponding displacement for this wavelet is

$$dz_s = \frac{ads}{x} \sin\left[\omega t - k\left(x + l_p\right)\right] = \frac{ads}{x} \sin\left[\omega t - k\left(x + s \sin \theta\right)\right] \tag{6.6.2}$$

and the total displacement is obtained by integrating (6.6.2) to obtain the contributions of all the wavelets emanating from all the elements of the slit, i.e.

$$z = \int_{-w/2}^{+w/2} dz_s = \int_{-w/2}^{+w/2} \frac{a}{x} \sin\left[\omega t - k\left(x + s \sin \theta\right)\right] ds \tag{6.6.3}$$

Following integration and rearrangement of terms we get

$$z = \frac{2a}{x} \sin\left[\omega t - kx\right] \int_0^{w/2} \cos\left(ks \sin \theta\right) ds = \frac{aw}{x} \frac{\sin\left(\frac{kw}{2} \sin \theta\right)}{\frac{kw}{2} \sin \theta} \sin\left[\omega t - kx\right] =$$

$$= A_0 \frac{\sin \beta}{\beta} \sin\left[\omega t - kx\right] = A \sin\left[\omega t - kx\right] \tag{6.6.4}$$

Thus, the important physical result is that the resultant displacement will also be a simple harmonic motion, determined by θ, but which will vary with position. The amplitude, $A = A_0 \frac{\sin \beta}{\beta}$, where $A_0 = aw/x$ and $\beta = \frac{kw}{2} \sin \theta = \frac{\pi w}{\lambda} \sin \theta$. Note that at $\beta = 0$, $A = A_0$, we observe a maximum value, and at $\beta = \pm\pi, \pm2\pi, \ldots, A = 0$, we observe minimum values. The intensity, I, on the screen is

$$I = A^2 = \frac{A_0^2 \sin^2 \beta}{\beta^2} \tag{6.6.5a}$$

(a)

(b)

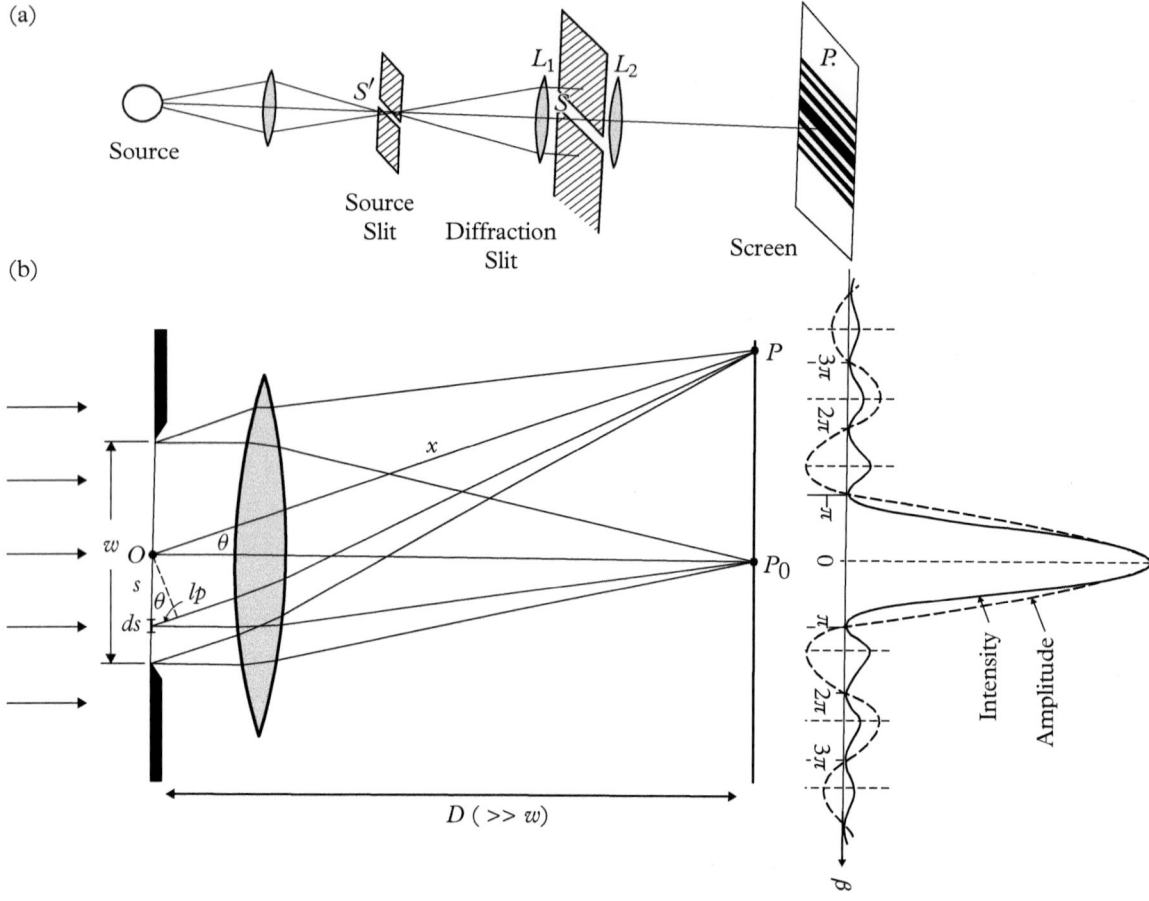

Figure 6.6.3 (a) Experimental arrangement to obtain the Fraunhofer diffraction pattern from a single diffracting slit. (b) Geometrical construction to determine the intensity distribution of the wave whose amplitude and intensity variations are also shown.

Both the amplitude and intensity variation, as a function of β, in units of radians, are also plotted in Figure 6.6.3b. Now, if the plane wave is incident at some angle, ϕ, then β is given in its more general form, i.e. $\beta = \frac{\pi w}{\lambda}(\sin\theta + \sin\phi)$, and includes both θ and ϕ.

Further, if the slit has a finite size, i.e. width, w, and height, h, then its amplitude $A_{w \times h} = \frac{awh}{x}\frac{\sin\beta}{\beta}\frac{\sin\gamma}{\gamma}$ where $\beta = \frac{\pi w}{\lambda}\sin\theta$ and $\gamma = \frac{\pi h}{\lambda}\sin\Omega$, where θ and Ω define the angles of the diffracted ray in two orthogonal planes. Its intensity, I, is given by

$$I = A_{w \times h}^2 = I_0 \left(\frac{\sin\beta}{\beta}\right)^2 \left(\frac{\sin\gamma}{\gamma}\right)^2 \tag{6.6.5b}$$

(a)

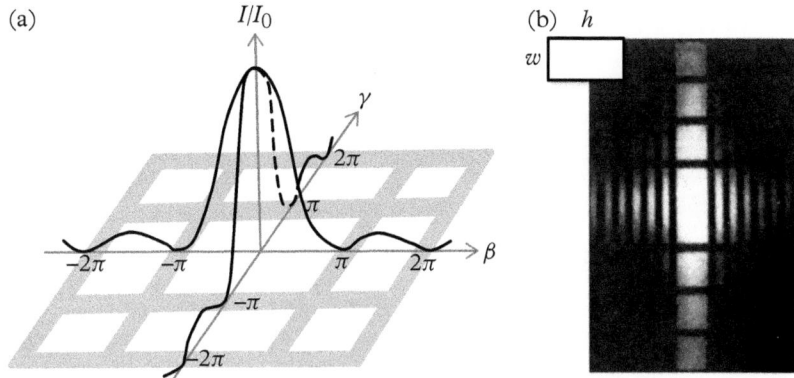

(b) *h*

w

Figure 6.6.4 (a) Fraunhofer diffraction from a square slit ($h = w$), showing the intensity variation along the two orthogonal directions. (b) An actual diffraction pattern of a rectangular slit with $h = 2w$; in the latter, the spacing in the horizontal and vertical directions are different and determined by the dimensions of the slit.

(a) Adapted from Hecht (2001). (b) Adapted from Jenkins and White (1981).

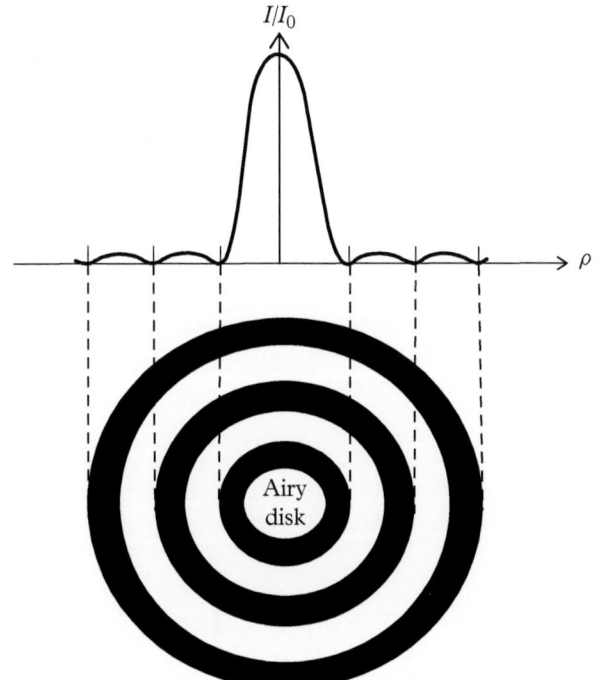

Figure 6.6.5 Fraunhofer diffraction from a circular aperture. The variable, $\rho = kR\sin\theta$, where R is the radius of the aperture, and k is the magnitude of the wave vector of the incident wave.

where $I_0 = \left(\frac{awh}{x}\right)^2$. Figure 6.6.4b shows the Fraunhofer diffraction pattern of such a rectangular aperture ($h = 2w$).

If instead of a rectangular slit, a circular aperture is used, a similar integration has to be carried out over the two-dimensional surface. The problem was originally solved by Airy[9] in 1835, with the solution obtained in terms of Bessel functions. Details can be found in any advanced book on optics (Fowles, 1989; Hecht, 2001), but the diffraction pattern obtained with such a circular aperture (Fig. 6.6.5) is also circularly symmetric, with a bright central disc—

[9] George B. Airy (1801–92) was an English mathematician and astronomer.

known as the Airy disc—surrounded by concentric bands of diminishing intensity. Alternatively, using a simple symmetry argument, the diffraction pattern can be generated by rotating the diffraction pattern obtained from a single slit (Fig. 6.6.3) about its axis. The angular radii of the finite disc are given by $\sin\theta = 1.22\,\lambda/D$, where $D = 2R$, is the diameter of the aperture.

6.6.2 Fraunhofer Diffraction from Double and Multiple Slits

Consider Figure 6.6.6 to observe the Fraunhofer diffraction pattern from a double slit, where the two slits of width, w, are separated by a distance, d. Following the derivation for the single slit, we can show that the displacement on the screen is given by

$$z = 2\frac{aw}{x}\frac{\sin\beta}{\beta}\cos\gamma\,\sin\left[\omega t - kx\right] \tag{6.6.6}$$

where $\beta = \frac{\pi w}{\lambda}\sin\theta$, $\gamma = \frac{\pi d}{\lambda}\sin\theta$, and $A_0 = \frac{aw}{x}$. The intensity is given by

$$I = 4A_0^2\frac{\sin^2\beta}{\beta^2}\cos^2\gamma \tag{6.6.7}$$

which shows positions of maxima, with values of $I_{\max} = 4A_0^2$ determined by $\cos^2\gamma$ at $\gamma = 0, \pi, 2\pi\ldots$, or equivalently when $d\sin\theta = n\lambda$, where n is an integer. Note that the factor, $(\sin\beta/\beta)^2$, derived earlier (6.6.5a) for the single-slit appears as an

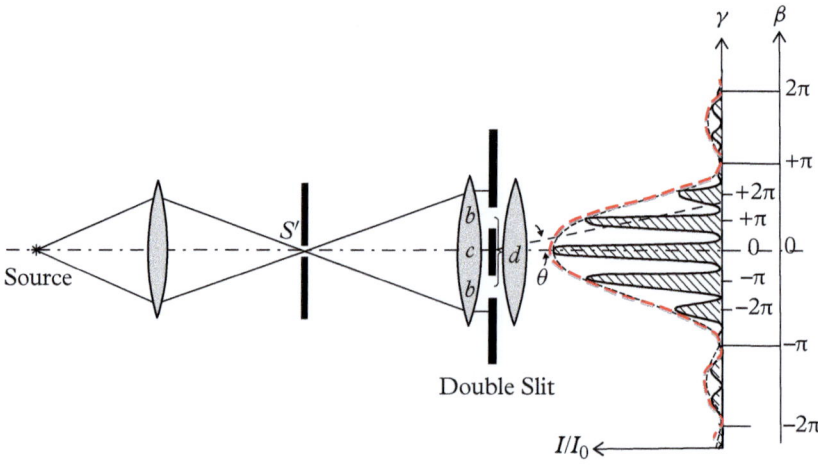

Figure 6.6.6 Fraunhofer diffraction from a double-slit aperture, including the intensity distribution. Note that the effect of the slit width, β, modulates the effect of their separation, γ.

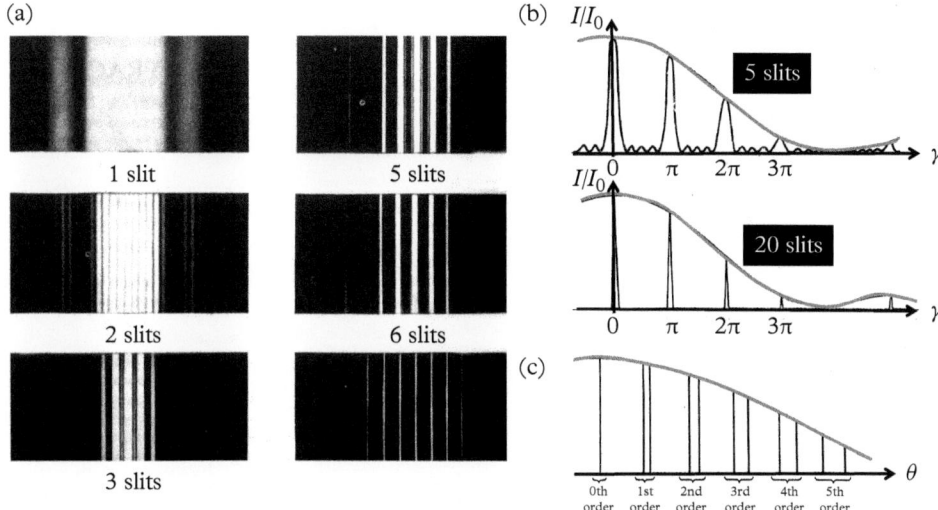

Figure 6.6.7 (a) Fraunhofer diffraction patterns from gratings with different number of slits. The sharpness of the *principal maxima* increases with the number of slits. (b) The intensity distribution plotted for 5 and 20 slits, (c) The separation of two different wavelengths as a function of diffraction order, $m = 0, 1, 2, 3, 4, \ldots$.

Adapted from Jenkins and White (1981).

envelope of the diffraction pattern. The resultant intensity, I, will be zero if either $\beta = \pi, 2\pi, 3\pi, 4\pi, \ldots$ or $\gamma = \pi/2, 3\pi/2, 5\pi/2 \ldots$, as shown in Figure 6.6.6.

Example 6.6.1: What happens to the diffraction pattern for a double slit if the width, w, is equal to half the spacing, d, i.e. $w = d/2$?

Solution: The angle at which the *maxima* of the diffraction occurs will be at $\sin^{-1}(n\lambda/d)$. However, the angle at which the *minima* of the envelope occurs will be at $\sin^{-1}(n\lambda/w) = \sin^{-1}(2n\lambda/d)$. Thus, every other *maximum* ($n = 2, 4, 6, \ldots$) in the diffraction pattern will be missing.

As the number of slits, N, is increased from two (Fig. 6.6.7), we can see that the sharpness of the principal maxima increases rapidly with the number of slits, and for $N = 20$, the lines become very narrow indeed. Further, secondary maxima appear as shown for $N = 3$ and $N = 5$ slits. Now we can generalize (without proof) our analysis to N slits, each of width, w, and separated by distance, d, to give an intensity

$$I_N = A^2 = A_0^2 \frac{\sin^2 \beta}{\beta^2} \frac{\sin^2 N\gamma}{\sin^2 \gamma} \tag{6.6.8}$$

Recall that $\beta = \frac{\pi w}{\lambda} \sin \theta, \gamma = \frac{\pi d}{\lambda} \sin \theta,$ and $A_0 = \frac{aw}{x}$. Further, when $N = 2$, we retrieve the result for the double slit, (6.6.7), and in the general case, maximum values, $I_N^{\max} = N^2 A_0^2$, are expected for $\gamma = 0, \pi, 2\pi.....$ The new term, $(\sin^2 N\gamma)/(\sin^2 \gamma)$, represents the interference effect for N slits. Note that

$$\lim_{\gamma \to m\pi} \frac{\sin N\gamma}{\sin \gamma} = \lim_{\gamma \to m\pi} \frac{N \cos N\gamma}{\cos \gamma} = \pm N \tag{6.6.9}$$

Thus, the maxima at $\gamma = 0, \pi, 2\pi.....$ correspond exactly in position to the maxima observed for a double slit as these conditions for γ also lead to the condition, $d \sin \theta = m\lambda$, where m is an integer. However, $I_N^{\max} = N^2 A_0^2$, and the intensity is proportional to the square of the number of slits. Further, for a given order, m, the angular separation (Fig. 6.6.8) between two spectral lines differing in wavelength, $\Delta \lambda$, is given by

$$\Delta \theta = \frac{m \Delta \lambda}{d \cos \theta} \tag{6.6.10}$$

The secondary maximum also follows a pattern that is related to the number of slits. As the number of slits increases, the amplitude of the secondary maximum decreases (Fig. 6.6.7).

Fraunhofer diffraction from a grating with multiple slits provides the basis for a physical understanding of the diffraction of X-rays and electrons by crystalline materials, to be discussed in detail in §7 and §8, respectively. A diffraction grating is in essence a one-dimensional crystal that has three distinct variables: slit spacing, slit width, and number of slits. The slit spacing, d, is equivalent to the lattice spacing, and determines the directions or overall geometry of the diffracted beams. The effect (β) of slit width, w, modulates the effect (γ) of their spacing. Analogously for X-rays, the atoms occupying the motif determine the observed intensities. Here, we assume that the diffraction is not dynamical and the diffracted beams do not interact with one another. Finally, the total number of slits determines the number and intensities of subsidiary diffraction peaks (satellites) on either side of the main diffraction peak. In fact, the number (intensities) of satellites increase (decrease) with the number of slits. For bulk crystals, the number of lattice planes involved in X-ray diffraction is very large, and as a result the satellite peaks have no observable intensity. However, thin films and multilayer materials with finite extent and a limited number (of the order of 100 or 1000) of lattice planes are an exception, and show well-defined satellite peaks that can be helpful in their analysis. Figure 7.9.12 shows a practical example.

6.6.3   Resolving Power of a Diffraction Grating

The angular width of a principal fringe (Fig. 6.6.6), defined as the separation between a peak and its adjacent minimum, is obtained by setting the change in $N\gamma$ equal to π, i.e. when $\Delta\gamma = \pi/N = \frac{\pi d}{\lambda}\cos\theta\,\Delta\theta$, or when

$$\Delta\theta = \frac{\lambda}{Nd\cos\theta} \qquad (6.6.11)$$

Thus $\Delta\theta$ is inversely related to N, and Figure 6.6.7c shows how, when N is very large, we observe sharp fringes in the diffraction pattern corresponding to different orders, $m = 0, \pm 1, \pm 2, ...$ From (6.6.10) and (6.6.11), we can show that the resolving power (RP) of the grating, defined as $\lambda/\Delta\lambda$, is given by

$$\text{R.P.} = \lambda/\Delta\lambda = Nm \qquad (6.6.12)$$

This is illustrated in Figure 6.6.8 and in Figure 6.6.7c. Note that the resolving power is not dependent on the size, w, of the individual grating (or ruling) or their spacing, d; it only depends on the total number, N, of gratings (rulings) and the order, m, of the diffraction. For various optical spectroscopies, discussed in §3, reflection gratings are typically used. They contain 600 lines/mm over a total width of 10 cm, giving a total of 60,000 lines, and a resolving power of $60,000m$, where m is the order of the diffraction. By suitably shaping the grooves, e.g. giving them a saw-tooth profile, the majority of diffracted light can be made to appear in one order, thus increasing the efficiency of the grating. For this to be successful, the spacing between the lines has to be very uniform, and moreover, its magnitude must be much smaller than the wavelength.

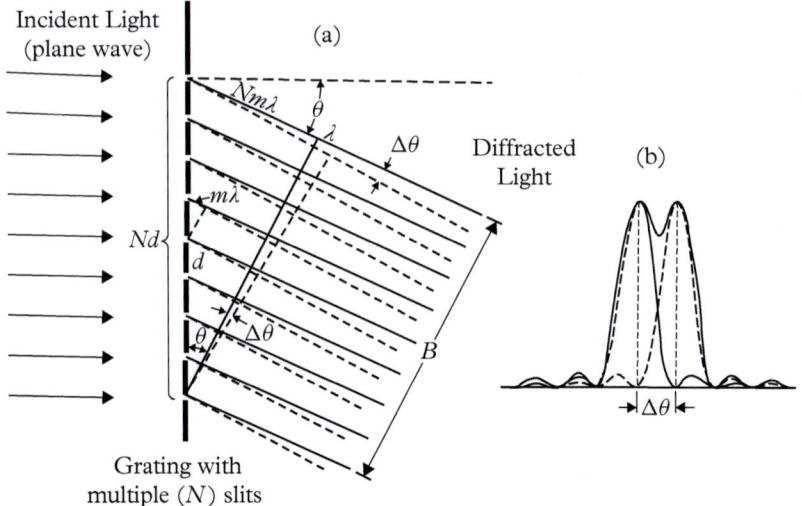

Figure 6.6.8 (a) A transmission diffraction grating arrangement to resolve spectral lines. (b) The angular separation of two spectral lines, $\Delta\theta$, just resolved by the diffraction grating.

Adapted from Jenkins and White (1981).

Example 6.6.2: The optical path through a transmission diffraction grating is shown in the following figure.

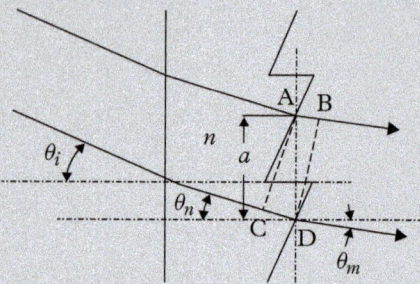

Show that the optical path length difference is *independent* of the refractive index of the grating.

Solution: From (6.5.2), $\sin \theta_i = n \sin \theta_n$, where n is the refractive index of the grating.

The optical path difference $= CD - AB = m\lambda = a\,(n \sin \theta_n - \sin \theta_m) = a\,(\sin \theta_i - \sin \theta_m)$, and does not depend on the refractive index.

6.6.4   Fresnel Diffraction

Fresnel diffraction can be easily understood by referring to Figure 6.6.9, where wave (1) starts at the bottom edge of the slit, and wave (2) starts at the middle of the slit. Note that the width, d, of the slit is much smaller than the distance, D, of the slit from the screen, i.e. $d \ll D$, and no lenses are required. The path difference, Δx, between rays (1) and (2), at the point, P, located at an angle, θ, is $\Delta x = (d/2) \sin \theta$, with an intensity minimum when $\Delta x = \lambda/2$. If we choose θ in such a way that $(d/2) \sin \theta = \lambda/2$, then every ray from the bottom half of the slit will be cancelled by another ray from the top half of the slit. Thus, the resultant intensity at this position, defined by θ, on the screen will remain zero and we get the first minimum. Similarly, even for a single slit, higher order minima are observed when $d \sin \theta = n\lambda$, where n is an integer.

Figure 6.6.9 Conditions for Fresnel diffraction. Note that, in reality, $d \ll D$.

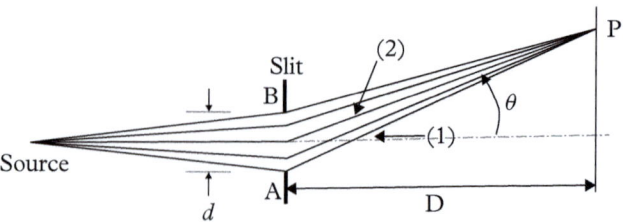

6.6.5   Fresnel Half-Period Zones

Consider the divergent spherical wave front, outlined by BCDE, traveling to the right (Fig. 6.6.10a). Following Huygens principle, every point on this spherical surface can be thought of as a source of secondary wavelets. To determine the effect of these wavelets at the point, P, of interest at a distance, OP $=$ b, from the center, we divide the wave front, BCDE, into circles, at distances, $s_1, s_2, s_3, ...$, along the arc, as shown, such that each of these circles are one half of a wavelength further from P, i.e. the circles are at distances $b + \lambda/2, b + 2\lambda/2, b + 3\lambda/2, ...$ from P. The areas, S_m, of the circles or zones, are all effectively equal, as can be seen in the simple construction shown in Figure 6.6.10b. The path difference, $l_p = HQP - HOP$, must be a multiple of $\lambda/2$ at the borders of the zone. From the sagittal formula,[10] for any zone with radius, s, we get

$$l_p = \frac{s^2}{2a} + \frac{s^2}{2b} = s^2 \frac{a+b}{2ab} \tag{6.6.13}$$

giving the radii, s_m, of the Fresnel zones as

$$m\frac{\lambda}{2} = s_m^2 \frac{a+b}{2ab} == \frac{s_m^2}{2}\left(\frac{1}{a} + \frac{1}{b}\right) \tag{6.6.14}$$

with the area of any one zone being the difference in the areas of two adjacent circles.

$$S_m = \pi\left(s_m^2 - s_{m-1}^2\right) = \pi\frac{\lambda}{2}\frac{2ab}{(a+b)} = \pi b\lambda\frac{a}{a+b} \tag{6.6.15}$$

In practice, the area of each zone is independent of m, and to first order a constant. Thus, by definition, each zone will send out successive wavelets that differ by a phase of π at P. Their amplitude, A_m, given by

$$A_m = k\frac{S_m}{d_m}(1 + \cos\theta) \tag{6.6.16}$$

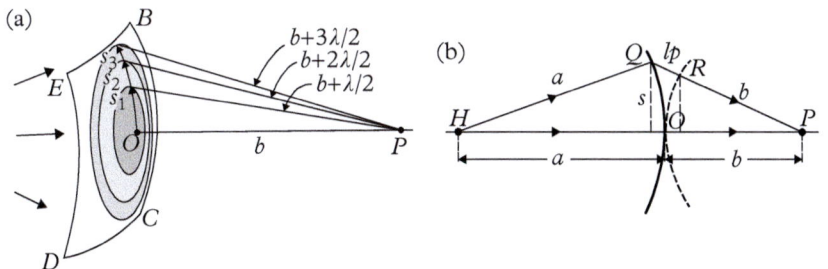

(a)

(b)

Figure 6.6.10 (a) A spherical wave front divided into half-period zones. (b) The path difference, Δ, at the point, P, from a distance, s, from the pole of the spherical wave.

Adapted from Jenkins and White (1981).

---

[10] The sagittal formula states that the distance, s, from the center of a circular arc of radius, r, to its base of length, l, is given by $r = \frac{s}{2} + \frac{l^2}{2s}$.

where k is a constant, $d_m = b + l_p$, is the distance of each zone to P, and θ is the angle at which the light leaves the zone, will add in a series with alternative signs to give a resultant amplitude, A, as

$$A = A_1 - A_2 + A_3 - A_4 + A_5 + ...(-1)^{m+1}A_m \qquad (6.6.17)$$

since successive zones differ by a phase of π. In (6.6.16), $(1 + \cos\theta)$ is an "obliquity" factor that causes successive terms in (6.6.17), to decrease very slowly. We can rearrange (6.6.17) as

$$A = \frac{A_1}{2} + \left(\frac{A_1}{2} - A_2 + \frac{A_3}{2}\right) + \left(\frac{A_3}{2} - A_4 + \frac{A_5}{2}\right) + ... \pm \frac{A_m}{2} \qquad (6.6.18)$$

and because the amplitude for neighboring zones are nearly equal, we get for odd m, the total amplitude at P, as

$$A = \frac{A_1}{2} + \frac{A_m}{2} \qquad (6.6.19a)$$

However, if m is even, we get

$$A = \frac{A_1}{2} - \frac{A_m}{2} \qquad (6.6.19b)$$

In summary, the resultant amplitude at P, is either half the sum (odd m) or half the difference (even m) of the contributions of *only the first* and *last zones*. Further, if m is very large, the entire spherical wave front when divided into zones results in $\theta \to 180°$ for the last zone. Thus, the obliquity factor $(1 + \cos\theta)$ in (6.6.16) becomes negligible and $A_m \to 0, A \to A_1/2$, i.e. the amplitude of the whole wave is just half of the first zone acting alone.

6.6.6   Diffraction by a Circular Aperture or Disc

For a spherical wave encountering a circular aperture (Fig. 6.6.11a) with radius, $r = s_1$, where s_1 was defined earlier (Fig. 6.6.10a), we get from (6.6.19a) for the total amplitude

$$A = \frac{A_1}{2} + \frac{A_1}{2} = A_1 \qquad (6.6.20a)$$

However, if $r = s_2$, we get

$$A = \frac{A_1}{2} - \frac{A_2}{2} \approx 0 \qquad (6.6.20b)$$

(a) (b)

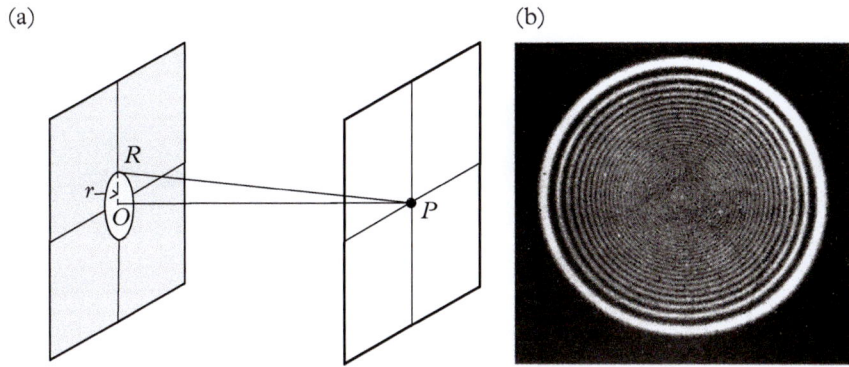

Figure 6.6.11 (a) Light passing through an open slit showing the geometric arrangement. (b) The Fresnel diffraction observed from a circular opening.

Adapted from Jenkins and White (1981).

Thus, the intensity will oscillate with the size of the hole. Also, the path difference will vary along the axis of the aperture and the intensity will vary as a function of distance, b.

From the previous, as a corollary, if the spherical wave is diffracted by a circular disc, the resulting amplitude will again be a summation, as discussed for the circular aperture, but since the first few zones are blocked, they are omitted from the summation. This leads to a bright spot in the center of the shadow image and the circular disc acts as crude lens.

6.6.7 Zone Plates and Their Applications in X-Ray Microscopy

Consider the special case of a half-period zone (§6.6.5), called a zone plate, where the transmitted light from every other half-period zone is blocked. As a result, either all the positive or all the negative terms in the summation (6.6.17) are removed. Either way, the amplitude, A, at P is increased many times. From (6.6.14), the radii of the zones, s_m, satisfies the condition $s_m \sim \sqrt{m}$ for any given a, b, and λ. Thus, in practice, by drawing circles with successive radii proportional to $\sqrt{m}$, where m is an integer, blackening alternative (either odd or even) zones, and allowing the light to pass through the rest, a zone plate can be constructed (Fig. 6.6.12a). Such a zone plate will produce an intense spot on its axis at a distance corresponding to the radii of the zones and the wavelength of the light used. The spot is so intense that the zone plate effectively functions as a lens. We illustrate this concept with some numbers.

Assume that the first five *odd* zones are exposed. The effective amplitude, (6.6.17), is $A = A_1 + A_3 + A_5 + A_7 + A_9 = 5A_1$. Without the zone plate, for large m, the whole wave gives an amplitude of $A_1/2$; thus, we can generalize and

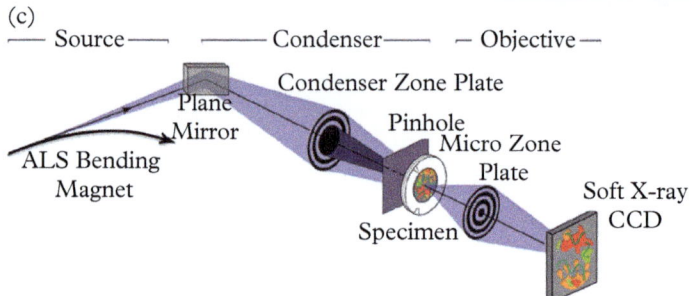

(b) $\Delta r = 25$ nm, D = 63 μm, N = 618 zones

(a)

(c)

Figure 6.6.12 (a) Zone plates. (b) Details of a zone plate fabricated for focusing an X-ray at a synchrotron beam line. Notice the period ($\Delta r \sim 25$ nm) and the number, $m = 618$, of zones. (c) A schematic of a transmission X-ray microscope with zone plates as condenser and objective for imaging.

Figures courtesy of the Advanced Light Source, Berkeley.

say that by exposing m zones, only odd or even, we get an amplitude at P that is $2m$ times larger than that without the zone plate. Moreover, the intensity at P is $4m^2$ times greater. In addition, from (6.6.14), the source and image distances follow the standard lens formula (6.8.3) as

$$\left(\frac{1}{a} + \frac{1}{b}\right) = \frac{1}{f} = \frac{m\lambda}{s_m^2} \qquad (6.6.21)$$

with the focal length, f, given by the value of b when $a \to \infty$, i.e.

$$f = \frac{s_m^2}{m\lambda} = \frac{s_1^2}{\lambda} \qquad (6.6.22)$$

In practice, such zone plates are lithographically fabricated of alternating rings of light (e.g. silicon) and heavy (e.g. tungsten) elements, and find great utility in focusing high-intensity X-rays such as in synchrotron beam lines. Figure 6.6.12b shows a typical zone plate fabricated by careful *e*-beam lithography and containing 618 zones. It is commonly used as a lens for a transmission X-ray microscope (Fig. 6.6.12c) in dedicated beam lines at synchrotron sources. Further details of zone plates and their use in imaging can be found in Atwood (2000).

Example 6.6.3: A zone plate has a focal length of 0.5 m for a wavelength of 500 nm. What is the radius of the first and sixteenth circles of the zone plate?

Solution: The radii, s_m, of the Fresnel zones can be calculated from (6.6.22), i.e. $s_m = (f\,m\,\lambda)^{1/2}$.

Thus, the radius of the first circle, $s_1 = (f\lambda)^{1/2} = (0.5 \times 500 \times 10^{-9})^{1/2} = 5 \times 10^{-4}$ m.

And the radius of the sixteenth circle, $s_{16} = 4\,s_1 = 2 \times 10^{-3}$ m.

6.7 Visually Observable: Characteristics of the Human Eye

We now move from diffraction to discuss optical imaging and microscopy. For any visual observation, we have to first understand the characteristics of the human data collection and image recording system. So, we begin with a brief summary of the characteristics of the human eye to answer the question: "what is visually observable?"

The human eye (Fig. 6.7.1) is a lens system that projects a positive image on a light sensitive surface (*retina*). The amount of light entering the eye is determined by a diaphragm (*iris*), which controls the light entering the eye through a variable hole (*pupil*). Two types of photo detectors are used to create black and white

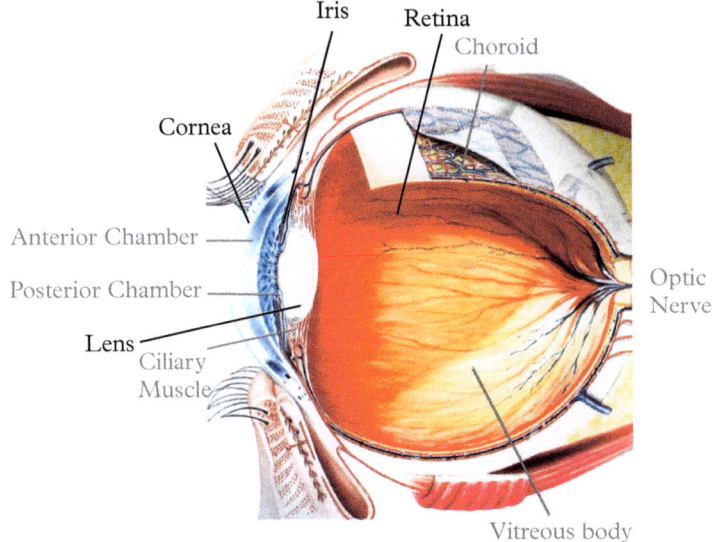

Figure 6.7.1 The human eye showing the principal optical components.

Adapted from *Anatomica*, Global Book Publishing (2005).

(*rods*) and color (*cones*) images. We have already remarked that the human eye is sensitive to wavelengths (visible spectrum) ranging from 380 nm to 700 nm, i.e. colors from violet to dark red (Fig. 1.3.2). However, the *peak sensitivity* of the human eye corresponds to green light with wavelength, $\lambda = 550$ nm. In practice, this is also one of the characteristic emission peaks ($\lambda = 546$ nm) of a mercury arc lamp (Fig. 5.2.1). Further, to take advantage of this peak sensitivity of the human eye, optical microscopes are focused using a green filter, and viewing screens for electron microscopes, both transmission and scanning, were traditionally coated with a green phosphor.

Another important characteristic of the human eye is its *integration time*, which is the time required for the retina to collect sufficient photons to form an image. Typically, this is ~0.1 s. In addition, the ambient light environment also affects the performance of the human eye. In total darkness the eye sees occasional flashes of light arising from random noise, but once it is adapted to the darkness, referred to as *dark adapted*, it achieves maximum sensitivity at low light levels and registers a good 50% of the incident photons! A reasonably good image can then be formed with ~100 photons collected by each picture element, or pixel. Thus, the human eye is a remarkable "instrument" and performs as well as the best available night vision system, with one exception being that the latter can integrate signals for times longer than 0.1 s.

Last, but most importantly, we are interested in determining the *resolution* of the eye, or its ability to separate two features that subtend the smallest angle at the eye, for a given pupil diameter, or aperture, at a specific distance of the objects from the eye. We address resolution in the next section.

6.8 Optical Microscopy

6.8.1 Resolution: Rayleigh and Abbe Criteria

Following the previous discussion of the Airy disc (Fig. 6.6.5), Abbe[11] showed that a *point source* subtending an angle, 2θ, at the lens will have an apparent diameter, ∂, on the screen given by

$$\partial = \frac{1.2\lambda}{n\sin\theta} \tag{6.8.1}$$

where λ is the wavelength of the light and n is the refractive index of the medium (Fig. 6.8.1a). Thus, an object with a diameter, D, will have a size $D + \partial$ on the screen. Rayleigh[12] then introduced a simple criterion for the resolution of the lens, by defining it as *the separation of two point sources where the maximum intensity from one source coincides with the minimum intensity of the other* (Fig. 6.8.1b). Hence, its resolution, $\delta = \partial/2$ and is given by

$$\delta = \frac{0.6\lambda}{n\sin\theta} \tag{6.8.2}$$

[11] Ernst Abbe (1840–1905) was a German physicist.

[12] John William Strutt, Lord Rayleigh (1842–1919) made fundamental discoveries in acoustics and optics that are basic to the theory of wave propagation. He received the Nobel prize in Physics (1904) and was cited for "investigations of the densities of the most important gases and for his discovery of argon in connection with these studies."

(a) (b)

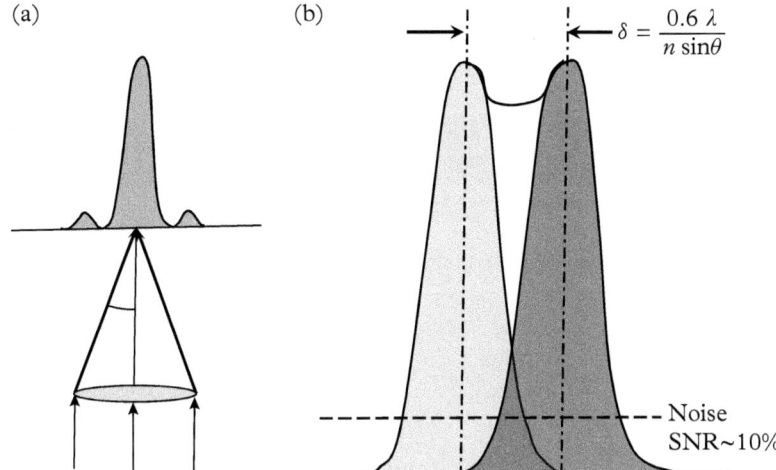

$$\delta = \frac{0.6\,\lambda}{n\,\sin\theta}$$

Noise
SNR~10%

Figure 6.8.1 (a) The image of a point object at infinity is given by the Abbe criterion. The width of the first peak is a function of the angular aperture, θ, of the lens and the wavelength of light, λ. (b) The Rayleigh criterion for resolution is given in terms of the image of two point-objects at infinity. The resolution, δ, is defined such that the angular separation of the two images gives the maximum of one peak at the position of the minimum of the other.

In addition, note that the signal intensity has to be well above background noise; typically signal to noise ratio, SNR ~ 10, is required for detection.

Example 6.8.1: The human eye has a normal aperture (*pupil*) of ~ 1.2 mm and it is physically difficult to focus it closer than ~ 15 cm (called the *near limit*). What is the resolution of the eye when green light ($\lambda = 550$ nm) is used?

Solution: For the eye, $\sin\theta = 0.6/150 \simeq 0.004$, and for green light, $\lambda = 550$ nm, in air ($n = 1$), we get a resolution $\delta_{eye} = \frac{0.6\times550}{1\times0.004} \simeq 82500$ nm $\simeq$ 82.5μm ~ 0.1 mm.

Example 6.8.2: What is the *optimal* magnification for an optical microscope (Fig. 6.8.5)?

Solution: The best optical systems are designed such that $n\sin\theta = 1$.

Thus, its resolution is $\delta_{microscope} \approx 0.6\lambda \simeq 330$nm. For the eye to be able to optimally see a magnified image, it should satisfy the condition $\delta_{eye} = \delta_{microscope}M$.

The optimal magnification for an optical microscope is $M_{opt} \simeq \delta_{eye}/\delta_{microscope} \simeq 0.1$ mm/330 nm ~ 300. In other words, an optical microscope with magnification, $M \sim 300 - 400$ will reveal all the information the human eye can resolve. Higher magnifications, M, do not bring any significant advantage, and using $M > 500$ for an optical microscope is often a waste of resources.

6.8.2 Geometric Optics and Aberrations

A brief discussion of elementary geometric optics is required to understand the operating principles of an optical microscope. As seen (Fig. 6.5.2), when light passes from a medium of low refractive index (e.g. air) to a medium of higher refractive index (e.g. glass) it is deflected, and the angles of transmission, for a given incidence angle, are determined by their respective refractive indices. Thus, a parallel beam of light, incident on a *convex* glass lens, is deflected such that the angle of deflection depends on the distance from the optic axis, and results in the light being focused at the focal point of the lens (Fig. 6.8.2a). The distance, f, of the focal point along the optic axis is proportional to the wavelength of light used. Further, simple geometric optics (Fig. 6.8.2b) determines the imaging characteristics and the magnification of the image. The distance of the object and image are simply related as

$$-\frac{1}{u} + \frac{1}{v} = \frac{1}{f}$$

(6.8.3)

and the magnification, M, is defined as the ratio $v/|u|$.

6.8.2.1 Lens Defects: Aberrations, Distortions, and Astigmatism

The lenses used in optical (and also other electromagnetic radiation) imaging do not give perfect images and are characterized by two important aberrations, *chromatic* and *spherical* (Fig. 6.8.3). Consider light emitted from the position P (Fig. 6.8.3a) with two different wavelengths, λ and $\lambda + \Delta\lambda$. As mentioned, the greater the wavelength, the greater is the focal length, and hence, these two wavelengths are brought to focus at two different points, Q and Q', on the optic axis. As a result, visible "white" light, consisting of a range of colors or wavelengths, are not brought to focus at a fixed point, but will be dispersed. Thus, point P, now imaged as a disc of radius $\Delta R/M$, where M is the magnification, defines the *chromatic aberration* coefficient

$$C_c \propto (\Delta\lambda/\lambda)/\Delta R$$

(6.8.4)

Note that in the case of electrons (see §9.2.2), $\Delta\lambda/\lambda$ is replaced by $\Delta v/v$, where v is the velocity, and (6.8.4) remains unchanged.

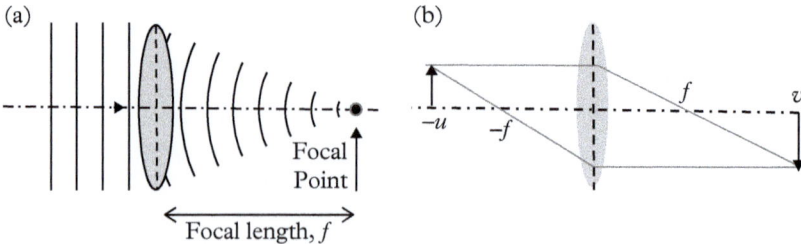

Figure 6.8.2 (a) A parallel beam of light incident on a convex lens converges on the focal point, at a fixed distance, f, along the optic axis. (b) Simple ray tracing showing image formation in a convex lens.

(a)

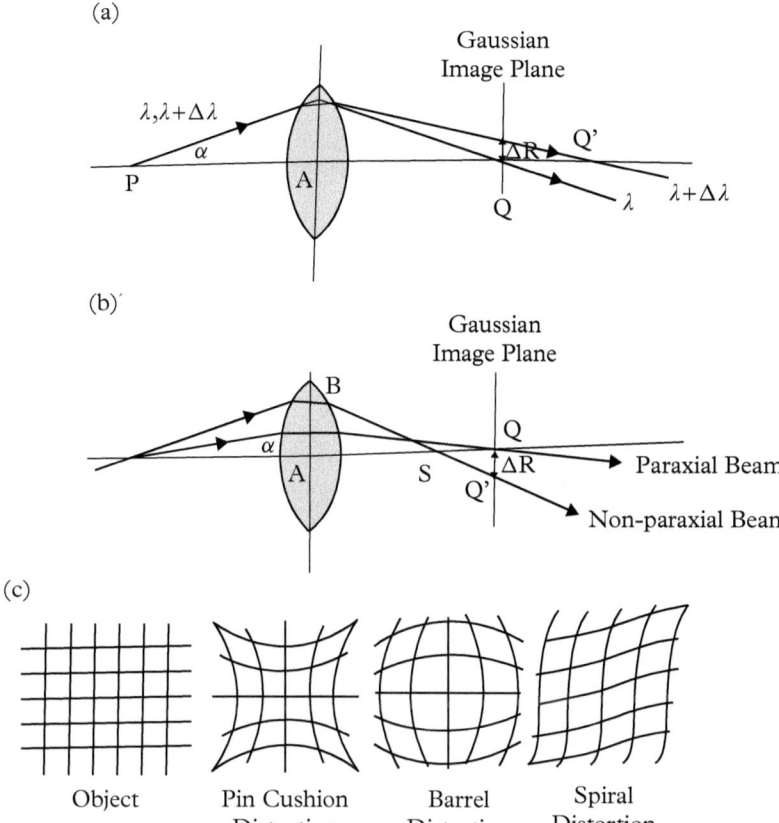

Figure 6.8.3 The two principal defects or aberrations commonly observed in both optical and electromagnetic lenses. (a) Chromatic aberration: rays leaving the same point, P, but with different wavelengths, λ and $\lambda + \Delta\lambda$ (or velocities, v and $v + \Delta v$) are brought to focus at different points, Q and Q'. As a result, the image of the point appears as a disc of radius, ΔR, on the Gaussian image plane. (b) Spherical aberration: Paraxial and non-paraxial rays are brought to focus at different points along the optic axis. (c) Other commonly observed distortions in both optical and electromagnetic lenses.

Adapted from Flewitt and Wild (2003).

The focal length of the lens varies across the lens (Fig. 6.8.3b). For paraxial[13] illumination, all rays originating from P form images on the image plane at Q. However, if the rays are not paraxial, the rays are bent further at the extremities of the lens, and displace the image by a distance, $\Delta R/M$, to Q'. Again, the point source, P, has an apparent radius $\Delta R/M$, which determines the *spherical aberration* coefficient, C_s, of the lens. Further, the resolution, δ, of the lens is now related to the spherical aberration coefficient as

$$\delta = \lambda^{3/4} C_s^{1/4} \qquad (6.8.5)$$

The spherical aberration coefficient, C_s, is particularly important in defining the performance of electromagnetic lenses used in electron microscopes and is discussed further in §9.2.2.

Spherical aberration of the lens causes the image magnification to vary in proportion to the cube of the distance from the optic axis and gives rise to three

[13] A paraxial ray makes a small angle (θ) with the optical axis of the system, and lies close to the axis throughout the system.

principal distortions. This is illustrated in Figure 6.8.3c, for point objects displaced from the optic axis. If the magnification of the image increases with distance, it gives rise to pincushion distortion. Alternatively, if the image decreases with distance, a barrel distortion is observed. Finally, if the image is subject to an angular rotation as a function of its distance from the optic axis, it causes a spiral distortion and produces sigmoidal-shaped images. For further details, see Flewitt and Wild (2003).

Finally, if the lens produces images without perfect axial symmetry, the image planes in two orthogonal directions differ from each other. In practice, this causes the vertical and horizontal components of the image to focus on different planes, and is known as *astigmatism*. As a result, no sharp image planes exist and there is a region of "confusion" between two sharply focused images. In optical systems, this defect is inherent in the manufacturing of the lens.

6.8.2.2   *Depth of Field and Depth of Focus*

The resolution possible for an object with its image in focus in the image plane, given by (6.8.2), depends on the numerical aperture, or $n\sin\theta$. Thus, the object can be displaced from its position without sacrificing the resolution (Fig. 6.8.4), and the distance it can be displaced along the optic axis and still remain in focus is called the depth of field, d_f. For a given resolution, δ, of the lens, it is given by

$$d_f = \delta \tan \alpha \qquad (6.8.6)$$

where α is the half-angle subtended by the objective aperture at the focal point. The depth of field determines how stable the specimen stage should be for optimal imaging. If $n \sin\alpha \sim 1$, then $\delta \sim 0.6\,\lambda \sim 300$ nm, and $d_f = \delta \tan \alpha \simeq 0.3 - 0.5$ μm, which is the accuracy required for the specimen positioning system.

Further, the distance over which the image remains in focus is called the depth of focus, D, and is given by

$$D = M^2 d_f \qquad (6.8.7)$$

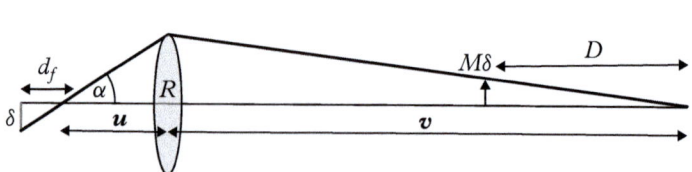

Figure 6.8.4 For a finite resolution, the depth of field, d_f, gives the range of positions that the object can occupy and still produce a focused image. Similarly, the focused image may be observed over a range of distances called the depth of focus, D.

where M is the magnification, and d_f is the depth of field. Note that D is not so critical as it is dependent on the square of the magnification, and even for $M = 100$, it is of the order of a millimeter.

6.8.3 The Optical Microscope

An optical or light microscope is a versatile tool, regularly used in metallography (§6.11), for examining, evaluating, and quantifying the microstructure of materials. Optical microscopes operate in either transmission or reflection modes, and the images produced have a typical resolution of ~250 nm. The reflection mode is particularly suited for examining opaque specimens of materials with the quality of the image enhanced by etching or polishing the surfaces of the specimens to be observed (see §6.11.1). Needless to say, care should be taken to ensure that the region of a specimen observed by any microscopy technique is truly representative of the sample.

6.8.3.1 Vertical Illumination in Reflection Geometry

A modern optical microscope (Fig. 6.8.5) has three components: the illuminating system, the imaging system, and the specimen stage which holds and moves the specimen in the X, Y, and Z directions, with mechanical stability consistent with the depth of field of the instrument. The imaging system consists of two lenses, one with very short focal length, called the *objective*, and the other, with a somewhat longer focal length, called the *eyepiece* (or *ocular*). The specimen or object (1) is placed just outside the focal point of the objective lens (Fig. 6.8.5b) and forms a real, magnified image at the first image plane (2), which becomes the object for the second lens, the eyepiece. The eyepiece magnifies (2) and forms a large virtual image (3), which then becomes the object for the eye to form a real final image (4) on the retina.

The specimen is illuminated with the condenser system, which is adjusted to uniformly illuminate the objective aperture, and focus an image of the source on the back focal plane of the objective lens. By eliminating any unwanted light reflected from the numerous surfaces of the lens, the illumination is further optimized. The condenser aperture is used to limit the amount of light, with improved contrast obtained by using a smaller aperture diameter. The brightest available source of light should be used, e.g. a mercury arc lamp or tungsten-halide discharge tube (§5.2.1.1). Appropriate filters can be used if monochromatic light, such as green wavelength light, is desired for viewing.

> **Example 6.8.3:** You have two positive lenses, both with a focal length of 20 mm, to make an optical microscope (one of them serves as the objective and the other as the eyepiece). If the *object* is positioned 25 mm from the objective (a) what is the separation between the lenses, and (b) what is the expected magnification of this microscope. Assume that an object at the focal point forms an image at a distance of 25 cm.

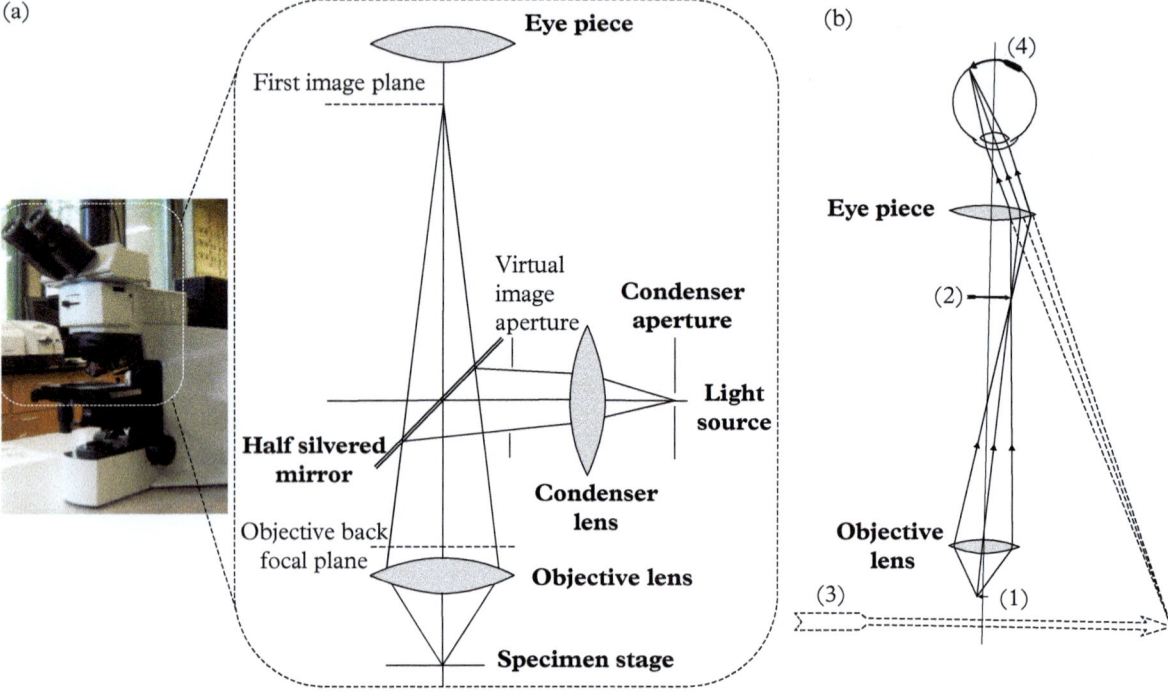

Figure 6.8.5 (a) An optical microscope and a schematic drawing of all its principal components in the reflection geometry and vertical illumination commonly used in metallography. (b) Ray diagram showing the image formation, including the eye of the viewer, in an optical microscope.

Solution:

(a) The intermediate image distance can be calculated from the lens formula (6.8.3), as $\frac{1}{f} = \frac{1}{u} + \frac{1}{v} = \frac{1}{20} = \frac{1}{25} + \frac{1}{v}$. Thus $v = 100$ mm. This is the distance from the objective lens to the intermediate image to which we need to add the focal length of eyepiece to get the separation of the two lenses: 100 mm + 20 mm = 120 mm.

(b) The magnification of the objective is $M_{obj} = v/u = 100/25 = 4\times$ (inverted image)

For the eyepiece, the object is at the focal point and the image is formed at a distance of 25 cm. Thus, the magnification of the eyepiece is $M_{ep} = 250/20 = 12.5\times$.

The total magnification of the microscope $M = M_{obj} \times M_{ep} = 4 \times 12.5 = 50\times$.

6.8.3.2 Direct, Oblique, and Dark Field Imaging

In addition to direct illumination discussed previously, which produces a bright field image, one can obtain an oblique or dark field image (Fig. 6.8.6a) by tilting the incidence angle. Oblique incidence can be obtained using off-axis illumination by simply shifting the condenser aperture (Fig. 6.8.6b). In dark field imaging, light incidence is such that no specularly reflected light enters the objective lens; in practice, instead of tilting the illumination, this is achieved by using an annular cone of light focused on the object plane. Dark field illumination is limited by low-image brightness but with an improvement in contrast (Fig. 6.8.6c).

6.8.3.3 Interference Contrast Microscopy

In its simplest form (Fig. 6.8.7a), a half-silvered cover slip is placed above the specimen, which splits the beam into two, half of which is reflected back from the cover slip, ray 1, and the other half, ray 2, is reflected from the specimen surface.

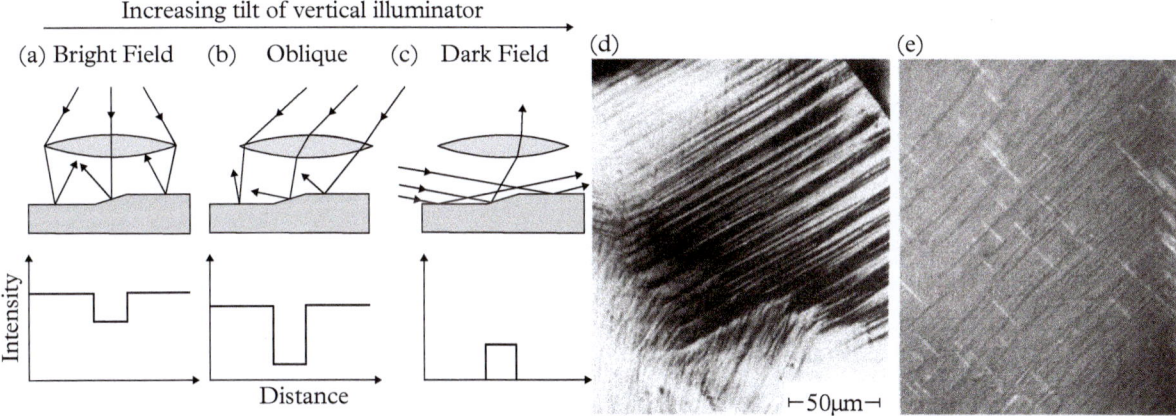

Figure 6.8.6 (a) Direct or bright field, (b) oblique, and (c) dark field illumination of a specimen, obtained by progressively tilting the illuminating beam. The reflected intensity distribution for each of these cases is also shown. Slip traces observed (d) in deformed β-phase by oblique incidence (note dark contrast), and (e) in α-phase (Bainite) by dark field imaging (note bright contrast), in a Cu–40%Zn alloy.

Adapted from Flewitt and Wild (2003).

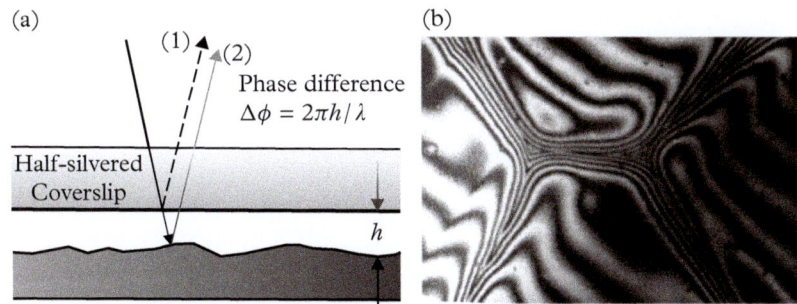

Figure 6.8.7 (a) Schematic representation of the two-beam interference method, and (b) image of a groove along a grain boundary.

Adapted from Brandon and Kaplan (2008).

The two beams, which were originally coherent, now have a phase difference depending on the additional path length ray 2 has to travel to the specimen surface and back. This phase difference leads to interference fringes (Fig. 6.8.7b) that can be interpreted in terms of local thickness variations.

Assume that the coefficient of reflection for the half-mirrored glass slip is R; then $(1-R)$ is the transmission coefficient. After reflecting from the specimen surface, again only $(1-R)$ of the intensity is transmitted by the slip; as a result $(1-R)^2$ of the original intensity is reflected from the specimen. For strong interference, the original reflected intensity, R, should be the same as the final transmitted intensity, $(1-R)^2$. Equating terms, we get $R = (1 - R)^2$, or $R = 1/\left(1 + \sqrt{2}\right) = 1/2.414 \simeq 0.41$. For the total path difference, $2h$, of the two beams to give rise to destructive interference (black fringe), it must be equal to $(2n + 1)\lambda/2$, where n is an integer and λ is the wavelength. Thus, when $(2n + 1)\lambda/2 = 2h$ is satisfied, dark fringes occur for height differences, h, of $\lambda/4, 3\lambda/4, 5\lambda/4....$ and successive fringes reflect height contours, $\Delta h = \lambda/2$. To detect local variations in specimen height or topography we need to detect changes in the positions of the fringes. If we assume that 10% shift in fringe separation can be detected, and using green light ($\lambda \sim 550$ nm), we can expect to resolve height differences with an accuracy of ± 25–30 nm. Further improvement in such two-beam interference microscopy can be achieved by using a drop of immersion oil between the coverslip and the specimen to effectively change the wavelength by the factor, n, the refractive index of the oil.

Interference contrast microscopy finds particular use in detecting details from optically transparent specimens that would otherwise lead to minimum contrast in a traditional optical microscope.

6.8.3.4   *Optical Microscopy with Polarized Light*

As seen in §1.3.4, a polarized light beam has its electric field vector component, **E**, aligned in a specific direction normal to the direction of propagation. Recall that when the direction of **E** is fixed in one plane but its magnitude varies along the propagation direction (Fig. 6.2.7) the light is called plane polarized. Further, following our discussion in §6.2.6, it is easy to see that if two plane polarized waves with their vectors inclined at some angle to one another are combined, it results in a third plane polarized wave with a vector sum of the original waves. Alternatively, a plane polarized wave can also be resolved into two orthogonal components, lying on arbitrary planes separated by 90° (Fig. 6.8.8b).

In a polarizing microscope, Figure 6.8.9a, the incident light is plane polarized by inserting a *polarizer* into the path of the condenser system. When this plane polarized light is reflected from an optically anisotropic surface, it undergoes a phase change and becomes elliptically polarized (see §6.2.6) and the electric vector now has a component orthogonal to the plane of polarization of the initial light. A second polarizer, called the *analyzer*, is placed between the objective lens and the eyepiece but with its polarization plane at 90° to the first polarizer; this

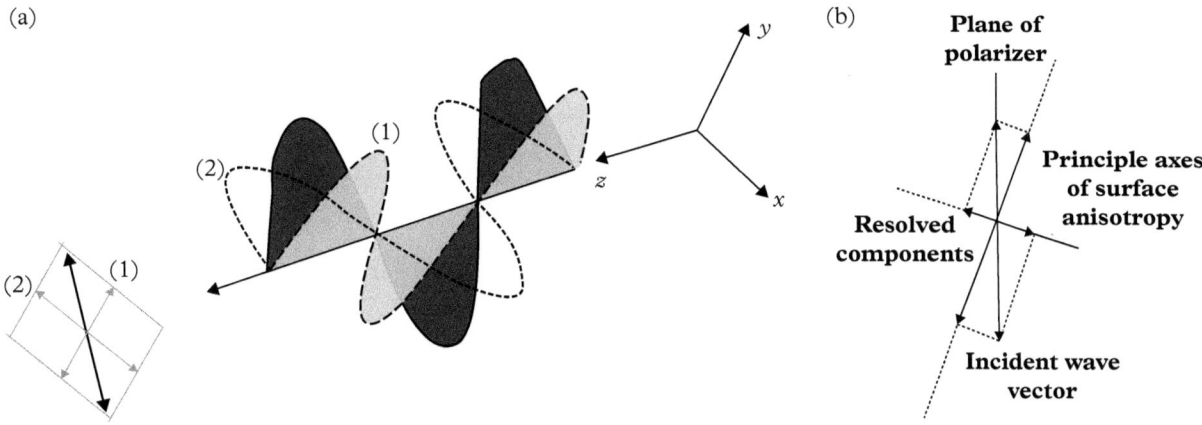

Figure 6.8.8 (a) Plane polarized light consisting of two beams in phase and polarized orthogonal to one another. (b) A plane polarized light beam can be resolved into arbitrary orthogonal components.

Adapted from Brandon and Kaplan (2008).

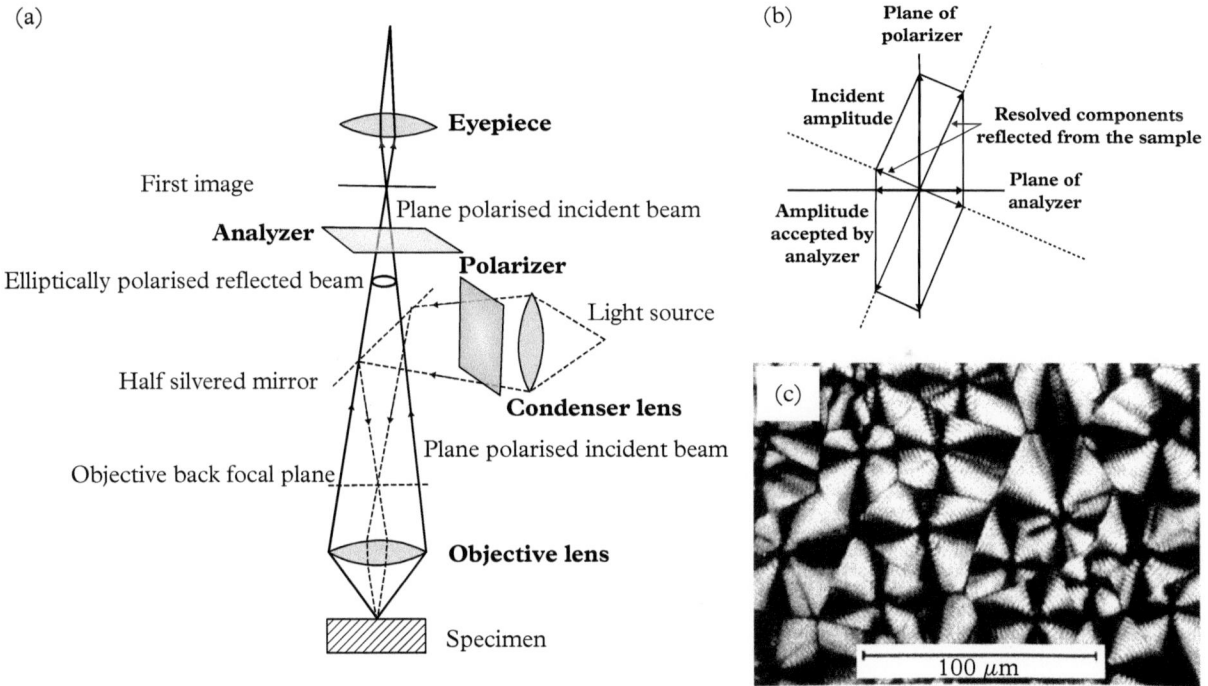

Figure 6.8.9 (a) An optical microscope with crossed-polarizer and analyzer used for imaging. (b) A plane polarized beam reflected from an optically anisotropic specimen surface will be elliptically polarized. An analyzer, polarized orthogonal to the analyzer, picks up the perpendicular component to form the image. (c) Details of the crystallization process of a polymer revealed by polarizing microscopy.

Adapted from Brandon and Kaplan (2008).

will detect the vector component perpendicular to the plane of polarization of the incident light (Fig. 6.8.9b). Thus, the phase change at the reflecting surface can be measured and specific features can be imaged, e.g. the crystallization process of polymers (Fig. 6.8.9c).

6.8.4   Confocal Scanning Optical Microscopy (CSOM)

CSOM, in which a finely focused beam of light is rastered over the specimen surface, is the light analog of scanning electron microscopy (§10). It is widely used in biological imaging. In conventional optical microscopy, discussed thus far, the entire field of view of the specimen is imaged. Moreover, the signal is also collected above and below the focal plane, which contributes significant blur to the final image degrading its contrast and sharpness (Fig. 6.8.11). In contrast, in CSOM, the illumination of the specimen is restricted to a single point at a specific depth of the specimen, which is then scanned to produce a complete image. Further, a confocal imaging aperture is introduced in the optical pathway, such that all light coming from regions above and below the focal plane of the microscope is attenuated and does not contribute to the image. Thus, the final image only contains the in-focus information.

Figure 6.8.10 shows the principle of confocal microscopy. The laser light from the source passing through the illuminating aperture, is reflected by the dichroic

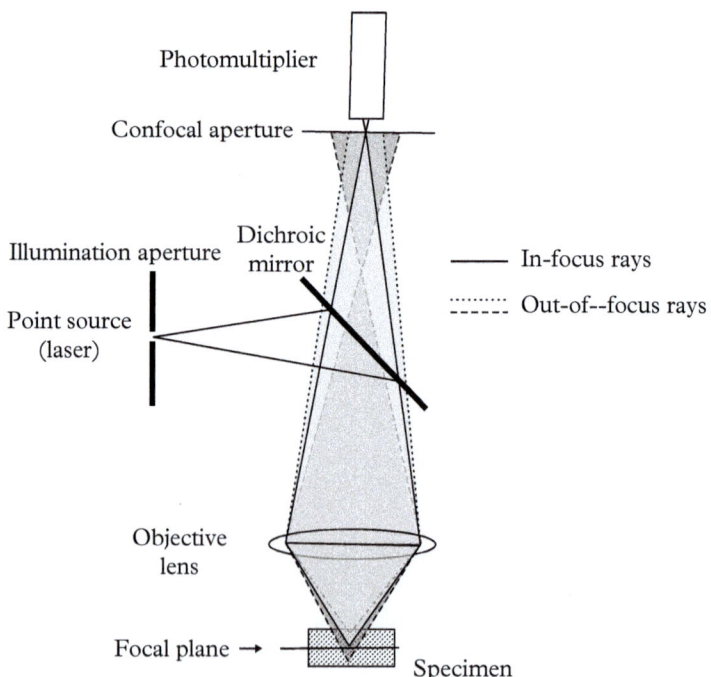

Figure 6.8.10 Schematic representation of a confocal microscope (see text for details).

Adapted from Sheppard and Shotton (1997).

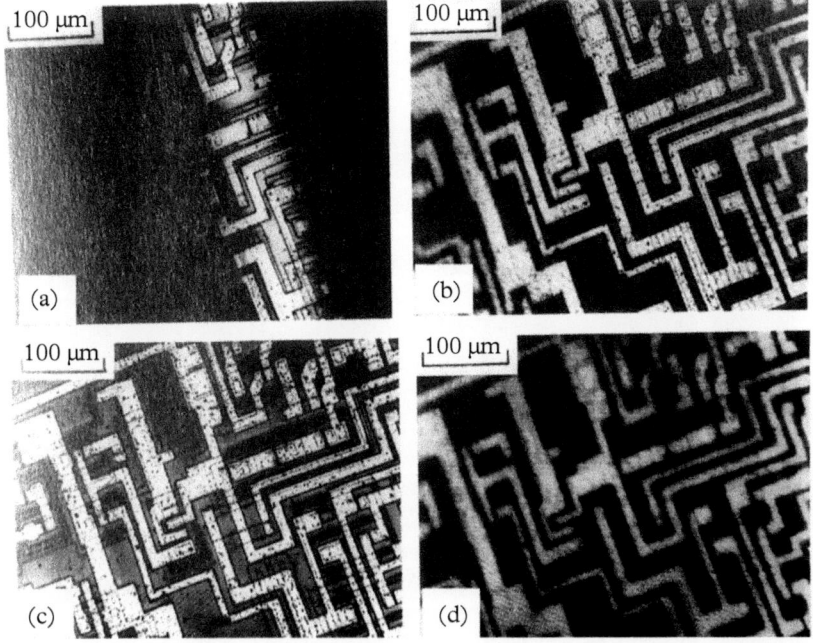

Figure 6.8.11 Comparison of confocal scanning optical microscope image of an electronic microcircuit with that obtained from a conventional full-field image. (a) In CSOM, a small region where only the central in-focus region is imaged, with all other regions appearing dark. (b) Conventional optical microscopy showing significant blur in the image. (c) The entire CSOM image of the specimen, all of it in focus. (d) Same as (c) but without the confocal optics; notice the significant blurring.

Adapted from [1].

mirror,[14] and then focused by the microscope objective lens to a diffraction limited spot at the focal plane of a three-dimensional specimen. *Reflected or fluorescent signal* is then produced from an imaging volume (voxel) at the focal depth, as well as from regions above and below it. However, only light from the in-focus voxel is allowed to pass through the confocal aperture to be detected; all other contributions from regions different from the focal plane are severely attenuated and do not contribute anything to the final image. Note that unlike conventional optical microscopes, which have a limited depth of field (§6.8.3.2), a CSOM can be operated as though it had an infinite depth of field by slowly and progressively changing its focal plane in depth for each scan of the specimen surface.

Figure 6.8.11a is a CSOM image, taken in reflection mode, of a small region of a tilted microcircuit using light of 633 nm wavelength. It is compared with a standard optical image (Fig. 6.8.11b) that shows significant blur, and a CSOM image (Fig. 6.8.11c) of the entire specimen showing all of it in focus.

CSOM combined with internal variations in autofluorescence is a versatile technique of particular importance in biology and the life sciences. Figure 6.8.12b illustrates this method of imaging through a sunflower pollen grain with *optical sections* gathered in 0.5 μm steps along the microscope optical axis. Pollen grains (20–40 μm in diameter) typically yield blurred images in a conventional fluorescence microscope (Fig. 6.8.12a); needless to say, they also lack information about internal structural details.

---

[14] A dichroic mirror shows significantly different reflection or transmission properties at two different wavelengths.

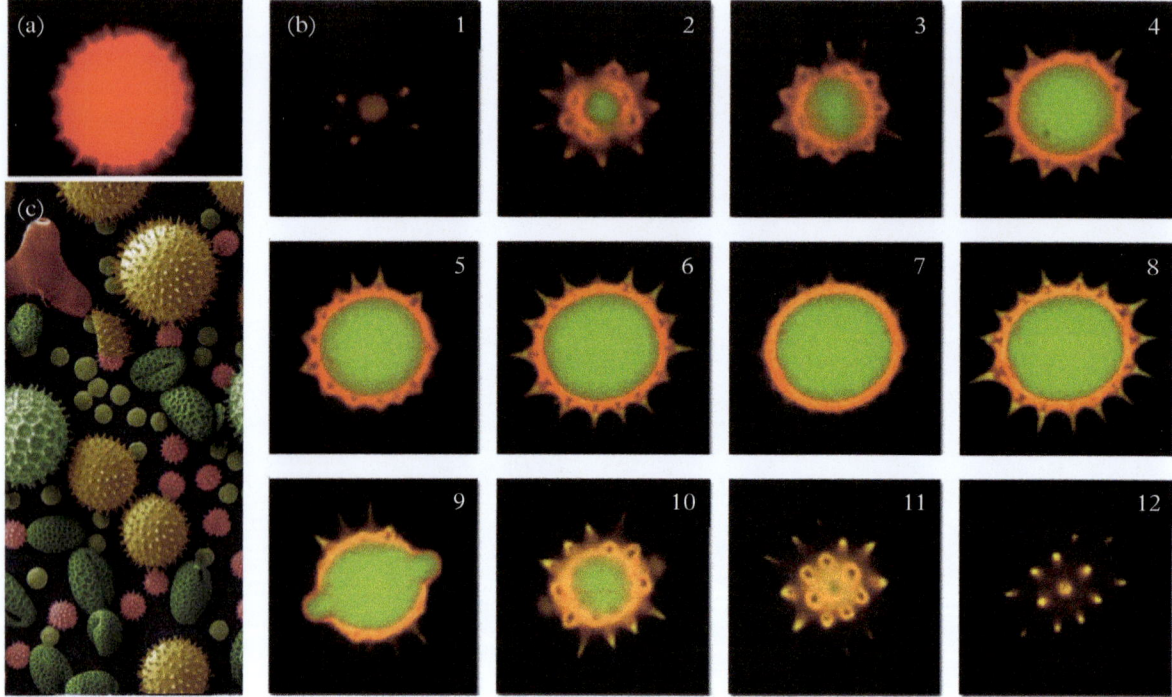

Figure 6.8.12 Images of a sunflower pollen grain using (a) conventional optical microscopy, and (b) CSOM sections (1–12) starting from the top in 0.5 μm steps along the optic axis using a dual argon-ion (488 nm; green fluorescence) and green helium/neon (543 nm; red fluorescence) laser system, showing its internal structure. (c) A false color scanning electron microscope (§10) image of the same pollens.

(a–b) Downloaded with permission from http://olympus.magnet.fsu.edu/primer/techniques/confocal/confocalintro.html. (c) Courtesy Marie Curie Institute.

In summary, CSOM excludes most of the light from the specimen that is not from the focal plane of the microscope. Thus, the image has better contrast than that of a conventional microscope and allows not only better observation of fine details but also helps build three-dimensional reconstructions of the specimen by assembling a series of images of thin slices taken along the optical axis. The excellent handbook by Sheppard and Shotton (1997) provides further details of CSOM.

6.8.5   Metallography

The simplicity and accessibility to optical microscopy makes it a widely used method to evaluate microstructures of materials. Some practical comments on specimen preparation and the application of optical microscopy are included here;

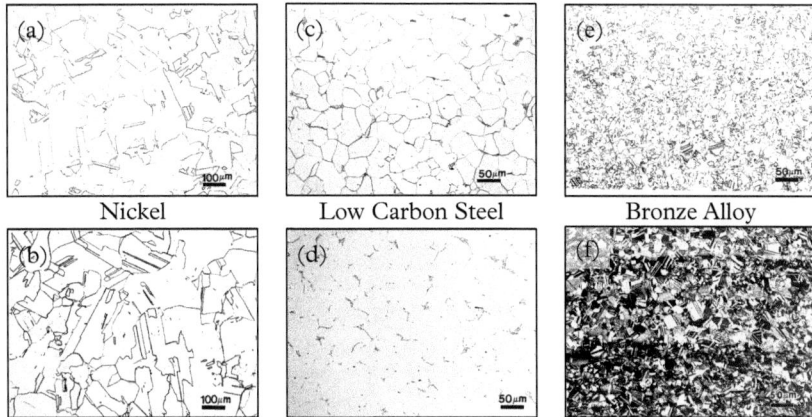

Nickel   Low Carbon Steel   Bronze Alloy

Figure 6.8.13 Effect of etching in metallography. The control of the degree of etching is important as shown for a 270-grade nickel etched with Kallings #2 reagent for (a) under-etched and (b) properly etched specimen under bright field imaging. Different etchants reveal different aspects of the microstructure shown here in bright field for a low-carbon steel specimen etched with (c) 2% Nital and (d) 4% Picral. Similarly, grain structures of a phosphor bronze alloy etched by (e) equal parts NH_4OH and 3% H_2O_2, and (f) Klemm's I tint etch.

All figures adapted from Williams, Pelton, and Gronsky (1991).

further details on metallography can be found in specialized monographs (Vander Voort, 1984).

In the examination of materials by optical microscopy, the preparation of specimens for observation is critically important and involves appropriate sectioning and polishing of their surface, followed by etching. For best results, it is best to examine specimens in the as-polished condition as it reveals certain microstructural features, such as inclusions, intermetallic phases, cracks, or porosity, which may be obscured by etching. Metal specimens are often etched with acid or base solutions to reveal the details of the microstructure. Initially, a general-purpose etchant (Table 6.8.1) is used, and is then followed by more specialized ones (Fig. 6.8.13).

Metallography is largely used to study opaque specimens; thus bright-field imaging in reflection mode is often used. However, optically anisotropic[15] specimens are best examined by cross-polarized illumination (Fig. 6.8.14a). Moreover, for examination with cross-polarized illumination, dark field illumination (Fig. 6.8.14b), or interference contrast imaging (§6.8.3.3), the specimens must be polished to a higher quality than what is required for bright field illumination to avoid seeing fine scratches in the images.

[15] Interacts with light differently in different crystallographic directions.

Table 6.8.1 Etchants commonly used in metallography.

Name of Etchant	Composition	Conditions for use	Suitable for etching
ASTM #30	Ammonia – 50 ml Hydrogen Peroxide (3%) – 100 ml DI water – 50 ml	Mix ammonia and water before adding peroxide. Use fresh. Swab surface for 5–45 s.	Copper, copper alloys, and Cu–Si alloys.
Adler Etchant	Copper ammonium chloride – 6 g Hydrochloric acid – 100 ml Ferric chloride, hydrated – 30 g DI water – 50 ml.	Combine all, immerse specimen for several seconds.	300 series stainless steels, Hastelloy, superalloys.
Carpenter's stainless steel etch	$FeCl_3$ – 7 g $CuCl_2$ – 2 g Hydrochloric acid – 100 ml Nitric acid – 4.9 ml Ethanol – 100 ml	Combine all, immerse specimen for several seconds at 20 °C.	Duplex and 300 series stainless steels.
Kalling's #2	$CuCl_2$–5 g, Hydrochloric acid – 100 ml, Ethanol – 100 ml	Combine all, immerse or swab specimen at 20 °C.	Duplex and 400 series stainless steel, Ni–Cu alloys and superalloys.
Keller's etch	Hydrochloric acid – 3 ml Nitric acid – 5 ml Hydrofluoric acid – 2 ml DI water – 190 ml.	Always use fresh etchant. Immerse specimen for 10–30 s.	Al and Ti alloys.

Reagent	Composition	Procedure	Applications
Klemm's reagent	Sodium thiosulfate, saturated – 250 ml. Potassium metabisulfite – 5g.	Etch for few seconds to minutes.	α–β brass, bronze, tin, cast-iron phosphides, ferrite, martensite and retained austenite.
Kroll's reagent	Nitric acid – 6 ml Hydrofluoric acid – 2 ml DI water – 190 ml.	Swab specimen for maximum 20 s.	Ti and its alloys.
Nital	Nitric acid – 1–10 ml Ethanol – 100 ml.	Immerse specimen for up to 1 min.	Iron, carbon and alloy steels, cast iron. Mn- alloys, magnetic alloys, Mg.
Picral	Picric acid – 2–4 g Ethanol – 100 ml.	Seconds to minutes. Prevent etchant from drying or crystallizing as it can explode!	Microstructures containing ferrite, carbide, pearlite, martensite, and bainite.

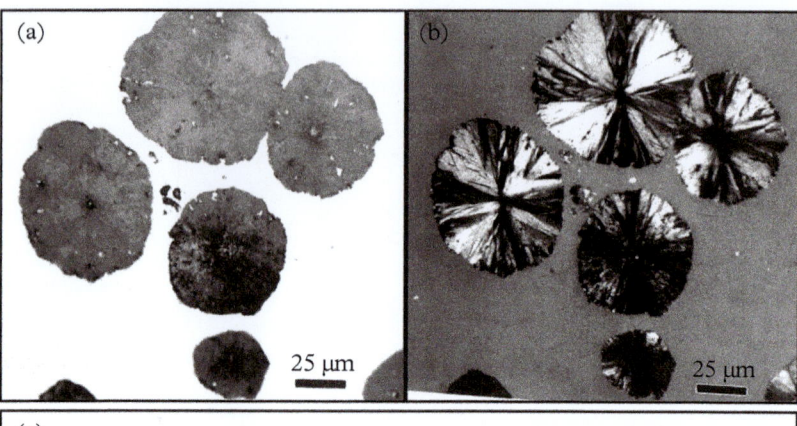

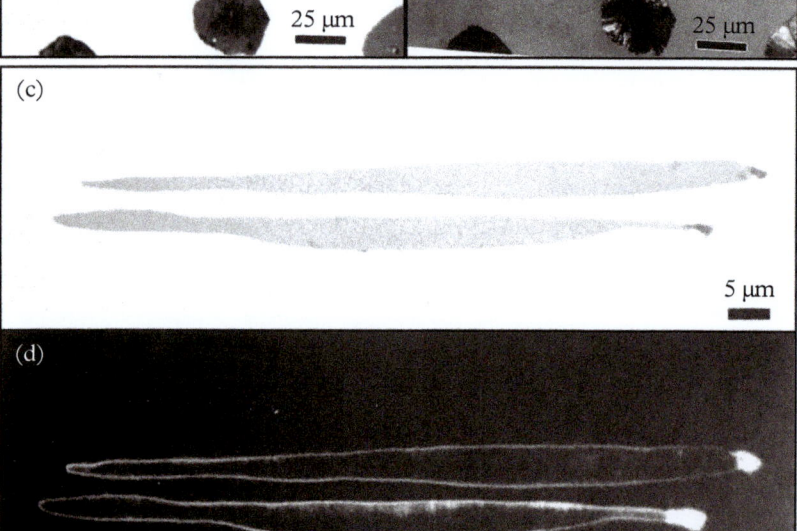

Figure 6.8.14 Metallography of annealed ductile iron containing graphite nodules in (a) bright-field at 400×, and (b) cross-polarized light microscopy. MnS with calcium–manganese sulfide tips shown in (c) bright field, and (d) dark field imaging.

All figures adapted from Williams, Pelton, and Gronsky (1991).

6.9   Ellipsometry

Ellipsometry is an optical measurement technique that is used to determine properties of specimens, largely thin films, based on changes in the polarization state of light reflected (transmitted) from (through) the specimen surface. The polarization change is measured in terms of an amplitude ratio, Ψ, and a phase change, Δ (§6.9.3). The technique is primarily used to determine optical constants and film thickness, but with additional sophistication in analysis it can also measure chemical composition, roughness, and crystallinity (see Fujiwara, 2007).

In general, the interaction of light with matter can be described in terms of a *complex refractive index, N,* which consists of the *index of refraction, n,* describing the phase velocity of the light in the medium as $v = c/n$, and the material dependent *extinction coefficient, K,* such that

$$N = n + iK \qquad (6.9.1)$$

Further, the optical properties of a material are described by the complex dielectric constant

$$\varepsilon = \varepsilon_1 + i\varepsilon_2 \qquad (6.9.2)$$

where $\varepsilon = N^2$, $\varepsilon_1 = n^2 - K^2$ and $\varepsilon_2 = 2nK$. Even though the light slows down as it enters a medium with higher index, its frequency remains unchanged, and its wavelength is shortened. The extinction coefficient, K, describing the loss of energy from the light to the medium, is given by

$$\mu = 4\pi K/\lambda \qquad (6.9.3)$$

where μ describes the attenuation of light (Beer's law) in the medium of thickness, t, introduced earlier, §1.3.5.1, as (1.3.11): $I = I_0 \exp(-\mu t)$.

This is also illustrated in Figure 6.9.1; however, when there is no absorption in the medium, $\mu = K = 0$. Furthermore, the complex dielectric constant and the complex refractive index are related through

$$n = \left[\frac{\varepsilon_1 + \left(\varepsilon_1^2 + \varepsilon_2^2\right)^{1/2}}{2} \right]^{1/2} \qquad (6.9.4)$$

(a)

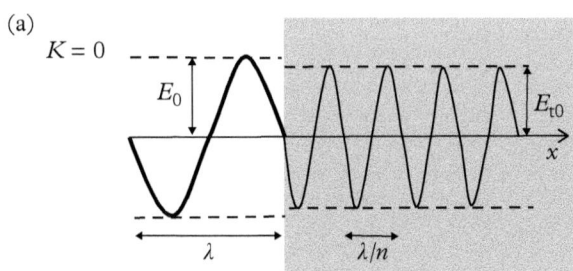

(b)

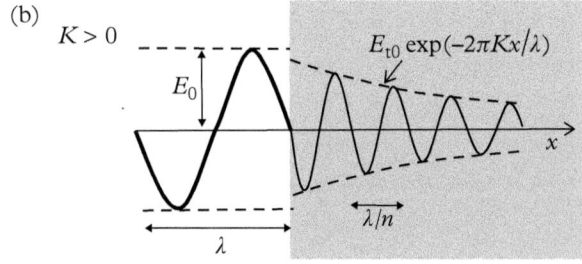

Figure 6.9.1 The propagation of an electromagnetic wave (only E-component shown) in (a) a transparent medium ($K = 0$), and (b) in an absorbing medium ($K > 0$). Note that in the latter, $E_0 > E_{t0}$, as some of the light is reflected at the interface.

Adapted from Fujiwara (2007).

$$K = \left[\frac{-\varepsilon_1 + \left(\varepsilon_1^2 + \varepsilon_2^2\right)^{1/2}}{2} \right]^{1/2} \qquad (6.9.5)$$

Finally, from K, (6.9.4), and using (6.9.3) the absorption coefficient, μ, can also be obtained.

6.9.1   p- and s-Polarized Light Waves, and Fresnel Equations of Reflection

In ellipsometry, we are interested in the reflection of polarized light at oblique incidence from a specimen surface (Fig. 6.9.2). It is common practice to classify the light into s- and p-polarization components, depending on the direction of oscillation of its electric field. In p-polarization (Fig. 6.9.2), the electric field of the light oscillates in the plane of incidence and reflection, whereas in s-polarization, the electric field oscillates in a plane orthogonal to this plane of incidence and reflection.

Figure 6.9.3a,b shows the propagation of the electric field, **E**, and magnetic induction, **B**, of p- and s-polarized light upon reflection from a surface. The ability of a surface to reflect a specific electromagnetic wave is characterized by the amplitude reflection coefficient. In general, upon reflection both **E** and **B** components of the light wave have to satisfy the boundary condition that their components parallel to the interface are continuous at the interface. Thus, the parallel components on the incidence side (throughout this section, we use subscripts i for incidence, r for reflection, and t for transmission) should be equal to that on the transmitted side. Hence, for the p-polarized light (Fig. 6.9.3a) we get

$$E_{ip} \cos\theta_i - E_{rp} \cos\theta_r = E_{tp} \cos\theta_t \qquad (6.9.6a)$$

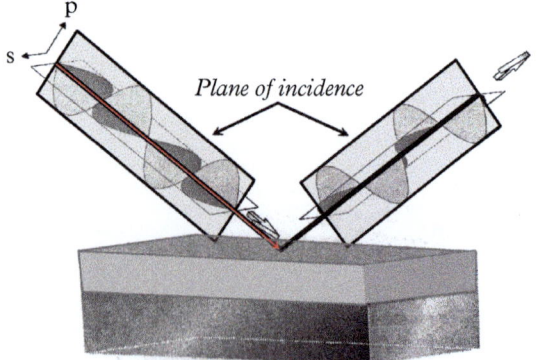

Figure 6.9.2 Reflection of s- and p-polarized light waves from a surface.

Adapted from Fujiwara (2007).

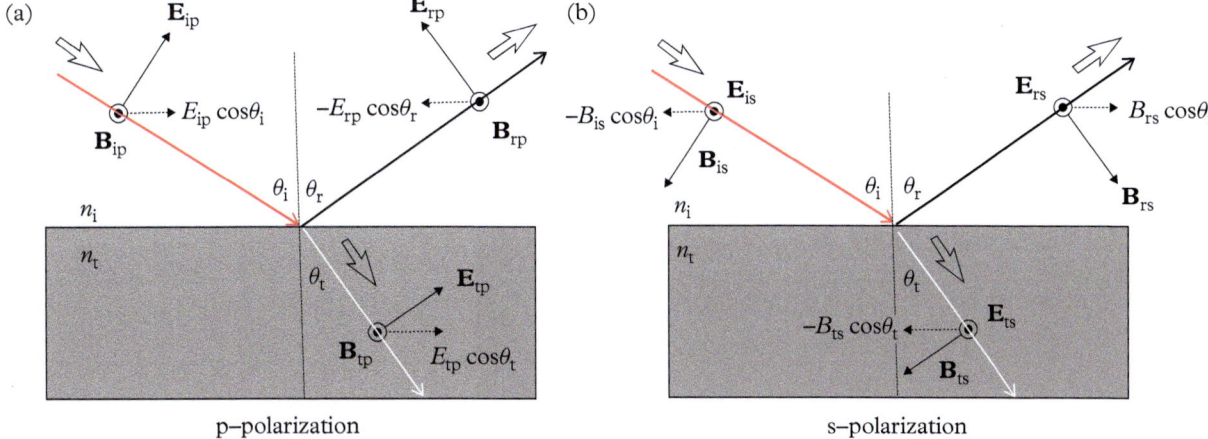

Figure 6.9.3 The electric field, **E**, and magnetic induction, **B**, for (a) p-polarization and (b) s-polarization, at the interface between two different materials.

Adapted from Fujiwara (2007).

$$B_{ip} + B_{rp} = B_{tp} \tag{6.9.6b}$$

Further, for a medium with refractive index, n, $E = vB = Bc/n$. Hence, (6.9.6b) can be rewritten as

$$n_i \left(E_{ip} + E_{rp} \right) = n_t E_{tp} \tag{6.9.6c}$$

By eliminating E_{tp} between (6.9.6a) and (6.9.6c), and setting $\theta_i = \theta_r$, we can obtain the amplitude reflection coefficient, r_p, for p-polarization

$$r_p = \frac{E_{rp}}{E_{ip}} = \frac{n_t \cos \theta_i - n_i \cos \theta_t}{n_t \cos \theta_i + n_i \cos \theta_t} \tag{6.9.7}$$

Similarly, for s-polarized light (Fig. 6.9.3b) the boundary condition at the interface gives

$$E_{is} + E_{rs} = E_{ts} \tag{6.9.8a}$$

$$-B_{is} \cos \theta_i + B_{rs} \cos \theta_r = -B_{ts} \cos \theta_t \tag{6.9.8b}$$

which leads to amplitude reflection coefficient

$$r_s = \frac{E_{rs}}{E_{is}} = \frac{n_i \cos \theta_i - n_t \cos \theta_t}{n_i \cos \theta_i + n_t \cos \theta_t} \tag{6.9.9}$$

The Fresnel equations for reflection (6.9.7; 6.9.9) are valid even if the refractive index, n, is replaced by the complex refractive index, N, where $\varepsilon = N^2$. It can then be shown that

$$r_p = \frac{\varepsilon_t N_{ii} - \varepsilon_i N_{tt}}{\varepsilon_t N_{ii} + \varepsilon_i N_{tt}} \qquad (6.9.10)$$

$$r_s = \frac{N_{ii} - N_{tt}}{N_{ii} + N_{tt}} \qquad (6.9.11)$$

where $N_{ii} = N_i \cos\theta_i$ and $N_{tt} = \left(\varepsilon_t - \varepsilon_i \sin^2\theta_i\right)^{1/2}$. This can also be written in polar coordinates as

$$r_p = |r_p| \, \exp(i\delta_{rp}) \qquad (6.9.12)$$

$$r_s = |r_s| \, \exp(i\delta_{rs}) \qquad (6.9.13)$$

In practice, in ellipsometry we measure the ratio of the reflection amplitude coefficients, r_p/r_s. Without proof, it is worth mentioning that the difference between r_p and r_s is maximized, to first order, for the angle of incidence equal to the Brewster angle, θ_B, defined as

$$\tan\theta_B = n_t/n_i \qquad (6.9.14)$$

At the air/glass interface, $n_t/n_i = 1.49$ and $\theta_B = \tan^{-1}(1.49) = 56°$. Further, θ_B depends on the wavelength, and for typical semiconductors, measurements are carried out at $\theta_i \sim 60\text{-}80°$.

Example 6.9.1: Fused glass has a refractive index, $n = 1.45$. For a film of this material placed in air and with Brewster angle incidence, calculate the reflection coefficient for both s- and p-polarized light.

Solution: The Brewster angle, (6.9.14), is $\theta_B = tan^{-1}(1.45) = 55.4° = \theta_i$. The angle of refraction is given by Snell law, (6.5.2),

$$\theta_t = sin^{-1}\left(\frac{n_i sin\theta_i}{n_t}\right) = sin^{-1}\left(\frac{1.0 \, sin55.4°}{1.45}\right) = 34.6°. \qquad (6.9.14)$$

Then, $r_p = \dfrac{E_{rp}}{E_{ip}} = \dfrac{n_t \cos\theta_i - n_i \cos\theta_t}{n_t \cos\theta_i + n_i \cos\theta_t} = \dfrac{1.45\cos(55.4) - 1.0\cos(34.6)}{1.45\cos(55.4) + 1.0\cos(34.6)} = 0.0001.$

And $r_s = \dfrac{E_{rs}}{E_{is}} = \dfrac{n_i \cos\theta_i - n_t \cos\theta_t}{n_i \cos\theta_i + n_t \cos\theta_t} = \dfrac{\cos(55.4) - 1.45\cos(34.6)}{\cos(55.4) + 1.45\cos(34.6)} = -0.3552.$

6.9.2   Optical Elements Used in Ellipsometry

To determine the polarization state of the light, optical components used in ellipsometry include polarizers (analyzers) and retarders. Typically, the former is used to extract linearly polarized light from unpolarized light, and the latter are used to convert linearly polarized light to circularly polarized light. Figure 6.2.6 shows three cases of the polarization state—linear, circular, and elliptical—typically encountered in ellipsometry as a function of the phase difference, $\Delta = \delta_x - \delta_y$, assuming the same amplitude, $E_{x0} = E_{y0}$ for the two phase components.

Recall from §4, Table 4.1.1, that a cubic crystal, such as NaCl, has its atoms arranged in a highly symmetric form with four threefold axes along the <111> directions. Light emanating from a point source within the material will propagate uniformly in all directions. Thus, it will be characterized by a single index of refraction. Alternatively, if the crystal belongs to the hexagonal, tetragonal, and trigonal systems, with uniaxial symmetry, the atoms are so arranged in the unit cell that the propagating light will encounter a different or asymmetric interaction depending on the direction of propagation. These materials are optically anisotropic and birefringent,[16] and there is only one direction (optic axis) about which the atoms are symmetrically arranged. Thus, these materials have two principal indices of refraction, $n_\perp$ and $n_\parallel$, perpendicular and parallel to the optic axis, respectively. The difference, $\Delta n = n_\parallel - n_\perp$, is a measure of the *birefringence*. Calcite is a good example of a uniaxial, optically active crystal.

Finally, crystal systems of even lower symmetry (orthorhombic, monoclinic, and triclinic) have two optic axes and are called biaxial. Such crystals have three principal indices of refraction. In practice, the birefringence of a biaxial crystal (e.g. mica) is the difference between its largest and smallest values of the indices of refraction.

6.9.2.1   Polarizer (Analyzer)

A standard polarizer is made of two prisms of a uniaxial calcite crystal (Fig. 6.9.4a). Such a polarizer (analyzer) is also known as a Glan–Taylor prism, and produces linearly polarized light along the optical anisotropy axis of the crystal.

6.9.2.2   Compensator (Retarder) and Photoelastic Modulator

In a typical ellipsometry set up, a compensator is placed either in front of the analyzer or after the polarizer. Its main function is to convert linear to circularly polarized light, and uses an anisotropic birefringent crystal that has two orthogonal axes along which light propagates with different velocities. The propagating light, initially linearly polarized at 45°, is then converted to left circularly polarization (Fig. 6.9.4b). For a thickness, d, of the compensator, this is accomplished by generating a phase difference, δ, between $\mathbf{E}_x$ and $\mathbf{E}_y$, due to differences in their velocities of propagation, where

$$\delta = \frac{2\pi}{\lambda}(n_e - n_o)d \qquad (6.9.15)$$

[16] Having two different refractive indices.

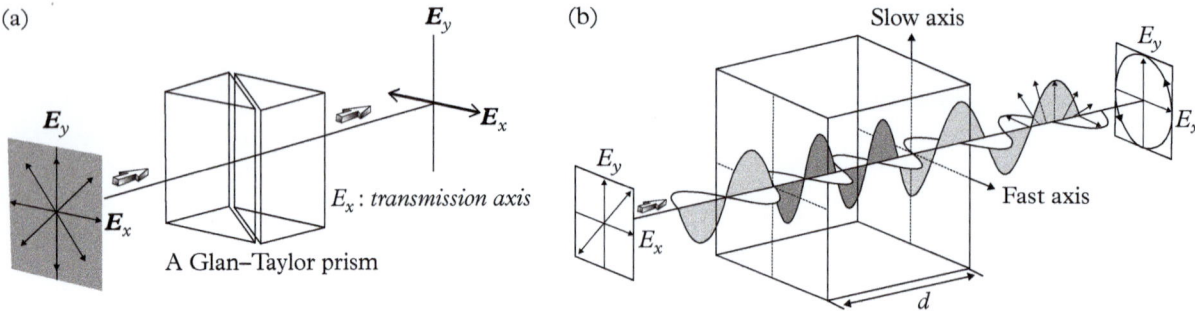

Figure 6.9.4 (a) A Glan–Taylor prism used as a polarizer (shown) or as an analyzer. (b) A compensator converts linearly polarized light to circularly polarized light by using a birefringent crystal that causes light to propagate with different velocities in orthogonal directions.

Adapted from Fujiwara (2007).

and n_e and n_o are the indices of refraction for the two components traveling with different velocities. Needless to say, δ depends on both thickness and wavelength, and for the special case of $\delta = \pi/2$, which corresponds to a wavelength difference, $\Delta\lambda = \lambda/4$, the compensator is known as a quarter wave plate.

In many isotropic materials, the application of stress leads to optically anisotropic behavior, called photoelasticity, and introduces birefringence behavior proportional to the applied stress and with optic axis aligned along the direction of stress. The phase shift, δ, in such photoelastic modulators varies continuously with time; details are beyond the scope of this book and can be found in more advanced texts (Fujiwara, 2007).

6.9.3   Ellipsometry Measurements

As mentioned, ellipsometry measurements monitor the change in polarization of p- and s-polarized light, $\mathbf{E}_{ip}$ and $\mathbf{E}_{is}$, optimally incident at the Brewster angle and reflected from the specimen, to obtain optical constants and film thickness. Figure 6.9.5 shows a typical measurement set up. Note that when $\theta = 0°$, the incident and reflected waves overlap completely; also, for the incident light (shown) linearly polarized at 45°, $E_{ip} = E_{is}$ is satisfied. As mentioned (§6.9.1), the amplitudes of the reflected light for p- and s-polarization differ significantly, and ellipsometry measures their amplitude ratio, Ψ, and the phase difference, Δ, between the s- and p-polarization. In the simplest case, Ψ is characterized by the refractive index, n, and the phase difference, Δ, represents the light absorption as described by the extinction coefficient, K. In other words, (n,K) can be determined from (Ψ, Δ) by applying the Fresnel equations (6.9.7; 6.9.9).

Figure 6.9.6 shows the measured ellipsometry spectra of SiO_2, for two different thickness, illustrating the variation of the amplitude ratio, Ψ, and phase lag, Δ, as a function of wave number.

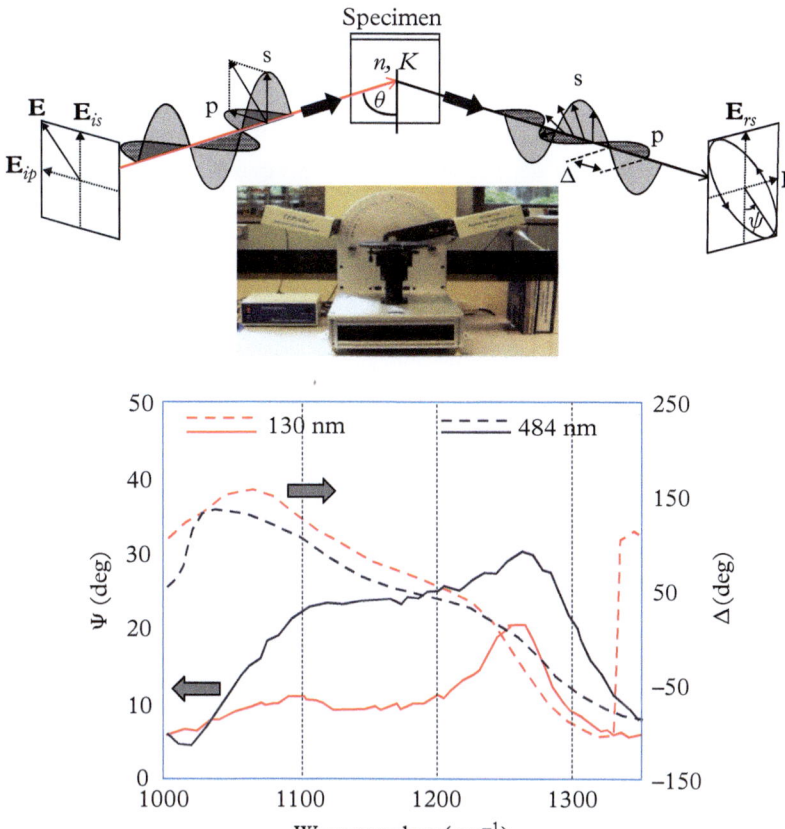

Figure 6.9.5 Top: Measurement principle of ellipsometry. The amplitude ratio, Ψ, and the phase lag, Δ, are measured and interpreted in terms of the optical constant, n and K, and other parameters of the thin film. Bottom: A commercial ellipsometer used for routine analysis of thin films.

Top: Adapted from Fujiwara (2007). Bottom: Courtesy University of Washington.

Figure 6.9.6 Ellipsometry spectra (Ψ and Δ) for two different thickness (130 nm and 484 nm) of SiO_2, as a function of wavenumber.

Adapted from [3].

In ellipsometry, the amplitude ratio, ρ, is related to Ψ and Δ by the following relationship

$$\rho = \frac{r_p}{r_s} = \tan\Psi \exp(i\Delta) \tag{6.9.16}$$

Now, using (6.9.9) and (6.9.7) we can rewrite (6.9.16) as

$$\rho = \frac{r_p}{r_s} = \tan\Psi \exp(i\Delta) = \frac{E_{rp}/E_{ip}}{E_{rs}/E_{is}} \tag{6.9.17}$$

In a simple case (Fig. 6.9.5), $E_{ip} = E_{is}$, and hence (6.9.17) is modified as

$$\rho = \frac{r_p}{r_s} = \tan\Psi \exp(i\Delta) = \frac{E_{rp}}{E_{rs}} \tag{6.9.18}$$

Further, in polar coordinates

$$\tan\Psi = \frac{|r_p|}{|r_s|} \text{ and } \Delta = \delta_{rp} - \delta_{rs}. \qquad (6.9.19)$$

In summary, ellipsometry measures the ratio of two values, and as such, is very robust, accurate, and reproducible. It is a model-based method, and after Ψ and Δ are experimentally determined, a model of the thin film is built, unknown optical constants and/or thickness parameters are varied, and Ψ and Δ values are calculated using the Fresnel equations (6.9.7) and (6.9.9). The calculated Ψ and Δ values, which best match the experimental ellipsometry data provide the optical constants and thickness parameters of the specimen. Depending on the material, thickness in the range of a few Å to a few micrometers can be measured by ellipsometry; however, for optically opaque materials, such as metals with strong absorption and shallow penetration of light, ellipsometry can routinely measure only optical constants, and not the thickness [2].

Summary

Propagation of light can be described as the simple harmonic motion of transverse waves, where the rate of change of its phase with time (distance) gives its angular frequency (wave vector). The linear superposition, or addition, of two or more SHWs produces a new SHW with a different amplitude and phase angle that can be derived from a simple geometric construction known as the phasor representation. Alternatively, it can be solved for analytically using complex notation and the related Argand[17] diagram. Additionally, combining waves that propagate on orthogonal planes can give rise to linear, elliptical, or spherical polarization, depending on their amplitudes and phase differences.

Two classical experiments inform our understanding of the behavior of electromagnetic waves. Huygens studied the propagation of light through a pin hole and, to explain the observed intensity distribution (which deviated significantly from geometric optics), he postulated that every point on an advancing wave front serves as a source of new spherical wavelets. The subsequent Young double-slit experiment demonstrated the principle of optical interference and diffraction. These experiments paved the way for our understanding of diffraction, which is now broadly classified as Fraunhofer and Fresnel diffraction. Fraunhofer diffraction is expected when the incident and diffracted waves are plane waves, i.e. the source and detector are significantly far away that the curvature of the wave front can be neglected, and is relevant to the understanding of the behavior and resolving power of diffraction gratings (§3.5.1) used in a wide range of spectroscopy measurements. Moreover, generalization of Fraunhofer diffraction to scattering by a three-dimensional arrangement of scatterers (Figure 1.4.7), such as atoms in crystals, forms the basis of diffraction methods (X-rays, §7, and electrons and neutrons, §8) in materials characterization. In contrast, Fresnel diffraction is observed when either the source or detector is close enough that the

[17] Jean-Robert Argand (1768–1822) was a Swiss/French amateur mathematician.

curvatures of the wave front becomes significant. From a practical point of view, Fresnel diffraction finds application in the design of zone plates used in focusing high-energy X-rays for microscopy at synchrotron radiation facilities.

Optical microscopy relies on visual observation and works best when the instrumentation is tailored to the characteristics of the human eye. The latter is sensitive to the visible spectrum ($\sim$380 nm $< \lambda <$ 700 nm), but with peak sensitivity for green light ($\lambda = 550$ nm). The Rayleigh criterion defines the resolution for any imaging system. For the human eye this resolution is $\sim$0.1mm at the optimal wavelength. Further, matching the resolution of the eye with that of an optical microscope, it can be shown that there is an upper limit (500$\times$) for the magnification of the microscope. Lenses, in general, used in all microscopy suffer from three major defects—spherical and chromatic aberrations, and astigmatism—that limit their performance.

Optical microscopes work in transmission or reflection modes; the latter is particularly useful for studying opaque specimens. In addition to direct illumination producing bright field images, microscopes can be operated under oblique and dark field imaging modes, or to produce interference contrast, or imaging with polarized light. A newer variation of optical imaging is confocal scanning microscopy, where the light (probe) is focused to a single point at a specific depth in the specimen, and then scanned to produce a complete image. Metallography, widely used to examine materials by optical microscopy, provides a wealth of information on the materials microstructure. It requires that the surfaces be mechanically polished, and if necessary, chemically etched to provide optimal contrast.

Finally, the polarization state of light reflected (or transmitted) from the surface of a specimen can be studied to obtain details of the optical properties of the material. Known as ellipsometry, this technique finds wide use in the characterization of materials, particularly thin films.

....

FURTHER READING

Attwood, D. *Soft X-Rays and Extreme Ultraviolet Radiation.* Cambridge: Cambridge University Press, 2000.

Brandon, D., and W. D. Kaplan. *Microstructural Characterization of Materials*, 2nd ed. Chichester: Wiley, 2008.

Flewitt, P. E. J., and R. K. Wild. *Physical Methods for Materials Characterization.* Boca Raton: IoP Press, 2003.

Fowles, G. R. *Introduction to Modern Optics*, 2nd ed. New York: Dover, 1989.

Fujiwara, H. *Spectroscopic Ellipsometry: Principles and Applications.* Hoboken: John Wiley & Sons, 2007.

Hecht, E. *Optics.* Reading: Addison Wesley, 2001.

Jenkins, F. A., and H. E. White. *Fundamentals of Optics.* Auckland: McGraw-Hill, 1981.

Sheppard, C. J. R., and D. M. Shotton. *Confocal Laser Scanning Microscopy.* Oxford: BIOS Scientific Publishers, 1997.

Vander Voort, G. F. *Metallography: Principles and Practice.* New York: McGraw-Hill, 1984.

Williams, D. B., A. R. Pelton, and R. Gronsky, eds. *Images of Materials.* Oxford: Oxford University Press, 1991.

···

REFERENCES

[1] Wilson, T., and D. K. Hamilton. "Dynamic Focusing in the Confocal Microscope." *Journal of Microscopy* 128, no. 2 (1982): 139–43.

[2] Rothen, A. "The Ellipsometer, an Apparatus to Measure Thicknesses of Thin Surface Films." *Review of Scientific Instruments* 16 (1945): 26.

[3] Downloaded from OSA Publishing (2019).

[4] Paterson, S. M., Y. S. Casadio, D. H. Brown, J. A. Shaw, T. V. Chirila, and M. V. Baker. "Laser Scanning Confocal Microscopy Versus Scanning Electron Microscopy for Characterization of Polymer Morphology: Sample Preparation Drastically Distorts Morphologies of Poly(2-Hydroxyethyl Methacrylate)-Based Hydrogels." *Journal of Applied Polymer Science* 127, no. 4 (2013): 4296–4304.

···

EXERCISES

A. Test Your Knowledge

Identify ALL the correct statements for each statement.

1. All light waves
   (i)   vibrate along the direction of propagation.
   (ii) vibrate in a plane normal to the direction of propagation.
   (iii) are transverse waves.

2. The addition of *any* two waves with displacements in orthogonal planes but propagating in the same direction
   (i)   *always* gives rise to polarized light.
   (ii) gives rise to linearly polarized light if their phase difference is $0°$.
   (iii) gives rise to circularly polarized light if their phase difference is $\pi/2$.

3. Huygens principle states that
   (i)   light travels in straight lines.
   (ii) light is bent when it hits an aperture.
   (iii) every point on the advancing wave front of light is a new source of light waves.

4. In the Young's double slit experiment, we observe
    (i) an interference pattern.
    (ii) a pattern that depends on the phase difference between two waves.
    (iii) intensity maxima that satisfy $m\lambda = d\sin\theta$.
5. The refractive index of a medium
    (i) is the ratio of the speed of light in vacuum to the speed of light in the medium.
    (ii) depends on the wavelength of light in the medium.
    (iii) can cause total internal reflection.
6. (i) Fresnel and Fraunhofer diffraction are the same.
    (ii) In Fresnel diffraction the source of light is at a finite distance.
    (iii) In Fraunhofer diffraction BOTH the source of light and the screen are at infinite distance.
    (i) In Fresnel diffraction a spherical wave front is incident on the object.
    (ii) In Fraunhofer diffraction a uniform plane wave is incident on the object.
7. For Fraunhofer diffraction through a slit
    (i) the spacings of the minima are proportional to the size of the slit.
    (ii) the spacings of the minima are *inversely* proportional to the size of the slit.
    (iii) the spacing of the observed minima in two perpendicular directions are the same for a *rectangular slit*.
8. An Airy disc
    (i) is light and floats in air.
    (ii) is the bright central disc observed in Fraunhofer diffraction when a _____ aperture is used.
    (iii) is always observed in Fresnel diffraction.
9. A zone plate
    (i) is used to focus X-rays.
    (ii) If used to focus X-rays, satisfy the object and image distance given by the standard lens formula, $\frac{1}{a} + \frac{1}{b} = \frac{1}{f}$.
    (iii) in practice is made of alternating rings of W and Si.
10. The human eye
    (i) is uniformly sensitive to all wavelengths of visible light.
    (ii) has a peak sensitivity to green light.
    (iii) has a variable integration time.
    (iv) has an integration time of 0.1 s.
    (v) when dark adapted can form images with 100 photons/pixel.
11. The light microscope to be used by human eyes
    (i) should have the highest magnification possible.
    (ii) should have an optimal magnification of 300–400×
    (iii) with a magnification, M > 1000, would not be particularly useful.

12. The best lenses have
    (i)   no defects.
    (ii) chromatic and spherical aberrations.
    (iii) a depth of field and a depth of focus that are of the same magnitude.

13. The optical microscope
    (i)   is a versatile tool for materials research and development.
    (ii) is regularly used in metallography.
    (iii) has *only two* components: the illumination and imaging systems.
    (iv) has only vertical illumination.

14. The optical microscope can produce images with
    (i)   oblique incidence.
    (ii) interference contrast.
    (iii) polarized light.

15. Confocal microscopy
    (i)   is the optical analog of an SEM.
    (ii) has an infinite depth of field.
    (iii) causes significant blurring of images.

16. In metallography
    (i)   the quality of images can be improved by suitable etching of the specimen.
    (ii) the quality of images can be improved by polishing, especially for dark field illumination.
    (iii) it is best to first examine the specimen after polishing but before etching.

17. Ellipsometry
    (i)   is an optical method to determine optical properties of materials.
    (ii) measures changes in the polarization of light upon reflection from a surface.
    (iii) involves s- and p-polarized light.
    (iv) when necessary, uses a half-wave plate.
    (v)   is a curve fitting method.

18. The resolving power of a grating
    (i)   depends on the size of the rulings.
    (ii) is independent of the spacing of the rulings.
    (iii) depends on the total number of rulings.
    (iv) depends on the diffraction order.

B.   Problems

1. In the **Fresnel diffraction** geometry if you place a small circular obstacle (opaque) why do you observe a bright spot at the center (along the optic axis) of the shadow?

2. Using **Fraunhofer diffraction**, a grating with a large number of slits can be used to measure the wavelength of an unknown radiation accurately. Why?

3. **Refraction:** Show that when an incident wave (ray) passes through a medium, such as glass, of well-defined thickness and limited by plane parallel slides, the emergent wave is parallel to the incident ray.

4. **Resolving power of a lens:** A lens has a diameter of 4 cm and a focal length of 40 cm. It is illuminated by a beam of monochromatic light with $\lambda = 560$ nm.

   (a) What is the radius of the central disc observed in the diffraction pattern at the plane of focus?

   (b) What is the resolving power of this lens at this wavelength?

5. **Resolving power of a grating:**

   (a) Sodium has two yellow lines with wavelengths, $\lambda_1 = 589.0$ nm and $\lambda_2 = 589.6$ nm in the visible spectrum. To what transitions from the energy levels of sodium, discussed in §2, do these lines correspond?

   (b) Suppose you have a grating with 20,000 lines and a length of 0.04 m. Can this grating resolve the two yellow lines of sodium?

6. The **Fraunhofer pattern** for an ideal grating with N slits was given as

$$I \approx A_0^2 \frac{\sin^2 \beta}{\beta^2} \frac{\sin^2 \gamma}{\gamma^2} \quad \text{where } \beta = (\pi w \sin \theta)/\lambda \text{ and } \gamma = (\pi d \sin \theta)/\lambda$$

$$(6.9.19)$$

Make a qualitative sketch of the intensity pattern for FIVE equally spaced slits with $d/w = 4$. Label several points on the x-axis with corresponding values of β and γ.

7. **Fresnel Diffraction:** Consider a small hole, 1 mm in diameter, on an opaque screen illuminated by light of $\lambda = 590.0$ nm. We define the farthest point of darkness (FPD) as the point at which only two Fresnel zones are within the aperture. Now:

   (a) Calculate the distance along the optic axis from the screen to the FPD.

   (b) Would the FPD distance change with order? How?

8. A given point is vibrating with **simple harmonic motion**. Its period is 5.0 s and its amplitude is 0.03 m. If the initial phase angle is $\pi/3$ radians, find

   (a) the initial displacement.

   (b) the displacement after 12.0 s.

   (c) Plot this in a graph.

9. **Ellipsometry:** Light of wavelength, $\lambda = 600$ nm, is incident on a quartz crystal film, placed in air, at an incident angle of 45°. Calculate the reflection coefficient for s- and p-polarized light. What will these coefficients be for a Brewster angle incidence?

10. **Addition of simple harmonic motion at right angles** in the y- and z-planes. Consider the two-component motion given by

$$\frac{y}{a_1} = \sin(\omega t - \alpha_1) \qquad (\text{Ex6.10.1})$$

and

$$\frac{z}{a_2} = \sin(\omega t - \alpha_2) \qquad (\text{Ex6.10.2})$$

(a) Expand equation (Ex6.10.1) to give (Ex6.10.3).
(b) Expand equation (Ex6.10.2) to give (Ex6.10.4).
(c) Multiply (Ex6.10.3) by $\sin \alpha_2$ to give (Ex6.10.5).
(d) Multiply (Ex6.10.4) by $\sin \alpha_1$ to give (Ex6.10.6).
(e) Subtract (Ex6.10.5) from (Ex6.10.6) to give (Ex6.10.7).
(f) Similarly, multiply (Ex6.10.3) by $\cos \alpha_2$ to give (Ex6.10.8).
(g) Multiply (Ex6.10.4) by $\cos \alpha_1$ to give (Ex6.10.9).
(h) Subtract (Ex6.10.9) from (Ex6.10.8) to give (Ex6.10.10).
(i) Now, square both (Ex6.10.7) and (Ex6.10.10) to give (Ex6.10.11) and (Ex6.10.12).
(j) Add (Ex6.10.11) and (Ex6.10.12) and show that this gives the solution

$$\sin^2(\alpha_1 - \alpha_2) = \frac{y^2}{a_1^2} + \frac{z^2}{a_2^2} - \frac{2yz}{a_1 a_2}\cos(\alpha_1 - \alpha_2) \qquad (\text{Ex6.10.2})$$

11. Compare the **resolution and depth of field** of an optical and a transmission electron microscope. Assume that we are using *blue light* for the optical microscope. For the electron microscope assume an acceleration voltage of 200 kV and the angle $\alpha = 5°$. Make any other relevant assumptions and state them clearly.

12. Consider the following two **waves propagating on orthogonal planes** $\mathbf{E}_x = \hat{\mathbf{i}}a_1 \sin(kz - \omega t)$ and $\mathbf{E}_y = \hat{\mathbf{j}}a_2 \sin(kz - \omega t + \delta)$ where δ is the relative phase difference between them.

(a) If $\delta = 0$ or $m\pi$, where m is an integer, show that the superposition of these two waves gives rise to linear polarization for both even and odd values of m.

(b) If $\delta = -\pi/2 + 2m\pi$, where $m = 0, \pm 1, \pm 2, \ldots$, and $a = a_1 = a_2$, what is the resultant wave? What happens to the resultant wave, at some point z_0, as a function of time? Hint: Use $t = 0$, and $t = kz_0/\omega$ to discuss your answer.

(c) If $\delta = \pi/2 + 2m\pi$, where $m = 0, \pm 1, \pm 2, \ldots$, and $a = a_1 = a_2$, what is the resultant wave at some point z_0, as a function of time? What is the difference between (b) and (c)?

13. Read the paper by Paterson et al. [4] and compare **laser scanning confocal microscopy** with scanning electron microscopy (§10), especially in the context of characterizing the morphology of hydrogels.

14. A thin film of copper, with refractive index, $n_{Cu} = 0.95$, placed in air ($n_{air} = 1.0$), is investigated with a polarized laser of wavelength, $\lambda = 550$ nm. For an incident angle, $\theta_i = 45°$, in air, calculate:
    (a) the transmitted angle, θ_t, in Cu.
    (b) the amplitude reflection coefficient for p-polarized (r_p) and s-polarized (r_s) light.
    (c) the amplitude ratio (ρ).
    (d) the difference $\Delta r = |r_s - r_p|$.
    (e) Now calculate and plot the variation of Δr as a function of θ_i for $0 < \theta_i < 90°$.
    (f) The **Brewster angle** is defined as the value of θ_I when Δr is a maximum. Determine the Brewster angle from your plot.
    (g) Compare with the estimate of the Brewster angle given by (6.9.14) and comment on the result.

7

X-Ray Diffraction

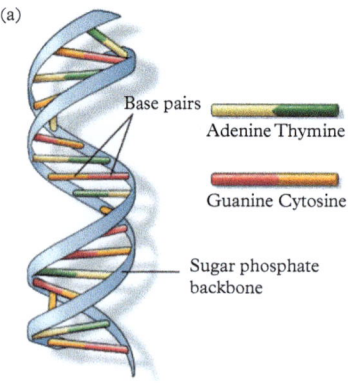

(a)

Base pairs

Adenine Thymine

Guanine Cytosine

Sugar phosphate backbone

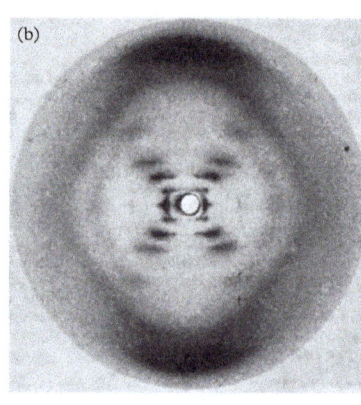

(b)

The "double helix" structure (a) of DNA, first proposed by Crick and Watson [1], is now universally accepted. However, this X-ray diffraction photograph (b) by Franklin and Gosling [2] of a DNA fiber containing many millions of aligned DNA strands was absolutely crucial in establishing the correctness of the structural model proposed by Crick and Watson. The two arms of spots are characteristic of the helical structure, and the angle between them represents the ratio of the width of the molecule to the repeat distance of the helix. The fourth spot along each arm is missing, which indicates that the two helices are intertwined.

(Figure credit: US National Library of Medicine).

Principles of Materials Characterization and Metrology. Kannan M. Krishnan, Oxford University Press (2021).
© Kannan M. Krishnan. DOI: 10.1093/oso/9780198830252.003.0007

7.1   Introduction

In this chapter, we turn our attention to one of the important and widely used methods of materials characterization: X-ray diffraction. Before one can tackle this subject in some detail, it is important to have a good understanding of the elementary principles of diffraction and crystallography introduced in §4. This includes concepts of the reciprocal lattice (§4.2), Bragg's law and the Laue condition (§4.3) for diffraction, as well as optics, especially the description of simple harmonic waves (§6.2) and optical diffraction (§6.6). All these topics have been presented earlier.

Here, we make our way systematically by first describing the interaction of X-rays with electrons, then their scattering by atoms to define the atomic scattering factor, followed by their scattering by the unit cell to develop the concept of structure factors, and finally, their relationship to the observed diffraction intensities. It then describes various methods of X-ray diffraction popular in laboratory settings for both powder and single crystal specimens.

The chapter describes the main factors influencing X-ray diffraction intensities, including temperature, absorption, multiplicity, and Lorentz polarization, and then introduces typical applications of X-ray diffraction in materials characterization: accurate measurement of lattice parameters, grain size, and lattice strain measurements; identification of crystallographic phases and the refinement of their structures; and order-disorder phase transitions. At the end, the chapter illustrates the versatility of X-ray diffraction with brief descriptions of *in situ* measurements in a synchrotron, and a miniaturized instrument sent to Mars by NASA to analyze the structure of minerals on the Martian surface.

It is important to emphasize that X-ray diffraction finds widespread use not only in materials sciences and engineering (MSE), but also in physics, chemistry, biology, geology, art conservation, etc., and as such it is an invaluable method. Further, a good understanding of X-ray diffraction will make it easy to appreciate related methods of electron and neutron diffraction, which are discussed in Chapter 8.

There are far too many books on X-ray diffraction, each treating the subject from a different perspective. In writing this chapter, I have particularly benefitted from Cullity (1978), de Graef and McHenry (2007), Giacovazzo et al. (2007), Hammond (2006), Klug and Alexander (1974), Schwartz and Cohen (1987), Warren (1990), and Woolfson (1997). Many more details that cannot be included in a single chapter are to be found in these comprehensive texts on X-ray diffraction, and the reader is encouraged to consult them as appropriate.

7.2   Interaction of X-Rays with Electrons

When an X-ray beam interacts with an atom two processes are possible. X-ray photons may be absorbed with either the emission of a photoelectron (§2.3) forming the basis of X-ray photoemission spectroscopy (§3.8), or following an inner-shell ionization the emission of a characteristic X-ray photon (§2.5.2.1) forming the basis of X-ray fluorescence spectroscopy (XRF). Alternatively, the X-ray beam may be scattered (§1.4.2), and such scattering may also include two different components.

First, we observe an unmodified scattered wave with the same wavelength as the primary beam of incident X-rays. This is referred to as Thomson[1] scattering and can be explained classically as follows. The primary X-ray beam is an electromagnetic wave (§6.2) with the electric vector, $\mathbf{E}$, at any point varying sinusoidally with time, and directed in a plane orthogonal to the direction of propagation. When this wave encounters an electron—a charged particle—the time-varying electric field of the incident wave will exert a force on the electron and set the latter also in oscillation around its mean position. This causes the electron to be accelerated and decelerated, and in classical electromagnetism such a repetitively accelerated charge will emit an electromagnetic wave. We refer to this absorption of the incident wave and its re-emission at the same wavelength as *scattering*. Further, we observe that this scattering of X-rays by electrons is *coherent*, by which we mean that there is a definite relationship between the phase of the scattered and incident X-ray waves. In the case of scattering of X-rays by electrons, this phase shift is $\lambda/2$; since all the electrons of the atom scatter with the same phase-shift, the scattered X-ray waves are coherent. Finally, even though the X-rays are scattered by an electron in all directions, the intensity of the scattered wave depends only on the angle of scattering.

Second, we also observe a modified scattered wave, with a wavelength *longer* than that of the incident X-rays. This Compton[2] scattering is incoherent with the incident wave (i.e. the phase relationship between them is random), and can only be explained in a quantum mechanical framework, including the principle of wave-particle duality introduced in §1.3.3. Now, the incident X-ray beam is considered to be a stream of X-ray quanta (photons), each of energy, $E_1 = h f_1$, and their interactions with the electron is best described in terms of billiard ball-like elastic collisions with a conservation of both energy and momentum. In this scenario, it can be shown that the change of wavelength of the scattered X-rays is independent of the wavelength of the incident radiation, but is dependent only on the scattering angle.

We now describe both the classical Thomson (coherent) and Compton (incoherent) scattering in detail, and discuss their relevance to X-ray diffraction from materials.

7.2.1   Thomson Coherent Scattering

We consider an electron at the origin, O, interacting with an *unpolarized* primary X-ray beam of intensity, I_0, traveling along the x-axis (Fig. 7.2.1). Our goal is to determine the intensity, $I_{2\theta}$, of the scattered wave at the position, P, at a distance, R, from the electron, and at an angle, 2θ, from the x-axis. Note that the point, P, is in the xy-plane. Further, since the incident beam is *unpolarized*, its electric field, $\mathbf{E}$,

[1] Sir J. J. Thomson (1856–1940) was an English physicist who received the 1906 Nobel prize in Physics and cited for "his theoretical and experimental investigations on the conduction of electricity by gases".

[2] A. H. Compton (1892–1962) was a US physicist who received the 1927 Nobel prize in Physics and cited for "discovering the effect named after him".

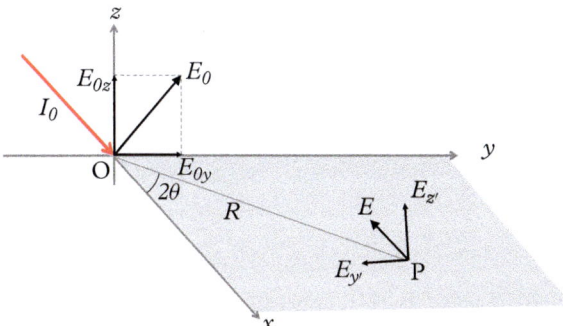

Figure 7.2.1 Classical scattering of an incident electromagnetic radiation of intensity, I_0, by an electron at O. The components of the electric vector of the scattered and incident radiations are shown.

can take all possible orientations in the yz-plane with equal probability. Without loss of generality, we first consider any one orientation, $\mathbf{E}_0$, with components E_{0y} and E_{0z}, as shown, and later average it over all directions.

We also know from the theory of electromagnetism that a charge, e.g. an electron, at position, O, subject to an acceleration, $\mathbf{a}$ (Fig. 7.2.2) will generate an electromagnetic radiation at any point, P, with an electric vector, $\mathbf{E}$, of amplitude

$$E = \frac{ea}{4\pi\varepsilon_0 Rc^2} \sin\alpha \tag{7.2.1}$$

where c is the velocity of light. The direction of E is always perpendicular to OP and lies in the plane of OP and $\mathbf{a}$.

When an X-ray beam is incident on the electron (Fig. 7.2.1), its electric field component will set the electron in oscillation, with acceleration components

$$a_z = \frac{E_{0z}e}{m} \tag{7.2.2a}$$

and

$$a_y = \frac{E_{0y}e}{m} \tag{7.2.2b}$$

As a result, at any point, P, applying (7.2.1), we find the electric vector components of the scattered wave as

$$E_{z'} = \frac{e^2}{4\pi\varepsilon_0 Rc^2 m} E_{0z} \tag{7.2.3a}$$

and

$$E_{y'} = \frac{e^2}{4\pi\varepsilon_0 Rc^2 m} E_{0y} \cos 2\theta \tag{7.2.3b}$$

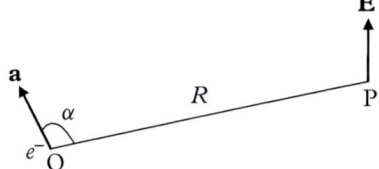

Figure 7.2.2 The relationship of a scattered electromagnetic wave at position, P, arising from the acceleration of an electron at O. Both vectors, $\mathbf{a}$ and $\mathbf{E}$, are in the same plane. Simply put, the component, $a \sin\alpha$, of the acceleration seen by an eye placed at the point, P, of observation will determine the electric field produced.

Then, the net amplitude, E, of the scattered radiation at the point, P, is given by

$$E^2 = E_{z'}^2 + E_{y'}^2 = \left(\frac{e^2}{4\pi\varepsilon_0 Rc^2 m}\right)^2 \left(E_{0z}^2 + E_{0y}^2\cos^2 2\theta\right) \tag{7.2.4}$$

Now, since the initial beam is unpolarized, we take directional averages, i.e.

$$\left\langle E_0^2\right\rangle = \left\langle E_{0y}^2\right\rangle + \left\langle E_{0z}^2\right\rangle \tag{7.2.5}$$

Further, since the y- and z-axis are equivalent, we get

$$\left\langle E_{0y}^2\right\rangle = \left\langle E_{0z}^2\right\rangle = \frac{1}{2}\left\langle E_0^2\right\rangle \tag{7.2.6}$$

Thus, (7.2.4) can be modified as

$$\left\langle E^2\right\rangle = \left\langle E_0^2\right\rangle \left(\frac{e^2}{4\pi\varepsilon_0 Rc^2 m}\right)^2 \frac{\left(1 + \cos^2 2\theta\right)}{2} \tag{7.2.7}$$

Since the observable quantity is the intensity, $I_{2\theta}$, of the scattered radiation we can set $\left\langle E_0^2\right\rangle = I_0$ and rewrite (7.2.7) to obtain the intensity of the classical scattering of an electromagnetic wave by a free electron as

$$I_{2\theta} = I_0\left(\frac{e^2}{4\pi\varepsilon_0 Rc^2 m}\right)^2 \frac{\left(1 + \cos^2 2\theta\right)}{2} = I_0\left(\frac{\kappa}{R}\right)^2 \frac{\left(1 + \cos^2 2\theta\right)}{2} \tag{7.2.8}$$

This is the classical Thomson scattering equation. The quantity, $\kappa \left(= \frac{e^2}{4\pi\varepsilon_0 c^2 m}\right)$, with the dimension of length, equals 2.82×10^{-15} m, and in classical electromagnetic theory is referred to as the radius of the electron. Note that the scattered intensity ratio, $I_{2\theta}/I_0$, at a distance of a few centimeters from the electron, is of the order of 10^{-26}. Even though this intensity ratio is very small, in practice the number of electrons even in a very small quantity (mg) of the specimen is $\sim 10^{20}$, and hence we get a measurable signal when all their contributions are combined. Further, even the lightest nucleus (a proton) has a mass $\sim 1,840$ times that of the electron. Since the scattered intensity is inversely proportional to the square of the mass, we expect only the electrons to effectively scatter the X-rays. In other words, even though the nucleus has a charge, because of its very high mass, it contributes negligibly to X-ray scattering. The factor $\left(1 + \cos^2 2\theta\right)/2$ in (7.2.8) is referred to as the polarization factor; this nomenclature is ironic as it arises because of our assumption of an unpolarized incident beam. Finally, as mentioned earlier, such Thomson scattering by electrons is coherent, with a uniform phase different of π between the scattered and incident radiation.

We are ultimately interested in the scattering of X-rays by atoms; in §7.3, we describe how the coherent scattering of the individual electrons are combined to describe the intensity of scattering by the atom, and later in §7.4, by the periodic arrangements of atoms in a crystalline unit cell.

Example 7.2.1: How can Thomson scattering be used to polarize and analyze X-rays?

Solution: We use the simple rule illustrated in Figure 7.2.2 and consider a beam of unpolarized X-rays incident along the X-axis on a scatterer (block of Carbon) at position, O, in the figure below.

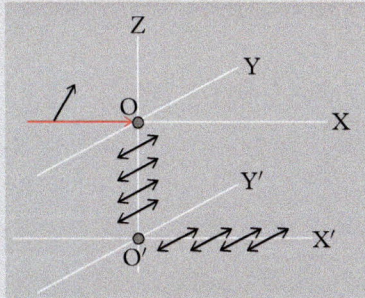

The electrons in the scatterer are accelerated in all directions in the Y–Z plane. However, if we place an eye at the point O′, we will only see the y-component of the electron acceleration. Thus, any X-rays scattered in the direction of O–O′, i.e. through a scattering angle of 90°, will be linearly polarized (i.e. produces a polarized beam).

Now, if we place another scatterer at O′, all its electrons will be accelerated only along the Y′ direction. If we place the eye at Y′ we will see no component of the Y′-acceleration, but at X′ we will see a maximum.

In summary, the first scattering of 90° at O produces a linearly polarized beam, and the second scattering at 90° at O′ serves as the analyzer. Adapted from Warren (1990).

7.2.2 Compton Incoherent Scattering

To describe Compton scattering, it is best to invoke the quantum mechanical principle of wave–particle duality (§1.3.3), and describe the elastic[3] interaction as a collision of an X-ray photon and an electron. Figure 7.2.3a shows the geometry of the scattering, with the incident photon traveling along AO, interacting with the electron at O, and after the collision moving along OP with a small change in wavelength, $\Delta \lambda$; the electron recoils and moves along OE.

[3] In some texts, Compton scattering is called an inelastic collision. Here, they are referring only to the fact that the energy of the incident and scattered photons is different, and ignore the kinetic energy of the recoil electron.

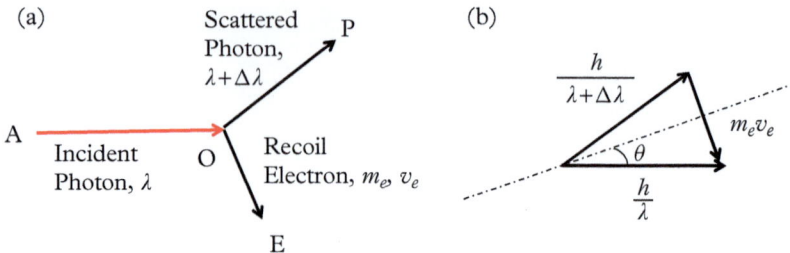

Figure 7.2.3 (a) The geometry of scattering of the X-ray photon and an electron. (b) The momentum vector diagram involved in Compton scattering.

For an elastic collision, the total energy is conserved. This gives

$$\frac{hc}{\lambda} = \frac{hc}{\lambda + \Delta\lambda} + \frac{1}{2}m_e v_e^2 \qquad (7.2.9)$$

or, by rearranging terms

$$\frac{hc}{\lambda^2}\Delta\lambda = \frac{1}{2}m_e v_e^2 \qquad (7.2.10)$$

Further, the momentum (Fig. 7.2.3b) is also conserved in the elastic collision. This gives

$$\frac{h}{\lambda}\sin\theta = \frac{1}{2}m_e v_e \qquad (7.2.11)$$

Eliminating v_e between the two equations (7.2.10) and (7.2.11) gives

$$\Delta\lambda = \frac{2h}{m_e c}\sin^2\theta = \frac{h}{m_e c}(1 - \cos 2\theta) = 0.024\,(1 - \cos 2\theta)\,\text{Å} \qquad (7.2.12)$$

where the physical constants have been replaced with their actual values.

Note that the change in the wavelength, $\Delta\lambda$, associated with Compton scattering is independent of the initial wavelength, and the maximum value of $\Delta\lambda$ is expected for $2\theta = \pi$, with a value, $\Delta\lambda = 0.048$, which is quite significant for the scattering of X-ray photons with $\lambda \sim 1$ Å. However, unlike Thomson scattering, the phase of the Compton-scattered wave has no relationship to the incident wave. As a result, it contributes *insignificantly* to the diffraction intensity; but it does contribute some background intensity in diffraction patterns. In general, both Thomson and Compton scattering take place, and one must resort to a complete Quantum Mechanical theory to accurately account for the coherent and incoherent scattering contributions. However, even in such a full Quantum Mechanical treatment, the total intensity of the scattered wave, per electron, taking both Thomson and Compton contributions into account, is closely equal to the value given by the classical Thomson equation (7.2.8).

7.3 Scattering by an Atom: Atomic Scattering Factor

Let us first consider the classical scattering of an X-ray beam by two electrons confined to a small volume in atomic orbit (Fig. 7.3.1). In the forward direction (Fig. 7.3.1a), $2\theta = 0$, both electrons will scatter the incident X-ray wave by an identical phase difference of $\lambda/2$. More importantly, at distances sufficiently far from the two electrons the path length of the wave scattered by the two electrons is the same; they arrive *in phase* and hence their *amplitudes* can be added directly to give the scattered wave amplitude. By extension, for an atom with Z electrons (only two are shown in the figure), the forward scattered wave amplitude is Z times the amplitude of the wave scattered by a single electron.

In any other direction, scattering of the incident X-ray by electrons in different positions introduce changes in the phase of the scattered wave because of their path length difference (such as A′BB′, for the two electrons shown in Figure 7.3.1b). The difference, typically less than a wavelength, results in partial coherence of the waves in the scattered direction, at any distance, R, far from the two electrons, causing a decrease in amplitude when compared to the scattered wave in the forward direction. We can represent the scattering conditions in vector form (Fig. 7.3.1c). For the point, P, at a distance, R, far from the electrons, the primary and scattered wave directions can be represented by the unit vectors, $\hat{\mathbf{s}}_0$ and $\hat{\mathbf{s}}$, respectively. Then, the total path length, $l_1 + l_2$, for the scattered plane wave from any electron at $\mathbf{r}_n$, observed at the position, P, is

$$l_1 + l_2 = \mathbf{r}_n \cdot \hat{\mathbf{s}}_0 + R - \mathbf{r}_n \cdot \hat{\mathbf{s}} = R - (\hat{\mathbf{s}} - \hat{\mathbf{s}}_0) \cdot \mathbf{r}_n \qquad (7.3.1)$$

Following Thomson scattering, the scattered wave in complex notation (see §6.2.4), with frequency, f, can be written as

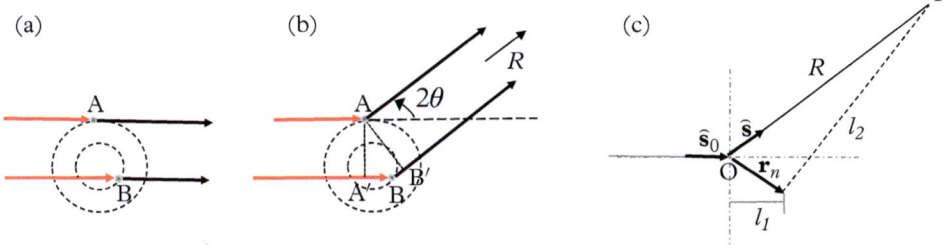

Figure 7.3.1 Scattering of an incident X-ray beam by two representative electrons in an atom in (a) forward scattering direction, and (b) a direction at an angle, 2θ, with a path difference, A′BB′, between the waves scattered by electrons at positions A and B. (c) The scattering geometry in vector form, showing the path length, $l_1 + l_2$, for the wave scattered by the electron at position, $\mathbf{r}_n$, and observed at position, P, at a distance, R, far from O.

$$E = E_0 \frac{\kappa}{R} \exp 2\pi i \left(ft - \frac{(l_1 + l_2)}{\lambda} \right) = E_0 \frac{\kappa}{R} \exp 2\pi i \left(ft - \frac{R - (\hat{\mathbf{s}} - \hat{\mathbf{s}}_0) \cdot \mathbf{r}_n}{\lambda} \right)$$

(7.3.2)

where $\kappa = \frac{e^2}{4\pi\varepsilon_0 c^2 m}$, as defined in (7.2.8). At position, P, the resultant electric field component, E_P, of the scattered wave, for n electrons can be written as the sum of the waves scattered from the individual electrons as

$$E_P = E_0 \frac{\kappa}{R} \exp 2\pi i \left[ft - \frac{R}{\lambda} \right] \sum_n \exp \frac{2\pi i}{\lambda} [(\hat{\mathbf{s}} - \hat{\mathbf{s}}_0) \cdot \mathbf{r}_n]$$

(7.3.3)

Now, instead of assuming that each electron is a localized scatterer of X-rays, we consider that each electron is spread out as a diffuse cloud of charge with a charge density, ρ, expressed in electron units per unit volume. Then, $\rho \, dV$ is the ratio of the charge in any volume unit, dV, normalized to the charge of one electron. Thus, for a single electron, $\int \rho \, dV = 1$. To get the contribution for the scattering from a single electron, we must integrate over the volume, dV, making sure that we account for the phase difference for scattering from each charge element dV. Then the sum in (7.3.3) is replaced by the integral

$$E_P = E_0 \frac{\kappa}{R} \exp 2\pi i \left[ft - \left(\frac{R}{\lambda} \right) \right] \int \exp \frac{2\pi i}{\lambda} [(\hat{\mathbf{s}} - \hat{\mathbf{s}}_0) \cdot \mathbf{r}] \rho dV$$

(7.3.4)

where the individual electrons at positions, $\mathbf{r}_n$, are replaced by charge elements ρdV at $\mathbf{r}$. The quantity represented by the integral

$$f_e = \int \exp \frac{2\pi i}{\lambda} [(\hat{\mathbf{s}} - \hat{\mathbf{s}}_0) \cdot \mathbf{r}] \rho dV$$

(7.3.5)

is called the electron scattering factor, and represents the ratio of scattering by an electron with distributed charge normalized by the scattering from a single electron in the classical theory.

As a first approximation, we will assume that the charge distribution for each electron in the atom is spherically symmetric, i.e. $\rho = \rho(r)$. Then, referring to the geometry in Figure 7.3.2, we can show that $(\hat{\mathbf{s}} - \hat{\mathbf{s}}_0) \cdot \mathbf{r} = 2 \sin \theta r \cos \phi$, and $dV = 2\pi r^2 \sin \phi d\phi dr$. We can now rewrite (7.3.5) as

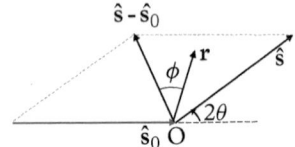

Figure 7.3.2 For the atom centered at O, the relationship between $(\hat{\mathbf{s}} - \hat{\mathbf{s}}_0)$ and $\mathbf{r}$.

$$f_e = \int_{r=0}^{\infty} \int_{\phi=0}^{\pi} \exp \left[i \frac{4\pi \sin \theta}{\lambda} r \cos \phi \right] \rho(r) 2\pi r^2 \sin \phi d\phi dr$$

(7.3.6)

We now set $s = \frac{\sin \theta}{\lambda}$, and carry out the integral over ϕ. Thus

$$f_e = \int_{r=0}^{\infty} r^2 \frac{\sin 4\pi sr}{sr} \rho(r) dr$$

(7.3.7)

Finally, for an atom with n electrons, we add their individual contributions to get

$$f_{atom} = \sum_n f_{e_n} = \sum_n \int_{r=0}^{\infty} r^2 \frac{\sin 4\pi sr}{sr} \rho_n(r) dr \qquad (7.3.8)$$

The simple number, f_{atom}, is called the *atomic scattering factor* and represents the ratio

$$f_{atom} = \frac{\text{ampitude of the wave scattered by the atom}}{\text{amplitude of the wave scattered by a single electron}} \qquad (7.3.9)$$

It can be computed if the radial distribution of the electron density in the atom, $\sum_n \rho_n(r)$ is known. Note that f_{atom} is a function of the parameter s, and hence it varies as a function of $\frac{\sin\theta}{\lambda}$. Figure 7.3.3a shows typical plots of f_{Au} and f_{Cu}. It is clear from the plot that for both elements, $f_{atom}(0) = Z$, where the number of electrons is equal to the atomic number, Z. In other words, for each element the dimensionless atomic scattering factor begins at Z for forward scattering ($\theta = 0$) and then decreases with scattering angle to a very low value for backward scattering ($2\theta = \pi$); it also decreases with decreasing wavelength, λ, of the incident X-rays.

Figure 7.3.3b shows the atomic scattering factor for O^{2-}, Ne, Si^{4+}, all three containing 10 electrons ($Z = 10$). Again, for all three ions, $f(0) = Z = 10$, but then the scattering factor falls off below 10 for larger values of scattering angles. However, the rate at which they fall with angle is dependent on the relative size of the atom or ion. Of the three, O^{2-} is the largest in size and hence the phase difference for scattering between the electrons is largest for large scattering angles, leading to the most destructive interference. At the other end, Si^{4+} is smallest in size, with Ne in between, and this is also reflected in the angular dependence.

Tables for atomic scattering factors for all the elements are readily available in two forms. In many textbooks, they are listed in tabular form as a function of $\frac{\sin\theta}{\lambda}$, typically in steps of 0.1, from 0 to 1.2. We do the same here in Table 7.3.1. Note that the tabulated values of f_{atom} are strictly valid only when the wavelength of the scattered radiation is much smaller than the absorption edge of the scattering atom; if the two values are close, a small correction called the *anomalous dispersion* correction, must be applied (see Warren, 1990). Alternatively, the atomic scattering factors are parametrized, as a function of $\frac{\sin\theta}{\lambda} (= s)$ in terms of curve fitting parameters, using four exponential functions. In this case, the scattering for any value of s can be computed from

$$f_{atom}(s) = \sum_{i=1}^{4} a_i \exp\left(-b_i s^2\right) + c \qquad (7.3.10)$$

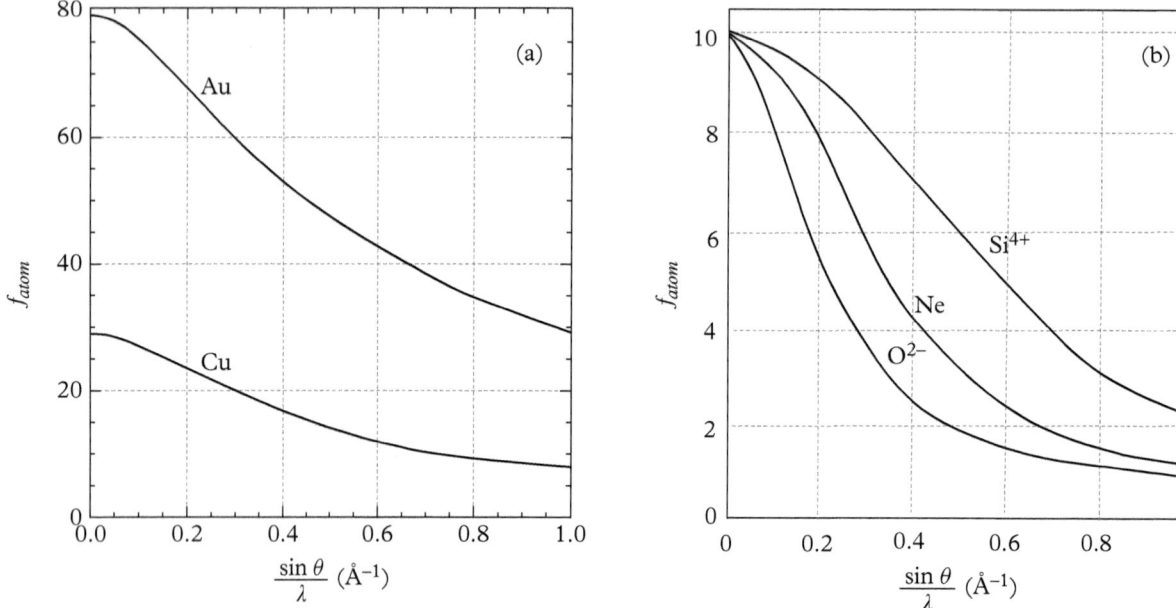

Figure 7.3.3 The atomic scattering factor as a function of $\frac{\sin\theta}{\lambda}$ for (a) two elements, Au and Cu, and (b) for three isoelectronic atoms and ions with 10 electrons.

where the wavelength is expressed in units of Å (= 0.1 nm), s is determined from the diffraction angle, θ_{hkl}, and the corresponding terms are tabulated, as the Cromer–Mann parameters [3], for some representative elements in Table 7.3.2.

We have determined the coherently scattered wave amplitude by considering that the electronic charge distribution in an atom is distributed in space and not localized at a specific point. In quantum mechanics, the wave function, Ψ, describes the state of the electron and is related to the electron distribution, ρ, simply as $\rho = |\Psi|^2$. In the case of a spherically symmetric potential, such as that experienced by an electron in an atom, the wave function can be shown to be a product of a radial component, $R(r)$, and an angular component, $Y(\theta, \phi)$, determining the shape of the orbital, i.e. $\Psi = R(r)\, Y(\theta, \phi)$. Thus, our assumption of spherical symmetry included in the electron distribution, $\rho(r)$, is not unreasonable, especially for a completely filled shell.

Further, electrons occupy discrete energy levels in atoms. Thus, Thomson scattering must result in no change in the energy level of the electron; in other words, the electron must be so tightly bound that no momentum is transferred to it upon impact. In contrast, Compton scattering must involve either well-defined atomic transitions between bound states or the complete ejection of the electron from the atom.

In summary, the complete quantum mechanical treatment shows that the intensities of the incoherent scattering become significant only for large values

Table 7.3.1 X-ray Atomic Scattering Factors (dimensionless numbers). Adapted from Warren (1990)

		$(\sin\theta)/\lambda$ (Å⁻¹)															
		0.0	0.1	0.2	0.3	0.4	0.5	0.6	0.7	0.8	0.9	1.0	1.1	1.2	1.3	1.4	1.5
H	1	1.0	0.81	0.48	0.25	0.13	0.07	0.04	0.02	0.02	0.01	0.01					
He	2	2.0	1.83	1.45	1.06	0.74	0.52	0.36	0.25	0.18	0.13	0.10	0.07	0.05	0.04	0.03	0.03
Li	3	3.0	2.22	1.74	1.51	1.27	1.03	0.82	0.65	0.51	0.40	0.32	0.26	0.21	0.16		
Be	4	4.0	3.07	2.07	1.71	1.53	1.37	1.20	1.03	0.88	0.74	0.62	0.52	0.43	0.37		
B	5	5.0	4.07	2.71	1.99	1.69	1.53	1.41	1.28	1.15	1.02	0.90	0.78	0.68	0.60		
C	6	6.0	5.13	3.58	2.50	1.95	1.69	1.54	1.43	1.32	1.22	1.11	1.01	0.91	0.82	0.74	0.66
N	7	7.0	6.20	4.60	3.24	2.40	1.94	1.70	1.55	1.44	1.35	1.26	1.18	1.08	1.01		
O	8	8.0	7.25	5.63	4.09	3.01	2.34	1.94	1.71	1.57	1.46	1.37	1.30	1.22	1.14		
F	9	9.0	8.29	6.69	5.04	3.76	2.88	2.31	1.96	1.74	1.59	1.48	1.40	1.32	1.25		
Ne	10	10.0	9.36	7.82	6.09	4.62	3.54	2.79	2.30	1.98	1.76	1.61	1.50	1.42	1.35	1.28	1.22
Na	11	11.0	9.76	8.34	6.89	5.47	4.29	3.40	2.76	2.31	2.00	1.78	1.63	1.52	1.44	1.37	1.31
Na⁺	11	10.0	9.55	8.39	6.93	5.51	4.33	3.42	2.77	2.31	2.00	1.79	1.63	1.52	1.44	1.37	1.30
Mg	12	12.0	10.50	8.75	7.46	6.20	5.01	4.06	3.30	2.72	2.30	2.01	1.81	1.65	1.54		
Al	13	13.0	11.23	9.16	7.88	6.77	5.69	4.71	3.88	3.21	2.71	2.32	2.05	1.83	1.69	1.57	1.48
Si	14	14.0	12.16	9.67	8.22	7.20	6.24	5.31	4.47	3.75	3.16	2.69	2.35	2.07	1.87	1.71	1.60
P	15	15.0	13.17	10.34	8.59	7.54	6.67	5.83	5.02	4.28	3.64	3.11	2.69	2.35	2.10	1.89	1.75
S	16	16.0	14.33	11.21	8.99	7.83	7.05	6.31	5.56	4.82	4.15	3.56	3.07	2.66	2.34		
Cl	17	17.0	15.33	12.00	9.44	8.07	7.29	6.64	5.96	5.27	4.60	4.00	3.47	3.02	2.65		
Cl⁻	17	18.0	16.02	12.20	9.40	8.03	7.28	6.64	5.97	5.27	4.61	4.00	3.47	3.03	2.65	2.35	2.11
A	18	18.0	16.30	12.93	10.20	8.54	7.56	6.86	6.23	5.61	5.01	4.43	3.90	3.43	3.03		
K	19	19.0	16.73	13.73	10.97	9.05	7.87	7.11	6.51	5.95	5.39	4.84	4.32	3.83	3.40	3.01	2.71
Ca	20	20.0	17.33	14.32	11.71	9.64	8.26	7.38	6.75	6.21	5.70	5.19	4.69	4.21	3.77	3.37	3.03
Sc	21	21.0	18.72	15.39	12.39	10.12	8.60	7.64	6.98	6.45	5.96	5.48	5.00	4.53	4.09	3.68	3.31
Ti	22	22.0	19.41	16.07	13.20	10.83	9.12	7.98	7.22	6.65	6.19	5.72	5.29	4.84	4.41	4.01	3.64
V	23	23.0	20.47	17.03	14.03	11.51	9.63	8.34	7.48	6.86	6.39	5.94	5.53	5.10	4.71	4.30	3.93
Cr	24	24.0	21.93	18.37	15.01	12.22	10.14	8.72	7.75	7.09	6.58	6.14	5.74	5.34	4.94	4.55	4.18
Mn	25	25.0	22.61	19.06	15.84	13.02	10.80	9.20	8.09	7.32	6.77	6.32	5.93	5.54	5.18	4.80	4.45
Fe	26	26.0	23.68	20.09	16.77	13.84	11.47	9.71	8.47	7.60	6.99	6.51	6.12	5.74	5.39	5.03	4.69
Co	27	27.0	24.74	21.13	17.74	14.68	12.17	10.26	8.88	7.91	7.22	6.70	6.29	5.91	5.58	5.23	4.90
Ni	28	28.0	25.80	22.19	18.73	15.56	12.91	10.85	9.33	8.25	7.48	6.90	6.47	6.08	5.75	5.41	5.09
Cu	29	29.0	27.19	24.63	19.90	16.48	13.65	11.44	9.80	8.61	7.76	7.13	6.65	6.25	5.90	5.57	5.25
Zn	30	30.0	27.92	24.33	20.77	17.42	14.51	12.16	10.37	9.04	8.08	7.37	6.84	6.42	6.07	5.73	5.43
Ga	31	31.0	28.65	24.92	21.47	18.26	15.38	12.95	11.02	9.54	8.46	7.64	7.05	6.58	6.21	5.88	5.58
Ge	32	32.0	29.52	25.53	22.11	19.02	16.19	13.72	11.68	10.08	8.87	7.96	7.29	6.77	6.37	6.02	5.72
As	33	33.0	30.47	26.20	22.69	19.69	16.95	14.48	12.37	10.67	9.34	8.32	7.57	6.98	6.54	6.17	5.86
Se	34	34.0	31.43	26.91	23.24	20.28	17.63	15.20	13.06	11.27	9.83	8.71	7.86	7.21	6.72	6.31	5.99

continued

Table 7.3.1 Continued

$(\sin\theta)/\lambda(\text{Å}^{-1})$		0.0	0.1	0.2	0.3	0.4	0.5	0.6	0.7	0.8	0.9	1.0	1.1	1.2	1.3	1.4	1.5
Br	35	35.0	32.43	27.70	23.82	20.84	18.27	15.91	13.78	11.93	10.41	9.19	8.24	7.51	6.95	6.51	6.16
Kr	36	36.0	33.44	28.53	24.40	21.34	18.82	16.54	14.44	12.57	10.97	9.66	8.62	7.81	7.19	6.70	6.31
Rb	37	37.0	34.11	28.97	24.75	21.29	18.55	16.30	14.47	12.94	11.66	10.58	9.65	8.84	8.14	7.53	6.99
Sr	38	38.0	35.06	29.83	25.51	21.96	19.15	16.84	14.96	13.39	12.07	10.95	9.99	9.16	8.44	7.80	7.24
Y	39	39.0	36.01	30.68	26.28	22.64	19.76	17.39	15.46	13.84	12.48	11.32	10.34	9.48	8.73	8.08	7.50
Zr	40	40.0	36.96	31.54	27.04	23.32	20.37	17.94	15.95	14.29	12.89	11.70	10.68	9.80	9.03	8.36	7.76
Nb	41	41.0	37.91	32.40	27.81	24.01	20.98	18.49	16.43	14.74	13.31	12.08	11.04	10.13	9.33	8.64	8.02
Mo	42	42.0	38.86	33.25	28.57	24.69	21.60	19.04	16.95	15.20	13.73	12.46	11.39	10.45	9.64	8.92	8.29
Tc	43	43.0	39.81	34.12	29.34	25.38	22.21	19.60	17.46	15.65	14.15	12.85	11.74	10.78	9.94	9.21	8.55
Ru	44	44.0	40.76	34.98	30.12	26.07	22.83	20.16	17.96	16.12	14.57	13.24	12.10	11.11	10.25	9.49	8.82
Rh	45	45.0	41.72	35.84	30.89	26.76	23.46	20.72	18.47	16.58	14.99	13.63	12.46	11.45	10.56	9.78	9.09
Pd	46	46.0	42.67	36.70	31.67	27.46	24.08	21.28	18.98	17.05	15.42	14.02	12.82	11.78	10.87	10.07	9.37
Ag	47	47.0	43.63	37.57	32.44	28.16	24.71	21.85	19.50	17.52	15.85	14.42	13.19	12.12	11.19	10.37	9.64
Cd	48	48.0	44.58	38.44	33.22	28.85	25.34	22.42	20.02	17.99	16.28	14.81	13.56	12.46	11.51	10.66	9.92
In	49	49.0	45.5	39.3	34.0	29.6	26.0	23.0	20.5	18.5	16.7	15.2	13.9	12.8	11.8	11.0	10.2
Sn	50	50.0	46.5	40.2	34.8	30.3	26.6	23.6	21.1	18.9	17.2	15.6	14.3	13.2	12.1	11.3	10.3
Sb	51	51.0	47.5	41.1	35.6	31.0	27.2	24.1	21.6	19.4	17.6	16.0	14.7	13.5	12.5	11.6	10.8
Te	52	52.0	48.4	41.9	36.4	31.7	27.9	24.7	22.1	19.9	18.0	16.4	15.1	13.8	12.8	11.9	11.0
I	53	53.0	49.4	42.8	37.1	32.4	28.5	25.3	22.6	20.4	18.5	16.8	15.4	14.2	13.1	12.2	11.3
Xe	54	54.0	50.3	43.7	37.9	33.1	29.2	25.9	23.2	20.9	18.9	17.2	15.8	14.5	13.4	12.5	11.6
Cs	55	55.0	51.3	44.5	38.7	33.8	29.8	26.5	23.7	21.3	19.4	17.7	16.2	14.9	13.8	12.8	11.9
Ba	56	56.0	52.3	45.4	39.5	34.5	30.4	27.0	24.2	21.8	19.8	18.1	16.6	15.3	14.1	13.1	12.2
Ta	73	73.0	68.6	60.4	53.1	46.9	41.7	37.3	33.6	30.4	27.7	25.4	23.3	21.6	20.0	18.6	17.4
W	74	74.0	69.5	61.3	54.0	47.6	42.3	37.9	34.1	30.9	28.2	25.8	23.7	21.9	20.3	18.9	17.7
Re	75	75.0	70.5	62.2	54.8	48.3	43.0	38.5	34.7	31.4	28.7	26.3	24.2	22.3	20.3	18.9	17.7
Os	76	76.0	71.5	63.1	55.6	49.1	43.7	39.1	35.3	32.0	29.1	26.7	24.6	22.7	21.1	19.6	18.3
Ir	77	77.0	72.4	64.0	56.4	49.8	44.4	39.7	35.8	32.5	29.6	27.1	25.0	23.1	21.4	19.9	18.6
Pt	78	78.0	73.4	64.9	57.2	50.6	45.0	40.3	36.4	33.0	30.1	27.6	25.4	23.5	21.8	20.3	18.9
Au	79	79.0	74.4	65.8	58.0	51.3	45.7	41.0	37.0	33.5	30.6	28.0	26.3	24.3	22.5	21.0	19.6
Hg	80	80.0	75.3	66.7	58.8	52.1	46.4	41.6	37.5	34.1	31.1	28.5	26.3	24.3	22.5	21.0	19.6
Tl	81	81.0	76.3	67.6	59.7	52.8	47.1	42.2	38.1	34.6	31.6	29.0	26.7	24.7	22.9	21.3	19.9
Pb	82	82.0	77.2	68.5	60.5	53.6	47.8	42.9	38.7	35.1	32.1	29.4	27.1	25.1	23.3	21.7	20.3
Bi	83	83.0	78.2	69.3	61.3	54.3	48.5	43.5	39.4	36.1	33.1	30.5	25.5	23.6	22.0	22.0	20.6
Th	90	90.0	85.0	75.6	67.1	59.6	53.3	47.9	43.3	39.4	36.1	33.1	30.5	28.3	26.3	24.5	22.9
U	92	92.0	86.9	77.4	68.7	61.1	54.7	49.2	44.5	40.5	37.1	34.0	31.4	29.1	27.0	25.2	23.6

Table 7.3.2 Cromer–Mann atomic scattering parameters for select elements and ions [3].

Element	Atomic #	a_1	a_2	a_3	a_4	b_1	b_2	b_3	b_4	c
C	6	2.3	1.02	1.589	0.865	20.844	10.208	0.569	51.651	0.216
O^{2-}	8	4.758	3.837	0	0	7.831	30.05	0	0	1.594
Na^+	11	3.256	3.936	1.3998	1.0032	2.6671	6.1153	0.2001	14.039	0.404
Si	14	5.7941	3.224	2.428	1.3215	2.5761	34.1775	0.8694	85.341	1.2314
Si^{4+}	14	4.4392	3.2035	1.1945	0.4165	1.6417	3.4376	0.2149	6.6537	0.7463
Cl^-	17	18.292	7.2084	6.5337	2.3386	0.0066	1.1717	19.5424	60.4486	16.378
Cr	24	10.6406	7.3537	3.324	1.4922	6.1038	0.392	20.2626	98.74	1.1832
Fe	26	11.9185	7.0485	3.3433	2.2723	4.8739	0.3402	15.933	79.034	1.40818
Cu	29	13.338	7.1676	5.6158	1.6735	3.5828	0.247	11.3966	64.8126	1.191
Cs^+	55	23.965	21.2204	9.7673	1.6155	0.2045	3.4388	23.4941	49.706	−2.567
Au	79	37.3027	14.9306	10.3425	2.0123	1.0081	6.5255	16.51	76.9117	14.3992

Adapted from Graef and McHenry (2007).

of s (> 1). However, for values of s (< 1) relevant for X-ray diffraction, the classical Thomson coherent scattering is a very good approximation of the observed intensities. Finally, in a crystal, the amplitudes of the coherent scattering from different atoms are added together; on the other hand, if the scattering is incoherent, it is only the intensities that add. As a result, we ignore the incoherent scattering and consider the coherent scattering as a very good approximation of the observed intensities.

Example 7.3.1: Calculate the scattering factor for a sphere of uniform charge density, ρ_0, of radius, R_0. How will the intensity of scattering vary as a function of R_0? Can you use this variation to determine the radius R_0?

Solution:

(a) The scattering factor for a sphere of uniform charge density is given by

$$f = \int_0^{\infty} 4\pi r^2 \rho(r) \frac{\sin sr}{sr} dr = \int_0^{R_0} 4\pi r^2 \rho_0 \frac{\sin sr}{sr} dr$$

$$= \frac{4\pi \rho_0}{(sR_0)^3} [\sin(sR_0) - sR_0 \cos(sR_0)]$$

(b) The intensity of scattering will be proportional to the square of the scattering factor $I \propto [\sin(sR_0) - sR_0 \cos(sR_0)]^2$ and will vary as a function of sR_0 as shown in the following figure.

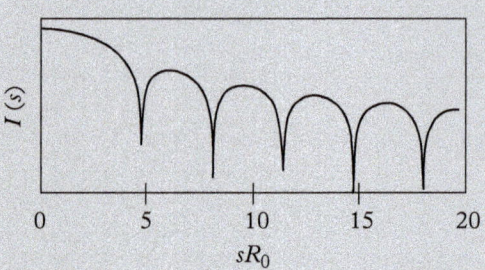

Further, the scattered intensity will be zero when

$$\sin(sR_0) - sR_0 \cos(sR_0) = 0 \text{ or } \tan(s_0 R_0) = s_0 R_0.$$

Thus, from successive minima in the scattering, which correspond to successive roots for s_0, we can determine R_0. Please note that often the symbol q is used in the scientific literature in place of $s = \frac{\sin\theta}{\lambda}$.

7.4   Scattering by a Crystal: Structure Factor

Now that we have described how the X-ray beam is scattered by an atom, our next step is to sum up the contributions from all the atoms in a crystal and specifically determine how they contribute to the diffracted intensity.

Strictly speaking, as the X-ray beam propagates in a crystal it is attenuated due to successive scattering by the atoms. Moreover, beams "reflected" by atomic planes deep in the crystal can be re-reflected, and it is possible that these re-reflected beams interfere destructively with the incident beam. As a result, one needs a "*dynamical*" theory to describe the true interaction of the X-ray beams in a crystal and predict the diffracted intensities. However, if the volume of the overall crystal is small, such dynamical effects can be ignored. Fortunately, this is the case for X-ray diffraction, and in practice a single-scattering model, also known as the kinematic theory, describes the intensities quite well. In contrast, for electron diffraction (§8), a dynamical theory is important.

To calculate diffraction intensities, we will assume that Bragg's law (§4.3.1) and the equivalent Laue condition (§4.3.2) are satisfied. Further, since the crystal can be considered to be a periodic repetition of the unit cell, we only need to determine the exact nature of the scattering from the atoms in the unit cell and their contributions to the diffracted intensity. We will also see that the intensities of the diffraction patterns depend on the nature—-centered or not—-of the unit cell, and the symmetry elements present in the crystal. In particular, symmetry elements, e.g. glide planes and screw axes, lead to *systematic absences* in intensities for diffraction from certain planes, even though they may satisfy the Bragg conditions. In such cases, the scattering from different atoms in the unit cell interfere destructively and result in zero observed intensities.

We follow, qualitatively, the same approach as in the previous section for determining the scattering from a single atom. Instead of determining the phase difference for scattering from electrons, we calculate the phase differences and amplitudes for scattering from all the atoms in the unit cell. We will see that the problem of scattering from a unit cell resolves into the addition of waves with different amplitudes and phases, but of the same frequency, following either the vector method introduced in §6.2.3, or in the general case, using the complex notation presented in §6.2.4. The resultant wave, with scattering contributions from all the atoms in the unit cell, contributing to diffraction by a specific set of planes, (hkl), is then described by the *structure factor*, F_{hkl}, and just like the atomic scattering factor, f_{atom}, (7.3.9), it can be defined as

$$|F_{hkl}| = \frac{\text{amplitude of the wave scattered by all the atoms in a unit cell}}{\text{amplitude of the wave scattered by a single electron}} \quad (7.4.1)$$

The corresponding diffraction intensity, I_{hkl}, is given by $|F_{hkl}|^2$, or strictly speaking, since it is a complex number, as $I_{hkl} = F^*_{hkl}F_{hkl}$, where F^*_{hkl} is the complex conjugate of F_{hkl}, with additional corrections for specific experimental factors to be described later in §7.10.

To begin, consider diffraction from a primitive lattice, with only one atom, scattering factor, f_0, at the origin (Fig. 7.4.1a). The incident and scattered X-ray beams satisfy the Bragg condition for the lattice planes (hkl) shown in the figure. Then, the beams scattered from all the atoms on the surface would be in phase, and by definition, the beams scattered from successive planes, d_{hkl}, $2d_{hkl}$ ... nd_{hkl}, would exhibit phase differences of 2π, 4π ... $2n\pi$. In other words, they would all be in phase, interfering constructively, to give a scattering amplitude that is a sum of the atomic scattering factor. Since we have only one atom in the unit cell, the structure factor for this trivial case is then $F_{hkl} = f_0$.

Next, consider the same primitive lattice, but with two atoms, one at the origin with scattering factor, f_0, and the other, with scattering factor, f_1, at a position, $\mathbf{r}_1$, with fractional coordinates (u_1, v_1, w_1) in the *unit cell* (Fig. 7.4.1b). Thus

$$\mathbf{r}_1 = u_1\mathbf{a} + v_1\mathbf{b} + w_1\mathbf{c} \quad (7.4.2)$$

where, $\mathbf{a}$, $\mathbf{b}$, and $\mathbf{c}$, are the unit cell vectors. Further, the incident beam, with unit vector, $\widehat{\mathbf{s}}_0$, and diffracted beam, $\widehat{\mathbf{s}}$, satisfy the diffraction condition for the lattice planes, (hkl), with inter-planar spacing, d_{hkl}, as shown. The path difference (Fig. 7.4.1b) for the beams scattered from the two atoms is given by

$$\text{P.D.} = \text{AC–BD} = \mathbf{r}_1 \cdot (\widehat{\mathbf{s}} - \widehat{\mathbf{s}}_0) \quad (7.4.3)$$

Since the diffraction satisfies the Laue condition

$$\mathbf{k} - \mathbf{k}_0 = \mathbf{g}_{hkl} = \frac{2\pi}{\lambda}(\widehat{\mathbf{s}} - \widehat{\mathbf{s}}_0) = h\mathbf{a}^* + k\mathbf{b}^* + l\mathbf{c}^* \quad (4.3.7)$$

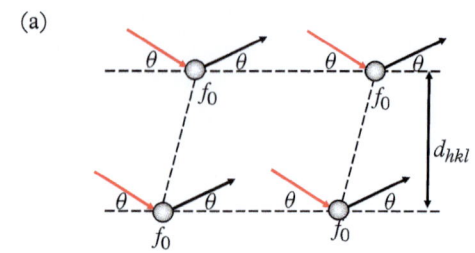

(a)

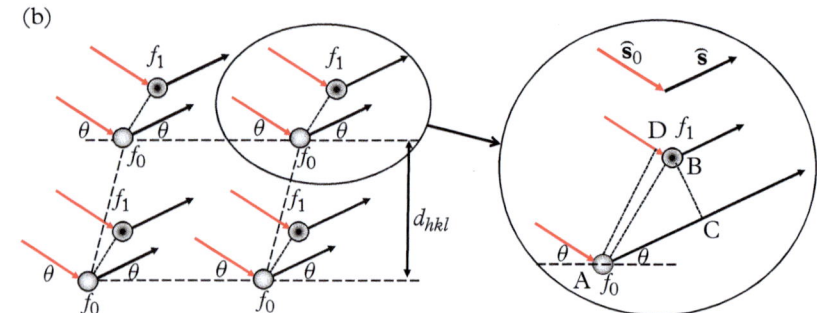

(b)

Figure 7.4.1 Bragg diffraction for lattice planes (*hkl*), from a primitive lattice with (a) one atom at the origin, and (b) two atoms with atomic scattering factors, f_0 and f_1, located in the unit cell.

Adapted from Hammond (2006).

Thus, we have

$$\widehat{\mathbf{s}} - \widehat{\mathbf{s}}_0 = \frac{\lambda}{2\pi}\left(h\mathbf{a}^* + k\mathbf{b}^* + l\mathbf{c}^*\right) \tag{7.4.4}$$

and the path difference, (7.4.3), is now

$$\mathbf{r}_1 \cdot (\widehat{\mathbf{s}} - \widehat{\mathbf{s}}_0) = \frac{\lambda}{2\pi}\left(h\mathbf{a}^* + k\mathbf{b}^* + l\mathbf{c}^*\right) \cdot (u_1\mathbf{a} + v_1\mathbf{b} + w_1\mathbf{c}) = \lambda\left(u_1 h + v_1 k + w_1 k\right) \tag{7.4.5}$$

because by definition $\mathbf{a} \cdot \mathbf{a}^* = \mathbf{b} \cdot \mathbf{b}^* = \mathbf{c} \cdot \mathbf{c}^* = 2\pi$, and $\mathbf{a} \cdot \mathbf{b}^* = 0$ etc.; see (4.2.4).

Since a path difference of λ is equal to a phase difference of 2π, we can write the phase difference, ϕ_1, between the waves scattered by the two atoms as

$$\phi_1 = 2\pi\left(u_1 h + v_1 k + w_1 k\right) \tag{7.4.6}$$

Following the graphical method for the superposition of waves, using the vector-phase diagram introduced in §6.2.3, where the vector lengths are proportional to the atomic scattering factors, f_0 and f_1, with a phase angle, ϕ_1, we obtain the resultant structure factor, F_{hkl} (Fig. 7.4.2a).

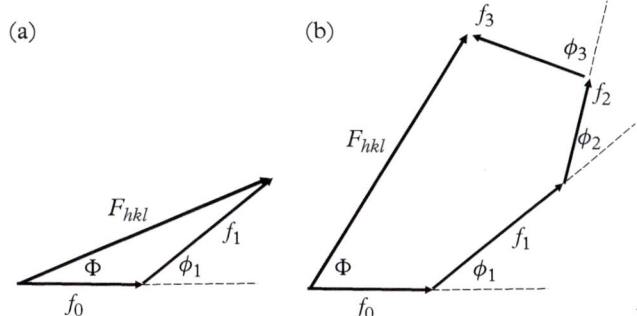

(a)

(b)

Figure 7.4.2 Graphical representation for the superposition of waves scattered by (*a*) two and (*b*) four atoms in the unit cell. The resultant wave has an amplitude given by the structure factor, F_{hkl}, and a phase difference, Φ, with respect to the incident beam.

We can readily extend this analysis to all n atoms, $\mathbf{r}_i$, each with fractional coordinates (u_i, v_i, w_i), and scattering phase angles, ϕ_i, occupying the unit cell as a summation. Since it becomes cumbersome to do this in vector form, as illustrated, even for four atoms (Fig. 7.4.2b), we employ the complex notation introduced in §6.2.4, and write the structure factor in general terms as

$$F_{hkl} = \sum_{i=1}^{n} f_i \exp i\phi_i = \sum_{i=1}^{n} f_i \exp 2\pi i \,(u_i h + v_i k + w_i k) \tag{7.4.7}$$

where the summation is over all atoms, n, in the unit cell. Since the structure factor is a complex number, the intensity of a particular diffraction peak is given by

$$I_{hkl} = F_{hkl}^* F_{hkl} \tag{7.4.8}$$

We now demonstrate the calculation of the structure factor using various examples.

7.5 Examples of Structure Factor Calculations

First it is important for the reader to review some important relations[4] of complex exponential functions that we encounter in calculating structure factors.

7.5.1 Face-Centered Cubic (FCC) Structure

The face-centered cubic unit cell (*International Tables*, Space group #225, $Fm\bar{3}m$) shown in Figure 4.1.11f is characterized by a basis of four atoms, with fractional coordinates, *uvw*, given by 000, ½ ½ 0, ½ 0 ½, and 0 ½ ½. Assuming that each of these four positions are occupied by an atom of the same kind, with atomic scattering factor, f, we can write the structure factor as

---

[4] $e^{i\phi} = \cos\phi + i\sin\phi$
$e^{i\phi} + e^{-i\phi} = 2\cos\phi$
$e^{in\pi} = (-1)^n = e^{-in\pi}$

$$F_{hkl} = \sum_{i=1}^{4} f_i e^{2\pi i(u_i h + v_i k + w_i l)}$$
$$= f\left[e^{2\pi i \cdot 0} + e^{2\pi i\left(\frac{h}{2}+\frac{k}{2}\right)} + e^{2\pi i\left(\frac{k}{2}+\frac{l}{2}\right)} + e^{2\pi i\left(\frac{l}{2}+\frac{h}{2}\right)} \right] \tag{7.5.1}$$

using the relationship $e^{in\pi} = (-1)^n$, we can rewrite (7.5.1) as

$$F_{hkl} = f\left[1 + (-1)^{h+k} + (-1)^{k+l} + (-1)^{l+h} \right] \tag{7.5.2}$$

It is now easy to see that when h, k, and l, are *either all even or all odd*, referred to from now on as *unmixed indices*, each of the four terms in F_{hkl}, (7.5.2), has a value equal to $+1$, and the structure factor becomes

$$F_{hkl} = 4f \quad \textit{unmixed indices} \tag{7.5.3a}$$

with $I_{hkl} = |F_{hkl}|^2 = 16 f^2$.

On the other hand, if the indices, h, k, and l, are *mixed*, i.e. two odd and one even or two even and one odd, the sum of the three exponents equals -1 and the structure factor is

$$F_{hkl} = 1 - 1 = 0 \quad \textit{mixed indices} \tag{7.5.3b}$$

with $I_{hkl} = 0$.

Example 7.5.1: For FCC Cu, $a_0 = 3.61$ Å, calculate the structure factor for the first three allowed reflections, using Cu Kα radiation ($\lambda = 1.54$ Å).

Solution: FCC crystals show non-zero values of the structure factor for only unmixed indices. See Table 7.5.1. The first three allowed reflections are, (111), (200), and (220).

From Table 4.1.2, for a cubic system $d_{hkl} = \dfrac{a_0}{\sqrt{h^2+k^2+l^2}}$.

The parameter $s = \dfrac{\sin\theta_{hkl}}{\lambda} = \dfrac{1}{2d_{hkl}}$.

We use the Cromer–Mann parameters for Cu, and calculate the scattering factors using (7.3.10).

Finally, the structure factor for unmixed indices is given by (7.5.3a). We now solve in the form of a table.

Reflection	$d_{hkl} = \dfrac{a_0}{\sqrt{h^2+k^2+l^2}}$	$s_{hkl} = \dfrac{\sin\theta_{hkl}}{\lambda} = \dfrac{1}{2d_{hkl}}$	$f_{hkl}(s) = \sum_{i=1}^{4} a_i \exp\left(-b_i s^2\right) + c$	$F_{hkl} = 4f$
(111)	2.08 Å	0.2404	22.04	88.19
(200)	1.805 Å	0.277	20.71	82.84
(220)	1.276 Å	0.3917	16.77	67.07

7.5.2 Body-Centered Cubic (BCC) Structure

The structure factor for a BCC structure (*International Tables*, Space group #229, $I\,m\bar{3}m$) (Fig. 4.1.11i) can be calculated considering that the same kind of atom is located at the coordinates, 000, and ½ ½ ½, within the unit cell, as

$$F_{hkl} = f\left[e^{2\pi i.0} + e^{2\pi i\left(\frac{h}{2}+\frac{k}{2}+\frac{l}{2}\right)}\right] = f\left[1 + e^{\pi i(h+k+l)}\right] = f\left[1 + (-1)^{h+k+l}\right]$$

(7.5.4)

Thus

$$\begin{aligned}F_{hkl} &= 2f \\ I_{hkl} &= 4f^2\end{aligned} \qquad \text{for } h+k+l = \text{even} \qquad (7.5.5a)$$

and

$$\begin{aligned}F_{hkl} &= 0 \\ I_{hkl} &= 0\end{aligned} \qquad \text{for } h+k+l = \text{odd} \qquad (7.5.5b)$$

Note that, in addition, the structure factor is independent of the size and shape of the unit cell. Thus, any reflection with $h+k+l = $ odd, will always be missing (= zero intensity) in a diffraction pattern for a body-centered cell, irrespective of whether it is cubic, orthorhombic, or tetragonal.

The intensities based on the structure factors for diffraction of different planes (*hkl*) in FCC and BCC crystal structures are compared in Table 7.5.1. Note that many of them do not give rise to a diffracted beam with an observable intensity because of centering extinctions.

When Bragg's law is satisfied, we can use the interplanar spacing, d_{hkl} (Table 4.1.2) for any crystal system, to determine the value of s, i.e. $s = \frac{\sin\theta_{hkl}}{\lambda} = \frac{1}{2d_{hkl}}$, and using the appropriate atomic scattering factors from the Cromer–Mann parametrization, we can calculate the ideal values of the structure factors and the diffracted intensity.

Table 7.5.1 Intensities for diffraction from different planes in BCC and FCC crystals.

(*hkl*)	100	110	111	200	210	211	220	221	300	311	222
BCC	0	$4f^2$	0	$4f^2$	0	$4f^2$	$4f^2$	0	0	0	$4f^2$
FCC	0	0	$16f^2$	$16f^2$	0	0	$16f^2$	0	0	0	$16f^2$

Example 7.5.2: Calculate the structure factor for some select reflections for Cr_{BCC} using $\lambda = 1.54$ and lattice parameter $a_0 = 2.884$ Å.

Solution: We solve this by applying (7.5.5), using values for the atomic scattering factor from Table 7.3.1, in the form of a table. Note that only $h+k+l = $ even, have observable intensities.

hkl	$s = \frac{\sin\theta_{hkl}}{\lambda} = \frac{1}{2d_{hkl}} = \frac{\sqrt{h^2+k^2+l^2}}{2a_0}(\text{Å}^{-1})$	f (Table 7.3.1)	F_{Cr} (7.5.5)
000	0	24	48
100	0.173	-	0
011	0.245	16.858	33.716
111	0.3	-	0
200	0.347	13.615	27.23
210	0.388	-	0
121	0.425	11.70	23.40
120	0.388	-	0
310	0.5482	9.43	18.86
311	0.575	-	0
320	0.625	-	0

7.5.3 Hexagonal Close Packed (HCP) Structure

We refer to the primitive hexagonal structure (Fig. 4.1.11e) with the unit cell containing two identical basis atoms, with atomic scattering factor, f, occupying positions with fractional coordinates, 000—the A layer, and 1/3 2/3 ½—the B layer. Substituting these two coordinates, we get the structure factor

$$F_{hkl} = f\left[e^{2\pi i.0} + e^{2\pi i\left(\frac{h}{3}+\frac{2k}{3}+\frac{l}{2}\right)}\right] = f\left[1 + e^{2\pi i\left(\frac{h}{3}+\frac{2k}{3}+\frac{l}{2}\right)}\right] \tag{7.5.6}$$

Let us consider two cases $hkl = 0002 \sim 00.2$, and $hkl = 10\bar{1}0 \sim 10.0$. We can see that

$$F_{00.2} = f\left[1 + 1\right] = 2f \tag{7.5.7a}$$

$$F_{10.0} = f\left[1 + e^{\frac{2\pi i}{3}}\right] = f\left[1 + \cos\frac{2\pi}{3} + i\sin\frac{2\pi}{3}\right] = f\left[0.5 + i0.866\right] \tag{7.5.7b}$$

The structure factor for the latter is not real but complex, and to calculate the intensity, we have to multiply F_{hkl} by its complex conjugate, F^*_{hkl}. We now rewrite (7.5.6) as

$$F_{hkl} = f\left[1 + e^{2\pi i\left(\frac{h}{3}+\frac{2k}{3}+\frac{l}{2}\right)}\right] = f\left[1 + e^{2\pi ip}\right] \tag{7.5.6}$$

where $p = \frac{h}{3} + \frac{2k}{3} + \frac{l}{2}$ to give, the general intensity

$$\begin{aligned}I_{hkl} &= f^2\left[1 + e^{2\pi ip}\right]\left[1 + e^{-2\pi ip}\right] = f^2\left[2 + e^{2\pi ip} + e^{-2\pi ip}\right] = f^2\left[2 + 2\cos 2\pi p\right] \\ &= f^2\left[2 + 2\left(2\cos^2\pi p - 1\right)\right] = 4f^2\cos^2\pi p\end{aligned}$$

$$\tag{7.5.7}$$

Expanding p, we get

$$I_{hkl} = 4f^2\cos^2\pi\left(\frac{h+2k}{3}+\frac{l}{2}\right) \qquad (7.5.8)$$

Now, we can readily see that $I_{hkl} = 4f^2$, when $\frac{h+2k}{3}+\frac{l}{2} = n$, where n is an integer. In addition, when $h+2k = 3m$, where m is an integer, and l is odd, we get $I_{hkl} = 0$. In practice, this leads to missing reflections, or absences, for various hkl such as 11.1, 11.3, 22.1, 22.3, etc.

7.5.4 Cesium Chloride (CsCl) Structure

The structure is illustrated in Figure 4.1.13a, with Cl^- located at 000 and Cs^+ located at ½½½. Using atomic scattering factors,[5] f_{Cl} and f_{Cs}, we get for the structure factor

$$F_{hkl} = f_{Cl}e^{2\pi i.0} + f_{Cs}e^{2\pi i\left(\frac{h}{2}+\frac{k}{2}+\frac{l}{2}\right)} = f_{Cl} + f_{Cs}e^{\pi i(h+k+l)} = f_{Cl} + f_{Cs(-1)}{}^{h+k+l}$$
$$(7.5.9)$$

Thus

$$F_{hkl} = f_{Cl} + f_{Cs} \quad \text{for } h+k+l = \text{even}$$
$$F_{hkl} = f_{Cl} - f_{Cs} \quad \text{for } h+k+l = \text{odd} \qquad (7.5.10)$$

Note that in both cases in (7.5.10), the structure factor is a real number. We now have two intensities

$$I_{hkl} = (f_{Cl} + f_{Cs})^2 \quad \text{for } h+k+l = \text{even}$$

$$I_{hkl} = (f_{Cl} - f_{Cs})^2 \quad \text{for } h+k+l = \text{odd} \qquad (7.5.11)$$

with the former intensities proportional to the square of the *sum* of the atomic scattering factors and called *fundamental* reflections, and the latter, substantially weaker reflections, with intensities proportional to the square of the *difference* of the atomic scattering factors, and called *superlattice* reflections.

Further, if Cs^+ and Cl^- are randomly distributed in the same BCC structure, then they would randomly occupy the two positions 000 and ½½½, and the atomic scattering factor for each site would be the average value, $\frac{f_{Cl}+f_{Cs}}{2}$, and satisfying the symmetry requirements the reflections with $h+k+l = $ odd would have zero intensity. In other words, the intensities would behave the same way as for a BCC crystal. However, if there were chemical order in the crystal, deviating from a random distribution of the two ions, then the superlattice reflection would show a finite intensity proportional to the degree of order, with maximum intensities observed for the ideal CsCl structure (see §7.11.4 for further details of chemical *order–disorder transitions*).

[5] Strictly speaking, we should use the scattering factors for ions and not atoms, but here we neglect the difference.

Example 7.5.3: The structure of sodium chloride, NaCl (Fig. 4.1.13b), can be interpreted as two interpenetrating FCC lattices, one occupied by Cl^-, and the other by Na^+. The Na^+ lattice is displaced with respect to the Cl^- lattice by the vector, $\tau = \frac{a_0}{2}(\hat{\mathbf{x}} + \hat{\mathbf{y}} + \hat{\mathbf{z}})$, where a_0 is the lattice parameter, giving atom locations

Cl^-	000	½ ½ 0	½ 0 ½	0 ½ ½
Na^+	½ ½ ½	00 ½	0 ½ 0	½ 0 0

Calculate the structure factor, F_{hkl}, and intensity, I_{hkl}, for NaCl.

Solution: Since both Cl^- and Na^+ occupy an FCC lattice, the structure factor for NaCl can be derived from that of the FCC structure, applying (7.5.2) to both the ions as

$$
\begin{aligned}
F_{hkl} &= f_{Cl}e^{2\pi i.0}\left[1 + (-1)^{h+k} + (-1)^{k+l} + (-1)^{l+h}\right] \\
&\quad + f_{Na}e^{2\pi i\left(\frac{h}{2}+\frac{k}{2}+\frac{l}{2}\right)}\left[1 + (-1)^{h+k} + (-1)^{k+l} + (-1)^{l+h}\right] \\
&= \left[f_{Cl} + f_{Na}e^{\pi i(h+k+l)}\right]\left[1 + (-1)^{h+k} + (-1)^{k+l} + (-1)^{l+h}\right] \\
&= \left[f_{Cl} + f_{Na}(-1)^{h+k+l}\right]\left[1 + (-1)^{h+k} + (-1)^{k+l} + (-1)^{l+h}\right] \quad (7.5.12)
\end{aligned}
$$

The first term corresponds to the contribution of the basis atoms of Cl^- at 000 and Na^+ at ½½½, and the second term is the contribution from the face-centering translations already encountered in §7.5.1. Then, the intensities for the structure are given by

$$
I_{hkl} = |F_{hkl}|^2 = \left[f_{Cl} + f_{Na}(-1)^{h+k+l}\right]^2\left[1 + (-1)^{h+k} + (-1)^{k+l} + (-1)^{l+h}\right]^2
$$
$$(7.5.13)$$

Thus, similar to the FCC structure, I_{hkl} (NaCl) $= 0$, for mixed indices. However, for unmixed indices

$$
F_{hkl} = 4\left[f_{Cl} + f_{Na}(-1)^{h+k+l}\right]
$$

and

$$
\begin{aligned}
I_{hkl} &= 16\left[f_{Cl} + f_{Na}(-1)^{h+k+l}\right]^2 \\
&= 16[f_{Cl} + f_{Na}]^2 \qquad \text{for } h+k+l = \text{even} \\
&= 16[f_{Cl} - f_{Na}]^2 \qquad \text{for } h+k+l = \text{odd}
\end{aligned}
$$

Thus, we can clearly see that the addition of another atom to the basis has decreased the intensities of some of the reflections. This analysis can be readily extended to the diamond structure (see Problem 7.1).

7.6 Symmetry and Structure Factor

7.6.1 Crystals with Inversion Symmetry

These crystals are also referred to as those with a center of symmetry at the origin. In these structures, for every atom at position, $\mathbf{r}$, with fractional coordinates uvw, there exists another atom of the same kind at position, $-\mathbf{r}$, with fractional coordinates $\overline{uvw}$, in the unit cell. Then, the structure factor can be split into two terms

$$F_{hkl} = \sum_{j=1}^{N} f_j e^{2\pi i(hu_j + kv_j + lw_j)} = \sum_{j=1}^{N/2} f_j e^{2\pi i(hu v_j + kv_j + lw_j)} + \sum_{j=1}^{N/2} f_j e^{-2\pi i(hu_j + kv_j + lw_j)}$$

$$= 2 \sum_{j=1}^{N/2} f_j \cos 2\pi \left(hu_j + kv_j + lw_j \right) \tag{7.6.1}$$

Thus, we can conclude that the structure factor for a *centrosymmetric crystal* is always real for all reflections.

7.6.2 Friedel Law

From the preceding, it follows that the diffraction pattern for centrosymmetric crystal[6] is also centrosymmetric. Now consider the reflections I_{hkl} and $I_{\overline{hkl}}$, i.e. two reflections that are on opposite sides of the direct beam, for a non-centrosymmetric crystal, in a primitive lattice with one atom at fractional coordinates uvw in the unit cell. Then, its structure factor, $F_{hkl} = f e^{2\pi i(hu + kc + lw)}$, gives an intensity

$$I_{hkl} = F_{hkl} F_{hkl}^* = f^2 e^{2\pi i(hu + kv + lw)} e^{-2\pi i(hv + kv + lw)}$$

$$= f^2 e^{2\pi i(hu + kv + lw)} e^{2\pi i(\overline{h}u + \overline{k}v + \overline{l}w)} = I_{\overline{hkl}} \tag{7.6.2}$$

Thus, even if a crystal does NOT have a center of symmetry, the corresponding diffraction pattern will be centrosymmetric. This rule, known as *Friedel law*, has important consequences. Recall that there are only 32 point groups possible in 3D crystallography (§4.1.7). As a result of Friedel law, these 32 point groups are reduced to 11 centrosymmetric groups, called Laue groups, in their diffraction patterns. This is particularly relevant for electron diffraction, especially in a transmission electron microscope (see §8.5). Further details on Friedel law can be found in Hammond (2006) and Giacovazzo (2002).

7.6.3 Systematic Absences

Certain symmetry elements, such as glide planes and screw axes, when present in a crystal structure give rise to systematic absences of certain specific reflections in diffraction. A glide plane is a symmetry element that simultaneously combines

[6] Georges Friedel (1865–1933) was a French mineralogist and crystallographer.

a reflection or mirror with a translation. Similarly, a screw axis is a rotation axis combined with a translation. We demonstrate the effect of a glide plane on the structure factor with the following example.

Example 7.6.1: Consider an *n*-glide mirror, parallel to the (001) plane going through the origin with a translation component, $\tau = \frac{1}{2}\left(\hat{\mathbf{x}} + \hat{\mathbf{y}}\right)$. Identify any systematic absences when this glide plane is present.

Solution: When an *n*-glide is present, for each atom at position *uvw*, there is an atom of the same kind at $u + 1/2, v + 1/2, \overline{w}$. The structure factor is then given by

$$
\begin{aligned}
F_{hkl} &= \sum_{j=1}^{N/2} f_j e^{2\pi i(hu_j + kv_j + lw_j)} + \sum_{j=1}^{N/2} f_j e^{2\pi i\left[h\left(u_j + \frac{1}{2}\right) + k\left(v_j + \frac{1}{2}\right) - lw_j\right]} \\
&= \sum_{j=1}^{N/2} f_j \left[e^{2\pi i(u_j h + v_j k)}\right]\left[e^{2\pi i w_j l} + e^{2\pi i\left(\frac{h+k}{2}\right) - w_j l}\right]
\end{aligned}
\tag{7.6.3}
$$

Then, for any reflection of the type, *hk0*, we get

$$
\begin{aligned}
F_{hk0} &= \sum_{j=1}^{N/2} f_j \left[e^{2\pi i(u_j h + v_j k)}\right]\left[1 + e^{\pi i(h+k)}\right] \\
&= \sum_{j=1}^{N/2} f_j e^{2\pi i(u_j h + v_j k)} \, 2\cos \pi \left(\frac{h+k}{2}\right) e^{\pi i\left(\frac{h+k}{2}\right)}
\end{aligned}
\tag{7.6.4}
$$

It is easy to see that $F_{hk0} = 0$, when $h + k = 2m + 1$, where *m* is an integer.

To satisfy this criterion, if such an *n*-glide plane were to be present, there will be systematic absences of *hk0* reflections, such as $210, 120, 320, 430, \ldots$ etc.

Similar absences can be calculated for other types of glide planes or screw axes. More details can be found in de Graef and McHenry (2007), and in Problem 7.6.

7.7 The Inverse Problem of Determining Structure from Diffraction Intensities

It is clear from §7.5 and §7.6 that, once the structure or the symmetry elements in the unit cell is known, in principle, we can calculate the diffraction intensities. In practice, we also need to correct them for the experimental factors, discussed in §7.10, that influence the diffraction intensities. However, because we are only measuring the intensities, we are unable to say anything meaningful about the phase, Φ, of the diffracted wave. Unfortunately, this phase information is critical to solve the inverse problem that is often of interest. Note that all vectors, F_{hkl}, such as in Figure 7.4.2b, as long as they have the same modulus will give the same intensity, $I_{hkl} = F^*_{hkl}F_{hkl}$, irrespective of their phase, Φ_{hkl}, or the direction in which F_{hkl} points. In other words, if we are given an unknown crystal structure, we need both the structure factor, F_{hkl}, and the phase, Φ_{hkl}, of the diffracted wave for several sets of planes, (hkl), to unequivocally identify the positions of the atoms. In special cases, such as the centrosymmetric crystal with the origin coinciding with the center of symmetry (§7.6.1), where the structure factor, F_{hkl}, is a real number with no imaginary component, the inverse problem can be solved. However, in general for non-centrosymmetric crystals, the inverse problem cannot be solved but close approximation can be made using advance techniques that are beyond the scope of this book, but can be found in specialized texts, e.g. Giacovazzo et al. (2002).

7.8 Broadening of Diffracted Beams and Reciprocal Lattice Points

Ideally, in a diffraction experiment satisfying Bragg's law, $\lambda = 2d_{hkl} \sin \theta_{hkl}$, the diffracted beam should appear as a sharp line with intensity, $I_{hkl} = |F_{hkl}|^2$, at a position, $2\theta_{hkl}$, defined by the relation between the diffracted and the incident or primary beam (Fig. 7.8.1). In practice, for crystals of finite thickness, the diffracted beam is broadened (Fig. 7.8.1) and its intensity is modified by experimental factors as discussed in §7.10. The broadening is conceptually similar to that discussed for Fraunhofer diffraction (§6.6.2) from a grating with a finite number of slits.

To derive an expression for the broadening of the diffracted beam, we consider a finite crystal with thickness, $t = md_{hkl}$, perpendicular to the set of diffracting

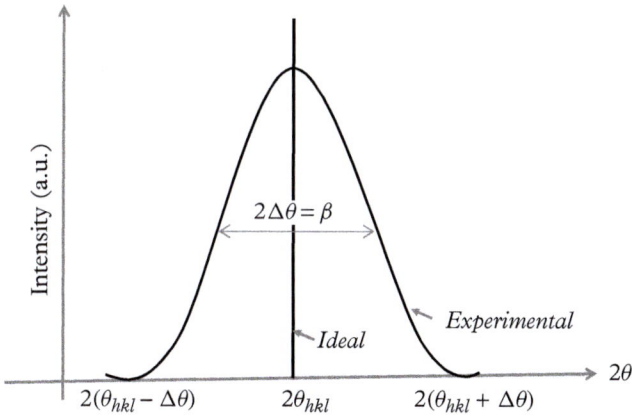

Figure 7.8.1 Broadening, β, of a diffraction peak can be due to a variety of factors including the crystallite or grain size, microstrain, and instrument factors (see §7.11.2).

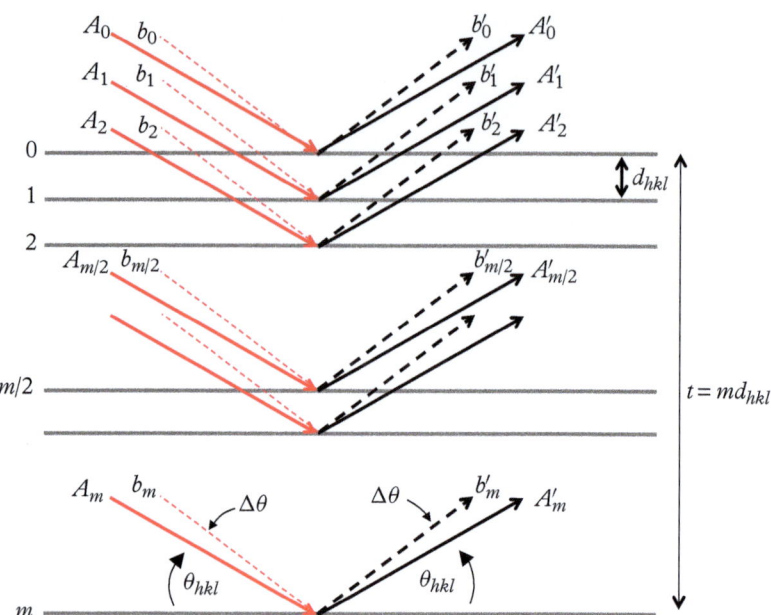

Figure 7.8.2 Bragg reflection from a crystal of thickness, t, measured perpendicular to the set of planes with spacing, d_{hkl}. Arrows represent the incident (A_n), and reflected (A'_n) beams at the exact Bragg orientation, θ_{hkl}, for the set of successive planes $n = 0,\ldots,m$; similarly, (b_n, b'_n) represent incident and reflected beams at a small deviation, $\Delta\theta$, from the Bragg angle.

Adapted from Hammond (2006).

planes, d_{hkl} (Fig. 7.8.2). The planes satisfy the exact diffraction condition at the Bragg angle, θ_{hkl}, for the incident beams, $A_0, A_1, \ldots A_{m/2} \ldots A_m$, with diffracted beams, $A'_0, A'_1, \ldots A'_{m/2} \ldots A'_m$, reflected from successive planes, $0 \ldots m/2 \ldots m$ (Fig. 7.8.2). Since Bragg's law is satisfied, the path difference between rays reflected from plane, 0, and successive planes in the crystal is $n\lambda$, where $n = 1,\ldots,m/2 \ldots \ldots m$; this leads to constructive interference for rays reflected from all planes, as expected for Bragg diffraction.

It is impractical to obtain a truly parallel primary beam, in spite of our best efforts at collimation (see §7.9.5.1), and a small angular deviation, $\Delta\theta$, of the incident beam is always to be expected. Now, consider the set, $b_0, b_1, \ldots b_{m/2} \ldots b_m$, of such primary beams, all with angular deviation, $\Delta\theta$, incident on successive planes with reflected beams, $b'_0, b'_1, \ldots b'_{m/2} \ldots b'_m$, also shown in the figure. Since $\Delta\theta \ll \theta_{hkl}$, the path difference between $b_0 b'_0$ and $b_1 b'_1$ is insignificantly different from λ, and a constructive interference between the two will persist. However, for consecutive planes, n, the path difference (P.D.) between $b_0 b'_0$ and $b_n b'_n$, will gradually increase with n, and at some critical depth, the path difference will be such that it will give rise to a destructive interference, i.e. P.D. $= \lambda/2$.

Let the first two such planes that show *constructive* interference be $n = 0$ and $n = m/2$; they then satisfy Bragg's law

$$\frac{m}{2}\lambda = 2\frac{m}{2}d_{hkl}\,\sin\theta_{hkl} \qquad (7.8.1)$$

For *destructive* interference from the same set of planes, at small angular deviation, $\theta_{hkl} + \Delta\theta$, we get from Bragg's law

$$\frac{m}{2}\lambda + \frac{\lambda}{2} = \frac{m}{2}d_{hkl}\,\sin(\theta_{hkl} + \Delta\theta) \tag{7.8.2}$$

When this condition, (7.8.2), is satisfied all subsequent *pairs* of lattice planes, 1 and $m/2 + 1$, 2 and $m/2+2, \ldots$, also satisfy the destructive interference criterion, such that the entire crystal as a whole will result in a destructive interference. When this happens, we expect to see zero intensity in the diffracted beam, thus defining the width of the diffracted beam on either side of the Bragg angle, θ_{hkl}. We can expand (7.8.2) to get

$$\frac{m}{2}\lambda + \frac{\lambda}{2} = 2\frac{m}{2}d_{hkl}\,[\sin\theta_{hkl}\cos\Delta\theta + \cos\theta_{hkl}\sin\Delta\theta] \tag{7.8.3}$$

For small angles, $\sin\Delta\theta = \Delta\theta$, $\cos\Delta\theta = 1$, $t = m\lambda$, and using (7.8.1), we can simplify (7.8.3) as

$$\frac{\lambda}{2} = t\cos\theta_{hkl}\,\Delta\theta \tag{7.8.4}$$

or

$$2\Delta\theta = \frac{\lambda}{t\cos\theta_{hkl}} = \beta \tag{7.8.5}$$

The relationship (7.8.5) that relates the broadening, β, of the diffracted X-ray beam to the thickness, t, of the crystal in a direction normal to the lattice planes, d_{hkl}, is known as the *Scherrer*[7] *equation*. Here, β is the angular width of the beam at half the maximum peak height, as defined in Figure 7.8.1. The Scherrer equation (7.8.5) is of great practical utility as it relates the peak broadening to the average crystallite or grain size in a specimen (see §7.11.2). However, in practice, to accommodate the overall shape of the crystal (7.8.5) is modified as

$$\beta = K\frac{\lambda}{t\cos\theta_{hkl}} \tag{7.8.6}$$

where K ($\sim$1) is an empirical constant that accounts for the crystallite shape.

It is also possible to include this broadening in the Ewald sphere construction (Fig. 4.3.2) by extending the infinitesimal reciprocal lattice point to form a "node" of finite size (Fig. 7.8.3). Now it can account for the fact that the diffracted beam satisfies the Laue condition (4.3.7), even when it is broadened by the angular range of $2\beta = 4\Delta\theta$. This is possible if and only if the reciprocal lattice point is broadened to a "node" of finite dimensions/length allowing it to intersect the reflecting sphere as the crystal rotates. Let Δg_{hkl} represent the extension of the reciprocal lattice point about its mean position.

[7] Paul Hermann Scherrer (1890–1969) was a Swiss physicist and the co-inventor of the Debye–Scherrer method (§7.9.3).

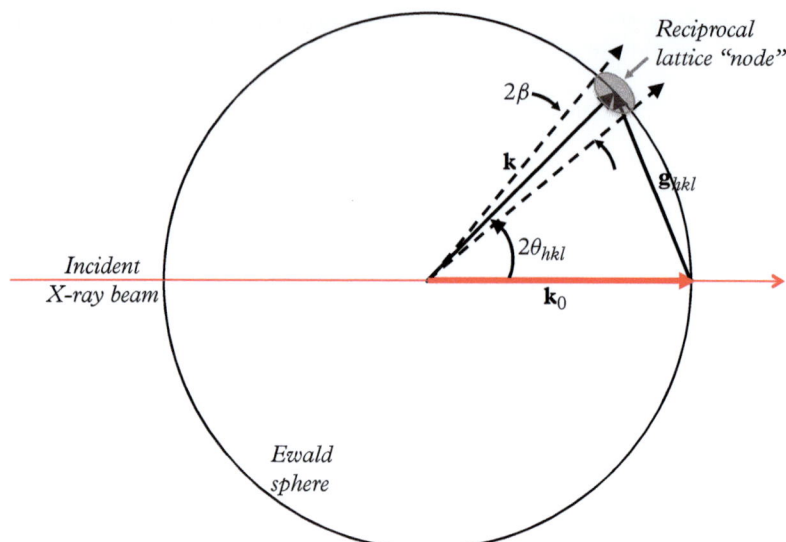

Figure 7.8.3 The Ewald sphere construction for a reflected beam broadened by 2β corresponding to the extension of the reciprocal lattice node by 2π/t, where t is the thickness of the crystal in a direction normal to the diffracting planes.

Since

$$| \mathbf{g}_{hkl} |= g_{hkl} = \frac{2\pi}{d_{hkl}} = \frac{4\pi \sin \theta_{hkl}}{\lambda} \tag{7.8.7}$$

we get

$$\Delta g_{hkl} = \Delta \left(\frac{4\pi \sin \theta}{\lambda} \right) = \frac{4\pi \cos \theta}{\lambda} \Delta \theta \tag{7.8.8}$$

and substituting for $\Delta\theta$ from (7.8.5), we get

$$\Delta g_{hkl} = \frac{4\pi \cos \theta}{\lambda} \frac{\lambda}{2t \cos \theta} = \frac{2\pi}{t} \tag{7.8.9}$$

Thus, the extension of the reciprocal lattice point is proportional (by a factor of 2π) to the inverse of the crystal dimension (thickness, t) in a direction normal to the diffracting planes. We can now generalize this argument to all other directions of the crystal and say that the reciprocal lattice nodes have a finite size that is inversely related to the shape and size of the crystal. For thin films, surfaces, or plate-like crystallites, the nodes become rods or streaks in a direction normal to the thickness. This is particularly important in electron diffraction in a transmission electron microscope (see Fig. 8.5.4) where electron transparent thin foils of specimens are used, and where the corresponding reciprocal lattice nodes are referred to as rel-rod (see §8 and §9). Further, as we have seen in Fraunhofer diffraction for a finite number of gratings (§6.6.2), the nodes in X-ray diffraction are also surrounded by subsidiary minima and maxima. If the crystal thickness is large, these maxima/minima can be ignored in X-ray diffraction. However, in thin film multilayers, where there are a finite number of layers, these secondary maxima and minima are also observed and can provide additional information about the specimen; this is discussed further in §7.9.4, and specifically in Figure 7.9.12.

> **Example 7.8.1:** Consider a powder diffraction experiment of polycrystalline iron using Cr Kα X-ray radiation.
>
> (a) What would be the first three most intense reflections?
> (b) What would be the peak broadening in degrees for these three reflections if the crystallites were (i) 1 mm and (ii) 50 nm, in diameter. Comment on the results.

Solution:

(a) We assume BCC iron with lattice parameter, $a = 0.2866$ nm. From Table 7.5.2, we determine that the three most intense reflections are (011), (002), and (211).

(b) The wavelength of Cr Kα is $\lambda = 0.2291$ nm. We apply Bragg's law to determine the angle of diffraction, and (7.8.6) to determine peak broadening (assuming $K \sim 1$), and solve in the form of a table.

hkl	d_{hkl} (Å)	θ_{hkl}	β(rad) $d = 1$mm	β(deg) $d = 1$mm	β(rad) $d = 50$ nm	β(deg) $d = 50$ nm
011	2.027	34.41°	2.78×10^{-7}	1.6×10^{-5}	0.0056	0.32°
200	1.433	53.07°	3.81×10^{-7}	2.2×10^{-5}	0.0076	0.44°
211	1.17	78.25°	1.125×10^{-6}	1.0×10^{-4}	0.0225	1.29°

Clearly, for the 1 mm crystallites, all reflections have negligible broadening. However, for the 50 nm crystallites, we see that the diffraction peaks would be rather broad with a measurable broadening of 0.32°–1.29°. Further, the broadening increases with the Bragg angle or smaller interplanar distances.

7.9 Methods of X-Ray Diffraction

Bragg's law (4.3.3) and its equivalent formulation in reciprocal space, the Ewald sphere construction (Fig. 4.3.2), and the Laue criterion (4.3.7) are the basis for the applications of X-ray diffraction in materials characterization. We have already seen the use of Bragg's law in X-ray analysis, especially in the form of wavelength dispersion spectroscopy (§2.5.1). In that case, using crystals of known inter-planar spacing, d_{hkl}, the intensities as a function of scattering angle, θ, were measured, and when Bragg's law is satisfied, the unknown wavelength, λ, is determined. Here, our interest is to use X-ray diffraction to determine the lattice parameter and structure of an unknown material, be it in crystalline, polycrystalline or amorphous form.

In general, there are two variables, θ and λ, that can satisfy Bragg's law for a crystal with a series of values of interplanar spacing, d_{hkl}. In a scattering experiment, for a given d_{hkl}, we can satisfy Bragg's law by either keeping θ fixed and varying λ, or keeping λ fixed and varying θ. The former, the Laue method, is traditionally applied to the study of single crystal specimens. The latter method, has a number of variations; two of them–the Debye–Scherrer method (fix λ and vary θ), with particular applicability to the study of powder specimens, and diffractometry (fix λ and rotate the crystal/specimen), are discussed here. Note that we use the Ewald sphere construction and the Laue criterion to discuss these methods; before we proceed, it would be helpful to review §4.3.

7.9.1   The Laue Method for Single Crystals

The geometry of the Laue method is straight forward and illustrated in Figure 7.9.1. A narrow beam of white radiation, such as the continuous spectrum from an X-ray tube (§2.4), is allowed to be incident on a fixed single crystal specimen. In the Laue method, either the forward diffracted beam is recorded in transmission (Fig. 7.9.1a), provided the specimen is sufficiently thin, or in the case of a bulk crystal, the backward diffracted beams are recorded by back reflection (Fig. 7.9.1b). In both cases an array of spots, or a diffraction pattern, is recorded (Fig. 7.9.1c,d). The common practice is to record the back-reflected pattern, as this allows the use of thicker, opaque specimens without the need for any specimen thinning procedures.

Details of Laue X-ray diffraction can be understood in terms of the reflecting or Ewald sphere construction (Fig. 7.9.2), introduced in §4.3.2. The incident, X-ray white radiation used in the experiment, can be considered to vary continuously from wavelengths λ_{min} to λ_{max}. In the figure, only two intermediate wavelengths, λ_1 and λ_2, such that $\lambda_{min} < \lambda_1 < \lambda_2 < \lambda_{max}$, are explicitly shown. Thus, only the reciprocal lattice points in the shaded area (that would be intersected by the spheres of one of the many Ewald spheres at intermediate wavelengths) contribute to the observable diffraction pattern. Hence, for the case shown in Figure 7.9.2, 44 nodes contribute to the Laue pattern.

In general, the Laue diffraction pattern produced in the laboratory is not easy to interpret for anything other than single crystal specimens. However, the advent of synchrotron radiation (§2.4.3) and powerful computers for interpretation have provided new avenues of application for the technique. In particular, the Laue method has been applied to resolve the structure of small molecules [4] and macromolecular crystallography [5], including dynamic, time-resolved measurements [6]. Finally, the simplicity of the experimental set-up makes for

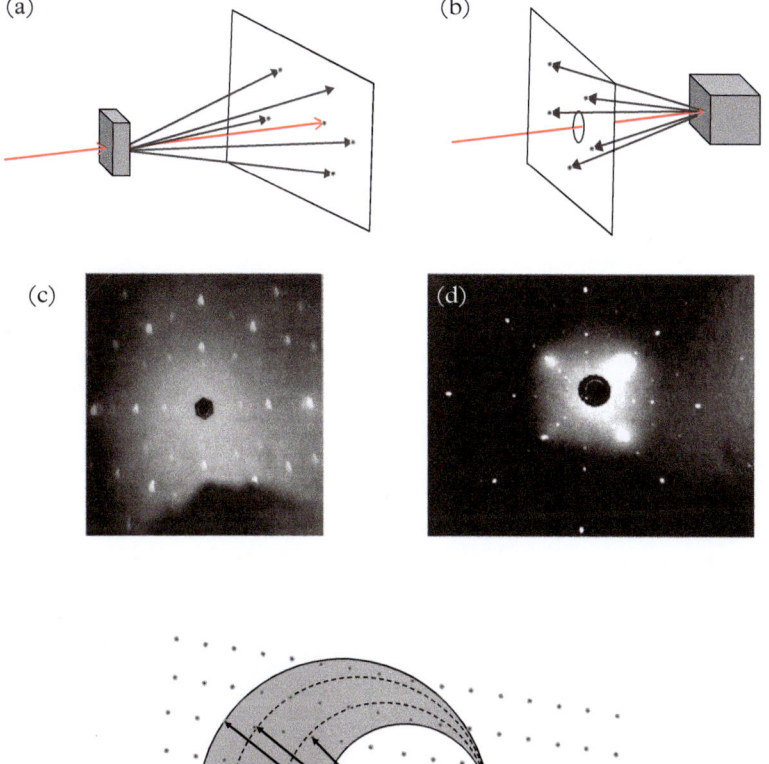

(a) (b) (c) (d)

Figure 7.9.1 The Laue method for single crystals in (a) the transmission, and (b) back-reflection, geometries. Examples of the back-reflection Laue pattern of an aluminum crystal with incident beam parallel to (c) [011] (note: shadow is from the goniometer holding the specimen), and (d) [001] clearly showing the fourfold symmetry of the intersecting zones.

Adapted from (c) Cullity (1978), and (d) Hammond (2006).

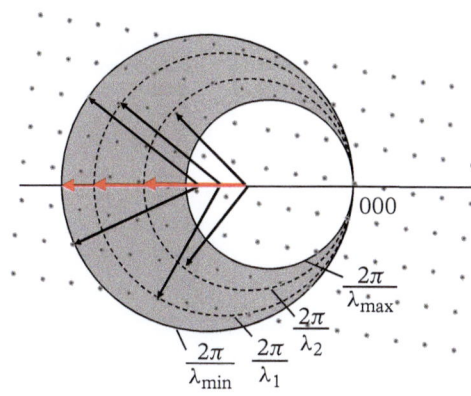

Figure 7.9.2 The Ewald sphere construction using "white" radiation over a range of X-ray wavelengths, $\lambda_{min} < \lambda < \lambda_{max}$, for diffraction from a single crystal specimen. All the nodes (total = 44) in the shaded region are observed in the Laue pattern, but only a few (six) are indicated by arrows.

a very portable Laue diffractometer, including remotely carrying out diffraction analysis of geological specimens on Mars (see §7.11.5).

7.9.2 Diffractometry of Powders and Single Crystals

Consider a powder specimen with millions of grains aligned in random orientations, each of which contains various lattice planes, (hkl), that under appropriate conditions of diffraction angle, $2\theta_{hkl}$, satisfy Bragg's law. In practice, as the direction of the incident beam is varied, there are two ways that the angle between the incident beam and the scattered beam can always be maintained at 2θ in a

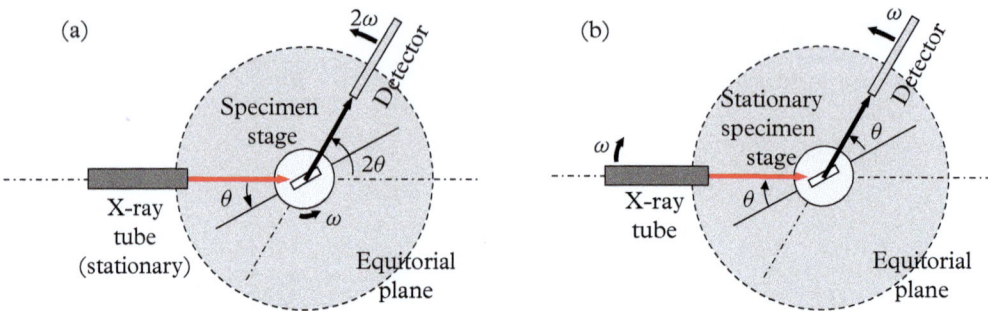

Figure 7.9.3 (a) The Bragg–Bretano geometry, also known as the θ–2θ scan in X-ray diffraction. Here the specimen is rotated at half the angular velocity of the detector, and the source is stationary. (b) A θ–θ scan; here the specimen is stationary and both the source and detector are rotated at the same angular velocity, ω, but in opposite directions.

diffractometer: (*i*) the X-ray tube or source is stationary and the detector and specimen rotate with angular velocities of 2ω and ω, respectively, in the same manner (Fig. 7.9.3a); this is known as the Bragg–Bretano geometry or commonly as the $\theta - 2\theta$ scan, and (*ii*) the specimen is stationary and both the source and detector are rotated towards each other at the same angular velocity (Fig. 7.9.3b); this is known as the $\theta - \theta$ scan. Both these arrangements, where the orientation of the X-ray source, the specimen, and the detector are well defined, ensure that Bragg's law is satisfied at all times.

Now, consider the $\theta - 2\theta$ diffractometry from the reciprocal space point of view (Fig. 7.9.4). From the Laue criterion (4.3.7), we expect to observe a diffracted beam with sufficient intensity when any reciprocal lattice point intersects the surface of the Ewald sphere. In a powder specimen, the grains can be considered to be randomly oriented. Hence, its reciprocal lattice can be derived from that of the equivalent single crystal by the free rotation about the center of the limiting sphere. Thus, each reciprocal lattice point or node will generate a spherical surface, which will then interact with the Ewald surface in the form of a cone of allowed reflections, for all individual grains that subtend an angle, 2θ, with the incident beam.

A diffractometer uses the equatorial geometry (Fig. 7.9.3), whereby the diffracted beam angles are measured in a plane defined by the incident X-ray beam and the rotation of the detector about an axis passing through the specimen as shown. As the angular relations between the detector, the polycrystalline powder specimen and the incident beam are varied systematically in the equatorial plane, either in the form of $\theta - 2\theta$ or $\theta - \theta$ scans, peaks corresponding to the various planes satisfying Bragg's law are observed. Figure 7.9.5 shows a typical example of a powder specimen of NaCl. Note that, in practice, a diffractometer detector will not detect X-rays from planes where normal does not lie in the equatorial plane,

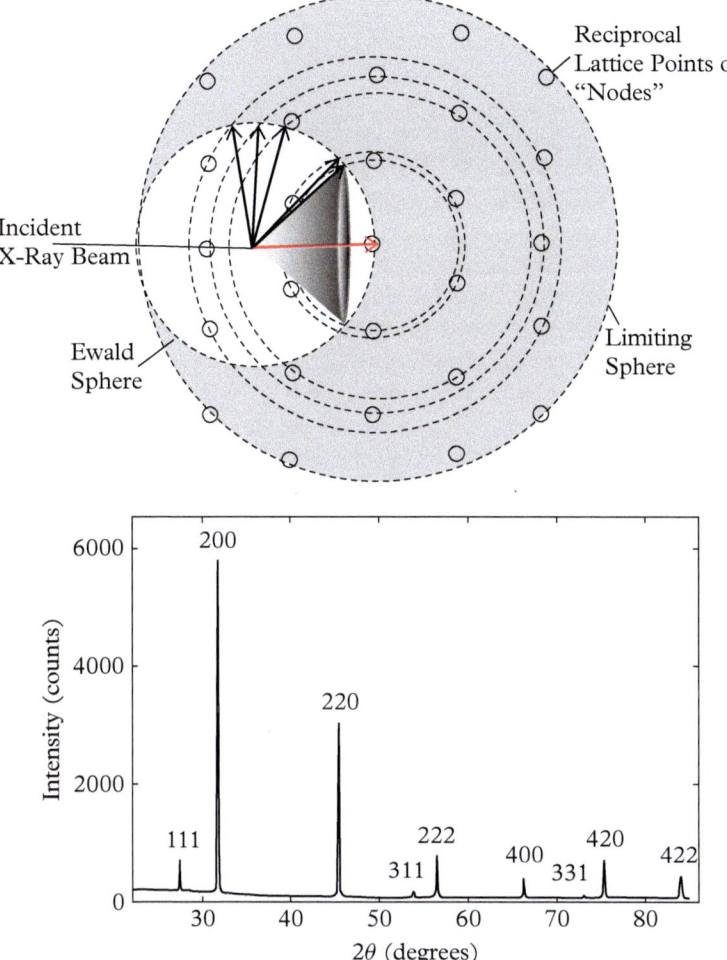

Figure 7.9.4 Diffraction from a powder or polycrystalline specimen is equivalent to rotating the reciprocal lattice about the center of the limiting sphere. Each reciprocal lattice vector now forms a sphere, which then intersects the Ewald sphere in a cone of allowed reflections, representing all the individual crystallites/grains that subtend an angle, $2\theta_{hkl}$, with the incident beam.

Figure 7.9.5 Powder diffraction data for NaCl. Notice that the $h + k + l =$ even reflection have higher intensity than those with $h+k+l =$ odd, consistent with the scattering factors for Na and Cl being in phase for the former and out of phase for the latter (see Example 7.11.1). Notice that the most intense reflections are 200, 220, and 222, in good agreement with powder data file shown in Figure 7.11.1.

even though they may satisfy the Bragg condition. In other words, the intersection of the Ewald sphere and the reciprocal lattice (sphere) is in the form of a cone (as shown for one reflection in Fig. 7.9.4), and we only measure one section in the equatorial plane. Further, only a small fraction of grains in the powder specimen will contribute to the observed pattern, as only planes with normal in the equatorial plane contribute to the diffracted intensities. As a result, it is necessary that the average powder size in the specimen be sufficiently small to give a statistically good sampling of all the possible lattice parameters in the specimen.

Large databases, called powder diffraction files (PDFs), for data sets with extensive entries numbering well over 200,000 different materials, are published by the *International Center for Diffraction Data*.[8] To identify any material, the $2\theta_{hkl}$

[8] http://www.icdd.com

values for the first three most intense X-ray diffraction peaks are compared with these tables, and generally, this is the first step in solving the crystal structure and phase of a new or unknown material (see §7.11.3).

Even though we have discussed powder diffractometers, single-crystal specimens may also be examined in a diffractometer by using a Eulerian cradle, also known as a four-circle goniometer, shown schematically in Figure 7.9.6, to mount and orient the specimen. As the name implies, there are four rotational motions available, three of which are associated with the crystalline specimen, and one with the detector/counter. The instrument has a main axis, normal to the equatorial plane containing the incident and diffracted beams passing through the crystal. There are two possible rotations about this main axis: the rotation of the detector, 2θ, to satisfy the Bragg diffraction condition for a particular reflection, and the rotation of the cradle, ω, as shown. Further, once the detector has been set at the Bragg angle, the scattering vector, $\mathbf{g}$, can be oriented to make the correct angle with the incident and diffracted beams by rotating the crystal about two additional axes, Φ and χ, also shown in the figure. Note that, if the cradle is *fixed* with its χ-plane perpendicular to the incident X-ray beam, the ω degree of freedom is redundant; such a set-up is also common and is known as a three-circle goniometer.

In practice, only two rotations are required to bring a reciprocal lattice node to the intersection of the Ewald sphere and the equatorial plane. Starting with $\chi = 0$, a rotation about Φ (or ω) can bring the required reciprocal lattice node on the Ewald sphere as shown in Figure 7.9.7a. A subsequent rotation, $\Delta\chi$, about the χ-axis can bring the node to the equatorial plane (Fig. 7.9.7b). Then, the diffracted beam can be measured by placing the detector at the required 2θ angle. Note that, in such arrangements, it is possible that the physical χ-circle may obstruct the scattered beam from reaching the detector; thus, the extra degree of

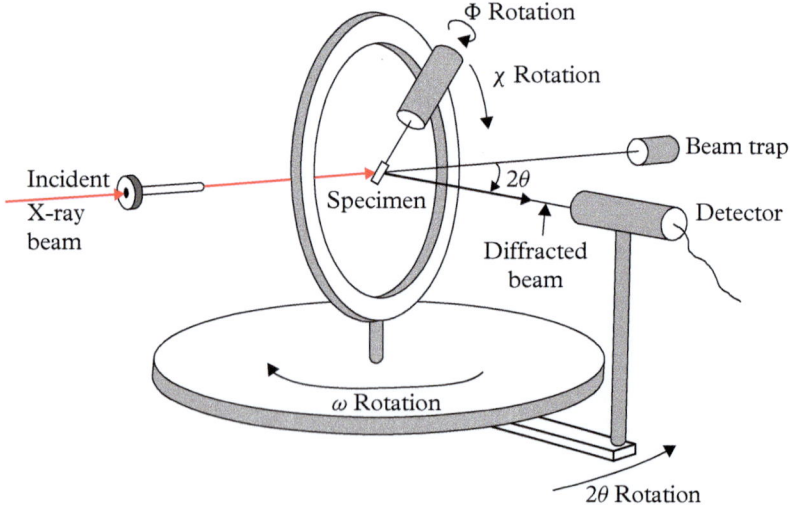

Figure 7.9.6 A four-circle diffractometer, also known as a Eulerian cradle, used for single crystal analysis. The detector rotates about the 2θ axis in the equatorial plane, and the specimen can be oriented in any way by a combination of three different rotation axes, ω, χ, and Φ.

Adapted from Woolfson (1997).

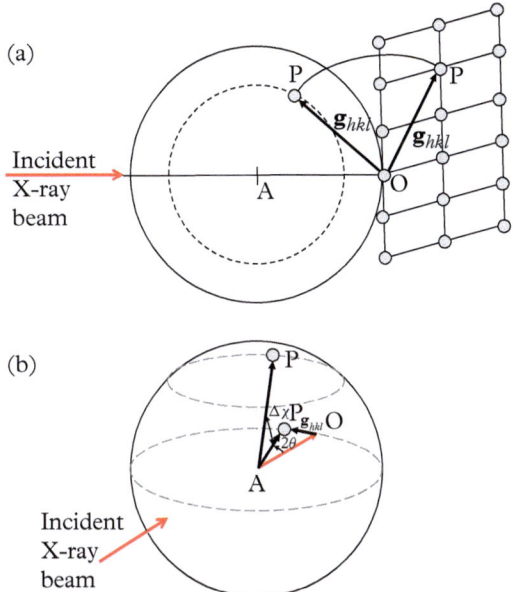

Figure 7.9.7 Typical method to bring a reciprocal lattice node to the intersection of the Ewald sphere and the equatorial plane. (a) First, the crystal is rotated about Φ until the node lies on the sphere, as shown in this projection looking down the instrument main axis. (b) Second, it is rotated about χ to bring the node to the equatorial plane, where it can be measured by diffraction as shown.

Adapted from Giagovazzo et al. (1992).

freedom afforded by the four-circle goniometer can prove to be of valuable utility. Most modern single-crystal diffractometers with four-circle Eulerian cradles are computer controlled with sophisticated software to carry out the tasks of crystal orientation, data collection, and solving the crystal structure. Finally, both powder and single-crystal diffractometers can be modified with heating or cooling units to carry out measurements at both high and low temperatures (see §7.11.4).

7.9.3 Debye–Scherrer Method for Powders

An earlier alternative to the powder diffractometer is the Debye–Scherrer camera that belongs to a family of transmission "cameras", where the specimen is placed in between the X-ray source and a stationary recording film. In this family, the Debye–Scherrer camera (Fig. 7.9.8a) is unique, because even for standard specimen to film distances of ∼8.0 cm, it records a wide range of 2θ angles as it registers both the forward- and back-scattered X-ray beams. It also has the distinct advantage of working with very small quantities of powder specimens, that consists of a large number of tiny crystallites—-typical volume of crystallite is 5 μm^3 and thus in a ∼1 mm^3 specimen, we expect ∼10^7 crystallites—-arranged in a random orientation.

The powder specimen is usually formed in the form of a needle-shaped cylinder, and housed in a thin-walled glass tube at the center of the camera. When a monochromatic, well-collimated X-ray beam is incident on the specimen in a Debye–Scherrer camera, the diffraction geometry is similar to that for the

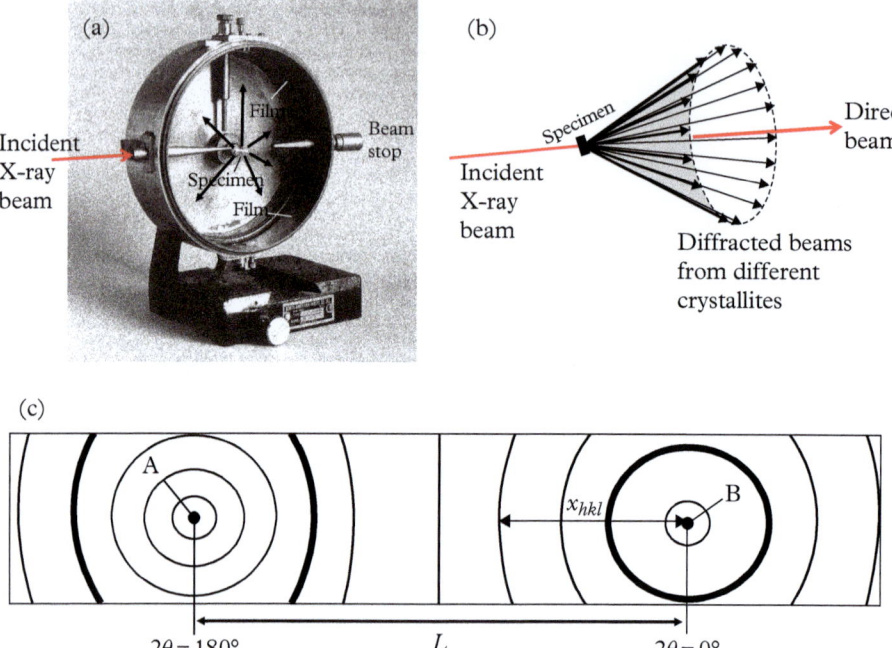

Figure 7.9.8 (a) The Debye–Scherrer camera, showing the incident beam, the beam stop, the specimen, the film, and the beams diffracted in all directions. (b) The locus of the diffracted beams from all the crystallites in a powder specimen for a given reflection, *hkl*. (c) A straightened photographic film, showing the rings that originate from intersection of the cone in (b); the position of the ring, x_{hkl}, normalized by the separation, L, of the incident, A, and beam-stop, B, positions give the angle θ_{hkl}.

(a) Adapted from Giagovazzo et al. (1992). (b) Adapted from Woolfson (1997).
(c) Adapted from de Graef and McHenry (2007).

diffractometer (Fig. 7.9.4) with the difference being the specimen is cylindrical instead of a flat thin film. For the particular crystal structure in such a powder specimen, each possible crystallographic plane, (*hkl*), that satisfies the Bragg condition, $2\theta_{hkl}$, is now present in large numbers at random orientations because of the small crystallite size. As a result, all the diffracted beams for the specific reflection, *hkl*, make an angle, $2\theta_{hkl}$, with the incident beam and collectively form a cone of semi-angle, $2\theta_{hkl}$, originating at the specimen as shown in Figure 7.9.8b. When each cone of diffracted beams for a specific reflection intersects the circular film mounted on the inside of the camera, a diffraction pattern of ring-segments is recorded; when the film is straightened out, it appears as in Figure 7.9.8c.

The two holes on the film correspond to the position (A) of the incident beam, or $2\theta = 180°$, and position (B) of the transmitted beam stop, or $2\theta = 0°$; the distance, L, between points A and B on the film is recorded. The value of $2\theta_{hkl}$ for any particular diffraction ring can be determined by measuring its distance, x_{hkl}, from position, B, and taking the ratio to give

$$\theta_{hkl} = \frac{180°}{2}\frac{x_{hkl}}{L} \qquad (7.9.1)$$

Using Bragg's law, each observed ring can be identified with a specific interplanar spacing, d_{hkl}. Comparing the list of observed spacing with the PDF (§7.11.3), the structure may be identified. However, a large number of diffracted beams will have similar 2θ values causing them to overlap making it difficult to resolve their intensities separately (see Exercise 7.4). As a result, a diffractometer is preferred for the analysis of powder specimens; moreover, for new and unknown materials, generally X-ray diffraction of single crystal specimens gives the best structural analysis (see §7.11.5). If this is not possible, sophisticated refinement (§7.11.3) of powder diffraction data gives good results.

7.9.4 Thin Films and Multilayers: Diffractometry, Reflectivity, and Pole Figures

As we have seen in §7.9.2 and in Figure 4.3.5), the standard or symmetrical Bragg–Bretano geometry (Fig. 7.9.3) provides the interplanar spacing of only those lattices planes parallel to the surface of a thin film specimen. This is achieved by orienting the incident and reflected beams, including all the required collimation, such that they make equal angles with the specimen surface (Fig. 7.9.9a). Figure 7.9.3 shows how, in practice, this geometry can be implemented in two ways by keeping the X-ray tube/source stationary and rotating both the specimen and detector, or alternatively, keeping the specimen/film stationary and rotating the source and detector in opposite direction but at the same angular rate. In this case, if the specimen is a single crystal film there will be only one peak; however, if the film is multilayer, such as the tri-layer film shown in Figure 7.9.10, there will be one (family of) peak(s) for each of the layers, and the substrate.

Figure 7.9.9b shows an alternative *asymmetric* arrangement, common in X-ray diffractometry of thin films. Here, the specimen is tilted at an angle, α, which can be varied, but the angle between the incident and reflected beam is always maintained at 2θ. As a result, the incident, $(\theta + \alpha)$, and reflected, $(\theta - \alpha)$, beams make different angles with the specimen surface, thereby making it possible to record diffraction from planes that are NOT parallel to the specimen surface. Note that the maximum angle of specimen tilt, $\alpha = \theta$, and using the Ewald sphere construction (Fig. 7.9.9c) we can now see the regions of reciprocal space, or the portion of the limiting sphere that can be sampled by the symmetric and asymmetric scattering geometries.

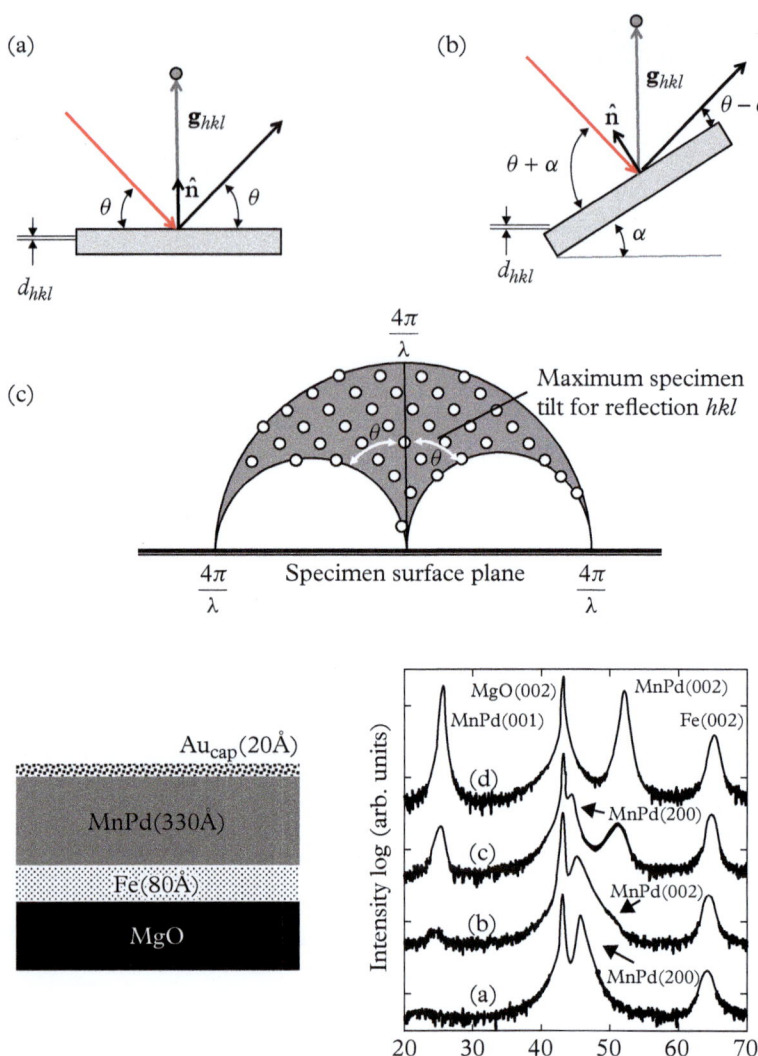

Figure 7.9.9 Diffractometer settings for single crystals. (a) A symmetric setting where all the reciprocal lattice vectors involved in diffraction are from planes with normal to the specimen surface. (b) An asymmetric setting where the reciprocal lattice vectors are inclined by an angle, α, with respect to the surface normal, $\hat{\mathbf{n}}$. (c) All the possible reflecting planes lie within the shaded region of the limiting sphere.

Figure 7.9.10 High angle, $\theta - 2\theta$ X-ray scan, using Cu Kα radiation, from the multilayer structure (shown on the left), grown at different temperatures: (a) 100 °C, (b) 200 °C, (c) 350 °C, and (d) 450 °C. The orientation of the tetragonal MnPd film changes from an a-axis normal to a c-axis normal when the growth temperature is increased from 100 °C to greater than 450 °C.

Adapted from [7].

We now present three distinct examples of diffraction from (1) a polycrystalline film, (2) a single crystal film subject to a Φ-scan, and (3) a multilayer film at both small and large angles.

(1) A *polycrystalline* film with many grains oriented in random directions, is mounted on a *four-circle* goniometer (Fig. 7.9.6) and aligned such that a specific reflection, *hkl*, is in the Bragg orientation, with a peak at $2\theta_{hkl}$ observed in X-ray diffraction. This implies that some of the polycrystalline grains are oriented in such a way that they satisfy the Bragg condition. Now slowly rotate the crystal

around the Φ-axis while maintaining the $\theta - 2\theta$ scattering geometry in the equatorial plane. Since the polycrystalline film is composed of small crystalline grains oriented at random, we expect to continue seeing a diffracted intensity from some of the grains, independent of the value of Φ. In other words, if we plot the intensity of the specific reflection, I_{hkl}, it should have the appearance of a circle, i.e. is the same for all values of Φ.

(2) Now consider a *single crystal* film, grown on another single crystal substrate. Again, set the scattering condition to satisfy Bragg's law for a specific reflection, *hkl*. The multiplicity and the angles between the families of planes {hkl} is well defined for a crystal with a given symmetry; thus, rotating the crystal about the Φ-axis will result in measurable intensity, I_{hkl}, only at those values of Φ consistent with the symmetry of the crystal. Such a plot of intensities for a given reflection, *hkl*, as both Φ and χ are varied is commonly referred to as a pole figure. We now illustrate this with a specific example.

Figure 7.9.11 shows pole figures for MgO films grown epitaxially on single crystal substrates of (a) $SrTiO_3$ (STO), a perovskite structure with $a \sim 3.90$ Å, and (b) $LaAlO_3$ (LAO), a perovskite structure with $a \sim 3.8$ Å. Both pole figures were measured for the MgO_{222} reflection set at $2\theta_{222} = 78.6°$. Note that in these plots, Φ is the angle by which the specimen is rotated about the surface normal, and χ is the angle by which the surface is tilted out of the equatorial plane. Thus, the three concentric circles represent χ values of $30°$, $60°$, and $90°$. In both figures, notice a ring of four MgO_{222} spots at $\chi \sim 55°$; further, there is an

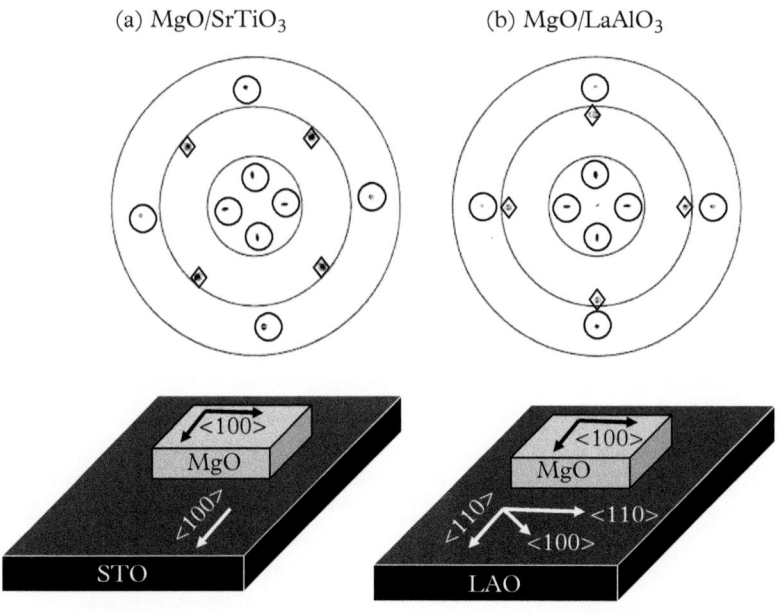

(a) MgO/SrTiO$_3$ (b) MgO/LaAlO$_3$

Figure 7.9.11 Pole figure measurements showing the variation of Φ around the circumference of the circle, and χ increasing radially, with three concentric circles representing $\chi = 30°, 60°$, and $90°$, respectively. (a) For MgO/SrTiO$_3$, the MgO$_{222}$ reflection at $2\theta_{222} = 78.6°$ ($\Diamond$), shows four peaks at $\chi = 55°$, consistent with the epitaxial relationship, MgO $\langle 100 \rangle$ $\|$SrTiO$_3$ $\langle 100 \rangle$, as illustrated here. The STO$_{(013)}$ reflections at $\chi = 18°$ and $72°$ are circled. (b) For MgO/LaAlO$_3$, the MgO$_{222}$ reflections occur at the same Φ angles, illustrating a different epitaxial relationship rotated by $45°$, i.e. MgO $\langle 100 \rangle$ $\|$LAO $\langle 110 \rangle$.

Adapted from [7].

inner ($\chi \sim 18°$) and outer ring ($\chi \sim 72°$) of spots that correspond to the substrate (013) reflections. However, there is one significant difference between the two figures. For the $SrTiO_3$ substrate (Fig. 7.9.11a), the MgO_{222} peaks occur midway in Φ angles between the STO_{013} peaks, whereas for the $LaAlO_3$ substrates, the MgO_{222} peaks occur at the same Φ angles as the LAO_{013} peaks. From this we can conclude [8] that the orientation relationship (Fig. 7.9.11) between the MgO films is$MgO \langle 100 \rangle \| SrTiO_3 \langle 100 \rangle$, and $MgO \langle 100 \rangle \| LAO \langle 110 \rangle$, and the relative lattice orientations for each case is as shown.

(3) *Multilayers* are sequences of very thin crystalline films of two or more materials (such as Co and Pt), grown (often) on single crystal substrates. Each of the distinct layers displays their own characteristic high-angle Bragg peak. Typically, as the layers are very thin, these Bragg peaks are broadened, and in principle, it is possible to use their broadening to estimate the thickness of the individual layers following the Scherrer equation (7.8.5). However, the repeat distance, Δ, of the multilayer or superlattice can also be determined from the location of the satellite reflections which occur on either side of the Bragg peaks. For a multilayer with long-range structural coherence, the repeat distance is given by

$$\Delta = \frac{\lambda}{2 (\sin \theta_2 - \sin \theta_1)} \tag{7.9.2}$$

where λ is the wavelength of the X-rays, and $2\theta_1$, $2\theta_2$ are the Bragg angles of adjacent satellite peaks. Converting the Bragg angles to nominal distances, we can rewrite (7.9.2) as

$$\frac{1}{\Delta} = \frac{1}{d_2} - \frac{1}{d_1} \tag{7.9.3}$$

Figure 7.9.12a shows an example [9] of a $(Co_{3\text{Å}}Pt_{15\text{Å}})_{30}$ multilayer grown on GaAs(111) with a $Ag_{200\text{Å}}$ buffer layer, using Cu Kα $\left(\lambda = 1.54 \text{ Å} \right)$ radiation. Peaks are observed at 2θ values of 30.5 (Pt, $n = -2$), 35.5 (Pt, $n = -1$), 38.2 (Ag), 40.6 (Pt, $n = 0$), and 45.6 (Pt, $n = 1$). Using (7.9.2), it is easy to see that the multilayer period, $\Delta \sim 18\text{Å}$ (see Problem 7).

At low angles ($2\theta = 1 - 5°$), we see additional Bragg reflections from the repeat period, Δ, of the multilayers, i.e. reflections that satisfy $\lambda = 2\Delta\sin\theta$. These measurements at low angles, generally termed as reflectivity measurements, show very fine oscillations that depend on the total thickness, T, of the multilayer film, where $T = n\Delta$ and n is the number of repeating units. The analysis of these fine oscillations is beyond the scope of this book but can be readily found in Parrat [10]. In addition, at slightly larger angles, satellite peaks are observed, from which using (7.9.2), we can also obtain the period of the multilayers. Again, for the same multilayer a low angle scan (reflectivity) is shown in Figure 7.9.12b. Two satellites are observed at $\theta \sim 2.5°$ and $\theta \sim 4.9°$, consistent with a period of 18.4 Å.

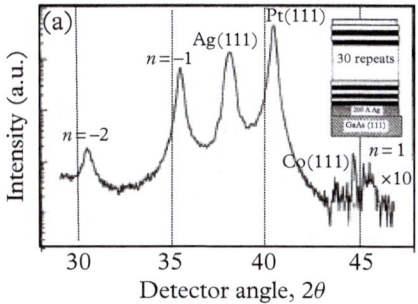 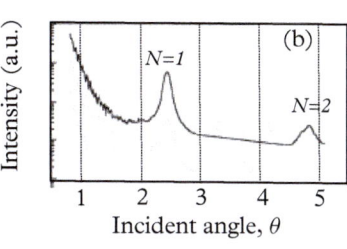

Figure 7.9.12 X-ray diffraction from a $(Co_{3Å}Pt_{15Å})_{30}$ multilayer grown epi-taxially on GaAs(111) with a $Ag_{200Å}$ buffer layer, using Cu Kα $\left(\lambda = 1.54\text{ Å}\right)$ radiation. (a) High-angle scan with the multilayer shown in the inset. (b) A low-angle scan showing the very fine oscillations corresponding to the total thickness ($\sim$540+200 Å) of the multilayer, and two satellites ($N = 1, 2$) determined by the multilayer period ($\sim$18 Å).

Adapted from [7].

7.9.5 Practical Considerations: Collimators and Monochromators

For X-ray diffraction, ideally, we require photons that correspond to a single wavelength. Moreover, all the diffraction methods discussed so far in this section require a narrow X-ray beam to be incident on the specimen.

In the laboratory, X-ray tubes (§2.4.1) generate characteristic X-ray lines (often, more than one) superimposed on a Bremsstrahlung background (Fig. 2.4.1). Appropriate absorption filters, as discussed in §2.4.2, are used to select a narrow wavelength out of the continuous X-ray spectrum produced by the X-ray tube (Fig. 2.4.4). In addition, the radiation from an X-ray tube is emitted in all directions; hence, methods to generate a narrow parallel beam, or collimation, in the direction of the specimen is required.

Alternatively, synchrotron sources (§2.4.3) produce high-intensity X-rays that are spatially very confined and pencil like (Fig. 2.4.5), but here too the wavelength of the radiation is over a broad continuous spectrum of energy. In this case, devices to select a narrow X-ray wavelength out of the synchrotron radiation, such as monochromators, are required.

We now provide a brief description of such collimators and monochromators.

7.9.5.1 Collimators

A collimator is typically used to produce a narrow, parallel, cylindrical beam of X-rays. In practice, a simple pinhole design (Fig. 7.9.13) consists of a narrow cylinder with two apertures (A_1 and A_2) to define the beam, and a third aperture (A_3)—the guard aperture, which does not define the beam but stops all radiation

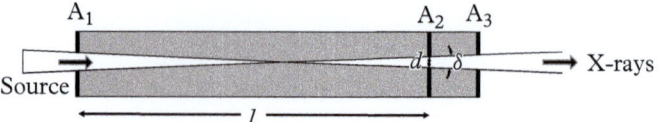

Figure 7.9.13 A pinhole collimator. The maximum angle of divergence, δ, depends on the diameter, d, of the defining aperture (A_2), and the separation, l, between the two apertures A_1 and A_2.

scattered by the second defining aperture (A_2). The apertures are normally circular in shape, but square and rectangular shapes can also be used. When cylindrical pinhole apertures are used in the laboratory, they are combined with either filters or monochromators to obtain a narrow wavelength of X-rays.

For the collimator shown in Figure 7.9.13, the divergence angle of the beam, δ, can be calculated, assuming small δ, from the dimensions of the cylindrical pinhole as

$$\tan\frac{\delta}{2} = \frac{d/2}{l/2} = \frac{d}{l} = \sin\frac{\delta}{2} = \frac{\delta}{2} \tag{7.9.4}$$

Then, $\delta = 2d/l$, where the divergence angle is in radians. The dimensions of a typical pinhole collimator are $d = 0.3 - 0.5$ mm, and $l = 40 - 50$ mm; thus, taking average values we get $\delta = 0.8/45 = 0.018$ radians $= 1.02°$.

7.9.5.2 *Monochromators*

We have seen earlier (Fig. 2.4.4) that a filter can be used to select a wavelength interval, out of the spectrum produced by an X-ray tube. The filter absorbs a range of unwanted radiation but allows the X-rays of interest for the diffraction experiment to go through. Alternatively, as we have seen in the design of wavelength dispersive spectrometers (§2.5.1.1), specific single crystals (Table 2.5.1) can be used to select X-rays of a specific wavelength. Applying the same principle of single crystal diffraction, we can also produce a narrow beam of X-rays within a very narrow range of wavelengths. Such devices, known as single crystal monochromators, are based on the simple application of Bragg's law, (4.3.3). When radiation of different wavelengths is incident on a single crystal, diffracted beams are produced in different directions depending on the diffraction angle (β) and the wavelength (λ) of the X-rays. In other words, a specific wavelength out of the X-ray spectrum can be selected by choosing a specific diffraction angle (β). In the simplest monochromator, a single crystal is cut in such a way that a major set of crystallographic planes parallel to one of the crystal faces satisfies the Bragg diffraction condition. To be truly effective the interplanar spacing of the monochromator, in the specific crystal orientation, should be in the range to produce a reasonable X-ray beam at the required wavelength. Moreover, to minimize the loss of intensity, a reflection that has a small scattering angle is chosen to minimize the contributions from the Lorentz polarization factor (Fig. 7.10.4).

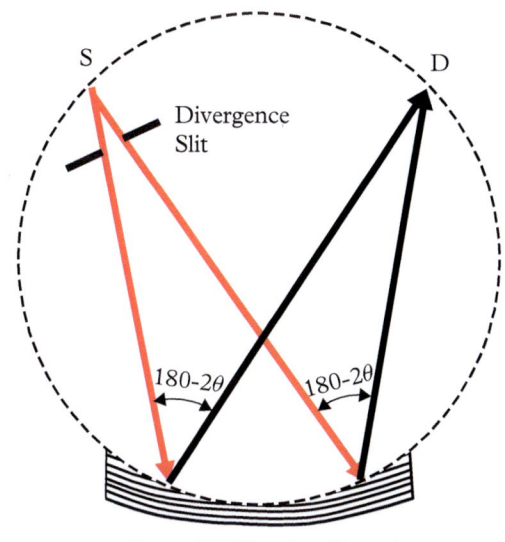

Curved Diffracting Crystal

Figure 7.9.14 A curved crystal diffracting monochromator. The source may be a point or a line, emitted by an X-ray tube; often it is polychromatic. The diffracting planes (*hkl*) of the curved crystal make an angle, θ, with respect to the incident and diffracted beams. The point D is now the focused monochromatic source for X-rays of a specific wavelength (e.g. $K\alpha_1$). See Figure 2.5.1.

A particularly clever design of a very efficient X-ray monochromator uses a curved diffracting crystal (Fig. 7.9.14). The polychromatic source produces characteristic lines ($K\alpha_1, K\alpha_2, K\beta$, etc.) above the continuum background (Fig. 2.4.1). A curved and shaped crystal is used that has a radius twice that of the focusing circle, but shaped such that the inner surface fits the focusing circle, as shown. The position of the source, S, the angle $(180-2\theta)°$, and the interplanar spacing, d_{hkl}, of the crystal are chosen to select a specific line, such as $K\alpha_1$, and bring the divergent beam from S, to a focus at the point, D, on the focusing circle. Further, the X-rays emerging from D are highly monochromatic and effectively serve as the source producing a narrow beam of X-rays.

7.10 Factors Influencing X-Ray Diffraction Intensities

In §7.4 we derived an expression for the structure factor, F_{hkl}, and the intensity, I_{hkl}, of a specific reflection, *hkl*, in X-ray diffraction. In practice, this ideal intensity is modified by a variety of experimental factors. We now discuss each of the four principal factors.

7.10.1 Temperature Factor

Earlier, while calculating structure factors (§7.4 and §7.5), we assumed that the crystal is a collection of atoms occupying fixed positions in the unit cells. However, the atoms vibrate about their mean or equilibrium positions with the amplitude of the vibration, typically of the order of one-tenth the lattice spacing (§5.3.5.1), increasing as the temperature increases.

As discussed in §3, the atoms in a crystal are bound to each other by various types of forces, and their equilibrium positions correspond to the energy minima. Hence, as their positions are perturbed, the atoms will return to their original positions in the form of oscillations. The collective vibration modes of all the atoms in a crystal are known as *phonons*. However, it is still reasonable to assume, as a first approximation, that the thermal motions of atoms in a crystal are independent of each other. In any case, these oscillations will effectively modify the electron density of each atom and alter their ability to scatter X-rays (or, for that matter, also electron and neutron probes). A larger vibration amplitude implies a larger *spatial extent* of the electrons, or alternatively, for the same number of electrons, an effectively lower electron density. This reduces the atomic scattering factor, f_T, at any temperature, T, as well as the diffraction intensity, $I_{hkl}(T)$, for a specific reflection. Typically, the period of the atomic vibration is much smaller than the time scale of an X-ray scattering experiment. As a result, only the time-averaged displacement of the atoms with respect to their equilibrium position is required to describe the effect of thermal motion on diffraction. Without going into the specific details (see Giacovazzo, 2002), we can simply state that to accommodate the effect of temperature, the atomic scattering factor is multiplied by an attenuation or damping factor:

$$ f_T\left(\frac{\sin\theta}{\lambda}\right) = f_0\left(\frac{\sin\theta}{\lambda}\right) e^{-B(T)\left(\frac{\sin\theta}{\lambda}\right)^2} = f_T(s) = f_0(s)e^{-B(T)s^2} = f_0(s)e^{-M} $$

$$(7.10.1)$$

Here, the subscript, 0, refers to the ideal value at 0 K. Further, the exponential factor, $B(T)$, at any temperature, T, is proportional to the mean square displacement, $\langle r_i^2 \rangle$, of the atom, i, in a direction normal to the reflecting planes. The exact value of $\langle r_i^2 \rangle$ is rather difficult to determine, but the dependence of $B(T)$ on temperature was first studied by Debye,[9] who also developed an analytical expression for f_T for atomic elements. The theory of X-ray scattering by lattice vibrations was further developed by Waller[10] in 1925, who also provided the definitive treatment of the subject. Thus, the factor, $B(T)$, are now known as the Debye–Waller factors. They are listed for some representative elements in Table 7.10.1, and a complete listing of them for a wider range of elements can be found in Peng et al. [11].

In practice, the Debye–Waller factors, $B(T)$, are not known accurately for most crystals; thus, experimental measurements are preferred, but if they are not available, elemental values are used as a first step. At room temperature they range from $\sim$10 Å^2 (the first column elements; Li, Na, K . . .), to $\sim$1–3 Å^2 (the second column elements; Be, Mg, . . .), to $\sim$0.3–0.7 Å^2 (for the remaining elements) in the periodic table. Needless to say, $B(T)$ depends on temperature, as summarized for select elements in Table 7.10.1. Finally, the *intensity* of a diffracted beam is now modified by the temperature factor, e^{-2M}, with respect to the value at 0 K,

[9] Peter J. W. Debye (1884–1966) was an American physical chemist who received the 1936 Nobel prize in Chemistry for his "contributions to the study of molecular structure."

[10] Ivar Waller (1888–1991) was a Swedish physicist.

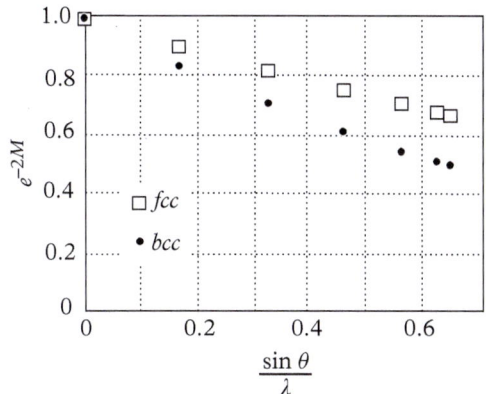

Figure 7.10.1 Variation of the temperature factor for FCC and BCC iron at 280°K as a function of the scattering parameter.

Table 7.10.1 Debye–Waller factors, $B(T)$, in Å^2, for some representative elements as a function of temperature. A more detailed list is in [12], from which this is taken.

T (K)	C (Dia)	Na (BCC)	Fe (BCC)	Fe (FCC)	Zn (HCP)	Au (FCC)
90	0.1300	2.0516	0.1493	0.1715	0.3512	0.1908
120	0.1311	2.7348	0.1332	0.2286	0.4683	0.2544
170	0.1336	3.8721	0.1886	0.3238	0.6632	0.3602
220	0.1370	5.007	0.2441	0.4189	0.8580	0.4659
270	0.1412	6.1389	0.2995	0.5140	1.0526	0.5714
280	0.1422	6.3648	0.3106	0.5330	1.0915	0.5925

and e^{-2M} decreases with scattering parameter, $\left(\frac{\sin\theta}{\lambda}\right)$; Figure 7.10.1 shows an example for Fe, both FCC and BCC, at 280°K.

7.10.2 Absorption or Transmission Factor

As X-rays travel through a crystal, both the incident and diffracted beams are absorbed. The absorption reduces the intensity of the X-rays traveling through the material by a factor proportional to the path-length, l, travelled and the linear absorption coefficient, μ, of the material:

$$I = I_0 e^{-\mu l} \qquad (2.4.2)$$

where I_0 is the incident and I the diffracted intensities. Strictly speaking, the path-length is dependent on the precise location of the scattering event (Fig. 7.10.2), as well as the incident and scattering angles.

Normally, we are interested in the scattered or transmitted intensity, which is defined by the transmission factor, $T(\theta) = I/I_0$, and to calculate it for the entire crystals of volume, V, we integrate (2.4.2) to get

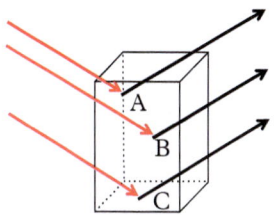

Figure 7.10.2 X-rays incident on three different points in the crystal with different primary and diffracted path lengths.

$$T(\theta) = \frac{1}{V} \int_v e^{-\mu(p+d)} dv \qquad (7.10.2)$$

where p and d, are the path lengths of the primary and diffracted beams, respectively, travelled in the crystal after being scattered at any volume element, dv. Recall (§2.4.2) that the mass absorption coefficients, $\mu_i^m = \mu_i/\rho_i$, are known (Table 2.4.1) for most elements, i, and it is easy to calculate the linear absorption coefficient, μ, for any material as

$$\mu = \rho \sum_i w_i \mu_i^m \qquad (7.10.3)$$

where ρ is its overall density, and w_i are the mass fractions of its constituent elements. Note that μ_i^m is smaller for lower atomic numbers and for shorter wavelengths. As a result, the absorption correction becomes more important for heavier elements and longer wavelengths.

An analytical evaluation of $T(\theta)$, (7.10.2), would depend on the specific beam path, which is a function of the reflection, hkl, and the overall shape of the crystal. Unfortunately, the integral, (7.10.2), cannot be calculated analytically even for the simplest crystal shapes, and hence, numerical methods are used [12]. Even though these results are beyond the scope of the book, we summarize the key results and trends. First, the transmission coefficient, $T(\theta)$, depends on the geometry of the diffraction method; as such, we only consider the two most common methods of diffractometry and the Debye–Scherrer camera. Second, the absorption is always greater for the low-θ reflections, and the difference between low-θ and high-θ reflections decreases as the linear absorption coefficient decreases.

In diffractometry, for a beam of fixed cross-section, when θ is small the area of illumination on the specimen surface is large, but its depth of penetration is small; alternatively, when θ is large, the area of illumination is small, but the depth of penetration is large. As a result, the volume of interaction between the X-ray beam and the specimen remains constant and independent of θ. Thus, $T(\theta)$ is independent of θ, and is a constant for a material:

$$T_{diffractometry} = \frac{1}{2\mu} \qquad (7.10.4)$$

Thus, for all reflections, hkl, the absorption reduces the intensity by the same factor, and if ratios of intensities are being considered, the absorption effect cancels out and it can be ignored.

In the Debye–Scherrer camera, a thin needle-like cylindrical specimen is used, and for this geometry $T(\theta)$ is difficult to calculate. However, the functional form of the dependence of absorption with angle, θ, mentioned earlier (i.e. the absorption is always greater for the low-θ reflections), is exactly opposite of the temperature factor (Fig. 7.10.1). Hence, it is reasonable to assume that the two effects of absorption and temperature cancel out when the Debye–Scherrer camera is used. Further details, specifically on absorption effects in diffraction, can be found in Giacovazzo (2002) and Cullity (1978).

7.10.3 Lorentz Polarization Factor

We have already seen (7.2.8) that when a totally unpolarized beam of X-rays is scattered in different directions, even by a single electron, its intensity is affected by the polarization factor, $P(\theta)$, given by

$$P(\theta) = \frac{1}{2}\left(1 + \cos^2 2\theta\right) \tag{7.10.5}$$

where θ is the Bragg angle. Note that, in theory, $0.5 < P(\theta) < 1.0$, but in practice, the variation is less substantial. Moreover, there are three other geometrical corrections that apply, which are combined with the polarization and together written as the Lorentz polarization factor, $L_P(\theta)$.

In §7.8 we saw that diffraction takes place not only at the exact Bragg orientation, but also when the orientation deviates from the exact angle. In the reciprocal formulation, or the Ewald sphere construction, diffraction is considered to have occurred when the Ewald sphere crosses a reciprocal lattice node, which, as mentioned (§7.8), has a finite volume in reciprocal space. Now, as the crystal is rotated (Fig. 7.10.3) at constant angular velocity, ω, if a specific reciprocal lattice node remains in diffracting position for a longer time, then its intensity can be higher. Depending on the method used, the time a node is in diffracting condition depends on the position and size of the node, as well as the velocity with which it is swept past the Ewald sphere.

In the simplest analysis, the Lorentz factor, $L_P(\theta)$, can be shown to be

$$L_P(\theta) = \frac{\omega}{v_n \lambda} \tag{7.10.6}$$

where v_n is the radial component of the velocity of the reciprocal lattice node. The linear velocity, v, of the point, P, in Figure 7.10.3 is

$$v = |\mathbf{g}|\,\omega \tag{7.10.7}$$

and its radial component, v_n, is

$$v_n = |\mathbf{g}|\,\omega \cos\theta \tag{7.10.8}$$

Now, when the Bragg condition is satisfied

$$|\mathbf{g}_{hkl}| = \frac{2\pi}{d_{hkl}} = \frac{2\pi}{\lambda} 2\sin\theta \tag{7.10.9}$$

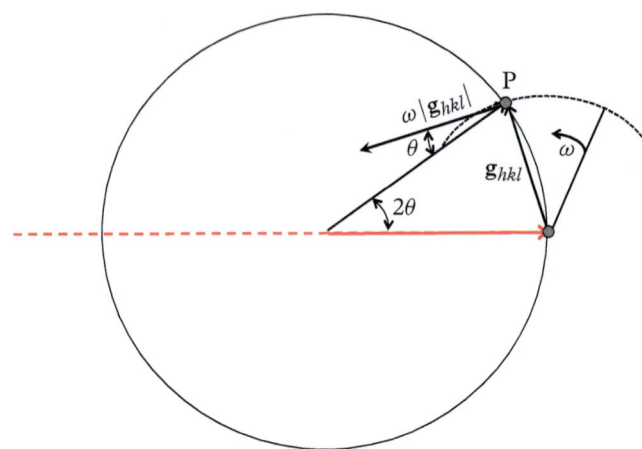

Figure 7.10.3 Lorentz correction, $L(\theta)$, for the rotation of a crystal with angular velocity, ω, about an axis normal to the plane defined by the incident and diffracted X-ray beams.

from which we get

$$v_n = 2\pi \frac{\omega}{\lambda} (2 \sin \theta \cos \theta) \tag{7.10.10}$$

and substituting in (7.10.6), we get

$$L_P(\theta) \approx \sin^{-1} 2\theta \tag{7.10.11}$$

Now, for a powder crystalline specimen, we have seen that the crystal diffracts X-rays in the surface of a cone, with their tips ending on the surface of the Ewald sphere with an opening angle of the cone of 2θ (Fig. 7.9.4 and Fig. 7.9.8). However, only a small fraction of the total intensity scattered by a family of lattice planes, {*hkl*}, is intercepted by the detector. For example, in a Debye–Scherrer camera of radius, R, the radius of the diffracted cone, r_{hkl}, at the point of intersection with the camera is $R \sin 2\theta$, with a total length of the diffraction line given by $2\pi r_{hkl} = 2\pi R \sin 2\theta$. Thus, the diffraction intensity per unit length of the diffraction line is proportional to $\sin^{-1} 2\theta$, which is the second trigonometric correction. Lastly, for a powder specimen with randomly oriented grains, the number of grains oriented favorably for diffraction with an angle 2θ is proportional to $\cos \theta$.

Combining all these three factors, we get the Lorentz-polarization factor

$$L_P(\theta) \approx \left(1 + \cos^2 2\theta\right) \left(\sin^{-1} 2\theta\right) \left(\sin^{-1} 2\theta\right) \cos \theta \approx \frac{1 + \cos^2 2\theta}{2 \sin^2 \theta \cos \theta} \tag{7.10.12}$$

where the proportionality constant has been dropped and only the angular dependence (Fig. 7.10.4) is included.

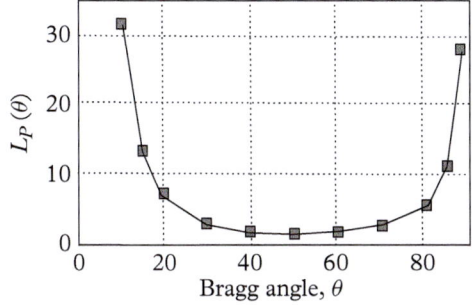

Figure 7.10.4 The variation of the Lorentz-polarization factor with the Bragg angle.

7.10.4 Multiplicity

This last correction factor takes into account the total number of variants of a specific family of lattice planes {*hkl*} that contribute to a given diffracted intensity, especially in a polycrystalline specimen with randomly oriented crystallites. The total number of planes in a given family, known as the multiplicity, P_{hkl}, depends on the specific crystal structure. For example, consider the {*hkl*} planes, where $h \neq k \neq l$. Then, $P_{hkl}^{cubic} = 48$, $P_{hkl}^{hexagonal} = 24$, $P_{hkl}^{tetragonal} = 16$, $P_{hkl}^{orthorhombic} = 8$, $P_{hkl}^{monoclinic} = 4$, and $P_{hkl}^{triclinic} = 2$. Thus, the intensity, I_{hkl}, must be multiplied by $P_{hkl}^{crystal}$ to give the observed intensity.

7.10.5 Corrected Intensities for Diffractometry and the Debye–Scherrer Camera

We can now include all these corrections factors: multiplicity, $P_{hkl}^{crystal}$, Lorentz-polarization, $L_P(\theta)$, transmission, $T(\theta)$, and Debye–Waller, $B(T)$, and write a general expression for the observed diffraction intensity as

$$I = I_{hkl} P_{hkl}^{cryst} L_P(\theta)\, T(\theta)\, e^{-2M} = |F_{hkl}|^2 P_{hkl}^{cryst} \frac{1+\cos^2 2\theta}{2\sin^2\theta\cos\theta}\, T(\theta)\, e^{-2M} \tag{7.10.13}$$

which simplifies for a polycrystalline specimen as

$$I_{diffractometry} = |F_{hkl}|^2 P_{hkl}^{cryst} \frac{1+\cos^2 2\theta}{2\sin^2\theta\cos\theta} \frac{1}{2\mu} e^{-2M} \tag{7.10.14}$$

for diffractometry, and

$$I_{D-S} = |F_{hkl}|^2 P_{hkl}^{cryst} \frac{1+\cos^2 2\theta}{2\sin^2\theta\cos\theta} \tag{7.10.15}$$

for Debye–Scherrer patterns.

Example 7.10.1: Powder diffraction data, using Mo Kα X-rays $\left(\lambda = 0.71 \text{ Å}\right)$, for two different samples are summarized in the table below. Assuming that the samples are cubic (FCC, BCC, diamond, or simple), index the observed reflections/peaks for both specimens. Find the lattice parameters and identify the structures of the two samples. How unique is your structural analysis?

| Peak # | Sample A | | Sample B | |
	2θ (Deg)	I/I_{max}	2θ (Deg)	I/I_{max}
1	19.6	100	20.2	100
2	22.7	44	28.7	15
3	32.3	25	35.4	27
4	38.1	28	41.1	9
5	39.8	7	46.2	14
6	46.3	5	50.9	3.5
7	50.8	26		
8	52.2	31		

Solution: We calculate the d-spacing for each peak. We then divide the d-spacing for each peak by that of the first peak, and then square it. For a cubic crystal, this gives us the ratio of the squares of the indices, i.e. $\left(\frac{d_1}{d_n}\right)^2 = \frac{h_n^2 + k_n^2 + l_n^2}{h_1^2 + k_1^2 + l_1^2}$. The results are shown in the table, which includes the consistent indexing and the related lattice parameter.

Peak #	2θ (Deg)	d_{hkl} (Å)	$(d_1/d_n)^2$	hkl	a (Å)
1	19.6	2.1	1	111	3.64
2	22.7	1.81	1.34 ($\sim$4/3)	200	3.62
3	32.3	1.28	2.69 ($\sim$8/3)	220	3.62
4	38.1	1.1	3.64 ($\sim$11/3)	311	3.65
5	39.8	1.04	4.08 ($\sim$12/3)	222	3.60
6	46.3	0.91	5.33 ($\sim$16/3)	400	3.64
7	50.8	0.83	6.4 ($\sim$19/3)	331	3.62
8	52.2	0.81	6.7 ($\sim$20/3)	420	3.62

Since all the indices are unmixed, this must be an FCC structure. We can develop a similar table for Sample B:

Peak #	2θ (Deg)	d_{hkl} (Å)	$(d_1/d_n)^2$	Simple Cubic			BCC	
				hkl	*a* (Å)		hkl	*a* (Å)
1	20.2	2.02	1	100	2.02		110	2.86
2	28.7	1.43	2.0	110	2.02		200	2.86
3	35.4	1.17	3.0	111	2.03		211	2.87
4	41.1	1.01	4.0	200	2.02		220	2.86
5	46.2	0.90	5.0	210	2.01		310	2.85
6	50.9	0.83	~6.0	211	2.03		222	2.88

It is possible to index the diffraction data both as a simple cubic with $a = 2.02$ Å and a BCC structure with $a = 2.86$Å. However, we also have the intensities, which we can analyze as follows:

Peak #	2θ (Deg)	Lorentz polarization factor (LPF)	Simple Cubic			BCC		
			hkl	Multiplicity, *m*	$\frac{I}{LPF \times m}$	hkl	Multiplicity, *m*	$\frac{I}{LPF \times m}$
1	20.2	31.06	100	6	0.54	110	12	0.27
2	28.7	14.87	110	12	0.08	200	6	0.17
3	35.4	9.45	111	8	0.34	211	24	0.12
4	41.1	6.8	200	6	0.25	220	12	0.11
5	46.2	5.22	210	24	0.13	310	24	0.11
6	50.9	4.2	211	24	0.09	222	8	0.10

We calculate the Lorentz polarization factor (7.10.12), and the multiplicity based on the crystal structure for each reflection. We then divide the observed intensity by the product of the Lorentz polarization factor and the multiplicity. If the fit is reasonable, the resultant value should be only proportional to the atomic scattering factor squared, and should decrease monotonically with the scattering angle. Thus, we can infer that Sample B is BCC (in this case, it happens to be Fe_{BCC}).

7.11 Applications of X-Ray Diffraction

X-ray diffraction is a versatile tool, used in materials characterization and analysis in a wide range of disciplines. We conclude this chapter with some typical applications of X-ray diffraction, both here on Earth and on Mars [13]!

7.11.1   Measurement of Lattice Parameters

How do we measure the lattice spacing, d_{hkl}, of a crystalline material most accurately by X-ray diffraction? As a first step, we define the resolving power, $\partial d/d$, of a diffraction experiment by taking the derivative of the Bragg equation, $\lambda = 2d \sin \theta$, for a *fixed* wavelength, λ, used in the experiment as

$$0 = 2 \sin \theta \partial d + 2d \cos \theta \partial \theta \qquad (7.11.1)$$

which gives the resolving power

$$\frac{\partial d}{d} = -\frac{\cos \theta}{\sin \theta} \partial \theta = -\cot \theta \partial \theta \qquad (7.11.2)$$

Thus, for a fixed value of $2 \ \partial\theta$, which defines the minimum separation between reflections that can be measured in a specific instrument, we can get the smallest resolving power when $\cot \theta$ has the smallest value possible. We know from the table of values of $\cot \theta$ that it has largest values for $\theta \sim 0°$, and then rapidly decreases as the Bragg angle, $\theta \to 90°$. Thus, where possible, reflections with large values of 2θ should be used for the most accurate measurements of lattice parameters. In diffractometry, physical limitations of bringing the detector close to the X-ray source restricts 2θ to be less than $150°$. However, large 2θ angles, especially in the back-reflection geometry, are routinely observed in Debye–Scherrer cameras, which is the principal factor driving their continued use today.

7.11.2   Crystallite or Grain Size and Lattice Strain Measurements

As shown earlier (Fig. 7.8.1), the broadening of the peak, β, can be related to the crystallite size by the Scherrer equation (7.8.6). However, the broadening observed in an experiment, β_{expt}, is also influenced by parameters such as the detector slit width, area of the specimen that is illuminated, whether the incident X-ray beam includes the $k_{\alpha 2}$ line or not, etc. To get around these instrumental factors, a similar X-ray diffraction pattern is measured using a large-grain material or a single crystalline material, whose contribution to the broadening can be ignored, and obtain only the broadening, β_{inst}, contribution from the instrument. The difference, $\beta = \beta_{expt} - \beta_{inst}$, can then give the true broadening from the crystallite or grain size through the Scherrer equation, (7.8.5). In practice, the accuracy of such measurements is affected by defects, such as small angle grain boundaries, and/or lattice strain in the material.

Strain in a material can be either on the macroscale or on the microscale. In the former, the whole specimen is subject to some tension or compression, and as a result the lattice spacing, d_{hkl}, increase or decrease in the direction of strain. In X-ray diffraction, the peaks shift in position, and by measuring them for the specimen in different positions or orientations, the magnitude and direction of the strain can be determined.

In contrast, if the strain is on the microscale, the magnitude and direction of the strain vary from crystallite/grain to crystallite/grain. Then, rather than observing a peak shift, we observe a broadening of the peaks in X-ray diffraction. To describe this mathematically, we use (7.11.2), but replace $\frac{\partial d}{d}$ by the elastic strain, ε. Thus,

$$\varepsilon = -\cot\theta\,\partial\theta \qquad (7.11.3)$$

Hence, the broadening, β_{strain}, from the strain, defined as the full width at half-maximum of the peak, $2\partial\theta$, is

$$\beta_{strain} = -\frac{2\varepsilon}{\cot\theta} = -2\varepsilon\tan\theta \qquad (7.11.4)$$

In other words, the broadening due to crystal size (Scherrer equation) varies as sec θ, whereas the broadening due to microstrain varies as tan θ. If the goal is to experimentally determine the microstrain, the instrument contribution to the broadening should be measured first and subtracted as described. Finally, if the Young modulus for the material is known, the measured microstrain, ε, can be interpreted in terms of the stress in the material.

7.11.3 Phase Identification and Structure Refinement

Powder diffraction is most commonly used for the quantitative and qualitative analysis of crystalline materials. From our discussions of powder diffraction so far, it is clear that the lattice spacing, d_{hkl}, and the relative intensities of the diffraction peaks, I_{hkl}, for a polycrystalline material is dependent on the nature of the material and its specific crystalline form. As such, the analysis of powder diffraction patterns is a widely used method for the identification of unknown materials. For this purpose, starting in the 1930s, specific files containing details of the diffraction patterns of a number of known materials have been compiled. Today, these data files known as the powder diffraction files (PDF), contain more than 893,400 standard entries, and the Joint Committee for Powder Diffraction Standards (JCPDS) distributes the collection of these files, in the form of a "card" for each standard material.

Figure 7.11.1. shows a typical example of a PDF card for NaCl annotated for the new user. Most of the information in the card is readily interpretable and the intensities are expressed with respect to the strongest line/peak, which is assigned an arbitrary value of 100. It is possible to match the information provided in such a "standard" card with the diffraction pattern of an unknown; for example, the card (Fig. 7.11.1) can be readily matched with the powder diffraction pattern for NaCl (Fig. 7.9.5). However, for simple materials such as NaCl even though the

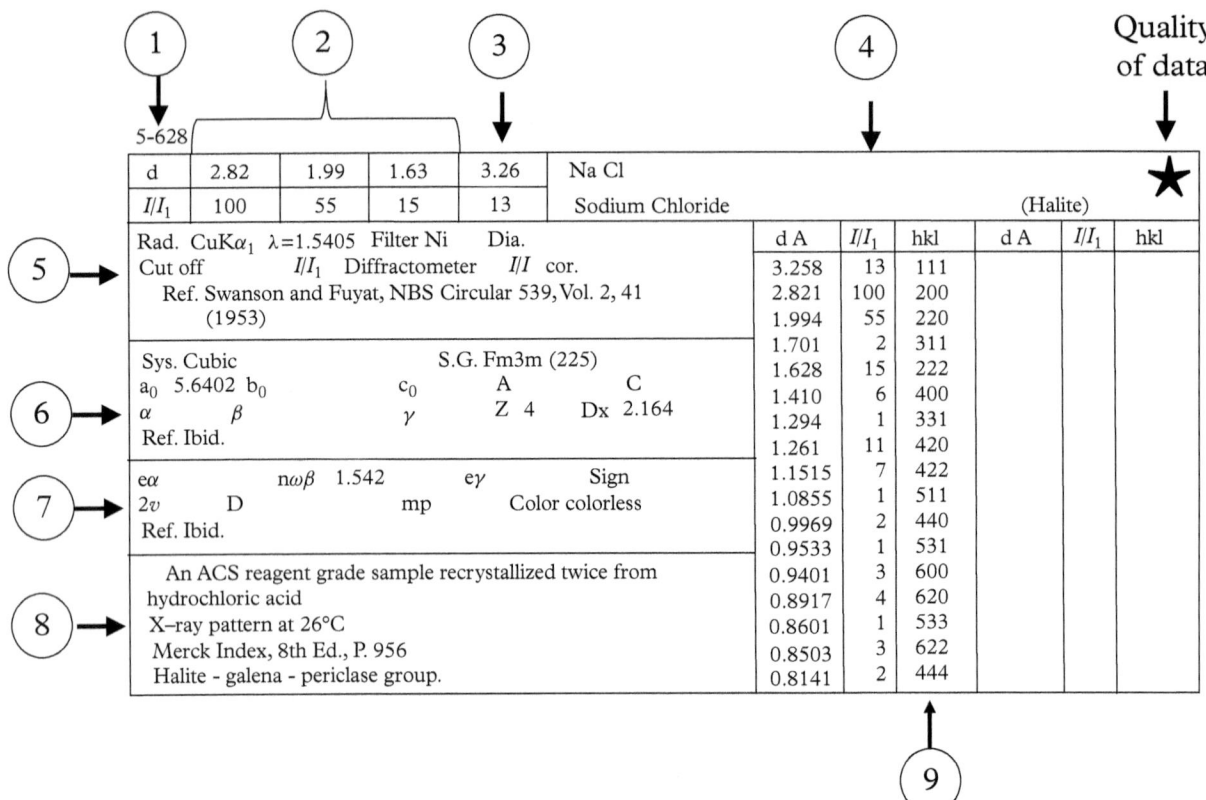

Figure 7.11.1 Annotated PDF card for NaCl: (1) File number. (2) Three strongest lines, corresponding to the 200, 220, and 222 reflections. (3) Lowest-angle line. (4) Chemical formula and name. (5) Description of diffraction method used. (6) Crystallographic data. (7) Optical and other data. (8) Description of specimen. (9) Details of the diffraction pattern. Reproduced from the JCPDS powder diffraction data file. See Figure 7.9.5 for the θ-2θ XRD scan of NaCl, which matches very well with this PDF file.

matching can be done manually by inspection, for more complex materials/phases, these days the matching is done largely using computer methods. Finally, if there are more than one crystallographic component or phases in the specimen, they can be simultaneously identified. However, a quantitative analysis is more complicated because the relationship between the intensities and the amount of a given phase in a mixture is nonlinear due to absorption effects. Details of such analysis of phase mixtures is beyond the scope of this book, but can be found in more advanced texts such as Klug and Alexander (1974).

For polycrystalline materials with large unit cell volumes and/or low symmetry, indexing of diffraction patterns is rather difficult because of the frequently overlapping peak intensities. However, if a rudimentary or imperfect structural model of the material is available, the observed diffraction intensity can be compared with that calculated based on the model. Such a least-squares refinement, originally pioneered by Rietveld [13], is now routinely used to refine structural details of powder specimens, with results comparable to that of single crystals.

Example 7.11.1: Show that the observed X-ray powder diffraction (Fig. 7.9.5) is that of NaCl. Assume Cu Kα (1.54Å) radiation was used.

Solution: We solve this as a table:

Peak position (2θ) in Figure 7.9.5.	27.5	31	46	54	56	67	73	75	84
Peak intensity	650	6100	3000	200	900	400	100	900	400
Relative intensity	10.7	100	49.1	3.3	14.75	6.6	1.6	14.75	6.6
$d = \frac{\lambda}{2\sin\theta}$	3.24	2.88	1.97	1.7	1.64	1.4	1.29	1.265	1.15

Both the peak positions (d-spacings) and the relative intensities match very well with the PDF card for NaCl (Figure 7.11.1). In fact, the indexing of all the observed peaks as show in Figure 7.9.5 is indeed correct.

In the Rietveld method, the entire profile of the powder pattern is fitted to a calculated pattern consisting of all the Bragg reflections, assuming each of them to be of a specific functional form (typically a Lorentzian or a Gaussian function, or a sum of such functions is used) and centered at their respective Bragg reflection position. These complex shapes for the peaks are required to accommodate the actual shapes of the X-ray emission lines when laboratory sources are used; however, if synchrotron sources are used a simple Gaussian shape may suffice. The fitting program minimizes the residual function

$$S = \sum_i w_i \{Y_{io} - Y_{ic}\}^2 \qquad (7.11.5)$$

where w_i is the least-squared weight, given by

$$w_i^{-1} = \sigma_i^2 = \sigma_{ip}^2 + \sigma_{ib}^2 \qquad (7.11.6)$$

and σ_{ip} is the standard deviation associated with the peak, and σ_{ip} is that associated with the background intensity, Y_{ib}. Y_{io} is the intensity observed at the ith step of $2\theta_i$, and Y_{ic} is the calculated intensity contributions from the Bragg reflection and background

$$Y_{ic} = k \sum_{hkl} m_{hkl} L_{hkl} |F_{hkl}|^2 G\left(\Delta\theta_{i_{hkl}}\right) + Y_{ib} \qquad (7.11.7)$$

where k is a scale factor, m is the multiplicity factor, F is the structure factor, $\Delta\theta_i = 2\theta_i - 2\theta_c$, where $2\theta_c$ is the calculated position for the Bragg reflection, and G is the profile function, all for the reflection hkl. Various parameters can be adjusted for refinement and these include details of the unit cell, atomic positions, the Debye–Waller thermal parameters and the parameters that define the peak–shape function, G. Figure 7.11.6 shows an example of Rietveld analysis.

7.11.4   Chemical Order–Disorder Transitions

Consider a simple binary alloy composed of elements A and B. At high temperatures, atoms of A and B are distributed randomly in the lattice, occupying each atomic site with the same probability. Such a solid solution is called a disordered alloy. Now, when the temperature is lowered, the atoms arrange themselves such that the A atoms occupy a certain lattice site in an orderly manner, and the B atoms occupy a different set of lattice sites. Thus, we have two sublattices in the crystal, one with atom A, and the other with atom B. Such a solid solution is known as *chemically ordered*, or a *superlattice*. Note that the superlattice must conform to one of the 14 Bravais lattices (§4.1.2); as such, the ordering is only a change in the lattice type, and calling it a superlattice, though common, is not strictly correct. In alloys, when this type of chemical order of atoms is observed over large distances compared to the unit cell, it is known as *long-range order*. Typically, an alloy disordered at high temperatures transitions into an ordered one as the temperature is lowered at a characteristic temperature, T_c, called the ordering temperature. Alloy systems that show such changes are said to exhibit order–disorder transitions. Classic examples of ordered alloys in the cubic system (Fig. 7.11.2) are referred to as the CuZn (BCC)-, CuAu (FCC)-, and Cu_3Au(FCC)-type ordered structures.

It is reasonable to expect alloys that exhibit order–disorder transitions to exist in some intermediate state, where the degree of order is incomplete. The degree of order can be quantitatively described by an order parameter, S, defined as

$$S = \frac{r_A - X_A}{1 - X_A} \tag{7.11.8}$$

where r_A is the fraction of A lattice sites occupied by the A atoms, and X_A is the fraction of A atoms in the alloy. For a completely random distribution of atoms in a disordered alloy, $r_A = X_A$, and $S = 0$, and for a perfect long-range ordered alloy, $r_A = 1$, and $S = 1$.

The presence of long-range order, and specifically, the order parameter, can be measured by X-ray, electron, or neutron diffraction. In the case of alloys, when there is negligible change in the unit cell size upon ordering, there will be no observable change in the position of the X-ray diffraction lines; however, ordering significantly affects the peak intensities. Additional reflections, referred to as superlattice lines (see §7.5.4), are often observed signaling a change in the

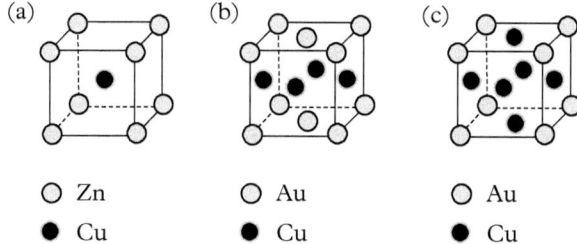

○ Zn   ○ Au   ○ Au

● Cu   ● Cu   ● Cu

Figure 7.11.2 Unit cells of ordered binary alloys with atom positions as indicated. (a) The CuZn (a *Cubic P* lattice, which transforms from the *Cubic I* lattice in the disordered phase). Note that this structure is similar to the CsCl structure discussed in §7.5.4 (b) The CuAu structure (a *Cubic P* lattice, but strictly speaking there is a tetragonal distortion in the ordered phase, and as such, it is a *Tetragonal P* lattice. The corresponding disordered phase is the *Cubic F* lattice). (c) The Cu$_3$Au structure (a *Cubic P* lattice, with a basis of 3Cu+Au). In the disordered phase, it is a *Cubic F* lattice.

Bravais lattice upon ordering. In addition, other types of diffraction lines, known as fundamental reflections, are observed in the same position, for both the ordered and disordered states of the alloy. Note that, in general, the superlattice reflections have a lower intensity compared to the fundamental. Nevertheless, the order parameter, S, can be obtained from experimentally measured diffraction data by a simple ratio of observed intensities:

$$S = \frac{F_f}{F_s} \sqrt{\frac{m_f (L_P)_f I_s}{m_s (L_P)_s I_f}} \tag{7.11.9}$$

where the subscripts s and f refer to the superlattice and fundamental lines, m is the multiplicity, (L_P) is the Lorentz polarization factor (7.10.12), F is the structure factor, and I is the observed intensity, often measured as the integrated area of the peak.

For the CuZn alloy (Fig. 7.11.2a) by calculations similar to those in §7.5.2 for a BCC crystal, we can show that the structure factors are

$$F_{CuZn} = f_{Cu} + f_{Zn} \quad \text{for } (h+k+l) \text{ even}$$

$$F_{CuZn} = f_{Cu} - f_{Zn} \quad \text{for } (h+k+l) \text{ odd} \tag{7.11.10}$$

Thus, the fundamental lines, $(h+k+l)$ even, are observed in both ordered and disordered alloys. However, the superlattice lines, $(h+k+l)$ odd, with intensities proportional to the degree of order, S, are only observed for the ordered alloys. As a first approximation, assuming all else is equal, we write the ratio of observed intensities, based on their respective structure factors, as

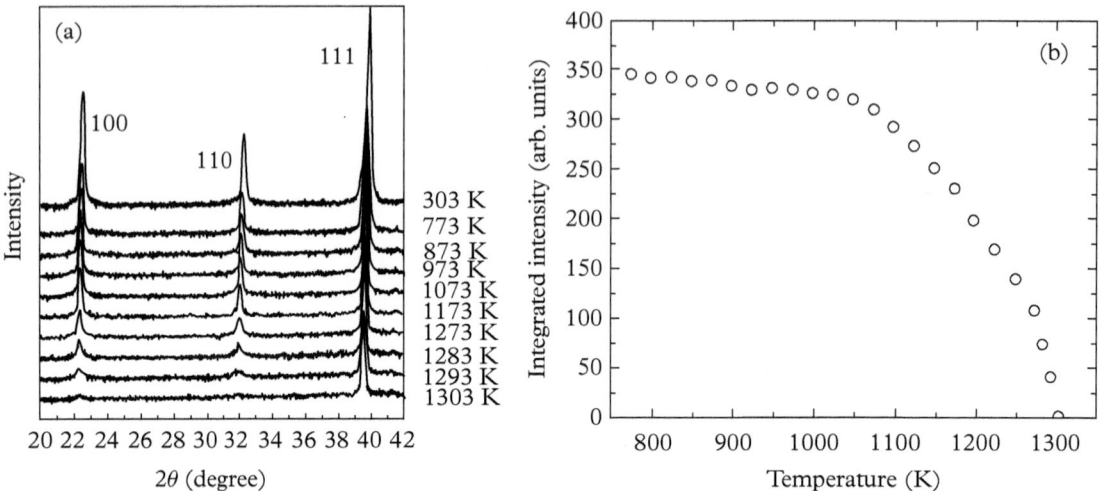

Figure 7.11.3 (a) Temperature dependence, from 300K to 1303K, of the X-ray diffraction from a specimen of $(Cu_{0.5}Fe_{0.5})\, Pt_3$. The fundamental (111) reflection and the superlattice (100) and (110) reflections are shown. (b) A plot of the integrated intensity of the superlattice (100) reflection as a function of temperature, showing the order-disorder transition occurs at $\sim$1300°K.

Adapted from [13].

$$\frac{I_s}{I_f} = \frac{|F_s|^2}{|F_f|^2} = \frac{|f_{Cu} - f_{Zn}|^2}{|f_{Cu} + f_{Zn}|^2} = \frac{|29 - 30|^2}{|29 + 30|^2} \approx 0.003\% \qquad (7.11.11)$$

As a result, the superlattice reflections in CuZn alloys are much too weak, compared to the fundamental ones, to be easily detected by X-ray diffraction. This is representative of a general problem with X-ray diffraction as the atomic scattering factors are proportional to the atomic number, and hence, the positions of atoms with similar atomic numbers in the unit cell are hard to determine. This problem is often overcome using neutron diffraction, where the scattering amplitudes are determined by the nucleus (and not the electrons) and the scattering factors vary irregularly with atomic number (see §8.8).

Based on this example, we can expect the ratios of intensities (in the forward direction) to be $\frac{I_s}{I_f} = \frac{50^2}{166^2} \approx 10\%$, and this ordering can be readily detected by X-ray diffraction. Figure 7.11.3a shows the X-ray diffraction peaks, both fundamental and superlattice reflections, in an alloy of Pt_3 ($Cu_{0.5}Fe_{0.5}$), having the Cu_3Au ordered structure, measured as a function of temperature. The temperature dependence of the integrated intensity of the superlattice (100) reflections (Fig. 7.11.3b), indicates an order–disorder transition temperature of $\sim$1300K (see Problem 7.7).

Example 7.11.2: Consider the Cu_3Au alloy shown in Figure 7.11.2c. For the completely disordered alloy, the probability that an Au atom occupies an atomic site in the FCC lattice is ¼, and for occupation by a Cu atom it is ¾ (this follows from the stoichiometry). For the ordered alloy, the positions, $u_nv_nw_n$, of the basis atoms are 000 for Au, and ½ ½ 0, 0 ½ ½, and ½ 0 ½, for the three atoms of Cu. Calculate the structure factor and intensities for the disordered and ordered alloy if $f_{Au} = 79$ and $f_{cu} = 29$ (see also Fig. 7.11.3).

Solution: The structure factor for the disordered alloy (see §7.5.1), is given by

$$F_{hkl} = 4 \times 1/4 \times f_{Au} + 4 \times 3/4 \times f_{Cu}$$

$$= f_{Au} + 3f_{Cu} \qquad \text{for } h, k, l \text{ all even or all odd}$$

$$F_{hkl} = 0 \qquad \text{for } h, k, l \text{ with mixed indices.}$$

For the ordered alloy the structure factor is given by (7.5.1), but by taking the specifics of chemical ordering into consideration, i.e.

$$F_{hkl} = f_{Au}e^{2\pi i.0} + f_{Cu}\left(e^{2\pi i\left(\frac{h}{2}+\frac{k}{2}\right)} + e^{2\pi i\left(\frac{k}{2}+\frac{l}{2}\right)} + e^{2\pi i\left(\frac{l}{2}+\frac{h}{2}\right)}\right)$$

Thus

$$F_{hkl} = f_{Au} + 3f_{Cu} = 166 \qquad \text{for } h, k, l \text{ all even or all odd (fundamentals)}$$

$$F_{hkl} = f_{Au} - f_{Cu} = 50 \qquad \text{for } h, k, l \text{ with mixed indices (superlattices)}$$

with corresponding intensities given by

$$I_{hkl} = |F_{hkl}|^2 = (f_{Au} + 3f_{Cu})^2$$

$$= (79 + 3 \times 29)^2 = 166^2 = 27556 \quad \text{for } h, k, l \text{ all even or all odd}$$

$$I_{hkl} = |F_{hkl}|^2 = (f_{Au} - f_{Cu})^2$$

$$= (79 - 29)^2 = 50^2 = 2500 \qquad \text{for } h, k, l \text{ with mixed indices.}$$

7.11.5   Short-Range Order (SRO) and Diffuse Scattering

From Figure 7.11.3b, we notice that above the critical temperature, T_c, the long-range order disappears, and the intensity of the superlattice reflection goes to zero. This suggests that the atomic arrangement in the alloy, for $T > T_c$, may be random. However, if you look carefully at the diffuse scattered intensity (Fig. 7.11.4) that forms the background of the diffraction pattern, it often shows a small peak,

Figure 7.11.4 The measured intensity from a single crystal of Cu$_3$Au, along the *h*00 direction for 405°C and 450°C. The short-range order peaks at 100 and 300 are normalized with respect to the fundamental 200 reflection.

Adapted from [14].

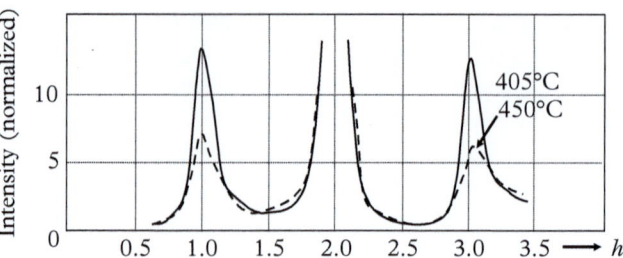

indicating that the material is not completely disordered and a correlation between nearest neighbors can persist.

For alloys consisting of a matrix and guest atoms, the guest atoms can interact either with the matrix atoms or with each other. Initially, for dilute alloys, the guest atoms mainly interact with the host/matrix atoms, but with increasing concentration, guest–guest interactions also become important. This gives rise to two competing possibilities; one, where unlike atoms prefer to be nearest neighbors, a tendency known as short-range order (SRO), and typically, observed in monovalent matrices. Alternatively, like atoms attract each other, a process known as clustering, and typically, observed in multivalent matrices.

Now, consider the Cu$_3$Au structure shown in Figure 7.11.2c. When the crystal is perfectly long-range ordered, an Au atom at 000 is surrounded by 12 Cu atoms at ½ ½ 0 (and equivalent sites, three shown in the figure). Similarly, every Cu atom is surrounded by four Au atoms and eight Cu atoms. Ideally, for $T > T_c$, this ordered arrangements breaks down and this atomic arrangement becomes truly random. Then, any given Au atom will have nine Cu atoms (= 3/4*12, based on the structure and composition) as nearest neighbors. However, above T_c it is observed that there are, on average, 11 Cu atoms around each Au atom. In fact, such SRO is a general effect and any solid solutions exhibiting long-range order also exhibits SRO above T_c, with the degree of SRO decreasing with increasing temperature.

Figure 7.11.4 shows X-ray diffraction intensities for Cu$_3$Au, taken at two different temperatures above $T_c = 390°C$. SRO maxima, i.e. 100 and 300 type reflection, are observed at the same positions as the superlattice reflections, with intensities ∼6–12 % of the fundamental 200 reflection. Notice that if a single crystal specimen is not used, other sources of diffuse scattering, primarily temperature, mask the SRO intensities. It is possible to quantify the degree of SRO using an appropriate parameter based on such measurements. Further details can be found in Schwartz and Cohen (1987).

7.11.6   *In Situ* **X-Ray Diffraction at Synchrotrons**

X-ray diffraction is an outstanding method for structure analysis, with the added advantage of substantial penetration (e.g. compared to electrons; see §8), allowing for studies of bulk materials and specimens in complicated environments. Furthermore, the high photon flux available in third-generation synchrotron sources (§2.4.3) makes it possible to collect X-ray diffraction data in relatively short time scales, and carry out *in situ* measurements (§1.5). Unlike the measurements of chemical order–disorder transitions described in §7.11.4, which strictly measure the thermodynamic *end states*, of particular importance here are measurements to understand *processes* such as phase transitions, crystal growth, and the role of buried interfaces in many natural processes. For example, in crystallization or thin film deposition, an atom transforms from being completely disordered in the liquid or gas phase to a crystalline form in the solid. To understand such processes as a function of thermodynamic variables, the probe should minimally interfere with the process of transformation. However, even though charged particles, particularly electrons, can be used for *in situ* measurements, they would have difficulty reaching buried interfaces, nor would they be able to penetrate gases under high pressure. That leaves X-rays and neutrons as potential probes. The latter lacks general availability of high flux sources (spallation sources, §5.2.3, are not common), leaving only X-ray photons at synchrotron sources as meaningful probes for such experiments. In addition, X-ray scattering satisfies the Born approximation (§8.2) and such kinematic scattering lends itself to relatively easy interpretation of diffraction data. Further practical details of implementing such *in situ* diffraction studies in synchrotrons are beyond the scope of this book, but interested readers are referred to the monograph edited by Ziegler et al. (2014), which gives a good introduction to such measurements, with emphasis on studies of thin film growth and deeply buried interfaces.

7.11.7 X-Ray Diffraction Measurements on Mars

We end this chapter with an interesting and practical example from geology: the first X-ray diffraction measurements of the crystal structure of various minerals from soils and drilled samples found on Mars conducted by the Curiosity rover [15]. In summary, these measurements reveal an abundance of primary basaltic minerals, amorphous components, and varied hydrous alteration products, including phyllosilicates [15]. As shown in Figure 7.11.5a, the X-ray diffraction instrument, called CheMin, is a shoe box-sized apparatus, based on a variation of the Debye–Scherrer camera in the transmission geometry. It contains a micro-focus X-ray source, a pinhole for collimation, a rotating specimen holder (which holds both the collected Martian samples, as well as standards from Earth, e.g. the beryl-quartz powder used for calibration Fig. 7.11.5c), and a CCD camera for detection of the scattered X-rays. All the specimens are powders, and the ring patterns observed on the CCD camera (Fig. 7.11.5c) are radially integrated to obtain the X-ray diffraction patterns (Fig. 7.11.5b). The specimens collected on Mars are put into Mylar or Kapton specimen holders, which produce minimal background; however, the Al light shield used to protect the CCD detector introduces distinct background peaks at $2\theta = 25.8°$ and $32.1°$ (Fig. 7.11.5d).

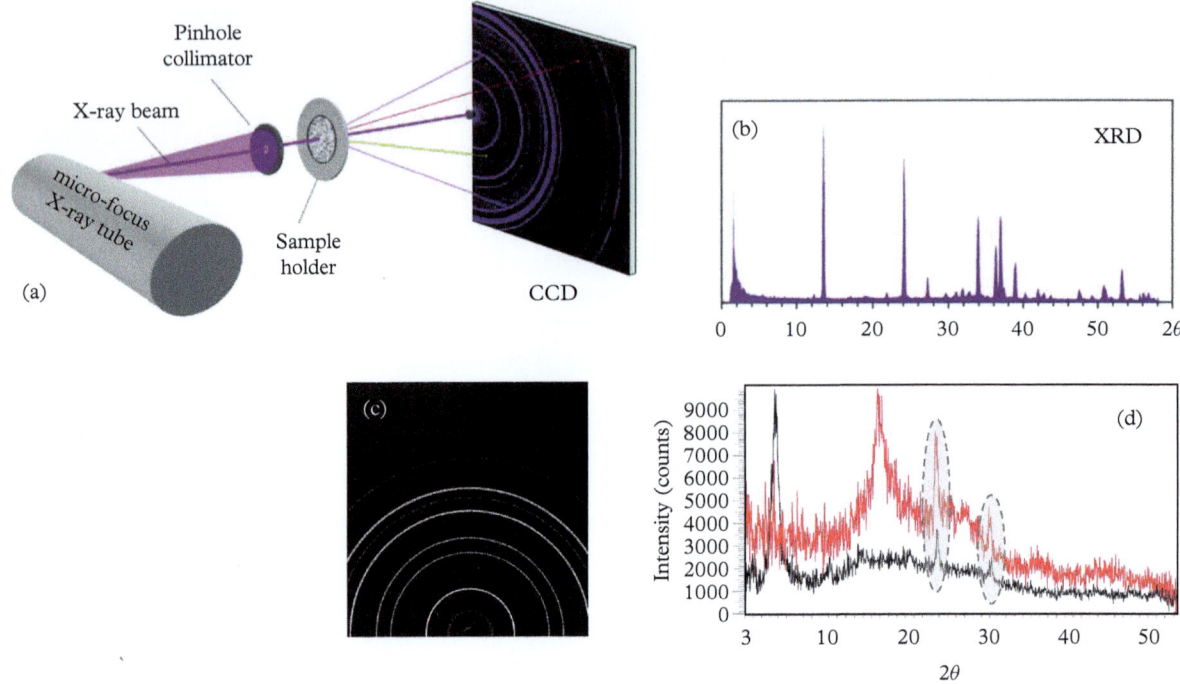

Figure 7.11.5 (a) A miniature diffractometer, CheMin, sent to Mars on the Curiosity rover, working on the transmission geometry and using an energy-discriminating CCD two-dimensional detector. (b) The two-dimensional image in (*a*) can be integrated to obtain the X-ray powder diffraction data. (c) A two-dimensional X-ray diffraction pattern of the beryl-quartz standard measured on Mars. (d) A comparison of the XRD patterns from empty Kapton (black) and Mylar (red) specimen holders. Circled peaks at 25.8° and 32.1° are due to scattering from the Al light shield on the CCD detector.

Adapted from [15].

A typical X-ray diffraction pattern (Fig. 7.11.6) integrated from the two-dimensional image (inset), recorded on the CCD camera, for Martian Rocknest is compared with the calculated plot, and refined with the Rietveld method [16]. Needless to say, the Rietveld refinement provides a very good fit to the observed data, with negligible difference or residue, except at $2\theta = 25.8°$, which corresponds to the background peak for scattering from the Al light shield for the CCD detector. Further details of the analysis of other Martian minerals, using CheMin, can be found in the original manuscript [15].

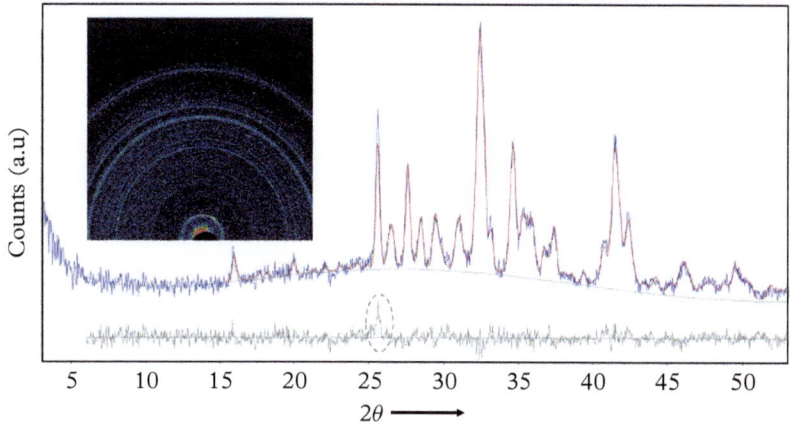

Figure 7.11.6 X-ray data from a specimen of the Rocknest aeolian bedform (dune) on Mars. Inset shows the two-dimensional XRD pattern. Observed (blue, integrated from the two-dimensional image in the inset) and calculated (red) plots from Rietveld refinement using data for Rocknest. The calculated background (black line) is inscribed at the base of the observed pattern and the gray pattern at the bottom is the difference plot (observed − calculated). The difference at ~25.8°, circled, is due to scattering from the Al light shield on the CCD.

Adapted from [15].

Summary

X-rays with a wavelength on the order of the atomic spacing in solids are perfectly suited for diffraction studies from a wide range of materials. Historically, X-ray diffraction has been instrumental in our understanding of the structure, crystallography, or arrangement of atoms and molecules in many important biological, geological, or technological materials.

X-rays are predominantly scattered by the electrons in solids. This can be elastic and coherent (Thomson) scattering with a well-defined phase shift of half the wavelength. Or, it can be inelastic and incoherent (Compton) scattering with the change in wavelength dependent only on the scattering angle. The net scattering of X-rays by an atom is given by the atomic scattering factor, which has the largest value corresponding to its atomic number (Z) for forward scattering ($\theta = 0°$) and decreases with increasing scattering angle; further, it also decreases with decreasing wavelength of the X-rays. Hence, the scattering factor is parameterized as a function of the variable, $s \left(= \frac{\sin \theta}{\lambda} \right)$, and is tabulated as the Cromer–Mann parameters for elements and ions.

Diffraction intensities from a crystal result from a summation of the contributions of all the atoms in the unit cell, taking their phases also into consideration. The amplitude of the diffracted wave, thus calculated for any reflection, is known as the structure factor, F_{hkl}, and its square gives the intensity. If the coordinates of all the atoms in the unit cell are known, the structure factor can be calculated for any diffracted reflection, corresponding to a specific set of lattice planes (hkl), as illustrated for a variety of crystal structures in §7.5.

For single crystals, the diffraction pattern reflects the underlying symmetry (point group) of the crystal. However, even if the crystal does not have a center of symmetry, the corresponding diffraction pattern will be centrosymmetric (Friedel law). Further, certain symmetry elements (glide planes and screw axes), if present in a crystal, give rise to systematic absences of reflections in diffraction. In polycrystalline materials, the diffracted beam is broadened inversely proportional to the thickness of the grain size (Scherrer equation).

Ideal X-ray intensities, given by the square of the structure factor, are modified by four principal experimental factors. The temperature factor accounts for the thermal vibration of the atoms about their equilibrium position in the unit cell. Known also as the Debye–Waller factors, they are tabulated for most crystals. As X-rays travel in a crystal, both the incident and diffracted beams are absorbed in a manner proportional to the path length and the linear absorption coefficient in the material. Further, an unpolarized beam of X-rays is scattered differently in different directions, even by a single electron; this and other angular dependence are accounted for by the Lorentz-polarization factor. Finally, the total number of variants (multiplicity) of a family of lattice planes have to be included, especially in polycrystalline materials.

Bragg's law and its equivalent in reciprocal space (Laue criterion and the Ewald sphere construction) are the basis of all X-ray diffraction measurements. There are two variables, θ and λ, that can satisfy the diffraction conditions for a crystal with a given lattice spacing, d_{hkl}. We can keep θ fixed and vary λ (Laue method) and apply it to the study of single crystals. Alternatively, we can keep λ fixed and vary θ, which has a number of variations for studies of powders (Debye–Scherrer method), both polycrystalline materials and single crystals (diffractometry), and even thin films and multilayers (reflectivity).

X-ray diffraction finds a wide range of applications in the analysis and development of materials. The representative examples discussed here include measurements of lattice strain, identification of phases and structure refinement, studies of chemical order–disorder transitions, and revealing the structure and function of biological molecules, including that of DNA.

. . . .

FURTHER READING

Cullity, B. D. *Elements of X-Ray Diffraction*. Boston: Addison-Wesley, 1978.

Giacovazzo, C., H. L. Monaco, G. Artioli, D. Viterbo, G. Ferraris, G. Gilli, G. Zanotti, and M. Catti. *Fundamentals of Crystallography*. Oxford: Oxford University Press, 2002.

de Graef, M., and M. McHenry. *Structure of Materials*. Cambridge: Cambridge University Press, 2007.

Hammond, C.*The Basics of Crystallography and Diffraction*. Oxford: Oxford University Press, 2006.

Klug, H. P., and L. E. Alexander. *X-Ray Diffraction Procedures for Polycrystalline and Amorphous Materials*. New York: Wiley, 1974.

Preuss, E., B. Krahl-Urban, and R. Butz. *Laue Atlas*. New York: Wiley, 1974.

Schwartz, L. B., and J. B. Cohen, *Diffraction from Materials*. 2nd ed. New York: Springer-Verlag, 1987.

Warren, B. E.*X-ray Diffraction*. New York: Dover, 1990.

Woolfson, M. M.*An Introduction to X-Ray Crystallography*. Cambridge: Cambridge University Press, 1997.

Ziegler, A., H. Graafsma, X. F. Zhang, and J. W. M. Frenken, eds. *In Situ Materials Characterization Across Spatial and Temporal Scales*, Ch. 2. New York: Springer, 2014.

. . . .

REFERENCES

[1] Watson, J. D., and F. H. C. Crick. "Molecular Structure of Nucleic Acids: A Structure for Deoxyribose Nucleic Acid." *Nature* 171 (1953): 737–8.

[2] Franklin, R. E., and R. G. Gosling. "Evidence for 2-Chain Helix in Crystalline Structure of Sodium Deoxyribonucleate." *Nature* 172 (1953): 156–7.

[3] Cromer, D. T., and J. B. Mann. "X-Ray Scattering Factors Computed from Numerical Hartree–Fock Wave Functions." *Acta Crystallographica Section A* 24, no. 2 (1968): 321–4.

[4] Wood, I. G., P. Thomson, and J. C. Mathewman. "A Crystal Structure Refinement from Laue Photographs Taken with Synchrotron Radiation." *Acta Crystallographica Section B* 39, no. 5 (1983): 543–7.

[5] Moffat, K., D. Szebenyi, and P. Bilderback. "X-Ray Laue Diffraction from Protein Crystals." *Science* 223 (1984): 1423.

[6] Moffat, K. "Time-Resolved Macromolecular Crystallography." *Annual Review of Biophysics and Biophysical Chemistry* 18 (1989): 309_32.

[7] Blomqvist, P., K. M. Krishnan, and D. E. McCready. "Growth of Exchange-Biased MnPd/Fe Bilayers." *Journal of Applied Physics* 95, no. 12 (2004): 8019.

[8] Stampe, P. A., and R. J. Kennedy. "X-ray Characterization of MgO Thin Films Grown by Laser Ablation on $SrTiO_3$ and $LaAlO_3$." *Journal of Crystal Growth* 191, no. 3 (1998): 478–82.

[9] Cho, N.-H., K. M. Krishnan, C. A. Lucas, and R. F. C. Farrow. "Microstructure and Magnetic Anisotropy of Ultrathin Co/Pt Multilayers Grown on GaAs (111) by Molecular-Beam Epitaxy." *Journal of Applied Physics* 72, (1992): 5799.

[10] Parratt, L. G. "Surface Studies of Solids by Total Reflection of X-Rays." *Physical Review* 95, no. 2 (1954): 359–69.

[11] Peng, L. M., G. Ren, S. L. Dudarev and M. J. Whelan. "Robust Parameterization of Elastic and Absorptive Electron Atomic Scattering Factors." *Acta Crystallographica Section A* 52, no. 2 (1996): 257–76.

[12] Lipson, H. In *International Tables for X-Ray Crystallography, Vol II: Mathematical Tables*, edited by J. S. Kasper and K. Londsdale, 291–315. Birmingham: The Kynock Press, 1959.

[13] Ahmad, E., M. Takahashi, K. Iwasaki, and K. Ohshima. "X-Ray Diffraction Study of Order–Disorder Phase Transition in CuMPt6 (M=3d Elements) Alloys." *Journal of the Physical Society of Japan* 78, no 1 (2009): 014601.

[14] Moss, S. C. "X-Ray Measurement of Short-Range Order in Cu3Au." *Journal of Applied Physics* 35 (1964): 3547.

[15] Bish, D., D. Blake, D. Vaniman, P. Sarrazin, T. Bristow, C. Achilles, P. Dera, S. Chipera, J. Crisp, R. T. Downs, J. Farmer, M. Gailhanou, D. Ming, J. M. Morookian, R. Morris, S. Morrison, E. Rampe, A. Treiman, and A. Yen. "The First X-Ray Diffraction Measurements on Mars." *International Union of Crystallography Journal* 1, no. 6 (2014): 514–22.

[16] Rietveld, H. M. "A Profile Refinement Method for Nuclear and Magnetic Structures." *Journal of Applied Crystallography* 2, no. 2 (1969): 65–71.

. . . .

EXERCISES

A. Test Your Knowledge

For each statement, choose all the right answers. Note that more than one response for each question may be correct.

1. To measure lattice parameters most accurately in XRD, we prefer
   (a) high Bragg angles with large 2θ values.
   (b) low Bragg angles.
   (c) back reflections in a Debye–Scherrer camera.
   (d) to use a diffractometer over a Debye–Scherrer camera.
2. A body-centered cell will have
   (a) zero intensity for $h + k + l =$ odd.
   (b) zero intensity for $h + k + l =$ even.
   (c) the same *hkl* reflections with zero intensity for cubic, tetragonal, and orthorhombic unit cells.
3. An accelerating electric charge at a point, O, will generate
   (a) no electromagnetic radiation.
   (b) an electromagnetic radiation in all directions at any other point, P.
   (c) an electromagnetic radiation in a very *specific* direction at any other point, P.
4. X-rays are scattered
   (a) equally by all components of an atom.

    (b) primarily by the electrons of an atom.

    (c) only by the nucleus of the atom.

5. The contributions of the nucleus to X-ray scattering is

    (a) negligible because it has a positive charge.

    (b) negligible because its mass is $\sim 10^3$ times that of the electron.

    (c) significant and should always be considered.

6. Compton scattering

    (a) is incoherent.

    (b) contributes insignificantly to the observed X-ray diffraction intensities.

    (c) contributes mainly to the background in X-ray diffraction.

7. The atomic scattering factor, f,

    (a) decreases at large scattering angles.

    (b) is derived from the assumption that the charge distribution in the atom is spherically symmetric.

    (c) equals the atomic number for forward scattering.

8. Diffraction intensities can be calculated using single-scattering models

    (a) for X-rays all the time.

    (b) for X-rays, if we assume that the volume of the overall crystal/specimen is small.

    (c) for electron beam incidence, but generally gives erroneous results.

9. The structure factor for an FCC crystal is

    (a) non-zero for mixed indices (hkl).

    (b) zero for mixed indices (hkl).

    (c) non-zero for un-mixed indices (hkl).

10. Comparing X-ray diffraction from BCC and FCC structures, we see that

    (a) all reflection (hkl) does not show a measurable intensity.

    (b) $I_{hk0} = 0$, for BCC.

    (c) FCC shows uniformly higher intensities than BCC.

11. The structure factor for

    (a) HCP crystals is always an imaginary number.

    (b) CsCl is always a real number.

    (c) CsCl would be the same as a BCC crystal if Cs and Cl are randomly distributed in the unit cell.

    (d) NaCl can be derived from that of an FCC structure.

12. A centrosymmetric crystal has

    (a) a point of inversion symmetry at the origin.

    (b) a structure factor that is always a real number.

    (c) a diffraction pattern that is also centrosymmetric.

13. The Friedel law

    (a) states that non-centrosymmetric crystals also have a centrosymmetric diffraction pattern.

    (b) reduces the 32 point groups in three dimensions to the 11 Laue groups.

    (c) has no relevance to diffraction in theory or practice.

14. In diffraction, the presence of glide planes
    (a) gives rise to systematic absences.
    (b) and/or screw axes do not make any difference.
    (c) and/or screw axes have well-defined absences.

15. In diffraction, since we only measure the intensities
    (a) we do not know the phase of the diffracted waves.
    (b) we can always determine the position of all atoms in the unit cell.
    (c) we cannot unequivocally determine the position of all atoms in the unit cell.
    (d) we can determine the position of all atoms in the unit cell of centrosymmetric crystals.

16. The broadening of a diffraction peak is
    (a) affected by crystallite or grain size.
    (b) affected by microstrain in the individual crystallite.
    (c) large for single crystal specimens.
    (d) described by the Scherrer equation.
    (e) affected by instrumentation factors.
    (f) always proportional to sec θ.
    (g) equivalent to extending the infinitesimal reciprocal lattice point to a finite-sized node.
    (h) not interpretable in the Ewald sphere construction.

17. A collimator is used to produce _____ of X-rays.
    (a) a monochromatic beam
    (b) a narrow parallel beam
    (c) only a circular shaped beam

18. A monochromator _____ of X-rays.
    (a) produces a select and narrow wavelength
    (b) has as its main function the production of a narrow, well-shaped beam
    (c) works by absorbing a certain range

19. What we call a superlattice diffraction peak _____ in order–disorder transitions.
    (a) is a misnomer
    (b) is the result of the generation of a new lattice, different from the 14 Bravais lattices,
    (c) reflects the change from one Bravais lattice to another

20. A miniaturized X-ray diffractometer
    (a) can be constructed based on the Debye–Scherrer camera design.
    (b) is too fragile to survive the journey to Mars.
    (c) produced exceptional structural data of the minerals on the surface of Mars.

21. Chemical order–disorder transitions in alloys
    (a) cannot be probed by XRD.
    (b) can be studied by XRD for ALL materials.

(c) can be studied by XRD if the superlattice reflection intensity is significant.

(d) cannot be studied by neutron diffraction.

(e) shows a characteristic transition temperature that can be determined by XRD.

22. The Powder Diffraction Files (PDFs) contain entries
   (a) for more than 30,000 separate materials.
   (b) on the crystal structure and the three principal XRD lines, with relative intensities, for all materials.
   (c) that are too complicated to interpret or most routine work.

23. For polycrystalline materials with large unit cells
   (a) indexing of powder diffraction data is easy.
   (b) indexing of powder diffraction data is difficult because of overlapping peaks in the spectrum.
   (c) refinement of XRD data using a model structure often gives results comparable to single crystal X-ray analysis.

24. A crystal monochromator
   (a) works on the principle of Bragg diffraction.
   (b) can only be used for laboratory X-ray tubes, but not for synchrotron radiation.
   (c) can be designed to minimize the contribution from the Lorentz-polarization factor.

B.   Problems

1. Following the example of NaCl, Example 7.5.3, derive the expression for the **structure factor of the diamond structure** that can be regarded as two interpenetrating FCC lattices shifted by the translation vector, $\boldsymbol{\tau} = \frac{1}{4} (\mathbf{x} + \mathbf{y} + \mathbf{z})$. Further, show that $I_{hkl} = 0$ for mixed indices, and $I_{hkl} = 0$ when $h + k + l = 4n + 2$, where n is an integer.

2. Using **Cromer–Mann parameters** (Table 7.3.2) for Cr, calculate the structure factor for the following reflections: $000, 001, 011, 020, 121, 130,$ and 321. Check your results with the data in Example 7.5.2.

3. Using the **Cromer–Mann parameters** (Table 7.3.2) for NaCl, show that the structure factors for the following reflections are

   $F_{000} = 112$

   $F_{110} = 0$

   $F_{002} = 85.7$

   $F_{111} = 18.5$

   $F_{220} = 72.8$

   Use $\lambda = 1.54$ Å and the lattice parameter for NaCl, $a_0 = 0.564nm$.

4. Consider a cubic unit cell of lattice parameter $a \sim 5$ Å, subject to an X-ray diffraction experiment using Cu Kα radiation.

   i) What is the volume of the unit cell in real space?

   ii) What is the volume of the reciprocal unit cell?

   iii) What is the total available volume within the limiting sphere?

   iv) From *i–iii*, approximately how many independent reflections do you expect to see in a typical Debye–Scherrer pattern?

   v) Including **Friedel law**, how many independent rings of intensities do you expect to see?

   vi) Comment on the implications of the number of observable rings on a Debye–Scherrer pattern. Think about the potential overlaps.

5. **Sphalerite**, or zinc blende (§4.1.6), with the chemical formula of ZnS, has an FCC structure with a lattice parameter $a = 0.541$ nm.

   i) Write down the coordinates for Zn and S, in the unit cell.

   ii) Calculate the structure factor for sphalerite

   iii) For what values of hkl would you see a nonzero intensity? If possible, write this down in some simple form.

   iv) Calculate the intensities for the top three diffraction lines. Compare your results with PDF file.

6. A **screw axis** is normally represented by a 4×4 transformation matrix. Consider the 4_1 screw axis, parallel to the *c*-axis, which moves an atom in any general position (x, y, z) to three other equivalent positions: $(-y, x, z+1/4)$, $(-x, -y, z+1/2)$, and $(y, -x, z+ \frac{3}{4})$.

   i) Write down the structure factor for these four positions.

   ii) Now consider only the reflections of the type $(00l)$ and rewrite the simplified structure factor.

   iii) Using the following mathematical relationships

$$\sum_j e^{\pi i \frac{il}{2}} = \frac{1 - e^{2\pi il}}{1 - e^{2\pi i\frac{l}{2}}}$$

   and

$$\lim_{l \to 4n} \frac{1 - e^{2\pi il}}{1 - e^{2\pi i\frac{l}{2}}} = 4$$

   show that reflections of the type $(00l)$, for $l \neq 4n$, are absent.

7. For the **Co–Pt multilayer** data, using Cu Kα radiation shown in Figure 7.9.12, calculate the period from both the low-angle and high-angle scans. Comment on the similarity and/or differences in the result.

8. For the **CuAu ordered structure** shown in Figure 7.11.2, calculate the structure factor for the fundamental and superlattice reflections. Assume that the ordered alloy remains cubic with no change in lattice parameter. What is the intensity ratio I_s/I_f? Is our assumption valid in practice?

9. **Order–disorder transition**: Consider the $(Cu_{0.5}Fe_{0.5})\,Pt_3$ alloy diffraction data shown in Figure 7.11.3.

   i) Calculate the structure factors and the ideal intensity ratios for the fundamental (111) and superlattice (100) reflections.

   ii) For the data at 303 K, approximate the integrated intensities, I_s and I_f.

   iii) Calculate the order parameter using (7.11.9).

   iv) The chemical order parameter, S, can be related to T_c and the chemical ordering energy, V_{order}, that stabilizes the ordered state, by the Bragg–Williams theory. Thus, near the ordering temperature

$$S^2 = \frac{3\,(T_c - T)}{T}$$
$$V_{order} = 4k_B T_c$$

   If possible, calculate T_c, from the experimental dependence of the order parameter near the critical temperature. Then, use the value of T_c to calculate the chemical ordering energy.

10. Suppose you are given a sample and told that it is orthorhombic. You are also given the lattice parameters a, b, and c. Now you are asked to determine to which of the **four orthorhombic lattices**—primitive, base-centered, body-centered, or face-centered, as shown in Figure 7.Pr.10—this sample belongs.

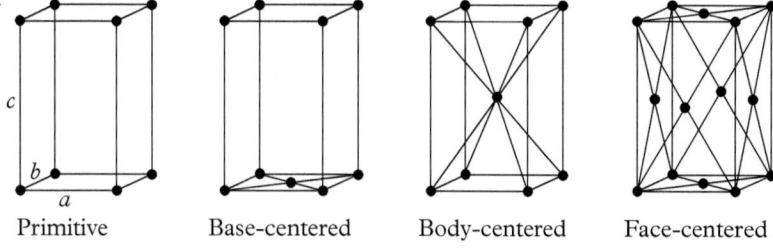

Primitive Base-centered Body-centered Face-centered

Figure 7.Pr.10 The four orthorhombic lattices.

Describe, in detail, how you would answer this question, including the indices of the peaks etc. that you will use in your analysis? Hint: Calculate the structure factors for each of the lattices.

11. $SrZrO_3$ has the **perovskite structure**.

i) What are the fractional coordinates (basis) of the three elements in a primitive cubic lattice.

ii) Show that the X-ray structure factor for the $(00l)$ reflection is given by $S_{00l} = f_{Sr} + (-1)^l f_{Zr} + \left[1 + 2(-1)^l\right] f_O$, where f are the atomic scattering factors.

iii) Neglecting the angular dependence of the atomic scattering factor, calculate the ratio $I_{(002)}/I_{(001)}$ of the X-ray diffraction intensities (Hint: Use Z).

Diffraction of Electrons and Neutrons

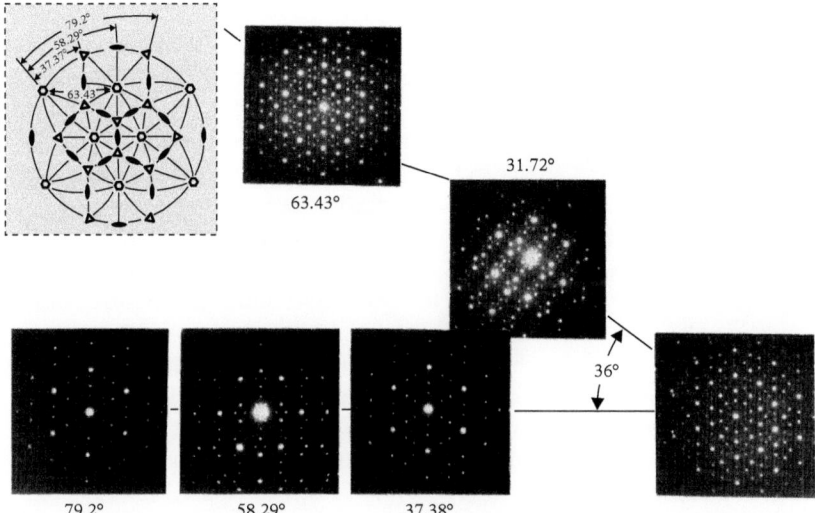

Transmission electron diffraction patterns taken from a single grain of the icosahedral phase in an Al-14at%Mn alloy and rotated through angles corresponding to the symmetry elements of the icosahedral group, $m\overline{3}\,\overline{5}$, as shown in the stereographic projection (*top left*). For his discovery of these phases, also known as quasicrystals, Professor D. Shechtman was awarded the Nobel Prize in Chemistry in 2011. Moreover, this electron diffraction work fundamentally altered the definition of a crystalline material. The International Union of Crystallography changed the definition of a crystal from "a substance in which the constituent atoms, molecules, or ions are packed in a regularly ordered, repeating three-dimensional pattern" to a broader one as "any solid having an essentially discrete diffraction pattern." For further details, see the 2011 Nobel Prize press release at https://www.nobelprize.org/prizes/chemistry/2011/press-release/.

Principles of Materials Characterization and Metrology. Kannan M. Krishnan, Oxford University Press (2021).
© Kannan M. Krishnan. DOI: 10.1093/oso/9780198830252.003.0008

8.1 Introduction

This chapter builds on the previous one and discusses diffraction of electrons and neutrons, with emphasis on the former. In general, because of their negative charge, the scattering of electrons by atoms is significantly stronger than that of X-rays. As a result, electron diffraction can be sensitive to surface crystallography, as well as probe the structure of very small volumes of materials in transmission using focused beams.

The chapter begins by discussing the reciprocal lattice nets of surfaces and the Ewald sphere construction (§4.3.2) for diffraction in two dimensions. We then discuss two important surface diffraction methods: low-energy electron diffraction (LEED) and reflection high-energy electron diffraction (RHEED), illustrating each one with an important application in materials science. This is followed by a discussion of the basic principles of transmission electron diffraction (TED) in the context of their implementation in a transmission electron microscope (TEM), typically operating at high acceleration voltages (>100 kV). We restrict our discussion to the kinematical theory of electron diffraction, mostly in the column approximation, and for completeness, briefly introduce the dynamical diffraction of electrons. Diffraction from perfect crystals is treated in this chapter, and studies of imperfect crystals are introduced in §9.5.2.

Practical implementation of TED includes three important methods: selected area diffraction (SAD), Kikuchi lines and patterns, involving both elastic and inelastic scattering, and convergent beam electron diffraction (CBED) using focused probes. These three are introduced and some typical applications of TED are presented. We conclude with a brief introduction to neutron scattering, including their use in studies of magnetic order. The emphasis in this chapter is on electron diffraction as they are readily encountered in most institutions and laboratories and unlike neutron sources that are generally available only at few national user facilities.

For those interested in pursuing these subjects further, a more detailed discussion of TED can be found in numerous texts at different levels of sophistication. The list of good sources is too long, but a partial selection in chronological order of publication is Hirsch et al. (1965), Thomas and Goringe (1980), Cowley (1984), Williams and Carter (1996), and Fultz and Howe (2013). Similarly, surface electron diffraction is discussed in Pendry (1974) and Vickerman and Gilmore (2009), and neutron scattering is discussed in Bacon (1962), Schwartz and Cohen (1987), and Willis and Carlile (2009).

8.2 The Atomic Scattering Factor for Electrons

Unlike X-rays, electrons interact with *both* the nucleus and the surrounding electron cloud of the atom; the nature of these two interactions is opposite in sign because of their respective charges. The scattering of an electron, considered as a particle with charge, e^-, and wavelength, λ $(=h/mv)$, by the nucleus, considered as a point charge, q $(=+Ze)$, can be analyzed based on their electrostatic interaction. This analysis is similar to the classic Rutherford scattering of positively charged α-particles $(\text{He}^{2+}, +2e)$ by the *atomic nuclei*, which was introduced in §5.3.5.3.

In the scattering of electrons by an atom, the α-particle in Rutherford scattering is replaced by the electron (Fig. 8.2.1) and the fraction of electrons, $d\sigma$, scattered into a solid angle, $d\Omega$, at an angle, 2θ, by the *nucleus*, is given by

$$|f_e(2\theta)| = \left|\frac{d\sigma}{d\Omega}\right|^{1/2} = \frac{Ze^2}{2mv^2\sin^2\theta} \tag{8.2.1}$$

where $f_e(2\theta)$ is the (complex) electron scattering amplitude or scattering factor. The derivation of (8.2.1) can be found in texts on modern physics, such as Sproull and Phillips (2015). In the first Born[1] approximation, i.e. assuming each incident electron is scattered *only once* within the atom, only the amplitude of $f_e(2\theta)$ is important for scattering and its phase is neglected. However, the phase component of f_e is important in high resolution phase contrast imaging in a TEM, as discussed in §9.2.7.3. Note that the Coulomb scattering depends on the product of two charges (and the scattering angle).

In addition to the positively charged nucleus, the electrons are also scattered by the negatively charged atomic electron cloud. It is logical to assume, as we have done in §7.3 for X-rays, that the scattering from all the surrounding electrons is given by the sum of their individual scattering, provided we account for the *phase factor* arising from the *spatial distribution* of the electronic charge cloud around the nucleus. The latter has been calculated for incident X-rays as the atomic scattering factor, f_{atom}, (7.3.8). Now, if we assume that the electron distribution around the nucleus is *spherically symmetric*, we can replace the electron cloud by a point charge, $-f_{atom}\,e$, at the center of the nucleus, and modify (8.2.1), to calculate the total electron scattering factor, f_e, for the atom as a sum of the nuclear and electron scattering contributions:

$$f_e = \frac{(Z - f_{atom})\,e^2}{2mv^2\sin^2\theta} = \frac{me^2\lambda^2}{2h^2\sin^2\theta}\,(Z - f_{atom}) \tag{8.2.2}$$

where the de Broglie wavelength, $\lambda = h/p$, (1.3.3), is used for the electron. Substituting the numerical values for the physical constants, we get

$$f_e\left(\frac{\sin\theta}{\lambda}\right) = f_e(s) = \frac{0.0239}{s^2}\,(Z - f_{atom}). \tag{8.2.3}$$

in units of Å, unlike f_{atom} for X-rays, which is dimensionless. Note that the scattering parameter, s, is defined as in §7, i.e. $s = \frac{\sin\theta}{\lambda}$. Strictly speaking, the relativistic mass of the electron increases with velocity as $m = \gamma\,m_0$, where m_0 is

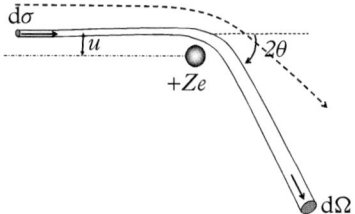

Figure 8.2.1 Rutherford scattering geometry of electrons with atomic nuclei, $+Ze$, and impact parameter, u. Electrons originating from an area $d\sigma$ are scattered through an angle 2θ into a solid angle, $d\Omega$. The scattering angle depends on the impact parameter; as u decreases, 2θ increases. For small angle scattering, particularly relevant for transmission electron diffraction, the differential scattering cross section, $d\sigma/d\Omega \propto \theta^{-4}$.

[1] Max Born (1882–1970) was a German physicist who received the 1954 Nobel prize in Physics and was cited for "fundamental research in Quantum Mechanics, especially in the statistical interpretation of the wave function."

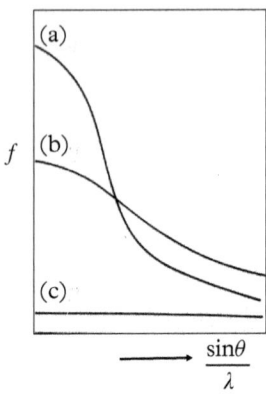

Figure 8.2.2 Schematic representation of the variation of the atomic scattering factors for (a) electrons, (b) X-rays, and (c) neutrons, as a function of the scattering parameter, $\frac{\sin\theta}{\lambda}$. Notice that neutron scattering has a negligible angular dependence.

its rest mass and γ is the relativistic mass correction factor. Thus, for electrons with energy, E (keV), it is necessary to multiply (8.2.3) by γ, given by

$$\gamma = \frac{1}{\sqrt{1 - (v/c)^2}} = 1 + \frac{E}{m_e c^2} \simeq 1 + \frac{E\,[keV]}{511} \qquad (8.2.4)$$

to obtain practically useful values of electron scattering factors. Note that the constant, 0.0239, in (8.2.3) is much larger than the scattering of X-rays by a single electron. Further, $(Z - f_{atom})/s^2 > f_{atom}$ for $s \leq 0.3 - 0.4$, and $(Z - f_{atom})/s^2 < f_{atom}$ for $s > 0.3 - 0.4$. Figure 8.2.2 shows the relative atomic scattering factors for X-rays, electrons, and neutrons, as a function of the scattering parameter, s. Most importantly, because $f_e > f_{atom}$, for values of s encountered in TED, the scattering of electrons by atoms is greater than that for X-rays by a large factor (10^3–10^4). This has a number of important implications for the diffraction of electrons by materials.

First, scattering effects from small volumes or very small quantities of materials can be readily detected, making it possible to undertake SAD (such as in a TEM, §8.6 and §9.2.7.1), and diffraction studies with electron probes focused on the nanoscale.

Second, electron beams are attenuated much more rapidly and are much less penetrating than X-ray beams in materials. As a result, electron beams are readily absorbed/scattered, even by air, and require vacuum columns for diffraction experiments.

Third, electron diffraction in transmission requires very thin specimens (typically, $\leq$100 nm thick) to avoid multiple scattering, and in the parlance of TEM, suitable specimens for observations are often referred to as thin foils.

Fourth, the decrease of diffracted intensities with the scattering angle, 2θ, is much more rapid for electrons compared to X-rays (Fig. 8.2.2). Combined with much shorter wavelengths (than X-rays), and a much larger radius for the Ewald sphere (Fig. 4.3.2) it results in most of the TED patterns being largely scattered in the *forward direction* and restricted to a much smaller angular range ($2\theta \leq \pm 4°$).

Fifth, for thicker specimens, electron diffraction is often carried out in a glancing geometry, but such electron diffraction patterns (RHEED) provide information only from the surface structure ($t \leq 10$ nm) of the specimen. RHEED is discussed in more depth in §8.4.3. Alternatively, surface (monolayer) structures can be studied by using low-energy ($\sim$100 eV) electrons, and LEED is discussed further in §8.4.1.

In general, electrons after scattering once can be easily diffracted multiple times by different lattice planes over short distances in the material. In contrast, the probability of X-rays being scattered multiple times is small and can be ignored. Hence, the simple *kinematical* approach where the intensity of the diffracted beam is considered to be proportional to the square of the structure factor, i.e. $I_{hkl} = |F_{hkl}|^2$, valid for X-rays, (7.4.8), may generally give erroneous results for electron diffraction intensities. Multiple scattering effects, or *dynamical* diffraction theory, is required to describe electron diffraction accurately; however, for very thin films, the kinematical theory will suffice as a good approximation. The dynamical theory of electron diffraction is introduced very briefly in §8.5.4, but a detailed discussion is well beyond the scope of this book. Interested readers are referred to more specialized texts, e.g. Hirsch et al. (1965), Cowley (1984), and Fultz and Howe (2013).

The discussion in §8.3 follows from the introduction to crystallography and diffraction presented in §4. In particular, we revisit the basics of the "crystallography" of surfaces and the related modification in two dimensions of the Ewald sphere construction.

Example 8.2.1: Calculate the atomic scattering factor for FCC Cu [111] reflection for electrons at 5 keV and 100 keV incidence. The lattice parameter $a_{Cu} = 3.62$ Å.

Solution: The interplanar spacing $d_{111} = 2.08$ Å, for Cu from Table 4.1.2. The deviation parameter $s_{111} = \frac{\sin\theta}{\lambda} = \frac{1}{2d_{111}} = 0.239$.

Using the value of s_{111}, we determine $f_{atom} = 22.094$, following (7.3.10) and the Cromer–Mann parameters for Cu in Table 7.3.2.

Applying (8.2.3), we get $f_e\,(s_{111}) = 2.884$ Å.

Applying the relativistic mass correction factor, (8.2.4), we get

$$f_e(5 \text{ keV}) = 2.912 \text{ Å}$$
$$f_e(100 \text{ keV}) = 3.448 \text{ Å}$$

8.3 Basics of Electron Diffraction from Surfaces

In Chapter 7, we described the use of X-ray diffraction in the bulk for three-dimensional crystal structure measurements. Since X-rays are only scattered weakly by the electrons surrounding the nucleus, they penetrate deeply into materials and allow us to probe the bulk crystal structure. Neutrons, as we will see in §8.8, interact even more weakly with solids, penetrate more deeply and with negligible angular dependence (Fig. 8.2.2). Hence, X-ray and neutron scattering are not the first choice for *surface structure* determination, although such measurements are possible using specialized geometries (see §7.8.4 on X-ray reflectivity). Thus, electrons that are strongly interacting, with mean free path lengths of the order of a few nanometers at energies of ~100 eV, are the logical choice for diffraction studies of surfaces.

Recall that in two dimensions there are only five lattice nets possible (Fig. 4.1.15) as opposed to the 14 Bravais lattices (Table 4.1.1) in three dimensions. We now begin a discussion of surface diffraction with a brief summary of the nomenclature used to describe surface lattices, followed by the Ewald sphere construction, now modified for two dimensions.

8.3.1   Surface Reconstruction, Surface Nets, and Their Notation

Ideally, we can identify a surface by referring to the terminating plane of the bulk material, i.e. Au(100), Fe(110), etc. Mathematically, we can describe the surface net as a translation vector

$$\mathbf{T} = m\mathbf{a}_s + n\mathbf{b}_s \tag{8.3.1}$$

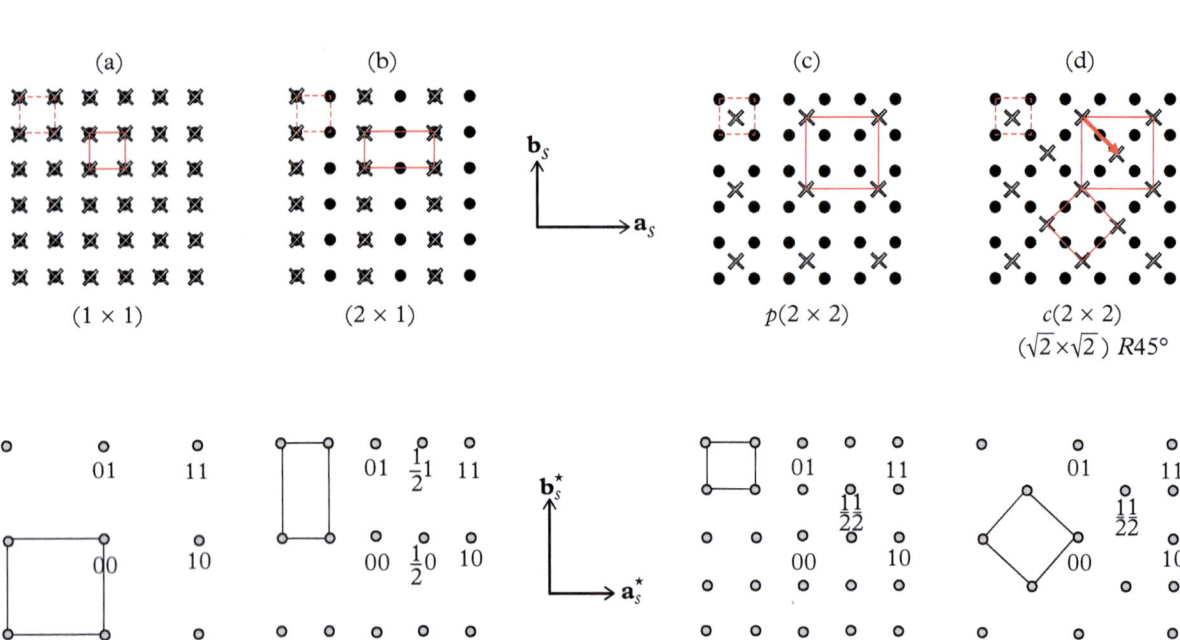

Figure 8.3.1 Examples of surface lattice structures (top row), where the filled dots represent the periodicity of the substrate and the crosses represent the surface lattice: (a) (1×1) reconstruction, (b) (2×1) reconstruction, (c) a primitive (2×2) reconstruction, and (d) a (2×2) reconstruction with an additional centering translation that give rise to a *multiple* cell—the alternative *unit* cell, also shown, is rotated by 45°. Both (c) and (d) show adsorbate atoms on the surface; if the adsorbate were to be oxygen, and the substrate an Au(100) surface, the nomenclature for (d) would be explicitly modified as Au(100) $\left(\sqrt{2} \times \sqrt{2}\right) R45° - O$. In all cases, the substrate lattice nets are shown by dashed lines and the surface lattice net is shown by continuous lines. For each reconstruction in the top row the corresponding reciprocal lattice, including the unit cell, is shown in the bottom row; typically, these patterns are observed in LEED, for normal electron incidence. See Figure 8.4.3.

Adapted from Woodruff and Delchar (1992).

where the subscript, s, refers to the surface, m and n are integers, and $\mathbf{a}_s$ and $\mathbf{b}_s$ are surface unit cell vectors of the crystal (or positions occupied by surface adsorbates). In reality, due to the break in symmetry at the surface, the atoms on the top-most monolayer of the surface can rearrange themselves into a new net. Physically, this surface reconstruction arises from a minimum energy configuration different from the bulk because the break in symmetry reduces the number of nearest neighbor atoms on the surface. This description also applies to other atoms *adsorbed* on surfaces. In the general case, the reconstructed surface unit-cell vectors, $\mathbf{a}_s$ and $\mathbf{b}_s$, are related to the translation vectors, $\mathbf{a}$ and $\mathbf{b}$, of the ideal bulk terminating layer as

$$\mathbf{a}_s = M\mathbf{a} \ \text{ and } \ \mathbf{b}_s = N\mathbf{b} \tag{8.3.2}$$

where M and N are integers. The reconstructed surface is then known as a $(M \times N)$ *reconstruction*. Further, if the surface net is rotated by ϕ, with respect to the underlying bulk lattice, the nomenclature is modified as $(M \times N)R\phi$ (Fig. 8.3.1d), where R indicates rotation. Moreover, if the surface net is best defined using a centering translation (Fig. 8.3.1d) instead of it being a primitive one (Fig. 8.3.1c) it is indicated explicitly as $c(M \times N)$. Finally, if the surface layer includes a specific adsorbed species, the adsorbate is explicitly indicated. Figure 8.3.1 provides some examples of surface nets, including reconstruction and positions occupied by the adsorbates, as well as the nomenclature used to describe them.

8.3.2   Reciprocal Lattice Nets and Ewald Sphere Construction in Two Dimensions

The reciprocal lattice construction for a two-dimensional surface lattice follows from our definition in three dimensions, (4.2.1), but in a simplified form as

$$\mathbf{a}_s^* = \frac{2\pi}{A} \left(\mathbf{b}_s \otimes \hat{\mathbf{n}} \right)$$

$$\mathbf{b}_s^* = \frac{2\pi}{A} \left(\hat{\mathbf{n}} \otimes \mathbf{a}_s \right) \tag{8.3.3}$$

where the area, $A = \mathbf{a}_s \cdot \left(\mathbf{b}_s \otimes \hat{\mathbf{n}} \right)$, and $\hat{\mathbf{n}}$ is a unit vector normal to the surface. Figure 8.3.1 also shows the corresponding reciprocal lattice nets (bottom row).

Diffraction processes from surfaces (Fig. 8.3.2a) can also be described by the Ewald sphere construction (Fig. 4.3.2) but appropriately modified for the two-dimensional periodicity parallel to the surface (Fig. 8.3.2b). As a consequence, only the wave vector component parallel to the surface is conserved with the addition of a reciprocal lattice vector. We follow our earlier convention for incident, $\mathbf{k}_0$, and diffracted, $\mathbf{k}$, beams, but then use superscript, $||$, and $\perp$ to indicate their components parallel and perpendicular to the surface.

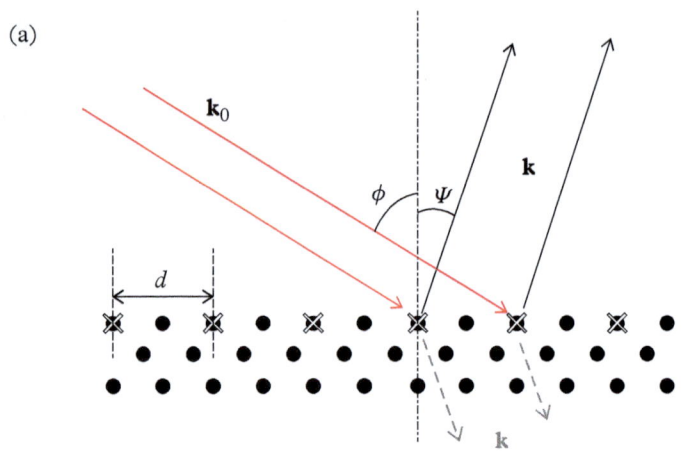

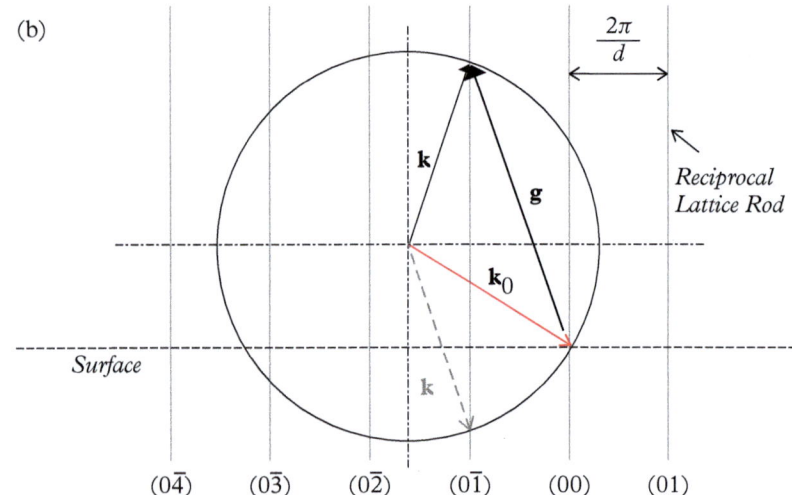

Figure 8.3.2 Schematic representation of surface diffraction in (a) real space, where ϕ is the angle of incidence, and ψ is the exit angle. Each cross (dot) represents a row of surface (bulk) atoms normal to the page. The corresponding Ewald sphere construction in reciprocal space, including the reciprocal lattice rods, is shown in (b); notice the two possibilities for the diffracted wave, with one, shown dashed, propagating into the crystal that is not observed.

Adapted from Vickerman and Gilmore (2009).

First, the conservation of energy gives

$$k_0^2 = k^2 \qquad (8.3.4)$$

or, using the parallel and perpendicular components, we get

$$\left(k_0^{\parallel}\right)^2 + \left(k_0^{\perp}\right)^2 = \left(k^{\parallel}\right)^2 + \left(k^{\perp}\right)^2 \qquad (8.3.5)$$

Similarly, the conservation of momentum gives

$$\mathbf{k}^{\parallel} = \mathbf{g}_{hk} + \mathbf{k}_0^{\parallel} \qquad (8.3.6)$$

where $\mathbf{g}_{hk} = h\mathbf{a}_s^* + k\mathbf{b}_s^*$ is a surface reciprocal lattice vector defined in (8.3.3). Strictly speaking, the perpendicular component, $k^\perp$, need not be conserved in the surface diffraction process. In addition, now that the diffraction conditions are dependent on a reciprocal lattice-net vector with only two components, we can also represent the diffracted beam with only two indices, (hk). Further, the Ewald sphere construction is modified (Fig. 8.3.2b). The incident and diffracted beams lie in three dimensions (Fig. 8.3.2a) but, most importantly, reciprocal lattice rods of infinite length perpendicular to the surface and passing through the reciprocal lattice points replace the reciprocal lattice of Figure 4.3.2. This is because the surface is a two-dimensional net with infinite periodic repeat distance *normal* to the surface. As the distance between adjacent reciprocal lattice points is inversely proportional to the corresponding distance in real space, we expect the reciprocal lattice points normal to the surface to be infinitesimally close to one another, and to be effectively perceived as a continuous line or rod. The Bragg diffraction condition is satisfied for every incident beam that emerges in a direction given by the intersection of the Ewald sphere with one of the reciprocal lattice rods (Fig. 8.3.2b). Compared to three dimensions, the presence of the reciprocal lattice rods for surfaces greatly relaxes the conditions for the observation of diffracted beams. In the bulk, a small change in incident electron energy ($\propto$ radius of the Ewald sphere), or the direction of the scattering wave-vector, $\mathbf{k}$, will results in the absence of many beams and the excitation of new ones. In contrast, for the case of surfaces, diffracted beams may occur at all energies provided the corresponding reciprocal lattice rod lies within the Ewald sphere. In addition, for each reciprocal lattice rod two diffraction beams can satisfy the Bragg diffraction conditions (Fig. 8.3.2b); however, only half of them are back-scattered and observed; the other half propagates forward into the bulk crystal and are not observed in surface diffraction.

Finally, the indexing of the diffracted beams from surfaces is, by convention, with reference to the *unreconstructed* substrate real and reciprocal lattice net. As a result, if the reconstructed surface, or the distributions of adsorbates, have larger periodicities, the corresponding *surface reciprocal lattice net* would be smaller than that for the bulk lattice and the additional reciprocal lattice points are indexed with fractional indices (Fig. 8.3.1c,d).

With this brief background of the diffraction of electrons from surfaces, we introduce two practical and commonly used surface diffraction methods involving low- (LEED) and high- (RHEED) energy incident electrons.

8.4 Surface Electron Diffraction Methods and Applications

We can see from (1.3.5) that electrons of energy $\sim$150 eV have a wavelength of $\sim$0.1 nm (1 Å), making them suitable for diffraction from crystalline materials with similar lattice parameters. Further, from the universal curve (Fig. 1.3.6) plotted for the mean free path length of electrons as a function of energy, we can see that the penetration depth is ~0.25 nm, of the order of a surface monolayer or two, for electron energies in the range 50–100 eV. This ensures that these low-energy electrons have good surface sensitivity in diffraction experiments. When electrons of such low energy, incident normal to a crystal surface, are elastically back-scattered, the resulting diffraction pattern forms the basis of the technique of LEED and is used for surface structure determination.

Alternatively, we can use high-energy electrons incident at grazing angles on the surface of a crystal. The shallow angle of incidence of the electrons results in a negligible component of their angular momentum normal to the surface. As a result, their penetration depth into the surface is very small, and their scattering from the surface in the reflection geometry forms the basis of RHEED.

These two techniques, LEED and RHEED, along with a representative application for each one, are introduced in the sections to follow.

8.4.1   Low-Energy Electron Diffraction (LEED)

Figure 8.4.1 shows a schematic diagram of an experimental set up for LEED measurements. The electron gun, described in §5.2.2, produces a beam of variable energy in the range ~20–300 eV, an energy spread of ~0.5–1.5 eV, and a current of ~1 μA. The beam is delivered to the specimen surface by a system of electrostatic focusing lenses. After interaction with the specimen surface, the back-scattered electrons travel in straight lines in the field-free region to a set of three grids surrounding the electron gun. The first grid is set at ground, the same potential as the specimen. The second set of grids are set to discriminate and eliminate any inelastically scattered electrons that have lost small amounts of energy, and allow only the nearly elastically scattered ones to go through. Such a discriminator to remove electrons that have lost more than a few eVs of energy is also known as a *retarding field analyzer*. The final grid is set at +5 kV to reaccelerate the diffracted electrons and have them impinge on a fluorescent screen and create the LEED pattern.

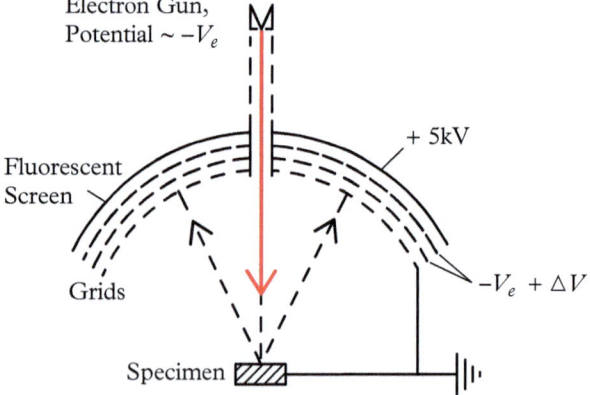

Figure 8.4.1 A schematic representation of the LEED apparatus. A low-energy beam is incident normal on the specimen and a hemispheric analyzer, with multiple grids for energy discrimination, detects the electrons diffracted from the surface in the backward direction.

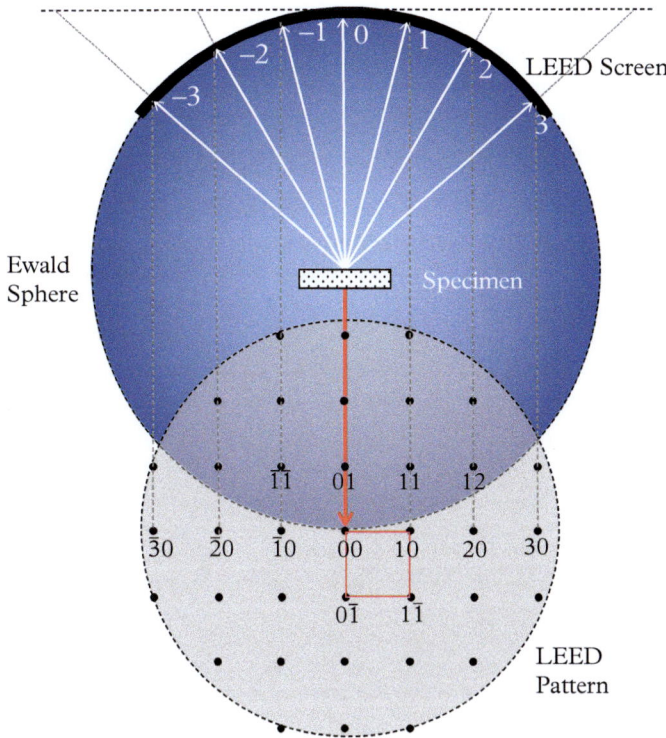

Figure 8.4.2 The Ewald sphere construction for LEED under normal incidence. A cubic (001) surface, a schematic of the two-dimensional LEED pattern, typically observed (see Fig. 8.4.3a). The diffracted beams, corresponding to a linear section of the reciprocal lattice, detected in the backward direction are indicated in the Ewald sphere construction.

Figure 8.4.2 shows the corresponding Ewald sphere construction for normal incidence. Comparing the two figures, it is clear that the LEED pattern observed is just a projection of the surface reciprocal lattice net, with a magnification determined by the energy of the incident electrons. Note that, strictly speaking, the position of the gun eliminates the possibility of recording the (00) beam for normal incidence. Further, LEED experiments are extremely sensitive to the cleanliness of the specimen surface; typically, experiments are carried out in an ultra-high vacuum (UHV) system ($\sim 10^{-10}$ torr, or $\sim 10^{-8}$ Pa) giving rise to atomically clean surfaces.

In a typical LEED experiment a portion of the reciprocal space, i.e. those electrons that are diffracted only in the backward direction are recorded. The most common data acquired is the variation of the intensity, $I(hk)$, as a function of position along a small portion of the Ewald sphere (Fig. 8.4.2) where h and k, are the diffraction indices. The positions of the spots and their relative intensities (e.g. Fig. 8.4.3) give information on the surface or over-layer unit lattice net size and shape. The exact positions of the atoms on the surface cannot be accurately determined from a visual inspection of the diffraction pattern, but instead, it requires an analysis of the diffraction intensities. The latter is more complicated to do as it requires that the intensity distribution, $I(V)$, or the variation of the intensity as a function of incident beam energy/voltage for a number of diffraction geometries, be measured and compared with model calculations (see Pendry, 1974). Other structural information, including surface phase transitions can be obtained by measuring the intensity variation, $I(T)$, as a function of temperature. Further, the kinetics of order–disorder transitions (§7.11.4) on surfaces can also be obtained by measuring the time evolution of diffraction intensities (Lagally, 1985). Alternatively, the intensity variation, $I(\alpha)$, of specific diffraction features as a function of adsorbate surface coverage, α, is very useful in adsorption studies. A simple example of the *adsorption* of different elements on a W(100) surface is presented in the next section. Last, but not least, LEED is commonly used to establish the cleanliness and surface structural order; the latter is also correlated with other physical properties such as lattice dynamics, surface diffusion kinetics, and electronic and magnetic properties. A quick summary of LEED can also be obtained in the *Encyclopedia of Materials Characterization* (Brundle, Evans, and Wilson, 1992), or in Amelinckx, Gevers, and van Landuyt (1978); for more detailed discussions, see Pendry (1974) and/or Lagally (1985).

Example 8.4.1: Plot the expected LEED pattern for the two cubic (110) surfaces shown in the following figure for (a) smooth and (b) stepped surfaces. For (b) assume that the coherence length is larger than the step width. Comment on what will happen if the step width is irregular.

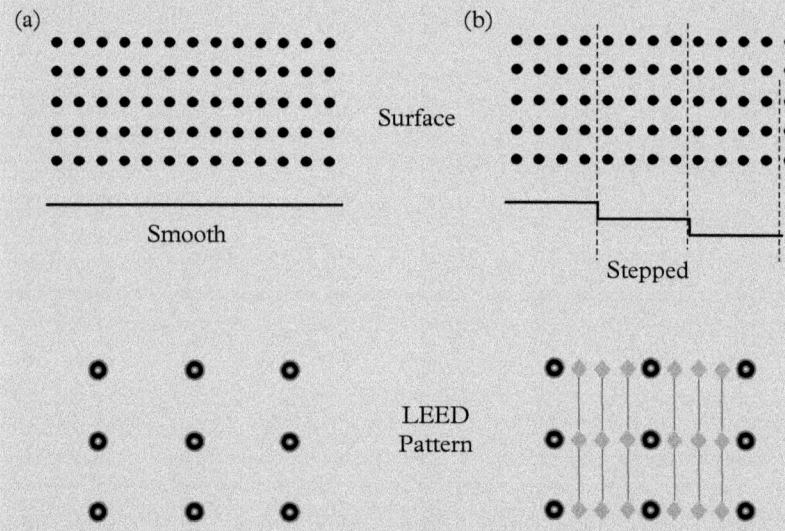

Solution: (a) The reciprocal lattice for the (110) surface can be calculated from (8.3.3). Note that the reciprocal lattice net is rotated by 90° with respect to the real lattice.

(b) Because of the periodic steps, the surface repeat unit is now enlarged by a factor of four leading to a LEED pattern that is roughly a (4×1) reconstruction. Strictly speaking the real repeat units should be $(4^2 + 1^2)^{1/2} = (17)^{1/2}$. If the step width is irregular, the additional spots will be streaked in the direction of the disorder.

8.4.2 Adsorption Studies on Surfaces Using LEED

Figure 8.4.3 shows a set of LEED patterns for various adatoms adsorbed on a W(100) surface. The diffraction pattern from a clean W(100) surface (Fig. 8.4.3a) corresponds to a surface with a square lattice in both real and reciprocal space in Figure 8.3.1a. When oxygen adatoms are adsorbed on the W(100) surface (Fig. 8.4.3b) additional spots in the half-order position, indicating a doubling of the surface lattice parameters for oxygen, consistent with the primitive $p(2 \times 2)$–O superstructure is seen; this corresponds to Figure 8.3.1c. In contrast, the LEED pattern (Fig. 8.4.3c) for the adsorption of hydrogen on the W(100) surface is nonprimitive. It is clearly a centered, $c(2 \times 2)$–H superstructure (Fig. 8.3.1d) with a rotation of 45° with respect to the substrate, W(100), surface. Finally, Figure 8.4.3d shows the LEED pattern for the co-adsorption of carbon monoxide and nitrogen on a W(100) surface. Clearly, the pattern corresponds to a (4×1)–N, CO reconstruction. Further, when compared to the (2×1) pattern in Figure 8.3.1b, this pattern confirms the presence of two domains of the (4×1) superstructure rotated by 90° with respect to each other.

From this brief illustrative example, we conclude that LEED provides a very good map of the reciprocal lattice net of the specimen surface monolayer, and allows the determination of the surface lattice vectors, $\mathbf{a}_s$ and $\mathbf{b}_s$, and

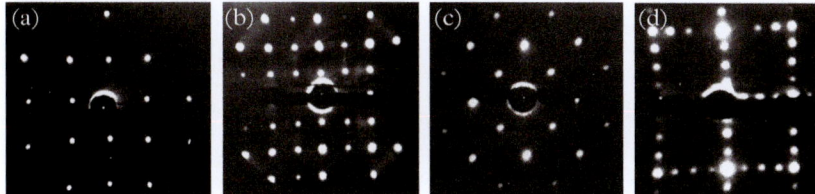

Figure 8.4.3 Photographs of LEED diffraction patters from (a) (1×1) or W(100) surface, (b) $p(2 \times 2)$–O reconstruction for a monolayer of oxygen atoms adsorbed on the surface, (c) a non-primitive $c(2 \times 2)$–H surface, and (d) two domains of (4×1)–N,CO reconstructed surfaces of adsorbed gases on W(100) surface.

Adapted from Amelinckx, Gevers, and van Landuyt (1978).

their relationship to the substrate lattice vectors, **a** and **b**. Further, LEED patterns provide some information about the lateral, inter-atomic interactions between the adsorbed atoms on the surface. For example, if there were no interactions, the adsorbed atoms would be distributed randomly on top of the atoms on the terminating layer of the substrate; however, if there were to be a lateral repulsion between neighboring adatoms, the tendency would be to form a $c(2 \times 2)$ superstructure, where each adatom is separated by at least one substrate surface atom (Fig. 8.3.1d; Fig. 8.4.3c).

8.4.3 Reflection High-Energy Electron Diffraction (RHEED)

Figure 8.4.4a shows a typical experimental arrangement for RHEED. The high-energy (5–100 keV), parallel beam is incident at grazing angle $\phi \sim 87$–$90°$ (Fig. 8.4.4b) and the elastically diffracted electrons are detected by a retarding field analyzer and then displayed on a fluorescent screen. If necessary, the intensity of a specific diffracted beam, useful for measuring RHEED oscillations (Fig. 8.4.8) can also be monitored as function of time or film growth.

The higher energy (smaller wavelength, λ, or larger wavevector, k_0) of the incident beam in RHEED, compared to the much smaller energy in LEED, makes a significant difference in the observed diffraction patterns. The Ewald sphere in RHEED (Fig. 8.4.5a) is now very large compared to the reciprocal lattice dimensions. As a result, its intersection with the reciprocal lattice rods, for an

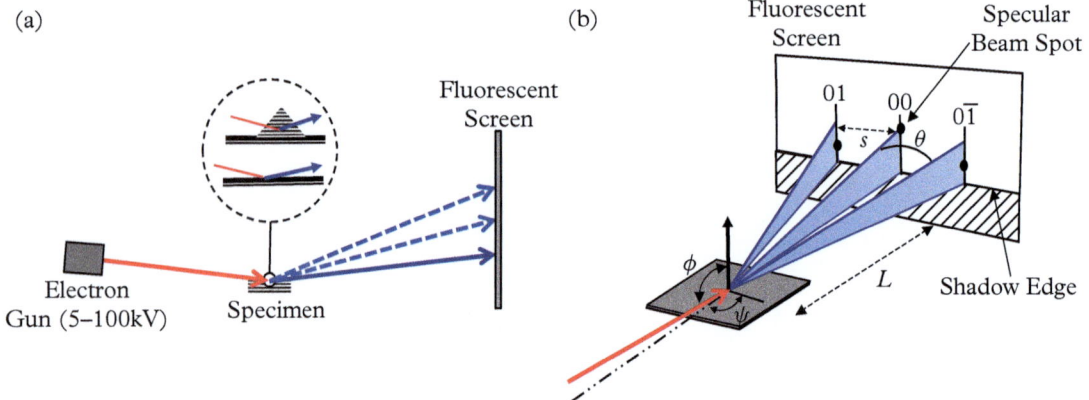

Figure 8.4.4 (a) A schematic of the experimental apparatus used for reflection high-energy electron diffraction in side view. (b) A three-dimensional view of the same, defining the incident angle, $\phi \sim 87$–$90°$, the azimuth angle, ψ, and the diffraction angle, θ, as well as other details of the experiment used in determining details of the surface lattice net from the RHEED pattern.

Adapted from [1].

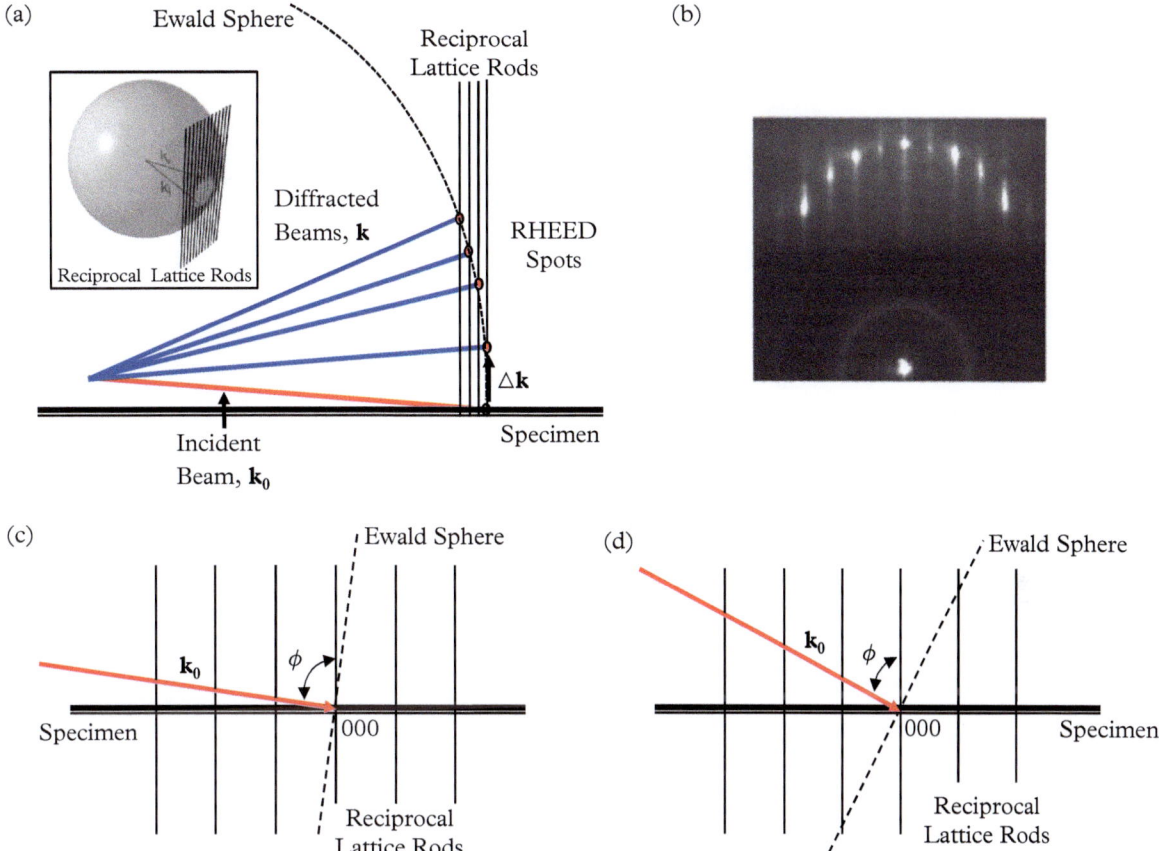

Figure 8.4.5 (a) The Ewald sphere and the reciprocal lattice rods involved in RHEED (shown in the inset). The scattering geometry, including the incident (k_0) and scattered (k) beams, and the associated scattering vector (Δk), is illustrated. (b) The ideal two-dimensional pattern from a Ge(001) surface. (c) For high-energy electrons incident at grazing angles, $\phi \sim 90°$, the Ewald sphere (with radius exaggerated here for clarity as a straight line) intercepts very few reciprocal lattice rods. (d) Reducing the angle of incidence allows few more reciprocal lattice-rods to be intercepted by the Ewald sphere.

(b) Obtained courtesy of Prof. Tsui, University of North Carolina, Chapel Hill. Adapted from Vickerman and Gilmore (2009).

ideal two-dimensional specimen surface, appears as a set of closely spaced dots (Fig. 8.4.5b). The corresponding RHEED pattern shows streaks (Fig. 8.4.6a). In practice, the radius of the Ewald sphere is so large that only a few spots/streaks are observed in a RHEED pattern. However, if the angle of incidence, ϕ, is changed additional diffraction conditions may be satisfied (Fig. 8.4.5c,d) as other lattice rods may be intercepted by the Ewald sphere. Thus, the arrangement of reciprocal lattice rods in three dimensions can be obtained. Such changes in ϕ are obtained by rocking the specimen, or alternatively by rotating it (changing the azimuthal angle, ψ, in Figure 8.4.4b) about its normal. However, in practice, the surface tends to have atomic scale roughness either due to incomplete coverage or surface defects, and now the surface is called quasi two dimensional.

The size of the surface unit mesh can be readily obtained from the RHEED pattern. From Figure 8.4.4b, with the separation, s, measured between the streaks of the pattern, and distance, L, between the specimen and the fluorescent screen, we obtain the diffraction angle, θ, as

$$\tan \theta = \frac{s}{L} \approx \sin \theta \tag{8.4.1}$$

for small θ. Further, for a cubic lattice ($a = b$), the surface diffraction satisfies the condition,

$$a \sin \theta = \left(h^2 + k^2\right)^{1/2} \lambda \tag{8.4.2}$$

where h and k are the diffraction indices. Then, substituting for $\sin\theta$ from (8.4.1), we get

$$a = \left(h^2 + k^2\right)^{1/2} \lambda \left(\frac{L}{s}\right) \tag{8.4.3}$$

and thus, allowing the lattice parameter, a, to be determined.

In RHEED, as well as in LEED, we are interested in knowing the *coherence length*/area involved in a measurement. By coherence length/area, we mean how large an area of the surface is involved in the production of a particular diffraction pattern/measurement. Both the energy spread of the incident beam and its angular divergence contribute to the coherence length of a RHEED pattern, and for most common experimental arrangements it is ~200 nm; typically, it is much smaller for LEED (5–10 nm). In other words, RHEED may not be able to detect any surface disorder if it occurs on a scale larger than the coherence length. However, for the small grazing angles used, RHEED is very sensitive to surface disorder on scales much smaller than the coherence length (Fig. 8.4.6 a,b). In addition to the quasi-two-dimensional surface arising from surface disorder causing streaks, the primary RHEED beam may pass through any surface protrusions or asperities (Fig. 8.4.6b). Such transmission of the incident beam will also be sensitive to the additional periodicity of the lattice planes normal to the surface. As a result, instead of streaks, spots appear in the RHEED pattern with their separation being inversely related to the inter-planar spacing *normal* to the surface. The contribution from the reflected and transmission features can be easily resolved in RHEED. When the incident angle, ϕ, is varied the transmission features will change in intensity but not in position; in contrast, the surface features will move continuously in position because the intersection of the Ewald sphere with a lattice rod will move up or down as ϕ is changed.

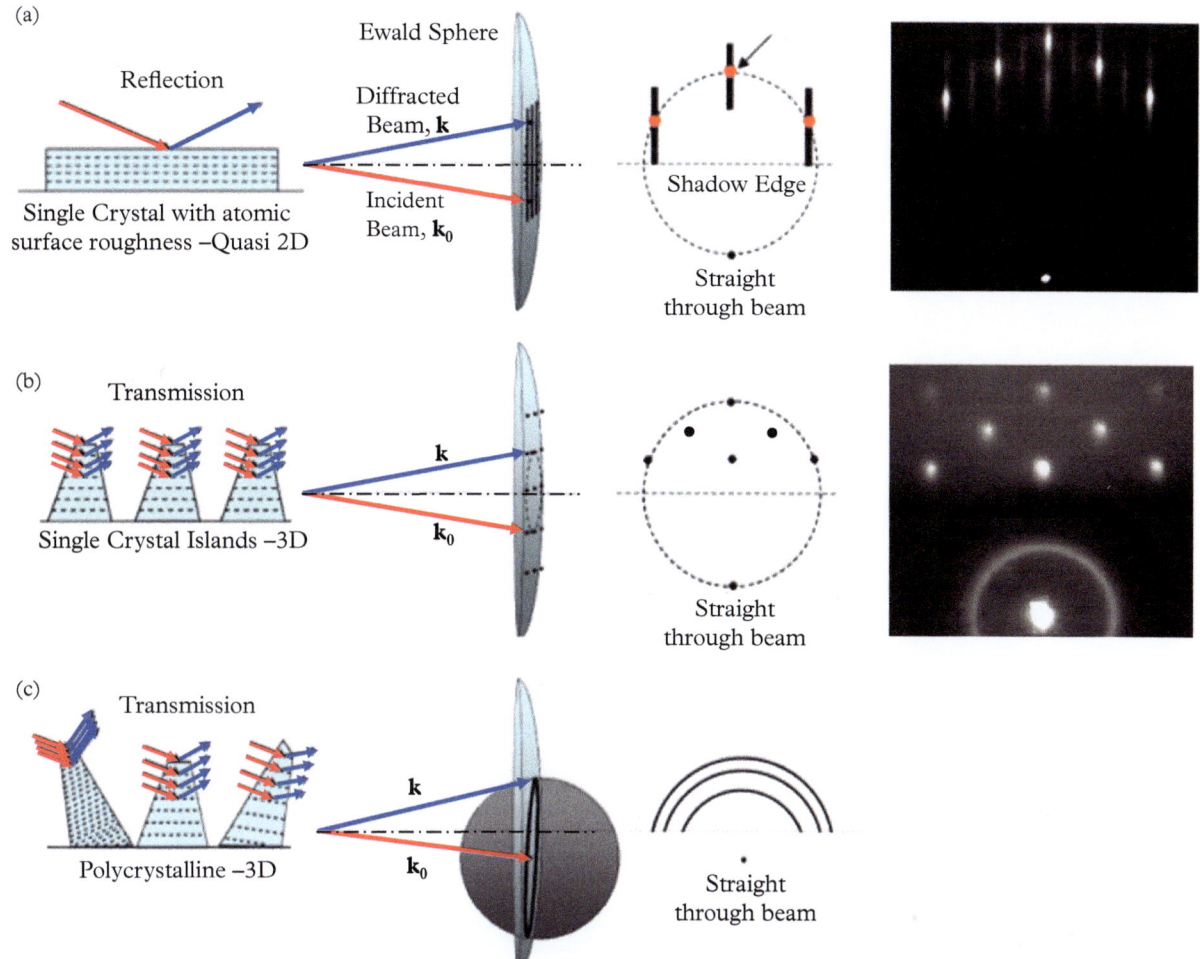

Figure 8.4.6 A schematic representation of three possible RHEED scattering geometries, depending on crystal surface morphology, with actual patterns obtained from metal layers grown on Ge(001). In each row, in sequence, from left to right we show a schematic of the film surface and indicate the incident and diffracted electrons in real space (left), the corresponding Ewald sphere construction, the resultant diffraction pattern (schematic) with both the shadow edge (dashed line) and the straight-through beam, which is the incident beam without hitting the substrate, and the observed RHEED pattern (right). The different morphologies are (a) a clean single crystal surface with atomic scale roughness, (b) single crystal surface with island growth, and (c) a polycrystalline film, where the grey sphere is the reciprocal structure for a polycrystalline crystal with randomly oriented grains.

Courtesy Prof. Tsui, UNC, Chapel Hill. Adapted from [2].

8.4.4   RHEED Oscillations: *In Situ* Monitoring of Thin Film Growth

Equilibrium growth morphologies of thin films can be predicted based on a simple free energy consideration [3] comparing the free energy of the substrate, $\gamma_{\text{substrate}}$, which exists before the film growth, with the sum of the free energies of the film surface, γ_{film}, and the film/substrate interface, $\gamma_{\text{interface}}$, which is established after the film has been deposited. If

$$\gamma_{\text{substrate}} > \gamma_{\text{film}} + \gamma_{\text{interface}} \qquad (8.4.4)$$

the first monolayer of the film will prefer to coat the entire substrate surface and a layer-by-layer growth is expected (Fig. 8.4.7i). However, once the first monolayer is deposited, further growth of the film will not encounter the substrate surface and the interface energy is now also different. Moreover, the lattice mismatch (if any) between the substrate and the film will give rise to a misfit strain energy, which will increase with growing film thickness. As a result, after the first layer, a modified form of growth may replace the continued layer-by-layer growth, where the next layer forms as islands. This is known as the Stranski–Krastanov (S–K) growth mode (Fig. 8.4.7ii). This mode not only allows the energetically favorable first layer of the film to remain exposed but also makes it possible for the subsequent islands to relieve any strain by lateral relaxation. Alternatively, if

$$\gamma_{\text{substrate}} < \gamma_{\text{film}} + \gamma_{\text{interface}} \qquad (8.4.5)$$

the film nucleates directly as a three-dimensional island (Fig. 8.4.7iii), and leaves as much of the low-energy substrate surface exposed as possible.

Figure 8.4.7 (a) Equilibrium growth mode determined by different surface and interface energies. (i) Layer-by-layer growth. (ii) Stranski–Krastanov (S–K) growth and (iii) island growth. (b) Non-equilibrium growth modes are a function of substrate temperature, evaporation rates, and step density (iv) step-flow growth and (v) island growth.

Adapted from Krishnan (2016).

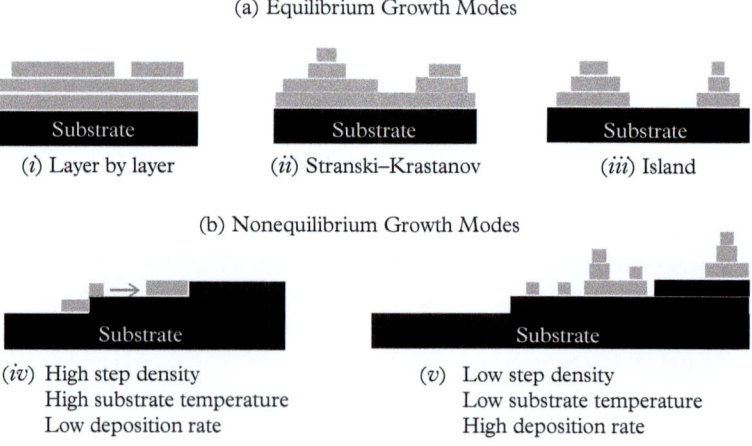

(a) Equilibrium Growth Modes

(*i*) Layer by layer   (*ii*) Stranski–Krastanov   (*iii*) Island

(b) Nonequilibrium Growth Modes

(*iv*) High step density
High substrate temperature
Low deposition rate

(*v*) Low step density
Low substrate temperature
High deposition rate

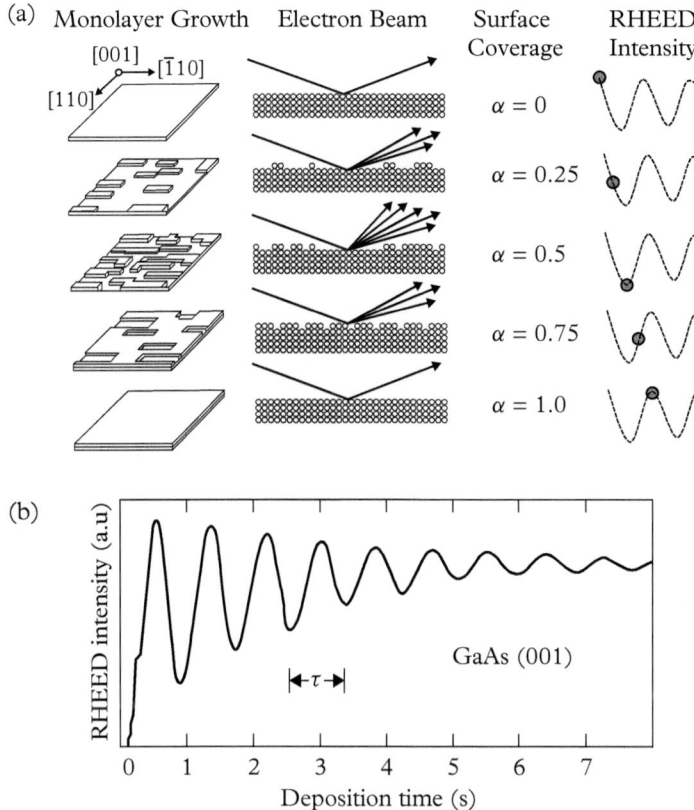

Figure 8.4.8 Variation of the intensity of the specularly reflected beam in RHEED as a function of the fractional surface coverage, α, for layer-by-layer growth. (a) The intensity oscillates from maximum ($\alpha = 0$), to a minimum ($\alpha = 0.5$), to a maximum ($\alpha = 1.0$) again. (b) RHEED oscillations with deposition time for films of GaAs(001) grown by molecular beam epitaxy (MBE), confirming the layer-by-layer growth morphology.

Adapted from [5].

Equilibrium growth of thin films generally requires high substrate temperatures ($T_g > 200\ °C$); however, nonequilibrium growth at lower temperatures ($T_g < 200\ °C$) and high deposition rates are often employed. In such conditions, the role of surface steps becomes critical in determining nonequilibrium growth morphology [4]. At high growth temperatures and/or low deposition rates, the deposited atoms have sufficient time and energy to diffuse to the nearest step followed by bonding at this location with high coordination number (Fig. 8.4.7iv). This is known as step-flow growth. However, at reduced growth temperatures, or high growth rates, or low step densities, the deposited atoms do not have sufficient time and energy to diffuse to the step edges. Instead, when they arrive on the substrate surface they nucleate spontaneously as islands, which continue to grow by incorporating new adatoms as they are deposited (Fig. 8.4.7v). Eventually, they coalesce into a smooth monolayer, but then nucleate again as islands forming a rough, incomplete, monolayer surface. This process is repeated as the film grows and results in alternating rough and smooth surfaces for each additional half a monolayer coverage. The most common way to monitor this type of growth is by RHEED, which shows a periodic oscillation of the intensity with every half-monolayer coverage (Fig. 8.4.8).

The scattering geometry is set up in such a way that the reflected electron beams from the upper terraces interfere destructively with the ones from the lower terraces. Thus ideally, this interference is out of phase, or completely destructive with zero intensity, for any half-monolayer coverage. Alternatively, it can be in phase, or constructive with maximum intensity, for any complete monolayer coverage. In practice, the beam intensity oscillates showing a minimum (maximum) for out-of-phase (in-phase) conditions, and these RHEED oscillations can not only be a very good measure of the growth morphology (Fig. 8.4.8), but compare very well with surface imaging methods (Fig. 8.4.9).

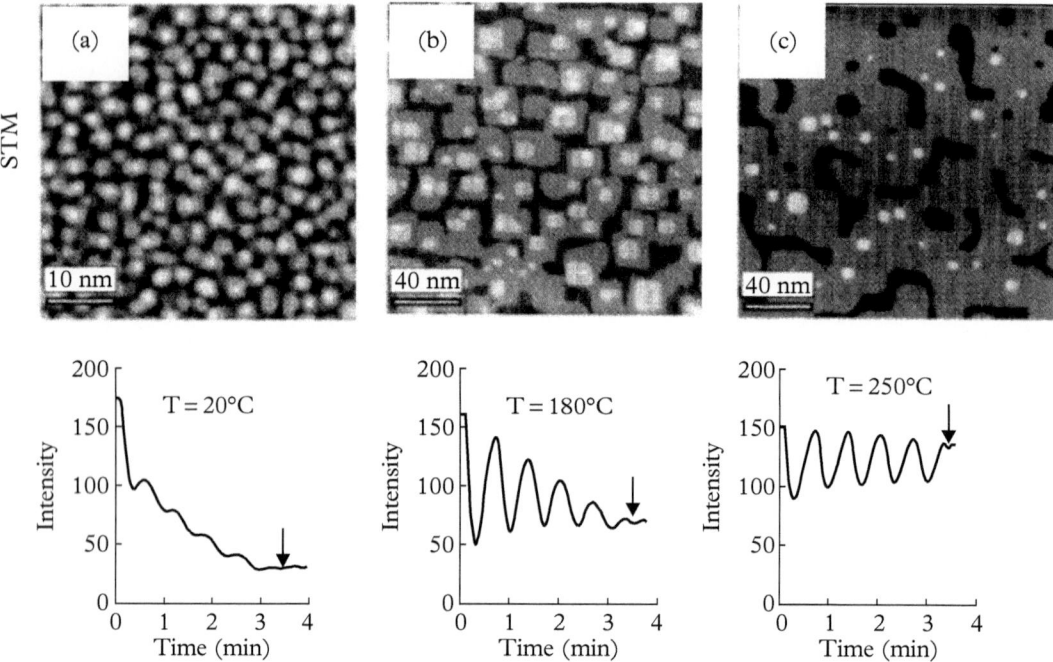

Figure 8.4.9 Images taken with a scanning tunneling microscope or STM (described in §11) of a Fe monolayer growing on an $Fe_{(100)}$ surface. Three different growth modes: (a) island, (b) Stranski–Krastanov and (c) near layer-by-layer growth modes, obtained at different temperatures are shown. STM images (top) of the growth surfaces correlate nicely with the RHEED oscillations (bottom).

Adapted from Krishnan (2016).

Example 8.4.2: The following figures represent RHEED patterns of the growth of Bi on graphite at (a) room temperature as a function of growth up to 16 MLs of Bi, and (b) after 8 ML growth at two different substrate temperatures. What can you say about the growth of Bi on graphite based on the RHEED data? Please look up Zayed and Esayed-Ali, *Phys. Rev. B* **72**, (2005): 205426, for details.

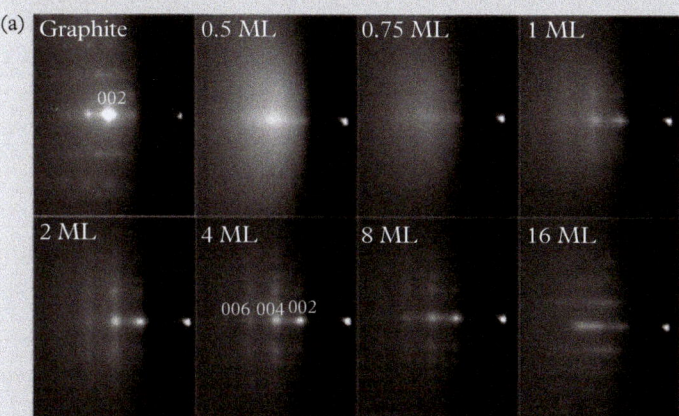

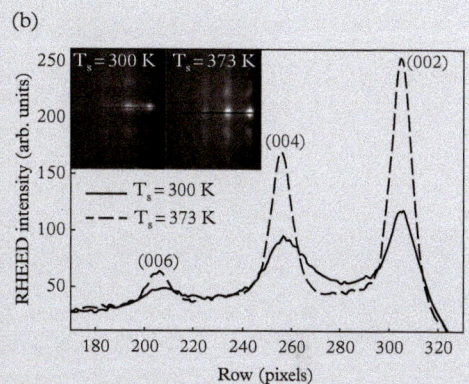

Solution: From (a), we can initially see the spot pattern of graphite. At 0.5 ML we can see the decay of the graphite spot intensity and the appearance of a diffuse background. As the thickness of Bi increases, we can see the appearance of a Bi spot with increasing intensity up to 8 ML. The spots can be indexed as the rhombohedral structure of Bi. There is no intensity oscillation in the intensity of the graphite spot, suggesting growth of three-dimensional islands. However, at 16 ML of Bi elongated RHEED streaks are visible. This suggests coalescence and formation of asymmetric shape crystallites.

(b) No change in the relative spot position is observed for the two different substrate temperatures, indicating no change in film growth orientation. However, narrower peaks are observed at higher temperature (373 K) compared to the room temperature (300 K) growth. This suggests that films grown at 373 K have an enhanced crystallite size and/or a higher degree of orientation order, when compared to 300 K growth.

8.5 Transmission High-Energy Electron Diffraction

We now describe the diffraction of high-energy (>100 keV) electrons, also referred to as *fast electrons* to distinguish them from the bound or itinerant electrons in the material, in the context of observations in a TEM. A TEM is a most versatile instrument (§9.2), routinely used in materials analysis, and which is capable of providing comprehensive diffraction, imaging, and spectroscopic information of thin foil specimens. Moreover, such information can be obtained in a TEM from small volumes, with excellent microstructural sensitivity, and at the highest spatial resolution.

The interaction of a fast electron beam with a specimen (see §5.3.2 and §9.3) is complex and includes elastic, inelastic, coherent, and incoherent scattering (§8.5.1), as well as secondary processes such as the emission of element-specific, characteristic X-rays (Fig. 5.3.5). In this section and the next, we focus on elastic scattering and the diffraction of electrons in a TEM, including some typical applications in materials research. Various imaging methods, the associated mechanisms of contrast in a TEM, and the two important analytical methods of energy-dispersive X-ray spectroscopy (EDXS) and electron energy-loss spectroscopy (EELS), implemented in a TEM, are discussed in §9. Note that a TEM, when it includes EDXS and/or EELS spectrometers, is also called an *analytical electron microscope* (AEM). After we discuss TEM/AEM in §9, we present details of the complementary technique of scanning electron microscopy (SEM) in §10.

We begin with elastic scattering (§5.3.2) as this interaction is fundamental to electron diffraction from crystals as well as the observation of image contrast in a TEM. As discussed in §8.1, fast electrons interact strongly with the atomic nuclei, in addition to the outer electrons, and the resultant Coulomb forces cause their elastic scattering. Typically, the angle of scattering, 2θ, is small (~ 1–$2°$), and in the process of scattering both kinetic energy and angular momentum are conserved. In other words, the fast electrons are scattered elastically without any loss of energy. Note that at high incident energies and large scattering angles, the transfer of energy from the fast electron to the nucleus does take place with multiple consequences, including back-scattering and sputtering (§5.3.2). The specific case of back-scattered electrons, in the context of image contrast in a SEM, is discussed in §10.3.3.

Similar to X-rays, the interaction of the fast electrons with a periodic crystal can be characterized by Bragg diffraction (Fig. 8.5.1a) and the associated Laue criterion (§4.3) in reciprocal space (Fig. 8.5.1b). However, unlike X-rays, the wavelengths of fast electrons, λ_e, is much smaller than the typical inter-planar spacing, d_{hkl}, in crystalline materials (Fig. 4.3.5). As a result, the radius of the Ewald sphere is very large with a very flat surface, and it intersects the reciprocal lattice in multiple zones (Fig. 4.3.4). As for the diffracted intensities on the *exit surface* of the thin foil (Fig. 8.5.1c) in the simplest column approximation (§8.5.4), it is assumed that the amplitudes, A_g, of the Bragg diffracted waves are small compared to that (A_0) of the *forward-scattered* beam. Since the fast electrons are scattered strongly by crystals, this approximation known as the *kinematical* theory of diffraction is only valid for thin foils, typically less than a few nm in thickness. Simply put, the physical principles of kinematical scattering (§8.5.3) are not very different from that described in terms of the structure factor for X-rays (§7.4), but

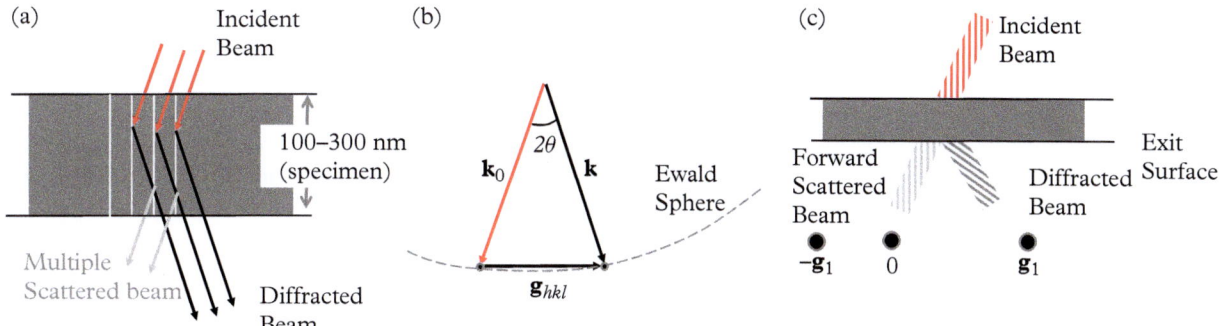

Figure 8.5.1 Diffraction of fast electrons in transmission through a thin foil specimen, illustrating (a) Bragg condition for the incident and diffracted beams in real space, (b) Ewald sphere construction and the Laue criterion in vector form, and (c) the column approximation (see text for details).

with appropriate modifications for (a) electron scattering factors (§8.2), different from those for X-rays, (b) the much smaller wavelength, λ_e, of the fast electrons, and (c) the shape of the object or thin foil *through which* it must pass.

For most crystals, the thin foil approximation is strictly not valid and the intensity is transferred from the forward-scattered beam to the diffracted beam as the fast electron propagates in the crystal. Further, it is highly probable that after propagating a few more nanometers into the foil, the diffracted beam transfers intensity back to the forward-scattered beam. This back and forth variation in the amplitudes and intensities of the forward-scattered and diffracted beams is referred to as *dynamical* diffraction. A comprehensive theory of dynamical diffraction of fast electrons requires a quantum mechanical solution of the Schrodinger equation in the periodic potential of the crystal. Further, since the fast electrons experience the periodic potential of the crystal, their wave functions must also reflect the symmetry of the crystals. Such solutions of the Schrodinger equation that reflect the translational symmetry are called Bloch waves (Kittel, 1986) and the theory of electron diffraction in terms of Bloch waves was originally developed by Bethe[2] [6].

The dynamical theory of electron diffraction also defines a characteristic distance in the material where the wave amplitude is transferred back and forth between the forward-scattered and diffracted beams. This distance, known as the *extinction distance*, ξ_g, for a specific diffracted beam (or reflection) defined by the reciprocal lattice vector, $\mathbf{g}$, is directly proportional to the volume of the unit cell, V_c, and inversely to the structure factor, $F(\theta_B)$, at the exact Bragg angle, θ_B, for that reflection as described in (8.5.19). We outline the rudiments of the dynamical theory of electron diffraction in §8.5.4, and describe a method to measure the extinction distance by CBED in §8.6.2.

[2] Hans A. Bethe (1906–2005) was a German–American physicist who received the Nobel Prize in Physics in 1967 and was cited for "his contributions to the theory of nuclear reactions, especially his discoveries concerning the energy production in stars."

8.5.1 Coherent, Incoherent, Elastic, and Inelastic Scattering

Recall our discussion of the propagation of light in transmission through a pin hole (Fig. 6.3.1) and a grating. When the spherical waves from any two openings arrive in phase (the path difference is an integral number of wavelengths) at a specific point on the screen, they constructively interfere, and appear bright (Fig. 6.4.1). At other points on the screen, the waves arrive out of phase, interfere destructively, and disappear. As a result, taking this interference process for the entire grating into consideration, we see alternative bright and dark lines on the screen (Fig. 6.6.7). The diffraction of electrons in transmission through crystals is analogous to the grating; however, the atomic arrangement is in three dimensions and a similar path difference argument for the waves scattered by each atom then predicts a regular "interference" or diffraction pattern of beams/spots instead of lines (see discussion on page 368).

In general, we say that the scattering from a set of different atoms located at positions, $\mathbf{r}_i$, in a material is *coherent* if the amplitude, A_{coh}, of the scattered wave is

$$A_{coh} = \sum_{\mathbf{r}_i} A_{\mathbf{r}_i} \tag{8.5.1}$$

where $A_{\mathbf{r}_i}$ is the amplitude of the scattered wave from each $\mathbf{r}_i$. The total coherent diffraction intensity measured for waves scattered from all sites is

$$I_{coh} = \left| \sum_{\mathbf{r}_i} A_{\mathbf{r}_i} \right|^2 \tag{8.5.2}$$

depending on the constructive or destructive interference of the different wave amplitudes, $A_{\mathbf{r}_i}$. Note that even though it is not explicitly indicated, the intensity depends on the relative phases of the scattered waves and the position, $\mathbf{r}_i$, of the scattering atoms. Furthermore, recall the Thomson model of the scattering of an X-ray wave by a bound electron (§7.2.1), where the electric field component of the X-ray wave drives the bound electron into oscillation. This "oscillator" then reradiates at the same frequency, but with a well-defined phase difference (of π for X-rays) between the scattered and incident waves. To reiterate, such scattering with a well-defined phase difference is also termed coherent.

Alternatively, the "oscillator" electron, may be coupled to another electron in the solid; further, we can reasonably assume that this coupling in a material, being quantum mechanical, has multiple degrees of freedom. Then, following interaction with the oscillator electron with multiple pathways of excitation, the phase of the scattered wave cannot be easily predicted, and such scattering is called *incoherent*. Thus, the scattering at all positions, $\mathbf{r}_i$, in the material does not preserve any phase relationship between the incident and scattered waves. Then, the total incoherent intensity

$$I_{incoh} = \sum_{\mathbf{r}_i} |A_{\mathbf{r}_i}|^2 \qquad (8.5.3)$$

is the sum of the intensities of the individual waves scattered from each position, $\mathbf{r}_i$. Further, the angular dependence of incoherent scattering for N atoms is not different from that of a single atom.

Apart from coherent and incoherent scattering, interactions can also be *elastic* or *inelastic* (§1.3.1). The former involves no difference in energy between the incident and scattered waves; in contrast, the latter always involves a change in energy. Typically, diffraction experiments involve elastic and coherent scattering, whereas spectroscopy measurements are generally due to inelastic and incoherent scattering.

8.5.2 Basics of Electron Diffraction in a Transmission Electron Microscope

The principles of the diffraction of waves by a periodic crystal lattice were introduced in §4.3, formulated both in real space (Bragg's law, Fig. 4.3.1) and in reciprocal space (Ewald sphere construction, and the Laue criterion, Fig. 4.3.2). In the latter formulation, diffraction maxima are excited when the Ewald sphere passes through a reciprocal lattice point. In §7, we used both real and reciprocal lattice approaches to describe the diffraction of X-rays. It was shown that, if the lattice is not primitive and contains more than one atom in the unit cell, not all points in reciprocal space permitted by the Laue criterion are observed in diffraction. Such systematic absences (§7.6), i.e. which reflections are allowed and which are not, are determined by the structure factor (§7.4). Section 7.5 shows examples of structure factors for many crystal structures calculated for X-ray scattering.

The diffraction of electrons in transmission can be understood quantitatively using the same formulism as that used for X-rays, provided we account for some principal differences in scattering between X-rays and fast electrons.

First, the wavelength of electrons, λ_e, for the range of energies encountered in TEM (>100 keV), is very small compared to typical interatomic distances in crystals. Hence, we expect the radius of the Ewald sphere to be very large (Fig. 4.3.4) when compared to the reciprocal lattice unit cell dimensions. As a result, the observed electron diffraction patterns are a *planar cross-section* of the reciprocal lattice of the crystal. Second, in calculating the structure factors, F_{hkl}, for diffraction intensities, the atomic scattering factors, f_e, for electrons, (8.2.3), corrected for relativistic effects, (8.2.4), must be used.

Third, in transmission through thin foils, we are always interested in the intensity of the scattered or diffracted beams at the *exit surface* of the specimen (Fig. 8.5.1c). Hence, the diffracted intensities are not only dependent on the structure factor (defined by the details of the unit cell) but also on the *external shape* of the specimen/crystal. The latter, called the *shape transform*, also modifies the shape of the reciprocal lattice points–one example discussed is the reciprocal lattice rods (Fig. 8.3.2) in surface diffraction—allowing for diffraction to be observed even when the Ewald sphere does not exactly coincide with a reciprocal lattice point, but passes in the near vicinity of such a point.

Fourth, and this bears repeating, since the fast electrons are strongly scattered by atoms, multiple scattering is routine and the intensity of the diffracted and forward-scattered beams dynamically oscillate with thickness.

8.5.3   Kinematical Theory of Electron Diffraction

We now calculate the diffraction intensities at the exit surface (Fig. 8.5.1c) for a fast electron beam incident on a crystalline thin foil. We consider two atoms, one at the origin, O, and the other at the general position, P, given by the vector, $\mathbf{r}$, in the crystal (Fig. 8.5.2).

For the incident beam, $\mathbf{k}_0$, we are interested in determining the intensity of the diffracted beam, $\mathbf{k}$. The path difference, Δl, for the two scattered waves emanating from the positions O and P, along the direction, $\mathbf{k}$, is

$$\Delta l = r \cos \alpha - r \cos \beta = r(\cos \alpha - \cos \beta) = r \left[\frac{\mathbf{k} \cdot \mathbf{r}}{|\mathbf{k}||\mathbf{r}|} - \frac{\mathbf{k}_0 \cdot \mathbf{r}}{|\mathbf{k}_0||\mathbf{r}|} \right] \quad (8.5.4)$$

However, for elastic scattering, $|\mathbf{k}| = |\mathbf{k}_0| = 2\pi/\lambda$, the phase difference between the two waves is

$$\Delta \phi = \Delta l \frac{2\pi}{\lambda} = (\mathbf{k} - \mathbf{k}_0) \cdot \mathbf{r} \quad (8.5.5)$$

Note that when the Bragg condition is satisfied, the Laue criterion gives $\mathbf{k} - \mathbf{k}_0 = \mathbf{g}_{hkl}$, where $\mathbf{g}_{hkl}$ is any reciprocal lattice vector of the crystal. Further, at the Bragg orientation, $\mathbf{g}_{hkl} \cdot \mathbf{r}_l = 2m\pi$, where m is an integer.

We use an approach similar to that used to calculate the structure factor, (7.4.7), to calculate the resultant amplitude, $A(\theta)$, of the wave scattered in the direction, $\mathbf{k}$, at an angle, θ (Fig. 8.5.2). It is given by

$$A(\theta) = \sum_i f_e^i(\theta) \exp[i(\mathbf{k} - \mathbf{k}_0) \cdot \mathbf{r}] \quad (8.5.6)$$

where f_e^i, (8.2.3), is the electron atomic scattering factor for atom, i. To calculate the amplitude of the diffracted beam at the exit surface, the summation in (8.5.6) has to be carried out over all atoms, i, in the thin foil along the path of the beam. We can simplify the summation by breaking down the coordinates of all the atoms, $\mathbf{r}$, into two parts, i.e. $\mathbf{r} = \mathbf{r}_i + \mathbf{r}_l$, where $\mathbf{r}_i$, $i = 1 \ldots n$, are the coordinates of all the atoms within the unit cell, and $\mathbf{r}_l$, $l = 1 \ldots N$, are the coordinates of all the unit cells in the three-dimensional crystal lattice encountered by the fast electron. Then

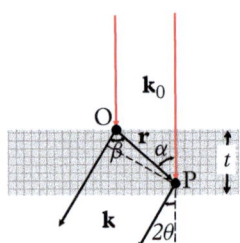

Figure 8.5.2 Transmission electron scattering geometry for a thin foil specimen of thickness, t, considering one atom, O, at the origin and the other, P, at a general position, $\mathbf{r}$. The incident, $\mathbf{k}_0$, and diffracted, $\mathbf{k}$, wave-vectors are also indicated.

$$A(\theta) = \sum_{i=1}^{n} f_e^i(\theta) \exp[i(\mathbf{k} - \mathbf{k}_0) \cdot \mathbf{r}_i] \sum_{l=1}^{N} \exp[i(\mathbf{k} - \mathbf{k}_0) \cdot \mathbf{r}_l] = F_{hkl} G \qquad (8.5.7)$$

The first term, F_{hkl}, in (8.5.7) is the structure factor discussed earlier for X-ray scattering in §7.4, but is now modified using the appropriate electron scattering factors, $f_e^i(\theta)$, discussed in §8.2. Further, the structure factor is only relevant when the Laue criterion, $\mathbf{k} - \mathbf{k}_0 = \mathbf{g}_{hkl}$, is exactly satisfied. Then, following (7.4.7), at the exact Bragg angle, θ_B, it can be calculated as

$$F_{hkl} = \sum_{i=1}^{n} f_e^i(\theta_B) \exp 2\pi i (hu_i + kv_i + lw_i) \qquad (8.5.8)$$

To get the total amplitude at the exit surface, the structure factor, F_{hkl}, is multiplied by the contribution from the lattice. The shape transform or the lattice amplitude, G, in (8.5.7) is determined by the overall shape of the crystal, and includes a summation over all N unit cells in the crystal in the path of the electron beam. We now assume that the intensity, I_0, of the primary beam remains unchanged as it propagates in the crystal, and we also allow for small deviations from the exact Laue condition, $\mathbf{k} - \mathbf{k}_0 = \mathbf{g}_{hkl}$, to contribute to the diffraction. This deviation is called the *excitation error*, $\mathbf{s}$ $(= s_x, s_y, s_z)$, and connects the reciprocal lattice vector, $\mathbf{g}_{hkl}$, to the Ewald sphere along the direction of the incident beam (Fig. 8.5.3a). Most importantly, the Laue criterion, (4.3.7), is now modified as

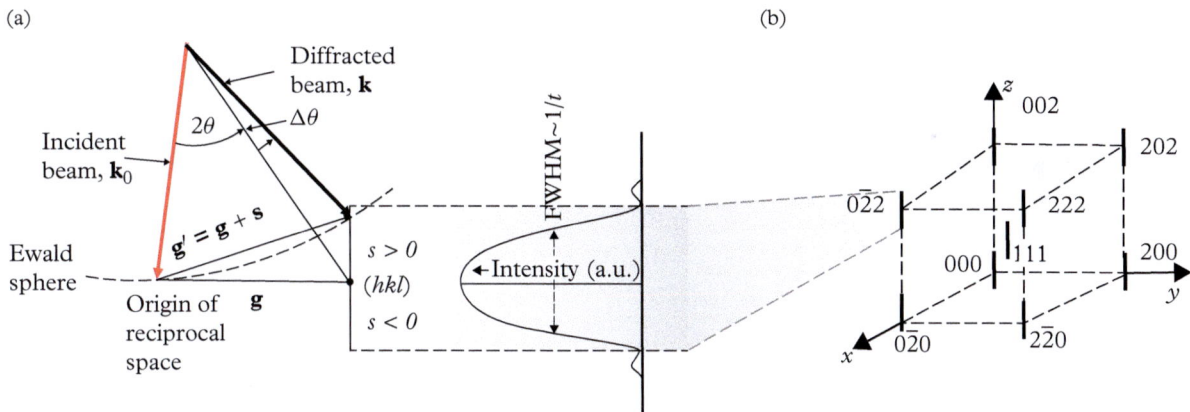

Figure 8.5.3 (a) The Ewald sphere construction for electron diffraction, including the definition of the excitation error, s. For a crystal of thickness, t, the intensity of the diffracted beam has a half-width of $1/t$, and subsidiary minima at n/t, where n is an integer (note: only two are shown). (b) Schematic illustration, for a thin foil, of the effective streaking of all reciprocal lattice points normal to the foil surface.

Adapted from Edington (1975).

$$\mathbf{k} - \mathbf{k}_0 = \mathbf{g}_{hkl} + \mathbf{s} \tag{8.5.9}$$

Substituting (8.5.9) in the expression for the lattice amplitude, G, i.e. the second summation term in (8.5.7), we get for the phase angle in the exponent

$$(\mathbf{k} - \mathbf{k}_0) \cdot \mathbf{r}_l = (\mathbf{g}_{hkl} + \mathbf{s}) \cdot \mathbf{r}_l = \mathbf{g}_{hkl} \cdot \mathbf{r}_l + \mathbf{s} \cdot \mathbf{r}_l \tag{8.5.10}$$

However, at the exact Bragg orientation, $\mathbf{g}_{hkl} \cdot \mathbf{r}_l = 2\pi m$, where m is an integer. When N is large, we can replace the summation in (8.5.7) with an integral for G. Then, for a bulk crystal in the shape of a parallelepiped of dimensions $N_1 a, N_2 b, N_3 c$, where a, b, c are the unit cell parameters, we get

$$G = \frac{1}{V_c} \int_0^{N_1 a} \int_0^{N_2 b} \int_0^{N_3 c} \exp 2\pi i \left(s_x x + s_y y + s_z z \right) dx dy dz$$

$$= \frac{1}{V_c} \frac{\sin \pi N_1 a s_x}{\pi s_x} \frac{\sin \pi N_2 b s_y}{\pi s_y} \frac{\sin \pi N_3 c s_z}{\pi s_z} \tag{8.5.11}$$

where $V_c = a \times b \times c$ is the volume of the unit cell. Incorporating the values of F_{hkl}, (8.5.8), and G, (8.5.11), for an incident beam of intensity, I_0, the intensity, I_{hkl}, of the diffracted beam is

$$I_{hkl} = I_0 |F_{hkl}|^2 G^2 \tag{8.5.12}$$

This result that the intensity is a product of the contributions from the structure factor and the shape transform, is important to get a good physical understanding of TED. The excitation errors, s_x, s_y, and s_z, are inversely proportional to the corresponding crystal dimensions, $N_1 a, N_2 b, N_3 c$, respectively. For a thin film with large physical extent in x- and y-directions, and with the z-axis normal to the film, we can set $s_x = s_y = 0$, and calculate the contribution of the lattice amplitude, G, to the diffracted intensity, I_G, from (8.5.11) as

$$I_G = |G|^2 = \frac{1}{V_c^2} \frac{\sin^2 \pi N_3 c s_z}{(\pi s_z)^2} \tag{8.5.13}$$

Note that this behavior is similar to the case of the diffraction of light, at a grating with M slits and separation, a, as discussed in §6.6.2. The total diffracted intensity, (8.5.12), for the case of transmission through a thin foil is then given by

$$I_{hkl} = I_0 |F_{hkl}|^2 I_G = I_0 |F_{hkl}|^2 \frac{1}{V_c^2} \frac{\sin^2 \pi N_3 c s_z}{(\pi s_z)^2} \tag{8.5.14}$$

Figure 8.5.3a shows the variation of the intensity with the excitation error, s_z, for any arbitrary reflection. Figure 8.5.3b shows the formation of streaks, or relrods, on all the reciprocal lattice points. From the scattering geometry shown in Figure 8.5.3a, and using Bragg's law, we can calculate the excitation error, s_z, as

$$s_z = \mathbf{g}_{hkl}\Delta\theta = \frac{2\pi}{d_{hkl}}\Delta\theta = \frac{4\pi}{\lambda}\sin\theta_{hkl}\Delta\theta \qquad (8.5.15)$$

In general, the diffraction intensity, I_{hkl}, in reciprocal space varies with the overall shape of the specimen. For a thin film of thickness, t, used in TED, we have $N_3 c = t$ in (8.5.14), and the intensity has the shape of a rod or spike normal to the thin film (Fig. 8.5.3a). Referred to as *relrods*, the lengths of the spikes are inversely proportional to the thickness of the foil (recall that in two dimensions, a surface gives rise to an infinitely long relrod; Fig. 8.3.2). The direction of the spikes is along the beam direction for normal incidence. Furthermore, this geometry—narrow relrods intersecting with the Ewald sphere—results in sharp diffraction intensities and, in fact, an observable diffraction pattern even at considerable deviation from the Bragg angle.

This description of the distribution of intensities in reciprocal space can be generalized for crystallites of different shapes (Fig. 8.5.4). Also known as *shape transforms*, they are particularly relevant for diffraction from small precipitates; for example, needle-shaped precipitates (Fig. 8.5.4c) give rise to reciprocal lattice points in the shape of plates/discs normal to their long axis. Overall, one can conclude that the intensity distribution of the reciprocal lattice points is always extended in a direction parallel to the smallest dimension of the precipitates; furthermore, the observed diffraction pattern also depends on the direction of incidence of the electron beam with respect to the orientation of the precipitates (Fig. 8.5.4d).

8.5.4 The Column Approximation, Dynamical Diffraction, and Diffraction from Imperfect Crystals

We now consider the two-beam diffraction condition (Fig. 8.5.5) where a parallel electron beam is incident on a thin foil and produces a forward scattered beam, $\mathbf{k}_0$, with intensity, I_0, and only one strongly diffracted beam, $\mathbf{k}$, with intensity, I_g, at the Bragg condition. Note that in reality, since the reciprocal lattice points have spikes and the Ewald sphere is rather flat (Fig. 8.5.4), several reflections ($2\mathbf{g}, 3\mathbf{g}, \ldots$) are also excited. The two beams (Fig. 8.5.5a) at the point, P, on the exit surface are produced, by multiple scattering events, only in the region defined by the triangle ABP. Further, the angle $A\hat{P}B = 2\theta_B$, or twice the Bragg angle. Then, for typical values of thickness, $t = 100$ nm, $\lambda_e = 3.7 \times 10^{-12}$ m (100 keV electrons), and $\theta_B \sim 0.5°$, the distance AB is only 2 nm. As a result, to calculate I_0 and I_g, the triangle ABP can be replaced by two columns, one for the forward-scattered and

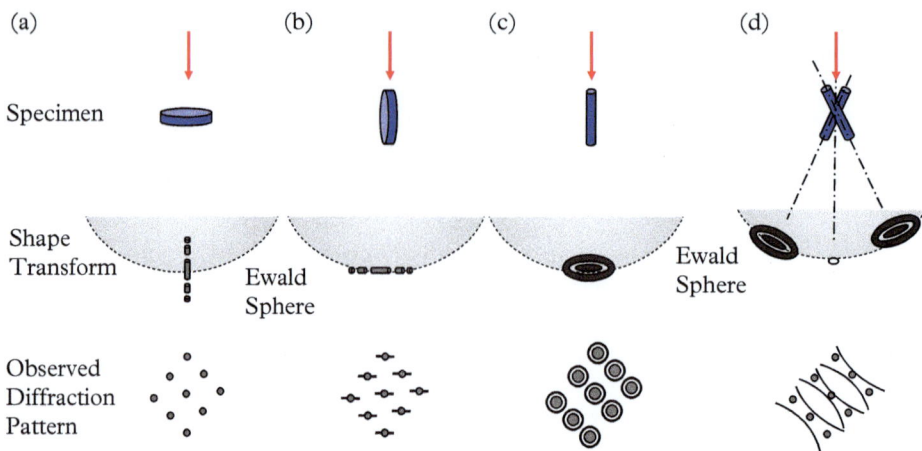

Figure 8.5.4 Effect of crystallite shape on the shape transform and the diffracted intensities in transmission for thin plates oriented (a) normal and (b) parallel to the beam, and for rods oriented (c) parallel and (d) inclined to the beam.

Adapted from Thomas and Goringe (1980).

Figure 8.5.5 The column approximation. (a) In the two-beam case, only the electrons—both forward scattered and diffracted—included in the narrow triangle, ABP, contribute to the intensity at P. For thin foils, t ~ 100 nm, the distance AB ~2 nm. (b) Thus, to calculate diffracted intensities, the forward-scattered and diffracted beams are approximated by columns along the beam directions.

Adapted from [7].

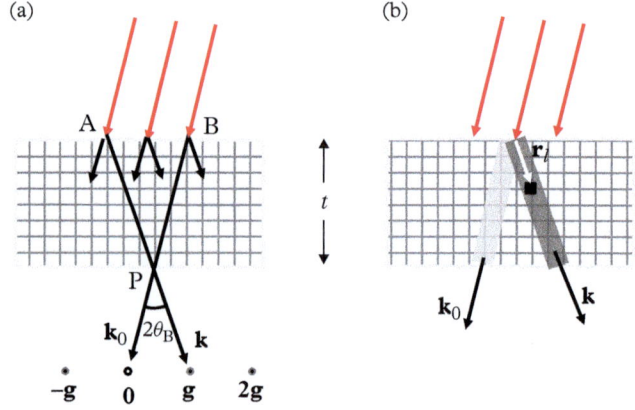

the other for the diffracted beam (Fig. 8.5.5b). Note that, to first approximation, the intensities of adjacent columns are independent of each other.

In this column approximation, the amplitude, A_g, of the diffracted beam at the exit surface is then obtained by integrating the contributions of each unit cell at a particular depth element along the column (Fig. 8.5.5b) over the thickness of the foil. In the kinematical theory, the lattice contribution to the intensity of a specific diffracted beam is obtained from (8.5.13) as

$$I_g \sim \frac{\sin^2 \pi t s}{(\pi s)^2} \qquad (8.5.16)$$

and depends on both the excitation error, s, and the thickness, t. This equation is best plotted as an amplitude-phase diagram (Fig. 8.5.6). The circle has a radius of $1/2\pi s$, and decreases rapidly as s increases, or in other words, as the specimen is tilted away from the exact Bragg orientation. However, for a fixed value of s, as the thickness, t, increases, the intensity oscillates between two maximal values proportional to $(\pi s)^{-2}$.

Furthermore, if the diffracted intensity is measured as a function of position, the image produced by such diffraction contrast has two very distinct characteristics. If the thin foil is wedge shaped, then the image will show a series of dark and bright bands reflecting the above intensity oscillation with thickness; such fringes are known as *thickness fringes* (Fig. 8.5.7). Along a particular fringe the specimen thickness is constant, and for a wedge-shaped specimen, the fringes are separated in thickness by a distance of $1/s$ between them. However, in practice, the thickness fringes are not visible at larger thicknesses and their spacing is not given accurately by (8.5.16). In addition, for thicker (large t) specimens, from (8.5.16), when $s = 0, I_g \sim t^2$; thus, I_g would become larger and larger with thickness, and the principal assumption of the kinematical theory that the diffracted intensity is small, i.e. $I_g << I_0$, is not valid. Thus, to reiterate, the kinematical theory is only valid for very thin crystals, and also when $s >> 0$, a condition frequently used in imaging defects.

Alternatively, if the thin crystalline foil is of uniform thickness, but is bent and not flat, the excitation error, s, now changes spatially with position. If the variation in the excitation error, s, is uniform, it will also produce an intensity distribution with position, and these are known as *bend contours* (Fig. 8.5.8).

The *dynamical theory* of electron diffraction overcomes the limitations of the kinematical theory, and is discussed in more specialized texts—see Hirsch

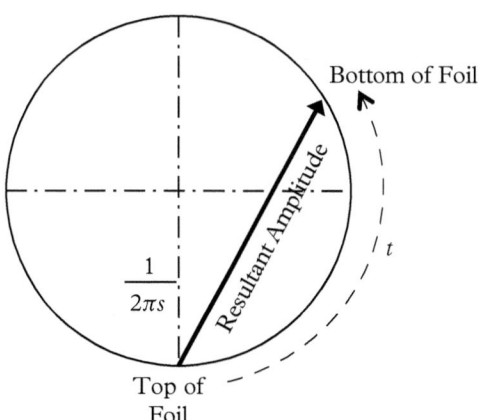

Figure 8.5.6 Amplitude–phase diagram for the lattice contribution to the intensity in the kinematical theory.

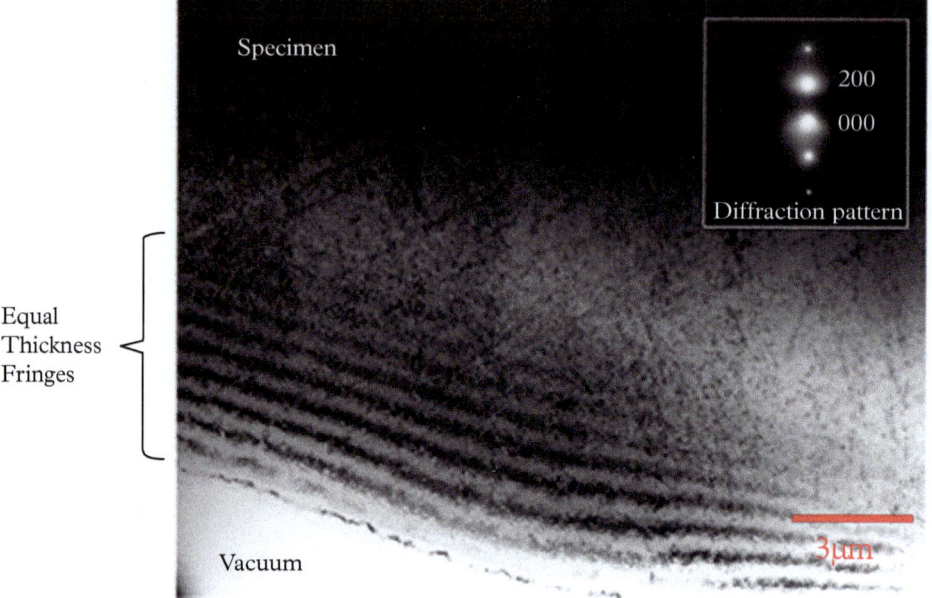

Figure 8.5.7 Bright field image of a AlCu alloy with the [200] reflections excited, showing thickness fringes.

Courtesy JEOL Ltd.

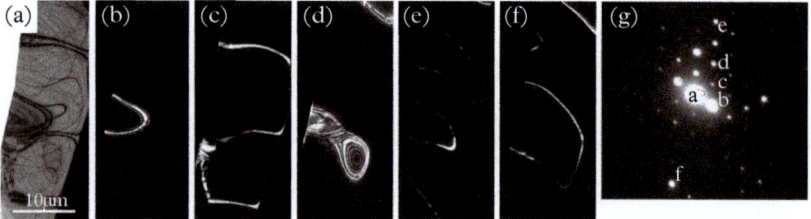

Figure 8.5.8 Bend contours observed in thin foil of mica. (a) Bright field image. (b)–(f) Dark field images formed by reflections indicated in (g) the selected area diffraction pattern from the same area shown in (a). Notice that in this bent crystal, the image shows a strong intensity in a dark field image at positions where the Bragg condition is satisfied.

Courtesy JEOL Ltd.

et al. (1965), Cowley (1984), Williams and Carter (1990), Reimer (1993), and Fultz and Howe (2013). Here, we limit ourselves to a very brief mention of the assumptions made in its formulation and present one of its key results. Unlike the kinematical theory, in the dynamical theory it is now assumed that as the electron propagates in the crystals, each thickness element scatters its intensity back and forth between the forward-scattered and diffracted beams. A system of coupled equations (known as the Howie–Whelan equations, [8]) describes the amplitudes of the diffracted, $A_g(z)$, and forward-scattered, $A_0(z)$, beams, as they propagate through the thickness, z, in the thin foil specimen, and when they are solved—as in the dynamical theory—correctly predict the intensities, I_g, of the diffracted beam. The solution, similar in form to (8.5.16), is given by

$$I_g \sim \frac{\sin^2 \pi t \bar{s}}{(\pi \bar{s})^2} \qquad (8.5.17)$$

where, the excitation error, s, is modified as

$$\bar{s} = \left(s^2 + \frac{1}{\xi_g^2} \right)^{1/2} \qquad (8.5.18)$$

and includes the *extinction distance*, ξ_g, given by

$$\xi_g \approx \frac{\pi V_c}{\lambda_e F_g(\theta_B)} \qquad (8.5.19)$$

Here, V_c is the volume of the unit cell, $F_g(\theta_B)$ is the structure factor for the specific reflection, and λ_e is the electron wavelength. ξ_g is the difference between two thickness extinctions, and as mentioned before is a characteristic distance in the material where the wave amplitude is transferred back and forth between the forward-scattered and diffracted beams. For TEM electrons of energies ~ 100–200 keV, and typical low-order reflections, the extinction distance, $\xi_g \sim 20 - 200$ nm.

So far, we have discussed perfect crystals, but most materials are imperfect and harbor various defects such as dislocations (§4.1.9.2) and stacking faults (§4.1.9.3). The analysis of defects, particularly dislocations, using diffraction contrast is discussed further in §9.5.2.

Example 8.5.1: Calculate the extinction distance for Cu_{111} for electron incidence at 100 keV.

Solution: The extinction distance is given by (8.5.19). Cu is FCC with $a_0 = 0.361$ nm. Thus, the volume of its unit cell, $V_c = (0.361)^3 = 0.047$ nm^3.
  The wavelength of 100 keV electrons is $\lambda = 0.0037$ nm.
  The interplanar spacing $d_{111} = 0.208$ nm (see Table 4.1.2 and Example 7.5.1)
    From Example 8.2.1, $f_e(100$ keV$) = 0.3448$ nm.
    From Example 7.5.1, the structure factor $F_{111} = 4f_e = 4 \times 0.3448 = 1.3792$ nm

> Applying (8.5.19), we get an *extinction distance*, $\xi_g \approx \frac{\pi V_c}{\lambda_e F_g(\theta_B)} = \frac{\pi \text{ x } 0.047}{0.0037 \text{ x } 1.3792} = 28.9$ nm.
>
> Note that the extinction distance depends on the lattice parameter (through the unit cell volume, V_c), the atomic number (through the structure factor, $F_{\mathbf{g}}$), and the acceleration voltage (through the electron wavelength, λ).

8.6 Transmission Electron Diffraction Methods

There are three principal types of electron diffraction patterns produced in a TEM: (a) SAD with parallel illumination, producing spot (single crystal grains) or ring (polycrystalline materials) patterns, (b) Kikuchi line patterns involving both inelastic and elastic scattering, particularly from thicker specimens, and (c) CBED patterns obtained from small areas of the specimen using a focused incident probe. These are discussed in the next three sections.

8.6.1 Selected Area Diffraction: Ring and Spot Patterns

For crystalline specimens, TED patterns simply correspond to a planar section of the reciprocal lattice (Fig. 4.3.4; Fig. 8.5.3) normal to the incident beam direction and recorded in the form of an image. Moreover, in a TEM the diffraction pattern is obtained from the area of interest in the specimen by using a selected area aperture of diameter, d_{SA}, placed in the first image plane of the objective lens (Fig. 9.2.9). The diffraction pattern for crystalline specimens is a magnified image of the reciprocal lattice array/spots on the surface of the Ewald sphere (Fig. 8.6.1). In the observed diffraction pattern, the measured distance, R_{hkl}, between the forward scattered beam (origin) and any diffracted beam (spot), can be related to the appropriate lattice spacing, d_{hkl}, for the specific reflection, $\mathbf{g}_{hkl}$, in terms of the camera length, L, as

$$\lambda_e L = R_{hkl} d_{hkl} \tag{8.6.1}$$

[3] Modern TEMs with digital cameras will often include/embed the camera constant, λL, when recording diffraction patterns. Moreover, intermediate lens astigmatism should be adjusted for round ring patterns, projector lens distortion affects outer rings, and therefore unless you have done the calibration yourself this should not be taken at face value.

As a result, to obtain accurate measurements of d_{hkl}, it is necessary to know the value of the camera length, L, which varies according to lens setting and does not correspond at all to the physical specimen-camera distance, for a given instrument (TEM) under specific experimental conditions of specimen position (height, z), objective lens excitation current etc.[3]

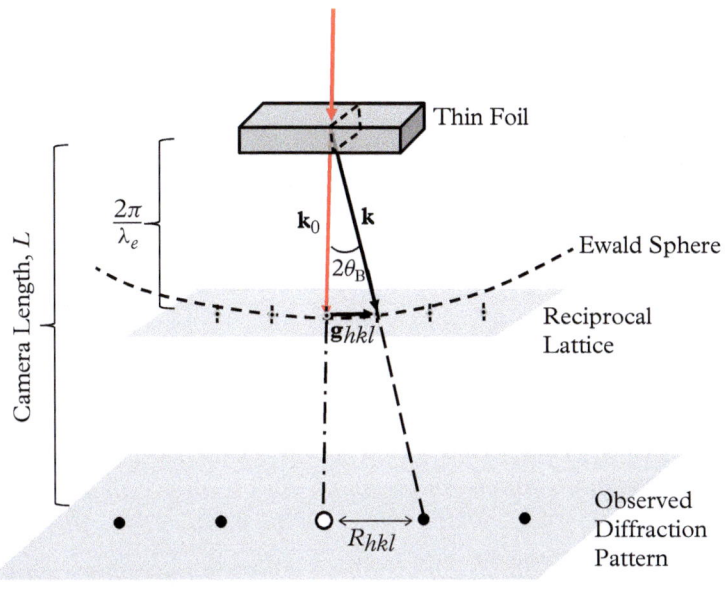

Figure 8.6.1 A schematic diagram of the formation of an electron diffraction pattern in transmission, with a camera length, L, that is calibrated for specific settings (lens excitations) in a TEM.

(a) (b) (c)

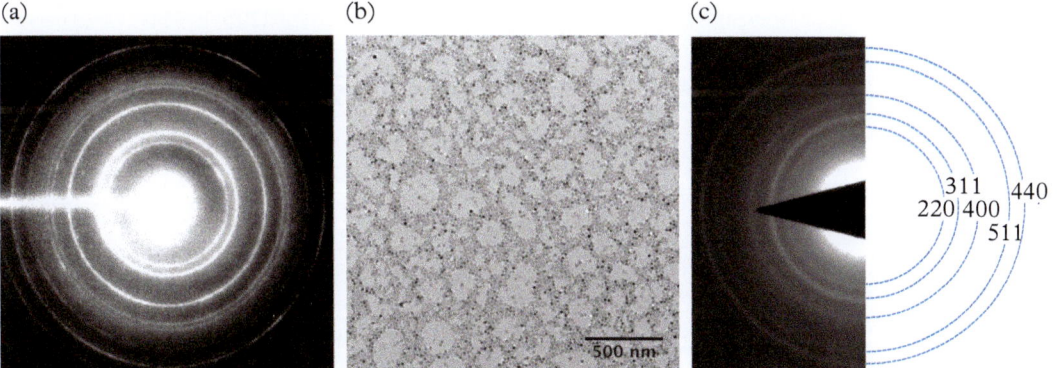

Figure 8.6.2 (a) A Debye–Scherrer type ring pattern obtained from a small-grain, polycrystalline copper specimen that can be used to determine the camera length, L. (b) Image of monodisperse iron oxide nanoparticles, ~25 nm in diameter, dispersed on an ultrathin carbon film. (c) Ring pattern from (b) that can be indexed as magnetite (Fe_3O_4).

Courtesy Dr. Ryan Hufschmid.

In practice, for a specific TEM, the camera length is best determined experimentally by measuring the Debye-Scherrer ring pattern (§7.9.3) from a *standard*, small-grain, polycrystalline material, such as an evaporated gold thin film (Fig. 8.6.2a). Here, all the radii, R_{hkl}, for the rings are measured accurately from the observed pattern, all inter-planar spacing, d_{hkl}, are known, the electron wavelength, λ_e, is calculated from the acceleration voltage, and then, L is determined accurately from (8.6.1).

For crystalline materials, the electron diffraction patterns for incident beam directions along low-index zone axis orientations (Fig. 4.1.10) reflect the crystallographic symmetry of its reciprocal lattice and as such, display highly symmetric patterns. Figure 8.6.3 shows examples of spot patterns for single crystals with FCC, BCC, diamond cubic, and HCP structures.

Example 8.6.1: Index the ring pattern in Figure 8.6.2a and determine the camera length, L, of the instrument. The manufacturer claims that $L = 65$ cm, for this setting. What is the error, if any? Assume 100 kV incident electrons.

Solution: The wavelength for 100 keV electrons, $\lambda = 0.037$ Å, and $a_{Cu} = 3.61$ Å. This is the classic FCC pattern following the sequence, two-together, one alone, two-together . . . , and the first five rings can be indexed as 111, 200, 220, 311, 222 . . . . . We now solve for the camera length, L, using (8.6.1), in the form of a table:

Reflection	111	200	220	311	222
Ring diameter measured (cm)	2.22	2.58	3.65	4.28	4.56
Ring radius, r (cm)	1.11	1.29	1.83	2.14	2.28
d_{hkl} (Å)	2.08	1.81	1.28	1.09	1.04
$\lambda L = r\, d_{hkl}$ (Å cm)	2.31	2.33	2.34	2.33	2.37
Camera length, L (cm)	62.43	62.97	63.24	62.97	64.05

Average value of the camera length, $L_{avg} = 63.35$ cm.

The error in the manufacturer's value is ~2.5%.

A characteristic feature of TED, because multiple scattering is prevalent, is *double diffraction*. As a consequence, a diffracted beam, $\mathbf{g}_1$, can be re-diffracted by another reflection, $\mathbf{g}_2$, either in the same or another crystal/grain, to produce an observable double-diffracted beam, $\mathbf{g}_1 + \mathbf{g}_2$, which may or may not be allowed for the crystal based on its structure factor. Such reflections, also indicated in Figure 8.6.3 for various zone axis orientations, display intensities that are not readily interpretable (see Example 8.6.1). Thus, specific reflections that are forbidden by the extinction rules in structure factor calculations, §7.5, are nevertheless observed in electron diffraction.

Finally, to analyze the crystal structure of an unknown specimen/material by electron diffraction, it is important to obtain sections of its reciprocal lattice in multiple directions of electron incidence. This is best done by tilting the crystal, for a fixed electron beam incidence direction, through a large angular range (± 45–$60°$), depending on the goniometer included as part of the specimen stage in the TEM) around any one crystallographic axis. Then, the diffraction patterns

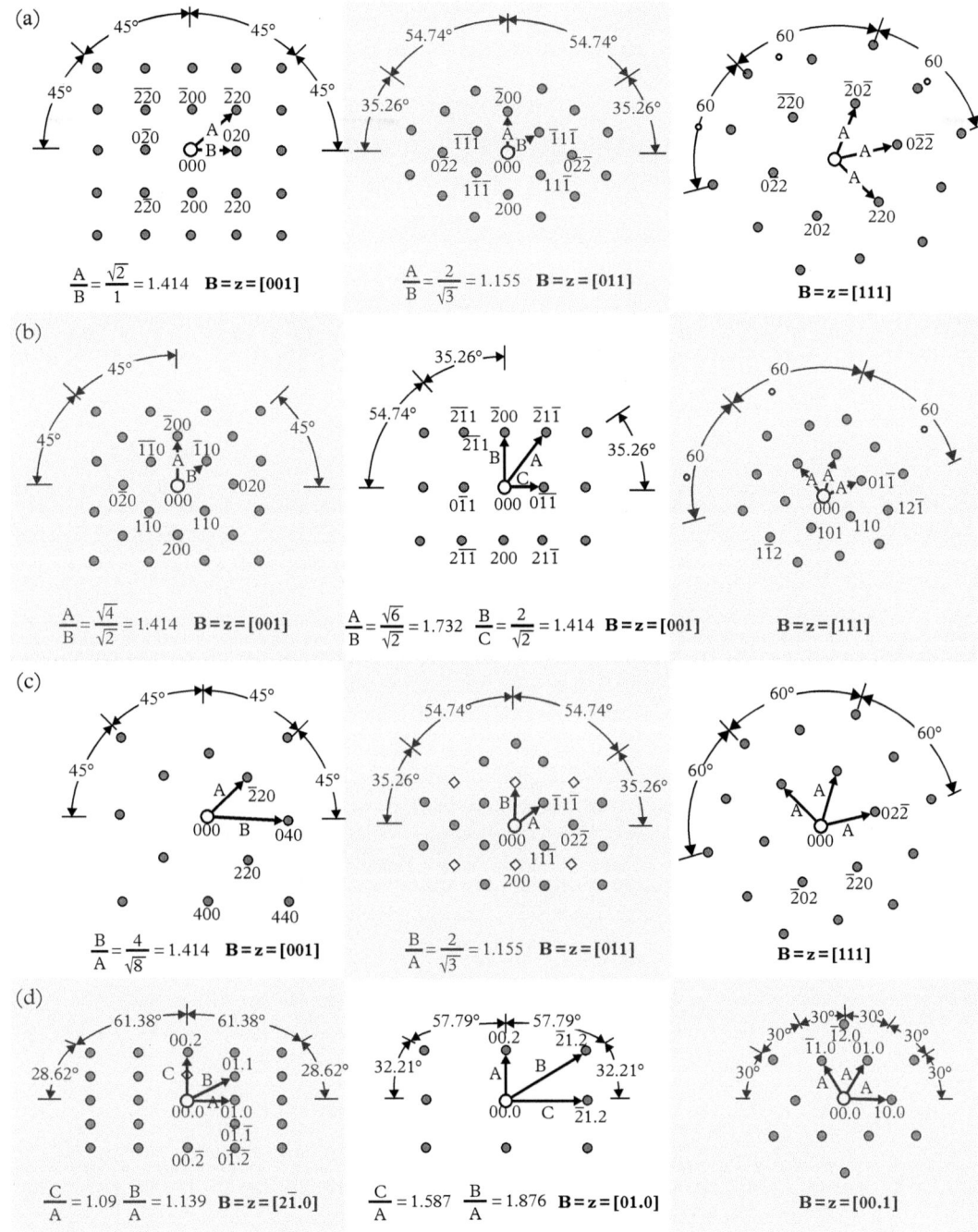

Figure 8.6.3 Single crystal spot transmission electron diffraction patterns for (a) FCC, (b) BCC, (c) Diamond cubic, and (d) HCP structures. Unfilled diamonds show reflections that appear due to double diffraction. The beam direction, **B**, and the zone axis, **z**, defined in §4.1.4, satisfying the cross-product rule, (4.1.8), are also indicated.

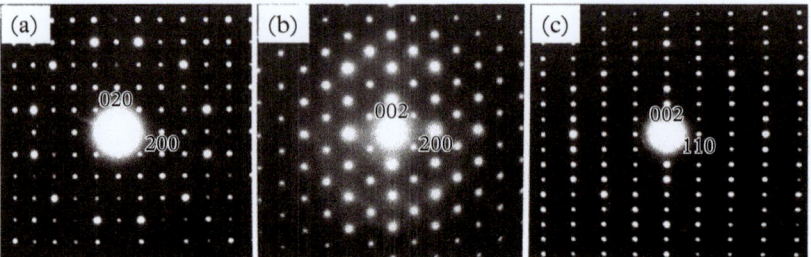

Figure 8.6.4 SAD patterns of $Fe_{14}Nd_2B$ along three different orientations. The halo around the forward-scattered (transmitted) and diffracted spots, as well as the faint background intensity observable in (b) are attributed to the presence of inelastically and incoherently scattered electrons.

Adapted from Shindo and Hiraga (1998).

are recorded in multiple zone axis orientations. Figure 8.6.4 shows an example of electron diffraction of a permanent magnet alloy, $Nd_2Fe_{14}B$, taken along three different zone axes orientations. Note that in Figure 8.6.4b, reflections of the type $h + l = 2n + 1$ are not observed; however, in both (a) and (c) they are present due to double diffraction.

Tilting of the crystals, from one zone axis orientation to another, where possible, is accomplished by following directions indicated by Kikuchi lines/patterns. The Kikuchi lines originate from a superposition of inelastic and elastic scattering of electrons in thicker crystals, and are discussed in the next section.

Example 8.6.2: Index the diffraction pattern of a specimen of silicon shown in the following image. The diameter of the spots corresponds to their intensity. Are the weak reflections, enclosed in the dotted circle, allowed from the structure factor for this structure? Then, explain their presence in the diffraction pattern.

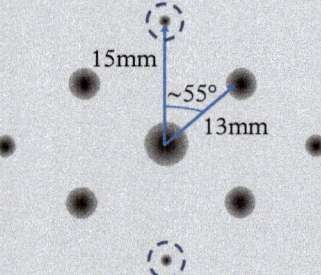

Solution: Silicon is a diamond cubic crystal structure. Comparing with Figure 8.6.3, we can easily index this as a $\mathbf{B} = \mathbf{z} = [011]$ pattern and assign the reflection as in the following figure.

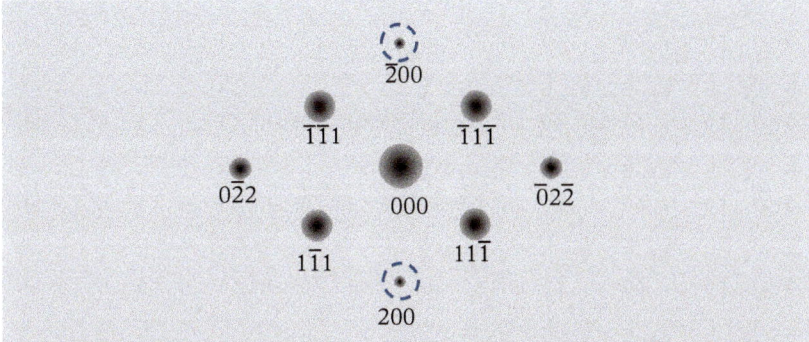

From Problem 7. 1, i.e. structure factor calculations for the diamond cubic structure, we know that $I_{hkl} = 0$, when $h + k + l = 4n + 2$, where n is an integer. Thus, the 200 reflections are forbidden for Si, and should be absent.

However, the 111-beam with structure factor, $F_{111} \neq 0$, can act as a new incident beam and be re-diffracted by the (111) planes. In other words, we can have double or dynamical diffraction, combining the two reflections, i.e. $(11\bar{1}) + (\bar{1}11) = (200)$, to faintly excite the 200 reflection. Thus, the 200 reflection is only "kinematically forbidden", and may be excited by dynamical diffraction. Unlike for X-rays, where single scattering dominates, double diffraction is a common feature in electron diffraction.

8.6.2 Kikuchi Lines, Maps, and Patterns

In crystalline thin foils that are reasonably thick, i.e. about half the penetration distance of the fast electron, and with low defect densities, the background intensity between the Bragg spots in TED displays bands of bright and dark lines. Occurring in pairs, these parallel bright and dark lines known as *Kikuchi lines*, named after their discoverer [9], are used to determine the sign and magnitude of the excitation error, **s**. They also define the sense of tilt of the specimen, reveal true crystal symmetry unlike the SAD patterns, and guide the experienced TEM operator in tilting a crystalline specimen from one zone axis orientation to another.

The underlying mechanism and geometry of the formation of Kikuchi lines can be understood from the diffraction of those electrons that have previously undergone inelastic scattering (Fig. 8.6.5a). A thin crystal is oriented such that the incident beam does not satisfy the Bragg law of diffraction for the planes (*hkl*), as shown in the figure. We focus our attention on the inelastically scattered electrons at any point, O, inside the crystal. Figure 8.6.5a shows the distribution of the inelastic electrons is forward peaked; moreover, we assume that the energy losses (<50 eV) are small compared to the incident electron energy ($\sim$100 keV), and the inelastically scattered electrons have the same wavelength, λ_e, as the incident

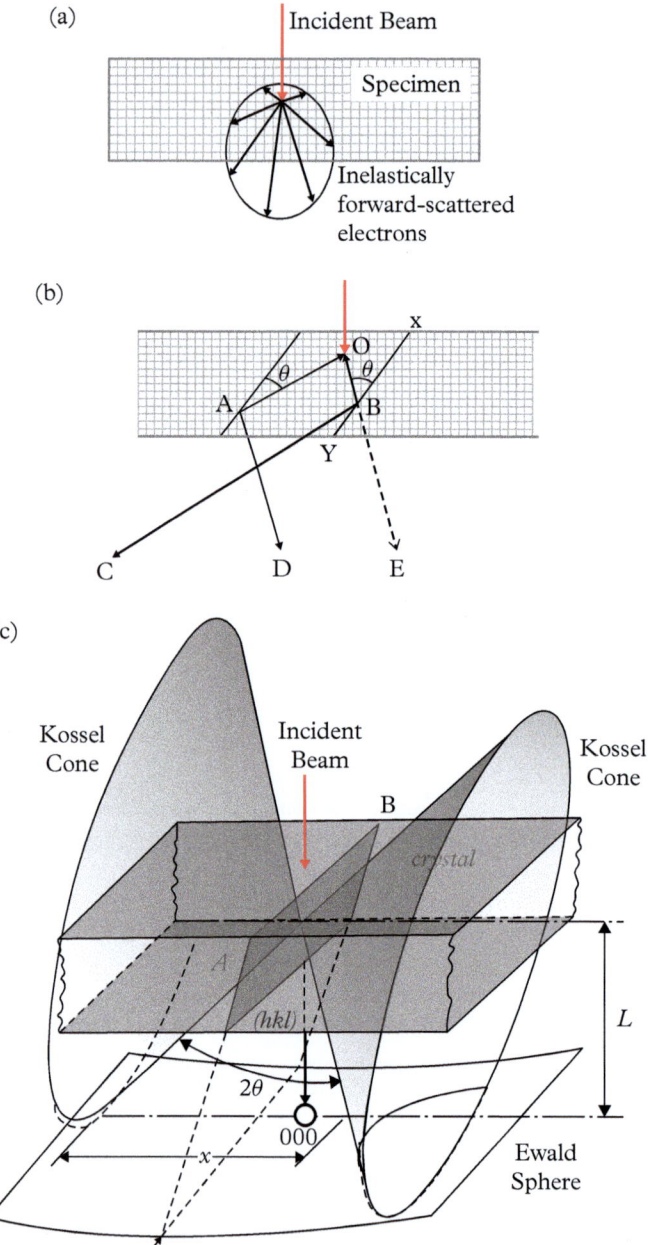

Figure 8.6.5 Schematic drawing illustrating the formation of Kikuchi lines in transmission electron diffraction of thin foils. (a) The distribution of inelastically scattered electrons, shown as a polar diagram, is forward peaked. (b) A fraction of the inelastically scattered electrons satisfy the Bragg condition for diffraction by a set of planes (hkl), which then form (c) the Kossel cones as shown; the intersection of the Kossel cones with the Ewald sphere at the plane of observation of the diffraction patterns appear as the Kikuchi lines.

Adapted from Edington (1975).

beam. Then, it is very likely that a fraction of the inelastically scattered electrons travels along the direction defined by the angle, θ_B, with respect to the planes (hkl) and satisfy Bragg's law (Fig. 8.6.5b). Since the electrons are now traveling along all directions before diffraction from the (hkl) planes, the diffracted beams will lie on two Kossel cones (Fig. 8.6.5c), with one pair of cones for each set of $\pm \mathbf{g}$ reflection. One of these cones corresponds to a deficit of intensity near the transmitted beam, and the other cone is indicative of an excess intensity further away from the transmitted beam. At the plane of observation of the diffraction pattern, the Kossel cones intersect the Ewald sphere along two hyperbolae (Fig. 8.6.5c). However, close to the forward scattered beam (optic axis for the TEM), and because of the small value of θ_B, they appear as pairs of dark and bright lines, separated by an angular distance of $2\theta_B$.

A unique and useful feature of the Kossel cones is that they behave as though they are fixed to the crystal. Now, when the crystal is tilted with respect to the incident beam the Kikuchi lines will move; in contrast, the Bragg diffraction spots are fixed in position and will only change in intensity, and that too, over a limited tilting range. In particular, at the exact Bragg condition, $s = 0$, one of the bright Kikuchi lines will pass through a Bragg diffraction spot and its complementary dark Kikuchi line will coincide with the transmitted or forward-scattered beam. In general, the position of the Kikuchi lines can be used to measure the exact value of the excitation error, $\mathbf{s_g}$, for any reflection, $\mathbf{g}$. Figure 8.6.6 shows schematically the diffraction geometry; for $s > 0$, the excess (bright) Kikuchi line is further away from $\mathbf{g}$, at a distance, x, as shown, and for $s < 0$, it is closer to the transmitted beam, O, than $\mathbf{g}$. If the camera length, L, is known, and the distances R_{hkl} and x (Fig. 8.6.6) are measured, from the scattering geometry we can show that

$$s \sim \frac{x \lambda_e}{R_{hkl}} g^2 \qquad (8.6.2)$$

In addition, for contrast analysis not only is the excitation error, s, an important parameter, but it is important to tilt the crystal such that a specific order of the reflection ($\mathbf{g}$, $2\mathbf{g}$, $3\mathbf{g}$ etc.) is excited or is in the exact Bragg orientation. This is also accomplished by using Kikuchi lines (Fig. 8.6.7).

Last, but not least, at the exact zone axis orientation, the Kikuchi lines can be drawn as perpendicular bisectors for every allowed reflection, $\mathbf{g}_{hkl}$. Figure 8.6.8a shows an example of the Kikuchi line construction for the [001] zone axis for an FCC crystal. The Kikuchi pattern from one zone axis can be extended to a second zone axis (pole) by following a pair of Kikuchi lines that are common to both poles (Fig. 8.6.8b). In this manner a complete Kikuchi map of the angular relationship between different poles, useful for "navigation" in reciprocal space while performing diffraction experiments in a TEM for any particular crystal

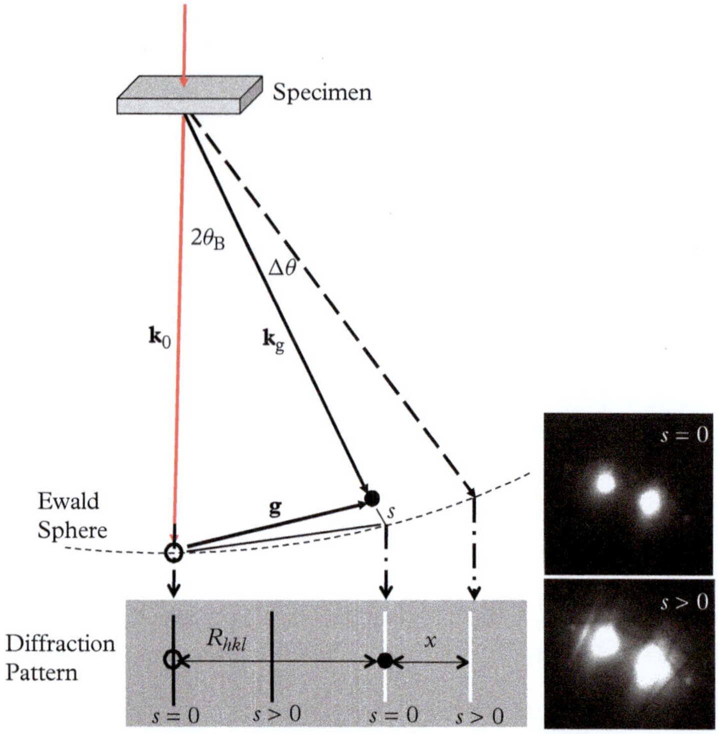

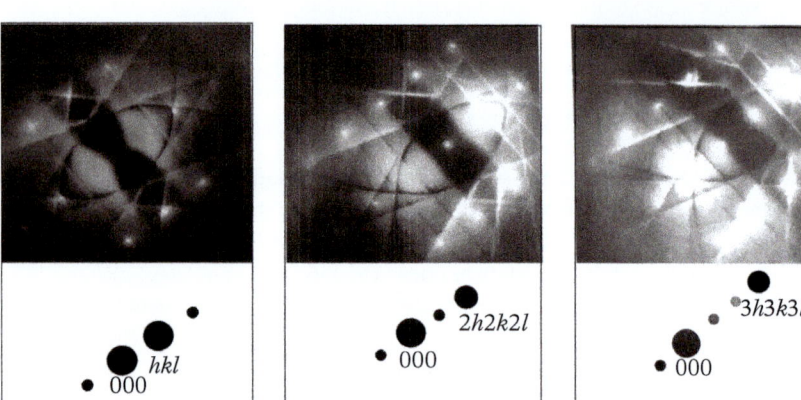

Figure 8.6.6 The Ewald sphere construction showing two different conditions of the excitation error, $s = 0$, and $s > 0$. In practice, this is achieved by tilting the crystal and monitoring the relative orientation of the Kikuchi lines with respect to the Bragg diffraction peaks, as shown on the right.

Figure 8.6.7 Illustrating the use of Kikuchi lines for tilting to an exact Bragg orientation; from left to right, conditions for exciting $\mathbf{g}_{hkl}$, $2\mathbf{g}_{hkl}$, and $3\mathbf{g}_{hkl}$, respectively.

Adapted from Williams, Pelton, and Gronsky (1991).

structure can be generated. Figure 8.6.8c shows an example of a Kikuchi map, for the FCC structure. Needless to say, similar Kikuchi maps for other crystal structures, such as BCC, HCP, etc., can be constructed or found in the literature; see, for example, Edington (1975), Thomas and Goringe (1980), and Williams and Carter (1990).

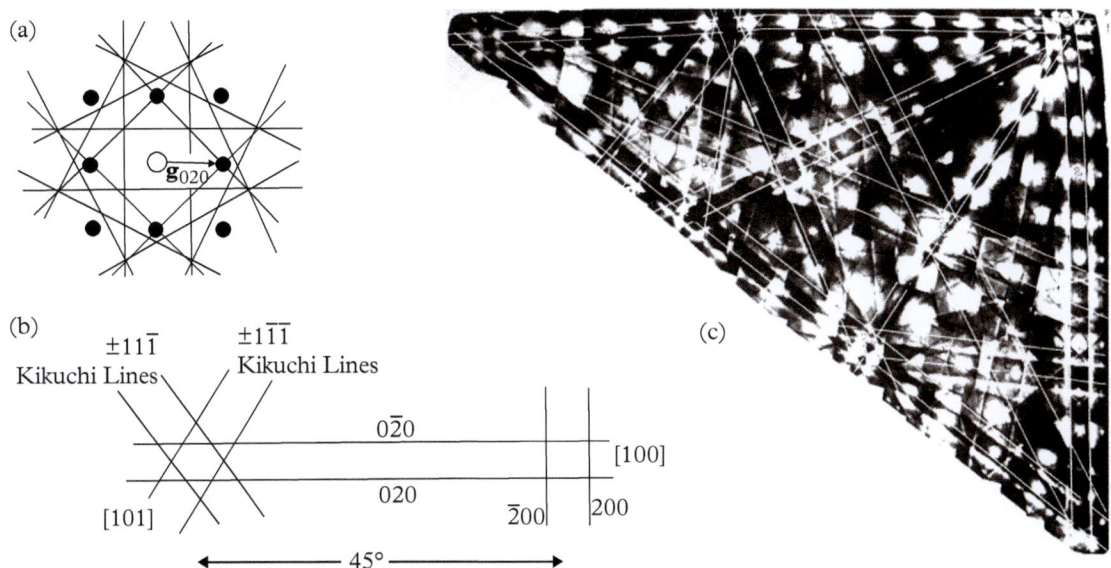

Figure 8.6.8 Construction of Kikuchi lines illustrated for the [001] zone axis, by drawing (a) the perpendicular bisectors for all allowed reflections. (b) By extending pairs of Kikuchi lines common to two poles with well-defined crystallographic orientations, a Kikuchi map such as the one for (c) FCC crystals can be established.

Adapted from Edington (1975).

8.6.3 Convergent Beam Electron Diffraction (CBED)

SAD in a TEM is implemented using an aperture in the first image plane (§9.2.7.1) in front of the diffraction or intermediate lens. However, the spherical aberration of the electromagnetic lens (see §9.2.2 and §9.2.9 for a detailed discussion of lens aberrations) limits the area selected for diffraction, using parallel illumination, to ~1 μm in diameter. To obtain diffraction from smaller areas of the order of 10 nm, a focused probe with a convergence semi-angle, α (Fig. 8.6.9) is used. If the convergence semi-angle is very small, $\alpha << \theta_B$ (Fig. 8.6.9a), a pattern of small spots with no discernible features inside any of them, for different Bragg vectors, $\mathbf{g}_{hkl}$, excited by a fixed incident wave vector, $\mathbf{k}_0$, similar to a SAD pattern (Fig. 8.6.10a) is obtained. For larger convergence semi-angles $\alpha \sim \theta_B$ (Fig. 8.6.9b) CBED patterns, with each spot now in the form of a disc (Fig. 8.6.10b,c), but with distinct features inside them, are observed. Each nonoverlapping disc in a CBED pattern (Fig. 8.6.10b) displays additional features, called higher-order Laue zone (HOLZ) lines. These dark lines within each disc correspond to allowed reflections at high scattering angles in the HOLZ rings (Fig. 8.6.10c,d shows a first order FOLZ ring). For even larger convergence semi-angles, $\alpha >> \theta_B$, we

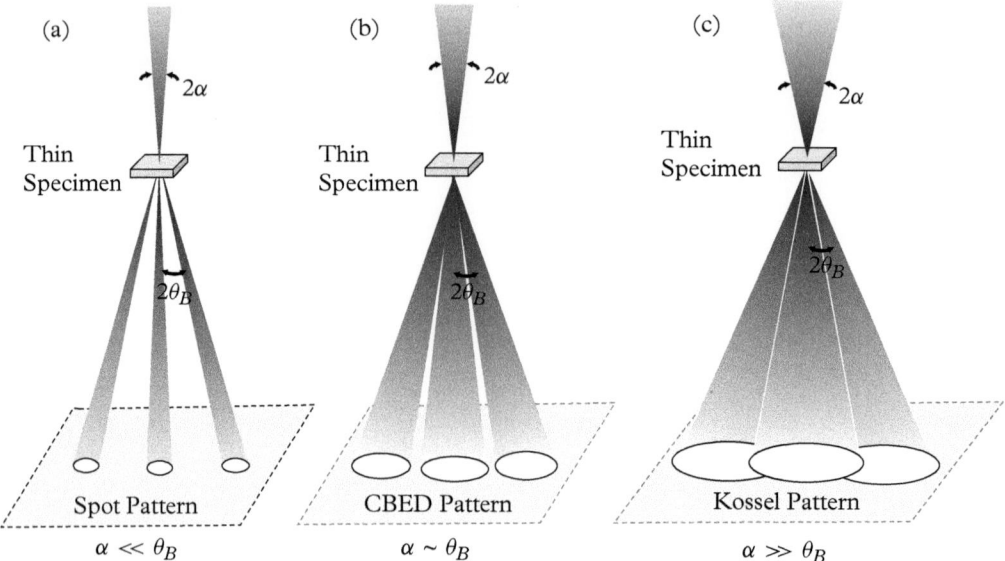

Figure 8.6.9 Effect of increasing beam convergence semi-angle, α, as compared to the Bragg angle, θ_B, on transmission electron diffraction patterns using a stationary probe, illustrating the formation of (a) micro diffraction pattern, (b) convergent beam electron diffraction, and (c) Kossel patterns.

observe a Kossel pattern of overlapping discs (Fig. 8.6.10d) with dark and bright line features similar to Kikuchi patterns. We shall restrict our discussion here to CBED patterns. Kossel patterns are discussed in Reimer (1993).

A convergent probe includes a large range of incident wave vectors, $\mathbf{k}_0$, within the illuminated cone (Fig. 8.6.9b) and therefore each point in a CBED disc corresponds to a particular incident wave vector. The intensity distribution within the forward-scattered disc and any diffraction disc represent the intensity variations with the excitation error, $\mathbf{s}$. Further, the intensity variation in a disc at the exact Bragg orientation (Fig. 8.6.11) reflects the variation in intensity with the modified excitation error, $\bar{s}$, defined in (8.5.18), and as plotted in Figure 8.5.3a.

A simple method has been developed to utilize the thickness fringes in CBED patterns (Fig. 8.6.11) to not only measure the thickness, t, of the specimen, but also the extinction distance, ξ_g, of the material. Each of the minima in the Bragg diffraction disc has a characteristic angular deviation, $\Delta\theta_i$, that can be measured. The corresponding excitation error, s_i, can be calculated from the relationship

$$s_i = \frac{\lambda_e}{d_{hkl}^2}\left(\frac{\Delta\theta_i}{2\theta_B}\right) \tag{8.6.3}$$

where all variables are known, or measured from Figure 8.6.11a. Further, the deviation parameter, s_i, is related to the extinction distance, ξ_g, simply as

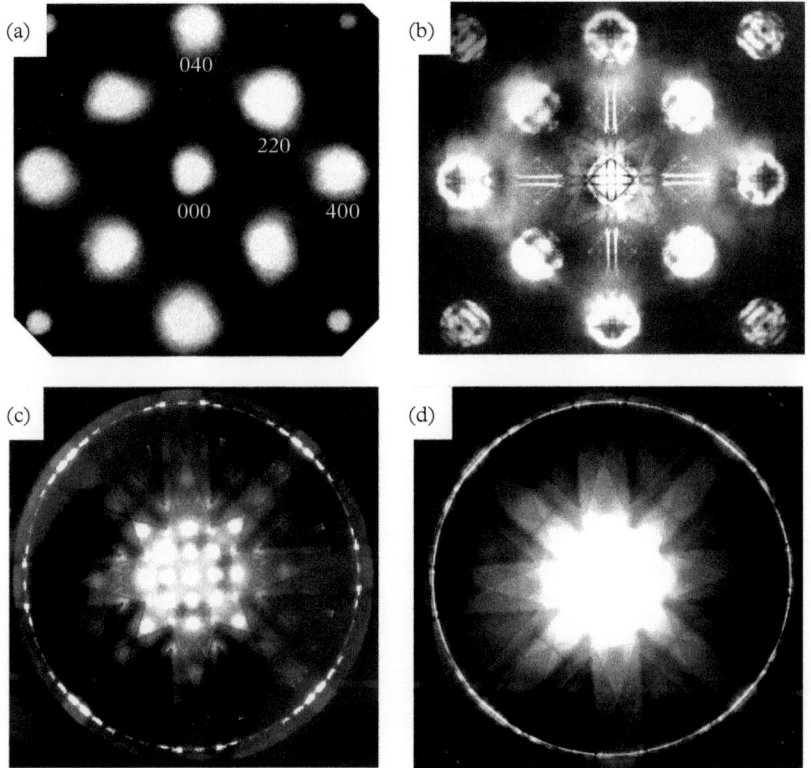

Figure 8.6.10 Diffraction patterns of a single crystal spinel, $MgAl_2O_4$, in the [001] zone axis orientation. (a) Selected area diffraction pattern obtained with parallel illumination (fixed $\mathbf{k}_0$); Notice the absence of the (200) type reflections, forbidden by the structure factor. (b) A convergent beam electron diffraction (CBED) pattern of the zero-order Laue zone (ZOLZ) showing dynamical contrast within the individual discs, including in the forbidden (200) reflections; in this case a medium convergence angle for the incident beam representative of a range of incident vectors, $\mathbf{k}_0$, is employed. (c) The entire CBED patterns showing both the ZOLZ and the first-order Laue zone (FOLZ) with individual disc features resolved. (d) A larger convergence angle results in the overlap of all the individual discs in the CBED pattern, but the Kikuchi lines remain visible and the pattern retains its overall symmetry—these are also known as Kossel patterns.

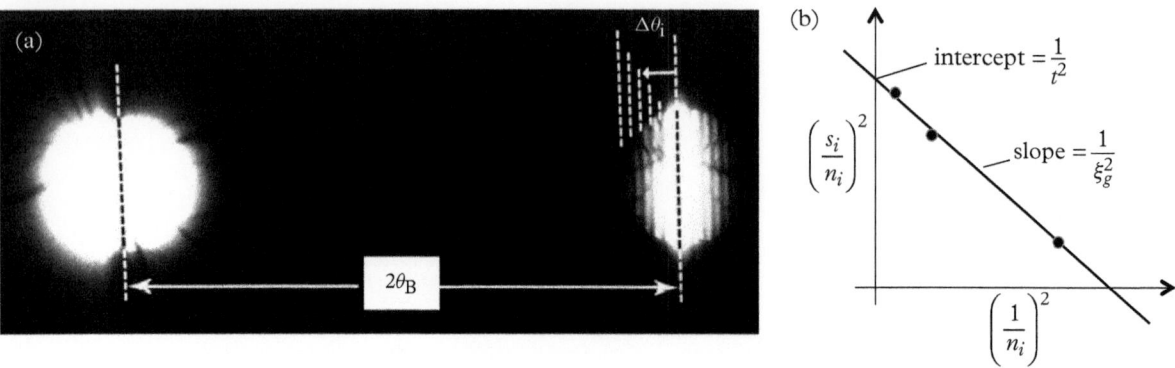

Figure 8.6.11 (a) Observation of thickness fringes in a convergent beam pattern at the exact ($s = 0$) Bragg orientation. (b) The fringe spacing can be measured and plotted to obtain both the thickness (intercept) and the extinction distance (slope) for that specific reflection.

Courtesy Chuck Echer.

$$\frac{s_i^2}{n_i^2} = -\frac{1}{\xi_g^2}\frac{1}{n_i^2} + \frac{1}{t^2} \tag{8.6.4}$$

where n_i is an integer. To determine t and ξ_g, it is common practice to plot $(s_i/n_i)^2$ versus $(1/n_i)^2$, assigning, $n_1 = 1$, $n_2 = 2$, etc. If the result is a straight line (Fig. 8.6.11b), then its slope is $-(1/\xi_g)^2$, and its intercept is $(1/t)^2$. If the result is not a straight line, the numbers, n_i, are reassigned different integer values, until a straight-line fit is obtained, and the slope and intercept is then determined to give values of the extinction distance and thickness, respectively.

As mentioned, if the diffraction pattern is recorded with a large collection angle along a low-index zone axis (Fig. 8.6.10c), higher order Laue zones, in the form of rings, are observed. Because of the large Bragg angles associated with the HOLZ reflections, they do not appear as discs/spots, but as lines. In addition, every bright line in a HOLZ ring has a corresponding dark line in the forward scattered beam (Fig. 8.6.10b), and can thus be indexed. These HOLZ lines in the forward scattered disc are sharp and appear, unlike Kikuchi lines, even in thin specimens. Moreover, since the HOLZ lines are both sharp and correspond to large Bragg angles, they are very sensitive to small changes in lattice parameters. Further, if the CBED pattern is carefully aligned along the exact zone axis (Fig. 8.6.10b), the symmetry of the pattern reflects the point group symmetry of the crystal. In additions, symmetries present in the forbidden reflections, such as in the (200) reflections in Figure 8.6.10b, is consistent with the space group of the specimen. CBED patterns are used to make elaborate analysis of the symmetry of the crystal at high spatial resolution that is defined by the size of the convergent probe incident on the specimen. Details of symmetry analysis by CBED can be found in [10].

8.7 Examples of Transmission Electron Diffraction of Materials

There are numerous applications of TED in materials characterization and analysis. Here we present select practical examples, mostly from nanoparticles and thin film heterostructures, to illustrate some basic applications; many more applications in metallurgy and ceramics are discussed in Thomas and Goringe (1980), Williams and Carter (1990), and Fultz and Howe (2013).

8.7.1 Indexing a Single Crystal Diffraction Pattern

Many applications require the indexing, or assigning specific reflections, $\mathbf{g}_{hkl}$, to the observed peaks in the intensities of the diffraction patterns. For a single crystal or large grain of the material, it is best done by tilting the crystal to a low-index zone axis orientation and recording the diffraction pattern (Fig. 8.6.4; Fig. 8.6.10a). Then, using the known camera length, L, and wavelength, λ_e, the distance, R_i, measured for each of the spots can be associated with a specific inter-planar spacing, d_i, using (8.6.1). Further, the ratios of any two values $\frac{R_i}{R_j} = \frac{d_j}{d_i}$, gives the ratios of possible inter-planar spacing. After this, using the measured set of ratios, d_j/d_i, and the symmetry of the zone axis pattern, and comparing them to either a table of allowed ratios or zone axis patterns (Fig. 8.6.3), one can make a first attempt at indexing the reflection as well as the zone axis orientation. The veracity of the indexing can be further cross-checked by using pairs of reflection and applying the zonal equation, (4.1.8), to them (Fig. 4.1.10). For example, Figure 8.6.10 is a [001] zone axis, and the three reflections indexed as (220), (400) and (040), are consistent with the cross-product of the zonal equation, (4.1.8). Methods for indexing more complicated diffraction patterns, often encountered in materials analysis, are discussed in all three specialized books mentioned earlier.

8.7.2 Polycrystalline Materials and Nanoparticle Arrays

TED from polycrystalline materials, including an assembly of randomly oriented nanoparticles, can be interpreted in a manner similar to X-ray diffraction from powder specimens (§7.9.2). If the specimen is truly polycrystalline and randomly oriented, its reciprocal lattice is a set of nested spheres, with each of its radii corresponding to a reciprocal lattice vector of the crystal. The intersection of these nested spheres with the relatively "flat" Ewald sphere in electron diffraction produces a set of rings in the SAD pattern (Fig. 8.7.1). The width of each ring, similar to a powder pattern in X-ray diffraction, gives an estimate of the average size of the grains or nanoparticles. Further, the intensity of each ring can be "circularly integrated" to produce a plot (Fig. 8.7.1g) of the intensity as a function of inverse lattice spacing, $1/d_{hkl}$, and then compared to the powder diffraction files (§7.11.3) to help identify the crystal structure.

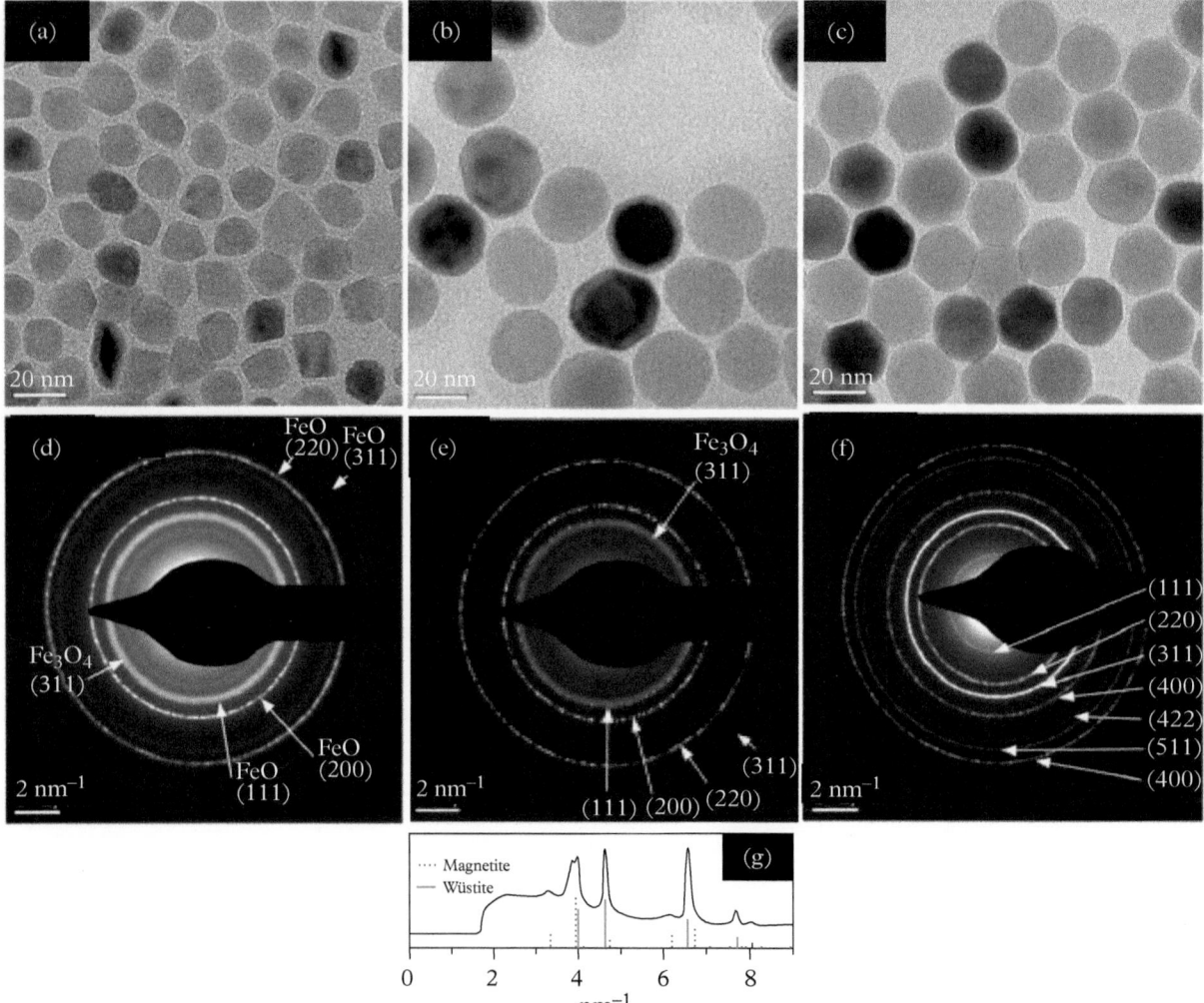

Figure 8.7.1 Time evolution of the crystal structure and phase purity of iron oxide nanoparticles synthesized by chemical routes in an organic solvent. At some intermediate stage (a, d) the nanoparticles are mainly FeO with a trace of Fe_3O_4. At the end of the synthesis (b, e) the nanoparticles are faceted, uniform in size, but uniformly *maghemite* (Fe_3O_4). Circular integration of the ring pattern produces a powder pattern (g) that matches very well with the PDF files for maghemite. However, (c, f), when a controlled oxidation step is introduced in the synthesis the nanoparticles have the most desirable phase, *magnetite* (Fe_3O_4).

Adapted from [11].

Figure 8.7.1 illustrates these ideas when applied to the synthesis of iron oxide nanoparticles by the decomposition of chemical precursors in organic solvents [11]. Figure 8.7.1a,d shows the particles at some intermediate stage of the synthesis; even though a faint ring indicates the presence of trace quantities of Fe_3O_4 (magnetite), it is mainly FeO. At the end of the synthesis, without an oxidation step (Fig. 8.7.1b,e,g) the nanoparticles are predominantly maghemite (Fe_2O_3), with the integrated intensity (Fig. 8.7.1g) matching very well with its powder diffraction file (PDF) data. Finally, when a controlled oxidation step is included, the nanoparticles attain their equilibrium shape and the SAD pattern is consistent with magnetite (Fe_3O_4) as confirmed by the indexed rings in Figure 8.7.1f.

8.7.3 Orientation Relationships Between Crystals or Phases

Another important application of electron diffraction in materials analysis is to determine the orientation relationship between two or more different crystals in a specimen, such as a precipitate distributed in a matrix, a thin film grown on a substrate, etc. A complete analysis requires the determination of the relationships of both crystallographic planes and directions in the two crystals.

Figure 8.7.2a shows the complex microstructure observed in Sm–Co permanent magnet alloys, doped with small amounts of Fe and Zr. It consists of three phases, hexagonal $SmCo_5$, rhombohedral Sm_2Co_{17}, and a thin platelet (Z) phase containing most of the Zr additions. The typical microstructure shown includes a network of fine ($\sim$200 nm in diameter) cells of the rhombohedral $Sm_2(Co,Fe)_{17}$, separated by cell walls ($\sim$5–20 nm thick) of a $Sm(Co,Cu,Zr)_5$ boundary phase, and a third platelet phase, containing Zr, running across many cells and cell boundaries. A SAD pattern from this microstructure is shown in Figure 8.7.2b, and it is indexed in Figure 8.7.2c. From the diffraction pattern, we can infer the orientation relationship between the phases, and conclude that all the phases are crystallographically coherent. This is important as these alloys are called pinning-type permanent magnets and their magnetic hardness arises from their ability to pin the motion of domain walls in a structurally coherent microstructure. For further details on the role of microstructure in determining the behavior of hard magnets, see Krishnan (2016).

8.7.4 Chemical Order in Materials

Here we illustrate studies of chemical order with the simple case of a bilayer thin film of Fe and MnPd, epitaxially grown on a single crystal $MgO_{[001]}$ substrate. MnPd is a chemically ordered $L1_0$ structure (you can visualize this as a variant of the CuAu structure; Fig. 7.11.2b) based on a face-centered tetragonal unit cell, and with lattice parameters (Fig. 8.7.3a). It can be grown epitaxially [12, 13], either

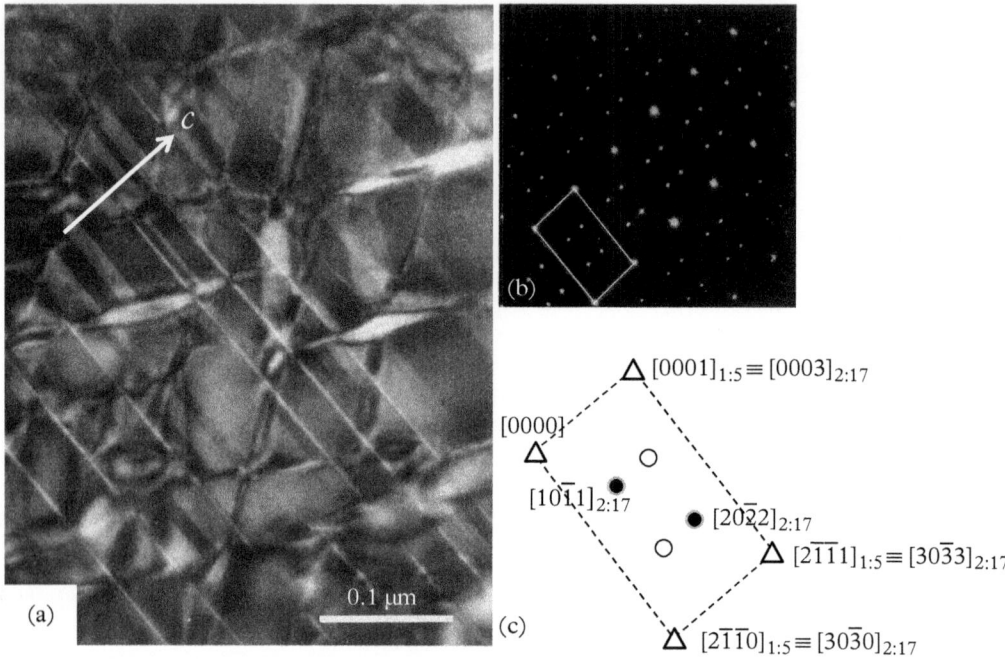

Figure 8.7.2 Orientation relationship between the different crystallographic phases, SmCo₅ and Sm₂Co₁₇, in the complex microstructure observed in Sm–Co permanent magnet alloys.

Courtesy of Dr. L. Rabenberg.

with the c-axis normal (Fig. 8.7.3a) or with the a-axis normal (c-axis in plane) to the film plane (Fig. 8.7.3c). When chemically ordered as mentioned, the Mn and Pd atoms arrange themselves in alternating layers and the inter-planar period doubles. Then, in a typical diffraction pattern (Fig. 8.7.3f), both the regular (002) reflections and the chemically ordered (001) reflections are visible; moreover, the relative intensity of the (001) reflection, compared to the (002) reflection, is a good indicator of the order parameter (§7.11.4), or the degree of chemical order. For a film grown with the c-axis normal, we can clearly see the chemical order, or alternating layers of Mn and Pd atoms perpendicular to the film normal, in the X-ray scan (Fig. 8.7.3b). However, it is not possible to say if an a-axis normal film is chemically ordered or not from an X-ray scan (Fig. 8.7.3d), as the scattering vector, $\mathbf{k} - \mathbf{k}_0$, is parallel to the film normal in a symmetric θ-2θ X-ray scan, but the ordering is in planes normal to the film surface (Fig. 8.7.3c). However, a TED pattern from a plan-view specimen with an in-plane scattering vector can register the chemical order in the a-axis normal film (see Figure 4.3.5 and related discussion). This is illustrated for chemically disordered (Fig. 8.7.3e) and ordered (Fig. 8.7.3f) films, where the latter clearly shows the (001) reflections representing

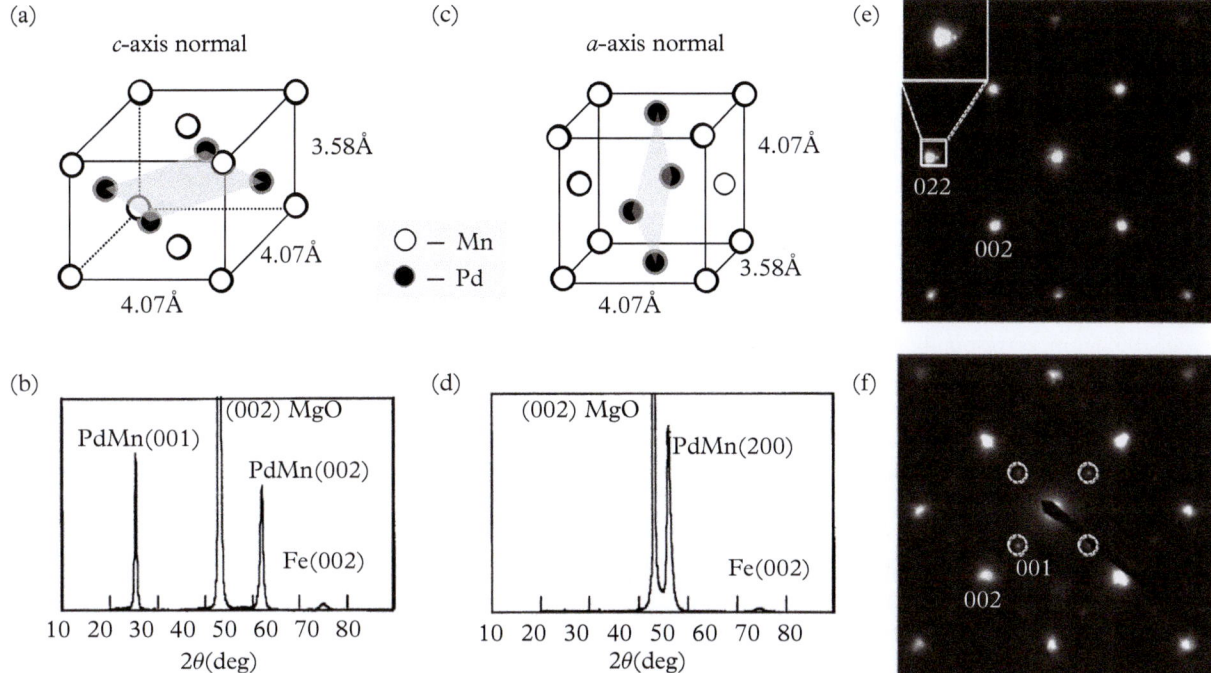

Figure 8.7.3 MnPd is a L1$_0$ ordered tetragonal structure that can be grown epitaxially either with (a) the *c*-axis normal, or (b) the *a*-axis normal to the film plane. In X-ray θ–2θ scans, the diffraction vector is normal to the film plane and thus it is only able to resolve chemical order (b) in the *c*-axis oriented film, with the observation of the MnPd (001) peak, but not in (d) the *a*-axis normal films. However, plan view electron diffraction of as grown (e) and annealed (f) specimens clearly show the chemical order in the *a*-axis normal films.

Adapted from [12, 13].

the chemical order, with their fourfold symmetry corresponding to the epitaxial growth of the layers on a (100) surface.

Figure 4.3.5 introduced the difference in scattering vectors conceptually between θ-2θ X-ray scans and plan-view electron diffraction. However, Figure 8.7.3 is a practical example illustrating the importance of selecting the appropriate technique with the right scattering geometry for a specific problem.

8.7.5 Diffraction from Long-Period Multilayers

We now consider an example of diffraction from a multilayer, with an artificial period of ~6.7 nm, and consisting of alternate layers of Mo (5.4 nm) and Si (1.3 nm), as clearly visible in the image of a cross-section specimen (Fig. 8.7.4a). The growth of individual layers, particularly Mo, is columnar with some texture. This is reflected in the angular spread of the SAD spot, as shown for Mo (011)

Figure 8.7.4 A cross-section specimen of a $[Mo_{5.4}Si_{1.3}]_n$ multilayer, grown on a single crystal Si substrate. (a) A bright field (BF) image, clearly showing the period of the bilayers, and the columnar growth of each of the layers. (b) The selected area diffraction pattern from the multilayer structure; the central region is blown up, and the fine spots indicate a periodicity of ∼6.7 nm corresponding to the periodicity of the Mo–Si bilayers observed in the BF image. How do we know that the two layers are Mo and Si? And, what is the atomic structure and is there intermixing of Mo and Si at the interface? These questions are answered by high spatial resolution spectroscopy and imaging methods in a TEM (Fig. 9.5.9) and discussed further in §9.5.3.

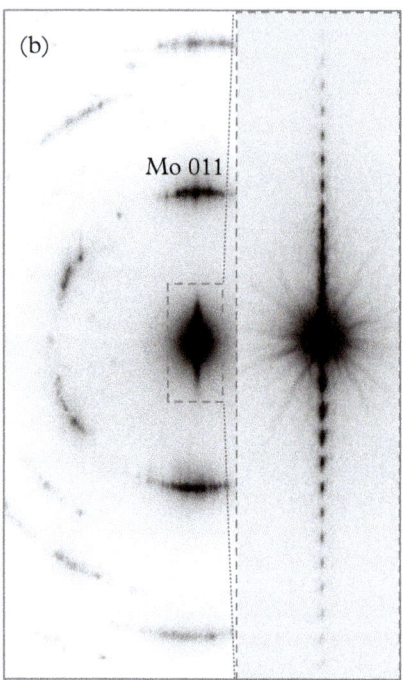

in Figure 8.7.4b. Further, the fine spacing of the superlattice reflections, shown in the magnified image of the central portion of the diffraction pattern, corresponds to the observed periodicity of ∼6.7 nm. Note that this analysis is straightforward as the Mo (011) reflection serves as an internal calibration, without knowing the cameral length, to measure the multilayer periodicity.

8.7.6　Twinning

A twin is a common phenomenon observed in materials and is important in determining their mechanical properties. Typically, a twin can be formed by shear, such that all atoms on one side of the twin boundary (or plane) appear to be in positions of mirror image with respect to the other (Fig. 8.7.5a). The twin planes depend of the crystal structure, and are {111} and {112} for FCC and BCC crystals, respectively. Alternatively, because the atoms are spherically symmetric, a twin can also be considered as arising from a 180° rotation about an axis normal to the twin plane. Since the reciprocal lattice points are simply related to the real lattice (§4.2), the reciprocal lattice for the twinned crystal defined by the indices (PQR), is related to the original crystal, defined by the indices, (pqr), by a simple

(a)

Twin
Plane

Matrix
Crystal

Twinned
Crystal

(b)

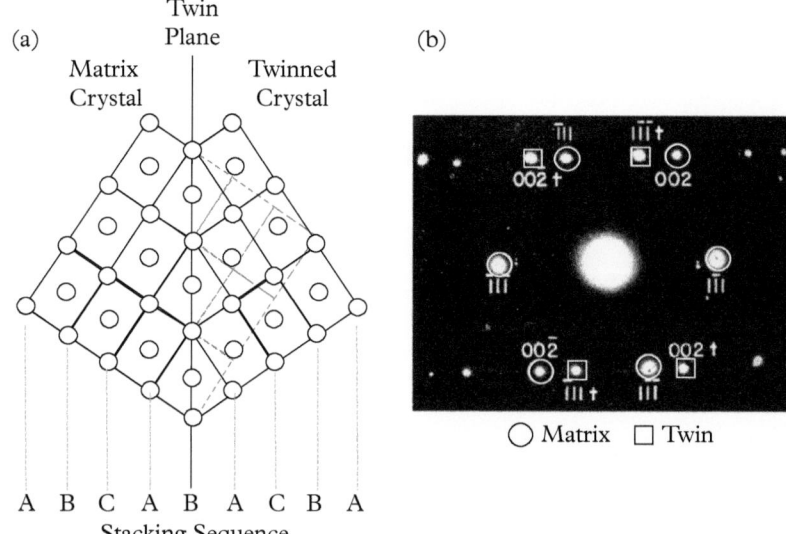

○ Matrix   □ Twin

A B C A B A C B A
Stacking Sequence

Figure 8.7.5 Twinning in an FCC crystal, showing (a) the position of atoms for a (110) section of a twin on $(1\bar{1}1)$; notice the reversal of the ABC stacking sequence of the matrix in the twin to ACB. (b) The diffraction pattern (indexed) observed for such twinning in an FCC crystal in the [110] zone axis pattern.

Adapted from Edington (1975).

transformation matrix, T_{hkl}, defined by rotation of 180° about the normal to the twin plane (hkl), as

$$(PQR) = T_{hkl}(pqr) \tag{8.7.1}$$

where the twinning matrix, T_{hkl}, depends on the crystal structure. For a cubic crystal, it is given by

$$T_{hkl} = \frac{1}{(h^2 + k^2 + l^2)} \begin{pmatrix} h^2 - k^2 - l^2 & 2hk & 2hl \\ 2hk & -h^2 + k^2 - l^2 & 2kl \\ 2hl & 2kl & -h^2 - k^2 + l^2 \end{pmatrix} \tag{8.7.2}$$

As mentioned, for an FCC crystal twinning occurs on the {111} plane. Using these values for hkl in (8.7.2), we get for the transformation matrix:

$$T_{111}^{FCC} = \frac{1}{(3)} \begin{pmatrix} -1 & 2 & 2 \\ 2 & -1 & 2 \\ 2 & 2 & -1 \end{pmatrix} \tag{8.7.3}$$

Figure 8.7.5b shows a diffraction pattern for an FCC crystal structure observed for the [110] zone axis, but twinned on $(1\bar{1}1)$; it is left as an exercise to the reader to show that this pattern is consistent with (8.7.3).

Example 8.7.1: The following figure is an electron diffraction pattern of a steel microstructure consisting of a matrix of tempered martensite (α-Fe, BCC, strong spots, $a_0 = 2.866$ Å), and particles of Fe_3C (cementite, orthorhombic, $a = 4.524$ Å, $b = 5.088$ Å, and $c = 6.741$ Å, weak spots). Index the pattern and establish the orientation relationship between the two phases. Assume a camera constant, $\lambda L = 46$ mm Å. Notice that α-Fe appears to consist of two twin-related variants.

Solution: Based on Table 4.1.2, we can calculate the d_{hkl} spacings for orthorhombic cementite as a table:

hkl	001	010	100	011	101	110	002	111	012	102	020	112	102
d_{hkl} (Å)	6.74	5.09	4.52	4.06	3.76	3.38	3.37	3.02	2.81	2.70	2.54	2.39	2.38

We can then index the diffraction pattern, using the camera constant, by measuring distances of the various reflections in the diffraction pattern, as follows:

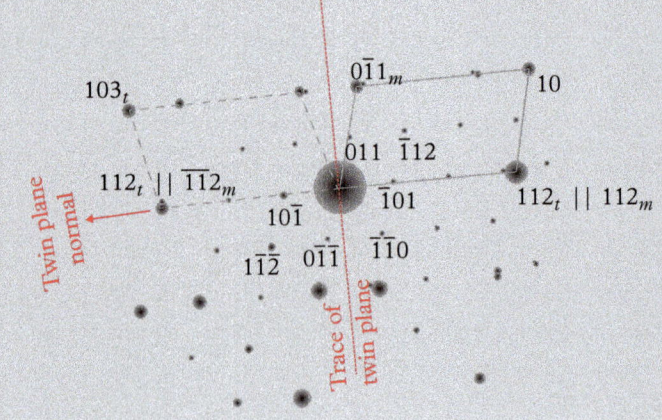

The α-Fe matrix (indexed as m) is shown as a solid line and its twin (indexed as t) orientation as the dotted line. The cementite reflections are indexed without a subscript.

This gives the orientation relationship:

$$(011)_{Fe_3C} \parallel (0\bar{1}1)_{\alpha-Fe} \text{ and} (\bar{1}01)_{Fe_3C} \parallel (112)_{\alpha-Fe}$$

Now to determine the zone axis, we take cross-products, (4.1.8), to give $[\bar{1}11]_{Fe_3C} \parallel [3\bar{1}\bar{1}]_{\alpha-Fe}$ (see Figure 4.1.10).

8.8 Interactions of Neutrons with Matter

Neutron-based methods require access to national user facilities and are not routinely encountered in materials characterization and analysis; however, for the sake of completeness, we briefly discuss the scattering of neutrons by atoms. Even though neutrons are much more massive than electrons, carry no charge, and are electrically neutral, they possess a magnetic moment that can be polarized and contribute to scattering by magnetic materials. The interactions of neutrons with the nucleus is very short range and of the order of the nuclear radius, which is $\sim 10^{-4}$ times the atomic radius. As a result, most materials are not able to stop thermal neutrons and they penetrate deep (typically mms) into the specimens; a consequence of this is the requirement of a substantially larger quantity of a material for neutron scattering experiments.

Example 8.8.1: Calculate the wavelength of a thermal neutron (rest mass, $m = 1.67 \times 10^{-27}$ kg) emerging from a reactor at room temperature (300 K) with an energy of 1.5 $k_B T$.

Solution: We assume that the energy of the neutron is entirely kinetic. Thus $\frac{1}{2}mv^2 = 1.5\, k_B T$. The wavelength, λ, is given by the De Broglie relation, $\lambda = h/p = h/mv = h/(3m\, k_B T)^{1/2}$. Substituting the values of m, h, k_B, and T, we get $\lambda = 1.456$ Å

8.8.1 Nuclear Interactions

A neutron beam interacts with an atom in two distinct ways: first, with the nucleus using short-range nuclear forces, and second, with the magnetic moment of the atom via the spin of the neutron.

Each atom of nuclear radius, R_N, presents an effective surface area of $4\pi R_N^2$, which is not penetrable by the incident neutron beam. Typically, R_N is small compared to the wavelength, λ_n, of the thermal neutrons and is given by $R_N = 1.5 \times A^{1/3} \times 10^{-13}$ m, where A is the atomic mass number (see §5.3.5.1). The quantity $4\pi R_N^2$ is called the scattering cross-section and gives the total intensity of the neutrons scattered by the nucleus in all directions; however, because $R_N \ll \lambda_n$, the nucleus behaves as a point-scattering object and the scattered intensities exhibits negligible angular, or $\left(\frac{\sin\theta}{\lambda_n}\right)$ dependence (Fig. 8.2.2). The intensity scattered into a unit solid angle is R_N^2 (because the total solid angle of a sphere is 4π), and its amplitude, or the scattering factor, is R_N.

In addition to the cross-section described, based on the physical scattering geometry the interaction of the neutrons can be thought to have a contribution similar to the Thomson scattering of X-rays with the outer electrons (§7.2.1). Thus, when the neutron is in very close proximity to the nucleus, we can think of them initially as forming a metastable nucleus–neutron "complex", which then decays to re-emit the neutron. At an appropriate energy, resonance effects can occur and contribute an additional term to the scattering factor. Considering the two—geometric and resonance—contributions, we can write the total scattering factor for the neutron as

$$f_n = R_N - \left(\frac{\gamma_N^R}{2k_n E_R}\right) \tag{8.8.1}$$

where $k_n = 2\pi/\lambda_n$ is the wave-vector of the thermal neutrons, E_R is the resonance energy to form the neutron-nucleus "complex", and γ_N^R is the resonance energy for the re-emission of the neutron. The second term in (8.8.1) is also independent of $\left(\frac{\sin\theta}{\lambda_n}\right)$, and so the total scattering does not have any angular dependence (Fig. 8.2.2). However, it is possible for the second resonance term to be larger than the first (examples are ^{1}H, ^{48}Ti, ^{62}Ni, ^{55}Mn), because the scattering is out of phase by π with respect to the geometric contribution. If and when this is the case, f_n can have negative values.

The values of f_n are measured experimentally to a high degree of precision; Table 8.8.1 compares the scattering factor for neutrons, electrons, and X-rays for hydrogen, copper, and tungsten. Note that unlike for electrons and X-rays, f_n varies nonmonotonically with atomic number. Hence, neighbouring elements in the

Table 8.8.1 Comparison of scattering factors for neutrons, electrons, and X-rays.

	f_n (10^{-14}) m	f_n (10^{-14}) m	f_e (10^{-14}) m	f_e (10^{-14}) m	f_X (10^{-14}) m	f_X (10^{-14}) m
$\frac{\sin\theta}{\lambda}$	0.1	0.5	0.1	0.5	0.1	0.5
^{1}H	−0.378	−0.378	4,530	890	0.23	0.02
^{63}Cu	0.67	0.67	51,100	14,700	7.65	3.85
W	0.466	0.466	118,000	29,900	19.4	12.0

Adapted from Schwartz and Cohen (1987).

periodic table may have appreciable differences in their neutron scattering factors, and can thus be easily resolved. Further, even light elements, such as hydrogen, have significant neutron scattering factors and can contribute significant intensity to neutron scattering. Thus, neutron scattering is more sensitive to the presence of hydrogen in the sample compared to X-rays and electrons. There are two other practical implications. First, since f_n does not depend on $\frac{\sin\theta}{\lambda_n}$, the nuclear scattering does not decrease significantly at high angles. As a result, data for reflections at high scattering angles can be collected, and structural details can be obtained at much higher accuracy than electrons or X-rays (see §7.11.1). Second, the typical energy of a neutron with $\lambda_n = 1$ Å is about 0.1 eV, which is comparable to the energy of phonons (thermal vibration modes) in the crystal. This can lead to inelastic scattering of neutrons.

8.8.2 Magnetic Interactions

A single neutron has a magnetic moment, $\mu_n = \frac{\gamma_n e h}{m_e c}$, where $\gamma_n = 1.9$ is the gyromagnetic ratio of the neutron, e and m_e are the charge and mass of the electron, and c is the velocity of light. The magnetic moment of the neutron can interact with unpaired electrons of magnetic atoms giving rise to additional neutron scattering. Similar to X-rays and electrons, this magnetic interaction occurs over a finite volume around the nucleus and also decreases with $\frac{\sin\theta}{\lambda_n}$.

The magnetic scattering amplitude for the neutron is given by

$$p_{mag} = \left(\frac{e^2}{m_e c^2}\right)\gamma_n S f_{mag} \sim 0.54 S f_{mag}\left(10^{-14}m\right) \tag{8.8.2}$$

where S is the electron spin quantum number for the atom, and the magnetic scattering factor, f_{mag}, is analogous to the atomic scattering factor, f_{atom}, in X-ray diffraction, but is determined only by the distribution of electrons with unpaired spins in the atom; further, f_{mag} is normalized such that $f_{mag} = 1$ for $\theta = 0$. Also, p_{mag} is of the same order as the nuclear scattering.

Figure 8.8.1a shows a typical neutron scattering geometry from a magnetic sample, defining the angle α between the scattering vector, $\mathbf{q}$, and the direction of spin orientation, $\mathbf{S}$. For an unpolarized neutron beam the nuclear and magnetic scattering intensities are additive and for any reflection hkl, we get

$$|F|^2 = |F|^2_{nuc} + \sin^2\alpha|F|^2_{mag} \tag{8.8.3}$$

If the specimen has only a single magnetic domain, then α is the same for all atoms; however, if there are multiple domains it has to be averaged over all spin orientations. The magnetic spins in both ferromagnetic and antiferromagnetic materials can be aligned by the application of external fields to satisfy the condition of either $\sin^2\alpha = 0$ or $\sin^2\alpha = 1$. If $|F|^2$ is measured under these two conditions, from (8.8.3), we can separately determine the contributions from the nuclear and magnetic scattering.

Figure 8.8.1 (a) Neutron scattering geometry from a magnetic sample indicating the neutron polarization, **p**, the scattering vector, **q**, and the sample magnetization, **s**; the latter two are separated by the angle, α, as shown. On any crystal lattice, different magnetic ordering such as (b) ferromagnetic, (c) antiferromagnetic, and (d) helimagnetic, are possible. In addition to the nuclear contribution (i), magnetic scattering contributes additional intensity (ii) for ferromagnetic order, additional peaks (iii) for the antiferromagnetic order, and satellites (iv) for helimagnetic order.

Adapted from Krishnan (2016).

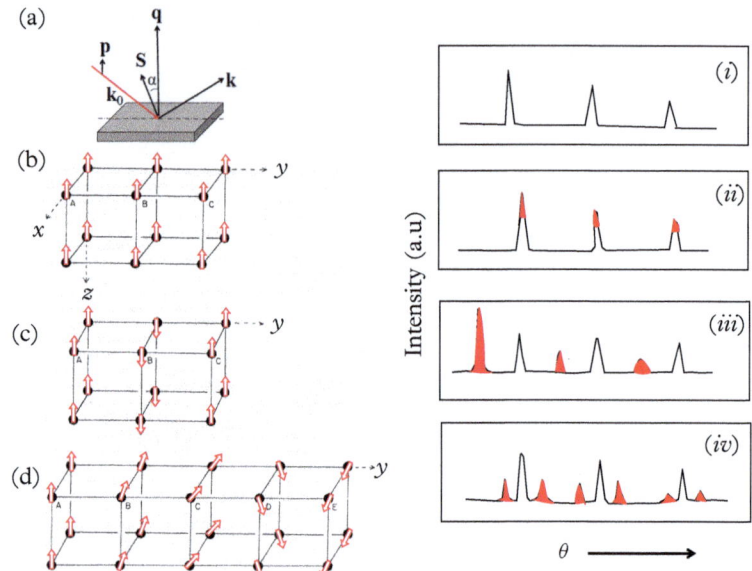

For our elementary discussion, we only consider monoatomic ferromagnets, where all spins are parallel, and antiferromagnets where the coupling between nearest neighbor spins is antiparallel. In materials where the magnetic unit cell coincides with the chemical unit cell (Fig. 8.8.1b) the nuclear contribution (Fig. 8.8.1i) and the magnetic contribution are additive (Fig. 8.8.1ii). In many antiferromagnetic materials, the magnetic unit cell is a multiple of the chemical cell (Fig. 8.8.1c) and then additional reflections from the purely magnetic scattering (Fig. 8.8.1iii), are observed. Lastly, for a helimagnetic structure formed on the same crystal lattice (Fig. 8.8.1d) additional satellites for each of the reflections (Fig. 8.8.1iv), are observed.

Finally, it is possible to obtain polarized neutrons using suitable single-crystal monochromators—a most common one is the alloy Fe_8Co_{92}, which produces 99.7% polarization. Either a constructive or destructive interference between the magnetic and nuclear scattering amplitudes for a given polarization can take place. In this way, a weak magnetic scattering in the midst of strong nuclear scattering can be detected. Further details of neutron diffraction can be found in Bacon (1962) or Willis and Carlile (2009).

8.8.3   *In Situ* Kinetic Studies Using Neutrons: Hydration of Cement

As demonstrated, neutrons interact with the nucleus and magnetic moments of the atoms with sufficient strength for the investigation of bulk specimens. Typically, a scattering probability of ~10% is targeted, and for a thermal neutron beam this requires about 1 billion atoms or a 1-cm thick layer of Al, but for hydrogenated samples this is reduced by a significant factor to thicknesses of ~ 0.1 mm. Moreover, the low absorption of neutrons deposits very little energy and does not perturb the specimen. Thermal neutrons range in energy from a few neVs up to eVs, corresponding to wavelengths from sub-Å to several mm, and suitable for studying subatomic to supramolecular length scales. Even though the wavelengths are comparable to those of X-rays, the equivalent thermal neutrons have much lower energies that match very well with energies of thermal excitations in materials, making it possible to carry out dynamic experiments. Furthermore, neutrons interacting with nuclei are sensitive to isotopes and deuteration (substituting deuterium for hydrogen, see §3.10.1) is routinely employed in neutron scattering studies of soft matter and biological samples. Finally, kinetic studies are carried out it two different ways: single shot experiments to investigate processes on the time scale corresponding to the counting times (seconds to hours) of neutron experiments. Alternatively, stroboscopic or pump-probe measurements, requiring repeatable processes, access much shorter time-scales down to nanoseconds.

An example, of the former is the time-resolved neutron-scattering study of the hydration of cement. Hydrogen exhibits an extraordinarily strong quasi-elastic scattering for neutrons allowing the possibility of distinguishing components with different mobility of free and bound water (Fig. 8.8.2a). Quantitative interpretation of the time evolution of this scattering, leads to the so-called bound water index (BWI), which can be used to describe the hydration process of cement. The

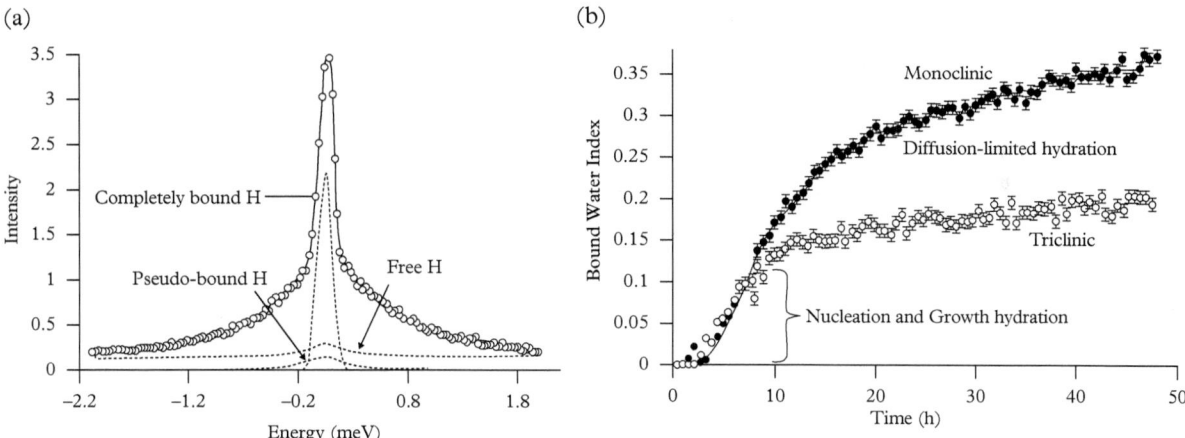

Figure 8.8.2 (a) The quasi-elastic scattering from hydrogen during the hydration of cement, showing components from the different mobility of free and bound water. (b) The time evolution of the bound water index for triclinic and monoclinic forms of tricalcium silicate, a principal component of cement.

(a) Adapted from [17]. (b) Adapted from [18].

variation of BWI with time (Fig. 8.8.2b) depends on the polymorph of tricalcium silicate that is used in the cement. A monoclinic modification enhances diffusion-controlled hydration, leading to the formation of a better-quality cement [18].

Further details of a range of *in situ* and kinetic studies using thermal neutrons can be found in the excellent review in Ziegler et al. (2014), Ch. 5.

Summary

Electrons interact both with the nucleus and the surrounding electron cloud of atoms. Thus, electron scattering factors include the sum of nuclear and electron scattering contributions, and are significantly greater than that for X-rays, especially in the forward-scattering direction. As a result, electron diffraction can be sensitive to surfaces and small nanoscale volumes of materials. Further, diffracted electron intensities decrease more rapidly with scattering angle, and in transmission are forward peaked. Moreover, electrons are multiply scattered even over short distances in materials; as a result, simple kinematic or single-scattering approaches valid for X-rays, need to be modified (dynamical scattering theory) to quantitatively account for the observed electron diffraction intensities.

Electrons with energies in the range of 50–100 eV have a wavelength of 0.1 nm and a penetration depth of ∼0.25 nm, making them suitable for studies of surfaces, including their reconstruction. In LEED, the electrons diffracted in the backward direction provide a record of a section of the reciprocal space of the two-dimensional surface. LEED is routinely used to study the distribution/arrangement of adsorbed atoms on clean surfaces. RHEED is used to probe surface crystallography. Particularly, in *in situ* experiments, it is used to monitor the mechanism of thin film growth by measuring the variation in intensity with growth time (RHEED oscillations).

In transmission, the interaction of high-energy (of the order of ∼100 keV) electrons with thin foil specimens is complex and involves elastic, coherent, inelastic, and incoherent contributions. The coherent and elastic components taken together constitute electron diffraction from crystals in transmission, and can be understood using the Ewald sphere construction introduced in §4. However, the wavelength of high-energy electrons is very small compared to typical inter-planar spacing in crystalline materials. As a result, the radius of the Ewald sphere is very large and its surface intersecting the reciprocal lattice is very flat, giving an electron diffraction pattern that is a planar cross-section of the reciprocal lattice. Furthermore, electron diffraction intensities observed in transmission through thin foils are products of the structure factor defined by the details of the unit cell, and a lattice amplitude factor determined by the *overall* size/shape of the specimen. This causes each reciprocal lattice point to have a finite size and shape, and leads to diffraction being observed even when the scattering geometry deviates from the exact Laue condition by a small excitation error.

In the column approximation of kinematic scattering, with only two *independent* beams—a forward-scattered one and a diffracted beam in the Bragg condition with very small intensity—it can be shown that the amplitude of the diffracted beam at the exit surface of the crystal oscillates with both the thickness and the excitation error. As a result, if a wedge-shaped specimen is used, its image will show a series of dark and bright bands (thickness fringes) that are separated in thickness proportional to the inverse of the excitation error. Alternatively, a bent foil of uniform thickness with lateral variations in the excitation error will show intensity variations (contrast) in the form of bend contours. The kinematic theory fails for crystals of large thickness, and a dynamical theory, involving coupled differential equations, where the intensity is scattered back and forth between the forward-scattered and diffracted beams as the electron propagates in the crystal is required.

Electron diffraction in a TEM can be broadly classified into three categories. SAD using parallel beam incidence, which produces spot (single crystalline grains) or ring (polycrystalline materials) patterns. In particular, incident beams directed along low-index zone axis orientations, reflect the symmetry of the crystal and can be easily identified. A significant feature of these patterns is double diffraction that leads to positive intensity even for reflections that are forbidden by the extinction rules in structure factor calculations. The second, Kikuchi lines and maps involving a combination of inelastic and elastic scattering, appear as a pair of dark and bright lines, and define the sense of tilt of the specimen, reveal true crystal symmetry in SAD patterns, and help guide the operator in tilting the specimen from one zone axis orientation to another. Finally, spherical aberration in electromagnetic lenses limit the application of SAD apertures to areas less than 1 µm in diameter. For diffraction from smaller areas, a focused or convergent probe is used to produce CBED patterns. If such a diffraction pattern is recorded with a large collection angle along a low-index zone axis, higher-order Laue zones in the form of rings are observed. CBED patterns can provide elaborate crystallographic or symmetry analysis of a material at high spatial resolution.

Unlike electrons, the interaction of neutrons with the nucleus of atoms is very short range ($\sim 10^{-4}$ times the atomic radius). In addition, it can interact with the magnetic moment of the atom via the spin of the neutron. The atomic scattering factor for neutrons has two contributions: a geometric one related to the scattering cross-section of the atomic nuclei, and a resonance contribution—not unlike Thomson scattering for X-rays—from the resonance effect arising from the decay of the neutron-nucleus complex. The overall scattering factor is a difference of these two terms and does not have any angular dependence. In addition, for some materials (^{1}H, ^{48}Ti, ^{62}Ni, ^{55}Mn) the second resonance term can be larger than the first cross-section term, giving rise to negative values of the scattering factor. Finally, magnetic scattering of neutrons finds particular use in studies of magnetic order (ferromagnetic, antiferromagnetic, spin canting, etc.) in materials.

....

FURTHER READING

Amelinckx, S., R. Gevers, and J. van Landuyt. *Diffraction and Imaging Techniques in Materials Science*, Vol II. Amsterdam: North-Holland, 1978.

Bacon, G. E. *Neutron Diffraction*. Oxford: Oxford University Press, 1962.

Brundle, C. R., C. A. Evans Jr., and S. Wilson, *Encyclopedia of Materials Characterization*. Washington, DC: Butterworth-Heinemann, 1992.

Cowley, J. M. *Diffraction Physics*. Amsterdam: North Holland, 1984.

de Graef, M., and M. McHenry. *Structure of Materials*. Cambridge: Cambridge University Press, 2007.

Edington, J. W. *Electron Diffraction in the Electron Microscope*. London: Macmillan, 1975.

Fuchs, E., H. Oppolzer, and H. Rehme. *Particle Beam Microanalysis: Fundamentals, Methods and Applications*. New York: VCH Publications, 1990.

Fultz, B., and J. Howe. *Transmission Electron Microscopy and Diffractometry of Materials*. Heidelberg: Springer, 2013.

Hammond, C. *The Basics of Crystallography and Diffraction*. Oxford: Oxford University Press, 2006.

Hirsch, P. B., A. Howie, R. B. Nicholson, D. W. Pashley, and M. J. Whelan. *Electron Microscopy of Thin Crystals*. Washington, DC: Butterworth, 1965.

Kittel, C. *Introduction to Solid State Physics*. New York: Wiley, 1986.

Krishnan, K. M. *Fundamentals and Applications of Magnetic Materials*. Oxford: Oxford University Press, 2016.

Lagally, M. G. *Methods of Experimental Physics—Surfaces*, edited by R. L. Park and M. G. Lagally. New York: Academic Press, 1985.

Pendry, J. B. *Low Energy Electron Diffraction*. New York: Academic Press, 1974.

Reimer, L. *Transmission Electron Microscopy*. Berlin: Springer, 1993.

Schwartz, L. B., and J. B. Cohen, *Diffraction from Materials*, 2nd ed. Berlin: Springer-Verlag, 1987.

Shindō, D., and K. Hiraga. *High-Resolution Electron Microscopy for Materials Science*. New York: Springer, 1998.

Sproull, R. L., and W. A. Phillips, *Modern Physics: The Quantum Physics of Atoms, Solids and Nuclei*, Third Edition. New York: Dover, 2015.

Thomas, G., and M. J. Goringe. *Transmission Electron Microscopy of Materials*. New York: Wiley, 1979.

Vickerman, J. C., ed. *Surface Analysis: The Principal Techniques*. Chichester: Wiley, 2000.

Vickerman, J. C., and I. S. Gilmore, eds. *Surface Analysis: The Principal Techniques, 2nd Edition*. Chichester: Wiley, 2009.

Williams, D. B., and C. B. Carter. *Transmission Electron Microscopy, Vol. II. Diffraction*. New York: Plenum Press, 1996.

Williams, D. B., A. R. Pelton, and R. Gronsky, eds. *Images of Materials*. Oxford: Oxford University Press, 1991.

Willis, B. T. M., and C. J. Carlile. *Experimental Neutron Scattering*. Oxford: Oxford University Press, 2009.

Woodruff, D. P., and T.A. Delchar. *Modern Techniques of Surface Science*. Cambridge: Cambridge University Press, 1992.

Ziegler, A., H. Graafsma, X. F. Zhang, and J. W. M. Frenken. *In-Situ Materials Characterization across Spatial and Temporal Scales*. Berlin: Springer, 2014.

....

REFERENCES

[1] Bölger, P., and P. K. Larsen. "Video System for Quantitative Measurements of RHEED Patterns." *Review of Scientific Instruments* 57, no. 7 (1986): 1363–7.

[2] Tang, F., T. Parker, G.-C. Wang, and T-M Lu. "Surface Texture Evolution of Polycrystalline and Nanostructured Films: RHEED Surface Pole Figure Analysis." *Journal of Physics D: Applied Physics* 40 (2007): R427.

[3] Joyce, B. A., P. J. Dobson, J. H. Neave, K. Woodbridge, J. Zhang, P. K. Larsen, and B. Bôlger. "RHEED Studies of Heterojunction and Quantum Well Formation during MBE Growth: From Multiple Scattering to Band Offsets." *Surface Science* 168, (1986): 423–38.

[4] Bauer, E. "Struktur und Wachstum duenner Aufdampfschichten." *Zeitschrift fur Kristallographie* 110 (1958): 372–94.

[5] Burton, W. K., N. Cabrera, and F. C. Frank. "The Growth of Crystals and the Equilibrium Structure of their Surfaces." *Philosophical Transactions of the Royal Society of London. Series A, Mathematical and Physical Sciences* 243, no. 866 (1951): 299–358.

[6] Bethe, H. "Theorie der Beugung von Elektronen an Kristallen." *Annalen der Physik* 87, (1928): 55–129.

[7] Takagi, S. "Dynamical Theory of Diffraction Applicable to Crystals with Any Kind of Small Distortion." *Acta Crystallographica* 15 (1962): 1311–12.

[8] Howie, A., and M. J. Whelan. "Diffraction Contrast of Electron Microscope Images of Crystal Lattice Defects. III. Results and Experimental Confirmation of the Dynamical Theory of Dislocation Image Contrast." *Proceedings of the Royal Society A, Mathematical, Physical, and Engineering Sciences* 267, no. 1329. (1962): 206–30.

[9] Kikuchi, S. "Diffraction of Cathode Rays by Mica." *Japanese Journal of Applied Physics* 4, no. 6 (1928): 271–4.

[10] Buxton, B. F., J. A. Eades, J. W. Steeds, and G. M. Rakham. "The Symmetry of Electron Diffraction Zone Axis Patterns." *Philosophical Transactions for the Royal Society of London. Series A, Mathematical and Physical Sciences* 81, no. 1301 (1976): 171–94.

[11] Kemp, S. J., R. M. Ferguson, A. P. Khandhar, and K. M. Krishnan. "Monodisperse Magnetite Nanoparticles with Nearly Ideal Saturation Magnetization." *RSC Advances* 6 (2016): 77452–64.

[12] Cheng, N., J.-P. Ahn, and K. M. Krishnan. "Epitaxial Growth and Exchange Biasing of PdMn/Fe Bilayers Grown by Ion-Beam Sputtering." *Journal of Applied Physics* 89, no. 11 (2001): 6597.

[13] Blomqvist, P., K. M. Krishnan, and D. E. McCready. "Growth of Exchange-Biased MnPd/Fe Bilayers." *Journal of Applied Physics* 95 (2004): 8019.

[14] Shull, C. G., W. A. Strauser, and E. O. Wollan. "Neutron Diffraction by Paramagnetic and Antiferromagnetic Substances." *Physical Review* 83 (1951): 333.

[15] Estrup, P. J., and E. G. McCrae. "Surface Studies by Electron Diffraction." *Surface Science* 25, no. 1 (1971): 1–52.

[16] Takayanagi, K., Y. Tanishiro, S. Takahashi, and M. Takahashi. "Structure Analysis of Si(111)-7 × 7 Reconstructed Surface by Transmission Electron Diffraction." *Surface Science* 164, no. 2–3 (1985): 367–92.

[17] Peterson, V. K., C. M. Brown, and R. A. Livingston. "Quasielastic and Inelastic Neutron Scattering Study of the Hydration of Monoclinic and Triclinic Tricalcium Silicate." *Chemical Physics* 326 (2006): 381–9.

[18] Peterson, V. K. "Studying the Hydration of Cement Systems in Real Time Using Quasielastic and Inelastic Neutron Scattering." In *Studying Kinetics with Neutrons*, edited by G. Eckold, H. Schober, and S. E. Nagler, 19–74. Springer Series in Solid State Physics **161**. Heidelberg: Springer-Verlag, 2010.

··

EXERCISES

A. Test Your Knowledge

For each statement, choose all the right answers. Note that more than one response, or none of them, may be correct.

1. Electron diffraction depends on scattering of the electron beam by
   (a) the positively charged nucleus.
   (b) the negatively charged electron cloud.
   (c) Coulombic interactions.
2. The atomic scattering factor for electrons
   (a) is the same as that for X-rays.
   (b) depends on the scattering angle.

(c) is ALWAYS larger than that for X-rays.

(d) does NOT depend on the atomic number.

3. The large atomic scattering factor for electrons

   (a) makes it possible to detect their diffraction from small volumes.

   (b) attenuates diffracted electron intensity very rapidly with thickness in transmission.

   (c) requires very thick specimens in transmission.

   (d) allows diffraction experiments to be done in air without vacuum.

   (e) rapidly decreases electron diffraction intensities with scattering angle.

4. The *kinematical* theory of electron diffraction in transmission

   (a) is the same as the discussion of X-ray diffraction in §7.

   (b) applies to only very thin crystals.

   (c) includes multiple scattering effects.

   (d) predicts the oscillation of diffraction intensities with thickness.

5. The reconstruction of a crystal surface

   (a) describes only the first atomic monolayer on the surface.

   (b) is due to the break in symmetry.

   (c) geometrically relates the surface unit cell to the unit cell of the underlying crystal.

6. Surface electron diffraction

   (a) also satisfies the Ewald sphere construction.

   (b) conserves the TOTAL electron momentum.

   (c) does NOT conserve the component of the electron momentum normal to the surface.

   (d) can be carried out only with low-energy (<200 eV) electrons.

7. In low-energy electron diffraction, the

   (a) observed pattern is a projection of the surface reciprocal lattice net.

   (b) magnification of the diffraction pattern cannot be changed.

   (c) experiments require very high surface cleanliness of the specimen.

8. Low-energy electron diffraction patterns

   (a) give information on the size and shape of the surface lattice net.

   (b) can be used to determine atom positions in the unit cell.

   (c) measured as a function of temperature can be used to study phase transitions.

9. In reflection high-energy electron diffraction, bulk features

   (a) are not observed.

   (b) can be separated from the surface features by varying the incidence angle.

   (c) produce spots rather than streaks.

10. RHEED oscillations observed in situ in thin film growth can

    (a) monitor growth thickness.

    (b) separate layer-by-layer growth from island growth.

    (c) be not correlated with surface imaging of growth.

11. In a transmission electron microscope
    (a) the high-energy incident electrons are also known as fast electrons.
    (b) diffraction, the Ewald sphere has a very large radius.
    (c) diffraction, we observe a planar cross-section of the reciprocal lattice.
12. The *extinction distance* in transmission electron diffraction
    (a) defines a thickness when you no longer see (extinct!) a diffraction pattern.
    (b) is a characteristic length associated with each reflection.
    (c) corresponds to a thickness where the wave amplitude is a constant.
13. Scattering of electrons from a set of atoms at positions, r_i, is
    (a) *coherent* if the resulting amplitude is the sum of the individual scattering amplitudes.
    (b) *incoherent*, if the phase of the scattered wave cannot be easily predicted.
    (c) *inelastic*, if there is no change in energy between the incident and scattered beams.
14. Transmission electron diffraction intensities at the exit surface of a thin foil are
    (a) proportional to the square of the structure factor.
    (b) not dependent on the shape of the crystal.
    (c) are affected by multiple scattering effects.
15. The *excitation error* in electron diffraction
    (a) defines the deviation from the exact Bragg condition.
    (b) affects the diffracted intensities.
    (c) can be used to orient the crystal at the exact Bragg condition.
    (d) varying spatially in a foil of uniform thickness produces *bend contours*.
16. A *ring pattern* in selected area diffraction
    (a) is produced by a polycrystalline specimen.
    (b) from a *standard* material can be used to calibrate the *camera length* of a TEM.
    (c) cannot be produced from an assembly of nanoparticles.
17. *Kikuchi lines* and patterns observed in a TEM
    (a) involve only elastic scattering.
    (b) involve only inelastic scattering.
    (c) are fixed to the crystal and move in position with specimen tilt.
18. Convergent beam electron diffraction patterns
    (a) arise from intermediate values of the beam convergence semi-angle.
    (b) can be used to measure local specimen thickness.
    (c) include HOLZ lines that reflect the *space group* symmetry of the crystal.
19. A beam of neutrons interacts with the specimen atoms
    (a) very strongly, requiring only small volumes of the specimen for diffraction experiments.

(b) in two different ways—with the nucleus and with its magnetic moment.

(c) with very little angular variation of the scattered neutron intensities.

20. The magnetic moment of thermal neutrons
    (a) is very weak.
    (b) interacts with the unpaired spins of magnetic atoms.
    (c) can help resolve magnetic and chemical order in crystals.

B.   Problems

1. Calculate the **electron wavelength**, λ_e, when accelerated through a potential of
   (a) 200 V, (b) 5 kV, (c) 20 kV, (d) 100 kV, (e) 200 kV, and (f) 1000 kV
   i) For each case calculate λ_e, both with and without relativistic corrections. Comment on your results.
   ii) Associate each of these energies with a specific technique (SEM, TEM, LEED, RHEED, etc.) that we have discussed so far.

2. Calculate the **atomic scattering factor** for FCC Au [111] for electrons accelerated through 5 kV and 200 kV potentials.

3. Consider the transmission electron diffraction of a **diamond structure** material.
   (a) Draw the [001] reciprocal lattice. Calculate the structure factor, F_{hkl}, for the *principal reflections* seen in electron diffraction and indicate which would be *strong* and which would be *weak* in intensity.
   (b) Now, draw the [011] reciprocal lattice. Calculate the structure factor, F_{hkl}, for the *principal reflections* seen in electron diffraction and indicate which would have zero structure factor, $F_{hkl} = 0$. Could any of these reflections with $F_{hkl} = 0$ show a measurable nonzero intensity? If so, how?

4. **Low-energy electron diffraction** patterns for NiO at different incident beam energies are shown in Figure 8.Pr.4. Index each of the patterns. Using the Ewald sphere construction, if possible, drawn to scale, explain the changes in the LEED patterns.

5. Calculate the **scattering factors** for plutonium (Pu) and hydrogen (H) for
   (*a*) X-rays
   (*b*) Electrons
   (*c*) Neutrons.
   Now, which of three above would be best to study the structure of the compound plutonium hydride? Why?

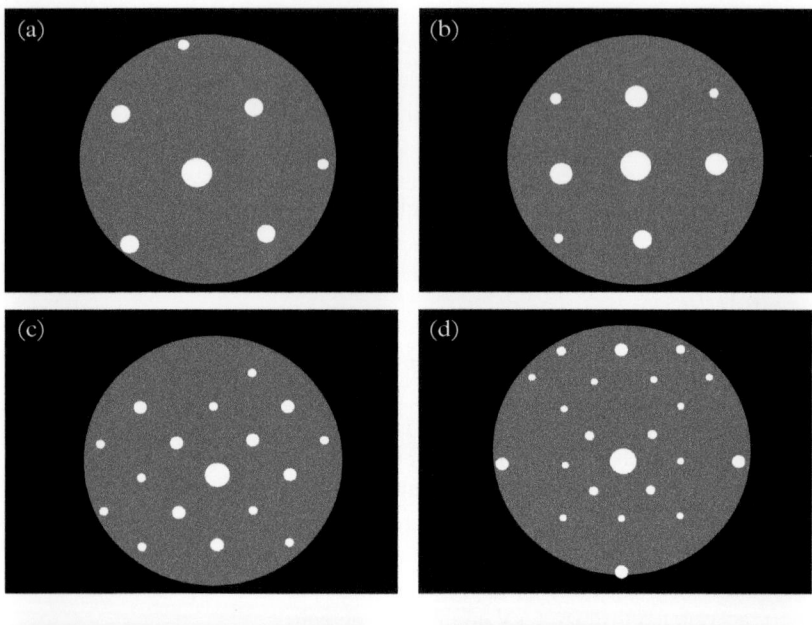

Figure 8.Pr.4 LEED patterns of NiO at different incident electron beam energies of (a) 150 eV, (b) 240 eV, (c) 450 eV, and (d) 550 eV.

Figure 8.Pr.7 Electron diffraction patterns for (a) FCC crystal, and (b) cubic, i.e. either an FCC or BCC structure.

6. Show that the radii, G_n, of a **higher-order Laue zone**, where n is its order, is given by

$$G_n = \sqrt{2\frac{2\pi}{\lambda}nH - n^2H^2} \qquad (8.8.3)$$

where H is the spacing of the reciprocal lattice planes normal to the electron beam.

7. **Indexing** transmission electron diffraction patterns shown in Figure 8.Pr.7.
   (i) Index the patterns.
   (ii) Label the low order reflections.
   (iii) Calculate the lattice parameter in (*b*) if the camera constant, $\lambda L = 60$ mm Å.

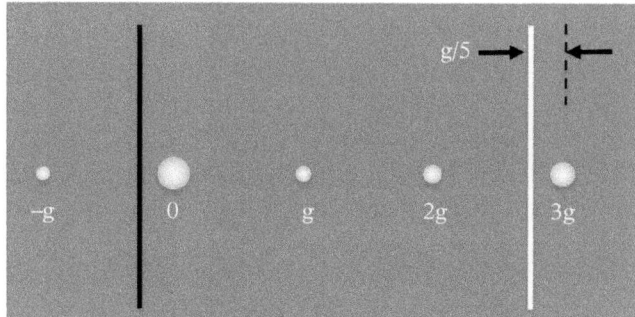

Figure 8.Pr.9 A tilted crystal with a row of spots and Kikuchi lines as shown.

Figure 8.Pr.10 The selected area diffraction pattern (drawn to scale) for the two-phase microstructure.

8. Can **thickness fringes** and **bend contours** be distinguished by transmission electron diffraction? How?

9. The **deviation parameter**, s, can be determined by tilting a crystal to show a row of spots in transmission electron diffraction with the Kikuchi lines appearing as shown in Figure 8.Pr.9. Calculate the value of s, for the reflection $3g$, as shown. State your assumptions.

10. The **orientation relationship** between a Fe (BCC, $a = 0.287$ nm) matrix, showing strong reflections, and a Fe_2TiSi (FCC, $a = 0.57$ nm) second phase, showing weak reflections, can be determined from the selected area diffraction pattern shown in Figure 8.Pr.10. Assume that $\lambda L = 3.15$ Å cm. Index the pattern and identify the orientation relationship between the two phases.

11. For the diffraction pattern in the [013] orientation of an FCC crystal
    (a) Draw the narrowest **Kikuchi bands**, intersecting the Kikuchi pole.
    (b) Index the Kikuchi lines.

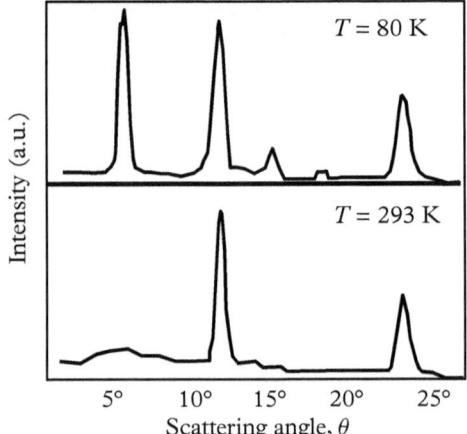

Figure 8.Pr.12 The selected area diffraction pattern for a twinned region in an FCC crystal.

Figure 8.Pr.13 Neutron diffraction for MnO above and below the Néel temperature.

Adapted from [14].

12. A selected area diffraction pattern for a **twinned FCC crystal** is shown in Figure 8.Pr.12.
    (a) Index the diffraction pattern for the original and twinned crystal.
    (b) Identify the orientation relationship between the two.
    (c) What is the boundary plane relating the two crystals?

13. The **neutron diffraction** pattern for antiferromagnetic MnO (rock Salt structure, $a = 4.43$ Å), at two different temperatures is shown in Figure 8.Pr.13.
    (a) Index the two diffraction patterns. Make any reasonable assumptions.
    (b) Explain the difference between the two patterns. (Hint: magnetic ordering).

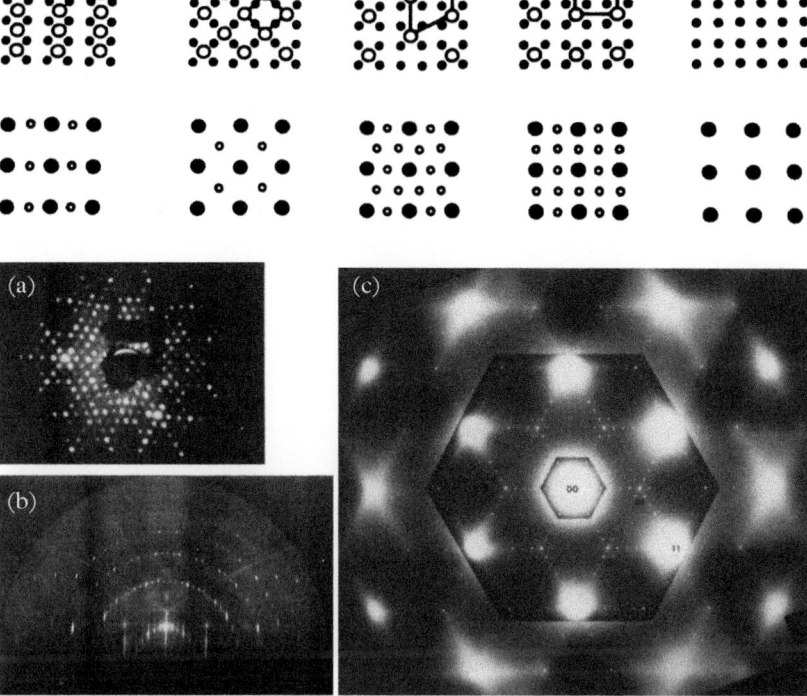

Figure 8.Pr.15 Surface atomic structure and the corresponding LEED patterns.

Figure 8.Pr.16 Si(111) surface studied by (a) LEED, (b) RHEED, and (c) TED.

14. For the **neutron diffraction** pattern shown (Fig. 8.Pr.13), if the kinetic energy of the thermal neutron emerging in thermal equilibrium is $2k_BT$, what is its temperature.

15. The real space (top) and LEED patterns (bottom) for various **overlayers** on the (100) surface of a crystals are shown in Figure 8.Pr.15. Label each of the patterns according to the notation described in §8.3.1.

16. The **surface reconstruction** of Si(111) has been extensively studied by (a) LEED, (b) RHEED, and (c) TED, as shown in Figure 8.Pr.16. Describe what you think is going on at the surface of Si(111).

    Read [15] and [16], from which these patterns are adapted, to get more details.

    What advantages, if any, do any one of these techniques have over the others?

    Compare these results with that obtained by STM, discussed in §11.

Transmission and Analytical Electron Microscopy

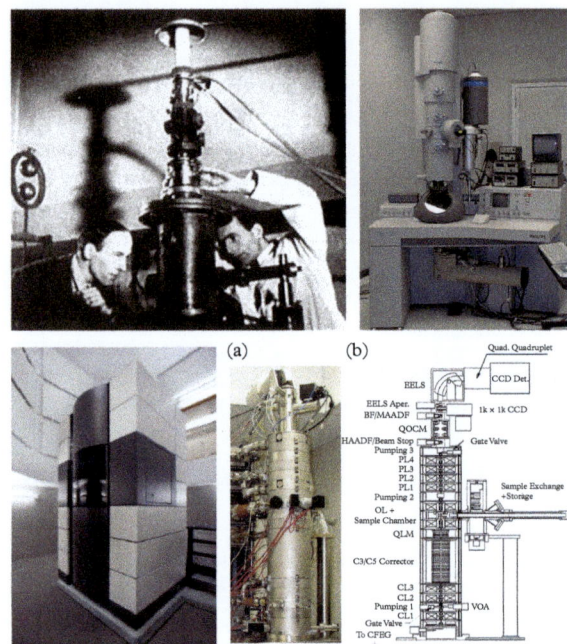

Top, left: The very first transmission electron microscope built in 1931 by Ruska and Knoll in Berlin.

Top, right: A commercial 200 kV analytical electron microscope, *circa* 2000, with a Schottky emission gun, an energy dispersive X-ray windowless detector, and an electron energy-loss spectrometer with parallel detection.

Bottom, left: A state-of-the-art (*circa* 2015) TEAM I double-aberration-corrected (scanning) transmission electron microscope (S/TEM) capable of producing images with 50 pm resolution. It includes a special high-brightness Schottky emission electron source, a source monochromator, a high-resolution GIF Tridiem energy-filter, CEOS hexapole-type spherical aberration correctors, and a chromatic aberration corrector. The illumination aberration corrector corrects coherent axial aberrations up to 4th order, as well as 5th-order spherical aberration, and sixfold astigmatism.

Photographs courtesy of the National Center for Electron Microscopy, Lawrence Berkeley National Laboratory, Berkeley, USA.

Bottom, right: The electron-optical column and schematic drawing of the aberration-corrected Nion UltraSTEM 100 scanning transmission electron microscope, operated in UHV, and capable of sub-Å resolution in imaging and rapid analysis with an atom-sized electron probe, and optimized EELS coupling optics.

Adapted from [6].

Principles of Materials Characterization and Metrology. Kannan M. Krishnan, Oxford University Press (2021).
© Kannan M. Krishnan. DOI: 10.1093/oso/9780198830252.003.0009

9.1 Introduction

Electron microscopes are of two types: transmission (TEM) and scanning (SEM). They are indispensable for materials characterization. We discuss TEM in this chapter and SEM in §10.

The optics of a TEM are similar to a conventional light microscope (§6.8.4) operated in transmission mode, while the SEM is similar to the confocal scanning optical microscope (§6.8.5). SEM and TEM differ in one significant way. Similar to an optical microscope, a standard TEM collects the image information simultaneously over the full magnified field of view, whereas the SEM collects the signal and forms the image sequentially, pixel by pixel (§1.4.4), with a minimum "dwell" time required to get a statistically significant signal at each pixel that constitutes the image. In addition, the configuration of a standard TEM is usually inverted with respect to an optical microscope, and the electron gun, which replaces the light source, is on the top and the image recording system is at the bottom. However, compared to optical microscopes, electron microscopes generally produce images with much higher magnifications and superior resolution arising from the much shorter wavelengths of electrons when accelerated through large potentials ($\sim$100 kV). Finally, TEM requires thin specimens (less than $\sim$0.1 μm or so, depending on the average atomic number of the sample and the acceleration voltage used), and preparing (§9.6) such electron transparent specimens (also referred to as *thin foils*) that are truly representative of the sample, without damaging the underlying microstructure or contaminating the specimen, is an "art" form and is a critical component of good TEM work.

The first electron microscope, pioneered by Ruska[1] and Knoll, appeared in the 1930s and present-day instruments have evolved significantly to routinely deliver a resolution of the order of 0.1 nm—well within the range of resolving interatomic distances in crystalline materials. In state-of-the-art TEMs, where aberrations of the electromagnetic lenses are corrected, the resolution can now approach $\sim$ 40–50 pm (Fig. 1.4.14). The basic components or building blocks of any transmission electron microscope (Fig. 9.1.1) can be broadly divided into four sections, depending on their function, as the electron source (§9.2.1), illumination (§9.2.3), image formation (§9.2.4), and magnification (§9.2.6). In addition, specimen handling (§9.2.5) and image recording (§9.2.10) components, complete the description of the instrument.

The *electron gun* can be either thermionic, field-emission (§5.2.2), or something in between, i.e. the Schottky emission gun, each with different characteristics of total current (I), current density (j), beam energy, E_0, and its energy spread, ΔE, as described in Table 5.2.2. On leaving the source, the electrons are brought to a cross-over of diameter, d_{co}, which then functions as a virtual source for the TEM, and as the beam enters the main microscope column it is further characterized by its divergence angle, α_{co} (Fig. 5.2.5). Note that a smaller d_{co} improves beam coherence and the resultant contrast in phase contrast imaging (§9.3.5). The *illumination* section consists of two or more *condenser lenses* (CL) to demagnify

[1] Ernst Ruska (1906–1988) was a German physicist and inventor of the electron microscope. Belatedly recognized in 1986 with the Nobel prize in Physics, he was cited "for his fundamental work in electron optics, and for the design of the first electron microscope."

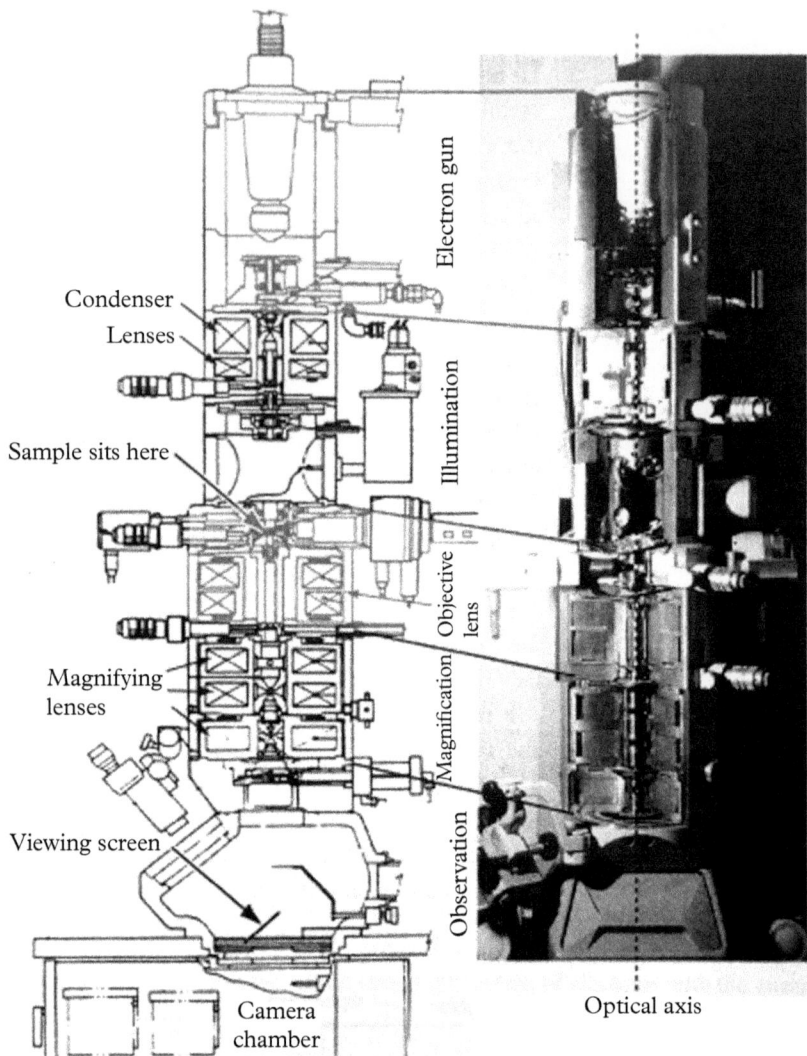

Condenser
Lenses

Sample sits here

Magnifying
lenses

Viewing screen

Camera
chamber

Electron gun

Illumination

Objective
lens

Magnification

Observation

Optical axis

Figure 9.1.1 A cross-sectional view illustrating the four sections—electron gun or source, illumination, imaging or objective lens, and magnification—of a typical transmission electron microscope.

Adapted from de Graef (2009).

the beam cross-over and to focus, shape, and orient/direct the electron beam onto the specimen surface, with a typical spot size of 1–2 μm, providing a suitable area of illumination to image the sample even at the highest magnifications. Note that in the TEM the electromagnetic lenses are adjusted by changing the lens currents to effectively change their focal length, and not by physically changing their positions as in an optical microscope (§6.8.4). Then, in the *imaging* section, after the beam interacts and is transmitted through the specimen, the first image is formed by the *objective lens* (OL), with its design and aberrations characteristics defining the true imaging performance of the transmission electron microscope.

The bottom or *magnification* section of the TEM consists of additional *intermediate lenses* (IL) that form the intermediate image from the first OL image, which is then magnified further by the *projector lenses* (PL) for final viewing or recording. Finally, since high-energy electrons have a limited mean free path length (§5.3.5.1) in air, the entire electron microscope column, including the specimen, is kept in high vacuum.

A TEM produces information of the specimen both in reciprocal (diffraction) and in real (imaging) space; we have already described the physics (§8.5) and methods of transmission electron diffraction in §8.6. To carry out such imaging and diffraction experiments, the thin-foil specimen is placed in the OL at the center of the electron-optic axis (*z*-axis), using a *specimen stage* that is mechanically quite complex. The commonly used *double-tilt* specimen stage allows for *x-y* movement to select the microstructural feature of interest, and includes a *goniometer* to tilt accurately about two orthogonal axes in the plane of the specimen to perform precise diffraction and crystallography experiments. In addition, it includes a modest range of motion along the optic axis (*z*-axis) of the beam, which allows specimen position to be adjusted (*eucentric height*) to minimize movement of the image on specimen tilt. Alternatively, some specimen stages, known as *tilt-rotation* holders (Fig. 9.2.8), also allow for the rotation of the specimen about the optic axis in combination with one or two axes of tilt.

Image contrast in a TEM is obtained in a number of different ways (§9.3). Electrons are elastically scattered or diffracted by thin specimens (§8.5) over small angles of the order of 1–2°. A set of variable apertures in the *back focal plane* of the OL limits the transmission of scattered electrons to a small angular range (~1 mrad). When this *objective aperture* is positioned to allow the directly transmitted beam to go through and form the image, it produces what is called a *bright field* (BF) image. Now all the scattered or diffracted beams are excluded and the contrast in the image arises from variations in density, thickness, and crystallographic orientation of the specimen. Alternatively, either the aperture can be positioned or the incident beam tilted to allow only a specific diffracted beam to go through the objective aperture and form a *dark field* (DF) image (§9.2.7). Finally, if a sufficiently large objective aperture is used, the transmitted and a select number of diffracted beams are allowed to go through, and together they interfere and form a *phase contrast* or high-resolution image of the atomic columns in the specimen (§9.3.5). Alternatively, high-resolution images of atomic columns in the specimen can also be formed in a TEM by scanning a highly focused beam of Ångstrom dimension (in this configuration it is known as scanning transmission electron microscopy, or STEM) over the thin foil surface and recording the transmitted electrons incoherently scattered through large angles (> 70 mrad). This method is called high-angle annular dark field (HAADF) or *Z-contrast* imaging (§9.3.4). Last but not least, the magnetic microstructures or domains in the specimen can be imaged (§9.3.6) using the deflection of the incident electron beam by the Lorentz forces as they pass through the specimen (Krishnan, 2016).

Furthermore, the already versatile TEM can be made even more so by incorporating additional components in the instrument. In a TEM with a highly coherent FEG source, the addition of an *electron biprism*—utilizing the electric fields generated by a micron sized conducting wire—in the optical column, splits the beam into two coherent beams. In the absence of a specimen, the self-interference of these two coherent beams will produce a set of fringes on the viewing screen. If the biprism is located on the optic axis above the specimen plane, and the specimen is placed on one side of the wire, such that it interacts only with one of the two coherent beams and modifies its phase, the specimen potential now alters this self-interference. This is the basis of *electron holography* (Tonomura, 1999), a rapidly developing technique of imaging in a TEM (§9.3.7), with particular application in the characterization of electronic and magnetic materials (see Krishnan, 2016, Ch. 8). Finally, the interaction of the electron beam with the specimen produces a variety of signals (Fig. 5.3.5) over different depths in the specimen (Fig. 10.2.6); some of these signals are discussed further in the context of a SEM in §10.2.3. In the optical column of a TEM it is also possible to install an energy dispersive X-ray spectrometer (§2.4) for the detection and quantification of characteristic X-rays (§9.4.2.5), and/or a post- or in-column magnetic sector electron detector (Fig. 9.4.8) for electron energy-loss spectroscopy (§9.4.2) in a TEM (Egerton, 1996; Reimer 1995). Such analytical electron microscopes (AEM) are very versatile and produce information (§9.4) on the physical, chemical, and magnetic microstructure of the specimen at the highest spatial resolution (Fig. 9.1.2).

In summary, a state-of-the-art AEM is a truly versatile instrument with a variety of beam-specimen interaction and signals, resulting in a wide range of techniques (Fig. 9.1.2). These include:

- conventional BF and DF imaging (§9.2.7)
- selected-area (§8.6.1) diffraction, convergent beam electron diffraction (§8.6.3), and microdiffraction (§9.4.4)
- high-resolution phase contrast imaging (§9.2.7.3 and §9.3.5)
- HAADF and *Z*-contrast imaging (§9.3.4)
- Lorentz microscopy—Fresnel, Foucault, and differential phase contrast imaging (§9.3.6)
- electron holography (§9.3.7)
- energy dispersive X-ray spectrometry (§9.4.3)
- electron energy-loss spectroscopy and energy-filtered imaging (§9.4.2)
- electron tomography (§9.5.1).

From this brief introduction, it may be obvious to the reader that a TEM is a sophisticated and versatile instrument capable of analyzing all aspects of the specimen microstructure. In this chapter, our goal is to present a comprehensive

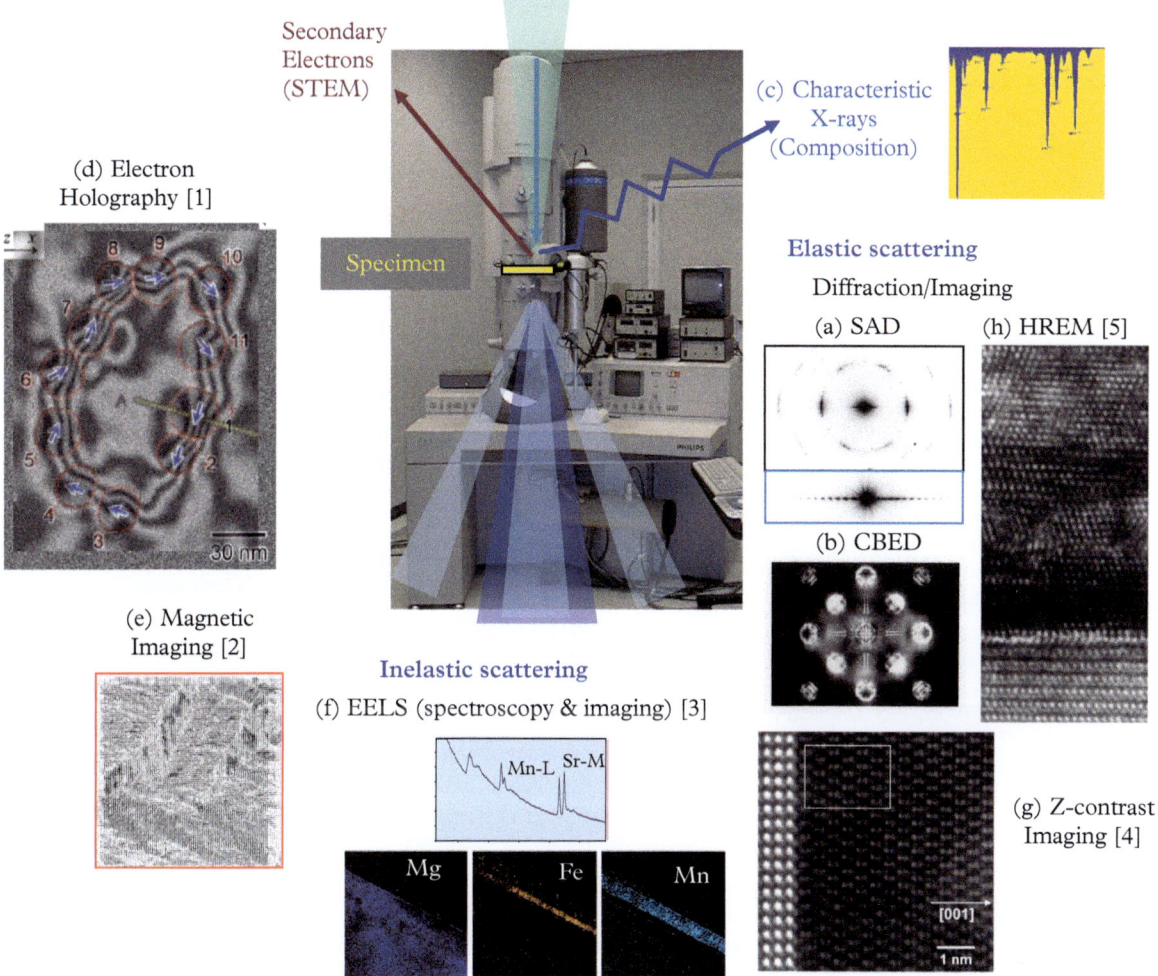

Figure 9.1.2 Transmission and analytical electron microscopy illustrating principal beam-specimen interactions, associated signals, and related techniques that include (a) selected area and (b) convergent beam electron diffraction, (c) energy dispersive X-ray spectrometry, (d) electron holography, (e) magnetic domain imaging, (f) electron energy-loss spectroscopy and energy-filtered imaging, (g) *Z*-contrast imaging, and (h) high-resolution electron microscopy. All these results are from the author's laboratory, some unpublished, and the others, as indicated, are adapted from [1–5]. See also Figure 5.3.5 and Figure 10.2.6.

introduction to the instrument, principal techniques implemented in it, and the rich palette of possibilities to characterize materials microstructures—structural, chemical, electronic, and magnetic—at the highest spatial resolution. The rest of the chapter is divided into five sections. The first, §9.2, is focused on the instrument, its principal components, and its different modes of operation.

The second, §9.3, describes beam-specimen interactions, relevant contrast mechanisms, and imaging methods that arise from mainly elastic interactions. The third, §9.4, deals with inelastic beam-specimen interactions, related spectroscopy methods, and broadly, analytical electron microscopy as implemented in transmission or scanning-transmission instruments. The chapter concludes with select applications of TEM (§9.5) and a brief section on specimen preparation methods (§9.6).

For more detailed and extensive discussion of specific topics, methods, and applications, as well as how they are implemented in a TEM, the reader should consult many of the publications and dedicated texts cited at the end of the chapter, including de Graef (2003), Egerton (1996), Fultz and Howe (2013), Goldstein and Joy (1979), Hirsch et al. (1965), Hren, Joy, Romig, and Goldstein (1986), Krishnan (2016), Reimer (1993, 1995), Shindo and Hiraga (1998), Spence (1981), Thomas and Goringe (1980), Tonomura (1999), and Williams and Carter (1996).

9.2 Elements and Operations of a Transmission Electron Microscope

We now describe the four principal sections of a TEM instrument: electron source, illumination, imaging, and magnification. This is followed by a description of the different diffraction and imaging modes emphasizing the principal contrast mechanisms. All these discussions involve simple geometric optics introduced in §6.8.2. We conclude the section with an introduction to scanning transmission electron microscopy (STEM), where we use focused probe illumination and raster the probe to form an image. The important principle of reciprocity, demonstrating the equivalence of TEM and STEM imaging, and which is often invoked in image interpretation, is then introduced. Finally, aberrations commonly observed in electromagnetic lenses and strategies for their correction are briefly outlined.

9.2.1 Electron Sources: Thermionic, Field, and Schottky Emission

Electron guns or sources are broadly of two types (Table 5.2.2): *thermionic emission*, which produces electrons of sufficient kinetic energy when heated to overcome the work function, Φ, and escape the solid tip, and *field emission*, which generates electrons by modifying the work function by applying an intense electric field to the tip such that the electrons can tunnel through (§11.2) from the solid to vacuum (Fig. 5.2.4b). The emission current is established by keeping the cathode at a negative potential, U_c, and the potential difference between the cathode and anode, U_0, draws the emitted electrons and accelerates them through the required acceleration voltage (see Fig. 5.2.5).

Thermionic guns, typically using W or LaB_6 filaments, have a larger cross-over diameter ($d_{co} \sim 10$–$40\ \mu m$) that limits the smallest probe that can be formed, generate high currents ($I \sim 5$–$100\ \mu A$), but low current densities ($j \sim 1 \times 10^4$–$2.5 \times 10^5\ A/m^2$) when focused, and a higher energy spread ($\Delta E \sim 1$–$2\ eV$), which affects the energy resolution in electron energy-loss spectroscopy (§9.4.2). Cold field-emission gun (FEG) sources have a smaller cross-over diameter ($d_{co} \sim 5$–$10 \times 10^{-3}\ \mu m$) and thus a higher current density ($j \sim 10^8 - 10^{10}\ A/m^2$), in spite of their significantly lower total current ($I \sim 1$–$10 \times 10^{-3}\ \mu A$). Typically, a cold FEG delivers a current density of a nA per nm^2 on the specimen; it also produces electrons with a narrower energy width ($\Delta E \sim 0.1$–$0.5\ eV$). The performance of the electron gun is characterized by its brightness, β, defined in units of current density per unit solid angle, which depends on the acceleration potential, U_0. Typically, at $U_0 = 100\ kV$, the FEG source is much brighter ($\beta \sim 2 \times 10^{12} - 10^{13}\ Am^{-2}sr^{-1}$), compared to a thermionic source with tips made of either W ($\beta \sim 1$–$5 \times 10^9\ Am^{-2}sr^{-1}$) or LaB_6 ($\beta \sim 2 \times 10^{10}\ Am^{-2}sr^{-1}$).

Thermal field emission guns, also known as Schottky emission guns or sources (SEG) that are increasingly common in electron microscopes, have characteristics in between thermionic and FEG sources. Typically, a FEG has higher brightness, a better energy resolution, and a lower total current, compared to a SEG. Thus, a FEG is preferable for highest resolution STEM imaging (§9.2.8 and §9.3.4) and energy dispersive X-ray microanalysis (§9.4.3), as well as fine structure analysis in electron energy-loss spectroscopy (EELS; §9.4.2). Moreover, in state-of-the-art electron microscopes where *spherical aberrations* are corrected (§9.2.9), the smaller energy spread minimizes the *chromatic aberration* and gives rise to superior spatial resolution. However, the total current of a FEG ($I \sim 10\ nA$) is much smaller than that of a SEG ($I \sim 100$–$200\ nA$). Thus, using a FEG it is difficult to illuminate large areas for magnifications less than $75,000\times$; further, when using a FEG with large area CCD cameras ($4\ k \times 4\ k$ pixels) for TEM imaging, long exposures are required for good statistics and to avoid shot noise arising from the discrete nature of the electron charge in the image. Finally, stability of the emission of a FEG requires "flashing" of the tip every few hours depending on the manufacturer, thus limiting its use for long exposure work, such as the long sequence of imaging required for electron tomography (§9.5.1), and which is typically carried out overnight.

Field emission, thermionic, and Schottky electron sources/guns have been discussed in detail in §5.2.2 and salient features useful for their selection are summarized in Table 5.2.2. As for deciding between a FEG and a SEG it often comes down to the specific use of the instrument, but as a simple rule of thumb, for imaging and best energy resolution a FEG is best, but if you were to do microanalytical work with good statistics (§9.4.2.5 and §9.4.3), a SEG may be better.

9.2.2 Electromagnetic Lenses

Electrons can be brought to a focus with both electrical and magnetic fields. The former, which forms the basis of electrostatic lenses, ends up having large aberration coefficients. Hence, they are now of diminished importance as lenses in electron microscopy, but are still used for electron beam deflection and in building spectrometers, often in conjunction with magnetic fields (e.g. Wien filter, §5.2.4.1). The latter, electromagnetic lenses, involve the action of magnetic fields with *axial symmetry* on the trajectory of fast electrons accelerated through kilovolt potentials. These lenses (Fig. 9.2.1) utilize soft iron pole-pieces with negligible coercivity to avoid hysteresis effects in lens operation. Further, unlike optical lenses made of glass, their focal length is variable and controlled by changing the lens current. We now discuss how a simple magnetic lens in the form of a solenoid focuses an electron beam.

An electron traveling in a magnetic field experiences a Lorentz force

$$\mathbf{F} = -e\mathbf{v} \otimes \mathbf{B} \tag{5.2.10}$$

and is deflected at right angles to the plane containing the magnetic field, $\mathbf{B}$, and the original direction of motion with velocity, $\mathbf{v}$. As a result, it will follow a helical

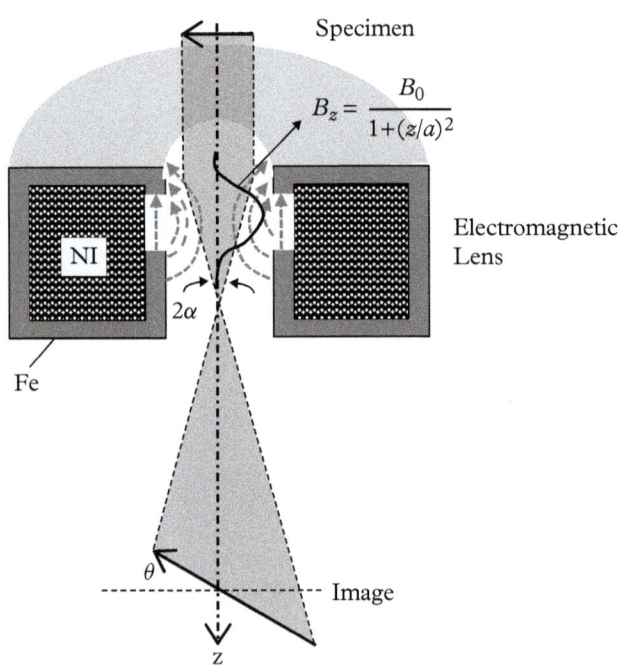

Figure 9.2.1 An electromagnetic lens with axially symmetric magnetic field. The component, B_z, has a maximum value, B_0, at the lens center and a full width at half maximum of $2a$. The image rotates by an angle, θ, depending on the particular lens design.

Adapted from Reimer (1993).

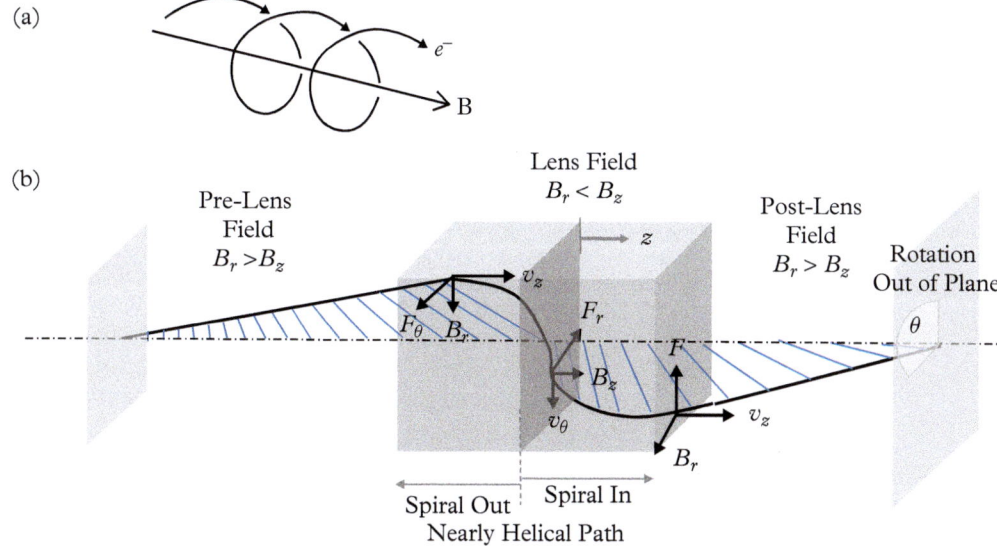

Figure 9.2.2 (a) The helical trajectory (exaggerated) of an electron in a magnetic field of induction, **B**, arising from the Lorentz force. (b) The electron trajectory through an electromagnetic lens causes both a rotation in the plane of travel by θ as well as a focusing of the electron along the optic axis at a position determined by the strength of the magnetic field.

Adapted from Fultz and Howe (2013).

path as shown in Figure 9.2.2a. It is best to write the equation of motion of the electron subject to the magnetic field of the lens in cylindrical polar coordinates. The electron velocity, $\mathbf{v} = v_r\hat{\mathbf{r}} + v_\theta\hat{\boldsymbol{\theta}} + v_z\hat{\mathbf{z}}$, in the lens field of induction, $\mathbf{B} = B_r\hat{\mathbf{r}} + B_\theta\hat{\boldsymbol{\theta}} + B_z\hat{\mathbf{z}}$, will experience a Lorentz field with components

$$\mathbf{F} = F_r\hat{\mathbf{r}} + F_\theta\hat{\boldsymbol{\theta}} + F_z\hat{\mathbf{z}} = -ev_\theta B_z\hat{\mathbf{r}} + -e\left(v_z B_r - B_z v_r\right)\hat{\boldsymbol{\theta}} + ev_\theta B_r\hat{\mathbf{z}} \qquad (9.2.1)$$

Note that for a lens with an axially symmetric field, $B_\theta = 0$, is assumed.

The electron trajectory is shown in Figure 9.2.2b. Initially, the electron travels in the plane of the paper with velocity, **v**, at an angle to the optic axis and experiences the *prefield* of the lens that is predominantly radial, i.e. $B_r \gg B_z$. As a result, it experiences a force, F_θ, pointing it out of the plane of the paper with a new velocity component, v_θ, causing it to have a spiral motion. Now, when the electron with both v_z and v_θ components enters the lens field with $B_z > B_r$, it now experiences a strong radial force, F_r, that focuses the electron beam. At the exact center ($z = 0$) of the symmetrical lens, $v_r = 0$, v_θ is a maximum, and the electron travels in a helical path with only v_θ and v_r components. As it traverses the exit half of the magnetic lens, because of a finite v_θ component, the electron is now further focused; in addition, B_r changes sign and v_θ starts to decrease. By symmetry, by the time the electron exits the post-field of the lens, $v_\theta = 0$, its spiral motion stops, and the electron comes to a focus along the optic axis. The field component, B_z, generated in the lens is directly proportional to the lens excitation current and hence the focal length decreases (v_θ increases) as the lens current is increased. Note that unlike optical lenses the path of the electron beam is rotated out of the plane of the paper through an angle, θ, in radians given by

$$\theta = \frac{0.15}{\sqrt{E}} \int\limits_{Axis} B_z dz \tag{9.2.2}$$

where E is the electron energy (eV). For 100 keV electrons passing through a field of $1~T\mu_0^{-1}$, over a distance of 0.5 cm in the lens, we get $\theta = \frac{0.15}{3.16\times10^2} \times 10^4 \times 0.5 = 2.4$ rad. As a result, either the *focusing of an image or a change in magnification*, both of which involve changing the electromagnetic lens field, is accompanied by a rotation of the image about the optic axis. In addition, the angle subtended by the electron beam on the optic axis in an electron microscope is very small, i.e. of the order of $1°$. Hence, compared to its angular spread, the path length of the electron beam is rather large and results in a very small effective numerical aperture, i.e. $n \sin \alpha \sim 10^{-2}$.

Earlier, in §6.8.2, we described the behavior of an optical lens in terms of geometric optics, with the assumption that the lens is sufficiently thin compared to the total optical path. We presented the simple relationship (6.8.3) between the focal length, and the position of the object and image. For electromagnetic lenses, this *thin-lens approximation* is strictly not valid; however, it is common practice to use geometric optics to illustrate the various imaging and diffraction modes of an electron microscope, and we shall also do the same in the sections that follow, knowing fully well that it is not strictly correct.

The diffraction limited resolution, δ_D(nm), of an electromagnetic lens can be obtained to first order using the Rayleigh criterion, (6.8.2), with a small ($\alpha \sim 0.57° \sim 10$ mrad) angular beam and the refractive index, $n = 1$, for vacuum, as

$$\delta_D = \frac{0.61\lambda}{n \sin \alpha} \approx \frac{0.61\lambda}{\alpha} \approx 60\lambda \tag{9.2.3a}$$

where we have set $\sin\alpha = \alpha$, for small α. We can write the wavelength, λ (nm) in terms of the electron energy, E (keV), corrected for relativistic effects as

$$\lambda = \frac{0.03878}{\sqrt{E\left(1 + 0.9785 \times 10^{-3}E\right)}} \tag{9.2.3b}$$

to give

$$\delta_D = \frac{0.61\lambda}{\alpha} = \frac{0.61}{\alpha} \frac{0.03878}{\sqrt{E\left(1 + 0.9785 \times 10^{-3}E\right)}} \tag{9.2.3c}$$

This suggests a better resolution is to be expected at higher voltages. As a result, many *high-voltage* electron microscopes (HVEMs) with acceleration voltage up to 3MV have been constructed, but most of them have found limited use because of significant *beam damage* of the specimen (Table 5.3.2; Fig. 5.3.8). Instead, to achieve ultimate resolution, *aberration corrected microscopes*, §9.2.9, operating at 60–300 kV, are currently used. An example of the design and development of such an aberration-corrected TEM is discussed in [6].

Similar to an optical lens (Fig. 6.8.3), an electromagnetic lens also shows both *spherical and chromatic aberration*. Electron beams originating at the same point but incident on the lens at different distances from the optic axis are brought to focus at different points on the optic axis (spherical aberration). As a result, the image of a point on the optic axis appears as a *disc of least confusion* on the Gaussian image plane. For spherical aberration this disc of least confusion limits the aberration-limited resolution, δ_S, which is related to the beam divergence, α, approximately as

$$\delta_S = 0.5 C_S \alpha^3 \tag{9.2.4}$$

where C_S is the *spherical aberration coefficient*, and a critical parameter that defines the imaging performance of the electromagnetic lens. We can see that for any lens with a given C_S, best performance is achieved when $\delta_D = \delta_S$, and observed at an optimal divergence angle, α_{opt}, obtained by setting (9.2.3a) equal to (9.2.4) and given by

$$\alpha_{opt} = \left(\frac{0.61\lambda}{C_S}\right)^{1/4} \sim 0.9 \left(\frac{\lambda}{C_S}\right)^{1/4} \tag{9.2.5}$$

The angle, α_{opt}, defined in practice by an aperture is determined by both the acceleration voltage (λ) and the lens characteristics (C_S).

Example 9.2.1: Calculate the diffraction limited resolution and the optimum divergence angle for an electromagnetic lens with $C_S = 0.6$ mm, in a TEM operated at 100 kV.

Solution: From (9.2.5), we get the optimal divergence angle

$$\alpha_{opt} = \left(\frac{0.6 \times 0.0037 \times 10^{-9}}{0.6 \times 10^{-3}}\right)^{1/4} = 8 \times 10^{-3} \text{ rad.}$$

From (9.2.3c), we get the diffraction limited resolution

$$\delta_D = \frac{0.61}{8 \times 10^{-3}} \frac{0.03877}{\sqrt{100}(1 + 0.9788 \times 10^{-3} \times 100)^{1/2}} = 0.28 \text{ nm.}$$

The electromagnetic lens also suffers from *chromatic aberration*, which brings electrons originating from the same point but of different energies to focus at different points on the optic axis. This is because higher energy electrons are less deflected by the magnetic field of the electromagnetic lens (Fig. 9.2.2). Again, the chromatic aberration is also defined by a disc of least confusion at the Gaussian image plane that limits the resolution, δ_C, to

$$\delta_C = C_C \frac{\Delta E}{E_0} \alpha \qquad (9.2.6)$$

where ΔE accounts for instabilities in both the source (gun) and lens currents, and the chromatic aberration constant, C_C, is of the order of the focal length, f, for weak lenses, and $C_C \sim 0.6f$, for strong lenses. For typical electron microscopes with a thermionic source and $E_0 = 100$ keV, $\Delta E \sim 1.5$ eV, $C_C = f = 2$mm, $\alpha \sim 10$ mrad, we get $\delta_C \sim 0.3$ nm. The spherical aberration of an electromagnetic lens can be corrected using pairs of magnetic hexapoles, and is discussed further in §9.2.9.

In addition, a cone of electron rays from a point, P, on the specimen at some distance, y, from the optic axis is generally focused as an ellipse at the point, P′, in the Gaussian image plane (GIP) (Fig. 9.2.3). This is because of the loss of axial symmetry in the lens resulting in a variation in the focal length for the two orthogonal axes, x and y, shown in the figure. This is called *astigmatism* (Fig. 9.3.20) and is an inherent feature in electron-optical lenses due to a residual asymmetry, or inhomogeneity, of the magnetization of the pole piece. Moreover, astigmatism is very sensitive to specimen effects including misalignment, asymmetry, and contamination. However, such twofold astigmatism is routinely corrected in a TEM by introducing additional sets of correction coils with variable magnetic fields, and placed orthogonal to each other about the optic axis. The current in the coils can be periodically adjusted during operation of the TEM to correct for astigmatism in the image (Fig. 9.3.20).

Finally, the depth of field, d_f, of an electromagnetic lens, is the same as that defined earlier for an optical lens (Fig. 6.8.4). It is the axial distance over which the specimen can be moved and yet produce a focused image with resolution, δ_S, and is given by

Figure 9.2.3 A simple ray diagram illustrating astigmatism in a lens arising from a loss of axial asymmetry. As a result, the focal length varies about the optic axis with two principal lines of foci, along orthogonal directions, as shown.

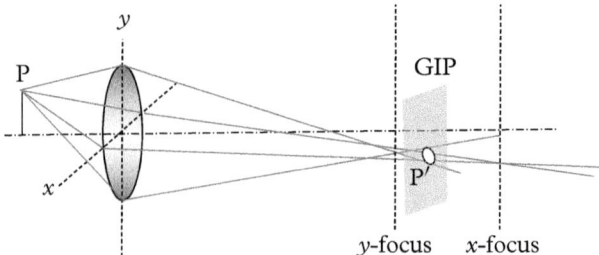

$$d_f = \delta_S/\alpha \qquad\qquad (9.2.7)$$

The depth of focus, D, is the range of distances over which the image appears in focus in the image plane of the lens (Fig. 6.8.4). For a convergence semi-angle, $\alpha \sim 10$ mrad, and magnification, M, such that the effective aperture $\alpha' = \alpha/M$ in the final image, the depth of focus is related to the depth of field:

$$D = \frac{\delta_S M}{\alpha'} = \frac{\delta_S M^2}{\alpha} = M^2 d_f \qquad\qquad (9.2.8)$$

which is very large (see Example 9.2.2). Now that we have a general idea of the characteristics of an electromagnetic lens we can move on and describe the different sections of a transmission electron microscope, starting with the illumination section.

Example 9.2.2: Is a typical TEM specimen with thickness less than 100 nm appropriate for imaging? Why? In addition, in a TEM the viewing screen and the image recording system are separated by 20 cm. For a magnification of 50,000×, can we both observe and record the image? Assume typical parameters for the TEM with a thermionic source.

Solution: For a typical TEM with $\alpha \sim 10$ mrad, $\delta_S \sim 0.3$ nm, we get $d_f = 0.3 \times 10^{-9}/10^{-2} = 0.3 \times 10^{-7}$ m $= 30$ nm.

The depth of focus is smaller than the specimen thickness and so it would be difficult to image it through its entire thickness. However, if $\alpha \sim 1$ mrad, $d_f = 300$ nm, and it should be possible to image it easily. The depth of focus, (9.2.8), for $\alpha \sim 10$ mrad, gives $D = M^2 d_f = \left(5 \times 10^4\right)^2 \times 30 \times 10^{-9} = 75$ m!

Clearly, the image will be in focus in both the viewing screen and the image recording system.

9.2.3 The Illumination Section

The illumination section (Fig. 9.1.1) transfers the electron cross-over produced by the gun (Fig. 5.2.5) with diameter, d_{co}, to the specimen. The cross-over serves as the object for the illumination system that consists of two or more CLs (called C1, C2, ...). Two illumination conditions—a broad parallel beam and a convergent beam focused on the specimen—commonly encountered in TEM are discussed next. Note that the beam is most coherent when the incident beam is parallel.

9.2.3.1 Parallel Beam Operation of a TEM

To illuminate the specimen with a parallel beam of electrons, uniformly over a region of several microns, and magnifications ranging from 20,000× to 100,000×, a system of two CLs (Fig. 9.2.4) is used. For a thermionic source, with a large

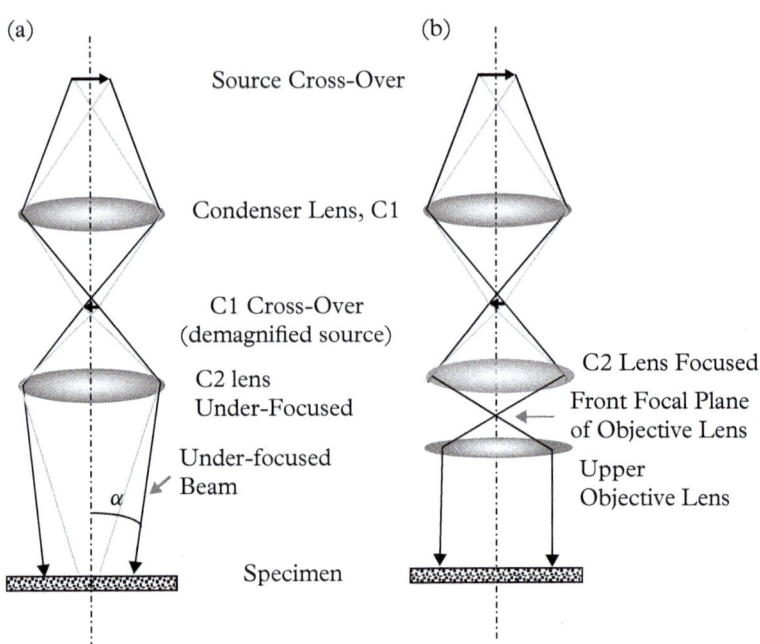

Figure 9.2.4 Ray diagram illustrating two ways of producing "parallel" illumination on the specimen using (a) an under-focused C2 lens, and (b) a focused C2 lens to produce an image of the source cross-over at the front focal plane of the upper objective lens. The corresponding diffraction patterns produced in the back focal plane of the objective lens (Fig. 8.6.9 and Fig. 8.6.10) show sharp diffraction discs (very small diameter that depends on α) for (a), and very narrow spots for (b).

cross-over diameter (Table 5.2.2) of several tens of micrometers, the first lens, C1, forms a *demagnified* image of the cross-over by an order of magnitude. On the other hand, for FEG sources where the cross-over diameter may be much smaller than desired for illumination, the lens C1 would be excited to magnify the cross-over. Then, in either case, the second condenser lens, C2, is excited to produce a substantially under-focused image of the C1 cross-over on the specimen, with a convergence semi-angle, $\alpha < 10^{-4}$ rads ($\sim 0.01°$; Fig. 9.2.4a), which for all practical purposes can be considered as a "parallel" beam. Such a parallel beam produces sharp diffraction patterns (very small discs) and the highest contrast in images. Note that typically, in this mode of imaging the C1 lens excitation remains unchanged (set at some value recommended by the manufacturer) and the C2 lens current determines the magnification (which is also related inversely to the area of the specimen illuminated by the beam). Alternatively, if a TEM is also used to generate a focused electron beam, e.g. in a STEM, additional control of the beam is made possible using the upper pole piece of the OL. In this arrangement, the C2 lens is focused to produce an image of the C1 cross-over at the front focal plane of the upper OL, which will then generate a true parallel beam on the specimen (Fig. 9.2.4b).

9.2.3.2   *Focused Probe Formation for Illumination and Scanning*

A focused probe (<5 nm in diameter) illumination is often required in a TEM to increase the local probe intensity and carry out analytical (spectroscopic) and diffraction measurements of the specimen with high spatial resolution. Because such a probe is incoherent and does not provide a useful image of the entire specimen, the beam has to be rastered or scanned over the specimen area; this type of operation/imaging is referred to as the STEM mode, which is discussed further in §9.2.8, along with the principle of reciprocity.

When a FEG source is available, the small gun cross-over diameter allows one to easily transfer the demagnified C1 cross-over image, as a focused probe onto the specimen surface with a focused C2 lens and aperture (Fig. 9.2.5a). However, if a thermionic source with a gun cross-over diameter of $\sim$10–20 µm is used, typical C1 and C2 lenses cannot completely demagnify the cross-over to the required probe size. In this case, the C2 lens is turned off and the upper pole piece of the OL is used as a third condenser lens (condenser-OL) to obtain a focused and convergent probe (Fig. 9.2.5b). Now, since the C2 lens is switched off, the convergence semi-angle, α, is proportional to the diameter of the C2 aperture.

Note that in practice, two sets of pre-specimen scan coils (Fig. 9.2.6) are included in the illumination section to translate (deflect) and tilt the beam. These double-deflection coils can be used to implement the DF mode (§9.2.7.2) of imaging by tilting the incident beam (Fig. 9.2.6b) such that only a specific Bragg diffracted electron beam is along the optic axis and enter the objective aperture to

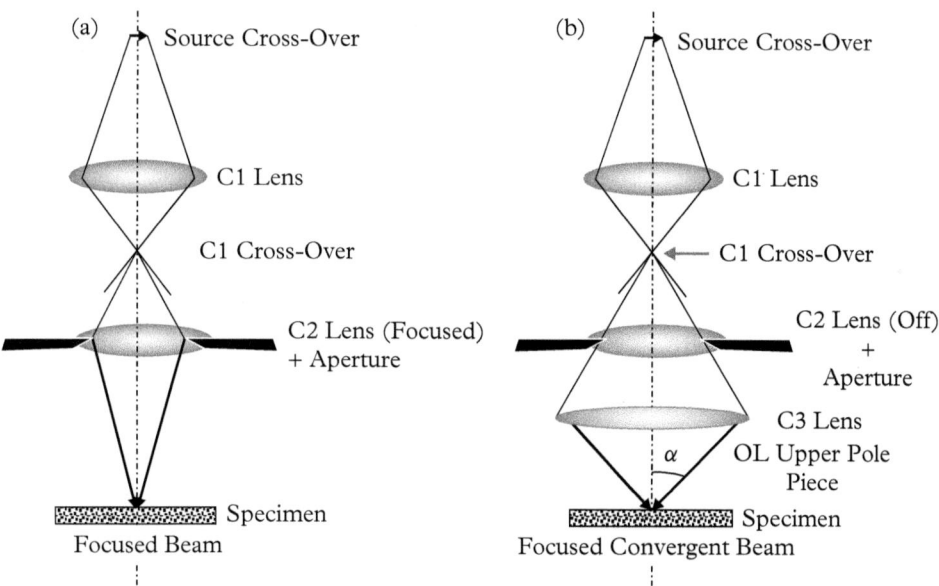

Figure 9.2.5 Ray diagram showing the illumination optics to obtain a focused probe on the specimen (a) using a focused C2 lens, particularly with a FEG source, and (b) for a thermionic source, turning C2 lens off and using the objective lens upper pole piece as a third condenser lens to obtain a highly convergent beam.

Figure 9.2.6 Two scan coils, upper (U) and lower (L), allow for (a) translation and (b) tilt of the electron beam. In addition, the convergent probe formed when the objective lens upper pole-piece is used as an additional condenser lens (C3), can be (c) scanned or (d) rocked by the same double-deflection coils.

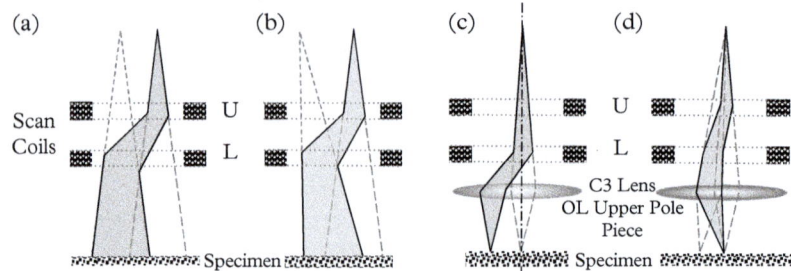

form the image. Alternatively, for best imaging in STEM the beam should always remain parallel to the optic axis as it is rastered over the specimen surface. The two scan coils accomplish this requirement (Fig. 9.2.6c); in addition, they help tilt the beam off axis such that it can be made to be incident on the specimen at any required angle (Fig. 9.2.6d), to meet the requirements of any specific diffraction condition (§9.2.7). Further practical details on the use of the scan coils can be found in Williams and Carter (1996).

9.2.4 The Imaging Section: Objective Lens and Aperture

The OL forms the first intermediate image of the specimen with a magnification of only 20–50×. However, for high-performance TEMs the OL should be of the highest quality and its astigmatism be so minimal that the stigmator is only required to correct the astigmatism from contamination of the aperture. Further, in TEM design and performance the spherical and chromatic aberration are most significant only for the OL. For all other lenses with high magnifications, M, the effective divergence angle, $\alpha = \alpha_0/M$. As demonstrated (9.2.4), the disc of least confusion for spherical aberration, $\delta_S \propto \alpha^3 \approx (\alpha_0/M)^3$, and hence it can be ignored for all other lenses used subsequently for higher magnification.

An objective aperture (OA) from the available set (typically 3–4 in number) can be inserted in the back focal plane of the OL. The radius, r_{OA}, of the aperture controls the divergence angle $\alpha_0 = r_{OA}/f_{OL}$, where f_{OL} is the focal length of the OL (Fig. 9.2.7a). The angle, α_0, prevents the passage of all electrons with an angular range, $\theta \geq \alpha_0$, through the aperture; as a result, the scattering contrast increases with decreasing aperture diameter. On the other hand, for high resolution phase contrast imaging (§9.2.7) the aperture size has to be increased significantly to accommodate the contribution of as much of the high spatial frequencies to the image as possible.

The integrity of the OA will affect the quality of the final image. It should be heat resistant and capable of tolerating the high current densities (~ 10 A/m^2) expected in the focal plane of the OL; generally, thin metal foils of Pt, or Pt–Ir alloy, with circularly punched holes are used.

(a) Electron Beam
From Condenser Lens

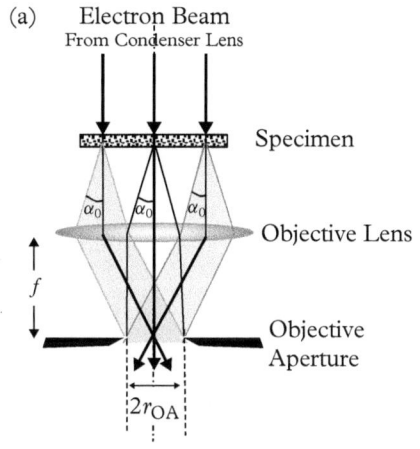

Specimen

Objective Lens

Objective
Aperture

$2r_{OA}$

f

α_0

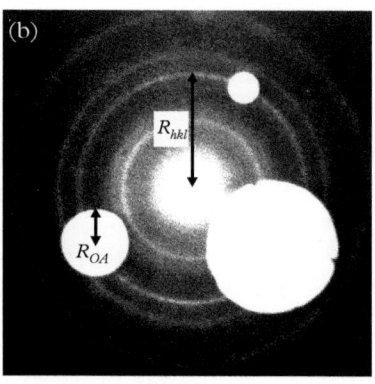

(b)

R_{hkl}

R_{OA}

The minimum imaging resolution, $\delta_{\min}$, of the OL is calculated by adding in quadrature the diameters of the discs of least confusion for diffraction, (9.2.3), spherical aberration, (9.2.4), and chromatic aberration, (9.2.6), at the optimal divergence angle, α_{opt}, calculated in (9.2.5). Thus

$$\delta_{\min}^2 = \delta_D^2 + \delta_S^2 + \delta_C^2 \qquad (9.2.9)$$

This is strictly valid only if all the broadening is of Gaussian shape. The effect of chromatic aberration is small and, to first order, can be neglected. Then using the expression for α_{opt}, (9.2.5), δ_D, (9.2.3a), and δ_S, (9.2.4), we can show that

$$\delta_{\min} \approx 0.75 C_S^{1/4} \lambda^{3/4} \qquad (9.2.10)$$

Notice that the best resolution is achieved for the smallest wavelength (the highest acceleration voltage), and the smallest spherical aberration coefficient.

Example 9.2.3: You are in the market to buy a TEM with the highest imaging resolution, and a limited budget. TEM1 operates at 100 kV with an OL $C_S = 1$ mm. TEM2 operates at 200 kV but to keep costs competitive, it is offered with the OL having a $C_S = 1.5$ mm. Which one would you buy?.

Solution: From (9.2.3b), the wavelengths for the two instruments are $\lambda_1 = 0.0037$ *nm*, $\lambda_2 = 0.00251$ nm.

We can use (9.2.10) to compare the two instruments. Then, $\delta_{\min}(1) = 0.5$ nm and $\delta_{\min}(2) = 0.29$ nm. Naturally, you will buy TEM2, which has a better resolution.

Example 9.2.4: Assume that Figure 9.2.7b shows the SAD pattern for polycrystalline gold, and the ring of interest corresponds to the (400) reflection. Calculate the angular divergence for the aperture shown.

Solution: The lattice parameter of gold, $a_{Au} = 4.078$ Å.
Thus $d_{400} = 1.019$Å. We assume 100 keV electron incidence; thus $\lambda = 0.037$ Å.
From Bragg's law, $\theta_{400} \sim \frac{\lambda}{2d_{400}} = 0.018$ rads. From Figure 9.2.7b, $R_{OA}/R_{400} = 0.28$.
Hence, $\alpha_{OA} = (R_{OA}/R_{400})\,\theta_{400} = 0.28 \times 0.018 = 0.0051 \approx 5$ mrads.

9.2.5   Specimen Handling and Manipulation

A TEM specimen is a disc or support grid, either 3.05 mm or 2.3 mm in diameter,[2] with some part of it suitably transparent to the high-energy electrons for imaging, diffraction, and analytical studies. It sits in a cup and is clamped in place using clamping rings, and this cup is part of a specimen holder that is inserted into the TEM stage. High mechanical stability of the stage is critical to avoid blurring of images at high magnifications due to specimen drift or vibrations during the time of image acquisition. In addition, high precision is required in both specimen translation at high magnifications and in orienting/tilting to achieve specific diffraction conditions (§8.5 and §8.6). The physical dimension and construction of the specimen holder is very important in a TEM as the specimen is located in the gap between the upper and lower pole pieces of the OL. Further, the objective aperture assembly must also be accommodated within the pole piece gap. Finally, to investigate crystalline materials by diffraction and imaging, it must be possible to tilt the specimen stage to ensure electron incidence at specific angles along well-defined crystallographic directions.

To meet these requirements, two different designs of specimen holders have been developed over time. The *side-entryholder*, in the form of a rod holding the specimen cup at one end with an electric motor attached for specimen tilt and rotation, is now quite standard (Figure 9.2.8). The range of tilt is determined by the size of the pole piece gap with the largest ones allowing ±60°. In addition, side-entry holders allow for specimen translation along the z-direction (optic axis) to ensure that the specimen position coincides with one tilt axis of the holder. As a result of this *eucentric* position, the specimen can be rotated about this axis without any perceptible lateral translation. It is easy to see that side-entry holders keep the specimen connected to the outside environment via the long rod, which may be an undesirable avenue for mechanical instability of the specimen. Recent innovations in design and development of computer-controlled specimen stages have come a long way in overcoming this limitation. Finally, side-entry holders result in an asymmetric bore of the OL and raise significant challenge in their design for ultimate resolution.

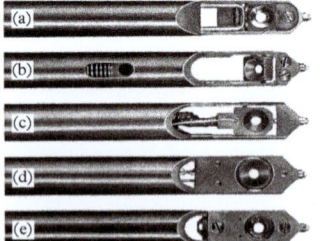

Figure 9.2.8 Examples of side-entry holders with (a) single-tilt, (b) double-tilt, (c) cooling, (d) heating, and (e) rotation features.

Adapted from Williams and Carter (1996).

[2] These sizes are purely historical and have no real scientific rationale.

The alternative, *top-entry holder*, is a cartridge that contains the specimen cup and is introduced from above through the bore of the OL pole piece. Such cartridges are particularly suitable for high-resolution electron microscopes (HREM), which are designed with an especially small gap to perform close to their resolution limit; hence, the range of tilt for such specimen holders is small and limited to ±25°.

In practice, side-entry holders are very versatile with many available designs (Fig. 9.2.8) that include (a) single tilt and tilt-rotation holders, (b) double-tilt holders with tilt axes fixed along two orthogonal directions, (c) low-background holders where the cup and clips that hold the specimen are made of Be, a low-atomic number element, which minimizes the generation of X-ray background signal for quantitative microanalysis (§9.4.3), (d) cooling holders for either liquid-N_2 or liquid-He specimen temperatures, and suitable for analytical TEM (§9.4), especially when combined with a double-tilt mechanism and a low-background Be cup, (e) heating holders that can go up to ~1,300 °C in conventional TEMs, and higher temperatures in HVEMs with larger pole-piece gaps, and (f) specialty holders (not shown) for mechanical straining, electron beam-induced current (EBIC), cathodoluminescence (CL; see §10.6.2 for a description of CL), and liquid studies.

Further details of specimen holders and stages, as well as many more practical details, are found in William and Carter (1996).

9.2.6 The Magnification Section

The first image formed by the OL is further magnified twice, first by a set of intermediate lenses (IL)—even though only one IL is shown in Figure 9.2.9—and then finally, by the projector lens (PL) as shown for the formation of the BF image in Figure 9.2.9. The excitation and the magnifications of the OL and the projector lens are generally constant, and the first intermediate lens typically has a magnification of 25×. The overall magnification of the image is a product of the magnifications of the three lenses (OL, ILs, and PL), and the variable magnifications of the other intermediate lenses; this gives a maximum magnification for the TEM of the order of $10^6 \times$. Moreover, the use of multiple intermediate lenses can help minimize, if not eliminate, image rotation, which is unavoidable in electron-optical lenses (Fig. 9.2.1). For low-magnification (~50–100×) operations, the OL is switched off, and the first IL is used as the OL to form the image. In addition, these set of intermediate lenses allow for the selection of different camera lengths, (8.6.1), in diffraction mode. Finally, the magnifications displayed by the instrument is often inaccurate; a calibration image obtained with a test specimen can help in establishing the magnification more accurately.

With multiple lenses used in imaging to ensure optimal performance of the TEM, it is imperative that the optic axis of all the lenses be aligned accurately. This is done systematically, starting by aligning the axis of the electron source with the overall illumination system, followed by the axis of the OL, and the

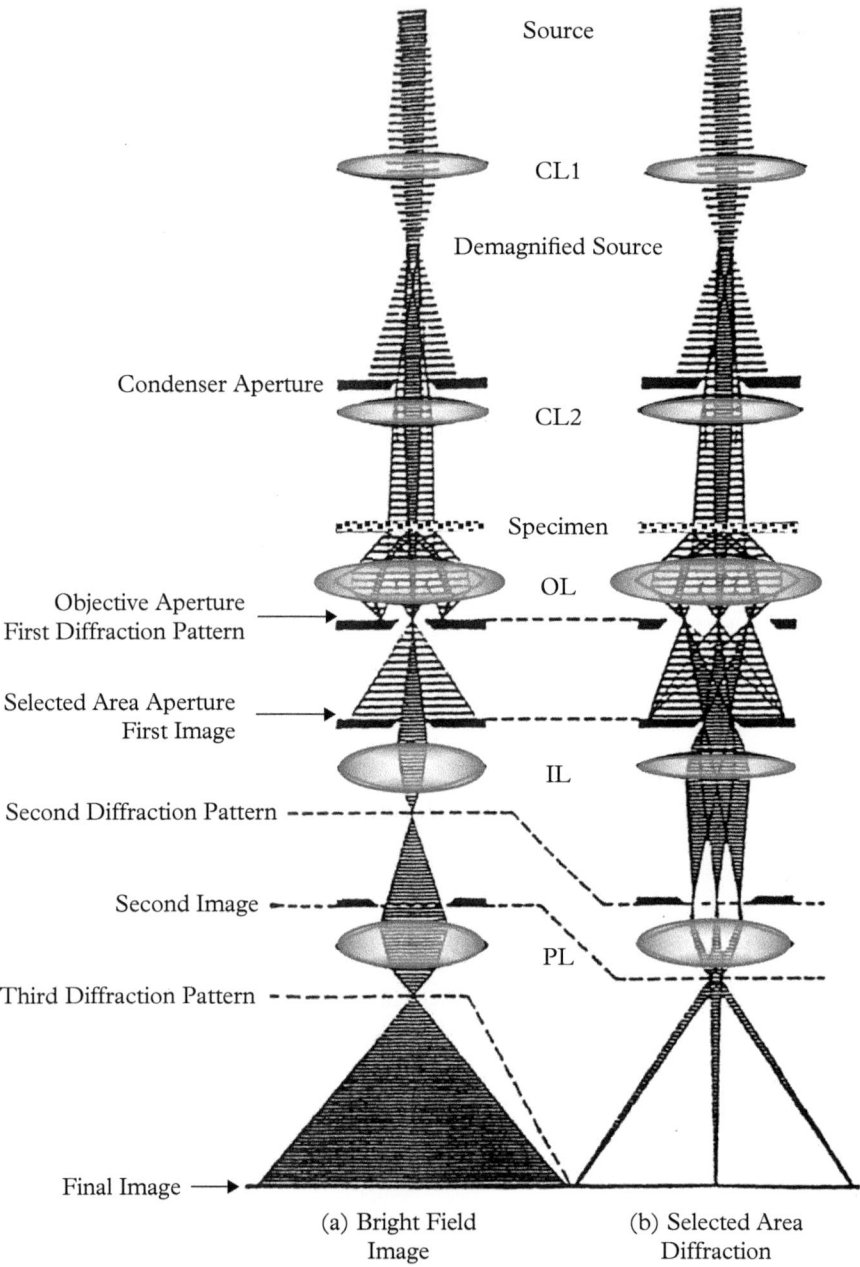

Figure 9.2.9 Ray diagrams illustrating the electron beam optics of a TEM for (a) bright field imaging, and (b) selected area diffraction.

CL, condenser lens; OL, objective lens; IL, intermediate lens; and PL, projector lens. Adapted from Reimer (1993).

magnification system. At every step the electron beam must be both tilted and shifted, using double deflection coils (similar to Figure 9.2.6), to ensure proper alignment. Further specifics on the alignment procedure can be obtained from the manufacturer of the particular TEM, and generic details are found in Williams and Carter (1996).

9.2.7 Imaging and Diffraction Modes

The basic operations of a TEM are two-fold and involve the projection of either an image or a diffraction pattern of the specimen onto the final viewing screen. This is mainly accomplished by the OL (Fig. 9.2.9), which takes the electrons emerging from the specimen and disperses them to create a diffraction pattern (Fig. 9.2.9b) in its back focal plane or an image (Fig. 9.2.9a) in the first image plane. The physical principles and methods of transmission electron diffraction are discussed in §8.5 and §8.6, respectively. Here, we concentrate on the role of the instrument in carrying out diffraction and imaging experiments.

9.2.7.1 Selected Area Diffraction (SAD) and Bright Field (BF) Imaging

To form an image (see ray diagram, Fig. 9.2.9a), the intermediate lens is adjusted such that the first image plane of the OL is the object plane for the intermediate lens; the projector lens that follows down the optical column then projects the image from the intermediate lens onto the final viewing screen. However, if the intermediate lens is adjusted such that the back focal plane of the OL serves as its object plane, a diffraction pattern (Fig. 9.2.9b) with contributions from all the electrons from the specimen is finally observed. The direct or forward-scattered beam of such a diffraction pattern often exhibits very high intensity; to reduce this intensity, and be able to restrict the diffraction pattern to a small area of the specimen, we insert a selected-area aperture in the first image plane of the OL. Alternatively, a beam blocker may also be used to protect the detector from the central beam. The resulting selected area diffraction (darker hatching in Figure 9.2.9b) overcomes these two difficulties, and also confines the diffraction pattern to the selected area. In practice, apertures cannot be fabricated with diameters less than 10.0 μm; thus, for a typical OL magnification of 20×, the minimum selectable area is only ~0.5 μm in diameter. In diffraction, to select areas smaller than that allowed by such selected area diffraction, a focused probe optics (Fig. 9.2.5) is used. This is referred to as microdiffraction and discussed further in §9.4.4.

Once the SAD pattern is observed on the viewing screen, images of the specimen can be formed from it in two different ways using either the bright, forward scattered beam (transmitted spot) or one (or many) of the beams diffracted by the specimen. The specific beam for forming the image is selected by inserting the objective aperture in the back focal plane of the OL (Fig. 9.2.9a and Fig. 9.2.10a) and then removing the selected area aperture. In BF imaging, the objective aperture *centered* on the optic axis, selects the forward-scattered or transmitted beam to form the image. Further, the size of the objective aperture determines the divergence semi-angle, α, of the beam (Fig. 9.2.7), which in turn controls the effect of the spherical, (9.2.4), and chromatic, (9.2.6) aberration on the resolution of the OL. Last, but not least, in this BF mode of imaging, the objective aperture provides mass-thickness (§9.3.2) or diffraction (§9.3.3) contrast, by restricting the electrons scattered to semi-angles, $\theta < \alpha$, and eliminating the background signal from electrons scattered at larger angles.

Alternatively, by allowing either one or more of the diffracted beams to pass through the objective aperture, images can also be formed. These are known as *dark field* (DF) and *phase contrast* imaging modes, respectively, and are discussed later.

Example 9.2.5: For the three-lens imaging microscope (Fig. 9.2.9) are the images and diffraction patterns inverted with respect to the image? How is this relationship dependent on the number, n, of lenses following the OL?.

Solution: Figure 9.2.9 also shows the ray paths for (a) BF imaging and (b) selected area diffraction. Each lens introduces an inversion and thus a three-lens system produces an inverted image with respect to the specimen. In contrast the diffraction pattern is not inverted by the OL, and so it is inverted with respect to the image (but not with respect to the specimen). However, if the number of lenses involved in forming the image and diffraction are different, this relationship will not hold.

In general, for n lenses *following the OL*, the final image will be inverted with respect to the specimen if n is even, while the diffraction pattern will be inverted with respect to the image if n is odd.

9.2.7.2 *Dark Field (DF) Imaging*

In the DF imaging, the objective aperture blocks the primary/transmitted beam, and the image is formed with the diffracted beam in one of two distinct ways. The objective aperture can be displaced (Fig. 9.2.10b) to select a specific diffracted beam to form the image. In this case, the electrons contributing to the image pass through the OL along trajectories that are off the optic axis, leading to significant aberrations and astigmatism, as well as difficulty in focusing the image. To overcome these problems, and avoid off-axis aberrations, the initial beam is tilted such that the diffracted beam contributing to the image propagates along the optic axis. This preferred mode is known as *centered* DF imaging (Fig. 9.2.10c) Finally, increasing the objective aperture size allows multiple Bragg diffracted beams to interfere and contribute to the final image (Fig. 9.2.10d), producing a fringe pattern representative of the crystal lattice planes. The underlying physics of such phase contrast is introduced in the next section.

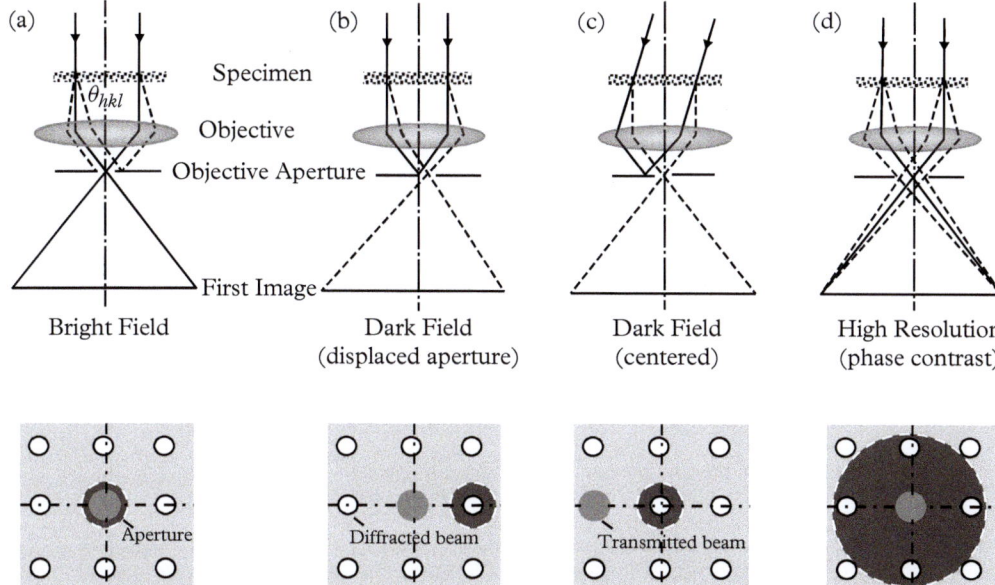

Figure 9.2.10 Ray optics (top) and focal plane of the OL where the OA is sited (bottom) for (a) bright field, (b) dark field, (c) centered or tilted dark field, and (d) phase contrast imaging modes. In each case, the positions of the primary/transmitted beam, the objective aperture, and the diffracted beams contributing to the image, are also indicated.

Adapted from Williams and Carter (1996).

Figure 9.2.11 compares BF, DF, and phase contrast images of a grain boundary in silicon nitride. The corresponding SAD patterns with the position of the objective aperture used in forming the image are also shown. Notice the fine resolution of the grain boundary in the DF image (Fig. 9.2.11b) and a quantitative measurement of its width ($\sim$12 Å) based on the lattice fringe periodicity of the parent phase on either side of the grain boundary (Fig. 9.2.11c).

9.2.7.3 *Phase Contrast Imaging and the Contrast Transfer Function*

Recall that a light beam traveling in vacuum and interacting with a transparent object (§6.5) of refractive index different from unity suffers a phase change relative to the wave in vacuum. Analogously, incidence of high energy electrons on regions of different electrostatic potential in a specimen, results in changes in their velocity and wavelength. If such a phase change is uniform, the direction of propagation of the wave does not change, but if there are regions of lateral variation in phase change, the electron is scattered from its original direction of propagation. As we have seen, such scattering angles are small ($\sim$10 mrad) for electrons and the electron barely undergoes a sideways movement (typically of the order of 1 Å) when it goes through a thin ($\sim$10 nm thickness) specimen as in

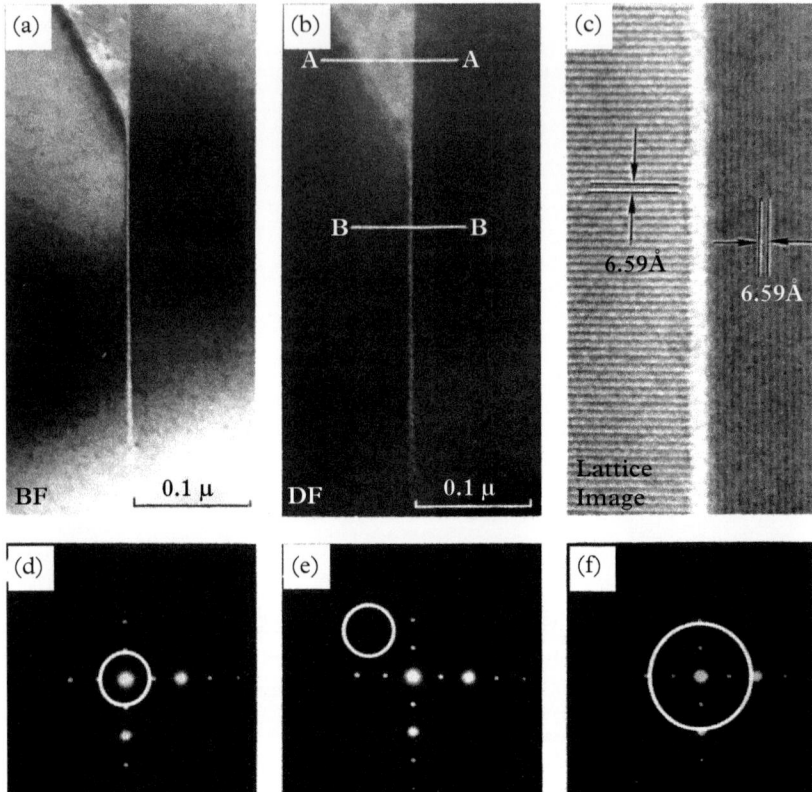

Figure 9.2.11 A grain boundary phase in silicon nitride shown in (a) bright field, (b) dark field, and (c) phase contrast modes of imaging. The corresponding SAD patterns and the positions of the OA forming the image are shown in (d), (e), and (f), respectively.

a TEM. To first order, for thin foils the phase change depends on the distribution of electrostatic potential down a straight path through the object. Thus, if the potential distribution of the specimen is given by the function, $\Phi(x, y, z)$, an electron traveling along the z-direction will undergo a phase change proportional to the projection of that potential on the x-y plane given by

$$\phi(x, y) = \int \Phi(x, y, z)\, dz \qquad (9.2.11)$$

The phase difference between a wave traveling in vacuum, and a wave traveling through the medium is given by the product of $\phi(x,y)$ and an interaction constant, σ, which determines the strength of the interaction. We can show that $\sigma = \pi/(\lambda E)$, where λ is the wavelength and E is the energy of the electron. Then, the effect of the phase change on the incident wave of amplitude, $A(x,y)$, results in an amplitude at the exit surface, $f(x,y)$, of $A(x,y)$ multiplied by a specimen transmission function, $T(x,y)$, given by

$$f(x,y) = A(x,y)\,T(x,y) = A(x,y)\exp\left[-i\sigma\phi(x,y)\right] \qquad (9.2.12)$$

This is known as the *phase object approximation* (POA). To reiterate, this approximation ignores the sideways scattering of the wave, as well as any inelastic scattering that may lead to absorption, and is only valid for thin specimens. This model can be further simplified for *very thin specimens*, by expanding the exponential as

$$f(x,y) = 1 - i\sigma\phi(x,y) \qquad (9.2.13)$$

where for simplicity, we have set $A(x,y) = 1$. The constant 1 (first term) now represents the direct beam unaffected by the object and produces a sharp peak in the back focal plane, and $i\sigma\phi(x,y)$ is the scattering function that describes the distribution of scattered amplitudes, and i indicates a phase change of $\pi/2$. In this case, we have what is referred to as the *weak phase object* (WPO) approximation. A pure phase object (9.2.12), when focused perfectly in an ideal microscope will generate no contrast because the image intensity $I = |f(x,y)|^2 = A^2(x,y) = 1$. The key idea in phase contrast imaging is to convert the phase contrast to an amplitude contrast by adding a phase of $\pm\pi/2$ to the scattered wave by taking advantage of the lens aberration and adjusting the defocus of the OL. When this is done carefully, such that only scattered or diffracted beams with a phase difference of $\pm\pi/2$ contribute to the image, the observed intensity distribution, $I(x,y)$, is given by:

$$I(x,y) = 1 + 2\sigma\phi(x,y) \qquad (9.2.14)$$

Now consider a WPO specimen in a TEM (Fig. 9.2.12) with the simplest case of only two beams going through: a forward scattered beam/wave which is negligibly attenuated with respect to the incident beam, and a Bragg scattered beam/wave, with very low intensity that has been scattered by the atoms and as a result has undergone a phase change. At the exit surface these two waves interfere, and the sum-wave will differ in amplitude from the forward scattered wave that provides the background intensity. Now, there are two extreme scenarios: a best case where the Bragg scattered wave is perfectly in phase with the forward scattered wave, and their amplitudes add yielding an image where the atomic columns appear bright (negative contrast) against a dark background. Alternatively, we can expect a scenario where the two waves are perfectly out of phase, their amplitudes subtract, and the atomic columns appear dark (positive contrast) against a bright background in the image. This is the simple physical principle behind phase contrast imaging but to carry out this imaging in practice, the operation of the microscope needs to be controlled to obtain appropriate contrast conditions in the image.

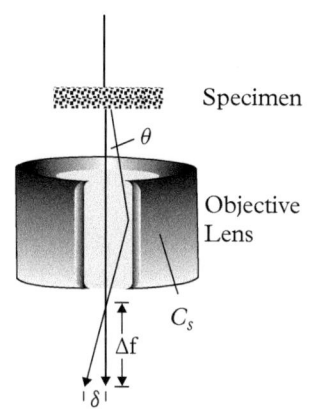

Figure 9.2.12 Ray diagram of a schematic OL, showing the displacement, δ, of a beam scattered through an angle, θ, due to both spherical aberration and the defocus of the lens.

Adapted from Williams, Pelton, and Gronsky (1991).

As mentioned, in addition to the specimen, the optics of the electron microscope, mainly the OL, introduces additional phase shifts to the beams that form the image. This is a direct consequence of the path difference between the off-axis, Bragg diffracted beam, and the forward-scattered, transmitted beam that travels down the optic axis. For a beam traveling at any general scattering angle, α, the spherical aberration, C_S, of the OL, causes a lateral displacement of the focal point by $C_S\alpha^3$, (9.2.4). Similarly, if the lens is defocused by Δf, it will also cause a displacement of $\alpha\,\Delta f$. The total displacement, $\delta(\alpha)$, is therefore the sum of these two terms:

$$\delta(\alpha) = C_S\alpha^3 + \alpha\,\Delta f \tag{9.2.15}$$

Recall that for a specific Bragg scattering angle, θ_{hkl}, the total scattering angle of the corresponding diffracted beam in the TEM is $\theta = 2\theta_{hkl}$ (Fig. 8.6.1). Hence, the total displacement, can be integrated over all angles, up to θ, to give the total path length difference, $D(\theta)$, between the forward scattered and any off-axis Bragg diffracted beam:

$$D(\theta) = \int_0^\theta \delta(\alpha)\,d\alpha = C_S\frac{\theta^4}{4} + \Delta f\frac{\theta^2}{2} \tag{9.2.16}$$

Now, for small Bragg scattering angles, $\lambda = 2d_{hkl}\sin\theta_{hkl} = 2d_{hkl}\theta_{hkl} = d_{hkl}\theta = \theta/g_{hkl}$. Hence, we can rewrite (9.2.5), dropping the subscripts, in terms of the magnitude, $g_{hkl} = 1/d_{hkl}$, of the reciprocal lattice vector[3], $\mathbf{g}_{hkl}$, as

$$D(\mathbf{g}) = C_S\frac{\lambda^4 g^4}{4} + \Delta f\frac{\lambda^2 g^2}{2} \tag{9.2.17}$$

The phase difference, $\phi(\mathbf{g})$, between the Bragg diffracted beam, $\mathbf{g}$, and the forward scattered beam, $\mathbf{g} = 0$, is given by

$$\phi(\mathbf{g}) = \frac{2\pi}{\lambda}D(\mathbf{g}) = \frac{\pi}{2}C_S\lambda^3 g^4 + \pi\,\Delta f\lambda g^2 \tag{9.2.18}$$

Clearly, the best way to control the phase difference and manipulate the contrast in the phase contrast image is to change the focus of the lens. Thus, by selecting the proper focus setting or defocus condition, and compensating the phase shift due to diffraction and spherical aberration, the operator can change the contrast in the image. Note that, in the general case, where many beams are selected by a larger objective aperture and contribute their phases simultaneously to the image, their total contribution can be quite complex and sophisticated simulations (Fig. 9.2.15) may be required to interpret the image.

[3] Earlier and throughout the book, we have defined, $|\mathbf{g}_{hkl}| = 2\pi/d_{hkl}$ (4.2.3b), but to be consistent with the conventions of the HREM literature, only in this section we define $|\mathbf{g}_{hkl}| = 1/d_{hkl}$.

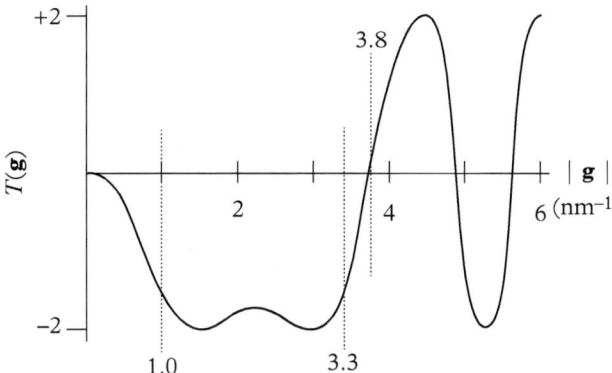

Figure 9.2.13 Contrast transfer function, $T(\mathbf{g})$, for a weak phase object (very thin specimen) as a function of reciprocal lattice vector, $\mathbf{g}$, drawn for $C_S = 1$ *mm*, $\lambda = 0.0025$ nm (200 keV), and $\Delta f = -58$ nm. This function is superimposed with axial symmetry over the electron diffraction pattern to determine which diffracted beams transfer with uniform phase and contribute to the phase contrast image.

Adapted from Williams, Pelton, and Gronsky (1991).

We are interested only in intensities that contribute to the contrast in the image and not the amplitudes. Therefore, following from (9.2.14) and (9.2.18), we can define the phase *contrast transfer function* (CTF), $T(\mathbf{g})$, for a weak phase object (Spence, 1981) as

$$T(\mathbf{g}) = 2\sin\phi(g) \qquad (9.2.19)$$

to account for the effect of the lens aberration and defocus on the amplitude distribution of the diffraction pattern. Note that this implies that the phase contrast in a TEM is oscillatory in nature as illustrated in the plot of $T(\mathbf{g})$ vs. $|\mathbf{g}|$ (Fig. 9.2.13). Further, a positive $T(\mathbf{g})$ produces negative contrast (bright image), and a negative $T(\mathbf{g})$ produces positive contrast (dark image). An alternative way to look at the transfer function, $T(\mathbf{g})$, is to consider it as a filter function that allows a range of spatial frequencies to be transferred by the lens to provide contrast in the image. Some important features of the plot are worth mentioning. The abscissa represents distances in reciprocal space or in the diffraction pattern, with larger values of $\mathbf{g}$ representing smaller features in the specimen that are transferred to the image. For the specific case shown in Figure 9.2.13, the function, $T(\mathbf{g})$, crosses the abscissa at $\mathbf{g}_1 = 3.8$ nm^{-1}, and then repeatedly as $\mathbf{g}$ increases; this first cross-over is sometimes considered as the resolution limit ($1/3.8 \sim 0.26$ nm) of the microscope. At higher values of $\mathbf{g}$, the transfer function is also modulated by other functions; we discuss this shortly. For the case shown in Figure 9.2.13, all beams between $g = 1$ and $g = 3.3$ nm^{-1} are transferred to the image with nearly identical phase—this

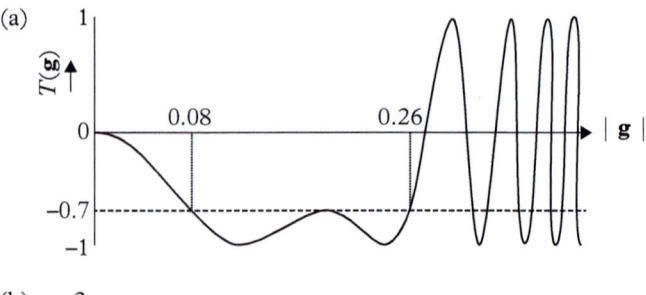

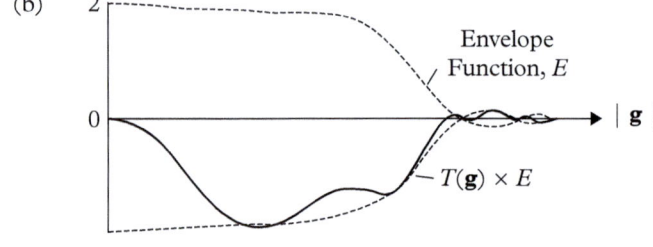

Figure 9.2.14 (a) Contrast transfer function for a microscope with $C_S = 2.2$ mm and $\Delta f = -100$ nm. (b) The same contrast transfer function now damped by an envelope function.

Adapted from Williams and Carter (1990).

corresponds to real space details ranging from 0.3nm to 1nm in the object—and produce black dots (atomic columns) against a bright background in the image.

In theory, the plot of $T(\mathbf{g})$ can extend to very large values of $\mathbf{g}$, but in practice, they are affected by envelope damping functions arising from the chromatic aberration and other factors (Fig. 9.2.14). As a result, irrespective of the focus, the envelope function is akin to a virtual aperture in the back focal plane of the OL and restricts the higher pass bands from contributing to the image. This sets another limit on the resolution of the microscope and is referred to as the *information limit* of the instrument.

In practice, an observed high-resolution phase contrast image varies with both the specimen thickness and the defocus of the OL. Hence, true interpretations of phase contrast images of crystalline materials require extensive computer simulations to match the observed and simulated images. Image simulations are beyond the scope of this book, and interested readers can find further details on all aspects of HREM in Buseck, Cowley, and Eyring (1988), and Spence (1981). Figure 9.2.15 is a typical example and shows a montage of simulated images of alumina with varying conditions of defocus and thickness. All the images represent the same structure, but under different phase interference conditions of thickness and defocus!

9.2.7.4 Scherzer Defocus and Resolution.

Using the Rayleigh criterion (Fig. 6.8.1) we have defined the microscope resolution as its ability to distinguish two closely spaced point objects. In phase contrast imaging, we define the resolution differently in terms of the CTF (Fig. 9.2.13). We require a flat response in the transfer function—in the earlier example, we

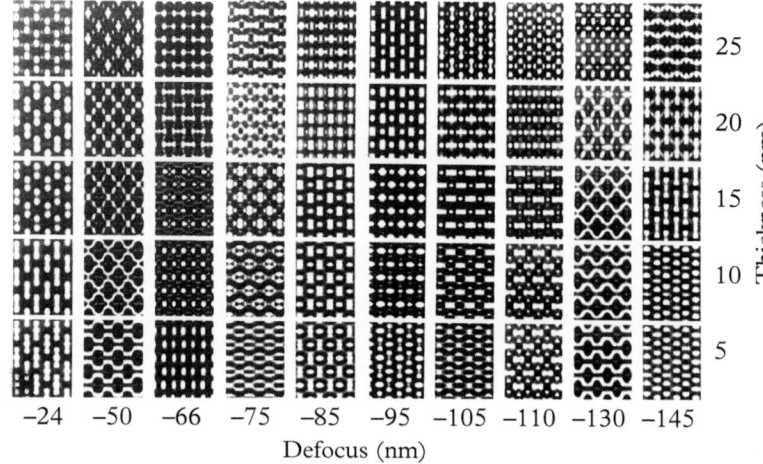

Thickness (nm)
25
20
15
10
5

−24 −50 −66 −75 −85 −95 −105 −110 −130 −145

Defocus (nm)

Figure 9.2.15 Computer simulated images of an alumina (Al_2O_3) crystal in the [011] orientation calculated for different values of defocus, Δf, and specimen thickness, t. Even though the crystal structure has not changed, the images look different because of changes in how the beams with different phases interfere. See also Figure 1.4.14.

Adapted from Williams, Pelton, and Gronsky (1991).

identified this as being between 1 nm^{-1} and 3.3 nm^{-1}—such that as many beams as possible are transferred though the optical/imaging system with identical phase. The best transfer function is the one with minimal number of zero crossings and such an optimization of the transfer function can be accomplished by balancing the effect of spherical aberration with a particular negative value of the defocus, now known after its discoverer as the Scherzer defocus.

The ideal transfer function curve is obtained when $\phi(\mathbf{g}) \sim -120°$, i.e. when $-120° < \phi < -60°$ we can get values of $\sin \phi$ close to 1. Further, for $\sin \phi$ to be nearly flat, $d\phi/dg = 0$. So, we find the value of Δf that satisfy $\phi(\mathbf{g}) \sim -120°$ and $d\phi/dg = 0$. Differentiating (9.2.18) we get

$$\frac{d\phi(\mathbf{g})}{dg} = 2\pi C_S \lambda^3 g^3 + 2\pi \Delta f \lambda g = 0 = C_S \lambda^2 g^2 + \Delta f \qquad (9.2.20)$$

Further, when $\phi = -120°$, (9.2.18) becomes

$$-\frac{2\pi}{3} = \pi C_S \frac{\lambda^3 g^4}{2} + \pi \Delta f \lambda g^2 \qquad (9.2.21)$$

We solve (9.2.20) and (9.2.21) for the optimal or Scherzer defocus

$$\Delta f_{Sch} = -\left(\frac{4}{3} C_S \lambda\right)^{1/2} = -1.155 (C_S \lambda)^{1/2} \qquad (9.2.22)$$

Further, at this value of the defocus by substituting in (9.2.18) we find that the cross-over, where $T(\mathbf{g}) = 0$, occurs at

$$g_{Sch} = 1.52 C_S^{-1/4} \lambda^{-3/4} \qquad (9.2.23)$$

The reciprocal of g_{Sch}, gives the resolution, r_{Sch}, of the TEM at the Scherzer defocus as

$$r_{Sch} = \frac{1}{1.52} C_S^{1/4} \lambda^{3/4} = 0.66 C_S^{1/4} \lambda^{3/4} \qquad (9.2.24)$$

which is a useful value to compare the performance of different TEMs. Note that (9.2.24) gives a value similar to (9.2.10), but the two are derived using different approaches.

Example 9.2.6: (a) What is the magnitude of the phase difference for a $\mathbf{g} = 200$ beam from a thin specimen of Cu using axial illumination, 200 keV electrons, $C_S = 3$mm, and Scherzer defocus.
(b) Can you image the 200 planes in the axial mode?

Solution: (a) The lattice parameter of Cu is 0.361 nm. For 200 kV electrons, $\lambda = 2.51$ pm.
For Cu, $g_{200} = 1/d_{200} = (2/0.361) \times 10^9$ m^{-1} = 5.54×10^9 m^{-1}.
From (9.2.22) we get $\Delta f_{Sch} = -1.155(C_S \lambda)^{1/2} = 10^{-7}$ m = 100 nm.
From (9.2.18), $\phi(\mathbf{g}) = \frac{2\pi}{\lambda} D(\mathbf{g}) = \frac{\pi}{2} C_S \lambda^3 g^4 + \pi \Delta f \lambda g^2$

$$= \frac{\pi}{2} 0.003 \left(2.51 \times 10^{-12}\right)^3 (5.54 \times 109)^4$$
$$- \pi 10^{-7} \left(2.51 \times 10^{-12}\right)(5.54 \times 109)^2$$
$$= 70.19 - 24.20 = 46 \text{ rad.}$$

(b) In the axial mode, the resolution is defined by setting the first zero in the phase CTF at the Scherzer defocus, (9.2.24), i.e.

$$r_{Sch} = 0.66\, C_S^{1/4} \lambda^{3/4} = 0.66(0.003)^{1/4} \left(2.51 \times 10^{-12}\right)^{3/4} = 0.31 \text{ nm}$$

However, for copper $d_{200} = 0.18$ nm.
Hence, the resolution limit is much larger than the spacing of the 200 planes in Cu, and so, they will not be imaged. (Note: we have neglected the phase factor).

In practice, the effect of the CTF on phase contrast image characteristics can be best illustrated using a very thin amorphous specimen, which by its very nature includes a projected potential with a continuous variation in spatial frequencies. Figure 9.2.16 shows images and corresponding Fourier transforms representing the spatial frequencies transferred to the image (CTF) of amorphous germanium taken on a JEOL 200CX microscope (200 keV, $C_S = 1.2$ mm) as a function of defocus. The minimum contrast occurs not at the Gaussian focus ($\Delta f = 0$)

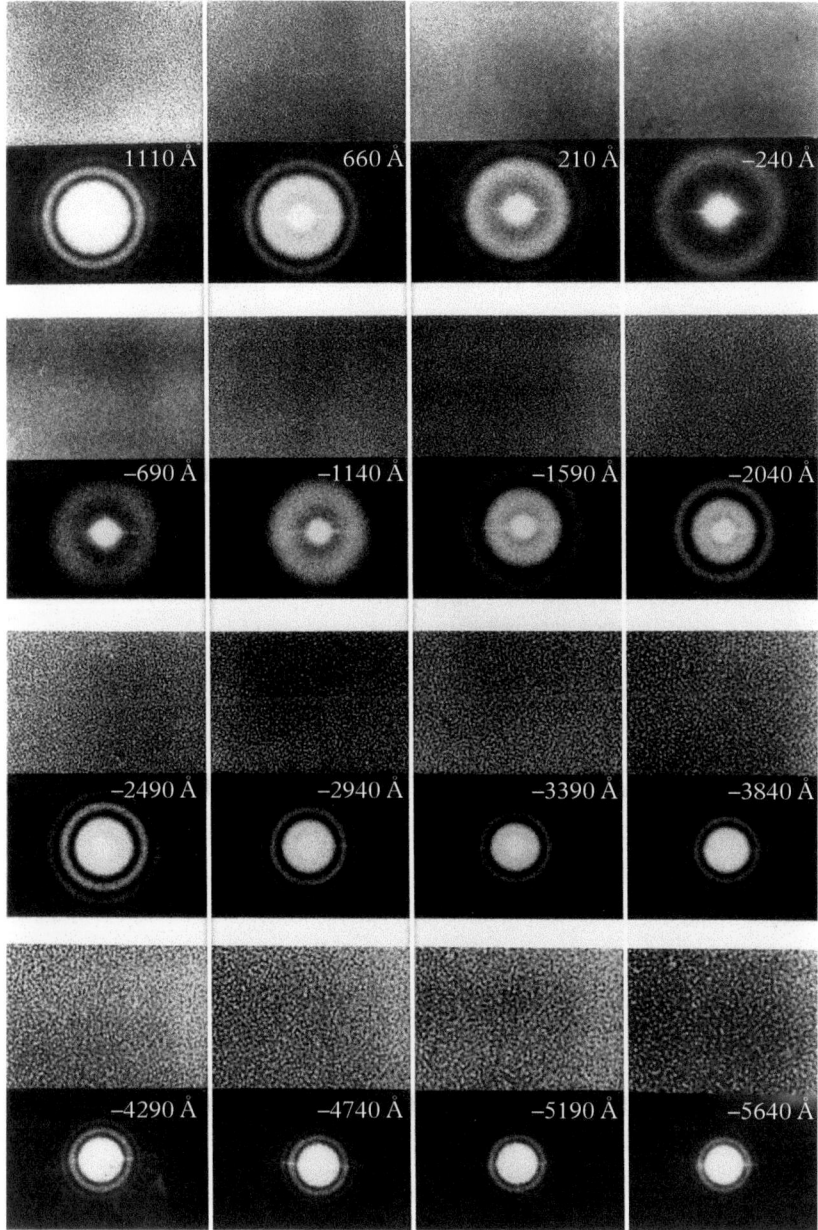

Figure 9.2.16 Through focus series images of amorphous Ge taken on a JEOL 200CX microscope. The corresponding Fourier transforms are included. These figures show the change in contrast and contrast transfer function with increasing defocus.

Adapted from Fultz and Howe (2013).

but at $\Delta f_{min} = -0.44(C_S \lambda)^{1/2}$, which for this microscope is at $\Delta f_{min} = -240\text{Å}$. On either side, as we increase (under-focus) or decrease (overfocus) the defocus, we see increased granularity in the image. At the Scherzer defocus, (9.2.22), $\Delta f_{Sch} = -690\text{Å}$, we see a broad uniform white ring in the Fourier transform. At larger defocus, larger spatial frequencies are transferred to the image—note alternating bright and dark rings—but at the same time, many zeros are included in the transfer function.

9.2.8   Scanning Transmission Mode and the Principle of Reciprocity

Imaging in a scanning transmission electron microscope[4] involves the formation of a focused probe (Fig. 9.2.5b) using the upper pole piece of the OL, and translating/rastering the probe over the specimen surface (Fig. 9.2.6c) to form the image. Throughout the scan the incident beam is maintained parallel to the optic axis using pairs of deflection coils, such that the beam always pivots about the front focal plane (FFP) of the upper pole piece of the OL (Fig. 9.2.17), i.e. the main probe-forming lens, and forms a demagnified image of the C1 cross-over on the specimen plane. The demagnification, $M = f_0/a$, where a is the distance of the upper OL pole piece from the C1 cross-over, and f_0 is the focal length of the OL. For typical values, $f_0 = 1$ mm, $a = 250$ mm, we get M = 1/250, and for a C1 cross-over diameter of 0.25 μm, for a highly excited lens, we get a probe diameter $d_p \sim 0.25 \times 10^{-6}/250 = 1$ nm, on the specimen surface. In reality, using FEG sources, probe diameters as small as 1 Å, with sufficient current to generate STEM images, can be obtained. Then, a diffraction pattern is formed on the back focal plane (BFP) of the OL, which is then transferred to the detector plane using intermediate and projector lenses (Fig. 9.2.17). Such a diffraction pattern is called *stationary* as the pattern does not move even when the beam is scanned because the BFP is conjugate with the FFP of the OL. However, instead of scanning, if we stop the probe at any point on the surface of the specimen, a convergent beam electron diffraction pattern, discussed in detail in §8.6.3, is formed on the BFP. A BF *electron detector* located on the optic axis, counts the number of electrons in the direct or forward scattered beam. For DF imaging, the diffracted electrons can be detected by the same BF detector by tilting the incident beam such that the diffracted electrons travel down the optic axis. A more common method of DF imaging is to use an annular (circular with a central hole) detector—with the central hole accommodating the BF detector—to detect the diffracted electrons. Note that the BF (either in TEM or STEM) and ADF imaging provide complementary contrast as illustrated in Figure 9.2.18.

To relate the image intensity in a STEM to that in a conventional TEM (CTEM) we introduce the important principle of *reciprocity*, which applies to the idealized situation of point emission and detection of electrons. This principle is based on the observation that the STEM instrument is optically

[4] A brief historical note is in order: Manfred von Ardenne [7] invented the STEM around the same time as Knoll and Ruska invented the TEM. The main problem with the STEM technique, in the early days, was noise as small probes were needed for high resolution for which the available current was limited. In fact, von Ardene quickly abandoned the STEM project in favor of Ruska's design of the TEM. Later, Albert Crewe incorporated a cold FEG source into the instrument, made it into a viable electron microscope [8, 9], and demonstrated the possibility of visualizing single atoms [10].

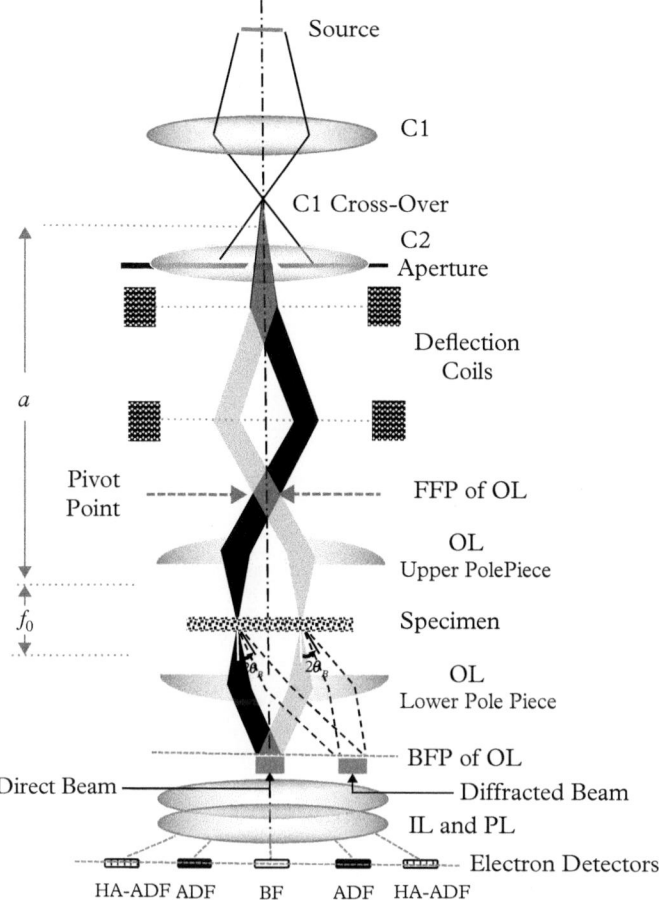

Figure 9.2.17 Ray diagram illustrating STEM optics. C2 is switched off, the OL upper pole-piece forms the probe, and the beam is rocked around the front focal point of the OL. The stationary diffraction pattern is formed in the back focal plane (BFP) of the OL and all electrons scattered through the same angle, $2\theta_B$, are brought to a focus at the same point on the BFP. The pattern may be detected directly or it may be transferred to the detectors by intermediate and projector lenses. Images are formed by the bright field (BF) detector collecting the direct or forward-scattered beam, the annular dark field (ADF) detector collecting electrons scattered or diffracted in different directions, and a high-angle HA-ADF detector to collect incoherently scattered electrons to form a Z-contrast image. Also see Figure 9.3.11.

similar to the CTEM, but with the important difference of the electrons traveling (hypothetically speaking) in the opposite direction (Fig. 9.2.19). The reciprocity principle may be stated as the following: *the amplitude of a radiation at point B, arising from a source at point A, is the same as the amplitude at point A, arising from a source at point B.* It applies to scalar fields and waves, as well as elastic scattering. For electron waves it can be extended to vector fields, including magnetic fields (lenses, deflector coils, etc.), provided the reversal of the electron beam direction is accompanied also by a reversal in the magnetic field direction. Extension of the theorem to finite (not point) sources and detectors can be made if we assume that they are incoherent, and the image intensity is then given by adding the intensities separately for all the points on the source and detector.

A STEM instrument provides a very flexible arrangement to detect both elastically and inelastically scattered electrons as well as a variety of other signals,

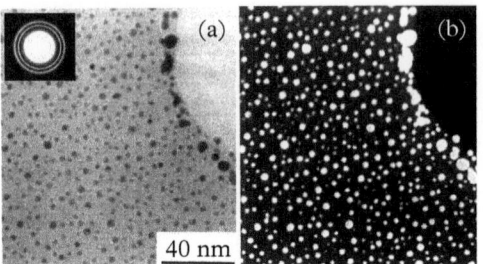

Figure 9.2.18 Images of Au islands on a C film. (a) BF image obtained by imaging the direct beam. (b) An annular dark field (ADF) image, obtained by blocking the forward-scattered beam in the diffraction pattern (inset in a) and collecting only the diffracted electrons. It is easy to recognize that the BF and ADF detectors provide complementary contrast. See also Figure 9.3.10.

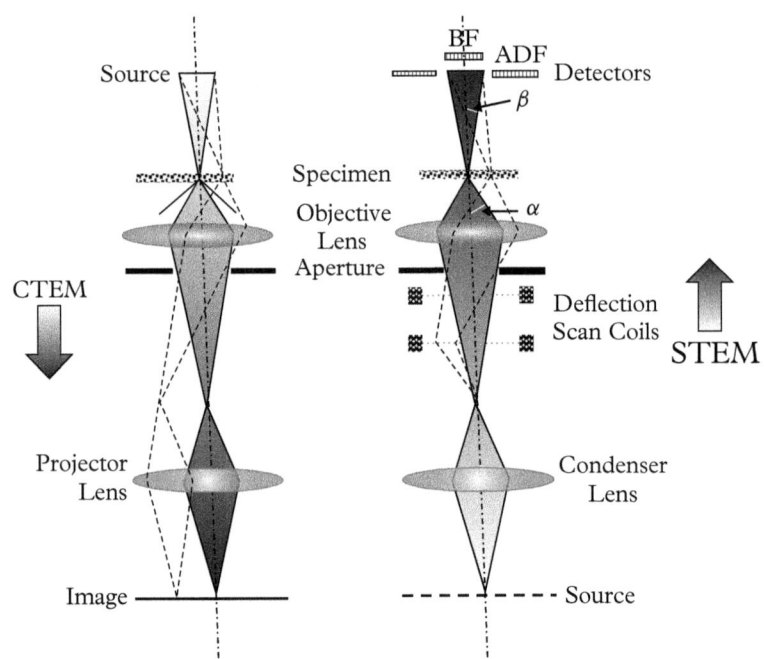

Figure 9.2.19 Ray diagrams illustrating the reciprocity theorem for a conventional (CTEM) and scanning (STEM) transmission electron microscopes. Note that as the arrow on either side indicates the direction of electron travel in the two cases are anti-parallel.

including X-rays and Auger electrons, generated by the interaction of the fast electron with the specimen. These are discussed in later sections of this chapter dealing with analytical electron microscopy (§9.4). Finally, a STEM image is formed without the use of lenses, and only the probe determines the final image resolution. The probe characteristics are determined by the spherical aberration of the OL, but the images in STEM are not affected by chromatic aberration to the significant extent that they are in a CTEM.

Example 9.2.7: A focused probe 10 nm in diameter is formed by a 200 kV STEM with a gun brightness of 5×10^9 A/m^2/sr. If the convergence semi-angle of the probe is 8 mrad, and aberrations can be neglected, how many electrons does the probe deliver on the specimen per second?.

Solution: The current, I, in the probe is given by

$$I = \beta \times \frac{\pi d^2}{4} \times \pi \alpha^2 = 5 \times 10^9 \times \frac{\pi \left(10 \times 10^{-9}\right)^2}{4} \times \pi \left(8 \times 10^{-3}\right)^2 = 7.9 \times 10^{-11} \text{A}$$

Then the number, N, of electrons per second $N = I/e$, where e is the charge of an electron. Thus $N = \frac{I}{e} = \frac{7.9 \times 10^{-11}}{1.6 \times 10^{-19}} = 4.9 \times 10^8$ electrons/sec.

9.2.9 Correction of Lens Aberrations

We have seen that the main imaging, or objective, lens suffers from spherical aberration with typical values of the coefficient, $C_S \sim 1$ mm. Hence, the resolution (9.2.10) of a 200 keV TEM is $\delta_{\min} = 0.262$ nm, which is two orders of magnitude larger than the wavelength ($\lambda_{200keV} \sim 0.0025$ nm) of the fast electrons. The resolution can be improved either by increasing the acceleration voltage to reduce the wavelength (but, this leads to additional problems of specimen damage— see §5.3.2.2), or by designing better lenses with smaller values of C_S. It is also possible to correct or completely compensate for the spherical aberration of an electron-optical lens by using two hexapole elements and two lens doublets, and accommodate the large image sizes typical in a TEM, as originally proposed by Rose [11–14], and as shown in Figure 9.2.20. With these correctors, C_S drops by three orders of magnitude to a typical value of $C_S \sim 3$ μm, but now, the chromatic aberration becomes significant. To minimize the chromatic aberration contribution, (9.2.6), the energy spread of the beam, ΔE, may be first reduced as much as possible. This is accomplished by incorporating a pre-specimen monochromator, which can reduce the inherent energy spread of the source from $\Delta E \sim 0.7$ eV (SEG) to $\Delta E \sim 0.2$ eV. However, any further correction of C_C will require a much more complex arrangement of electron optical components. Aberration corrected microscopes have made spectacular advances in atomic scale imaging with chemical sensitivity [4, 15–18]; see Figure 1.4.14 for a comparison of images of the same material taken with a traditional electron optics ($C_S = 0.6$ mm) and an aberration corrected TEM. Note that in addition to chromatic aberration, higher order lens aberrations that become significant as the spherical aberration goes down are also fully or partially corrected by an aberration corrector. Further details can be found in [11–14].

Figure 9.2.20 The Oxford JEOL JEM2200FS microscope. The two shiny additions to the column are the CEOS aberration correctors. The upper one corrects the probe-forming lens (STEM) and the lower one corrects the image-forming lens (TEM).

Courtesy of Andrew McKnight.

9.2.10 Image Recording and Detection of Electrons

For completeness, it must be mentioned that images and diffraction patterns in a TEM, traditionally recorded by photographic plates, are now recorded with a CCD camera, image plate [19], and for STEM detectors, using a combination of a scintillator and photo multiplier tube (PMT); see Figure 10.2.2 for a detector with similar design used in a SEM. In the latter case, ∼2,000 photons are generated by each electron incident on the scintillator, a parabolic mirror then reflects the generated photons towards the PMT, with about 100 photons/electron reaching the cathode of the PMT. Ideally, this high level of efficiency can make single electron counting feasible. Further details of image recording devices can be found in Reimer (1993), and the implementation of the electron detector in a commercial STEM instrument is discussed in [6]. Finally, it is worth mentioning that recent developments in microelectronics, have allowed the design of direct single electron detectors with fine pixels, and fast readouts, which are sufficiently radiation hard for practical use in electron microscopy [106,107], particularly for resolving molecular structures by cryo-imaging.

9.3 Beam–Solid Interactions, Contrast Mechanisms, and Imaging Methods

The interactions of an electron beam with a specimen can be broadly classified (see §8.5.1 for a detailed discussion) as elastic or inelastic, when electrons are considered as particles, and when considering their wave nature, as coherent (same wavelength and phase) or incoherent (different wavelengths and/or no phase relationship). In elastic scattering no energy is transferred from the beam to the solid and the energy, E_{el}, of the electron leaving the thin foil specimen on the exit side is the same as its original energy, E_0, i.e. $E_{el} = E_0$. Needless to say, no energy is transferred if the electron passes through the specimen without any interaction; in TEM such electrons are included in the direct or forward-scattered beam, which contains all electrons that pass through the thin specimen in the direction of the incident beam. Elastic scattering includes electrons deflected by the Coulomb interactions with the positive nuclear charge ($+Ze$), as discussed earlier in §8.2, losing negligible energy and contributing to the diffraction signals (§8.5) in a TEM. Alternatively, the incident electrons can transfer energy to the specimen thereby losing their energy and emerging from the exit surface with energy $E_{el} < E_0$. The electron energy-loss spectrum, $I(\Delta E)$, where $\Delta E = E_0 - E_{el}$, of inelastically scattered electrons (§9.4.1) emerging in the forward direction (Fig. 9.3.11) can be monitored to provide information on the electronic structure and chemical composition of the specimen (§9.4.2). Further, the energy transferred from the beam to the specimen raise the latter to an excited state; the subsequent de-excitation processes produce a variety of signals such as characteristic X-rays, Auger and secondary electrons, and CL. All these signals can be monitored by attaching suitable detectors to an electron-optical column (Fig. 9.1.2) and such a TEM (discussed further in §9.4) is called an analytical electron microscope (AEM).

9.3.1 Elastic Interactions

As a starting point, we treat the electrons as particles, ignoring their wave-like behavior, and describe the incoherent scattering of electrons by atoms. The electron beam penetrating the electron cloud of the atom with nuclear charge, $+Ze$, at a distance, r, experiences a Coulomb force, F, in the classic Rutherford scattering model (Fig. 8.2.1 and Fig. 9.3.1) given by

$$F = \frac{Ze^2}{4\pi \varepsilon_0 r^2} \tag{9.3.1}$$

Clearly, the force, F, increases with decreasing r, and increasing atomic number, Z. It is also logical to expect the scattering angle, θ, to increase with F, and there is a small, but finite possibility that the electrons can also be back-scattered (see §10.3.3). Further, in passing through the specimen, an electron may be scattered once (single scattering), a few times (plural scattering), many times (multiple scattering), or not at all (unscattered)! So, to describe the probability of an electron undergoing a scattering event we use either an *interaction cross-section*, σ_{int}, or the distance it travels between two interactions in the specimen, known as the *mean free path length*, λ_{mfp}.

In physical terms, the interaction cross-section, σ_{int}, is the effective area presented by the atom to the electron for a specific interaction event. Further, dividing the cross-section by the total area provides a measure of the interaction probability. Moreover, the cross-section can be defined by an effective radius, r_{int}, such that $\sigma_{int} = \pi r_{int}^2$. For elastic scattering

$$r_{elast} = \frac{Ze}{U_0 \theta} \tag{9.3.2}$$

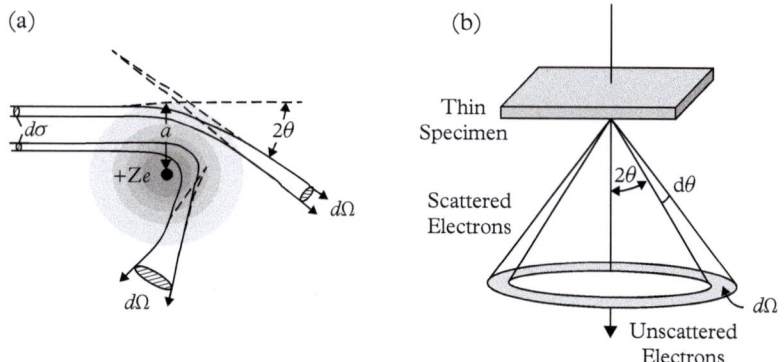

(a)

(b)

Thin Specimen

Scattered Electrons

Unscattered Electrons

Figure 9.3.1 (a) Elastic scattering of an electron by an atom with nuclear charge, $+Ze$, in the particle model showing the differential cross section, $d\sigma / d\Omega$, for two different cases. (b) Scattering of an electron by a single atom through a semi-angle, θ, with an increment in solid angle, $d\Omega$, observed for a change $d\theta$ in the scattering angle.

(a) Adapted from Reimer (1993).

where U_0 is the accelerating potential (Volts) and θ is the scattering angle. Clearly, the likelihood of scattering increases for larger atomic number elements, but decreases with acceleration voltage and scattering angle. The total interaction cross section, σ_T, includes both elastic and inelastic contributions:

$$\sigma_T = \sigma_{elast} + \sigma_{inelast} \tag{9.3.3}$$

For a specimen containing N atom/volume, we can write the total scattering cross section, Q_T, as

$$Q_T = N\sigma_T = \frac{N_0 \sigma_T \rho}{A} \tag{9.3.4}$$

where N_0 is the Avogadro number, A is the atomic mass, and ρ is the density of the specimen. Further, for a thin specimen of thickness, t, the probability of a scattering event is

$$P = Q_T t = \frac{N_0 \sigma_T}{A}\rho t \tag{9.3.5}$$

where the quantity, ρt, is known as the mass thickness. Note that the mean free path length, typically tens of nanometers in a TEM, is related inversely to the scattering cross-section, i.e. $\lambda_{mfp} = 1/Q_T$. For a specimen of thickness, t, the probability, P, of interaction is $P = t/\lambda_{mfp} = tQ_T$, and the intensities of the incident, I_0, and transmitted, I_{trans}, electrons are simply related as

$$I_{trans} = I_0 \exp(-P) = I_0 \exp\left(-\frac{N_0 \sigma_T \rho t}{A}\right) \tag{9.3.6}$$

Thus, the intensity of the unscattered or transmitted part of the beam decreases exponentially with mass thickness. Further, σ_{elast}, which is a significant part of σ_T, (9.3.3), is closely related to the scattering factor, $f_e(\theta)$, (8.2.4), and is strongly dependent on the wavelength and the atomic number.

Strictly speaking, we are interested in describing the angular distribution of electrons scattered by the atom. To do so, we define a differential scattering cross-section, $\frac{d\sigma}{d\Omega}(\theta)$ as the scattering of a parallel incident beam of electrons passing through the specimen element, $d\sigma$, into a cone of solid angle, $d\Omega$, at an angle, 2θ. The differential scattering cross-section (Fig. 9.3.1b) is important as it is related to the measured quantity of the atomic scattering factor, $f(\theta)$, described earlier in the context of electron diffraction, as

$$f(\theta) = \left|\frac{d\sigma}{d\Omega}\right|^{1/2} \tag{8.2.1}$$

Rutherford first determined the differential scattering cross-section for the atomic nucleus, ignoring the scattering by the outer electrons, as

$$\frac{d\sigma}{d\Omega}(\theta) = \frac{e^4 Z^2}{16 E_0^2 \sin^4\left(\frac{\theta}{2}\right)} \tag{9.3.7}$$

The solid angle, 2Ω, is simply related to the scattering angle, 2θ, i.e., $\Omega = 2\pi(1-\cos\theta)$. Thus

$$\frac{d\sigma}{d\Omega} = \frac{1}{2\pi \sin\theta} \frac{d\sigma}{d\theta} \tag{9.3.8}$$

By substituting (9.3.8) in (9.3.7), integrating from 0 to π, and substituting the values for the physical constant we get the cross-section, σ_{atom}, for the incoherent scattering by a single atom:

$$\sigma_{atom} = 1.62 \times 10^{-24} \frac{Z}{E_0} \cot^2\left(\frac{\theta}{2}\right) \tag{9.3.9}$$

Following (9.3.5), we can calculate the probability of atomic scattering for a specimen of thickness, t, as

$$Q_{atom} t = 1.62 \times 10^{-24} \frac{N_0 \rho t}{A} \frac{Z}{E_0} \cot^2\left(\frac{\theta}{2}\right) \tag{9.3.10}$$

Note that, so far, we have ignored the scattering from the electron cloud, which because of its negative charge will reduce the differential scattering contribution. This is called screening and if this effect is also included, it can be shown (Reimer, 1993) that the differential scattering cross-section, in terms of the Bohr radius, $a_0 = \varepsilon_0 h^2 / \pi m_e e^2$, (2.2.2), and λ_{rel}, the relativistic corrected wavelength, (8.2.6), can be written as

$$\frac{d\sigma}{d\Omega}(\theta) = \frac{\lambda_{rel}^4 Z^2}{64\pi^4 a_0^2 \left[\sin^2\left(\frac{\theta}{2}\right) + \left(\frac{\theta_{screen}}{2}\right)^2\right]^2} \tag{9.3.11}$$

where

$$\theta_{screen} = \frac{0.117 Z^{1/3}}{E_0^{1/2}} \tag{9.3.12}$$

accounts for the electron screening effects. This form (9.3.11) of the differential scattering cross-section is widely used in electron diffraction, particularly in the calculation of atomic scattering factors using the relationship (8.2.3).

Figure 9.3.2 (a) The screened relativistic Rutherford cross-section as a function of scattering angle for three elements. (b) A Monte Carlo simulation for two materials, low Z and high Z, showing most electrons propagating through a thin specimen are forward scattered.

(a) Adapted from Williams and Carter (1996).

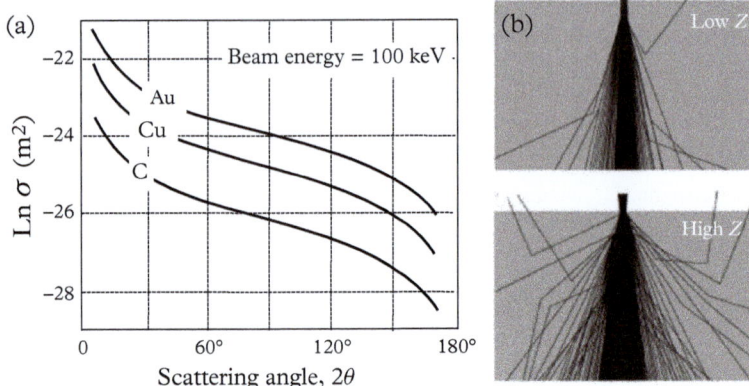

Figure 9.3.2a shows the variation of the screened relativistic Rutherford cross-section as a function of scattering angle for three different elements. The cross-section decreases by six orders of magnitude as 2θ varies from 0 to π. Further, Monte Carlo simulations for materials with either low or high average atomic number (Fig. 9.3.2b) show that most electrons are either not scattered or scattered through very small angles in the forward direction. The probability of scattering through larger angles and the resultant beam broadening increases with Z. The intensity of the direct beam is reduced due to scattering or deflection of the beam from the forward direction, which depends on Z, allowing us to distinguish different materials. Now, with this simple description of elastic scattering, we can discuss the three principal contrast mechanisms in the TEM that are closely linked to the formation of images: mass-thickness, diffraction, and phase contrast.

9.3.2 Mass–Thickness Contrast

The mass-thickness scattering contrast in a TEM occurs from the incoherent Rutherford scattering of electrons, discussed in the previous section. The probability of such scattering (9.3.10) depends strongly on the atomic number and the mass-thickness of the specimen, and as it is heavily forward-peaked, it contributes significantly for small angles ($\leq 5°$) of scattering, especially for amorphous and biological materials. In contrast, for crystalline materials, Bragg scattering angles for high energy electrons are small ($<3° \sim 50$ mrads) and diffraction (coherent scattering) competes favorably in providing image contrast. However, for very large scattering angles (>70 mrads), the contribution of coherent scattering is negligible, and the weak incoherent scattering, with dependence only on the atomic number, Z, can provide image contrast. Imaging contrast obtained with such high-angle scattered electrons, also known as Z-contrast imaging, is discussed further in §9.3.4.

We now ignore diffraction contrast (§9.3.3), and discuss mass-thickness contrast, which is particularly applicable to incoherent scattering by nonperiodic structures such as amorphous materials, polymers, and biological materials. We can simply surmise that the likelihood of the Coulomb interaction of a fast electron with an atom in the specimen increases with its atomic charge or atomic number. Thus, elements with heavier mass will scatter more strongly than lighter elements. Further, the strength of scattering depends on the number of atoms encountered by the electron in its path through the specimen. Therefore, for a given mean free path length, thicker specimens will scatter more than thinner specimens, and will appear darker in the forward scattered image. These two effects taken together, and referred to as mass-thickness contrast when implemented in a TEM, are shown schematically in Figure 9.3.3. Here, in the BF image, only the direct beam contributes to the image, and electrons scattered by thicker and/or higher mass areas are blocked by the objective aperture, and appear darker when compared to thinner and/or lower mass areas.

Mass-thickness contrast can be controlled by the size of the objective aperture and the energy (or the wavelength) of the electron beam. If a larger objective aperture is used, more electrons are included in the BF image, increasing the overall image intensity but decreasing the contrast in the specimen between areas that scatter and those that do not scatter the electron beam. On the other hand,

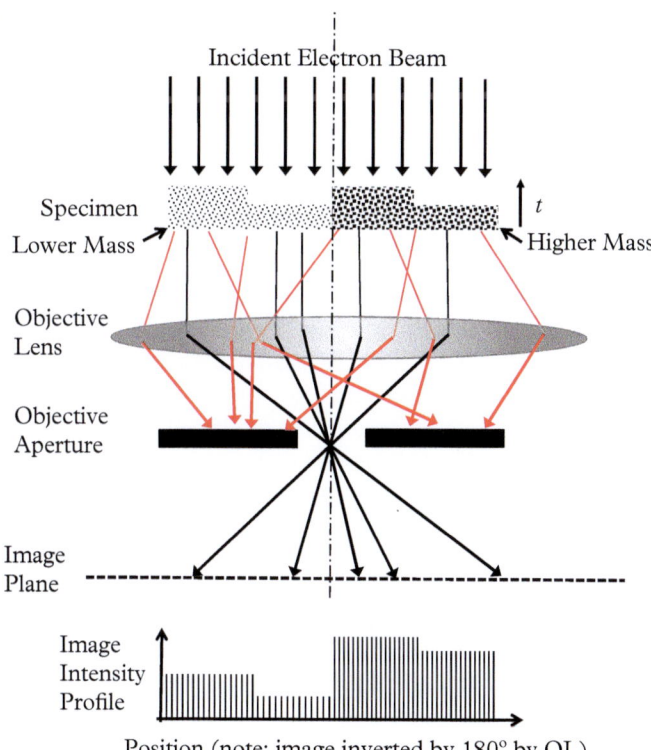

Figure 9.3.3 Schematic illustration of mass-thickness contrast in a BF image. Thicker and higher mass areas of the specimen scatter electrons away from the forward direction and are blocked by the objective aperture, producing darker contrast in the image. Such contrast is used in biology, by staining regions of interest with heavy metals, such as uranium or osmium, to enhance contrast.

Figure 9.3.4 TEM images of HT-1080 cells with fluorescently tagged superparamagnetic iron-oxide nanoparticles (SPIONs) and (a,b) imaged by confocal fluorescence microscopy indicating the internalization of the SPIONs in the cell. The cells were also embedded with resin, stained with osmium, and ultra-thin sectioned for TEM imaging, with mass contrast illustrating images (a,c) before and (b,d) after internalization with location indicated by arrows. Further magnified imaging by TEM confirms encapsulation of SPIONs in (e) cytoplasmic vesicles including endosomes, and (f) lysosomes.

Adapted from [28].

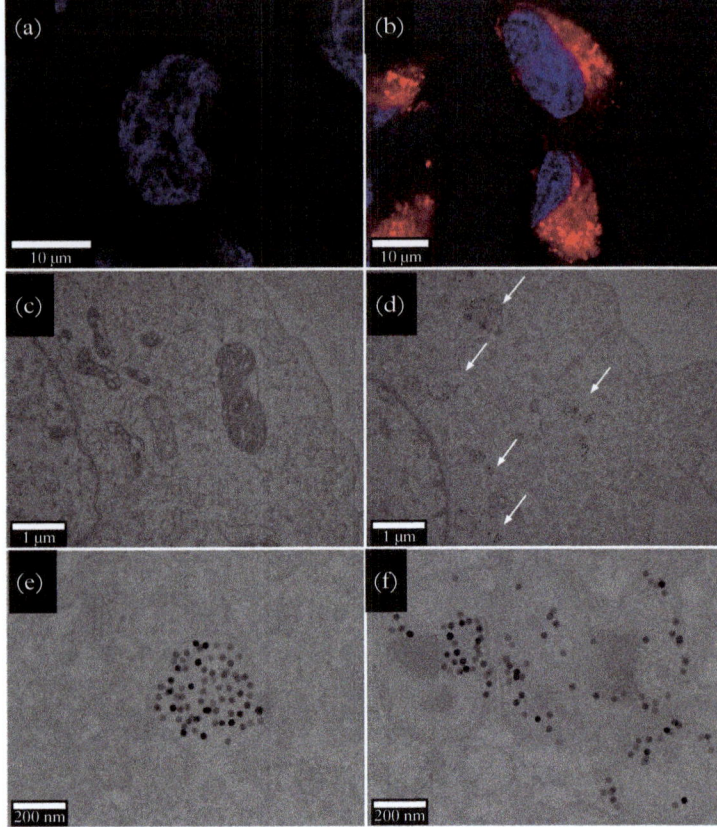

if the energy (keV) of the electron is increased, all three, i.e. its wavelength, the scattering angle, and the cross-section, will decrease. Then, more electrons are included in a given objective aperture, decreasing the contrast, even though the overall brightness of the image increases.

Such mass-thickness contrast arising from incoherent elastic scattering exists in all materials, and is the most important image formation mechanism in noncrystalline materials; however, in crystalline materials diffraction contrast dominates. In general, biological specimens do not generate much contrast, either mass-thickness or diffraction, because they are typically composed of carbon, nitrogen, and oxygen—atoms with similar low atomic numbers with no long-range order or periodicity. Thus, tissues are negatively stained (make them appear darker) with heavy metals, such as osmium, uranium, or lead. In particular, osmium preferentially binds to the unsaturated bond in lipid bilayers and crosslinks it with a neighboring lipid, and thus osmium is used to enhance contrast of membranes. Figure 9.3.4[5] shows a typical example of the imaging of cells, stained with Os, and with internalized SPIONs. Similarly, polymer materials are also stained with

[5] Biological samples can also be studied by Cryo-TEM, by plunging them in liquid ethane to form vitreous ice. In fact, Dubochet, Frank, and Henderson received the Nobel prize in Chemistry in 2017 for "developing cryo-electron microscopy for the high-resolution structure determination of biomolecules in solution".

heavy metal oxides to selectively penetrate specific parts of the "microstructure," locally change the atomic number, and enhance mass-thickness contrast.

9.3.3 Diffraction Contrast

For crystalline materials, electron diffraction in a TEM (§8.5) is determined by the crystal structure and the specimen orientation or tilt. At specific (Bragg) angles of diffraction we observe *coherent elastic scattering* (Fig. 8.6.1) and we can use such diffraction to create contrast in a TEM image. Figure 9.2.10 shows how, by using a parallel incident beam (in practice, this may also require an under-focused C2 lens, Fig. 9.2.4a), a BF image (Fig. 9.2.10a) is formed by placing the objective aperture around the forward-scattered (direct) beam in the back focal plane of the OL; similarly, a DF image (Fig. 9.2.10b,c) is formed by selecting any of the diffracted beams to form the image. However, unlike mass-thickness contrast, to get good diffraction contrast in both BF and DF images, the specimen has to be tilted to the right orientation.

Recall that diffraction by a specific set of planes, (hkl), results in the strong excitation of the $\mathbf{g}_{hkl}$ beam (Fig. 8.6.6) in transmission electron diffraction. Hence, areas that appear bright in a DF image of a crystalline specimen using a specific reflection correspond to regions where those planes are oriented to satisfy the Bragg diffraction condition. If we tilt the crystal such that only one specific reflection is strong (in addition to the transmitted beam) in diffraction, and all other reflections are weakly excited, we enhance the contributions to the DF image from only one specific lattice planes. Such an orientation is called the *two-beam condition* (Fig. 8.6.7) and is most desirable for diffraction contrast imaging. From any zone axis pattern (Fig. 8.6.4) the specimen can be tilted slightly in different directions, to obtain a number of two-beam conditions, such that in each orientation a different reflection (and the transmitted or forward scattered beam) is strongly excited.

In §8.6.2 we discussed the use of Kikuchi lines to set up diffraction conditions accurately by controlling the excitation error, s, and setting up the Bragg condition ($s = 0$) as accurately as possible (Fig. 8.6.6). Earlier, we presented the method of centered DF imaging (Fig. 9.2.10c), where the incident beam is tilted such that the specific diffracted beam of interest travels down the optic axis to form the image. In practice, we wish to obtain a strong two-beam diffraction condition to form a centered DF image. If the incident beam is tilted to bring a specific reflection, $\mathbf{g}_{hkl}$, down the optic axis, it often becomes weak in intensity, and instead, the $3\mathbf{g}_{hkl}$ reflection becomes strong (Fig. 9.3.5b); this is known as the *weak beam* imaging condition [108], often used to image dislocations with good contrast (Fig. 9.5.8b). However, if we tilt the beam such that the originally weak $\mathbf{g}_{\bar{h}\bar{k}\bar{l}}$ reflection is now down the optic axis (Fig. 9.3.5c) we achieve the required, but strong, two-beam condition. Diffraction contrast is often used to image and analyze defects. Best image contrast of defects in the two-beam condition is obtained with small positive excitation errors, $s > 0$ (Fig. 9.3.6).

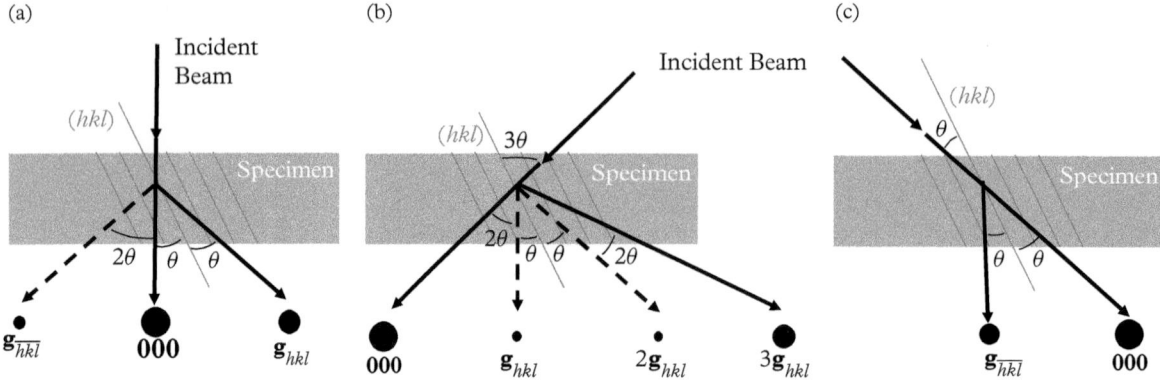

Figure 9.3.5 Schematic illustration for setting up optimal two-beam diffraction conditions for DF imaging. (a) Standard condition. (b) Incident beam tilted through $2\theta_{hkl}$ such that the diffracted beam, $\mathbf{g}_{hkl}$, is down the optic axis. In this case the diffracted beam is weak as indicated by the size of the diffraction spot. (c) By tilting the beam by $-2\theta_{hkl}$, i.e. in the opposite direction, a strong higher intensity $\mathbf{g}_{\overline{hkl}}$ reflection is now along the optic axis, which makes for a good centered dark field image.

Figure 9.3.6 Variation in diffraction contrast as a function of the excitation error: (a) $s = 0$, (b) $s > 0$, but small, and (c) $s > 0$, but large. The best contrast is observed in (b), even though the defect appears narrower in (c).

Adapted from Williams and Carter (1996).

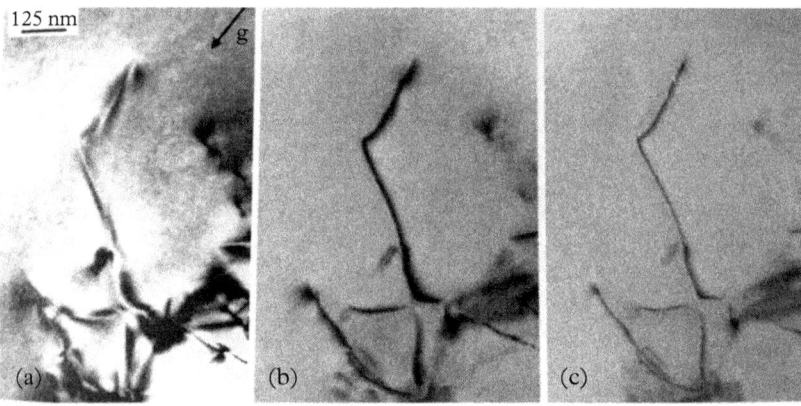

Diffraction contrast is widely used to resolve the microstructure of practical materials. Broadly speaking such microstructure features include changes in local orientation without changes in composition (grains, twins, precipitates, etc.), lattice defects (dislocations, stacking faults, etc.), and multiple-phase systems with additional changes in composition and/or structure, including their interfaces. Figure 9.3.7 shows diffraction contrast images of twins (see §8.7.6) formed in copper as a result of explosive deformation.

A classic application of diffraction contrast is to determine the Burgers vector, **b**, of perfect dislocations, especially in isotropic materials. In particular, this

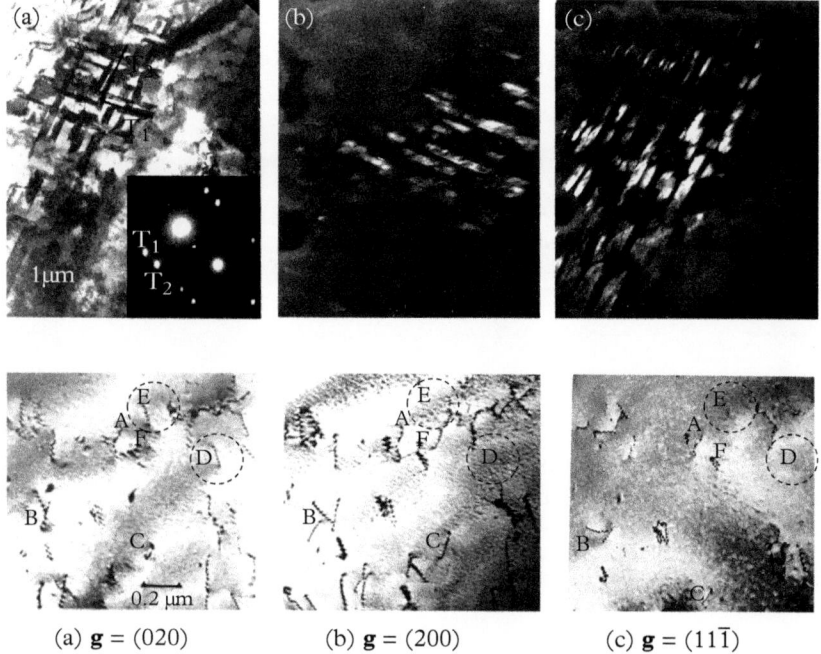

Figure 9.3.7 Twins in copper. (a) Bright field image, with (inset) the selected area diffraction pattern indexed. (b, c) Dark field images using the twin spots T_1 and T_2, clearly identifying the twins by their contrast reversal.

Adapted from Thomas and Goringe (1979).

Figure 9.3.8 Images of the dislocations in aluminum taken with different operative reflections. Note that the dislocations marked D and E (encircled) can be observed only for $\mathbf{g} = (020)$. Applying the $\mathbf{g} \cdot \mathbf{b} = 0$ criterion gives the Burgers vector $\mathbf{b} = \frac{1}{2}[011]$ for these dislocations.

Adapted from Edington (1975).

analysis takes advantage of the feature of dislocation images that make them invisible in BF and DF images when the condition $\mathbf{g} \cdot \mathbf{b} = 0$ is satisfied (§9.5.2). This is illustrated in Figure 9.3.8 for dislocations in aluminum. However, most materials are elastically anisotropic and the simple invisibility criterion cannot be applied; more sophisticated analysis is required and the interested reader may consult Edington (1975), Thomas and Goringe (1979), or Williams and Carter (1996) for details.

Diffraction contrast is observed in both TEM and STEM images, even though in the latter case, the contrast is a lot poorer. To get good diffraction contrast images in TEM, we need a parallel incident illumination, a strong two-beam diffraction condition, and allow only one of the two beams, either transmitted (BF) or diffracted (DF) beam, to pass through the OA and form the image. In STEM, as we have seen, we require a large collection angle to generate sufficient signal intensity in the detector. However, by the reciprocity principle, we require this collection angle in STEM to be as narrow as possible to approximate the parallel illumination condition in the TEM. Hence, diffraction contrast in STEM is only possible for small collection angles, which naturally reduces the image intensity and makes it noisy (Fig. 9.3.9). Some of this limitation can be overcome using a high brightness FEG source, but diffraction contrast is not the forte of STEM.

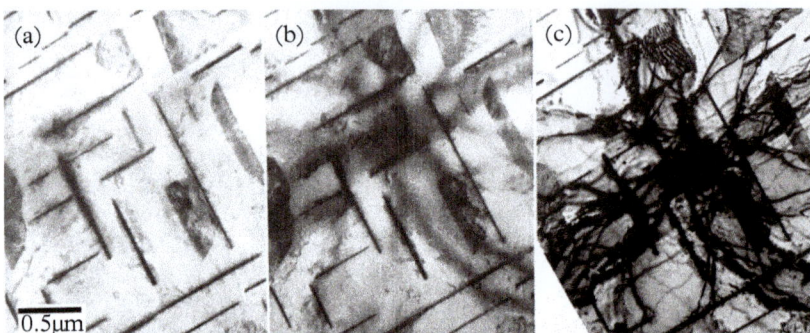

Figure 9.3.9 Comparison of diffraction contrast in STEM and TEM images. The same region of an Al-4 wt% Cu specimen is imaged in bright field STEM under (a) normal and (b) narrower collection angles; in the latter case, some diffraction contrast (bend contours) is visible, and (c) the corresponding TEM BF image. In all images, the Cu-rich precipitates show excellent mass-thickness contrast.

Adapted from Williams and Carter (1996).

9.3.4 High-Angle Incoherent Scattering: *Z*-Contrast Imaging

The Coulomb interaction of the fast electron beam with the positive potential $(+Ze)$ of the atomic nucleus is strong enough to scatter the electrons into high angles $(>3°)$. In fact, the electrons can even be back-scattered, and then detected for imaging in a scanning electron microscope (SEM; §10.3.3). The Rutherford cross-section, (9.3.11), applicable for such scattering, depends strongly on the atomic number $(\sim Z^2)$, and offers the possibility of chemical contrast in these images. In other words, areas and particles of high-Z elements scatter strongly and produce bright contrast in images recorded with electrons scattered into high angles (Fig. 9.3.10); in particular, this effect is used in HAADF imaging in a STEM as schematically illustrated in Figure 9.3.11.

Note that if a small detector is used, the collection angle of the HAADF detector has to be large enough to avoid detection of Bragg scattered or diffracted electrons. However, if we use a detector large enough to collect many Bragg reflections, then we get an incoherent image [110, 111]. This ensures that the detector collects purely incoherent but elastically scattered electrons that are described well by the Rutherford scattering model, (9.3.11). Optimally, such a detector would collect electrons scattered—incoherently—through angles larger than 70 mrad ($\sim 8°$). As a result, the intensities of scattering from individual atoms can be added together and the images can be interpreted directly in terms of the

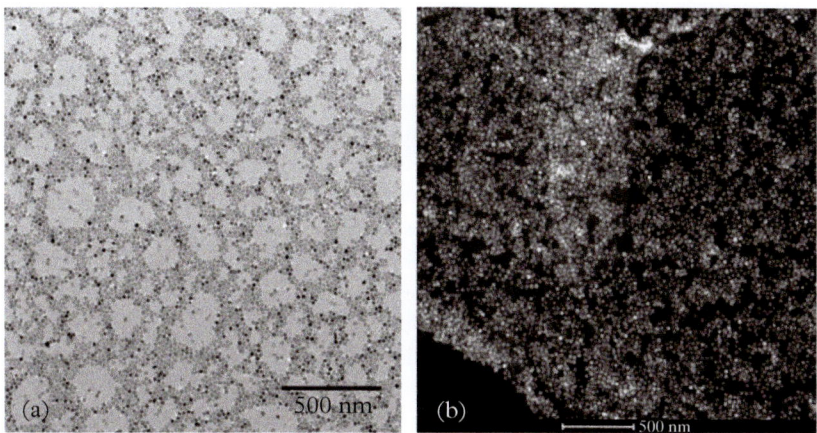

Figure 9.3.10 Images of iron oxide nanoparticles drop cast on ultra-thin carbon films mounted on Cu grids at nearly identical magnifications. (a) A bright field TEM image showing *dark contrast* for the particles with strong diffraction contrast—notice how some particles are darker because they are oriented crystallographically for strong Bragg scattering. (b) A dark field HAADF–STEM image, with contribution from only incoherent scattering over sufficiently large angles (>70 mrads) such that the particles provide bright contrast. Compare with Figure 9.2.18, where the DF image in the STEM, collected with an ADF detector, includes strong contributions from diffracted electrons, which by definition has significant coherent scattering contributions.

Images courtesy Dr. R. Hufschmid.

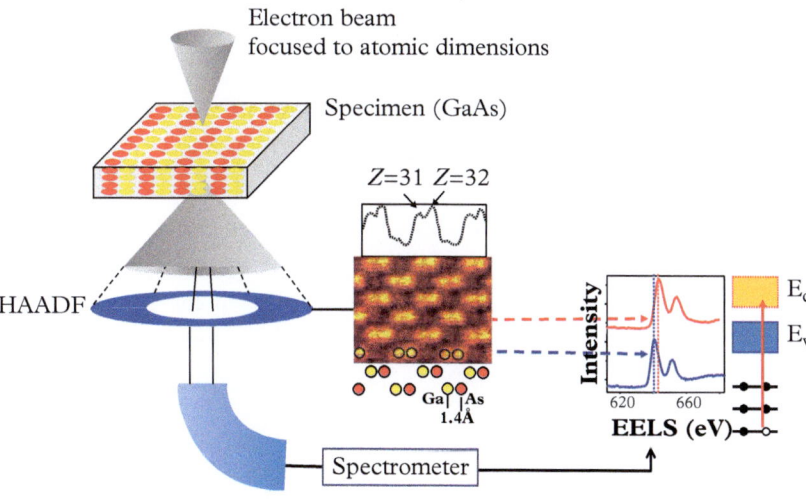

Figure 9.3.11 Schematic illustration of a STEM instrument with a HAADF detector for atomic resolution, *Z*-contrast imaging, and an EELS spectrometer for simultaneous chemical and electronic structure measurements of the specimen.

GaAs image courtesy of Professor S. J. Pennycook.

atomic species (*Z*) and their position. Hence, the use of the name, *Z*-contrast imaging, to describe this technique.

Now, we briefly discuss the achievable resolution in such *Z*-contrast imaging. Typically, the spatial extent of an atomic scattering potential is ~0.01–0.03 nm, and modern dedicated STEM instruments produce probe sizes of the order of 0.07–0.20 nm, with the lower limits achieved in aberration corrected instruments operating at 200 kV and generating about 30 pA of current. As a first-order approximation, a vertical column of atoms encountered by the probe in a thin foil specimen can be considered as a δ-function potential. An ultimate STEM, using *Z*-contrast imaging with HAADF detectors, can produce an image resolution that is a convolution of the atomic potential with the spatial distribution of the electron current in the probe. Hence, for aberration-corrected STEM instruments with sub-Å probe diameters, individual atomic columns can be resolved (Fig. 9.3.12).

The success of such aberration corrected STEM instruments in producing atomic resolution images, has generated interest in using lower acceleration voltages to avoid knock-on damage, or atomic displacements, of specimens (§5.3.2.2) during imaging. However, at lower voltage, chromatic aberration becomes the limiting factor for spatial resolution in *Z*-contrast imaging in a STEM [20]. To overcome this limitation, various monochromator designs have been incorporated in an aberration corrected STEM [20–23].

It is logical to see that the use of a HAADF detector also allows the placement of an energy-loss spectrometer along the optic axis of the microscope, in place

Figure 9.3.12 (a) HAADF *Z*-contrast image of a Co-doped TiO$_2$ film, grown epitaxially on LaAlO$_3$ substrate. The raw data, shown in the inset, has been smoothed to remove noise. Notice the bright contrast from the heavier La in the substrate on the left. (b) Shows a schematic of the TiO$_2$ anatase lattice, with grey and red spheres showing Ti and O atoms, respectively, along the projection of the image.

Adapted from [24].

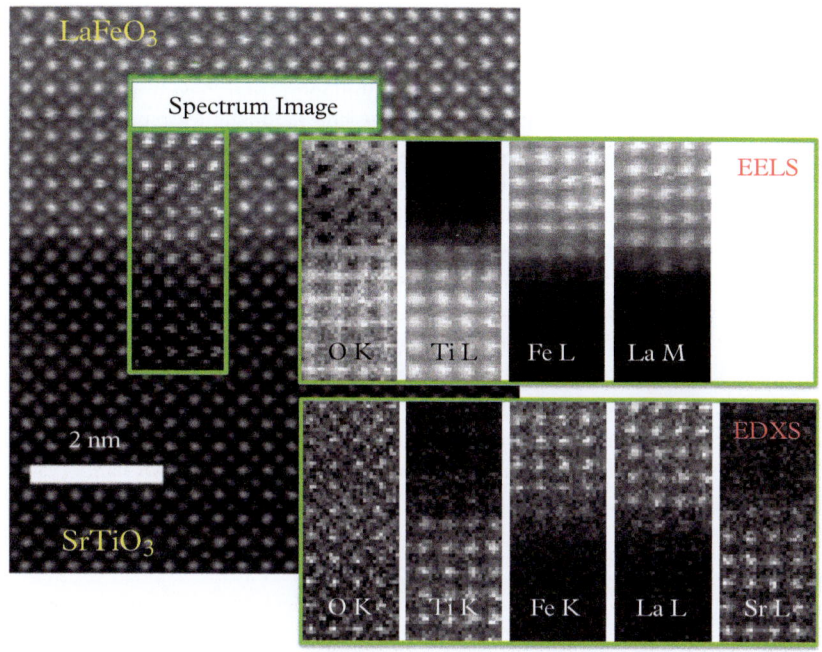

Figure 9.3.13 Simultaneously acquired atomic resolution, Z-contrast images, EELS, and the much noisier energy dispersive X-ray spectra (EDXS) maps from a $LaFeO_3/SrTiO_3$ interface.

Adapted from [25], which is also an excellent review of STEM with a historical perspective.

of a BF detector, to collect the inelastically scattered EELS (Fig. 9.3.11). By synchronizing the EELS acquisition with the raster scan of the probe, each pixel in the high-resolution STEM image can be associated with a specific EELS signal, providing simultaneous structural, chemical, and bonding information of the specimen, all at atomic resolution. EELS is discussed further in §9.4.2, but Figure 9.3.13 illustrates the power of this technique.

In summary, a HAADF detector in a STEM uses an Å-size probe and collects incoherently scattered electrons at large angles (~70–150 mrads) to produce high-quality Z-contrast images at atomic resolution. By the reciprocity theorem, to accomplish the same in a TEM will require equivalent large angles of beam convergence, which is rather difficult, if not impossible. Even when the so-called hollow cone illumination [109] is used in a TEM, we can only have convergence semi-angles of a few mrads, making it impossible to obtain pure Z-contrast images. It is inevitable at the small convergence semi-angles prevalent in the TEM that some diffraction contrast is included in the image.

9.3.5 High-Resolution Electron Microscopy (HREM): Phase Contrast Imaging in Practice

Phase contrast images are best formed when a thin crystal is oriented such that the incident electron beam is parallel to a low-index zone axis and a fixed number of diffracted beams are selected by the objective aperture to form the image

(Fig. 9.2.10d). Generally, the illumination is called *axial* if the transmitted or forward-scattered beam is along the optic axis, but if it makes an angle with the optic axis, it is called *off-axis* illumination. Such phase contrast images produce a pattern of bright and dark lines or dots arising from the interference of the diffracted beams with the direct beam and reflecting the periodicity of the crystal lattice orthogonal to the projection direction.

In the simplest form of phase contrast imaging, the interference of any two beams passing through the back focal plane of the OL and selected by the objective aperture gives rise to *lattice fringes* in the image. Such lattice fringes are formed even when the incident beam is not exactly aligned along a specific crystallographic direction of the specimen, and are observed over a wide range of lens defocus and specimen thickness values. In fact, in a specimen containing nanoparticles, or grains, with a Debye–Scherrer ring diffraction pattern, it will be impossible to align the beam parallel to a specific crystallographic direction in *all* nanocrystals. Nevertheless, even in specimens with a distribution of nanocrystal orientations, lattice fringes and even individual atomic columns can be observed (Fig. 9.3.14) without any careful alignment of the diffraction conditions, provided

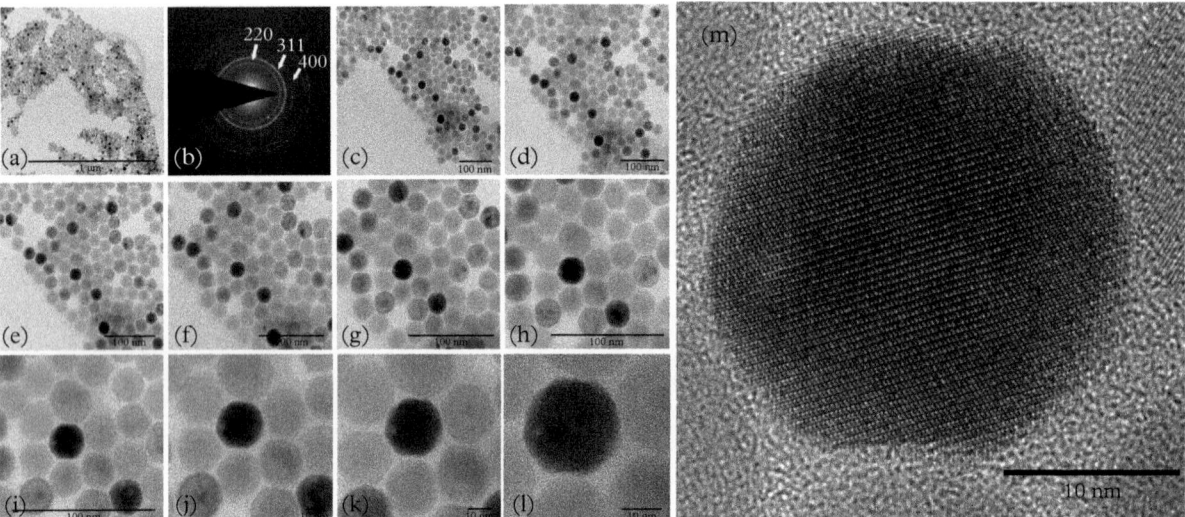

Figure 9.3.14 (a) A low-magnification image of a high purity and monodisperse iron oxide nanoparticles dispersed on a carbon grid. (b) A selected area diffraction pattern from the same region showing a ring pattern corresponding to the magnetite phase. (c)–(h) Images of the same region of the nanoparticles taken at increasingly higher magnifications. (i)–(l) phase contrast images, centered on the same particle in strong diffraction contrast, at increasing magnifications. (m) A high-resolution phase contrast image of a particle, aligned to the [310] zone axis, taken on a C_S-corrected 300 kV TEM, at a defocus of 213 nm showing lattice fringes of individual atomic columns.

Courtesy Dr. R. Hufschmid [26].

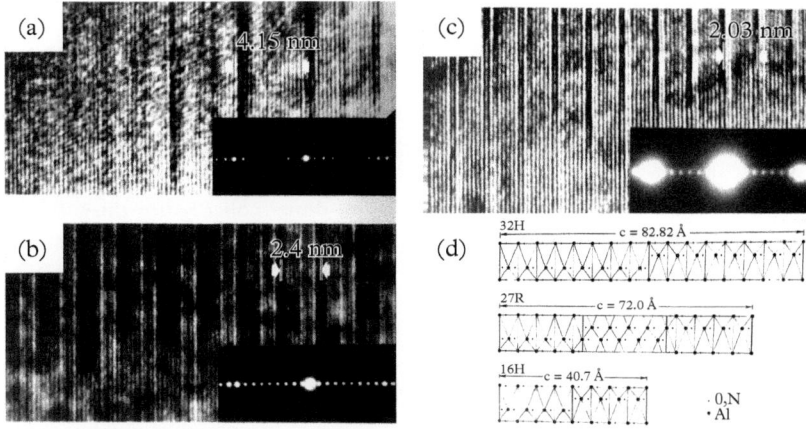

Figure 9.3.15 Polytypes in the Al-O-N system are formed by the insertion of a chemically distinct layer into another chemically distinct structure. One-dimensional lattice images observed in polytype structures of (a) 32H where the unit cell consists of two blocks, each containing sixteen layers, (b) 27R with three blocks of nine-layer repeats, and (c) 16H with two blocks of eight-layer repeats. (d) A schematic of the projections of the unit cell on the (110) planes of the three polytype structures. The 001-diffraction row for each structure is shown in the inset in (a)–(c).

Adapted from [27].

the reflections excited correspond to lattice spacing larger than the resolution of the microscope. Generally, since the diffraction conditions are not exactly specified, lattice fringes provide information only about the morphology, shape, and select inter-planar spacing in such nanocrystals.

Next, we have what are known as *one-dimensional structure images*, typically observed in long-period unit cell crystals. By tilting the orientation of the specimen, the incident beam can be made orthogonal to the long-axis of the crystal unit cell. In this case a *systematic row* of diffracted spots in one dimension, symmetric about the transmitted beam, can be excited. Phase contrast images with this systematic row of diffraction spots can provide additional information, albeit in one dimension, about the crystalline specimen. Figure 9.3.15 is such a one-dimensional structure image of *polytype*[6] structures (see Verma and Krishna, 1966) formed in the ternary Al-O-N system.

Strictly speaking, an HREM image contains no more structural information than the underlying diffraction pattern. If the incident beam is parallel to a specific crystallographic axis, typically of low index, and satisfies the Bragg condition in two dimensions, the diffraction pattern will reflect the symmetry and lattice parameters of the crystal system and unit cell, respectively. In this case, a phase contrast image formed by the interference of the transmitted and diffracted beams

[6] Forms of a crystalline substance that only differ in one dimension of the unit cell.

will produce a *two-dimensional lattice fringe* image, representative of the unit cell but not specifically the location of the individual atomic columns within it. Figure 9.3.16 shows such a two-dimensional image of β-Si_3N_4, with the incident beam along [110], and including (002) and (111) reflections, in addition to the transmitted beam. This image is rich in detail and includes many types of defects including dislocations, stacking faults, as well as twin and tilt grain boundaries. From a practical point of view, since such two-dimensional lattice images are formed with a small number of diffracted beams, the images can be formed over a

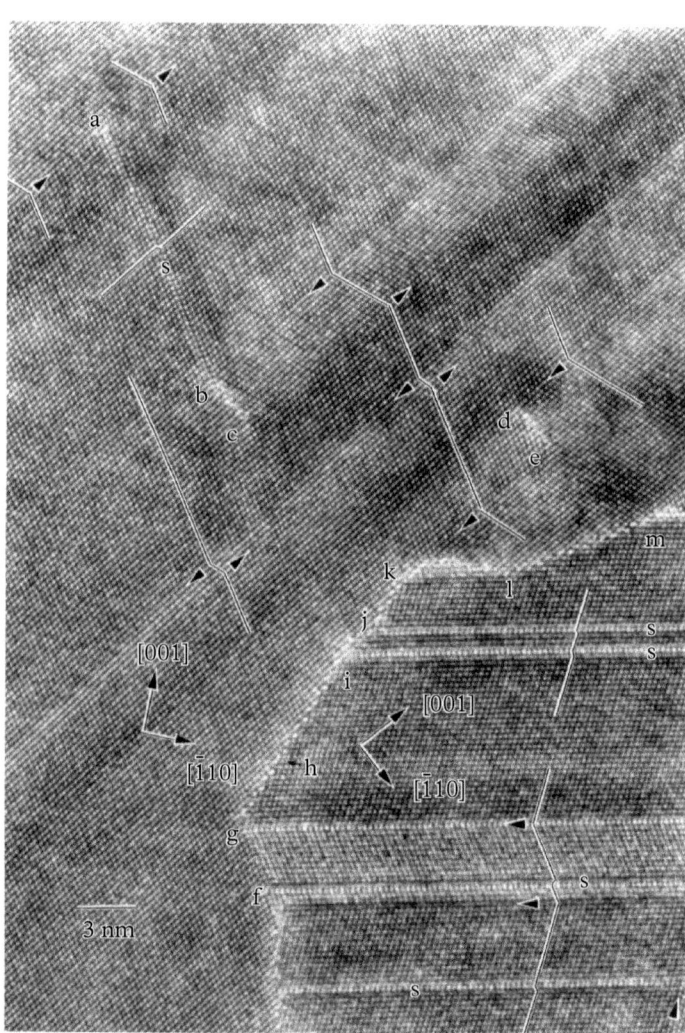

Figure 9.3.16 Two-dimensional lattice image of β-Si_3N_4 with incident beam parallel to [110]. Image shows a number of defects including dislocations (b–c, d–e), stacking faults (s), tilt grain boundaries (f through m), and twin boundaries indicated by arrowheads.

Adapted from Shindō and Hiraga (1998).

wide range of defocus (including the Scherzer value) and thickness. Again, from these images it is not easy to say whether the atomic columns appear as black or white dots in these images. However, to analyze defects and their structural details, these types of images should be taken as close to the Scherzer defocus condition as possible; moreover, very thin specimens (~5–10 nm) will facilitate direct interpretation of the images of defects.

The most sophisticated images in HREM are those that represent atomic positions as bright or dark regions in a *two-dimensional structure image*. However, true structure images are obtained only in very thin specimens (~8–10 nm thick), where the amplitude of the various diffracted beams are proportional to the thickness. Figure 9.3.17 shows the amplitude of various diffracted beams in β-Si$_3$N$_4$; up to 7 nm in thickness they all show a linear dependence with thickness. The corresponding HREM images (Fig. 9.3.18) simulated for various defocus values show best structure images are found around the Scherzer defocus (48.7 nm). Further, the optimal thickness to be used depends on the size of the unit cell, even if they are of similar densities. Figure 9.3.19 illustrates that the structure image of α-Si$_3$N$_4$ can be best obtained for specimens twice the thickness of β-Si$_3$N$_4$.

In practice, recording of optimal HREM images require that the TEM is corrected for astigmatism and specimen drift is avoided during the time of image acquisition. Both can be accomplished by recording a high magnification image of the amorphous carbon layer, typically found at the edge of the specimen. A Fourier transform or diffractogram of the image will give a good indication of the misalignment, astigmatism, and specimen drift (Fig. 9.3.20). Astigmatism typically causes the diffractogram to look elliptical instead of circular, whereas drift causes the disappearance of parts of the rings along the drift direction. Such astigmatism can be corrected in a TEM using the pair of stigmators—a set of two magnetic quadrupole lenses, one above the other, and rotated by 45°—positioned above the condenser lens, C1, to produce a circular incident beam, and above the OL to produce a focused, minimum-contrast image.

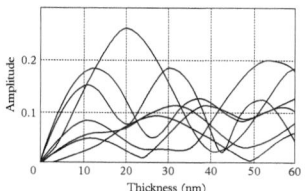

Figure 9.3.17 Dependence of the amplitudes of different diffracted beams as a function of specimen thickness calculated for a 400 kV TEM.

Adapted from Shindō and Hiraga (1998).

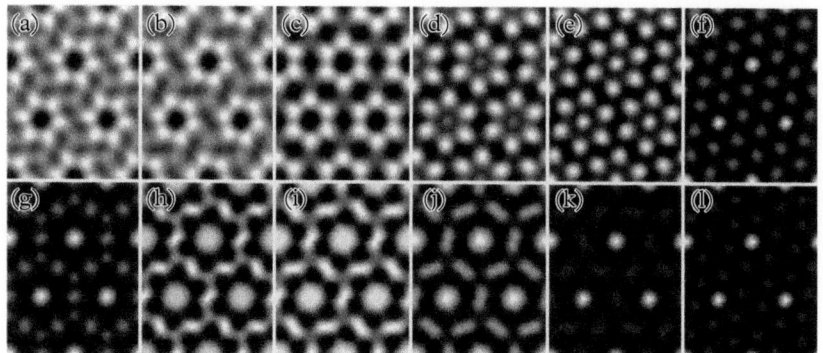

Figure 9.3.18 Simulated high-resolution images of a thin (3 nm) specimen of β-Si$_3$N$_4$, for [001] incidence and 400 kV, as a function of defocus varying from −40 nm to +70 nm in steps of 10 nm. The structure images (Fig. 9.3.19) appear close to Scherzer defocus ($\Delta f = 30 - 50$ nm). Compare with Figure 1.4.14.

Adapted from Shindō and Hiraga (1998).

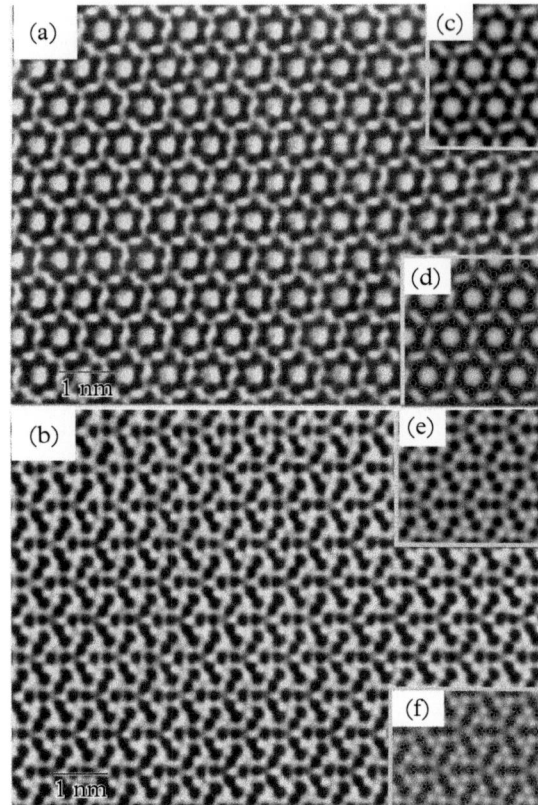

Figure 9.3.19 Structure images of (a) β-Si_3N_4, and (b) α-Si_3N_4. The corresponding simulated images, assuming a thickness of 3 nm and defocus, $\Delta f = 45$ nm, with and without the atom positions superimposed are in (c, d) and (e, f), respectively.

Adapted from Shindō and Hiraga (1998).

Once the astigmatism and specimen drift are eliminated, a good HREM image can be recorded, provided steps are taken to ensure that the microscope is well aligned, an optimal thin area of the specimen has been identified, and the optics are set up with the required diffraction pattern. Finally, it is required to set the optimum defocus. To do so, again an amorphous region of the specimen is used to first find the focus condition that produces a *minimum contrast condition*, i.e. when nearly all details in the image of the amorphous material disappear.[7] In this case, the CTF is close to zero over a wide range of spatial frequencies and occurs at a defocus, $\Delta f_{mc} = -0.44(C_S\lambda)^{1/2}$. For a given microscope, with a given C_S, the minimum contrast occurs at a specific value of defocus, e.g. for a JEOL 4000EX microscope, it is at $\Delta f_{mc} = -18$ nm. From here, a specific number of defocus steps would take one to the Scherzer defocus, which for the same microscope is at $\Delta f_{Sch} = -49$ nm. To avoid any errors in defocus setting, it is common practice to record images over a range of focus values (through focus series).

HREM structure images have two significant limitations: they provide very little information on the atomic number of the elements in the specimen, and they reveal only a projection of the atomic columns in the crystal structure

[7] Fresnel fringes can also be used for focusing. Typically, a white fringe means under-focus, and so raising the specimen or increasing objective lens strength will get the image into focus.

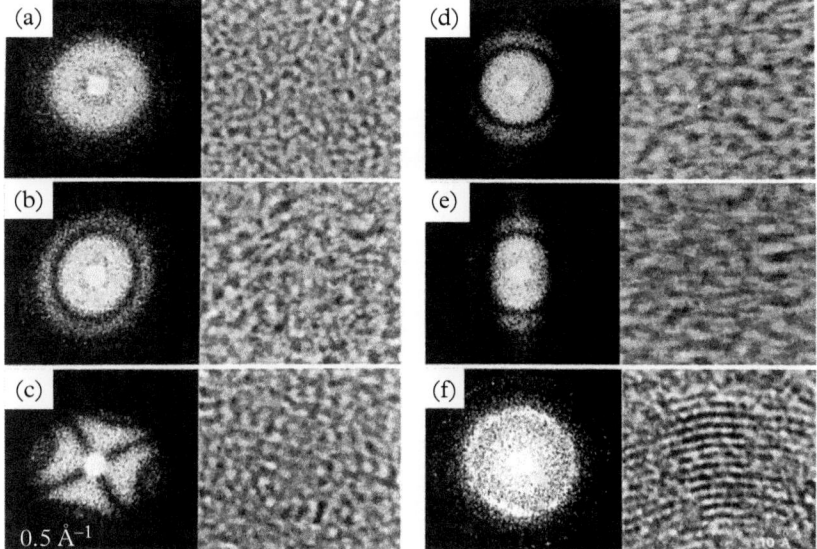

Figure 9.3.20 Images of an amorphous carbon film and their diffractograms under different imaging conditions using a 300 kV HREM. (a) Well-aligned, no drift, astigmatism corrected. (b) Some astigmatism. (c) Significant astigmatism. (d) No astigmatism but with small drift (0.3 nm). (e) No astigmatism but with larger drift. (f) Well-aligned, no drift, and no astigmatism with graphite fringes of 0.344 nm spacing.

Adapted from Williams and Carter (1996).

along the incident beam direction. Moreover, they require very thin specimens for direct structural interpretations, and in many cases where there may be significant surface reconstruction (§8.3.1), thicker specimens may be necessary to reveal the true structural details. In the latter case, inelastic scattering of electrons in the specimen will deplete the elastic wave field, which in turn, will affect the phase contrast images for specimens of thickness greater than a few tens of nanometers. In §9.4 we briefly introduce inelastic scattering and most importantly, present two important spectroscopy methods that arise from inelastic scattering that complement imaging and, with the incorporation of appropriate spectrometers, provide quantitative chemical and electronic structure information of the specimen in an (analytical) electron microscope.

Example 9.3.1: A thin film of NiO ($a = 4.18$ Å) includes small precipitates of Ni ($a = 3.52$ Å) that are perfectly aligned, i.e. $(001)_{NiO} \parallel (001)_{Ni}$ and $[100]_{NiO} \parallel [100]_{Ni}$. In a TEM experiment a diffraction pattern is taken such the beam direction is along the [001] direction of the crystals. Then a two-beam lattice image is also produced.

    (a)   Draw the (001) zone axis pattern for NiO and Ni.

    (b)   What would the diffraction pattern look like if double diffraction is included?

    (c)   Describe the two-beam lattice image including double diffraction.

Solution: We refer to the following figure to answer the questions.

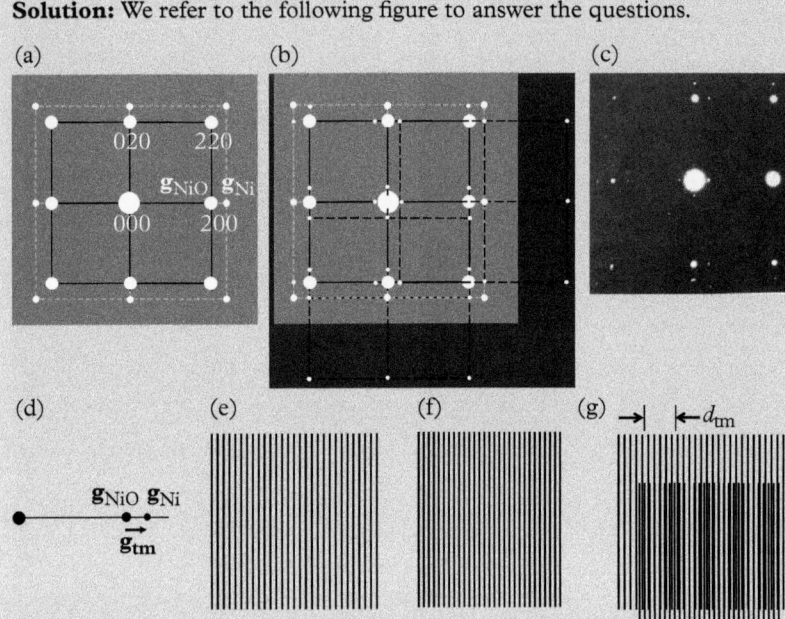

(a) The (001) zone axis pattern for the two crystals, with the reflections indexed, are drawn to scale. NiO has the larger lattice parameter and since it is the matrix has brighter intensity (diameter of the circles).

(b) In double diffraction, each diffracted spot of the precipitate (Ni) acts as an incident beam for the matrix phase (NiO). In other words, the two diffraction patterns are convoluted. A simple way to visualize this is to separate several matrix patterns (two shown) displaced by diffraction vectors of the precipitate. This generates all the double diffraction patterns.

(c) An experimental diffraction pattern, adapted from Williams and Carter (1996), from perfectly aligned Ni and NiO, agrees very well with (b).

(d) Lattice fringes are produced by the interference of transmitted and diffracted beams. Similarly, we can also produce an interference between two beams that are diffracted from the matrix and precipitate. Such fringe patterns are called Moiré patterns and the simplest kind, called a translational Moiré, involves lattice planes that are parallel, but with different spacings. The two reciprocal lattice vectors, $\mathbf{g}_{NiO}$ and $\mathbf{g}_{Ni}$, when perfectly aligned will effectively generate another spacing defined by $\mathbf{g}_{tm} = \mathbf{g}_{Ni} - \mathbf{g}_{NiO}$, with spacing given by $1/\mathbf{g}_{tm}$ in the lattice image.

(e) The lattice image of NiO, and (f) the lattice image of Ni, showing the (200) lattice planes to scale.

(g) The translational Moiré pattern produces an additional period, d_{tm}, where $d_{tm} = \frac{1}{\mathbf{g}_{tm}} = \frac{d_{NiO}}{1 - \frac{d_{Ni}}{d_{NiO}}} = \frac{2.09}{1 - \frac{1.76}{2.09}} = 13.24$ Å, as illustrated in the figure. Note that Moiré patterns can also be produced by the rotation of one crystal with respect to the other.

9.3.6 Magnetic Contrast: Lorentz Microscopy

The two conventional modes of Lorentz[8] microscopy, Fresnel[9] and Foucault[10] imaging, are now well established. In Fresnel imaging, a small *defocus*, Δz, is introduced, and contrast arises wherever there is a spatially varying *in-plane component* of magnetic induction. Fresnel contrast can be understood based on a classical ray diagram (Fig. 9.3.21a). The electrons on passing through the

[8] Hendrik Antoon Lorentz (1853–1928) was a Dutch physicist who shared the Nobel prize (1902) in Physics with P. Zeeman and was cited for "their research into the influence of magnetism upon radiation phenomenon."

[9] Augustin-Jean Fresnel (1788–1827), French physicist who contributed to early developments in optics.

[10] Jean-Bernard-Leon Foucault (1819–1868) was a French physicist.

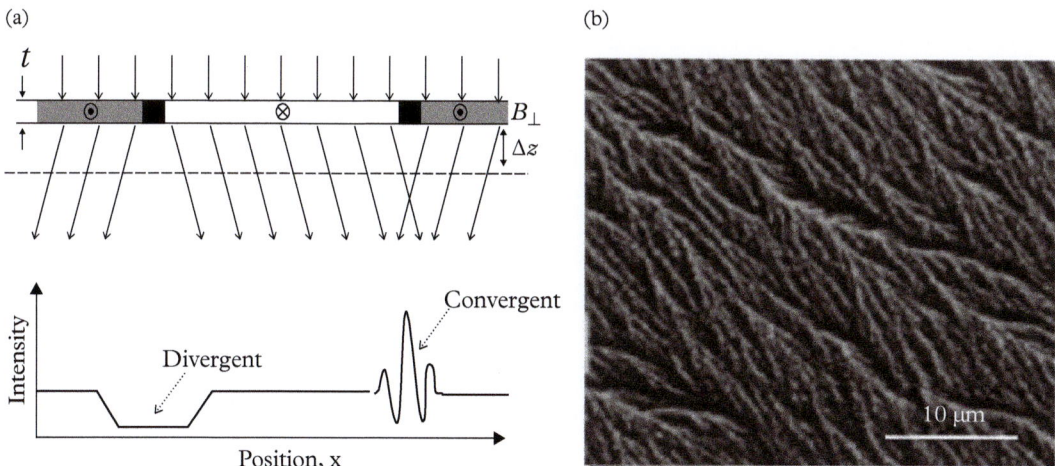

Figure 9.3.21 (a) Schematic illustration of the contrast in a Fresnel image, obtained with a defocus, Δz, of two ferromagnetic domains separated by a 180° domain wall. Note that even if there is a net deflection by the magnetic specimen, in a focused bright field image it will not show any magnetic contrast. (b) Example of a Fresnel image of domains in a 30-nm thick cobalt film. The magnetization ripple contrast allows for interpretation of the direction of magnetization in the interior of the domains [29].

specimen are deflected by the Lorentz force arising from its in-plane induction integrated over its thickness. In the example, the two domains with antiparallel directions of induction will deflect the beam in opposite directions. At any plane not coincident with the specimen plane, defined by the defocus, Δz, there will be regions beneath the domains where the electron intensity will be enhanced or reduced above a uniform background signal. Hence, domain walls are revealed as narrow dark or bright bands, on a uniform background with no contrast from the interior of the domains (Fig. 9.3.21b). Based on this simple description of Lorentz deflection, Fresnel images of domain walls arising from both the convergent (bright) and divergent (dark) electrons should be uniform and devoid of any additional fine structure. However, the narrow bright band corresponding to the convergent wall reveals additional fringes that arise from the interference of the waves from the neighboring regions of uniform but opposing directions of magnetic induction. Details of the intensity distribution of the fringes, if carefully analyzed, using models of the magnetization distribution in the walls, combined if necessary, with micromagnetic simulations, can provide additional information on any induction variation within the wall itself.

Fresnel imaging is characterized by operational simplicity, high contrast, and no directional preference for imaging domain walls. However, since there is no contrast from the interior of the domains, in the absence of magnetization ripple[11] it is difficult to determine the local direction of induction.

In contrast to Fresnel imaging, the Foucault mode is an in-focus method of domain imaging and relies on a deliberate shifting of the objective aperture to generate magnetic contrast (Fig. 9.3.22a). If the objective aperture is positioned in the back focal plane such that scattering vector components, $k_x < 0$, are obstructed and $k_x > 0$ are allowed to transmit unchanged, an image as shown will be recorded. In this simple illustration, using a classical description, only those electrons with positive value of the deflection angle, $\beta_x > 0$, contribute to a bright image. Thus, domains that deflect electrons in a well-defined direction, selected by the position of the objective aperture, will appear as high-intensity, bright areas in the image. All other electrons scattered outside the objective aperture will appear as zero intensity. Thus, manipulating the position of the objective aperture allows the imaging of domains and the qualitative determination of the induction along any given direction (Fig. 9.3.22b).

Fresnel and Foucault methods are complementary; the former determines the location of domain boundaries and the latter provides information on the directions of in-plane induction inside specific domains. However, under standard operation both methods do not provide a *quantitative* description of the spatial variation of the induction in the specimen.

The alternative method of *differential phase contrast* (DPC) *imaging* is best implemented in a STEM equipped with a large circular electron detector that is split into four quadrants (Fig. 9.3.18a). In the STEM, a small probe of electrons, often from a coherent source, is scanned across the specimen and the scattered electrons are detected to form the image in a time-sequential manner. The

[11] Magnetization ripple is a characteristic fluctuation of magnetization always oriented perpendicular to the average magnetization direction of the domain.

(a)

(b)

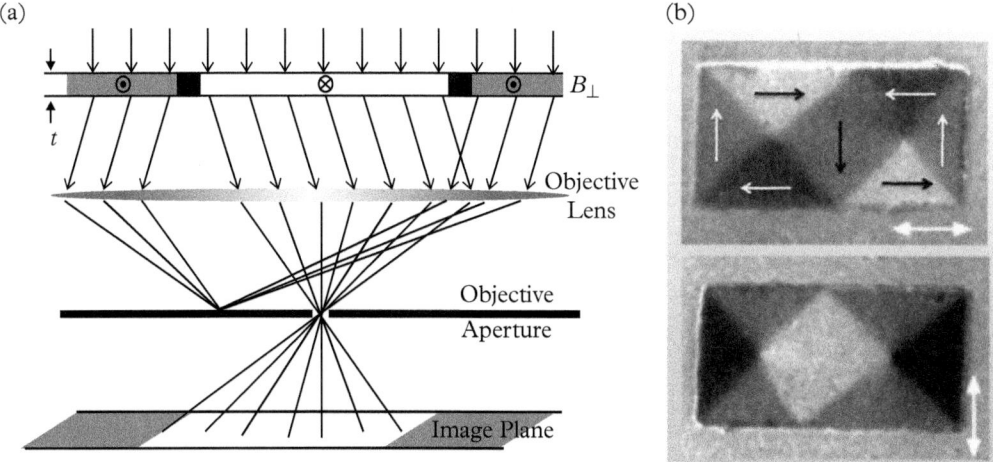

Figure 9.3.22 (a) Schematic illustration of the image formation in a TEM in the Foucault mode. Notice that the interior of the domains is imaged. (b) Foucault images of a micron-scale magnetic element along two perpendicular scattering directions allow the unambiguous determination of the magnetization direction in the domains.

Images courtesy of Professor J. N. Chapman.

detector, of fixed size, is positioned in the far-field with respect to the specimen. A set of post-specimen lenses between the specimen and detector ensure both a variable cameral length and the collection of electrons over a wide angular range of scattering. In this arrangement, the DPC image is the difference in signal between the two halves or opposing quadrants of the detector [30, 31].

Figure 9.3.23a shows a classical representation of the formation of DPC images and its relationship to the magnetic structure. The focused probe is scanned across the specimen, and at each point it is deflected by the local Lorentz angle:

$$\beta_x(x) = \frac{e\lambda}{h} \int_{-\infty}^{+\infty} B_y(x, z)\, dx = B_y(x) \frac{e\lambda t}{h} \qquad (9.3.13)$$

where $B_y(x)$ is the average in-plane induction along the trajectory of the electrons. It then passes through the descan coils before arriving at the detector to ensure that, in the absence of a specimen, the beam is centered and the number of electrons arriving at each sector of the detector is the same and independent of the position of the probe on the specimen. Thus, if the probe undergoes a deflection, β, at any point on the specimen as shown, the electrons falling on the opposing sectors of the detector would be different. The difference in signal would be proportional to β in both magnitude and direction. Further, by scanning the probe over the specimen surface, the deflection at every position, $\beta(x,y)$, can

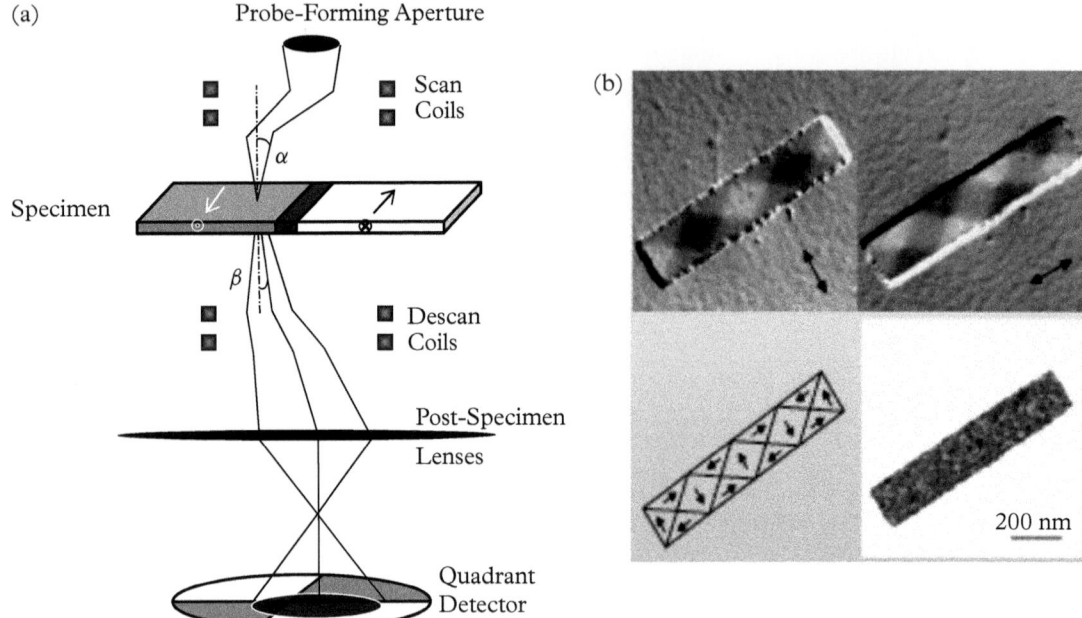

Figure 9.3.23 (a) Schematic arrangement of the implementation of the differential phase contrast imaging in a STEM. The intensities measured in each of the four quadrants of the detector provides a quantitative map of the magnitude and direction of the beam deflection at each position, proportional to the in-plane induction, as the probe is scanned across the specimen surface. (b) Representative DPC images of a cobalt magnetic element. Two sets of difference images, a vector map of the magnetization and the sum of the intensities in the four quadrants, equivalent to a conventional TEM image are shown.

Images courtesy of Professor J. N. Chapman.

be mapped. A quadrant detector is used to resolve both components of the in-plane induction (Fig. 9.3.23a). Since this is an in-focus method, if the signal from the four quadrants is added together, instead of taking the difference, a normal BF image of the specimen, revealing any non-magnetic features (such as defects, grain boundaries, etc.) of interest, is also produced. If a STEM with a quadrant detector is not available, a DPC image can also be produced in a conventional TEM. Using reciprocity arguments, we see [32] that DPC contrast maps of the in-plane induction can be obtained in a conventional TEM by digitally combining a series of Foucault images taken with small increments of the beam tilt in two orthogonal directions.

From a practical point of view, DPC imaging in a STEM can provide resolutions of the order of ~10 nm. It is very sensitive and is able to detect small quantities such as a few layers of Fe. However, all TEM imaging requires electron transparent specimens; often, this entails thinning of the specimen and this procedure may affect the domain structures to be observed. Also, typically electron-optical lenses generate strong magnetic fields that may affect/alter the magnetic structure of the specimen during imaging. There are three ways of avoiding the effect of the fields generated by the excitation of imaging lenses— turn the OL off or move the specimen away from the lens. Both these methods degrade the resolution of the instrument. Alternatively, dedicated field-free lenses specially designed for magnetic imaging can be installed. Finally, in DPC imaging due to the contribution from microstructural features the scattered pattern may not be symmetric and may sometime lead to pronounced contribution to the phase contrast image. To suppress the effect of such scattering, a ring detector (essentially an octant detector), instead of a quadrant, is used to exclude the contribution from the central beam [33].

9.3.7 Electron Holography

Even though electron holography was originally proposed [33,34] to correct for microscope aberrations as a means to achieve superior resolution, its usefulness for magnetization measurements [34], recognized immediately afterwards, is emphasized here. The applications of electron holography to magnetic materials are based on the recording of an interference pattern from which the amplitude and phase of the object is reconstructed. Magnetic materials are strong phase objects and the phase shift of electrons passing through a magnetic specimen is proportional to the flux enclosed. For example, two beams starting from a coherent point source, passing through a thin ferromagnetic specimen (Fig. 9.3.24a) at two different points, P_1 and P_2, and then brought to the same observation point on the image plane, would undergo a phase difference, $\Delta\phi$, due to the enclosed flux given by

$$\Delta\phi(x) = \frac{et}{\hbar}\int_{x_1}^{x_2} B_y(x)\,dx \qquad (9.3.14)$$

where x_1, x_2 are the positions of P_1, P_2, respectively, and $B_y(x)$ is the average value of the in-plane induction of the specimen along the electron trajectory. Naturally, if P_1 and P_2 were to lie on the same magnetic line of force in the film, the phase difference would be zero. Therefore, magnetic lines of force may be directly observed as contour maps of the electron wave propagation by electron holography. This simple interpretation is valid provided the phase shift is entirely magnetic in origin.

In practice, an *electron biprism* (typically, a charged wire placed between two earthed plates) is required to split the incoming electron wave into two virtual coherent sources and form the interference pattern. Even though the phase difference between any two arbitrary points in the specimen plane can be defined in this way, in the most commonly implemented form of electron

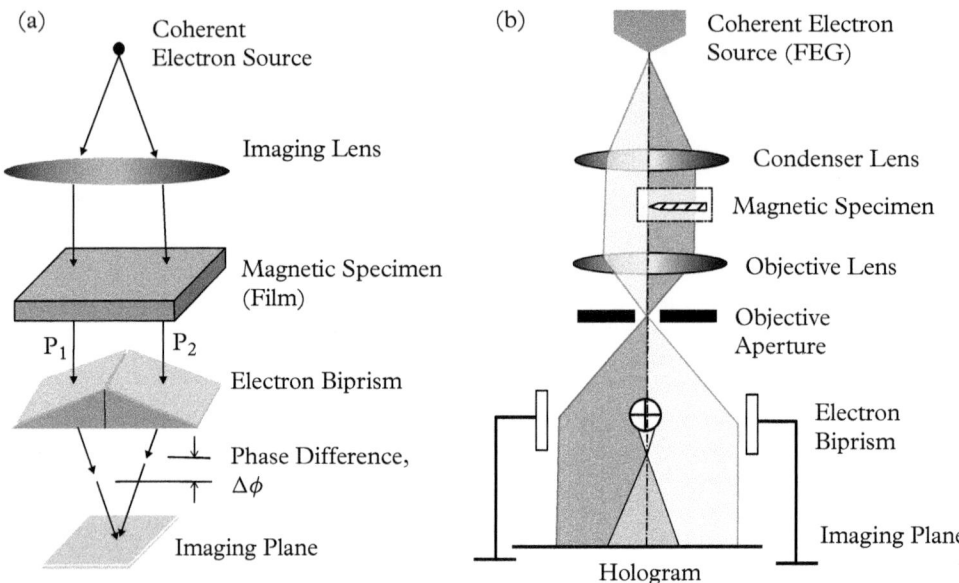

Figure 9.3.24 (a) Principle of electron holography. Two beams from a coherent source converging at the same observation point after passing through a magnetic specimen, with in-plane induction as shown by arrow, would undergo a phase difference proportional to the enclosed flux. (b) Schematic drawing of the experimental arrangement for off-axis holography imaging in a TEM. A charged wire serves as the biprism splitting the beam into a reference wave (light) and a wave that passes only through the specimen (dark) and causes a phase change. The interference of the two waves produces the hologram.

Adapted from Krishnan (2016).

holography, a vacuum or a reference wave that avoids the specimen completely is made to overlap with the wave scattered by the specimen to form the hologram (Fig. 9.3.24b). This configuration is termed off-axis or side-band holography and is one of at least twenty different ways identified to implement electron holography [35] in a TEM. Finally, the magnetic phase image is reconstructed from the hologram by on-line digital processing methods [37] to achieve truly quantitative electron holography. The resolution of the reconstructed image depends on the spacing of the fringes in the hologram. For magnetic imaging, a large number of fringes are desirable, and as large as 500 fringes have been reported [36]. To accomplish this, electron sources with high spatial coherence are desirable. High brightness, spatially coherent, field emission gun sources [38], now routinely available on commercial TEMs, greatly facilitate the implementation of electron holography. The principles of off-axis electron holography are discussed next; however, the monograph by Tonomura (1999) and a recent review [116] are comprehensive and contain more details of all aspects of electron holography.

Figure 9.3.24b shows the experimental set-up for holography in a conventional TEM with a coherent FEG source. An electrostatic biprism is used to overlap the object (interacting with the specimen placed such that it covers about half the field of view) and reference wave (the other non-interacting half) to form the interference pattern. Ideally, the biprism would be rotatable, in plane, to facilitate the alignment of the interference fringes with the microstructural features of interest in the specimen. Quantitative analysis of phase shifts is carried out by acquiring data with a linear detector having a large dynamic range such as a CCD camera. The intensity distribution in the hologram, using coherent reference and object wave functions, Ψ_r and Ψ_0, respectively, is given by

$$I(x,y) = |\Psi_r(x,y)|^2 + \left|\Psi_0(x,y)\right|^2 + |\Psi_r(x,y)\|\Psi_0(x,y)| \left[e^{i(\phi_1-\phi_2)} + e^{-i(\phi_1-\phi_2)}\right]$$
$$= A_r^2 + A_0^2 + 2A_rA_0 \cos \Delta\phi \qquad (9.3.15)$$

where ϕ is the phase angle and A is the amplitude of the waves. When this hologram is recorded, a series of cosine fringes are superimposed on a normal TEM BF image (Fig. 9.3.25a). A Fourier transform of the hologram (Fig. 9.3.25b) reveals two complex conjugate side bands that, in addition to the fundamental cosine frequency, contain additional intensity distributions that include complete information on the amplitude and phase of the waves. The separation of the side bands can be controlled by the relative angular deviation of the object and reference waves or, in practice, the voltage applied to the biprism. Only one of the side bands is necessary for the reconstruction process.

To reconstruct the hologram, one of the side bands is extracted, re-centered. and then its inverse Fourier transform is calculated. The resulting image is complex and using the real (r) and imaginary (i) data sets, the phase (ϕ) and amplitude (A) can be calculated as

$$\phi = \tan^{-1}\left(\frac{i}{r}\right)$$
$$A = \sqrt{r^2 + i^2} \qquad (9.3.16)$$

This is shown in Figure 9.3.25c,d. In one dimension, the change in phase of the object wave incident in the $\hat{z}$ direction, on a specimen in the x-y plane, is given by

$$\phi(x) = C_E \int V(x,z)\, dz - \frac{e}{\hbar}\int B_\perp(x,z)dxdz \qquad (9.3.17)$$

where, $V(x,z)$ is the mean inner potential and $B_\perp$ is the component of the magnetic induction perpendicular to both $\hat{x}$ and $\hat{z}$. Here, C_E is given by

$$C_E = \frac{2\pi}{\lambda E}\frac{E+E_0}{E+2E_0} \qquad (9.3.18)$$

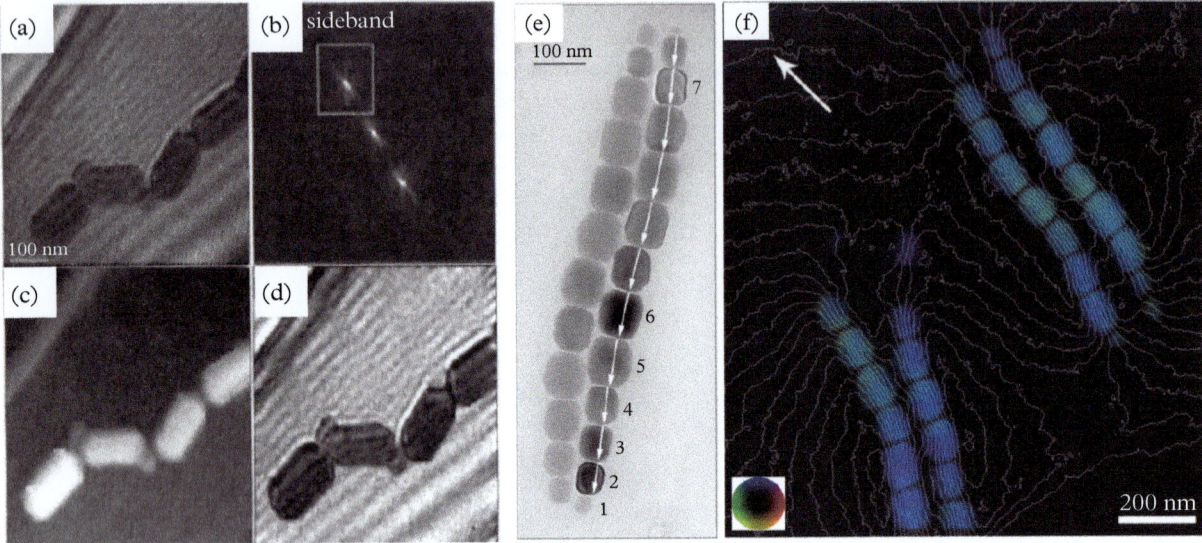

Figure 9.3.25 (a) Off-axis hologram for a chain of magnetosomes. (b) Fourier transform of the hologram showing the side bands, one of which is used for the reconstruction. (c) The phase image, and (d) the amplitude image. (e) Bright field image of a double chain of magnetite magnetosomes, with white arrows representing [111] crystallographic directions. (f) Magnetic phase contours measured using off-axis holography from two sets of magnetosomes, after magnetizing parallel and antiparallel to the arrow. The color wheel shows the direction of magnetic induction. Contours of spacing 0.25 radians can be seen.

(a–d) Adapted from [39]. (e–f) Adapted from [117].

where, λ is the wavelength, E is the kinetic energy, and E_0 is the rest-mass energy of the electron. Assuming that both V and $B_\perp$ do not vary with $\hat{z}$, neglecting any stray magnetic and electric fields, we can simplify (9.3.17) as

$$\phi(x) = C_E V(x) t(x) - \frac{e}{\hbar} \int B_\perp(x) t(x) dx \qquad (9.3.19)$$

Differentiating with respect to x, we get

$$\frac{d\phi(x)}{dx} = C_E \frac{d}{dx} [V(x) t(x)] - \frac{e}{\hbar} B_\perp(x) t(x) \qquad (9.3.20)$$

If the specimen is uniform in thickness and composition, the first term in (9.3.20) is negligible, and the in-plane induction is proportional to the phase gradient. However, in typical TEM specimens showing significant and rapid thickness variations, the mean inner potential contribution, $V(x) t(x)$, can dominate and complicate the determination of the magnetic contribution. However, time reversal operation of the electron beam—conceptually simple by physically inverting the specimen [40], but difficult to implement in practice—can be used to eliminate the effect of the mean inner potential. Currently, a phase sensitivity of $\pi/50$ can be routinely achieved in electron holography. Thus, magnetic induction on the spatial scale of 5 nm can be detected [41]. Further improvement in spatial resolution may require thicker specimens, larger intrinsic specimen magnetization, or longer data acquisition times.

9.4 Analytical Electron Microscopy (AEM) and Related Spectroscopies

In addition to elastic scattering, which gives rise to the diffraction and imaging methods in a TEM (§9.3), inelastic interactions of the electron beam with the inner-shell, valence, or conduction electrons of the specimen provide information about the electronic structure and chemistry of the solid. The main inelastic processes (Fig. 9.4.1) can lead to intra-band, inter-band, and plasmon excitations, and the energy transferred to the specimen can be subsequently emitted in the form of secondary (§10.2.3) electrons, light, or CL (§10.6.2), or following inner-shell ionization, by the emission of characteristic X-rays (§2.3.2), or Auger (§2.3.3) electrons. Such inelastic scattering is significant; in fact, for elements with atomic number smaller than Cu, the total single electron inelastic scattering exceeds the total elastic scattering.

The incident electron beam of energy, E_0, loses energy characteristic of the process of excitation, and the energy it transfers to the specimen can be observed as an energy loss, ΔE, of the incident electron reducing its kinetic energy to $E_0 - \Delta E$. A magnetic sector electron spectrometer (Fig. 9.4.8) mounted at the end of the electron column can record this electron energy loss-spectrum (EELS) and provide a wealth of information about the specimen; note that the underlying physics of EELS is similar to XAS (§3.9). In a STEM using a highly focused, and where possible, coherent probe, EELS is often combined with micro-diffraction (§9.4.5) or Z-contrast imaging (Fig. 9.3.11) to provide both crystallographic and electronic structure information at sub-nm spatial resolution. Note that in the EELS literature there is sometimes a tendency to separate the electronic structure from the crystal structure, but this is quite arbitrary as the two are strongly related in a solid.

The addition of an energy dispersive X-ray spectrometry (EDXS) detector (§2.5.1.1), typically mounted on the electron optical column to detect and quantify the characteristic X-rays, completes the capabilities of such an AEM. An important advantage of EELS over EDXS is its collection efficiency (Fig. 9.4.2). Beam electrons inelastically scattered by inner-shell ionization processes are concentrated within small scattering angles in the forward direction

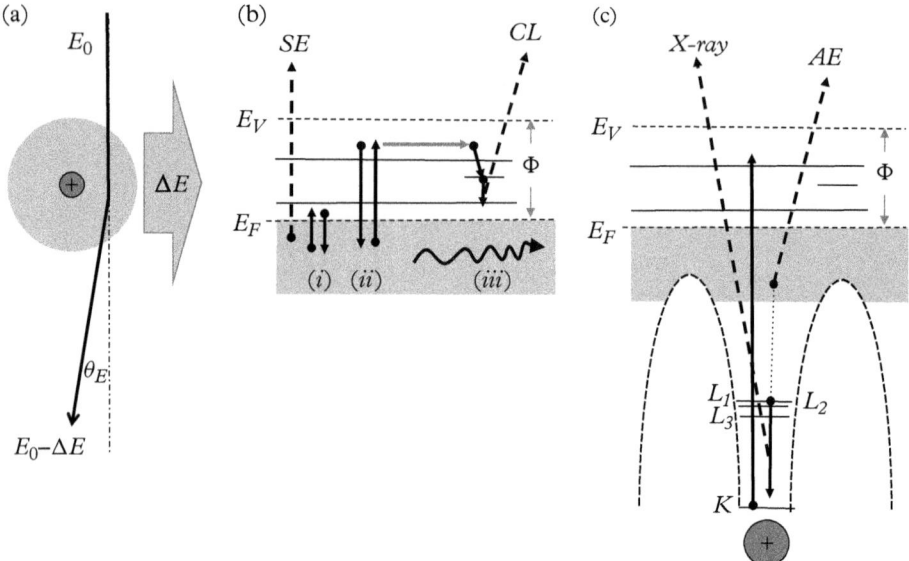

Figure 9.4.1 Inelastic interactions of the electron beam with the specimen. (a) The related energy loss, ΔE, results in scattering through very small angles, θ_E, and is measured in electron energy-loss spectroscopy (EELS). (b) If ΔE is small in magnitude it leads to (i) intra-band and (ii) inter-band transitions, as well as (iii) plasmon excitations that are collective oscillations of the valence band. If the excited electrons have energies greater than the work function, Φ, they leave as secondary electrons (*SE*). Inter-band excited electrons can recombine by the emission of light (*CL*) or as a radiation-free phonon (not shown) generation. (c) At much larger energy losses, inner shell ionizations take place and the de-excitation process that follows produces either characteristic X-rays or Auger electrons (*AE*). Detecting the former is similar to EPMA (§2.5.2.2), and the latter is typically not detected in an analytical electron microscope. Note that for labeling purposes this figure shows one-electron atomic orbitals even though in practice EELS measures differences in total energy between many-electron states in the solid. See Egerton (1996) for a more detailed discussion.

and the EELS spectrometer, with a typical entrance aperture semiangle, $\beta < 20$ mrad, will collect them with efficiencies of the order of ~50 %. On the other hand, characteristic X-ray quanta are emitted isotropically in all directions and only a small solid angle (~0.1–1 sr) are detected by the EDXS detector, which results in very inefficient X-ray collection (or 0.8–8 % of all emitted X-rays) and requires long counting times to establish good statistics. In addition, since EELS observes the primary inelastic scattering event, unlike the secondary de-excitation process in EDXS, it can provide information on the wide range of possible inelastic interactions. These are introduced in the next section.

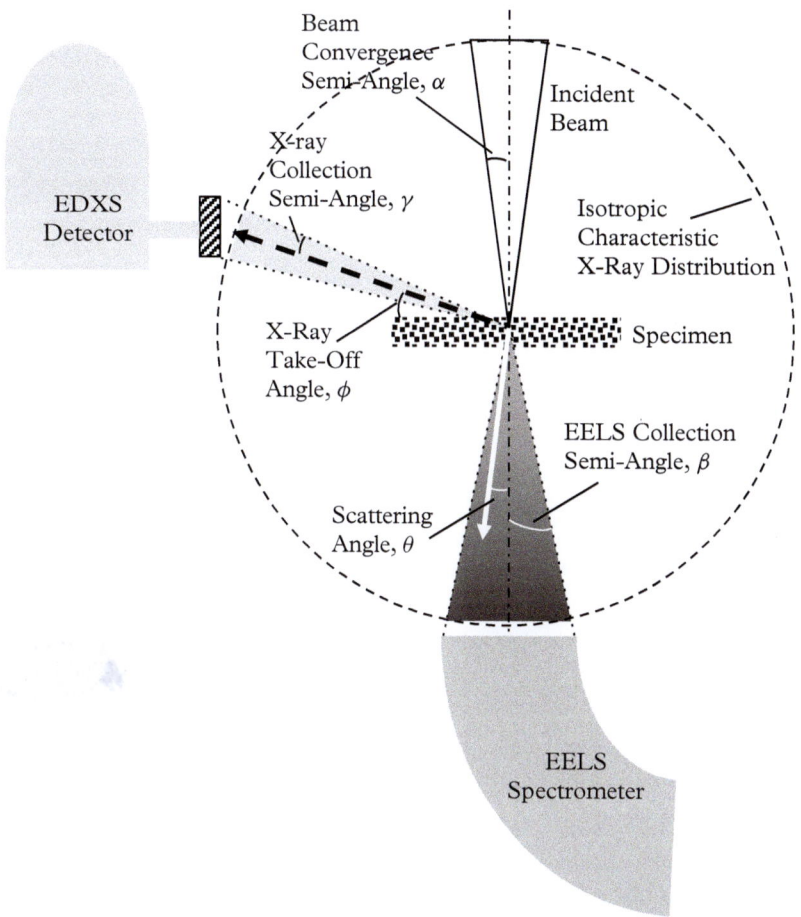

Figure 9.4.2 Comparison of the spatial distribution and collection efficiencies of the beam electrons inelastically scattered in the forward direction that constitute the EELS signal, and the isotropically distributed characteristic X-ray signals, arising from de-excitation process, only a small fraction of which is collected by the EDXS detector. The figure also illustrates the experimental arrangements of EELS and EDXS in a TEM, and the scattering angle, θ, the electron-beam convergence semi-angle, α, and the EELS spectrometer acceptance semi-angle, β, as well as the X-ray take-off angle, ϕ, and the collection semi-angle, γ.

9.4.1 Inelastic Scattering and Spectroscopy

Inelastic scattering processes arise from the interaction of the beam[12] electrons with the electrons in the solid (Fig. 9.4.1). It conserves the *total* energy and momentum, and part of the kinetic energy of the incident beam electrons is converted to excitations of the electrons in the solid. Needless to say, the primary electrons lose the same energy and scatters through small angles in the forward direction (Fig. 9.4.2). For our discussion, we divide the inelastic scattering and the associated energy losses into three broad categories.

At very low energy losses (20 meV–1 eV), we expect excitations of oscillations in molecules and gases, as well as phonon excitations in solids. Recall that the energy widths of the primary beam (Table 5.2.2) ranges from ∼1eV for

---

[12] We also refer to beam electrons interchangeably as incident, primary, or fast electrons.

thermionic guns to 0.1–0.5 eV for field emission and Schottky sources. Hence, the observation of such very low energy-losses in a TEM requires monochromatization of the primary beam, using in-column energy filters (such as the Ω-filter mentioned in §9.4.2.6). We shall not discuss this energy regime here, but further details can be found in the monograph by Ibach and Mills (1982) and the recent developments in high-energy resolution monochromators and spectrometers with sub-10 meV resolution [112] and sub-Å spatial resolution [113].

At slightly larger energy losses, $\Delta E \sim$ 1–50 eV, intra-band and inter-band excitations of the outer atomic electrons, and collective longitudinal oscillations of the valence or conduction electron bands, known as plasmons (Fig. 9.4.1b) with a broad maximum in energy-loss are observed. To first order, these discrete surface and volume plasmon losses, ΔE_p, are proportional to the square root of the electron concentration, n, in the band, i.e. $\Delta E_p \propto n^{1/2}$. Note that there is a clear distinction between the underlying physics of energy lost to plasmon excitations and to inner-shell ionization. In the former case, the initial state of the electrons in the band (see §3.2) occupies a range of energies but in the latter case they occupy a sharp energy level (Fig. 9.4.1c). As a result, unlike inner-shell ionization, the energy loss to collective excitation of the band electrons cannot be treated in terms of simple atom-like behavior but requires the treatment of the solid as a many electron system. This is difficult to do in quantum mechanics, and to overcome this difficulty it is common practice to describe plasmon excitations phenomenologically in terms of the dielectric behavior of the solid. Further details of this approach can be found in Raether (1980), Reimer (1995), and Egerton (1996).

The cross-section for plasmon excitation for a primary electron of energy, E_0, per unit solid angle, is given [42] per band-electron per unit volume as

$$\frac{d\sigma_p}{d\Omega} = \frac{1}{2\pi a_0} \frac{\theta_p}{\theta^2 + \theta_p^2} \tag{9.4.1}$$

where $a_0 = 0.529$ Å is the Bohr radius (§2.2.1), the characteristic plasmon scattering angle, $\theta_p = \Delta E_p / 2E_0$, the element of solid angle, $d\Omega = 2\pi \sin\theta d\theta$, and θ $(0 < \theta < \pi)$ is the scattering angle (Fig. 9.4.2) measured with respect to the trajectory of the incident beam. Typical values for aluminum for $E_0 = 100$ keV are $\Delta E_p \sim 15$ eV, $\theta_p = 0.15$ mrad. This implies that the energy-lost electrons associated with plasmon scattering are highly forward peaked, and hence we can set $d\Omega = 2\pi \sin\theta d\theta = 2\pi\theta d\theta$, and integrate (9.4.1) to obtain the total plasmon cross-section:

$$\sigma_p = \int d\sigma_p (\theta) = \frac{\theta_p}{2\pi a_0} \int_0^{\theta_1} \frac{2\pi\theta}{\theta^2 + \theta_p^2} d\theta \tag{9.4.2}$$

Based on typical scattering angles in a TEM, we can safely set the upper limit of the integration, $\theta_1 = 0.2$, and multiply by the factor $\frac{n_c A}{\rho N_0}$, where n_c is the number of conduction band electrons per atom, A is the atomic weight of the elements in the specimen, ρ is its density, and N_0 is the Avogadro number, to obtain a total plasmon scattering cross-section per atom/m^2 as

$$\sigma_p = \frac{n_c A}{2\rho N_0 a_0} \theta_p \left[\ln \left(\theta_p^2 + (0.2)^2 \right) - \ln \theta_p^2 \right] \qquad (9.4.3)$$

Since $\Delta E_p \propto n^{1/2}$, in plasmon scattering the electron beam loses energy in quantized units that are detected as strong peaks in EELS. Moreover, the wavelength of the longitudinal plasmon oscillations is of the order of ~100 interatomic spacings and the associated EELS signal is not localized. In other words, the spatial resolution of plasmon excitations in EELS measurements is determined by the plasmon wavelength (several nanometers) and not by the size of the focused probe.

Last, but not least, inelastic scattering can lead to inner-shell (K, L, M, N, and O) ionization and the excitation of an electron either to an unoccupied state or to the continuum, well above the Fermi level. The EELS spectrum, $d\sigma/dE$, for the ionization of a core electron with binding energy, E_c, will show an increase in intensity at energy loss $\Delta E = E_c$, followed by a long tail for $\Delta E > E_c$, classically referred to as a *saw-tooth shaped curve*. Other shapes such as a delayed maximum, "white lines," etc., can also be observed (Fig. 9.4.5). In addition, for the energy region up to 50 eV above the onset of the edge, $E_c < \Delta E < E_c + 50$ eV, the *energy loss near edge structure* (ELNES), similar to NEXAFS (§3.9.2), which depends on the binding state of the core electron and its atomic environment, can be observed. Further, interference effects between the outgoing excited electron wave and the electron wave back-scattered from neighboring atoms, known as *extended energy-loss fine structure* (EXELFS), is observed in the form of long-range oscillations superimposed on the underlying edge shape. With analysis similar to EXAFS (§3.9.2), EXELFS gives information on the nearest neighbor distances and coordination number (Fig. 3.9.7).

The cross-section, Q, for inner-shell ionization, attributed to Bethe [43], is given by

$$Q = \frac{\pi e^4 b_s n_s}{E_0 E_c} \log \left(c_s \frac{E_0}{E_c} \right) = \frac{6.51 \times 10^{-20} b_s n_s}{U E_c^2} \log \left(c_s U \right) \qquad (9.4.4a)$$

where n_s is the number of electrons in the specific shell, E_c is the ionization energy, b_s and c_s are constants appropriate for the specific shell, and $U = E_0/E_c$ is the overvoltage. The ionization cross-section, Q, varies with over-voltage (Fig. 9.4.3a) and the ionization probability is significant if $U > 5$. As the voltages used in TEM

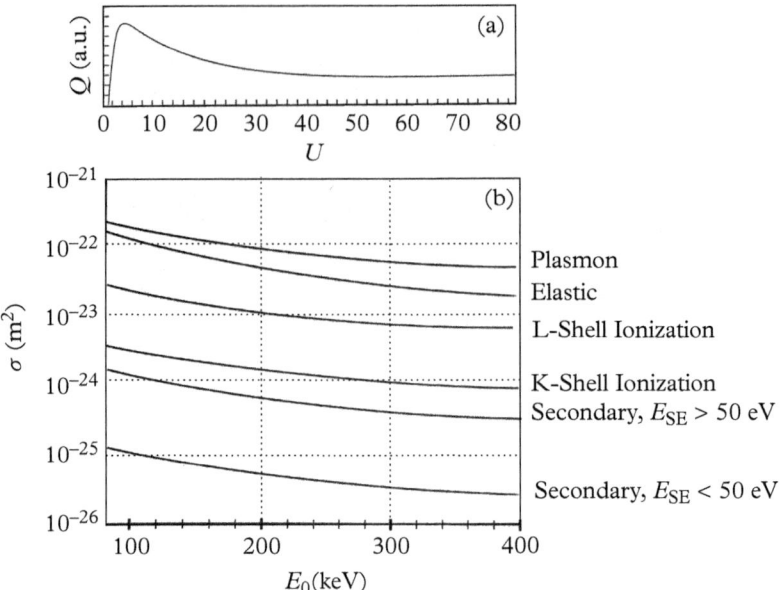

Figure 9.4.3 Inelastic scattering cross-sections. (a) Variation of the ionization cross-section, Q, as a function of the overvoltage, U ($= E_0/E_c$). Above a critical value, $U \geq 5$, the ionization is most probable. (b) The cross-sections for various inelastic scattering processes in aluminum as a function of acceleration voltage. The elastic cross-section is included for comparison.

Adapted from Joy, Romig, and Goldstein (1986).

are significantly high ($E_0 > 100$ keV), relativistic effects become important, and (9.4.4a) is modified as

$$Q = \frac{\pi e^4 b_s n_s}{\left(\frac{m_0 v^2}{2}\right) E_c} \left\{ \log\left(c_s \frac{m_0 v^2}{2 E_c}\right) - \log\left(1 - \beta^2\right) - \beta^2 \right\} \qquad (9.4.4b)$$

where

$$\beta = \frac{v}{c} = \left\{ 1 - \left[1 + \frac{E_0}{511} \right]^{-2} \right\}^{1/2} \qquad (9.4.5)$$

and E_0 is expressed in keV. The value of the constants, b_s and c_s, has been the subject of considerable discussion in the literature and are often adjusted for specific experiments. For low beam energies, $E_0 < 20$ keV, and small overvoltages, $4 \leq U \leq 25$, for K shells, $b_s = 0.9$ and $c_s = 1.05$ [44]. For higher beam energies, $5.5 \leq U \leq 25$, and higher atomic number elements, e.g. Ni, very different values of $b_s = 1.05$ and $c_s = 0.51$ are recommended [45]. Numerous other formulas to describe the inelastic cross-sections are also available [46, 47]. Figure 9.4.3b shows the cross-section for various inelastic scattering processes in Al as a function of the incident beam energy.

Finally, a very important quantity that is involved in calculating scattering contrast is the ratio, γ, of the total inelastic scattering to the total elastic scattering cross-section:

$$\gamma(Z) = \frac{\sigma_{inelastic}}{\sigma_{elastic}} \approx \frac{20}{Z} \qquad (9.4.6)$$

Figure 9.4.4 plots the inverse of this ratio, $1/\gamma(Z)$, is plotted in along with experimentally measured values; clearly $\sigma_{inelastic}$ is much larger than $\sigma_{elastic}$ for atoms with $Z < 20$ (compare $\gamma_{Pt} = 0.25$ with $\gamma_C = 3.3$).

In summary, inelastic scattering transfers energy from the incident electron beam to the specimen. This generates a variety of signals, principally EELS, which provides a wealth of information about the specimen. The de-excitation processes, following inner shell ionization, produce characteristic X-rays that can be detected using EDXS and quantified to obtain compositional information. These two methods, as implemented in a TEM, are discussed in the sections that follow.

9.4.2 Electron Energy-Loss Spectroscopy (EELS) in a TEM

The experimental arrangement for EELS, including the important angles involved, is schematically illustrated in Figure 9.4.2. The incident beam of intensity, I_0, and energy, E_0, is inelastically scattered as it passes through the specimen into the analyzing spectrometer, which is mounted at the end of the electron-optical column (Fig. 9.1.3). In the process of inelastic scattering, it suffers an energy-loss, ΔE, and emerges from the specimen with an energy, $E = E_0 - \Delta E$, and intensity distribution, $I(E)$, or $I(\Delta E)$. The spectrometer (Fig. 9.4.8) magnetically disperses $I(E)$ such that all electrons emerging with a specific loss, $\Delta E = E_0 - E$, are brought to a focus at a specific point on the detector (§9.4.2.4). The entrance aperture to the spectrometer can be made narrow and positioned at any scattering angle, θ (Fig. 9.4.2). Then, the EELS spectrum collected as a function of the angle, θ, includes in addition to the energy-loss, the momentum distribution of inelastically scattered electrons, and can be used to provide information on the complex dielectric function of the materials (see Raether, 1980). However, here we restrict our discussion to the case where the spectrometer integrates all electrons scattered up to a maximum collection angle, β (Fig. 9.4.2); by such angular integration the momentum information is sacrificed to gain the advantage of simple interpretation of the EELS spectrum in terms of the chemical and physical properties of the specimen. Now, the ratio $I(E)/I_0$ at any energy, E, will be proportional to the ratio of the inelastic scattering cross-section, $\sigma_{inelastic}(\Delta E, \beta)$, representing the probability of detecting an incident electron subject to a specific energy-loss, ΔE, and scattered through a maximum angle, β, in the forward direction. The constant of proportionality, N, is equal to the number of atoms per unit area contributing to the observed scattering event. As we have seen in the last section, the inelastic scattering cross-section is related to specific physical interactions between the electron beam and the specimen, and if a standard specimen with a known concentration, N, is

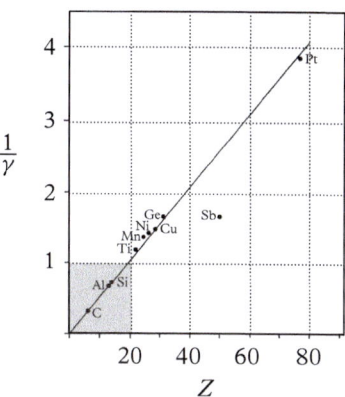

Figure 9.4.4 Measured ratios of the elastic-to-inelastic scattering cross-section, $1/\gamma$, as a function of atomic number, Z. Note the importance of inelastic scattering for $Z < 20$.

Adapted from Reimer (1995).

used, the appropriate cross-section can be determined [48–50]. Alternatively, if the scattering cross-section is known, the concentration, N, can be determined from the measured intensity and quantitative information about the specimen composition, also known as microanalysis, can be obtained (§9.4.2.5).

Figure 9.4.5 shows a representative EELS spectrum of an oxide material. It can be broadly divided into three energy regions: no-loss (NL), low-loss, and core-loss, and we shall now discuss them individually.

9.4.2.1 *The No-Loss Region*

The no-loss region of the EELS spectrum (Fig. 9.4.5) contains its most visible feature, the zero-loss peak (ZLP), and includes three types of electrons: (i) those that are not scattered on passing through the specimen, (ii) elastically scattered electrons, and (iii) those that generate a phonon excitation with energy-loss, $\Delta E < 1$ eV. The energy width (full width at half maximum, FWHM) of the

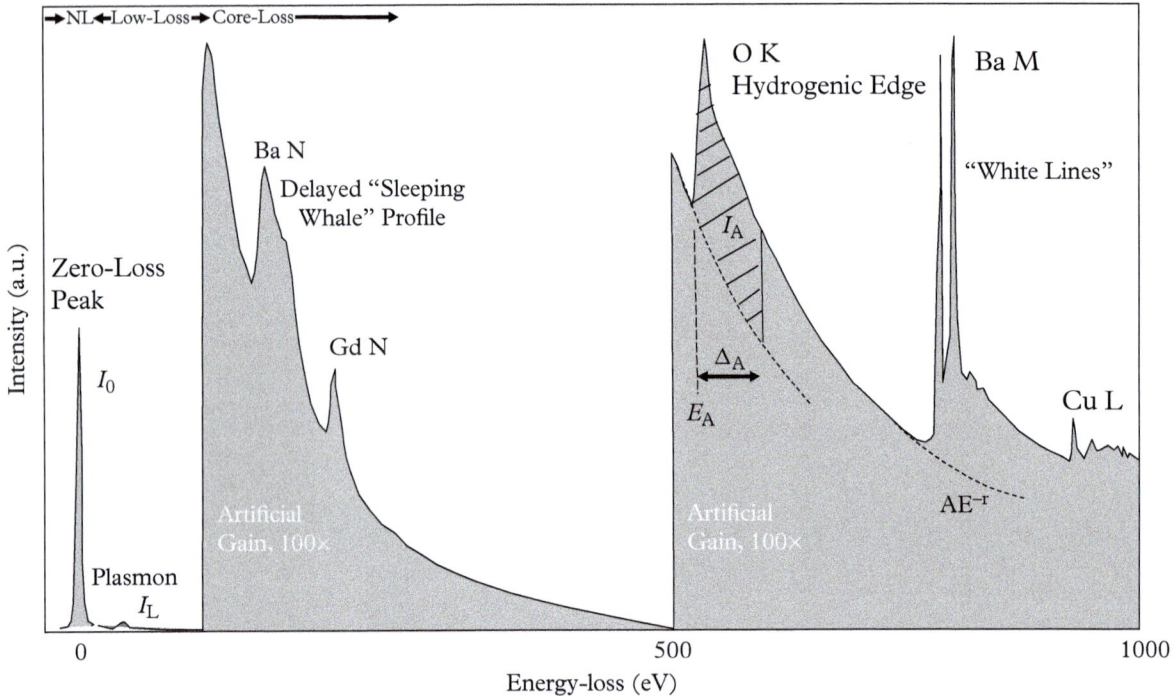

Figure 9.4.5 Electron energy-loss spectrum of $GdBa_2Cu_3O_{7-\delta}$, illustrating various parameters (I_0, Zero loss peak intensity, I_L, low loss intensity, E_A, edge onset energy, Δ_A, integration energy window, I_A, integrated intensity above background for element A), and the background model AE^{-r}, described in the text. The figure also shows the plasmon loss and three types of edge shapes typically observed in EELS. Note the artificial gain applied in the display.

Adapted from [51].

ZLP is a measure of the energy resolution of the experimental arrangement and is a function of both the energy spread of the source, including instabilities of the acceleration potential, and the energy resolution of the spectrometer. Taken together, typically the FWHM ~ 0.2–1.5 eV, and depends on the type of source/gun as well as the stability of the monochromator used in the TEM.

The angular distribution of the unscattered electrons—no interaction with the specimen—can be expected to be the same as that of the focused incident beam, i.e. a cone of angular range ~ 1–5 mrad wide (Fig. 9.4.6a). The elastically scattered electrons, typically deflected by the positive nuclear charge, and given by the screened Rutherford cross-section (9.3.11), form a broad angular distribution (~ 30 mrad) for amorphous materials (Fig. 9.3.2a). However, in crystalline materials, Bragg diffraction modifies this angular distribution with

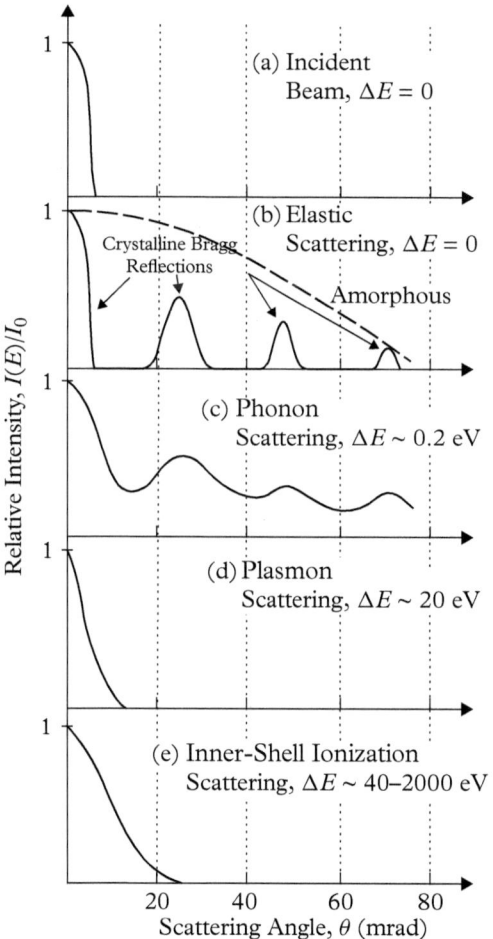

Figure 9.4.6 Angular distribution of (a) incident and unscattered electrons, (b) elastically scattered electrons (Rutherford cross-section) including the Bragg diffracted beams in crystalline materials, (c) phonon scattered electrons, which broaden the individual Bragg scattered beams, (d) plasmon scattered electrons, and (e) electrons scattered by inner-shell ionization with large energy losses.

Adapted from Joy, Romig, and Goldstein (1986).

peaks corresponding to the Bragg angles (Fig. 9.4.6b). Note that the broad angular distribution of elastically scattered electrons impacts our ability to efficiently collect the inelastically scattered electrons, especially for thick specimens. Finally, the electrons that excite phonons, or vibration modes in the specimen, also contribute to the ZLP. They lose small amounts of energy ($\Delta E \sim 0.2$ eV) but they have a broad angular distribution. Typically, in crystalline materials it is of the order of 5 mrad, and centered around each of the Bragg diffracted beams (Fig. 9.4.6c).

9.4.2.2 The Low-Loss Region

This energy region of the EELS spectrum (Fig. 9.4.5) extends from the edge of the ZLP up to an energy loss, $\Delta E < 40$ eV, and includes the plasmon peak, especially in metals, as its principal feature. The integrated signal intensity, I_L, in this region is about 5–10 % of the intensity, I_0, of the ZLP. In practice, since most ($\sim$80 %) of the electron beam passing through a thin specimen is either unscattered or elastically scattered, the *total* inelastically scattered electron intensity, I_T, is essentially the sum of the no-loss and low-loss regions, i.e. $I_T = I_0 + I_L$. As we see shortly (9.4.8), this has implications in specimen thickness measurements.

The energy losses in this region are a result of the direct electrostatic interactions of the electron beam with atomic electrons. It involves interactions or excitations with various bound states (Fig. 9.4.1b) including (i) intra-band, and (ii) inter-band transitions. However, the most prominent feature in this energy region is due to plasmon excitations, observed predominantly in metals, and attributed to the collective excitation of the delocalized electron "gas" or band. This excitation is not localized and occurs over a volume of several nanometers; further, the plasmon oscillations are rapidly damped with a typical lifetime (see §2.3.1) of 10^{-15} s. The energy loss, ΔE_p, of the beam generates a plasmon of angular frequency, ω_p, given by

$$\Delta E_p = \hbar\omega_p = \hbar \left(\frac{ne^2}{\varepsilon_0 m} \right)^{1/2} \tag{9.4.7}$$

where n is the free electron density in the band. The characteristic scattering angles, $\theta_p = \Delta E_p/2E_0$, for plasmon scattering are rather small (<0.1 mrad); as mentioned before, beam electrons suffering plasmon losses are forward peaked with cut-off angles, $\theta_c \sim 100\theta_p \approx 5 - 10$ mrad (Fig. 9.4.6d). The plasmon mean free path length, $\lambda_p \sim 100$ nm, for $E_0 = 100$ keV, such that in an average specimen most of the electrons transmitted lose energy by plasmon excitations. In fact, in thicker specimens, multiple plasmon excitations at energy losses $m\Delta E_p$ (m is an integer) are often encountered.

Finally, as alluded earlier, specimen thickness information can be obtained from the energy loss spectrum. The intensity in the EELS spectrum falls off so rapidly that $I_T = I_0 + I_L$ can be safely assumed. Then, the local thickness, t, of the specimen can be related to the mean free path, λ_p, of plasmon excitation

Table 9.4.1 Plasmon loss data at 100 keV for some elements.

Material	ΔE_p (calc) (eV)	ΔE_p (expt) (eV)	θ_p(mrad)	θ_c(mrad)	λ_p (calc) (nm)
Li	8.0	7.1	0.039	5.3	233
Be	18.4	18.7	0.102	7.1	102
Al	15.8	15.0	0.082	7.7	119
Si	16.6	16.5	0.090	6.5	115
K	4.3	3.7	0.020	4.7	402

Adapted from Egerton (1986).

simply as

$$t = \lambda_p \ln (I_T/I_0) \tag{9.4.8}$$

and hence t can be found if λ_p is known. Table 9.4.1 summarizes typical parameters for plasmon excitations of some metals at 100 keV.

9.4.2.3   The Core Loss Region

At higher energy losses, $\Delta E > 50$ eV, one observes edges corresponding to the interaction of the beam electrons with the deeply bound core electrons of the specimen. The intensity in this region is very much lower than in the no-loss and low-loss regions, and falls so rapidly with increasing energy-loss that an artificial gain is required to see the underlying features (Fig. 9.4.5).

When the incident beam interacts with an atom in the specimen, a minimum energy corresponding to the binding energy, E_c, of the core electron is required for its ionization. Thus, the onset of the edge in the energy-loss spectrum corresponds to the ionization energy of the core electron. Moreover, as the binding energies of the core electrons are a unique function of the atomic number, the position of the onset of the edge in EELS can be used to unequivocally identify the elements that constitute the specimen and quantify its chemical composition. In addition, in the process of ejecting the core electron to the continuum, the beam electron can impart varying amounts of energy, up to a maximum of $E_0 - E_c$, to it. However, the probability of doing so decreases with increasing energy loss. As a result, the overall shape of an ideal core edge in EELS (see O K edge in Figure 9.4.5) is a sharp onset followed by a smooth decay, extending with decreasing probability to higher energy losses, and this tail often results in significant edge overlaps in multi-element specimens.

Moreover, preceding edges and plasmon losses can impart considerable background to subsequent edges observed at higher energy losses. In practice, the background in EELS has been found to have the form $A(\Delta E)^{-r}$, where A is a constant and the exponent, r, is usually in the range $3 < r < 5$; both A and r are obtained by curve-fitting (Fig. 9.4.5; see [52, 53] and Egerton, 1986). The problem is compounded in thicker specimens due to plural scattering, but again they can be deconvoluted to obtain the single-scattered spectrum [54, 55].

The intensity of a core-edge in EELS varies as a function of the energy loss, ΔE, and the scattering angle, θ. Hence, the cross-section is written in the form of a second differential:

$$\frac{d^2\sigma_{inelastic}}{d\Omega d(\Delta E)} = \frac{e^4}{(4\pi\varepsilon_0)^2 E_0 \Delta E}\left(\frac{1}{\theta^2 + \theta_E^2}\right)\left(\frac{d\mathcal{F}}{dE}\right) \qquad (9.4.9)$$

Here, the characteristic angle for inelastic scattering, $\theta_E = \Delta E/2E_0 = 1.33$ mrad for $\Delta E = 532$ eV (O K edge) and beam energy, $E_0 = 200$ keV, $\mathcal{F}$ is the generalized oscillator strength (GOS) that describes the response of an atom when a specific energy and momentum are supplied by the collision of a beam electron, and $d\mathcal{F}/dE$, is the GOS per unit energy loss. Strictly speaking, to calculate (9.4.9) accurately, knowledge of the atomic wave functions and the band structure of the solid are required. For small scattering angles, θ, and energy-loss edge energy, E_c, the contribution of the GOS is a constant. Then, the angular distribution of inner-shell excitations is of the form $\left(\theta^2 + \theta_E^2\right)^{-1}$, with a maximum value for the forward-scattered direction ($\theta = 0$). Hence, it is safe to assume that the average inelastic scattering angle is of the order of ~ 5 mrad, which is much smaller than the characteristic angle, θ_0 ($=\frac{\lambda Z^{1/3}}{2\pi a_0}$), for elastic scattering (Fig. 9.4.6). In addition, the cross-section shows a long tail and even though a majority of the inelastically scattered electrons are forward peaked, some of them are scattered at angles $\theta >> \theta_E$. In practice, this results in not all the energy-lost electrons of the characteristic edge being detected, as finite collection angles ($\beta < 50$ mrad) are used to minimize electron-optical aberrations; in fact, both the shape of the spectrum and the intensities are affected by these finite apertures. Moreover, even though the inner-shell ionization signal will increase as the collection angle, β, is increased, eventually at some critical value, $\beta \sim (2\theta_E)^{1/2} \sim 50$ mrad, the contribution from the GOS will tend to zero. Nevertheless, a substantial amount of the signal can be collected for $\beta = 20$ mrad. Finally, for large energy losses, $\Delta E >> E_b$, the contribution of the GOS is small in the forward direction ($\theta = 0$), but increases to a maximum value (peak) at $\theta \sim (\Delta E/E_0)^{1/2}$; known as the Bethe ridge, this effect is discussed in more advanced texts such as Egerton (1986) and Reimer (1993).

Integrating (9.4.9) from $\Delta E = E_c$ to $\Delta E = E_0$, gives the total cross-section for the inner-shell ionization, which is typically of the order of 10^{-24} m^2/atom for a light element, such as carbon at 100 keV. Compared to plasmons, this is at least two orders of magnitude weaker, and often requires the introduction of an artificial gain in displaying the core-loss spectrum (Fig. 9.4.5). Finally, a typical spectrometer collects the signal over a finite angular range (β) and energy window (Δ_A) (Fig. 9.4.5). Therefore, in carrying out microanalysis (§9.4.2.5) only a fraction of the total cross-section, known as the partial cross-section, $\sigma_{inelastic}(\beta, \Delta_A)$, is appropriate and used for quantification.

When a small collection aperture ($\beta \sim 25$ mrad) is used, the predominant transitions observed in an energy-loss spectrum are the ones governed by the diploe selection rules (§2.3.1); in that sense, EELS spectra are very similar to X-ray absorption spectra (discussed earlier in §3.9). Standard spectroscopic nomenclature is used to label the core edges, i.e. K (1*s*), L_1 (2*s*), L_2 ($2p_{1/2}$), L_3 ($2p_{3/2}$), M_1 (3*s*), M_2 ($3p_{1/2}$), M_3 ($3p_{3/2}$), M_4 ($3d_{3/2}$), M_5 ($3d_{5/2}$),..., etc. In practice, only a few of these transitions can be recorded with ease and the relevant edges are those corresponding to an initial state of maximum *l* for a given *n*[56]. Hence, transitions originating in 2*p*, 3*d*,... initial states are an order of magnitude greater in intensity than those from 2*s*, 3*s*, 3*p*,..., etc.

Three basic edge shapes (Fig. 9.4.5) are broadly observed; as the wave functions of the core electrons undergo little change upon forming a solid, a simple atomic model can predict their general shapes [57, 58]. They are: (a) "saw tooth" profile such as those for hydrogen-like wave functions for K-shell edges; experimentally observed edges, such as O K in Figure 9.4.5, conform to this general shape but with some additional fine structure near the edge onset; (b) a "sleeping whale" profile, as seen for the Ba $N_{4,5}$ edge in Figure 9.4.5, or a delayed maximum observed ~20 eV above the ionization edge, usually resulting from large centrifugal barriers due to the $l'(l' + 1)$ term in the radial Schrodinger equation, and commonly observed for the $L_{2,3}$ edge for the third period elements Na–Ar; and (*c*) "white lines", arising from distinct spin-orbit split levels and typically observed for the 3*d* transition metals (Fig. 3.9.3). As discussed in §3.9.2, $L_{2,3}$ edges probe the *d*-symmetry portion of the final state wave functions, and can become large and narrow with sharp threshold peaks in a solid with high density of unoccupied *d* states. Similar effects can be observed in M edges as shown for Ba $M_{4,5}$ (Fig. 9.4.5). The L_3/L_2 ratio is often different from the statistical value of 2.0 based on initial state occupation (Fig. 9.4.7) and can be used to determine the oxidation state of the transition metal [59].

Superimposed on the broad edge shapes are the fine structures due to solid-state effects. At the edge threshold, one can measure a displacement of the onset, or "chemical shift". In particular, positive chemical shifts are observed with increasing oxidation states (Fig. 9.4.7) because oxidation removes valence electrons, which leads to reduced screening of the nuclear field and a deepening of the potential well of the initial state. In EELS, the ionization edge threshold is a function of both the initial state as well as the position and nature of the vacant states around the Fermi level. The energy-loss near edge structure (ELNES), much like NEXAFS discussed in §3.9.2, and observed ~3–50 eV above the ionization edge, can be interpreted in the first approximation using a simple one electron transition model between the initial state and a vacant final state as

$$I(\Delta E) = T(\Delta E) N(E) \tag{9.4.10}$$

where $T(\Delta E)$, the transition probability, is a slowly varying function of energy loss, ΔE, and $N(E)$ is the density of final states. This simple model is reasonable for the interpretation of ELNES, provided $N(E)$ is defined to include the

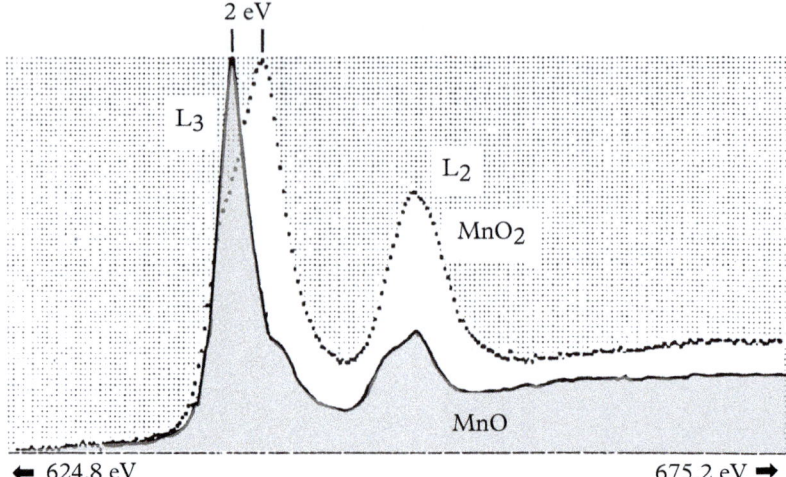

Figure 9.4.7 The $L_{3,2}$ edge for Mn^{2+} and Mn^{4+} illustrating both the chemical shift (2 eV) and the change in L_3/L_2 ratio with the oxidation state.

Adapted from [51].

following ideas: (a) dipole selection rules apply, i.e. $N(E)$ is interpreted as a symmetry projected density of states clearly distinguishing the K and L edges; (b) core level states are highly localized, i.e. $N(E)$ is a local density of states determined for that particular lattice site and reflecting the local symmetry; and (c) in reality, $N(E)$, is a joint density of initial and final states and broadening, based on the lifetime of the core hole and the final states, is incorporated. For more advanced readers, the theory and analysis of unoccupied electron states in ELNES and XANES is discussed at length in Fuggle and Inglesfield (1992).

Finally, long range oscillations of weak intensity similar to EXAFS (Fig. 3.9.6), superimposed on the high-energy tail of a core-loss ionization edge, gives information on the nearest neighbor distances and coordination number for that specific element in the specimen. Such extended energy-loss fine structure (EXELFS) data is analyzed in a manner very similar to EXAFS as discussed in §3.9.2.

9.4.2.4 The EELS Spectrometer and Signal Detection

The magnetic spectrometer, also known as a magnetic prism because of its similarity to an optical prism that disperse white light into its frequency components, spatially disperses electrons according to their kinetic energy. In the spectrometer electrons traveling down the optic axis (z-direction) of the TEM (Fig. 9.4.8) encounter an orthogonal magnetic field of induction, B, along the y-direction. The electrons will experience a constant Lorentz force, $F_L = evB$, in the region of the magnetic field and travel in a circular orbit of radius, R, which can be determined by simply balancing the Lorentz force with the centripetal force:

$$R = \frac{\gamma m_0 v}{eB} \qquad (9.4.11)$$

Here, $\gamma = \left(1 - v^2/c^2\right)^{-1/2}$ is the relativistic correction factor, and m_0 is the rest mass of the electron. The spectrometer is designed such that the electrons emerge from the magnetic region with a total angle of deflection, $\phi = 90°$. Electrons that have suffered inelastic scattering in the specimen end up with different energies and velocities as they enter the spectrometer. Specifically, electrons undergoing larger energy losses will have smaller velocities, v, and smaller radius, R. Hence, they will leave the magnetic field region of the spectrometer with a larger deflection that is proportional to their energy loss. In addition, the magnetic sector behaves as a lens and brings all electrons of the same energy leaving the same *object point* to focus on the same point on the image plane of the spectrometer. Thus, electrons losing a specific amount of energy, ΔE, compared to the primary beam of energy, E_0, are brought to focus at a different point displaced by a distance, $\Delta x = D. \Delta E$, on the image plane, where

$$D = \frac{2R}{E_0} = 2\frac{\gamma m_0 v}{E_0 eB} = \frac{\Delta x}{\Delta E} \tag{9.4.12}$$

is called the dispersion of the spectrometer. For a microscope operating at $E_0 = 100$ keV, and an electron orbit radius, $R = 0.2$ m, we get a typical dispersion $D = 4$ µm/eV. These dispersed electrons can be collected, in principle, in two different ways: serial and parallel detection (Fig. 9.4.8).

A narrow slit of width, d_s, placed in the image plane of the spectrometer will allow only electrons of a specific and narrow energy loss to pass through to the detector. By ramping the magnetic field in small steps electrons of different energy (loss) can be detected serially by the detector. In this case, the minimum energy width, δE_d, that can be detected by the slit would be $\delta E_d = d_s/D$, where D is the dispersion (9.4.12). Overall, the energy resolution of the spectrometer, δE, goes in quadrature with all its contributions, which include the energy spread, δE_0, of the source before the beam reaches the specimen, the energy resolution, δE_{so}, of electron-optical components, and the spatial resolution of the electron detector given by the slit width, δE_d. Thus

$$(\delta E)^2 = (\delta E_0)^2 + (\delta E_{so})^2 + (\delta E_d)^2 \tag{9.4.13}$$

Alternatively, the entire spectrum can be detected in parallel using photodiode arrays (PDAs), as in the original design, or more recently replaced with CCD or direct electron detection cameras (Fig. 9.4.8b). If a PDA is used, the spatial resolution of the detector, defined by the interdiode spacing of the photodiode array, is ~25 µm. For a typical spectrometer with a dispersion $D = 4$ µm/eV, calculated earlier, a PDA can only resolve ~6 eV. To better the energy resolution, it is necessary to magnify the spectrum before projecting it onto the detector plane. A simple electron-optical lens can be used for magnification, but as we have seen before (§9.2.2), this will lead to magnification-dependent rotation of the image. To overcome this, a quadrupole lens system for magnification and

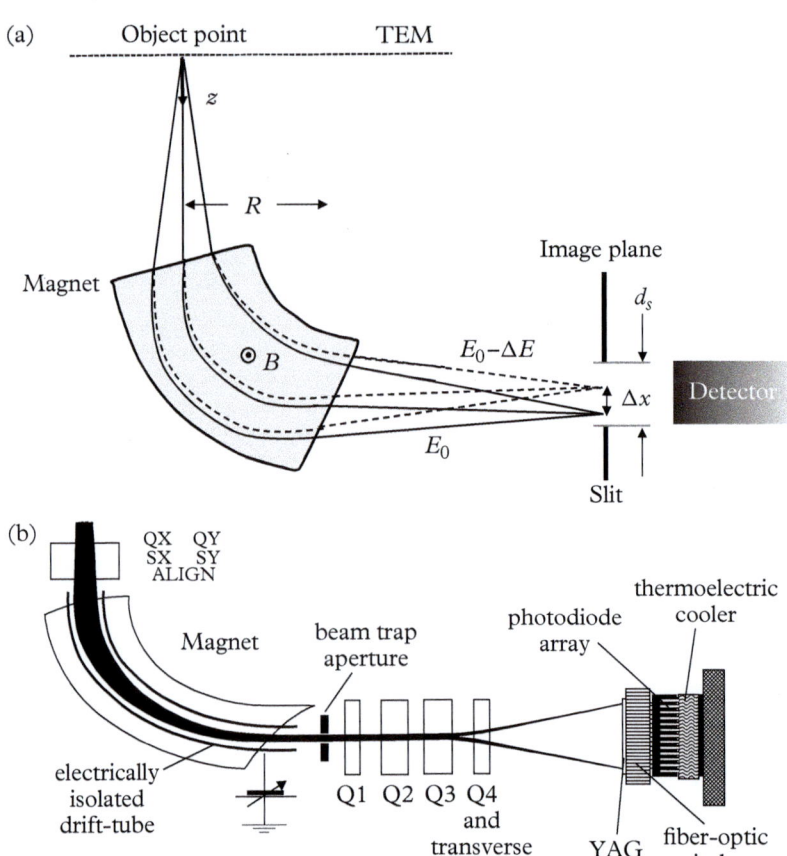

Figure 9.4.8 (a) Schematic layout of the EELS spectrometer for serial detection. Electrons leaving the object point from the specimen with the same energy are brought to focus at the same point on the image plane of the spectrometer. The dispersion, $\Delta E/\Delta E$, of the spectrometer depends on the magnetic field and is typically a few μm/eV. A slit of narrow width, d_s, is used for energy selection. (b) The original design of a commercial parallel detection EELS system showing the pre-spectrometer focusing and alignment coils (QX, QY, SX, SY), quadrupole lens array (Q1–Q4), and position sensitive detector, which is a YAG scintillator couple to a photodiode array (PDA). Nowadays, the PDA is replaced with CCD cameras or direct electron detection cameras.

Adapted from Egerton (1986).

rotation-free focusing of the spectrum on the detector plane is used [60]. Note that in both cases, good energy resolution requires that an electron-beam cross-over of small diameter be placed at the spectrometer object plane. This is best accomplished by positioning either a low-magnification image (image coupling, or seeing a diffraction pattern on the viewing screen, as the spectrometer looks at final projector cross-over) or the central diffracted beam (diffraction coupling, or seeing an image on the viewing screen) at the object point of the spectrometer. Finally, it is important to mention that serial spectrometers are no longer available commercially.

9.4.2.5   *Microanalysis Using Inner-Shell Ionization Edges*

In addition to the analysis of near-edge fine structure (ELNES) to obtain bonding and chemical information, EELS is used for quantitative microanalysis, especially for light elements ($Z < 11$), where it has significant advantage in the TEM over the alternative energy-dispersive X-ray spectroscopy (EDXS) method (§9.4.3).

However, the detection of hydrogen in a solid is complicated because other low-loss features (ZLP and plasmons) often obscure the H^{1s} edge. Moreover, in metallic hydrides, the $1s$ electron is often incorporated in the conduction band, resulting in a shift of the host Fermi level. Nevertheless, shifts in the plasmon energy observed in hydrides have been interpreted in terms of the composition in a variety of metal hydride systems [61, 62]. However, such interpretation is difficult as it involves understanding the modification of the band structure due to the addition of hydrogen, and often leads to detection limits one order of magnitude worse than the simple EELS microanalysis (described later) using core edges for $Z > 3$.

For a specimen that is thin enough to avoid plural scattering, typically 20–50 nm in thickness, the quantification procedure is straightforward and involves measurement of the area of the appropriate ionization edge, after background subtraction. There are a number of methods available in the literature (Egerton, 1996; Williams and Carter, 1996) to obtain such an integrated intensity from the energy-loss spectrum, but we present the two-area method of background fitting to the power law form AE^{-r}, and determining the constants A and r. Consider the ideal spectrum (Fig. 9.4.9) showing the energy regions and integrated intensities required for background fitting and quantification. The background region, $E_1 - E_2$, over which the fitting is desired, is divided into two equal halves with measured integrated intensities, I_1, and I_2, as shown. Then we can obtain the power law constants as

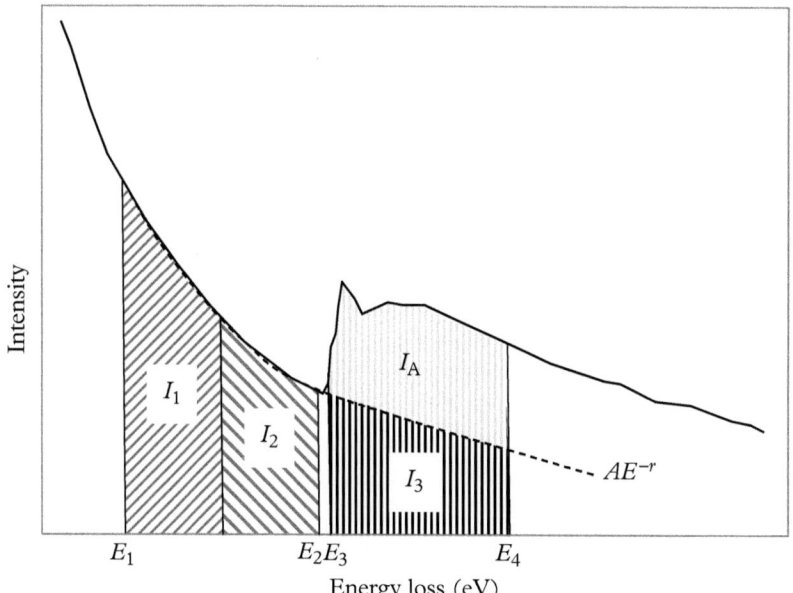

Figure 9.4.9 A portion of an ideal energy-loss spectrum showing the regions needed for fitting the background, its power-law (AE^{-r}) extrapolation under the edges, and the integrated peak intensity.

$$r = \frac{2 \log (I_1/I_2)}{\log (E_2/E_1)} \tag{9.4.14}$$

and

$$A = \frac{(I_1 + I_2)(1 - r)}{E_2^{1-r} - E_1^{1-r}} \tag{9.4.15}$$

Once these constants are known, the background can be extended into the region, $\Delta_A = E_4 - E_3$, under the edge, and the required integrated intensity, $I_A(\beta, \Delta_A)$ for the element A, can be obtained from the gross integrated intensity, I_3, as

$$I_A(\beta, \Delta_A) = I_3 - A\frac{E_4^{1-r} - E_3^{1-r}}{1 - r} \tag{9.4.16}$$

The area, I_A, or the counts for an element A, is a product of the incident current, $\mathcal{J}_0$, the number of atoms, N_A, per unit volume of the specimen, the thickness, t, of the specimen, and the total ionization cross-section, σ_A, for the excitation of the appropriate inner-shell by the incident electrons. However, the entrance aperture to the spectrometer limits collection of inelastically scattered electrons to angles less than β, and hence only a fraction, $I_A(\beta)$, of the core-loss signal is measured. Moreover, in most microanalysis situations with multi-element specimens, to avoid edge overlap, the intensity, $I_A(\beta, \Delta_A)$, over a limited energy-loss window, Δ_A, following the ionization edge, E_A, is measured (Fig. 9.4.9). Then the absolute concentration, N_A, of the element A is

$$N_A = \frac{I_A(\beta, \Delta_A)}{G\mathcal{J}_0\sigma_A(\beta, \Delta_A)t} \tag{9.4.17}$$

where G is any artificial gain incorporated in the spectrum, and $\sigma_A(\beta, \Delta_A)$ is the partial ionization cross-section corresponding to the scattering angle, 2β, and inner-shell losses between energies E_A and $E_A + \Delta_A$. If plural scattering can be avoided or corrected for by appropriate deconvolution methods, $\mathcal{J}_0$ can be replaced with the area under the zero-loss peak, I_0 (Fig. 9.4.5).

Often, only relative abundances of two elements, A and B, are of interest. Then, (9.4.17) gives

$$\frac{N_A}{N_B} = \frac{I_A(\beta, \Delta_A)}{I_B(\beta, \Delta_B)} \frac{\sigma_B(\beta, \Delta_B)}{\sigma_A(\beta, \Delta_A)} \tag{9.4.18}$$

where it is assumed that the data for the two edges are measured under identical experimental conditions of illumination, specimen thickness and scattering angles, and the artificial gains, G_B and G_A are the same. Then, the accuracy in the analysis is largely determined by errors arising from the removal of the background using the AE^{-r} form, and the correctness of the partial ionization cross-section, either measured or calculated, used in the analysis.

Two methods of calculating partial ionization cross-sections are generally in use. An approximate, but easily programmable model, SIGMAK, for K-shell edges based on hydrogen-like wave functions and scaled to account for the nuclear charge, along with a screening constant independent of Z, has shown good agreement with experimental measurements [63]. For L-shells, the equivalent SIGMAL [64], uses an additional empirical factor to match experimental data as the simple treatment for screening is inaccurate [65]. Both SIGMAK and SIGMAL can also be obtained by a parametric approximation by writing the partial ionization cross-section for the element A, with edge onset at E_A (eV), and energy window, Δ_A (eV) as

$$\sigma_A (\beta, \Delta_A) = \sigma \cdot f(\beta) \cdot g(\Delta_A) \qquad (9.4.19)$$

where the saturation cross-section, σ, is given by

$$\sigma = \frac{F \ln \left(\frac{0.055 E_0}{E_A} \right)}{E_0 \cdot E_A} \ \text{m}^2/\text{atom} \qquad (9.4.20)$$

and E_0 is the incident beam energy in eV, and the constant, $F = 1.60 \times 10^{-17}$ m^2/atom for K-edges, and $F = 4.51 \times 10^{-17}$ m^2/atom for L-edges. The characteristic inelastic angle, θ_E^A, for the average energy-loss

$$\theta_E^A = \frac{E_A + \Delta_A/2}{2E_0} \qquad (9.4.21)$$

can be used to calculate the second term, $f(\beta)$, in (9.4.19) as

$$f(\beta) = \frac{\log \left(1 + (\beta/\theta_E^A)^2 \right)}{\log \left(2/\theta_E^A \right)} \qquad (9.4.22)$$

where, β is the spectrometer acceptance semi-angle in radians. The last term, $g(\Delta_A)$, in (9.4.19) has the form

$$g(\Delta_A) = 1 - \left[\frac{E_A}{E_A + \Delta_A} \right]^{s-1} \qquad (9.4.23)$$

and the exponent s is given by

$$s = H - 0.334 \ln \left(\beta/\theta_E^A \right) \qquad (9.4.24)$$

with the constant H $= 5.31$ for K-edges, and H $= 4.92$ for L-edges.

The cross-sections can also be calculated more accurately using Hartree–Slater (HS) wave-functions [58, 66], assuming that the element is in atomic form, and neglecting solid-state or exciton effects. In the calculations of partial ionization cross-sections both methods agree to within 5 % for K-shell and 10% for L-shell edges. Alternatively, we can measure the partial ionization cross-sections using standard specimens and two systematic measurements, one for K- and L-edges [67], and the other for M-edges [68], and both are available in the literature. However, some of the experimentally measured cross-sections show large variations [65], and some have been collected with large collections angles ($\beta \sim 50$ mrad) that may be susceptible to errors due to lens aberration effects [69]. Finally, using the integrated edge-to-background ratios, minimum detectable mass (MDM) has been defined for EELS microanalysis; typically, using current technology, MDM $\sim 10^{-19}$ g. Further details can be found in Joy, Romig, and Goldstein (1986), Egerton (1996), and Williams and Carter (1996).

Example 9.4.1: An unknown compound shows two edges at (A) 188 eV and (B) 284 eV in an energy loss spectrum. Following Figure 9.4.9, for edge A, the energies are $E_1 = 163$ eV, $E_2 = 186$ eV, $E_3 = 188$ eV, and $E_4 = 258$ eV. The corresponding intensities are $I_1 = 36829$, $I_2 = 28226$, and $I_3 = 107112$ counts. Similarly, for edge B, the energies are $E_1 = 240$ eV, $E_2 = 280$ eV, $E_3 = 284$ eV, and $E_4 = 354$ eV, and the intensities are $I_1 = 26189$, $I_2 = 20132$, and $I_3 = 53273$ counts. Identify the elemental edges A and B, and determine their chemical ratio. Make any appropriate assumptions.

Solution: We first make the reasonable assumption that these spectra were taken in a TEM with $E_0 = 100$ keV, and with a typical spectrometer acceptance semi-angle $\beta = 3$ mrad.

The integration window $\Delta = E_4 - E_3 = 70$ eV is the same for both edges. Looking at the binding energies for the core electrons (Table 2.2.2), we can determine that A is the Boron (B) K-edge, and B is the Carbon (C) K-edge. We can calculate the parametrized SIGMAK partial ionization cross-sections using (9.4.19–9.4.24). Thus

$$\sigma_K^B (\beta = 3 \text{ mrad}, \Delta = 70 \text{ eV}) = 5.84 \times 10^{-25} \text{ m}^2/\text{atom}$$
$$\sigma_K^C (\beta = 3 \text{ mrad}, \Delta = 70 \text{ eV}) = 8.33 \times 10^{-26} \text{ m}^2/\text{atom}$$

Next, we have to analyze the spectra and determine the integrated intensities. In order to do this, we have to do the power-law fitting of the backgrounds. Using (9.4.14) and (9.4.15) we determine, A and r for the two edges. For the B K-edge and C K-edge we get

$$r_B = 4.03, A_B = 3.017 \times 10^{12} \text{ and } r_C = 3.412, A_C = 1.982 \times 10^{11}.$$

Then, we can extrapolate the background under the respective edges, and using (9.4.16) determine the integrated intensities in each edge as

$$I_B \, (\beta = 3 \text{ mrad}, \Delta = 70 \text{ eV}) = 28128$$
$$I_C \, (\beta = 3 \text{ mrad}, \Delta = 70 \text{ eV}) = 12034$$

We can now determine the relative ratios of the two elements using (9.4.18) as

$$\frac{N_B}{N_C} = \frac{I_B \, (\beta, \Delta_A)}{I_C \, (\beta, \Delta_B)} \frac{\sigma_C^K \, (\beta, \Delta_B)}{\sigma_B^K \, (\beta, \Delta_A)} = \frac{28128 \times 8.33 \times 10^{-26}}{12034 \times 5.84 \times 10^{-25}} = 0.3334$$

In other words, this compound is BC_3. In fact, such a compound was newly synthesized and analyzed by EELS many years ago [114].

9.4.2.6 Energy Filtered Imaging

In an electron microscope, we can combine the chemical information in EELS with the spatial information of the microstructure at a resolution better than 1 nm [70, 71]. Since this technique is based on EELS, energy filtered transmission electron microscopy (EFTEM) can be a powerful method to study light elements ($3 < Z < 11$), and as such it is particularly applicable to the study of *biological materials* where the elements, C, N, and O are predominant. The technique uses an imaging filter attached to a TEM that makes it possible to create images only with electrons that have lost a specific energy. The simplest application of EFTEM is to filter out unwanted inelastically scattered electrons to improve contrast and interpretability of both TEM images and diffraction patterns (Fig. 9.4.10). It can also be used to selectively image the specimen with inelastically scattered electrons that have lost a specific energy; we now discuss this further.

The imaging filter focuses the image onto an energy-dispersive plane, where an energy "window" selects only those electrons that have lost a specific energy, $\Delta E \pm \delta E / 2$, where $E = E_0 - \Delta E$ is the selected energy, and δE is the energy width of the electrons used in imaging. There are two commercially designed image filters available: the Gatan image filter (GIF) [73], mounted at the end of the electron-optical column, and the in-column Zeiss omega filter (Ω-filter) [74]. The GIF is similar to the EELS spectrometer with parallel detection (Fig. 9.4.8b), but with the addition of a set of quadrupole and sextupole lenses to focus the image onto a CCD camera. The Ω-filter is mounted in-column, and includes a set of four magnetic sectors that bend the electron beam in the shape of Ω (hence the name) to form an energy dispersed diffraction pattern at the exit plane of the filter along the original optic axis of the TEM. Then, an aperture can select the required energy range for the image.

From an information point of view, the EELS spectrum can be considered as an additional energy dimension to the image. Thus, we have a three-dimensional data set for the specimen, combining a conventional spatial image $I(x,y)$ with the spectroscopic information $I(E)$, at each point, (x,y), as shown schematically in

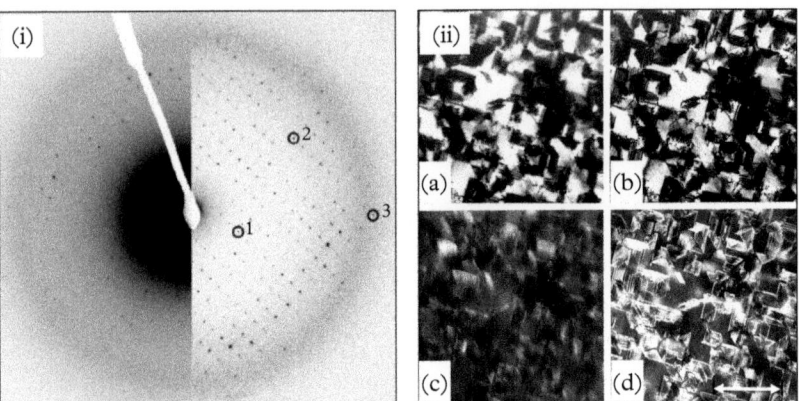

Figure 9.4.10 (i) Comparison of electron diffraction patterns from a thin crystal of F41 (P2$_1$, $a = 51.8$ Å, $b = 36.5$ Å, $c = 118.7$ Å, $\beta = 90.8°$) without (left) and with (right) energy filtering. The background intensity near the beam stop is almost 100 times and overall, 5.5 times more on the left. (ii) Microtwins in an epitaxially grown film of Ag on NaCl with (a) unfiltered and (b) zero-loss filtered bright field images, and (c) unfiltered and (d) zero-loss filtered dark field images obtained by shifting the objective aperture.

(i) Adapted from [72]. (ii) Adapted from Reimer (1995).

Figure 9.4.11a. Note that this approach can also be used to obtain data sets in higher dimensions, for example, by adding the time dimension such as in four-dimensional electron microscopy; see §9.5.5 and Zewail (2008). The acquisition of such a comprehensive three-dimensional data set can be carried out in two different ways, depending on whether a conventional TEM or a dedicated STEM is used. In the former, EFTEM, images covering a specific energy-loss range selected by the energy-selecting slit are collected in parallel by the CCD camera to form two-dimensional images (Fig. 9.4.11b); the complete three-dimensional data set is obtained by acquiring a sequence of images using different energy windows (horizontal sectioning of the data cube). In the latter, STEM-EELS (Fig. 9.4.11c), the energy-selecting slit is removed and the entire energy loss spectrum is collected in parallel at each pixel (vertical sectioning of the data cube). The EFTEM typically provides images of larger areas ($10^5 – 10^7$ pixels) compared to STEM-EELS, which contains about 10^4 (100×100) pixels and takes about 10^3 s of acquisition time ($= 10^{-1}$ s dwell time/pixel). Figure 9.4.12 shows the complementarity of the two methods; note that for a given dose the signal is always higher for STEM-EELS compared to EFTEM, an important factor in imaging biological materials that are highly dose sensitive.

To obtain quantitative images representative of the chemical composition of the specimen, the *three-window elemental mapping* method is often used. Figure 9.4.13 shows a schematic of the method. Three images are acquired, two pre-edge to

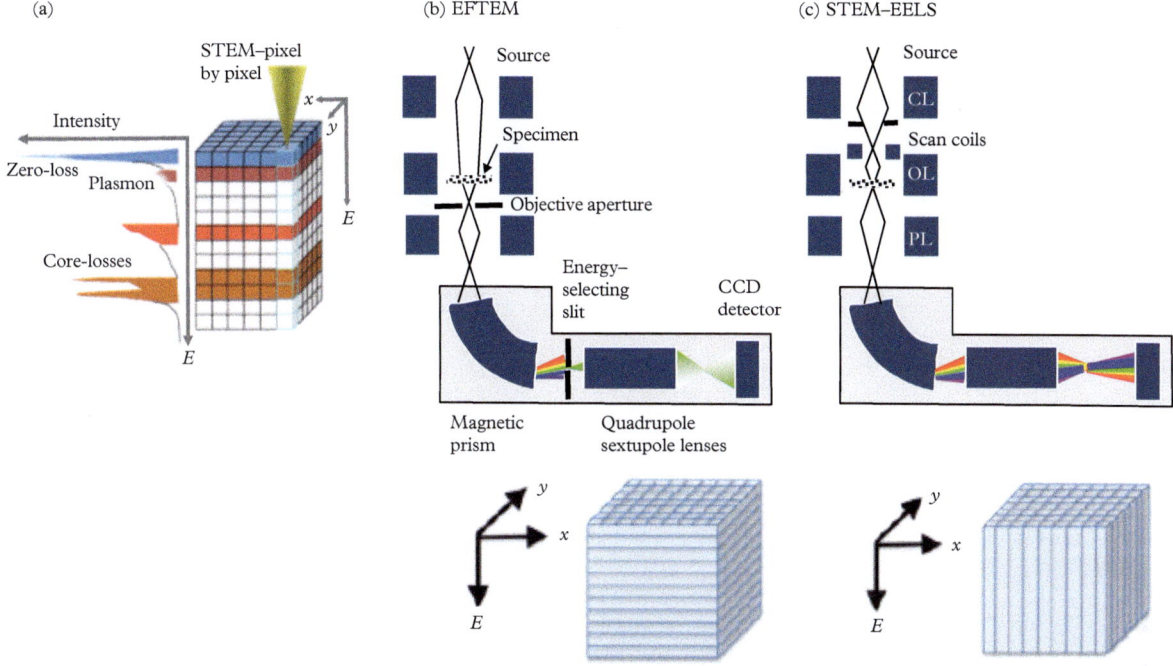

Figure 9.4.11 (a) Schematic of the 3D data set cube obtained by spectroscopic imaging using energy-loss spectroscopy; x and y are spatial coordinates, and $E = E_0 - \Delta E$ is the energy axis. Comparison of compositional imaging in (b) a conventional energy-filtered transmission electron microscope (EFTEM) fitted with a post-column energy filter (an in-column energy filter will also work), and energy-selecting slit. (c) In a scanning transmission electron microscope (STEM-EELS), the slit is removed and the EELS spectrum is collected in its entirety at each pixel. In these two cases the 3D data sets (x, y, E) are built up differently either by (b) horizontal or (c) vertical sectioning of the (x, y, E) cube as shown.

Adapted from [77].

estimate the background under the edge by extrapolation, and one post-edge. By subtracting the extrapolated background pixel by pixel from the post-edge image, the chemical map is formed. An alternative to this method is the *jump ratio* method, which divides the post-edge image by the pre-edge image. Such a jump ratio image provides *qualitative* images of changes in the spectrum at the edge of interest, but it is not quantitative (see §1.4.4. for basics of digital imaging).

Figure 9.4.14 shows EFTEM images of a thin film standard with varying thickness of Mn layers, sandwiched between PdMn layers, using both the three-window and jump ratio methods. It is clear that EFTEM can provide sub-nm resolution in these images. Further information on energy-filtered imaging can be found in review articles [75, 76], and in the dedicated monograph on the subject edited by Reimer (1995).

Figure 9.4.12 Comparison of EFTEM and STEM-EELS imaging modes showing the availability of total number of pixels and the number of electrons per pixel. Different beam currents, I, and acquisition times, *t*, are shown by diagonal lines. Assuming 3-nm width pixels, the total image width is shown on the right, and the total electron dose is shown on top. Note that for a given dose, the signal is always higher for STEM-EELS.

Adapted from [77].

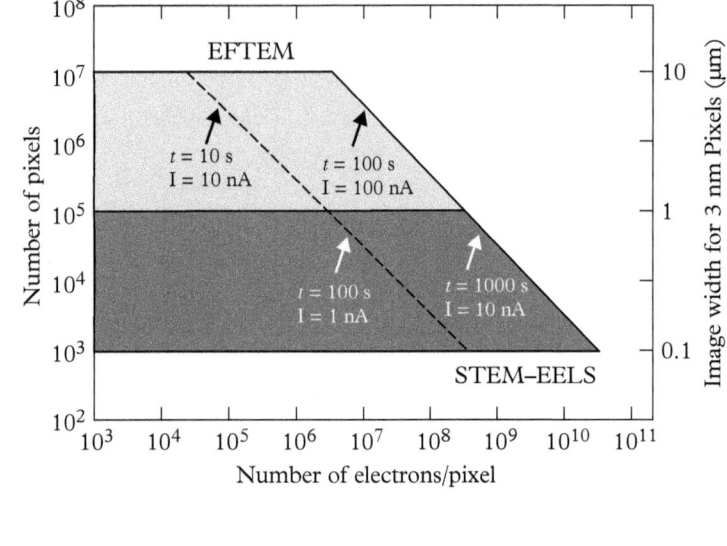

Figure 9.4.13 Illustration of the three-window technique for a multilayer specimen of $Pr_xCa_{1-x}MnO_3$, with $x = 1, 0, 0.25, 0.5, 0.75, 1$, respectively, for each layer grown on a substrate of $SrTiO_3$. Two pre-edge images are used to extrapolate the background, following a fit to the AE^{-r} model. The post-edge image is acquired and the extrapolated background image is subtracted to give the Pr elemental map image, in agreement with the Pr content in the layers.

Adapted from [75].

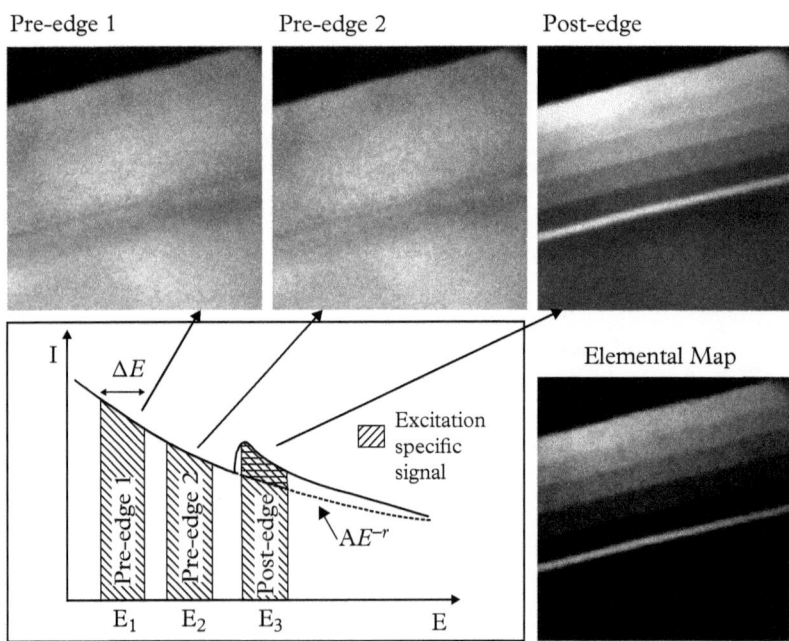

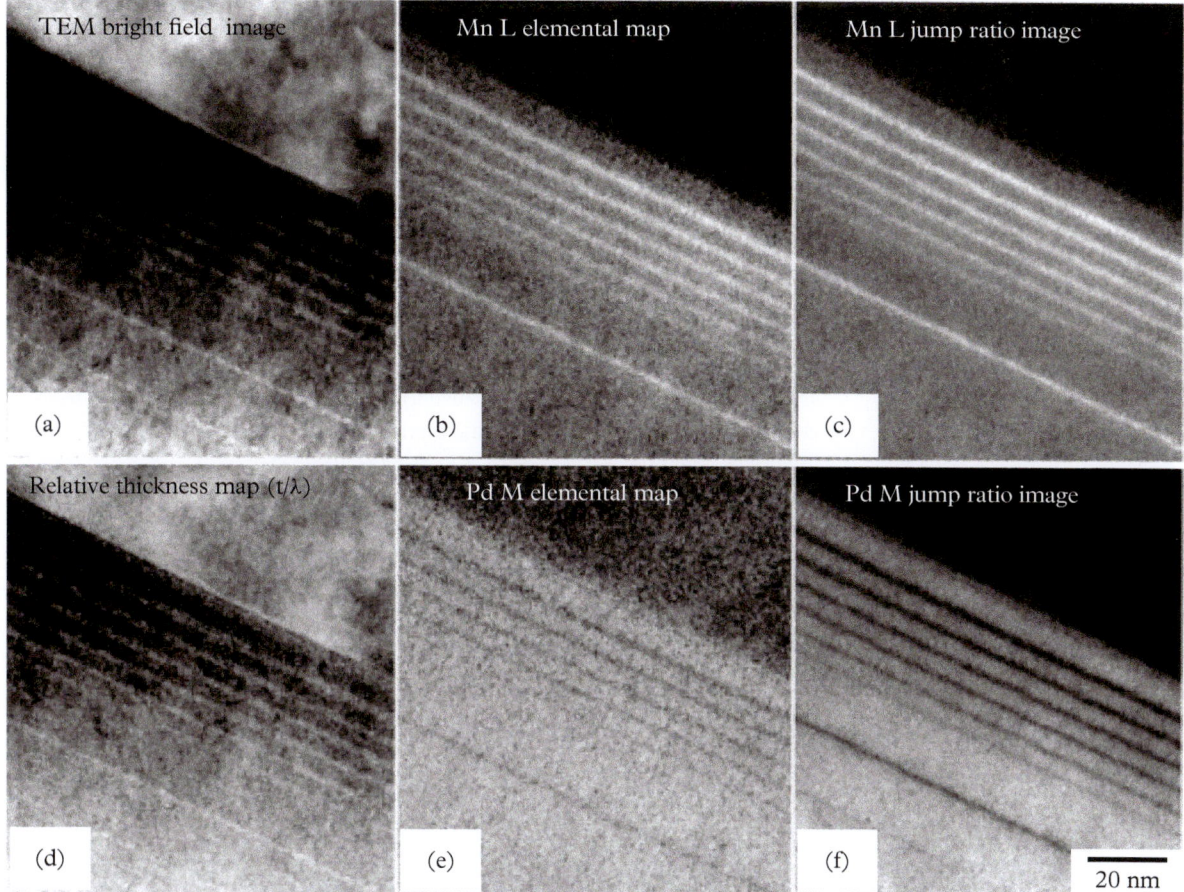

Figure 9.4.14 EFTEM images of a $Mn_t(PdMn)_{3.7}$ multilayer specimen with $t = 2.6, 2.2, 1.8, 1.4, 1.0,$ and 0.47 nm showing (a) BF image, (b) Mn three-window elemental map, (c) Mn jump-ratio map, (d) relative thickness (t/λ) map, (e) Pd three-window elemental map (Pd $M_{4,5}$), and (f) Pd jump-ratio map. Note that even the thinnest Mn layers are clearly resolved, confirming sub-nm spatial resolution.

Adapted from [71].

9.4.3 Quantitative Microanalysis with Energy-Dispersive X-Ray Spectrometry

Following inelastic scattering and an inner-shell ionization event, the specimen responds by one of two possible de-excitation processes involving either the emission of a characteristic X-ray photon or an Auger electron (§2.3). As mentioned, characteristic X-rays can be detected in a TEM by incorporating an energy dispersive X-ray detector (§2.5.1.2), positioned at a well-defined take-off angle,

ϕ, to collect a small fraction of the total X-ray emission (Fig. 9.4.2). The X-ray spectrum detected by the EDXS detector (Fig. 9.4.15) can be used to measure the characteristic X-ray intensities, above the background signal, $I_A, I_B, \ldots,$ of the various elements, A, B, ..., in the specimen and quantify them in terms of the chemical composition, i.e. mass concentration, $C_A, C_B, \ldots,$ using either a simple ratio method or by using appropriate thin film standards.

We begin by making the simple assumption that in an electron-transparent, thin foil TEM specimen, very few incident electrons are back-scattered, and most of the beam electrons follow a trajectory roughly equal to the thickness, t, of the specimen at the point of analysis (note that in a typical wedge-shaped TEM specimen, t can vary from point to point). We also assume that, on average, the beam electrons lose very little energy (5 eV/nm) on traversing through the specimen. Then, for a specimen satisfying the thin film criteria (above), for any element A, the characteristic X-ray intensity, I_A, measured over a specific time interval is given by

$$I_A = \kappa C_A \omega_A Q_A \frac{t}{A_A} \eta_A \qquad (9.4.25)$$

where κ is a proportionality constant, dependent on instrumental parameters including the incident beam intensity and the detector geometry, ω is the fluo-

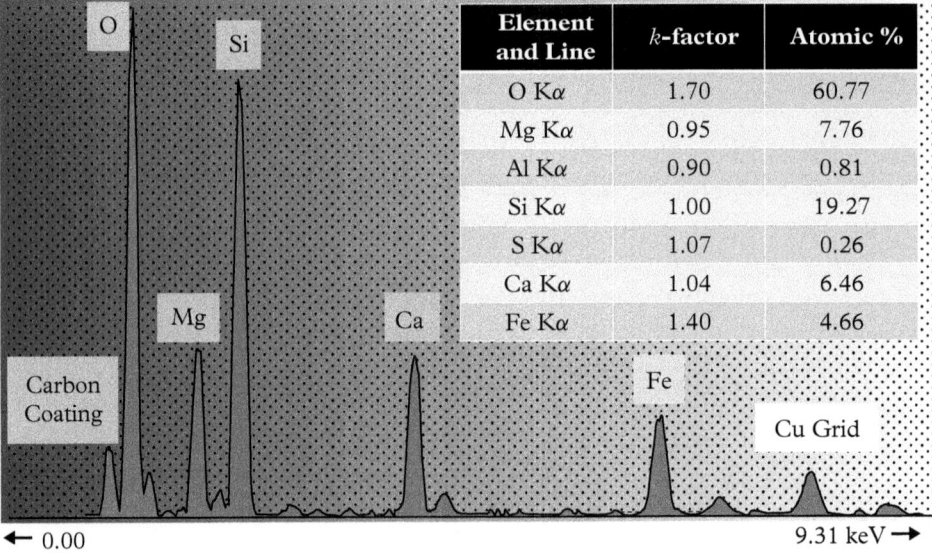

Element and Line	k-factor	Atomic %
O Kα	1.70	60.77
Mg Kα	0.95	7.76
Al Kα	0.90	0.81
Si Kα	1.00	19.27
S Kα	1.07	0.26
Ca Kα	1.04	6.46
Fe Kα	1.40	4.66

← 0.00   9.31 keV →

Figure 9.4.15 EDXS spectrum and microanalysis using experimentally measured k-factors of a thin film glass standard (K411) obtained from NIST. A 200 kV AEM, with a LaB$_6$ source and an ultra-thin window detector were used [79].

rescent yield for characteristic X-rays (Fig. 1.4.5), Q is the cross-section for a particular inner-shell (K, L, M, . . . , etc.) ionization given by (9.4.4a) that varies with the incident beam energy, E_0, and the binding energy, E_c, of the core electron, through the overvoltage, U ($=E_0/E_c$), A is the atomic weight and η is the efficiency of detection for characteristic X-rays of a specific energy; in all cases, the subscript refers to a specific element A. As mentioned (§5.1.1), the EDXS detector has various layers—the Be or ultra-thin window (if used) layer, the Au surface layer, and the Si dead layer—each of different thickness where the characteristic X-rays can be absorbed. Then, the efficiency, η, of detection is given by the sum of the absorption contribution, (2.4.2), of these individual layers as

$$
\eta_{\mathrm{A}} = \left\{ \exp\left[-\left(\frac{\mu}{\rho}\right)_{\mathrm{A}}^{\mathrm{Be(UTW)}} \rho_{\mathrm{Be(UTW)}} X_{\mathrm{Be(UTW)}} - \left(\frac{\mu}{\rho}\right)_{\mathrm{A}}^{Au} \rho_{Au} X_{Au} - \left(\frac{\mu}{\rho}\right)_{\mathrm{A}}^{Si} \rho_{Si} X_{Si} \right] \right\}
$$
$$
\left\{ 1 - \exp\left[\left(\frac{\mu}{\rho}\right)_{\mathrm{A}}^{Si} \rho_{Si} Y_{Si} \right] \right\} \tag{9.4.26}
$$

where the thickness of the layers, $X_{\mathrm{Be(UTW)}}$, X_{Au}, and X_{Si}, as well as the thickness of the active silicon layer in the detecting crystal, Y_{Si}, vary from detector to detector, and are not easily available. However, since the intensities $I_{\mathrm{A}}, I_{\mathrm{B}}, \dots$, for different elements are obtained simultaneously under the same conditions in a given EDXS spectrum, a simpler method of ratios [78] can be applied. Thus, the ratio of intensities:

$$
\frac{I_{\mathrm{A}}}{I_{\mathrm{B}}} = \frac{\left(\frac{Q\omega}{A}\right)_{\mathrm{A}} \eta_{\mathrm{A}}}{\left(\frac{Q\omega}{A}\right)_{\mathrm{B}} \eta_{\mathrm{B}}} \left(\frac{C_{\mathrm{A}}}{C_{\mathrm{B}}}\right) \tag{9.4.27}
$$

is proportional to the mass concentration ratio, and where the proportionality constant, κ, and thickness t, cancel out. This is generally written as

$$
\frac{C_{\mathrm{A}}}{C_{\mathrm{B}}} = k_{\mathrm{AB}} \frac{I_{\mathrm{A}}}{I_{\mathrm{B}}} \tag{9.4.28}
$$

and k_{AB} is referred to as the Cliff–Lorimer k-factors, or simply as the k-factors, for elements A and B. In a binary alloy, we have the additional relationship, $C_{\mathrm{A}} + C_{\mathrm{B}} = 1$, which allows us to solve for the concentrations. Similarly, in a ternary system, we have

$$
\frac{C_{\mathrm{A}}}{C_{\mathrm{B}}} = k_{\mathrm{AB}} \frac{I_{\mathrm{A}}}{I_{\mathrm{B}}}
$$
$$
\frac{C_{\mathrm{C}}}{C_{\mathrm{B}}} = k_{\mathrm{CB}} \frac{I_{\mathrm{C}}}{I_{\mathrm{B}}}
$$
$$
C_{\mathrm{A}} + C_{\mathrm{B}} + C_{\mathrm{C}} = 1 \tag{9.4.29}
$$

which can be solved for the concentrations provided k_{AB} and k_{CB} are known.

There are two ways to determine the k_{AB} factors: either by calculating the cross-sections and using (9.4.27), or using a *standard specimen* with known concentrations, C_A, C_B . . . , by measuring the intensities, I_A, I_B . . . , and determining k_{AB} . . . from (9.4.28). It is common practice to standardize the k_{AB}-factors with respect to Si. Then

$$k_{AB} = \frac{k_{ASi}}{k_{BSi}} = \frac{k_A}{k_B} \tag{9.4.30}$$

where, in the end, the subscript, Si, is understood and not explicitly included.

The k-factors for various inner shell excitations have been measured experimentally; Figure 9.4.16 shows a careful set of measurements for a 200 kV instrument, using an ultra-thin window detector. Using these k-factors and (9.4.20) modified for more elements, the spectrum (Fig. 9.4.15) for the glass standard obtained from NIST,[13] was analyzed with good agreement. Note that most commercial detectors come with a software package for data acquisition and analysis that includes the ability to calculate the ionization cross-sections and k-factors using in-built parameters for the detector. This is adequate for most common microanalysis applications; however, to make the most accurate analysis, there is no substitute to the measurement of k-factors for a particular EDXS detector, using standards of known composition. In addition, to correct for the absorption of the characteristic X-rays in the specimen (§2.5.2.3) accurate measurement of the thickness, t, at each analysis point is necessary; if the standard is a crystalline material, the CBED method (Fig. 8.6.11) can be used to measure the local

[13] https://www.nist.gov/publications/ status-microanalysis-standards-national- institute-standards-and-technology-nist

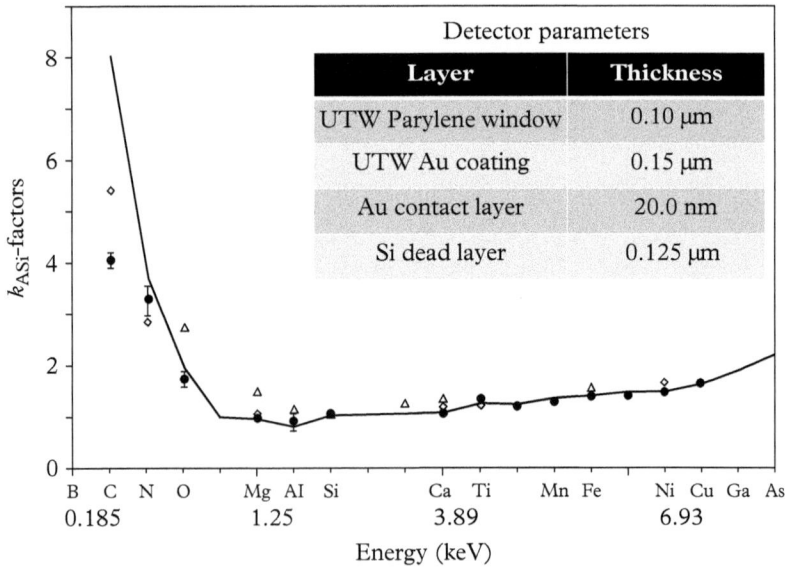

Figure 9.4.16 Experimentally measured k-factors for the K lines for all the principal elements up to Cu, using an UTW detector. The inset shows the measured thicknesses of the principal layers of the detector required for accurate microanalysis.

Adapted from [79].

Detector parameters	
Layer	**Thickness**
UTW Parylene window	0.10 μm
UTW Au coating	0.15 μm
Au contact layer	20.0 nm
Si dead layer	0.125 μm

specimen thickness, t. Details of such accurate k-factor measurements, including the corrections for local specimen thickness, are given in [79]. Recently, a new quantitative thin-film X-ray analysis procedure, termed the ζ-factor method and which overcomes the two principal limitations of the Cliff–Lorimer method—(1) use of pure-element rather than multi-element thin-specimen standards, and (2) built in X-ray absorption correction with simultaneous thickness measurements— has been proposed. Details can be found in [115].

The X-ray counting statistics follow Gaussian behavior; hence the standard deviation, $\sigma = \sqrt{N}$, where N is the number of counts in the peak above background. For 3σ level of confidence, the error in the measurement is $3\sqrt{N}$ with a relative error of $3N^{-1/2}$. Thus, for the ratio method of microanalysis, the relative error in the measurement of the ratio C_A/C_B, is the sum of the errors of I_A, I_B, and k_{AB}. In practice, accumulating $\sim$10,000 counts in each peak will guarantee a relative error of 3% in the measured concentration. Another criterion important in microanalysis is the minimum mass fraction (MMF) that can be detected. Statistically, it can be expressed as

$$MMF = \left[\left(\frac{P}{B} \right) \times P \times \tau \right]^{-1/2} \qquad (9.4.31)$$

where P is the elemental counting rate, P/B is the peak to background ratio, and τ is the counting time. To improve MMF, all three terms should be maximized. To improve P, it is best to increase the current density in the electron beam/probe, which is limited by contamination rate, beam sensitivity of the specimen (§5.3.2), and the stability of the stage. Further, for a given detector location (take-off angle, ϕ), the specimen should be tilted such that the X-ray signals are maximized and minimally absorbed in the specimen. And, where possible, the highest beam energy available should be used (why?). For more practical details on EDXS microanalysis, interested readers should consult Williams and Carter, Volume IV (1996).

Example 9.4.2: (a) Using the STEM probe described in Example 9.2.7, we carry out an EDXS microanalysis of a thin film, 30 nm thick, of iron with Cr impurity. The beam rapidly contaminates the specimen in about 10 seconds. Assume that the background is 2% of the Fe peak, and the Cr peak can be detected if its intensity is at least five times the statistical variation of the background. If the minimum detectable mass of Cr is 1 at %, what should be the minimum number of counts of Fe Kα per second to make this analysis possible?

(b) How does the minimum detectable concentration of Cr change if the probe diameter is increased to 100 nm?

(c) What will happen if the specimen thickness is doubled to 60 nm, and the probe diameter is kept at 10 nm?

Solution: (a) Assume that the we detect N counts of Fe Kα per second. Then, for 10 seconds, the background is $0.02 \times 10 \times N = 0.2\,N$ counts, with a statistical variation of $(0.2\,N)^{1/2}$. For Cr to be detected, the intensity of Cr Kα, should be at least $5 \times (0.2N)^{1/2}$ counts. We assume that the cross-sections for Fe and Cr are the same. Then, for a minimum detectability of 1 at % we have $\frac{5 \times (0.2N)^{1/2}}{10N} = \frac{1}{100}$.

Solving for N we get $N = 500$ counts of Fe Kα per second.

(b) When the probe diameter is increased by a factor of 10, the total counts of iron will increase by a factor of 100. Thus in 10 seconds, before contamination stops the experiment, we will count a total of $500 \times 10\ \text{sec} \times 100 = 500,000$ counts of Fe Kα. This will have $0.02 \times 500,000$ counts in the background, with a statistical variation of $(0.02 \times 500,000)^{1/2} = 100$. Thus, for Cr to be detected, we should have a minimum of $5 \times 100 = 500$ counts. Then, the minimum detectability of Cr would be $500/500000 = 0.1$ at % of Cr.

(c) If the specimen thickness is doubled, assuming there is no multiple scattering, the detection limit will improve by a factor of $2^{1/2}$, when compared to (a). Thus the minimum detectability of Cr would be $1/2^{1/2} = 0.707$ at %.

Example 9.4.3: Quantitative analysis combining EDXS and EELS: There are significant difficulties in the analysis of light elements, such as B, in a heavy rare-earth matrix. The microstructure of a Fe–Nd–B alloy, shown in the following figure (a), contains an Fe–Nd matrix and a second phase (RE) containing B. The quantitative analysis of B by EDXS is difficult, and the analysis of Nd by EELS is difficult because the cross-section for the M-edge is not readily available.

A 100 kV TEM equipped with an EDXS and EELS spectrometer is used for all analysis. We use 100 eV energy windows for the analysis of all elements, and a collection half-angle $\beta = 13.5°$ for EELS analysis.

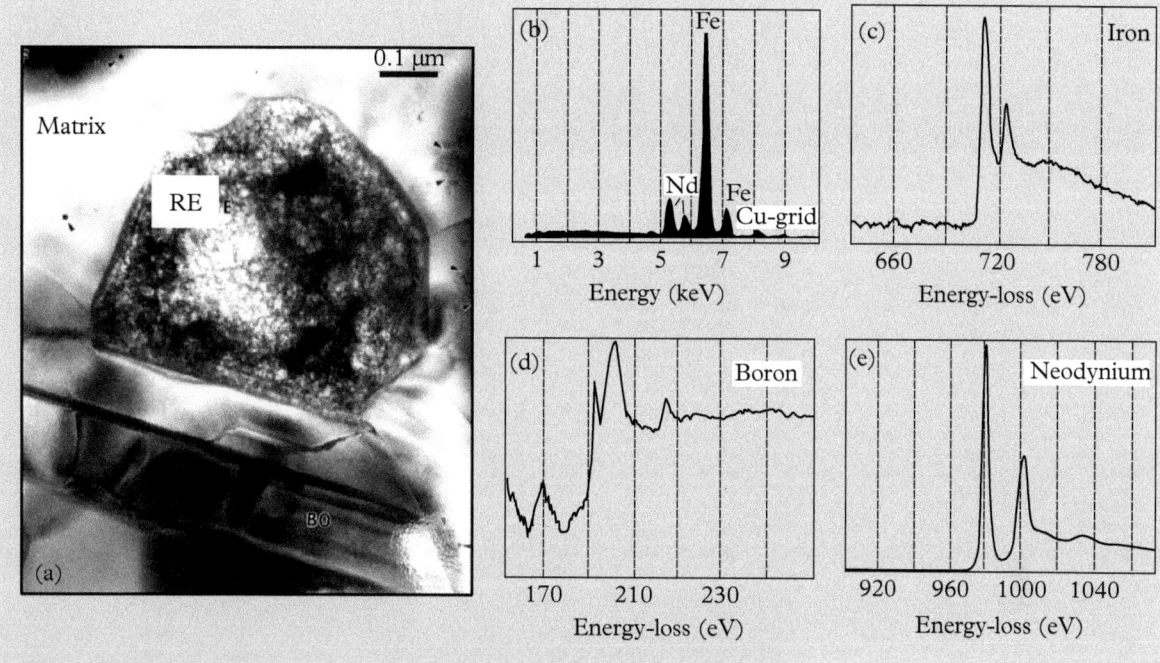

(a) An EDXS spectrum as shown in Figure (b) is acquired from the matrix of known composition. For this particular detector the k-factors, $k_{FeSi} = 12.6$ and $k_{NdSi} = 1.8$, and the measured intensities for the peaks are $I_{Fe} = 3,150$ and $I_{Nd} = 450$ counts, what is the composition of the matrix phase?

(b) We collect an EELS spectrum from the matrix and observe an Fe L-edge and a Nd M-edge. We calculate a cross section $\sigma_{Fe\text{-}L} = 2.75 \times 10^{-21}$ cm^2/atom. The measured intensities in the edges, similar to the ones shown in (c) and (e) are $I_{Fe\text{-}L} = 8.6 \times 10^6$ and $I_{Nd\text{-}M} = 4.88 \times 10^6$ counts, respectively. What is the cross-section for the M-edge of Nd?

(c) We can also calculate the B K-edge cross-section as $\sigma_{B\text{-}K} = 9.85 \times 10^{-20}$ cm^2/atom. We now have all the required cross-sections for the EELS analysis of the unknown Fe–Nd–B phase. If the measured intensities are $I_{B\text{-}K} = 9.998 \times 10^5$, $I_{Fe\text{-}L} = 4.189 \times 10^3$ and $I_{Nd\text{-}M} = 1.30 \times 10^4$ counts, respectively, what is its composition?.

Solution: (a) We determine the k-factor $k_{FeNd} = k_{FeSi}/k_{NdSi} = 12.6/1.8 = 7.0$. From (2.5.7), for the unknown atomic concentrations, C_{Fe} and C_{Nd}, we get $I_{Fe}/I_{Nd} = k_{FeNd} (C_{Fe}/C_{Nd}) = 3150/450 = 7 (C_{Fe}/C_{Nd})$. Thus $C_{Fe}/C_{Nd} = 1$, and assuming that this is a pure binary alloy, $C_{Fe} + C_{Nd} = 1$. Hence, $C_{Fe} = C_{Nd} = 0.5$. Thus, the matrix phase is an equiatomic alloy FeNd.

(b) We apply (9.4.18) to calculate the unknown cross-section for the Nd M-edge. Thus

$\frac{N_{Fe}}{N_{Nd}} = 1 = \frac{I_{Fe-L}}{I_{Nd-M}} \frac{\sigma_{Nd-M}}{\sigma_{Fe-L}}$. Hence, $\sigma_{Nd-M} = \frac{I_{Nd-M}}{I_{Fe-L}} \sigma_{Fe-L} = \frac{4.88}{8.6} 2.75 \times 10^{-21} = 1.56 \times 10^{-21}$ cm^2/atom.

(c) For the unknown, we apply (9.4.18) for pairs of elements as follows

$$\frac{N_{Fe}}{N_{Nd}} = \frac{I_{Fe-L}}{I_{Nd-M}} \frac{\sigma_{Nd-M}}{\sigma_{Fe-L}} = \frac{4.189 \times 10^3}{1.30 \times 10^4} \frac{1.56 \times 10^{-21}}{2.75 \times 10^{-21}} = 0.1828$$

Similarly, $\frac{N_{Fe}}{N_B} = \frac{I_{Fe-L}}{I_{B-K}} \frac{\sigma_{B-K}}{\sigma_{Fe-L}} = \frac{4.189 \times 10^3}{9.998 \times 10^5} \frac{9.85 \times 10^{-20}}{2.75 \times 10^{-21}} = 0.15$

and $\frac{N_{Nd}}{N_B} = \frac{I_{Nd-M}}{I_{B-K}} \frac{\sigma_{B-K}}{\sigma_{Nd-M}} = \frac{1.30 \times 10^4}{9.998 \times 10^5} \frac{9.85 \times 10^{-20}}{1.56 \times 10^{-21}} = 0.8225$

We also know that $N_{Fe} + N_{Nd} + N_B = 1$. We can then calculate the concentrations in the alloy as $N_{Fe} = 7.6$ at %, $N_{Nd} = 41.7$ at%, and $N_B = 50.7$ at %.

9.4.4 Microdiffraction

Even though microdiffraction involves elastic scattering, we discuss it here as it requires using a fine probe incident beam, such as that encountered in an AEM or a dedicated STEM. We motivate the use of microdiffraction, often encountered in the TEM as convergent beam electron diffraction (§8.6.3), using two simple arguments.

First, as seen in §9.7.2.1 and Figure 9.2.9, a selected area aperture of diameter, $\varphi \sim 10$ μm, located in the first image plane of the OL, is used to select the area of the specimen for diffraction with an incident parallel beam. Then, the selected area on the specimen has a nominal diameter of φ/M, where M is the magnification of the OL, which is usually of the order of 20–50 times. This gives a minimum selectable area on the specimen of $\sim$0.2–0.5 μm. Many materials' microstructures, such as grain boundary phases, nanoparticles, thin film heterostructures, semiconductor device components, etc., with feature sizes smaller than 0.2–0.5 μm require microdiffraction methods with focused beams for effective structural analysis. Unlike broad beam methods, e.g. selected area diffraction, that provide information averaged over larger volumes, microdiffraction methods provide local structural information defined largely by the size of the probe.

Second, the spherical aberration of the OL also contributes to the size of the actual specimen area included in the diffraction. Even though the selected area diffraction aperture is conjugate to the specimen, there is a discrepancy or displacement, Δd, between the area selected in the image plane and where exactly in the specimen the diffracted beams that pass through the aperture originate from. To a good first approximation (Hirsch et al., 1977), Δd is given by

$$\Delta d = C_S \theta^3 + (\Delta f) \theta \tag{9.4.32}$$

where C_S (~1–6 mm) is the spherical aberration coefficient of the OL, θ is the scattering angle of the electron beam, and Δf is the defocus of the lens. Thus, Δd can be significant for large Bragg angles, θ, and the associated diffracted beams may not originate from the same area as that selected by the aperture. Note that Δf may be positive or negative and since $C_S \theta^3$ is always positive, the overall error in selecting the area for diffraction can be minimized by choosing Δf carefully; however, this approach is tedious and not commonly used.

Example 9.4.4 You are imaging an orthorhombic crystal ($a = 8$ Å, $b = 7$ Å, $c = 6$ Å) using a 200 keV TEM with $C_S = 1.2$ mm. What is the displacement between the area selected and the exact area from which a (111) diffracted beam arises, if the defocus is -4000 Å?

Solution: From Table 4.1.2, with $(hkl) = (111)$, $d_{111} = 3.9587$ Å. From (9.2.3b), λ (200 keV) $= 0.025$ Å. Thus $\theta = 0.1809$. Hence, from (9.4.32), we get $\Delta d = 7.03$ μm. This displacement is quite large!

For these two reasons, microdiffraction methods, primarily CBED (introduced in §8.6.3) using a focused probe ~1–10 nm in diameter, and a convergence semi-angle, α, of the order of the Bragg angle, i.e. $\alpha = \theta_B$, are preferred for structural characterization of microstructural features in the sub-μm length scale. Other salient applications of CBED in materials characterization include the following.

1. As we have seen in X-ray and EELS microanalysis, absolute quantification of concentration, (9.4.26), requires determining the local specimen thickness, t. As discussed earlier (§8.6.3; Fig. 8.6.11), using the two-beam CBED method, the fringes seen in the discs for crystalline materials can be used to obtain the local thickness. This same convergent probe used for microanalysis, without any change in the TEM/STEM operating conditions, can be used for the thickness measurements.

2. The radii of the higher-order Laue zones (HOLZ; Fig. 8.6.10c) can be simply related to the crystal periodicity along the beam direction. For a HOLZ diameter, G_n, in units of Å^{-1}, measured from the CBED pattern, where n is the order of the ring, it can be shown that

$$G_n = \sqrt{2\frac{2\pi}{\lambda}nH - n^2 H^2} \qquad (9.4.33)$$

where $H = 2\pi/d_{uvw}$ is the spacing of the reciprocal lattice along the electron beam direction. Neglecting the second order term, the period, d_{uvw}, of the crystal in the beam direction is then given by

$$d_{uvw} = \frac{8\pi^2}{G_1 \lambda} \qquad\qquad (9.4.34)$$

where G_1 is the diameter of the first-order Laue zone (FOLZ) ring. Note that the unit cell parameters, including angles (Fig. 8.6.3) normal to the beam direction can be obtained from the analysis of the ZOLZ pattern.

3. Under favorable conditions the space group of the crystals can be determined from CBED patterns [80]. Such analysis involves details of the contrast within the discs of the CBED pattern, which requires somewhat thicker crystals. However, for thin crystals, an elegant method using a STEM unit and rocking the electron beam produces large-angle CBED patterns without overlapping of neighboring discs [81, 82].

In addition to §8, further details of electron microdiffraction, including CBED, can be found in specialized texts on TEM, including Williams and Carter (1996), Joy, Romig, and Goldstein (1986), and Reimer (1993).

9.5 Select Applications of TEM

9.5.1 Electron Tomography

A TEM image is essentially a two-dimensional projection of the thin foil specimen along the direction of the primary beam. Thus, features within the depth of the specimen cannot be separately resolved. Stereo pairs can be recorded to resolve them if they are discrete features, but if they are continuously varying functions of mass or density they cannot be easily resolved. One way to overcome these limitations and obtain three-dimensional information with true depth sensitivity is by electron tomography. By *tomography* we mean the reconstruction of a three-dimensional image of the object based on a series of two-dimensional projected images of the specimen acquired in different directions. The processing or reconstruction of the three-dimensional image from a set of two-dimensional projected images, critical for the successful implementation of tomography, involves two principal tasks:

1) alignment of the two-dimensional projected images with respect to each other, such that they can be referred to a common three-dimensional coordinate system describing the object to be reconstructed, and

2) computation of the three-dimensional image based on a reconstruction algorithm using one of the many approaches available.

The technique of X-ray computed axial tomography (CAT), now common in diagnostic medicine, firmly established the principles of tomography as it is used today [83]. In this method, a thin fan-like beam is used to illuminate a slice of the object; the source and a linear detector placed behind the object are moved together to obtain a series of one-dimensional images/projections, which are then reconstructed to form a two-dimensional image. Finally, to obtain a three-dimensional image, such two-dimensional image sets along different directions are recorded and reconstructed.

Electron tomography in a TEM considers each image to be included in the reconstruction as a projection of the object along the primary beam direction. This important assumption is justified by the large depth of focus (9.2.8), which is so much larger than the diffraction-limited image resolution (9.2.3a). Moreover, since the specimen is finite, the full structural image at a predetermined resolution can be obtained by recording a series of projected images by sequentially tilting the specimen about a single axis over the largest range of angles possible (preferably 180°) and maximizing the number of images acquired in the series (Fig. 9.5.1a). The relationship between the resolution, δ, and the increments in the tilt angle, $\Delta\alpha$, are related to the diameter, D, of the object, and the number, N, of equally spaced projections included in the reconstruction is given by [85]:

$$\delta = D\Delta\alpha = \pi D/N \qquad (9.5.1)$$

Thus, for the reconstruction of an object 25 nm in diameter (a typical nanoparticle, say), with a resolution of 0.25 nm (= 2.5 Å) will require 100 projections and a tilt increment of 1.8°. However, the TEM specimen geometry (extended in x- and y-, but limited to the thickness in the z-direction) can be better sampled by nonequal tilt angle increments [86].

In practice, the difficulty in tilting the specimen over the full angular range causes distortions in the reconstructed image due to the missing data (Fig. 9.5.2) [87]. Second, in acquiring the large number ($\sim$100) of projected images required for reconstruction, the specimen is subject to a large electron dose. To minimize the dose and exposure time, automated acquisition routines, especially with high precision, computer-controlled goniometer specimen stages are used [88–90]. Finally, as mentioned, each individually projected image has to be accurately aligned, i.e. tilted to a common tilt axis by changing the spatial position (x-y) and the rotational (θ) orientation of the specimen. To do so, a series of sequential cross-correlations [91], and least-squares tracking of fiducial markers [92] are employed.

For reconstruction (Fig. 9.5.1b), a real space back-projection method [93] is widely used. This involves projecting each two-dimensional image along the original tilt angle, and the superposition of these back-projected images gives the reconstructed image. However, the single tilt-axis geometry favors the sampling of low-frequency components of the image over the high frequency ones. This error is overcome by using a "weighting filter", an approach that is referred to as the weighted back-projection method (Fig. 9.5.1b). Alternatively, iterative reconstruction methods [94, 95]—based on the principle that a reprojection of the three-dimensional reconstructed object along the original projection angle/direction

(a) Data acquisition: projection of 2D images along the beam direction

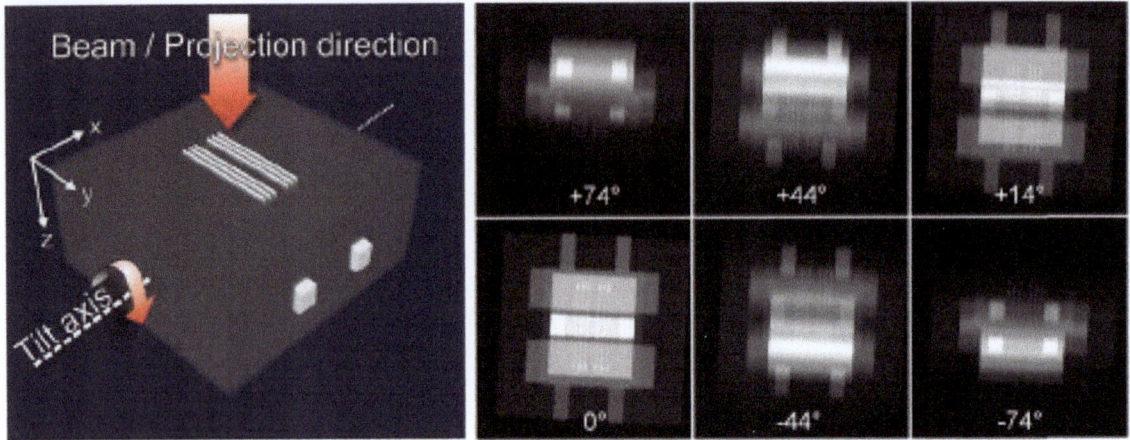

(b) Tomographic reconstruction using consecutive 2D slices

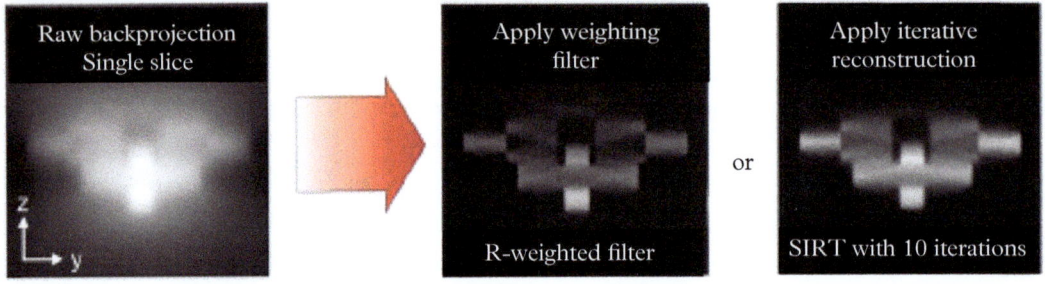

(c) 3D visualization and segmentation

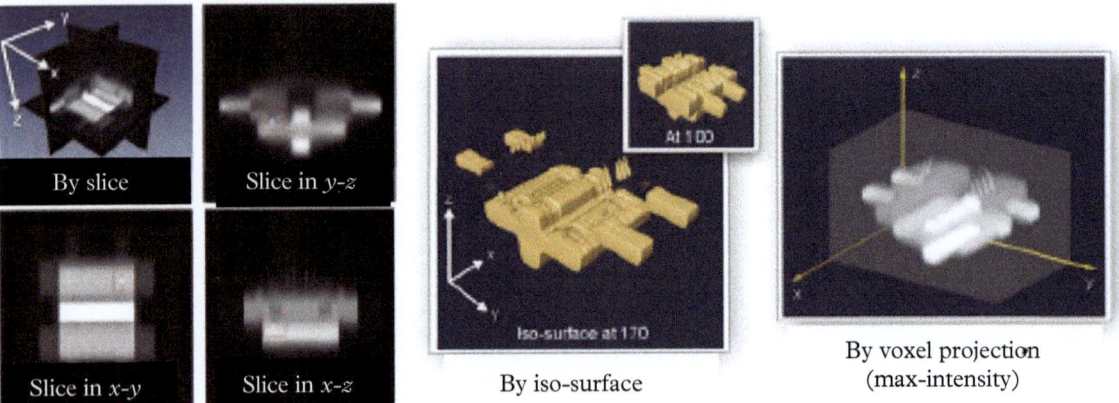

Figure 9.5.1 The method of electron tomography. (a) The object is sampled by obtaining a tilt series of projected images, which are then aligned to a common tilt axis. (b) The three-dimensional image is reconstructed from the two-dimensional projected images using a back-projection algorithm. The raw back-projection shows significant blurring, which is overcome by appropriate frequency weighting or using an iterative approach to the reconstruction. (c) The reconstructed three-dimensional image of the object is visualized in a number of different ways including slicing, surface rendering, or voxel projection.

Adapted from [84].

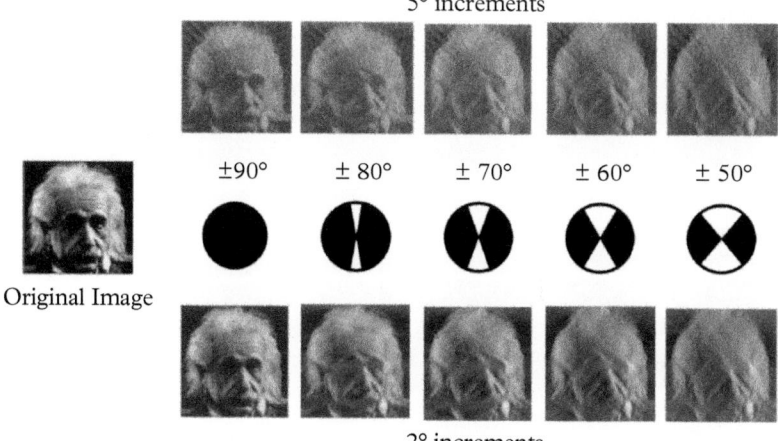

5° increments

Original Image

2° increments

Figure 9.5.2 Illustration of the effect of tilt range (missing data shown as a wedge in the middle row) and tilt increments (top: 5° increments, and bottom: 2° increments) on the reconstructed image. The reconstructed image is almost identical to the original when the tilt range is ±90° and tilt increment is 2°; however, when tilt range is ±50° and tilt increment is 5°, there is no similarity.

Adapted from [90].

should be identical to the original image—are used, particularly for noisy and under-sampled images [96].

The key projection requirement for tomographic reconstruction is that the intensity in the image show a monotonic relationship with thickness of the specimen. For biological [90] and amorphous specimens, *mass-thickness contrast* (§9.3.2) satisfies this requirement. For crystalline materials, especially high-Z, such as heterogeneous catalyst particles, tomographic reconstruction has been successfully demonstrated using a set of tilt-series images obtained with a high angle annular dark-field (HAADF) detector (§9.3.4) in a dedicated STEM. The intensity in such Z-contrast images is linearly dependent on depth [97], and as mentioned (9.3.11), the differential cross-section is proportional to Z^2. Electron tomography with chemical sensitivity has also been demonstrated using projected elemental maps, obtained with an EFTEM (§9.4.2.6), along different direction. Figure 9.5.3 illustrates EFTEM tomography of nanoscale precipitates in intermetallic alloys.

Electron tomography is widely used in biology with particular emphasis on the reconstruction of images of single particles of complex macromolecules [90]. In fact, electron tomography, as a means to construct three-dimensional images from two-dimensional projections, owes a significant part of its development to the biological community. This is because electron tomography bridges a critical resolution gap between methods providing atomic resolution images and other microscopic methods, such as confocal light microscopy (§6.8.5) that allow imaging of living cells. Even though preparation of biological specimens involves embedding the cellular structures in vitrified ice, the resolution of ~2–5 nm achievable in electron tomography, allows the possibility of imaging individual macromolecules to obtain information about their spatial relationship inside the cells.

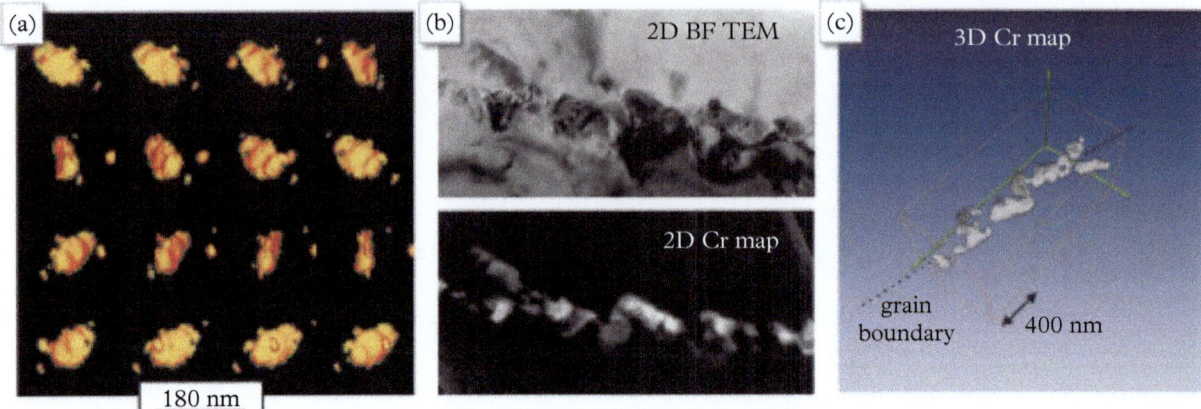

Figure 9.5.3 Electron tomography of precipitates in intermetallic alloys using EFTEM images. (a) A series of surface rendered images following reconstruction of yttria precipitates in NiAl. (b) Images of grain-boundary carbide precipitates in 316 stainless steel showing a BF image (top) and a Cr-EFTEM map (bottom). Diffraction effects obscure the shapes of the carbide precipitates in the BF image, whereas the Cr-EFTEM map shows unambiguous contrast only from the precipitates; the latter are more suitable for 3D tomographic reconstruction. (c) The reconstructed tomographic image (voxel projection) showing the precise shape and orientation of the precipitates with respect to the matrix.

Adapted from [84].

The total mean free path, Λ_{tot}, combining elastic and inelastic scattering in vitrified ice is ~100 nm at $E_0 = 100$ keV. Typically, biological specimens (TEM sections) are several 100 nms in thickness, and so multiple scattering is always observed. Further, biological specimens (cells) and vitrified ice are largely composed of light elements ($Z < 10$), and hence, inelastic scattering in the forward direction dominates over elastic scattering (§9.4.1). This leads to a strong inelastic background in the images, and the associated noise causes significant blurring and obscures finer details in the images of macromolecules. As a result, EFTEM have been extensively used in biological imaging to filter out the majority of the inelastic scattering and using only the zero-loss peak to improve contrast in the images. However, the associated large doses required for EFTEM mapping limits its use in tomographic imaging. Many strategies including use of lower acceleration voltages to reduce knock-on damage (§5.3.2.2), the use of a FEG source with a narrow primary energy spread, and a liquid-He specimen stage to reduce the radiation damage to the specimen, are currently being employed [90].

In spite of these dose concerns, by combining EFTEM with electron tomography, three-dimensional distribution of elements, such as P, were obtained in thin sections of the nematode *Caenorhabdites elegans*, prepared by high-pressure freezing and plastic embedding [98]. The sections were sufficiently thin to allow jump-ratio imaging (see §9.4.2.6 and Fig. 9.4.13) using energy losses above and below the P $L_{2,3}$ edge. Subsequently, EFTEM projection maps (P jump ratio

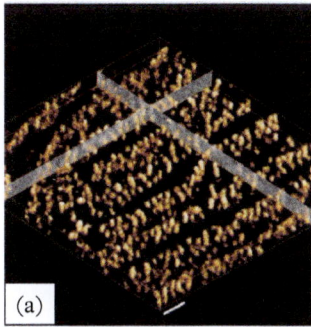
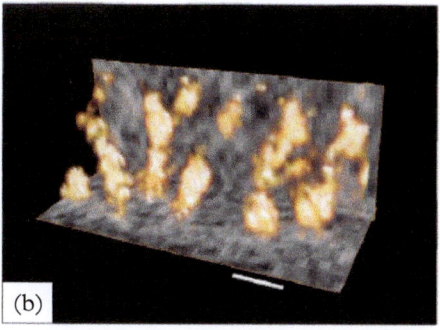

Figure 9.5.4 Topographic reconstruction of phosphoros distribution in a section of the *C. elegans* cell. (a) Volume rendered image of rows of ribosomes along stacks of endoplasmic reticulum membranes. (b) Higher magnification image of the phosphoros distribution shows individual ribosomes located at different heights within the section. Scale bar = 20 nm in both images.

Adapted from [98].

images) were taken at 5° intervals over an angular tilt range of ±55° for two orthogonal tilt directions. Colloidal gold particles were used as fiduciary markers to align the images. A reconstructed three-dimensional image of the phosphoros distribution (Fig. 9.5.4) in the surface rendering, clearly shows features 15–20 nm in diameter that were identified as ribosomes distributed along the stacked membranes of the endoplasmic reticulum and in the cytoplasm. Surprisingly, these specimens were able to withstand a high electron dose of $\sim 10^7 e/\text{nm}^2$ in the process of image alignment, acquisition, and tomographic reconstruction.

Complementing this, correlative imaging methods using dual function probes, such as Fluorogold^{TM},[14] hold much promise to identify specific bound elements and biological molecules in cellular structures [77]. For example, Fluorogold labels can detect specific proteins in the fluorescence optical microscope, followed by the localization of the same proteins in a TEM using the electron-dense gold labels [99], combined with electron tomography to image them in three dimensions.

9.5.2 Analysis of Defects: Dislocations and Stacking Faults

One of the most well-developed applications of TEM is to investigate the diffraction contrast associated with images of dislocations (§4.1.9.2), stacking faults (§4.1.9.3), and other defects such as grain boundaries, interfaces, etc.

In defective crystals the positions of atoms are slightly displaced (by the vector, $\mathbf{R}$) from their original sites (defined by $\mathbf{r}$) in the ideal crystal (Fig. 9.5.5); thus, their positions are given by $(\mathbf{r} + \mathbf{R})$. Then, following (8.5.10), for the atoms in a crystal that are misoriented by $\mathbf{s}$, from the exact Bragg condition in diffraction,

[14] Hydroxystilbamidine (trade name FluoroGold) is a fluorescent dye that emits different frequencies of light when bound to DNA and RNA.

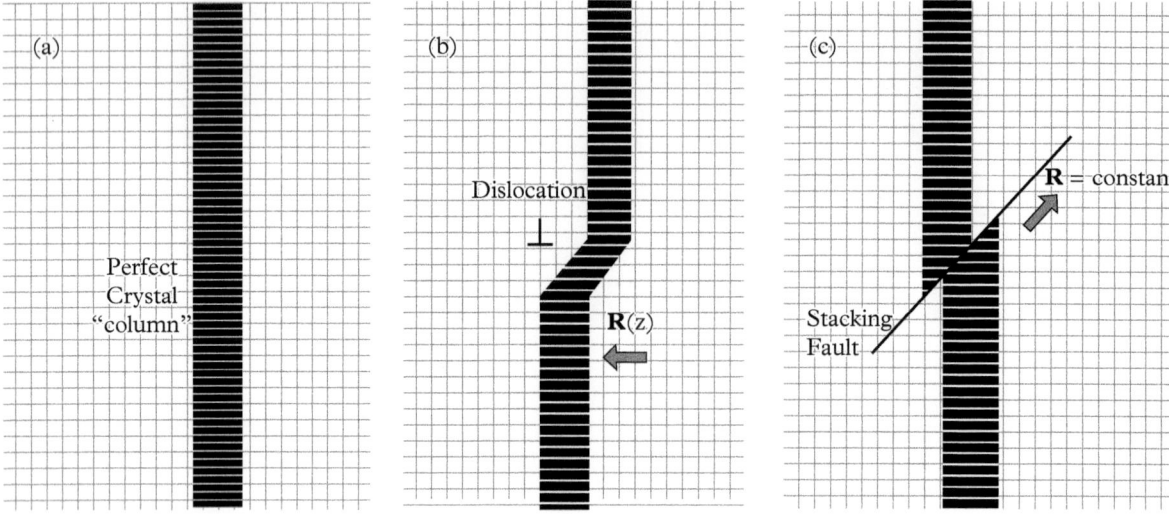

Figure 9.5.5 The effect of defects on a columnar beam in (a) a perfect crystal, and a crystal with (b) a dislocation with the lattice displacement varying along z, and (c) a stacking fault with a constant lattice displacement in the lower half of the thin foil.

Adapted from Fuchs, Oppolzer, and Rehme (1990).

and are defective with an additional displacement, $\mathbf{R}$, the phase angle, ϕ_g, for any reflection, $\mathbf{g}$, is given by

$$\phi_g = (\mathbf{g} + \mathbf{s}) \cdot (\mathbf{r} + \mathbf{R}) = [(\mathbf{g} \cdot \mathbf{r}) + (\mathbf{g} \cdot \mathbf{R}) + (\mathbf{s} \cdot \mathbf{r}) + (\mathbf{s} \cdot \mathbf{R})] \tag{9.5.2}$$

The term $(\mathbf{s} \cdot \mathbf{R})$ is a product of two small vectors and can be neglected. The term $(\mathbf{g} \cdot \mathbf{R})$ is an integer and does not affect the phase change. The two remaining terms, $(\mathbf{g} \cdot \mathbf{R}) + (\mathbf{s} \cdot \mathbf{R})$, added together contribute to the total phase shift for either any lattice defect, or misorientation from the Bragg angle, θ_B, respectively. Further, at the exact Bragg condition, $\mathbf{s} = 0$, the scalar product, $(\mathbf{g} \cdot \mathbf{R})$, dominates the phase angle. Thus, for specific Bragg reflections, $\mathbf{g}$, where the contrast of the lattice defect satisfies the criterion

$$\mathbf{g} \cdot \mathbf{R} = 0 \tag{9.5.3}$$

its image disappears, and under these conditions, both the direction and magnitude of $\mathbf{R}$ can be determined. For stacking faults (Fig. 8.5.8c), $|\mathbf{R}| = $ constant; then, when the condition $(\mathbf{g} \cdot \mathbf{R}) = 0$ is satisfied, the stacking faults become invisible. In the case of dislocations (Fig. 8.5.8b), $\mathbf{R}$ is a function of position, $\mathbf{r}$, and specifically for thin foils, it is a function of depth, $\mathbf{R}(z)$. Further, for a pure screw dislocation, by definition the Burgers vector, $\mathbf{b}$, is parallel to the displacements,

R, for all atoms (Fig. 4.1.20b), which are all along the dislocation line defined by the unit vector, $\hat{\mathbf{u}}$. Then, **R** can be replaced by **b**, and when the condition $\mathbf{g}\cdot\mathbf{b}$ = 0 is satisfied, pure screw dislocations become invisible. In the general case, the displacement, **R**, in an isotropic solid for a general (mixed) dislocation, where **b** is not parallel to $\hat{\mathbf{u}}$, is given by

$$\mathbf{R} = \frac{1}{2\pi}\left(\mathbf{b}\alpha + \frac{1}{4(1-\nu)}\left\{\mathbf{b}_e + \mathbf{b}\otimes\hat{\mathbf{u}}\left[2(1-2\nu)\ln|r| + \cos 2\alpha\right]\right\}\right) \quad (9.5.4)$$

where $\mathbf{b}_e$ is the edge component of the Burgers vector, ν is the Poisson ratio, and polar coordinates (α) are used. Then, for a pure edge dislocation $\mathbf{b} = \mathbf{b}_e$, and $(\mathbf{g}\cdot\mathbf{R})$ involves two terms, i.e. $(\mathbf{g}\cdot\mathbf{b}_e)$ and $(\mathbf{g}\cdot\mathbf{b}_e \otimes \hat{\mathbf{u}})$. The second cross-product term arises from the buckling of the glide plane by the presence of the edge dislocation, and can complicate the analysis of dislocations with an edge component, as they may not go completely out of contrast. Nevertheless, the $(\mathbf{g}\cdot\mathbf{b}) = 0$ criterion can be generalized to all dislocations and stated as a rule; i.e. if the Burgers vector, **b**, of the dislocation is perpendicular to the active diffraction vector, **g**, no diffraction contrast will be observed from the dislocation and it will be invisible. This is illustrated schematically in Figure 9.5.6. Further if two Bragg reflections $\mathbf{g}_1$ and $\mathbf{g}_2$, satisfy

$$\mathbf{g}\cdot\mathbf{b} = 0, \text{then } \mathbf{g}_1 \otimes \mathbf{g}_2 \| \mathbf{b}.$$

In practice, the dislocations become invisible even when $\mathbf{g}\cdot\mathbf{b} < 1/3$. In addition, the analysis and determination of **b** is more complicated because the dislocations may not be invisible, even when $\mathbf{g}\cdot\mathbf{b} = 0$, if there is a contribution of the term $\mathbf{g}\cdot\mathbf{b}_e \otimes \hat{\mathbf{u}} \neq 0$. Further details of such analysis can be found in Edington (1976). Table 9.5.1 shows the values of various $\mathbf{g}\cdot\mathbf{b}$ combinations for perfect dislocations in FCC crystal. Orientations such as [110] are particularly useful because it gives access to **g** vectors of the form (002), $(1\bar{1}1)$, $(2\bar{2}0)$, $(1\bar{1}3)$, etc. Figure 9.5.7 shows

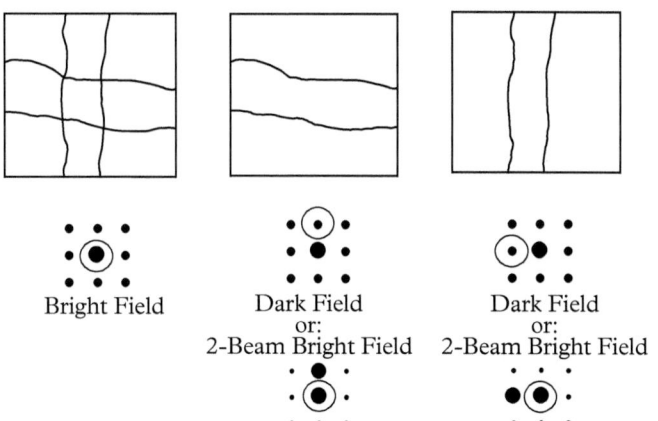

Bright Field

Dark Field
or:
2-Beam Bright Field

Dark Field
or:
2-Beam Bright Field

Figure 9.5.6 Schematic illustration of the diffraction contrast from dislocations, with **b** and lines in the plane of the image.

Adapted from Fultz and Howe (2013).

Table 9.5.1 Values of **g·b** for perfect dislocations in FCC crystals.

Plane of dislocation	$_b\backslash{}^g$	$\bar{1}11$	$\bar{1}\bar{1}1$	$11\bar{1}$	002	$0\bar{2}0$	$2\bar{2}0$
$(1\bar{1}1)$ or $(1\bar{1}\bar{1})$	$\frac{1}{2}[110]$	0	0	1	0	−1	0
$(1\bar{1}\bar{1})$ or $(11\bar{1})$	$\frac{1}{2}[101]$	1	0	0	1	0	1
$(1\bar{1}1)$ or $(11\bar{1})$	$\frac{1}{2}[011]$	0	1	0	1	−1	−1
(111) or $(11\bar{1})$	$\frac{1}{2}\left[1\bar{1}0\right]$	1	−1	0	0	1	2
(111) or $(1\bar{1}1)$	$\frac{1}{2}\left[10\bar{1}\right]$	0	−1	1	−1	0	1
(111) or $(\bar{1}11)$	$\frac{1}{2}\left[0\bar{1}1\right]$	1	0	−1	1	1	1

Adapted from Thomas and Goringe (1980).

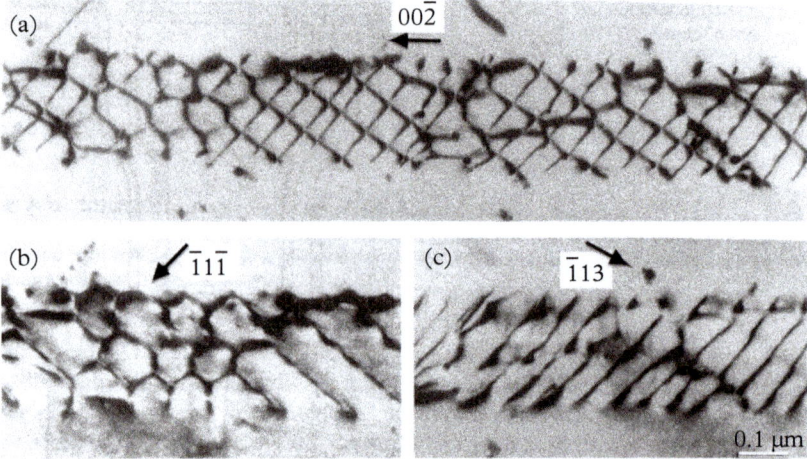

Figure 9.5.7 Bright field images in the two-beam condition of edge dislocations in a TiAl alloy. The corresponding **g** vectors are shown.

Adapted from Fultz and Howe (2013).

g·b analysis of edge dislocations in TiAl. It can be seen that the dislocations are primarily edge type with $\mathbf{b} = \frac{1}{2}(110)$.

Sharper images of dislocations, at better resolution (Fig. 9.5.8) can be obtained by weak-beam dark-field imaging using the **g**-3**g** diffraction condition (Fig. 9.3.5b). Further details of weak beam imaging can be found in [100].

Diffraction contrast of stacking faults is not discussed here, but similar analysis including establishing whether the stacking faults are intrinsic or extrinsic, as well as practical details and pitfalls, is found in Edington (1976), Williams and Carter (1996), and Fultz and Howe (2013).

9.5.3 Thin Films and Multilayers: An Example

We now revisit the TEM analysis of the thin film $(\mathrm{Mo}_{5.4\mathrm{nm}}\mathrm{Si}_{1.3\mathrm{nm}})_n$ multilayer, introduced in §8.7.5, and answer some of the question raised earlier to provide a complete picture of its structure, composition, and microstructure. The BF image (Fig. 9.5.9a) of a cross-section specimen of the multilayer clearly shows a substrate

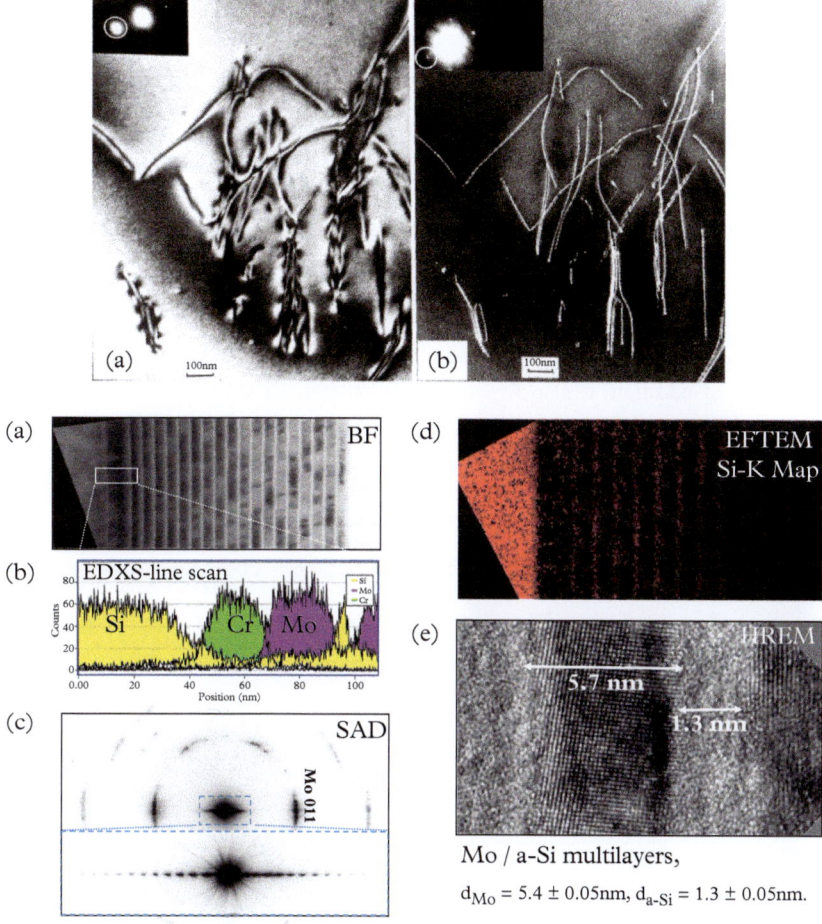

Figure 9.5.8 Dislocations in silicon imaged with a (a) strong 220 diffraction beam (inset), and (b) weak-beam 220 dark field, **g-3g**, condition. Notice the significantly better resolution in the weak-beam image.

Adapted from [100], which also describes weak-beam imaging in detail.

Figure 9.5.9 Comprehensive analysis of a nominal $[Mo_{5.4nm} Si_{1.3nm}]_n$ multilayer cross-section specimen using a 200 kV analytical electron microscope with Schottky emission gun, EDXS detector, and a post-column image filter. (a) Bright field image, (b) X-ray line scan, (c) selected area diffraction pattern, (d) energy-filtered image using the Si K edge, and (e) a high-resolution phase contrast image shows that the Mo layer in reality is slightly thicker (5.7 nm) than expected.

on which we can observe a thick first layer, followed by two layers that together repeat with a periodicity of ∼7 nm. Moreover, the thicker of the two repeating layers shows diffraction contrast within the layer, indicative of columnar growth, but the thinner layer shows no contrast, suggesting that it may be amorphous. Further, the selected area diffraction pattern (Fig. 9.5.9c) reveals additional specifics about the crystallographic structure of the layers. First, it should be noted that the diffraction pattern is comprised of two parts: reflections from the crystalline Mo layer, and superlattice reflections, shown in the magnified part of the central portion of the diffraction pattern with very fine spacing that correspond to the overall multilayer period. The Mo reflection can be indexed as (011), and shows some texture (formation of arcs) of the columnar grains; moreover, the $Mo_{(011)}$ reflection serves as an internal calibration. The fine superlattice spacing are 1/37.8 of the $Mo_{(011)}$ reflection, and hence the superlattice period is d_{011} (∼ 0.1767 nm) × 37.8

= 6.68 nm. Finally, the HREM phase contrast image (Fig. 9.5.9e) confirms the overall periodicity of the multilayer (which is to be expected), the high crystalline quality of the Mo layer as well as the amorphous nature of the Si layer, and most importantly that the Mo/a–Si interface is not sharp but atomically intermixed.

What are the multilayers made of? To answer this question, we can turn to energy dispersive X-ray microanalysis using a 0.5 nm probe. This is possible because the microscope (Fig. 9.1.2) is equipped with a high-current density Schottky emission gun and a windowless energy dispersive X-ray detector. X-ray spectra were obtained by positioning the probe at different points along a straight line starting at the substrate and through the multilayer stack, and the characteristic X-ray intensities were plotted as a function of position (Fig. 9.5.9b). The analysis clearly revealed alternating layers to be Mo (5–5.5 nm) and Si (1–1.5 nm); in addition, it also revealed an unknown Cr underlayer used in the growth that was not known before the TEM analysis. Finally, an EFTEM image (Fig. 9.5.9d) using Si K-edge ($\Delta E \sim 1800$ eV) confirms the silicon distribution. It also confirms that EFTEM can obtain images with resolution of the order of 1 nm, even at such high energy-losses as the Si K-edge (~ 1740 eV loss).

9.5.4 TEM in Semiconductor Manufacturing: Metrology, Process Development, and Failure Analysis

Continued technological scaling of metal oxide semiconductor field effect transistors (MOSFETs), ubiquitous in integrated circuits (ICs), presents significant challenges in (a) determining precisely the size and extent of the ultrathin layers and interfaces that make up the individual transistor (*metrology*), (b) development of alternative high-*k* dielectrics to replace silicon dioxide as the gate oxide used to isolate the gate electrodes from the source-drain channel (*process development*), and (c) identifying the root cause of a failure site indicated by electrical measurements (*failure analysis*) [101]. The narrowest feature in an IC is the gate oxide, or the thin dielectric layer that forms the basis of the field-effect device structure. Silicon dioxide is widely used and the physical thickness of this gate is of the order of 1 nm; however, alternative gate oxides, such as HfO_2, ZrO_2, etc., which can be thicker than 1 nm because of their higher dielectric constants, are in development. Nevertheless, in manufacturing of ICs, accurate measurements of the thickness of the gate oxide layer are necessary as even a 10 % decrease in its thickness will change the leakage current by one order of magnitude [101]. Moreover, a thickness of 1 nm of the oxide layer corresponds to about five silicon atoms spanning the layer [102]; of these five, at least two of the atoms will be at the silicon–oxide interface with electrical properties expected to be different from those in the bulk oxide. It is easy to recognize that such rapidly shrinking device features below the nanometer scale will push the required metrology well beyond the resolution limits of even the best SEMs, which is of the order of 5–7 nm (§10.2.4). Further, in manufacturing, process anneals at $\sim 1,000°C$

are involved; hence, it is important to study not only the stability but also the interactions between the dielectric layer and the silicon channel of the device at elevated temperatures. Generally, most electrical failures of the device to be resolved require the physical and chemical characterization of defects buried deep inside the structure. SEM only provides image contrast based on surface topography and/or difference in the atomic weights of the elements distributed on the surface. In practice, a buried defect will not change the surface topography, and by extension the contrast in imaging with an SEM, unless a decorative etch is used; this is also a destructive process and hence is not applicable. The last resort to address these three issues in semiconductor manufacturing is a TEM, which has the required resolution and is also a versatile instrument providing structural, crystallographic, and chemical information (Fig. 9.1.2). However, the most important key to addressing these problems with a TEM is the development of site-specific specimen preparation of the device/wafer using a focused ion beam instrument [103]. Specifically, the advanced dual column FIB-SEM instruments (Fig. 10.7.3) with separate high-resolution drift-free ion beam and electron-beam columns allow for the simultaneous imaging and sectioning of the specific defective portion of the IC device into an electron transparent plan-view or cross-section TEM specimen (Fig. 10.7.4c). Methods of TEM specimen preparation are discussed briefly in Section 9.6), and we now present some representative examples of MOSFET structures studied by TEM [101].

Once the proper TEM specimen is prepared, conventional BF imaging (Fig. 9.5.10) can be used to evaluate the profile and critical dimensions of the components of the MOSFET, including the poly-Si gate, nitride spacer, active Si trench depth, etc. Such measurements of the critical dimensions are often used to tailor the etch processes to improve the gate profile, including tapering and footing, and ensure that the device meets the performance specifications.

Again, if a damage-free ultra-thin (thickness <20 nm) TEM specimen is prepared, HREM imaging can directly measure details of the nm-scale gate oxide layer. In this case, it is important to ensure that the specimen is prepared from a device region where the gate oxide is free of overlapping features. Figure 9.5.11a

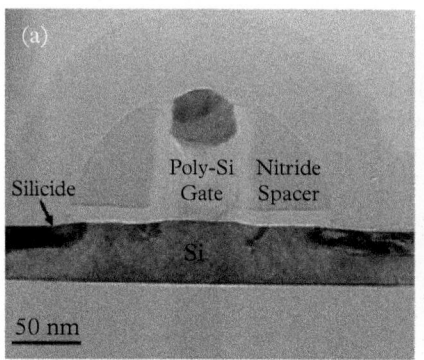 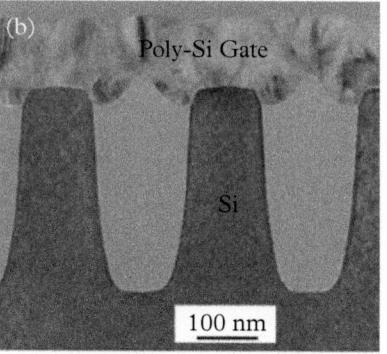

Figure 9.5.10 TEM bright field image of (a) poly-Si gate structure, and (b) poly-Si over active and trench areas.

Adapted from [101].

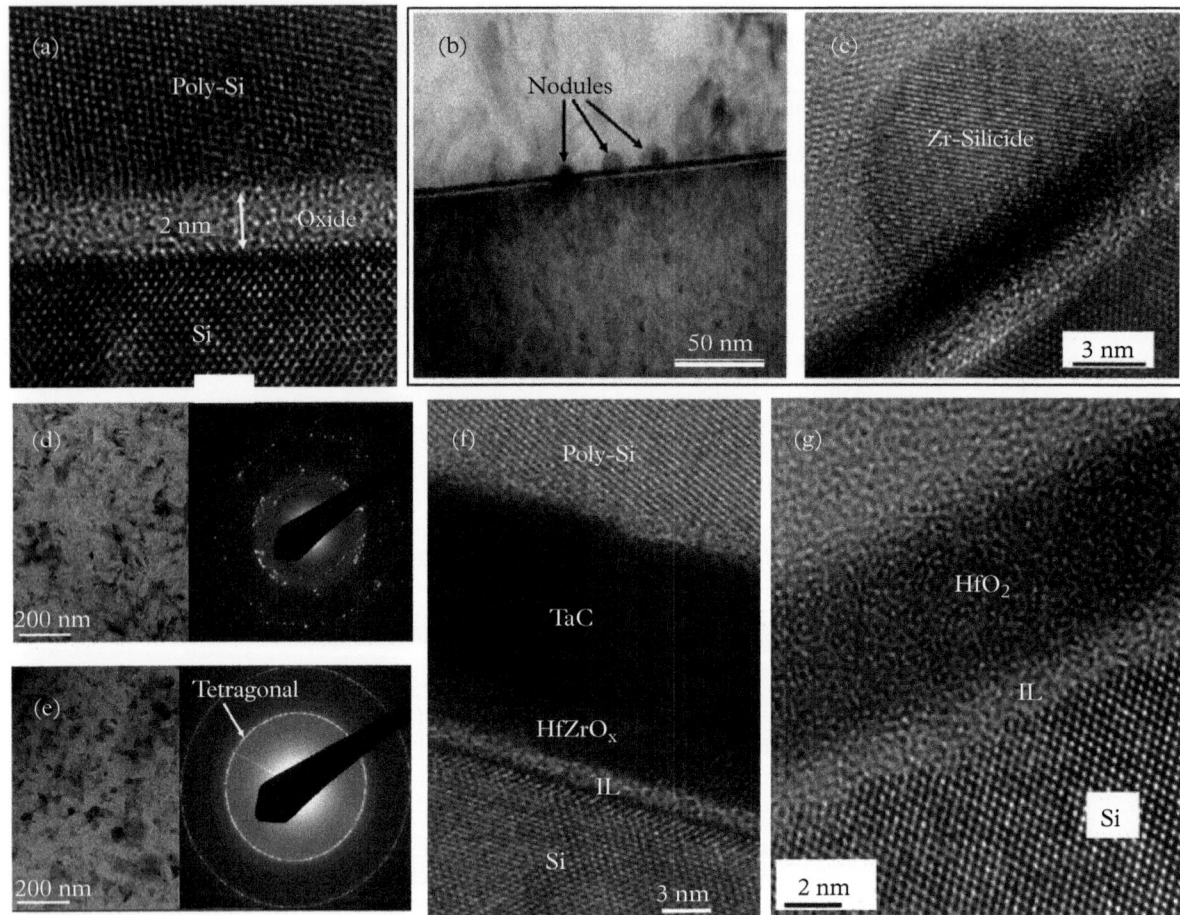

Figure 9.5.11 The gate dielectric. (a) A HREM image of a cross-section specimen showing the details of the poly-Si/SiO$_2$/Si interface and the 2 nm thick dielectric layer. (b) Low-magnification image showing the formation of nodules at the poly-Si/ZrO$_2$ gate interface. (c) HREM confirms that the nodules are crystalline Zr–silicide. Plan view BF images show (d) formation of a low dielectric monoclinic phase in a pure HfO$_2$ film, but (e) formation of a high-k dielectric tetragonal phase in HfZrO$_x$ films. (f) HREM of HfZrO$_x$ gate dielectric film, 3 nm thick, with a 1-nm interfacial layer between the Si channel and the gate oxide. (g) Amorphous HfO$_2$ film, 6 nm thick, with a 1.2-nm amorphous intermediate layer between Si and the HfO$_2$ layer.

Adapted from [101].

shows an HREM image in the <110> orientation of the Si substrate, of the poly-Si/SiO$_2$/Si structure, and a 2-nm thick layer of the gate oxide. Note that the Si/SiO$_2$ interface is smooth atomically; however, if there were to be a step of half the Si unit cell at the interface, the measurement of the local SiO$_2$ thickness of 2 nm will differ by ~15%.

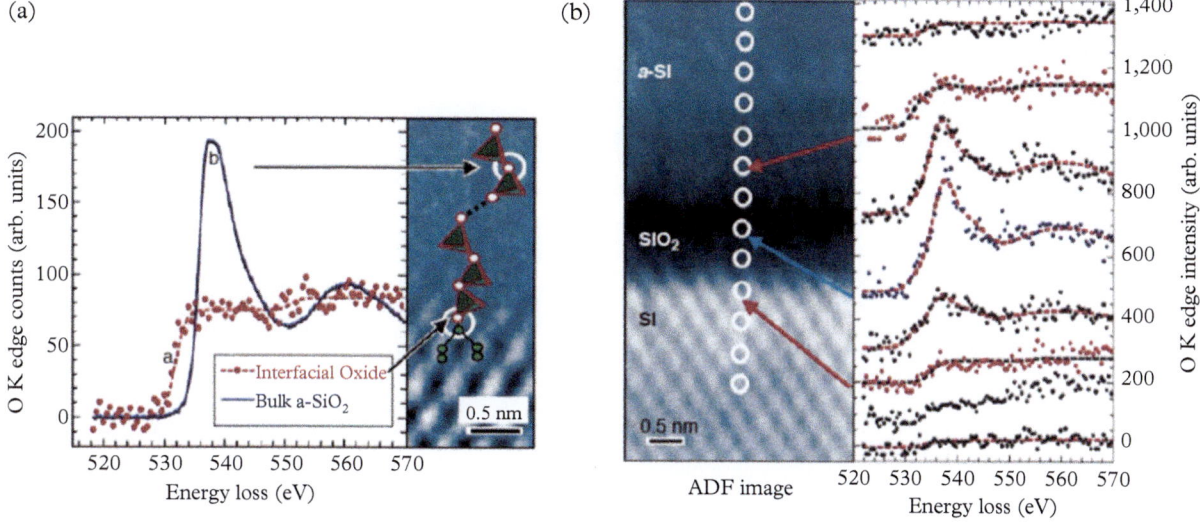

Figure 9.5.12 (a) High-resolution EELS spectrum of the O K edge of bulk amorphous silicon (marked as b) and from the O atoms at an atomically smooth interface (marked as a). The figure (right) is a dark field image of the interface obtained with an annular detector. (b) Similar analysis of the amorphous-Si/SiO$_2$/Si gate stack, showing the annular dark field image (left) and the EELS O K edge spectra (right) obtained from individual atomic columns (typically 100 atoms in depth) as indicated by arrows. The spectra were fit to a combination of bulk-like and interface-like spectra obtained from (a). From this it was inferred that bulk like behavior of the oxide is preserved only if the thickness of the SiO$_2$ layer is greater than 0.7 nm.

Adapted from [102].

Further details of the chemical composition and electronic structure, at atomic resolution, across a 1-nm thick SiO$_2$ gate oxide layer has been measured using EELS in a dedicated STEM [102]. The fundamental question is "how thick must the SiO$_2$ layer be before its bulk electrical properties can be obtained?" Based on this study (Fig. 9.5.12), we conclude that a fundamental limit of 0.7 nm can be placed on the usable silicon dioxide gate dielectric before it breaks down.

Several materials, particularly ZrO$_2$ and HfO$_2$, are being considered as gate dielectric materials to replace SiO$_2$, primarily to reduce gate leakage with scalable equivalent oxide thickness. Initial attempts found that the poly-Si/ZrO$_2$ gate interface degrades with high temperature anneal and forms nodules of Zr-silicide (Fig. 9.5.11b,c). Moreover, a film of HfO$_2$ deposited on silicon forms the monoclinic, low dielectric phase (Fig. 9.5.11d), but the mixed compound HfZrO$_x$, forms the desired high-k tetragonal phase (Fig. 9.5.11e). Thus, the latter is an attractive candidate as a replacement gate dielectric material. In fact, cross-section TEM analysis of HfZrO$_x$ dielectric structures (TaC metal gate, and poly-Si cap; Fig. 9.5.11f) shows that they are polycrystalline, ~3 nm thick, with an interfacial oxide layer 1 nm thick. In addition, no interactions between the HfZrO$_x$ layer and

Figure 9.5.13 TEM bright field image of cobalt-silicide/poly-Si/Si/ insulator stack, showing spiking and voiding in the Co–silicide layer. Relatively uniform Ni–silicide region in the source/drain region of a PMOS device.

Adapted from [101].

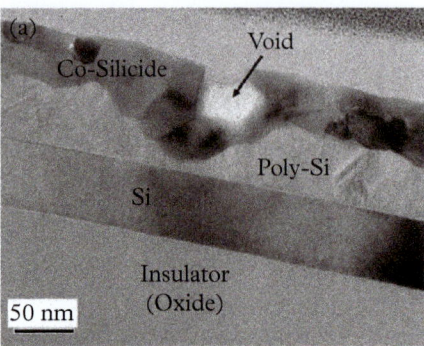
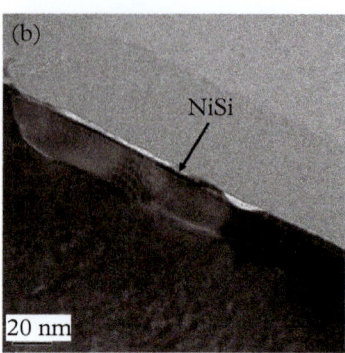

the Si channel was observed even after 1,000°C anneals, confirming good thermal stability of the HfZrO$_x$ layer. In fact, further development has made it possible to obtain gate oxides of amorphous HfO$_2$ with an intermediate amorphous layer (Fig. 9.5.11g). Finally, the narrow line width of 65 nm CMOS technology has limited the integration of Co-silicide lines due to spiking and void formation (Fig. 9.5.13a). On the other hand, Ni-silicide (Fig. 9.5.13b) has emerged as a promising candidate.

Finally, TEM methods are critical for failure analysis of defects during the IC fabrication process. A variety of defects (Fig. 9.5.14) are observed. These include observations of dislocations introduced during the process (Fig. 9.5.14a), as well as stacking faults (Fig. 9.5.14b) in the Si substrate, both observed in planar micrographs. Electrical shorts (Fig. 9.5.14c) between different circuit elements are common failures arising from different causes; here, a Co–Si conducting layer, only a few nanometers wide is the culprit. Sometimes, a via can fail due to a high resistance interface because of the presence of an interfacial layer (Fig. 9.5.14d) between the via and a Cu metal line. Figure 9.5.14e shows an example of a gate dielectric breakdown due to the presence of a gate-substrate short, and Figure 9.5.14f shows a magnified image of the substrate damage caused by the gate oxide breakdown. A Z-contrast STEM image of a failing transistor shows a bright contrast (higher Z) for the element protruding under the gate; EDXS analysis confirms that it is cobalt–silicide (Fig. 9.5.14g).

Transmission electron microscopy, in all its variations, is increasingly an essential metrology tool for semiconductor IC manufacturing. Advances in specific site TEM specimen preparation by focused ion beam milling has largely enabled the application of TEM in semiconductor manufacturing. Specimen preparation is discussed very briefly in §9.6.

9.5.5   Dynamic Measurements in a TEM

As discussed, a TEM provides information on the microstructure (structural, chemical, magnetic, or morphological) of a specimen at unprecedented levels of

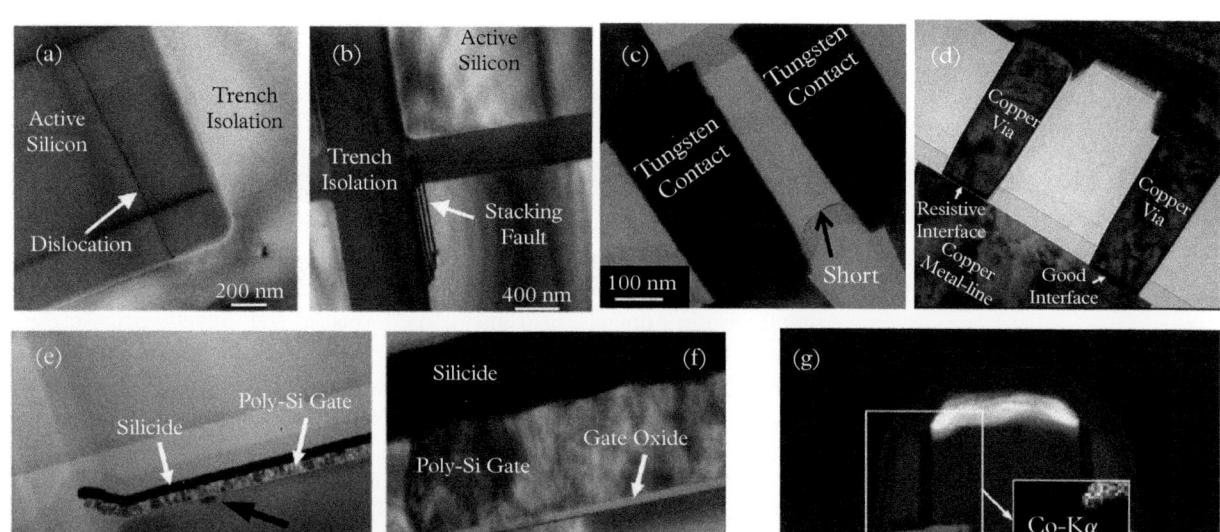

Figure 9.5.14 Failure analysis of semiconductor devices by transmission electron microscopy. Planar specimens reveal defects such as (a) dislocations and (b) stacking faults in silicon. In addition, failures can be attributed to (c) unwanted shorts between contact, and (d) poor resistive interfaces. Sometimes gate dielectric breakdown is caused by (e) a gate-substrate short, which can also cause (f) significant damage. (g) *Z*-contrast image shows that the failure of this transistor is dues to a short arising from a protruding Co–silicide layer.

Adapted from [101].

resolution. However, sustained and continuous development of the instrument and tailored specimen holders has resulted in substantial progress in studying materials under specific and controlled dynamic environments. Such measurements can be broadly divided into two categories.

The first one, *in situ* TEM, treats the microscope as a versatile workstation, instead of just as an imaging or spectroscopic tool. *In situ* TEM is broadly defined [119] as "the *real-time* observation of a specimen in a TEM while some form of *stimulus*, or external field, is applied *directly* to it." Typical external stimulus that are applied in a TEM include heating, cooling, gas and liquid reactions, electric and magnetic fields, mechanical forces (both tensile and compressive), and ion beam radiations. Often, these studies are dependent on the development of specially designed specimen holders and an appropriate TEM that can accommodate them, as well as fast image recording systems. As mentioned (§7.11.6), in the context of X-ray diffraction, such *in situ* studies go well beyond understanding initial and final states of the specimen/material subject to a stimulus, but includes the elucidation of the intermediate steps (and its evolution in real time) of the material. In short, such *in situ* methods convert the TEM into a miniaturized laboratory to dynamically interrogate chemical reactions, structural transitions, evolution of physical properties, etc., in real time and high spatial and temporal resolution.

The second class of dynamic measurements, called ultrafast TEM, typically involves an optical (laser) pump and an electron probe. A reaction in the TEM specimen is triggered, usually with a pulsed-laser irradiation, and a probing electron beam with a well-defined time delay is used to create and record a distinct signal (diffraction pattern or image) from the specimen. Even though this is a simple idea in principle, instrumental performance and limitations make real-life implementation of ultrafast TEM difficult and challenging. In particular, the major departure from the normal function of the TEM as a source of continuous electrons to one that generates a short electron pulse in response to an external trigger with sufficient intensity/signal to be recorded is the primary challenge. Needless to say, ultrafast electron detection methods also need significant development to make more progress.

Further details of these emerging methods can be found in a special report on dynamic *in situ* methods [119] as well as in the dedicated monograph edited by Ziegler et al. (2014).

9.6 Preparation of Specimens for TEM Observations

The preparation of electron transparent specimens is often the most critical and time-consuming part of transmission electron microscopy. Various methods have been developed for different materials and types of samples. However, in all cases, good specimens for TEM observations and measurements need to be thinned down to <150 nm for conventional BF/DF imaging (can be thicker if a high-voltage microscope with E_0 >1 MeV is used), <20 nm for EELS and EFTEM to avoid plural scattering (§9.4.2), and <10 nm for high-resolution phase contrast imaging (§9.3.5).

For bulk samples, specimen preparation involves a number of steps, starting with cutting a disc 3 mm in diameter and thickness ~0.1 mm to fit into the TEM specimen holder (Fig. 9.2.8). Such a disc may be punched out from a ductile metallic sheet, trepanned from a bulk ceramic, cut from a bar, or machined from a larger section. The metal or ceramic disc section is then ground down, if necessary, to the required thickness (~0.1 mm). Often, a dimple grinder is also used to produce a spherical dimple at the center of the 3 mm disc (Fig. 9.6.2a). This mechanically pre-thinned or dimpled disc is subsequently thinned very carefully to the thickness required for final observation using different methods depending on the material in question. For example, electrochemical polishing or electrolyte thinning (Fig. 9.6.1) is used for metals and alloys, chemical polishing/etching is used for semiconductors, and ion beam milling—a universal

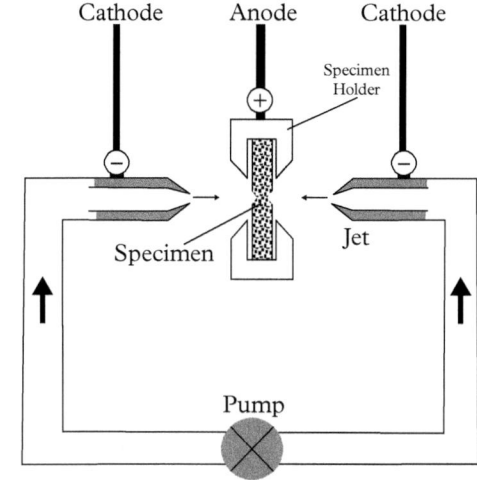

Figure 9.6.1 Schematic drawing of a double-jet electropolishing system.

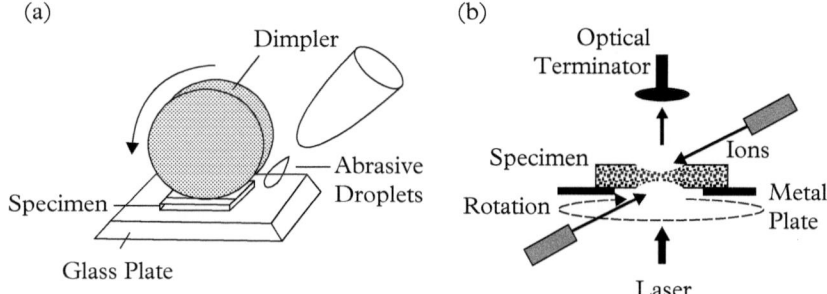

Figure 9.6.2 Preparation of TEM specimen by ion milling requires two steps: (a) dimple grinding to ~50 μm thickness, followed by (b) ion milling to perforation.

Adapted from Leng (2013).

method—is used for all materials (Fig. 9.6.2b). In addition, the focused ion beam tool (§10.7.2) is used to make site-specific specimens of complex structures and devices.

For powders and nanoparticles, specimen preparation begins with dispersing them stably in a suitable liquid or solvent. A drop of this dispersion, typically ~0.2 μL, is placed on an ultrathin carbon film mounted on a 3 mm TEM grid; after the solvent evaporates, the particles can be observed in a TEM. Unfortunately, this method is challenging because the particles often agglomerate on the grid, especially on the meniscus as the drop dries. To overcome this problem, it is better to spray the dispersion over the carbon grid using a commercial nebulizer, atomizer, or ink-jet printer, which forms tiny droplets a few micrometers in diameter and each containing only a few particles, depending on the concentration.

For polymers and biological tissues, specimens are generally made with a microtome that uses a glass or diamond knife to cut sequences of very thin sections of the sample (Fig. 9.6.3). These sections can be studied individually, or in sequence to create a three-dimensional structure of the biological tissue.

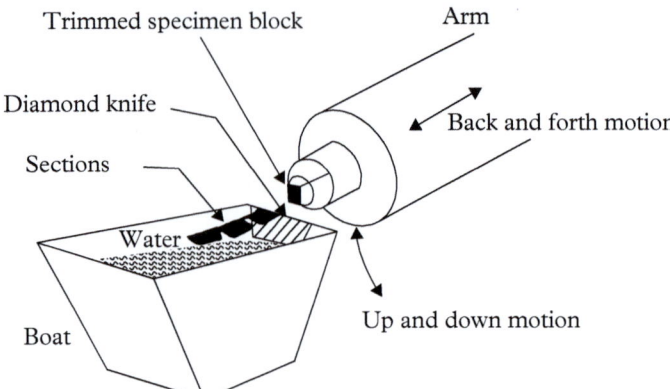

Figure 9.6.3 Schematic illustration of ultramicrotomy.

Adapted from Leng (2013).

It is important to ensure, regardless of the method used, that in the course of specimen preparation its morphology, microstructure, and chemistry remain unchanged, the final specimen maintains true fidelity to the original microstructure under investigation, and contamination is avoided. In other words, due diligence is required to avoid any artifacts arising from the specimen preparation process. We now briefly outline various method of specimen preparation commonly used in TEM analysis.

9.6.1 Chemical and Electrochemical Polishing

This long-established method is discussed in detail in Goodhew (1972), with focus on metals/alloys and some semiconductors. Even though electropolishing is used for metals/alloys, and chemical polishing for semiconductors, they both typically use a polishing solution that contains a mixture of an oxidizer to form an oxide layer on the surface and an etching agent that removes the oxide. For example, to thin silicon, a mixture of HNO_3 (to oxidize) and HF (to etch or dissolve the oxide layer) is used [104]. In the case of electropolishing, a simple electrolytic cell is constructed (Fig. 9.6.1) where the specimen is used as the anode. In order to avoid preferential polishing at the edges of the specimen, it is generally coated with an insulating paint, but a small area in the center of the 3-mm disc is left open for electropolishing. A pump forces the electrolytes through a nozzle directed at the specimen, and when the formation of the hole is detected with a laser and photocell, the polishing is terminated. It is important to terminate the etching immediately after the hole is detected; otherwise, the specimen would be etched too rapidly and render it poor for imaging. Needless to say, every material requires its own etchants, maintained at a specific solution temperature (both heating and cooling may be required, depending on the sample material) and at the appropriate voltage. A list of electropolishing conditions for different materials can be found in Thomas and Goringe (1980).

9.6.2 Ion-Beam Milling

In this method, a beam of energetic inert-gas ions (typically Argon) accelerated to several kV is directed at the specimen at oblique incidence to locally remove materials by sputtering. Electrical conductivity is not required for ion milling; thus, the method is universal and works for all inorganic materials. Figure 9.6.2b shows the general procedure of ion milling of a previously mechanically ground and dimpled specimen. Note that the specimen is mounted on a metallic ring holder to protect the edge of the disc, and is constantly rotated during ion milling about an axis normal to the specimen to create a symmetric dimple on the surface. Again, a laser beam and light sensor, placed on opposite sides of the specimen, are used to detect the formation of the hole and terminate the milling. Always, a compromise has to be made between obtaining the highest thinning rates and minimizing damage to the specimen due to the ion beam. Typically, the sputtering rate increases rapidly with ion energy (up to $\sim$10 keV) and shows a maximum at an angle of incidence, $\psi = 20$–$25°$. The ion beam can also rapidly raise the temperature of the specimen; to avoid specimen heating during ion milling it is often cooled to liquid nitrogen temperatures.

9.6.3 Ultramicrotomy and Preparation of Biological Materials

This method is commonly used to prepare thin ($\sim$100 nm) sections of polymeric materials or biological tissues (Fig. 9.6.3). First, the specimen is trimmed with a glass knife before it is sliced with the diamond knife. Ultramicrotomy can also be used to prepare specimens of powders by embedding them in acrylic or epoxy resin, allowing the resin to cure, and then slicing them with the glass or diamond knife. For fine control of slice thickness, a piezoelectric drive can be used to advance the specimen between slices.

Biological specimens are prepared by ultramicrotomy for confocal and TEM imaging (Fig. 9.3.4). A typical procedure used to prepare such specimens is described in detail in [28].

9.6.4 Preparation of Cross-Section Specimens

Cross-sectional specimens, such as those shown in Figure 9.4.13, have to be prepared for thin films, multilayers, and layered materials with particular interest in characterizing one or more buried interfaces.

The method is summarized schematically in Figure 9.6.4. Rectangular sections are cut from the wafer and glued together to form a block greater than 3 mm in thickness. A rod $\sim$2.8 mm in diameter is trepanned from this block. The rod is then inserted and glued into a metallic tube with an outer diameter of 3 mm. This tube assembly, with the cross-section assembly inside, is then sliced into thin disc

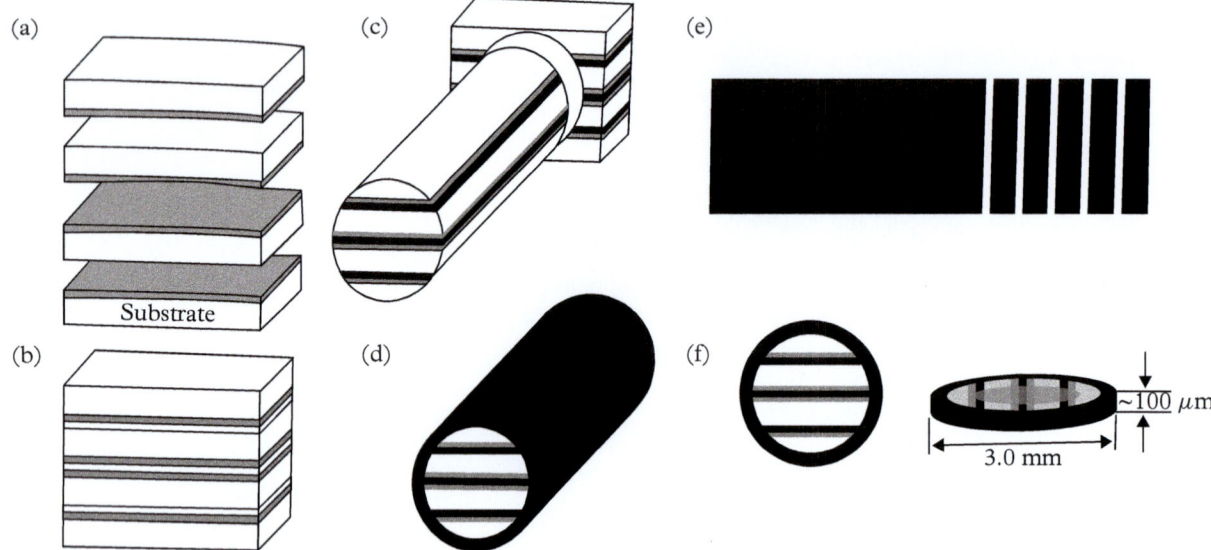

Figure 9.6.4 Schematic illustration of preparing cross-section specimens of thin films. It involves (a) cutting rectangular sections from the wafer, (b) gluing them together to make a block larger than 3 mm in thickness, (c) trepanning the block to produce a rod ~2.8 mm in diameter, (d) inserting the rod and gluing it into a tube of 3.0 mm outer diameter, (e) cutting the tube into thin sections, and (f) mechanically thinning the sections to ~100 μm in thickness and dimpling it in the center. The sections are then ion-milled to perforation.

Adapted from Brandon and Kaplan (2008).

sections. Subsequent mechanical thinning, dimpling, and ion milling (§9.6.2) is followed to prepare the TEM specimen. A review of this method, with particular emphasis on semiconductor specimens, can be found in [105].

9.6.5   Focused Ion-Beam (FIB) Milling

With the development of dual-beam (separate ion and electron beam columns) instruments (§10.7.2), a focused ion beam can be used to prepare thin films sections for TEM observations in a very specific site-selective manner [120]. Two methods are commonly used: the H-bar and the lift-out methods (Fig. 9.6.5a,b).

The H-bar method (Fig. 9.6.5a) is relatively straightforward to implement. After mechanical thinning and polishing a section of the materials of interest, a protective coating of platinum to avoid damage from the high energy Ga ion beam, is deposited on the surface of the region to be thinned. Then, the Ga ion beam is rastered over two regions of the surface, which are typically separated by about 1 μm, as shown. The ion beams cut trenches in the specimen, leaving behind a bridge, ~1 μm thick, in the center (thus the name H-bar), Subsequently, using lower-energy ions the bridge is thinned further to the required electron transparency, followed by a final ion polish at 2 keV, to remove any surface

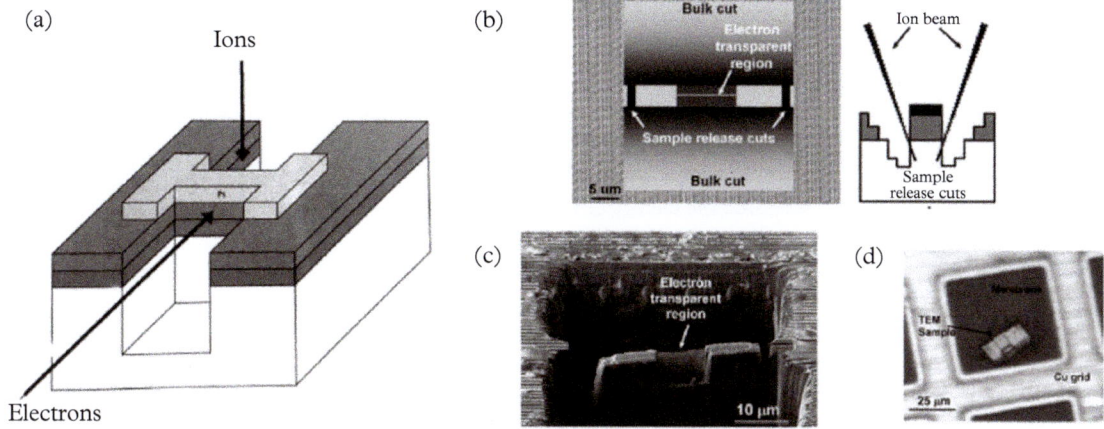

Figure 9.6.5 Use of a focused ion beam tool for TEM specimen preparation. (a) The H-bar method involves cutting two trenches on either side of a bridge as shown. The specimen is viewed with the electron beam orthogonal to the ion beams used for cutting. (b) The lift-out method follows the same procedure as the H-bar method (c), but finally a U-cut is made to release the specimen, and using a manipulator to place it on a TEM grid (d). See also Figure 10.7.4c.

(a) Adapted from Brandon and Kaplan (2008). (b–d) Adapted from [101].

damage. The specimen is rotated by 90° and viewed edge on in the TEM as shown.

The alternative lift-out method (Fig. 9.6.5b–d) is more precise than the H-bar method. It requires a dual-beam FIB and a nanomanipulator (not shown) to pick up the specimen and deposit it on a TEM grid. The first part of the procedure is similar to the H-bar method. However, after the thin film specimen is established as the bridge, a U-cut underneath the specimen is made with the ion beam to release it, and the specimen is then lifted off with the nanomanipulator (often, the specimen is attached to the nanomanipulator by depositing a thin layer of Pt) and deposited on a TEM grid.

Finally, recent developments in FIB-based TEM sample preparation techniques, particularly through recipe-driven TEM sample preparation, combined with TEM/STEM image acquisition and critical dimension metrology on an offline work station, can lead to faster turnaround time of TEM analysis [121–124]. From the industrial point of view, this is the key to its successful application in process development, yield enhancement, manufacturing, and failure analysis of semiconductor ICs.

Summary

TEM has significantly advanced our imaging capabilities well beyond optical microscopy that reached its far-field diffraction limit in the early part of the twentieth century. The wavelength of electrons decreases with increasing beam voltage and this was exploited by TEMs over a period of 50 years or so, to show a steady increase in their resolving power. This trend could not be sustained because of the significant radiation damage of specimens that limits the use of TEMs at higher voltages. Fortunately, recent developments in correcting electron-optical aberrations have shown a reliable pathway to improving resolution without increasing the energy of the fast electron beam. In fact, resolution of sub-Å image features has been demonstrated with aberration-corrected lenses, and there is still hope for further improvement as these microscopes generally operate far from their ultimate diffraction limit.

Elastic scattering of the fast electron by the specimen atoms is the most important interaction that generates contrast in TEM images. The small angle scattering of the fast electrons by the nuclei of the atoms in the specimen determines the scattering contrast, and the image is formed by the interception or transmission of the scattered electron by the objective aperture. The incoherent Rutherford scattering, particularly applicable to nonperiodic structures and biological materials, contributes to the mass contrast. On the other hand, coherent elastic, or Bragg, scattering contributes to diffraction contrast, and images are made with either the transmitted (BF) or the diffracted beam (DF) selected by the objective aperture. Such diffraction contrast finds ready use in the imaging and analysis of defects in crystalline materials.

Understanding high-resolution phase contrast (HREM) requires the use of wave-optical theory of imaging. In the focal plane of the OL, where the diffraction pattern is observed, each observed reflection, θ_{hkl}, corresponds reciprocally to a period, d_{hkl}, in the specimen, or to a spatial frequency, $q = 1/d_{hkl}$, which then determines the amplitude distribution, $F(q)$ of the exit electron wave. Further, a phase shift that depends on the scattering angle, the aberration coefficient (C_S) of the lens, and the degree of defocus, is applied to $F(q)$ as an exponential phase factor. Finally, the image intensities observed may be expressed as a CTF for the different spatial frequencies. More importantly, the CTF characterizes the effect of the instrument on image formation, and does not depend on any particular specimen. The envelope function, arising from chromatic aberration and other factors, further dampens the CTF and sets the information limit of the instrument. Finally, optimum imaging and resolution in BF is observed at the Scherzer defocus, at which a broad band of spatial frequencies is imaged with positive phase contrast.

Magnetic microstructure in materials is studied by measuring the deflection of the fast electron beam by the in-plane induction of the specimen. This is done by recording either a defocused image (Fresnel mode), or selecting those electrons that have been deflected in a specific direction using either an aperture (Foucault mode), or a quadrant/segmented detector (differential phase contrast). In principle, these signals are proportional to the first or second derivative of the phase shift of the electrons that have passed through the specimen. The alternative—off-axis electron holography—provides a direct measurement of the phase shift of the electron wave. Both electrostatic potential and magnetic flux density around a specimen contribute to the observed phase shift. Typically, an electrostatic biprism is used to split a highly coherent beam; the resultant interference of an electron wave that has passed through the specimen (object wave) with another part of the same wave propagating only through vacuum (reference wave) forms the holographic interference pattern, from which the in-plane induction distribution is reconstructed.

In a STEM, a focused beam, which can be smaller than the spacing between atomic columns, is scanned across the specimen. The elastically scattered intensity at high angles, which can be recorded by an *annular detector*, when the beam is on and off the atomic column, forms the Z-contrast image. If the collection angles are substantially larger than the beam convergence semi-angle (typically, three or more times), the imaging can be considered as *incoherent*, especially for thin specimens. Then, such STEM-HAADF can be considered to completely eliminate diffraction and phase contrast, with the observed contrast being monotonic with thickness and sensitive to the composition (i.e. proportional to $Z^{1.8}$, where Z is the atomic number, and asymptotically trending towards Z^2, the Rutherford scattering limit). Inelastically scattered electrons, that form the basis of EELS, are forward peaked and can be collected by simultaneously placing a spectrometer on the optic axis. Such an ADF/EELS geometric arrangement provides simultaneous structural, chemical, and electronic structure information at the highest spatial resolution.

EELS in a TEM probes the local electronic structure and composition. It is forward-peaked and records the energy lost by the fast electron in the process of exciting a core electron to unoccupied local states while leaving behind a core hole. Such energy losses in transmission are characterized by excitation with frequencies ranging from the near-infrared to hard X-ray regions. The analysis of the loss spectrum provides information on the *unoccupied* local density of states, partitioned by site symmetry, nature of the chemical species, and the angular momentum of the final state. If the probe is small enough to focus on a *single column* of atoms, the site-specific information is readily interpretable. The unique binding energy of the core-level electron provides the chemical specificity. Both the angular momentum of the initial state and the dipole selection rules determine the final state angular momentum. The shape or fine structure of the EELS spectrum, specifically close in energy-loss to the edge onset, is influenced by the nature of the *core hole*. When it is well screened, the spectrum reflects the ground state density of states; however, if there exists strong core hole coupling, excitonic effects not only influence the shape of the spectrum, but also make it possible to use it as a fingerprint of the local charge and coordination chemistry. Energy-lost electrons can also be used to form images either by recording them pixel by pixel as a sequence of spectra (spectrum imaging), or by choosing a particular energy-loss (window) and forming an image with electrons that have lost that specific energy (EFTEM). In addition, following the formation of the core hole, de-excitation processes can involve the emission of a characteristic X-ray photon.

Incorporating an energy dispersive X-ray detector in the TEM column allows the detection of these X-rays, providing another method to obtain compositional information, including chemical maps of the specimen at high spatial resolution.

A series of (S)TEM images of the specimen, recorded every one or two degrees about a tilt axis and over as large a specimen tilt range as possible can be back-projected to reconstruct a three-dimensional image. This process of electron tomography works best when the recorded signal is a monotonic function of thickness or density of the specimen; this projection criterion is readily satisfied by the mass contrast employed in imaging biological or amorphous materials. Diffraction contrast does not satisfy the projection criterion; however, the incoherent scattering observed in HAADF-STEM is monotonic with thickness and is ideal for tomographic reconstruction, as has been demonstrated for a number of nanoscale objects in catalysis, metallurgical precipitates, and even biogenic magnetite crystals. Generally, the resolution in electron tomography is equal to the angular increment in the tilt series multiplied by the size of the object.

TEM has found wide use in the characterization of both biological and physical materials, with applications ranging from fundamental studies to routine metrology in the semiconductor industry. However, specimen preparation is the ultimate limitation in TEM, with many recent developments, including focused ion beam milling. In all cases, the challenge is to prepare a thin foil specimen that is truly representative of the microstructure of the sample, and one that is not damaged/altered in the process of preparing the thin specimen for (S)TEM observation. Further, to achieve ultimate resolution (say, in a STEM), an optimal specimen should be thin enough to prevent any broadening of the probe, but thick enough to provide a bulk-like scattering signal.

....

FURTHER READING

Brandon, D., and W. D. Kaplan. *Microstructural Characterization of Materials*, 2nd ed. Hoboken: Wiley, 2008.

Buseck, P. R., J. M. Cowley, and L. Eyring, eds. *High-Resolution Transmission Electron Microscopy and Associated Techniques*. Oxford: Oxford University Press, 1988.

Graef, M. *Introduction to Conventional Transmission Electron Microscopy*. CUP, 2009.

Edington, J. W. *Practical Electron Microscopy in Materials Science*. London: Macmillan, 1976.

Egerton, R. *Electron Energy-Loss Spectroscopy in the Electron Microscope*. Plenum Press, 1996.

Fuchs, E., H. Oppolzer, and H. Rehme. *Particle Beam Microanalysis: Fundamentals, Methods and Applications*. New York: VCH Publications, 1990.

Fuggle, J. C., and J. E. Inglesfield, eds. *Unoccupied Electronic State: Fundamentals of XANES, EELS, IPS and BIS*. Berlin: Springer, 1992.

Fultz, B., and J. Howe. *Transmission Electron Microscopy and Diffractometry of Materials*. Heidelberg: Springer, 2013.

Goodhew, P. F. *Specimen Preparation in Materials Science*. Amsterdam: North-Holland/Elsevier, 1972.

Hirsch, P. B., A. Howie, R. B. Nicholson, D. W. Pashley, and M. J. Whelan. *Electron Microscopy of Thin Crystals*. New York: Krieger, 1977.

Hren, J., J. I. Goldstein, and D. C. Joy. *Introduction to Analytical Electron Microscopy*. New York: Plenum Press, 1979.

Ibach, H., and D. L. Mills. *Electron Energy-Loss Spectroscopy and Surface Vibrations*. New York: Academic Press, 1982.

Joy, D. C., A. D. Romig Jr., and J. I. Goldstein, eds. *Principles of Analytical Electron Microscopy*. New York: Plenum Press, 1986.

Krishnan, K. M. Chapter 8. *Fundamentals and Applications of Magnetic Materials*. Oxford: Oxford University Press, 2016.

Leng, Y. *Materials Characterization*, 2nd ed. Weinheim: Wiley, 2013.

Raether, H. *Excitation of Plasmons and Interband Transitions by Electrons*. New York: Springer, 1980.

Reimer, L. *Transmission Electron Microscopy*. Berlin: Springer, 1993.

Reimer, L., ed. *Energy-Filtering Transmission Electron Microscopy*. New York: Springer, 1995.

Shindō, D., and K. Hiraga. *High-Resolution Electron Microscopy for Materials Science*. New York: Springer, 1998.

Spence, J. C. H. *Experimental High-Resolution Electron Microscopy*. Oxford: Oxford University Press, 1981.

Thomas, G., and M. J. Goringe. *Transmission Electron Microscopy of Materials*. New York: Wiley, 1979.

Tonomura, A. *Electron Holography*. Berlin: Springer, 1999.

Verma, A. R., and P. Krishna. *Polymorphism and Polytypism in Crystals*. New York: Wiley, 1966.

Williams, D. B., and C. B. Carter. *Transmission Electron Microscopy. Vol. 1–IV*. New York: Plenum Press, 1996.

Williams, D. B., A. R. Pelton, and R. Gronsky. *Images of Materials*. Oxford: Oxford University Press, 1991.

Zewail, A. H. *Physical Biology: From Atoms to Medicine*. London: Imperial College Press, 2008.

Ziegler, A., H. Graafsma, X. F. Zhang, and J. W. M. Frenken. *In Situ Materials Characterization across Spatial and Temporal Scales*. Berlin: Springer, 2014.

....

REFERENCES

[1] Takeno, Y., Y. Murakami, T. Sato, T. Tanigaki, H. S. Park, D. Shindō, R. M. Ferguson, and K. M. Krishnan. "Morphology and Magnetic Flux Distribution in Superparamagnetic, Single-Crystalline Fe_3O_4 Nanoparticle Rings." *Applied Physics Letters* 105, no. 18 (2014): 183102.

[2] Verbist, K., E. C. Nelson, T. C. Anthony, J. A. Brug, and K. M. Krishnan. "Lorentz Transmission Electron Microscopy in a Standard CM200FEG." *Proceedings of the 14th International Congress on Electron Microscopy*, 503–4. Cancun, Mexico, August 31–September 4, 1998.

[3] Krishnan, K. M., C. Nelson, C. J. Echer, R. F. C. Farrow, R. F. Marks, and A. J. Kellock. "Exchange biasing of permalloy films by Mn_xPt_{1-x}: Role of composition and microstructure." *Journal of Applied Physics* 83 (1998): 6810.

[4] Griffin Roberts, K., M. Varela, S. Rashkeev, S. T. Pantelides, S. J. Pennycook, and K. M. Krishnan. "Defect-Mediated Ferromagnetism in Insulating Co-doped Anatase TiO2 Thin Films." Physical Review B 78 (2008): 014409.

[5] Farrow, R. F. C., S. S. P. Parkin, R. F. Marks, K. M. Krishnan, and N. Thangaraj. "Quenching of Giant Magnetoresistance by Interface Roughening and Alloying in Annealed $[(Ni_xFe_{1-x})_yAu_{1-y}]$/Au Multilayers." *Applied Physics Letters* 69 (1996): 1963–5.

[6] Krivenek, O. L., G. J. Corbin, N. Dellby, B. F. Elston, R. J. Keyse, M. F. Murfitt, C. S. Own, Z. S. Szilagyi, and J. W. Woodruff. "An Electron Microscope for the Aberration-Corrected Era." *Ultramicroscopy* 108 (2008): 179–95.

[7] von Ardene, M. "Das Elektronen-Rastermikroskop. Praktische Ausführung." *Z. Tech. Phys* 19 (1938): 407–16.

[8] Crewe, A. V. "Scanning Electron Microscopes: Is High Resolution Possible?" *Science* 154 (1966): 729–38.

[9] Crewe, A. V., J. Wall, and L. M. Welter. "A High-Resolution Scanning. Transmission Electron Microscope." *Journal of Applied Physics* 39 (1968): 5861–8.

[10] Crewe, A. V., J. Wall, and J. Langmore. "Visibility of Single Atoms." *Science* 168 (1970): 1338–40.

[11] Rose, H. "Outline of a Spherically Corrected Semiaplantic Medium-Voltage Transmission Electron Microscope." *Optik* 85 (1990): 19–24.

[12] Haider, M., H. Rose, S. Uhlemann, E. Schwan, B. Kabius, and K. Urban. "A Spherical-Aberration-Corrected 200 kV Transmission Electron Microscope." *Ultramicroscopy* 75 (1998): 53.

[13] Haider, M., S. Uhlemann, E. Schwan, H. Rose, B. Kabius, and K. Urban. "Electron Microscopy Image Enhanced." *Nature* 392 (1998): 768–9.

[14] Lanio, S., H. Rose, and D. Krahl. "Test and Improved Design of a Corrected Imaging Magnetic Energy Filter." *Optik* 73 (1986): 56–68.

[15] Batson, P. E., N. Delby, and O. L. Krivanek. *Nature* 418 (2002): 617.

[16] Muller, D. A., L. Fitting Kourkoutis, M. Murfitt, J. H. Song, H. Y. Hwang, J. Silcox, N. Dellby, and O. L. Krivanek. "Atomic-Scale Chemical Imaging of Composition and Bonding by Aberration-Corrected Microscopy." *Science* 319 (2008): 1073–6.

[17] Krivanek, O. L., M. F. Chisholm, V. Nicolosi, T. J. Pennycook, G. J. Corbin, N. Dellby, M. F. Murfitt, C. S. Own, Z. S. Szilagyi, M. P. Oxley, S. T.

Pantelides, and S. J. Pennycook. "Atom-By-Atom Structural and Chemical Analysis by Annular Dark-Field Electron Microscopy." *Nature* 464 (2010): 571–4.

[18] Thorel, A., J. Ciston, T. P. Bartel, C-Y. Song, and U. Dahmen. "Observation of the Atomic Structure of ß'-SiAlON Using Three Generations of High-Resolution Electron Microscopes." *Philosophical Magazine A* 93 (2013): 1172–81.

[19] Mori, N., T. Oikawa, T. Katoh, J. Miyahara, and Y. Harada. "Application of the 'imaging plate' to TEM image recording." *Ultramicroscopy* 25 (1988): 195–201.

[20] Krivanek, O. L., J. P. Ursin, N. J. Bacon, G. J. Corbin, N. Dellby, P. Hrncirik, M. F. Murfitt, C. S. Own, and Z. S. Szilagyi. "High-Energy-Resolution Monochromator for Aberration-Corrected Scanning Transmission Electron Microscopy/Electron Energy-Loss Spectroscopy." *Philosophical Transactions of the Royal Society A: Mathematical, Physical and Engineering Sciences* 367 (2009): 3683–97.

[21] Bell, D. C., C. J. Russo, and G. Benner. "Sub-Angstrom Low-Voltage Performance of a Monochromated, Aberration-Corrected Transmission Electron Microscope." *Microscopy and Microanalysis* 16 (2010): 386–92.

[22] Essers, E., G. Benner, T. Mandler, S. Meyer, D. Mittmann, M. Schnell, and R. Höschen. "Energy Resolution of an Omega-Type Monochromator and Imaging Properties of the MANDOLINE Filter." *Ultramicroscopy* 110 (2010): 971–80.

[23] Tiemeijer, P. C., M. Bischoff, B. Freitag, and C. Kisielowski. "Using a Monochromator to Improve the Resolution in TEM to Below 0.5 Å. Part I: Creating Highly Coherent Monochromated Illumination." *Ultramicroscopy* 114 (2012): 72–81.

[24] Griffin Roberts, K., M. Varela, S. Rashkeev, S. T. Pantelides, S. J. Pennycook, and K. M. Krishnan. "Defect-Mediated Ferromagnetism in Insulating Co-doped Anatase TiO_2 Thin Films." *Physical Review B* 78 (2008): 014409.

[25] Pennycook, S. J. "Advances in Scanning Transmission Electron Microscopy." *Microscopy and Analysis* September (2012): 59–65.

[26] Hufschmid, R. PhD Thesis, University of Washington, 2019.

[27] Krishnan, K. M., et al. *Materials Research Society Symposium Proceedings* 60, (1986): 211.

[28] Teeman, E., C. Shasha, J. E. Evans, and K. M. Krishnan. "Intracellular Dynamics of Superparamagnetic Iron Oxide Nanoparticles for Magnetic Particle Imaging." *Nanoscale* 11 (2019): 7771.

[29] Donnet, D. M., K. M. Krishnan, and Y. Yajima. "Domain Structures in Epitaxially Grown Cobalt Thin Films." *Journal of Physics D: Applied Physics* 28 (1995): 1942–50.

[30] Chapman, J. N., P. E. Batson, E. M. Waddell, and R. P. Ferrier. "The Direct Determination of Magnetic Domain Wall Profiles by Differential Phase Contrast Electron Microscopy." *Ultramicroscopy* 3 (1978): 203–14.

[31] Daykin, A. C., and A. K. Petford-Long. "Quantitative Mapping of the Magnetic Induction Distribution Using Foucault Images Formed in a Transmission Electron Microscope." *Ultramicroscopy* 58 (1995): 365–80.

[32] Chapman, J. N., I. R. McFadyen, and S. McVittie. "Modified Differential Phase Contrast Lorentz Microscopy for Improved Imaging of Magnetic Structures." *IEEE Transactions on Magnetics* 26 (1990): 1506–11.

[33] Gabor, D. "Microscopy by Reconstructed Wave-Fronts." *Proceedings of the Royal Society A: Mathematica, Physical and Engineering Sciences* 197 (1949): 454–87.

[34] Gabor, D. "Microscopy by Reconstructed Wave Fronts: II." *Proceedings of the Physical Society. Section B* 64 (1951): 449–69.

[35] Cowley, J. M. "Twenty Forms of Electron Holography." *Ultramicroscopy* 41 (1992): 335–48.

[36] Tonomura, A., T. Matsuda, H. Tanabe, N. Osakabe, J. Endo, A. Fukuhara, K. Shinagawa, and H. Fujiwara. "Electron Holography Technique for Investigating Thin Ferromagnetic Films." *Physical Review B* 25 (1982): 6799.

[37] de Ruijter, W. J., and J. K. Weiss. "Detection Limits in Quantitative Off-Axis Electron Holography." *Ultramicroscopy* 50 (1993): 209–20.

[38] Tonomura, A., T. Matsuda, and J. Endo. "Spherical-Aberration Correction of an Electron Lens by Holography." *Japanese Journal of Applied Physics* 18 (1979): 1373.

[39] Dunin-Borkowski, R. E., M. R. McCartney, R. B. Frankel, D. A. Bazylinski, M. Pósfai, and P. R. Buseck. "Magnetic Microstructure of Magnetotactic Bacteria by Electron Holography." *Science* 282 (1998): 1868–70.

[40] Tonomura, A., T. Matsuda, J. Endo, T. Arii, and K. Mihama. "Holographic Interference Electron Microscopy for Determining Specimen Magnetic Structure and Thickness Distribution." *Physical Review B: Condensed Matter and Materials Physics* 34 (1986): 3397.

[41] Lichte, H., H. Bauzhof, and R. Huhle. In *Electron Microscopy* 98, edited by H. A. Calderon Benavides and M. José Yacaman, 559. Bristol: IOP, 1998.

[42] Ferrel, C. R. "Angular Dependence of the Characteristic Energy Loss of Electrons Passing Through Metal Foils." *Physical Review* 101 (1956): 554.

[43] Bethe, H. A. "Zur Theorie des Durchgangs schneller Korpuskularstrahlen durch Materie." *Annalen der Physik* (Leipzig) 397, no. 3 (1930): 325–400.

[44] Pockman, L. T., D. L. Webster, P. Kirkpatrick, and K. Harworth. "The Probability of K Ionization of Nickel by Electrons as a Function of Their Energy." *Physical Review* 71 (1947): 330.

[45] Powell, C. J. "Cross Sections for Ionization of Inner-Shell Electrons by Electrons." *Reviews of Modern Physics* 48 (1976): 33.

[46] Inokuti, M. "Inelastic Collisions of Fast Charged Particles with Atoms and Molecules—The Bethe Theory Revisited." *Reviews of Modern Physics* 43 (1971): 297.

[47] Eusemann, R., H. Rose, and J. Dubochet. "Electron Scattering in Ice and Organic Materials." *Journal of Microscopy* 128 (1982): 239–49.

[48] Egerton, R. F. "K-shell Ionization Cross-Sections for Use in Microanalysis." *Ultramicroscopy* 4 (1979): 169–79.

[49] Leapman, R. D., P. Rez, and D. F. Mayers. "*K, L,* and *M* Shell Generalized Oscillator Strengths and Ionization Cross Sections for Fast Electron Collisions." *Journal of Chemical Physics* 72 (1980): 1232.

[50] Krishnan, K. M., and C. J. Echer. "Measurements of Ionization Cross-Sections for EEL Microanalysis Under Well-Defined Scattering Conditions." In *Microbeam Analysis 1991: Proceedings of the 26th Annual Conference of the Microbeam Analysis Society,* edited by D. G. Howitt, 259–62. San Jose, California, August 4–9 1991. San Francisco: San Francisco Press, 1991.

[51] Krishnan, K. M. "Electron Energy-Loss Spectroscopy Fundamentals and Applications in the Characterization of Minerals." In *Spectroscopic Characterization of Minerals and Their Surfaces,* edited by L. M. Coyne, S. W. S. McKeever and D. F. Blake, 54–74. Washington, DC: ACS Publications, 1990.

[52] Trebbia, P. "Unbiased Method for Signal Estimation in Electron Energy Loss Spectroscopy, Concentration Measurements and Detection Limits in Quantitative Microanalysis: Methods and Programs." *Ultramicroscopy* 24 (1988): 399–408.

[53] Krishnan, K. M., and M. T. Stampfer. *Proceedings of the Annual Meeting of the Electron Microscopy Society of America* 46 (1988): 538.

[54] Egerton, R. F., B. G. Williams, and D. T. G. Sparrow. "Fourier Deconvolution of Electron Energy-Loss Spectra." *Proceedings of the Royal Society A: Mathematics, Physical and Engineering Sciences* 398 (1985): 395–404.

[55] Swyt, C. R., and R. D. Leapman. In *Microbeam Analysis,* edited by A. D. Romig and J. I. Goldstein, 45. San Francisco: San Francisco Press, 1984.

[56] Colliex, C. In *Advances in Optical and Electron Microscopy,* edited by V. Cosslett and R. Bauer, 65. London: Academic Press, 1984.

[57] Manson, S. T. "Inelastic Collisions of Fast Charged Particles with Atoms: Ionization of the Aluminum L Shell." *Physical Review A* 6, no. 3 (1972): 1013.

[58] Leapman, R. D., P. Rez, and D. F. Meyers. "*K, L,* and *M* Shell Generalized Oscillator Strengths and Ionization Cross Sections for Fast Electron Collisions." *J. Chem. Phys.* 72 (1980): 1232.

[59] Rask, J. H., B. A. Miner, and P. R. Buseck. "Determination of Manganese Oxidation States in Solids by Electron Energy-Loss Spectroscopy." *Ultramicroscopy* 21 (1987): 321–6.

[60] Krivanek, O. L., C. C. Ahn, and R. B. Keeney. "Parallel Detection Electron Spectrometer Using Quadrupole Lenses." *Ultramicroscopy* 22 (1987): 103–15.

[61] Stephens, A. P., and L. M. Brown. "Observation by Scanning Transmission Electron Microscopy of Characteristic Electron Energy-Losses Due to Hydrogen in Transition Metals." *Institute of Physics Conference Series* 52 (1980): 341–2.

[62] Zaluzec, N. P., T. Schober, and D. G. Westlake. "Application of EELS to the Study of Metal–Hydrogen Systems." *Proceedings of the 39th Annual Meeting of the Electron Microscopy Society of America*, edited by G. W. Bailey, 194. Baton Rouge: Claitor, 1981.

[63] Egerton, R. F. "K-shell Ionization Cross-Sections for Use in Microanalysis." *Ultramicroscopy* 4 (1979): 169–79.

[64] Egerton, R. F. *Proceedings of the 39th Annual Meeting of the Electron Microscopy Society of America*, edited by G. W. Bailey, 198–9. Baton Rouge: Claitor, 1981.

[65] Egerton, R. F. In *Scanning Electron Microscopy*, Vol. 2, edited by O. Johari, 505. O'Hare: AMF, 1984.

[66] Ahn, C. C., and P. Rez. "Inner Shell Edge Profiles in Electron Energy Loss Spectroscopy." *Ultramicroscopy* 17 (1985): 105–15.

[67] Malis, T., K. Rajan, J. M. Titchmarsh, and C. Weatherly. In: *Intermediate Voltage Electron Microscopy*, edited by K. Rajan, 78. Mahwah: Philips Electron Optics Publishing Group, 1987.

[68] Hofer, F., P. Golob, and A. Brunegger. AEM-workshop. Manchester: UMIST, 1987.

[69] Ahn, C. C., and O. L. Krivanek. *EELS Atlas: A Reference Collection of Electron Energy Loss Spectra Covering All Stable Elements*. Warrendale: Gatan, Inc., 1983.

[70] Kurata, H., S. Moriguchi, S. Isoda, and T. Kobayashi. "Attainable resolution of energy-electing image using high-voltage electron microscope." *Journal of Electron Microscopy* 45 (1996): 79–84.

[71] Grogger, W., B. Schaffer, K. M. Krishnan, and F. Hofer. "Energy-Filtering TEM at High Magnification: Spatial Resolution and Detection Limits." *Ultramicroscopy* 96 (2001): 481–9.

[72] Yonekura, K., S. Maki-Yonekura, and K. Namba. "Quantitative Comparison of Zero-Loss and Conventional Electron Diffraction from Two-Dimensional and Thin Three-Dimensional Protein Crystals." *Biophysical Journal* 82 (2002): 2784–97.

[73] Gubbens, A. J., B. Kraus, O. L. Krivanek, and P. E. Mooney. "An Imaging Filter for High-Voltage Electron Microscopy." *Ultramicroscopy* 59 (1995): 255–65.

[74] Egle, W., A. Rilk, J. Bihr, and M. Menzel. "Microanalysis in the EM 902: Tests on a New TEM for ESI and EELS." In *42nd Annual Proceedings of the Electron Microscope Society of America*, edited by G.W. Bailey, 566–7. San Francisco: San Francisco Press, 1984.

[75] Verbeeck, J., D. van Dyck, and G. van Tendeloo. "Energy-Filtered Transmission Electron Microscopy: An Overview." *Spectrochimica Acta* B59 (2004): 1529–34.

[76] Pennycook, S. J., and C. Colliex. "Spectroscopic Imaging in Electron Microscopy." *MRS Bulletin* 37 (2012): 13–18.

[77] Aronova, M. A., and R. D. Leapman. "Development of Electron Energy-Loss Spectroscopy in the Biological Sciences." *MRS Bulletin* 37 (2012): 53–62.

[78] Cliff, G., and G. W. Lorimer. "The Quantitative Analysis of Thin Specimens." *Journal of Microscopy* 103 (1975): 203–7.

[79] Krishnan, K. M., and C. J. Echer. In *Analytical Electron Microscopy*, edited by D.C. Joy, 99. San Francisco: San Francisco Press, 1987.

[80] Buxton, B. F., J. A. Eades, John Wickham Steeds, and G. M. Rackham. "The Symmetry of Electron Diffraction Zone Axis Patterns." *Proceedings of the Royal Society A: Mathematics, Physical and Engineering Sciences* 281 (1976): 171–94.

[81] Eades, J. A. "Zone-Axis Patterns Formed by a New Double-Rocking Technique." *Ultramicroscopy* 5 (1980): 71–4.

[82] Tanaka, M., R. Saito, K. Ueno, and Y. Harada. "Large-Angle Convergent-Beam Electron Diffraction." *Journal of Electron Microscopy* 29, no. 4 (1980): 408–12.

[83] Cormack, A. M. "Representation of a Function by Its Line Integrals, with Some Radiological Applications." *Journal of Applied Physics* 34 (1963): 2722.

[84] Weyland, M., and P. A. Midgley. "Electron Tomography." *Materials Today* 7, no. 12 (2004): 32–40.

[85] Crowther, R. A., L. A. Amos, J. T. Finch, and A. Klug. "Three-Dimensional Reconstructions of Spherical Viruses by Fourier Synthesis from Electron Micrographs." *Nature* 222 (1970): 421–5.

[86] Saxton, W. O., W. Burmeister, and M. Hahn. "Three-Dimensional Reconstruction of Imperfect Two-Dimensional Crystals." *Ultramicroscopy* 13 (1984): 57–70.

[87] Koster, A. J., R. Grimm, D. Typke, R. Hegerl, A. Stoschek, J. Walz, and W. Baumeister. "Perspectives of Molecular and Cellular Electron Tomography." *Journal of Structural Biology* 120 (1997): 276–308.

[88] Dierksen, K., D. Typke, R. Hegerl, and W. Baumeister. "Towards Automatic Electron Tomography II. Implementation of Autofocus and Low-Dose Procedures." *Ultramicroscopy* 49 (1993): 109–20.

[89] Dierksen, K., D. Typke, R. Hegerl, A. J. Koster, and W. Baumeister. "Towards Automatic Electron Tomography." *Ultramicroscopy* 40 (1992): 71–87.

[90] Koster, A. J., H. Chen, J. W. Sedat, and D. A. Agand. "Automated Microscopy for Electron Tomography." *Ultramicroscopy* 46 (1992): 207–27.

[91] Frank, J., and B. F. McEwen. "Alignment by Cross-Correlation." In *Electron Tomography: Three-Dimensional Imaging with the TEM*, edited by J. Frank, 205–13. New York: Plenum Press, 1992.

[92] Lawrence, M. C. "Least-Squares Method of Alignment Using Markers." In *Electron Tomography: Three-Dimensional Imaging with the TEM*, edited by J. Frank, 197–204. New York: Plenum Press, 1992.

[93] Radermacher, M., T. Wagenknecht, A. Verschoor, and J. Frank. "Three-Dimensional Reconstruction from a Single-Exposure, Random Conical Tilt Series Applied to the 50S Ribosomal Subunit of *Escherichia coli.*" *Journal of Microscopy* 146 (1987): 113–36.

[94] Gordon, R., R Bender, G T Herman. "Algebraic Reconstruction Techniques (ART) for Three-Dimensional Electron Microscopy and X-Ray Photography." *Journal of Theoretical Biology* 29 (1970): 471–81.

[95] Gilbert, P. "Iterative Methods for the Three-Dimensional Reconstruction of an Object from Projections." *Journal of Theoretical Biology* 36 (1972): 105–17.

[96] Weyland, M., and P. A. Midgley. "Extending Energy-Filtered Transmission Electron Microscopy (EFTEM) into Three Dimensions Using Electron Tomography." *Microscopy and Microanalysis* 9, no. 6 (2003): 542–55.

[97] Howie, A. "Image Contrast and Localized Signal Selection Techniques." *Journal of Microscopy* 117 (1979): 11–23.

[98] Leapman, R. D., E. Kocsis, G. Zhang, T. L. Talbot, and P. Laquerriere. "Three-Dimensional Distributions of Elements in Biological Samples by Energy-Filtered Electron Tomography." *Ultramicroscopy* 100 (2004): 115–25.

[99] Robinson, J. M., and T. Takizawa. "Correlative Fluorescence and Electron Microscopy in Tissues: Immunocytochemistry." *Journal of Microscopy* 235 (2009): 259–72.

[100] Ray, I. L. F., and D. J. H. Cockayne. "The Dissociation of Dislocations in Silicon." *Proceedings of the Royal Society of London. Series A, Mathematical and Physical Sciences* 325 (1971): 543–54.

[101] Rai, R. S., and S. Subramanian. "Role of Transmission Electron Microscopy in the Semiconductor Industry for Process Development and Failure Analysis." *Progress in Crystal Growth and Characterization of Materials* 55 (2009): 63–97.

[102] Muller, D. A., T. Sorsch, S. Moccio, F. H. Baumann, K. Evans-Lutterodt, and G. Timp. "The Electronic Structure at the Atomic Scale of Ultrathin Gate Oxides." *Nature* 399 (1999): 758–61.

[103] Overwijk, M. H. F., F. C. van den Henvel, and C. W. T. Bulle-Lieuwma. *Journal of Vacuum Science & Technology B: Microelectronics and Nanometer Structures Processing, Measurement, and Phenomena* 11 (1993): 2021.

[104] Booker, G. R., and R. Stickler. "Method of Preparing Si and Ge Specimens for Examination by Transmission Electron Microscopy." *British Journal of Applied Physics* 13, (1962): 446–9.

[105] Bravman, J., and R. Sinclair. "The Preparation of Cross-Section Specimens for Transmission Electron Microscopy." *Journal of Electron Microscopy Technique* 1 (1984): 53–61.

[106] Faruqi, A. R. "Direct Electron Detectors for Electron Microscopy." *Advances in Imaging and Electron Physics* 145, (2007): 55–94.

[107] Faruqi, A. R., and G. McMullan. "Direct Imaging Detectors for Electron Microscopy." *Nuclear Instruments and Methods in Physics Research Section A* 878 (2018): 180–90.

[108] Cockayne, D. J. H. "Weak-Beam Electron Microscopy." *Annual Review of Materials Science* 11 (1981): 75–95.

[109] Saxton, W. O., W. K. Jenkins, L. A. Freeman, and D. Smith. "TEM Observations Using Bright Field Hollow Cone Illumination." *Optik (Jena)* 49 (1978): 505–10.

[110] Nellist, P. D., and S. J. Pennycook. "Incoherent Imaging Using Dynamically Scattered Coherent Electrons." *Ultramicroscopy* 78 (1999): 111–24.

[111] Jesson, D. E., and S. J. Pennycook. "Incoherent Imaging of Thin Specimens Using Coherently Scattered Electrons." *Proceedings of the Royal Society of London A* 441 (1993): 261–81.

[112] Krivanek, O. L., T. C. Lovejoy, M. F. Murfitt, G. Skone, P. E. Batson, and N. Dellby. "Towards Sub-10 meV Energy Resolution STEM-EELS." *Journal of Physics Conference Series* 522 (2014): 012023.

[113] Batson, P. E., N. Dellby, and O. L. Krivanek. "Sub-Ångstrom Resolution Using Aberration Corrected Electron Optics." *Nature* 418 (2002): 617–20.

[114] Krishnan, K. M. "Structure of Newly Synthesized BC_3 Films." *Applied Physics Lett*ers 58 (1991): 1857.

[115] Watanabe, M., and D. B. Williams. "The Quantitative Analysis of Thin Specimens: A Review of Progress from the Cliff–Lorimer to the New Zeta-Factor Methods." *Journal of Microscopy* 221 (2006): 89–109.

[116] Kovacs, A., and R. Dunin-Borkowski. "Magnetic Imaging of Nanostructures Using Off-Axis Electron Holography." In *Handbook of Magnetic Materials*, Vol 27, edited by E. Brück, 59–153. Amsterdam: Elsevier, 2018.

[117] Simpson, E. T., T. Kasama, M. Pósfai, P. R. Buseck, R. J. Harrison, and R. E. Dunin-Borkowski. "Magnetic Induction Mapping of Magnetite Chains in Magnetotactic Bacteria at Room Temperature and Close to the Verwey Transition Using Electron Holography." *Journal of Physics Conference Series* 17, no. 1 (2005): 108.

[118] Dahmen, U., and K. H. Westmacott. "Observations of Pentagonally Twinned Precipitate Needles of Germanium in Aluminum." *Science* 233 (1986): 875–6.

[119] Sharma, R., P. A. Crozier, and M. M. J. Treacy. "National Science Foundation Report: Dynamic *in situ* Electron Microscopy as a Tool to Meet the Challenges of the Nanoworld." The Buttes, Tempe, Arizona. January 3–6, 2006.

[120] Giannuzzi, L. A., and F. A. Stevie. "A Review of Focused Ion Beam Milling Techniques for TEM Specimen Preparation." *Micron* 30, no. 3 (1999): 197–204.

[121] Kang, H. H., J. F. King, O. D. Patterson, S. B. Herschbein, J. P. Nadeau, and S. E. Fuller. "High Volume and Fast Turnaround Automated Inline TEM Sample Preparation for Manufacturing Process Monitoring." *36th ISTFA Conference Proceedings* (2010): 102–7.

[122] Rai, R., E. Chen, Y. Zhang, D. Nedeau, Y. Chen, W. Zhao, S. K. Lim, Z-H. Mai, and J. Lam. "Automated TEM Sample Preparation from Smaller Device Structure Regions of Semiconductor ICs using Inline Dual-Beam CLM+ and TEMLink 150." *Microscopy and Microanalysis* 19, Suppl 2 (2013): 900–1.

[123] Ugurlu, O., M. Strauss, G. Dutrow, J. Blackwood, B. Routh Jr., C. Senowitz, P. Plachinda, and R. Alvis. "High-Volume Process Monitoring of FEOL 22nm FinFET Structures Using an Automated STEM." *SPIE Proceedings* 8681 (2013): 868107–120.

[124] Tan, H., W. Weng, R. Rai, C. Kang, L. Dumas, I. Brooks, and A. Katnani. "Advanced Industrial S/TEM Automation and Metrology: Boundary of Precision." *ASMC Proceedings* (2018): 131–5.

. . . .

EXERCISES

A. Test Your Knowledge

For each statement, choose all the right answers. Note that more than one response, or none, may be correct.

1. A *standard* TEM
   a) collects image information simultaneously over the full magnified field of view.
   b) can be broadly divided into four sections based on their function.
   c) *always* has a thermionic source.
   d) produces information of the specimen *only* in real space.
2. The first electron-beam cross-over
   a) is formed immediately after leaving the source.
   b) does not function as the virtual source for the TEM.
   c) affects the beam coherence.
3. A goniometer is used to
   a) tilt the specimen accurately about two orthogonal axes.
   b) adjust the specimen height to be in the *eucentric* position.
   c) rotate the specimen in plane.
4. Image contrast in a TEM is observed due to
   a) mass-thickness.
   b) local changes in specimen surface topography.
   c) diffraction.
   d) interference of electron beams scattered in different directions.

5. In a TEM, a biprism
   a) is used for electron holography.
   b) splits the beam into two coherent beams.
   c) is a complicated device that is difficult to install.

6. A state-of-the-art analytical electron microscope include capabilities for
   a) CBED.
   b) EDXS.
   c) EELS.
   d) Auger electron spectroscopy.
   e) Lorentz microscopy with a DPC detector.

7. Thermionic guns
   a) *only* use LaB_6 filaments.
   b) are operated at very low temperatures to avoid heating.
   c) generate low total current and high current densities.
   d) are *always* preferred over FEG sources.

8. The lenses commonly used in a TEM
   a) are electromagnetic in nature.
   b) are not electrostatic because of large aberration coefficients.
   c) are electromagnetic and designed with no axial symmetry.
   d) have only a fixed focal length.
   e) both deflect and focus the electron beam.

9. An electromagnetic lens
   a) strictly speaking, cannot be considered as a thin lens.
   b) function can be illustrated by geometric optics.
   c) resolution is diffraction limited.
   d) does not show any chromatic aberration.
   e) shows spherical aberration that can be corrected by pairs of magnetic sextupoles.

10. For electromagnetic lenses
    a) best performance is observed at an optimal divergence angle.
    b) astigmatism is not an issue.
    c) the depth of field is related to the aberration-limited resolution.

11. The depth of focus for an electromagnetic lens typically used in a TEM is of the order of
    a) millimeters.
    b) centimeters.
    c) meters.

12. A parallel beam illumination in a TEM is
    a) difficult to set up.
    b) achieved by an under-focused C2 lens excitation.
    c) achieved by a focused C2 lens excitation.

13. In STEM
    a) a focused C2 lens is used to produce the focused probe on the specimen.

    b) the C2 lens is switched off and the OL upper pole piece focuses the probe.

    c) scan coils are used to keep the beam parallel to the optic axis as it is rastered.

14. The best resolution in an OL is achieved for
    a) the lowest acceleration voltage.
    b) the smallest wavelength of the electron beam.
    c) the smallest spherical aberration coefficient.

15. A side-entry specimen holder
    a) is the same as a top-entry holder in design.
    b) allows for a larger range of tilt compared to the top-entry holder.
    c) is susceptible to mechanical vibrations.
    d) can be adapted for a wide range of temperatures.

16. Imaging
    a) in BF involves selecting the forward scattered beam with the OA.
    b) in DF involves selecting one of the diffracted beams by shifting the OA.
    c) in DF can involve tilting the incident beam.
    d) in HREM is a combination of all of the above.

17. In phase contrast imaging
    a) HREM images are formed by the interference of the transmitted and diffracted beams.
    b) the phases of the waves for imaging can be controlled by lens defocus.
    c) the contrast transfer function determines which diffracted beams contribute to the image.

18. The contrast transfer function
    a) is a filter function that determines which spatial frequencies contribute to the image.
    b) is affected by the envelope function, which serves as an aperture in the BFP of the OL.
    c) when combined with the envelope function determines the information limit.

19. The Scherzer defocus
    a) corresponds to a CTF with a minimum number of zero crossings.
    b) corresponds to a particular negative value of the defocus.
    c) and the corresponding Scherzer resolution can be used to compare imaging performances of different TEMs.

20. In STEM
    a) the ADF *and* HAADF detectors are used for *routine* dark field imaging.
    b) the HAADF detector is used for *Z*-contrast images.
    c) the image intensities can be related to TEM images by the reciprocity principle.

21. The interaction of an electron beam with a specimen
    a) is classified as elastic or inelastic when the electrons are considered as particles.
    b) is classified as coherent or incoherent when the electrons are considered as waves.
    c) generally transfers no energy to the specimen.
22. Elastic scattering
    a) of electrons by atoms is incoherent.
    b) cross-section depends on the acceleration potential.
    c) intensity is inversely related to the scattering angle.
23. Rutherford cross-section for elastic scattering
    a) decreases by ~ 6 orders of magnitude as the scattering angle varies by $180°$.
    b) shows a Z^2 dependence at very large angles in the forward direction.
    c) includes no corrections for screening effects.
24. Mass-thickness contrast
    a) is applicable to incoherent scattering from nonperiodic structures.
    b) applies specifically to biological specimens.
    c) will make thicker specimens appear brighter in the forward-scattered beam.
25. Diffraction contrast
    a) arises from coherent elastic scattering.
    b) requires the specimen to be tilted to the right orientation.
    c) is used to analyze defects such as dislocations.
    d) is not observed in STEM.
26. *Z*-contrast imaging _____ distribution of the electrons in the probe.
    a) ... depends on high-angle incoherent scattering ...
    b) ... requires a HAADF detector with a large collection angle that avoids ...
    c) ... has a resolution given by the convolution of the atomic potential and the spatial ...
27. Phase contrast imaging
    a) produces lattice fringes *only*.
    b) produces HREM images with *more* information than the underlying diffraction pattern.
    c) should be carried out close to Scherzer defocus for easy interpretation.
28. Recording of optimal HREM images requires that the
    a) TEM is corrected for astigmatism.
    b) specimen drift is avoided.
    c) microscope is well aligned.
    d) microscope be operated in the minimum contrast condition.
29. Magnetic contrast can be obtained
    a) by Fresnel imaging.

b) in Foucault mode.

c) using a quadrant detector in DPC mode.

d) by electron holography.

30. Electron holography is used to

a) correct microscope aberrations.

b) image the in-plane specimen induction.

c) reconstruct images from interferograms formed by the reference and object waves.

31. Inelastic scattering of the electron beam

a) involves interactions with inner-shell, valence, and conduction band electrons in the specimen.

b) conserves energy and momentum of the electron beam.

c) conserves *total* energy and momentum of the system.

d) *only* involves intra-band and inter-band processes.

32. In an analytical electron microscope

a) characteristic X-rays are emitted isotropically in all directions.

b) energy-loss electrons are observed only in the forward direction.

c) energy-loss electrons are typically detected with an efficiency $\sim$50%.

33. Plasmon resonance

a) leads to collective excitation of the valence band in semiconductors.

b) can be simply understood in terms of atom-like transitions.

c) is described phenomenologically in terms of the dielectric behavior of the solid.

d) can give information on the electron density of the band.

34. A transmission EELS spectrum

a) is broadly divided into no-loss, low-loss, and core-loss regions.

b) can be quantified to provide compositional information.

c) includes fine structure that is interpreted in terms of bonding and electronic structure.

35. The no-loss region of the EELS spectrum includes electrons that are

a) unscattered.

b) elastically scattered.

c) *photon* scattered.

36. The low-loss region of the EELS spectrum

a) results from direct electrostatic interactions of the electron beam with atomic electrons.

b) includes excitations arising from inter-band and intra-band transitions.

c) provides no information on the thickness of the specimen.

37. EELS core-loss edges

a) have three basic shapes.

b) display fine structure due to solid state effects.

c) show *positive* chemical shift with increasing oxidation state.

d) show *negative* chemical shift with increasing oxidation state.

e) show an intensity above background proportional to the concentration.

38. To carry out good microanalysis with EELS
    a) it is preferable to have thin specimens and avoid _____ scattering effects.
    b) edge overlaps must be avoided.
    c) requires accurate partial ionization cross-sections.

39. True or false:
    a) EFTEM and STEM-EELS collect the three-dimensional data set in an *identical* manner.
    b) Quantitative chemical maps in energy filtered imaging are obtained with the three-window method.
    c) The jump ratio method provides quantitative maps.
    d) EFTEM can provide images with sub-nm spatial resolution.

40. X-ray microanalysis (EDXS) in a TEM
    a) requires standards.
    b) is possible without standards.
    c) is best done at the highest acceleration voltage.
    d) is routinely analyzed with the Cliff–Lorimer method.
    e) is generally not subject to the absorption correction.

41. Selected area diffraction
    a) typically has a minimum selectable area of 0.2–0.5 micrometers.
    b) has a significant error in selecting a specific area for *very small* Bragg angles.
    c) can provide structural information from small areas better than CBED.

42. Electron tomography
    a) reconstructs a three-dimensional image based on a number of two-dimensional projections.
    b) has a resolution directly proportional to the number of two-dimensional projections included in the reconstruction.
    c) is not used to reconstruct images with chemical sensitivity.
    d) commonly uses the real-space back projection reconstruction method.
    e) is only used in biology to study macromolecules.

43. TEM is used in semiconductor manufacturing for
    a) metrology.
    b) process development.
    c) failure analysis.

B. Problems

1. Calculate the **diffraction limited resolution** and the optimum divergence angle for an electromagnetic lens with $C_S = 0.6$ mm in a microscope operated at 100 kV.

2. Consider a **lens with spherical aberration**, δ_S, and the diffraction limited resolution, δ_D. Assume that its effective resolution, δ_{eff}, can be

computed as $\delta_{eff}^2 = \delta_S^2 + \delta_D^2$. Show that the optimal beam divergence angle, α_{opt}, is given by (9.2.5). Now show that the effective resolution at this optimal angle is given by $\delta_{eff}(\alpha_{opt}) \approx 0.5\,(C_S\lambda^3)^{1/4}$.

3. Is a typical **TEM specimen** with thickness <100 nm appropriate for imaging? Why? In a TEM the viewing screen and the image recording system are separated by 20 cm. For a magnification of 50,000×, can we both observe and record the image? Assume typical parameters for the TEM with a thermionic source.

4. Assume that the **SAD pattern** for polycrystalline gold is shown in Figure 9.2.7b, and the ring of interest corresponds to the (400) reflection. Calculate the angular divergence for the aperture shown.

5. (a) Calculate in the form of a table, and plot the optimal **convergence semi-angle and the Scherzer radius** as a function of acceleration voltage from 100 to 1000 kV, for two different values of $C_S = 1$ mm and 3 mm, respectively. (b) If you wish to resolve the {222} planes of NaCl, what voltages are required for these two values of C_S?

6. What are the relative advantages and disadvantages of **EDXS and EELS**, the two common analytical methods implemented in a TEM?

7. The **electronic structure** of Au is given in the following figure.

	n	l	j	# of e^-	E_c (keV)
K	1	0	1/2	2	80.72
L_I	2	0	1/2	2	14.35
L_{II}	2	1	3/2	2	13.73
L_{III}	2	1	3/2	4	11.92
M_I	3	0	1/2	2	3.42
M_{II}	3	1	1/2	2	3.15
M_{III}	3	1	3/2	4	2.74
M_{IV}	3	2	3/2	4	2.30
M_V	3	2	5/2	6	2.21

(a) If the **electron transitions** satisfy the selection rules $\Delta n \geq 1$, $\Delta l = \pm1$, and $\Delta j = \pm1$ or 0, draw all the allowed transitions following the ionization of (i) a K-shell, (ii) any one of the three L-shells.

(b) Calculate the values of the Au K and L X-ray emission spectra, and label all the transitions (hint: refer to §2).

(c) What would be the relative intensities of the principal K-lines. Note that the probability of a transition occurring is related to (i) the number of electrons in the initial shell, and (ii) screening of the electrons between the initial and final states.

(d) Which of these X-ray lines cannot be detected by a typical EDXS detector? Why not?

(e) Which of these lines cannot be excited by a 100 keV TEM? Why not?

8. (a) Calculate and plot the SIGMAK and SIGMAL **cross-sections** for Ni–L and O–K edges, using (9.4.19–9.4.24), as a function of the integration window, $0 < \Delta_A < 400$ eV, for two collection semi-angles $\beta = 13.5$ mrad and $\beta = 4.5$ mrad. Assume a 200 keV incident beam.

(b) The measured intensities for a specimen of Ni–O are given in the following table.

Δ_A (eV)	50	75	100	135
$I_{O\text{-}K}$	209551	330617	451699	508260
$I_{Ni\text{-}L}$	168380	283720	412994	554477

(i) Calculate the relative concentration (Ni/O) for the specimen, assuming $\beta = 4.5$ mrad, for the four integration windows shown.

(ii) Repeat (i) assuming $\beta = 13.5$ mrad using the same data set for intensities.

(iii) Comment on the importance of determining experimental parameters such as β in EELS quantification.

9. An unusual **HREM image** (Fig. 9.Pr.9) of a pentagonally twinned Ge precipitate in an Al matrix with the beam direction along <110> Ge and <100> Al is shown in the following figure. Read the original paper and explain this image, which is adapted from [118].

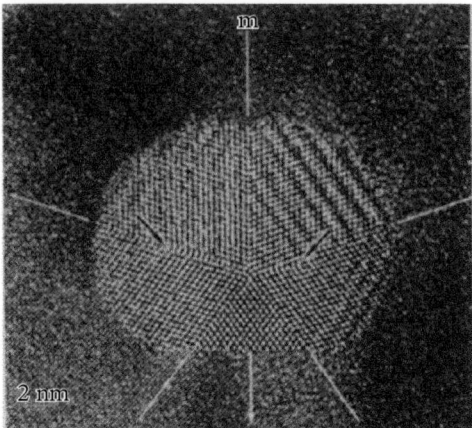

10. Three different possibilities to form **moiré patterns** (Fig. 9.Pr.10) from regions containing a dislocation are shown in the following:

(a) lattice image of a region containing an extra half plane (dislocation) and a regular lattice.

(b) small rotation of the regular lattice leading to the formation of a rotational moiré.

(c) small change in the spacing of the regular lattice.

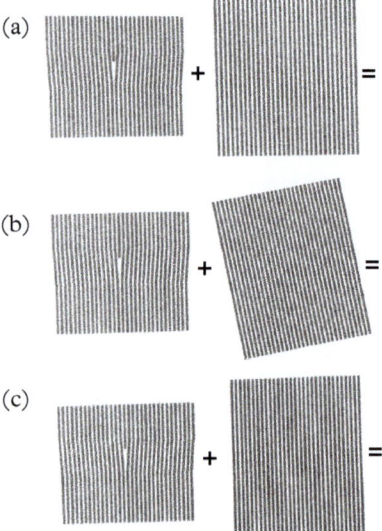

What would be the resultant moiré pattern for each case (hint: see Example 9.3.1)?

Comment on how easy it would be to interpret such a defect using these moiré patterns.

11. Figure 9.Pr.11 shows experimental EELS and EDXS spectra from a specimen.

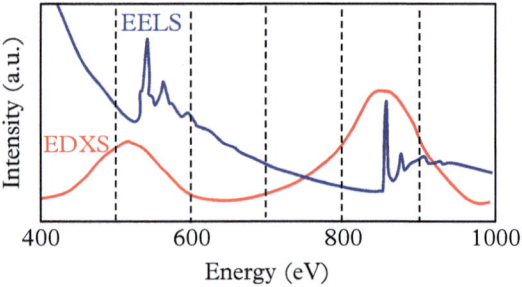

(a) Identify the elements in the unknown specimen.

(b) These spectra illustrate the physical principles of the two techniques very well. Discuss them in detail, highlighting the differences and similarities.

Scanning Electron Microscopy

10

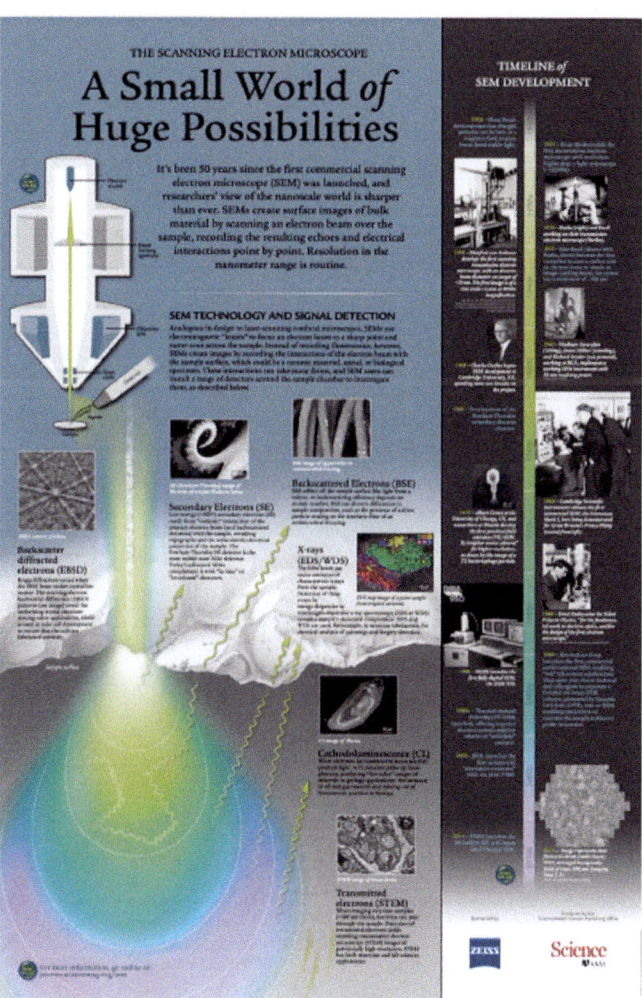

Poster produced by the American Association for the Advancement of Science in 2015 to mark the 50th anniversary of the introduction of the first commercial SEM. Downloaded with permission from posters.sciencemag.org/sem.

Principles of Materials Characterization and Metrology. Kannan M. Krishnan, Oxford University Press (2021).
© Kannan M. Krishnan. DOI: 10.1093/oso/9780198830252.003.0010

10.1 Introduction

A scanning electron microscope (SEM) is a versatile tool that is both widely and routinely used in materials characterization and analysis to generate *morphological*, *chemical*, and *structural* information, at high resolution, from a specimen. Analogous to a confocal scanning optical microscope (§6.8.5), an SEM uses electromagnetic lenses (§9.2.2) to focus and raster an electron beam over the surface of the specimen. The interaction of the electron beam with the specimen produces a range of signal radiations—secondary and back-scattered electrons, characteristic X-rays, beam-induced current, ultraviolet and visible light—that are then detected by various detectors placed appropriately in the specimen chamber to maximize their detection and form images. Image resolutions down to the nanometer scale can be achieved. Recent developments of the environmental SEM allow for the study of wet specimens, of particular importance in the life sciences. Further, the incorporation of a separate focused ion beam (FIB) column in an SEM has made it possible to mill or micromachine a specimen to desired shapes at the nanometer length scale.

Details of the instrument, the various signals produced by electron–specimen interactions, and their detection, as well as their application in materials characterization are presented here. For further details on all aspects of SEM, with particular emphasis on physical principles, the reader should consult a comprehensive monograph, such as Reimer (1985) and/or Goldstein et al. (2003).

10.2 The Scanning Electron Microscope

10.2.1 The Instrument

Figure 10.2.1 shows a commercial SEM equipped with a thermionic electron source, operating at 5–25 kV, and used routinely for teaching and research. However, it is also common to find SEMs with field emission and Schottky sources as guns; recall that electron sources—thermionic, field-emission, and Schottky—were discussed earlier in §5.2.2 and §9.2.1, and their performance characteristics, particularly their total current and current density, as well as brightness, were summarized in Table 5.2.2.

The key design principle of an SEM is to demagnify the current crossover (the virtual source) of the electron gun (see Fig. 5.2.6) and produce a small electron probe at the specimen surface. This is accomplished in an SEM by the condenser and objective (probe forming) lenses. The condenser lens design is normally different for thermionic and field emission guns. The main advantage of a field emission gun is its overall 2–3 orders of magnitude higher brightness and the small crossover diameter ($\sim$10 nm) of the source. As a result, only one and not two condenser lenses (Fig. 10.2.2) is required to form a small ($\sim$1 nm) probe with sufficient current. Note that field emission sources suffer from

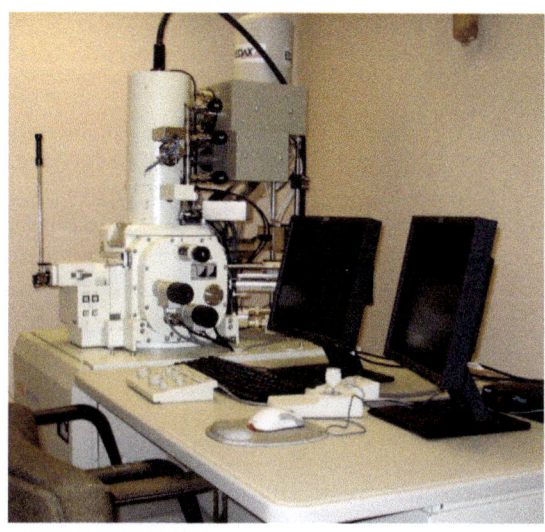

Figure 10.2.1 A commercial SEM with an energy dispersive X-ray detector.

Image courtesy University of Washington.

short-term current fluctuations (∼2–5%) and the overall emission current also drifts with time.

The condenser system containing one or more lenses is used to adjust the diameter of the electron beam by strengthening (broader beam) or weakening (narrower beam) its excitation. This is followed by a selectable aperture (a metal plate with holes of various sizes) placed between the condenser and the final probe-forming (objective) lens. After passing through the condenser lens, the electron beam illuminates the aperture plate. Depending on the size of the hole, the aperture allows a part of the electron beam to reach the objective lens (Fig. 10.2.3). When the excitation of the condenser lens is increased, the electron beam broadens significantly on the aperture plate, and for a given aperture diameter, the number of electrons reaching the objective lens, also known as the probe current, decreases. In contrast, for a weaker condenser lens excitation, there is less broadening of the beam on the aperture, which results in a larger beam current. In other words, the excitation of the condenser lens allows the operator to control *both* the probe *diameter* and probe *current*. Moreover, for many practical applications, it is important to have control over the electron beam diameter, aperture, and current, but these critical parameters are mutually dependent through the gun brightness (see §10.2.4 for further details). Finally, the objective lens is used for focusing the image, and this lens, in practice, determines the final diameter of the electron probe. In general, a high-quality objective lens with minimal aberrations (see §9.2.2) is required to produce high-quality images.

For thermionic emission, the virtual source has a diameter, $d_c \sim 20$–40 μm, and for a Schottky emission gun, $d_c \sim 10$ nm. Thus, the electron-optical system of an SEM consists of one (Schottky) or two (thermionic source) condenser lenses and a final probe-forming lens, that in combination reduce the probe to a

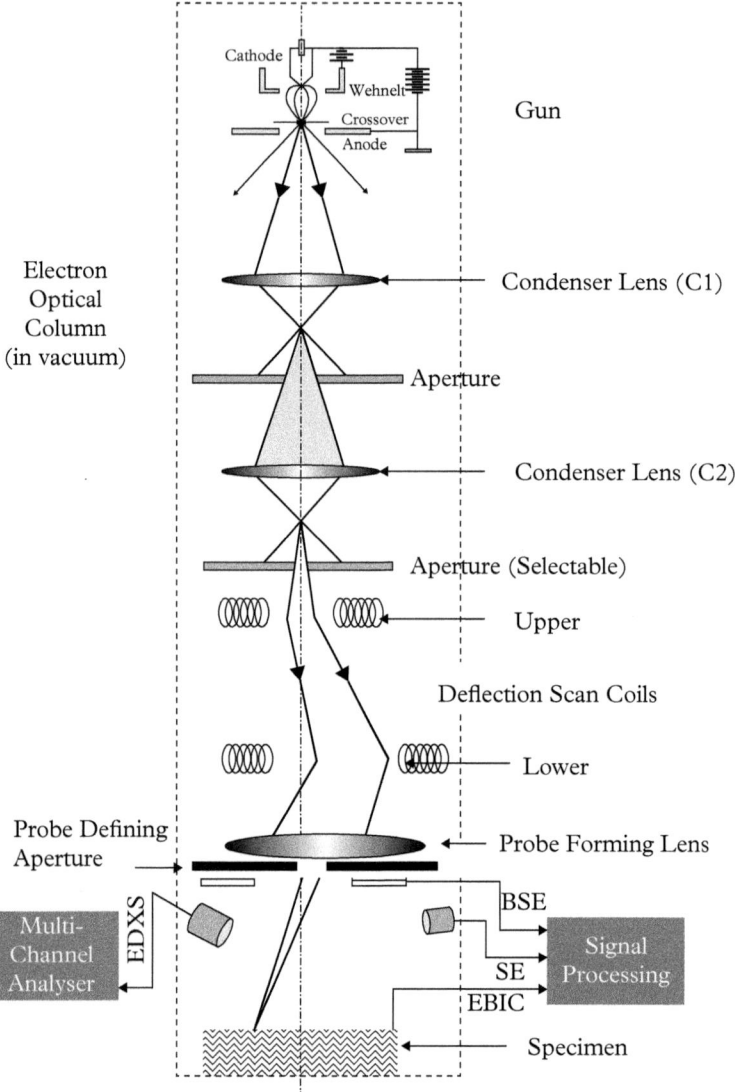

Figure 10.2.2 Schematic representation of an SEM (not to scale), with a two-condenser lens system, including the ray optics and the main components that provide different contrasts in imaging. Notice the large working distance, d_w (~5–40mm), between the final probe forming lens and the specimen.

well-defined spot of 1–2 nm diameter for the Schottky source, or 8–10 nm diameter for a thermionic source. This probe is then rastered over the specimen surface using a pair of beam deflection coils, and the electrons (secondary or backscattered) are collected as a function of position and amplified in the detector to produce an image. The magnetic deflection coils move the beam using the Lorentz force, $\mathbf{F} = q\mathbf{v} \otimes \mathbf{B}$ (Fig. 10.2.4). Normally two sets of coils are used to keep the beam close to the optic axis and minimize spherical aberration effects. Note

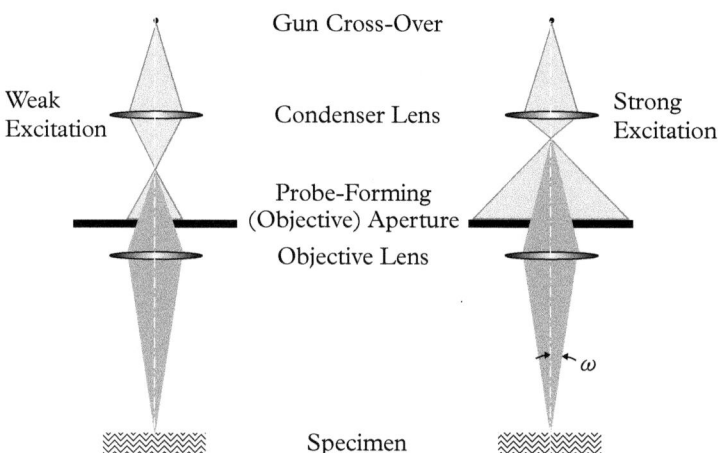

Gun Cross-Over

Weak Excitation

Condenser Lens

Strong Excitation

Probe-Forming (Objective) Aperture

Objective Lens

ω

Specimen

Figure 10.2.3 Formation of an electron probe by a condenser and objective lens, illustrating the role of the probe-forming aperture and the condenser lens excitation.

that the final probe-forming lens operates at a relatively large working distance to enable the accommodation of various detectors (Fig. 10.2.2) required for collecting the different signals (quanta of radiation, or particles) and to generate contrast with high enough efficiency. As a result, the spherical aberration (§9.2.2) of the probe-forming lens increases, which, in turn, limits the size of the smallest probe of the incident electrons that can be generated.

In practice, one is interested in varying and selecting an appropriate probe size and probe current. However, they cannot be varied independently and they are also affected by the selectable aperture of the last probe-forming lens. Typically, apertures with semi-angle, $\alpha \sim 0.005$ rad, are used for routine high-resolution work. If we wish to increase the depth of field, §10.2.5, smaller apertures, reduced by one or two orders of magnitude, or higher acceleration voltages (§10.3.4) are desirable. Overall, the resolution of an SEM is determined by (a) the spherical aberration (designed to be as small as possible) of the final probe-forming lens, (b) the smallest diameter probe, which is determined by the electron-optics, and (c) the size of the final (selectable) aperture. In practice, the achievable combination of good resolution and large enough depth of field ($\sim$4–40 mm) makes SEM a versatile tool for materials characterization and analysis.

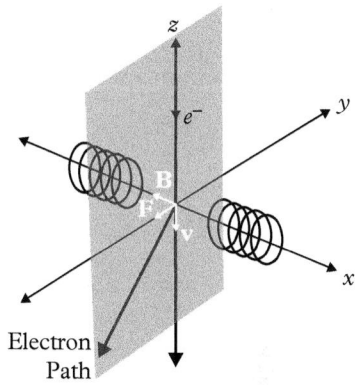

Figure 10.2.4 Beam deflection coils use the Lorentz force, $\mathbf{F} = q\mathbf{v} \otimes \mathbf{B}$, to control the trajectory of the electrons.

Example 10.2.1: An SEM column is to be built with three lenses (two condenser and one objective). Assume that each lens produces a crossover image at a distance, $L_i \gg f_i$, in *front of the next lens*. Assume that each of the three lenses have the same demagnification, M_i, and the same focal lengths $f_i \sim 10$ mm. What should be the minimum length of the total electron-optical column for a probe diameter, $d_0 = 10$ nm and a cross-over diameter, $d_c = 50$ μm?

Solution: For each lens, the position of the image will be given by (6.8.3), i.e. $\frac{1}{u} = \frac{1}{f_i} - \frac{1}{L_i}$, or $u = \frac{f_i L_i}{L_i f_i} \approx f_i$. In other words, the image will be near the lens focus with a demagnification, $M_i = f_i/L_i$. Thus, for a three-lens system, the electron probe diameter, d_0 will be related to the crossover diameter, d_c, as $d_0 = M_1 M_2 M_3 d_c = M_i^3 d_c$. Then,

$$M_i = \left(\frac{d_0}{d_c}\right)^{1/3} = \left(\frac{10 \times 10^{-9}}{50 \times 10^{-6}}\right)^{1/3} = \left(\frac{1}{5000}\right)^{1/3} \approx \frac{1}{17} = \frac{f_i}{L_i}.$$

Thus, $L_i = 17 f_i = 170$ mm.

For a three-lens system, the column length should be greater than $3L_i = 510$ mm ≥ 0.5 m

10.2.2 The Everhart–Thornley Electron Detector

The widely used Everhart–Thornley electron detector (Fig. 10.2.5) consists of a Faraday cage, a scintillator, a light guide, and a photomultiplier. It is often optimized for secondary electron detection, and the Faraday cage is kept at a positive potential of +250 V to attract low-energy secondary electrons, which travel through large deflections as shown; it can also be kept at a negative potential (–50 V) to repel or screen out the secondary electrons, and detect only the higher energy (keVs) back-scattered electrons that travel in straight lines. The scintillator, kept at a large positive potential (+12 kV), accelerates the electrons entering the Faraday cage and converts them into photons. These photons now travel through the light guide to the photomultiplier with a signal gain of $\sim 10^6$, which is then further amplified, if necessary, to display as an image with appropriate contrast (Fig. 10.3.8).

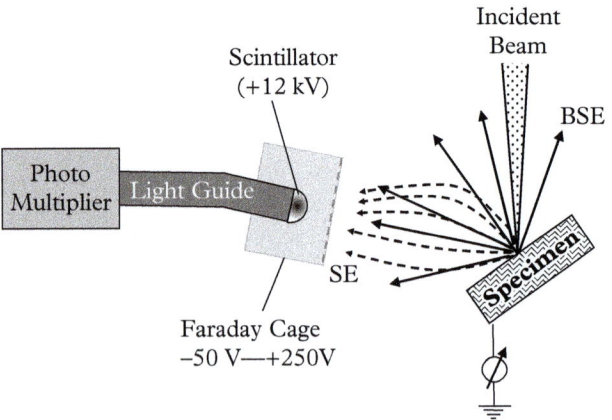

Figure 10.2.5 The Everhart–Thornley (E–T) detector used for collecting secondary and back-scattered electrons.

10.2.3 Beam–Solid Interactions and Signals

In general, the complex interaction of an electron beam with any specimen gives rise to multiple methods of image contrast. If a bulk specimen were to be illuminated in a conventional SEM with an electron beam, typically of lower energy ($\sim$5–30 keV) compared to a transmission electron microscope (TEM) (typically $\geq$100 keV), the beam-specimen interactions (Fig. 10.2.6a) would produce among other emissions, back-scattered and secondary electrons. Back-scattering, arising from the interactions of the electrons with the atomic nuclei (§5.3.2), can be elastic (scattered electrons retain their primary energy) or inelastic (10–20% lower in energy). The trajectories of the high-energy back-scattered electrons are almost straight lines; however, as they emerge from the specimen, they are also sensitive to the magnetic induction distribution inside the specimen. Alternatively, secondary electrons with energies much lower than that of the incident beam are emitted from the excitations of atoms by the primary electron probe. The trajectories of these secondary electrons can be effectively controlled by the electrostatic fields of the E–T detector (Fig. 10.3.6). They are also influenced by the stray

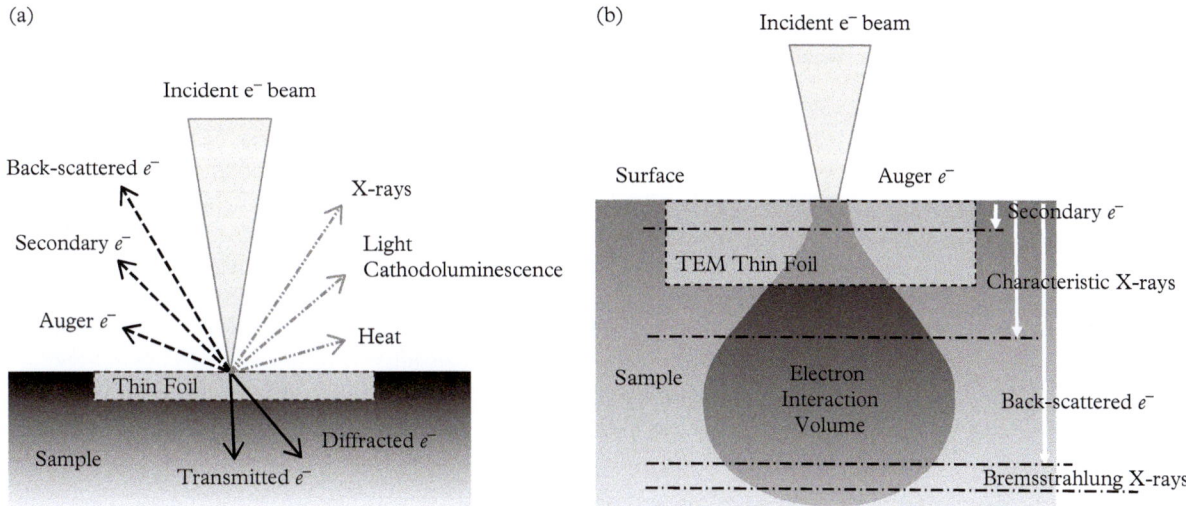

Figure 10.2.6 Beam-solid interactions in a TEM and SEM. (a) Interactions of a high-energy primary electron beam with a solid, illustrating the variety of signals (electrons and photons) produced. Representative thicknesses (5–50 nm) of electron transparent thin foils relevant to TEM imaging (§9) are indicated. (b) Different signals are produced at different depths and provide different depth sensitivity for secondary (<5–50 nm), Auger (surface), and back-scattered (15 μm at 200 keV) electrons. It is important to recognize that thinning a specimen for observation in a TEM may alter its intrinsic microstructure; this is particularly problematic for domain observations, as it changes the magnetostatic interactions in the specimen, and careful attention should be paid to avoid such artifacts of specimen preparation.

Adapted from Krishnan (2016).

Figure 10.2.7 Cathodoluminescence in semiconductors following high-energy electron incidence. Following excitation of an electron from the valence band to the conduction band, relaxation can occur via a number of defect states in the band-gap, to emit light in the visible or UV/IR portion of the electromagnetic spectrum.

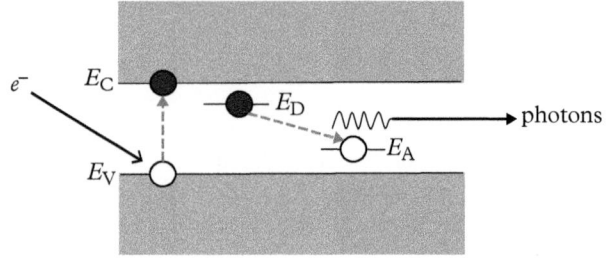

magnetic fields above a ferromagnetic specimen. As a result, two distinct magnetic contrast mechanisms (§10.5) in an SEM, referred to as Type I and Type II, based on the detection of secondary and back-scattered electrons, respectively, using collectors with directional sensitivity and appropriate energy selectivity have been developed. In addition, the secondary electrons are also spin-polarized and the state of their polarization depends on the magnetization direction of the specimen at the location from which they emerge. Detection of the spin-polarization of secondary electrons in a spatially resolved fashion leads to another important method (SEMPA) of imaging magnetic contrast (§10.5.2).

Furthermore, cathodoluminescence (CL) due to recombination of electron–hole pairs in the specimen created by the energetic incident electrons, can also be detected in an SEM. This luminescence phenomenon can be explained using a simple band picture of the solid (Fig. 10.2.7). An insulating or semiconducting material can be visualized (Fig. 3.2.1) as having a valence band and a conduction band separated by a band gap (§3.2). When fast electrons of sufficiently high energy are incident on this material, electrons from the valence band may be promoted to the higher-energy conduction band. When these excited electrons attempt to return to their ground state (valence band), they may be trapped temporarily (on the order of microseconds) by intrinsic (structural defects) or extrinsic (impurity) defect states in the band gap (Fig. 10.2.7). As the electrons vacate these traps and return to the valence band, the energy difference is emitted as radiation, typically in the visible spectrum ($\lambda \sim 380$–700 nm), with some photons also observed in the ultraviolet or infrared part of the electromagnetic spectrum (Fig. 1.3.2). Such luminescence due to the recombination of electron–hole pairs created by energetic electrons is called CL (§10.6.2).

There are several possible pathways in which the excited electrons can interact with the defect-states (traps) to produce luminescence. Once the electrons are excited to the conduction band, they may not encounter a defect-state, E_D, and fall back to the valence band; alternatively, they may move randomly through the crystal structure until a defect-state is encountered. From that defect state, the electron might return to the ground state (valence band) or it may encounter multiple defect states in the gap, such as E_A, to transition into and emit photons with wavelengths dependent on the energy differences. However, the intensity of the CL is generally a function of the density of the defect-states in the material.

10.2.4   The Incident Probe Size and Spatial Resolution

The resolution of the SEM is determined by the smallest probe, d_p, with sufficient current that can be formed on the image plane. It is also determined by the size of the final aperture and the working distance. To first order, it is assumed that the current density is uniform over a circle defined by the ideal probe diameter, d_p (nm), given by:

$$d_p = \left(\frac{4i_p}{\pi^2 \beta \alpha^2}\right)^{1/2} = \left(\frac{4i_p}{\pi^2 \beta}\right)^{1/2} \alpha^{-1} = A_0 \alpha^{-1} \qquad (10.2.1)$$

where i_p is the total current (A) in the probe, β is the gun brightness (A cm^{-2} sr^{-1}), which relates the electron current density, j_p (A cm^{-2}) on-axis, and α, the beam semi-convergence angle (mrad), as $j_p = \pi \beta \alpha^2$, and $j_p = i_p / (\pi d_p^2 / 4)$. The brightness depends on the type of gun and is tabulated in Table 5.2.2; a cold FEG is typically 10^3 times brighter than a thermionic source. The diameter of the probe can be decreased by increasing the convergence angle, α; however, this will also increase the spherical aberration. Hence, an optimal diameter of the probe, d_{opt}, including the spherical aberration coefficient, C_S (mm), (see §9.2.2), is given by:

$$d_{opt} = A C_S^{1/4} \left(\frac{i_p}{\beta} + \lambda^2\right)^{3/8} \qquad (10.2.2)$$

where $A \sim 1$, is a constant, and λ is the wavelength of the electron that depends on the acceleration voltage. Figure 10.2.8 illustrates the variation of the probe diameter as a function of (a) the acceleration voltage, (b) the probe current, and (c) the convergence semi-angle. Notice the minimum probe diameter occurs at an optimal value of the convergence semi-angle.

The ultimate spatial resolution of an SEM is determined by the smallest probe focused on the specimen surface with sufficient current in it to obtain a meaningful signal (see §10.2.6). One simple approach to increase the resolution (decrease the probe size) is to increase the gun acceleration voltage. In some cases, where specimen damage or charging (§10.4.1) is a concern, lower acceleration voltages are used even though this may lead to a loss of resolution.

Example 10.2.2: (a) What is the effective electron probe diameter including spherical aberration, chromatic aberration, and the diffraction error? Assume Gaussian distribution where the corresponding diameters (Hint: see §9.2.2) can be added in quadrature? (b) Now, making appropriate

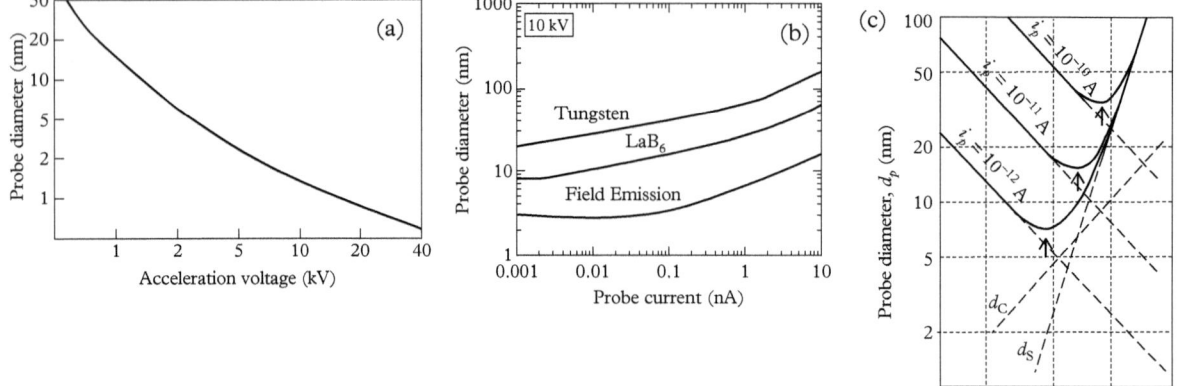

Figure 10.2.8 The resolution of the SEM is largely determined by the diameter of the electron beam, which depends on the acceleration voltage. It is also a function of the aberration of the probe-forming lens and any diffraction effects, as well as the probe current. (a) The probe diameter as a function of kV, for $C_S = 3$ mm, $C_C = 4.2$ mm, $i_p = 1$ pA, and a LaB_6 cathode. (b) The probe diameter as a function of the probe current, at 10 kV, and for three different sources. (c) The probe diameter as a function of the convergence semi-angle for three different probe currents and $E = 20$ keV, $\Delta E = 1$ eV, $\beta = 7 \times 10$ A cm^{-2} sr, $C_S = 50$ mm, and $C_C = 20$ mm; in each case, the minimum probe diameter, indicated by arrows, occurs at an optimum aperture. The diameters due to spherical, d_S, and chromatic, d_C, aberration are also indicated.

Adapted from Reimer (1985).

assumptions about the magnitude of these individual contributions, find an expression for the optimal semi-convergence angle, α_{opt}, which minimizes the overall probe diameter.

Solution:

(a) From §9.2.2, for a probe with semi-convergence angle, α, the diffraction limited diameter, (9.2.3a), is $d_D = 0.6\lambda\alpha^{-1}$

The spherical aberration limited diameter, (9.2.4), is $d_S = 0.5\ C_S\ \alpha^3$

The chromatic aberration limited diameter, (9.2.6), is $d_C = C_C\ \alpha\ \Delta E/E_0$

The ideal probe diameter, (10.2.1), is $d_p = A_0\ \alpha^{-1}$

The effective probe diameter, d_{eff}, is given by adding these in quadrature as:

$$d_{eff}^2 = d_p^2 + d_D^2 + d_S^2 + d_C^2 = \left[A_0^2 + (0.6\lambda)^2\right]\alpha^{-2}$$
$$+ 0.25\ C_S^2\ \alpha^6 + [C_C\ \Delta E/E_0]^2\alpha^2.$$

(b) To determine, α_{opt}, we make the following reasonable approximations:

We assume that the constant $A_0 \gg \lambda$, thus neglecting the diffraction error. For typical operating voltages of 10–20 keV, $\Delta E/E_0$ is very small, and the chromatic aberration term can be neglected. Then $d_{eff}^2 = A_0^2 \, \alpha^{-2} + 0.25 \, C_S^2 \, \alpha^6$, and the optimal value, α_{opt}, can be found by minimizing d_{eff}, by setting $\partial d_{eff}/\partial \alpha = 0$. Thus

$$\alpha_{opt} = (4/3)^{1/8}(A_0/C_S)^{1/4}.$$

10.2.5 Depth of Field

Even at low magnifications, the SEM offers a significantly larger depth of field, compared to optical microscopes, as discussed later. The depth of field, d_f, is determined by the convergence semi-angle, α, of the incident electron beam (Fig. 10.2.9). It is defined as the depth range, or distance, over which the specimen can be displaced without blurring the image. Alternatively, it defines the vertical range where the loss of resolution is smaller than the line spacing in the scanned image.

Now, consider a beam with a fixed convergence semi-angle α (Fig. 10.2.9) related to the diameter, d_A, of the final probe forming aperture, and the working distance, d_W, as:

$$\tan \alpha = \frac{d_A}{2d_W} \qquad (10.2.3)$$

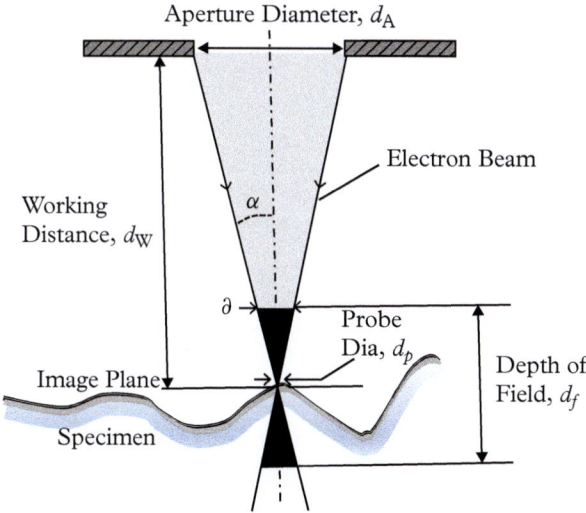

Figure 10.2.9 The convergence of the electron beam with the semi-angular aperture, α, determined by the final aperture diameter, d_A, and the working distance, d_W, of the instrument, gives the probe diameter, d_p, on the image plane. The disk of confusion, ∂, is the acceptable resolution, and d_f is the depth of field of the SEM.

Then, for the range of axial shift, defining the depth of field, d_f, any blurring of the image is measured by the "disk of confusion", ∂, such that:

$$d_f = \frac{\partial}{\tan \alpha} \tag{10.2.4}$$

For small α, the depth of field simplifies as:

$$d_f = \partial/\alpha = \frac{\partial}{\tan^{-1}\left(\frac{d_A}{2d_W}\right)} \tag{10.2.5}$$

In practice, ∂ is the *achievable resolution* of the SEM. Figure 10.2.10 plots the relationship between d_f and ∂ for an SEM and compares it to an optical microscope using light of wavelength, $\lambda = 500$ nm. It is clear that, even at low magnifications, the SEM offers a significantly larger depth of field.

In summary, a smaller aperture size, d_A, and a longer working distance, d_W, gives a larger depth of field, d_f. This is illustrated in Figure 10.2.11 with a set of secondary electron images of an appropriate test specimen taken with different combinations of the working distance and limiting aperture diameters (see also Fig. 10.3.14).

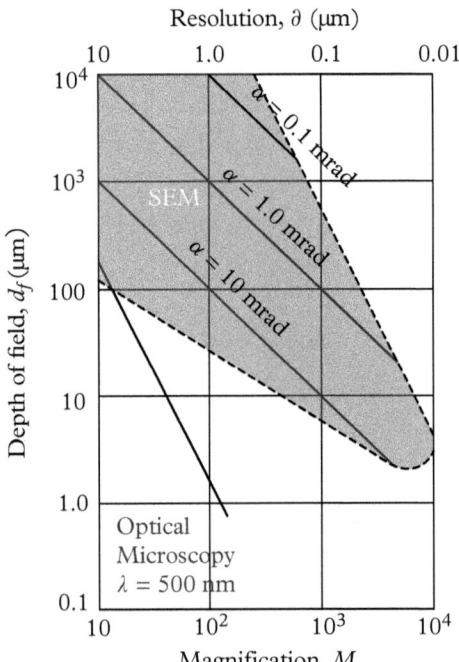

Figure 10.2.10 The relationship between the depth of field, d_f, the magnification, M, and the resolution, ∂, of a scanning electron microscope. The highlighted area with dotted boundary, is the resolution limit based on Figure 10.2.8c.

Adapted from Reimer (1985).

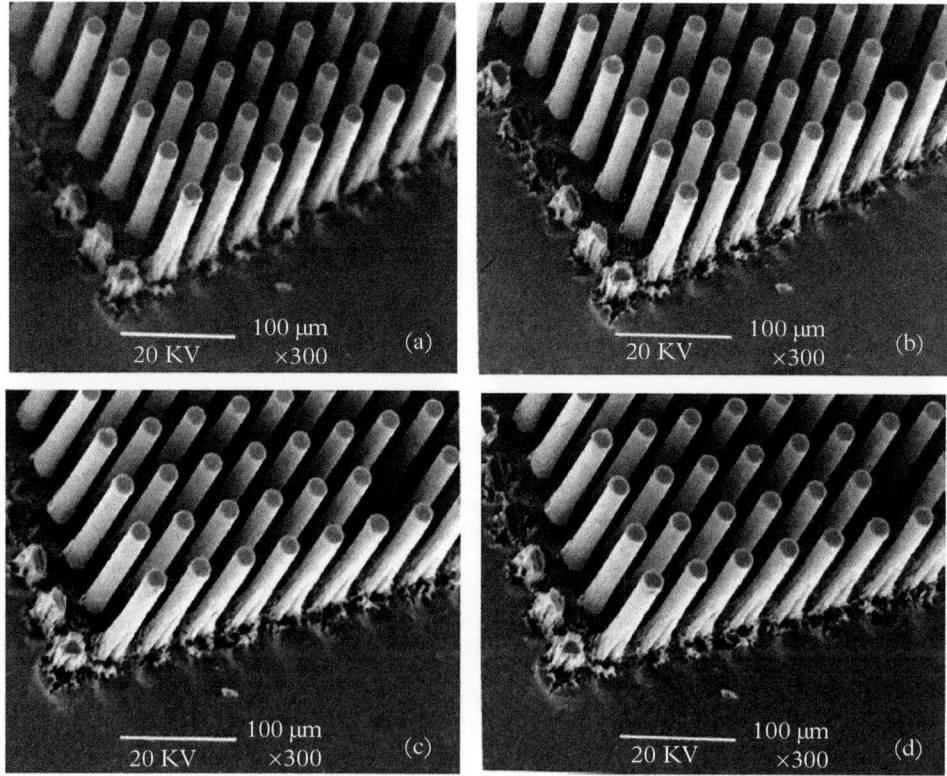

Figure 10.2.11 The effect of different combinations of the final aperture size, d_A, and the working distance, d_W, which also determines the semi-convergence angle, α, on the depth of field, d_f: (a) $\alpha = 11.2°$, $d_W = 25$ mm, $d_A = 170$ μm, (b) $\alpha = 3.28°$, $d_W = 25$ mm, $d_A = 50$ μm, (c) $\alpha = 5.81°$, $d_W = 48$ mm, $d_A = 170$ μm, and (d) $\alpha = 1.71°$, $d_W = 48$ mm, $d_A = 50$ μm.

Adapted from Leng (2013).

Example 10.2.3: Derive a simple expression for the dotted boundary, showing the limits of the relationship between the depth of field, d_f, and the resolution, ∂, in Figure 10.2.10. (Hint: assume that the resolution of an SEM behaves similar to a light microscope with $n = 1$).

Solution: The depth of field is related to the resolution, (10.2.4), as $d_f = \partial / \mathrm{Tan}\,\alpha$.

Assuming that the relationship of the resolution to the semi-convergence angle is similar to the corresponding one for a light microscope, we get (with $n = 1$)

$$\partial \sim \lambda / \sin \alpha.$$

But $\tan \alpha = \sin \alpha / \cos \alpha = \sin \alpha / \left(1 - \sin^2 \alpha\right)^{1/2}$. Then, the depth of field is

$$d_f = \partial / \tan \alpha = \partial \left[\left(\frac{\partial}{\lambda}\right)^2 - 1 \right]^{1/2}$$

which describes the limits in Figure 10.2.10.

10.2.6   Noise and Contrast in Imaging

Consider two adjacent regions in an SEM image, producing intensities I_1 and I_2, with standard deviation, σ_1 and σ_2, respectively (Fig. 10.2.12). The degree to which we can separately resolve the two regions is defined as image contrast. We now present a simple mathematical estimate for the contrast.

To statistically resolve the two regions in the SEM image, the intensities should satisfy:

$$I_1 + 3\sigma_1 \leq I_2 - 3\sigma_2 \tag{10.2.6}$$

We define the contrast, C, in any image as the ratio of the signal difference between the two regions divided by the original signal, i.e.:

$$C = \frac{I_2 - I_1}{I_1} \tag{10.2.7}$$

or:

$$I_2 = I_1 \left(1 + C\right). \tag{10.2.8}$$

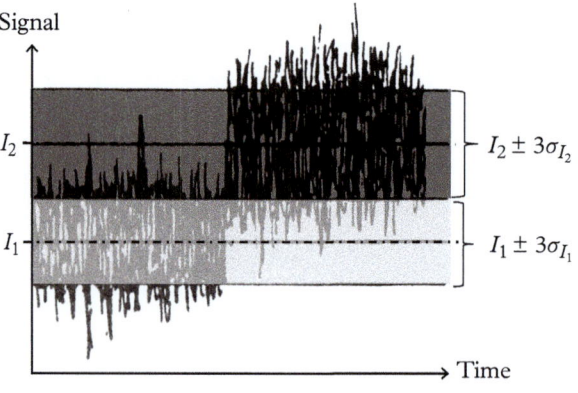

Figure 10.2.12 Two adjacent regions in an SEM with different intensities and noise used to determine the acceptable contrast in the image.

Then, from (10.2.6):

$$I_1 C \geq 3 (\sigma_1 + \sigma_2) \qquad (10.2.9)$$

If the noise in the intensity signal is random, $\sigma = \sigma_1 = \sigma_2$, and:

$$C \geq \frac{6\sigma}{I} \qquad (10.2.10)$$

which is a conservative estimate of the required contrast in a resolvable image. Note that the simple criterion, (10.2.10), is not just particular to SEM but defines the required contrast in all imaging methods. (See Problem 3 at the end of this chapter).

10.2.7 Elastic and Inelastic Scattering, and Beam Broadening

Figure 10.2.13 illustrates the beam-broadening in a specimen of a polymer, PMMA, done as an etching experiment in an SEM, and shows the eventual pear shape of the beam. High-energy electrons incident on the specimen undergo both elastic and inelastic scattering, which, in turn, determine the spatial extent of the beam in the specimen. For thick or bulk specimens imaged in an SEM, the inelastic scattering dominates. The inelastically scattered electrons will eventually reach thermal equilibrium ($E = k_B T$) and be absorbed by the specimen.

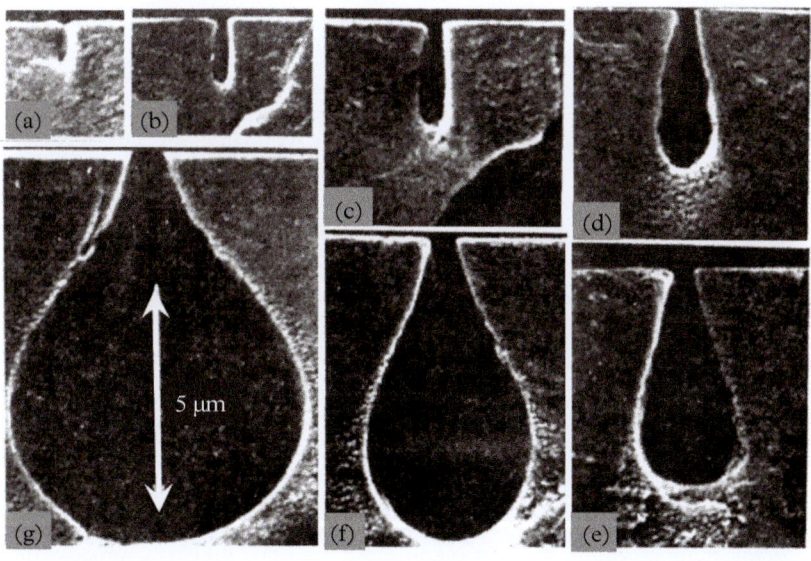

Figure 10.2.13 Cross-section image of an etching experiment on polymethyl methacrylate (PMMA) for a 20 keV incident beam (diameter <1 μm) at a fixed dose as a function of time. The interaction volume starts initially (a) as a thin cylinder, but eventually (g) becomes pear-shaped in the specimen with an extent of several μm.

Adapted from [14].

Figure 10.2.14 Monte Carlo simulation of 200 electrons propagating in a specimen of aluminum, at 30 keV incidence. The trajectories of electrons that are back-scattered define the diffusion depth, x_D, and those that achieve thermal equilibrium and are eventually absorbed in the solid define the penetration depth, x_R.

Adapted from Brandon and Kaplan (2008).

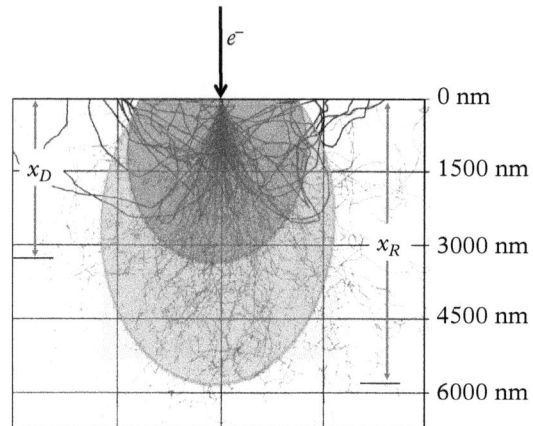

Figure 10.2.15 Illustration of the trajectory envelopes of a point incident beam, defining the diffusion and penetration depths in an SEM and their variation with specimen atomic number and beam energy.

Adapted from Brandon and Kaplan (2008).

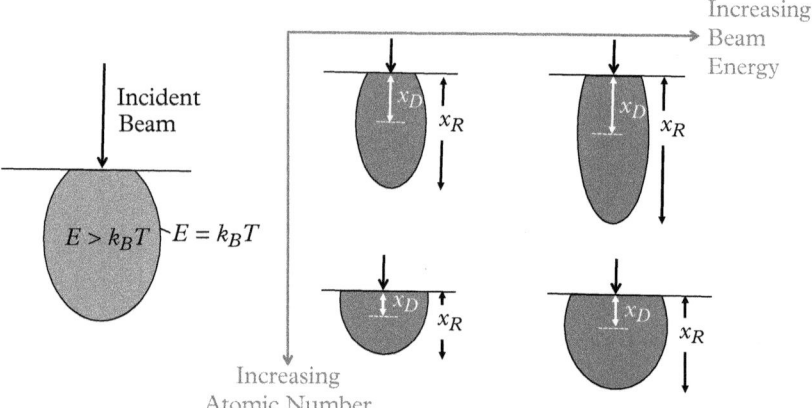

To describe these interactions quantitatively, we define two characteristic lengths as illustrated by the Monte Carlo simulation in Figure 10.2.14: (a) the diffusion depth, x_D, beyond which the electrons are randomly scattered and move in any direction, and (b) the penetration depth, x_R, at which the electron energy is reduced to $E = k_B T$, and become indistinguishable from the specimen electrons. These simulations ignore crystallographic effects, such as lattice anisotropy and channeling effects; the latter scatters the incident electrons preferentially along specific crystallographic directions, and is discussed further in §10.4.2.

Both x_D and x_R, decrease with increasing atomic number, Z, and decreasing incident beam energy, E_0 (Fig. 10.2.15). The shape of the envelope of the cumulative electron trajectories remains unchanged with increasing incident beam energy; however, the shape of the envelope changes significantly with atomic number. The diffusion depth, x_D, defined by the back-scattering is more sensitive to atomic number, Z, than the penetration depth, x_R. Thus, the lateral spread of the beam, roughly proportional to $(x_D\text{-}x_R)$, is affected by Z, and overall, the envelope has a less elongated profile with increasing Z. See also the earlier discussion of beam broadening of incident electrons in §5.3.2.1. Also, when interpreting images and the composition of the specimen, it is important to recognize that different signals (secondary and back-scattered electrons, and characteristic X-rays) are produced at different depths (Fig. 10.2.6b).

10.3 Image Contrast in a Scanning Electron Microscope

The two most important signals for image contrast in an SEM are secondary and back-scattered electrons. By convention, electrons with energies $E_{SE} < 50$ eV are called secondary electrons, and electrons with energies $E_{BSE} > 50$ eV are called back-scattered electrons. Further, it is common practice to define their respective yields as:

$$\eta = i_{BSE}/i_p \tag{10.3.1}$$

and:

$$\delta = i_{SE}/i_p \tag{10.3.2}$$

where i_{BSE}, i_{SE}, and i_p, are the back-scattered, secondary, and incident probe currents, respectively.

We can also define the total electron yield, σ, per unit incident probe current as:

$$\sigma = \eta + \delta \tag{10.3.3}$$

and typical plots of total outgoing and incident currents, as a function of incident electron energy, are shown in Figure 10.3.1. At sufficiently high incident beam energies, if $\sigma > 1$, then more electrons leave the specimen than those incident, and if the specimen were an insulator it will become positively charged. On the other hand, if $\sigma < 1$, it will become negatively charged and will affect the stability of the beam for subsequent imaging. By selecting an optimal acceleration voltage, E_1 or E_2, which results in zero net current flow, the charging of the specimen by the incident beam can be eliminated (Fig. 10.3.1).

Example 10.3.1: An SEM has a current density of 3.0×10^5 A/m, an ideal probe diameter of 2 nm, and a secondary electron current of 5×10^{-13} A. Assuming there is no charging of the specimen, what would be the back-scattered electron current?

Solution: The total probe current is related to the current density as $j_p = i_p / (\pi d_p^2 / 4)$. Thus,

$$i_p = j_p \left(\pi d_p{}^2 / 4 \right) = 3.0 \times 10^5 \times 3.14 \times \left(2 \times 10^{-9} \right)^2 / 4 = 9.42 \times 10^{-13} \text{ A}.$$

Since there is no charging, applying (10.3.3), we have

$$\sigma = \eta + \delta = \frac{i_{BSE}}{i_p} + \frac{i_{SE}}{i_p} = 1 \text{ or}$$

$$i_{BSE} = i_p - i_{SE} = (9.42 - 5) \ 10^{-13} \text{ A} = 0.442 \text{ pA}.$$

10.3.1   Factors Influencing Secondary Electron Emission

A typical spectrum of secondary electron energies (Fig. 2.6.6) shows a maximum at a peak energy, E_P, and a distribution with a full width at half maximum, ΔE_{FWHM} (Fig. 10.3.2). Both E_P and ΔE_{FWHM} depend upon the material, and typically, ΔE_{FWHM} is smaller for insulators when compared to metals.

Figure 10.3.3 shows the dependence of the secondary electron yield, δ, on the incident electron energy. As beam energy is increased, more secondary electrons are created; however, the secondary electrons are created at increasing depths with increasing beam energy and gradually the number of secondary electrons observed decreases. This is because the secondary electrons can only leave the

Figure 10.3.1 The two points of zero current flow, E_1 or E_2, where the incident current flow is equal to the outgoing total (combined secondary and back-scattered electrons) electron current, are recommended operating points that prevent electrostatic charging of the specimen, thereby improving the quality of the image. See Figure 10.3.3 and Figure 10.3.4.

Adapted from Brandon and Kaplan (2003).

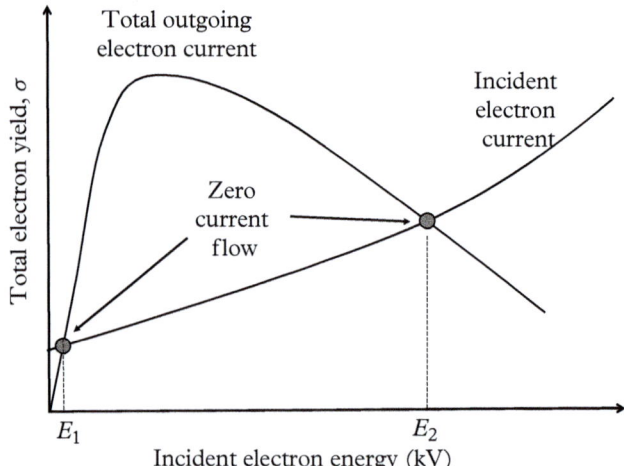

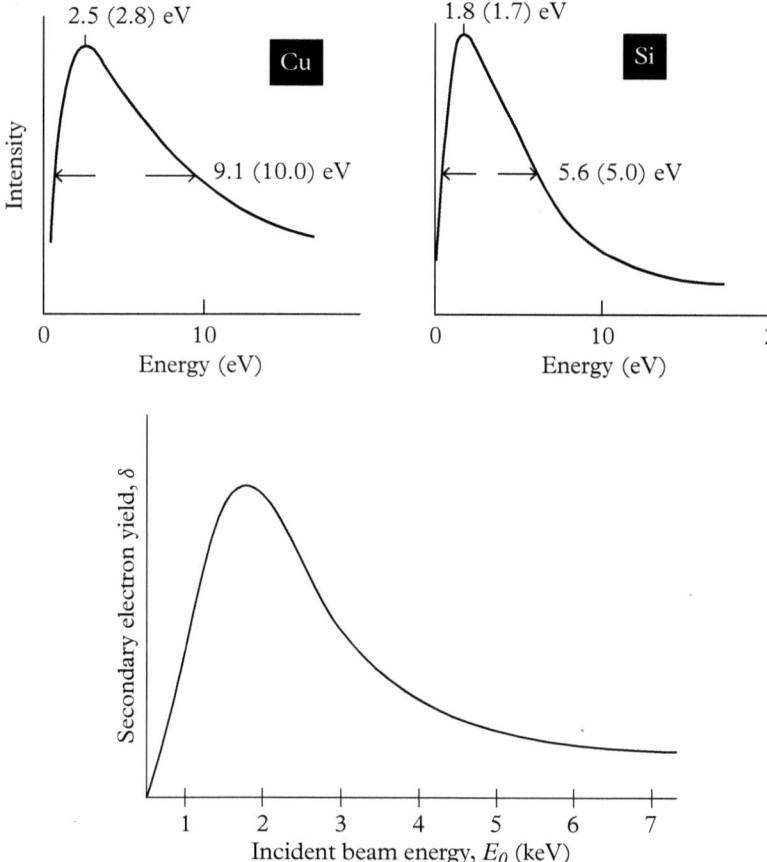

Figure 10.3.2 Theoretical predictions of the intensity distribution of secondary electrons as a function of energy for two different elements, Cu and Si. The peak position, E_P, in their energy distribution is independent of the acceleration voltage. Experimental values observed for peak position, E_P, and widths, ΔE_{FWHM}, are in parenthesis.

Adapted from [1].

Figure 10.3.3 Effect of incident beam energy, E_0, on secondary electron yield, δ.

specimen from a small escape depth of the order of 0.5–1.5 nm for metals and 10–20 nm for insulators (see Fig. 1.3.6). Thus, the yield, δ, rises sharply with beam energy but then decreases (Fig. 10.3.3).

The secondary electron yield, δ, and the back-scattered electron yield, η, increase with the average atomic number of the specimen (Fig. 10.3.4) for incident beam energy of 25–30 keV. As the average atomic number of the specimen is increased, more back-scattered electrons are generated, and they in turn, generate more secondary electrons. In general, for high-Z elements, more secondary electrons are created. The Z-dependence of secondary electron yield is more pronounced at lower beam energies.

The yield of secondary electrons also depends on the work function, Φ, of the specimen. In practice, Φ is a function of the composition, surface atomic structure, and surface adsorption/contamination. However, the effect of work function may be ignored if the specimen has a layer of surface hydrocarbon

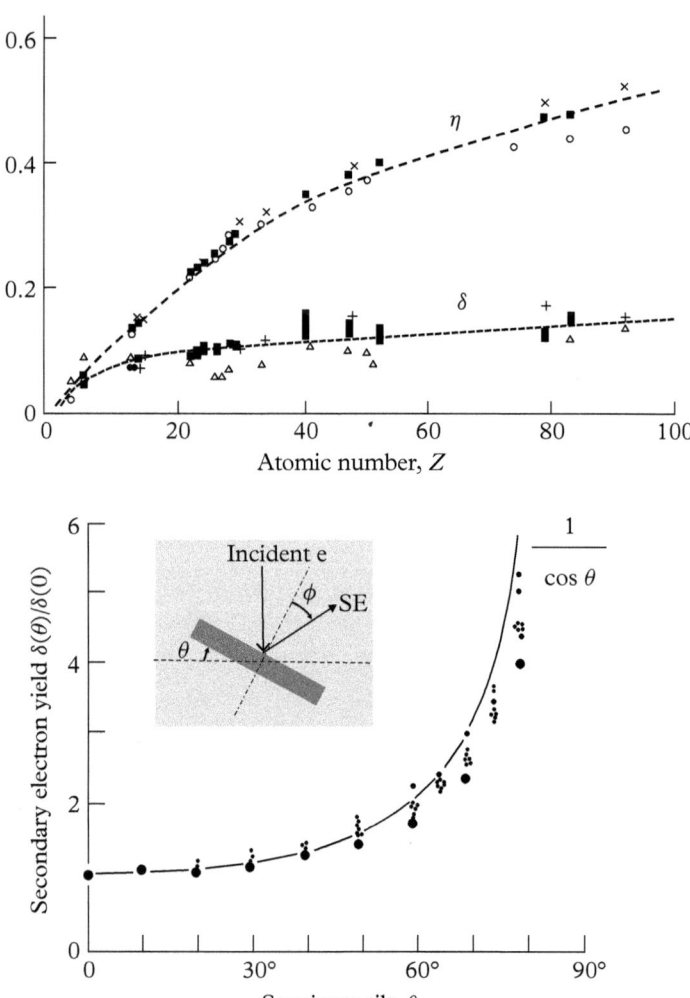

Figure 10.3.4 The variation of the secondary, δ, and back-scattered, η, electron yields as a function of average atomic number, Z, of the specimen for beam energies in the range 25–30 keV.

Adapted from Reimer (1985).

Figure 10.3.5 Secondary electron yield data as a function of specimen tilt, θ, for a medium atomic number element (Cu). Notice the good agreement with $1/\cos \theta$. The experimental geometry is shown in the inset.

Adapted from Reimer (1985).

contamination, which is often the case.[1] The secondary electron yield also increases with increasing specimen tilt angle θ (Fig. 10.3.5, inset), and this is an important factor in topographic contrast. To first order, for medium atomic number elements, the effect of specimen tilt can be approximated by a sec θ dependence. Further, the angular distribution of secondary electron emission is dependent on $\cos \phi$, where ϕ is the direction of emission with respect to the surface normal of the specimen. Thus, we can write the dependence of the secondary electron yield, $d\delta/d\Omega$, on the angles θ and ϕ as:

[1] Contamination can arise from both the sample and the vacuum system of the SEM. See [16] for further details.

$$\frac{d\delta\,(\theta, Z)}{d\Omega} \propto \delta\,(0, Z)\ \sec \theta\ \cos \phi \qquad (10.3.4)$$

For more advanced readers, further details can be found in Reimer (1985).

Example 10.3.2: What would be the SEM back-scattering yield for the high-T_c superconductor, $YBa_2Cu_3O_7$, assuming that $\eta = \sum_i \eta_i c_i$, where c_i are the mass fractions?

Solution: The mass fractions are: Y(0.133), Ba(0.412), Cu(0.286), and O(0.168). From Figure 10.3.4, the back-scattering yields for the individual elements based on their atomic numbers can be interpolated: Y(0.32), Ba(0.39), Cu(0.28), and O(0.10). Then the back-scattering yield for $YBa_2Cu_3O_7$ is 0.30.

10.3.2 Topographical Contrast in Secondary Electron Imaging

There are a number of mechanisms that provide topographical contrast in secondary electron imaging.

10.3.2.1 Surface-Tilt Contrast

This is the dependence of the secondary electron yield, δ, on the tilt angle, θ, (Fig. 10.3.5) of the surface element of the specimen. This is similar to reflected light effects due to orientation in a light microscope. Further, this is also subject to the trajectory effect where electrons from surfaces oriented towards the detector are collected preferentially (Fig. 10.3.6a), but this can be overcome partially by applying a bias voltage to collect secondary electrons with high efficiency.

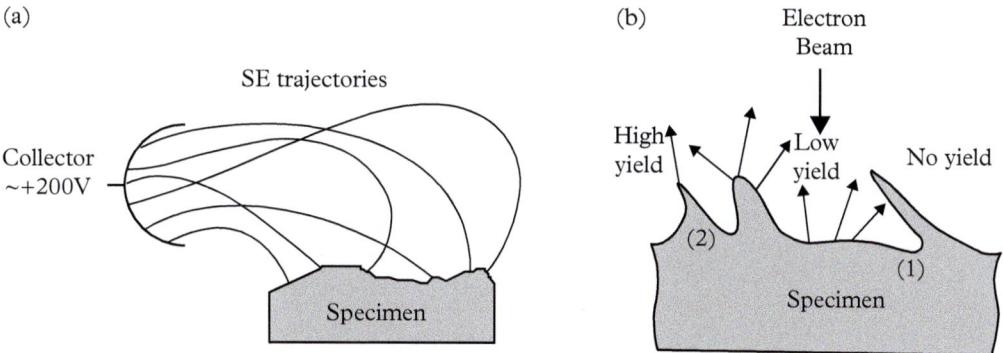

Figure 10.3.6 (a) A small bias field can increase collection of SE. (b) SE emission depends on surface irregularities. Note that there is no yield from region (1) due to shadowing effects, and a high yield from region (2) due to the number effect.

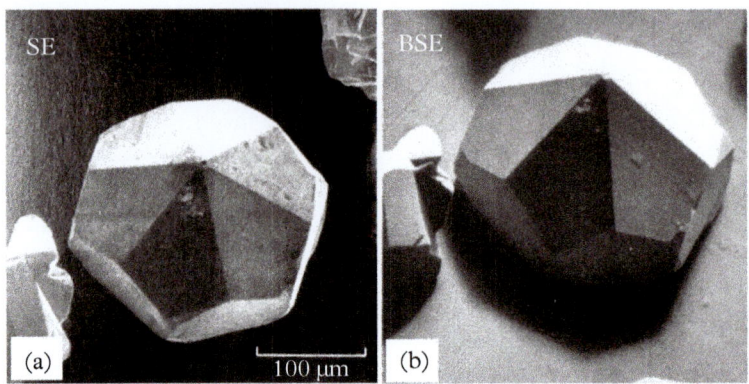

Figure 10.3.7 Surface tilt and shadow contrast observed in (a) secondary electron and (b) back-scattered electron images in an SEM from a single crystal. Notice the enhanced shadow in the BSE image.

Adapted from Reimer (1985).

10.3.2.2   Shadow Contrast

This arises from the suppression of the secondary electron emission from points behind protuberances (Fig. 10.3.6b). Figure 10.3.7 illustrates the effect of surface tilt and shadow contrast for both secondary and back-scattered electrons.

10.3.2.3   The Effect of the Average Atomic Number

This is the effect of the nature of the material, on the secondary electron yield, δ, including its increase with increased back-scattering (Fig. 10.3.4) as a function of atomic number, Z.

10.3.2.4   Number Effect

This is a manifestation of the diffusion length of the secondary electrons in the materials. As a result, more electrons escape from sharp edges of the specimen (Fig. 10.3.6b).

10.3.2.5   Charging Effect

This is discussed in detail in §10.4.1.

10.3.2.6   External Magnetic Fields

These occur from ferromagnetic specimens that can produce Type I magnetic contrast sensitive to the magnetic microstructure or the arrangement of domains. This is discussed further in §10.5 and in greater detail in Krishnan (2016).

SEM is routinely used in the analysis of the failure of engineering materials. Figure 10.3.8 illustrates the failure modes in three distinct classes of engineering materials: metals, ceramics, and composites, which are observed using secondary electrons.

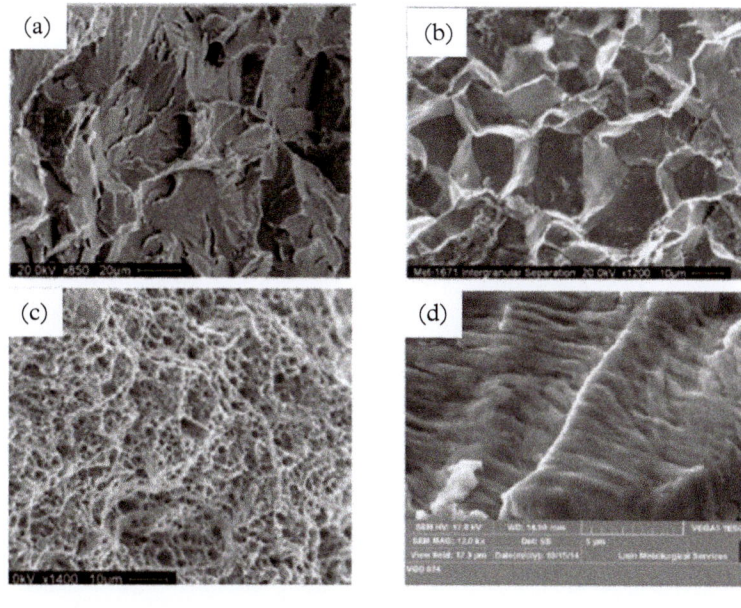

Figure 10.3.8 The greater depth of field of an SEM allows high resolution imaging of rough surfaces and can be useful in analyzing the failure of engineering materials. (a) Transgranular cleavage fracture caused by impact overload of a low-ductility material. (b) Intergranular fracture with rock-like appearance highlighting microstructural precipitates that weakened grain boundaries. (c) Ductile fracture showing dimpled cup and cone surface arising from a tensile overload of a ductile material. (d) Beach marks indicative of failure due to fatigue induced cyclic stress.

Adapted from [15].

Example 10.3.3: What is the observed contrast in secondary electron imaging between two areas with a difference, $\Delta\theta$, in the local tilt angle of the surface?

Solution: From (10.3.4), and Figure 10.3.5, the secondary electron yield, δ, can be written as $\delta = \delta_0 \sec\theta$, where δ_0 is the value at zero tilt. Then, the difference in secondary yield with angle can be found by differentiating this, i.e. $\Delta\delta = \delta_0 \sec\theta \tan\theta \Delta\theta$.
To first order, the observed contrast, (10.2.7), $C \sim \Delta\delta/\delta = \tan\theta \Delta\theta$.

10.3.3 Angular Dependence of Back-Scattered Electrons and Topographic Information

For normal incidence of the electron beam of intensity, i_p, a polar intensity plot (Fig. 10.3.9a) gives the back-scattered electron intensity distribution. Its angular dependence is:

$$i_{BSE} = Ai_p \cos\psi \qquad (10.3.5)$$

where ψ is the angle of detection, defined with respect to the incident beam direction, as shown. If the specimen is tilted at an angle, θ (Fig. 10.3.9b), the back-scattered electron intensity is forward peaked and is described by:

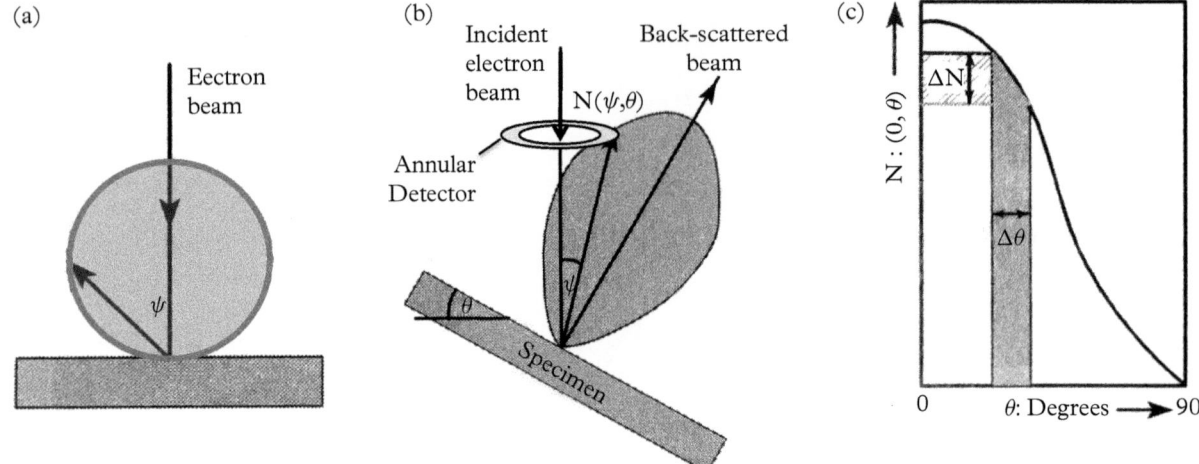

Figure 10.3.9 The BSE yield (a) depends on the angle, ψ, of detection with respect to the incident beam shown here as a polar intensity plot for normal incidence, and (b) is forward peaked for an incident angle, θ, defined by the specimen tilt. (c) For a fixed annular detector, the intensity, $N\,(\psi = 0, \theta)$, decreases with increasing specimen tilt, θ. Further, by detecting a narrow window, ΔN, regions of specific tilt, $(\theta, \Delta\theta)$, in the specimen can be imaged.

Adapted from Flewitt and Wild (2003).

$$i_{BSE} = \frac{AZ^{1/2}i_p}{9\,(1 + \cos\theta)} \tag{10.3.6}$$

where A is a constant, i_p is the incident beam current, and Z is the specimen atomic number.

The back-scatter electron yield, η, increases with the specimen tilt, θ, with a maximum obtained when $\psi = \theta$, i.e. the specimen normal is oriented towards the detector. However, unlike normal incidence, now the angular variation is more rapid around the peak value. Further, for an annular detector placed as shown in Figure 10.3.9b around the incident beam, the number of BSE detected, $N\,(\psi = 0, \theta)$, will decrease as θ increases (Fig. 10.3.9c), which also provides additional *topographical* contrast. Specifically, imaging over a small voltage window using a voltage discriminating amplifier will detect a narrow range, ΔN, of the back-scattered electron yield and this corresponds to a narrow range of tilt, $\Delta\theta$ (Fig. 10.3.9c). This approach can be used to produce grey-scale images, resolving regions of a specific tilt in the specimen. If multiple detectors are used (Fig. 10.3.10), the relative surface orientation variation in the specimen can be unambiguously determined by interpreting the contrast variation in terms of local changes in orientation. Figure 10.3.11 shows a typical example of using back-scattered electron yield to image local change in the orientation of the grains in a polycrystalline specimen.

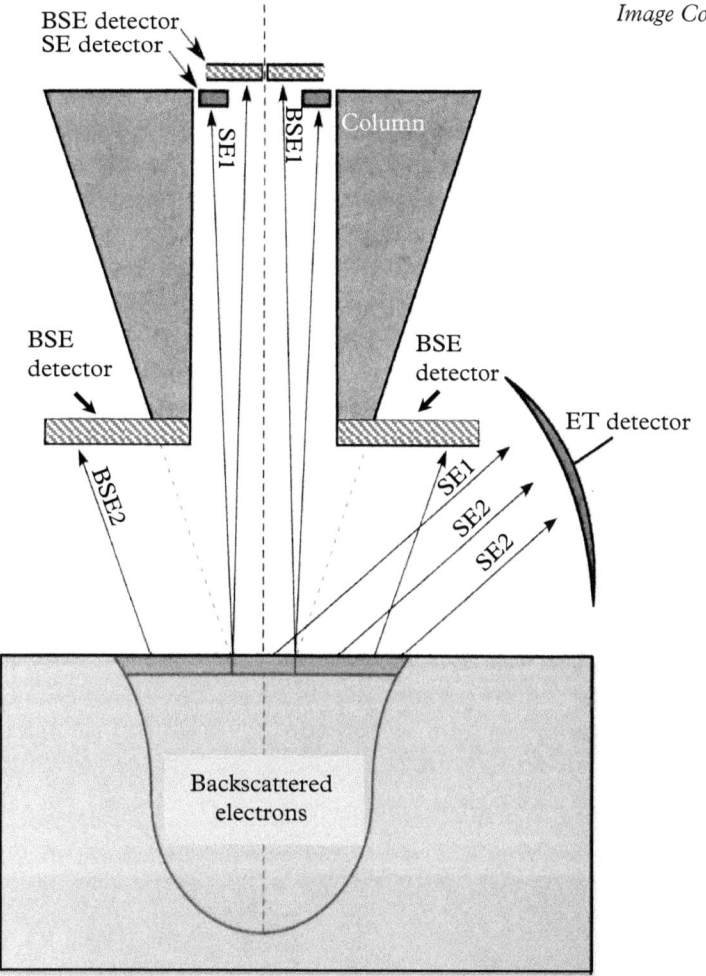

Figure 10.3.10 The placement of SE and BSE detectors in an SEM. SE1 and BSE1, placed in the optical column to detect electrons with very few inelastic scatterings, and produce the highest resolution images. The ET detector collects all SE2, as does the second BSE2 detector, to produce lower resolution images.

Adapted from Brandon and Kaplan (2003).

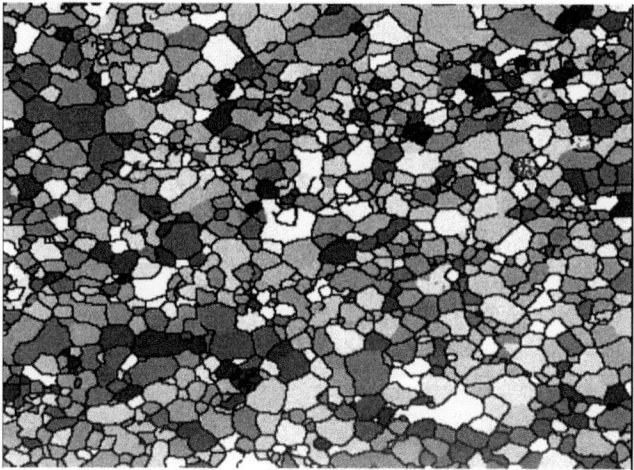

Figure 10.3.11 Back-scattered electron image of a polycrystalline specimen. The grey scale indicates variations in the orientation of the grains.

Adapted from Williams, Pelton, and Gronsky (1991).

10.3.4 Comparison of SEM Images with Different Operating Parameters

One of the principal advantages of imaging in an SEM is the ability to rapidly take images at different magnifications, while maintaining focus throughout the series, using a constant objective lens excitation and working distance (Fig. 10.3.12). This is because the magnification in an SEM image is the ratio of the width in the display (fixed at 10 cm, say) divided by the width of the area scanned by the electron beam, which can be easily varied by altering the scan coil currents.

It is also instructive to compare SEM images from the same area of a specimen under different imaging conditions to elucidate the effect of important parameters, i.e. acceleration voltage, electron beam size, and working distance of the specimen, on the contrast. First, we look at the effect of acceleration voltage (Fig. 10.3.13). Comparing secondary electron images at 5 kV and 25

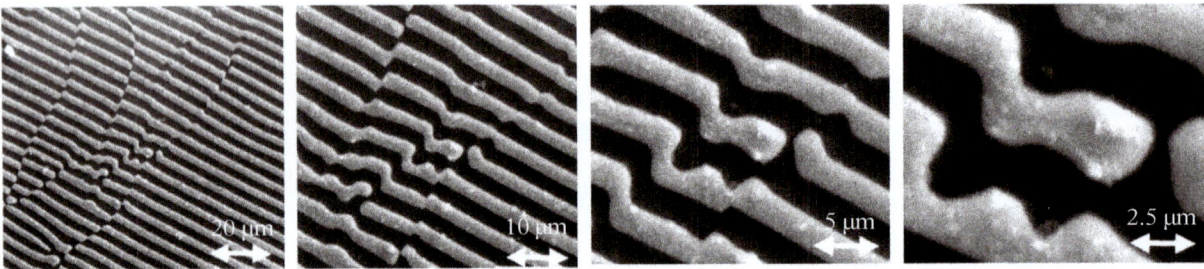

Figure 10.3.12 A series of images of a eutectic Cu–Al alloy taken at different magnifications by scanning smaller and smaller areas of the specimen using the electron beam.

Adapted from Goldstein et al. (2003).

Figure 10.3.13 Comparison of secondary electron images (a and b) shows improved resolution with increasing acceleration voltage, and better surface topographical features (c and d) with decreasing acceleration voltage.

Courtesy B. Cheney (SJSU).

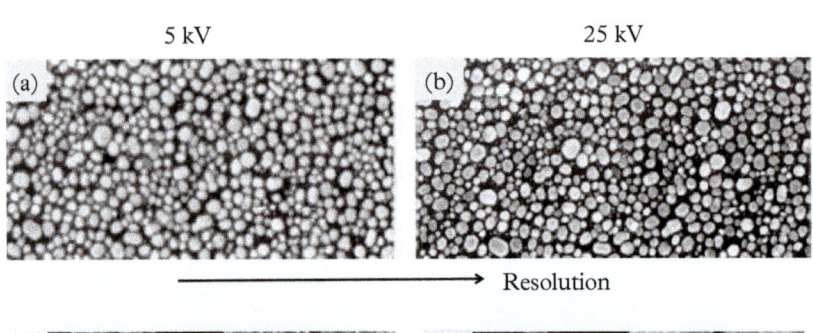
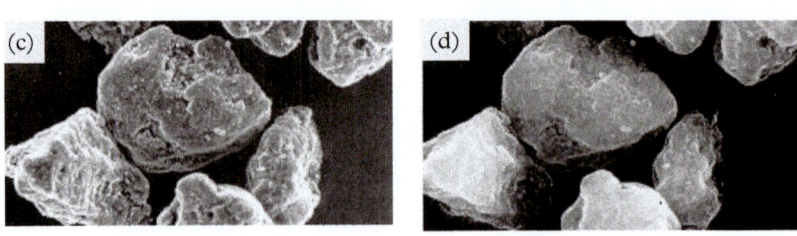

Depth of Field →

Working Distance

Spot Size

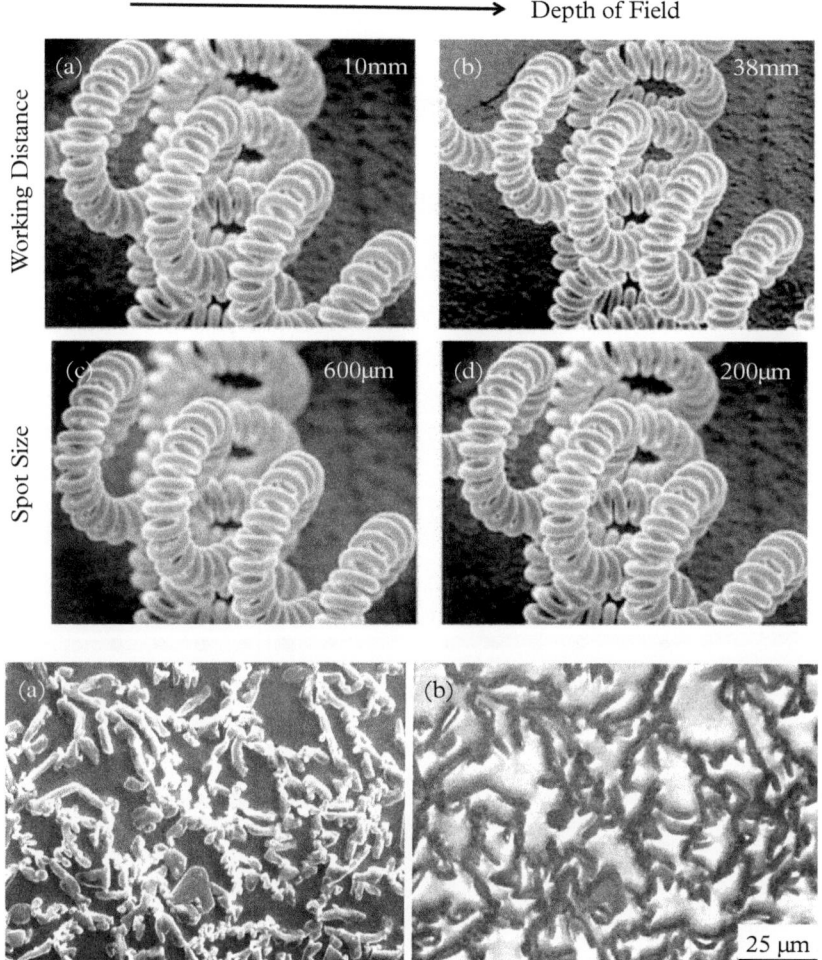

Figure 10.3.14 The depth of field in secondary electron imaging improves (compare a and b) with increased working distance, and smaller probe size (compare c and d).

Courtesy B. Cheney (SJSU).

Figure 10.3.15 Secondary (a) and back-scattered (b) electron micrographs, taken in an SEM at 25 kV, of alumina particles dispersed on a nickel substrate.

Adapted from Brandon and Kaplan (2003).

kV, we can conclude that the image resolution is enhanced at higher voltages (b vs. a), but surface topographical features are better resolved at lower voltages (c vs. d). Similarly, one can obtain a better depth of field in the secondary electron images (Fig. 10.3.14) when a smaller probe diameter (c and d) or a larger working distance (a and b) are used.

Finally, we compare the secondary and back-scattered images from the same area of a specimen. We have already seen an example of the images of a faceted single crystal (Fig. 10.3.7), which clearly shows an enhanced shadow contrast in the back-scattered electron image. In Figure 10.3.15, it is clear that the secondary electron images (a) provide better spatial resolution; however, Figure 10.3.16b shows elemental sensitivity with contrast dependent on atomic number—a principal advantage—of the back-scattered electron images.

Figure 10.3.16 Comparison of (a) secondary and (b) back-scattered scanning electron micrographs, at 25 kV, of a nickel alloy. Notice the compositional sensitivity in the BSE image.

Adapted from Leng (2013).

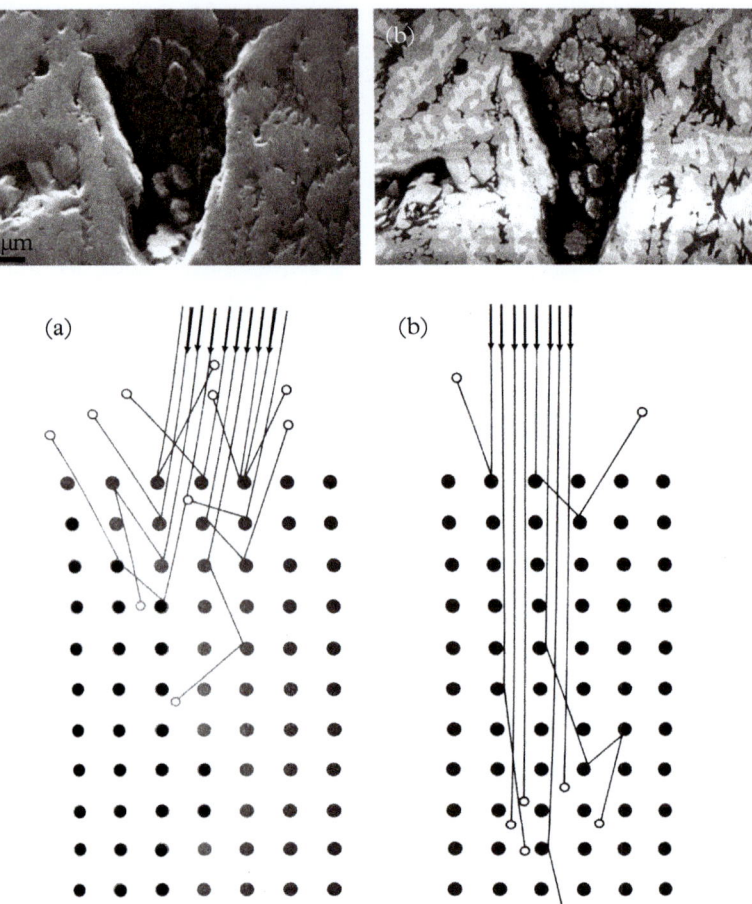

Figure 10.4.1 Electron channeling (b), with preferential penetration along certain crystallographic directions, leads to additional contrast, representative of the crystal symmetry, in the back-scattered electron signal in a crystalline material. Compare with (a) the random orientation.

Adapted from Williams, Pelton, and Gronsky (1991).

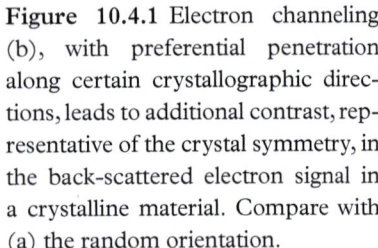

10.4 Channeling and Electron Back-Scattered Diffraction Patterns (EBSD)

In addition to the atomic number, or Z-dependence, observed even in amorphous materials, the back-scattered electron yield for a crystalline material also depends on the relative orientation of the incident beam with respect to the lattice. At any arbitrary angle of incidence (Fig. 10.4.1), the incident beam has a high probability of encountering an atomic nucleus and being scattered back from the lattice. However, if the incident angle is such that the beam travels along one of the high symmetry directions of the crystal (also, see Fig. 5.4.6, for a similar discussion of ions), then the incident electron can "channel" or penetrate deep into the lattice before encountering an atomic site (nucleus) to be back-scattered; in other words, it has a lower back-scattering probability.

To first order, the angles, θ, where such channeling occurs are close to those that satisfy the simple Bragg law of diffraction, (4.3.1). For an SEM operating at 20 kV, the wavelength of the electron is $\lambda \sim 0.01$ nm. In a typical crystal with $d_{hkl} \sim 0.3$ nm, the Bragg angle $\theta_{hkl} \sim 1\text{-}2°$. Thus, if the angle of incidence is varied by $1\text{-}2°$, the BSE yield will vary. Moreover, in an SEM, we have to consider the two-dimensional variation in crystallographic or beam orientation. In other words, we can use the "channeling" effect to map the crystallographic symmetry of the lattice *about the beam direction*. Such patterns are called electron back-scattered diffraction patterns (EBSD). Finally, to record EBSD patterns, the beam direction has to be varied and this is achieved by observing the specimen at low magnifications, since under such scans the incidence angle of the beam varies only by a few degrees over the size of the specimen. Figure 10.4.2 and Figure 10.4.3 show two examples of EBSD patterns.

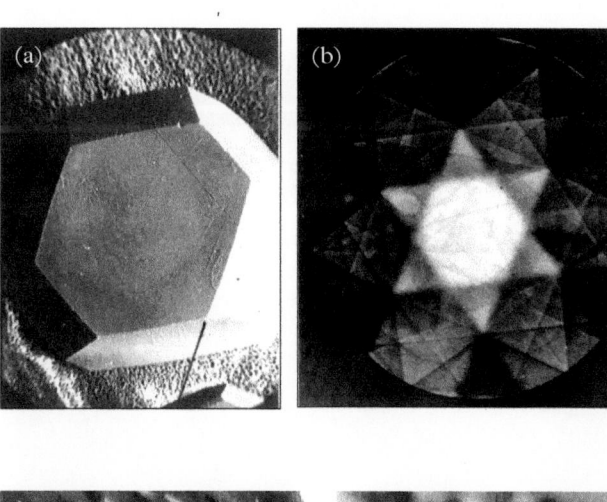

Figure 10.4.2 (a) A secondary electron image of an artificial diamond crystal, showing well-developed facets $\sim$300 μm in size. (b) An EBSD patterns showing the symmetry of the facet to be (111).

Adapted from Williams, Pelton, and Gronsky (1991).

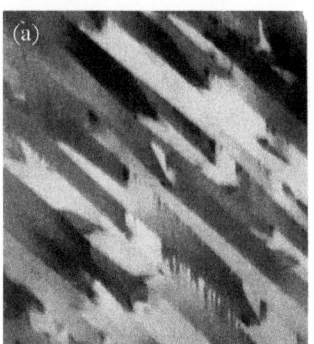 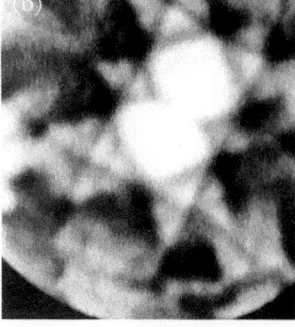

Figure 10.4.3 A silicon film regrown on an amorphous SiO_2 layer. (a) Back-scattered electron image showing the grain structure due to channeling contrast. (b) A selected area EBSD pattern shows that the regrown film has an average {110} orientation, but the individual grains are randomly misaligned.

Adapted from Williams, Pelton, and Gronsky (1991).

10.5   Imaging Magnetic Domains

10.5.1   Type I and Type II Magnetic Contrast

Figure 10.5.1a shows the geometric arrangement for observing magnetic contrast and image domains in an SEM. The secondary electron yield, δ, depends strongly on the angle of incidence, θ_i, (Fig. 10.3.5), as well as the incident beam energy, E_0, (Fig. 10.3.3), but shows only a secondary dependence on the atomic number, Z, of the material (Fig. 10.3.4). On the other hand, the yield of back-scattered electrons, η, is dependent on the incident beam energy, E_0, which also controls the depth of penetration, increases rapidly with Z, and slightly with θ_i. The collection efficiency of the detector, Ω, depends on its relative position, ψ, with respect to the point of emission from the specimen, and the size of its collection aperture, ξ. Thus, in general, the SEM signal, S, defined as the number of electrons at each position of the incident beam, collected per unit time, is given by the product

$$S = N\left(E_0, \theta_i, Z\right)\ \Omega\left(\psi, \xi\right) \tag{10.5.1}$$

where N is equal to δ for secondary, and equal to η for back-scattered electrons.

Magnetic contrast in an SEM arises in two ways: one that depends on variation in Ω (Type I) and the other that depends on variations in η (Type II). In Type I contrast, the stray field distribution, H_{stray}, outside the ferromagnetic specimen causes a local change in ψ and thus a change in collection efficiency, Ω. From

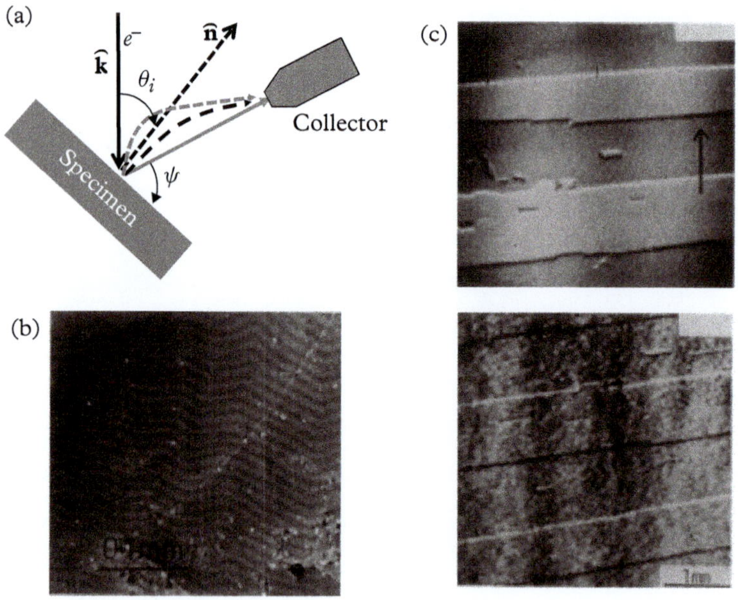

(a)

(b)

(c)

Figure 10.5.1 (a) Specimen-collector geometry in an SEM for magnetic imaging. (b) Stripe domains in an epitaxial garnet film observed with Type I contrast. (c) Domain wall contrast in Type II images in a 3% Si-Fe, $E_0 = 200$ kV back-scattered image with (top) $\theta_i = 45°$ and $\psi = 60°$ and (bottom) $\theta_i = 0°$ & $\psi = 45°$. The arrow indicates the direction of the specimen tilt and the detector is on the right side of each micrograph.

Adapted from [2].

considerations of the Lorentz force on the secondary electrons, maximum contrast is to be expected when domain walls are aligned along the symmetry axis of the detector. Further, to produce the maximum number of secondary electrons for Type I contrast, the primary beam energy has to be optimized to be less than 10 keV. Type II contrast is based on the variations of the back-scattered electron yield, η, across the specimen. Local variations in the induction, $\mathbf{B}$, of the specimen effectively result in a change in θ and thus a change in the back-scattered yield, η. The contrast, C_{mag}, between adjacent domains can be written [54] as

$$C_{mag} = \frac{\mathcal{F}(\theta, E_0)}{\overline{S}} \left\{ \left(\hat{\mathbf{k}} \otimes \hat{\mathbf{n}} \right) \cdot \Delta \mathbf{B} \right\} \tag{10.5.2}$$

where, $\mathcal{F}$ is an appropriate function describing the dependence on incident angle and incident energy, $\overline{S}$ is the average signal, and $\Delta \mathbf{B}$, represents the difference in flux density between the two domains. In general, for maximum back-scattered signal the highest available beam energy should be used. Note that the specimen has to be tilted to ensure that $\hat{\mathbf{k}} \otimes \hat{\mathbf{n}} \neq 0$, otherwise no contrast is observed.

Further, if the specimen is rotated about $\hat{\mathbf{n}}$, the normal direction, the contrast is zero if $\left(\hat{\mathbf{k}} \otimes \hat{\mathbf{n}} \right) \cdot \Delta \mathbf{B} = 0$. Therefore, the domains exhibit zero contrast if there is no change in magnetization parallel to the specimen tilt axis. The contrast is proportional to $\Delta \mathbf{B}$, provided both θ and E_0 remain constant. In general, the contrast is low and varies from 0.1% (20 keV) to 1% (200 keV); therefore, large beam currents are required. Figure 10.5.1b and Figure 10.5.1c show examples of Type I and Type II contrast, applied to magnetic materials. Further details about these methods can be found in books (see Reimer, 1988; Newbury, 1986) or reviews [2,3] of magnetic contrast in an SEM.

The advantage of this method is its simplicity. A standard SEM can be used, as is, without any modifications, provided it has the required secondary and back-scattered detectors in the optimal geometry. Further, these techniques require no special specimen preparations. However, the spatial resolution of these methods is limited: typically Type I is about 1 μm and Type II ranges from 1 μm (20 keV) to 2 μm (200 keV). However, the Type II method provides a large information depth of ~10–15 μm at 200 keV. Hence, the contrast is less sensitive to the surface and more representative of the bulk magnetic structure.

10.5.2   Scanning Electron Microscopy with Polarization Analysis (SEMPA)

SEMPA [4], also known as Spin-SEM [5], is a technique for directly imaging the magnetic microstructure of surfaces using the spin polarization of secondary electrons (Fig. 10.5.2a). It reflects the spin density of the material that, in turn, is proportional to the magnetization of the sample. By measuring the secondary electron polarization, as a focused electron probe is rastered over the surface of interest, an image of the magnetic domain structure can be generated. Note that

(a)

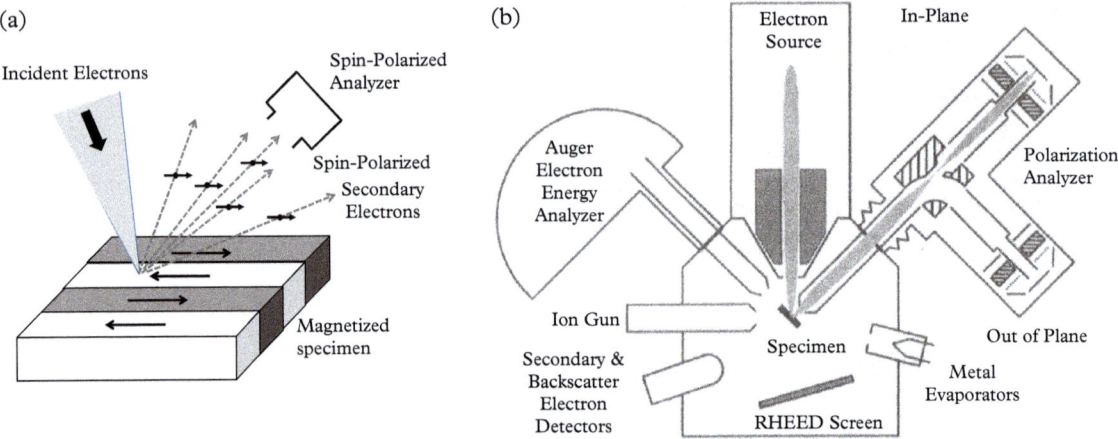

(b)

Figure 10.5.2 (a) The principles of SEMPA. As the incident electron beam is scanned over a ferromagnetic surface, it produces polarized secondary electrons, proportional to the specimen magnetization and direction, that are then detected and analyzed. (b) The microscope column showing the experimental arrangements, including the surface preparation and analysis facilities, and the set of two polarization analyzers. The Mott polarization analyzer is based on the asymmetric scattering of the spin-polarized electrons by a gold foil, which is then detected by the quadrant detector.

Adapted from [4, 6].

secondary electrons, in general, have a short mean free path in the solid; typical sampling depths are 0.5–1 nm for a transition metal [6]. For this reason, good SEMPA images require atomically clean surfaces devoid of contamination. Hence, specimen cleaning (e.g. ion sputtering) and surface characterization (e.g. Auger spectroscopy, §2.6.2) facilities should be integrated into the SEMPA system (Fig. 10.5.2b).

For simple $3d$ transition metals, e.g. Fe, Co, and Ni, the degree of spin polarization of secondary electrons is 28%, 19%, and 5%, respectively [7,8]. Moreover, the polarization is independent of the incident beam conditions [9] and can be interpreted directly. However, even though the secondary electron polarization is proportional to the magnetization, the proportionality constant is difficult to determine; as a result, SEMPA often measures the direction of magnetization and the *relative* magnitude of the moment.

The essential SEMPA system consists of a generic electron-optical column, such as an SEM, to form and raster a narrow electron probe, appropriate vacuum and surface preparation facilities to ensure a clean specimen surface, spin-polarization detectors, and data acquisition hardware and software (Fig. 8.4.1b). The detectors include transport optics to collect and accelerate most of the secondary electrons to the spin analyzer. Spin polarization analysis is performed using the common Mott-type spin analyzer (Fig. 10.5.3). The electron

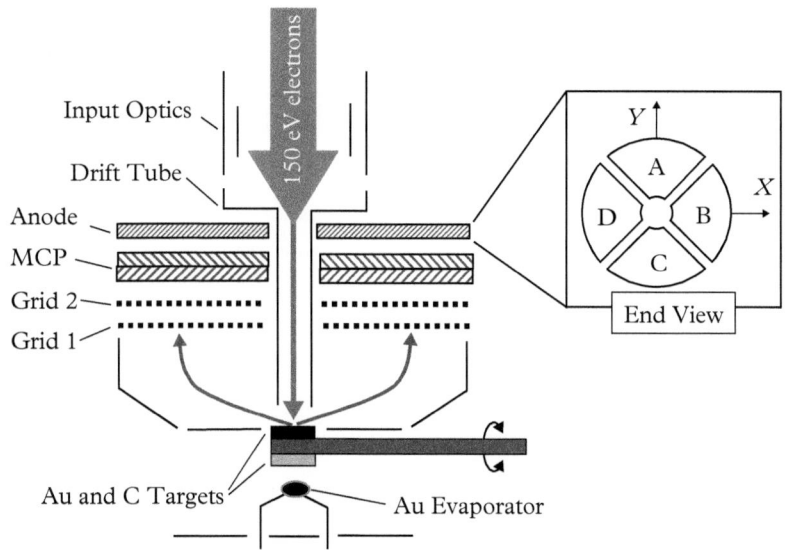

Figure 10.5.3 The Mott analyzer for measuring the spin polarization of the secondary electrons is based on the asymmetric scattering of the spin-polarized electrons by a gold foil which is then detected by the quadrant detector.

Adapted from [4].

polarization detected in this analyzer depends on the asymmetry of the spin-orbit scattering from an amorphous Au foil, which is then measured by a quadrant detector, similar to the one discussed in §9 for differential phase contrast (DPC) imaging in a TEM.

At any given time, two orthogonal components of the transverse spin polarization are measured. Hence, two analyzers are required to completely resolve all the components of the magnetization (Fig. 10.5.2b). However, since shape anisotropy [see Krishnan (2016), Ch. 6] often renders the magnetization to lie in the plane for most specimens analyzed in an SEMPA, the out of plane component of the magnetization, M_z, is seldom observed. Thus, from the observed components of the magnetization, say M_x and M_y, the magnitude $|\mathbf{M}|$, and local direction, θ, of the resultant magnetization can be simply derived:

$$|\mathbf{M}| = \sqrt{M^2{}_x + M_y^2} \qquad (10.5.3)$$

$$\theta = \tan^{-1}\left(\frac{M_y}{M_x}\right). \qquad (10.5.4)$$

Figure 10.5.4 shows a (100) surface of a Fe single crystal, with the measured components of the magnetization, M_x and M_y, the overall topography of the surface which is the total secondary electron intensity, I, and the magnitude, $|\mathbf{M}|$. Note that the magnitude of the magnetization is uniform, as expected, except at the domain walls. Also, 180° walls appear darker than 90° walls because of averaging of the magnetization from adjacent domains with a probe diameter comparable to the domain wall width.

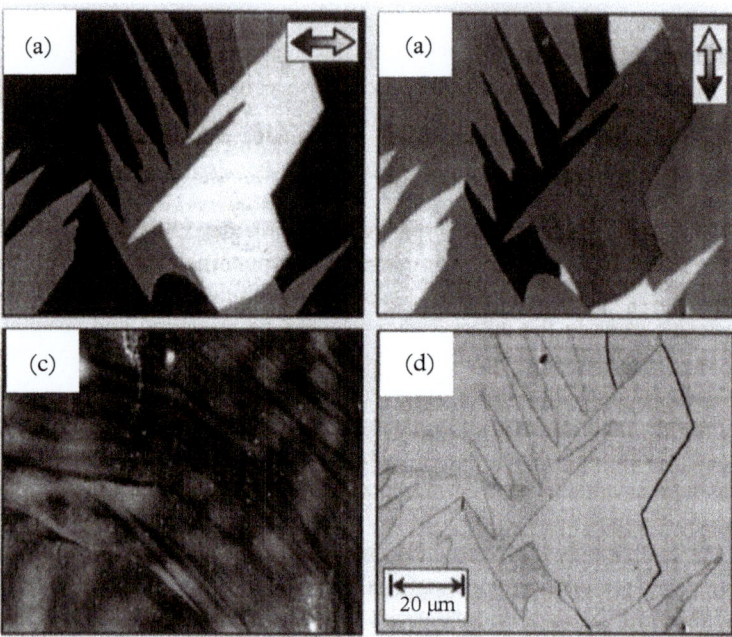

Figure 10.5.4 SEMPA images of a surface of a Fe(100) single crystals. (a) and (b) Images with polarization along two orthogonal directions. (c) The topographic or sum image. (d) Image of the magnitude, $|\mathbf{M}|$, of the magnetization, clearly delineating the 90° and 180° domain walls.

Adapted from [10].

SEMPA has been particularly important in the study of surface magnetization microstructures because of the very shallow escape depth (< 1–2 nm) of secondary electrons (Fig. 1.3.6) and its sensitivity to very small quantities of magnetic materials (as small as 1,000 atoms of Fe). It has also made seminal contributions in various aspects of surface magnetism, and particularly, in determining the nature and period of oscillatory magnetic coupling in metallic heterostructures [see Krishnan (2016), Ch. 10]. Further details of SEMPA and its applications can be found in the excellent review by Unguris [11].

10.6 Probing Sample Composition and Electronic Structure

10.6.1 Basics of X-Ray Microanalysis in an SEM

We have seen that a high-energy (5–30 keV) incident electron beam in an SEM can result in inner-shell ionization of most elements in the specimen. Further, the subsequent de-excitation processes can give rise to characteristic X-ray intensities from the elements present in the specimen as discussed in §2. The characteristic X-ray intensities reflect the composition of the specimen and can be interpreted as follows.

Consider a multicomponent specimen of elements, i, with atomic weights, A_i, atomic fractions, a_i, and mass fractions, m_i. Then, for any element, x, the atomic and mass fractions are inter-related as:

$$m_\mathrm{x} = \frac{a_\mathrm{x} A_\mathrm{x}}{\sum a_i A_i} \tag{10.6.1}$$

In quantitative X-ray microanalysis, we wish to determine the atomic, a_i, and/or mass, m_i, fractions of any multi-element specimen, based on the analysis of their characteristic X-ray intensities. To first order, we assume that the total number of X-ray counts, N_x, for element, x, is proportional to m_x. Then, for the same total electron flux, $i_\mathrm{p}\tau$, where τ is the counting time, if the total X-ray counts, N_x^s, from a number of standards each with known mass concentration, m_x^s, are measured for a binary alloy, a linear calibrations curve (Fig. 10.6.1) can be expected. Then, the number of X-ray counts, N_x^u, measured for our unknown specimen can be simply interpreted in terms of its concentration, m_x^u, as:

$$k_\mathrm{x} = \frac{N_\mathrm{x}^u}{N_\mathrm{x}^s} \simeq \frac{m_\mathrm{x}^u}{m_\mathrm{x}^s} \tag{10.6.2}$$

where k_x is independent of parameters such as the ionization cross-section, fluorescence yield, the detector/collection efficiency, etc., and $m_\mathrm{x}^s = 1$ for a pure single-element standard. In practice, the calibration curves, measured in binary alloy systems, deviate from linearity (Fig. 10.6.1).

This observed deviation from linearity requires three principal corrections of the data for analysis (§2.5.2.3).

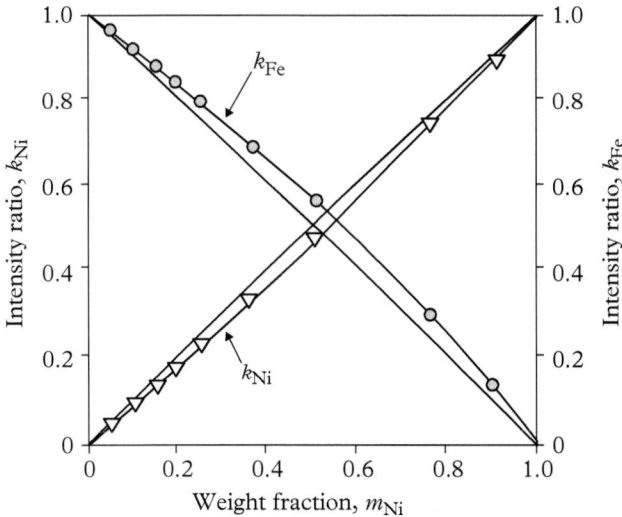

Figure 10.6.1 X-ray microanalysis calibration curves for a Fe-Ni binary alloy, showing the intensity ratios, k_Fe and k_Ni, as a function of the weight fraction, m_Ni, of Ni. The convex shape of k_Fe is the result of the fluorescence of Fe X-ray quanta by Ni X-rays, and the concave shape of k_Ni is due to the absorption of Ni X-rays by Fe atoms.

Adapted from Reimer (1985).

1. The atomic number (Z) correction, which takes into consideration the fraction of back-scattered electrons with exit energies larger than the ionization energies, E_x^I, for the specific shells of the elements, x, of interest. For a detailed analysis this requires a knowledge of the back-scattering energy spectrum, $d\eta/dE_B$, where E_B is the back-scattered electron energy. Further, this correction also includes the maximum path length of the incident electrons before their energy becomes smaller than E_x^I, as at this stage they are no longer able to produce any further X-rays.

2. The X-ray absorption (A) correction, which accounts for the attenuation of the X-rays, (2.5.4), defined by the mass attenuation coefficient, μ/ρ, of the specimen.

3. The fluorescence (F) correction, which accounts for the photoionization of atoms by any X-ray quanta of higher energy generated by other atoms in the multi-element specimen.

For the binary Fe-Ni calibrations standards shown in Figure 10.6.1, since the atomic numbers of Fe and Ni are close, the first correction can be ignored. Second, since Ni $K\alpha$ radiation fluoresces Fe atoms, we observe an enhancement in the Fe $K\alpha$ signal, and a convex shape in the k_{Fe} curve. Moreover, since the Ni $K\alpha$ X-ray quanta can all be potentially absorbed by the Fe atoms, a concave shape in the k_{Ni} curve is observed. As a good first approximation these concave/convex shapes can be fitted to a hyperbola:

$$\frac{m_x}{1 - m_x} = \alpha \frac{k_x}{1 - k_x} \tag{10.6.3}$$

where $\alpha < 1$ for the convex (Fe) shape, and $\alpha > 1$ for the concave (Ni) shape. For the data in Figure 10.6.1, fitting such hyperbolas gives $\alpha = 0.79$ for Fe, and $\alpha = 1.15$ for Ni. We can extend this analysis for multi-element specimens, and modify (10.6.2) and (10.6.3) to give:

$$k_x = \frac{\alpha_x m_x}{\sum \alpha_i m_i} \tag{10.6.4}$$

with α_i being calculated theoretically. Further details of this approach can be found in Reimer (1985).

In practice, a single standard, not necessarily a pure element, but with known concentrations, m_x^s, is prepared. Then, the unknown concentration, m_x^u, is given by correcting the factor, k_x, (10.6.2), as:

$$m_x^u = k_x m_x^s ZAF \tag{10.6.5}$$

where Z, A, and F, are multiplicative factors (§2.5.2.3) that correct for the effects of atomic number, absorption and fluorescence, respectively. As we can expect, these correction factors depend on the actual concentration in the specimen and hence they are applied iteratively. Different computer routines, with different approximations for the *ZAF* corrections are available, and the user should be aware of the details before using them. Again, see Reimer (1985) for further details.

The error analysis for X-ray microanalysis, both for energy dispersive and wavelength detection, follows Poisson statistics. Thus, if we collect N_x counts for a specific characteristic X-ray peak intensity in a specimen, the associated standard deviation, σ_x, for the measurements is $\sigma_x = \sqrt{N_x}$. Then, the relative standard error, ε_x, is given by $\varepsilon_x = \sigma_x / N_x = N_x^{-1/2}$. Thus, for a standard error of 1%, $N_x \geq 10^4$ counts are required in each characteristic X-ray peak; this is easily achievable in WDS (§2.5.1.1) but requires very long integration times for energy dispersive detection.

Using this simple statistical model, we can also estimate the analytical sensitivity, which is the smallest difference, $\Delta m = m_1 - m_2$, between two mass fractions, with total counts, N_1 and N_2, for the same element (note that we have dropped the subscript, x). If Δm is small, $N_1 \simeq N_2 = N$ and $\sigma_1 = \sigma_2 = \sqrt{N}$. Applying the law of error propagation, we can expect N_2 to be different from N_1 with 99.7% level of confidence if:

$$N_2 - N_1 > 3\sqrt{\left(\sigma_2^2 + \sigma_1^2\right)} \simeq 3\sqrt{2N} \simeq 4\sqrt{N} \qquad (10.6.6)$$

Hence, we can conclude that:

$$\frac{\Delta m}{m} = \frac{N_2 - N_1}{N} = \frac{4N^{1/2}}{N} = 4N^{-1/2} \qquad (10.6.7)$$

In practice, if we wish to have an analytical sensitivity $\Delta m/m \approx 1\%$, we will need $N = 1.6 \times 10^5$ counts in the X-ray peak of the element of interest. We can extend this discussion to also obtain an estimate of the minimum detectable limit, m_{min}, for any element, x, with counts N_x^P in the peak, that is slightly larger than the counts, N^B, in the adjacent continuous background. Then, $N_x^P - N^B << N^B$ and from (10.6.6), we get:

$$N_x^P - N^B > 3\sqrt{\left(\sigma_P^2 + \sigma_B^2\right)} \simeq 3\sqrt{N_x^P} = 3\sqrt{N^B} \qquad (10.6.8)$$

where σ_B has been neglected. Then, we can estimate the minimum detectable limit as:

$$m_{min}{}^x = \frac{N_x^P - N^B}{N^B} = \frac{3\sqrt{N^B}}{N^B} = \frac{3}{\left(i_p\tau\right)^{1/2}} \frac{n_B^{1/2}}{n_s} \qquad (10.6.9)$$

Here, n_B and n_s, are the counts obtained from the continuous background and a pure-element standard per unit flux, $i_p\tau$, i.e. $N = n i_p \tau$. In practice, n_s is determined by the trace element to be detected, and n_B is given by the matrix. Thus, the detection limit for heavy elements in a light element matrix can be quite low (few parts per million), especially if wavelength dispersive X-ray detection is used.

Example 10.6.1: X-ray microanalysis data for an Al-Cu alloy, using pure elemental standards for Cu $K\alpha$ and Al $K\alpha$, is shown in the following table.

X-ray line	Z_i	A_i	F_i	k_i
Cu $K\alpha$	1.21	1.0	1.0	0.0248
Al $K\alpha$	0.995	1.039	1.0	0.938

(a) Comment on the values of the Z A F correction data. Hint, use Table 2.4.1.

(b) What is the composition (wt%) of the alloy?

Solution:

(a) We comment on each of the correction factors individually.

   (i) $Z_{Al} = 0.995$, is very small. This suggests a small difference between the average atomic number of the standard (pure element) and the alloy specimen. Hence, we can infer that the alloy is Al-rich with very little Cu and $Z_{Cu} = 1.21$, is significant. It is consistent with the small number of backscattered electrons in the Al-matrix.

   (ii) $A_{Cu} = 1$. This is because the mass absorption coefficient, Table 2.4.1 for Cu $K\alpha$, i.e. $\left(\frac{\mu}{\rho}\right)_{Cu}^{Cu\,K\alpha} = 51.54\ \text{cm}^2/\text{g}$ and $\left(\frac{\mu}{\rho}\right)_{Al}^{Cu\,K\alpha} = 50.23\ \text{cm}^2/\text{g}$ are small. Similarly, $\left(\frac{\mu}{\rho}\right)_{Al}^{Al\,K\alpha} = 385.7\ \text{cm}^2/\text{g}$, is small, and even though $\left(\frac{\mu}{\rho}\right)_{Cu}^{Al\,K\alpha} = 5377\ \text{cm}^2/\text{g}$ is substantially larger, $A_{Al} = 1.039$, because there is very little Cu in the alloy.

   (iii) Cu $K\alpha$ radiation has more than enough energy to excite Al $K\alpha$ X-ray photons, but because the fluorescence is very inefficient, $F_{Al} = F_{Cu} = 1$.

(b) We apply (10.6.5) but since elemental standards were used, $m^s_{Al} = m^s_{Cu} = 1$.

   Then $m^u_{Al} = Z_{Al}A_{Al}F_{Al}k_{Al} = 0.995 \times 1.039 \times 1.0 \times 0.938 = 0.97$ or 97 wt% Al and $m^u_{Cu} = Z_{Cu}A_{Cu}F_{Cu}k_{Cu} = 1.21 \times 1.0 \times 1.0 \times 0.0248 = 0.03$ or 3 wt% Cu.

   In summary, this is an Al-3wt%Cu alloy (adapted from Goldstein et al., 2003).

10.6.2 Cathodoluminescence

Figure 10.6.2 shows a dedicated FEG-SEM for CL, introduced earlier in Figure 10.2.7. It combines a special micro-positioning stage and a parabolic mirror that focuses the weak CL signal. Figure 10.6.2e–h shows examples of the application of this instrument to the measurement of CL in nanoscale structure, which are discussed in the accompanying caption.

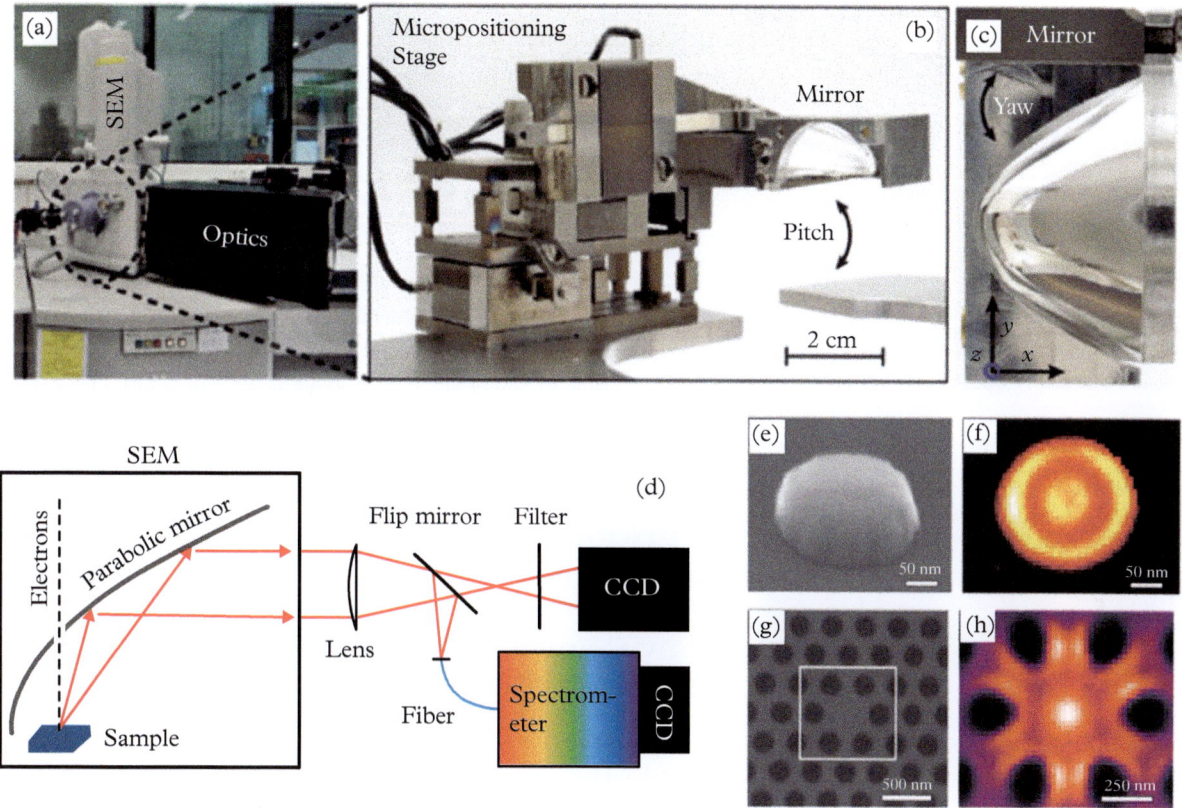

Figure 10.6.2 Details of (a) the dedicated field emission gun scanning electron microscope for cathodoluminescence studies. It includes (b) a micropositioning stage, (c) a focusing mirror, and (d) collection optics for two-dimensional signal detection. (e) SEM image of a silicon nanodisc (diameter 200 nm, height 100 nm) on a 300 nm-thick SiO_2 layer. (f) Two-dimensional cathodoluminescence excitation map of a 320 nm-diameter silicon disk at peak wavelength $\lambda_0 = 475$ nm. (g) SEM image of a photonic crystal cavity etched in a 200 nm-thick Si_3N_4 membrane. (h) Two-dimensional cathodoluminescence excitation map of the cavity, area shown by the white square in (g) for $\lambda_0 = 650$ nm.

Adapted from [12].

We have discussed the basic principles of CL in semiconductors in §10.2.3. In practice, such luminescence effects are observed in many materials and minerals and can be divided further into the following materials-related factors.

1. Intrinsic luminescence, which is characteristic of the host lattice. It can be due to non-stoichiometry (vacancies), structural imperfections (poor ordering in the crystal, etc.), and impurities (non-activators, see (3), below, that distort the lattice).

2. Extrinsic luminescence, which arises from impurities in the structure. The impurities, which are the most common source of CL, generate luminescent centers and are commonly transition, rare-earth, and actinide elements.

3. Activators are substitutional elements that, in trace quantities, promote CL, up to a certain point linearly with concentration in a material. However, with further increase in concentration, the CL will decrease (i.e. self-quench) and this will depend on crystallographic structure of the material. Some important activators include Mn^{2+}, Cr^{3+}, Fe^{3+}, Ti^{4+}, and rare earth elements.

4. Sensitizers or co-activators, which are ions (such as Pb^{2+}) that absorb energy, re-emit it to an activator (such as Mn^{2+}) and enhance the CL response of the activator.

5. Quenchers inhibit or eliminate CL in a material. The most important quencher is typically Fe^{2+}, but other $3d$ transition metal ions, such as Co^{2+}, Ni^{2+}, and Fe^{3+}, can also quench the CL.

10.7 Variations of Scanning Electron Microscopy

10.7.1 Environmental Scanning Electron Microscopy (ESEM)

There are many situations in materials characterization where specimens have to be examined under a gaseous environment; in addition, biological specimens are best observed in hydrated form. These and various other observations, such as in situ crystal growth and process of corrosion, can benefit from the use of an ESEM, which has a special differential pumping mechanism that separates the specimen (now in a gaseous environment) from the rest of the electron-optical column that is maintained under the required high vacuum. Such differential pumping is accomplished using two apertures, 100–200 μm in diameter, to create an intermediate space that is pumped independently to separate the specimen chamber from the microscope column (Fig. 10.7.1). This allows the observation of specimens in a hydrated state, even though the water/gas molecules will scatter the electron beam.

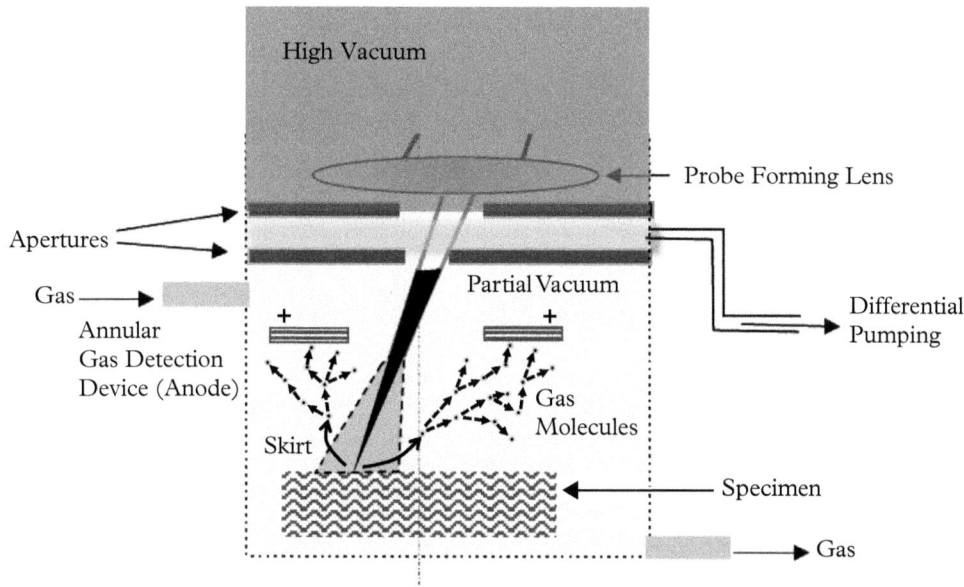

Figure 10.7.1 The lower part of a conventional SEM (Fig. 10.2.2) is modified with differential pumping between two apertures to form an ESEM. The specimen chamber can now be at substantially higher pressures to accommodate gases or specimen hydration. The probe is broadened in the form of a skirt due to scattering with the gas molecules, but the central core of the probe remains unaltered and generates both secondary and back-scattered electrons without significant change in resolution. A novel annular detector that serves also as an anode detects the avalanche of electrons (shown) generated by collisions of the signal electrons (SE or BSE) with the gas molecules.

Consider normal pressure, $P_0 \sim 1.013 \times 10^5$ Pa, and temperature, $T_0 \sim 273$ K; then, the density of the gas is:

$$\rho = \frac{M}{V_m} \tag{10.7.1}$$

where, M is the molecular weight, and $V_m = 2.24 \times 10^6$ m^3, is the molar volume. The mass-thickness, x, at a distance, t, in the specimen chamber from the final aperture, for a gas pressure, P, at temperature, T, is:

$$x = \rho t = \frac{M}{V_m} \frac{P}{P_0} \frac{T_0}{T} t \propto Pt \tag{10.7.2}$$

At 20 keV energy, the total mean free path length of electrons in water is $\sim 2 \times 10^{-9}$ m, and as the electron beam penetrates this distance in water, a fraction ($1/e \sim 37\%$) is not scattered, but the remaining 63% are both elastically and inelastically scattered through small angles. As a result, as the incident electrons travel a distance, x_t, of the order of a millimeter, 63% of the electrons suffer significant beam broadening, but the remaining 37% will remain tightly bound in the electron probe (Fig. 10.7.1); the exact size will be determined by the details of the electron optics. Thus, when $x_t = x$ in (10.7.2), we get $Pt = 2 \times 10^3$ Pa/mm; in other words, a reasonable working range is $P \sim 2 \times 10^3$ Pa, and $t = 1$ mm, which is quite appropriate for water molecules ($P_{H_2O} \sim 1.8 \times 10^3$ Pa at $T = 300\,K$). Figure 10.7.1 shows the nature of the electron probe in a hydrated environment, with the central beam relatively unchanged and a skirt of scattered electrons.

To record images in an ESEM, traditionally it was considered best to use back-scattered electrons combined with a scintillation detector. Recent developments include a gaseous detection device (GDD) that was developed to detect both secondary and back-scattered electrons. In the GDD detector arrangement (Fig. 10.7.1), the specimen acts as a cathode, and the positively charged GDD is placed above the specimen and below the final aperture. The emitted signal electrons, both SE and BSE, are rapidly attracted towards the anode, but in their trajectory, they generate a large number of additional electrons due to a collision cascade (or avalanche effect) with the gas molecules, leading to a near 100% signal detection efficiency. Further details of GDD and ESEM can be found in Stokes (2008).

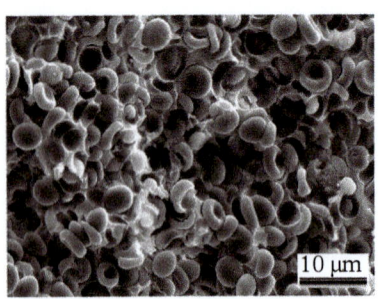

Figure 10.7.2 A, SEM image of red blood cells (erythrocytes) in the lungs, taken at 10 keV and 2000×.

Figure 10.7.2 shows a typical example of an ESEM image of red blood cells. Note that when insulating specimens are imaged in the ESEM, charging effects can be prevented because the positive gaseous ions present in the chamber neutralize the negative charge building up on the specimen. As a result, an added bonus of ESEM is that no specimen coating is necessary, even for insulating samples.

10.7.2 Combined Focused Ion-Beam (FIB) and Scanning Electron Microscope

An instrument with many structural similarities to an SEM, but using ion beam optics instead of an electron beam, can also be constructed (Fig. 10.7.3b). More than an imaging tool, such a FIB machine has been used extensively in the semiconductor industry for microfabrication and micromachining (see Fig. 10.7.4). A most important component of a FIB is the liquid ion source, which operates by the principle of field evaporation of the liquid metal around a sharp tungsten tip maintained at a very high positive potential. As discussed in §5.2.4.1, because of the much larger mass of ions compared to electrons, electrostatic lenses are favored for the focusing and guiding of ions; this is unlike SEMs, where electromagnetic lenses are commonly used.

A particular instrument of great practical utility is a combined dual beam FIB–SEM machine with *two columns* under vacuum, one for ions (FIB) and the other for electrons (SEM), both pointing to the same specimen area (Fig. 10.7.3). Such an instrument allows for routine imaging with the SEM column, combined with

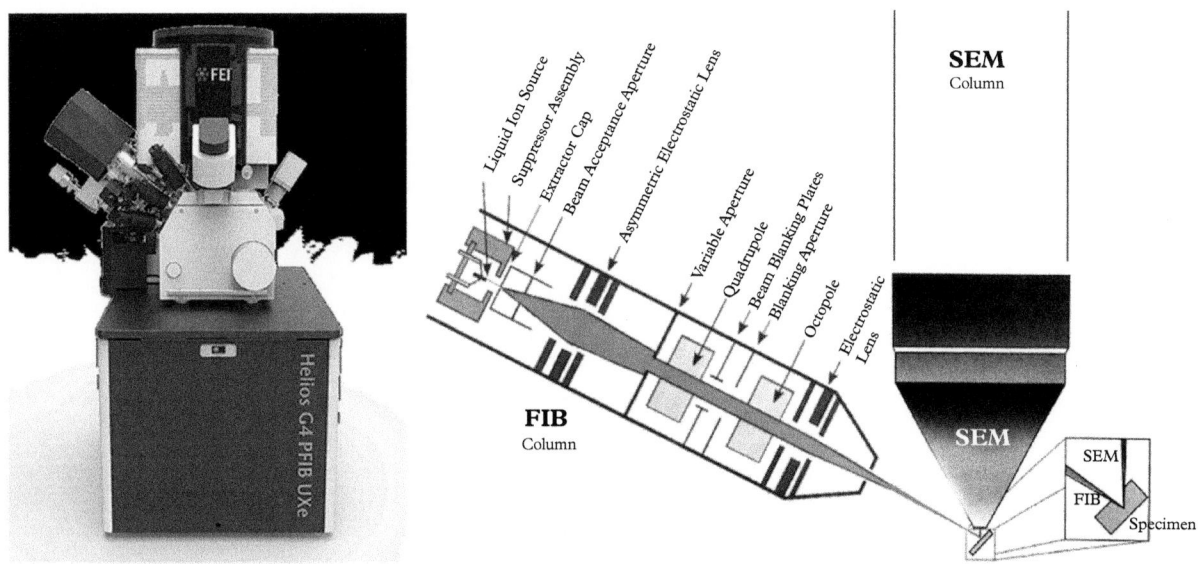

Figure 10.7.3 (a) A commercial dual beam FIB-SEM instrument. (b) A schematic representation of a generic FIB-SEM instrument with both ion- and electron-column oriented towards the same specimen area.

Adapted from Brandon and Kaplan (2008).

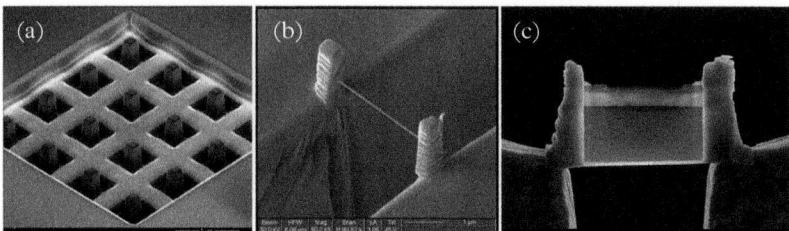

Figure 10.7.4 Examples of work done with a dual beam FIB-SEM dual-beam instrument. (a) Automated FIB preparation of pillars with well-defined shapes for applications such as atom probe or micro mechanical testing. (b) A platinum nanowire deposited and etched with the FIB and imaged with the SEM. (c) Dual-beam in situ 30 kV SEM view of a GaN ultra-thin electron-transparent specimen prepared for transmission electron microscopy observations, in less than an hour by FIB, and using a 1 keV FIB polishing to allow for less than 2 nm amorphized layers on each side of the lamella.

Downloaded from FEI.com

novel fabrication possibilities using the ion beam. This dual instrument has various features:

1. Micromachining, or micromilling, using the FIB to remove appreciable quantities of the materials, and mimicking conventional mechanical operations (turning, milling, etc.), but on the nanoscale. The electron beam, or the SEM, is used to monitor and tailor the milling process. Figure 10.7.4a shows an example of a microfabricated array of nanoscale pillars.

2. Gas-assisted etching using small amounts of reactive gas injected into the chamber through a capillary tube in close proximity to the specimen. The gas then reacts with the specimen and forms volatile species that are then removed by pumping. Some elements (such as Cu) have anisotropic response to the reactive gases and the resultant "ion-milling" leads to rough surfaces of the specimen. Subsequent reactive etching with other gases can be used to form a smooth surface.

3. Gas deposition of nanoscale structures can be accomplished by delivering precursor gases, through capillary needles. The precursors are chosen such that they decompose under the FIB, leading to deposition only where the ion or electron beams are localized. As a result, it is possible to build complex shapes; Figure 10.7.4b shows a simple example of a fabricated Pt nanowire.

4. Cross-sectioning of a specimen for subsequent examination by an SEM or TEM. Figure 10.7.4c shows a good example of an electron-transparent cross-section specimen of a GaN semiconductor for observation by a high-resolution TEM (§9.6.5).

5. Layer-by-layer serial sectioning of a specimen, using the FIB, and imaging with one of the detectors of the SEM. Even though this is a destructive technique, it helps to build a complex three-dimensional image of the specimen structure or composition.

10.8   Preparing Specimens for SEM

To carry out SEM observations, the specimen must meet the following simple conditions before being loaded on to the specimen stage.

1. The surface of the specimen to be studied must be clearly exposed.

However, if the internal structure of the specimen is of interest, a cross-section must be prepared. For mechanically hard materials, the specimen is simply fractured to obtain a cross-section. If the specimen changes characteristics at low temperatures, i.e. it is soft at room temperatures but becomes hard at low temperature, the material is typically frozen with liquid nitrogen and then fractured. For single crystals, such as semiconductors, fracturing along specific cleavage planes produces high quality flat specimens. Alternatively, if the material is soft, such as a polymer or a biological material, it can be cut with an ultra-microtome (which is also used to prepare thin sections for TEM observations, see §9.6.3) using a diamond blade. Metals, alloys, and mineral specimens can be prepared by mechanical polishing (see also §6.8.5), with abrasives gradually changed from coarse to fine, to finally produce a mirror-like surface. Finally, a FIB (§10.7.2) can also be used to produce cross-section specimens.

   2. The specimen must be mounted firmly on the stage.

In addition, it must make a good electrical contact to the specimen stage. For bulk specimens this is accomplished by using conducting paste. Powders and particles can be observed after dusting them on a conducting paste or a conducting double-sided adhesive tape.

   3. The specimen must have good conductivity to prevent charging effects.

For non-conducting materials, the simplest solution is to apply a highly conducting metallic thin film (~10 nm thick noble metal, typically Au, because it is highly stable and has a large secondary electron yield) as a coating on the specimen.

Summary

An SEM generates a range of morphological, structural, and chemical information about the specimen, based on image contrast arising from various electron–specimen interactions (Fig. 10.2.6a). The most widely used imaging contrast is based on detecting the secondary electrons, which can be collected efficiently and over a large bandwidth (10 MHz) by the Everhart–Thornley detector. Alternatively, the back-scattered electrons with the same energy as the incident beam, which are minimally affected by electrostatic fields and travel in straight lines, can also be detected. In both secondary and back-scattered electron imaging, the interpretation of images is simple and straightforward as the contrast is similar to that seen in optical microscopes (§6.8.4) with reflected light.

An important contrast mechanism in imaging back-scattered electrons is their sensitivity to the average atomic number of the specimen, by which different regions varying in composition can be spatially resolved. The back-scattered intensity is also sensitive to the orientation of the incident beam with respect to the crystal lattice. Thus, channeling or crystal orientation effects can provide both additional contrast, and local crystallographic information.

Ferromagnetic materials, with high internal magnetic induction, **B**, can alter the trajectories of both the secondary (Type I) and back-scattered (Type II) electrons. The contrast that this generates can be used to image magnetic domains in regular SEM instruments. However, specialized SEM instruments optimized for polarization analysis (SEMPA), using a Mott analyzer, can provide higher resolution magnetic contrast images.

Following inner shell ionization, the characteristic X-rays emitted by the specimen can also be detected by installing an energy dispersive X-ray detector in the SEM column and elemental maps of the specimen can be routinely obtained. In addition to characteristic X-ray photons, the incident electron bombardment can also stimulate the emission of ultraviolet and visible light from the specimen. A sensitive detector can detect such CL. Further, by using a parabolic mirror to focus the light emitted by the specimen, the sensitivity of CL measurements in an SEM can be significantly enhanced.

A less commonly used mode of contrast generation in an SEM is electron beam-induced current (EBIC); this is not discussed here. Other recent developments include the ESEM, with differential pumping to allow the specimens to be viewed under gaseous environments and/or under hydration. The latter is particularly useful to image biological specimens. Another variation is an SEM combined with a separate FIB column, both pointing towards the same area of the specimen. The FIB is used for top-down nanofabrication, including the preparation of electron-transparent TEM specimens (§9.6.5).

....

FURTHER READING

Brandon, D., and W. D. Kaplan. *Microstructural Characterization of Materials*. 2nd ed. Hoboken, NJ: Wiley, 2008.

Flewitt, P. E. J., and R. K. Wild. *Physical Methods of Materials Characterization*. Bristol: Institute of Physics Press, 2003.

Goldstein, J., D. Newbury, D. Joy, C. Lyman, P. Echlin, E. Lifshin, L. Sawyer, and J. Michael. *Scanning Electron Microscopy and X-Ray Microanalysis*. Berlin: Springer-Verlag, 2003.

Hubert, A., and R. Schäfer, *Magnetic Domains*. Berlin: Springer-Verlag, 2000.

Krishnan, K. M. *Fundamentals and Applications of Magnetic Materials*. New York: Oxford University Press, 2016.

Leng, Y. *Materials Characterization*. New York: Wiley-VCH, 2013.

Newbury, D. E., D. C. Joy, P. Echlin, C. E. Fiori, and J. I. Goldstein. *Advanced Scanning Electron Microscopy and X-Ray Microanalysis*, p. 147. New York: Plenum, 1986.

Ozawa, L. *Cathodoluminescence: Theory and Applications*. New York: Wiley-VCH, 1990.

Reimer, L. *Scanning Electron Microscopy*. Berlin: Springer-Verlag, 1995.

Stokes, D. J. *Principles and Practice of Variable Pressure Environmental Scanning Electron Microscopy*. Chichester: Wiley, 2008.

Williams, D. B., A. R. Pelton, and R. Gronsky. *Images of Materials*. New York: Oxford University Press, 1991.

· ·

REFERENCES

[1] Amelio, G. F. "Theory or the Energy Distribution of Secondary Electrons." *Journal of Vacuum Science and Technology* 7, (1970): 593.

[2] Jones, G. A. "Magnetic Contrast in the Scanning Electron Microscope: An Appraisal of Techniques and Their Applications." *Journal of Magnetism and Magnetic Materials* 8, (1978): 263–85.

[3] Tsuno, K. "Magnetic Domain Observation by Means of Lorentz Electron Microscopy with Scanning Technique." *Reviews in Solid State Science* 2 (1988): 623–58.

[4] Scheinfein, M. R., J. Unguris, M. H. Kelley, D. T. Pierce, and R. J. Celotta. "Scanning Electron Microscopy with Polarization Analysis (SEMPA)." *Review of Scientifiic Instruments* 61, (1990): 2501.

[5] Koike, K., and K. Hayakawa. "Observation of Magnetic Domains with Spin-Polarized Secondary Electrons." *Applied Physics Letters* 45, no. 5 (1984): 585–6.

[6] Unguris, J., R. J. Celotta, and D. T. Pierce. "Magnetism in Cr Thin Films on Fe(100)." *Physical Review Letters* 69, (1992): 1125.

[7] Kisker, E., W. Gudat, and K. Schröder. "Observation of a High Spin Polarization of Secondary Electrons from Single Crystal Fe and Co." *Solid State Communications* 44, no 5 (1982): 591–5.

[8] Hopster, H., R. Raue, E. Kisker, G. Güntherodt, and M. Campagna. "Evidence for Spin-Dependent Electron-Hole-Pair Excitations in Spin-Polarized Secondary-Electron Emission from Ni(110)." *Physical Review Letters* 50, (1983): 70.

[9] Penn, D. R., S. P. Apell, and S. M. Girvin. "Theory of Spin-Polarized Secondary Electrons in Transition Metals." *Physical Review Letters* 55, no 5 (1985): 518–21.

[10] Unguris, J., R. J. Celotta, and D. T. Pierce. "Magnetism in Cr Thin Films on Fe(100)." *Physical Review Letters* 69 (1992): 1125.

[11] Unguris, J. "Scanning Electron Microscopy with Polarization Analysis and its Applications". In *Magnetic Imaging and Its Applications to Materials*, edited by Y. Zhu and M. de Graef, 167–93. New York: Academic Press, 2001.

[12] Coenen, T., B. J. M. Brenny, E. J. Vesseur, and A. Polman. "Cathodoluminescence Microscopy: Optical Imaging and Spectroscopy with Deep-Subwavelength Resolution." *MRS Bulletin*, 40 (2015): 359–65.

[13] Romano-Rodriguez, A., and F. Hernandez-Ramirez. "Dual-Beam Focused Ion Beam (FIB): A Prototyping Tool for Micro and Nanofabrication." *Microelectronic Engineering* 84, no 5–8 (2007): 789–92.

[14] Everhart, T. E., R. F. Herzog, M. S. Chang, and W. J. DeVore. In *Proceedings of the Sixth International Conference on X-Ray Optics and Microanalysis,*

edited by G. Shinoda, K. Kohra, and T. Ichinokawa, 81. Tokyo: University of Tokyo Press, 1972.

[15] Adapted from *Fractography*. American Testing Services (June, 2012).

[16] Postek, M. T., A. E. Vladár, and K. P. Purushotham. "Does Your SEM Really Tell the Truth? How Would You Know? Part 2." *Scanning* 36, no. 3 (2014): 347–55.

. . . .

EXERCISES

A. Test Your Knowledge

For each one, identify ALL possible correct statements.

1. The wavelength of electrons _____ the acceleration voltage in a TEM/SEM.
    a) decrease with
    b) increase with
    c) is independent of
2. Resolution in an SEM is determined by the
    a) spherical aberration of the final probe forming lens.
    b) smallest probe that can be formed independent of the probe current.
    c) size of the final selectable aperture.
3. Resolution in an SEM depends on
    a) how much we magnify the image.
    b) how the lens configuration demagnifies the virtual source (crossover) of the gun.
    c) the acceleration voltage.
4. All electron lenses have
    a) spherical aberrations.
    b) chromatic aberrations.
    c) a depth of field and a depth of focus.
5. An SEM requires a *source* that
    a) has a large energy spread to image all kinds of materials.
    b) has a high temporal stability.
    c) always has a very high current density.
6. When compared to an optical microscope, an SEM has a much
    a) better resolution.
    b) better depth of field.
    c) higher magnification.
7. Resolution of an SEM is determined by the
    a) diameter of the probe.
    b) chromatic aberration of the condenser lens.
    c) spherical aberration of the final probe-forming lens.

8. The Everhart–Thornley detector
   a) is normally optimized for secondary electron detection.
   b) detects electrons by converting them into photons.
   c) cannot detect back-scattered electrons.

9. The probe size in an SEM
   a) decreases with increasing convergence angle.
   b) depends on the type of gun (thermionic or field emission).
   c) determines the depth of penetration of the incident electrons in the specimen.

10. Beam broadening in an SEM specimen is
    a) due to elastic scattering.
    b) due to inelastic scattering.
    c) independent of the sample atomic number.
    d) dependent on the acceleration voltage.

11. Secondary electron image intensity in an SEM
    a) depends on the work function of the specimen.
    b) decreases uniformly with incident beam energy.
    c) is directly proportional to the incident beam current.

12. Back-scattered electron images in an SEM depend
    a) only on topographical contrast.
    b) only on atomic number contrast.
    c) on both (a) and (b).

13. Back-scattered electrons in an SEM
    a) have a depth resolution defined by the diffusion distance, xD, in the specimen.
    b) do not have any angular dependence.
    c) can give information about the topography of the specimen.

14. EBSD patterns in an SEM
    a) give crystallographic information in two dimensions.
    b) have different yields with 1–2° variation in the angle of incidence.
    c) arise mainly from elastic scattering.

15. Magnetic contrast in an SEM can be obtained
    a) by measuring the polarization of the secondary electrons.
    b) from the effect of the stray fields from ferromagnetic specimen on the trajectory of the secondary electrons.
    c) and give domain images.

16. Electronic transitions and de-excitation processes produce signals in the SEM that can give information on _____ of the specimen.
    a) chemical composition
    b) semiconductor band gaps
    c) surface topography

B. Problems

1. Does the **working distance** of the final probe-forming lens from the specimen affect the image? How?

2. Consider the following statement:
    "The **secondary electron (SE) yield** shows an enhancement with the average atomic number of the specimen, especially at lower incident beam energies."
    (a) Is this statement correct?
    (b) If yes, why? If no, why not? Please explain briefly.

3. (a) A typical **thermionic emission source** used in an SEM has a total current of 1.6 μAmps. If 10% of the electron current is included in the probe, how many electrons are incident on the specimen in 10 s?
    You are using this SEM to image a zone-plate made of alternating layers of C_6 and W_{74}. Make the following assumptions:
    (i) The secondary electron (SE) yield is 0.2% for C and 0.05% for W.
    (ii) The back-scattered electron (BSE) yield is 0.1Z%, where Z is the atomic number.
    (iii) The efficiency of detection for both electrons is 10%
    (iv) The standard deviation, $\sigma = \sqrt{N}$, where N is the counts.
    (b) How many secondary electrons of C and W will be detected in 1 s?
    (c) How many BSE of C and W will be detected in 1 s?
    (d) What will be the contrasts across a W/C boundary for secondary and back-scattered electrons?
    (e) Is the contrast statistically significant in each case?
    (f) Which method, SE or BSE, would be better to image the W/C boundary? Give a brief explanation.

4. An SEM has three lenses—Condenser lens 1 (focal length = 10 mm, demagnification = 15), Condenser lens 2 (focal length = 15 mm, demagnification = 12), and an objective lens (focal length = 17 mm, demagnification = 20). If the electron gun has an initial cross-over diameter of 40 μm, what should the minimum **electron column length** be to have a final probe size of 15 nm? State your assumptions clearly.

5. The **microanalysis** of a geological specimen of pyroxene, using *stoichiometric oxide* standards is summarized in the following table:

Element	Standard used	Z_i	A_i	F_i	k_i
Si	SiO_2	0.973	1.106	0.999	0.496
Al	Al_2O_3	0.989	1.180	0.992	0.00959
Cr	Cr_2O_3	1.01	1.019	0.985	0.00561
Fe	FeO	1.136	1.008	1.00	0.0830
Mg	MgO	0.977	1.208	0.996	0.120
Ca	CaO	1.034	1.033	0.997	0.199
O					

a) Calculate the mass fraction of the metal in *each standard.*
b) Calculate the mass fraction of all the metals in the pyroxene
c) Assuming that the remaining mass is oxygen, calculate it.
d) Calculate the *atomic fraction* of all the elements in the pyroxene
e) Is this pyroxene stoichiometric?

6. For each of the **SEM images** shown in Figure 10.Pr.6, answer the related questions:

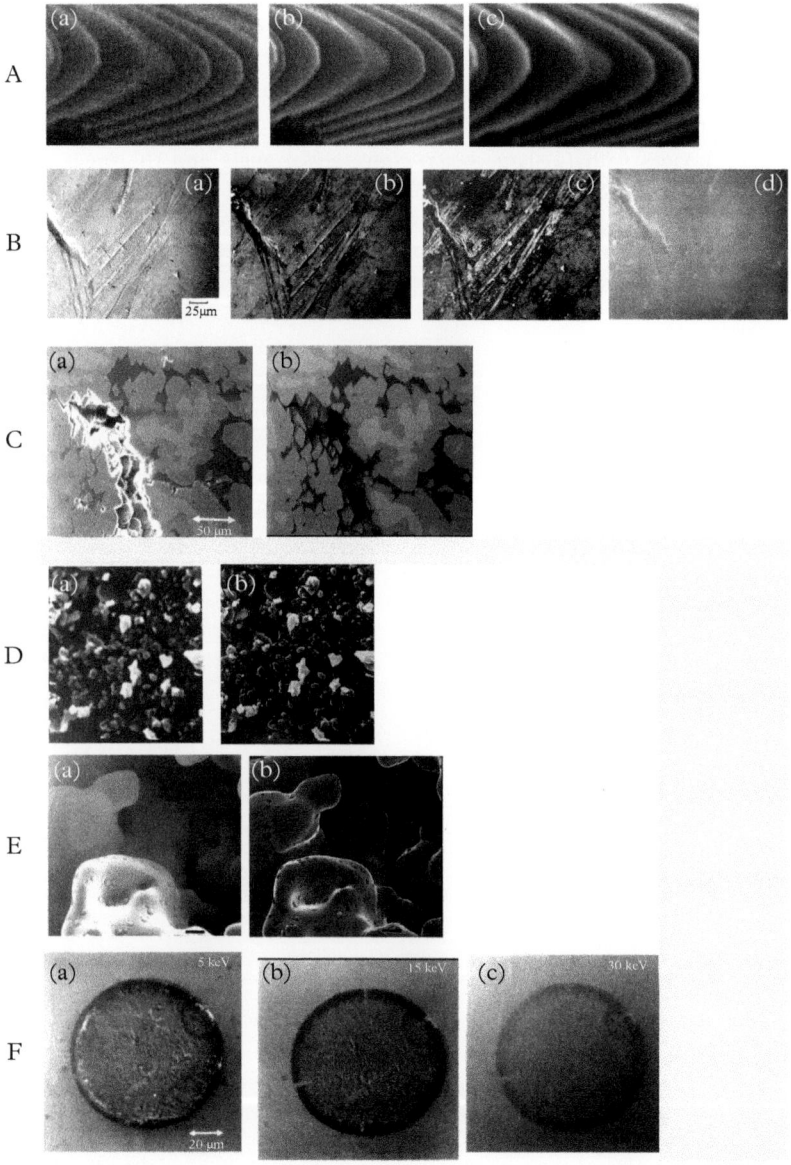

Figure 10.Pr.6 All images adapted from Goldstein et al. (2003).

a) Images of a stainless-steel metal filter with (a) smallest probe and minimum current, (b) small probe with adequate current, and (c) high current and a large probe size. How do the images change with these parameters?

b) Images of a platinum surface at (a) 20 keV, (b) 10 keV, (c) 5 keV, (d) 1 keV, with a positively biased E–T detector. What are the differences in the image with kV?

c) A nickel specimen with multiple phases using a (a) positively biased E–T detector, and (b) a solid-state BSE detector. What features are seen in each image?

d) Image with (a) typical lens defect, (b) corrected—what lens defect is this?

e) Images with (a) large beam convergence angle and (b) small beam convergence angle. What aspect of the image is affected by the beam convergence?

f) Detail of a surface oxide growth on copper (a) 5 kV, (b) 15 kV, and (c) 30 kV. What happens to the surface detail with acceleration voltage?

Scanning Probe Microscopy

11

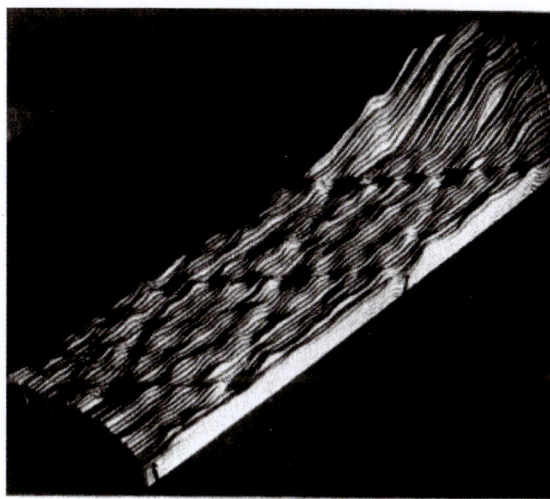

The first published image of Si (7×7) surface reconstruction obtained by scanning tunneling microscopy (STM) [3]. Binnig and Rohrer, the inventors of STM, received the 1986 Nobel prize in Physics.

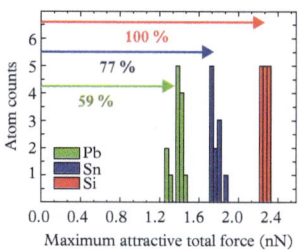

Atomic fingerprinting using dynamic, frequency modulated, atomic force microscopy (FM-AFM) imaging. Starting with a regular AFM image (left), the short-range chemical forces were normalized (middle) to distinguish (right) individual atoms of Si (red), Sn (blue) and Pb (green) in this ternary surface alloy [20].

Principles of Materials Characterization and Metrology. Kannan M. Krishnan, Oxford University Press (2021).
© Kannan M. Krishnan. DOI: 10.1093/oso/9780198830252.003.0011

11.1 Introduction

Scanning probe microscopy (SPM) is a family of techniques that is used to *map*, *measure*, and *manipulate* a surface or its properties at atomic or near-atomic resolution. In SPM, a sharp tip is scanned over the surface and either the *tunneling current* or *forces* arising from one of many possible interactions, depending on the specific nature of the tip and specimen surface, is measured as a function of tip position and plotted as an image. If such a tip-surface interaction is *near-field*, the limitations to image resolution discussed earlier for far-field techniques, such as optical microscopy (§6) and scanning electron microscopy (§10), which is of the order of half the wavelength of the photons or electrons, can be overcome. As this chapter shows, the geometric shape and size of the tip or probe, and the nature of its local interaction determines the resolution in SPM.

The field of SPM started with the invention [1] of the scanning tunneling microscope (STM) by Binnig and Rohrer.[1] Tunneling is a quantum mechanical transport phenomenon and cannot be understood in terms of classical physics concepts of drift or diffusion. In classical physics a particle, such as an electron, with energy smaller than a potential barrier will never be able to cross the barrier (Fig. 11.1.1a). However, in quantum mechanics where the wave-particle duality (§1.3.3) applies, the wave nature of the electrons gives it a finite probability to "tunnel" through to the other side of the barrier (Fig. 11.1.1b).

Such tunneling across artificial junctions/barriers has been a well-studied mechanism of transport in solid-state electronics (Mayergoyz, 2016). However, in developing the STM, Binnig and Rohrer demonstrated for the first time that such

[1] Gert Binnig and Heinrich Rohrer received the 1986 Nobel prize in Physics for "their design of the scanning tunneling microscope." They shared the prize with Ernst Ruska (1906–1988), who was recognized, rather belatedly, for "his fundamental work in electron optics, and for the design of the first electron microscope."

(a) Classical Mechanics

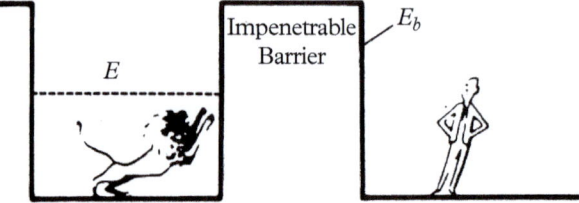

Figure 11.1.1 Tunneling through a potential barrier illustrating the difference between (a) classical and (b) quantum mechanical descriptions. In quantum mechanics the electron (shown here as a lion!) of energy, E, has a finite probability to tunnel through the barrier, $E_b > E$.

Adapted from Weisendanger (1994).

(b) Quantum Mechanics

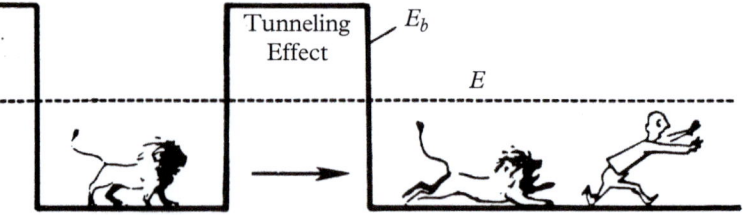

tunneling can take place in a controlled and predictable manner across a vacuum gap, between a sharp metal tip/needle and a surface, provided the tip is scanned at a distance of less than ~ 1 nm from the surface. This is because the tunneling current in an STM decays over a length scale of an atomic radius and flows from the terminating atom at the apex of the tip to a single atom on the surface (§11.3.1), or vice versa. Thus, atomic resolution is integral to the physics underlying the tunneling contrast mechanism in a STM. However, to achieve this resolution in practice, STM requires highly sensitive *piezoelectric*[2] *actuators* (§11.6) to move the tip orthogonally in three directions with control over atomic dimensions. It also requires high *mechanical stability* (no vibrations) of the experimental set-up. When both these criteria are satisfied, as in contemporary STM design and construction (Fig. 11.3.1), an electronic controller can be used to adjust the tip–surface separation such that either a constant/preset tunneling current with variable separation/height, or a constant height with variable tunneling current, is maintained throughout the scan. Then the separation or current, respectively, recorded as a function of lateral position, can be displayed at atomic resolution as a microscopic image of the surface. Strictly speaking, an STM maps a constant *local density of states* (§3.2) at the surface, which may be different from the surface *topography*. Moreover, it can also study the electronic structure of the surface by stopping the probe at any specific location and studying the tunneling current as a function of tip–surface voltage; this mode is known as scanning tunneling spectroscopy (STS, §11.3.2). Finally, the STM can also be used to manipulate or move atoms on surfaces as discussed in §11.3.3.

All scanning probe microscopes are variants of the STM described (Fig. 11.1.2). In all cases, a local probe/tip senses a *short-range interaction*, be it a current or force, with the surface. The probe is scanned using piezoelectric scanners and the strength of the local interaction—the tunneling current, or a force—is recorded as a function of *lateral position*, and plotted as an image.

The most important variant of the SPM is the *scanning force microscope* (SFM), commonly known as the *atomic force microscope* (AFM), which has the distinct advantage, unlike the STM, of being able to image nonconducting surfaces [2]. An SFM, as the name implies, measures the force between the tip and the surface by mounting the tip on a microfabricated cantilever that behaves as a force sensor. Either the deflection of the cantilever or the change in its dynamic response, as a result of the local tip–surface forces, is recorded to form the image. SFMs take advantage of various possible tip–surface interactions or forces, with specific designs optimized for sensing each one (Fig. 11.1.2).

In the remaining sections of this chapter, we present a comprehensive introduction to the instrumentation, physical principles, and applications of SPM. Our emphasis mainly is on STM and scanning/AFM. Further details on SPMs can be found in numerous, well-written textbooks and monographs, including Chen (1993), Wiesendanger (1994), Meyer, Hug, and Bennewitz (2004), and Eaton and West (2010).

[2] Piezoelectric materials show an electric polarization on the application of stress. This linear coupling between electrical and mechanical energies is described as a tensor, $P_i = d_{ijk}\sigma_{jk}$, where P_i is the induced polarization, and σ is the applied stress. Strictly speaking, here we are interested in the inverse piezoelectric effect, $x_{ij} = d_{kij}E_k$, where the same coefficients, d_{ijk}, are used to relate the applied electric field, E, to the induced strain, x.

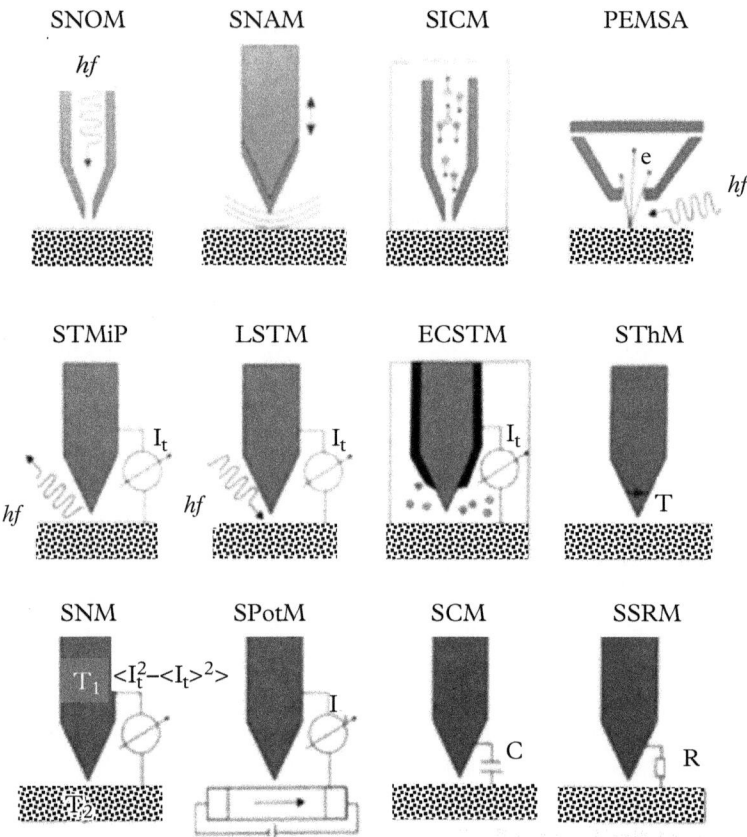

Figure 11.1.2 Various designs of scanning probe microscopes; all these methods are characterized by a local probe, registering a local signal/interaction, as it is scanned over the surface to form an image. Top (l to r): Scanning near-field optical microscope (SNOM), Scanning near-field acoustic microscope (SNAM), Scanning ion conductance microscope (SICM), and Photoemission microscope with scanning aperture (PEMSA). Middle (l to r): STM with inverse photoemission (STMiP), Laser scanning tunneling microscope (LSTM), Electrochemical STM (ECSTM), and Scanning thermal microscope (SThM). Bottom (l to r): Scanning noise microscope (SNM), Scanning tunneling potentiometry (SPotM), Scanning capacitance microscope (SCM), and Scanning spreading resistance microscope (SSRM). One of these (SThM) is discussed further in this chapter (§11.8.3), but detailed discussions of the rest can be found in Meyer, Hug, and Bennewitz (2004), from which this figure is adapted.

11.2 Physics of Scanning Tunneling Microscopy (STM)

11.2.1 Elastic Tunneling Through a One-Dimensional Barrier

The basic physical principles of operation of an STM can be understood from the simple model of elastic tunneling through a one-dimensional potential barrier of width, s (Fig. 11.2.1). By simply matching the wave function (see Wiesendanger (1994) for a detailed derivation) on either side of the barrier, it can be shown that the transmission coefficient, T, defined as the ratio, $T = j_t/j_i$, where j_t is the transmitted (region 3) and j_i ($= \hbar k/m$) is the incident (region 1) current density, is given by

$$T = \frac{16k^2\chi^2}{\left(k^2 + \chi^2\right)^2}e^{-2\chi s} \qquad (11.2.1)$$

where $k = \sqrt{2m_e E}/\hbar$, and the decay rate, χ, is given by

$$\chi = \frac{\sqrt{2m_e\left(E_0 - E\right)}}{\hbar} \qquad (11.2.2)$$

Here, a strongly attenuating barrier, with $\chi s \gg 1$, is assumed. It is now easy to see that the transmission coefficient, T, is dominated by the factor $\exp(-2\chi s)$. As such, T is dependent on the product of the barrier width, s (Å), and the effective barrier height, $(E_0\text{-}E)$ in eV, but not on the exact shape of the barrier. Note that if we make the reasonable assumption that the effective barrier height is of the order of the work function, Φ, and use Au ($\Phi \sim 5$ eV) as an example, we can see that a change in barrier width by 1 Å leads to one order of magnitude change in the barrier transmission (see Problem 1). This very high sensitivity of the tunneling current to the barrier width is the underlying motivation for STM.

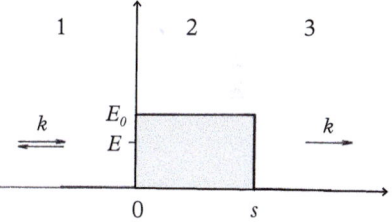

Figure 11.2.1 A one-dimensional potential barrier of height, E_0, and width, s. An electron of energy, E, and wave vector, k, impinging on the barrier in region 1, has a finite probability of tunneling *elastically* to region 3.

Example 11.2.1: A one-dimensional square potential barrier has a height of 10 eV and a width of 3 nm, and electrons with kinetic energy of 7 eV are trying to tunnel through this barrier.

(a) What would be the decay rate of these electrons?

(b) Would this be a strongly attenuating barrier?

(c) What is the transmission coefficient for the tunneling of these electrons?

(d) What will be the new transmission coefficient for the same electrons if the barrier width is reduced to 1 nm?

Solution:

(a) The decay rate (11.2.2) is

$$\chi = \left[2 \times 9.1 \times 10^{-31} \times (10 - 7)\left(1.6 \times 10^{-19}\right)\right]^{1/2} / \left(1.05 \times 10^{-34}\right)$$
$$= 8.90 \times 10^9 \text{ m}^{-1}.$$

(b) The attenuating barrier for a barrier width, $s = 3$ nm, is $\chi s = 8.9 \times 10^9 \times 3 \times 10^{-9} = 26.7 \gg 1$. Hence, we can assume that this is a strongly attenuating barrier.

(c) The transmission coefficient, (11.2.1), can be written as $T = T_0 \exp(-2\chi s)$, where by rearranging terms, it can be shown (the derivation is left as an exercise to the reader) that $T_0 = 16E(E_0 - E)/E_0^2 = 16 \times 7 \times 3/10^2 = 3.36$.
Thus, for a barrier width of 3nm, $T = 3.36 \exp(-2 \times 26.7) = 2.16 \times 10^{-23}$, which is a very small number, indeed.

(d) If the barrier width is 1 nm, the new transmission coefficient is

$$T = 3.36 \exp\left(-2 \times 8.9 \times 10^9 \times 1 \times 10^{-9}\right) = 6.25 \times 10^{-8}.$$

In other words, reducing the barrier height by a factor of three, or a small change in the barrier height, increases the transmission coefficient by 10^{15}, i.e. leads to a very large change in the transmission coefficient.

11.2.2 Quantum Mechanical Tunneling Model of the STM

In STM, the tunneling barrier width, s, is given by the vacuum gap, z, between the tip and the specimen surface, and the work function, Φ, replaces the barrier height, E_0. However, in addition to the effective barrier height, $(\Phi - E)$, and the tip–surface separation, z, the tunneling current, I_T, also depends on the applied bias voltage, V_b, and the local density of states, $N_s(E_F)$, of the specimen at the Fermi level, and is given by

$$I_T \propto V_b N_s(E_F) \exp\left[-2\frac{\sqrt{2m_e(\Phi - E)}}{\hbar} z\right] \propto V_b N_s(E_F) e^{-1.025\sqrt{\Phi}z} \quad (11.2.3)$$

This can be best understood by considering the energy level diagram of the tip and the specimen, starting with them being far apart to be electronically

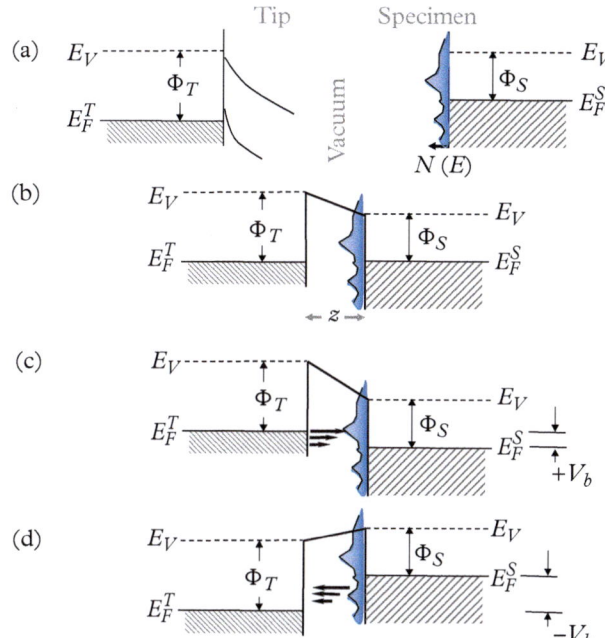

(a)

(b)

(c)

(d)

Figure 11.2.2 Energy level diagram for the tip (left) and the specimen (right) in a scanning tunneling microscope. (a) The tip and specimen, when far away are electronically independent. The work function, Φ, the vacuum level, E_V, and the Fermi level, E_F, for both the tip and specimen, as well as the density of states, $N(E)$, for the specimen are shown. (b) The tip and specimen are in equilibrium, i.e. $E_F^T = E_F^S$, and are separated by a small vacuum gap, z. (c) Positive specimen bias voltage, $+V_b$, with electrons tunneling from the tip to the specimen. (d) Negative specimen bias, $-V_b$, with electrons tunneling from the specimen to the tip.

Adapted from Wiesendanger (1994).

independent (Fig. 11.2.2a). When the tip is brought close to the specimen surface with a small vacuum gap, z, they are in equilibrium and their Fermi levels are aligned equal (Fig. 11.2.2b). Upon applying a bias voltage, V_b, to the specimen the energy levels will undergo a rigid shift in energy by the amount $|eV_b|$. This can be downward (Fig. 11.2.2c) for positive polarity of the bias, or upward (Fig. 11.2.2d) for negative polarity of the bias. Thus, for *positive specimen bias* (Fig. 11.2.2c) the tunneling current flows from the tip to the specimen, and probes its local *unoccupied* density of states as a function of bias voltage. On the other hand, for *negative bias* (Fig. 11.2.2d), the electrons tunnel from the *occupied states* of the specimen to the unoccupied states of the tip. Hence, by varying the sign (polarity) and magnitude of the bias voltage, both the occupied and unoccupied local density of states of the specimen can be probed. This is called STS and is discussed further in §11.3.2.

Example 11.2.2: An STM tip is used to probe an Au surface with a bias voltage of 2V. If the tip specimen height changes from 1 Å to 2 Å, what will be the percentage change in the tunneling current?

Solution: The work function for Au is $\Phi = 5$ V and $V_b = 2$ V. Applying (11.2.3), we calculate the percentage change as

$$\frac{I_T^2 - I_T^1}{I_T^1} \times 100 = \frac{e^{-1.025\sqrt{2\Phi}} - e^{-1.025\sqrt{\Phi}}}{e^{-1.025\sqrt{\Phi}}} \times 100 = \left(1 - e^{-1.025\sqrt{2\Phi} - \sqrt{\Phi}}\right) 100$$

$$= (1 - 0.387)\, 100 = 61\%$$

which is truly substantial.

11.3 Basic Operation of the Scanning Tunneling Microscope

Figure 11.3.1 shows a schematic diagram of the STM. It consists of a sharp metal tip that is brought in close proximity to a surface, creating an atomic-scale vacuum gap, using the *xyz* piezo drives. A bias voltage applied between the tip and the

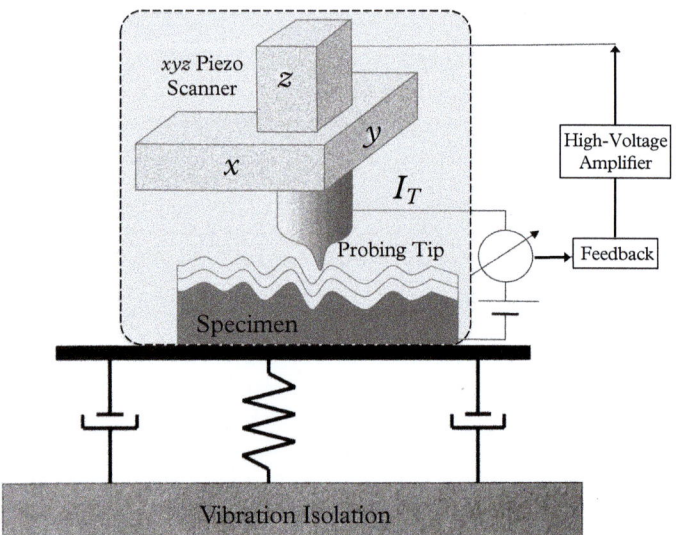

Figure 11.3.1 Schematic representation of a scanning tunneling microscope. The *xyz* piezo-scanner moves the tip in a controlled manner over the specimen surface. The feedback loop is used to keep a constant tunneling current, and the height, *z*, is recorded as a function of the scanned coordinates, *x* and *y*, to give the image. Note that a high voltage amplifier to drive the piezo scanner and good vibration isolation are required for atomic-resolution imaging. The entire microscope portion, enclosed in the dashed line, is kept under UHV to avoid surface contamination.

Adapted from Meyer, Hug, and Bennewitz (2004).

specimen results in electrons tunneling through the vacuum gap. Typically, the tunneling current is in the nA to pA range. This current signal is amplified and used as the input voltage for the feedback loop to the piezo drives, which adjusts the tip height to maintain a constant tunneling current as the tip is scanned over the surface. The tip height (or, in reality, the feedback voltage to the piezo drive) is plotted as a function of position to form the image. This imaging method is called the *constant current* mode. Alternatively, the tip can be moved at *constant height* and the variations in the tunneling current measured to form the image. These two modes of STM imaging are discussed in §11.3.1. The tip can also be physically stopped at any position, (x_i, y_i), and the bias voltage, V_b, or tip–surface separation, z, can be varied up or down to measure the variation of a local property such as the barrier height (work function). This spectroscopic mode is discussed in §11.3.2. Finally, an STM can also be used to manipulate atoms on a surface, by picking them up individually with the tip and moving them by translating the tip position; this is discussed in §11.3.3.

11.3.1   Imaging

We assume, without any loss of generality, that the specimen is fixed on the stage and the STM tip is scanned in the *x-y* plane, even though there are some STMs that do it the other way around. Anyway, for the purpose of our discussion the two arrangements are equivalent, and we now discuss the two common modes of STM imaging.

In the *constant height mode* (Fig. 11.3.2a), the tip does NOT move in the vertical (z) direction as it is scanned laterally (*x-y*) over the specimen surface. The tunneling current across the vacuum gap is recorded, which is inversely related to the local tip–surface distance, and varies as the surface profile of the specimen. However, the dependence of the tunneling current on the surface profile is not linear, and depends exponentially on the tip–surface separation as described by (11.2.3). In addition, the current is also dependent on the electronic structure, or the density of states, of the specimen (Fig. 11.2.2).

From an experimental point of view, the *constant height mode* works well for flat specimens, where the roughness or corrugation is smaller than the average tip–surface distance. Further, two practical issues are encountered: there is always a vertical drift of the tip due to thermal effects and often the overall specimen plane is not parallel but inclined to the lateral scanning plane (*x-y*) of the tip. To work around these issues, the feedback system is turned on and operated at low sensitivity to keep the average current at a preset value. The low sensitivity ensures that the feedback mechanism does not interfere with the fine variations of the current caused by local changes in the surface profile; at the same time, it is fast enough to recognize the *inclination* of the specimen surface.

In the *constant current mode* (Fig. 11.3.2b), the feedback system mentioned earlier is actively engaged and used to control the vertical position of the tip in such a way that a constant tunneling current (preset value) is always maintained.

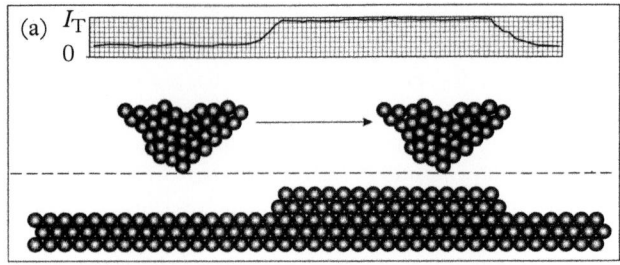

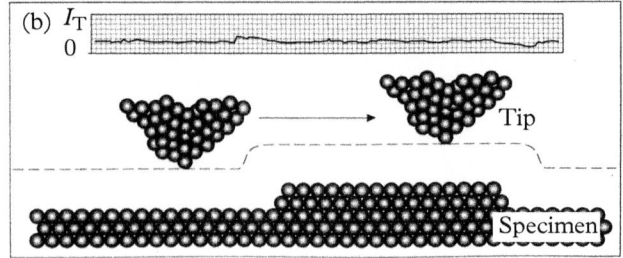

Figure 11.3.2 Two modes of STM imaging include (a) constant height mode, and (b) constant current mode. The tunneling current in both cases is shown in the inset (plot). The tunneling takes place between the atom at the apex of the tip and an atom on the surface of the specimen. Note that the topography measured is a *convolution* of the real surface with the physical dimension of the tip.

Adapted from Colton et al. (1998).

Then, the tip follows the surface profile assuming a constant energy barrier height. In practice, the voltage applied to the piezo drive in the z-direction controls the vertical tip position, and this voltage (plotted, as shown) is proportional to the surface profile. Note that in this mode, the scanning speed is limited by the reaction time of the feedback and scanning system.

The constant height mode is used for atomically flat specimens, as the time constant of the feedback system is not significant in this mode of operation. However, if the surface roughness/corrugation is significantly larger than the tip–surface distance, there is a very high probability of the constant-height tip crashing into the specimen; in that case, the constant current mode is preferred.

One of the very first demonstrations to highlight the capabilities of the STM was the study of the 7×7 reconstruction[3] of the Si(111) surface, which was resolved by direct real-space imaging [3], using the *constant current* mode (Fig. 11.3.3). The original recorded image of two 7×7 unit cells is shown in relief in Figure 11.3.3a; the dark lines of minima highlighting the boundaries of the rhombohedral 7×7 unit cells are clearly seen. A top view of the relief is shown in Figure 11.3.3b, where the sixfold rotational symmetry is obvious. The crosses indicate the adatom positions[4] that agree with an adatom model as shown in Figure 11.3.3c.

In addition to the tunneling current, $I_T(x,y)$, its derivative, $dI_T/dz(x,y)$, often referred to as barrier height or work function signal, is recorded simultaneously. Taking the derivative of (11.2.3), it can be shown that

$$dI_T/dz \propto -I_T\sqrt{\Phi} \qquad (11.3.1)$$

[3] Surface reconstruction and related nomenclature was introduced in §8.3.

[4] An atom that lies on a crystal surface is called an adatom. It can be thought of as the opposite of a surface vacancy.

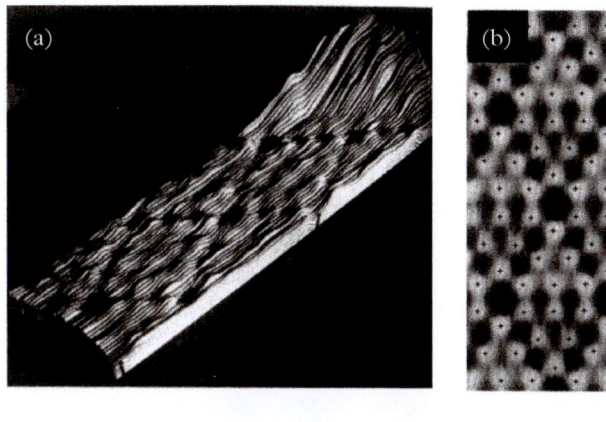
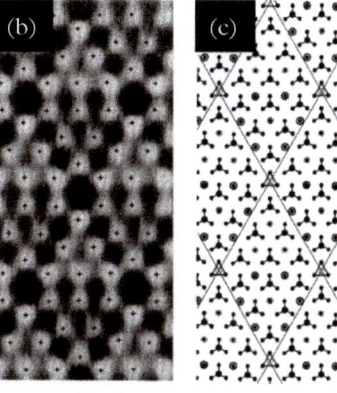
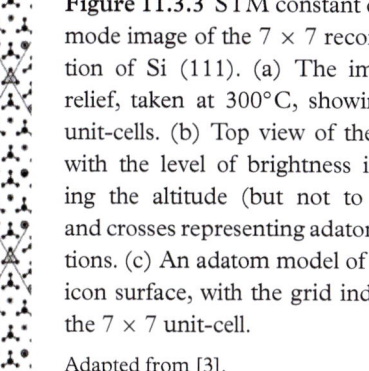

Figure 11.3.3 STM constant current mode image of the 7×7 reconstruction of Si (111). (a) The image in relief, taken at 300°C, showing two unit-cells. (b) Top view of the relief, with the level of brightness indicating the altitude (but not to scale), and crosses representing adatom positions. (c) An adatom model of the silicon surface, with the grid indicating the 7×7 unit-cell.

Adapted from [3].

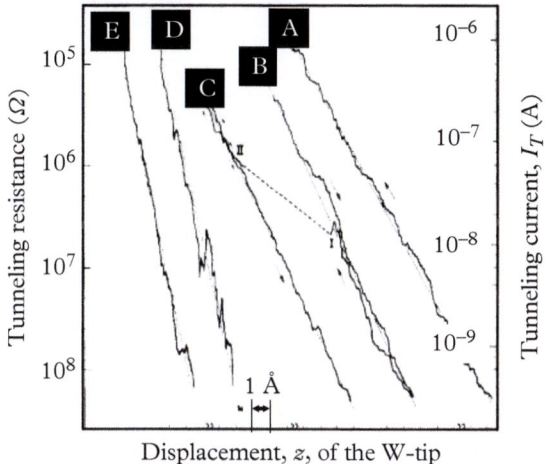

Displacement, z, of the W-tip

Figure 11.3.4 The variation of the tunneling current as a function of the W-tip displacement in a Pt specimen. The derivative of the tunneling current gives a work function, $\Phi \sim 0.6$–0.7 eV, for a normal surface (curves A, B, C). However, when the surface was repeatedly cleaved *in situ* in vacuum (curves D, E), a significantly steeper slope gives a value of $\Phi \sim 3.2$ eV.

Adapted from [4].

Thus, the *square* of the normalized derivative signal, $(dI_T/I_T)/dz$, is proportional to the height of the tunneling barrier, or to a first approximation, the work function, Φ. However, in practice the values of $[(dI_T/I_T)/dz]^2$ that are measured are smaller than theoretical values because of surface contamination. Alternatively, measuring this derivative signal gives an indication of the cleanliness of the tip, and is often used as a check before imaging studies. Figure 11.3.4 shows one of the first such measurements of the barrier height distribution in a Pt specimen using a tungsten tip [4]. Note that here the tunneling current (11.2.3) is fitted in the form $\ln I_T = -A\sqrt{\Phi}\, z + B$, and the *slope* is interpreted in terms of the *work function*.

11.3.2 Tunneling Spectroscopy

We have seen (Fig. 11.2.2) how the *local electronic density of states* of the specimen contributes to the tunneling current. Note that the current increases significantly if the bias voltage, V_b, allows electrons to tunnel to a maximum of the unoccupied density of states locally in the specimen. In addition, electrons at the highest energy—close to the Fermi level—experience the smallest effective tunneling barrier, as indicated by the magnitude of the arrows in Figure 11.2.2 c,d. This is also known as the barrier transmission coefficient, $T(E,eV_b)$, which is a function of the energy, E, of the electron and the bias voltage, V_b. Thus, to first order, a typical tunneling experiment would yield a tunneling current that is a product of the tunneling coefficient, $T(E, eV_b)$, and the local density of states, N(E), of the tip and specimen.

Figure 11.3.5 is a schematic illustration of a spectroscopic tunneling experiment with the tip biased positively by +1 V, allowing the electrons to tunnel from the specimen to the tip. For the purpose of discussion, Figure 11.3.5a shows the densities of states by different curves for the specimen (slow, or low frequency wiggles) and tip (fast, or high frequency wiggles). Figure 11.3.5b illustrates the expected tunneling current for both positive and negative bias voltage. We see that

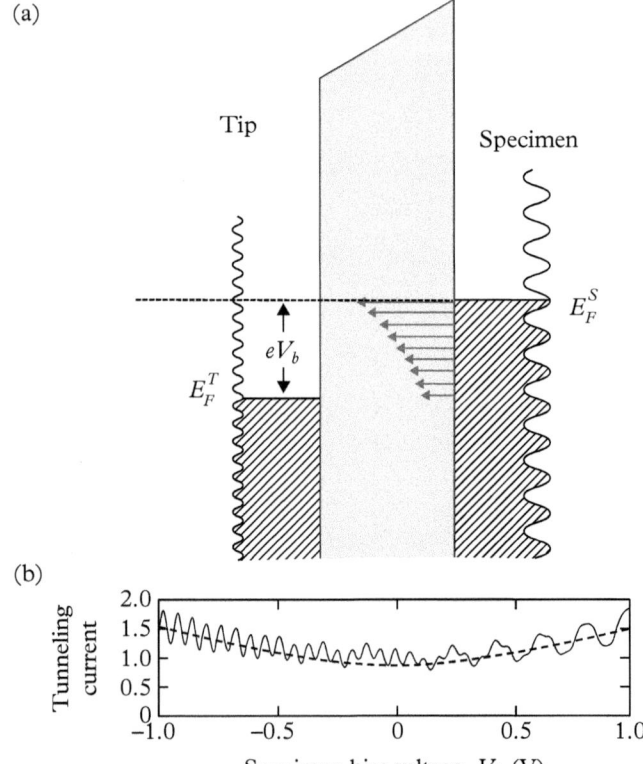

Figure 11.3.5 Schematic illustrations of the principles of tunneling spectroscopy using an STM. (a) The density of states (DOS) of the specimen (slow wiggles) and the tip (fast wiggles), as well as the applied positive bias voltage, V_b, leading to a tunneling current flow from the specimen to the tip, is shown. (b) The resultant tunneling spectrum, as a function of bias voltage, represents the DOS of the tip for negative bias, and the DOS of the specimen for positive bias. This signal is superimposed on a background (dashed line) arising from the transmission coefficient, $T(E,eV_b)$, that increases monotonically, independent of the sign, with the *magnitude* of the bias voltage.

Adapted from Wiesendanger (1994).

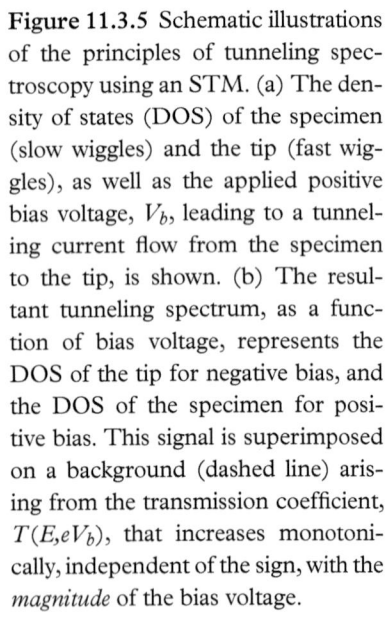

the transmission coefficient increases monotonically as a function of bias voltage and contributes a smooth background, on which the electronic structure (local density of states) is superimposed. We also see (Fig. 11.3.5b) that the resultant spectrum shows fidelity to the density of states of the specimen (slow wiggles) for positive bias voltage, and that of the tip (fast wiggles) for negative bias.

In practice, the simplest method to obtain spectroscopic information in STM is to sequentially record topographical images at different bias voltages using the constant current method (§11.3.1). For analysis of the spectra, we assume that at any specific bias voltage only the electronic states at the Fermi level of the specimen and/or tip contribute to the tunneling current; in other words, the interference of the local geometry/topography with the electronic structure is neglected. More sophisticated experimental arrangements for spectroscopy, where the effect of topography can be removed, are discussed in Wiesendanger (1994) and Chen (1993).

Again, one of the first applications of STS was the measurement of the bands of the Si(111) 2 × 1 reconstructed surface [5], using the differential conductivity, dI/dV, to directly measure the surface DOS. Data were obtained over the energy range -4eV $- +4$ eV, relative to the Fermi level, and showed the expected structure of two π-bonded surface bands and a new surface resonance 2.3 eV above the Fermi level. For those interested, further details can be found in [5].

11.3.3 Manipulation of Adsorbed Atoms on Clean Surfaces

There are two common ways to manipulate adsorbed atoms on surfaces using an STM tip. This includes both vertical (normal to the surface) and lateral (along the surface) manipulation (Fig. 11.3.6).

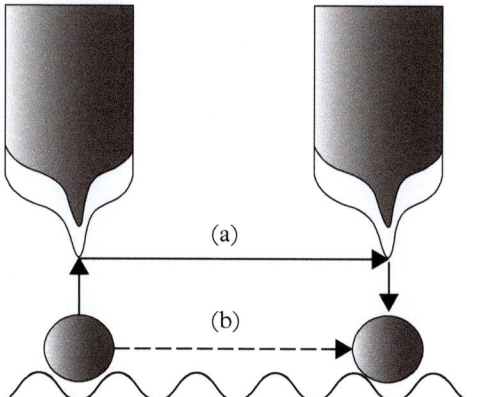

Figure 11.3.6 Ways of manipulating adsorbed atoms on clean surfaces (a) Vertical manipulation followed by lateral motion of the tip. (b) Lateral manipulation that requires overcoming the resistance from the surface to the adatom motion, typically in the $K\Omega - M\Omega$ range.

Adapted from Meyer, Hug and Bennewitz (2004).

In the case of vertical manipulation, the tip is brought into contact or near contact with the adsorbed atom. The atom is then transferred from the surface to the tip by the application of a voltage pulse of appropriate magnitude and sign. For example, in the case of Xenon atoms, the adsorbed atoms move in the same direction as the tunneling current by a process called *heat assisted electromigration* [6]. If necessary, once the atom is transferred to the tip, the latter can be retracted from the surface and moved laterally by changing the tip position. Then, a reverse voltage pulse can be applied to release the adsorbed atom to the surface as illustrated in path (a) of Figure 11.3.6. Alternatively, by adjusting its vertical position the tip can be set to form a weak bond with the atom and then moved laterally, path (b), to "push" the adsorbed atom from one location to another. The threshold resistance to move the atom depends on both the atomic species and the nature of the specific surface. Typical values are 5 MΩ for Xe on Ni(110), 200 KΩ for CO on Pt(111), and 20 KΩ for Pt adatoms on Pt(111).

Such atomic manipulation has provided an elegant platform to study the confinement of electrons in nanoscale dimensions [7–9]. The simple quantum mechanical (Rae, 1992) model of electrons confined in a one-dimensional potential well of length, L (Fig. 11.3.7), also known as the "*particle in a box*" model, predicts that the energy levels of the electron, for an infinitely high potential, $V_0 \to \infty$, are discrete and given by

$$E_n = \frac{h^2}{8m_e L^2} n^2 \tag{11.3.2}$$

where $n = 1, 2, 3, \ldots$. The corresponding wave functions are

$$\psi_n = \sqrt{\frac{2}{L}} \sin \frac{n\pi x}{L} \tag{11.3.3}$$

for $0 < x < L$, and $\psi = 0$, for regions outside the potential well. However, if V_0 is finite, then the number, n, of wave functions within the potential well is finite; moreover, the probability for the wave function to extend outside the well is also finite with exponentially decreasing amplitude as a function of distance from the boundary.

These fundamental ideas of quantum mechanics were spectacularly demonstrated by confinement of electrons in two-dimensional boxes, or quantum corrals, created by moving and positioning adsorbed atoms on surfaces. In one of the first such experiments [5, 7], 48 Fe atoms were positioned in the form of a circle on a Cu(111) surface (Fig. 11.3.8a). The itinerant electrons of Cu confined within this "circle" behave as though they are a two-dimensional *electron gas* and form a standing wave pattern. Moreover, STM measurements of the electronic structure in the quantum corral, are in good agreement with calculations (Fig. 11.3.8b).

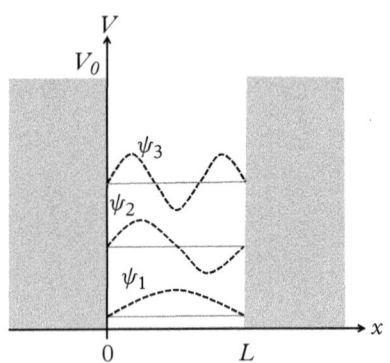

Figure 11.3.7 A one dimensional potential well of height, V_0, and length, L. The first three allowed wave functions are also shown.

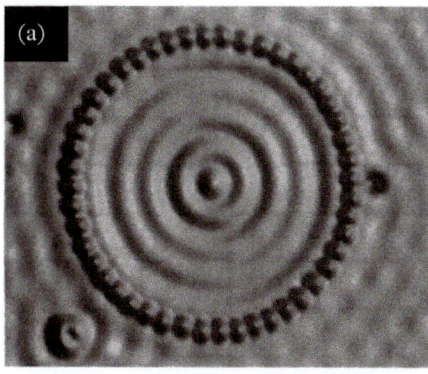
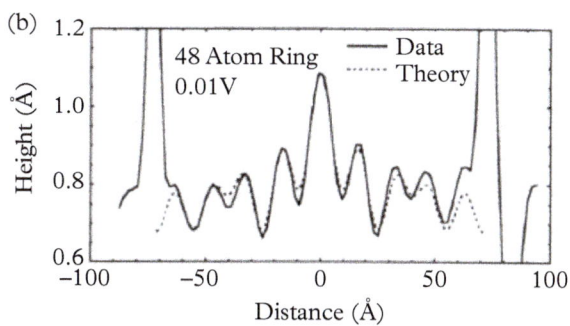

Figure 11.3.8 Spatially resolved STM image of the eigenstates of a two-dimensional electron gas in a quantum corral. (a) 48-atom Fe ring, with average diameter ~142.6 Å, constructed on Cu(111) enclosing a defect-free region of the surface. (b) Cross-section of the image data (solid line) and fit to theory (dotted line).

Adapted from [7].

11.4 Physics of Scanning Force Microscopy

The STM is a versatile imaging/spectroscopy tool, but it is restricted only to the study of conducting surfaces. Moreover, it was recognized very early in the development of the STM that whenever the tip–specimen distance was small enough for the flow of the tunneling current, significant forces would act simultaneously on the tip. It was speculated that if these atomic-range forces could be sensed, then an "atomic" force microscope, which can be applied to nonconducting materials, may be a possibility. This was indeed realized with the invention of the SFM [2] (Fig. 11.4.1). Here, the tip height is controlled such that the *force* between the sharp tip and the specimen during the scan is maintained at a constant value, allowing the measurement of the topography of the surface irrespective of its conducting properties. To measure the local force, the tip is mounted on a cantilever, which serves as a force sensor (§11.4.2). In principle, either the deflection of the cantilever or its dynamic response to the tip–specimen forces can be measured by a variety of methods. Strictly speaking, the total force combining the different contributions of the tip and the cantilever is measured (§11.4.3). Both short- and long-range forces should be considered; for example, the short-range van der Waals interaction is confined to the mesoscopic tip, but the long-range electrostatic interactions largely affect the cantilever. In addition to the *contact mode* described above, the SFM is also operated in two other modes: the *dynamic mode*, also known as noncontact AFM that provides true atomic resolution [10], and the *tapping mode*, also known as the *intermittent contact mode* that is widely used for imaging in air (§11.5). We now discuss these aspects of SFM and conclude this chapter by describing some applications of this versatile

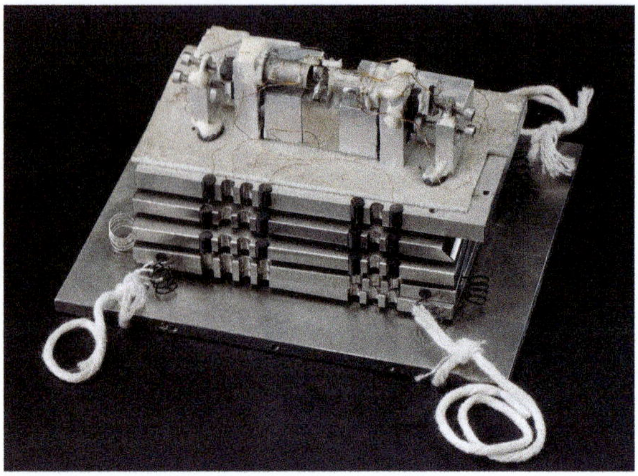

Figure 11.4.1 The first atomic force microscope built by Binnig et al. [2], on display at the Science Museum, London. The original "cantilever" used was a strip of gold foil with a small diamond tip glued to it.

Image downloaded from http://sciencemuseum.org.uk.

Figure 11.4.2 (a) A SFM tip and cantilever with important dimensions indicated. One common way to measure the displacement of the cantilever is to use an optical lever and a split photo-detector as shown. Both (*b*) normal forces, and (*c*) lateral forces can be measured.

(a) Adapted from Meyer, Hug, and Bennewitz (2004). (b, c) Adapted from Colton et al. (1998).

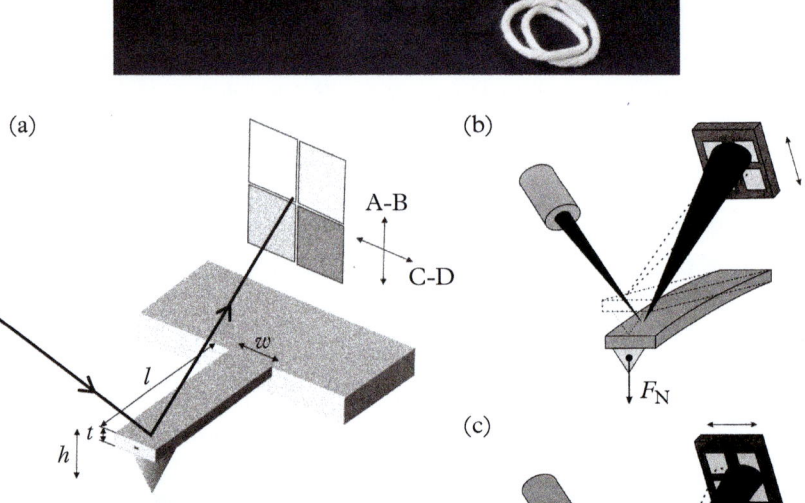

technique in characterizing both physical and biological materials. Advanced readers should consult specialized books for more information, including Meyer, Hug, and Bennewitz (2004), and Eaton and West (2010).

11.4.1 Mechanical Characteristics of the Cantilever

Unlike the STM, the central element of a SFM is the spring that measures the forces between the tip and surface. To measure normal tip–specimen forces it is important that the force sensor be rigid in two dimensions but relatively flexible or soft in its motion in the third dimension. A cantilever beam satisfies these requirements very well, and hence a microfabricated cantilever (Fig. 11.4.2) is normally used as the force detector. The force constant, k_f, also known as *stiffness*, for normal bending of a cantilever is

$$k_f = \frac{Ewt^3}{4l^3} \qquad (11.4.1)$$

where E is the Young modulus[5] of elasticity of the material, and w is the width, t is the thickness, and l is the length of the cantilever (Fig. 11.4.2a). The resonance frequency, f_0, of this cantilever is

$$f_0 = \frac{0.162t}{l^2} \sqrt{\frac{E}{\rho}} \qquad (11.4.2)$$

where ρ is its mass density. Note that by measuring the resonance frequency, as well as l and w, which can be obtained by SEM imaging, the thickness, t, of the cantilever can be determined. Cantilevers are commonly microfabricated of silicon; thus $\rho_{Si} = 2330$ kg/m^3, $E_{Si} = 169$ GPa and

$$f_0^{Si} = \frac{t}{l^2} \frac{1}{7.23 \times 10^{-4}} \qquad (11.4.3)$$

The important properties of a SFM cantilever are its stiffness, k_f, its resonance frequency, f_0, and its variation with temperature, $\partial f_0 / \partial T$, its quality or Q factor,[6] as well as its composition.

For SFM applications, the resonance frequency should be significantly above ambient (building) vibrations (1–100 Hz), or sound frequencies (1–10 kHz). Further, the typical spring constant for atoms in a solid is $k_{atom} \sim m_{atom}\omega^2$, where $m_{atom} \sim 5 \times 10^{-26}$ kg, and ω is of the order of phonon frequencies ($\sim 10^{13}$ Hz). For optimal SFM performance, the stiffness, k_f, of the cantilever should be of the same order of magnitude as k_{atom}. Typical values are $k_f \sim 0.1$–100 N/m. Further, to obtain atomic sensitivity, the thermal motion of the cantilever should be less than 1 Å. Thus, to avoid any thermal effects in SFM measurements, $k_f \langle \Delta z^2 \rangle \geq k_B T$. At room temperature, this gives $k_f \geq 0.4$ N/m. To achieve such spring constants, we can see from (11.4.1) that the cantilevers must have dimensions in the micrometer range. Moreover, the resonance frequency (11.4.2) of a cantilever depends inversely on its physical size.

If lateral forces are to be measured the torsional spring constant, k_t, of the cantilever given by

$$k_t = \frac{Gwt^3}{3h^2l} \qquad (11.4.4)$$

where G is the shear modulus and h is the length of the tip, is important.

In addition to the beam (Fig. 11.4.3b), triangular cantilever geometries (Fig. 11.4.3a) are widely used for topographic imaging. This is because their high

[5] $\sigma = E\varepsilon$, where σ is the stress, ε is the strain, and E is the Young modulus in units of GPa (10^9 N/m^2).

[6] The Q factor defines the resonance behavior of an oscillator. Typically, it is defined as the ratio of the resonance frequency, f_0, to its resonance width, Δf_0, i.e. $Q = f_0/\Delta f_0$. The Q factor (a dimensionless quantity) depends on the damping mechanisms present in the cantilever; for cantilevers operated in air, Q is a few hundred, but in vacuum, Q reaches hundreds of thousands.

torsional stiffness prevents them from being affected by lateral forces during the measurements. Details of the calculation of spring constants for triangular cantilevers can be found in [12].

Cantilevers are typically made of single crystalline silicon (beam) or silicon nitride (triangular) and manufactured by microfabrication processes in a clean room. In the case of silicon, the pyramidal tip points along <100> and has a cone angle of ~50°; the cone angle is etched further to a tip radius of ~10 nm. The silicon is highly doped to prevent charging effects. A silicon tip that is removed from the wafer has a coating of native silicon oxide, which is removed by etching with HF or by sputtering in a vacuum (Fig. 11.4.3c). The lateral resolution of an AFM can be improved by sharpening the tip, and many techniques are available to do so (Fig. 11.4.3d–g). See also Example 11.5.1.

In the next section we discuss different approaches to monitor the motion of the cantilever as it functions as a deflection sensor of the tip.

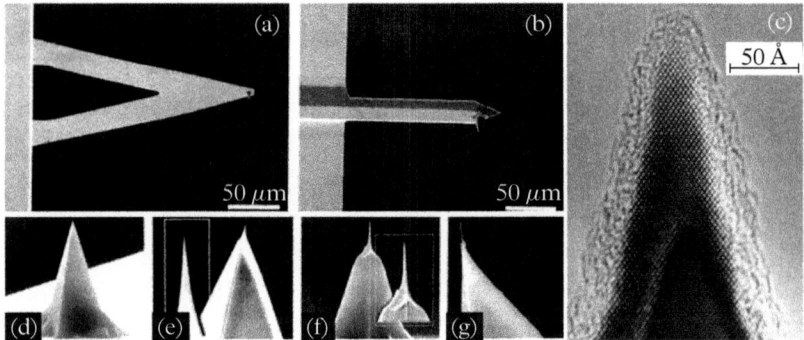

Figure 11.4.3 Scanning electron images of (a) triangular (Si_3N_4) and (b) beam (Si) shaped cantilever probes. (c) HR-TEM images of a very sharp silicon tip with the native oxide removed by etching; the amorphous coating is polymerized hydrocarbon—notice the crystal structure remains true to the bulk even at the very tip. Different types of sharpened tips: (d) standard Si tip, (e) electrochemically etched super-sharp Si tip, (f) ion-milled, high aspect ratio Si tip, and (g) a tip with a carbon nanotube attached.

(a–c) Adapted from [13]. (d–g) Adapted from Eaton and West (2010).

Example 11.4.1: A Si cantilever has the dimensions: $l = 0.225$ mm, $w = 0.038$ mm, $t = 0.006$ mm, and $h = 0.125$ mm. What is its bending force constant, torsional force constant, and resonance frequency? What is the maximum thermal motion of the tip at room temperature?

Solution: For silicon, the Young modulus, $E = 1.69 \times 10^{11}$ N/m^2, Shear modulus, $G = 0.68 \times 10^{11}$ N/m, and density $\rho = 2330$ kg/m^3.

Then, from (11.4.1), the bending force constant

$$k_f = \frac{Ewt^3}{4l^3} = \frac{\left(1.69 \times 10^{11}\right)\left(0.038 \times 10^{-3}\right)\left(0.006 \times 10^{-3}\right)^3}{4\left(0.225 \times 10^{-3}\right)^3} = 30.45 \, \text{N/m}.$$

The thermal motion of the tip $\left(\Delta z^2\right) \leq \frac{k_B T}{k_f} = \frac{25 \times 1.6 \times 10^{-22}}{30.45} = 1.32 \times 10^{-22}$
Thus $(\Delta z) \approx 1.14 \times 10^{-11}$m.
From (11.4.3), the resonance frequency $f_0^{Si} = \frac{t}{l^2} \frac{1}{7.23 \times 10^{-4}} = 163.9$ kHz.
From (11.4.4), the torsional spring constant
$$k_t = \frac{Gwt^3}{3h^2l} = \frac{(0.68 \times 10^{11})(0.038 \times 10^{-3})(0.006 \times 10^{-3})^3}{3(0.125 \times 10^{-3})^2 \, (0.225 \times 10^{-3})} = 52.92 \, \text{N/m}.$$

11.4.2 Cantilever as a Force Sensor

In AFM/SFM it is important to measure the forces between a sharp tip and the specimen surface. This geometry results in very high pressure (force/area) on the sharp tip. In addition to preventing the tip from breaking, the AFM has to be optimized for measuring very weak forces. To accomplish this a number of force sensor designs (Fig. 11.4.4) have been developed and implemented for AFM/SFM. The most common method used in the majority of commercial instruments is the *optical beam deflection* method. In this design, a light beam is deflected from the back side of the cantilever and its deflection is monitored using a four-segment, position sensitive, photodiode detector (Fig. 11.4.2a). This widely used design can measure very small movements in both vertical deflection (Fig. 11.4.2b) and torsional rotation (Fig. 11.4.2c) of the cantilever, providing very high normal and lateral force sensitivity.

Figure 11.4.4d shows an alternative way to detect the cantilever deflection by using the cantilever as the moving mirror in a Michelson interferometer (introduced earlier in Figure 3.5.1). This technique is very sensitive and is limited only by the thermal noise of the cantilever; moreover, potentially it can be implemented in the limited confines of an ultra-high vacuum (UHV) chamber and can also operate at low temperatures. However, this interferometry technique has met with only limited success because the probe can jump between interference fringes while scanning.

In the first AFM built by Binnig et al. [2], an STM tip positioned on the back side of the cantilever (Fig. 11.4.4a) was used to measure the motion of the cantilever via the variations in the associated tunneling current. Even though this technique is a viable force sensor, its implementation is very difficult and it suffers the added complexity of accounting for the forces between the STM tip and the cantilever.

Other methods integrate the sensor and the actuator (cantilever), creating significant advantages for dynamic measurements. It is also advantageous to

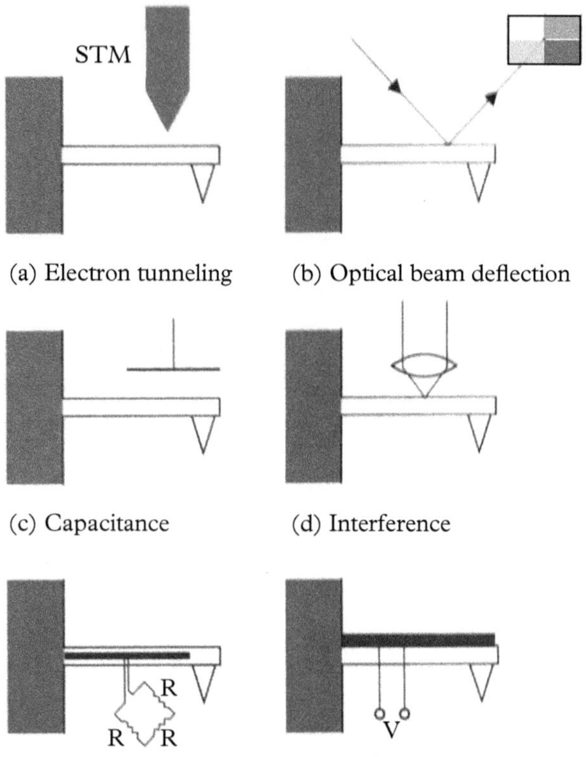

(a) Electron tunneling   (b) Optical beam deflection

(c) Capacitance   (d) Interference

(e) Piezoresistance   (f) Piezoelectricity

Figure 11.4.4 Schematic represen-tation of various deflection sensors designed for scanning force microscopy.

Adapted from Colton et al. (1998).

implement such deflection sensors in UHV environments as they do not require additional positioning of the sensor elements *in situ* in the chamber. For example, the cantilever and a counter electrode can be microfabricated as an integral part of the force sensor, and the cantilever deflection can be monitored by any changes in the capacitance between them (Fig. 11.4.4c). Such a technique is capable of fast measurements but the forces between the cantilever and the counter electrode may affect the performance.

In dynamic AFM, it was recognized quite early in the development of the technique that the requirements for the force sensor were similar to that of the time-keeping elements (quartz tuning forks) manufactured in the billions of numbers for low-cost watches. In principle, these tuning forks are cheap piezoelectric sensors and by attaching metallic tips to one of the prongs of the tuning fork (while keeping the other one fixed) they were modified to serve as AFM force sensors (Fig. 11.4.4f). Alternatively, high-quality self-sensing force sensors with a piezoresistive element integrated into the back side of the cantilever have been developed (Fig. 11.4.4e). Here, the resistance changes with the bending of the cantilever and it can be monitored as a force sensor.

11.4.3 Tip–Specimen Forces Encountered in an SFM

Except for the force sensor replacing the tunneling tip, the SFM is not conceptually different from an STM. The component of the tip–specimen force, F_{t-s}, in a direction normal to the surface that is relevant for SFM, has both short- and long-range contributions. In principle, for an SFM operating in a vacuum to achieve atomic resolution, the short-range (< 1 nm) chemical forces must dominate over the van der Waals, and the long-range (<100 nm) electrostatic, and/or magnetic forces between the tip and specimen. In addition, when operated in air, layers of water vapor or hydrocarbons may adhere to the tip or specimen and create capillary forces that also need to be considered.

In general, the energy, E_{t-s}, for all these tip–specimen interactions contributes to the force, i.e. $F_{t-s} = -\partial E_{t-s}/\partial z$, and the associated tip–specimen spring constant, $k_{t-s} = -\partial F_{t-s}/\partial z$. In SFM/AFM, the imaging signal is based on either F_{t-s} or some entity derived from F_{t-s}, depending on the mode of operation (§11.5).

The short-range forces can be treated akin to a chemical bond between atoms at the extremity of the tip and the specimen surface, with forces modeled in terms of the overlap of their electron wave-functions and the repulsion between their positively charged atomic cores. If the overlap of the electron wave-functions reduces their total energy, a bonding-like condition is realized and leads to attractive tip–specimen forces. Alternatively, repulsive forces can also arise from the Pauli exclusion principle, when there is a strong overlap of the wave-functions. The additional Coulombic repulsion of the ion cores also contribute over short distances if it is inadequately screened by the outer electrons. Early in the development of SFM, attempts were made to describe these short-range forces using Morse or Leonard–Jones potentials, but they were found to have significant shortcomings. More sophisticated models tailored to the specifics of the tip–specimen materials system being used have now been developed, with reasonably good results [11].

Most importantly, the attractive short-range forces are in the range of 0.5–1.0 nN per interacting atom. Moreover, at a distance of 0.5 nm from the surface, the short-range forces are of a magnitude comparable to the long-range forces between the tip and specimen. This tip–specimen vertical distance (0.5 nm) is typical for the noncontact mode of SFM operation §11.5.2, especially if atomic resolution in imaging is required. A state-of-the-art example of the application of AFM with chemical sensitivity is discussed in §11.8.1.

The van der Waals interaction arises from the fluctuations in the diploe moment of atoms and their mutual polarization. For short distances, the van der

Waals force is given by $F_{vdW} \propto 1/r^7$ for $r < 5$ nm, which reduces to $F_{vdW} \propto 1/r^8$ for $r > 5$ nm. To first order, approximating the tip–specimen geometry as a hemisphere (tip) in the proximity of a semi-infinite body (specimen), the van der Waals force (Israelichivli, 1991) is given by

$$F_{vdW} = \frac{A_H R}{6 d_{t\text{-}s}^2} \tag{11.4.5}$$

where A_H is the material-dependent Hamaker constant [27], R is the tip radius, and $d_{t\text{-}s}$ (~ 0.5 nm) is the tip–specimen separation. The Hamaker constant is of the order of 10^{-19} J, with an enhancement sometimes as large as 30 times for metals compared to insulators. For typical SFM conditions, we get $F_{vdW} \sim 2$ nN for commercial silicon tips and $F_{vdW} \sim 7 - 10$ nN for etched metal tips with $R \sim 100$ nm. This magnitude of F_{vdW} is a significant hindrance to achieving atomic resolution in scanning force microscopy. However, F_{vdW} can be significantly reduced by immersing the cantilever in water [14].

In addition, electrostatic forces, F_{el}, become important if the specimen and tip are both conductors and have an electrostatic potential difference, ΔU. Here, $\Delta U = U_{bias} - U_{cpd}$, typically ~ 1 V, where U_{bias} is the tip–specimen bias voltage and U_{cpd} is the contact potential difference arising from the different work functions of the tip and specimen. We can then show [15] that

$$F_{el} = \frac{\pi \varepsilon_0 R}{d_{t\text{-}s}} (\Delta U)^2 \tag{11.4.6}$$

with $F_{el} \sim 1.6$ nN for standard tip and specimen parameters; here $\varepsilon_0 = 8.854 \times 10^{-12}$ Fm^{-1} is the permittivity of free space. Alternatively, if the tip and specimen are ferromagnetic, the tip will experience a magnetic force, F_{mag}, given by

$$F_{mag} = -\mu_0 \int \mathbf{M}_{tip} \left(x', y', z' \right) \frac{\partial}{\partial z'} \left(\mathbf{H}_{sample} + t \right) dV' \tag{11.4.7}$$

where $\left(x', y', z' \right)$ is a coordinate frame attached to the tip, $\mathbf{t}$ (x,y,z) is the position of the tip in the laboratory frame of reference, $\mathbf{M}_{tip}$ is the magnetization of the tip, $\mathbf{H}_{specimen}$ is the field arising from the specimen, and the integral is over the volume, V', of the magnetic tip. We discuss magnetic force microscopy (MFM), using ferromagnetically coated tips in §11.8.2.

Finally, in ambient conditions, capillary forces, F_{cap}, arising from water vapor condensation, especially if the tip radius, $R < 100$ nm, have to be considered. The water molecules condensing between the tip and specimen will from a meniscus (Fig. 11.4.5). The capillary force can be estimated as

$$F_{cap} = \frac{4\pi R \gamma \cos\theta}{1 + \frac{d_{t\text{-}s}}{R(1 - \cos\phi)}} \tag{11.4.8}$$

where γ is the surface tension of water ($\sim$0.074 N/m, at 20 °C), θ is the contact angle, and ϕ is the angle of the meniscus. Now the *maximum value* of the capillary force is $F_{cap} = 4\pi R\gamma \cos\theta$, which is significantly larger ($\sim$90 nN) than the corresponding van der Waals force. The effect of the capillary force can be substantially reduced, if not eliminated, by coating the tip and specimen surface with hydrophobic amphiphilic molecules. As mentioned earlier, immersing the cantilever in water also can eliminate the capillary forces entirely. On a positive note, capillary forces can be exploited for nanoscale lithography, and §11.8.5 discusses the related technique of dip-pen nanolithography (DPN).

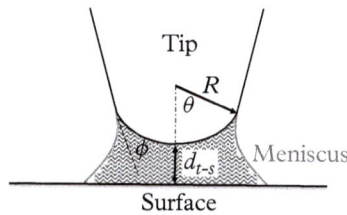

Figure 11.4.5 Schematic representation of the formation of the meniscus between the tip and specimen by the condensation of water vapor, which leads to capillary forces.

Example 11.4.2: Calculate the magnitudes of the van der Waals force for commercial silicon and etched metal tips. Also, calculate the electrostatic and the maximum capillary forces for typical scanning probe parameters.

Solution: For van der Waals force, we apply (11.4.5) with the following parameters: $A_H = 10^{-19}$ J, $R \sim 30$ nm and $d_{t\text{-}s} \sim 0.5$ nm for Si, with the radius $R \sim 100$ nm for etched metals. Thus, we get $F_{vdW} = 2$ nN for Si, and $F_{vdW} = 6.67$ nN for etched metals.
To calculate the electrostatic forces, we apply (11.4.6) with $R \sim 30$ nm, $d_{t\text{-}s} \sim 0.5$ nm, and $\Delta U \sim 1$ V, and get $F_{el} = 1.67$ nN.
The maximum capillary force, for $R \sim 100$ nm and $d_{t\text{-}s} \sim 0.5$ nm, is given by the numerator of (11.4.8), i.e. $F_{cap} = 4\pi R\gamma \cos\theta \sim 90$ nN.

11.5 Operation of the Scanning Force Microscope

The wide range of operating modes developed in scanning force microscopy [16], can be broadly classified as *static* and *dynamic*, where the bending (former) of the cantilever, or changes in vibrational properties (latter) are measured. In the static mode, the cantilever tip is in "contact" with the surface, and the tip–specimen distance is controlled to maintain a constant bending of the cantilever. Thus, the topography of the specimen, subject to a constant force on the cantilever, is recorded. Alternatively, the tip in "contact" is scanned at a constant distance from the surface and the variation in the force recorded as the image.

In contrast, the dynamic mode measures changes in the vibrational properties of the cantilever due to the local tip–specimen interactions. These properties include its resonance frequency, the amplitude of oscillation, and the phase difference between the induced excitation and resultant oscillations of the cantilever. Further, dynamic modes are differentiated based on feedback parameters, i.e. amplitude, frequency, or phase modulation, which are used to sense or control the tip–specimen distance [17]. Dynamic modes include both contact and noncontact modes of operation; however, the most common method is the intermittent contact or tapping mode of operation. In addition to topography, the tapping mode provides information on the physical and chemical properties of the surface while causing minimal damage during the scanning. Note that SFM modes can provide information beyond surface topography and properties by recording various signals (Fig. 11.1.1) as a function of distance or other parameters. Such measurements, known as force spectroscopy, provide wide-ranging information on tip–specimen interactions.

It is important to understand the difference between the interaction forces in noncontact and contact modes of operation. The tip–specimen interaction in the noncontact mode (Fig. 11.5.1a) consists of attractive long- and short-range contributions, with the latter making a significant contribution to the total force. Because the short-range forces between the outermost atoms of the tip and specimen are significant, atomic resolution imaging in noncontact mode

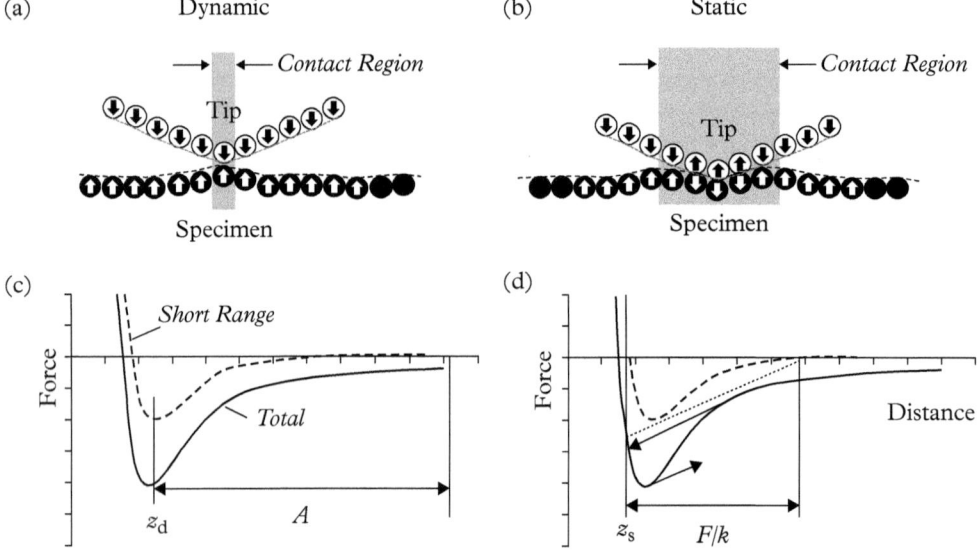

Figure 11.5.1 Interaction forces and contrast formation in (a, c) dynamic (noncontact), and (b, d) static (contact) modes in SFM. In (a) and (b) the arrows indicate the direction of forces on the atoms of the tip and specimen; note that the contact region, highlighted in grey, laterally extends over a region of several atoms in (b). The corresponding force-distance curves, including the total (continuous) and short-range (dashed) forces, are shown for (c) noncontact, and (d) contact modes. In (c), the amplitude range, A, of the oscillating tip used in dynamic measurements is indicated. In (d), arrows indicate the points where the tip jumps into and out of contact, with the slope of the straight dotted line indicating the stiffness, k, of the cantilever.

Adapted from Meyer, Hug, and Bennewitz (2004).

is possible without any surface damage. Alternatively, in the contact mode the situation is complex as some of the atoms of the tip and specimen are in repulsive contact (Fig. 11.5.1b). Even though the total force, $F_{t\text{-}s}$, may be attractive, the local deformation of the surface and the transition between attractive and repulsive forces may lead to a contact region that spans several atoms; as such, atomic resolution in contact mode is not possible. Static, contact mode topographic imaging is discussed in §11.5.1.

To further get an idea of the static and dynamic modes of operation, we have to consider the force–distance behavior as the tip is brought close to surface. Note that, typically, static and dynamic modes use cantilevers that are mechanically soft and rigid, respectively. In the dynamic mode (Fig. 11.5.1c), the position, z_d, in the force–distance curve corresponding to Figure 11.5.1a indicates a situation where the short-range attractive forces are a maximum and form a significant component of the total force. Further, the amplitude range, A, of the oscillations of the tip is always maintained such that the closest distance of approach of the tip to the surface prevents it from being trapped at the specimen surface. In this scenario, as shown, the total force is attractive even when the short-range force becomes repulsive. Dynamic, noncontact SFM imaging is discussed further in §11.5.3.

The force–distance curve for the contact mode (Fig. 11.5.1d) indicates a position, z_s, for the tip that is on the repulsive side of the short-range force curve. Further, as the tip is brought close to the surface in the static mode, it is characterized by an instability (indicated by the left arrow in the figure) known as "*jump to contact*," which arises from the elastic bending of the cantilever. Similarly, as the tip is retracted from the surface, it suffers another instability (indicated by the right arrow), known as "*jump off contact*." These instabilities can be understood with reference to the slope of the dashed line in the figure, which represents the spring constant, k, of the cantilever. Approaching the specimen, when the derivative of the total force curve becomes larger than k, the tip jumps to a contact position where the short-range repulsive force balances the attractive force. Similarly, upon retraction of the tip, the jump off occurs when the total force derivative becomes equal to the spring constant.

We now discuss some important modes of SFM operation in more detail.

11.5.1 Static Contact Mode for Topographic Imaging

The very first mode developed for SFM, static contact mode uses static measurements of the *deflection* of the cantilever to create a topographic image of the surface. In this mode, the tip is first brought into contact with the surface and its height adjusted to be in the short-range repulsive force regime (Fig. 11.5.1b). Overall, the tip position defines a force equilibrium between the long-range attractive force, F_{att}, of the tip and surface, and the short-range repulsive force, F_{rep}, of the tip apex and specimen, and external force, F_{ext}, due to the cantilever stiffness (Fig. 11.5.2a). Strictly speaking, in this regime, a combination of cantilever bending and local specimen compression will occur, and the set point height is chosen to minimize the resulting force and any related damage to the specimen surface or the tip but, at the same time, to avoid jump out-of-contact of the tip.

Once the vertical position of contact is established, the topographic image can be obtained by operating the scan in two different ways (Fig. 11.5.2b). In the constant force method, also known as the constant deflection method, the microscope feedback system is used to keep the cantilever deflection at a set value determined by the operator. From Hooke's law, the force, F, applied by the probe to the surface is $F = -k \times D$, where k is the spring constant of the cantilever and D is its vertical displacement. As a result, a probe with a small spring constant (a soft cantilever) will maximize the vertical deflection for a given force. As the tip (or specimen) is scanned, the short-range repulsive forces are very sensitive to the tip–specimen distance, and hence, images of constant repulsive forces can be interpreted in terms of the surface topography. Alternatively (Fig. 11.5.2b, bottom), the feedback system can be turned off and an image, scanned at constant height, can be recorded. Now the image comes entirely from the deflection of the cantilever and not the voltage applied to the z-piezo drive. Hence, this measurement is dependent on the specific calibration of the cantilever deflection.

The resolution of the contact mode SFM, depends on the lateral extent of the contact region (Fig. 11.5.1b) of the tip-apex with the specimen. The contact diameter, d_c, assumed to increase with the applied force, F, due to the elastic deformation of the tip and specimen (so-called Hertz model), is given by

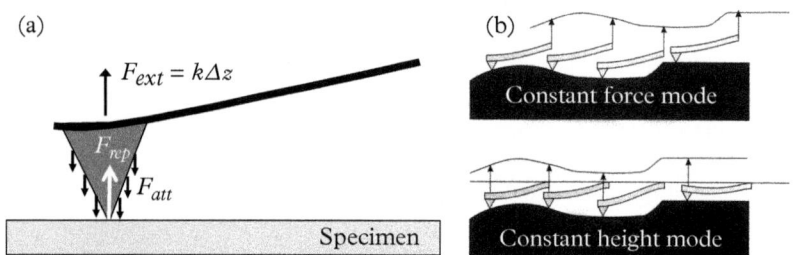

Figure 11.5.2 (a) Force balance affecting the cantilever displacement in the contact mode include the short-range tip-apex-surface repulsion, F_{rep}, the long-range tip–specimen attractive force, F_{att}, and the external force, F_{ext}, due to the cantilever spring constant. (b) Two different scanning modes are used in the contact mode for topographic imaging: the *constant force* or *constant displacement* mode (top) plotting the voltage applied to the feedback system to keep the cantilever deflection at a set value, and the *constant height* mode (bottom) where the feedback loop is turned off and the imaging signal is based on the cantilever deflection rather than the voltage applied to the z-piezo drive.

$$d_c = 2(DRF)^{1/3} \qquad\qquad (11.5.1)$$

where R is the tip radius, $D = (1 - v_t^2)/E_t + (1 - v_s^2)/E_s$, and v and E are the Poison ratio and Young modulus of the tip (t) and specimen (s), respectively. Again, for typical values of $R \sim 90$ nm, $E_t = E_s = 1.5 \times 10^{11}$ N/m^2, $v_t = v_s = 0.3$, we get $d_c \sim 2$–10 nm, for $F \sim 1$–100 nN, which is valid for ambient atmospheres. Thus, these contact diameters limit lateral resolution in contact SFM. However, if the force can be reduced, such as in high vacuum, to 0.1–10 nN, we can expect a smaller contact diameter ($d_c \sim 1$–4 nm). Alternatively, in liquids where the long-range force can be significantly reduced to ~ 10 pN, atomic resolution may be possible. Note that in this mode the tip is always in contact with the surface, and the ensuing short-range repulsive force may damage either the specimen surface, or the tip, during the scanning process. Since the tip is always in contact, in addition to the normal force, the tip and/or specimen may also experience lateral forces. Last, but not least, since the nature of the specimen surface may affect the result, this method could also probe the local nature/properties of the specimen.

Example 11.5.1: An alternative model, adapted from [46], of a contact mode operation of the AFM is shown in the following figure. The cantilever is represented by a spring of force constant, k, and the tip is represented as a sphere of radius, R (typically ~ 50 nm), and in the absence of external forces is separated by a distance, Z, from the specimen surface. The forces acting on the tip cause a deflection, δ (negative), from equilibrium, such that the actual separation is $D = Z + \delta$, as shown in (a).

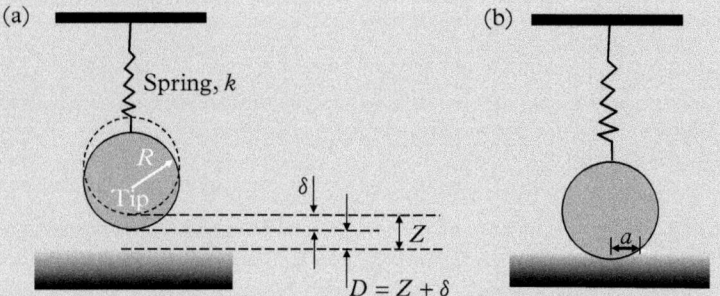

(i) Neglecting short-range forces, the equilibrium position satisfies the condition given by $\frac{A_H R}{6(Z+\delta)^2} + k\delta = 0$, where A_H is the Hamaker constant. This equation can be scaled such that it depends only on one parameter, $l = (A_H R/k)^{1/3}$. Further, at a distance, $D_{\text{crit}} = l/3^{1/3}$, the tip becomes unstable and snaps into contact with the surface as shown in (b). Calculate the scaling parameter, l, and the critical distance, D_{crit}, for typical values of tip parameters.

(ii) When the tip snaps into contact, Figure (b), assume that only the specimen is deformed. Then, the characteristic resolution scale will be defined by the region of contact which will have a minimum radius, $\mathcal{R} = \left(\frac{R^2 A_H}{8KD_{min}^2} \right)^{\frac{1}{3}}$ where $\frac{1}{K} = \frac{3}{2} \left(\frac{1-\nu^2}{E} \right)$, and the minimum distance between the sphere and surface is $D_{min} = 1.5$ Å. Calculate $\mathcal{R}$ for typical tip parameters. How can the resolution be improved?

Solution: Typical parameters for the tip are $E = 1.5 \times 10^{11}$ N/m^2, $R = 50$ nm, $\nu = 0.3$, $k = 0.4$ N/m and $A_H = 10^{-19}$ J. Thus $\frac{1}{K} = \frac{3}{2} \left(\frac{1-0.3^2}{1.5 \times 10^{11}} \right) = 9.1 \times 10^{-12}$.

(i) The scaling parameter $l = (A_H R/k)^{1/3} = 3.45$ nm and the critical distance $D_{crit} = l/3^{1/3} = 2.4$ nm.

(ii) The minimum radius of contact $\mathcal{R} = \left(\frac{(50 \times 10^{-9})^2 10^{-19}}{8(1.1 \times 10^{11})(0.15 \times 10^{-9})^2} \right)^{1/3} = 2.84$ nm, which will be the best achievable resolution. There are two ways to improve the resolution: 1) by reducing the radius, R, of the tip, either by developing "super tips" of smaller radius ($\sim$10 nm) or by sharpening existing tips, and 2) by reducing the Hamaker constant, by immersing the tip and specimen in different liquids.

11.5.2   Lateral Force Microscopy

In contact mode, the vertical deflection of the cantilever, indicated by the difference in the signal between the top and bottom half of the split photodiode detector (Fig. 11.4.2b) is used as the feedback signal. In addition, it is possible to compare the signals from the left and right sides of the split photodiode (Fig. 11.4.2c). If we do the latter, we will obtain information about the lateral deflection of the cantilever, which can be correlated with the mechanical interaction, particularly friction, of the probe with the specimen surface. Thus, this technique is referred to as lateral force microscopy (LFM) or as friction force microscopy (FFM). Strictly speaking, the lateral bending and the vertical deflection of the cantilever are coupled, and thus such scans provide information on both the specimen topography and the materials properties; in particular, the friction depends on the slope of travel of the tip.

The lateral deflection is always different in two different directions (e.g. forward and backward scans; Fig. 11.5.3) Even on perfectly flat and homogeneous specimens, the two signals will differ in magnitude. In general, changes in slope will affect forward and backward lateral scans in opposite fashion, and changes in materials friction properties will exhibit greater or smaller difference in forward and backward scans.

Orientation of the probe   Lateral deflection signal

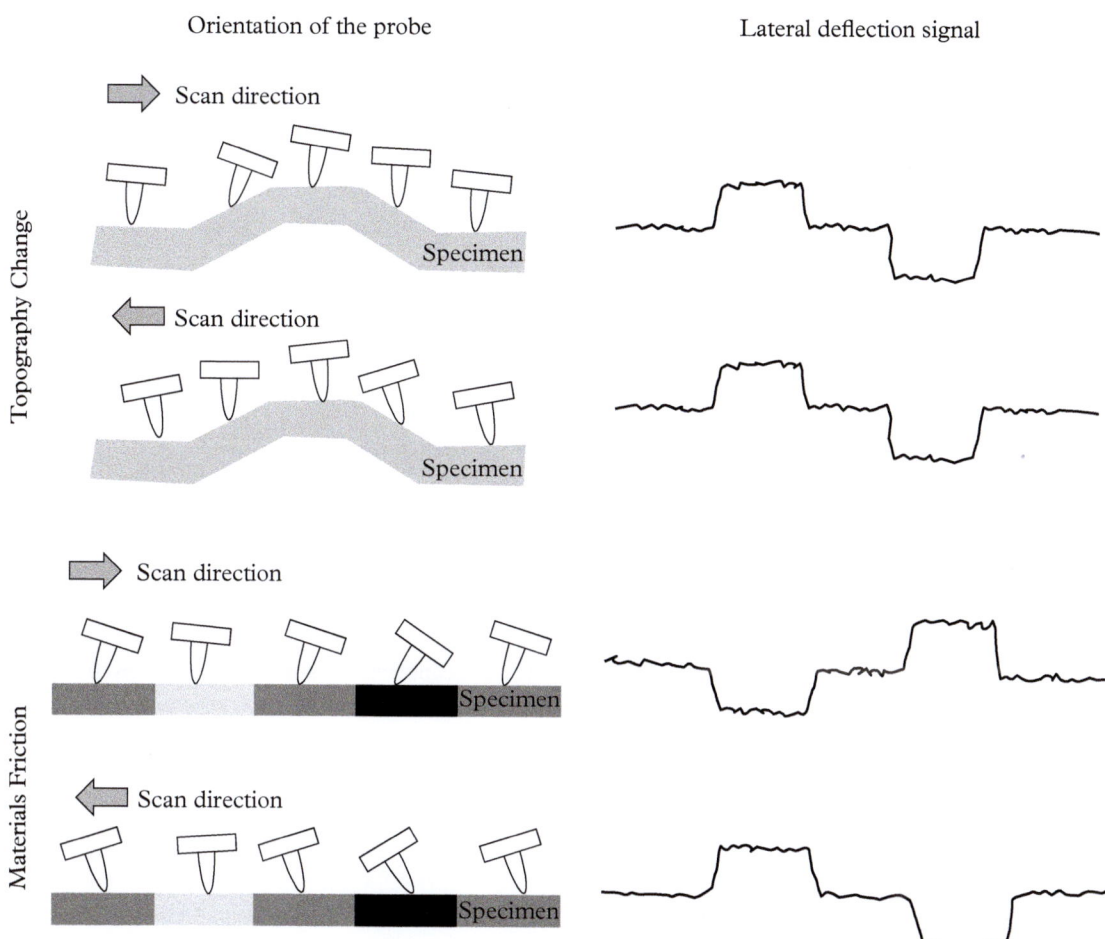

Figure 11.5.3 Lateral signal recorded from a specimen with only (top) topographical, and (bottom) materials friction variations. In the lower figure, darker colors represent higher friction. If the lateral deflection signal for the forward scans is subtracted from the reverse scan, it will show a constant value for the topography changes. However, such subtraction for materials friction, will give a significant difference variation and is indicative of local properties.

Adapted from Eaton and West (2010).

11.5.3 Dynamic Noncontact Modes of Atomic Force Microscopy

In dynamic mode, the amplitude, the resonance frequency, and the phase shift of the oscillation link the dynamics of the vibrating cantilever with a sharp tip to the local tip–surface interactions. In fact, any of the three parameters can be used as the feedback parameter to locally sense and image the topography and other properties of the surface.

In the amplitude-modulated atomic force microscopy (AM-AFM), a stiff cantilever with a sharp tip is mounted on an actuator and externally excited to oscillate at a fixed amplitude, A_{drive}, and a drive frequency, f_{drive}, that is close to but different from its resonance frequency, f_0, given by (11.4.2). As the tip approaches and interacts with the specimen, it causes a change in both its amplitude and phase (with respect to the driving signal) of oscillation. The oscillation amplitude is used as the feedback parameter to measure the topography. In addition, the phase shift variation could be used to map the variation in materials properties. Unfortunately, the change in amplitude as a result of tip–specimen interaction only occurs over a time scale of $\tau_{AM} = 2Q/f_0$, where Q is the quality factor. In vacuum, Q factors can be quite large ($\sim 10^5$) and the AM-AFM mode can be rather slow. To overcome this, the frequency modulated AFM (FM-AFM) was invented (Fig. 11.5.4).

Tapping or intermittent-contact mode (IC-AFM) is a dynamic mode that allows the probe to touch the specimen briefly to experience a repulsive tip–specimen interaction. In this mode the feedback is usually based on the amplitude modulation. Moreover, unlike in contact mode where the lateral forces cause problems, here they are eliminated as the tip moves perpendicular to the specimen as it is scanned. Further, since the cantilever withdraws the tip away from the specimen, any capillary forces from water vapor or other contamination on

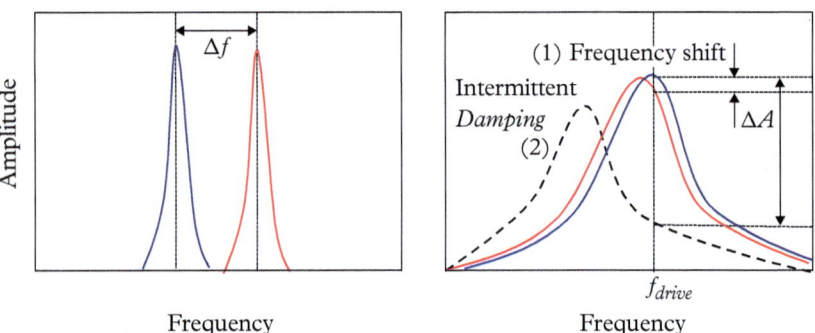

Figure 11.5.4 Two dynamic non-contact modes are used to detect the tip–specimen interaction in atomic force microscopy. (a) In frequency modulation (FM-AFM), the shift in frequency, Δf, when the cantilever is driven at its resonance frequency, caused by the tip–specimen interaction is detected. (b) In amplitude modulation (AM-AFM), the change in amplitude, ΔA, at a fixed frequency is detected. One variant of AM-AFM is the intermittent contact (IC-AFM) or tapping mode. In this case, the decrease in amplitude, ΔA, can arise from a frequency shift (1), or the damping (2), when the tip touches the surface.

Adapted from Meyer, Hug, and Bennewitz (2004).

the specimen surface affects it minimally. In IC-AFM, both the change in the amplitude as well as the phase (delay) of the probe is recorded (Fig. 11.5.5). The latter is sensitive to materials properties (Fig. 11.5.6).

AM-AFM imaging also provides the opportunity to map variations in the composition, friction, viscoelasticity, and adhesion properties of the specimen surface. The method is called phase contrast imaging and it measures the phase lag of the tip with respect to the drive signal, while the feedback keeps the amplitude constant. Figure 11.5.6 illustrates the difference in contrast obtained by standard topography and phase contrast obtained in AM-AFM imaging of a thin film of hydrogenated diblock copolymer (PEO-PB). The topography image shows no specific features; however, the phase contrast image shows that the copolymers

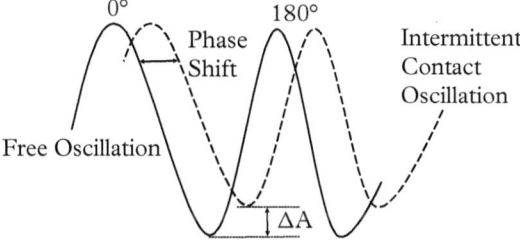

Figure 11.5.5 The effect of intermittent contact (tapping mode) on the oscillation of the cantilever includes a reduction in amplitude, ΔA, due to the repulsive contact with the specimen surface and a phase shift determined by the properties of the material.

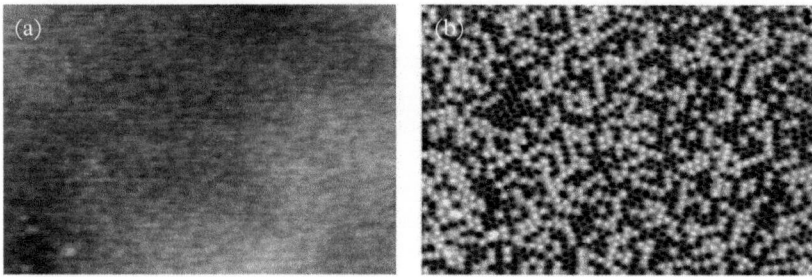

Figure 11.5.6 AM-AFM images of a block copolymer mesophase. (a) Topography, and (b) phase contrast image. In (b) individual spheres (12 nm diameter), which are either crystalline (light) or non-crystalline (dark) PEO micelles, are resolved. The long edge of the image is 1 mm and the maximum height variation detected in (a) was 10 nm.

Adapted from [17].

organize into spherical PEO micelles with crystallization occurring individually for each sphere.

Alternatively, in FM-AFM the cantilever is kept oscillating at its resonance frequency with a fixed amplitude. The local force gradient, F', between the tip and specimen causes a change in the effective cantilever stiffness, i.e. $\Delta k = F'$. Now, the fractional change, Δf, in the cantilever resonance frequency, f_0, can be related to the fractional change in the cantilever stiffness as

$$\frac{\Delta f}{f_0} = \frac{\Delta k}{2k} = \frac{F'}{2k} \tag{11.5.2}$$

The spatial dependence of the frequency shift, Δf, is the source of contrast in the FM-AFM image. The important advantage of this method over AM-AFM is that the change in resonance frequency occurs very fast within a single oscillation over a time scale $\tau_{FM} = 1/f_0$.

Initially, both AM-AFM and FM-AFM were conceived to be noncontact modes of operation and the cantilever was maintained at sufficient distance from the specimen to obtain a net attractive force between the tip and specimen. Note that most AFM experiments in UHV are performed in FM mode, whereas experiments in air or in liquids are performed in the AM mode.

11.6 Scanning Force Microscopy Instrumentation

The principal operation of an SFM is shown in the block diagram of Figure 11.6.1a, and the components of the stage, which is the heart of the instrument, are highlighted in Figure 11.6.1b. The stage includes the specimen holder, X-Y specimen positioning platform, the coarse vertical approach mechanism (Z motor), the specimen x-y-z scanner, and usually, an optical microscope to help position the tip over the microstructural feature of interest in the specimen. The propensity of the stage to vibrate mechanically varies inversely as its physical size, and for high resolution it is constructed to be as small as possible. In addition, the single most important factor that influences the vertical resolution of an AFM is the rigidity of the *mechanical loop* of the instrument. By the mechanical loop, we mean all the elements (x-y-z scanner, X-Y specimen stage, Z motor, and the probe) that are required to hold the probe at a precise and fixed distance from the surface.

The x-y-z scanners, which control the fine movement of the probe, are normally constructed of amorphous lead–barium–titanate (PBT) or lead–zirconium–titanate (PZT) piezoelectric ceramics materials and are optimized either for maximum expansion with respect to the applied voltage or for linear expansion-voltage characteristics. Generally, one of the two factors mentioned is optimized at the expense of the other.

A piezoelectric ceramic preserves its volume as it changes its geometry on the application of an electric voltage. Based on this simple idea, two standard

(a)

(b)

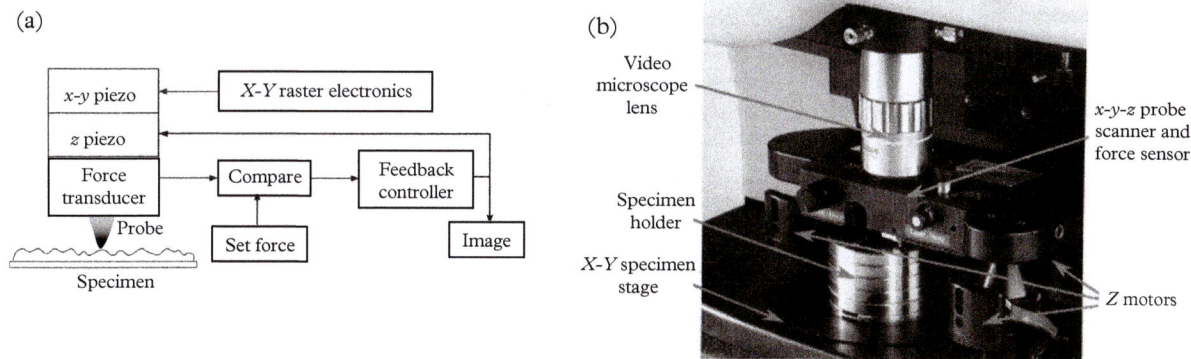

Figure 11.6.1 (a) Block diagram of the AFM in operation. (b) A photograph of an AFM stage with principal components identified.

Adapted from Eaton and West (2010).

configurations of piezoelectric scanners are designed to move the probe (or specimen) accurately in an AFM. The widely used tube scanner (Fig. 11.6.2a), with inner and outer electrodes, is compact, easy to fabricate, and produces precise movements, especially for small scan ranges. In addition, the clear optical path through the center of the tube makes for easier probe positioning; however, the scans suffer from nonlinearity. Alternatively, the tripod scanner (Fig. 11.6.2b) is the simplest three-dimensional scanner design and allows for independent control of motion of all three axes. In practice, all piezoelectric drives suffer from nonlinear behavior of *hysteresis* and *creep* as a function of applied voltage (see §11.7.2). In addition, an AFM/SFM instrument requires force sensors (the cantilever arrangements discussed in §11.4.2), appropriate electronics for generating voltage ramps corresponding to the x-y positioning of the probe, feedback control to drive the z-piezoelectric ceramic, and appropriate signal collection for amplitude, frequency, or phase modulation; these are discussed in detail in Eaton and West (2004).

Finally, from a practical point of view, the image profile will depend on the shape of the probe, with a sharper probe often giving a better horizontal resolution, and many techniques (Fig. 11.4.3d–g) are used to obtain sharp probe tips.

11.7 Artifacts in Scanning Probe Microscopy

11.7.1 Probe Artifacts

As mentioned in Figure 11.3.2, SPM images are a convolution of the specimen topography with the shape of the tip of the probe. In simple terms, all specimen features with a radius of curvature smaller than that of the probing tip are not properly imaged (Fig. 11.7.1a). If the probe tip is blunt, images of convex features,

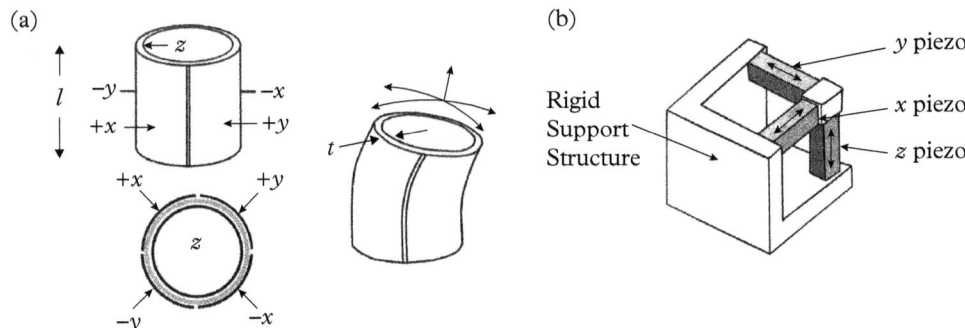

Figure 11.6.2 The two types of piezoelectric scanners used commonly in AFM. (a) The tube scanner, with length, l, and thickness, t, moves in the x-y-z directions using four electrodes outside for x-y motion and an inner electrode for z-motion. The movement of the piezodrives are $\Delta x = \Delta y \propto \frac{l^2}{t} V$ and $\Delta z \propto \frac{l}{t} V$. (b) The simplest tripod scanner with $\Delta x, \Delta y, \Delta z \propto V$. In both cases, V is the applied voltage.

Adapted from Eaton and West (2010).

such as particles, would be larger than expected in the lateral dimension (Fig. 11.7.1b); on the other hand, concave features, such as holes on a flat surface, may appear narrower than they actually are (Fig. 11.7.1c). Moreover, if the features have a higher aspect ratio than the probing tip, they are also not imaged properly and instead produce repeating images of an inverted probe or its side walls (Fig. 11.7.1d).

In addition, probes often get contaminated, and if such "dirty" probes are used they show repeating patterns in the images (Fig. 11.7.2a). Sometimes, when the tip is broken it can end up having a double tip; in this case, each feature is doubled and shows a false "twin" in the image (Fig. 11.7.2b). If this were to happen, it is best to replace with a fresh probe and check its performance using resolution standards available from national metrology laboratories or commercial SPM vendors.

> **Example 11.7.1:** For a probe geometry with cross-section described as an upside-down triangle of semi-angle, θ_0, the diameter of a particle measured, d_{meas}, is related to its actual diameter, d, by a convolution term, i.e.
>
> $$d_{meas} = d \left(\cos \theta_0 + \sqrt{\cos^2 \theta_0 + (1 + \sin \theta_0) \left(-1 + \left(\frac{\tan \theta_0}{\cos \theta_0} \right) + \tan^2 \theta_0 \right)} \right)$$
>
> Calculate the actual diameter for a particle with $d_{meas} = 100$ nm for
>
> (a) $\theta_0 = 30°$

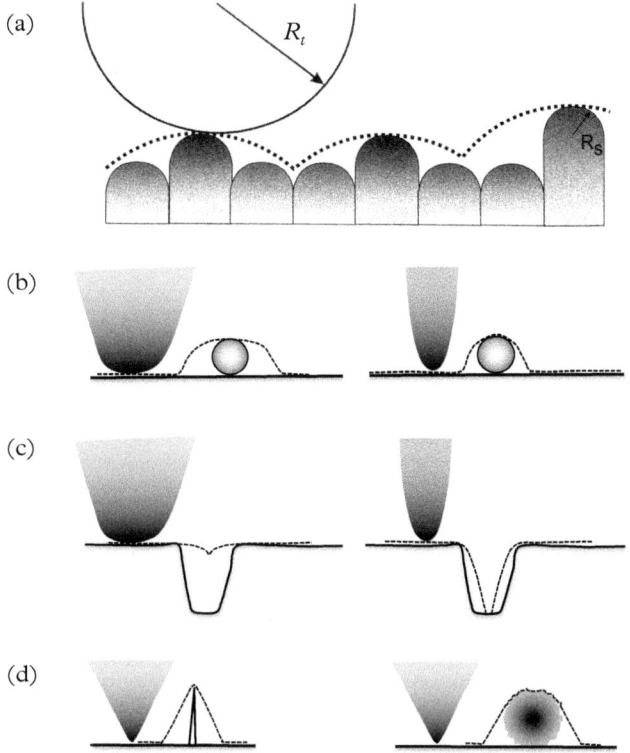

Figure 11.7.1 Probe tip artifacts: (a) Image of a standard surface of asperities with radius of curvature, R_s, much smaller than the tip radius, R_t. The profile measured (dotted line) is not representative of the surface but corresponds to an inverted probe tip of radius R_t. (b) Images (dotted lines) of a convex surface (particles) obtained by using two different tip radii. The height is accurate but the feature appears much broader when a probe of larger radius is used. (c) Images (dotted lines) of a concave surface (holes) appear less wide and less deep with the larger probe. (d) Images of features with high aspect ratio resemble images of the inverted probe (left) or its side walls (right).

(a) Adapted from Meyer, Hug, and Bennewitz (2004). (b–d) Adapted from Eaton and West (2010).

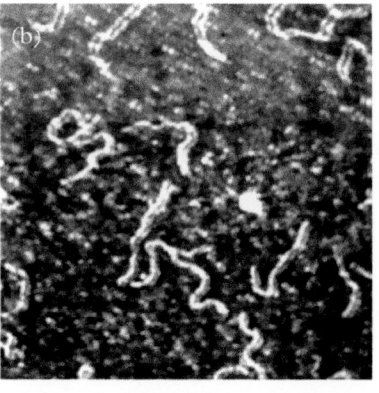

Figure 11.7.2 Examples of tip artifacts (a) Repeating images due to dirty or broken tips. (b) Images of DNA molecule with a broken tip with each molecule showing a false "twin" image next to it.

Adapted from Eaton and West (2010).

(b) $\theta_0 = 60°$.

Solution: The convolution term for

(a) is 1.732. Thus $d = 100/1.732 = 57.7$ nm

(b) is 3.732. Thus $d = 100/3.732 = 26.8$ nm.

Clearly, the sharper tip (a) requires a smaller convolution correction, but it is still substantial.

11.7.2 Instrument Artifacts

The two major artifacts of piezoelectric scanners are hysteresis and creep. Piezoelectric materials exhibit *hysteresis* in the displacement versus voltage behavior, and when a saw tooth voltage is applied, the position of the actuator can deviate by as much as 15% between forward and backward movements. In addition, when an instantaneous voltage is applied and maintained, the piezoelectric material may not only move to the new position, but may continue to move in the same direction even when the voltage remains unchanged. This movement is called *creep*. Even though the duration of creep is short, it distorts the scanned image; in practice, by waiting long enough for the piezo position to stabilize before imaging, this artifact can be minimized. Alternatively, position sensors have been integrated in the cantilever force sensor to measure the true motion of the tip relative to the specimen, and make real-time scan corrections [18]. Unfortunately, such hardware correction systems offer poor signal to noise, making it difficult to achieve high resolution. In that case, software corrections that are based on empirical approximations are used.

In addition, high-resolution AFM scanners also suffer from specimen *drift*. To avoid this artifact, specimens are often glued to the specimen support. Even then, thermal expansion of the specimen can appear as a movement of the specimen. By changing the scan direction, the nature of the drift can be clarified. In addition, by repeated measurement of a specimen feature the drift velocity can be determined; typically, this can be as much as nm/s at room temperature, but can be reduced by one order of magnitude to Å/s by lowering the operating temperature.

11.8 Select Applications of Scanning Force Microscopy

SFM or AFM is now well established and successfully applied in many fields. We now review some representative examples involving high-resolution AFM imaging (§11.8.1), measurement of local magnetic (§11.8.2) and thermal (§11.8.3) properties of materials, applications in the life sciences (§11.8.4), and lithography on the nanoscale (§11.8.5). A comprehensive discussion of all aspects of the subject can be found in more specialized textbooks, including Meyer, Hug, and Bennewitz (2004) and Eaton and West (2010).

11.8.1   Atomic Fingerprinting in Frequency Modulated Atomic Force Microscopy

As mentioned earlier, under UHV conditions, with excellent vibration isolation, and noncontact operation, FM-AFM can be used to obtain atomic resolution imaging. However, its development as a true metrology tool that is capable of chemically identifying individual atoms on the surface requires true innovation and careful calibration of the local chemical forces [20]. Recall that in FM-AFM, the atomic-scale interaction of the apex of the tip with the specimen surface atoms (Fig. 11.8.1a) involving short-range chemical and long-range van der Waals forces leads to a change in its resonant frequency. Such variation is measured as a function of the tip position and plotted as a three-dimensional atomic image of the surface (Fig. 11.8.1b).

Normally, it is difficult to establish the chemical identity of the individual atoms on the surface, as the forces measured above individual atoms (such as Si, Sn, and Pb, in the surface alloy shown in Figure 11.8.1) vary from scan to scan and from tip to tip. This is understandable as the precise atomic shape and composition of the apex of the tip (Fig. 11.8.1a) may change over time and differ from tip to tip. Even though the atomic structure of the tip is impossible to determine and control reproducibly, Sugitomo et al. [20] discovered a clever way to address the problem and demonstrated the possibility of atomic fingerprinting in FM-AFM imaging.

By repeatedly measuring a large number of force-displacement curves at atomic resolution for the three species (Si, Sn, and Pb), they observed that the strongest interaction was always between the tip and Si atoms. So, they normalized the maximum attractive force for individual atoms by the maximum force measured for Si atoms, and in this way, they created a distinctive fingerprint for the three different atoms, i.e. Si ~100%, Sn ~77%, and Pb ~55% (Fig. 11.8.1c). Furthermore, using atomistic modeling [21, 22] of the interaction between the tip and surface, they developed a standard procedure [23] now used to interpret atomic-scale AFM imaging. First, they confirmed plausible atomic structures for the tip and then calculated the forces for the three different species that matched well with the experimental variation in their individual force curves. Now, armed with such fingerprinting, they used the measurement of the chemical forces as the basis of atomic recognition in the FM-AFM images of a surface alloy of Si, Sn, and Pb atoms (Fig. 11.8.1d).

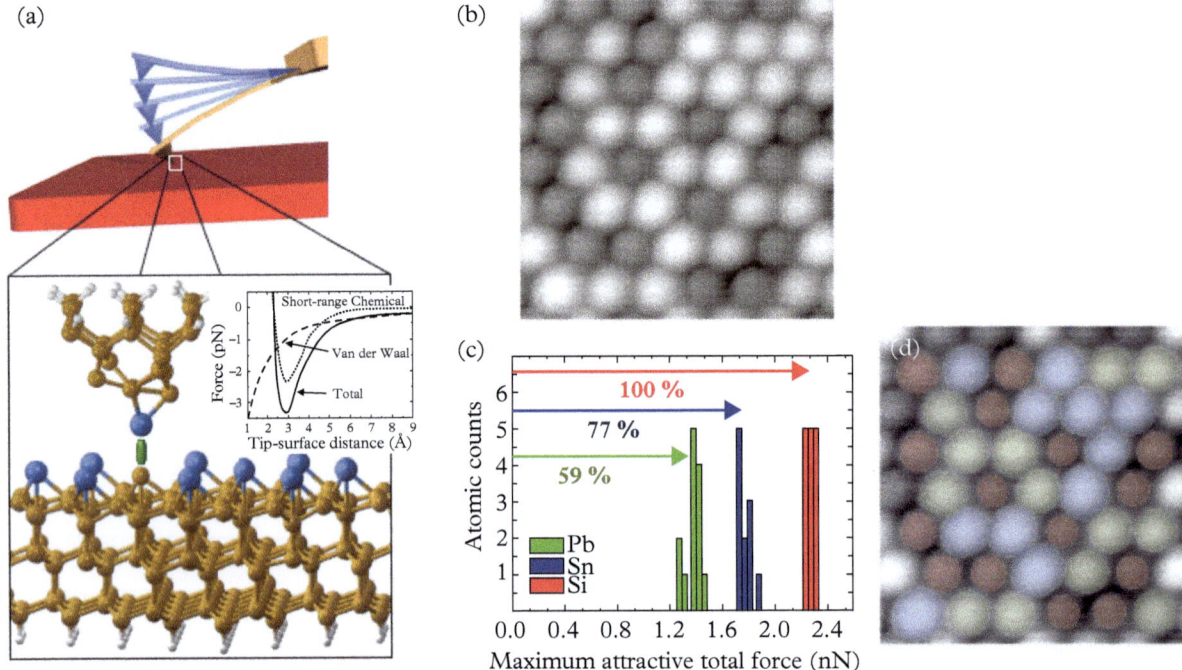

Figure 11.8.1 Atomic fingerprinting with frequency modulated atomic force microscopy. (a) Operation of the microscope in dynamic mode, illustrating the onset of the chemical bond between the atoms at the apex of the tip and the surface. The inset shows the contributions to the force-distance curves from the chemical and van der Waals forces. (b) The atomic resolution dynamic mode image of the surface obtained without resolving the chemical species. (c) Calibration of the attractive force for the three atomic species, normalized with that of Si. (d) The same image as in (b) but now color coded to identify the local chemical composition at atomic resolution.

Adapted from [20].

11.8.2 Magnetic Force Microscopy (MFM)

As mentioned earlier (11.4.7), the spatial variation of the stray magnetic fields on the surface of a specimen can be measured directly in an SFM, provided the tip is coated with a ferromagnetic thin film. In practice, this can be easily accomplished by coating a standard silicon tip with a thin magnetic film; typically, Co or alloys of Co–Cr, Co–Ni, etc., are used. This may have two detrimental consequences: one, it increases the radius of the tip leading to a deterioration of the resolution; and two, it may possibly soften the surface of the tip, mechanically speaking, leading to an increased wear rate. Moreover, when the tip and specimen are in contact, the magnetic forces are one order of magnitude smaller than the other tip–specimen forces, which are dominated by the shorter-range van der Waals interactions. However, magnetic forces are reasonably long range, and hence it is

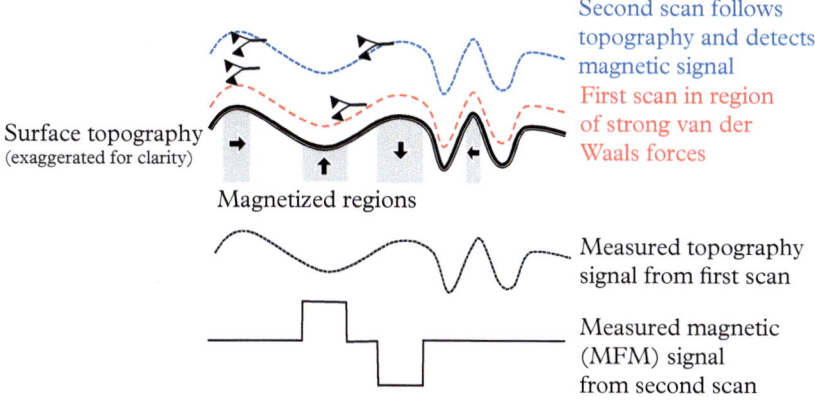

Second scan follows topography and detects magnetic signal

First scan in region of strong van der Waals forces

Surface topography (exaggerated for clarity)

Magnetized regions

Measured topography signal from first scan

Measured magnetic (MFM) signal from second scan

Figure 11.8.2 Schematic of one approach to implement magnetic force microscopy involving two scans. The first topography scan is carried out close to the surface in the region of strong van der Waals forces. The second scan uses the results from topography scan to keep the probe at constant height, but at sufficiently larger tip–specimen distances where only the magnetic force dominates to obtain the MFM image.

advantageous to measure them with the tip positioned 5–15 nm distant from the surface, to avoid interference from the other forces.

The key to successful MFM imaging is to separate the magnetic signal from the total signal, which includes the topographical variation (Fig. 11.8.2). In practice, the separation can be achieved by carrying out two scans. In the first scan, the tip is positioned close to the surface, where the van der Waals image dominates, to obtain a topographic image. Then, for the second scan, the tip–specimen distance is increased to sense only the magnetic forces, but the set point based on the first scan is varied so that the tip follows the topography of the surface (Fig. 11.8.3). Now, magnetic contrast in the MFM image can be obtained by measuring either the amplitude, frequency, or phase change in the cantilever oscillations as a function of position. Alternatively, similar information can be obtained by intermittent contact, or tapping mode AFM, where the momentary contact of the tip with the surface monitors the surface topography and the noncontact portion of the scan monitors the magnetic forces (Fig. 11.8.3). MFM is reviewed in [24]. The technique compares very well with other modes of magnetic imaging; for further information, see Krishnan (2016).

The spatial resolution in MFM depends on the magnetized part of the probe and the tip–specimen distance. Best lateral resolution can be achieved by keeping the magnetized part of the tip to the smallest volume possible, consistent with a detectable magnetic signal. Typically, this is achieved by keeping only a small particle of magnetized material at the apex of the tip. In practice, this condition is obtained by coating a standard tip (Fig. 11.8.4a) with the magnetic material, masking a small region of the tip apex, and then etching away the magnetic coating in the remaining portions of the tip. Such a super tip (Fig. 11.8.4b,c) will produce superior resolution MFM images (Fig. 11.8.4 d–g).

For completeness, it must be mentioned that there are other interesting developments in magnetic imaging that combine the imaging power of magnetic resonance imaging (MRI) and the sensitivity of the AFM to create a hybrid

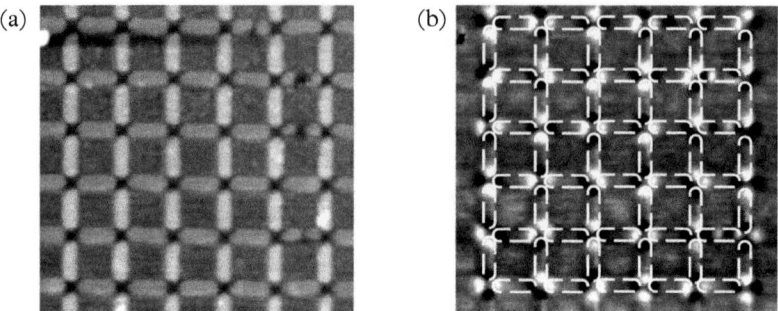

Figure 11.8.3 Images of a magnetic array of Fe nanoelements, each of height ~420 nm and width ~120 nm, and fabricated by e-beam lithography. (a) An AFM image in the tapping mode, showing the excellent geometric pattern of the array. (b) A MFM image with a lift height of ~50 nm, showing one of the possible magnetic "spin ice" configuration in this frustrated lattice system. Note that each element, shown by its outline, has a single domain configuration and is characterized by a pair of black and white "dots" corresponding to their magnetic polarity.

Adapted from [44] and [47].

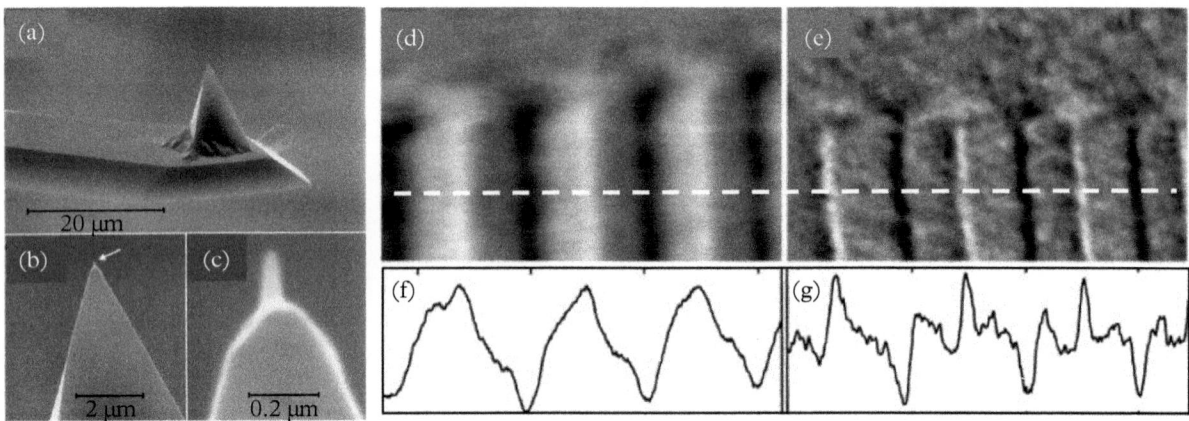

Figure 11.8.4 Improved resolution obtained in MFM by optimizing the magnetic coating on the cantilever tip. Scanning electron micrographs of (a) a standard fully coated tip, and (b) low magnification and (c) high magnification of a super-tip prepared by masking and etching the coated tip. MFM images taken of a hard magnetic disk surface with recording tracks at 2 μm periodicity with (d) the standard tip and (e) the super tip. The corresponding line scans, show the superior resolution of the super-tip (g), compared to the standard tip (f).

Adapted from [24].

technique that can resolve single atomic spins. Further details of this technique, known as magnetic resonance force microscopy (MRFM), can be found in [25].

11.8.3   Scanning Thermal Microscopy (SThM)

As a consequence of the second law of thermodynamics, all *irreversible* processes involving energy interactions with the surroundings require that some of the energy be dissipated in the form of heat. However, the transport of thermal energy can be difficult to control but the scanning thermal microscope (SThM) overcomes this difficulty by controlling the flow of energy near points of contact, enabling investigation of thermal transport phenomenon at small length scales.

Figure 11.8.5 shows a schematic diagram of a wire thermocouple AFM probe, which is the heart of an SThM. The probe is made of two thermocouple wires, electrochemically etched to form sharp tips, and then bonded together by capacitor discharge to form a thermocouple junction as well as an AFM tip. A reflecting metal strip is then attached across the two thermocouple wires for reflection sensing using a laser beam. Even though the smallest junction obtained (diameter ∼15 μm) is too large for topographical imaging, the junction often contains a sharp point that can effectively work as an imaging tip. Typically, the thermocouple probe is then calibrated by bringing it in contact with a copper surface maintained at uniform temperature.

The application of this SThM was first demonstrated with topographical (Fig. 11.8.6a) and thermal (Fig. 11.8.6b) images of a metal-semiconductor field effect transistor (MSFET) in operation. Note that in the thermal image the source and drain are cooler than the semiconductor. This is because the metal electrodes are of lower electrical resistance. In addition, the heating is highest under the gate with the narrowest electrical channels. Finally, the drain side was observed to be

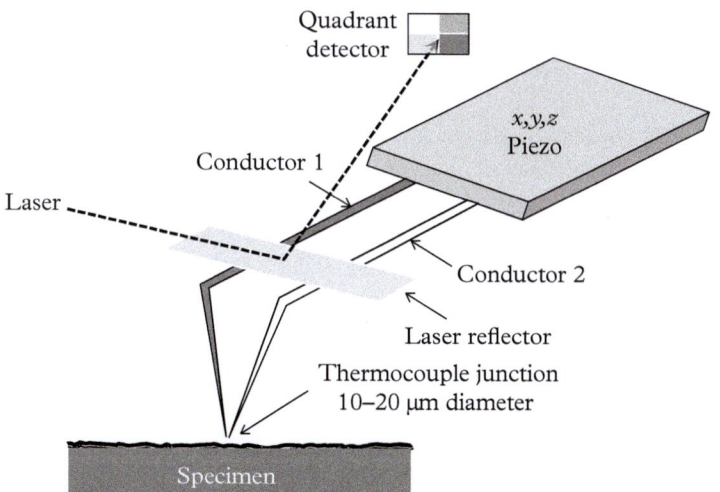

Figure 11.8.5 Schematic illustration of a cantilever thermocouple probe, formed by a two-wire junction, used for scanning thermal microscopy (SThM). In ambient conditions, the heat transfer through the liquid meniscus dominates.

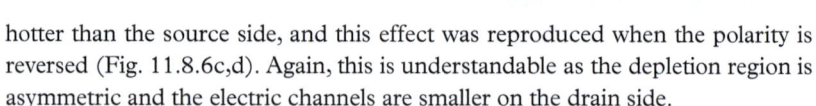

Figure 11.8.6 Scanning thermal microscopy images of a *n*-GaAs metal-semiconductor field effect transistor (MESFET). (a) Topography and (b) thermal images obtained using a wire thermocouple cantilever probe. Images obtained when the source and drain polarities are reversed (c, d), show that the hot spot is consistently on the drain side of the gate.

Adapted from [26].

hotter than the source side, and this effect was reproduced when the polarity is reversed (Fig. 11.8.6c,d). Again, this is understandable as the depletion region is asymmetric and the electric channels are smaller on the drain side.

This is but a simple discussion to illustrate the early development of SThM. Subsequent advances in probe fabrication, including single wire thermocouple probes, Schottky diode tips that can serve as both heaters and temperature sensors, and thin films temperature sensors deposited on commercial SiN_x cantilever probes, have contributed to a steady improvement of SThM. These developments are reviewed comprehensively in [26] and are also discussed in Meyer, Hug, and Bennewitz (2004).

11.8.4 Applications of Atomic Force Microscopy in the Life Sciences

Biological processes generally occur in a liquid environment and depend critically on the environment (pH and concentration of ions) and temperature. Moreover, biological building blocks (cells, membranes, molecules, etc.) demonstrate significant changes in structure and behavior when removed from the liquid environment and dried. The AFM, with its ability to image or measure specimens in physiologically relevant conditions—buffer solutions, body temperature (37°C), and varying pH—is ideally suited for studies in biology and the life sciences; see [27] and Morris, Kirby and Gunning (1999) for detailed discussions. These studies can be broadly classified into five areas.

11.8.4.1   *Imaging of Biomolecules*

Of the four important classes of biomolecules—carbohydrates, lipids, nucleic acids, and proteins—proteins have been extensively studied by AFM because of their importance in disease and other biological processes. In fact, a particular area where AFM excels over TEM or other crystallographic methods is in the study of protein complexes under realistic biological conditions, i.e. in water or in buffer solution. A specific example [28, 29] that illustrates the versatility of AFM is the study of *E. coli* chaperonins, GroEL and GroES, and their complexes that play a critical role in protein folding processes. Combining different imaging modes, including contact and high-speed tapping modes, it has been shown that the GroES ring sometimes forms a lid for the GroEL, adding a height of ∼5 nm for GroEL/ES complex and affects the kinetics of the association or disassociation of the two entities (Fig. 11.8.7).

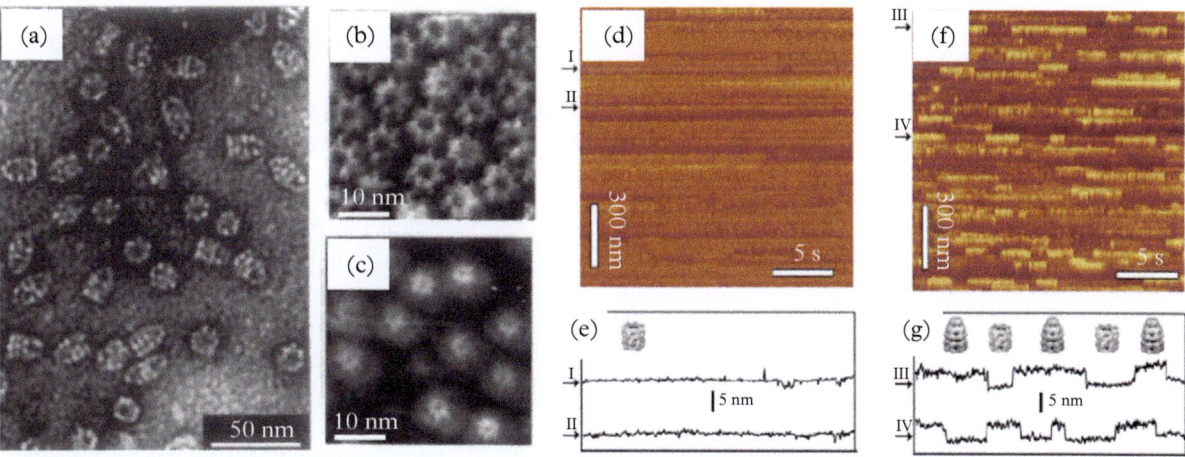

Figure 11.8.7 Images of GroEL and GroES chaperonins and their complexes. (a) TEM image of GroEL/ES complex via negative staining; nearly all GroEL show one or two GroES bound to them. (b) High resolution AFM image of the surface structure of the apical domains of the heptamer with a circular opening and seven elongated domains radiating outward. (c). Image of the GroEL/ES complex. Adapted from [28]. Association and dissociation of the GroEL/ES complex. (d) High-speed IC-AFM scans of GroEL in buffer solution without any GroES. (e) Height fluctuations of GroEL shows uniform traces. (f) Addition of GroES to GroEL in the buffer solution shows repeat variations in height. (g) The height varies between two values, differing by 3.6 ± 1 nm, consistent with the height difference between GroEL and GroEL/ES complex, indicating rapid association and disassociation.

Adapted from [29].

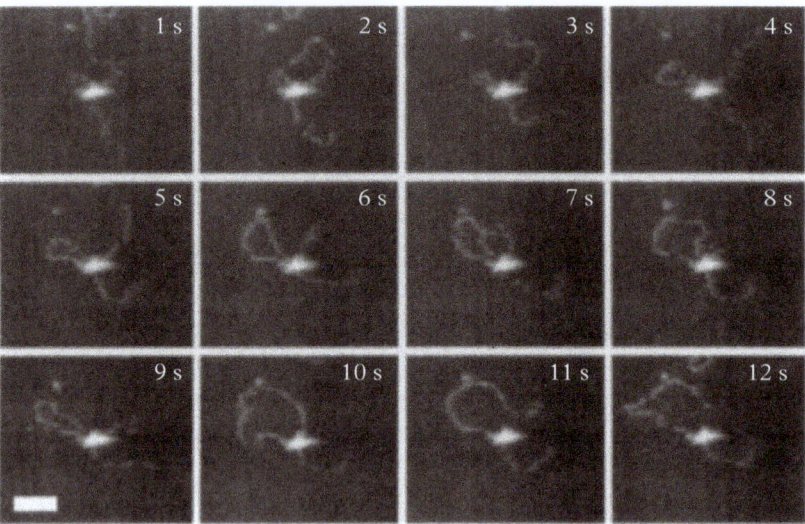

Figure 11.8.8 Studies of DNA translocation and looping in real time by fast-scan AFM. Time-lapse images of an EcoP15I-DNA complex obtained at 1 frame per sec. The elapsed time is shown in each image. Translocation and formation of an extruded loop between 1 and 10 s, before release of the loop at 11s.

Adapted from [30].

Another example is the study of DNA molecules, which are imaged by depositing a drop of solution on a freshly cleaved mica substrate. Note that both mica and DNA molecules are negatively charged under normal conditions and require positively charged cations (e.g. Ni^{2+} or Mg^{2+}) to create a bridge between the substrate and the DNA molecule to immobilize the latter, and allowing for its imaging in air or in solution. Further, controlling the ionic concentration in the solution allows the DNA molecules to move in two dimensions while ensuring that they remain affixed on the surface for imaging. This has allowed observations [30] of protein–DNA interactions in real time (Fig. 11.8.8).

11.8.4.2 *Imaging of Lipid Membranes*

It is well known (Karp, 2003) that plasma membranes, or lipid bilayers, separate the *intra*cellular components from the *extra*cellular environment of cells. Further, the membrane allows selective molecules to go in and out of the cell while blocking all other unwanted substances. Model phospholipid bilayers can be prepared by Langmuir–Blodgett methods, deposited on substrates, and the resultant flat surfaces and their phase separation [31] imaged [32] by AFM (Fig. 11.8.9).

11.8.4.3 *Imaging of Bacterial Cells*

AFM images complement the widely used optical imaging studies of bacteria. While the optical methods provide statistically significant data on the morphology of cells and genes, AFM provides unique details of cellular properties using force spectroscopy. In particular, well-known bacterial such as *E. coli* [33], *Bacillus* [34], and *Salmonella* [35], have been studied by immobilizing them on a surface (Fig. 11.8.10). Furthermore, a subject of medical relevance is the interaction of antibiotics with bacteria; an example [36] of AFM imaging of the response of

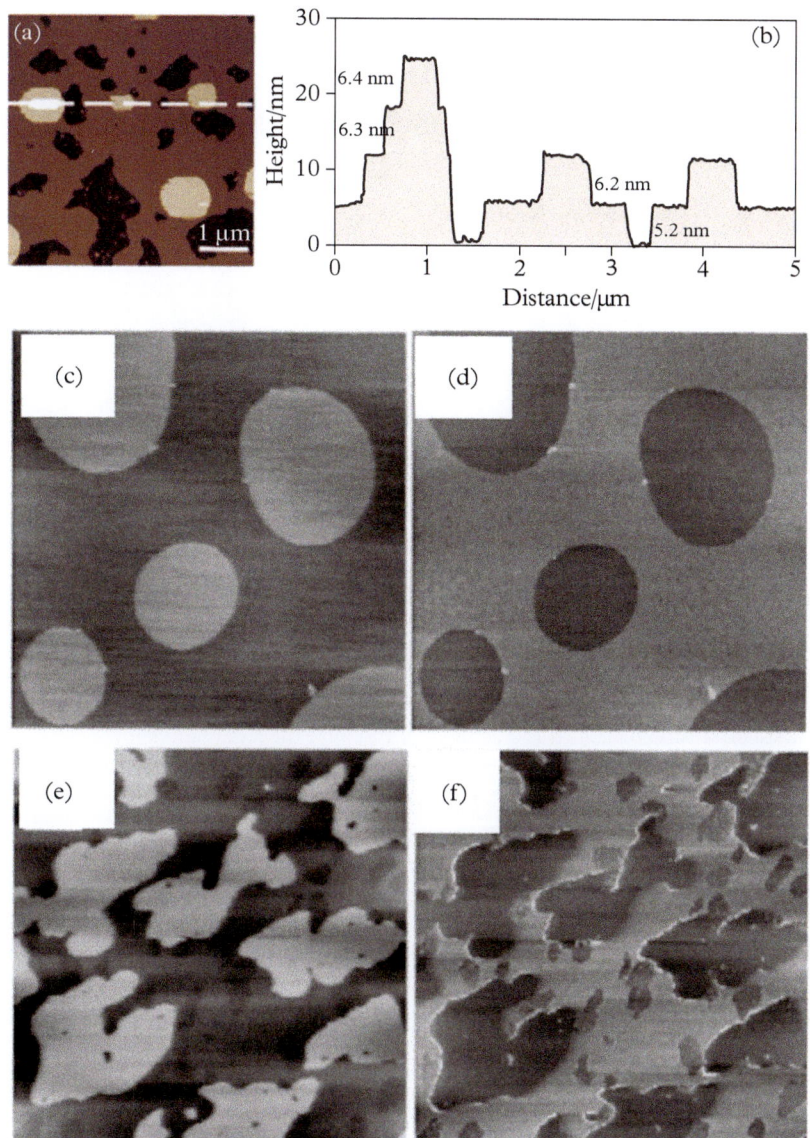

Figure 11.8.9 Imaging of bilipid layers. (a) Tapping mode (IC-AFM) image of multiple lipid bilayers on mica under a fluid. (b) Thickness of the bilayers range from 5.2 ± 0.1 nm adjacent to mica, to 6.2–6.4 ± 0.1 nm for the subsequent layers. (c, e) Topography, and (d, f) friction images of monolayers in air (c, d) and bilayers under water (e, f) of a mixture of saturated distearoyl-phosphatidylethanolamine (DSPE) and dioleoyl-phosphatidylethanolamine (DOPE).

(a–b) Adapted from [31]. (c–f) Adapted from [32].

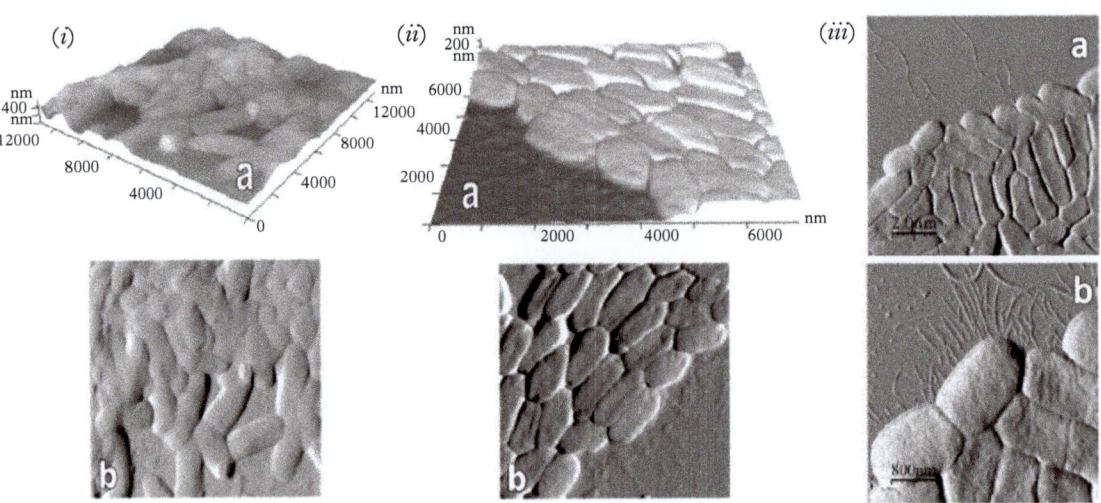

Figure 11.8.10 *Escherichia coli* bacterial cells dispersed on mica and imaged by AFM in contact mode applying a force of 1nN showing (a) 3D height and (b) deflection mode images. (i). Image of bacteria in liquid cultivation medium (ii) Image of bacteria in air. (iii) High-resolution AFM images of pili-like fimbrial structures and flagella in the curli mutant MAE14 after 4 h at two different scan sizes.

(i–ii) Adapted from [33]. (iii) Adapted from [35].

bacteria to antibiotic treatment, showing a significant reduction in stiffness after treatment, is illustrated in Figure 11.8.11.

11.8.4.4   *Studies of Mammalian Cells*

In addition to imaging cells under physiological conditions, AFM can be used for nanoindentation or nanomechanical response studies of live cells. This allows differentiation between diseased and healthy cells as the mechanical stiffness of a cell decreases upon tumor invasion and metastasis. The difference in stiffness between healthy and cancerous cells measured by AFM (Fig. 11.8.12) has been proposed as a way to diagnose cancer [37].

11.8.4.5   *Biological Force Spectroscopy*

Last, but not least, force distance curves of AFM cantilevers are used to study the strength of specific antigen–antibody interactions. In fact, the binding between avidin and biotin is now used as the standard in such measurements because of both its strength and specificity [38]. A variant of such force spectroscopy is to measure [39, 40] unfolding dynamics of proteins (Fig. 11.8.13). The protein is adsorbed onto a surface, often by covalent attachment to gold through cysteine residues engineered at the C-terminus. When the cantilever is lowered onto the surface one of the protein molecules attaches to it, typically by nonspecific adsorption. Next, when the cantilever is raised away from the surface by a piezo

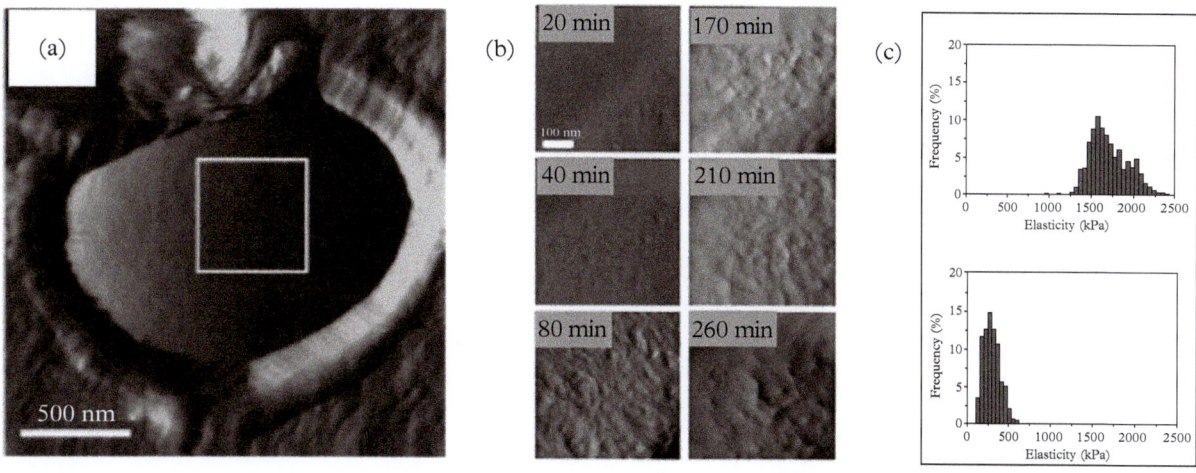

Figure 11.8.11 Direct AFM observation of bacterium *Staphylococcus aureus* cell wall digestion by the antibiotic Lysostaphin. (a) Low-resolution deflection image of a single *S. aureus* cell trapped in a pore of the polycarbonate membrane recorded in phosphate buffered saline (PBS) solution. (b) Imaging of lysostaphin-treated cells at high resolution. A series of deflection images recorded on top of a single cell after incubation with 16 μg/mL Lysostaphin for 20, 40, 80, 170, 210, and 260 min is shown. (c) The effect on cell wall stiffness with (top) no, and (bottom) 80 minutes of treatment.

Adapted from [36].

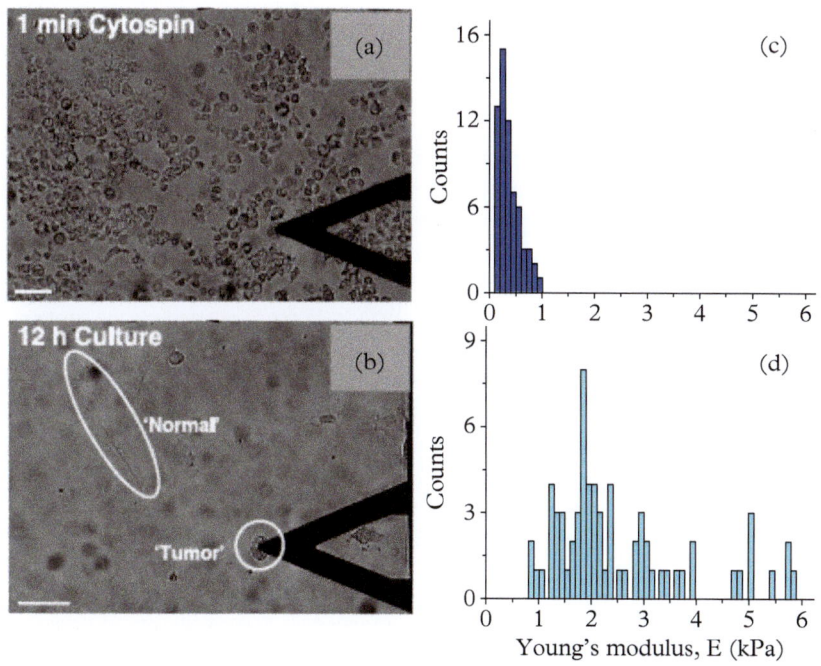

Figure 11.8.12 Can the AFM diagnose cancer cells? Optical image of (a) the cantilever and a mixture of normal and cancer cells and (b) 12 h *ex vivo* cultured cells demonstrating the round, spherical morphology of visually assigned tumor cells ("tumor") and the large, flat morphology of presumed benign mesothelial ("normal") cells; scale bar = 30 μm. Histograms of cell elastic stiffness for (c) tumor, and (d) normal reactive mesothelial cells.

Adapted from [37].

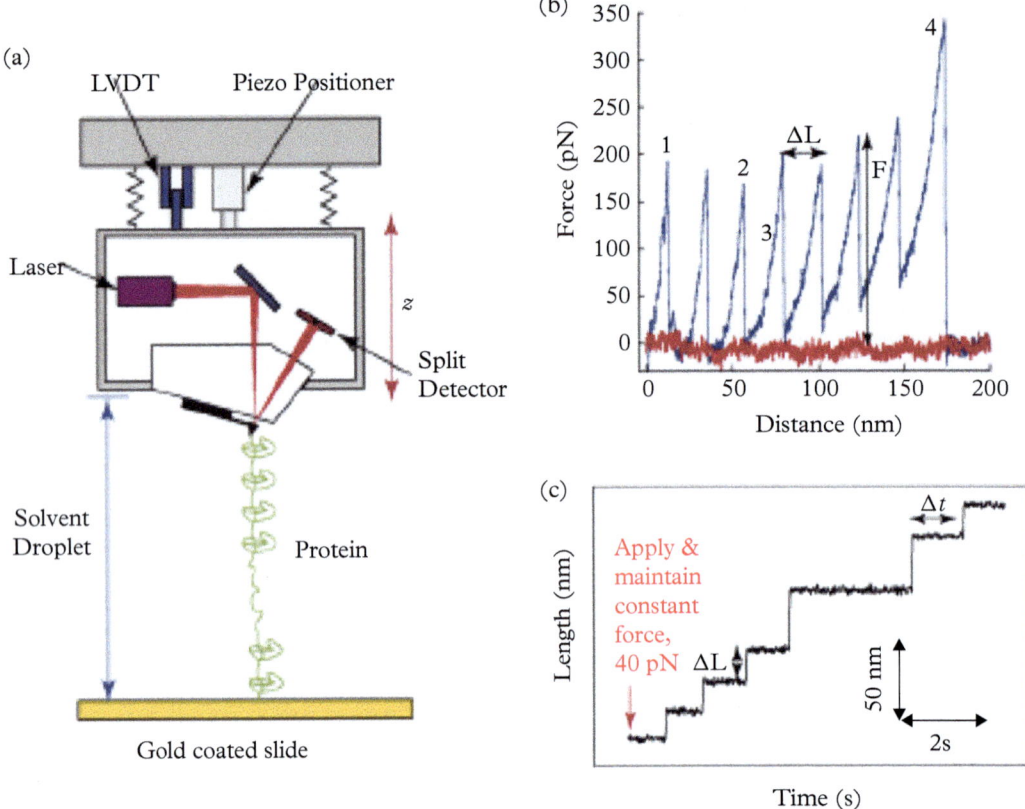

Figure 11.8.13 Use of the AFM to study protein folding and unfolding in the (b) constant velocity, and (c) constant force modes. (a) Schematic diagram showing the operation of the AFM. (b) A "typical" AFM trace with the instrument used in *constant velocity* mode to determine the unfolding force at a given pulling speed. The red trace represents the cantilever deflection during the approach to the surface, and the blue trace during the retraction. A sequence of peaks is observed in the force-distance curve. The first peak (1) reflects detachment of the tip and/or protein from the surface. Force is exerted on the protein until one domain unfolds (2). The unfolded polypeptide chain is then stretched (3) until another domain unfolds. The distance between unfolding events (marked as ΔL) is characteristic of the unfolding of a domain of ~90 amino acids. The domains unfold at a force, F~200 pN, as determined from the height of the unfolding peaks. Finally, the protein detaches from the cantilever (4) when all the domains have unfolded. (c) A "typical" AFM trace showing the same protein being pulled while the AFM is used in *constant force* mode. At the time marked by the red arrow, the force was switched from –20 pN to 40 pN to initiate the measurement. At this force, the unfolding steps (ΔL) have a mean size of ~20 nm (the two larger steps correspond to two domains unfolding simultaneously). In this experiment, the time between unfolding events (Δt) can also be determined.

Adapted from [39].

stage, a force is exerted on the protein. The position of the cantilever and the associated force it exerts is determined using a laser and a split photodiode detector. The linear voltage differential transformer (LVDT) monitors displacement in the z-direction.

11.8.5 Dip-Pen Nanolithography (DPN)

As shown schematically in Figure 11.4.5, the narrow-gap capillary formed between the AFM tip and the specimen, when an experiment is conducted in air, condenses water from the ambient atmosphere. Moreover, this is a dynamic problem, as that water will either be transported from the substrate to the tip, or vice versa, depending on the humidity and the substrate wetting characteristics. Initially, such condensed moisture was considered a nuisance for high-resolution

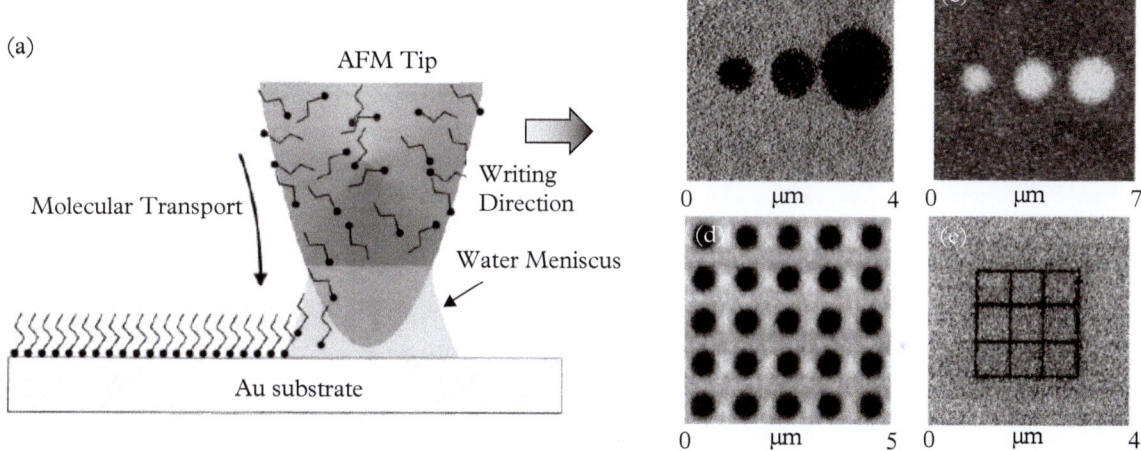

Figure 11.8.14 (a) Schematic representation of DPN. A water meniscus form between the AFM tip coated with 1-octadecanethiol (ODT) and the Au substrate. The size of the meniscus, which is controlled by relative humidity, affects the ODT transport rate, the effective tip-substrate contact area, and DPN resolution. Lateral force SFM images of DPN patterns written on Au substrates. (b) AFM tip, coated with ODT held in contact with the substrate for 2, 4, and 16 min (left to right) with relative humidity ∼45%, and the image recorded at a scan rate of 4 Hz. (c) Dots of 16-mercaptohexadecanoic acid created by holding the coated tip in contact with the substrate for 10, 20, and 40 s (left to right); relative humidity ∼35%. Comparing (b) and (c), we conclude that the transport properties of 16-mercaptohexadecanoicacid and of ODT differ substantially. (d) An array of ODT dots (∼20 s contact each) generated by DPN. (e) A molecule-based grid of lines 100 nm wide and 2 mm in length, written in a time of 1.5 minutes.

Adapted from [41].

AFM imaging. However, this liquid transport mechanism has been transformed into a nanolithography method [41], using the AFM tip to function as a nib (as in a fountain pen), initially coated with molecules (the ink), having a specific affinity for a substrate surface such as gold (the paper), to write patterns—using the x-y piezo drives—of these specific molecules on the sub-micron length scale. If the molecules, first coated on the tip, are chosen such that upon transport they anchor themselves by chemisorption on to the substrate surface, stable surface structures/patterns can be formed.

This method, first demonstrated using 1-octadecanethiol (ODT) molecules and a gold substrate, and schematically illustrated in Figure 11.8.14a, has been termed dip-pen nanolithography (DPN) by its inventors [41]. The uniform diffusion properties of two different inks are illustrated in Figure 11.8.14, for ODT (b), and an alkanethiol derivative, 16 mercaptohexa-decanoic acid (c), proving that this is a general technique applicable to a variety of "inks", and capable of writing different arrays of dots (d) and patterns (e). The resolution of DPN depends on the grain size of the substrate (similar to the texture of paper affecting conventional writing with a pen), and moreover, the self-assembly and chemisorption of the molecules on the surface can be used to control "ink" diffusion. Over the years, DPN patterns have been created with a number of biological molecules including proteins, peptides, and DNA, with applications of such arrays in proteomic and genomic testing [42].

Summary

SPM involves scanning a fine probe/tip very close to a specimen surface and measuring either the quantum-mechanical tunneling current (STM) or a force (SFM) based on one of many possible tip–surface interactions. In SPM, both the tunneling current and the force can be measured at high spatial resolution, as a function of tip position, using piezoelectric actuators to move the tip in three orthogonal directions. The tunneling current across the vacuum gap between the tip and surface provides either an image at atomic resolution (microscopy) or details of the local electronic structure of the surface (spectroscopy) measured as a function of specimen bias voltage. The tunneling tip can also be used to manipulate adsorbed atoms on a clean surface. The STM can be operated either in a constant current mode using a feedback loop to the piezo stage to control the tip height, or in a constant height mode (requires atomic flat specimens) and measuring the variable tunneling current.

In the SFM, a force sensor—a sharp tip mounted on a vibrating cantilever—replaces the tunneling tip of the STM. Short range (<1 nm) chemical and a variety of long-range (<100 nm) forces (van der Waals, electrostatic, magnetic, capillary, etc.), depending on the nature of the tip and the specimen, are encountered in the SFM. A SFM also functions in multiple modes. In the constant force mode, the tip height is controlled such that the force between the specimen surface and the tip is maintained a constant, allowing the measurement of surface topography irrespective of its conducting properties. Alternatively, in the dynamic mode, changes in the vibrational properties (resonance frequency, amplitude of oscillation, and phase difference between the cantilever oscillations and the drive signal) of the cantilever—which functions as the force sensor—due to the local tip–surface interactions are measured. Thus, the feedback parameters to control the tip–surface distance and form the image can be the frequency, amplitude, or phase modulation. Dynamic imaging includes both contact and noncontact modes, but the most common method is the intermittent contact or tapping mode.

SPM images are a convolution of the shape of the tip/probe with the specimen topography. In addition, the tip can be contaminated or broken, and this also appears as artifact in the image. Further, hysteresis and creep of the piezoelectric scanners can affect the accuracy in the positioning of the tip.

SPMs come in a number of designs to measure a large range of properties. They include, among others, optical, acoustic, conductance, electrochemical, capacitance, thermal, and magnetic interactions between the tip and surface. These methods find a wide range of applications both in the physical and life sciences. Scanning probe instruments can also be used to carry out nanoscale lithography.

. . . .

FURTHER READING

Chen, C. J. *Introduction to Scanning Tunneling Microscopy*. Oxford: Oxford University Press, 1993.

Colton, R. J., A. Engel, J. Frommer, H. E. Gaub, A. A. Gewirth, R. Guckenberger, W. Heckl, B. Parkinson, and J. Rabe, eds. *Procedures in Scanning Probe Microscopies*. Chichester: John Wiley and Sons, Ltd., 1998.

Eaton. P., and P. West. *Atomic Force Microscopy*. Oxford: Oxford University Press, 2010.

Israelichivli, J. *Intermolecular and Surface Forces*. New York: Academic Press, 1991.

Krishnan, K. M. *Fundamentals and Applications of Magnetic Materials*. Oxford: Oxford University Press, 2016.

Karp, G. *Cell and Molecular Biology*, 3rd ed. New York: Wiley, 2003.

Mayergoyz, I. *Quantum Mechanics for Electrical Engineers*. Hackensack: World Scientific, 2016.

Meyer, E., H. J. Hug, and R. Bennewitz. *Scanning Probe Microscopy: The Lab on a Tip*. Heidelberg: Springer-Verlag, 2004.

Morris, V. J., A. R. Kirby, and A. P. Gunning. *Atomic Force Microscopy for Biologists*. London: Imperial College Press, 1999.

Rae, A. I. M. *Quantum Mechanics*. Bristol: IOP Publishing, 1992.

Wiesendanger, R. *Scanning Probe Microscopy and Spectroscopy*. Cambridge: Cambridge University Press, 1994.

REFERENCES

[1] Binnig, G., H. Rohrer, Ch. Gerber, and E. Weibel. "Surface Studies by Scanning Tunneling Microscopy." *Physical Review Letters* 49 (1982): 57.

[2] Binnig, G., C. F. Quate, and Ch. Gerber. "Atomic Force Microscope." *Physical Review Letters* 56 (1986): 930.

[3] Binnig, G., H. Rohrer, Ch. Gerber, and E. Weibel. "7 × 7 Reconstruction on Si(111) Resolved in Real Space." *Physical Review Letters* 50 (1986): 120.

[4] Binnig, G., H. Rohrer, Ch. Gerber, and E. Weibel. "Tunneling Through a Controllable Vacuum Gap." *Applied Physics Letters* 40 (1982): 178.

[5] Strocio, J. A., R. M. Feenstra, and A. P. Fein. "Electronic Structure of the Si(111)2 × 1 Surface by Scanning-Tunneling Microscopy." *Physical Review Letters* 57 (1986): 2579.

[6] Eigler, D., C. P. Lutz, and W. E. Rudge. "An Atomic Switch Realized with the Scanning Tunnelling Microscope." *Nature* 352 (1991): 600–3.

[7] Crommie, M. F., C. P. Lutz and D. M. Eigler. "Confinement of Electrons to Quantum Corrals on a Metal Surface." *Science* 262 (1993): 218–20.

[8] Crommie, M. F., C. P. Lutz, and D. M. Eigler. "Imaging Standing Waves in a Two-Dimensional Electron Gas." *Nature* 363 (1993): 524–7.

[9] Heller, E. J., M. F. Crommie, C. P. Lutz, and D. M. Eigler. "Scattering and Absorption of Surface Electron Waves in Quantum Corrals." *Nature* 369 (1994): 464–6.

[10] Giessibl, F. J. "Atomic Resolution of the Silicon (111)-(7×7) Surface by Atomic Force Microscopy." *Science* 267 (1995): 68–71.

[11] Giessibl, F. J. "Advances in Atomic Force Microscopy." *Reviews of Modern Physics* 75 (2003): 949–83.

[12] Sader, J. E. "Parallel Beam Approximation for V-Shaped Atomic Force Microscope Cantilevers." *Review of Scientific Instruments* 66 (1985): 4583.

[13] Marcus, R., T. S. Ravi, T. Gmitter, K. Chin, D. Liu, W. J. Orvis, D. R. Ciarlo, C. E. Hunt, and J. Trujillo. "Formation of Silicon Tips with <1 nm Radius." *Applied Physics Letters* 56 (1990): 236.

[14] Ohnesorge, F., and G. Binnig. "True Atomic Resolution by Atomic Force Microscopy Through Repulsive and Attractive Forces." *Science* 260 (1993): 1451–6.

[15] Law, B. M., and F. Rieutord. "Electrostatic Forces in Atomic Force Microscopy." *Physical Review B* 66 (2002): 035402.

[16] Friedbacher, G., and H. Fuchs. "Classification of Scanning Probe Microscopies." *Pure and Applied Chemistry* 71 (1999): 1337–57.

[17] Garcia, R., and R. Pérez. "Dynamic Atomic Force Microscopy Methods." *Surface Science Reports* 47 (2002): 197–301.

[18] Griffith, J. E., and D. A. Grigg. "Dimensional Metrology with Scanning Probe Microscopes." *Journal of Applied Physics* 74 (1993): R83.

[19] Erlandsson, R., L. Olsson, and P. Moartensson. "Inequivalent Atoms and Imaging Mechanisms in Ac-Mode Atomic-Force Microscopy of Si(111)7 × 7." *Physical Review B* 54 (1996): R8309–12.

[20] Sugimoto, Y., P. Pou, M. Abe, P. Jelinek, R. Pérez, S. Morita, and Ó. Custance. "Chemical Identification of Individual Surface Atoms by Atomic Force Microscopy." *Nature* 446 (2007): 64–7.

[21] Sugitomo, Y., P. Pou, Ó. Custance, P. Jelinek, S. Morita, R. Pérez, and M. Abe. "Real topography, atomic relaxations, and short-range chemical interactions in atomic force microscopy: The case of the $\alpha-$Sn/Si$(111)-(\sqrt{3} \times \sqrt{3})$R30° surface." *Physical Review B* 73 (2006): 205329.

[22] Jelinek, P., H. Wang, J. P. Lewis, O. F. Sankey, and J. Ortega. "Multicenter Approach to the Exchange-Correlation Interactions in *Ab Initio* Tight-Binding Methods." *Physical Review B* 71 (2005): 235101.

[23] Hoffmann, R., L. N. Kantorovich, A. Baratoff, H. J. Hug, and H.-J. Güntherodt. "Sublattice Identification in Scanning Force Microscopy on Alkali Halide Surfaces." *Physical Review Letters* 92, (2004): 146103.

[24] Hartmann, U. "Magnetic Force Microscopy." *Annual Review of Materials Science* 29 (1999): 53–87.

[25] Rugar, D., R. Budakian, H. J. Mamin, and B. W. Chui. "Single Spin Detection by Magnetic Resonance Force Microscopy." *Nature* 430 (2004): 329–32.

[26] Majumdar, A. "Scanning Thermal Microscopy." *Annual Review of Materials Science* 29 (1999): 505–85.

[27] Parot, P., Y. F. Dufrêne, P. Hinterdorfer, C. Le Grimellec, D. Navajas, J-L. Pellequer, and S. Scheuring. "Past, Present and Future of Atomic Force Microscopy in Life Sciences and Medicine." *Journal of Molecular Recognition* 20 (2007): 418–31.

[28] Mou, J. X., S. T. Sheng, R. Y. Ho, and Z. F. Shao. "Chaperonins GroEL and GroES: Views from Atomic Force Microscopy." *Biophysical Journal* 71 (1996): 2213–21.

[29] Viani, M. B., L. I. Pietrasanta, J. B. Thompson, A. Chand, I. C. Gebeshuber, J. H. Kindt, M. Richter, H. G. Hansma, and P. K. Hansma. "Probing Protein–Protein Interactions in Real Time." *Nature Structural & Molecular Biology* 7 (2000): 644–7.

[30] Crampton, N., M. Yokokawa, D. T. F. Dryden, J. M. Edwardson, D. N. Rao, K. Takeyasu, S. H. Yoshimura, and R. M. Henderson. "Fast-Scan Atomic Force Microscopy Reveals that the Type III Restriction Enzyme EcoP15I is Capable of DNA Translocation and Looping." *Proc. Nat. Acad. Sci. of the USA* 104, (2007): 12755–60.

[31] Connell, S. D., and D. A. Smith. "The Atomic Force Microscope as a Tool for Studying Phase Separation in Lipid Membranes." *Molecular Membrane Biology* 23 (2006): 17–28.

[32] Dufrêne, Y. F., and G. U. Lee. "Advances in the Characterization of Supported Lipid Films with the Atomic Force Microscope." *Biochimica et Biophysica Acta—Biomembranes* 1509 (2000): 14–41.

[33] Bolshakova, A., O. I. Kiselyova, A. S. Filonov, O. Y. Frolova, Y. L. Lyubchenko, and I. V. Yaminsky. "Comparative Studies of Bacteria with an Atomic Force Microscopy Operating in Different Modes." *Ultramicroscopy* 86 (2001): 121–8.

[34] Plomp, M., and A. J. Malkin. "Mapping of Proteomic Composition on the Surfaces of Bacillus Spores by Atomic Force Microscopy-Based Immuno-labeling." *Langmuir* 25 (2009): 403–9.

[35] Jonas, K., H. Tomenius, A. Kader, S. Normark, U. Römling, L. M. Belova, and O. Melefors. "Roles of Curli, Cellulose and BapA in Salmonella Biofilm Morphology Studied by Atomic Force Microscopy." *BMC Microbiology* 7 (2007): 70.

[36] Francius, G., O. Domenech, M. P. Mingeot-Leclerc, and Y. F. Dufrêne. "Direct Observation of *Staphylococcus aureus* Cell Wall Digestion by Lysostaphin." *J. Bacteriology* 190 (2008): 7904–9.

[37] Cross, S. E., J-S. Jin, J. Tondre, R. Wong, J. Rao, and J. K. Gimzewski. "AFM-Based Analysis of Human Metastatic Cancer Cells." *Nanotechnology* 19 (2008): 384003.

[38] Moy, V. T., E. L. Florin, and H. E. Gaub. "Intermolecular Forces and Energies Between Ligands and Receptors." *Science* 266 (1994): 257–9.

[39] Forman, J. R., and J. Clarke. "Mechanical Unfolding of Proteins: Insights into Biology, Structure and Folding." *Current Opinion in Structural Biology* 17 (2007): 58.

[40] Oesterhelt, F., D. Oesterhelt, M. Pfeiffer, A. Engel, H. E. Gaub, and D. J. Müller. "Unfolding Pathways of Individual Bacteriorhodopsins." *Science* 288 (2000): 143–6.

[41] Piner, R. D., J. Zhu, F. Xu, S. Hong, and C. A. Mirkin. "'Dip-Pen' Nanolithography." *Science* 283 (1999): 661–3.

[42] Ginger, D. S., H. Zhang, and C. A. Mirkin. "The Evolution of Dip-Pen Nanolithography." *Angewandte Chemie International Edition*, 43 (2004): 30–45.

[43] Hamaker, H. C. "The London–van der Waals Attraction Between Spherical Particles." *Physica* 4 (1937): 1058–72.

[44] Parakkat, V. M., K. Xie, and K. M. Krishnan. "Tunable Ground State in Heterostructured Artificial Spin Ice with Exchange Bias." *Physical Review B* 99 (2019): 054429.

[45] Poggi, M. A., A. W. McFarland, J. S. Colton, and L. A. Bottomley. "A Method for Calculating the Spring Constant of Atomic Force Microscopy Cantilevers with a Nonrectangular Cross Section." *Anal. Chem.* 77 (2005): 1192–5.

[46] Hutter, J. L., and J. Bechhoefer. "Manipulation of van der Waals Forces to Improve Image Resolution in Atomic-Force Microscopy." *Journal of Applied Physics* 73 (1993): 4123.

[47] Parakkat V. M., G. M. Macauley, R. L. Stamps, and K. M. Krishnan, "Configurable artificial spin ice with site-specific local magnetic fields", *Physical Review Letters* 126 (2021): 017203.

. . . .

A.   Test Your Knowledge

1. Scanning probe microscopy can _____ a surface.
    (a) map the topography of
    (b) measure specific properties of
    (c) manipulate adsorbed atoms on
2. A scanning tunneling microscope maps
    (a) *only* the surface topography.
    (b) the local density of states of the specimen.
    (c) the local density of states of the tip AND specimen.
3. A tunneling electron
    (a) with energy less than a potential barrier has no probability of crossing it.
    (b) will ALWAYS provide a predictable signal between a tip and surface in STM.
    (c) will give an STM signal if the tip is scanned at a distance <1 nm from the surface.
4. The *resolution* of a scanning probe microscope depends on the
    (a) physical aspects of the tip, i.e. its size and shape.
    (b) nature of the interaction between the tip and surface.
    (c) speed at which the tip is scanned.
    (d) distance between the tip and the surface.
5. An SPM gives atomic resolution images if and only if it
    (a) includes a highly sensitive piezoelectric actuator.
    (b) has a high stability mechanical loop.
    (c) measures the tunneling current.
6. All scanning probe microscopy methods
    (a) are variants of the STM.
    (b) measure a short-range interaction.
    (c) map the variation of a force or current laterally across the surface.
    (d) provide images that are a convolution of the specimen topography and the shape of the tip.
7. In a scanning force microscope, the cantilever
    (a) functions as a force sensor.
    (b) deflection or its dynamic response is measured.
    (c) cannot measure local properties such as friction.
8. In STM and related spectroscopy
    (a) the barrier height is approximated by the work function.
    (b) the tunneling current depends on the applied bias voltage.
    (c) we can probe the *occupied* DOS of the specimen.
    (d) we cannot probe the *unoccupied* levels of the specimen.

9. Typically, in
   (a) STM, the tunneling currents is in the nA–pA range.
   (b) SFM, the stiffness of the cantilever is in the 1–100 N/m range.
   (c) SFM, capillary forces are $\sim$90 nN.

10. In STM operation
   (a) constant height mode moves the tip normal to the surface.
   (b) constant current mode measures surface topography subject to a preset current.
   (c) constant height mode is affected adversely by the overall specimen inclination.
   (d) constant height mode is best for atomically flat specimens.
   (e) the derivative of the tunneling current with height gives estimate of the work function.

11. An SFM
   (a) *requires* a non-conducting surface.
   (b) measures local forces on a tip.
   (c) can operate in contact, noncontact, and tapping modes.
   (d) can measure specimens in liquid environments.
   (e) can only measure specimens in air.

12. A cantilever used in SFM
   (a) serves as a force sensor.
   (b) is normally microfabricated of highly doped silicon.
   (c) shows variable performance depending on its stiffness, resonant frequency, and its quality factor.
   (d) should have a resonant frequency comparable to sound.
   (e) should have a thermal motion range > 1 Å.
   (f) typically has dimensions in the micrometers range.

13. Cantilever deflections in SFM can be measured
   (a) by the optical lever method.
   (b) using a Michelson interferometer, especially in UHV.
   (c) by monitoring the resistance changes in special piezoresistive cantilevers.

14. Atomic resolution in a scanning *force* microscope is achieved
   (a) only when chemical forces are comparable to the van der Waals forces.
   (b) in contact or static mode.
   (c) in dynamic noncontact mode.

15. Dynamic mode SFM imaging
   (a) is amplitude modulated in liquids or air.
   (b) is frequency modulated in vacuum.
   (c) is phase modulated to map variations in properties.

16. In contact mode SFM
   (a) the tip position defines a force equilibrium.
   (b) both cantilever bending and specimen compression are expected.
   (c) a set point is chosen to avoid jump out-of-contact of the tip.

(d) a mechanically hard cantilever is used.

(e) the spatial resolution is independent of the tip–surface contact diameter.

17. For magnetic force microscopy it is best to use _____ mode of operation.

(a) contact

(b) noncontact

(c) tapping

18. All piezoelectric scanners used in SFM

(a) suffer from hysteresis and creep artifacts.

(b) will always return the probe to the same position on the surface after driving it in a circle.

(c) may continue to move the tip even when the applied voltage remains unchanged.

19. A scanning *force* microscope is ideally suited for applications in the life sciences because it

(a) images under physiological conditions relevant for biology.

(b) can distinguish between normal and diseased cells using nanomechanical measurements.

(c) can measure forces relevant to protein folding and unfolding mechanisms.

20. Dip-pen nanolithography

(a) uses the water meniscus as a chemical transport mechanism.

(b) can only write patterns at a resolution determined by the grain size of the substrate.

(c) find applications in proteomic and genomic testing.

B. Problems

1. **Scanning tunneling microscopy**: An electron of energy, $E = 1$ eV, is tunneling through a barrier of energy $E_0 = 5$ eV. Calculate

   a) its wavevector, k.

   b) its tunneling decay rate, χ.

   c) Now show that the transmission coefficient, T, changes by one order of magnitude when the barrier width, s, changes by 1 Å.

2. A **commercial Si tip** for magnetic force microscopy has dimensions $l = 225$ μm, $t = 3$ μm, $w = 41$ μm, $h = 12.5$ μm. What is its force constant and resonance frequency?

3. **Cantilevers with nonrectangular (trapezoid) cross-sections** [45]: Some commercial tips have non-rectangular cross-sections with a trapezoid, with two different widths, w_a on the side of the tip, and w_b, for the opposite surface, being common. In this case, the force constant is

given by $k = 3EI/t^3$, where the second moment of the beam, $I_{trapezoid} = \frac{t^3 (w_a^2 + 4w_a w_b + w_b^2)}{36(w_a + w_b)}$.

Calculate the force constant for the following dimensions: $l = 225$ μm, $t = 3$ μm, $w_a = 24$ μm and $w_b = 41$ μm. Compare with the result of Problem 2.

4. The **work function**, Φ, in a scanning tunneling microscopy experiment. If the tunneling current, $I_T = 10^{-9}$ A, at a tip separation of 2 Å, and $I_T = 10^{-8}$ A, at a tip separation of 1 Å, determine the work function for this experiment. Assume that the proportionality constant in (11.3.1) is 1.

5. Consider a **scanning force microscopy** experiment conducted in air with the following parameters: Tip–specimen distance, $d_{t-s} = 5$ Å, Hamaker constant, $A_H = 5 \times 10^{-19}$ J, tip radius $R = 50$ nm, and electrostatic potential difference, $\Delta U = 1.5$ V.

   a) Calculate the van der Waals force as a function of tip–specimen distance.

   b) From *a)* estimate the distance at which the van der Waals force falls to half its value from the tip surface.

   c) From *b)* state if the van der Waals force can lead to atomic resolution images?

   d) What is the electrostatic force between the tip and specimen?

   e) What is the capillary force between the tip and specimen if the contact angle, $\gamma = 60°$.

   f) Compare all the forces and comment on which forces dominate SFM imaging and why.

   g) Assuming that the tip is made of silicon and the specimen is a **metal** with Young modulus $E = 2.0 \times 10^{11}$ N/m^2, and the Poison ratio of the tip and specimen are the same:

      (i) Calculate the contact diameter, d_c, if the SFM is operated in contact mode

      (ii) Now calculate d_c if the specimen is a **polymer** with $E = 0.09 \times 10^{12}$ N/m^2.

6. What type of SPM can be used to measure the **work function** of a material? How will you do the measurement and process the data? What other measurement technique, discussed earlier in §3, can also measure the work function? How do these two techniques compare?

7. Can a scanning probe microscope be used to measure the variation of **friction** on the surface of a specimen? If so, how?

8. A SFM **cantilever**, made of silicon, with the following dimensions: length, $l = 220$ μm, width, $w = 40$ μm, and height, $h = 125$ μm, has a torsional spring constant of 55.94 N/m. Calculate (a) the thickness of the cantilever, (b) its stiffness, and (c) its resonance frequency.

Summary Tables

12

The three tables in this chapter provide an easily accessible comparative summary of the key points and features of the major characterization methods discussed in the text. The techniques are classified into three broad groups/methods: spectroscopy/chemical, diffraction/scattering, and imaging. For each technique, a concise description of the method and its use is followed, in bullet form, by its salient details and other characteristics (resolution, sensitivity, etc.), as well as specimen requirements that determine its practice. A general estimate of cost and space requirements are also included, but needless to say, this is only a snapshot for comparison and is definitely subject to change with time.

Principles of Materials Characterization and Metrology. Kannan M. Krishnan, Oxford University Press (2021).
© Kannan M. Krishnan. DOI: 10.1093/oso/9780198830252.003.0012

Table 12.1 Spectroscopy and Chemical Methods.

Acronym and Section	Method and Use	Salient Details	Sensitivity, Limits, Resolution, etc.	Specimen	Cost and Space Requirements
AES §2.6.2	AUGER ELECTRON SPECTROSCOPY Determine elemental composition and the bonding of atoms of inorganic materials, especially with surface sensitivity.	• Electrons in and electrons out • All elements except H and He detected • Nondestructive method except when used with depth profiling by sputtering • Analysis quantitative with standards (semi-quantitative otherwise) • Weak features, often a derivative spectrum is used for analysis • Can be used for imaging—called scanning auger microscopy	• Sensitivity ~100 ppm for most elements but depends on the matrix • Depth probed ~2 nm. • Lateral resolution ~30 nm for Auger analysis (larger for imaging)	Specimen should be vacuum-compatible (typically inorganic materials)	• Cost: ~$100K–$1M • Space: ~3 m × 4 m
CL §10.6.2	CATHODO-LUMINESCENCE Quantitative analysis and distribution of defects and impurities in luminescent materials using electron probes	• Electrons in, photons out • Not an element-specific method • Nondestructive but electron beam may cause defects • Depth profiling by varying energy of incident electrons	• Detection limit $\geq 10^{18}$ at/m^2 • Depth probed ~10 nm–several μm, depending on incident beam energy • Lateral resolution ~1 μm • Standards required for quantification, that too, with difficulty	Solid, vacuum-compatible specimen	• Cost: ~$25K–$250K • Added to a SEM or TEM

EELS §9.4.2	ELECTRON ENERGY-LOSS SPECTROSCOPY		TEM-EELS		
	ELECTRON ENERGY-LOSS SPECTROSCOPY • In transmission: Quantitative measurement of the local specimen concentration, its electronic and chemical structure (ELNES), and nearest-neighbor atomic spacing (EXELFS). • In reflection: High-resolution measurement of vibration of atoms and molecules, and *intra-band* transitions with surface sensitivity	• Electrons in, electrons out • Nondestructive • Detect all elements, but H and He only under special conditions • Transmission EELS probes core-levels and *unoccupied density of states* • All transitions satisfy *dipole selection rules* • Quantitative analysis ~5–20% without standards; 1–2% with standards • Hydrogenic and Hartree–Slater cross-sections used for quantification, but best results obtained with experimentally measured cross-sections • Energy-filtered imaging possible • HREELS (reflection) probes vibrations in the 50–4000 cm^{-1} range	• Depth probed ~20 nm • Lateral resolution ~0.5–10 nm • Detection limit ~10^{-21} g • Energy resolution ~0.3–3 eV • MFP for single scattering ~60–100 nm HREELS (Reflection) • Depth probed ~0.05–3 nm • Lateral resolution ~100 nm–50 μm • Detection limit ~0.1 ML • Energy resolution ~10–100 meV	• Specimens must be solid and electron transparent (~10–100 nm) for TEM-EELS • Reflection HREELS best with single crystal, atomically flat, conducting specimens. Must be UHV compatible	• Cost: TEM-EELS (does not include TEM) spectrometer ~$100K Imaging filter ~$200–$300K • Cost: HREELS spectrometer ~$100K

continued

Table 12.1 Continued

Acronym and Section	Method and Use	Salient Details	Sensitivity, Limits, Resolution, etc.	Specimen	Cost and Space Requirements
EDXS §2.5.1.2 §9.4.3 §10.6.1	ENERGY DISPERSIVE X-RAY SPECTROMETRY Microanalysis of a specimen in a TEM, SEM, or EPMA system.	• Electrons/photons in, photons out • All elements with $Z \geq 5$ detected • Nondestructive method • Analysis quantitative with standards but standard-less analysis also possible • Can be used for compositional mapping when included in a SEM, EPMA, or STEM • Simultaneous acquisition with EBSD possible in a SEM	• Detection limits $\sim$ 200 ppm for isolated peaks of elements with $Z \geq 11$, but higher, 1–2 wt% for lower Z elements. • Depth sampled depends on specimen Z and kV, in the range 20 nm–µm. • Lateral resolution 0.5 µm in an SEM or EPMA, but 1–2 nm in a STEM using electron transparent thin foils. • Energy resolution $\sim$100 eV	• Specimen size determined by instrument—3 mm dia for TEM • Size limited by specimen stage in SEM for analysis of powders, solids, and composites	Cost: $\sim$\$50K–\$150K (does not include SEM or TEM)
EPMA §2.5.2.2	ELECTRON PROBE MICROANALYSIS Quantitative nondestructive microanalysis of a specimen using a dedicated electron probe	• Electrons in, photons out • Incident electron probe of energy 5–30 keV used • Includes WDXS and EDXS • Sophisticated instruments combine simultaneous X-ray detection with SEM imaging characteristics • Chemical mapping possible • Electron beam can damage specimen	• Analysis of micron-sized volumes at ppm levels • $Z > 3$ can be analyzed • Accuracy and detection limits of analysis depends on WDXS or EDXS • Requires ZAF correction for correct analysis	Same as for SEM	• Cost|: \$300K–\$800K • Space: 3 m × 2 m × 2 m

EXAFS §3.9.2 **EXTENDED X-RAY ABSORPTION FINE STRUCTURE** Inter-atomic distance and coordination number with element specificity.	• X-rays in, X-rays out • All elements with $Z \geq 3$ • Nondestructive method • Surface sensitive • Depth profiling achieved with glancing incident angle • Optimally done in a synchrotron beam line • Similar to extended energy-loss fine structure (EXELFS). Both data can be analyzed in the same manner	• Sensitive to structural distribution of atoms within ~0.5 nm of the X-ray absorber atom. • Measurement accuracy: 1–2% for interatomic distance, and 10–25% for coordination number. • Detection limit ≥ 100 ppm in bulk specimens.	Any specimen—solid, liquid, gas, including biological materials—can be analyzed.	• Cost: ~$300K for laboratory facility • Synchrotron beam line end station > $1M
FTIR §3.5.2 **FOURIER TRANSFORM INFRARED SPECTROSCOPY** Vibrational frequencies measured to identify chemical bonds, their concentrations, and surrounding environments.	• Photons (IR) in, photons (IR) out • Nonzero change in dipole moment with vibration mode required • Strictly, not an element-specific method but all elements can be analyzed. • Nondestructive • Chemical bonding information, particularly identification of functional groups possible • Standards required for quantification	• Detection limits $<10^{13}$ bonds/cm^3 • Lateral resolution ~0.5 cm–20 μm. • Group frequency region, 1250–4000 cm^{-1}, shows strongest absorption • Fingerprint region 650–1,250 cm^{-1}, shows common single bond stretching and bending • Spectral resolution ~1 cm^{-1}	• Any specimen—solid, liquid, gas, including biological materials—can be analyzed. • Vacuum not required.	• Cost: ~$50K–$150K (~$20K–if non-FT) • Space: Tabletop to ~2 m × 2 m

continued

Table 12.1 Continued

Acronym and Section	Method and Use	Salient Details	Sensitivity, Limits, Resolution, etc.	Specimen	Cost and Space Requirements
ICP-MS §5.4.4	INDUCTIVELY COUPLED PLASMA MASS SPECTROMETRY High sensitivity elemental and isotope analysis	• Use plasma or laser ablation to generate sample ions, and analyze them with a mass spectrometer • All elements with $Z \geq 3$. • Destructive method, original specimen cannot be retrieved • Quantitative analysis with good accuracy • ICP-OES (optical emission spectroscopy), a variant, measures intensities of characteristic light emissions of plasma for quantification	• Accuracy ~0.2% isotope, ≤5% quantitative. • Detection limit ≤ 1 ppb for most elements, improves with increasing Z • Depth sensitivity ~1–10 μm, per laser pulse, and lateral resolution ~20–50 μm, if laser is used for sample ablation.	Specimen must be a solution or digestible in a solvent.	• Cost: ~$150K–$750K • Space: ~2.5 × 2.5 m.
IPES §3.8	INVERSE PHOTOEMISSION SPECTRSCOPY Probes unoccupied density of states	• Electrons in, photons out • Two modes 1) Fixed electron energy and tunable photon detector 2) Variable incident energy and fixed-energy photon detector, known as Bremsstrahlung isochrome spectroscopy (BIS) • Accessible energy, $E_F < E < E_V$	Energy resolution ~25 meV	Specimen must be vacuum-compatible and solid	Cost: ~$150–$200K

| LEISS
§5.4.2 | **LOW-ENERGY ION SCATTERING SPECTROSCOPY**
Analysis of surface monolayers and detailed depth profiling up to ~10 nm from the surface. | • Ions ($^3\mathrm{He}^+$, $^4\mathrm{He}^+$ or $^{20}\mathrm{Ne}^+$) in, ions out
• Collison of inert ion beam follows simple laws of conservation of momentum with their energy loss used to identify the ions struck
• Can detect all elements
• Nondestructive but specimen can be damaged if subject to depth profile experiment
• Surface analysis technique | • Detection limit ranges from ~0.5% for C, to 50 ppm for heavy metals.
• Sensitivity ~0.01 monolayer
• Lateral resolution ~150 μm | • Specimen must be vacuum-compatible and solid | • Cost: ~\$25K–\$150K
• Space: ~ 3.0 × 3.0 m. |
| PIXE
§5.4.5 | **PARTICLE-INDUCED X-RAY EMISSION**
Rapid analysis of composition in multielement thin specimens | • Particles (ions) in, X-rays out
• Rapid analysis. All elements with $Z \geq 3$ detected
• Nondestructive method.
• Analysis quantitative (2–10% accuracy with standards)
• Best for thin specimens and surfaces | • Detection limit ~ 0.1–100 ppm because of much lower background compared to electron excitation.
• Depth probed ≤10 μm.
• Lateral resolution 5 μm–2 mm.
• Minimal beam broadening | • Can analyze all specimens—solid, liquids, and gases
• Specimen need not be in vacuum | • Cost: ~\$1M, includes an accelerator for 2-MeV He^+ ions.
• Space: ~10 m^2. |

continued

Table 12.1 Continued

Acronym and Section	Method and Use	Salient Details	Sensitivity, Limits, Resolution, etc.	Specimen	Cost and Space Requirements
Raman §3.6	RAMAN SPECTROSCOPY Measure vibrational frequencies of chemical bonds and identification of unknown compounds.	• Photons in, photons out • Nonzero change in polarizability with vibration mode required • Can analyze all elements but not element specific • Typically nondestructive, unless specimen is susceptible to damage from the high-intensity laser beam • Qualitative method • Imaging possible in a Raman microscope	• Detection limited to ~100 nm of the specimen surface. • Lateral resolution ~1 μm, requires lasers with *microfocus*.	• Can analyze all specimens—solids, powders, thin films, liquids, and gases • Specimen heating may be a problem	• Cost: ~\$100K–\$300K. • Space: ~1.5 m × 2.5 m
RBS §5.4.1	RUTHERFORD BACK-SCATTERING SPECTROSCOPY Quantitative depth profiling of thin films and multilayers.	• Ions in, ions out • Kinematic scattering assumed • Quantitative stopping power, associated with continuous loss of energy, to be known for analysis • Nondestructive depth-profiling method but specimens may suffer ion-implantation ($\sim10^{13}$ He atoms implanted)	• Detection limit $\sim10^{16}$–10^{20} at/m^2 or 1–10 at% for $Z \leq 8$, but 1–100 ppm for high-Z elements. • Lateral resolution ~1–4 mm. • Depth resolution ~15–20 nm • Maximum depth that can be analyzed ~2 μm, but ~20 μm if H$^+$ is used	Can analyze only solid, vacuum-compatible specimens.	• Cost: ~\$0.5M–\$1.0M. • Space: ~3 m × 8 m

SIMS §5.4.3	**Secondary Ion Mass Spectrometry** Analyze composition and trace element quantification as a function of depth	• Ions in, ions out • Can analyze all elements and isotopes • Destructive method—specimen removed by sputtering for depth profile • Low dose ($\leq 5 \times 10^{16}$ atoms/m^2)—Static SIMS—probes surface (1–2 ML) • Higher dose—Dynamic SIMS—destroys surface integrity and used for depth analysis ranging from nm to µm • Quadrupole or time-of-flight mass spectrometers used for analysis	• Mass range $\leq$1000 amu (quadrupole) and $\leq$10,000 amu (TOF) • Detection limit $\sim 10^{18}$–10^{22} atoms/m^3. • Lateral resolution ~50 nm–2 µm	Solid, vacuum-compatible specimens.	• Cost: ~\$0.5M–\$1.5M. • Space: ~3 m × 5 m
UPS §3.8	**Ultraviolet Photoelectron Spectroscopy** Electronic structure of gas molecules, solid *surfaces*, and adsorbates on surfaces	• Photons (UV) in, electrons out • Depth probed ~0.2–10 nm • Physics same as XPS, but much lower photon energies used • Can determine the work function	• Lateral resolution ~1–5 mm • Energy resolution ~25 meV	Vacuum compatible specimens.	• Add on to an XPS machine cost: ~\$30K • Space: ~3 m × 3 m

continued

Table 12.1 Continued

Acronym and Section	Method and Use	Salient Details	Sensitivity, Limits, Resolution, etc.	Specimen	Cost and Space Requirements
UV-Vis §3.4.2	ULTRAVIOLET AND VISIBLE SPECTROSCOPY Quantitative measurements of absorbance of light in dilute solutions of materials	• Light absorption in the visible (700 nm–380 nm) and ultraviolet (~380 nm–180 nm) result from excitations of electronic states • Absorption bands typically broad	• Requires ~ few mg for standard analysis • Broad range of solvent choices, including water, with quartz being a suitable window material • Range $\lambda \sim 180$–700 nm	Gives meaningful information about molecules containing conjugated systems and chromophores	• Cost: ~$2K–$10K • Space: ~ tabletop
WDXS §2.5.1.2	WAVELENGTH DISPERSIVE X-RAY SPECTROMETRY Measuring characteristic X-ray intensities using X-ray diffraction, and perform microanalysis of materials. Implemented in an EPMA.	• Electrons in, photons out • All elements with $Z \geq 5$ detected • Nondestructive method • Analysis quantitative with standards but standard-less analysis also possible • Can be used for compositional mapping when included in an EPMA • $\lambda_{max} = 1.9\ d$ of the crystal. Multiple crystals required to detect wide spectral range	• Detection limits ~200 ppm for isolated peaks of elements with $Z \geq 11$, but higher, 1–2 wt % for lower Z elements. • Depth analyzed depends on specimen Z and kV, in the range 20 nm–μm. • Lateral resolution 0.5 μm in EPMA • Energy resolution ~3–5 eV	Specimen size limited by stage for analysis of powders, solids, and composites.	• Cost: ~$100K–$150K • Space: same as EPMA
XAS §3.9.1	X-RAY ABSORPTION SPECTROSCOPY Probes core levels, *unoccupied* density of states (XANES), and nearest-neighbor distance and coordination number (EXAFS)	• X-rays in, X-rays out • Physics similar to EELS • Transitions satisfy dipole selection rules • Element specific measurements	Lateral resolution ~0.5 mm	Vacuum compatible solid	Cost: ~$400K for the end station (in addition to a synchrotron)

Technique	Principle	Applications	Quantitative analysis	Specimen requirements	Cost / space
XRF §2.5.2.1	• X-rays in, fluorescent X-rays out • Probes core levels • Analyze all elements with Z ≥ 4 • Characteristic X-rays can be analyzed either by WDS or EDXS • Best for bulk or thick specimens (compare with PIXE)	• Identify elements present in a specimen and determine composition • Total Reflection XRF is used in semiconductor industry to detect trace element contamination of wafers in production	• Accuracy in chemical analysis ± 1% • Depth analyzed ~10 μm but substantially less (~10–50 Å) in total reflection mode • Detection limit ~0.1 at % for bulk XRF, and ~10^{13} at/m² in TRXRF mode	• Specimen size ~3–5 cm • TRXRF requires a specular surface. • Vacuum not required but preferred • Specimen can be solid or liquid	• Cost: ~$50–300K for bulk XRF and ~$300K–$600K for TRXRF • Space: 2 m × 3 m.
XPS §3.8 Determine electronic state and composition in the top, 1–5 nm, layer of the specimen. Angle-resolved measurements (ARPES) used to determine band structure	• X-rays in, electrons out • Probes core levels and *occupied* density of states • Analyze all elements with Z ≥ 3 • Nondestructive method • Chemical analysis is semiquantitative, but can be quantitative with standards • Cannot detect trace quantities • Imaging possible with a Photoemission electron microscope (PEEM) implemented in a synchrotron beamline • X-ray photoelectron diffraction provides structural information of single crystals		• Depth resolution ~1–5 nm • Lateral resolution ~1–5 mm	• Vacuum compatible specimens, preferably with flat surfaces • Size of specimen depends on instrument	• Cost: ~$00K–$1M • Space: ~3 m × 4 m

Table 12.2 Diffraction and Scattering Methods.

Acronym and Section	Method and Use	Salient Details	Sensitivity, Limits, Resolution, etc.	Specimen	Cost and Space Requirements
EBSD §10.4	ELECTRON BACK-SCATTERED DIFFRACTION Gives crystallographic details about the microstructure of a specimen	• Electrons in, electrons out • Nondestructive • Implemented in a SEM • Diffraction pattern characteristic of the crystal structure and orientation • Signal obtained from a few tens of nms • Back-scattered electrons modulated by Kikuchi bands • Simultaneous acquisition with EDXS • Large specimen tilts (>70°) converts EBSD patterns to RHEED patterns	Large solid angle ±25°	Solid specimens to be cut, mounted with resin, and polished.	
Ellipsometry §6.9	ELLIPSOMETRY Characterize optical and dielectric properties, and measure thickness of thin films	• Photons in, photons out • Nondestructive and contact-free • Measures change in polarization of light upon reflection or transmission	• Spectral range ~380–700 nm (basic), 193–1000 nm (advanced) • Angle can be fixed (65°) or vary: Horizontal: 45°–90°, Vertical: 20°–90°	• Solid thin film specimens • Liquids can be measured with special cells	• Cost ~$10–20K • Space: tabletop

LEED §8.4.1	**LOW-ENERGY ELECTRON DIFFRACTION** Analyzing structure, crystallography, and cleanliness of surfaces	• Electrons in, electrons out. • Analyze all elements, but not element specific • Surface images can be obtained with additional lenses (LEEM) with typical resolution ~10 nm • Charging is a limiting factor. • Resolve atom position and surface step heights with 0.01 nm accuracy	• Depth probed ~0.4–2 nm. • Detection limit ~0.1 ML. • Surface length for detecting disorder ~20 nm • Lateral beam size ~0.1 mm (standard), ~10 μm (specialized)	Single crystals of conductors and semiconductors, but insulators can also be studied with specialized instruments.	• Cost: ~$75K. • Space: ~8 m²
Neutron Scattering §8.8	**NEUTRON SCATTERING AND DIFFRACTION** Studies of atomic and magnetic arrangements in materials, emphasizing structure determination and refinement.	• Neutrons in, neutrons out. • Analyze all elements, but not element specific • Particularly sensitive to H, but V is poorly detected • Nondestructive. • Weak interaction; hence, larger quantity of material required	• Poor lateral resolution. • Structural accuracy: ~10^{-4} nm in determining atom position • Phase identification for concentrations $\geq$1% molar	Requires crystalline specimens of substantial volume	Cost of access to user facilities varies

continued

Table 12.2 Continued

Acronym and Section	Method and Use	Salient Details	Sensitivity, Limits, Resolution, etc.	Specimen	Cost and Space Requirements
RHEED §8.4.3	REFLECTION HIGH-ENERGY ELECTRON DIFFRACTION Monitoring surface structure, particularly in thin film growth. Distinguish between two-dimensional and three-dimensional features.	• Electrons in, electrons out • Analyze all elements, but not element specific • Use intensity oscillation with thickness to monitor thin film growth	• Depth probed ~0.2–10 nm • Lateral resolution ~200 μm × 4 mm	Best for single crystals, conductors, and semiconductors.	• Cost: ~$50K–$200K. • Space: typically attached to a UHV thin film growth chamber
TED §8.6.3	TRANSMISSION ELECTRON DIFFRACTION Measure atomic structure, crystallography, strain etc. in thin foils	• Electrons in, electrons out • High-energy electrons (>100 keV) • Broadly classified as selected area diffraction (SAD, parallel beam) and convergent beam electron diffraction (CBED) • Nondestructive (except for specimen preparation), but long exposure with a focused probe can cause beam damage	• Lateral resolution is instrument dependent, best ~1 Å (focused probe) • Accuracy: Lattice parameters to four significant figures using CBED	Solid conductors or coated insulators	• Cost: ~$300K–$2M. • Space: 3 m × 3 m

| XRD §7 | X-ray Diffraction Crystallography, including atomic arrangements in the unit cell. Identification of crystalline phases. Strain, orientation and crystallite size determination | • X-rays in, X-rays out
• Analyze all elements, but low Z elements (particularly H) difficult. Not element specific
• Nondestructive for most materials
• Satisfies Bragg's law (real space) or the Laue criterion (reciprocal space) | • Depth ~few μm, but depends on the material
• Monolayer sensitivity with synchrotron radiation
• Detection limit ~3% in two-phase mixtures
• Lateral resolution is poor, but with microfocus ~10 μm | • Any material
• Analysis can be done in air | • Cost: $50K–$400K
• Space: ~8 m^2 |

818

Table 12.3 Imaging Methods.

Acronym and Section	Method and Use	Salient Details	Sensitivity, Limits, Resolution, etc.	Specimen	Cost and Space Requirements
APT §1.4.3	Atom Probe Tomography Three-dimensional imaging of a fine tip specimen with sub-nm resolution	• Destructive, removes atoms by successive evaporation • Evaporation with either voltage or laser (Nd: YAG, 1064 nm; or Nd:YVO$_4$, 355 nm) pulses	• Best resolution ~0.2 nm, typical 0.3 nm • Depth inversely related to laser wavelength	• Sharp, needle-shaped solid specimen (tip radius ≤100 nm) • Best prepared by FIB methods	Cost: ~$0.5M–$1M
CSOM §6.8.5	Confocal Scanning Optical Microscopy Optical imaging using a spatial pin-hole to block out-of-focus light in image formation	• Photons in, photons out • Operate in reflection mode • Scan either the specimen or the beam	Resolution ~0.2 μm (x, y), 0.5 μm (vertical)	All specimens, including biological ones	
Electron Holography §9.3.7	Electron Holography in a TEM Quantitative evaluation of the magnetic induction or electric potential of a thin foil specimen at high resolution	• Electrons in, electrons out • Preparation of electron transparent specimen may damage microstructural features • Requires HV condition. • Requires a coherent source (FEG or SEG)	• Spatial resolution ~20 nm (best ~5 nm) • Thickness integrated (≤100 nm) signal • Image acquisition time ~50 msec–10 sec	Specimens should be electron transparent and vacuum compatible	• Cost: ~$1M–$2M • Space: requires vibration isolation area ~3 m × 4 m

Electron Tomography §9.5.1	Technique for reconstructing three-dimensional images from a series of two-dimensional projections (images)	• Electrons in, electrons out • Consists of acquisition of the projection series, tomographic image reconstruction, and final image viewing and segmentation • Requires true projection, i.e. intensity must show monotonic relationship to some function of thickness or density. • Particularly suited for studying amorphous, polymer, and biological materials	• Electron transparent specimens of inorganic materials and biological specimens.	• Resolution ~10 nm • Requires large specimen tilt range single axis ~ ±70°, dual axis ~ ±50°	• Cost: ~$1M–$2M • Space: requires vibration isolation area ~3 m × 4 m
HREM §9.3.5	Phase contrast images of thin foils formed by interference of the direct transmitted beam with one or more diffracted beams, selected by an aperture	• Electrons in, electrons out • Image is a function of both defocus and thickness • Often requires simulations for interpretation • Resolution defined in terms of the contrast transfer function • Best transfer function, with minimum number of zero crossings, occurs at the Scherzer defocus • Correction for spherical and chromatic aberration possible	• Thin specimen ≤20 nm	• Resolution $\propto C_S^{1/4} \lambda^{3/4}$, improves with smaller λ (higher keV) and smaller C_S • Typically, resolution ~0.1 nm	

continued

Table 12.3 Continued

Acronym and Section	Method and Use	Salient Details	Sensitivity, Limits, Resolution, etc.	Specimen	Cost and Space Requirements
Lorentz Microscopy §9.3.6	LORENTZ MICROSCOPY IN A TEM Imaging magnetic domains in a TEM	• Electrons in, electrons out • Three possible methods: Fresnel and Foucault (F&F), and Differential Phase Contrast (DPC) • Qualitative (F&F) and quantitative (DPC) measurement of in-plane induction	• Spatial resolution ~10–50 nm (F&F), ~2–10 nm (DPC) • Thickness integrated, $t \leq 100$ nm. • Image acquisition time ~50 msec–60 sec.	Electron-transparent specimen.	
OM §6.8	OPTICAL OR LIGHT MICROSCOPY Method of choice for the first visual observation of a specimen and used for metallography	• Photons in, photons out • Nondestructive • Reflection and transmission geometry • Different imaging modes include using cross-polarized light, differential interference contrast, dark field, and phase contrast	Resolving power ~0.20 μm using white light	• All solid, liquid, and biological specimens • Image quality in metallography enhanced by polishing/etching surface	• Cost: ~$2.5K–$50K • Space: ~ Tabletop

SEM §10	**SCANNING ELECTRON MICROSCOPY** High-magnification imaging and chemical analysis, including mapping	• Electrons in, electrons out (imaging) • Electrons in, photons out (chemical analysis) • Nondestructive, but specimens may be subject to beam damage • Incident beam energy ~0.5 keV–30 keV	• Magnification range ~10×–300,000× • Lateral resolution ~1–50 nm for secondary electrons • Depth sampled varies from nm to μm, depending on kV, sample and analysis mode • Large depth of field, 10–10,000 μm	• Specimens may need a conducting coating to avoid charging effects • Size 0.1 mm–~10 cm	• Cost: $100K–$300K • Space: ~2 m × 3m
SEMPA §10.5.2	**SCANNING ELECTRON MICROSCOPY WITH POLARIZATION ANALYSIS.** Maps local magnetization and domain structure of surfaces	• Electrons in, electrons out • Typically uses a Mott analyzer to detect secondary electron polarization • Best to have electrostatic lenses • Requires UHV environment • Quantitative evaluation of magnetization distribution	• Spatial resolution ~150 nm (best ~20 nm) • Depth of information ~1–2 nm • Time for image acquisition ~1–100 min	• Specimens should be vacuum compatible • A clean surface is required	• Cost: $300K–$500K • Space: ~2 m × 3 m.

continued

Table 12.3 Continued

Acronym and Section	Method and Use	Salient Details	Sensitivity, Limits, Resolution, etc.	Specimen	Cost and Space Requirements
SFM §11.4	SCANNING FORCE MICROSCOPY Imaging in air, vacuum, or solution, and measuring surface forces with excellent resolution and accuracy	• Nondestructive • Cantilever used as force sensor • Probe physically interacts with the specimen • Nothing to focus and zero depth of field! • Imaging in ambient and liquid environments • Various forces can be measured • Contact, noncontact, and tapping modes	• Vertical resolution ~0.01 nm • Lateral resolution ~ atomic–0.1 nm • Typical force measurement ~1 nN • Maximum scan area 100×100 μm • Time for image ~ 2–5 min	• Conducting or insulating • Biological materials • Specimen can be in liquid or in air • Vibration-free environment required	• COST: $75K (air)–$200K (UHV) • Space: Instrument ~0.2 m × 0.2 m sits in 0.6 m × 0.6 m × 0.6 m vibration isolator
STM §11.2	SCANNING TUNNELING MICROSCOPY Imaging surface topography at atomic resolution, and measuring the local electronic structure in spectroscopy mode	• Nondestructive • Tunneling current probes both occupied and unoccupied density of states depending on bias • Can determine work function • Contaminated tips can create artefacts • Requires high vacuum	• Vertical resolution ~0.001 nm • Lateral resolution ~ atomic scale	• Flat, conducting specimen. • Insulators can be viewed with conducting coating • Vibration-free environment	• Cost: ~$150K including vacuum system • Space: ~ table-top

Technique	Description		Specimen
TEM §9	**TRANSMISSION ELECTRON MICROSCOPY** Observing the structural/crystallographic, chemical and magnetic microstructure of a specimen, with diffraction (SAD, CBED), imaging (mass, diffraction, and phase contrast), and spectroscopy (EELS, EDXS) at the highest possible lateral resolution	• Electrons in, electrons out • Elastic scattering (imaging and diffraction) and inelastic scattering (EDXS and EELS) • Magnetic contrast (Lorentz microscopy and electron holography) possible • Reciprocity theorem relates standard TEM and STEM modes • Characteristics depend on source/gun parameters • EM lenses corrected for aberrations in high-end machines	• Resolution ~0.3–5 Å • Convergence semi-angle, $\alpha \sim 10^{-2}$ rad, scattering semi-angle, $\beta \sim 15$ mrad	• Thin specimen (≤ 100 nm) • Preparing electron transparent foils is a challenge	• Cost: ~$0.5M–$2M and more! • Space: requires vibration isolation area ~3 m × 4 m
XTM §6.6.7	**X-RAY TRANSMISSION MICROSCOPY** Generate microscopic images of a thin specimen by scanning it in a focused X-ray beam	• Photons in, photons out • Use zone plates for X-ray focusing • Use dichroism for magnetic contrast • Element-specificity in imaging	Resolution using zone plates ~10 nm	Thin specimens, $t \leq 20$ nm	
Z-Contrast §9.3.4	**Z-CONTRAST IMAGING IN A STEM** Image formed by large angle Rutherford scattering with element sensitivity ($\propto Z^2$)	• High-angle incoherent scattering • Combines a STEM with a high-angle annular dark field detector to collect electrons incoherently scattered over large angles • Scattering angles ~70–150 mrads	Can resolve single atomic columns using sub-Å probes and aberration-corrected lenses	Thin electron-transparent specimens	

823

Index

(*fn* denotes foot-note)

Table of Values

Name	Symbol	Value	SI units	CGS units
Avagadro's Number	N_A	6.022×10^{23} mol^{-1}	—	—
Boltzmann's Constant	k_B	1.38062	$\times 10^{-23}$ J K^{-1}	$\times 10^{-16}$ erg K^{-1}
Bohr Magneton	μ_B	9.274	$\times 10^{-24}$ Am^2 or JT^{-1}	$\times 10^{-21}$ erg G^{-1}
Bohr's radius for H	a_0	5.292	$\times 10^{-11}$ m	$\times 10^{-9}$ cm
Electron Charge	e	1.6022	$\times 10^{-19}$ C	4.8032×10^{-10} esu
Electron rest mass	m_e	9.109	$\times 10^{-31}$ kg	$\times 10^{-28}$ g
Electric permittivity of free space	ε_0	8.854	$\times 10^{-12}$ F m^{-1}	1
Magnetic flux quantum	$\Phi_0 = h/(2e)$	2.0678	$\times 10^{-15}$ Tm^2	
Magnetic permeability of free space	μ_0		$4\pi \times 10^{-7}$ [Hm^{-1}]	1
Planck's constant	h	6.626	$\times 10^{-34}$ J s	$\times 10^{-27}$ erg s
Reduced Planck's constant	$\hbar = h/2\pi$	1.0546	$\times 10^{-34}$ J s	$\times 10^{-27}$ erg s
Proton rest mass	m_p	1.6726	$\times 10^{-27}$ kg	$\times 10^{-24}$ g
Velocity of light in free space	c	2.9979	$\times 10^{8}$ m s^{-1}	$\times 10^{10}$ cm s^{-1}
1 electron volt	eV	1.6022	$\times 10^{-19}$ J	$\times 10^{-12}$ erg
Atomic mass unit	amu	1.6606	$\times 10^{-27}$ kg	$\times 10^{-24}$ g

Periodic Table of the Elements

Key

Atomic Number → 24
Cr crystal structure → **Cr** bcc
Atomic Weight → 52.00
Electronegativity (Pauling) → 1.6
Energy(eV) to remove one electron → 6.76
Electron (outer) configuration → $3d^54s$

Legend:
- ▦ Nonmetal
- ▨ Intermediate
- ☐ Metal

IA	IIA												IIIA	IVA	VA	VIA	VIIA	VIIIA
1 **H** hcp 1.01 / 2.1 / 13.595 / $1s$																		2 **He** hcp 4.00 / — / 24.58 / $1s^2$
3 **Li** bcc 6.94 / 1.0 / 5.39 / $2s$	4 **Be** hcp 9.01 / 1.5 / 9.32 / $2s^2$												5 **B** rhom 10.81 / 2.0 / 8.30 / $2s^22p$	6 **C** diam 12.01 / 2.5 / 11.26 / $2s^22p^2$	7 **N** hcp 14.01 / 3.0 / 14.54 / $2s^22p^3$	8 **O** rhomb 16.00 / 3.5 / 13.61 / $2s^22p^4$	9 **F** mono 19.00 / 4.0 / 17.42 / $2s^22p^5$	10 **Ne** fcc 20.18 / — / 21.56 / $2s^22p^6$
11 **Na** bcc 22.99 / 0.9 / 5.14 / $3s$	12 **Mg** hcp 24.31 / 1.2 / 7.64 / $3s^2$	IIIB	IVB	VB	VIB	VIIB		VIII		IB	IIB		13 **Al** fcc 26.98 / 1.5 / 5.98 / $3s^23p$	14 **Si** diam 28.09 / 1.8 / 8.15 / $3s^23p^2$	15 **P** ortho 30.97 / 2.1 / 10.55 / $3s^23p^3$	16 **S** ortho 32.06 / 2.5 / 10.36 / $3s^23p^4$	17 **Cl** ortho 35.45 / 3.0 / 13.01 / $3s^23p^5$	18 **Ar** fcc 39.95 / — / 15.76 / $3s^23p^6$
19 **K** bcc 39.10 / 0.8 / 4.34 / $4s$	20 **Ca** fcc 40.08 / 1.0 / 6.11 / $4s^2$	21 **Sc** hcp 44.96 / 1.3 / 6.56 / $3d4s^2$	22 **Ti** hcp 47.87 / 1.5 / 6.83 / $3d^24s^2$	23 **V** bcc 50.94 / 1.6 / 6.74 / $3d^34s^2$	24 **Cr** bcc 52.00 / 1.6 / 6.76 / $3d^54s$	25 **Mn** bcc 54.94 / 1.5 / 7.43 / $3d^54s^2$	26 **Fe** bcc 55.85 / 1.8 / 7.90 / $3d^64s^2$	27 **Co** hcp 58.93 / 1.8 / 7.86 / $3d^74s^2$	28 **Ni** fcc 58.69 / 1.8 / 7.63 / $3d^84s^2$	29 **Cu** fcc 63.55 / 1.9 / 7.72 / $3d^{10}4s$	30 **Zn** hcp 65.41 / 1.6 / 9.39 / $3d^{10}4s^2$		31 **Ga** ortho 69.72 / 1.6 / 6.00 / $4s^24p$	32 **Ge** diam 72.64 / 1.8 / 7.88 / $4s^24p^2$	33 **As** rhom 74.92 / 2.0 / 9.81 / $4s^24p^3$	34 **Se** Mon 78.96 / 2.4 / 9.75 / $4s^24p^4$	35 **Br** Mon 79.90 / 2.8 / 11.84 / $4s^24p^5$	36 **Kr** fcc 83.80 / — / 14.00 / $4s^24p^6$
37 **Rb** bcc 85.47 / 0.8 / 4.18 / $5s$	38 **Sr** fcc 87.62 / 1.0 / 5.69 / $5s^2$	39 **Y** hcp 88.91 / 1.2 / 6.5 / $4d5s^2$	40 **Zr** hcp 91.22 / 1.4 / 6.95 / $4d^25s^2$	41 **Nb** bcc 92.91 / 1.6 / 6.77 / $4d^35s^2$	42 **Mo** bcc 95.94 / 1.8 / 7.18 / $4d^55s^2$	43 **Tc** hcp (98) / 1.9 / 7.28 / $4d^55s^2$	44 **Ru** hcp 101.07 / 2.2 / 7.36 / $4d^75s^2$	45 **Rh** fcc 102.91 / 2.2 / 7.46 / $4d^85s^2$	46 **Pd** fcc 106.4 / 2.2 / 8.33 / $4d^85s^2$	47 **Ag** fcc 107.87 / 1.9 / 7.57 / $4d^{10}5s$	48 **Cd** hcp 112.41 / 1.7 / 8.99 / $4d^{10}5s^2$		49 **In** fct 114.82 / 1.7 / 5.78 / $5s^25p$	50 **Sn** diam 118.71 / 1.8 / 7.34 / $5s^25p^2$	51 **Sb** Rhom 121.76 / 1.9 / 8.64 / $5s^25p^3$	52 **Te** hex 127.60 / 2.1 / 9.01 / $5s^25p^4$	53 **I** ortho 126.90 / 2.5 / 10.45 / $5s^25p^5$	54 **Xe** fcc 131.30 / — / 12.13 / $5s^25p^6$
55 **Cs** bcc 132.91 / 0.7 / 3.89 / $6s$	56 **Ba** bcc 137.33 / 0.9 / 5.21 / $6s^2$	57 **La**	72 **Hf** hcp 178.49 / 1.3 / 7.0 / $4f^{14}5d^26s^2$	73 **Ta** bcc 180.95 / 1.5 / 7.88 / $5d^36s^2$	74 **W** bcc 183.84 / 1.7 / 7.98 / $5d^46s^2$	75 **Re** hcp 186.20 / 1.9 / 7.87 / $5d^56s^2$	76 **Os** hcp 190.23 / 2.2 / 8.7 / $5d^66s^2$	77 **Ir** fcc 192.20 / 2.2 / 9.0 / $5d^9$	78 **Pt** fcc 195.08 / 2.2 / 8.96 / $5d^96s$	79 **Au** fcc 196.97 / 2.4 / 9.22 / $5d^{10}6s$	80 **Hg** rhom 200.59 / 1.9 / 10.43 / $5d^{10}6s^2$		81 **Tl** hcp 204.38 / 1.8 / 6.11 / $6s^26p$	82 **Pb** fcc 207.19 / 1.8 / 7.41 / $6s^26p^2$	83 **Bi** rhom 208.98 / 1.9 / 7.29 / $6s^26p^3$	84 **Po** cub (209) / 2.0 / 8.43 / $6s^26p^4$	85 **At** (210) / 2.2 / — / $6s^26p^5$	86 **Rn** fcc (222) / — / 10.74 / $6s^26p^6$
87 **Fr** (223) / 0.7 / — / $7s$	88 **Ra** bcc (226) / 0.9 / 5.28 / $7s^2$	89 **Ac**																

RARE EARTHS

| 57 **La** hcp 138.91 / 1.1 / 5.61 / $5d6s^2$ | 58 **Ce** fcc 140.12 / 1.3 / 6.91 / $4f5d6s^2$ | 59 **Pr** hcp 140.91 / 1.2 / 5.76 / $4f^36s^2$ | 60 **Nd** hcp 144.24 / 1.2 / 6.31 / $4f^46s^2$ | 61 **Pm** hex (145) / 1.2 / — / $4f^56s^2$ | 62 **Sm** rhom 150.35 / 1.2 / 5.60 / $4f^66s^2$ | 63 **Eu** bcc 151.96 / 1.2 / 5.67 / $4f^76s^2$ | 64 **Gd** hcp 157.25 / 1.2 / 6.16 / $4f^86s^2$ | 65 **Tb** hcp 158.92 / 1.2 / 6.74 / $4f^96s^2$ | 66 **Dy** hcp 162.5 / 1.2 / 6.82 / $4f^{10}6s^2$ | 67 **Ho** hcp 164.93 / 1.2 / 6.01 / $4f^{11}6s^2$ | 68 **Er** hcp 167.26 / 1.2 / 6.10 / $4f^{12}6s^2$ | 69 **Tm** hcp 168.93 / 1.2 / 6.18 / $4f^{13}6s^2$ | 70 **Yb** fcc 173.04 / 1.2 / 6.25 / $4f^{14}6s^2$ | 71 **Lu** hcp 174.97 / 1.2 / 5.43 / $4f^{14}5d6s^2$ |

ACTINIDES

| 89 **Ac** fcc (227) / 1.1 / 5.17 / $6d7s^2$ | 90 **Th** fcc 232.04 / 1.3 / 6.08 / $6d^27s^2$ | 91 **Pa** bct 231.04 / 1.5 / 5.89 / $5f^26d7s^2$ | 92 **U** ortho 238.03 / 1.7 / 6.05 / $5f^36d7s^2$ | 93 **Np** ortho (237) / 1.3 / 6.19 / $5f^46d7s^2$ | 94 **Pu** mono (244) / 1.3 / 6.06 / $5f^67s^2$ | 95 **Am** hcp (243) / 1.3 / 5.99 / $5f^77s^2$ | 96 **Cm** hcp (247) / 1.3 / 6.02 / $5f^76d7s^2$ | 97 **Bk** hcp (247) / 1.3 / 6.23 / $5f^97s^2$ | 98 **Cf** hcp (251) / 1.3 / 6.30 / | 99 **Es** hcp (252) / 1.3 / 6.42 / | 100 **Fm** (257) / 1.3 / 6.50 / | 101 **Md** (258) / 1.3 / 6.58 / | 102 **No** (259) / 1.3 / 6.65 / | 103 **Lr** (262) / 1.3 / — / |